THE BIOLOGY OF DINOFLAGELLATES

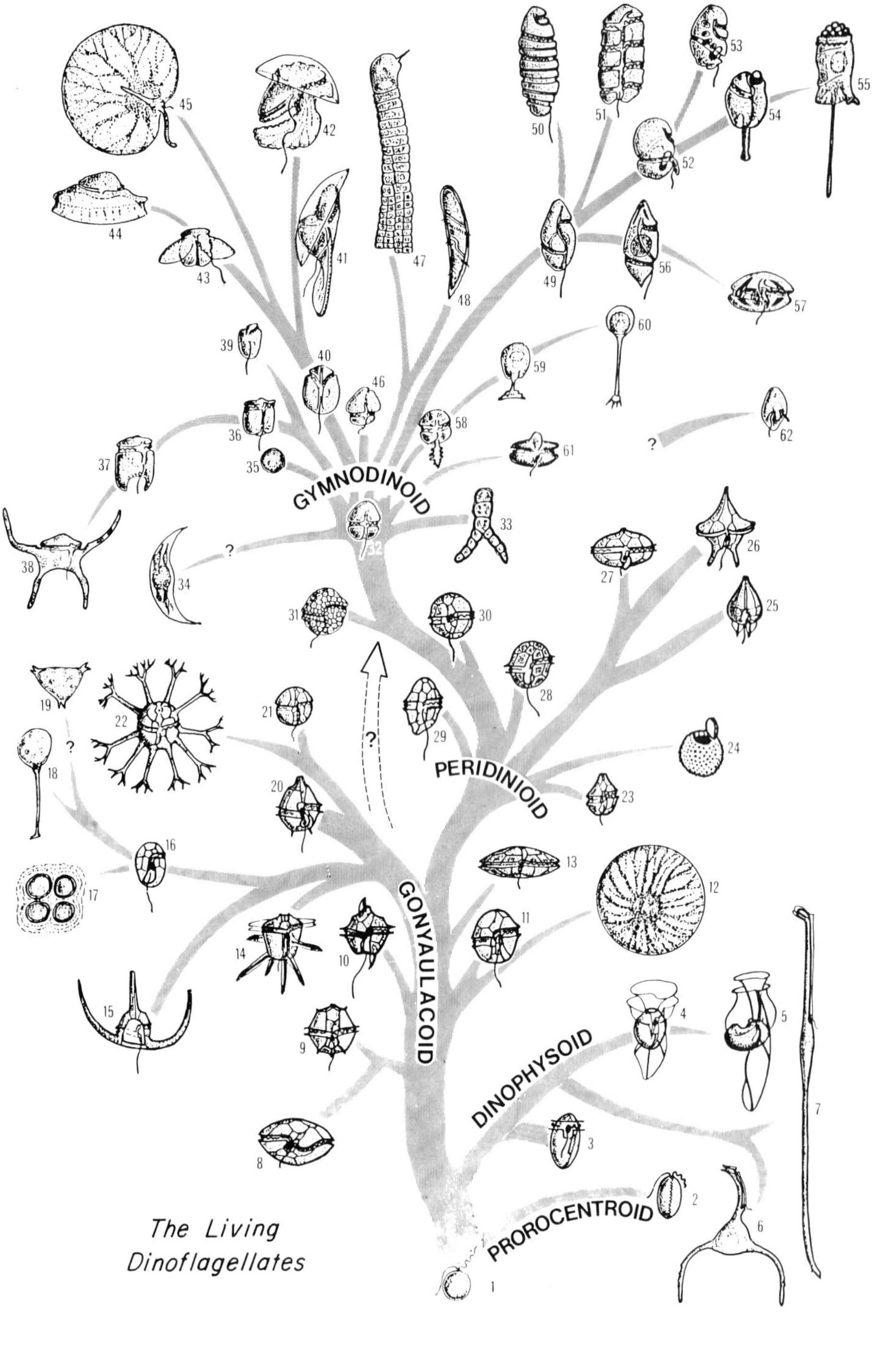
GYMNODINOID
PERIDINIOID
GONYAULACOID
DINOPHYSOID
PROROCENTROID
The Living
Dinoflagellates

Frontispiece. Probable relationships among extant dinoflagellates.

In this tree the ancestral dinoflagellate is assumed to have been desmokont, resembling present-day desmomonads. A case can also be made for the group to have arisen from a gymnodinoid form, possibly like *Oxyrrhis*, in which case the arrangement would be somewhat the inverse of the present arrangement. Only a quarter of the living genera are shown in the tree.

1, Desmomonad ancestor, resembling *Desmomastix* or *Pleromonas*. 2, *Prorocentrum*. 3, *Dinophysis*. 4, *Ornithocercus*. 5, *Parahistioneis*, *Histioneis*. 6, *Triposolenia*. 7, *Amphisolenia*. 8, *Gambierdiscus* (also *Ostreopsis*, *Coolia*). 9, *Triadinium* (= *Heteraulacus*). 10, *Pyrodinium*. 11, *Protogonyaulax* (also *Alexandrium*/*Gessnerium*, *Fragilidium*). 12, *Pyrocystis*. 13, *Pyrophacus*. 14, *Ceratocorys*. 15, *Ceratium*. 16, *Hemidinium* (possibly also *Crypthecodinium*). 17, *Gloeodinium*. 18, *Stylodinium*. 19, *Tetradinium* and other Phytodinialians. 20, *Gonyaulax*. 21, *Palaeophalacroma*. 22, *Cladopyxis*. 23, *Scrippsiella*. 24, *Thoracosphaera*. 25, *Podolampas* (and *Blepharocysta*). 26, *Protoperidinium*. 27, *Diplopsalopsis*, *Dissodium*, *Oblea* and other diplopsaloids. 28, *Peridinium*. 29, *Heterocapsa* (= *Cachonina*). 30, *Glenodinium*. 31, *Woloszynskia*. 32, *Gymnodinium*. 33, *Dinothrix*, *Dinoclonium*. 34, *Dissodinium*. 35, *Symbiodinium*, *Zooxanthella*. 36, 37, *Amphidinium* e.p. 38, *Brachydinium*. 39, 40, *Amphidinium* e.p. 41, *Kofoidinium*. 42, *Pomatodinium*. 43, *Cymbodinium*. 44, *Craspedotella*. 45, *Noctiluca*. 46, *Katodinium*. 47, *Haplozoon*. 48, *Blastodinium*. 49, *Gyrodinium*. 50, *Cochlodinium*. 51, *Polykrikos*. 52, *Proterythropsis* (and *Warnowia*). 53, *Nematodinium*. 54, *Erythropsidinium*. 55, *Greuetodinium* (= *Leucopsis*). 56, *Plectodinium*. 57, *Actiniscus*. 58, *Protoodinium*. 59, *Oodinium*. 60, *Chytriodinium*. 61, *Ptychodiscus*. 62, *Oxyrrhis*.

BOTANICAL MONOGRAPHS · VOLUME 21

THE BIOLOGY OF DINOFLAGELLATES

EDITED BY

F. J. R. TAYLOR

Professor, Departments of
Oceanography and Botany,
University of British Columbia

BLACKWELL SCIENTIFIC PUBLICATIONS

OXFORD LONDON EDINBURGH

BOSTON PALO ALTO MELBOURNE

Editorial offices:
Osney Mead, Oxford, OX2 0EL
(*Orders*: Tel. 0865 240201)
8 John Street, London, WC1N 2ES
23 Ainslie Place, Edinburgh, EH3 6AJ
52 Beacon Street, Boston
Massachusetts 02108, USA
667 Lytton Avenue, Palo Alto
California 94301, USA
107 Barry Street, Carlton
Victoria 3053, Australia

First published 1987

Typeset, printed and bound in Great Britain by William Clowes Limited, Beccles and London

DISTRIBUTORS

USA and Canada
Blackwell Scientific Publications Inc
PO Box 50009, Palo Alto
California 94303
(*Orders*: Tel. (415) 965-4081)

Australia
Blackwell Scientific Publications
(Australia) Pty Ltd
107 Barry Street,
Carlton, Victoria 3053
(*Orders*: Tel. (03) 347 0300)

British Library
Cataloguing in Publication Data

The Biology of dinoflagellates.—(Botanical monographs; v. 21)
1. Dinoflagellata
I. Taylor, F.J.R. II. Series
593.1'8 QL368.D6

ISBN 0-632-00915-2

CONTENTS

CONTRIBUTORS

DONALD M. ANDERSON, Woods Hole, Oceanographic Institution, Woods Hole, Mass. 02543, USA.

JEAN & MONIQUE CACHON, Université de Nice, Laboratoire de Protistologie Marine, Station Zoologique, 06230 Villefranche-sur-Mer, France.

JOHN D. DODGE, Department of Botany, Royal Holloway College, University of London, Egham, Surrey TW20 0EX, England.

MALTE ELBRÄCHTER, Biologische Anstalt Helgoland, Litoralstation List 2282 List auf Sylt, Hafenstrasse 3, West Germany (FRD).

GREGORY GAINES, Department of Oceanography, University of British Columbia, Vancouver, BC V6T 1W5, Canada.

DAVID K. GOODMAN, Arco Gas and Oil Company, P.O. Box 2819, Dallas, Texas 75221, USA.

CLAUDE GREUET, Université de Nice, Laboratoire de Biologie Animal et Cytologie, Campus Universitaire Valrose, 06034 Nice, France.

PAMELA J. KANETA, Haskins Laboratories of Pace University, 41 Park Row, New York, N.Y. 10038, USA.

MICHAEL LEVANDOWSKY, Haskins Laboratories of Pace University, 41 Park Row, New York, N.Y. 10038, USA.

LOIS A. PFIESTER, Department of Botany and Microbiology, 770 Van Fleet Oval, University of Oklahoma, Norman, Oklahoma 73019, USA.

UTSA POLLINGHER, Kinneret Limnological Laboratory, P.O. Box 345, Tiberias, Israel 14–102.

BARBARA PRÉZELIN, Department of Biological Sciences, University of California, Santa Barbara, California 93106, USA.

PETER J. RIZZO, Texas A & M University, Department of Biology, College Station, Texas 77843, USA.

YUZURU SHIMIZU, Department of Pharmacognosy, College of Pharmacy, University of Rhode Island, Kingston, Rhode Island 02881, USA.

BEATRICE M. SWEENEY, Department of Biological Sciences, University of California, Santa Barbara, California 93106, USA.

F. J. R. 'MAX' TAYLOR, Departments of Oceanography and Botany, University of British Columbia, Vancouver, BC V6T 1W5, Canada.

ROBERT K. TRENCH, Department of Biological Sciences, University of California, Santa Barbara, California 93106, USA.

NANCY W. WITHERS, Box 661, School of Medicine, Medical University of South Carolina, 171 Ashley Avenue, Charleston, South Carolina, 29401, USA.

This book is dedicated to the memory of Trygve Braarud (1903–1985), pioneering Norwegian phytoplanktologist, a student of dinoflagellates in all their aspects and a truly gentle man.

PREFACE

For many years dinoflagellates have languished as an obscure group among the miscellaneous collection of pigmented flagellates claimed both by botanists (as algae) and zoologists (as protozoans), but only desultorily treated by both. For reasons that are not obvious the study of dinoflagellates was primarily by protozoologists until the 1940s, whereas the field is now dominated more by phycologists (and palynologists). The series in which this volume appears is an indication of that, and so is the widespread, although not yet exclusive, use of botanical classification. However, the editor is a card-carrying protistologist!

Dinoflagellates first intruded upon my consciousness as annoyingly mobile members of the Indian Ocean plankton. When preserved they revealed the details of their complicated suits of armour (the naked ones having disintegrated in the excessively strong 5–7% formalin in use at the time), their architecture much more plastic than the geometric forms of the diatoms—their rivals as primary producers—and radiolarians.

Shortly afterwards, they made their presence known in a much more dramatic fashion when one of their species killed all the marine life in a nearby bay during a massive 'red tide' that was also luminous at night. No sooner had I arrived in British Columbia than there was a human fatality as a result of eating shellfish that had fed on a toxin-producing species. After such insistent attention-seeking it was impossible not to focus on this extrovert group. Further awareness of their unusual cell biological and biochemical features (see Chapter 1), possible primitiveness and usefulness as microfossils, made me grateful to have stumbled on such interesting creatures and surprised that they have been neglected for so long.

It can now truly be said that dinoflagellates are a group whose time has come, with meetings exclusively devoted to them (in Colorado Springs, in 1978, Tübingen in 1981 and Egham, UK in 1985) and to the fostering of communication between those studying the living forms (neontologists) and the growing army studying their fossilized remains. The literature on them is growing at a gratifying, if alarming rate, increasing the necessity for reviews and books to cope with it. Two books on the group have appeared: that by Sargeant in 1974, a slim but useful volume focused primarily on the fossil forms and one on special topics by Spector in 1984. The present work is intended to provide a review of the general biology of living forms, with a summary of the fossil record and its applications also included.

A recurring problem in works of this type is the choice of name in data allocation when the name of a species or genus, particularly a well known one, has been changed relatively recently. Such changes, albeit mandatory by the

Codes of Nomenclature, are annoying and confusing to non-taxonomists. Here, the majority of authors have used the most recent names, with the altered name in parentheses. The reasons for such changes are explained as briefly as possible in the Taxonomic Appendix. However, a few authors have preferred to use only the earlier, better known names.

Dinoflagellate studies are now so diverse that it is highly doubtful that a single author could provide a balanced, adequate summary and I thank the team of authors who have contributed so valuably to this volume, particularly those in rapidly developing areas who have patiently resigned themselves to the ponderous process involved in a work of this type.

As a result of the editor contributing rather more to the body of the text than is usual (or wise, and unanticipated), the assistance of Leslie Fenn Christian (general text) and Elizabeth du Fresne (index) is gratefully acknowledged. Finally, behind nearly every book like this is a family that has had to put up with the irreplaceable loss of hundreds of hours of family time: regrets and thanks to Andrew, Belinda, Christopher and Margaret for their part-time father/husband.

F. J. R. 'Max' Taylor

CHAPTER 1
GENERAL GROUP CHARACTERISTICS; SPECIAL FEATURES OF INTEREST; SHORT HISTORY OF DINOFLAGELLATE STUDY

F. J. R. TAYLOR
Departments of Oceanography and Botany, University of British Columbia, Vancouver, BC, Canada V6T 1W5

1 GENERAL DINOFLAGELLATE CHARACTERISTICS

Dinoflagellates are one of a dozen or so groups of predominantly unicellular, eukaryotic, flagellated organisms that possess both photosynthetic and non-photosynthetic members. These nutritionally versatile groups are referred to informally as 'phytoflagellates'. Similar groups in which all the members are non-photosynthetic are termed 'zooflagellates'. Neither of these assemblages appears to be monophyletic (Taylor 1978) and for this reason their formal use is discouraged. Cox (1980) has edited a volume summarizing phytoflagellate biology, with a chapter on dinoflagellates by Steidinger & Cox (1980).

By the late 1970s approximately 2000 living, and 2000 fossil species had been described. Roughly 40–60% of the living species are photosynthetic.* Because both photosynthetic and non-photosynthetic dinoflagellates are known, because they can swim, and because many have cell walls, they have been claimed by both botanists and zoologists, each emphasizing the 'plant' or 'animal' features. Botanists group them with the 'algae' and zoologists with the 'protozoa' and both have produced classification schemes for dinoflagellates. In fact, phytoflagellates demonstrate very nicely the inadequacy of the application of concepts developed from the study of the multicellular members of two more

* Exact numbers are not known because some of the species are poorly defined and some descriptions do not indicate if the species are photosynthetic or not. In formalin-preserved material the cells are usually bleached, and in some cases storage bodies may have been confused with chloroplasts.

recently evolved lineages, the Metazoa and Metaphyta, to a distantly-related group which evolved in its own special way. A useful, informal term which avoids the 'animal' or 'plant' non-issue is 'protist', meaning a unicellular eukaryotic organism.

There is no single feature which unites all dinoflagellates. Rather, assignment to the group is made because of the possession of one or more of a suite of characters illustrated diagrammatically in Fig. 1.1.

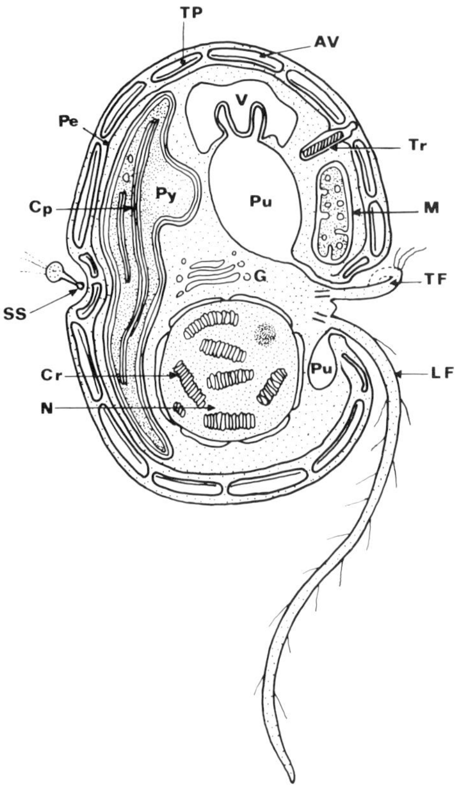

Fig. 1.1. Diagrammatic longitudinal section through the motile cell of a generalized dinoflagellate. The ventral side is to the right (slightly modified from Taylor, 1980a). AV, amphiesmal (thecal) vesicle; Cp, chloroplast, surrounded by a triple-membraned envelope; Cr, permanently condensed chromosome; G, Golgi apparatus; LF, longitudinal flagellum; M, mitochondrion with tubular cristae; N, nucleus (dinokaryon); Pe, pellicle; Pu, pusule; Py, pyrenoid; SS, striated strand (=paraxial rod) of the transverse flagellum; TF, proximal part of the transverse flagellum; TP, thecal plate within an amphiesmal vesicle; TR, trichocyst.

(a) The presence of a *dinokaryon*, i.e. a nucleus containing chromosomes which do not decondense during interphase and which appear fibrillar under electron microscopy (Chapter 3), containing very little basic protein (Chapter 4). Nucleosomes are absent. The nuclear envelope remains intact during mitosis and the spindle is extranuclear.

(b) The motile cell, or *mastigote*, possesses a single layer of flattened vesicles just beneath the cell membrane. There are two dissimilar flagella, one laterally directed and the other beating posteriorly. Their structure is unique (Chapters 2 and 3).

(c) A wall may be present or absent. If present it consists either of cellulose plates, located within the peripheral vesicles (and therefore intracellular), or a continuous layer beneath or within the vesicles in the mastigote, or forming a cyst wall which may be intra- or extracellular (Chapter 3). The latter may be strengthened with a resistant material resembling sporopollenin. On excystment the cell exits via an opening, the *archeopyle* which bears a fixed relationship to other cell structures (Chapter 2).

(d) Special, modified vacuole-like structures termed *pusules*, of unknown function, usually two per cell, opening by canals to the flagellar bases (Chapter 3).

(e) In photosynthetic cells, the presence of *peridinin*, complexed with chlorophyll (Chapter 5), as a major xanthophyll. Its presence is significant, but not its absence.

(f) The presence of unique sterols: *dinosterol* and *amphisterol* (see Chapter 9).

Other features common to dinoflagellates, but not unique to them, include the presence of chlorophyll a and c_2 (c_1 is usually absent), the production of starch and oils as reserves, mitochondrial cristae with circular cross-sections, and a triple-membraned envelope surrounding the chloroplast. Rod-like, ejectile bodies—*trichocysts*—occur in several phytoflagellate groups, but in detailed structure (Chapter 3) those of dinoflagellates are unique.

This is not a complete characterization of the group. There are many unusual features that occur in too few representatives to make them useful at the group level, such as the bioluminescence system (Chapter 7), or toxin production (Chapter 8), or extraordinary eyes (Chapter 3). As noted earlier, some dinoflagellates may lack one or several of these features. For example, in the large dinoflagellate *Noctiluca* the chromosomes disperse during most of the life cycle but are of typical appearance when they appear at sporulation; and in some parasites, such as *Syndinium*, the chromosomes never condense visibly. Some dinoflagellates produce siliceous, or calcareous skeletal elements, e.g. *Actiniscus* or the calcified cysts of *Scrippsiella* and *Thoracosphaera*. The mitotic spindle is intranuclear in *Oxyrrhis*.

2 SPECIAL FEATURES OF INTEREST IN DINOFLAGELLATE BIOLOGY

Few, if any, other protist groups can offer such a wide range of topics of unusual research interest although, until recently, dinoflagellates attracted little attention except among ecologists. The latter knew them as important members of the *phytoplankton*, the microscopic photosynthesizers contributing significant quantities of food for animals in both marine and freshwater habitats. Occasionally they occur in great abundance ('blooms'), causing the water to turn a reddish-brown colour, sometimes with harmful consequences to the inhabitants of the water or consumers of shellfish or fish, including humans. Such outbreaks, with or without harmful effects, are colloquially called *red tides*. The ecological significance of dinoflagellates is discussed further in Chapter 11.

In the 1930s it was discovered that some red tide dinoflagellates produce potent neurotoxins (of still unknown function) and since then many other toxins have been identified, including several concentrated by shellfish and others producing *ciguatera* fish poison. These are described in Chapter 8. Another subject of special interest is the luminescence produced by several of the marine species, of which the best known are *Noctiluca scintillans*, *Pyrodinium bahamense*, *Gonyaulax polyedra* and several species of *Pyrocystis*. The production of luminescence by *G. polyedra* has proven to be an excellent system for the basic study of cell rhythms and both these aspects are summarized in Chapter 7.

More recently, dinoflagellates have aroused the interest of phycologists, protozoologists, evolutionists, and cell biologists due to their extraordinary combination of apparent primitiveness (nuclear features) with elaborate morphological developments ('arms', 'wings', horns, domes, medusa-like shapes, differentiated multicellular forms, coenocytes) and cytological structures ('eyes', nematocysts, etc.). No student of dinoflagellates can make the elementary error of confusing primitiveness with structural simplicity. At the time when histones (basic nuclear proteins) were in vogue as controllers of gene expression it was obvious that dinoflagellates accomplished gene control perfectly well without normal histones, or, as we now know, nucleosomes (see Chapters 3, 4). The chromosomes remain permanently condensed. Mitosis is extraordinary, with a spindle formed externally to the permanently closed nuclear envelope ('cryptomitosis'; details in Chapter 3).

Dinoflagellates are major parasites of marine invertebrates and fish but, despite the publication of a major monograph by Edouard Chatton (1920, creator of 'pro-' and 'eucaryote', 1938: see section 3 of this chapter) they have not attracted the attention they deserve. The review by the Cachons here (Chapter 13) may help to bring them wider attention.

After Kawaguti (1944) and Freudenthal (1962) recognized that the numerous brown symbionts, termed 'zooxanthellae', were mostly dinoflagellates, it became a source of amazement that, not only were all those in reef-building corals, as well as giant clams, jellyfish, and anemones, dinoflagellates, but they seem to involve only a few species and their role in tropical reef ecology is considerable (see Chapter 12).

In the organic remains of palynological preparations from ancient rocks and sediments, dinoflagellate cysts have graduated from being a minor supplement to fossil pollen as stratigraphic and paleoecological indicators, to major tools in their own right, increasingly valued by petroleum geologists (Chapter 15). Many more dinoflagellate taxonomists can be found in the laboratories of petroleum companies than in academic institutions and it is no longer possible for either neontologists or palynologists to ignore each others' knowledge.

3 EARLIER STUDIES ON DINOFLAGELLATES

In this section an attempt has been made to describe major developments in the study of dinoflagellates up to the 1960s. Developments after that time are the subject for the remainder of this book. It gives an idea of the chronological development of knowledge and the principal contributors to the field. Further information on the development of particular aspects is contained in the introductory section of later chapters.

Some dinoflagellates are not only large enough to see with the naked eye, but are also luminescent (*Noctiluca* and *Pyrocystis*) and therefore it is not surprising that they were among the first to be observed. There is a description (quoted in full by Tappan 1980) by Henry Baker in his classic *Employment for the Microscope* (1753) which, contrary to the opinion of Kofoid & Swezy (1921, p. 412), unequivocally seems to refer to *Noctiluca*, even though he did not illustrate or name it. It was later illustrated by Slabber (1778). Other early references to possible or probable observations on *Noctiluca* were summarized by C. G. Ehrenberg in one of the earliest works on marine luminescence, *Das Leuchten des Meeres* (*Lights of the Sea*, 1834). Because of their unusual forms *Noctiluca* and other 'Cystoflagellates' were not recognized as dinoflagellates. Stein (1883) and Pouchet (1885a, b) made a reasonable case for their true identity but they were only formally included in 1921 by Kofoid & Swezy.

Wholly microscopic dinoflagellates appear first in the earliest general, systematic work to be published on protists: *Animalcula infusoria fluviatilia et marina ...* (*Freshwater and Marine Infusoria*) by the Dane, O. F. Müller, posthumously published in 1786. In this classic work he included Latin descriptions and figures of two freshwater species: *Bursaria* (now *Ceratium*) *hirundinella* and *Vorticella* (now *Glenodinium*) *cincta*, whose diagnoses had been published earlier (O. F. Müller 1773) and one marine species, *Cercaria* (also now *Ceratium*) *tripos* which had been listed by name only in 1777. According to Bergh (1881a, b) one of the two figures provided for *V. cincta* by Müller (1786, fig. 6) is of *Peridinium tabulatum*. No flagella were shown in these early, crude drawings, but Müller did refer to the presence of 'undique ciliis minutissimus' and described the movement of *C. tripos*.

The first generic name which is still in use for a dinoflagellate is *Ceratium*, created by F. von Paula Schrank (1793) for two species, one of which is no longer considered a dinoflagellate, and he added a third species, *C. macroceras*, in 1802. Nitszch (1817) correctly realized that *Cercaria tripos* O.F.M. belonged to Schrank's genus and made the transfer. Shortly after, Bory de St Vincent (1824) transferred two other of Müller's species to new genera: *Hirundinella quadricuspis* (for *B. hirundinella*: the practice of retaining the species name on transfer to another genus had not yet become established) and *Tripos muelleri* (for *C. tripos*), changes which were not accepted.

In 1830 G. A. Michaelis published a pioneering study on microscopic sources of luminescence in the Baltic Sea, showing their biological origin for the first time. Some of the sources that he illustrated, as seen on the surface of a filter, were 'eine Volvox', which was evidently a species of *Protoperidinium* (probably *P. depressum*) and two 'Cercarien', one of which was *Ceratium fusus* and the other *Prorocentrum micans*. The latter is not known to cause bioluminescence. A further source which he named to species, was *Cercaria tripos*. He illustrated the longitudinal flagellum for the first time, but thought it was double.

The genus *Peridinium* was added by the great taxonomic microscopist, Christian Gottfried Ehrenberg, based on the freshwater species *P. cinctum* (1830/32), only the first of many species he described over the next half century, including microfossils from flint flakes and chalks. In 1836 he presented a paper to the Berlin Academy in which he illustrated a species of fossil *Peridinium*: *P. pyrophorum*, as well as some spiny cysts which became known as 'xanthidia' for many years (published in 1838). Many of the latter are now known to also be dinoflagellate cysts (*cf.* the more extensive outline of the history of fossil dinoflagellate study by Sarjeant 1974). In 1838 he also published his massive, classic work *Die Infusionsthierchen als vollkommene Organismen* (*Infusion Animalcules as Complete Organisms*). In its 532 pages and 64 large-format plates he brought together the descriptions of hundreds of protists, many of which were new or recently described by him. The drawings became known for their realistic portrayal of external form, although his conception of internal structure was totally wrong. He placed those he recognized as dinoflagellates in the new group Peridineae, loosely termed 'Kranzthierchen' (wreath-encircled animals, referring to the erroneous impression that the girdle groove contained a ring of cilia, an error which was not corrected until nearly the end of the century). He recognized only four dinoflagellate genera, all of his own creation: *Peridinium* (into which he sank *Ceratium*); *Glenodinium*, differing by the presence of an eyespot ('Augenkranzthierchen'); and two genera *Chaetotyphla* and *Chaetoglena* which were soon recognized to be euglenoids and sunk into *Trachelomonas* (Dujardin 1841; Saville-Kent 1880–82). There were other dinoflagellates in his work, however. *Gymnodinium fuscum*, later to be made the type of the genus *Gymnodinium* Stein (1878), was included as a species of *Peridinium*, having been first described in an earlier paper. *Prorocentrum* was hiding in with the cryptomonads. *Dinophysis* was added later as a genus of walled ciliate (Ehrenberg 1840)!

Ehrenberg's extensive micropaleontological observations were summed up in another massive work: *Mikrogeologie* (1854). A second species of *Peridinium* had been added to the fossil record, there were more xanthidia, and the siliceous endoskeletons of several species of *Actiniscus* were described long before they were found to be made by dinoflagellates still alive today (Schütt 1891). Sarjeant (1974) has recounted the circumstances surrounding the obscure publication of the important fossil genus *Spiniferites* by Mantell (1850), an English micropaleontologist greatly influenced by Ehrenberg. The importance of Ehrenberg's findings to the history of phytoplankton ecology has been briefly evaluated by Taylor (1980b).

Because of his reluctance to take advantage of advances in microscopy, and his fertile imagination and fixed ideas, Ehrenberg tarnished his reputation by insisting that 'Infusionsthierchen' were simply scaled down animals, including those now known to be photosynthetic, such as diatoms and desmids. He was convinced that they had all the usual Metazoan organs, including mouths; chloroplasts were thought to be ovaries, the nucleus was a prostate gland or testis and the eyespot an eye. Although he recognized some as lacking intestines (grouped as the Anentera and including dinoflagellates), he thought vacuoles were multiple stomachs and his theory thus became known as the 'Polygastrica Theory'. The credit for debunking this view, by careful observation and lengthy argument, is given to Felix Dujardin (1841), Professor of Zoology at Rennes, who also described the peculiar dinoflagellate *Oxyrrhis* as a cryptomonad and reinstated *Ceratium*.

Further freshwater species were added by the Swiss, Maximillian Perty (1852), whose ideas on internal structure and function were nearly as odd as those of Ehrenberg and whose illustrations were much worse. He thought the two layers he could see (wall and interior) were two cells, one inside the other, and he thought that whole cells could arise from the chloroplasts. He called the latter 'Blastien' and, in this recognition of them as self-reproducing bodies, he almost anticipated the present view of their origin (see Taylor (1980c) for a historical review of the events leading up to the general acceptance of the Serial Endosymbiosis Theory for the origin of the eukaryotic condition), but he attributed too much to them and had no idea of their real function. Bailey (1853) thought dinoflagellates were embryonic stages of annelid worms.

One of the earliest papers on marine red tides was that by H. J. Carter (1858) on a dinoflagellate discolouring the sea off Bombay. He named it *Peridinium sanguineum*, but it is not reasonably referrable to any modern taxon. He showed that the green pigment present was similar, if not identical to chlorophyll and he described cyst formation, with division in the cysts. Carter provided a number of references to earlier accounts, such as those by Charles Darwin, and an early outbreak in Iceland in 1694, as possibly being due to dinoflagellates because of their appearance, behaviour and/or nocturnal luminescence, some of which resulted in fish kills (see Chapter 11 here). He even proposed that the First Plague of Egypt may have been caused by a dinoflagellate, although it could also have been due to cyanobacteria or some other bacteria.

The first tropical freshwater dinoflagellates were described from Egypt, principally from the Nile, by L. R. Schmarda (1854). In 1855 G. J. Allman (the discoverer of trichocysts in ciliates) found that a species of *Peridinium* was responsible for the brown discoloration of ponds in Phoenix Park, Dublin. He saw the chromosomes of the nucleus but thought these were striae on its surface and saw cyst formation. Confusingly, he thought cilia covered the whole surface of the cell, an observation which was to be erroneously used as further support for a relationship between dinoflagellates and ciliates.

In the same year the great German zoologist, Johannes Müller, described cells of the common tropical marine genus *Pyrocystis* in a paper on radiolarians, but he thought the former were encysted cells of *Noctiluca* and it was only much later (Wyville-Thomson in Murray 1876) that their true nature was realized.

In 1859, Édouard Claparède and Johannes Lachmann, two students of Johannes Müller, published the results of their observations on marine protists from Bergen, Norway. They translated Ehrenberg's Kranzthierchen as 'Cilioflagellates', considering them to be intermediate between flagellates and ciliates. They added the genus *Amphidinium*, recognized *Prorocentrum* as a dinoflagellate and provided a key for the five genera they placed in the group. The mode of division and cyst formation in *Peridinium* was included in the second part of their work (Claparède & Lachmann 1861).

Andrew Pritchard (1861) provided descriptions of all the species known to that date. All the dinoflagellates were assigned to *Peridinium* (including *Ceratium*) and *Glenodinium*. However, he missed the two species of *Dinophysis* that Ehrenberg had described in 1839, *Prorocentrum* and *Oxyrrhis* were included with the cryptomonads, and *Noctiluca* with the ciliates!

Only minor developments occurred over the next 20 years, including K. M. Diesing's (1866) creation of the genus *Gonyaulax* for peridinia (which he thought, like Bailey, were primitive worms), in which the girdle had a knee-like bend (hence the name). An earlier generic name of his, *Heteraulacus* (Diesing 1850) has been replaced by *Triadinium* (Dodge 1981) but also needs replacement (see Appendix). Warming (1875) demonstrated that the dinoflagellate theca was composed of cellulose.

In 1881 Karl Brandt correctly interpreted the small brown spheres in colonial radiolarians and some coelenterates as symbiotic protists to which he gave the generic name *Zooxanthella*. The latter name came to be informally applied, as 'zooxanthellae', to refer to any brown, photosynthetic symbionts associated with animals. Shortly afterwards, Patrick Geddes (1882) produced a further valuable contribution on zooxanthellae (which he attributed to his genus *Philozoon*). The dinoflagellate nature of most of these was only convincingly demonstrated by the observation of the motile cells in the mid-twentieth century (Kawaguti 1944) and their culture by Fruedenthal (1962).

A major contribution was provided by the Dane, Rudolph Bergh in 1881 (in an abbreviated Danish form, 1881a; and in a more extensive German text, 1881b). In addition to a useful historical summary and a systematic section in which he introduced *Diplopsalis*, *Protoceratium* and *Protoperidinium*, he became the first author to offer a reasoned, extensive phylogenetic exposition. Later (1884) he provided a concise review of the four major works, by Stein, Pouchet, Kelbs and Gourett, which all appeared in 1883 (see below).

The coenocyte *Polykrikos* had been seen in 1868 by Uljanin but he concluded that it was a metazoan larva, perhaps a turbellarian. Otto Bütschli (1873), after more careful examination, concluded that it was an unusual ciliate, and Bergh (1881b) clarified its dinoflagellate affinities. Saville-Kent's (1880–82) renowned *Manual of the Infusoria* contains relatively little of note concerning the dinoflagellates, which he included as only one of several groups in the Cilioflagellates. However, like Pritchard (1861), he published descriptions of all the forms which had been described to that date. The Polish protozoologist, Leon Cienkowski (Tsenkovsky), who did most of his work in Russia, described *Exuviaella* in 1881. Its distinction from *Prorocentrum* is no longer maintained. He had published two short papers on *Noctiluca* previously.

The year 1883 was a notable one in dinoflagellate history because four major works appeared, two of which were primarily cytological and two taxonomically significant. The first of the cytological papers was by Georg Klebs, he focused attention on the contents of freshwater dinoflagellates, recording 'diatomin' as the pigment in the pigmented internal bodies. More importantly, he was the first to assert that the propulsive structure in the girdle was an undulating flagellum ('ein schraubig gewundenes Band'), rather than the ring of cilia seen, through the power of suggestion, by so many before him. He shares honours with Pouchet (below) for the first observations on dinoflagellate chromosomes as bodies within the nucleus and he recognized the peculiarity of the nucleus of the group.

The work of greatest systematic importance to dinoflagellate biology in the nineteenth century was published by Friedrich Ritter von Stein, Professor of Zoology in Prague, 1883, as one of his series *Der Organismus der Infusionsthiere*. Earlier, in 1878, he had described *Gymnodinium* and *Hemidinium* in the first section on flagellates, but now he described eighteen more genera (fifteen of which are still in use), including *Cladopyxis*, *Heterocapsa*, *Amphidoma*, *Ceratocorys*, *Oxytoxum*, *Podolampas*, *Ptychodiscus* and the peculiar dinophysoids *Amphisolenia*, *Citharistes*, *Histioneis* and *Ornithocercus* ('einer der wunderbarsten und seltsamsten Thierformen, die mir je vorgekommen sind', p. 26). He also asserted the dinoflagellate character of *Prorocentrum* and *Noctiluca*. This rich harvest of strange new forms came particularly from tropical material that he ingeniously obtained from the gut contents of various preserved invertebrates (particularly salps, ascidians and medusae) brought back by others from the Mediterranean, Fiji and Samoa. At this time, when nets were coarse and sampling only superficial, zooplankton provided an excellent source of concentrated microplankton, including diatoms and radiolaria. He supplemented this with his own collections from Kiel and Helgoland (where rough weather limited his observations).

More significantly, Stein was the first observer to use thecal plate characteristics as taxonomically useful features and, where he could, he described these in remarkable detail. He recognized four latitudinal series:

Frontalen, Basalen (of the epitheca and hypotheca) and Endplatten, plus the girdle and sulcus, and named some special plates: the Rautenplatte (first apical) and Mundplatte (the ventral depression of *Ceratium*). His terminology is no longer used, but it provided the basis from which Kofoid (1907, 1909) developed his system. The theca, of closely fitting plates, led Stein to name the group the 'arthrodelen Flagellaten' to replace Cilioflagellates, since *Prorocentrum* and *Noctiluca* clearly lacked cilia (he did not know of Kleb's observations), but this never caught on.

In the same year Paul Gourret (1883) also described many warm water forms from the Mediterranean near Marseilles, figured less accurately than those of Stein and, because his work was published shortly after Stein's, he had the misfortune to lose priority in the naming of *Podolampas* and none of his new genera survived. However, many of his figures of tropical ceratia are recognizable and useful, he offered a variety of cytological and life-cycle details (although he thought many forms were larvae); had views on phylogeny and concluded his work with a review of Pouchet's (1883) paper.

C.- H.- Georges Pouchet produced a numbered series of five contributions on the biology of dinoflagellates between 1883 and 1892 (plus an earlier note on the dinoflagellate affinities of *Noctiluca*; 1882) which are notable for their careful cytological observations (he used osmic acid as a primary fixative) and the variety of source material. Operating out of the museum in Paris and the Laboratory of Maritime Zoology and Physiology at Concarneau (the latter being the earliest established coastal marine laboratory) he examined material from the Arctic, the Atlantic, and the Mediterranean. In 1894 he produced a general account of plankton collected by 'La Manche' on a cruise to Spitzbergen. Apart from a penchant for placing all his figures upside down, relative to later convention (and locomotion), he provided a wealth of cytological detail, surpassed only by Schütt in the nineteenth century. He described various new forms, including *Peridinium polyedricum* (now *Triadinium*), *Glenodinium obliquum* (which is almost certainly *Coolia monotis* Meunier) and short chains of cells , resembling *Protogonyaulax tamarensis* (Pouchet, 1883, fig. 36). He saw the vermicelli-like (his adjective) chromosomes in the nucleus and realized that dinoflagellates could be recognized by their nuclear features (1885a, p. 35: ('Il offre chez les Peridiniens . . . une uniformité remarquable dans sa constitution'; see also 1885b, pl. 26, fig. 26). He saw nuclear cyclosis, now thought to be associated with meiosis, in *Ceratium* (1885b) as well as paired cells of *C. fusus*, and he discovered the ocellus ('organe oculaire') in *Warnowia* (known for many years as *Pouchetia*). Because of his cytological techniques he was able to study athecate species in greater detail than his predecessors, including nematocysts and food vacuoles in *Polykrikos* (1887). Pouchet (1884) also described the first parasitic dinoflagellate, later named *Oodinium pouchetii* (Lemm.) Chatton, an ectoparasite on the appendicularian *Oikopleura*, although this was not accepted by Bütschli.

The three monumental protozoological volumes by J. A. Otto Bütschli in Bronn's *Klassen und Ordnüngen des Thier-Reichs* (1889), were for many years the 'Bible of Protozoologists' (to use Clifford Dobell's term). Characteristically, the general biology of dinoflagellates was reviewed very thoroughly by Bütschli, without producing many innovations. However, the name we use most frequently for the group derives from his 'Dinoflagellaten' (most authors before that using Ehrenberg's Peridineae, or Claparède and Lachmann's Cilioflagellates) and he was the first to produce tabulation diagrams for various genera (apical and antapical views), with plate terminology derived from Stein (1883). He also included a discussion of the phylogeny of the group, with a phyletic tree fairly similar to that of Bergh.

Bütschli failed to realize the significant difference between the ocellus and more common eye-spots, as have many more recent authors. Furthermore, he omitted *Erythropsis* (now *Erythropsidinium*), one of the most elaborate ocelloid-containing dinoflagellates. It had been described from Sorrento in Italy by Richard Hertwig (1884) while he was teaching at Bonn University. Although the latter did not recognize the dinoflagellate nature of this unusual organism, he was correct in considering it to be related to 'Infusoria'. He was unfortunate to be the subject of repeated, scathing attacks by Carl Vogt, who was convinced that it must have been a vorticellid ciliate fixed in the act of digesting a medusan eye. Kofoid & Swezy (1921) have detailed the lamentable history of this controversy. The structure of this extraordinary organism is discussed in Chapters 2 and 3 here.

Miscellaneous observations on freshwater dinoflagellates had been slowly accumulating, and in 1891 A. J. Schilling provided the first comprehensive account of that community (in his doctoral dissertation at the University of Basel and in a separate publication). He later wrote the authoritative section on dinoflagellates for Adolph Pascher's *Süsswasserflora* . . . (1913).

The most outstanding of the early observers (although little heralded in his own time) was undoubtedly Franz Schütt, Professor of Botany at Kiel (briefly) and then at Greifswald. After a few small, but significant papers on cyst formation (1887), pigments (1890) and the discovery of the cells producing the mysterious, siliceous, radiating skeletons seen many years before in fossil material and named *Actiniscus* by Ehrenberg (which he illegally renamed *Gymnaster* in 1891), he took part in the 'Plankton-Expedition', organized by Victor Hensen of Kiel (the 'father of quantitative plankton ecology' (Taylor 1980b)). The cruise, which used the vessel 'National', and followed a figure-of-eight track in the Atlantic Ocean, encompassed both boreal and tropical waters. Schütt's observations led to one of the earliest expositions on quantitative phytoplankton methodology (1892a), the first general text on marine phytoplankton ecology (*Das Pflanzenleben der Hochsee*, 1892b) and a monographic report on the dinoflagellates of the Plankton Expedition (1895). The latter, illustrated by more realistic drawings than any of his predecessors (and many to follow), is packed with unprecedented information on the cell contents of pelagic dinoflagellates, mostly derived from observing them in their living state while he was on board. Many of these features, such as the organization of the *pusules* (named by him) or extracellular extensions, could not be studied in fixed material and Schütt made the most of his opportunity, producing a work which is still the most detailed study on the total biology of tropical marine dinoflagellates. His term *amphiesma*, for the peripheral complex of the motile cell, was recently revived (Loeblich 1970) and seems to be particularly useful for the recognition of another feature which is highly characteristic of dinoflagellates, additional to the cellulose wall elements. Schütt's monograph also provides a chronological bibliography which lists the increasingly numerous minor papers which were appearing towards the end of the century and which cannot be detailed here.

Around the turn of the century Schütt (1900a, b) engaged in a lively dispute with George Karsten, who had also studied dinoflagellates, over whether dinoflagellate and diatom walls grew by intussusception (as in higher plants) or secondary apposition. He supported the latter, particularly because of his elegantly illustrated studies on *Ornithocercus*, where the lists grow out from the surface and later, when fully extended, become distally reticulated. Schütt believed that this occurred with additions of material by 'Aussenplasma' that extended, pseudopodia-like, through the thecal pores. He had seen such pseudopodia-like protrusions in *Podolampas* and other non-photosynthetic species. We now know that both observers were right, although Schütt didn't realize that the dinoflagellate theca is intracellular all the time, lying within amphiesmal vesicles, and thus can be added to fairly readily at any point (usually marginal). Taylor (1971a, 1973a) studied wall development in *Ornithocercus* further, in the first application of scanning electron microscopy to the group. Karsten described the phytoplankton collected by A. F. W. Schimper on the 'Valdivia' during the German Deep-Sea Expedition, including many figures of dinoflagellates, but these are rather stylized and lacking in detail (Karsten 1907). In this work he noted the presence of a 'Schattenflora' (shade flora) consisting largely of dinoflagellates.

Several careful taxonomic studies were published early in this century by the Norwegian botanist (and ex-schoolteacher) Eugen Jørgensen, who did most of his work at the Bergen Museum. His monograph on *Ceratium* (1911) is a classic and he later published several studies on Mediterranean forms (1920, 1923). His countryman, H. H. Gran, with whom he carried out more detailed work on diatoms, is much better known for his important publications on marine phytoplankton ecology. This is also true for the Swede, Per T. Cleve, although both described a few dinoflagellates in the course of their ecological studies in the late nineteenth and early twentieth centuries.

In 1912 Klebs produced another significant general biological work, this one on those forms, chiefly from freshwater, that live principally in a cyst-like, but photosynthetically active stage, e.g. *Cystodinium*. It contains a general discussion of cyst formation in dinoflagellates from a much broader viewpoint than that usually encountered.

The present terminology for gonyaulacoid and peridinioid plates and their short-hand, superscript notation (see Chapter 2 for details) was introduced by an American, Charles Atwood Kofoid, Professor of Zoology at Berkeley, California, in a paper on the tabulation of *Ceratium* (Kofoid 1907). His system was much more fully discussed in a paper on the theca of *Protoperidinium steinii* (Kofoid 1909) and he included details of the sulcal plates which had been unresolved in *Ceratium*. Kofoid's original observations were made chiefly on a cruise of the USS Albatross to Japan, with Alexander Agassiz, and at the Scripps Institution of Oceanography, which he helped found.

Kofoid became one of the major figures in dinoflagellate biology in the early twentieth century. His massive taxonomic works (e.g., on *Gonyaulax*, 1911), usually with co-authors (Kofoid & Swezy 1921; Kofoid & Skogsberg 1928; Kofoid & Adamson 1933) are renowned, not only for the exhaustive detail, but for the elegance of their illustrations (which is ironic because Kofoid was an indifferent artist and hired others to do his artwork). In the earlier phase of his work he was fortunate to have Josephine (Rigden) Michener as a collaborator, many of the original observations being made by her (Taylor 1971b, 1972).

Other taxonomic contributions of the early twentieth century were those by Marie V. Lebour (1925), mainly on waters around the British Isles, Erich Lindemann's revision of the section on dinoflagellates in Engler & Prantl's *Natürlichen Pflanzen-familien* (1928), a study on the Polish Baltic by Jadwiga Woloszynska (1929), and Josef Schiller's massive compilation of descriptions of virtually all the living dinoflagellate species in two parts of Rabenhorst's *Kryptogamen-Flora* (1933, 1937). The latter, although generally uncritical, is an indispensible aid to the taxonomy of the living species. Schiller lived in Vienna, but did most of his earlier observations in the Adriatic Sea. After the Second World War he worked on freshwater microflora. The Dane, Ove Paulsen, who earlier (1908) prepared the dinoflagellate section for *Nordisches Plankton*, wrote a critique of these taxonomic works. It was published posthumously in 1949 (edited by the diatomologist Jules Gröntved) and contains much valuable discussion of problems in dinoflagellate taxonomy that are still of current interest.

E. Zederbauer (1904) claimed to have made the first unequivocal observations on sexuality in dinoflagellates while he was studying *Ceratium hirundinella* and other phytoplankton in an alpine lake. In early morning samples he found pairs of cells joined by their ventral areas and he saw early zygotic cyst (zygospore) formation. However, many authors did not accept this as good evidence for sexuality (e.g. Paulsen 1949) because he did not describe nuclear events in detail. R. Lauterborn (1895) had previously provided a detailed description of mitosis in the same species. Further observations on *C. hirundinella* sexuality and the subsequent fate of the cysts were made by Huber & Nipkow (1922,

1923), who also provided details of the excystment process. Thanks to the seasonal varving of the Lake Zurich sediments they studied, they were able to determine that cysts of *Peridinium cinctum* could remain viable for 16 years or more. Von Stosch (1964) has summarized early observations on sexuality with particular reference to *Ceratium* (see Chapter 14 here).

A major contributor to the biology of freshwater dinoflagellates was E. Lemmermann (1910). He summarized his many observations in a contribution to the Flora of Mark Brandenburg. Geza Entz Jr. (1926) published a very detailed study of *Peridinium borgei.* Another important freshwater contribution was the monograph on *Peridinium* by M. Lefèvre (1928), which was not simply taxonomic, containing a great deal of general biological detail.

Lefèvre also had an interest in fossil dinoflagellates and later (1932, 1933) described bodies which he considered to be silicified thecae in Tertiary marine sediments from Barbados. He created the genus *Peridinites* for these. At almost the same time Georges Deflandre (1933), who became one of the most prolific authors on fossil dinoflagellates, found similar Tertiary fossils in New Zealand diatomaceous deposits. The study of fossils had been limited to minor observations (summarized by Sarjeant 1974) for many years, only commencing in earnest with the work of Walter Wetzel in the 1920s and in the 1930s by Deflandre, Alfred Eisenack and Otto Wetzel (no relative of Walter), much of this focused on Jurassic and Upper Cretaceous material.

Knowledge of the parasitic dinoflagellates was restricted to a few miscellaneous observations until their study was undertaken by the eminent protozoologist and parasitologist Édouard Chatton, Professor of the Sorbonne and director of the marine laboratories at Banyuls-sur-Mer, Villefranche-sur-Mer, and Sète. His considerable contributions were collected in a volume on parasitic dinoflagellates (Chatton 1920) and a summary of his work to 1937 (Chatton 1938). He also authored the chapter on dinoflagellates in Grassé's *Traité de Zoologie* (1952), although it only appeared posthumously, with revisions by Deflandre. André Lwoff (1948) wrote an obituary for him.

A significant work, which also appeared posthumuously in 1952, was that of Berthe Biecheler. She had been a student with Chatton at Sète in the 1930s. Here she used the dinoflagellates of the Etang de Thau, a rich lagoon next to the laboratory on the Mediterranean coast (systematically studied earlier by Pavillard) to aquire a great deal of information concerning many aspects of their biology, principally cytology, behaviour and nutrition. Tragically, she died of Hodgkin's disease in 1939 at the age of 38 , after publishing only a few short notes. Chatton undertook the preparation of her observations for publication, which finally appeared, under her name only, 13 years later and after his death. André Lwoff wrote a preface to this work, a mine of information whose value is only gradually being realized by non-French biologists (Chatton incorporated many of her observations in the Traité chapter).

Another biologist linked with the laboratory at Sète was Jules Pavillard,

Professor of the Institute of Botany at Montpelier, whose interests were more taxonomic. After a classic study on the Etang de Thau he worked chiefly on the open Mediterranean community. His major works consist of a study of the dinoflagellates of the Golfe du Lion (1909), one on material collected by the Prince of Monaco in the Atlantic (1931) and a seasonal, multi-depth study off Monaco (1937).

Much of the history of dinoflagellate study in this century has been a gradual, steady accumulation of taxonomic and distributional knowledge which will not be detailed here. Many of the references appear in later chapters of this volume. A few major contributions resulted from extensive oceanographic cruises, such as Nicolaus Peters' (1928) analysis of dinoflagellate distributions in the South Atlantic from 1925 to 1927, using material collected by the German research vessel 'Meteor', followed by the studies of E. Steeman Nielsen (1934, 1939a, b) on *Ceratium* distribution in the Pacific and Indian Oceans, using samples from the Danish 'Dana' Expedition collected from 1928 to 1930. The latter observer became much better known as a result of his development of the ^{14}C isotope method of measuring primary productivity, but his studies on *Ceratium* are still valuable sources of information (see Chapter 11), on vertical as well as horizontal distributions.

The material collected on the seventh cruise of the 'Carnegie' in 1929 served as the basis for the morphological and distributional studies of Herbert W. Graham on a variety of dinoflagellates (1942) and later, assisted by Natalia Bronikovsky, on *Ceratium* (1944). Graham's detailed and painstaking observations continued the pioneering studies of the sulcal plates begun by Li Sun Tai and Tage Skogsberg (1934) on dinophysoids and Tohru Abé (1936) on peridinoids. Only the more recent light microscopy of the Argentinian, Enrique Balech, rivals the fine detail of Graham's work, modern authors increasingly using scanning electron microscopy (SEM) to confirm these earlier observations. For example, Taylor (1973b, 1976a, b) supplemented light microscopy with SEM in a report on the dinoflagellates collected during the International Indian Ocean Expedition in 1963–64 by the American R. V. 'Anton Bruun', the most recent of the major expeditions to yield broad-scale distributional material.

The culture of dinoflagellates, a crucial requirement for physiological/autecological studies, lagged considerably behind that of several other protist groups, such as diatoms or ciliates. The first species to be cultured was the alga-associated heterotroph *Crypthecodinium cohnii* (although it was named *Gymnodinium fucorum* by its cultivator, E. Küster, 1908). He obtained it by letting pieces of the brown alga *Fucus* rot in sea water, a method still in use today. Other early successes with a few dinoflagellates, including one freshwater species, were recorded by Albert Barker (1935) of Hopkins Marine Station, who was the first to culture a variety of marine planktonic species, including *Prorocentrum*, *Heterocapsa*, *Dinophysis* and a species of *Ceratium*, and to conduct experiments on their requirements, optima and tolerances. For example, he found their temperature optima to be surprisingly high, considering the temperatures in which they are found, and that they preferred higher light intensities than some of the other organisms which he cultured. Their nitrate requirements seemed to be very low.

Shortly after this, the first dinoflagellate was linked with toxicity by Hermann Sommer and his colleagues at the Hooper Foundation in San Francisco, when they showed a correlation between the abundance of *Protogonyaulax* (*Gonyaulax*) *catenella* in the coastal plankton in 1933–35 and levels of paralytic shellfish poison (PSP) in mussels (Sommer *et al.* 1937). They were able to confirm this by feeding *P. catenella* to shellfish and eventually, by extracting the poison from the dinoflagellate. Since then, many other dinoflagellates have been found to produce toxins affecting both humans and marine fauna (see Chapter 8).

The 'modern period' of dinoflagellate study, the results of which are described in this book and therefore need little review here, essentially began in the 1950s. At this time the first electron microscopic observations were made on dinoflagellates: flagella (Pitelka & Schooley 1955), theca (Fott & Ludvik 1956; Braarud *et al.* 1958), and chromosomes (Grell & Wohlfarth-Bottermann 1957; Grassé & Dragesco 1957). The peculiar nature of the chromosomes, also evident in cytochemical studies (see Chapters 3 and 4 here) excited the interest of cell biologists in dinoflagellates, leading to much further study.

Cultures came into wider ecological use, at first chiefly in Norway by Trygve Braarud, his colleagues and students in Oslo (e.g. Braarud 1945; Braarud & Pappas 1951; Hasle & Nordli 1951), to determine tolerances and optima, and by Estella de Sousa e Silva in Portugal (e.g. Silva 1959) for cytological purposes. In his review on earlier physiological studies, most of which also began in the 1950s, Loeblich (1966) attributed this progress to the development of improved media by Luigi Provasoli, Michael Droop, Francis Haxo, Beatrice Sweeney and others, details of which he provided (see also Provasoli *et al.* 1957; McLachlan 1973; Loeblich 1975).

As a result of cultures, significant life-cycle connections were demonstrated. In 1962 H. D. Freudenthal cultured the 'zooxanthellae' from the jellyfish *Cassiopeia*, unequivocally establishing their dinoflagellate nature, particularly evident in their mastigote morphology and opening the way to their physiological study (see Chapter 12).

Cysts have been seen periodically during studies on natural populations, but most of these cysts have been of the ephemeral 'temporary' type, produced asexually under stress by ecdysis in species possessing a pellicular layer, e.g., several species of *Gonyaulax* (Schütt 1887, 1895; Taylor 1962, 1967a) or by release of a gelatinous material (Klebs 1883; Schütt 1895). Even the studies of resting cysts, such as the classic work of Huber & Nipkow (1922, 1923) on *Ceratium hirundinella* referred to earlier (see also Chapter 14), did not yield much information which appeared useful to those studying fossil cysts, since these cysts did not appear to fossilize*. Braarud (1945), Nordli (1951) and Braarud & Pappas (1951) observed the formation of spiny cysts in various cultures which provided clues to the interpretation of some of the fossil types, but there was little communication between those studying living forms with the palynologists studying fossils.

Sarjeant (1974) has described the rapid expansion in knowledge of fossil dinoflagellates after the work in the 1920s and 1930s by the Wetzels, Eisenack and Deflandre. As with the work on living forms, knowledge grew particularly rapidly from the 1950s onward, with the description of a great many organic-walled microfossils ('OWMs'), some of an obvious dinoflagellate nature, others apparently not, and a large number of spiny 'hystrichospheres' of unknown affinities. It was the careful, precise observation of William R. Evitt of Stanford University, which led to the view that many of the latter are also dinoflagellates, most probably cysts, based on detailed features of the excystment aperture (Evitt 1961, 1967) and the arrangement of the processes which suggested a positional correspondence with the plates of the theca. Cysts corresponding to known fossils were seen forming in thecae which had settled onto recent sediments by Martine Rossignol (1963) and Evitt & Davidson (1964). Finally David Wall (1966) and Wall & Dale (1967a, b) succeeded in hatching out motile cells of several species of *Gonyaulax* from cysts collected from sediment near Woods Hole, Massachusetts, the cysts being of the hystrichospherid type, thus confirming Evitt's contentions. Non-dinoflagellate OWMs were termed *acritarchs* and given an artificial classification by which various types—some of which may also be dinoflagellates but show insufficient signs to go by—are named and classified. More details on this important branch of dinoflagellate study are provided in Chapter 15.

* *Pseudoceratium pelliferum* Gocht, from the Lower Cretaceous, is very similar to the granular-walled cyst of *Ceratium hirundinella* recently described by Chapman *et al.* (1982); see also Wall & Evitt (1975).

4 REFERENCES

ABÉ, T. H. (1936) Notes on the protozoan fauna of Mutsu Bay. III. Subgenus *Protoperidinium*: Genus *Peridinum. Sci. Rep Tôhoku. Univ.*, Ser. 4 Biol. **11**, 19–48.

ALLMAN, G. J. (1855) Observations on *Aphanizomenon flosaquae* and a species of Peridinea. *Q.J. microsc. Sci.*, **3**, 21–25.

BAILEY, J. W. (1953) Notes on new species and localities of microscopical organisms. *Zool. Rept. Pacific Railroad Survey* (Williamson Exped.) 5(4), 97 pp, 26 pls also in *Smithsonian Soc. Trans.—Zool. Rept. Individ. Expeds.* (B) 10, (767) (1859), 4, 1–4, pp 97, pls 26. Washington.

BAKER, H. (1753) Of luminous water insects. In: *Employment for the Microscope*, pp 399–403, R. Dodsley, London.

BARKER, H. A. (1935) The culture and physiology of the marine dinoflagellates. *Arch. Mikrobiol.*, **6**, 157–181.

BERGH, R. S. (1881a) Bidrag til Cilioflagellaternes Naturhistoire. Dansk Vidensk. Medd. NatHistFor Kjob., Ser. 4, **3**, 60–76.

BERGH, R. S. (1881b) Der Organismus der Cilioflagellaten. Eine phylogenetische Studie. *Morph. Jb* **2** (1), 73–86, pl. 5.

BERGH, R. S. (1884) Neue Untersuchungen über Cilioflagellaten. *Kosmos*, **1**, 384–390.

BIECHELER, B. (1952) Récherches sur les Péridiniens. *Bull. biol.* Suppl. 36, 1–149.

BORY DE ST VINCENT, J. B. (1824) Encyclopédie méthodique, ou par ordre de matière, par une societé de gens de lettres. Histoire naturalle des Vers, Coquilles, Mollusques et Zoophytes par Bruguière, Lamouroux, De Lamarck, P. G. Deshayes et Eud. Deslongchamps. Vols 4 & 5 (488 pls) (1791–1832). Paris.

BRAARUD, T. (1945) Morphological observations on marine dinoflagellate cultures (*Porella perforata, Gonyaulax tamarensis, Protoceratium reticulatum*). *Avh. Norske VidenskAkad. Oslo, Math.-Naturv. Kl., 1944*, **11**, 1–18, pls 1–4.

BRAARUD, T. (1956) Haaken Hasberg Gran (1870–1955). *J. Cons. int. Explor. Mer.* **21**(2), 122–124.

BRAARUD, T., MARKALI J. & NORDLI E. (1958) A note on the thecal structure of *Exuviaella baltica* Lohm. *Nytt Mag. Bot.*, **6**, 43–45.

BRAARUD, T. & PAPPAS, I. (1951) Experimental studies on the dinoflagellate *Peridinium triquetrum* (Ehrb.) Lebour. *Avh. Norske Vidensk Akad i Oslo I. Mat: Naturv.* Kl., 1951, **2**, 1–23.

BRANDT, K. A. H. (1881) Ueber das Zusammenleben von Thieren und Algen. *Verh. physiol. Ges. Berl.*, 570–574.

BÜTSCHLI, O. (1873) Einiges über Infusorien. *Arch. mikr. Anat.*, **9**, 677–678.

BÜTSCHLI, O. (1889) Protozoa, In: *Klass. U. Ordn. des Thierreichs* 1 (Ed. H. G. Bronn), pp. 1089–2032, pls 56–79. Leipzig & Heidelberg.

CARTER, H. J. (1858) Note on the red colouring matter of the sea round the shores of the island of Bombay. *Ann. nat. Hist.* Ser. 3, **1**, 258–262.

CHAPMAN, D. V., DODGE, J. D. & HEANEY, S. I. (1982) Cyst formation in the freshwater dinoflagellate *Ceratium hirundinella*. *J. Phycol.*, **18**, 121–129.

CHATTON, E. (1920) Les Péridiniens Parasites. Morphologie, reproduction, éthologie. *Arch. Zool. exp. Gen.*, **59**, 1–475, pls 1–18.

CHATTON, E. (1938) *Titres et Travaux Scientifiques* (1906–1937) pp. 1–405. Sottano, Sete.

CIENKOWSKI [TSENKOVSKY], L. (1881) Otchet' 'o byelomorskoy ekshursii 1880 (Bericht über Exkursionen ins weisse Meer) *Trudy Sankt-Petersburgskago Obschchestra Estestroispytateley*, **12** (1), 130–171, pls 1–3.

CLAPARÈDE, E. & LACHMANN, J. (1859) Études sur les Infusoires et les Rhizopodes. Ordre III. Infusoires Cilioflagellés. *Mem. Inst. Genevois* **6**, mém. **1**, 392–412, pls 19, 20.

CLAPARÈDE, E. & LACHMAN, J. (1861) Études sur les Infusoires et les Rhizopodes. *Mem. Inst. Genevois*, **7**, 1–291.

COX, E. R. (Ed.) (1980) *Phytoflagellates*, Developments in Marine Biology 2, pp. 1–473. Elsevier/North Holland, New York, Amsterdam & Oxford.

DEFLANDRE, G. (1933) Note préliminaire sur un Péridinien fossile *Lithoperidinium oamaruense* n.g. n.s.p. *Bull. Soc. zool. Fr.*, **58**, 265–273.

DIESING, K. M. (1850) *Systema Helminthum*, 1, pp. 679. C. Gerold's Sohn, Vienna.

DIESING, K. M. (1866) Revision der Prothelminthen. *Sitzber. math. nat. Klasse Kgl. Akas. Wiss. Wien*, **52**, 287–401.

DODGE, J. D. (1981) Three new generic names in the Dinophyceae: *Herdmania*, *Sclerodinium* and *Triadinium* to replace *Heteraulacus* and *Goniodoma*. *Br. phycol. J.*, **16** (3), 273–280.

DUJARDIN, F. (1841) Histoire naturelle des Zoophytes Infusoires. *Suites A Buffon*. pp. 684, pls 22. Roret, Paris.

EHRENBERG, C. G. (1832, separate 1830) Beiträge zur Kenntniss der Organisation der Infusorien und ihrer geographischen Verbreitung, besonders in Sibirien. *Abh. preuss. Akad. Wiss.* 1830, 1–88, pls 1–8.

EHRENBERG, C. G. (1834) Das Leuchten des Meeres. Neue Beobachtungen nebst Uebersicht der Hauptmomente der geschichtlichen. Entwicklung dieses merkwürdigen Phänomens.

Abh. k. Akad. Wiss. Berlin, 1834, 411–575, pls 1, 2.
EHRENBERG, C. G. (1838, separate 1837) Über das Massenverhaltniss der jetzt lebenden Kiesel-Infusorien und über ein neues Infusorien-Conglomerat als Polirschiefer von Jastraba in Ungarn. *Abh. preuss. Akad. Wiss.*, *1836*, 109–135, pl 1–2.
EHRENBERG, C. G. (1840) Über jetzt wirklich noch zahlreich lebende Thier-Arten der Kreideformation der Erde. *Verh. preuss. Akad. Wiss. Berlin* 1839, 152–159.
EHRENBERG, C. G. (1854) *Mikrogeologie das Erden und Felsen schaffende Wirken des unsichtbaren Kleinen selbstandigen Lebens auf der Erde.* pp. 374, pls 1–40. Leopold Voss, Leipzig.
ENTZ, G., Jun. (1926) Beiträge zur Kenntniss der Peridineen. I. Zur Morphologie und Biologie von *Peridinium Borgei* Lemmermann. *Arch. Protistenk.*, **56**, 397–446, 1 pl.
EVITT, W. R. (1961) Observations on the morphology of fossil dinoflagellates. *Micropal.*, **7**, 385–420, pls 1–9.
EVITT, W. R. (1967) Dinoflagellate studies II. The archeopyle. *Stanford Univ. Publs. Geol. Sci.*, **10** (3), 1–83, pl 1. 1–11.
EVITT, W. R. & DAVIDSON, S. E. (1964) Dinoflagellate studies I. Dinoflagellate cysts and thecae. *Stanford Univ. Publs. Geol. Sci.*, **10** (1), 1–12, pl 1.
FOTT, B. & LUDVIK, L. (1956) Uber den submikroskopischen Bau de Panzers von *Ceratium hirundinella. Preslia*, **28**, 278–280.
FREUDENTHAL, H. D. (1962) *Symbiodinium* gen. nov. and *Symbodinium microadriaticum* sp. nov., a zooxanthella: Taxonomy, life cycle and morphology. *J. Protozool.*, **9** (1), 45–52.
GEDDES, P. (1882) On the nature and functions of the 'yellow cells' of radiolarians and coelenterates. *Proc. R. Soc. Edinb.*, **II**, 377–396.
GOURRET, P. (1883) Sur les Péridiniens du Golfe de Marseille. Ann. Mus. Hist. nat. Marseilles, Zool. 1, Mem. 8, 1–114, pls 1–4.
GRAHAM, H. W. (1942) Studies in the morphology, taxonomy, and ecology of the Peridiniales. *Scient. Results Cruise VII Carnegie, Biol. Ser.*, **3**, 1–129.
GRAHAM, H. W. & BRONIKOVSKY, N. (1944) The genus *Ceratium* in the Pacific and North Atlantic Oceans. *Scient. Results Cruise VII Carnegie, Biol. Ser.* **5**, 1–209.
GRASSÉ, P.-P. (1952) *Traité de Zoologie. Anatomie, systématique, biologie*, **1**, pp. 1071, pl. 1. Masson & Cie, Paris.
GRASSÉ, P.-P. & DRAGESCO, J. (1957) L'ultrastructure du chromosome des Péridiniens et ses conséquences génétiques. C.R. Acad. sci. Paris, **245**, 2447–2452.
GRELL, K. G. & WOHLFARTH-BOTTERMANN, K. E. (1957) Licht und elektronenmikroskopische Untersuchungen an den Dinoflagellaten *Amphidinium elegans* n. sp. *Z. Zellforsch. mikrosk. Anat.*, **47**, 7–17.
HASLE, G. R. & NORDLI, E. (1951) Form variation in *Ceratium fusus* and *tripos* populations in cultures and from the sea. *Avh. norske VidenskAkad. Oslo* 1. *Mat-Naturv. Klasse.*, **4**, 1–25.
HERTWIG, R. (1884) *Erythropsis agilis:* Eine neue Protozoa. *Morph. Jahrb.*, **10**, 204–212, pl. 6.
HUBER, G. & NIPKOW, F. (1922) Experimentelle Untersuchungen über die Entwicklung von *Ceratium hirundinella* O.F.M. *Z. Bot.*, **14**, 337–371.
HUBER, G. & NIPKOW, F. (1923) Experimentelle Untersuchungen über Entwicklung und Formbildung von *Ceratium hirundinella. Flora, Jena*, **116**, 114–215.
JØRGENSEN, E. (1911) Die Ceratien. Eine Kurze Monographie der Gattung *Ceratium* Schrank. *Int. Rev. Hydrobiol.*, **4** (Biol. Supple. Heft 1): 1–124, pls 1–10.
JØRGENSEN, E. (1920) Mediterranean Ceratia. *Rep. dan. oceanogr. Exped. Medit.* **1** (Biol. J. 1), 1–110.
JØRGENSEN, E. (1923) Mediterranean Dinophysiaceae. *Rep. dan. oceanogr. Exped. Medit.*, **2**, (Biol. J. 2), 1–48.
KARSTEN, G. (1907) Das indische Phytoplankton. *Wiss. Ergeb. deutsch. Tiefsee Exped.*, **2**, 221–548, pls 34–54.
KAWAGUTI, S. (1944) On the physiology of reef corals. VII. Zooxanthella of the reef corals is

Gymnodinium sp., Dinoflagellata; its culture *in vitro*. Palao Trop. biol. Stud., **2** (4), 675–679.

KLEBS, G. (1883) Ueber die Organisation einiger Flagellaten gruppen und ihre Beziehungen zu Algen und Infusoria. *Untersuch bot. Inst. Tübingen*, **1**, 233–362, pls 2, 3.

KLEBS, G. (1912) Über Flagellaten- und Algen-ahnliche Peridineen. *Verh. Naturh.-med. Ver. Heidelb.*, **11** (4), 369–451, pl. 10.

KOFOID, C. A. (1907) The plates of *Ceratium* with a note on the genus. *Zool. Anz.*, **32**, 177–183.

KOFOID, C. A. (1909) On *Peridinium steini* Jorgensen, with a note on the nomenclature of the skeleton of the peridinidae. *Arch. Protistenk.*, **16**, 25–47, pl. 2.

KOFOID, C. A. (1911) Dinoflagellata of the San Diego region—IV. The genus *Gonyaulax*, with notes on its skeletal morphology and a discussion of its generic and specific characters. *Univ. Calif. Publ. Zool.*, **8** (4), 187–286, pls 9–17.

KOFOID, C. A. & ADAMSON, A. M. (1933) The dinoflagellata: The family Heterodiniidae of the Peridinioidae. *Mem. Mus. comp. Zool. Harv*, **54**, 1–136, pls 1–22.

KOFOID, C. A. & SKOGSBERG, T. (1928) The Dinoflagellata: The Dinophysoidae. *Mem. Mus. comp. Zool. Harv.*, **51**, 1–766, pls 1–31.

KOFOID, C. A. & SWEZY, O. (1921) The free living unarmoured Dinoflagellata. *Mem. Univ. Calif.*, **5**, 1–562, pls 12.

KÜSTER, E. (1908) Eiene Kultivierbar Peridineen. *Arch. Protistenk.*, **II**, 351–362.

LAUTERBORN, R. (1895) Kern- und Zelltheilung von *Ceratium hirundinella*, O.F.M. *Zeitschr. wiss. Zool.*, **59**, 167–190, pls 12, 13.

LEBOUR, M. V. (1925) *The Dinoflagellates of Northern Seas*, pp. 250, pls 35. Mar. Biol. Ass., UK, Plymouth.

LEFÈVRE, M. (1928) Monographie des espèces d'eau douce du genre Peridinium. *Archs. Bot. Mém.*, **2**, Mem. 5, 1–210, pls 1–6.

LEFÈVRE, M. (1932) Sur la présence de Péridiniens dans un dépot fossile des Barbades. *C.R. hebd. Séanc. Acad. Sci., Paris*, **194**, 2315–2316.

LEFÈVRE, M. (1933) Les *Peridinites* des Barbades. Annls Cryptog. exot., **6**, 215–229.

LEMMERMANN, E. (1910) Klasse Peridiniales. In: *Kryptogamenflora der Mark Brandenburg und angrenzender Gebiete, III Algen I* (Schizophyceen, Flagellaten, Peridineen), (Ed. E. Lemmermann), pp. 563–682. Gebrüder Borntraeger, Leipzig.

LINDEMANN, E. (1928) Abteilung Peridineae (Dinoflagellatae). In: *Die Natürlichen Pflanzenfamilien nebst ihren Gattungen und wichtigeren Arten insbesondere den Nutzpflanzen*, 2. (Ed. A. Engler & K. Prantl), pp. 3–104, Wilhelm Englemann, Leipzig.

LOEBLICH, A. R. III (1966) Aspects of the physiology and biochemistry of the Pyrrhophyta. *Phykos*, **5** (182), 216–255.

LOEBLICH, A. R. III (1970) The amphiesma or dinoflagellate cell covering. *Proc. north American Paleont. Convention, Chicago, September 1969*, **2**, Part G. pp. 867–929. Allen Press, Lawrence, Kansas.

LOEBLICH, A. R. III (1975) A seawater medium for dinoflagellates and the nutrition of *Cachonina niei*. *J. Phycol.*, **2** (1), 80–86.

LWOFF, A. (1948) La vie et l'oeuvre d'Edouard Chatton. *Arch. Zool. exp. Gen.*, **85**, 121–137.

MANTELL, G. A. (1850) *A pictorial atlas of fossil remains, consisting of coloured illustrations selected from Parkinson's 'Organic remains of a Former World', and Artist's 'Antediluvian Phytology.'* pp. 208, pls 74. Henry G. Bohn, London.

MCLACHLAN, J. (1973) Growth media—marine. In: *Handbook of Phycological Methods* (Ed. J. R. Stein), pp. 25–51. Cambridge University Press, Cambridge.

MICHAELIS, G. A. (1830) *Ueber das Leuchten der Ostsee nach eigenen Beobachtungen*. pp. 52 pls 2. Perthes & Besser, Hamburg.

MÜLLER, J. (1855) Ueber *Sphaerzoum* und *Thalassicolla*. *Monatsber. Berl. Akad.*, 229–253.

MÜLLER, O. F. (1773) *Vermium terrestrium et fluviatilum seu animalium infusoriorum, helminthicorum et testaceorum, non marinorum secincta historia*. 1 (1) pp. 135 + [32 unnumbered pages]. Martini Hallager, Hauniae et Lipsiae.

MÜLLER, O. F. (1786) *Animalcula infusoria fluviatilia et marina, que detexit, systematice descripsit et ad vivum delineari curavit Otho Fridericus Müller, regi daniae quondam a consiliis conferentaie, plurium que academiarum et societatum scientiarum sodalis, sistit opus hoc posthumam quod cum tabulis aeneis L. in lucem tradit vidua ejus nobilissima, cur Othonis Fabricii.* pp. [56] + 367, pls 50. Nicolai Molleri, Hauniae.

MURRAY, J. (1876) Preliminary reports to Professor Wyville Thomson, F.R.S., Director of the civilian scientific staff, on work done on board the 'Challenger'. *Proc. R. Soc.*, **24** (170), 471–544, pls 20–24.

NITZSCH, C. L. (1817) Beitrage zur InfusorienKunde oder Naturbeschreibung der Zerkarien und Bazillarien. *Neueu Schriftr. naturf. ges. Halle*, **3** (1) pls 6.

NORDLI, E. (1951) Resting spores in *Goniaulax* [sic] *polyedra* Stein. *Nytt. Mag Naturvid.*, **88**, 207–212.

PAULSEN, O. (1908) Peridiniales. In: *Nordisches Plankton, Botanischer Teil* (Ed. by K. Brandt and C. Apstein), XVIII, pp. 1–124. Lipsius & Tischer, Kiel & Leipzig.

PAULSEN, O. (1949) Observation on dinoflagellates. *Biol. Skr.*, **6** (4), 1–67.

PAVILLARD, J. (1909) Sur les Péridiniens du Golfe du Lion (22 Note). *Bull. Soc. bot. France*, **54**, 225–231.

PAVILLARD, J. (1931) Phytoplancton (Diatomées, Péridiniens) provenant des campaignes scientifiques du Prince Albert 1er de Monaco. *Résult. Camp. sci. Prince Albert*, **1**, 82, 1–208.

PAVILLARD, J. (1937) Les Péridiniens et Diatomées pélagiques de la mer de Monaco pendant les années 1912, 1913 et 1914. *Bull. Inst. océanogr. Monaco*, **738**, 1–56.

PERTY, M. (1852) *Zur Kenntniss Kleinster Lebensformen nach Bau, Funktionen, Systematik, mit Specialverzeichniss der in der Schweiz beobachteten.* pp. 228, pls 17. Jent & Reinert, Bern.

PETERS, N. (1928) Die Perideenbevolkerung der Weddellsee mit besonderer Berücksichtigung der Wachstums- und Variationsformen. *Int. Rev. Hydrobiol. Hydrogr.*, **21**, 17–146.

PITELKA, D. R. & SCHOOLEY, C. N. (1955) Comparative morphology of some protistan flagella. *Univ. Calif. Publ. Zool.* 61 (2), 79–128, pls 15–26.

POUCHET, G. (1882) Sur l'évolution des Péridiniens et les particularites d'organisation que les rapprochent des Noctiluques. *C.R. Acad. sci. Paris*, **95**, 794–796.

POUCHET, G. (1883) Contribution a l'histoire des Cilioflagellées. *J. Anat. Physiol.*, **19**, 399–455, pls 18–22.

POUCHET, G. (1884) Sur en Péridinien parasite. *C.R. Acad. sci. Paris*, **98**, 1345–1346.

POUCHET, G. (1885a) Nouvelle contribution à l'histoire des Péridiniens marins. *J. Anat. Physiol.*, **21**, 28–88, pls 2–4.

POUCHET, G. (1885b) Troisième contribution à l'histoire des Péridiniens. *J. Anat. Physiol.*, **21**, 525–534, pl. 26.

POUCHET, G. (1887) Quatrième contribution à l'histoire des Péridiniens. *J. Anat. Physiol.*, **23**, 87–112, pls 9, 10.

POUCHET, G. (1892) Cinquième contribution a l'histoire des Péridiniens. *J. Anat. Physiol.*, **28**, 143–150, p. 11.

POUCHET, G. (1893) Sur le polymorphisme du *Peridinium acuminatum* Ehr., *C.R. Acad. Sci. Paris*, **117**, 703–705.

POUCHET, G. (1894) Histoire Naturelle. In: *Voyage de 'La Manche' a l'ile Jan-Mayen et au Spitzberg (Jullet-Aout, 1892). Nouv. Arch. Missions Sci. Lit.* **5**, 154–217, pl. 22.

PRITCHARD, A. (1861) *A history of Infusoria, including the Desmidiaceae and Diatomaceae.* 4th edn., pp. 968, pls 40. Whittaker & Co., London.

PROVASOLI, L., MCLAUGHLIN, J. J. A. & DROOP, M. R. (1957) The development of artificial media for marine algae. *Arch. Mikrobiol.*, **25**, 392–428.

ROSSIGNOL, M. Apercus sur le développement des Hystrichosphères *Bull. Mus. nat. Hist. nat. Paris* sér 2, 35(2), 207–212, pls 1, 2.

RUSSELL, F. S. (1972) Obituary. Dr. Marie V. Lebour. *J. mar. Biol. Ass. UK*, **52**, 777–778.

SARJEANT, W. A. S. (1974) *Fossil and Living Dinoflagellates*, 182 pp. Academic Press, London.

SAVILLE-KENT, W. (1880–82) *A Manual of the Infusoria*. 3 vols. **1**, 1–472; **2**, 473–913; **3**, pls 1–50.

SCHILLER, J. (1933, 1937) Dinoflagellatae (Peridineae) in monographischer Behandlung. In: *Kryptogamen-Flora von Deutschland, Osterreich und der Schweiz* (Ed. L. Rabenhorst), 10 [3], (1) pp. 1–617, (2) pp. 1–590, Akad. Verlages, Leipzig.

SCHILLING, A. J. (1891) Die süsswasser Peridineen. *Flora alg. Bot. Zeitschr.* **74**, 220–299, pls 8–10.

SCHILLING, A. J. (1913) Dinoflagellatae (Peridineae) In: *Die Süsswasser-flora Deutschlands, Osterreichs und der Schweiz* (Ed. A. Pascher) 3, pp. 66, Gustav Fischer, Jena.

SCHMARDA, L. K. (1854) Zur Naturgeschichte Aegyptens. *Denkschr. Wiener Akad.* **7**, 1–28, pls 7.

SCHRANK, F. von Paula (1793) Microscopische Wahrnehmungen. *Der Naturforscher*, **27**, 26–37, pl. 3.

SCHRANK, F. von Paula (1802) *Briefe naturhistorischen, physikalischen und okonomischen Inhaltes an Herrn.* pp. 374–376. B.S. NAU, Erlangen.

SCHÜTT, F. (1887) Ueber die Sporenbilding mariner Peridineen. *Ber. dt. bot. Ges.*, **5**, 364–374, pl. 18.

SCHÜTT, F. (1890) Ueber Peridineenfarbstoffe. *Ber. dt. bot. Ges.*, **8**., 9–32, pls 1, 2.

SCHÜTT, F. (1891) Sulla formazione scheletrice intracellulare di un dinoflagellato. *Neptunia*, **1** (10), 407–426, pl. 3.

SCHÜTT, F. (1892a) *Analytische Plankton-Studien. Ziele, Methoden und Anfangs-Resultate der quantitativ-analytischen Plankton Forschung.* pp. 117, Lipsius & Tischer, Kiel & Leipzig. [also published as part of vol. 1 of Neptunia.].

SCHÜTT, F. (1892b) Das Pflanzenleben der Hochsee. In: *Reisebeschreibung der Plankton-Expedition* (Ed. O. Krümmel), pp. 243–314. Lipsius & Tischer, Kiel & Leipzig.

SCHÜTT, F. (1895) Peridineen der Plankton-Expedition. *Ergebn. Plankton-Expedition der Humboldt-Stiftung*, **4**, M, a, A, 1–170, pls 27.

SCHÜTT, F. (1900a) Centrifugale und simultane Membranverdickungen. *Jahr. wiss. Bot.*, **35** (3), 467–534, pl. 1.

SCHÜTT, F. (1900b) Die Erklärung des centrifugalen Dickenwachsthums der membran. *Bot. Z.* (1900), **16/17**, 1–30.

SILVA, E. de S. (1959) Some observations on marine dinoflagellate cultures. I. *Prorocentrum micans* Ehr. and *Gyrodinium* sp. *Notas Est. Inst. Bid. mar.*, **21**, 239–254.

SLABBER, M. (1778) *Natuurkundige Verlustigengen, behelzende microscopishe Waarneemingen van In- en Uitlandse, Water- en Land-Dieren* (Haarlem, Basch), pp. 166, pls 18. 1st edn. dated 1771.

SOMMER, H. WHEDON, W. F., KOFOID, C. A. & STOHLER, R. (1937) Relation of paralytic shellfish poison to certain plankton organisms of the genus *Gonyaulax*. *Arch. Path.*, **24** (5), 537–559.

STEEMANN NIELSEN, E. (1934) Untersuchungen über die Verbtreitung, Biologie und Variation der Ceratien im Südlichen Stillen Ozean. *Dana Rep.*, **4**, 1–64.

STEEMANN NIELSEN, E. (1939a) Die Ceratien des Indischen Ozeans und der Ostasiatischen Gewasser mit einer allgemeinen Zusammenfassung über die Verbreitung des Ceratien in den Weltmeeren. *Dana Rep.*, **17**, 1–33.

STEEMANN NIELSEN, E. (1939b) Über die vertikale Verveitung der Phytoplanktonen im Meere. *Int. Rev. ges. Hydrobiol. Hydrogr.*, **38**, 421–440.

STEIDINGER, K. A. & COX, E. R. (1980) Free-Living Dinoflagellates. In: *Phytoflagellates* Developments in Marine Biology 2 (Ed. E. R. Cox), pp. 407–432. Elsevier/North Holland, New York.

STEIN, F. Ritter von (1878) *Der Organismus der Infusionsthiere nach eigenen Forschungen in systematischer Reihenfolge bearbeitet.* Abteilung III. Die Naturgeschichte der Flagellaten oder Geisselinfusorien. I Hälfte. Den noch nicht abgeschlossenen allgemeinen Teil nebst

Erklärung der sammtlichen Abbildungen enhaltend. pp. 154, pls 22. Wilhelm Engelmann, Leipzig.

STEIN, F. Ritter von (1883) *Der Organismus der Infusionsthiere nach Eigenen Forschungen III.* Abteilung II Hälfte. Die Naturgeschichte der Arthrodelen Flagellaten. pp. 1–30, pls 25. Wilhelm Engelmann, Leipzig.

STOSCH, H. A. von (1964) Zum problem der sexuellen Fortpflanzung in der Peridineengattung *Ceratium.*—*Wiss. Meersunters*, **10**, 140–152.

TAI, L.-S. & SKOGSBERG, T. (1934) Studies on the Dinophysoidae, marine armored dinoflagellates, of Monterey Bay, California. *Arch. Protistenk.*, **82**, 380–482, pls 11–12.

TAPPAN, H. (1980) *The Paleobiology of Plant Protists*, pp. 1028. W. H. Freeman & Co., San Francisco.

TAYLOR, F. J. R. (1962) *Gonyaulax polygramma* Stein in Cape waters: a taxonomic problem related to developmental morphology. *J. S. Afr. Bot.*, **28**, 237–242.

TAYLOR, F. J. R. (1971a) Scanning electron microscopy of thecae of the dinoflagellate genus *Ornithocercus. J. Phycol.*, **7**, 249–258.

TAYLOR, F. J. R. (1971b) Notes on some 'Kofoidiana'. *Phycologia*, **10** (1), 143–144

TAYLOR, F. J. R. (1972) Unpublished observations on the thecate stage of the dinoflagellate genus *Pyrocystis* by the late C. A. Kofoid and Josephine Michener. *Phycologia*, **II**, 47–55.

TAYLOR, F. J. R. (1973a) Topography of cell division in the structurally complex dinoflagellate genus *Ornithocercus.* J. Phycol., **9**, 1–10.

TAYLOR, F. J. R. (1973b) General features of dinoflagellate material collected by the 'Anton Bruun' during the International Indian Ocean Expedition. In: *The Biology of the Indian Ocean.* [Ecological Studies 3] (Ed. B. Zeitschel), pp. 155–169, pls 1–3. Springer-Verlag, Berlin.

TAYLOR, F. J. R. (1976a) Dinoflagellates from the International Indian Ocean Expedition. *Bibl. Bot.*, **132**, 1–234, pls 46.

TAYLOR, F. J. R. (1976b) Flagellate phylogeny: a study of conflicts. *J. Protozool.*, **23**, 28–40.

TAYLOR, F. J. R. (1978) Problems in the development of an explicit phylogeny of the lower eukaryotes. *BioSystems*, **10**, 67–89.

TAYLOR, F. J. R. (1980a) On dinoflagellate evolution. *BioSystems*, **13**, 65–108.

TAYLOR, F. J. R. (1980b) Phytoplankton ecology before 1900: supplementary notes to the 'Depths of the Ocean.' In: *Oceanography—the Past* (Ed. M. Sears & D. Merriman), pp. 509–521. Springer-Verlag, New York.

TAYLOR, F. J. R. (1980c) The stimulation of cell research by endosymbiotic hypotheses for the origin of eukaryotes. In: *Endocytobiology, Endosymbioses and Cell Biology* (Eds W. Schwemmler & H. E. A. Schenk), Vol. 1, pp. 917–942. Walter de Gruyer & Co. Berlin.

WALL, D. (1966 separate 1965) Modern Hystrichospheres and dinoflagellate cysts from the Woods Hole region. *Grana Palynol*, **6** (2), 297–314.

WALL, D. & DALE, B. (1967a) 'Living Fossils' in Western Atlantic plankton. *Nature* (*Lond.*), **211**, 1025–1026.

WALL, D. & DALE, B. (1967b) The resting cysts of modern marine dinoflagellates and their palaeontological significance. *Rev. Palaeobot. Palynol.*, **2**, 349–354, pl. 1.

WALL, D. & EVITT, W. D. (1975) A comparison of the modern genus *Ceratium* Shrank, 1793, with certain Cretaceous marine dinoflagellates. *Micropal.*, **21**, 14–44.

WARMING, E. (1875) Om nogle ved Danmarks Hyster levende Bacterier. *Vidensk. naturhist. Foren. Kjobenhan, 1875*, 1–414.

WOLOSZYNSKA, J. (1929) Dinoflagellatae polskiego bałticyu i błot nad piaŝnica. (Dinoflagellatae der Polnischen Ostsee sowie der an Piasnica gelegenen Sümpfe.) III. *Archiv. Hydrob. Ichtyologie*, **3** (3–4), 153–278, pl. 15.

ZEDERBAUER, E. (1904) Geschlectliche und ungeschlechtliche Fortplanzung von *Ceratium hirundinella. Ber. dt. bot. Ges.*, **22**, 1–8, pl. 1.

CHAPTER 2
DINOFLAGELLATE MORPHOLOGY

F. J. R. TAYLOR
Departments of Oceanography and Botany, University of British Columbia, Vancouver, BC, Canada V6T 1W5

1 CELL FORM

Within the dinoflagellates there has been a considerable diversification of form (see Frontispiece), including highly bizarre and elaborate modifications which represent some of the most sophisticated examples of differentiation among the protists (Taylor 1980). These can be considered to be modifications of a basic form (Fig. 2.1).

1.1 Basic form

The basic form of the swimming cell, termed the *mastigote*, is ovoid to roundly pyriform (pear-shaped), the anterior end usually more pointed than the posterior, the posterior often indented and lobed to some degree. The paired flagella arise from the side, one beating sideways around the cell, the other beating backwards. This flagella arrangement is known as the *dikokont* condition (Fig. 2.3).

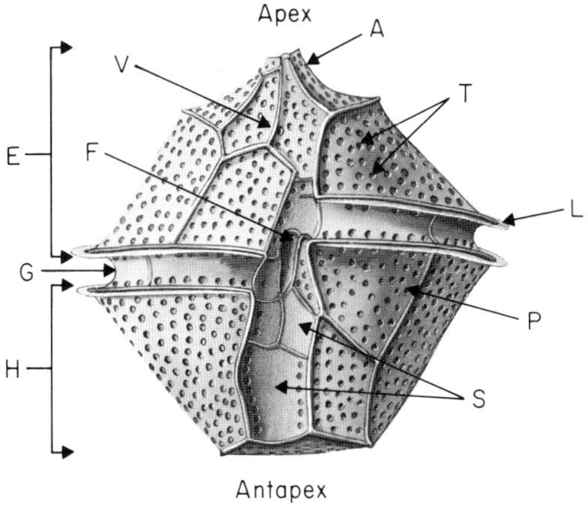

Fig. 2.1. Ventral view of the theca of *Gonyaulax polyedra* (× 1000). From Kofoid (1911), with permission (labelling additional). A, apical horn; E, epitheca; F, flagellar pore; G, girdle (cingulum); H, hypotheca; L, girdle list; P, plate suture; S, sulcal plates; T, trichocyst pores; V, ventral pore (some gonyaulacoids).

In a few dinoflagellates both flagella arise from the anterior end. This is the *desmokont* condition (Fig. 2.2). In desmokonts, e.g. *Prorocentrum*, the cells are ovoid, often flagellated, with no surface grooves, and may be broadest at the anterior or posterior end.

In a few dinoflagellates, such as *Oxyrrhis* or *Torodinium*, the flagella both arise posteriorly and are *opisthokont* (Fig. 2.4).

In dinokonts the flagella beat in surface furrows or grooves. The *transverse flagellum*, which beats latitudinally around the cell lies within the *girdle* groove (or *cingulum*). The proximal part of the *longitudinal flagellum* lies in the *sulcus*, the distal part projecting posteriorly beyond the cell, sometimes for a considerable distance. The girdle runs circumferentially around the cell, the sulcus only indenting the posterior surface of one side.

The girdle essentially divides the cell into an anterior part, the *episome* and a posterior part, the *hyposome* ('soma' = body) (Fig. 2.3). These terms of Kofoid (1909a, b) have been replaced by 'epicone' and 'hypocone' by some authors (e.g. Lebour 1925). If a theca is present the equivalent terms are *epitheca* and *hypotheca* respectively (Fig. 2.1). An anterior extension of the sulcal groove onto the episome, sometimes not continuous with the sulcus, is often present on athecate forms. It was termed the *acrobase* by Biecheler (1934a, b; see Fig. 2.5 here).

In the mastigote stage, if cell wall plates are present, they are invariably intracellular, lying within a row of vesicles, the *amphiesmal vesicles*. The outer surface is covered by the cell membrane, although this is not apparent under light microscopy and preparation of the wall for scanning electron microscopy often results in its loss.

The cell surface may be smooth, or raised in a variety of bumps, ridges, reticulations or processes, detailed below (Section 3.3). Delicate organic scales have been observed on the surface of species of *Oxyrrhis* and *Heterocapsa*. None were observed in a search of eleven other genera (Morrill & Loeblich 1981a).

1.2 Amphiesmal reticulum (argyrome)

With light or scanning electron microscopy the amphiesmal vesicles are not usually visible unless they contain thecal plates. However, when Biecheler (1934a, 1952) impregnated cells with colloidal silver (the Chatton–Lwoff technique) she observed a reticulated peripheral pattern which she termed the 'argyrome'. It seems to be a visible manifestation of the amphiesmal vesicles and trichocyst membranes (Fig. 2.5). In *Gyrodinium pavillardii* the amphiesmal vesicles are unusually rectangular, the girdle containing a higher density of regularly shaped vesicles resembling a brick road. The *acrobase*, looping up anticlockwise over the anterior end is clearly visible (and was named from such preparations by her). Apart from the latter, her other terminology for regions of reticulation do not seem particularly useful since they are simply the vesiculated regions of the episome (acromere, prosomere), the girdle (mesomere) and the hyposome (opisthomere). The exit pores of the trichocysts appear as small circles at the reticular (vesicle) junctions.

Other techniques can also reveal the reticulum. Bursa (1958) illustrated the pattern for *Woloszynskia limnetica* using iodine-stained material. The vesicular pattern can also be seen in some scanning electron microscopy preparations (e.g. Loeblich & Sherley 1979). Netzel & Dürr (1984) have illustrated a variety of vesicular patterns, which they term the *corticotype*.

1.3 Orientation

Orientation, relative to swimming direction, plays an important part in the descriptive terminology of dinoflagellates.

In desmokonts the pole directed towards the swimming direction is *anterior* (rarely apical), the opposite pole is *posterior*, and the broad surfaces of the cell are *lateral*. On close examination there is a difference between the lateral sides of the theca, if present, leading to the recognition of left and right sides (see Section 3).

In dinokonts the side from which the flagella arise is *ventral*, the opposite surface is *dorsal*, with *left* and *right* sides recognized by zoological convention, as in humans. The anterior pole is *apical* and the posterior *antapical*.

Most commonly dinokonts are illustrated from the ventral side because this view provides the most information. The left and right sides usually differ and care must be taken to avoid optical reversal, a common error resulting from the type of optical system or from focusing through the cell and drawing the far side.

1.4 Flagella

The two flagella of dinoflagellates are differentiated in a characteristic manner. One is ribbon-like and thrown into short-period waves (3–4 per 10 μm: Fig.

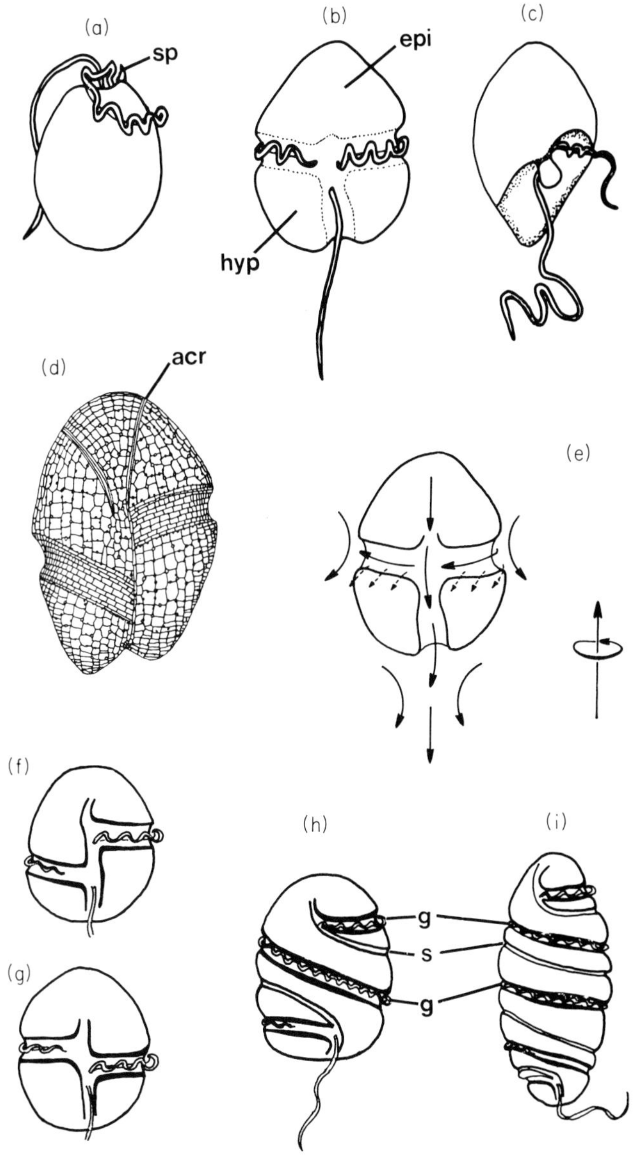

Fig. 2.2. Basic flagellar arrangements: (a) desmokont (e.g. *Prorocentrum*); (b) dinokont (e.g. *Gymnodinium*); (c) opisthokont (e.g. *Oxyrrhis*). (a) and (b) redrawn from Taylor (1980). epi, episome; hyp, hyposome: sp, spine. (d) The amphiesmal reticulum in *Gyrodinium pavillardii* as revealed by silver impregnation ('argyrome'). Redrawn from Biecheler (1952). acr, acrobase. (e) Water flow associated with the girdle and sulcus, from observations by Gaines & Taylor (1985). Arrows to the right indicate the direction of cell movement during swimming. (f)–(g) Girdle displacement: (f) left hand (descending); (g) right handed (ascending). (h)–(i) Torsion (left handed): (h) Two turns (*Cochlodinium clarissimum*); (i) Four turns (*C.augustum*). Redrawn from Kofoid & Swezy (1921). g, girdle (cingulum); s, sulcus. (Figures not to scale.)

2.3). One of its edges is contracted by a banded fibre, the 'striated strand'. This edge is invariably closest to the cell surface (Taylor 1975a). The axoneme (propulsive component) runs along the longer, outer edge. In dinokonts this flagellum is termed the *transverse* flagellum, but a flagellum of similar form is also present in desmokonts and will be given the same name here.

The second, *longitudinal* flagellum is not thrown into short period waves and can be simple, rounded in cross-section or ribbon-like (not as wide as the other). This is particularly so in *Ceratium* in which it can perform complex movements, including very rapid contraction (see Chapter 10), but it can be seen in *Gymnodinium* species as well. This flagellum usually beats posteriorly. In some species a curious loop is present close to the base (Herman & Sweeney 1977).

Very fine hairs are present, arising from the axonemal edge of the transverse flagellum. In *Oxyrrhis*, which has both flagella resembling the longitudinal type, very small, ovoid scales may also occur on one of the flagella (Clarke & Pennick 1972) and the hairs may be of different lengths.

In desmokonts such as *Prorocentrum*, the transverse (undulating) flagellum is tightly coiled over the cell surface, invariably curving initially over the left (least excavated) valve, but it is not attached except at its origin. Its waveform is like that in dinokonts (Fig. 2.3f). Dodge & Bibby (1973) and Soyer *et al.* (1982) found that its ultrastructure is essentially similar to that in dinokonts. The former also illustrated the details of roots associated with the two parallel basal bodies of the flagella of *Prorocentrum*, each shown to arise from a separate pore in the anterior platelet field. Biecheler (1952) observed that the transverse flagellum arose from the pore nearest the spine (the narrower pore) and the longitudinal from the larger, more distal pore in *Prorocentrum micans*. Parke & Ballantine (1957) and Loeblich (1976) asserted that the reverse was the case in the species they studied (both authors also reversing the 'left' and 'right' designations from that of earlier authors).

Subsequently, Loeblich *et al.* (1979) maintained that both flagella arose from the larger, distal pore. However, in their scanning electron micrographs in which two flagella are shown arising from the large pore, both flagella are of the longitudinal flagellar type, the transverse being absent. Soyer *et al.* (1982) observed triflagellated cells in which two longitudinal and one transverse were present. Taylor (1980) suggested that they may be zygotes (two longitudinal flagella are present on dinokont hypnozygotes). Recently Honsell & Talarico (1985) have provided good evidence for both normal flagella arising from the larger pore in *P. minimum*.

In dinokonts the flagella beat in the girdle and sulcal grooves: depressions in the cell surface or channels formed by 'lists' projecting from the surface. The transverse flagellum originates on the ventral side (by definition) and lies in the cingulum. It invariably winds to the cell's left and waves always proceed from the base to the tip (Gaines & Taylor 1985) as in most flagella, although Abé (1981) has reported reversals of beats in *Protoperidinium*. The striated strand

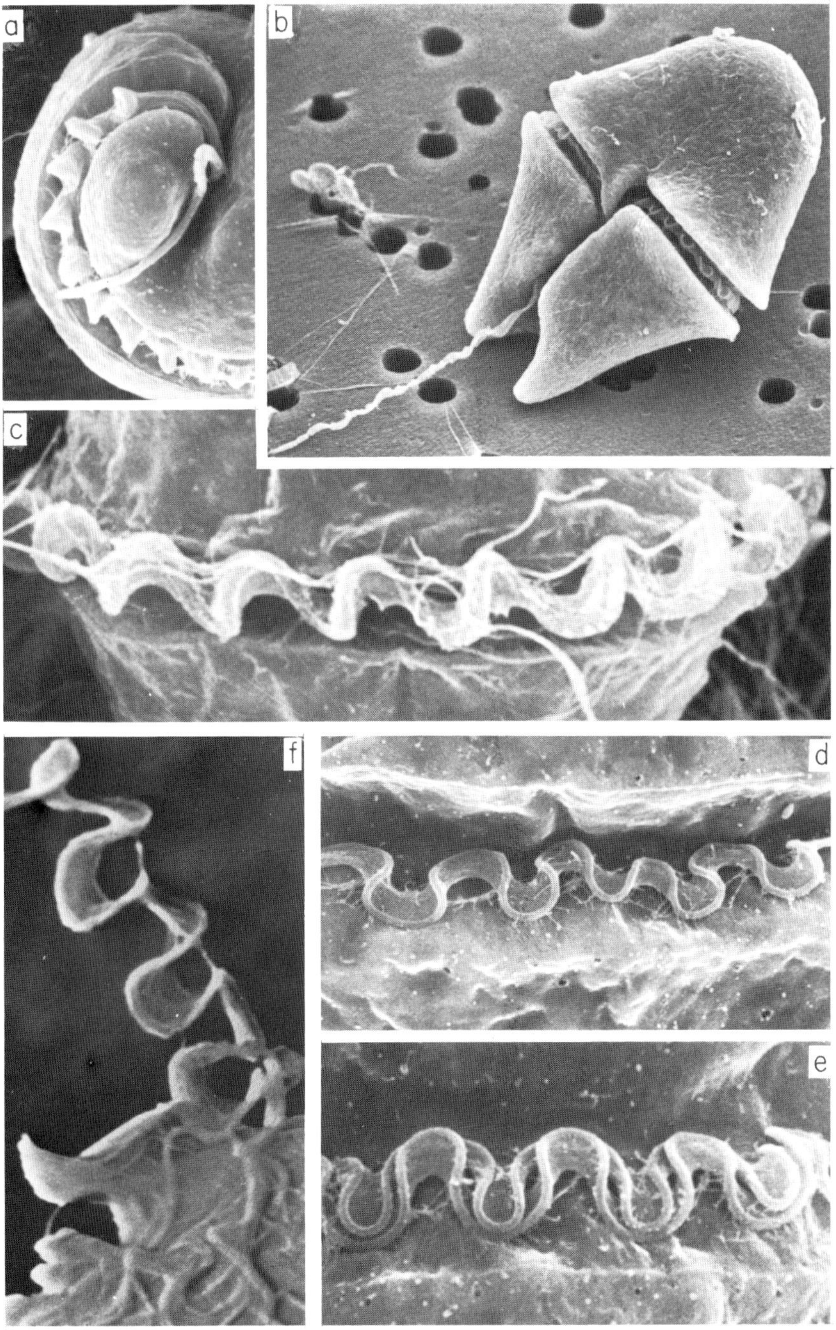

Fig. 2.3. Dinoflagellate flagella seen with scanning electron microscopy (SEM). (a) Antapical view of *Cochlodinium polykrikoides* (× 2500). (b) *Gymnodinium sanguineum* (× 800). (c) The transverse flagellum of *Heterocapsa* sp. (× 5300). (d)–(e) *Gyrodinium uncatenum*, single and double transverse flagella (× 4000). (f) The 'transverse flagellum' of *Prorocentrum micans* (× 6200). (a)–(e) courtesy of G. Gaines.

edge lies deepest within the girdle and its contraction may serve to hold the flagellum within the furrow as well as to allow propulsive waves to pass down the outer edge. The distal end of the flagellum may be attached to the cell (Leadbeater & Dodge 1966), although this is unusual.

Scanning electron microscopy and serial sectioning with transmission electron microscopy (refs in Chapters 3 and 10) have shown that the axoneme does not spiral around the striated strand as earlier models had suggested. It may appear to undulate back and forth in an arc termed 'hemi-helical' by LeBlond & Taylor (1976), but this is probably a species-specific fixation artefact. In live cells the axoneme coils in a spiral path (Berdach 1977; Berman & Roth 1979; Gaines & Taylor 1985), but remains continuously external to the striated strand (see Fig. 10.6). The ribbon has to be extraordinarily distensible and contractile in order to permit the latter, and Rees & Leedale (1980) found that it could invaginate in pouch-like intrusions within the ribbon to accommodate the subsequent stretching.

The longitudinal flagellum provides some forward propulsion and steers the cell in an oar-like fashion (Chapter 10), whereas the transverse flagellum produces both a rotatory and forward thrust. The beat of the transverse flagellum causes a forward thrust as the wave advances along the flagellum (LeBlond & Taylor 1976). Recent observations on cell rotation have shown unequivocally that the cells turn in the same direction as the beat, i.e. the cell is pulled around rather than pushed (Gaines & Taylor 1985), presumably due to the action of the hairs.

Planozygotes and planomeiocytes can be recognized, at least initially, by the presence of two longitudinal flagella which usually beat in synchrony, in a 'ski-track' pattern. Two transverse flagella may also be present, but are more difficult to see, e.g. in *Gyrodinium uncatenum* (Fig. 2.3e). Further information on flagella in dinoflagellate sexuality can be obtained from Beam & Himes (1980) and Chapter 14 here.

1.5 Major variations in form

ASYMMETRY, GIRDLE DISPLACEMENT AND TORSION

Although a few dinoflagellates are almost bilaterally symmetrical, the majority show distinctive asymmetry to a greater or lesser extent. As Kofoid (1912) noted, there is a tendency for the left side of the body to be enlarged relative to the right, usually shown in the posterior lobes or antapical horns or spines. The left sulcal list is invariably more developed than the right in dinophysoids, this being most extreme in *Ornithocercus*, *Parahistioneis* and *Histioneis* (Fig. 2.4c, d).

Reduction of the right horn is extreme in the subgenus *Amphiceratium* of *Ceratium*, but in a few, such as *C. dens* (Fig. 2.9a) the left antapical is reduced to a small, acute stump while the right antapical horn is well developed.

Commonly, the ends of the girdle do not meet one another in exact alignment and are said to be *displaced*. Not only is one end more anterior than the other, but *torsion* may cause them to overlap to a greater or lesser extent.

Displacement is said to be *left-handed* (=descending) or *right-handed* (=ascending) depending on whether the left (proximal) or right (distal) end is more anterior respectively (Fig. 2.2f, g). The degree of displacement is expressed in *girdle widths* and is measured from the top edge of the most anterior end to the top edge of the other. Left-handed displacement is much more common than right-handed, the latter being present only in a few *Protoperidinium* species and their relatives (*Diplopsalis*, etc.). LeBlond & Taylor (1976) suggested that displacement might redirect flow from the girdle, but they assumed a flow

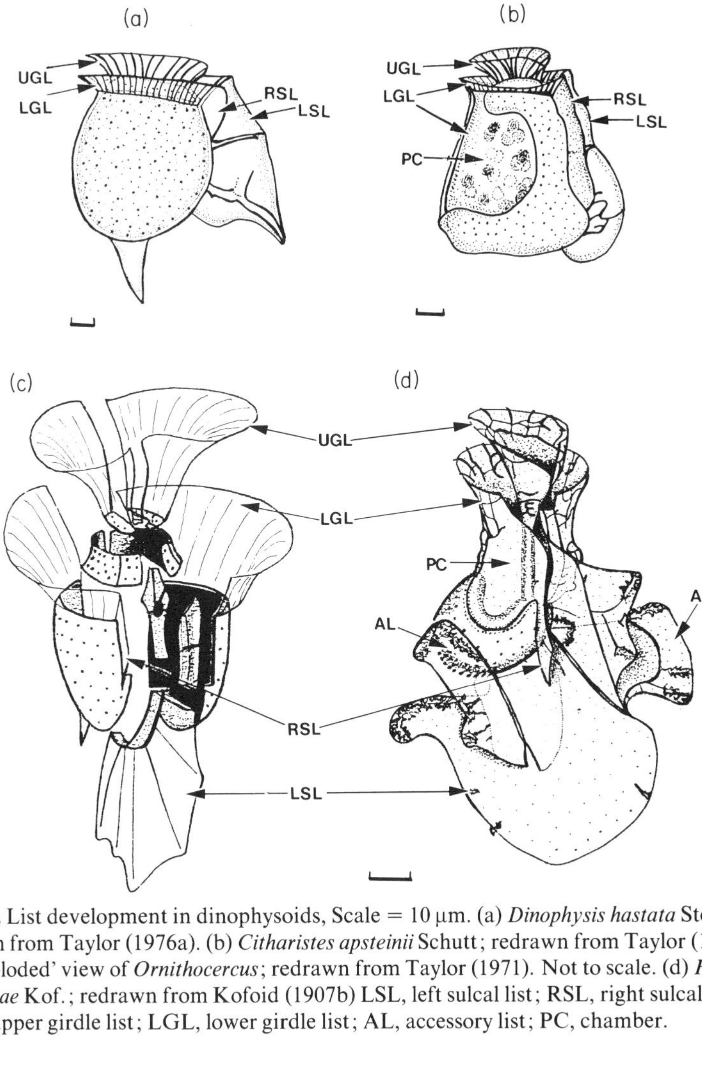

Fig. 2.4. List development in dinophysoids, Scale = 10 μm. (a) *Dinophysis hastata* Stein; redrawn from Taylor (1976a). (b) *Citharistes apsteinii* Schutt; redrawn from Taylor (1976a). (c) 'Exploded' view of *Ornithocercus*; redrawn from Taylor (1971). Not to scale. (d) *Histioneis josephinae* Kof.; redrawn from Kofoid (1907b) LSL, left sulcal list; RSL, right sulcal list; UGL, upper girdle list; LGL, lower girdle list; AL, accessory list; PC, chamber.

opposite to that now known to occur and so the commonality of left-handed displacement is still a mystery.

Torsion is almost invariably in the same direction, the proximal end extending further in a right apical direction, the distal end extending to the left antapical, causing the sulcus to incline between them. Such torsion can be readily seen in *Gyrodinium* and in *Gonyaulax* among the thecate genera. It is most extreme in the athecate genera *Warnowia* (=*Pouchetia*), *Nematodinium* and *Cochlodinium*. The greatest torsion, four turns, is found in *C. brandtii* and *C. augustum* (fig. 2.2i). Dorso-ventral asymmetry is also common. For example, in *Cladopyxis* (Fig. 2.6e) the processes lie in a single plane which is inclined at an angle to the girdle, again higher on the dorsal side than the ventral (Taylor 1976a).

FLATTENING

Flattening to varying degrees may be found in all the major planes: dorso-ventral, lateral or apico-antapical. Some extreme examples of dorso-ventral flattening can be seen in the brackish water species *Kryptoperidinium* (*Glenodinium*) *foliaceum* and in members of the section *Archaeceratium* of *Ceratium* such as *C. praelongum* (Fig. 2.6d), *C. gravidum* or *C. cephalotum*, and the subgenus *Platydinium* of *Heterodinium*, where the cells seem to show a 'leaf-like' adaptation to enhance chloroplast illumination (Taylor 1980). This may also influence sinking path.

Both lateral and dorso-ventral flattening are common in sand-dwelling dinoflagellates (Chapter 11) where they seem to be adaptations to life in the interstitial spaces between the sand grains. Phagotrophic cells which move over solid surfaces with the ventral surface downwards, the sulcal area being the site of ingestion (Chapter 6), are also commonly dorso-ventrally flattened.

Sessile (attached) protists are often flattened and in dinoflagellates this can be seen in some thecate species which live attached to the surface of macroalgae, sand or even animals. Examples of this are *Gambierdiscus toxicus* and species of *Ostreopsis*, which are flattened apico-antapically. A few planktonic taxa such as *Dissodium* spp., *Pyrophacus* spp. and *Ptychodiscus noctiluca*, are also apico-antapically flattened.

Some of the most elaborately complex noctilucoid dinoflagellates show either lateral (e.g. *Kofoidinium*, Fig. 2.7d) or apicoantapical flattening into leaf-like forms (*Petalodinium*, *Scaphodinium*, *Abedinium*, Fig. 2.7c) or extraordinary medusoid forms (*Craspedotella*, Fig. 2.7b; *Pratjetella*=*Leptodiscus*, Cachon & Cachon 1969).

ELONGATION

In several genera the cell body has undergone considerable elongation in the apico-antapical axis (see also horn formation, below). This is seen in a moderate way in the dinophysoid genus *Oxyphysis*, or in *Centrodinium* in which the cells are also laterally flattened. It is most extreme in the subgenus *Amphiceratium* of *Ceratium*, in which the apical and left antapical horns are greatly extended, the right antapical horn being reduced to a vestige, the whole cell resembling a long, curved needle with a slightly swollen cell body.

In the dinophysoid genera *Dinophysis* (a few species, such as *D. miles*), *Triposolenia* and *Amphisolenia* the cells may also be greatly extended, but in these the girdle is located at the extreme anterior end, raised up from the flagellar pore and cell body by a long 'neck' (Figs 2.5a, c, 2.6b), the antapical part of the body being drawn out into one or more (two in *Dinophysis* and *Triposolenia*) long, narrow extensions. Curvature is again evident. Kofoid (1906) determined the sinking path for *Triposolenia* (Fig. 2.5d) which seems to follow the zig-zag principle described under 'Asymmetry' and 'Flattening'. If swimming is continuous such a feature would be useless. Another possibility is that, like horns, elongation acts to increase the effective diameter of the cells, reducing grazing pressure from small grazers. Most of these extremely elongate forms are

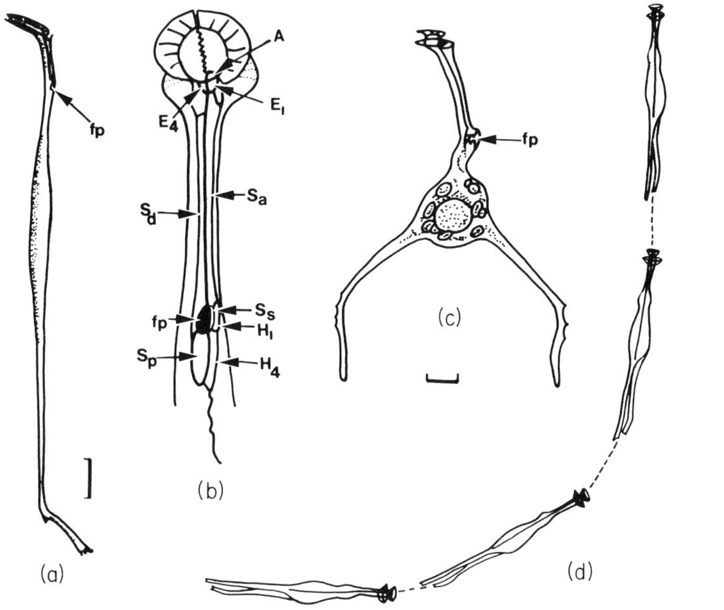

Fig. 2.5. (a) *Amphisolenia asymmetrica* Kofoid; redrawn from Taylor (1976a). Scale = 50 μm. (b) Sulcal plates of *Amphisolenia bidentata* redrawn from Balech (1977). Not to scale. A, apical platelet; E_1, E_4, ventral epithecal platelets; fp flagellar pore; H_1, H_4, ventral hypothecal plates; S_a, anterior sulcal; S_d, right sulcal; S_p, posterior sulcal; S_s, left sulcal. (c) *Triposolenia bicornis*, side view. fp, flagellar pore; redrawn from Taylor (1976a). Scale = 10 μm. (d) Sinking path of *Triposolenia* determined from models by Kofoid (1906). The curvature of the antapical horns produces the deflection in path. Redrawn from Kofoid (1906). Not to scale.

tropical, oceanic species (as are those with the longest horns), although *Ceratium fusus*, which is abundant in arctic cold-temperate waters, also shows the same modification to a moderate extent.

CELL EXTENSION: LISTS, HORNS AND SPINES

In many dinoflagellates the cell body may be extended into *horns*, a multiple form of body elongation, or the theca may possess ridges or flanges: *lists*, or elongate, narrow thecal extensions: *spines*.

The classical example of horn formation in dinoflagellates is the genus *Ceratium* (Fig. 2.9a–e) in which the apex is drawn out into an apical horn and there are two antapical horns (left and right), usually directed laterally. Horns are often formed by more than one plate. In *C. cornutum* each antapical horn involves two plates, although the tips are covered by only one of the plates (Happach-Kasan 1982).

A peculiar variation occurs in *Brachydinium*, in which four horns arise like arms and legs from the cell body (Fig. 2.10a). The wall, although continuous (pellicular?—see Section 3.1) is flexible and the horns move in a slow, waving fashion (J. Cachon in Léger 1971), completing a circle in 1–2 minutes. A closely related genus, *Asterodinium*, has an additional fifth horn arising from the episome, and *Microceratium* has only three.

Spines (solid or narrow, hollow wall extensions) are rare on dinoflagellate thecae, although common on cysts. Thecal spines are most developed in *Ceratocorys* (most extreme in *C. horrida* Fig. 2.6f), *Podolampas* (one or two antapical spines), *Micracanthodinium* (long, unbranched) or *Cladopyxis* (Fig. 2.6e). In *Prorocentrum* a small, tooth-like spine is developed to a greater or lesser extent, adjacent to the smaller anterior pore. Spines may form from the corners of plates (e.g. *Ceratocorys*) or from the centre.

Lists are flanges which arise from the margin of plates, most commonly along the girdle and sulcal edges where they may channel the flow more strongly. Upper and lower girdle lists (UGL, LGL) and left and right sulcal lists (LSL, RSL) may be present. In dinophysoids, list development has been taken to

---

Fig. 2.6. Scanning electron micrographs. (a) The periflagellar platelets of *Prorocentrum rhathymum* (= *P. mexicanum*?). Plate designations are those used by Taylor (1980). The dark arrow indicates the narrower flagellar pore from which the transverse flagellum appears to originate; the white arrow indicates the larger flagellar pore from which the longitudinal flagellum arises. The large fin invariably arises from 'a' and a small flange is often present on 'h'. Photomicrograph courtesy of G. Gaines (× 15 000). (b) *Triposolenia truncata*, left side. Photomicrograph courtesy of M.C. Carbonell (× 600). (c) Antapical view of *Triadinium* (*Heteraulacus*) *polyedricum*. Courtesy of M. C. Carbonell (× 560). (d) *Ceratium praelongum*, ventral view. Courtesy of M. C. Carbonell (× 200). (e) *Cladopyxis brachiolata*, three-quarter right side. From Taylor (1980), with permission (× 580). (f) *Ceratocorys horrida*, right side. Original (× 400).

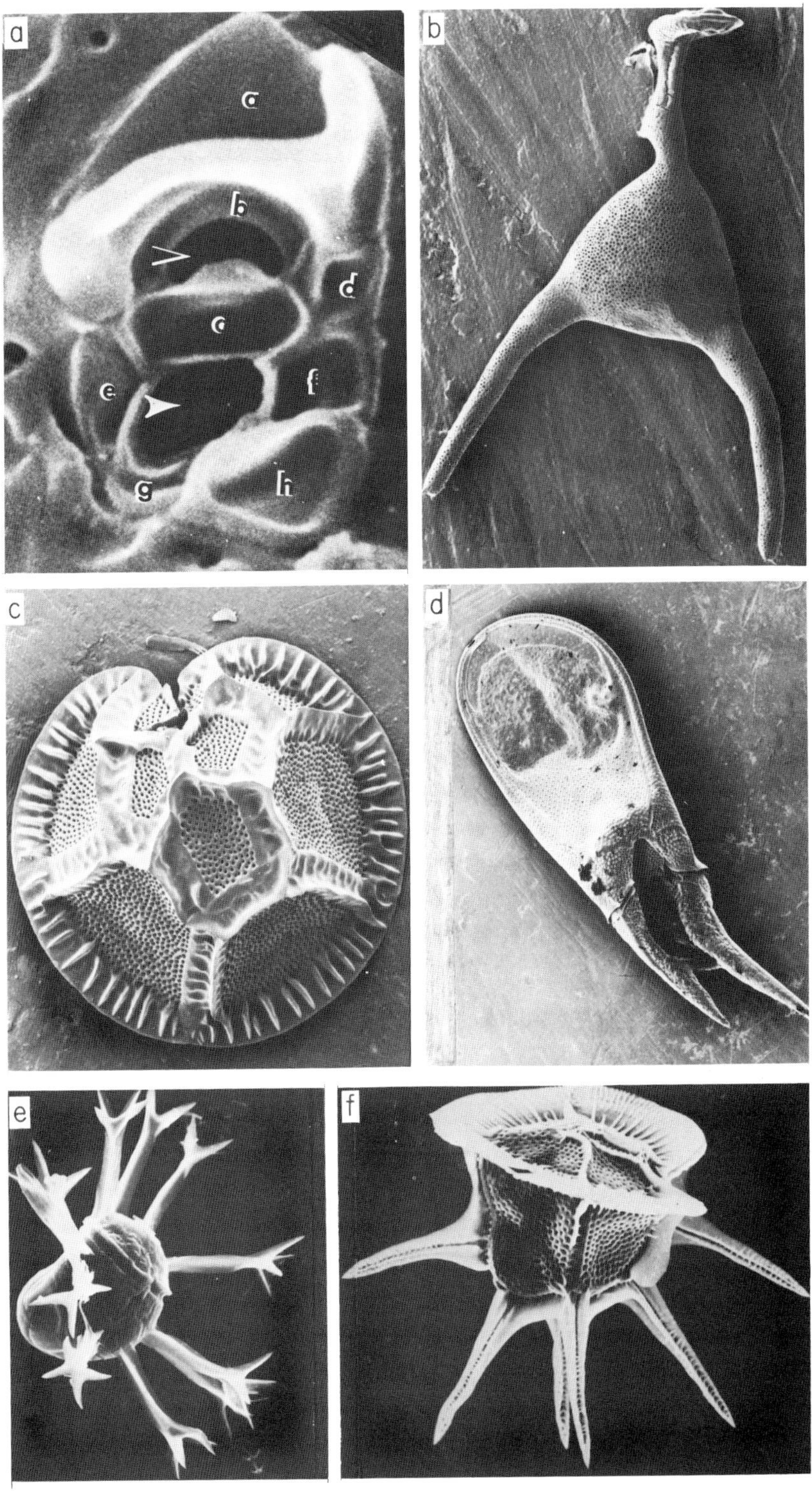
a
a
b
c
d
e
f
g
h
b
c
d
e
f

bizarre extremes. Taylor (1980) has summarized the gradient in development which can be seen, from least in *Dinophysis* to most in *Ornithocercus* and *Histioneis* (Fig. 2.4a–d). Both the girdle and sulcal lists can be elaborated. In *Ornithocercus* and *Parahistioneis* the girdle lists are greatly extended and project forwards. The UGL forms a complete flange, but it has a notch on its ventral edge directly above the sulcus, which may serve to channel water over the sulcus, possibly a feeding adaptation in this non-photosynthetic genus (Taylor 1971). Between the girdle lists numerous ectosymbiotic cyanobacteria ('blue-green algae'), termed *phaeosomes*, are usually present (see Chapter 12 and the review on planktonic symbioses by Taylor 1982). The RSL is relatively small and ends on the ventral side whereas the LSL is very extensive and keel-like, made of three components, and terminates on the dorso-antapical side (Figs. 2.4c and 14.3). The lists are strengthened by ribs and reticulations which become more elaborate as the cell gets older.

In *Histioneis* the girdle lists project forward (apically) much more (Fig. 2.4d). The LGL is collar-like, making the girdle cavity into a chamber with transparent walls that sometimes have lateral pouches. The UGL is shaped like a trumpet because the epitheca has been reduced to a minute disc, its anterior end flaring to semi-close the chamber made by the LGL. This cavity is termed the *phaeosome chamber* because it usually contains coccoid cyanobacteria and it seems to have evolved its unusual shape in order to house them (Taylor 1980).

The LSL of *Histioneis* is more elaborate than that of *Ornithocercus* and may have lateral accessory flanges, the whole effect being reminiscent of an orchid or pitcher-plant. The cell body is extended dorsally in a slipper-like shape.

Citharistes (Fig. 2.4b) produces a more enclosed phaeosome chamber. The cell body is bent into a C-shape and the epitheca shifted so that the walls of the chamber are made entirely by the body and the LGL, with only a small opening at the apical end.

The function(s) of the horns and spines is unknown. It is evident that they increase the surface area/volume ratio (and membrane surface area because the wall is formed internally), they increase the maximum cell dimensions and they increase 'form resistance': the increase in drag resistance in a particular direction. Similar structures are found in both thecae and cysts, and in other groups, some of which, such as diatoms, lack motility in the plankton.

It has been suggested that the longer, more slender horns of tropical, oceanic ceratia (relative to those in temperate or polar waters) are an adaptation to resist sinking in lower viscosity, warmer water (Gran 1912; Braarud 1962) or to lower nutrient levels (Peters 1932, with reference to phosphate). Steemann Nielsen (1939) and Böhm (1976) have shown that the data on which the 'flotation' arguments were based do not support the proponents' cases and, in any case, 'sinking' is a difficult concept to apply to the mastigote (swimming) stage since it is not known how continuously planktonic dinoflagellates swim (see Chapter 10). In the case of cysts it might be much more applicable (Sarjeant *et al.* in prep.), slower sinking resulting in greater transport/dispersal. The nutrient-uptake argument for the mastigote is greatly strengthened by observations on autotomy (self-cutting) of horns in *Ceratium* (see Section 5.2).

Oceanic dinoflagellates may be subjected to strong grazing pressure from zooplankton. An increase in maximum dimension would make it difficult for tintinnids (ciliated microzooplankton) to feed on them, but would increase their 'filterability' and availability to larger zooplankton. The data needed to evaluate this aspect are not yet available.

TENTACLE, PEDUNCLE AND PISTON

Some dinoflagellates, particularly in the related genera *Gymnodinium* and *Gyrodinium*, possess a small, movable, finger-like protrusion from the sulcus region, arising between the flagellar bases (e.g. in *Gyr. lebourae* (Lee 1977) and *Gym. sanguineum*). This is termed a peduncle. Its functions are not fully understood, but it appears to be sensory and is involved in feeding in some phagotrophic species such as *Gym. fungiforme* (see Chapter 6). It is easily overlooked and may be much more widespread than it appears. A similar structure was observed on the mastigote of *Symbiodinium* (*Zooxanthella*) *microadriaticum* by Loeblich & Sherley (1979).

Tentacles are more elongate and obvious structures, found in *Noctiluca*, *Pavillardia*, *Pronoctiluca* and *Spatulodinium*. At least in *Noctiluca*, the tentacle clearly plays a function in feeding, the large cell (Fig. 2.8a, b) waving the tentacle slowly back and forth, wiping it across the cytostomal groove, cells and other particles occasionally becoming trapped on extremely thin threads on the tentacle (Soyer 1970; Lucas 1982; Nawata & Sibaoka 1983). The other tentaculoid forms are also non-photosynthetic and the tentacle is probably related to feeding in their cases (Chapter 6).

The piston is a more sophisticated cell extension, resembling a tentacle, arising from the posterior of *Erythropsidinium* (= *Erythropsis*) and *Greuetodinium* (= *Leucopsis*), but capable of extraordinarily rapid and extensive elongation (15 or more times the body length) and contraction. Its ultrastructure is described in Chapter 3 and its probable function in Chapter 6.

BLADDER FORMS AND OTHERS

One of the adaptations recognized by Schütt (1892) as a method to reduce sinking rate in phytoplankton, particularly in non-motile forms such as diatoms, was an inflation in size (combined with a thin wall), in which the greater part of the cell consists of a vacuole filled with fluid. This modification, termed the 'bladder type' by Gran (1912), can be recognized in *Pyrocystis*, in which a pelagic cyst-like phase (termed a 'vegetative cyst' because it is photosynthetically active), surrounded by a smooth, continuous wall containing both cellulose and

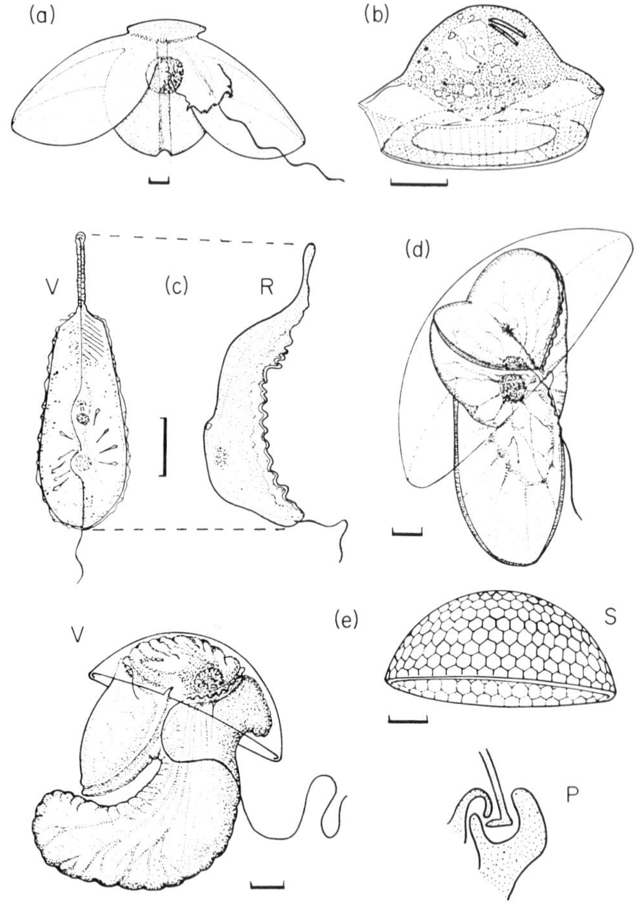

Fig. 2.7. Highly modified, noctilucoid species. All scales = 20 μm. (a) *Cymbodinium elegans*. Redrawn from Cachon & Cachon (1969). (b) *Craspedotella pileolus*. Redrawn from Cachon & Cachon (1969). (c) *Abedinium* (*Leptophyllus*)*dasypum*. V, ventral view; R, right side. Redrawn from Cachon & Cachon (1964, 1969). (d) *Kofoidinium pavillardii*, ventral view. Redrawn from Cachon & Cachon (1967). (e) *Pomatodinium impatiens*: V, ventral view; S, shell ('coque'); P, detail of the finger-like processes which hold the shell and rotate it. Redrawn from Cachon & Cachon (1967).

sporopollenin (Swift & Remsen 1970), becomes greatly distended by a vacuole. The cells are spherical, fusiform (Fig. 7.2) or lunate (also in *Dissodinium*; Elbrächter & Drebes 1978). Motile cells occur for a brief period; less than 1 hour. Although some may be athecate, the mastigotes of several species have a theca resembling *Protogonyaulax* or *Gessnerium* (Pincemin *et al.* 1981, 1982).

Although *Noctiluca* possesses a single flagellum (Fig. 2.8a) it is ineffectual in locomotion and the cell is effectively non-motile. However, like *Pyrocystis*, it is greatly expanded by a fluid vacuole, the cytoplasm forming a very thin peripheral layer and radiating strands with a mass containing the nucleus and

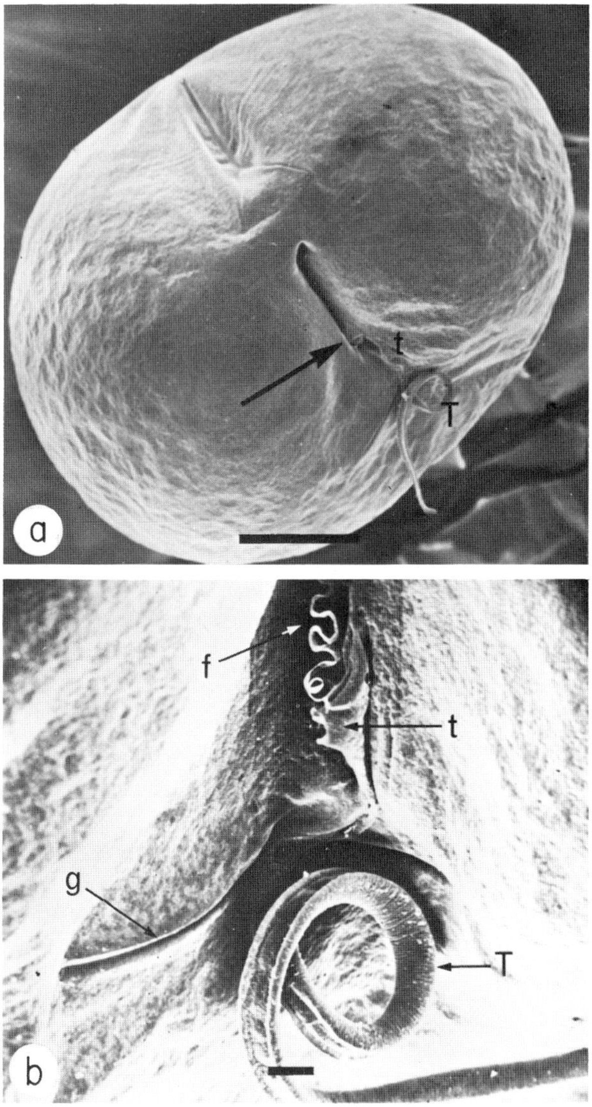

Fig. 2.8. Scanning electron micrographs of *Noctiluca scintillans*. (a) Whole cell. The arrow indicates the entrance to the cytostomal groove. T, tentacle; t, 'tooth' (possibly a modified transverse flagellum). Scale = 100 μm. (b) Detail of the tentacle (T) base, tooth (t), single flagellum (f) and superficial groove (g). Scale = 10 μm. Both micrographs from Lucas (1982), with permission.

cytostome. The ionic composition of the vacuolar fluid was analysed by Kessler (1966). He found that a buoyancy of 0·0036–0·0046 g cm^{-1} was due to the selective accumulation of Na^+ relative to K^+, a high concentration of light NH_4^+ ions and exclusion of heavier divalent ions such as SO_4^{2-}. The pH of the sap was 4·35. The cells can actively vary their buoyancy. The composition of *Pyrocystis* vacuolar sap is not known in the same detail, but NH_4^+ and NO_3^- can be stored in it if ample external NO_3^- is available (Bhovichitra & Swift 1977).

The Noctilucales can be subdivided into three families (zoological subfamilies): the Noctilucaceae, Kofoidiniaceae and Leptodiscaceae. The second of these, like the first, shows great expansion of the cell through increased vacuolar volume, starting from an *Amphidinium*-like stage but culminating in bizarre cells with a flattened keel, and an expanded episome, around the edge of which runs the cingulum. Both *Kofoidinium* (Fig. 2.7d) and *Pomatodinium* (Fig. 2.7e) produce a delicate, dome-shaped external structure, held by finger-like projections on the dorsal and ventral sides (Cachon & Cachon 1967). Most extraordinarily, the dome can be rotated, possibly aiding in concentrating food particles at the cytostome which lies at the centre of such a vortex, but this is not known for certain. The hyposome of *Pomatodinium* also shows strong contractile movements (Cachon & Cachon-Enjumet 1966).

Members of the Leptodiscaceae show a variety of leaf-like, butterfly-like or medusa-like flattened forms (see previous section and Fig. 2.7a–e).

Many parasitic dinoflagellates are almost unrecognizable as members of this group during most of their life cycle, but pass through a motile stage which betrays their origin (Chapter 13 and Chatton 1920).

2 GROWTH STATES

2.1 Chain formation

In dinoflagellates, chains are found principally in the gonyaulacoids, such as *Ceratium* spp., *Pyrodinium bahamense* var. *compressum*, *Peridiniella catenata*, *Protogonyaulax catenella*, and *Gessnerium monilatum*. A few gymnodinoids, such as *Gymnodinium catenatum*, *Cochlodinium catenatum* or *Cochl. polykrikoides* also form chains. In *Ceratium vultur* (Fig. 2.9d), up to twenty cells may form a chain. In *P. catenella* (Fig. 2.11a) it may be as many as sixty-four (F. Taylor, unpubl. obs.) although chains of eight, four or two are more common. Steidinger & Williams (1970) observed seventy cells in a chain of *Gessnerium* (*Gonyaulax*) *monilatum*. Division is oblique, as usual, and the individuals change alignment without separating. The absence of chain formers in peridinioids is probably due to their mode of division (eleutheroschisis), which involves contraction within the parent theca and division in the naked state, the parent wall being discarded (Chapter 14). However, *Protoperidinium denticulatum* forms pairs by a mode of division which is not known.

In dinophysoids the daughter cells often adhere by a remnant of the megacytic growth zone: the dorsal megacytic bridge (DMB; Taylor 1973a, b), e.g. in *Ornithocercus* and a few species of *Dinophysis*. In *D. miles* the cells may

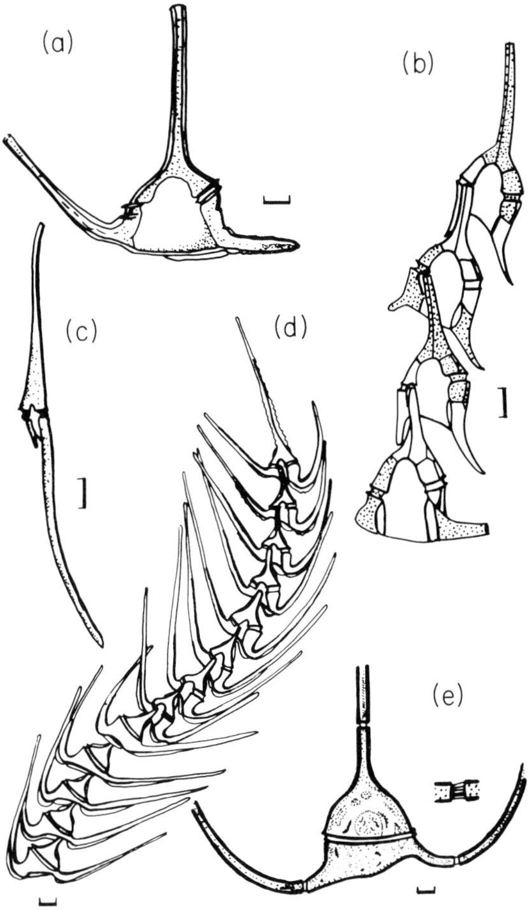

Fig. 2.9. *Ceratium* species. Scales = 20 μm. (a) *Ceratium dens*, an Indo-west Pacific species with an unusual left antapical horn. Redrawn from Taylor (1976). (b) Heteromorphic, four-celled chain of *C. tripos*. Unstippled plates are those most recently formed. Redrawn from Kofoid (1909b). (c) *C. fusus*, in which the right antapical horn is greatly reduced. Redrawn from Taylor (1976). (d) A chain of *C. vultur*. The torsion exhibited is not always present. Redrawn from Kofoid (1909b). (e) Autotomy in *Ceratium* sp. (inset of the groove which forms in the thecal plates of the horns: see text for details). Redrawn from Kofoid (1908).

remain laterally attached by DMBs through several divisions, forming a rosette of eight cells or more (Weber van Bosse 1901).

In *Ceratium*, chain formation involves a very precise alignment of the apical horn of the more posterior individual with a notch (the chain channel: Kofoid 1909b) at the distal end of the girdle on the right side of the ventral depression (Fig. 2.9b). In this genus there is no realignment after division, the contact being oblique. Apical horn length varies within individuals in a chain, being much longer in the anteriormost individual and shorter and more or less equal in the other members of the chain. In most chain-forming *Ceratium* species the antapical horns are similar in all members of the chain. However, in *Ceratium tripos* the individuals in a chain can be markedly different to one another in the length and orientation of the horns (Lohmann 1908; Kofoid 1909b). Anterior individuals have short, posteriorly directed antapical horns and the most posterior individual has normally directed (lateral, upward-curving) antapical horns (see Fig. 2.9b). Note that the epitheca of the anteriormost, and the hypotheca of the posteriormost, were originally part of the cell ancestral to all members of the chain. Kofoid (*loc. cit.*) interpreted this to be a form of mutation (from *C. tripos* to *C. californiense*), Lohmann (1908) suggesting that it may be a prelude to sexual reproduction. In view of most recent knowledge on sexuality in *C. horridum* (von Stosch 1972, and Chapter 14 here), the latter is most likely.

In species in which both single and chained individuals occur, such as *Peridiniella* (*Gonyaulax*) *catenatum* (Balech 1977) and *Pyrodinium bahamense* (Steidinger *et al.* 1980), single individuals are proportionately longer (apex to antapex) than those in the chains.

The factor(s) governing chain length, and its function in dinoflagellates, are unknown. Cultures of chain-forming species rarely form chains as long as those in the field, or they may not form chains at all.

On the basis of *Ceratium* material from the Atlantic Ocean, Schubert (1937) concluded that chain formation was a feature of populations in high salinity waters. If there is a correlation with salinity in dinoflagellates it must be relative or interactive with temperature or nutrients, for *Peridiniella catenatum* is a cold-water brackish species. Reproductive rate may also influence chain formation.

2.2 Coccoid forms

The term 'coccoid' is usually used to denote single or colonial, non-motile algal cells, not necessarily spherical (Fritsch 1935). Coccoid forms may be permanently, or only temporarily non-motile. Unlike resting cysts which are dormant, coccoid cells are actively photosynthetic and reproductive in the non-motile (amastigote) state. Like resting cysts, they are usually surrounded by a continuous wall and, alternatively, can be termed *vegetative cysts* (i.e. cysts which are fully active metabolically and reproductively), as opposed to resting cysts. Similarly, they are a major life-cycle stage, rather than a briefly transient state like an ecdysal (pellicle) cyst.

Nearly all the coccoid dinoflagellates are photosynthetic and most are benthic, freshwater species, but there are some ecologically important marine exceptions. In freshwater, the genera living predominantly in the coccoid state are *Hypnodinium*, *Phytodinium* (both simple, unattached cells), *Cystodinedria* (including *Phytodinedria*), *Dinastridium* (including *Bourrellyella*), *Cystodinium* (Fig. 2.10e) and *Tetradinium* (Fig. 2.10f) which show varying degrees of attachment to surfaces and/or spine development (reviewed by Bourrelly 1970).

Dinococcus (= *Raciborskia*) is now thought to be a life-cycle stage of *Cystodinium* (Pfiester & Lynch 1980). *Stylodinium* (Fig. 2.10f) is similar but develops a long terminal stalk by which it is attached to vegetation and has a motile cell with a tabulation resembling *Crypthecodinium*. *Cystodinium*, *Cystodinedria* and *Stylo-*

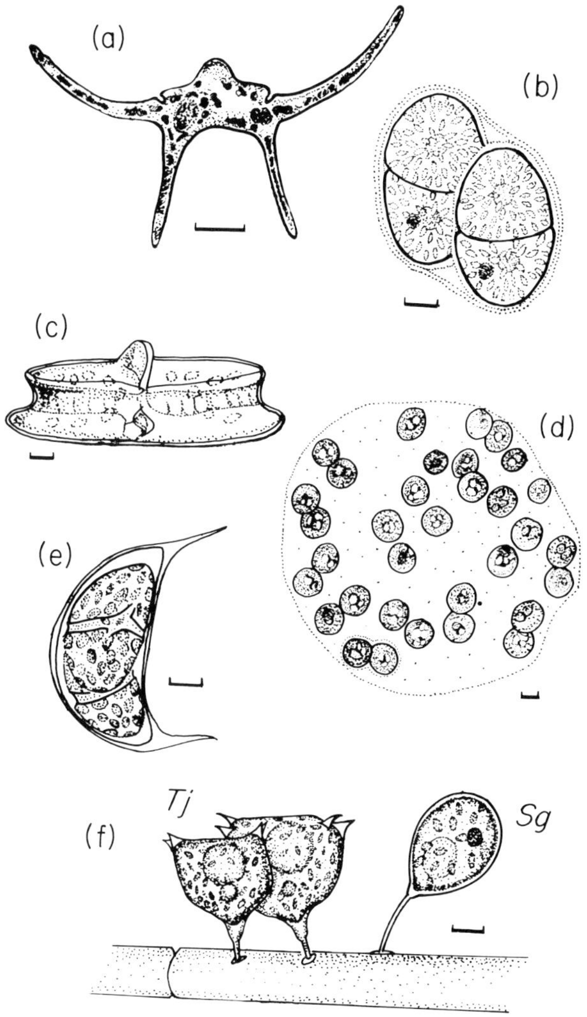

Fig. 2.10. Scales = 10 μm. (a) *Brachydinium capitatum*, a mastigote with a flexible pellicle. The 'arms' can move slowly. Flagella omitted. Redrawn from Taylor (1963). (b) *Gloeodinium montanum*. Dividing cells in mucilage. Redrawn from Thompson (1949). (c) *Ptychodiscus noctiluca* (= *carinatus*), another pelliculoid species. Flagella omitted. Redrawn from Taylor (1976). (d) *Gloeodinium marinum*; a mucilaginous colony. Drawn from a photomicrograph in Taylor (1976). (e) *Cystodinium iners*. Division stage. Redrawn from Thompson (1949). (f) A pair of cells of *Tetradinium javanicum* (*Tj*) and *Stylodinium globosum* (*Sg*) attached to a filament of *Oedogonium*. Redrawn from Thompson (1949).

dinium have been reported to have amoeboid stages which parasitize filamentous green algae (Section 2.4) and it is possible that others may also be shown to be parasitic. *Thaurilens* is a brackish water genus with a spiny, coccoid form similar to the freshwater genera above.

Cystodinium shows at least a superficial resemblance to members of the marine genera *Dissodinium* and *Pyrocystis* (see section on bladder forms). The former consists of two species, *D. pseudocalani*, and *D. pseudolunula*, which parasitize copepod eggs (see Elbrächter & Drebes 1978; and Chapter 13), having both spherical and crescentic, non-photosynthetic vegetative cysts. The latter species also has a thicker-walled, crescentic state (formerly named *Pyrocystis margalefii*) which is believed to be a resting stage by Drebes (1981).

Symbiotic species are also usually coccoid within the host (except for motile *Amphidinium* cells inside the turbellarian worm *Amphiscolops* and the zooxanthellae of the foraminiferans *Globigerinoides ruber* and *G. sacculifer*: see Taylor 1982 and Chapter 12). *Zooxanthella* and *Symbiodinium* live as spherical, continuous-walled cells within their numerous hosts, and divide in this state to produce more of their kind, but also produce transient gymnodinoid swarmers.

2.3 Palmelloids

Palmelloid cells are colonial, coccoid cells embedded in a common mucilaginous matrix. Three genera of dinoflagellates have adopted this habit: *Desmocapsa*, *Gloeodinium* (Fig. 2.10b, d) and *Rufusiella* (separated from others formerly assigned to *Ourococcus*, a genus of the green algae). Although most species are in fresh or brackish water (cf. Bourrelly 1970) *D. gelatinosa* and *Gl. marinum* (Fig. 2.10d) occur in the marine plankton. The latter can form colonies of up to sixty-four cells (Bouquaheux 1971; Taylor 1976a, b). In *Rufusiella* mucilage is produced on only one side to produce stratified stalks and they form a reddish intertidal zone on arctic shorelines.

The motile cell of *Desmocapsa* is desmokont, whereas that of *Gloeodinium montanum* resembles the dinokont genus *Hemidinium nasutum* (and the two may be conspecific).

2.4 Amoeboid forms

Amoeboid stages have been found in several members of the Dinococcales. Pascher (1916) described *Dinamoebidium* (= *Dinamoeba*) *varians*, a marine phagotroph which reproduces in a cyst, passing through a brief gymnodinoid stage before becoming amoeboid once more. He noted similarities between the life-cycle and cyst of *D. varians* and the freshwater dinococcalean *Cystodinium bataviense* (= *Dinococcus oedogonii*), although amoeboid stages in the latter and in *Stylodinium sphaera* and *Cystodinedria inermis*, two other dinococcaleans, have only been described recently (Pfiester & Popovsky 1979; Pfiester & Lynch 1980; Popovsky & Pfiester 1982).

In several dinoflagellates pseudopodial extensions have been seen to arise from the sulcal region, probably in association with feeding (see Chapter 6 for details). In *Katodinium spirodinioides* rhizopodia, used in feeding on diatoms, arise from the girdle and hypocone (Popovsky 1982). This is also the case in the benthic, putative dinoflagellates assigned to *Protaspis*. Members of *Protoperidinium* and related genera use a pseudopodial 'feeding veil' to surround food for extracellular digestion (Gaines & Taylor 1984).

2.5 Filamentous dinoflagellates

True filamentous dinoflagellates, in which the cells are arranged linearly, in direct contact with each other, are extremely rare. There are two basic types, probably not closely related to each other. All are marine species.

In the first type the photosynthetic cells have continuous external cell walls, the filaments being formed by wall-to-wall contact. In *Dinothrix* (Pascher 1914) very short, branching filaments of not more than ten cells are formed (Fig. 2.11e). The cells can also become separated by mucilage to form a palmelloid stage in *Dinoclonium* (Pascher 1927). Algal-like, wall-to-wall branching filaments are also formed, the branches tapering (Fig. 2.11d). In both genera some cells of the filaments can produce athecate, gymnodinoid, motile 'swarmers': mastigotes.

Quite a different type of filamentous development is found among some non-photosynthetic, parasitic dinoflagellates. In many, the process of sporogenesis involves the production of numerous sporocytes which adhere temporarily in a filamentous or tissue-like way before giving rise to flagellated swarmers, e.g. *Sphaeripara* (=*Neresheimeria*), *Amoebophrya* (see Chapter 13). However, in *Haplozoon* (Fig. 2.11c) the filament-like state is more pronounced. the multicellular form can actively writhe, the cells lacking walls. In *H. axiothellae* (Siebert 1973) the multicellular state begins with a single cell, the trophocyte, which bears a spine and embeds itself in a gut-cell of a marine annelid (which is ironic, since *Haplozoon* resembles a tapeworm, scaled down to a lower level of organization!). Repeated divisions result in cells termed gonocytes. Quadrinucleate sporocytes are shed from the distal end, giving rise to gymnodinoid swarmers (Shumway 1924).

Electron microscopy (Siebert & West 1974) has revealed that each cell has a typical amphiesmal region consisting of a single layer of vesicles containing thin, cryptic thecal platelets. Further details of this genus, *Sphaeripara*, *Atlanticellodinium* and related taxa can be found in Chapter 13.

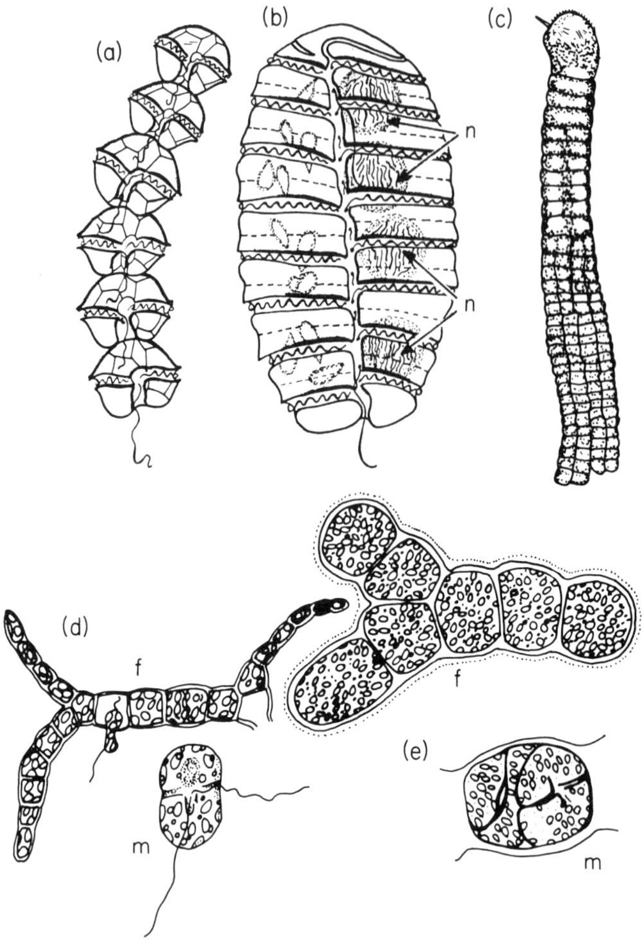

Fig. 2.11. (a) A chain of *Protogonyaulax* (*Gonyaulax*) *catenella*. Original. (b) A coenocytic cell of *Polykrikos schwartzii*. n, nuclei. Redrawn from Grassé (1952). (c) The multicellular parasite *Haplozoon delicatulum*. Redrawn from Grassé (1952). (d) A filament (f) and mastigote (m) of *Dinoclonium conradii*. Redrawn from Grassé (1952). (e) A filament (f) and mastigote formation (m) of *Dinothrix paradoxa*. Redrawn from Grassé (1952).

2.6 Coenocytes

A few dinoflagellates are coenocytic, i.e. possess multiple nuclei (Zool.: syncytial). The classic example of this is the athecate genus *Polykrikos* (Fig. 2.11b). The external morphology resembles that of a number of mastigotes connected together, each unit possessing paired flagella, a girdle, sulcus, episome and hyposome. The sulci run together, forming a continuous longitudinal furrow. The units are fused internally and in most cases, the number of nuclei is a half or a quarter of the number of individuals which appear to make up the exterior of the coenocyte.* Long, ribbon-like roots, visible with light microscopy, run to two sets of basal bodies and flagella ('mastigonts') from each nucleus. The coenocyte divides transversely at the centre (Morey-Gaines & Ruse 1980).

Kofoid & Rigden (1912) described a peculiar organism, *Gonyaulax series*, which resemble a chain of *Protogonyaulax catenella* in which all the cells are fused and there is a gradient in size from the largest in the centre towards either end. As it has not been seen since it is not known if this is its usual state or if it is a teratological (aberrant) form.

3 CELL WALLS OF MOTILE CELLS

Cell walls, semi-rigid or rigid peripheral coats, may be developed in the motile or non-motile phase, in one or in neither. There are significant differences between the wall formed in different stages of the life cycle.

In the motile (mastigote) phase the cell walls seem to be invariably intracellular, i.e. lying internal to the cell membrane (=plasmalemma).† The usual wall in this phase is the *theca*. This term usually refers to the composite, predominantly cellulosic, multipartite wall formed of tightly-fitting plates like a suit-of-armour (see below), lying within amphiesmal (=thecal) vesicles, as described in Chapter 3. In those with a great many vesicles it is the honeycomb of vesicles themselves, or some underlying layer, which becomes the primary strengthening formation of the cell periphery, similar to the situation in ciliates.

3.1 Pellicular layer

Some dinoflagellates have the ability to shed their theca and the outer amphiesmal region when under stress (*ecdysis*: Section 5.1) and emerge as a non-motile cell surrounded by a membrane (Section 5.2 and Chapter 3) and a thin, continuous wall. The layer which forms the wall of the resulting (asexually produced) cyst, is present in the motile cell, beneath the theca. Although such a layer was unwittingly seen by many early authors in species in which it is strongly developed, it was Loeblich (1970) who clearly identified it in the mastigote, restricting the old term *pellicle* to this layer. He recognized it as a continuous fibrous layer lying internal to the vesicular layer of the amphiesma (later reported to form within the vesicles by fusion of elements, in *Heterocapsa*: Morrill & Loeblich 1983; Morrill 1984), not losing its integrity when exposed to cellulase or other solvents of the thecal layer. Subsequently, it appears that

* The genus *Pheopolykrikos* was created for cells resembling *Polykrikos* but possessing chloroplasts and an equal number of nuclei to flagellar pairs. Some authors combine this genus with *Polykrikos*.

† The peculiar noctilucoids *Kofoidinium* and *Pomatodinium* produce a dome-like structure (Fig. 2.7d, e) which seems to be extra-cellular.

cellulose or related compounds may be a constituent of the pellicular layer, or even the prime constituent in freshwater *Ceratium* species. The usual chemical resistance of the pellicular layer appears to be due to the presence of a material similar to sporopollenin, a poorly-defined constituent of pollen walls and dinoflagellate resting cysts (Section 4.1). Pores may penetrate the layer (Loeblich 1970), particularly flagellar pores, and its surface may show varying degrees of ridging or reticulation, but it is often smooth.

Gymnodinoid cells do not usually preserve well in formalin concentration greater than 2% (Taylor 1978). However, some gymnodinoids also have pellicular development (see below) in the mastigote phase and preserve relatively well under such circumstances. Cells in which the pellicular layer is the principal strengthening structure of the motile stage can be said to be pelliculate.

Morrill & Loeblich (1981b) surveyed representatives of twenty genera for which cultures were available, principally using staining and acetolysis (the latter to detect sporopollenin-like material). They found a pellicular layer to be present in fifteen genera, only three of which (*Gloeodinium*, *Pyrocystis* and *Symbiodinium*) had an obvious, continuous wall layer. Many of these were known to produce ecdysal cysts, e.g. *Heterocapsa* spp., *Protogonyaulax tamarensis*. In some strains the layer was variably present or resistant. No sign of a pellicular layer was found in any prorocentroid or *Amphidinium* spp.

In all these, other than the coccoid genera, a pellicle is only cryptically present. However, in others not examined by Morrill & Loeblich because cultures were not available, a pellicular layer is evident in the mastigote phase, judging by light microscopy. Examples of such genera are the noctilucoids *Kofoidinium* and *Noctiluca*; gymnodinoids such as *Balechina* (=subgenus *Pachydinium* of *Gymnodinium*), *Ptychodiscus* (Fig. 2.10c) *Berghiella*, *Lophodinium*, *Sclerodinium*, *Herdmania*, *Brachydinium* (Fig. 2.10a) and *Asterodinium*. The rugose internal wall of *Achradina* (Fig. 2.16e and Section 3.10) appears to be a modified pellicular layer with silica impregnation. In these and other taxa the pellicular layer appears to be the principal strengthening element of the periphery of the *motile* stage. In several there can be an opening: in the sulcal area of *Ptychodiscus* and in the episome of *Achradina*.

Because it is formed in the same region of the cell as a cyst wall, often contains the same material and may even be a cyst wall precursor, it is not surprising that the pellicle has been confused with a resting cyst wall in the past, especially in preserved material in which flagella have been lost. For example, the mastigote of *Ptychodiscus* was considered to be a cyst by Cleve (who named it *Diplocystis*: see Boalch (1969) for observations on living cells). Loeblich (1970) and Tappan (1980) believed that the ridged wall of *Lophodinium* is a cyst rather than the pellicle of a motile cell. However, Carty & Cox (1985) have confirmed its formation by the mastigote, the wall dissociating into numerous hexagonal thecal plates. When *Brachydinium* (Fig. 2.10a) was first found in preserved material it was not realized that the cell was motile when alive and it was assigned to a coccoid order. It is now known to be a mastigote with a flexible pellicular wall.

In other pelliculate genera there is no doubt that there is a pellicle in the mastigote phase. In the noctilucoids mentioned above its presence has been confirmed with transmission electron microscopy (TEM) and it is likely that others (e.g. in the Leptodiscaceae) are also pelliculate. The wall of the fossil genus *Dinogymnium* is very similar to the recent genus *Balechina*, raising the strong probability that the latter is a fossilized pelliculate theca rather than a cyst (Taylor 1980). Other, larger species currently referred to the genus *Gymnodinium*, but with longitudinal lines on their surface, may also be pelliculate.

When a well-developed pellicular theca is present the wall may not cleave directly, the cells having to escape the pellicle during reproduction (pellicular ecdysis). This may be the reason for the archeopyle-like aperture in *Achradina* (or in *Dinogymnium*). In *Heterocapsa* the pellicle does not cleave (Morrill & Loeblich 1981b).

3.2 Thecal (tabular) layer

The term *theca* has been used conventionally to refer to the multipartite, cellulose wall of the motile cell (Kofoid 1907a, b), or for the entire peripheral complex: membranes, vesicles, trichocysts, plates, regardless of whether cellulose plates are present or not. Here it is restricted to the skeleton made up of the thecal plates. Schütt's term *amphiesma* is used for the entire peripheral complex. Netzel & Dürr (1984) prefer the term *cortex* for the entire complex. Further discussion is provided by Loeblich (1970), Taylor (1980), Morrill & Loeblich (1983) and Dodge (Chapter 3).

A dinoflagellate lacking cellulose plates in its amphiesmal (thecal) vesicles is said to be *athecate* (informally: naked or unarmoured) and *thecate* (armoured) if it has them. Although each thecal plate is located within a vesicle, the plates abut one another tightly, forming a firm peripheral skeleton lying just beneath the cell membrane. The plate junctions are termed *sutures*.

The plates may be thick, or so thin that they cannot be seen with the light microscope. There is generally an inverse relationship between the number and the thickness of the plates (Taylor 1980). Many species thought to be athecate have since been shown to possess delicate thecae (e.g. in the genus *Gymnodinium*) although there are many genuinely athecate dinoflagellates. In the latter (e.g. *Oxyrrhis*) it is the vesicles themselves which seem to act as peripheral strengthening elements, like a membranous honeycomb, with further support from microtubules and fibrous material (Chapter 3). Loeblich & Morrill (1979) have suggested that some of the extremely thin structures thought to be cryptic thecal plates in transmission electron micrographs of some species may be homologous to a membrane precursor to wall formation (see Chapter 3A.2) rather than to the plates themselves and may fuse to form the pellicle (Morrill & Loeblich 1983).

3.3 Thecal composition and ornamentation

Thecal plates usually consist primarily of cellulose (Loeblich 1970) or, in the case of *Peridinium cinctum*, a non-cellulosic glucan (Nevo & Sharon 1969). A detailed analysis of the wall of *Heterocapsa* (*Cachonina*) *niei* found that it consisted of 15% ash. The organic component consisted of 84% glucan, 3·5% chloroform–methanol–extractable lipid, 2·3–3·9% protein, with an unidentifiable residue of 8·2–9·8% (Loeblich 1970), the latter being interpreted as part of the pellicular fraction. Further tests, such as staining with iodine plus potassium iodide, followed by sulphuric acid, or the zinc-chlor-iodide reaction, confirmed the presence of cellulose.

The plates are nearly always penetrated by pores, termed *trichocyst pores* because it is assumed that most, if not all, are potential exit apertures through the theca for trichocysts. A 'one pore:one trichocyst' hypothesis is difficult to test, but seems to be reasonable.

The plate surface is finely fibrillar and can appear smooth under light microscopy, e.g. several *Prorocentrum* species. In some, such as *P. minimum*, the surface is covered by fine spines, termed *denticulae*. The pores may lie in concave depressions of the thecal surface, as in many dinophysoids including *Ornithocercus*, or in areas surrounded by ridges. Such pore areas are termed areolae. Many thecae are *reticulate*, the ridges forming a network (Fig. 2.16c). Another common element consists of linear longitudinal ridges termed *striae*, seen in many *Gonyaulax* species. Lists (flange- or wing-like outgrowths) are usually supported by ribs.

Maturational changes have been observed. The newly-formed theca often lacks ornamentation other than the pores (which is also the case for the internal thecal surface), e.g. in *Ornithocercus magnificus* (Taylor 1973a, b). As the wall thickens, areolar depressions become more evident until the pores lie at the base of strongly marked pits. In *Pyrodinium bahamense* var. *compressum* (Steidinger *et al.* 1980) the new wall surface lacks denticles except around the pores (unpubl. obs.), but the mature surface is strongly denticulated. In gonyaulacoids such as *Pyrodinium*, *Gonyaulax* or *Ceratium* in which the parent theca is shared between the daughters, the immature, new wall moiety can be recognized by this, at least initially.

When entire new walls are formed, e.g. after ecdysal excystment, the whole theca may undergo maturational changes. Pores and sutures are present from the beginning. In *Gonyaulax reticulata* and *G. polygramma* the first ornamentation consists of linear striae, with reticulation developing later (Taylor 1962; Steidinger 1968). In *G. polyedra* it consists of rings, termed 'pore rings' by Dürr & Netzel (1974), again followed by reticular development. Kalley & Bisalputra

(1971) concluded that in *Scrippsiella* (*Peridinium*) *trochoidea* the initial thecal unit was continuous and sutures formed progressively, but it now appears that the micrographs which led to this conclusion were of the pellicular surface (after ecdysis, possibly with the inner thecal membrane still present) rather than of the theca (Morrill & Loeblich 1981a, b). In cysts there is a terminology for the distribution of ornamentation relative to the paraplates (Section 4.2) which could be applied to thecae as well.

3.4 Intercalary growth zones

Increasing size of the theca is accommodated by growth of the plates at one or more of the plate margins. The marginal growth zones, which are usually striated (the striae at right angles to the suture) are termed *intercalary bands* (IBs). Trichocyst pores are absent from them. In prorocentroids and dinophysoids the single intercalary band, associated with the sagittal suture, is termed the *megacytic growth zone* (MGZ), 'megacytic' because the cell is at maximum volume when it is fully formed. It is invariably widest distal to the flagellar platelet region, with little detectable in the immediate vicinity of the flagellar pores, e.g. *Ornithocercus* (Taylor 1973a, b).

There is also little or no intercalary growth in the sulcal region of gonyaulacoids and peridinioids. In some (all?) gonyaulacoids intercalary growth is usually from only one of the adjoining plates, e.g. *Gonyaulax*, *Ceratocorys*, whereas both adjoining plates contribute to the intercalary zone in *Protoperidinium* spp. (Fig. 2.12). Depending on the species and growth state of IBs between, plates are not all of equal width. In many there is a decrease in width of band away from the cingulum, but there are no IBs between the cingular and the adcingular (pre- and postcingular) plates and the width may vary greatly within the same plate series. Boltovskoy (1979) noted that there appeared to be a correlation between the number of cingular plates (and hence, sutures) and the presence of IBs in the adcingular series. Thus, in *Protoperidinium depressum* only two of the sutures between precingular plates have IBs, their positions corresponding to the IBs within the cingular series of three plates. In contrast, *Peridinium willei* has four

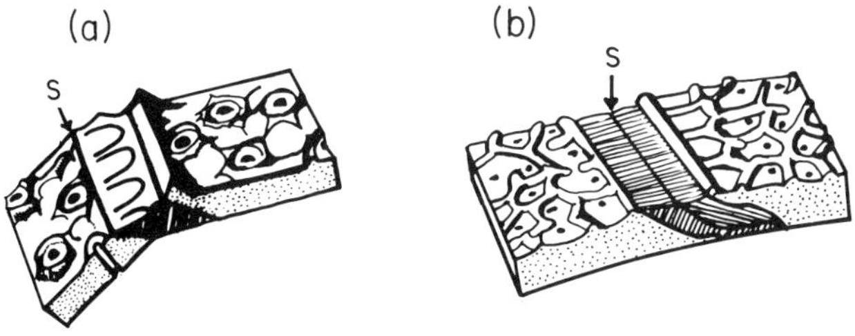

Fig. 2.12. Intercalary plate growth. (a) Additional material (hatched) produced by the overlapping margin only; e.g. *Gonyaulax polyedra*; s, suture. Redrawn from Gocht (1979). (b) Additional material contributed by both adjoining plates, e.g. *Protoperidinium grande*; s, suture. Redrawn from Gocht & Netzel (1974).

large adcingular IBs and two narrower ones in the precingular series, again corresponding roughly to the position of the six cingular IBs. He used this to argue that the IB pattern on the fossil *Palaeoperidinium* had six, rather than the three cingular plates present on modern *Protoperidinium* species, the latter being the genus thought to be most closely related to *Palaeoperidinium*. Unfortunately, this ingenious approach cannot be used with confidence because many modern *Protoperidinium* species produce six precingular IBs, even though they only have three cingular plates (with two cingular IBs). Peters (1927, 1929) discussed the development of IBs and illustrated them for several *Protoperidinium* species (then placed in *Peridinium*) in which this can be seen.

In sexual, thecate species the female cell, to which the male has suddenly added its volume, accommodates this by the production of very wide IBs (e.g. *Per. cinctum* (Dürr 1979b); *Per. volzii* (Pfiester & Skvarla 1979) and see Chapter 14 here). Pfiester has suggested that such extremely wide IBs may be used as one indicator of the zygotic condition.

The surface expression of IBs can be used to infer imbrication patterns (see next section and Fig. 2.12), even from that seen on the surface of some cysts. The development of pseudo-IBs on cyst walls, if not due to physical impression (a hypothesis once in vogue, but now not favoured), is surprising, because IBs are brief, transient growth features.

3.5 Plate overlap (imbrication)

In well developed gonyaulacoid and peridinioid thecae the margins of adjacent plates seldom simply abut one another. Instead the margins are bevelled and one plate usually overlaps the other (see Fig. 2.12 and Chapter 3 for ultrastructural details). Loeblich (1970) noted that the overlapped and underlapped margins of plates on *Gonyaulax polyedra* differ in surface appearance, the overlapped margin bearing a ridge lacking on the underlapped margin. Later Dürr & Netzel (1974) and Dürr (1979a) established the imbrication (overlap) pattern for the entire theca of *G. polyedra* (Fig. 2.13a) Gocht & Netzel (1974) and Dürr (1979b) doing the same for *Protoperidinium grande* and *Peridinium cinctum* respectively (Fig. 2.13b, c). Happach-Kasan (1982) has provided it for *Ceratium cornutum* and others are being added as awareness of its usefulness grows.

The imbrication pattern appears to be consistent, both in general trend, and in detail for a species. Secondly, overlapping margins can be recognized from surface features such as ridges, although this varies in different genera or species. Generally there is a more pronounced marginal ridge on the overlapping, rather than the underlapping edge.

Essentially there are two gradients in overlap: from dorsal to ventral and from cingulum to pole (Fig. 2.13d). The mid-dorsal adcingular plates (see p. 61) overlap all their neighbours (except the cingulum) and the mid-ventral plates are overlapped by all those around them, probably being the reason for their tendency to sink inwards into a depression such as the sulcus. Similarly, the apical pore complex is surrounded on all sides by overlapping neighbours. Various names have been suggested for the mid-dorsal plates which overlap all their neighbours; 'keystone plate' seems to be as good as any, one existing on the epitheca and the other on the hypotheca.

Gocht (1979, 1981) has been able to determine the imbrication patterns for several fossil cysts by using clues from surface topography. In *Hystrichogonyaulax cladophora*, an Upper Jurassic species, the overlap is shown with astonishing clarity, particularly considering that the cyst wall consists of only a single moiety.

Because the imbrication pattern seems to be very conservative it can be a useful aid in determining plate homologies. The keystone plate and the rhomb

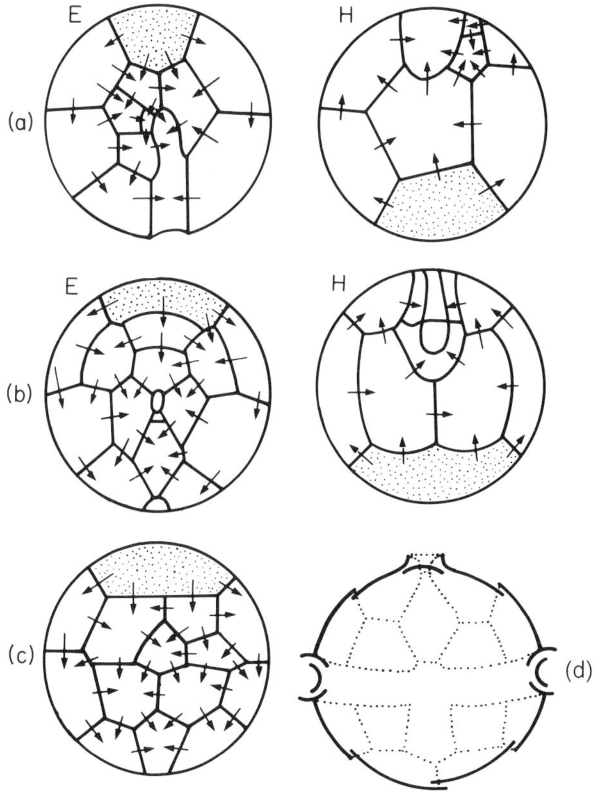

Fig. 2.13. Patterns of plate overlap (imbrication). 'Keystone' plates stippled. (a) *Gonyaulax polyedra*. E, epitheca; H, hypotheca. Redrawn from Dürr & Netzel (1974). (b) *Protoperidinium grande*. E, epitheca; H, hypotheca. Redrawn from Gocht & Netzel (1974). See also Fig. 2.18(g). (c) *Peridinium cinctum* epitheca. Redrawn from Dürr (1979b). See also Fig. 2.18(f). (d) Generalized scheme of plate overlap from the girdle to the poles in a peridinioid theca. Original.

plate can be recognized, when displaced, because of their overlap and underlap features respectively. Girdle plates may overlap, underlap or abut (see Netzel & Dürr 1984).

3.6 Tabulational types

Within the thecate dinoflagellates most taxa are characterized by their tabulation alone. Five basic types can serve to discuss tabulation: *prorocentroid, dinophysoid, gymnodinoid, gonyaulacoid,* and *peridinioid* (Taylor 1980). A sixth, *woloszynskoid,* has been distinguished from gymnodinoid by Netzel & Dürr (1984).

PROROCENTROID

This is found in the genus *Prorocentrum* (which includes *Exuviaella*) and probably in *Mesoporos* (= *Porella, Dinoporella*). The theca consists of two large plates, termed 'valves', and a small cluster around the flagellar pores: the 'periflagellar platelets' (Figs 2.6a and 2.14a). There are two larger pores at the anterior end and the valves usually have trichocyst pores. Flagellar insertion is discussed in Section 1.4.

The valves meet at the 'sagittal suture' which often has a jagged appearance. One valve is more excavated by the periflagellar pore field than the other and this was designated as the right valve by Bütschli (1885), the other being the left valve. Some recent papers have reversed this convention.

The least number of periflagellar platelets seems to be eight (Fig. 2.14a). Many early authors did not observe them due to their small size.* The largest number recorded so far is twelve in *P. micans*. Although there is some variability

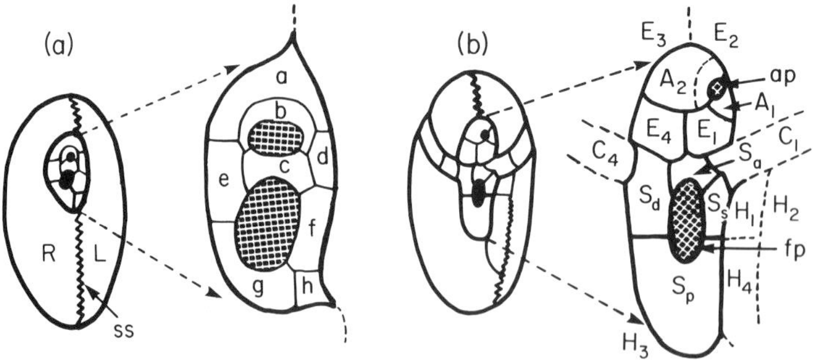

Fig. 2.14. (a) Basic prorocentroid organization: left, whole theca (three-quarter apical view); right, periflagellar platelets. Redrawn from Taylor (1980). (b) Basic dinophysoid organization: left, ventral view of whole theca; right, detail of the sulcal region. Redrawn from Taylor (1980).

* The platelets are most readily observed by staining (3·5 g chloral hydrate in 1 ml hydroiodic acid with a few iodine crystals added; can be diluted further with hydroiodic acid), scanning electron microscopy revealing the ridges and depressions, but not the sutures (Fig. 2.6a).

in the latter species, and our knowledge of these platelets is based on only a few species, there appears to be some predictability to the pattern. Taylor (1980) has labelled the platelets using lower case letters. Often a flange-like spine is formed from *a*. Platelet *b* may be inclined relative to the others. The pores are always separated by only one platelet, *c*.

In *Mesoporos* there is a fairly prominent pore in the centre of each valve. According to Braarud (1945), who cultured *M. perforatus*, the central pore is not as large as that illustrated by Lebour (1925). Details of the periflagellar platelets of this genus are not known, although Braarud observed two flagellar pores, and the valve indentation is the same as in *Prorocentrum*.

Some desmokonts, such as *Adinomonas*, *Haplodinium* and *Desmomastix*, appear to have simpler walls consisting of only one or two parts. They have been seen only once, more than 50 years ago, and require modern re-examination to determine their true wall condition.

DINOPHYSOID

In the dinophysoids the theca is fundamentally divisible into two lateral halves, separated by a jagged suture corresponding to the sagittal suture of prorocentroids. However, a girdle and sulcus are superimposed on this laterality and the plates in the pore area are larger and more readily seen (Fig. 2.14b).

As a group the dinophysoids are striking in their often bizarre, morphological elaborations involving the body form or, a particular dinophysoid feature, list elaborations (see Section 1.5 and Taylor 1980). However, the thecal tabulation is relatively conservative (Balech 1980; Taylor 1980), typically consisting of eighteen or nineteen plates. In the past, three different systems for designating the plates have been used. Those of Tai & Skogsberg (1934) and Balech (1967, 1980) differ in small but significant ways. Abé (1967) attempted to use the Kofoid system (Section 2.3.6). Here the Balech system is used for dinophysoids.

The epitheca consists of two large, left and right dorsal epithecal plates (E2, E3). A small, triangular, ventral part of the epitheca located above the sulcus, is subdivided into two small left and right ventral epithecal plates (E1, E4) which, like the larger ones, contribute to the upper girdle list, and two or three small apical platelets (A1, A2, A3) surrounding an apical pore, which in dinophysoids is only slightly bigger than the surrounding trichocyst pores and is ventrally displaced relative to gonyaulacoids and peridinioids (see below). In *Metaphalacroma* the first apical platelet seems to be missing (Balech 1980); in *Dinophysis* there are two apical plates and three in *Latifascia* (=*Heteroschisma*). Earlier observations of a single apical platelet in *Ornithocercus* (designated P by Norris 1969 and Taylor 1971) were probably incomplete, *Ornithocercus* and *Histioneis* now being regarded as having the same tabulation as *Dinophysis* by Balech (1980).

There are four cingular plates (C_{1-4}), the dorsal ones being larger than the ventral pair. The sulcus consists of four sulcal plates, designated by Balech in the same manner as those of gonyaulacoids and peridinioids: i.e. *Sa*, anterior sulcal; *Ss*, left (sinister) sulcal; *Sd*, right (dexter) sulcal; *Sp*, posterior sulcal; all surrounding a single, large flagellar pore from which both flagella arise. *Sp* is the largest of these. In those dinophysoids in which the girdle is at the anterior end of a long neck like a golf club (*Amphisolenia*, *Triposolenia*), the flagella arise from the body at a remote distance from the girdle and *Sa* and *Sd* are very elongated (Fig. 2.5a–c).

The hypotheca resembles the epitheca in having two large dorsal left and right hypothecal plates (H_2, H_3) and two smaller hypothecal plates (H_1, H_4) which are usually aligned one above the other to the left and below the sulcus. Their affinity to left and right halves, respectively, is revealed by the division (sagittal) suture passing between them. Each of these ventral hypothecal plates contributes to the left sulcal list which is very large in *Ornithocercus* (Fig. 2.4c) and *Histioneis* (Fig. 2.4d). *Latifascia* has a large, triangular H_1. *Citharistes* (Fig. 2.4b), with its unusual C-shaped body forming a dorsal 'phaeosome chamber', has two slender, additional plates, termed dorsal intercalaries by Balech (1980), which lie adjacent to the sagittal suture within the chamber.

GYMNODINOID

Until recently it was assumed that all the 'unarmoured dinoflagellates' (Kofoid & Swezy 1921) lacked a theca or other wall structure, as their name implied. The presence of a reticular cortical pattern could be detected with the use of staining (the 'argyrome' of Biecheler; see Section 1.2), but it was only through electron microscopy that the relationship between the amphiesmal vesicles and the reticular pattern became apparent. *Gymnodinium* is the archetypal 'naked dinoflagellate' and thus it came as a surprise when the type species, *G. fuscum*, was found to possess extremely delicate, plate-like structures within the vesicles (Dodge & Crawford 1969). Loeblich & Morrill (1979) have argued that these fine plates, which are osmiophilic, correspond to the transient 'thecal membrane' which acts as a precursor structure during thecal formation in more heavily thecate species (see Chapter 3). However, more well developed, but still thin platelets have been seen in other *Gymnodinium* species (Schnepf & Deichgräber 1972) and there seems to be a continuum from extremely thin platelets, invisible under normal light microscope observations, to clearly visible in *Woloszynskia* (whose separation from *Gymnodinium* may therefore be a matter of degree) and *Crypthecodinium*. Few authors have paid much attention to the pattern, apart from Biecheler (1952) and Netzel & Dürr (1984) and so its variations are not well known. The acrobase (Fig. 2.2d) usually extends up to the anterior pole and loops around it anticlockwise. However, clockwise apical loops are also known and some gymnodinoids have a crest-like differentiation that does not contact the sulcus, termed a *crista*.

By reducing thecal plate thickness while retaining peripheral strength, gymnodinoids have been able to use their plasticity to evolve into many elaborate forms (see Frontispiece), such as those found in the derivative groups, the noctilucoids (which have a well developed pellicular layer) and warnowiaceans.

GONYAULACOID AND PERIDINIOID

These two groups, which typify the classical 'armoured dinoflagellates', having a well developed girdle and sulcus, share the presence of five complete latitudinal series of plates (*apicals*, *precingulars*, *cingulars*, *postcingulars*, *antapicals*) plus one non-latitudinal series—the *sulcals*. Additional plates are termed *intercalaries* (anterior or posterior on the epi- or hypotheca respectively). An *apical pore complex* (*APC*) is usually present at or near the apical pole. In most taxa a single rhomboidal plate extends from the APC to the sulcus. It is traditionally assigned to the apical series.

Despite this shared pattern, which led to them being classified in a single order, the Peridiniales, there are differences in the thecal pattern, overlap, mode of division and cyst types which suggest that they form distinct lineages and Taylor (1980) has separated an order Gonyaulacales from the Peridiniales.

Details of these patterns and plate terminology (the Kofoid System) are given in the sections which follow and only some general comments will be given here.

In these types we see an approximation to a more radial pattern in which most of the plates (other than the sulcals) are more similar in size than in the

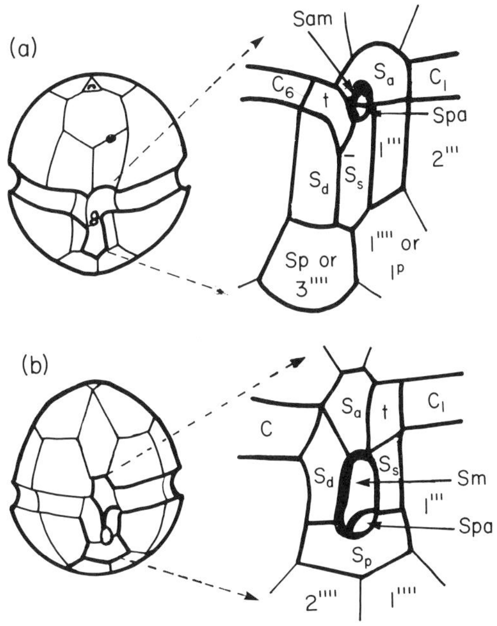

Fig. 2.15. (a) Basic gonyaulacoid organization: left, ventral view of whole theca (e.g. *Triadinium*); right, detail of the plates of the ventral region (Balech notation). Redrawn from Taylor (1980). (b) Basic peridinioid organization: left, ventral view of whole theca (e.g. *Protoperidinium*); right, the ventral region (Balech notation). Redrawn from Taylor (1980) and Balech (1980).

prorocentroids or dinophysoids. The body form is more conservative but thecal tabulation has undergone a greater diversification. The epithecal pattern shows the greatest diversity although the primary series are maintained. In particular, within the peridinioids the anterior intercalaries have become a major feature, constituting a series which is almost complete in *Glenodinium* (four anterior intercalaries).

The cingular (girdle) plates are relatively conservative, usually ranging from three (*Protoperidinium*) to six (*Peridinium*). A small platelet at the proximal end of peridinioids and distal end of gonyaulacoids is termed 'transitional' (*t*) because it could be considered either a cingular or a sulcal. In several genera, such as *Crypthecodinium* or *Hemidinium*, and some species of *Ceratium*, the girdle is incomplete. In the genera *Podolampas* (Fig. 2.16d) and *Blepharocysta*, the girdle depression has been lost entirely but the plates appear to be still present, perhaps fused to the postcingulars.

The flagella arise from between a cluster of smaller platelets as they do in prorocentroids and dinophysoids. Kofoid (1909a) termed this the 'ventral area', but most recent authors simply refer to the sulcal region, even if these platelets are not depressed in a furrow. The flagella usually arise close together. In peridinioids, such as *Peridinium*, *Protoperidinium* or *Dissodium*, there is a relatively large, distinct flagellar pore in the posterior part of the sulcus (Fig. 2.15b) in which there are one or two very delicate platelets. In gonyaulacoids such as *Protogonyaulax* or *Gonyaulax*, the flagella arise immediately below the anterior sulcal plate (Fig. 2.15a). The number of sulcal plates varies from four to eight, seven being most common (with some difference of opinion as to whether some marginal plates should be included or not; see Section 3.8).

Larger pores, or pore complexes, located near the apex of the cell in non-prorocentroid dinoflagellates, are termed *apical pores*. In dinophysoids they consist of a simple pore, barely bigger than others on the theca. In gonyaulacoids and peridinioids they are usually much more evident and situated closer to the centre of the epitheca (Fig. 2.16a, b).

The structure referred to by early authors as an apical pore covered by an 'apical closing plate' has been resolved to be an apical pore complex (APC), usually consisting of two plates: a peripheral apical platelet (Postek & Cox 1976) or *pore plate* (Dodge & Hermes 1981), surrounding an inner closing

---

Fig. 2.16. (a)–(e) Scanning electron micrographs, (f) light micrograph. (a) The apical pore complex (APC) of *Triadinium* (*Heteraulacus*) *polyedricum*. The apex is covered by only a single pore plate (Po). In most gonyaulacoids the inner portion is separated as a distinct cover plate (Pi). Original (× 6700). (b) The APC of *Protoperidinium grande*. Original (× 6500). (c) Reticulation on the ventral, epithecal plate surfaces of *P. grande*. Original (× 5000). (d) A partially disconnected theca of *Podolampas bipes* (ventral view). Courtesy of M. C. Carbonell (× 700). (e) The pellicle of *Achradina pulchra*. Courtesy of S. Honjo (× 1300). (f) The nuclear capsule of *Plectodinium mineatum* (= *nucleovolvatum*). The cell has disintegrated. Light micrograph. Original (× 1000).

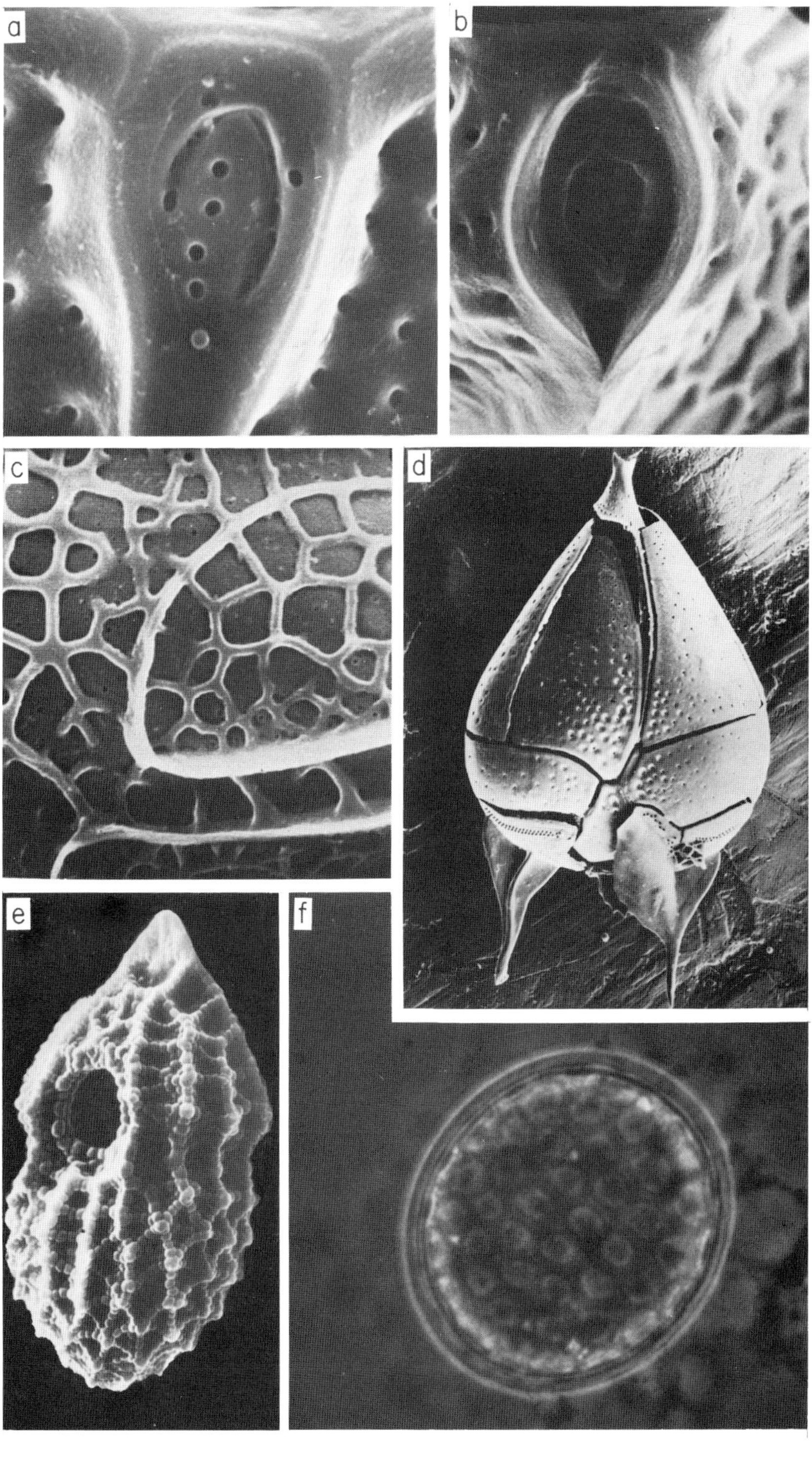
a
b
c
d
e
f

platelet or *cover plate* (Dodge & Hermes 1981), which may be only incompletely separated from the other in more primitive gonyaulacoids. The principal apertures are either a row of pores lying within a curved groove (gonyaulacoids) or a large opening (peridinioids).

The function(s) of the APC is not evident. There is clearly a battery of trichocysts at this anterior end, but it is also possible that it has a sensory role.

In chain-forming dinoflagellates the apical region is in direct contact with the antapex or, in the case of *Ceratium*, the girdle of the next most anterior individual. It is difficult to see if there is cytoplasmic continuity between individuals but, if so, the apical pore complex would be involved. In some species there is an additional pore, adjacent to the cover plate, which is absent from the anteriormost individual, e.g. in the thecate pairs of *Pyrocystis* cf. *fusiformis* (Pincemin *et al.* 1981). In *Protogonyaulax catenella* (Whedon & Kofoid 1936) and *Pyrodinium bahamense* (Steidinger *et al.* 1980) a posterior pore is developed in the posterior sulcal plate of all individuals in a chain, except the posteriormost. It seems to correspond in alignment to the apical pore of the individual posterior to it.

Some peridinioids lack an APC, e.g. the subgenus *Cleistoperidinium* of the genus *Peridinium*. Although the marine genus *Protoperidinium* (included in *Peridinium* for many years) usually has an APC it can be lacking, e.g. in *Protop. steidingerae* (Balech 1979).

Other enlarged pores may occur on the thecal surface. In many gonyaulacoids there is a *ventral pore* which usually lies close to the dextro-anterior of the first apical plate or its homologue. Its function is unknown. In *Ceratium carolinianum* there is an unusual digitiform depression on the dorsal side of the cell (Bourrelly & Couté 1976), and one is also present in *Ceratocorys horrida* (unpubl. obs.).

3.7 Plate designation systems

Several different systems have been proposed to give individual plates a designatory number or letter. Although some of these persisted into the 1920s (e.g. Lefèvre 1928) it was the system of Kofoid (1907a, 1909a) which came into universal use for peridinioids and gonyaulacoids. Over the years it has been slightly modified, particularly by Balech. In Kofoid's System the plates immediately anterior to the girdle are the *precingulars*, except for the mid-ventral plate (the 'rhomb' plate) when it extends to the apex. It and the row surrounding the apex are the *apicals*. If any plates occur between these and the precingulars they are designated as *anterior intercalaries*. The latter are often not a complete series and may even be interrupted (e.g. *Peridiniella*). The apical pole is usually recognizable by the apical pore complex (APC) with a single (P) or double (Po + Pi) plate covering. In *Protoperidinium* a minute platelet, designated the 'x' plate (Balech 1980) or preapical platelet, is located between the apex of the first apical and the APC.

The rows of plates immediately posterior to the girdle, other than those in the sulcus, are the *postcingulars* and those at the antapex are the *antapicals*. Any plates additional to these on the hypotheca and not within the sulcus are usually termed *posterior intercalaries*. Recently Balech (1980) has proposed excluding any plates contacting the sulcus from the latter. Following this course, which may lead to confusion due to different usages, the traditional posterior intercalary (1^{p}) of *Gonyaulax* would become the first antapical plate.

The girdle consists of a row of plates known as *cingulars* (or girdle plates) and those that line the sulcal depression are *sulcals*. Sometimes the pre- and postcingulars have been referred to collectively as the *adcingulars* (Gocht & Netzel 1974).

The individual plates in all series except the sulcals are numbered consecutively, starting with the plate closest to the midventral position and counting off anticlockwise (from left ventral to dorsal to right ventral).

Kofoid (1909a) used a superscript notation to designate each series in conjunction with the number of the plate in question: apicals$'$, anterior intercalaries^{a} (originally$^{\circ}$), precingulars$''$, postcingulars$'''$, posterior intercalaries^{p} and antapicals$''''$. Barrows (1918) suggested superscript *g* for those in the girdle, although C is now also used (Balech 1908). Most recent authors use Balech's terminology and designations (Sa, Sp, etc) for the sulcal platelets (see Fig. 2.15a, b). Apical pore platelets are indicated by P.

Kofoid realized that the number of plates in each series is a useful adjunct to descriptions and he used the *plate formula* as a shorthand means of doing this. It consists of listing the number of plates in each series, together with their series notation, beginning at the apex and concluding with the antapicals, e.g. P, $4'$, 3^{a}, $7''$, 3C, 6S, $5'''$, 0^{p}, $2''''$, which is the formula for the genus *Protoperidinium*. If there are no intercalaries the zero is often omitted, and queries are used when the number is unknown. The very small apical closing platelets, and the '*x*' platelet of *Protoperidinium*, have not been included in formulae by most authors although they are as distinguishable, as are some of the smallest sulcals.

The Kofoid System has also been used to refer to the 'paratabulation' of cysts (Section 4.1, Chapter 15; Evitt *et al.* 1977; Evitt 1985).

3.8 Problems with the Kofoid System

The Kofoid System has been widely adopted because it is usually simple to apply. However, it is not always so and there are some problems which make it a poor system for comparative purposes. The heart of the matter is that the system has no provision for the determination of plate *homology*. If one is comparing the tabulation of genus A with genus B it is essential to compare homologous structures. However, because plates are numbered simply from left to right, the fourth precingular of one may not be the fourth precingular of the other, for example. One subdivision (or fusion) of a plate will cause all those 'downstream' of it to change their numbers.

A greater difficulty arises from differing attributions of plates to series by various authors. For example, apical plates are usually defined as those in contact with the apical pore complex (APC) but if an APC is lacking, as it is in many freshwater *Peridinium* species, or cysts, this is difficult to apply. What if a plate does not touch, but is very close (Fig. 2.17b)? In fact, the 'first apical plate' of many marine *Protoperidinium* and *Scrippsiella* species does not contact the pore plate. It is separated from it by a small, elongate platelet (Fig. 2.18b, e), overlooked for many years, termed the *x* plate by Balech (1980) or the preapical plate or, because it often lies in a furrow, the canal plate (Fine & Loeblich 1976). This problem can be solved by tacit agreement to stretch the Kofoidean rules, or to classify the platelet as part of the APC, but arbitrariness is once more evident. If two epithecal patterns are identical, but the APC is in a different location (Fig. 2.18b, c), it alters the number of apical plates, an apical becoming an anterior intercalary plate.

There are also problems in deciding whether a plate is a sulcal or not: e.g. if it extends far onto the epitheca (in *Heterocapsa* the plate immediately anterior to the sulcus has been variously interpreted as first or eighth precingular, or the anterior sulcal plate), or where the sulcus and the cingulum meet, or when a plate is on the edge of the sulcal repression, or is in the depression in one species and not in another.

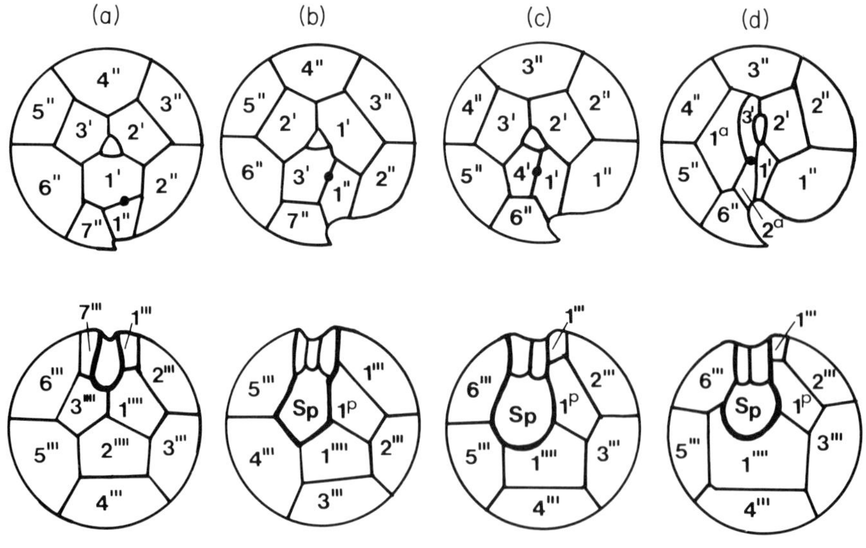

Fig. 2.17. Strict application of Kofoidean plate designations to gonyaulacoid genera: epithecae (upper row); hypothecae (lower row); sulcus (heavier line). (a) *Triadinium polyedricum*. (b) *Pyrodinium bahamense* (the plate marked 1″ is commonly interpreted as 1′). (c) *Protogonyaulax tamarensis*. (d) *Gonyaulax spinifera*.

Taylor (1976) illustrated the plates of *Triadinium* (= *Heteraulacus*), *Pyrodinium* and *Protogonyaulax*, applying the same designations to those considered to be homologous to *Gonyaulax* regardless of whether or not they touched the apical pore complex, were recessed into the sulcal depression, contacted the sulcus, or not. Here, to illustrate the weakness of the Kofoid System, they are numbered in Fig. 2.17a–c according to strict classical interpretation. Note that in all these genera there are equivalent plates, but that slight shifts in position result in their being considered to have quite different tabulations. This also obscures their relatedness. Graham (1942) for example, considered that *Triadinium* occupies an isolated position, although it seems clearly related to *Pyrodinium* and other gonyaulacoids (Taylor 1979, 1980). Besada *et al.* (1982) have similarly used *Gonyaulax* as the arbiter to determine plate homologies in *Gambierdiscus*, using Kofoidean designations. They did not justify why *Gonyaulax*, rather than another genus (e.g. *Triadinium*), should serve as the homology model. Peridinioid tabulation is illustrated in Fig. 2.18.

Eaton (1980) has redefined the epithecal series (with reference to cysts); proceeding from the cingulum toward the apex they are precingulars, apicals and apical-closing plates, to avoid the problem of not being sure where the apex is on many cysts. The latter results in most anterior intercalaries being redesignated as apicals, and the apicals as apical-closing plates.

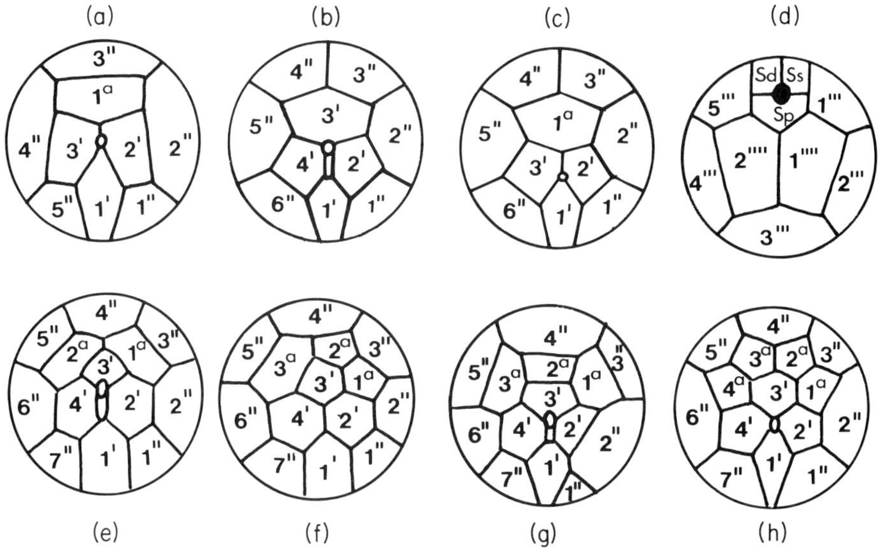

Fig. 2.18. Kofoidean plate designations for peridinioid taxa. (a) *Diplopsalis lenticula*, epitheca. In some, plate 3″ is subdivided as in (c). (b) *Peridiniopsis* sp. Note how the shift in APC position alters the tabulation compared with (c). (c) *Diplopsalis* sp. (d) Peridinioid hypotheca (the suture between 1⁗ and 2⁗ is missing in some members). (e) *Protoperidinium* (subgenus *Archaeperidinium*), e.g. *P. abei*. (f) *Peridinium cinctum*. (g) *Protoperidinium elegans*. (h) *Glenodinium* (*Sphaerodinium*) *cinctum*.

The potential for confusion in these and other 'modified Kofoid Systems' is evident, as stressed by Evitt (1985), and the modifications are entirely arbitrary. Instead of resorting to such modifications it seems best to recognize homologous plates by another system, retaining the Kofoid System for routine descriptions.

3.9 Plate homology determination

Before discussing the details of intergeneric homology determination in gonyaulacoids and peridinioids, there is a broader question of homology that requires consideration, i.e. the gross relationship between the plates of prorocentroids, dinophysoids and gonyaulacoids/peridinioids.

Dinophysoids show several resemblances to prorocentroids, even though they have more plates and a girdle and sulcus are present: the theca is separable into two primary halves separated by a serrated sagittal suture along which intercalary growth occurs. The right half is more invaded by the sulcus than the left, much the same as the relative excavation of the right and left valves of prorocentroids. The transverse flagellum is directed towards the left side. A second, much smaller pore is present on the ventral side of dinophysoids: the apical pore. These and other smaller features, strongly suggest that the left and right lateral halves of dinophysoids are probably homologous to the left and right valves of prorocentroids respectively. The dinophysoid sulcus, possibly with the inclusion of the small epithecals and apicals in the triangular area above, may correspond to the periflagellar platelet region of prorocentroids, the cingulum having developed normal to the axis of the two flagellar bases (Fig. 2.20b).

Possible homologies between these and gonyaulacoids and peridinioids are less apparent. Although the two valves of prorocentroids are usually looked on as lateral, the flagella being apical, this is largely determined by their direction of swimming (Fig. 2.19). One can also look on them as having an upper and lower half, like peridinioids, for example, but swimming in a ventral direction. The fission line (sagittal suture) may also be conservative and therefore aid in this task.

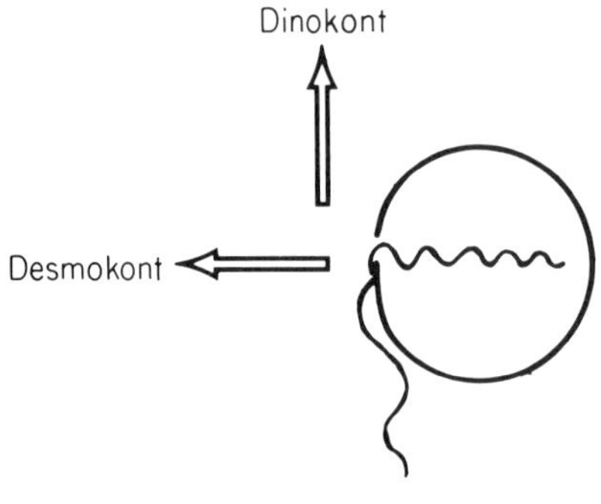

Fig. 2.19. Orientation of the cell during swimming in dinokonts and desmokonts.

Loeblich (1976) suggested that the most excavated valve corresponded to the epitheca and the less excavated valve to the hypotheca (he reversed the conventional left/right terminology). In his interpretation the cingulum would then lie roughly in the plane of the sagittal suture (Fig. 2.20a).

To the present author a different interpretation seems more reasonable.* The fission plane, in dinokont dinoflagellates that share the parent theca, invariably passes obliquely between the flagellar bases, dividing the cell into antero-sinistral and postero-dextral moieties (Fig. 14.2). Like dinophysoids, the sulcus lies more on the right than the left side of this plane. It is suggested here that it is the halves of the theca separated by the fission line that correspond to prorocentroid valves, rather than the halves separated by the cingulum, and that the left, least excavated valve is homologous to the antero-sinistral thecal half and the more excavated right valve corresponds to the postero-dextral thecal half (Fig. 2.20b), the periflagellar area once again corresponding to the sulcus. The direction of transverse flagellar beat (initially towards the cell's left side) is consistent with these interpretations.

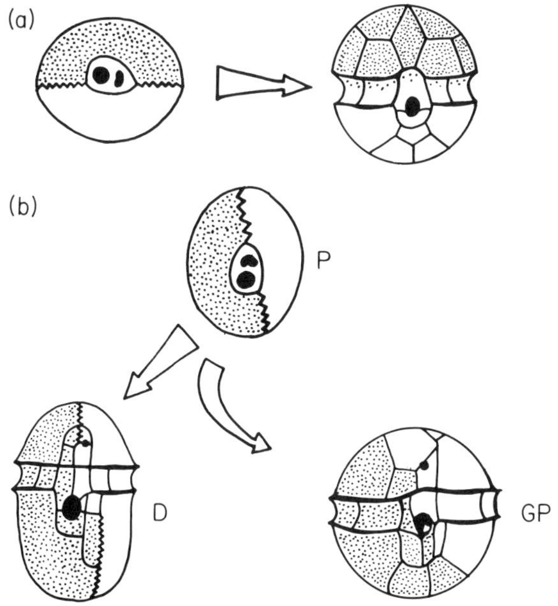

Fig. 2.20. (a) Gross homology between prorocentroid (left) and peridinioid (right) thecae implicit in the interpretations of Loeblich (1976) and Tappan (1980). The most excavated prorocentroid valve corresponds to the epitheca (all? cingulars?). (b) Gross homologies suggested as more probable here. The most excavated (right) valve of prorocentroids (P) is homologous to the right half of dinophysoids (D) and the postero-dextral moiety of gonyaulacoids (demarcated by the fission line) and peridinioids (GP).

*Initially presented at the First International Phycological Congress, 1982. Partly also proposed by Honsell & Talarico (1985).

Difficulties arise in determining individual plate homologies, chiefly as a result of apparent plate enlargement, reduction, displacement, subdivision or fusion. Displacement, which derives from the first two, can result in the allocation of homologous plates to different plate series. Examples of both apparent plate subdivision and fusion can be found in living taxa.

The simplest way to determine homologies is to find living thecal or fossil cyst tabulations which show apparently intermediate conditions. Often it is possible to construct remarkable morphological continua using living forms only (Taylor 1980). Fossils, not surprisingly, can also provide excellent clues, although the paratabulation on the cyst is often incomplete.

Bujak *et al.* (1980) have suggested that a plate or paraplate which is assigned to a series other than that dictated by strict Kofoidean criteria because of presumed homology, be termed a *homologue* and designated notationally by an asterisk, e.g. 1′* would indicate a plate believed to be homologous with a first apical plate even though it is not in a position which meets the Kofoidean criteria for a first apical plate. While this may be a step in the right direction, it still runs the risk of confusion with strict Kofoidean interpretations and offers no guidelines for homology determination.

Another approach is to develop a basic tabulational model and designation system which is less influenced by series attributions and from which more specialized tabulations can be derived. The first of these was developed by this author (first published in Brugerolle & Taylor 1979†) and used to compare toxic gonyaulacoids (Taylor 1979). It was detailed in a broader context (Taylor 1980) and has been extensively applied to fossil cysts by Evitt (1985). All tabulations used are first normalized to a sphere and obvious distortions removed, bearing in mind the common modifications described in Section 1.5, disregarding whether the plates lie within surface depressions (cingulum, sulcus) or not.

The model pattern is radially symmetrical, with an almost identical pattern on the epitheca and hypotheca (Fig. 2.21). It can be thought of as having six primary sectors (like segments of an orange) with three *apical polar plates* (A, B, C), and three *antapical polar plates* (X, Y, Z) corresponding to alternating sectors, and six *equatorial plates* (E1–E6 of Taylor (1979) or a–f of Evitt (1985)); Greek symbols were also proposed but are awkward to use and have been discarded. The remaining plates in the basic model are the *pre-equatorials* (1–6) and *post-equatorials* (I–VI). Subdivisions are recognized by anterior, posterior, median, dexter (right) and sinister (left) abbreviations as they would be oriented if on the ventral surface of the cell. The method allows for a maximum of nine subdivisions for each primary model plate (Fig. 2.21i).

The sector labelled 1 or I is that corresponding to the flagellar insertion in gonyaulacoids (the conventional first apical of gonyaulacoids being 1s, a subdivision of 1). The system was named the Taylor System (TS) by Evitt (1985)

† First presented orally at the conference "Marine Plankton and Sediments", Kiel, 1974.

to clearly distinguish it from the Kofoid System (KS). It is not intended to replace the latter for routine description, but to be used for purposes of comparison.

The relationship between the polar plate arrangement and the flagellar insertion (indicated by an inverted V in Fig. 2.21d–f) offers a distinction in symmetries. For convenience these are termed 'Y' or 'A' symmetry, depending on whether the polar plates can be reduced to these letters in the mind's eye. Thus, gonyaulacoids have a 'YA' symmetry (epithecal and hypothecal type, epitheca first), whereas most peridinioids (*sensu* Taylor 1980) have an 'AY'

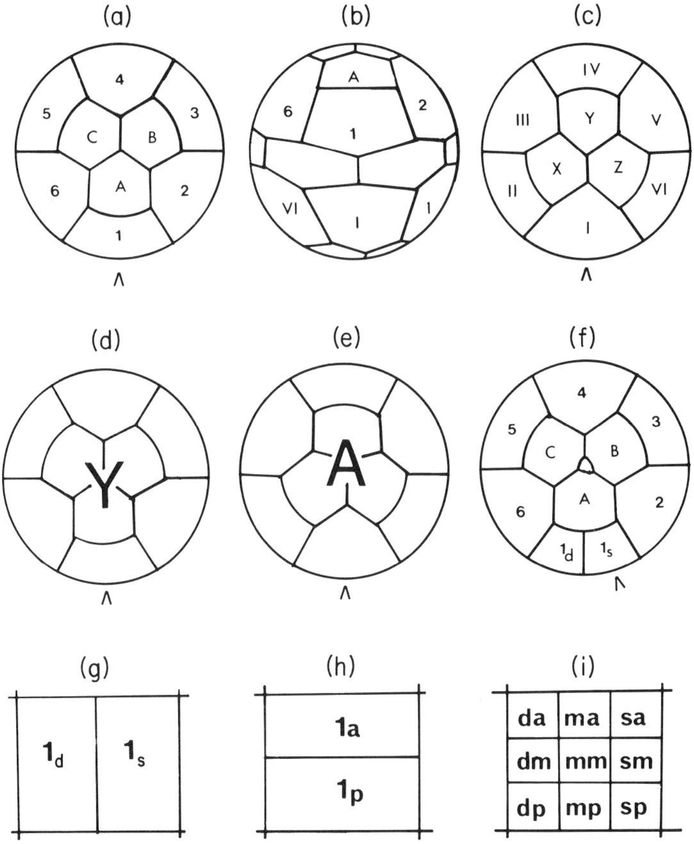

Fig. 2.21. The homology system for gonyaulacoids suggested by Taylor (1980). (a) epitheca; (b) ventral view; (c) hypotheca; (d), (e) the Y and A symmetries indicated by the three polar plates relative to the flagellar insertion; (f) the basic gonyaulacoid pattern in which 1_s becomes the Kofoidean first apical plate; (g), (h), (i) designations for plate subdivisions (o and i are used for outer and inner, as in the APC of most gonyaulacoids). Evitt (1985) has substituted *i, u, v* and *h* for *s, d, a* and *p* respectively.

symmetry. Inherent in the Taylor model is a difference in symmetry between the epitheca and hypotheca, but distortion could result in 'YY' or 'AA' symmetries.

Although it was found that the model reflected the basic symmetries, both of gonyaulacoids and peridinioids, the TS tabulational designations were initially provided only for gonyaulacoids. On further reflection it seems most probable that the peridinioid first apical plate (KS: 1′) corresponds to the gonyaulacoid plate 2 (TS). Thus, the two large antapicals in peridinioids would be Y and Z. The anterior intercalaries (KS) may be subdivisions of either the anterior polar plates or the pre-equatorials (TS) or, like the apical pore plate (P: KS and TS), they may have arisen by new plate formation *between* the other series. Such plates are referred to here as *intertabular* plates. In *Protoperidinium* there are two or three anterior intercalaries (Fig. 2.18e, g) (KS: 1^a–3^a), in *Glenodinium* (*sensu stricto*) four (Fig. 2.18h), and in gymnodinoids and the fossil suessoids one or more complete anterior intercalary series is present. The answer to the origin of the intercalaries lies in a study of probable intermediate forms. Unfortunately, many fossil peridinioids do not exhibit much tabulation. The determination of the nearest living forms should also help come to conclusions as to plate origins.

Eaton (1980) developed a model of cyst paratabulation (first present at the Penrose Dinoflagellate Conference in Colorado in 1979) which was derived by almost the reverse process of that by Taylor. His basic model (ES) has three series on both the epitheca and hypotheca (Fig. 2.22a, b), with seven plates (the greatest number in the material he was comparing) in each except for a single one at the antapical pole. As noted earlier, he named his series differently from the Kofoid System: those on the epitheca, proceeding from the cingulum to the apical pole, being the precingular, apical and apical closing series; those on the hypotheca being the postcingulars, antapicals and on antapical closing plates.

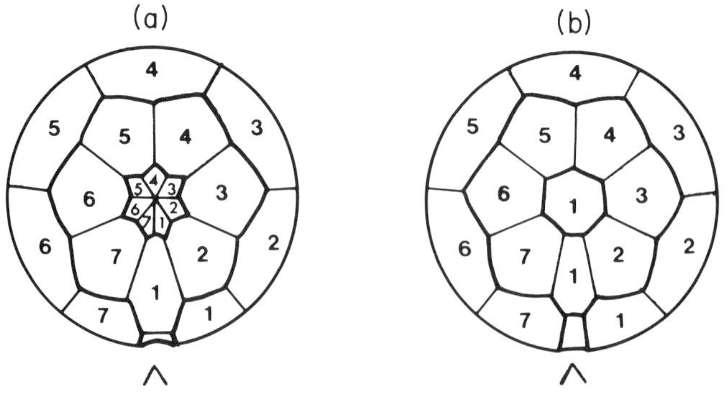

Fig. 2.22. The homology system of Eaton (1980) in which plate fusion is a major theme. Note the reversal in direction of numbering in the epitheca/epicyst (a) and hypotheca/hypocyst (b).

This is a radically different use of the Kofoidean terms 'apical' and 'antapical'. The epithecal plates are numbered serially from the midventral position, counterclockwise when seen from the apex in the usual Kofoidean fashion, and counterclockwise, when seen from the antapex, on the hypotheca (which is the reverse of the KS). Because most tabulations have fewer plates than his model, real plates are viewed as the *sum* of particular model plates, rather than as subdivisions in the TS.

There are other aids to homology determination. The imbrication pattern (Section 3.5) allows recognition of displaced mid-dorsal pre- and post-cingulars (TS:4, IV; ES:4), the first apical (TS:1s; ES:2ap.1) and the most overlapped antapical/posterior sulcal (TS:Z). The great similarity between the imbrication patterns of gonyaulacoids and peridinioids suggests that their differences in symmetry could be due to a primary distortional influence, rather than a drastic shift in flagellar insertion. On this and other grounds Happach-Kasan (1982) believes that both gonyaulacoids and peridinioids can be derived from the same basic ancestor with four apical plates.

In many gonyaulacoids there is often a 'ventral pore' which is almost invariably associated with suture 1s/A (TS) when present and the pore plate is asymmetrically elongated, 'pointing' to suture A/B in which the peridinioid APC may be located. The fission line seems to be relatively conservative in gonyaulacoids (peridinioids usually shed the whole parent theca).

3.10 Other internal skeletal elements

Some dinoflagellates possess internal structures which have some supportive function. Perhaps the best known of these are the paired, five-rayed, star-like elements formed deep within cells of the athecate genus *Actiniscus* (= *Gymnaster*) (Fig. 2.23b). As noted in Chapter 1, these elements were seen first in the fossil record and they are still commonly observed in sediments from the Cretaceous to the present (e.g. Orr & Conley 1976). Occasionally four such elements have been seen in *A. pentasterias*, two smaller than the other. This seems to be a pre-division state.

Paired siliceous spicules are also present in the apical end of *Plectodinium mineatum* (= *P. nucleovolvatum* (Biecheler 1934b), see Taylor, 1980) but these are much less substantial. A single, branched skeleton, also apparently siliceous, in the form of an inverted Y, with acute branched apices, is present in *Dicroerisma* (Fig. 2.23a).

Hovasse (1943) suggested a dinoflagellate affinity for the ebriids, non-photosynthetic flagellates with an internal, branching, siliceous skeleton resembling that of *Dicroerisma*, principally because their nuclei appear to have continuously condensed chromosomes. The latter have not yet been examined with electron microscopy and so this allocation must remain tentative although

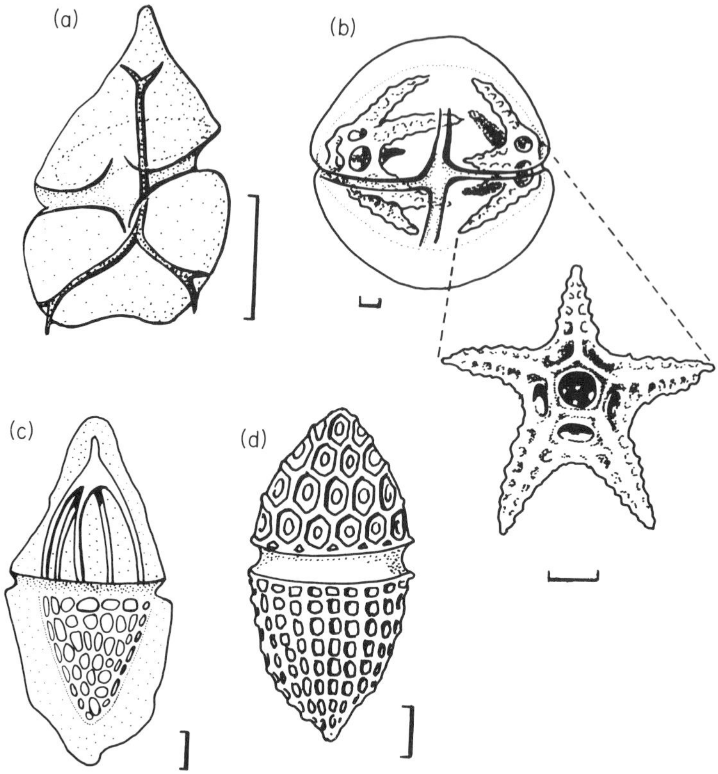

Fig. 2.23. Internal skeletal elements. Scale = 10 μm. Flagella omitted. (a) *Dicroerisma psilonereiella*, an athecate species with an internal, branching skeleton reminiscent of an ebriid developmental stage. Redrawn from Taylor & Cattell (1969). (b) *Actiniscus* (*Gymnaster*) *pentasterius*. Top, whole cell (redrawn from Schütt 1895); bottom, one of the star-shaped, internal elements, drawn from the SEM micrographs of Orr & Conley (1976). (c) *Monaster rete*. Redrawn from Schütt (1895). (d) *Amphitholus* (*Amphilothus*) *elegans*. Redrawn from Schütt (1895).

Loeblich (1982) has classified the Ebriophycidae as a subclass of the Dinophyceae. Only two living genera are known: *Ebria* and *Hermesinum*.

Internal, basket-like skeletons of unknown material were seen by Schütt (1895) in *Monaster* and *Amphitholus* (Fig. 2.23c, d). It is possible that they may be more pellicle-like in nature, showing some similarities to *Achradina* (Section 3.1). Tappan (1980) suggested that they might be cysts.

In addition to the more obvious internal skeletal elements described above, several genera exhibit degrees of strengthening of the nuclear envelope. Slight indications of this can be found in *Protoperidinium,* but it is particularly evident in some *Gymnodinium*, *Gyrodinium* and the genera considered to be gyrodinioid derivatives by Taylor (1980), i.e. *Polykrikos*, *Cochlodinium*, and members of the Warnowiaceae, such as *Erythropsidinium*, *Warnowia* and *Nematodinium*. With light microscopy the nuclear envelope may appear to have a double layer (Kofoid & Swezy 1921). Some of these have been studied ultrastructurally by Greuet (1972) who found a fibrous 'perinuclear capsule', penetrated by pores, to be present. Interestingly, perinuclear capsules are also present in *Actiniscus* and *Plectodinium*, (also suggested to be gyrodinioid derivatives by Taylor 1980), additional to their siliceous elements. The perinuclear capsule of *Plectodinium* is very strongly developed and can be easily seen with light microscopy, having a basket-like, reticular appearance (Biecheler, 1934a, b; and Fig. 2.16f). A link between dinoflagellates and radiolarians has been suggested by Hollande *et al.* (1962), partly on the basis of the presence of a capsule and siliceous elements in these dinoflagellates.

Perinuclear skeletal elements may be transient in the life cycle. Taylor (1969) observed that delicate platelets form around the nucleus of *Gonyaulax pacifica* just prior to encystment (presumably asexual) and can be seen in the temporary cyst after ecdysis (Fig. 2.24b).

A variety of perforated discs of unknown composition were observed in the cytoplasm of *Pavillardinium kofoidii* by Kofoid (1907b). Tappan (1980) has noted the similarity of these discs and others seen in a gymnodinoid species to fossils assigned to the genera *Rocella* and *Pseudorocella*, now thought to be a diatom and chrysophyte respectively, suggesting that they might be remnants of ingested food.

'Discoasters' are stellate microfossils, somewhat similar to coccoliths in form, composed of calcium carbonate (extensively reviewed by Tappan 1980). No living forms have been discovered and post-Pleistocene occurrences of loose discoasters are thought to be due to resuspension of fossil sediments. Bursa (1971) observed cell-like bodies with discoasters adhering to their surfaces on the underside of the ice in the Canadian Arctic. He attributed these to dinoflagellates, and lumped them with *Actiniscus*, also present in his material, incorrectly referring to the latter as discoasters although they are siliceous. Most subsequent authors have viewed Bursa's dinoflagellate discoasters as artefactual aggregates of scoured sedimentary material, but R. Horner (pers. comm.) has also seen discoaster-like structures adhering to Arctic dinoflagellates.

4 CYST MORPHOLOGY

4.1 General wall features

Cyst stages, in which the cell loses its flagella and becomes surrounded by a continuous wall, can be simple or highly complex in their morphology. Because the walls of resting cysts (= hypnospores, zygospores) preserve well in sediments the fossil record seems to consist for the most part of cyst walls and the taxonomy of fossil dinoflagellates is based entirely on cyst features (Chapter 15).

The simplest cyst coverings consist of a gelatinous layer produced

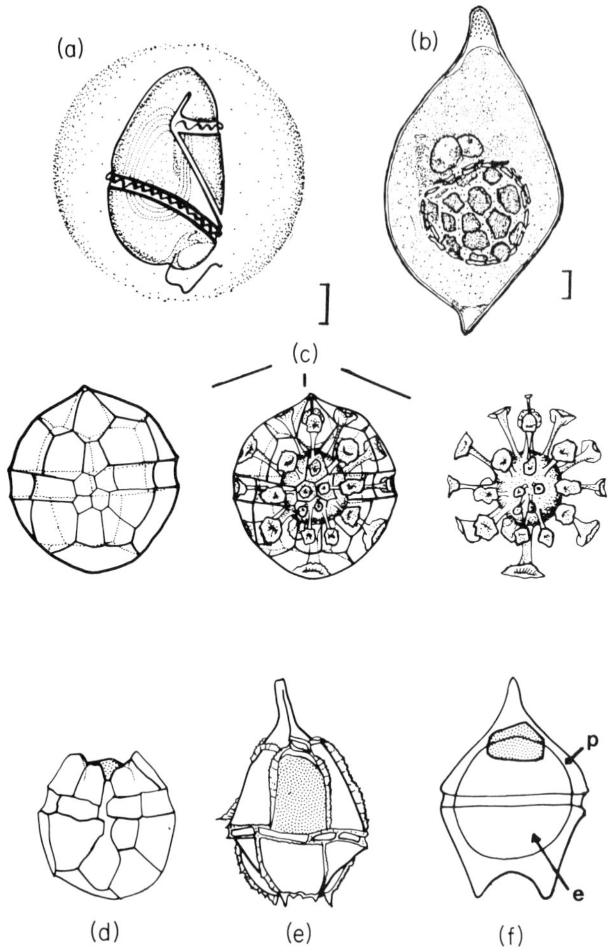

Fig. 2.24. Dinoflagellate cysts (see also Figs 15.3, 15.4). (a) *Cochlodinium pirum*: an exogenous, gelatinous cyst in which the cell can retain its flagella for 14 hours or more. Redrawn from Kofoid & Swezy (1921). (b) The temporary cyst produced by *Gonyaulax pacifica*, following apical ecdysis. The nucleus is surrounded by delicate platelets (not seen in other species). Redrawn from Taylor (1976). (c) The relationship between the thecal plates and the processes of a chorate cyst, inferred by Evitt (1963) from the cyst of *Hystrichosphaeridium*. The 'one process–one plate' relationship is only true for some spiny cysts. The tabulation is probably slightly more gonyaulacoid than shown in the diagram. Left, hypothetical theca; centre, cyst within the theca; right, detached cyst. Redrawn from Evitt (1963). (d) Apical archeopyle (detachment of the entire apical paraplate series). Redrawn from Evitt (1964). (e) Dorsal archeopyle (stippled), e.g. the 3″ paraplate of *Gonyaulacysta jurassica*. Redrawn from Evitt (1964). (f) Dorsal anterior intercalary archeopyle in a cavate cyst, e.g. *Deflandrea* sp. p, pericoel; e, endocoel. Redrawn from Sarjeant (1982).

exogenously, presumably by release from mucocysts, in rapid response to stress. This is found in many athecate genera such as *Gymnodinium* and *Gyrodinium* (Fig. 2.24a). Schütt (1895) illustrated many cells surrounded by a gelatinous envelope in his study on living plankton material. Klebs (1912) referred to such cells as 'Schleimcysten' (slime cysts) to distinguish them from those with a more substantial wall ('Hautcysten'). The outer surface may take on a more substantial appearance with time, or be covered with debris from the cell's surroundings, particularly in sand-dwelling species. Klebs (1883) found that this type of envelope in *Gymnodinium fuscum* stains dark blue with methyl green and also stains strongly with methyl violet, both dyes which stain trichocyst and mucocysts in other protists. Division can also take place in this type of cyst.

Most cysts are surrounded by more substantial walls. The wall (phragma) may consist of a single layer. This is most common in temporary ecdysal cysts, e.g. *Ceratium hirundinella* 'smooth cysts' (Chapman *et al.* 1982), and several species of *Gonyaulax* (Taylor 1969, 1976a, b), and it probably corresponds to the pellicular layer of the mastigote (Section 3.1). Most commonly the walls of resting cysts consist of two layers (Jux 1968a, b). Cellulose is commonly present in the cysts of living species, although it is absent from fossil cysts.

One or both of the layers is often strengthened with more resistant material, which accounts for their good sedimentary preservation. The strengthening material resembles *sporopollenin*, a fatty-acid-containing material which occurs in pollen and resists concentrated hydrochloric and sulphuric acids, acetic anhydride, 50% hydrofluoric acid (all used in the concentration and preparation of fossil cysts from sedimentary materials), nitric acid, 50% sodium hypochlorite, potassium hypochlorite, potassium hydroxide and ammonium hydroxide (Eisenack 1963; Evitt & Davidson 1964). A similar material has also been found in the pellicular layer of some mastigote-phase cells (Loeblich 1970), supporting the view that it is this layer which may differentiate into one of the layers of the resting cyst wall, as well as being able to serve as the immediate source of the temporary cyst wall (Taylor 1980).

In *Peridinium cinctum* there are three wall layers, each covered by a membrane (Dürr 1979c), with chitin in the outer layer (exospore) and sporopollenin in the inner endospore layer (see Chapter 14). In cysts of *Woloszynskia apiculata* von Stosch (1973) recognized four layers, with a peripheral *perispore* that disintegrates as the cyst develops, additional to an *exo-*, *meso-* and *endospore*. The exo- and the endospore are acid-resistant, the mesospore apparently being a source of mucilage which expands the exospore outwards during cyst maturation. Acid resistance is not a guarantee of preservation in sediments. Von Stosch found that the thick endospore dissolved during germination in this species.

Some cyst walls have an external, crystalline, calcium carbonate layer additional to the organic layer(s). *Scrippsiella trochoidea* (= *Peridinium trochoideum*) has an outer layer consisting of numerous calcitic, rod-like crystals (Wall *et al.* 1970). *Thoracosphaera heimii* occurs chiefly as a calcareous sphere but produces a motile peridinioid stage (Tangen *et al.* 1982). *Scrippsiella*-like paratabulation has been found on calcareous microfossils termed 'calcispheres', sometimes only on the inner surface of the cyst wall (Fütterer 1976; Keupp 1979, 1980). It seems likely that many of the calcispheres, formerly assumed to be coccolithophorids, are dinoflagellate cysts (Chapter 15). Calcitic cysts have only been found in peridinioid taxa with *Scrippsiella*-like tabulation so far and appear to form a distinct lineage (Bujak & Davies, 1983).

Although siliceous skeletal elements are known from motile stages of several species (Section 3.10) silicified cyst walls are known primarily from some fossil cysts (e.g. *Peridinites*) and in these it is not certain whether silicification occurred before, during or after fossilization (Tappan 1980). The strongly developed pellicle of *Achradina*, which may be homologous with a cyst wall although formed in the motile stage, appears to have some silica content (S. Honjo, pers. comm.) and silica granules are present in the wall of zygotic cysts of the freshwater species *Ceratium hirundinella* (Chapman *et al.* 1981; and Fig. 3.23 here).

Palynologists recognize several major categories of cyst types. If the cyst forms close to the thecal wall and does not subsequently develop extensive projections it is said to be *proximate* (Fig. 2.24f). Often such cysts reflect the shape of the parent thecate stage and some may lack ornamentation. If the processes or ridges exceed 20% of the shortest diameter of the central body, the cyst is *chorate* (Sarjeant 1982, and Fig. 2.24c here). Intermediate types, i.e. with surface structures extending 10–20% of the shortest diameter are *proximochorate* (Fig. 2.24e). Cysts with a cavity or cavities between their wall layers are said to be *cavate* (Fig. 2.24f). These in turn have been modified or subdivided by various authors, e.g. Norris (1978), Stover & Evitt (1978): for further discussion and illustration of examples see Chapter 15. The terminology which has been used for other cyst features has also become elaborate (see Sarjeant 1982 and Evitt 1985).

The cyst may show a division into two regions by a girdle-like feature, in which case the part thought to correspond spatially to the epitheca is the *epicyst* (=epitract) and that of the hypotheca the *hypocyst* (=hypotract). Many features which show a superficial resemblance to thecal structures, but which are expressed in the cyst morphology, use the same term as that used for the theca but with the prefix '*para*'-, hence *paratabulation*, *parasutures*, *paraplates*, *paracingulum*, *parasulcus*.

The paraplates are assigned to the same series as those of the theca (apicals, precingulars, etc.) and paratabulation formulae follow the Kofoid convention and superscripts.

When the wall consists of a single layer, that layer is referred to as the *autophragm*. When two layers are present the outer is the *periphragm*, the inner is the *endophragm*, the space between them is the *pericoel* and the inner cavity is the *endocoel* (Fig. 2.24f). Sometimes a third layer, the *mesophragm*, may occur between it and the endophragm, e.g. in *Deflandrea*, or an elaboration of honeycomb-like extensions of the periphragm may produce an additional outer layer, the *ectophragm* with *ectocoel* cavities, e.g. in some species of *Wetzeliella*. The latter is often discontinuous. Because of differential preservation, it is not possible at present to equate these layers directly with the peri-, exo-, meso- and endospore of living cysts, although there is clearly some relationship which will be elucidated with time. In some genera, a further outer layer appears to have been made up of adherant debris, probably originating as a mucilage layer. It is termed a *calyptra* in fossil cysts.

4.2 Surface features

The surfaces of cyst walls range from completely smooth, devoid of any features other than general shape and the excystment aperture (e.g. some *Protoperidinium*, *Protogonyaulax*, *Gessnerium*, *Fragilidium* and *Pyrophacus* cysts), to highly elaborate, equal to or greater in complexity than the theca.

A paratabulation may be expressed in the form of ridges which, if raised like flanges, are referred to as *crests* or *septa*. It may also be visible because of zones of less elaborate structure than on the rest of the surface. Such zones on the cyst, which may even mimic striated intercalary bands, are termed *pandasutural zones*. Patterning within the paraplate region is *intratabular*. If it is close to, and parallel with the margins, it is *penitabular*. If processes or the surface structures arise from the junction of paraplates they are *gonal*, and *intergonal* if they occur between the junctions. Sarjeant (1974, 1982) has collated the terminology for numerous types of surface ornamentation and processes. The distribution of the latter may also be used to determine a paratabulation, particularly if they outline the paraplates or arise from the centres of them. However, in others there appears to be no readily determinable relationship between the process distribution and paratabulation, or processes may arise from only some of the paraplates.

Scanning electron microscopy has revealed that in some fossil cysts the ornamentation, or paratabulational details, are on the *inner* surface of the phragma. This is the case in *Palaeoperidinium pyrophorum* (Gocht & Netzel 1976) and some calcified cysts of the genus *Pithonella* (Keupp 1981).

In some fossil cysts, notably *Dinogymnium* (May 1976), pores resembling flagellar and trichocyst pores have been found. This has revived the possibility that a few fossils correspond to the pellicle of the motile, rather than a dormant, sedentary phase (Taylor 1980 and Chapter 15).

4.3 Archeopyle

The aperture which forms during excystment is the *archeopyle* (earlier termed the pylome), the piece or pieces which separate off forming the *operculum*. When two layers are penetrated there can be a *periarcheopyle* and an *endoarcheopyle*. The operculum may be hinged (*adnate*, formerly called attached), or come off completely (*free*). Evitt (1967, 1984) developed a terminology for the numerous types of archeopyle, depending principally on position and the paraplates involved. An operculum is *simple* if it separates as only one piece and *compound* if separated into more than one piece. If only one paraplate is involved it is *monoplacoid*, and *polyplacoid* if more than one paraplate contributes to the operculum. A simple formula was developed to refer to the archeopylar type: A, I, P, C and H refer to its apical, intercalary, precingular, cingular or hypocystal position respectively. Parentheses surrounding the letter(s) indicate that more than one paraplate appears to be involved in a simple operculum (Norris 1978),* e.g. (3A) indicates an apical archeopyle involving three paraplates. (AI) indicates that the archeopyle involves both the apical and intercalary series (and several paraplates). In some cysts the entire epicyst may detach (AP), or merely the apical series (A, Fig. 2.24d). A plus sign was used to indicate the presence of a parasuture on the operculum between the series, e.g. A + P, but this has been dropped as it is superfluous. The number of opercular pieces separated off by such parasutures are indicated by numbers, e.g. 4A3P. A *combination* archeopyle involves more than one series. Many authors refer to the cuts through the wall associated with the formation of the archeopyle as sutures, because they truly separate the wall into pieces rather than being lines, ridges, areas or depressions corresponding in position to thecal sutures. As Evitt *et al.* (1977) noted, the opercular cuts usually correspond to thecal sutures in position and so 'parasuture' is just as appropriate. Recently Evitt (1985) has expanded the formulae by adding details of the paraplates involved below the main line.

In taxa in which both ecdysis and excystment occur within the life cycle, there is usually no positional correlation between the sutures involved in thecal plate separation in the former and the parasutures which form the archeopyle. For example, *Gonyaulax digitale* and *G. spinifera* undergo apical ecdysis, but their cysts have a dorsal P archeopyle in the cyst (like that in Fig. 2.24e) corresponding to the third precingular plate in position, a very common gonyaulacoid type. This corresponds to the epithecal 'keystone' (most overlapping) plate on the theca (Fig. 2.13a), which may have significance in the evolution of the archeopyle (Gocht & Netzel 1976). Epicystal intercalary archeopyles (I) are the commonest form in peridinioid taxa, e.g. *Wetzelliella*, *Deflandrea* (Fig. 2.24f). At present the only living genus thought to have a hypocystal archeopyle is *Pyrophacus* (cyst name *Tuberculodinium*) (Wall & Dale 1971). Boltovskoy (1979) found that in some *Peridinium* species there was a good positional agreement between the archeopyle and the ecdysal aperture (which he also termed an archeopyle, incorrectly).

* Formerly indicated as a bar above the letter(s).

The operculum is usually missing when cysts are recovered from sediments, although separated opercula may be recognized in the material, even occasionally within the cyst, and some remain attached in a hinge-like manner. In cysts in which paratabulation, or a regular, paraplate-related arrangement of spines, is not present, the correspondence between the archeopyle and the plates/paraplates involved can still be determined by ingenious interpretation of the angularity of the archeopyle margins, although the rounding-off of the corners in a 'reduced' archeopyle can make this a hazardous exercise. Some authors have attempted to reconstruct almost entire tabulations on the basis of archeopylar shape. Evitt's (1985) extensive, very clear exposition of cyst morphology may be consulted to see illustrations of all the features described here.

5 PHENOTYPIC VARIABILITY

Within a species the phenotype can vary due to a number of factors. Some of these are evident, such as growth, division, or encystment. Less evident or less understood, but of great potential interest, are environmental influences on form. For example, form may vary in the same region at different times of the year or may differ between populations of the same species in different regions, such as tropical and temperate, or neritic and oceanic forms. Finally, there are teratological aberrations ('freaks') which, although rare in the natural environment, may be more common in cultures and may be informative about the potential for change or give insights into morphogenesis. These aspects are considered further below. Variability in thecal pattern (tabulation) is also dealt with (Section 5.4).

5.1 Effects of division

CELL DIVISION

Because of the obliquity of the division plane (least in prorocentroids and dinophysoids) daughter cells can differ significantly in their proportions, particularly of epi- and hypotheca, immediately after division (see Chapter 14). Usually, this disparity is made up during the division process, the cells adhering until proportions are approximately restored. However, signs of the disproportionate sharing of maternal components can be seen in many dinoflagellates following division. For example, in the dinophysoid genus *Ornithocercus*, the antero-sinistral daughter receives only two parts of the large left sucal list, the postero-dextral daughter receiving the whole of the right sulcal list and the largest section of the left sulcal list (Taylor 1973a, b and Fig. 14.3). Some regrowth occurs while the cells adhere by a remnant of the intercalary growth zone, the dorsal megacytic bridge (DMB), but is only completed after separation.

In *Ceratocorys horrida* one daughter receives the four antapical spines, and the other the dorsal and ventral spines (Taylor 1976a).

The surface ornamentation of the thecal plates can be different between the maternally inherited, mature half of the theca and the newly-formed, immature half in those dinoflagellates which share the parent theca during cell division (desmoschisis). In those that do not (eleutheroschisis, common in peridinioids) both daughter cells form a complete new theca and look evenly developed over their thecal surface while undergoing maturational changes.

5.2 Ecdysis

This term, 'to become naked', was applied first to dinoflagellates by Kofoid (1908). It refers to the complete shedding of the theca by some gonyaulacoid and peridinioid species in response to environmental stress and/or cyst formation. Dürr & Netzel (1974) have provided ultrastructural evidence that the outer part of the vesicular membrane breaks away from the inner part, so that the new outer membrane is the collective inner moieties of the amphiesmal vesicles. Morrill (1984) has studied the process in *Heterocapsa*.

Because the theca lies internal to the cell membrane, within the amphiesmal vesicles, this process necessitates the shedding of all membranes external to the thecal plates, including the flagella (whose membrane is continuous with the cell membrane).

Usually the cell escapes from a predictable part of the theca, most commonly the apex in *Gonyaulax* ('apical ecdysis') or the left side between the girdle and the precingulars ('lateral ecdysis'), e.g. in *Scrippsiella*. Only those species that which have formed a pellicular layer (Section 3.1) beforehand are able to do this, the inner vesicular membrane and the pellicle forming the periphery of the ecdysal ('pellicle', 'temporary') cyst which results (Fig. 2.24b). A new theca forms beneath the pellicular layer on the one or two motile cells which arise from this asexual cyst.

5.3 Autotomy

This process of 'self cutting' is a means of rapid alteration of the horn length in many marine species of *Ceratium*, being most common in long-horned members of the sections Macroceros and Amphiceratium. Kofoid (1908) first applied the term to dinoflagellates in a lengthy descriptive study which determined the essential manifestations of the process, but not its function.

Shedding of the distal portions of all three horns, or the apical only, or the two antapical horns, is heralded by the development of grooves in the outer wall surface near the base of the horns (Fig. 2.9e). These are not at a fixed point but are proportionate distances away from the cell body. Horns which have been previously shortened may be further shortened by formation of an autotomy groove closer to the base. Presumably the cytoplasm is retracted immediately prior to shedding, although this part of the process has not been studied in detail. The appearance of the cell can be dramatically changed. *C. ranipes* var. *palmatum* loses its 'hand-like' distal parts completely.

Regeneration may follow autotomy. Regenerated distal horn sections can be recognized, at least initially, by the relative thinness of the regenerated wall (Kofoid 1908 and several specimens illustrated by Taylor 1976a). Occasionally the regenerated moieties may not be in regular alignment with the basal parts, e.g. in *C. vultur* var. *vultur* f. *recurvum*.

Kofoid (1908) showed that proportionality was usually maintained after autotomy with the exception of the apical horn. If the right antapical horn is shorter than the left, as is often the case in *Ceratium*, this is still the case after autotomy. Similarly, proportionality is maintained during regeneration.

Because the horns were thought to be important in the regulation of sinking, it was natural to assume that autotomy might be related to water density. Kofoid (1908) found that most autotomized individuals were from deeper levels (below 150 m) off California and calculated, using Ostwald's equation, that short-horned individuals would have a sinking property at 12°C similar to long-horned individuals at 20°C. The present author found that individuals of *C. horridum* and *C. macroceros* isolated into natural sea water enriched with Provasoli's 'ES' addition (McLachlan 1973) and kept at approximately the same temperature as that from which they were isolated, invariably autotomized overnight, regardless of other manipulations (light/dark).

L. Brand (pers. comm.) has preliminary indications that horn development may be related to trace metal levels. Although not yet experimentally verified, his observations raise interesting possibilities and are worth recording here. *C. hexacanthum* has a form in which the left antapical horn forms a long coil. In culture this long, coiled horn only developed in low concentrations of trace metal mix (Guillard f/50 medium), the horns being shorter in higher concentrations (f/2 or f/10). Similarly, in *Ceratium ranipes*, more 'fingers' were formed in cultures with the lower concentration of metal mix.

5.4 Seasonal changes

In dinoflagellates which occur in the plankton through most of the year it has been noticed that distinct seasonal changes in form occur, although the reasons for this are less evident. Pollingher (Chapter 11B) has summarized the well documented seasonal changes of form in the freshwater species *Ceratium hirundinella*, where there is a variation both in cell length and number of antapical horns. In marine ceratia, early studies—summarized by Steemann Nielsen (1934) and Böhm (1976)—indicated that winter and neritic forms had larger cell bodies than summer and oceanic forms. Nordli (1957) found that in *Ceratium fusus* oceanic forms clearly had longer horns than neritic forms.

Nielsen (1956) carried out a careful study of the seasonal variations in form in *C. tripos* at a single locality, the Anholt Knob light ship in the eastern Kattegat near Denmark (Fig. 2.25). He found the shortest cell lengths occurred in late summer or early autumn, the longest being in March–April, the maximum body diameter being positively correlated with cell length. In general these changes could be correlated best with temperature. Lopez (1955) found similar variations in *C. tripos* off Castellon in Spain, concluding that temperature had an influence on the abundance of individuals but not directly on horn length; 'small cells grow best in summer and the longer ones in winter.'

5.5 Thecal plate variability

Because thecal plate patterns have been used so extensively in the taxonomy of thecate dinoflagellates it is important to determine how conservative they are. Despite this there have been relatively few detailed studies of this aspect.

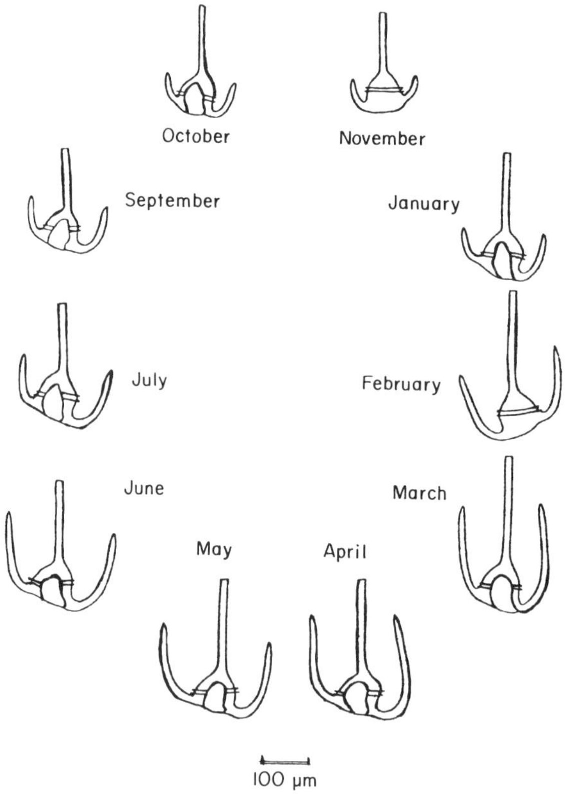

Fig. 2.25. Seasonal variation in the form of *Ceratium tripos*, observed in the eastern Kattegat by Nielsen (1956).

Several principles established by Barrows (1918) during his study of thecal variability in *Protoperidinium*, still hold true. He found that dinophysoids were more constant in tabulation than peridinioids. The hypothecal pattern is more stable than the epithecal pattern. He knew of no cases in which four sutures met precisely at one point. Within the genus he studied, he identified two dorsal and two ventral regions in which most of the variability occurred, these sutural variations being the basis for Jörgensen's (1912) subdivisions ('*ortho*', '*meta*', '*para*' first apical plates; '*quadra*', '*penta*', '*hexa*' second anterior intercalaries; see fig. 15.5) which he supported. He was wrong however, in thinking that the sulcal plates would be too variable to be useful. They have turned out to be extremely conservative, their study being pioneered by Abé (1936). Barrows believed that fewer plates were more primitive than many and that new plates arose by small platelets splitting off from the corners of larger plates, and subsequently expanding, rather than by the splitting of plates into more or less equal halves.

From an analysis of epithecal patterns in various peridinioid genera Taylor (1980) suggested that the upper left quadrant might be a 'hot spot' in which new plates have formed, this being the principal region in which one finds small, asymmetrically located platelets.

Diwald (1939), who concentrated more on biometrics and the recognition of particular taxa in *Protoperidinium*, provided examples both of plate splitting (the first apical) and apparent sutural loss, both being very rare events. Subdivision of the first apical plate was also seen in *Scrippsiella* (*Peridinium*) *faeroense* by Fine & Loeblich (1976).

Cultures, particularly clonal ones (i.e. started from a single cell), are instructive in this regard. Taylor (1975b) illustrated the variation in a clonal culture of *Protogonyaulax tamarensis* (Fig. 2.26). In this species the thecal pattern is easy to see with light microscopy after ecdysis occurs. Many intact hypothecae can be found, although the epitheca often disintegrates. Of 63 epithecae examined, 60 were normal for the species, one had both fewer apical and precingular plates and two had an extra apical plate. Of 450 hypothecae, 444 were relatively normal, three others having only five instead of six postcingulars, two having the correct number of postcingulars but with the sutures in abnormal locations, and four had supernumery antapicals (some had more than one abnormality). Only one clone was examined and so the ratio of abnormal to normal, which seems high here, may not be typical.

As Abé (1981) has noted, although one can find both fewer or more plates in each series, with the latter more common, the integrity of the latitudinal series generally remains intact. He greatly doubted the mirror-image variations, including girdle displacement reversals, reported by Mangin (1911), most of which do seem to be due to optical reversal by the observer. Variation in the degree of girdle displacement does occur within limits, and slight changes in sutural position can alter the shape of the first apical plate in *Protoperidinium*.

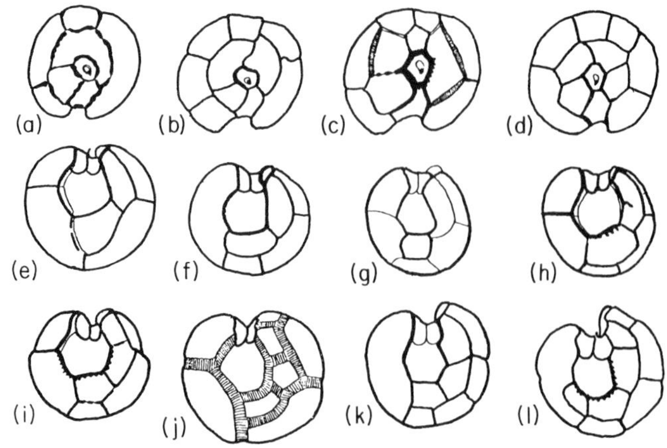

Fig. 2.26. Variability in the thecal plates of a culture of *Protogonyaulax tamarensis*. (a)–(d) epithecae ((b) is the usual number of plates); (e)–(i) hypothecae ((h) closest to the usual pattern). From Taylor (1975).

Cao Vien (1981) has studied the occurrence of doubling or absence of the third apical plate in a culture of *Peridinium bipes*. He found that, after division, one daughter cell always retained the abnormality and the other was normal (this species does not share the theca division). He interpreted his results in terms of mutational events. Further details of plate variability are provided by Netzel & Dürr (1984).

6 CONCLUSION

The diversity of form in dinoflagellates is great and the previous pages of this chapter can only provide an outline. Despite the wealth of information on form, however, our knowledge of the basic factors operating in dinoflagellate morphogenesis is still marginal. Information from ultrastructure, light microscopy, genetics, physiology, ecology and the fossil record are all needed to examine this important problem.

Peripheral microtubules undoubtedly play an important part in dinoflagellate morphogenesis, as they do in other protists. However, in addition, all the swimming cells (mastigotes) of dinoflagellates that have been examined ultrastructurally have a single peripheral layer of amphiesmal vesicles. Usually, each vesicle tightly abuts its neighbours so that their individual nature is not obvious, but Dodge's interpretation of them as individual sacs (Chapter 3) seems to be reasonable. In gymnodinoids their honeycomb-like reticulum may be the primary support for the cell periphery, but a pellicular layer may also be supportive. Bursa (1966) appreciated the importance of the reticular layer in dinoflagellate morphogenesis, but lacked the electron microscopical information to form a coherent picture.

In thecate dinoflagellates the cellulose plates are formed by the discharge of material into the amphiesmal (thecal) vesicles from below. It is the vesicles which form the primary tabular template. The thecal pattern has to be a reflection of the vesicular pattern, the sutures being the boundaries of adjacent vesicles. Presumably it is for this reason that the tabulation pattern exhibits surface tension features (e.g. the '3-suture junctions', polygonality, straight edges), particularly in those with many vesicles. The thecal plate thickness tends to be inversely related to the total number of vesicles and one finds greater deviation from the fundamental membrane-tensional patterns in those with the fewest plates (prorocentroids, dinophysoids). The greater the number of vesicles the more subequal in size they are.

Incomplete sutures are almost never seen in thecae and the few reports of them are dubious. Morrill & Loeblich (1981a, b) believe that the apparent presence of such in *Scrippsiella* (*Peridinium*) *trochoidea*, illustrated by Kalley & Bisalputra (1971), was due to mistaken observation of the pellicular, rather than thecal surface. However, in cysts, in which the wall is one unit (except for the archeopyle), the paratabulation may be an incomplete expression of the tabulation, and partially expressed parasutures are common.

It is a puzzle as to why the cyst wall should show any signs of the thecal tabulation and especially of transient growth states like intercalary bands. The earlier view that such surface markings are impressions of the inner surface of the theca under which the cyst wall formed (in proximate cysts) has given way to the view that nearly all are genetically-determined properties, intrinsic to the cyst. Even if some are shown to be impressional there is still much of the cyst wall which is relatively independent of the theca and, like the structures of larval forms, must contain features of adaptive significance to a dormant, benthicly deposited existence, even though a non-motile habit permits more elaborate wall ornamentation. If Morrill & Loeblich (1983) are correct in asserting that the pellicle results from a fusion of elements within the vesicles, and cyst walls form in the same manner, this would explain the presence of paratabulation patterns.

What determines the locality of a suture? There are signs that, apart from the innate tendency of membranous vesicles covering a surface to equalize in size (which may impart some of the radial symmetry), there are regions in the cell where sutures are more likely to be than others. Furthermore, marginal plate growth may be largely a cytoplasmic, rather than nuclear-regulated feature. For example, cells infected by the dinoflagellate *Amoebophrya ceratii* (Fig. 13.7), usually become greatly enlarged as the parasite grows within them, the latter digesting the host nucleus. The present author has observed that in greatly enlarged, infected cells of *Protogonyaulax catenella*, the thecal plates of the latter still abut one another, this expansional growth presumably being a mechanical response to the parasite growth. The enlargement of the planozygote theca (Chapter 14) may be a similar phenomenon.

7 REFERENCES

ABÉ T. H. (1936) Notes on the protozoan fauna of Mutsu Bay III. Subgenus Protoperidinium. *Sci. Rep. Tohoku. Univ.* Ser. 4. Biol. II, 19–48.

ABÉ T. H. (1967) The armoured dinoflagellata. II. Prorocentridae and Dinophysidae (B)—*Dinophysis* and its allied genera. *Publ. Seto mar. biol. Lab.* **15** (1), 37–78.

ABÉ T. H. (1981) *Studies on the family Peridiniidae: an unfinished monograph of the armoured Dinoflagellata.* pp. 1–409. Academia Scientific Book Inc., Tokyo.

BAILEY J. W. (1850) Microscopical observations made in South Carolina, Georgia and Florida. *Smithson. Contr. Knowl.* **2** (8), 1–48, pls. 1–3.

BALECH E. (1967) Dinoflagelados nuevos o interesantes del Golfo de Mexico y Caribe. *Rev. Mus. Argent. Cienc. nat. 'Bernardino Rivadavia', Hidrobiol.* **2**, 77–126.

BALECH E. (1977) Structure of *Amphisolenia bidentata* Schroder. *Physis* (Buenos Aires) **37**, 25–32.

BALECH E. (1979) Dinoflagelados. Campaña Oceanografica Argentina Islas Orcadas 6/75. *Servicio hidrogr. Naval Armada Argentina* **655**, 1–76.

BALECH E. (1980) On thecal morphology of dinoflagellates with special emphasis circular [*sic*] and sulcal plates. *An. Centro Cienc. mar. limnol., Univ. nat. autón. Mexico* **7**, 57–68.

BARROWS A. L. (1918) The significance of skeletal variations in the genus *Peridinium. Univ. Calif. publ. Zool.* **18**, 397–478.

BEAM C. A. & HIMES M. (1980) Sexuality and meiosis in dinoflagellates. In *Biochemistry and Physiology of Protozoa, Vol. 3* (ed. M. Levandowsky & S. H. Hutner), pp. 171–206. Academic Press, New York.

BERDACH J. T. (1977) *In situ* preservation of the transverse flagellum of *Peridinium cinctum* (Dinophyceae) for scanning electron microscopy. *J. Phycol.* **13**, 243-251.

BERMAN T. & ROTH I. L. (1979) The flagella of *Peridinium cinctum* fa. *westii*: *in situ* fixation and observations by scanning electron microscopy. *Phycologia* **18**, 307–311.

BESADA E. G., LOEBLICH L. A. & LOEBLICH A. R. III (1982) Observations on tropical, benthic dinoflagellates from ciguatera-endemic areas: *Coolia, Gambierdiscus,* and *Ostreopsis. Bull. mar. Sci.* **32**, 723–735.

BHOVICHITRA M. & SWIFT E. (1977) Light and dark uptake of nitrate and ammonium by large oceanic dinoflagellates: *Pyrocystis noctiluca, Pyrocystis fusiformis* and *Dissodinium lunula. Limnol. Oceanogr.* **22**, 73–83.

BIECHELER B. (1934a) Sur le réseau argentophile et la morphologie de quelques Péridiniens. *C. R. Soc. biol.* **115**, 1039.

BIECHELER B. (1934b) Sur un Dinoflagellé a capsule perinucléaire, *Plectodinium* n. gen. *nucleovolvatum* n. sp. et sur les relations des Péridiniens avec les Radiolaires. *C. R. Acad. Sci. Paris* **198**, 404–406.

BIECHELER B. (1952) Recherches sur les Péridiniens. *Bull. biol.* Suppl. **36**, 1–149.

BOALCH G. T. (1969). The dinoflagellate genus *Ptychodiscus* Stein. *J. mar. Biol. Ass. U.K.* **49**, 781–784.

BÖHM A. (1976) Morphologische Studien im Bereich der Pyrrhophyta. Das Problem Form und Selektion. *Bibl. Phycol.* **22**, 1–119.

BOLTOVSKOY A. (1979) Estudio comparativo de las bandas intercalares y zonas pandasuturales en los generos de dinoflagelados *Peridinium* s.s., *Protoperidinium* y *Palaeoperidinium. Limnobios* (9), 325–332.

BOURRELLY P. (1970) *Les Algues d'eau Douce.* Initiation à la Systématique. **3**, pp. 1–512. N. Boubeé & Cie, Paris.

BOURRELLY P. & COUTÉ A. (1976) Observations en microscopie électronique à balayage des *Ceratium* d'eau douce (Dinophycées). *Phycol.* **15**, 329–338.

BOUQUAHEUX F. (1971) *Gloeodinium marinum* nov. sp. Péridinien dinocapsale. *Arch. Protistenk.* **113**, 314–321.

BRAARUD T. (1945) Morphological observations—marine dinoflagellate cultures. *Avh. Norske*

Vidensk. Akad. Oslo **11**, 1–18.

Braarud T. (1962) Species distribution in marine phytoplankton. *J. Oceanogr. Soc. Japan., 20th Anniv. Vol.*, 628–649.

Brugerolle G. & Taylor F. J. R. (1979) Taxonomy, cytology and evolution of the Mastigophora. *Protozoological Actualities 1977* (eds Hutner S. H. & Bleyman L. K.), pp. 15–28. Proc. 5th Int. Congr. Protozool.

Bujak J. P. & Davies E. H. (1983) Modern and fossil Peridiniineae. *Amer. Ass. Stratigr. Palynol.* **13**, 1–202.

Bujak J. P., Downie C., Eaton G. L. & Williams G. L. (1980) *Dinoflagellate Cysts and Acritarchs from the Eocene of Southern England.* Special Papers in Palaeontology 24.

Bursa A. (1958) The freshwater dinoflagellate *Woloszynskia limnetica* n. sp. Membrane and protoplasmic structures. *J. Protozool.* **5**, 299–304.

Bursa A. S. (1966) Ectoplasm as a morphogenetic factor in the dinoflagellate *Woloszynskia limnetica. Verh. int. Ver. Limnol.* **16** (3), 1589–1594.

Bursa A. S. (1971) Morphogenesis and taxonomy of fossil and contemporary Dinophyta secreting discoasters. *Proc. 2nd Planktonic Conference Rome 1970*, 129–143. Edizioni Tecnoscienza, Rome.

Bütschli O. (1885) Dinoflagellata. In *Protozoa* (1880–1889) *Klass. u. Ordn. des Thierreichs.* 1 (ed. H. G. Bronn), pp. 906–1029. C. F. Winter'sche Verlagshandlung, Leipzig & Heidelberg.

Cachon J. & Cachon-Enjumet M. (1964) *Leptospathium navicula* nov.gen.nov.sp. et *Leptophyllus dasypus* nov.gen.nov.sp., Péridiniens *Noctilucidae* (Hertwig) du plancton néritique de Villefranche-sur-Mer. *Bull. Inst. océanogr. Paris* **62** (1292), 1–12.

Cachon J. & Cachon-Enjumet M. (1966) *Pomatodinium impatiens* nov.gen.nov.sp. Péridinien Noctilucidae Kent. *Protistologica* **2** (1), 23–30.

Cachon J. & Cachon M. (1967) Contribution a l'étude des Noctilucidae Saville-Kent. Les Kofoidininae Cachon J. et M., évolution morphologique et systématique. *Protistologica* **3** (4), 427–444.

Cachon J. & Cachon M. (1969) Contribution a l'étude des Noctilucideae Saville-Kent. Évolution morphologique, cytologie, systematique. *Protistologica* **5**, 11–33.

Cao Vien (1981) Sur le déterminisme génétique de deux variations morphologiques de la plaque 3′ du *Peridinium bipes* Stein, Péridinien d'eau douce. *C. R. Acad. Sci. Paris* **292**, 897–901.

Carty S. & Cox E. R. (1985) Observations on *Lophodinium polylophon. J. Phycol.* **21**, 396–401.

Chapman D. V., Dodge J. D. & Heaney S. I. (1982) Cyst formation in the freshwater dinoflagellate *Ceratium hirundinella. J. Phycol.* **18**, 121–129.

Chapman D. V., Livingstone D. & Dodge J. D. (1981) An electron microscope study of the excystment and early development of the dinoflagellate *Ceratium hirundinella. Br. phycol. J.* **16**, 183–194.

Chatton E. (1920) Les Péridiniens parasites. Morphologie, reproduction, ethologie. *Arch. Zool. exp. gen.* **59**, 1–475.

Clarke K. J. & Pennick N. C. (1972) Flagellar scales in *Oxyrrhis marina* Dujardin. *Br. phycol. J.* **7**, 357–360.

Claparède E. & Lachmann J. (1859) Etudes sur les Infusoires et les Rhizopodes. Ordre III. Infusoires Cilioflagellés. *Mem. Inst. Genevois* **6**, mém. 1, 392–412, pls. 19, 20.

Diwald K. (1939) Ein Beitrag zur Variabilität und Systematik der Gattung Peridinium (nach dem auf den osterreichischitalienischen Terminfahrten der 'Najade' in den Jahren 1911–1914 gesammelten Material). *Arch. Protistenk.* **93**, 121–184.

Dodge J. D. & Bibby B. T. (1973) The Prorocentrales (Dinophyceae) I. A. comparative account of fine structure in the genera *Prorocentrum* and *Exuviaella. Bot. J. Linn. Soc.* **67**, 175—187.

Dodge J. D. & Crawford R. M. (1969) The fine structure of *Gymnodinium fuscum. New Phytol.* **68**, 613–818.

DODGE J. D. & CRAWFORD R. M. (1970) A survey of thecal fine structure in the Dinophyceae. *Bot. J. Linn. Soc.* **63**, 53–67.

DODGE & HERMES (1981) A revision of the *Diplopsalis* group of dinoflagellates (Dinophyceae) based on material from the British Isles. *Bot. J. Linn. Soc.,* **82**, 15–26.

DREBES G. (1981) Possible resting spores of *Dissodinium pseudolunula* (Dinophyta) and their relationship to other taxa. *Br. phycol. J.* **16**, 207–215.

DÜRR G. (1979a) Elektronenmikroskopiche Untersuchungen am Panzer von Dinoflagellaten. I. *Gonyaulax polyedra. Arch. Protistenk.* **122**, 55–87.

DÜRR G. (1979b) Elektronenmikroskopische Untersuchungen am Panzer von Dinoflagellaten. II. *Peridinium cinctum. Arch. Protistenk.* **122**, 88–120.

DÜRR G. (1979c) Elektronenmikroskopische Untersuchungen am Panzer von Dinoflagellaten III. Die Zyste von *Peridinium cinctum. Arch. Protistenk.* **122**, 121–139.

DÜRR G. & NETZEL H. (1974) The fine structure of the cell surface in *Gonyaulax polyedra* (Dinoflagellata). *Cell Tiss. Res.* **150**, 21–41.

EATON G. L. (1980) Nomenclature and homology in peridinialean dinoflagellate plate patterns. *Paleontology* **23** (3), 667–688.

EISENACK A. (1963) Hystrichosphären. *Bio. Rev.* **38**, 107–139.

ELBRÄCHTER M. & DREBES G. (1978) Life cycles, phylogeny and taxonomy of *Dissodinium* and *Pyrocystis* (Dinophyta). *Helgoländer wiss. Meeresunters* **31**, 347–366.

EVITT W. R. (1963) Arrangement and structure of processes in *Hystrichosphaeridium* and its allies. *Pollen et Spores* **4**, 343–344.

EVITT W. R. (1964) Observations on the morphology of fossil dinoflagellates. *Micropal.* **7**, 385–420.

EVITT W. R. (1967) Dinoflagellate studies II. The archeopyle. *Stanford Univ. Publs. Geol. Sci.* **10** (3), 1–83, pls. 1–11.

EVITT W. R. (1985) Sporopollenin dinoflagellate cysts: their morphology and interpretation. *Amer. Ass. stratigr. Palynol.* Foundation, pp. 1–333.

EVITT W. R. & DAVIDSON S. E. (1964) Dinoflagellate studies I. Dinoflagellate cysts and thecae. *Stanford Univ. Publ. Ser.* **10**, 1–12.

EVITT W. R., LENTIN J. K., MILLIOUD M. E., STOVER L. E. & WILLIAMS G. L. (1977) Dinoflagellate cyst terminology. *Geol. Surv. Pap., Can.* 76–24, 1–11.

FINE K. E. & LOEBLICH A. R. III (1976) Similarity of the dinoflagellates *Peridinium trochoideum, P. faeroense* and *Scripsiella sweeneyae* as determined by chromosome numbers, cell division studies and scanning electron microscopy. *Proc. biol. Soc. Washington* **89**, 275–288.

FRITSCH F. E. (1935) *The Structure and Reproduction of the Algae*, Vol. I., pp. 1–791. Cambridge University Press, Cambridge.

FÜTTERER D. (1976) Kalkige Dinoflagellaten ('Calciodinelloideae') und die systematische Stellung der Thoracosphaeroideae. *N. Jb. Geol. Paläont. Abh.* **151** (2), 119–141.

GAINES G. & TAYLOR, F. J. R. (1984) Extracellular digestion in marine dinoflagellates. *J. Plankt. Res.* **6**, 1057–1061.

GAINES G. & TAYLOR F. J. R. (1985) Form and function of the dinoflagellate transverse flagellum. *J. Protozool.* **32**, 290–296.

GOCHT H. (1979) Correlation of overlapping system and growth in fossil Dinoflagellates. *N. Jb. Geol. Paläont. Abh.* **157** (3), 344–365.

GOCHT H. (1981) 'Direkter' Nachweis der Platten-Überlappung bei *Hystrichogonyaulax caldophora.* (Dinoflagellata, Oberjura). *N. Jb. Geol. Paläont. Mh.* H.3, 149–156.

GOCHT H. & NETZEL H. (1974) Rasterelektronenmikroskipische Untersuchungen am Panzer von *Peridinium* (Dinoflagellata). *Arch. Protistenk.* **116**, 381–410.

GOCHT H. & NETZEL H. (1976) Relief structures of the Cretaceous dinoflagellate *Palaeoperidinium pyrophorum* (Ehr) compared with thecal structures of recent *Peridinium* species. *N. Jb. Geol. Paläont. Abh.* **152** (3), 380–413.

GRAHAM H. W. (1942) Studies in the Morphology, Taxonomy, and Ecology of the Peridiniales.

Biology III. *Carnegie Inst. Washington Publ.* **542**, 1–129.
GRAN H. H. (1912) Pelagic Plant Life. In *The Depths of the Ocean.* (eds J. Murray & J. Hjort), pp. 307–386. Macmillan and Co., London.
GRASSÉ P.-P. (1952) *Traité de zoologie. Anatomie, systématique, biologie*, 1, pp. 1–1071, Masson & Cie, Paris.
GREUET C. (1972) Intervention de lamelles dans la formation de couches squelettiques au niveau de la capsule perinucléaire de péridiniens Warnowiidae. *Protistologica* **8**, 155–168.
HAPPACH-KASAN C. (1982) Studies on the construction of the theca of *Ceratium cornutum. Arch. Protistenk.* **125**, 181–207.
HERMAN E. M. & SWEENEY B. M. (1977) Scanning electron microscopic observations of the flagellar structure of *Gymnodinium splendens* (Pyrrophyta, Dinophyceae). *Phycologia* **16**, 115–118.
HOLLANDE A., CACHON J. & CACHON-ENJUMET M. (1962) Considération sur les affinités entre Dinoflagellés et Radiolaires. *C. R. Acad. Sci. Paris*, 2069–2071.
HONSELL G. & TALARICO L. (1985) The importance of flagellar arrangement and insertion in the interpretation of the theca of *Prorocentrum* (Dinophyceae). *Bot. mar.* **28**, 15–21.
HOVASSE R. (1943) Nouvelles recherches sur les flagellés a squelette siliceux: ébriidés et silicoflagellés fossiles de la diatomite de Saint-Laurent-La-Vérnede (Gard.) *Bull. Biol. Fr. Belg.* **77**, 271–284.
JØRGENSEN E. (1912) Bericht über die von der schwedischen hydrographisch-biologischen Kommission in der schwedischen Gewässern an den Jahren 1909–1910, eingesammelten Planktonproben. *Svenska Hydrog. Biol. Komm.* **4**, 1–20.
JUX U. (1968a) Über den Feinbau der Wandung bei *Cordosphaeridium inodes* (Klumpp, 1953). *Palaeontographica Abt. B.* v. **122**, 48–54, pls. 13–14.
JUX U. (1968b) Über den Feinbau der Wandung bei *Hystrichosphaera bentori* Rossignol 1961. *Palaeontographica Abt.* **123**, 147–152, pl. 31.
KALLEY J. P. & BISALPUTRA T. (1971) *Peridinium trochoideum*: the fine structure of the thecal plates and associated membranes. *J. ultrastr. Res.* **37**, 521–531.
KESSELER H. (1966) Beitrag zur kenntnis der chemischen und physikalischen Eigenschaften des Zellsaftes von *Noctiluca miliaris. Veroff. Inst. Meeresforsch. Bremerhaven*, 357–368.
KEUPP H. (1979) Lower Cretaceous Calcisphaerulidae and their relationship to calcareous dinoflagellate cysts. *Bull. Centre Rech. explor.-prod. Elf-Aquitaine* **3**, 651–663.
KEUPP H. (1980) *Pithonella patriciacreeleyae* Bolli 1974, eine kalkige Dinoflagellaten-Zyste mit internen Paratabulation (unter-kreide, Speeton/SE-England) *N. Jb. Geol. Paleontol. Mh.* **9**, 513–524.
KEUPP H. (1981) Calcareous dinoflagellate cysts of the Boreal lower Cretaceous (Lower Hauterivian to Lower Albian). *Facies* **5**, 1–190.
KLEBS G. (1883) Ueber die Organisation einiger flagellaten Gruppen und ihre Beziehungen zu Algen und Infusoria. *Untersuch bot. Inst. Tübingen* **1**, 233–362, pls. 2, 3.
KLEBS G. (1912) Uber Flagellaten-und Algen-ähnliche Peridineen. *Verh. naturh.-med. Ver. Heidelb.* **11**, 369–451, pl. 10.
KOFOID C. A. (1906) On the significance of the asymmetry in *Triposolenia. Univ. Cal. Publ. Zool.* **3**, 127–133.
KOFOID C. A. (1907a) Dinoflagellata of the San Diego Region III. Descriptions of new species. *Univ. Calif. Publ. Zool.* **3**, 299–340.
KOFOID C. A. (1907b) New species of dinoflagellates. *Bull. Mus. Comp. Zool. Harvard* **6**, 163–207.
KOFOID C. A. (1908) Exuviation, autotomy and regeneration in *Ceratium. Univ. Calif. Publ. Zool.* **4**, 345–386.
KOFOID C. A. (1909a) On *Peridinium steini* Jørgensen, with a note on the nomenclature of the skeleton of the Peridinidae. *Arch. Protistenk.* **16**, 25–47, pl. 2.
KOFOID C. A. (1909b) Mutations in *Ceratium. Bull. Mus. Comp. Zool. Harvard* **52**, 211–257, pls 1–4, 5 figs. in text.

KOFOID C. A. (1911) Dinoflagellata of the San Diego region, IV. The genus *Gonyaulax*, with notes on its skeletal morphology and a discussion of its generic and specific characters. *Univ. Cal. Publ. Zool.* **8** (4), 187–286 + 9 pls.

KOFOID C. A. (1912) Significance of certain forms of asymmetry of the dinoflagellates. *Proc. Seventh Int. Zool. Congr. 1907*, 928–931.

KOFOID C. A. & RIGDEN E. J. (1912) A peculiar form of schizogony in *Gonyaulax*. *Bull. Mus. Comp. Zool. Harvard* **54**, 335–348, pls. 1–2.

KOFOID C. A. & SWEZY O. (1921) The free-living unarmored Dinoflagellata. *Mem. Univ. Calif.* **5**, 1–562.

LEADBEATER B. & DODGE J. D. (1966) The fine structure of *Woloszynskia micra* sp. nov., a new marine dinoflagellate. *Brit. Phycol. Bull.* **3**, 1–17.

LEBLOND P. H. & TAYLOR F. J. R. (1976) The propulsive mechanism of the dinoflagellate transverse flagellum reconsidered. *BioSystems* **8**, 33–39.

LEBOUR M. V. (1925) *The Dinoflagellates of Northern Seas*. pp. 250, pls 35. Mar. Biol. Ass. UK, Plymouth.

LEE R. E. (1977) Saprophytic and phagocytic isolates of the colourless heterotrophic dinoflagellate *Gyrodinium lebouriae* Herdman. *J. mar. biol. Ass. UK.* **57**, 303–315.

LEFÈVRE M. (1928) Monographie des espèces d'eau douce du genre *Peridinium*. *Archs. Bot. Mém.* **2**, Mem. 5, 1–210, pls 1–6.

LÉGER G. (1971) Les populations phytoplanctoniques A. Géneralités et premier séjour (21–27 Février) *Bull. Inst. océanogr. Monaco* **69**, 1412 (A, B), 1–42.

LOEBLICH A. R. III (1970) The amphiesma or dinoflagellate cell covering. *Proc. North American Paleont. Convention, Chicago, September 1969*, **2**, Part G, pp. 867–929. Allen Press, Lawrence, Kansas.

LOEBLICH A. R. III (1976) Dinoflagellate evolution: speculation and evidence. *J. Protozool.* **23**, 13–28.

LOEBLICH A. R. III (1982) Dinophyceae. In *Synopsis and Classification of Living Organisms*. Vol. (ed. S. P. Parker), pp. 101–115). McGraw-Hill Co., New York.

LOEBLICH A. R. III & MORRILL L. C. (1979) Dinoflagellate cell wall structure and formation. 37th *Ann. Proc. Electron Microscopy Soc. Amer.*, 184–187.

LOEBLICH A. R. III & SHERLEY J. L. (1979) Observations on the theca of the motile phase of free-living and symbiotic isolates of *Zooxanthella microadriatica* (Freudenthal) comb. nov. *J. mar. biol. Ass. UK.* **59**, 195–205.

LOEBLICH A. R. III, SHERLEY J. L. & SCHMIDT R. J. (1979) The correct position of flagellar insertion in *Prorocentrum* and description of *Prorocentrum rhathymum* sp. nov. (Pyrrhophyta) *J. Plankton Res.* **1**, 113–120.

LOHMANN H. (1908) Untersuchungen zur Festellung des vollständigen Gehaltes des Meeres an Plankton. *Wiss. Meeresunters Abt. Kiel* **10**, 129–370, pls 9–17.

LOPEZ J. (1955) Variación alométrica en *Ceratium tripos*. *Inv. Pesq.* **2**, 131–159.

LUCAS I. A. N. (1982) Observations on *Noctiluca scintillans* (Macartney) Ehrenb. (Dinophyceae) with notes on an intracellular bacterium. *J. Plankton Res.* **4**, 401–409.

MANGIN L. (1911) Sur l'existence d'individus dextres et senistres chez certains Péridiniens. *C. R. Seanc. Acad. Sci. Paris* **153**, 27–32.

MCLACHLAN J. (1973) Growth media—marine. In *Handbook of Phycological Methods: culture methods and growth measurements* (ed. J. Stein), pp. 25–52. Cambridge University Press, Cambridge.

May F. E. (1976 Dinoflagellates: fossil motile-stage tests from the Upper Cretaceous of the northern New Jersey coastal plain. *Science* **193**, 1128–1130.

MOREY-GAINES G. & RUSE R. H. (1980) Encystment and reproduction of the predatory dinoflagellate, *Polykrikos kofoidi* Chatton (Gymnodinioles). *Phycologia* **19** (3), 230–232.

MORRILL L. C. (1984) Ecdysis and the location of the plasma membrane in the dinoflagellate *Heterocapsa niei*. *Protoplasma* **119**, 8–20.

MORRILL L. C. & LOEBLICH A. R. III (1981a) A survey for body scales in dinoflagellates and a

revision of *Cachonina* and *Heterocapsa* (Pyrrophyta). *J. Plankt. Res.* 3 (1), 53–65.
MORRILL L. C. & LOEBLICH A. R. III (1981b) The dinoflagellate pellicular wall layer and its occurrence in the divison Pyrrhophyta. *J. Phycol.* **17**, 315–323.
MORIL L. C. & LOEBLICH A. R. III (1983) Ultrastructure of dinoflagellate amphiesma. *Int. Rev. Cytol.***82**, 151–180.
NAWATA T. & SIBAOKA T. (1983) Experimental induction of feeding behavior in *Noctiluca miliaris*. *Protoplasma* **115**, 34–42.
NETZEL H. & DÜRR G. (1984) Dinoflagellate cell cortex. In *Dinoflagellates* (ed. D. Spector), pp. 43–105. Academic Press, Orlando.
NEVO Z. & SHARON N. (1969) The cell wall of *Peridinium westii*, a non-cellulosic glucan. *Biochim. Biophys. Acta.* **173**, 161–175.
NIELSEN J. (1956) Temporary variations in certain marine *Ceratia*. *Oikos* **7**, 256–272.
NORDLI E. (1957) Experimental studies on the ecology of *Ceratia*. *Oikos* **8**, 200–265.
NORRIS D. R. (1969) Thecal morphology of *Ornithocercus magnificus* (Dinoflagellata) with notes on related species. *Bull. Mar. Sci.* **19**, 175–193.
NORRIS G. (1978) Phylogeny and a revised supra-generic classification for Triassic-Quaternary organic-walled dinoflagellate cysts (Pyrrhophyta). *N. Jb. Geol. Palaönt. Abh.* **155**, 300–317.
ORR W. N. & CONLEY S. (1976) Silceous dinoflagellates in the northeast Pacific rim. *Micropaleont.* **22** (1), 92–99.
PARKE M. & BALLANTINE D. (1957) A new marine dinoflagellate: *Exuviaella mariae—lebouriae* n. sp. *J. mar. biol. Ass. UK.* **36**, 643–650.
PASCHER A. (1914) Ueber Flagellaten und Algen. *Ber. Deutsch. Bot. Ges.* **32**, 136–160, 430.
PASCHER A. (1916) Ueber eine neue *Amoeba—Dinamoebe* (*varians*) mit dinoflagellatenartigen Schwärmen. *Arch. Protistenk.* **36**, 116–117, pls. 7–9.
PASCHER A. (1927) Die braune Algenreihe aus der Verwandtschaft der Dinoflagellaten (Dinophyceen). *Arch. Protistenk.* **58**, 1–58, figs. 1–38.
PAULSEN O. & OSTENFELD C. H. (1911) Marine phytoplankton from the East-Greenland Sea IV General Remarks on the microplankton. *Meddr. Grønland* **43**, 319–325.
PAVILLARD J. (1937) Les Péridiniens et Diatomées pélagiques de la mer de Monaco pendant les années 1912, 1913 et 1914. *Bull. Inst. océanogr. Monaco* **7**, 1–56.
PETERS N. (1927) Das Wachstum des Peridiniumpanzers. *Zool. Anz.* **73**, 143–148.
PETERS N. (1929) Beitrage zur Planktonbevölkerung der Weddellsee. *Int. Rev. ges. Hydrobiol. Hydrogr.* **21**, 17–146.
PETERS N. (1931) Die Bevölkerung des Südatlantischen Ozeans mit Ceratien. *Wiss. Ergebn. dtsch. Atl. Exped. 'Meteor' 1925–1927*, **12** Biol. 1, 1–69.
PFIESTER L. A. & LYNCH R. A. (1980) Amoeboid stages and sexual reproduction of *Cystodinium bataviense* and its similarity to *Dinococcus* (Dinophyceae). *Phycol.* **19**, 178–183.
PFIESTER L. A. & POPOVSKY J. (1979) Parasitic, amoeboid dinoflagellates. *Nature* (*Lond.*) **279**, 421–424.
PFIESTER L. A. & SKVARLA (1979) Heterothallism and thecal development in the sexual life history of *Peridinium volzii* (Dinophyceae). *Phycologia* **18**, 13–18.
PINCEMIN J. M., GAYOL J. & SALVANO P. (1981) Observations on the thecate stage of the dinoflagellate *Pyrocystis* cf. *fusiformis* (clones NOB2 and III): variations in morphology and tabulation. *Arch. Protistenk.* **124**, 271–282.
PINCEMIN J. M., GAYOL P. & SALVANO P. (1982) *Pyrocystis lunula*: Dinococcidae giving rise to a biflagellated thecate stage. *Arch. Protistenk.* **125**, 95–107.
POPOVSKY J. (1982) Another case of phagotrophy by *Gymnodinium helveticum* Penard f. *achroum* Skuja. *Arch. Protistenk.* **125**, 73–78.
POPOVSKY J. & PFIESTER L. A. (1982) The life-histories of *Stylodinium sphaera* Pascher and *Cystodinedria inernis* Geitler. *Arch. Protistenk.* **125**, 115–128.
POSTEK M. T. & COX E. R. (1976) Thecal ultrastructure of the toxic marine dinoflagellate *Gonyaulax catenella*. *J. Phycol.* **12**, 88–93.

REES A. J. J. & LEEDALE G. F. (1980) The dinoflagellate transverse flagellum: three dimensional reconstructions from serial sections. *J. Phycol.* **16**, 73–83.

SARJEANT W. A. S. (1974) *Fossil and Living Dinoflagellates.* 182 pp. Academic Press, London.

SARJEANT W. A. S. (1982) Dinoflagellate cyst terminology: a discussion and proposals. *Can. J. Bot.* **60**, 922–945.

SARJEANT W. A. S., LACALLI T. & GAINES G. (1986) The cysts and skeletal elements of dinoflagellates: speculations on the causes for their development and morphology (in prep.).

SCHNEPF E. & DEICHGRÄBER G. (1972) Über den Feinbau von Theka, Pusule und golgi-apparat bei dem dinoflagellaten *Gymnodinium* spec. *Protoplasma* **74**, 411–425.

SCHUBERT, K. (1937) Die Ceratien der Deutschen Antarktischen Expedition auf der 'Deutschland' 1911/1912. *Int. Rev. Hydriobiol.* **34**, 373–431.

SCHÜTT F. (1892) *Analytische Plankton-Studien. Ziele, Methoden und Anfangs—Resultate der quantitativ-analytischen Planktonforschung.* pp. 117. Lipsius & Tischer, Kiel & Leipzig. [Also published as part of Vol. 1 of Neptunia].

SCHÜTT, F. (1895) Peridineen der Plankton-Expedition. *Ergebn. Plankton-Expedition der Humboldt-Stiftung* 4, M, a, A, 1–170, pls. 27.

SHUMWAY W. (1924) The genus *Haplozoon* Dogiel. Observations on the life history and systematic position. *J. Parasitol.* **11**, 59–74.

SIEBERT A. E. Jr (1973) A description of *Haplozoon axiothellae* n. sp., an endosymbiont of the polychaete *Axiothella rubrocincta. J. Phycol.* **9**, 185–190.

SIEBERT A. E. Jr & WEST J. A. (1974) The fine structure of the parasitic dinoflagellate *Haplozoon axiothellae. Protoplasma* **81**, 17–35.

SOURNIA A. (1982) Form and function in marine phytoplankton. *Biol. Rev.* **57**, 347–394.

SOYER M.-O. (1970) Les ultrastructures liées aux fonctions de relation chez *Noctiluca miliaris* S. (Dinoflagellata). *Z. Zellforsch* **104**, 29–55.

SOYER M.-O., PREVOT P. & DE BILLY F. (1982) *Prorocentrum micans* E., one of the most primitive dinoflagellates: 1. The complex flagellar apparatus as seen in scanning and transmission electron microscopy *Protistologica* **18**, 289–298.

STEEMANN NIELSEN E. (1934) Untersuchungen über die Verbreitung, Biologie und Variation der Ceratien im südlichen Stillen Ozean. *Dana Rep.* **4**, 1–64.

STEEMANN NIELSEN E. (1939) Die Ceratien des Indischen ozeans und der Ostasiatischen Gewässer mit einer allgemeinen Zusammenfassung über die verbreitung des Ceratien in den Weltmeeren. *Dana Rep.* **17**, 1–33.

STEIDINGER K. A. (1968) The genus *Gonyaulax* in Florida waters. I. Morphology and thecal development in *Gonyaulax polygramma* Stein, 1883. *Florida Bd. Conserv. Mar. Lab.* **1** (4), 1–5.

STEIDINGER K. A. & COX E. R. (1980) Free-living dinoflagellates. In *Phytoflagellates*, Developments in Marine Biology 2 (ed. E. R. Cox), pp. 407–432. Elsevier/North Holland, New York.

STEIDINGER K. A., TESTER L. S. & TAYLOR F. J. R. (1980) A rediscription of *Pyrodinium bahamense* var. *compressa* (Böhm) stat. nov. from Pacific red tides. *Phycologia* **19**, 329–337.

STEIDINGER K. A. & WILLIAMS J. (1970) *Dinoflagellates.* Memoirs Hourglass Cruises, **2**, 1–251. (Mar. Res. Lab. Fla. Dept. Nat. Res.).

STOSCH H. A. VON (1972) La signification cytologique de la 'cyclose nucleaire' dans le cycle de vie des dinoflagellés. *Mém. Soc. Bot. Fr.*, 201–212.

STOSCH H. A. VON (1973) Observations on vegetative reproduction and sexual life cycles of two freshwater dinoflagellates, *Gymnodinium pseudopalustre* Schiller and *Woloszynskia apiculata* sp. nov. *Br. Phycol. J.* **8**, 105–134.

STOVER L. E. & EVITT W. R. (1978) Analysis of pre-Pleistocene organic-walled dinoflagellates *Stanford Univ. Publs Geol. Sci.* **15**, 1–300.

SWIFT E. & REMSEN C. C. (1970) The cell wall of *Pyrocystis* spp. (Dinococales). *J. Phycol.* **6**,

79–86.
TAI L. & SKOSBERG T. (1934) Studies on the Dinophysoidae, marine armored dinoflagellates of Monterey Bay, California. *Arch. Protistenk.* **82**, 380–482, pls. 11–12.
TANGEN K., BRAND L. E., BLACKWELDER P. L. & GUILLARD R. R. L. (1982) *Thoracosphaera heimii* (Lohmann) Kampner is a dinophyte: observations on its morphology and life cycle. *Mar. Micropaleont.* **7**, 193–212.
TAPPAN H. (1980) *The Paleobiology of Plant Protists.* pp. 1028. W. H. Freeman & Co., San Francisco.
TAYLOR F. J. R. (1962) *Gonyaulax polygramma* Stein in Cape waters: a taxonomic problem related to developmental morphology. *J. S. Afr. Bot.* **28** (3), 237–242.
TAYLOR F. J. R. (1963) *Brachydinium*, a new genus of the Dinococcales from the Indian Ocean. *J. S. Afr. Bot.* **29** (2), 75–78.
TAYLOR F. J. R. (1969) Perinuclear structural elements formed in the dinoflagellate *Gonyaulax pacifica* Kofoid. *Protistologica* **5**, 165–167.
TAYLOR F. J. R. (1971) Scanning electron microscopy of thecae of the dinoflagellate genus *Ornithocercus. J. Phycol.* **7**, 249–258.
TAYLOR F. J. R. (1973a) Topography of cell division in the structurally complex dinoflagellate genus *Ornithocercus. J. Phycol.* **9**, 1–10.
TAYLOR F. J. R. (1973b) General features of dinoflagellate material collected by the 'Anton Brunn' during the International Indian Ocean Expedition. In *The Biology of the Indian Ocean.* [Ecological Studies 3] (ed. B. Zeitschel), pp. 155, 169, pls 1–3. Springer-Verlag, Berlin.
TAYLOR F. J. R. (1975a) Non-helical transverse flagella in dinoflagellates. *Phycologia* **14**, 45–47, figs 1–8.
TAYLOR F. J. R. (1975b) Taxonomic difficulties in Red tide and paralytic shellfish poison studies: the 'Tamarensis Complex' of *Gonyaulax. Environ. Letters* **9** (2), 103–119.
TAYLOR F. J. R. (1976a) Dinoflagellates from the International Indian Ocean Expedition. *Bibl. Bot.* **132**, pp. 1–234, pls. 1–46.
TAYLOR F. J. R. (1976b) Flagellate phylogeny: a study of conflicts. *J. Protozool.* **23**, 28–40.
TAYLOR F. J. R. (1978) Problems in the development of an explicit phylogeny of the lower eukaryotes. *BioSystems* **10**, 67–87.
TAYLOR F. J. R. (1979) The toxigenic gonyaulacoid dinoflagellates. In *Toxic Dinoflagellate Blooms*, Vol. I. Developments in Marine Biology. (eds D. L. Taylor & H. H. Seliger), pp. 47–56. Elsevier/North Holland, Amsterdam.
TAYLOR F. J. R. (1980) On dinoflagellate evolution. *BioSystems* **13**, 65–108.
TAYLOR F. J. R,. (1982) Symbioses in marine microplankton. *Ann. Inst. océanogr., Paris* **58** (suppl.), 61–90.
TAYLOR F. J. R. & CATTELL S. A. (1969) *Dicroerisma psilonereiella* gen. et sp. n., a new dinoflagellate from British Columbia coastal waters. *Protistologica* **5**, 169–172.
THOMPSON R. H. (1949) Immobile Dinophyceae. I. New records and new species. *Amer. J. Bot.* **36**, 301–308.
WALL D. & DALE B. (1971) A reconsideration of living and fossil *Pyrophacus* Stein, 1883 (Dinophyceae). *J. Phycol.* **7** (3), 221–235.
WALL D., GUILLARD R. R. L., DALE B., SWIFT E. J. & WATABE N. (1970) Calcitic resting cysts in *Peridinium trochoideum* (Stein) Lemmermann, an autotrophic marine dinoflagellate. *Phycologia* **9**, 151–156, figs. 1–10.
WEBER VAN BOSSE A. (1901) Études sur les algues de l'Archipel Malaisien. III. Note préliminaire sur les resultats algologiques de l'expediation du Siboga. *Ann. Jard. bot. Buitenzorg* **2**, 126–141.
WHEDON W. F. & KOFOID C. A. (1936) Dinoflagellata of the San Francisco region I. On the skeletal morphology of two new species—*Gonyaulax catanella* and *G. acatanella. Univ. Cal. publ. Zool.* **41**, 25–34.

CHAPTER 3
DINOFLAGELLATE ULTRASTRUCTURE AND COMPLEX ORGANELLES

JOHN D. DODGE
Botany Department, Royal Holloway College, Egham, Surrey, UK

C. GREUET
Université de Nice, Laboratoire de Biologie Animal et Cytologie, Campus Universitaire Valrose, 06034 Nice, France

A GENERAL ULTRASTRUCTURE

JOHN D. DODGE

1 INTRODUCTION AND GENERAL STUDIES

The ultrastructure of dinoflagellates was comprehensively reviewed in 1971 (Dodge 1971b) and in the account which follows most emphasis will be placed on work which has appeared subsequent to that date. A number of aspects of the ultrastructure have been considered by Loeblich (1976) and F. Taylor (1980) in connection with their discussions of dinoflagellate evolution and by Steidinger & Cox (1980). From the large number of studies which have now been made it is clear that dinoflagellates possess many unique structural features which may enable a positive identification of an organism as a member of this group on the basis of a single electron micrograph (Fig. 3A. 1) (Dodge 1976; see also Chapters 1 and 2). All have a distinctive nucleus, amongst the free-living species all have characteristic flagella and cell-covering (amphiesma or theca), many have an unusual organelle called the pusule and may have trichocysts. Chloroplasts are present in the symbiotic species and also in many of the free-living organisms. Certain species have organelles which are very elaborate for unicellular organisms; these include the ocelloids, nematocysts and 'piston' (see Chapter 3B).

Free-living species which have been the subject of fairly intensive investigation include the following;

Prorocentrales: *Prorocentrum mariae-lebouriae* (Faust 1974)
P. micans and *P. triestinum* (Dodge & Bibby 1973)

Gymnodiniales: *Amphidinium carterae* (Dodge & Crawford 1968)
Gymnodinium micrum (Leadbeater & Dodge 1966)
G. simplex (Dodge 1974)
G. breve (= *Ptychodiscus brevis*) (Steidinger *et al.* 1978)
G. fuscum (Dodge & Crawford 1969a)
Gyrodinium lebourae (Lee 1977)
Nematodinium armatum (Mornin & Frances 1967)
Oxyrrhis marina (Dodge & Crawford 1971b)
Aureodinium pigmentosum (Dodge 1967)
Woloszynskia coronata (Crawford & Dodge 1974)

Peridiniales: *Gonyaulax polyedra* (Schmitter 1971)
Peridinium cinctum (Spector & Triemer 1979)
P. westii (Messer & Ben-Shaul 1969)
Ceratium hirundinella (Dodge & Crawford 1970a)
C. tripos (Wetherbee 1975a, b, c)

Of the symbiotic species most work has been carried out on:

Symbiodinium (= Zooxanthella) *microadriaticum* (D. L. Taylor 1968; Kevin *et al.* 1969)

Endodinium chattonii (D. L. Taylor 1971a)

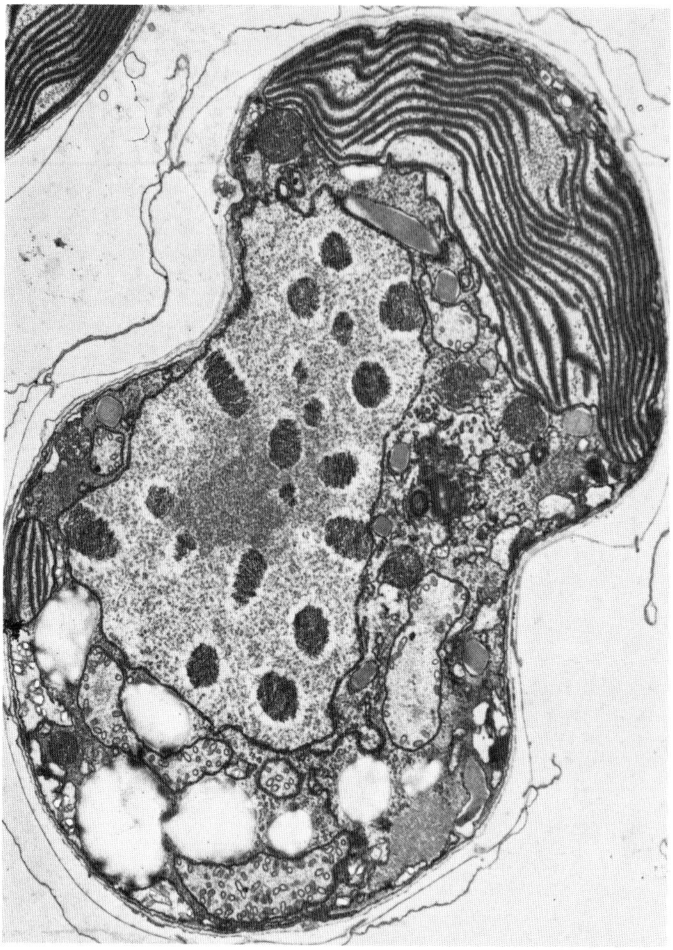

Fig. 3A.1. Longitudinal section of *Glenodinium hallii* showing the arrangement of the main organelles (× 11 250).

Amphidinium klebsii (symbiotic form) (D. L. Taylor 1971b).

Parasitic dinoflagellates have been the subject of a number of studies, e.g. *Haplozoon axiothellae* (Siebert & West 1974), *Oodinium* (Cachon & Cachon 1971), *Apodinium* (Cachon & Cachon 1973) (see Chapter 13).

Aspects of phagotrophy, one of the three nutritional modes in this group, have been reported from *Oxyrrhis marina* (Dodge & Crawford 1974) and *Gymnodinium fungiforme* (Spero & Morée 1981) (see Chapter 6).

2 THE CELL COVERING

Free-living motile dinoflagellates are bounded by an elaborate covering which has been variously termed the theca (Leadbeater & Dodge 1966) or amphiesma

(Loeblich 1970). The basic structure of this, as described by Dodge & Crawford (1970b), consists of a continuous outer membrane which bounds the flagella and runs completely over the cell surface except where there are pores. Under this there is a single layer of flattened vesicles which are usually appressed at their edges. In 'naked' dinoflagellates the vesicles may appear empty or contain only some dark-staining material (Fig. 3A. 2a, b) but in 'armoured' species the thecal plates each lie inside such a vesicle (Fig. 3A. 2b). Beneath the vesicles there are a variable number of microtubules which seem to be differently arranged in various species. Additionally, in species which readily shed their theca, e.g. prior to division, there may be an extra membrane beneath the vesicles and also an amorphous layer termed the pellicle (Loeblich 1970). Morrill & Loeblich (1981b) have surveyed the range of species with pellicles and found that in a number of them it contains what is thought to be sporopollenin and is presumably very resistant. Cellulose is also often present in thick pellicles.

There has been some disagreement as to which of the various membranes of the thecal system is the functional plasma membrane. Some authors have maintained that as the outer membrane is continuous with that around the flagella it must be the plasma membrane (Dodge 1973; Wetherbee 1975a; Dürr 1979a, b, c) whilst others have suggested that as plasmolysis or injury causes the protoplasm to round up beneath the thecal plates, then the plasma membrane must be situated there (Kalley & Bisalputra 1971; Steidinger *et al.* 1978). The thecal membranes of *Gonyaulax polyedra* have been studied by freeze-fracture (Sweeney 1976) and in the membrane associated with the pellicle it was found that on the protoplasmic fracture face there were about 6000 7 nm particles per μm^2. However, on the external face of the vesicle membrane the number and size of particles changed during the 24-hour cycle, being about 780 at 06.00 hours and 1530 at 18.00 hours. The relationship between this fact and the diurnal rhythm of the species has yet to be determined.

Several workers have reported the presence of delicate organic scales (Fig. 3A. 2d) on the outer surface of the thecal membranes (Clarke & Pennick 1976; Pennick & Clarke 1977; Morrill & Leoblich 1981a) in the genera *Oxyrrhis*, *Heterocapsa* (incl. *Cachonina*, and *'Glenodinium'*. In *Oxyrrhis* these structures are flattened discs with a spiral pattern but in the other organisms they tend to have a roughly triangular shape and are composed of a delicate plate bearing spines arranged in the form of three interlocking hexagons. As yet, nothing is known about the formation of the scales.

The contents of the thecal vesicles prove a very variable feature of these organisms. Since it was first shown that even 'naked' dinoflagellates have the vesicles (Leadbeater & Dodge 1966) there have been several reports of a thin layer of dark-staining material within them (see Fig. 3A, 2a) (e.g. Dodge 1974) but this material has not been identified. A number of dinoflagellates have delicate smooth plates within the vesicles. The early reports for *Aureodinium* (Dodge 1967) and *Katodinium* (Dodge & Crawford 1970) were based on TEM

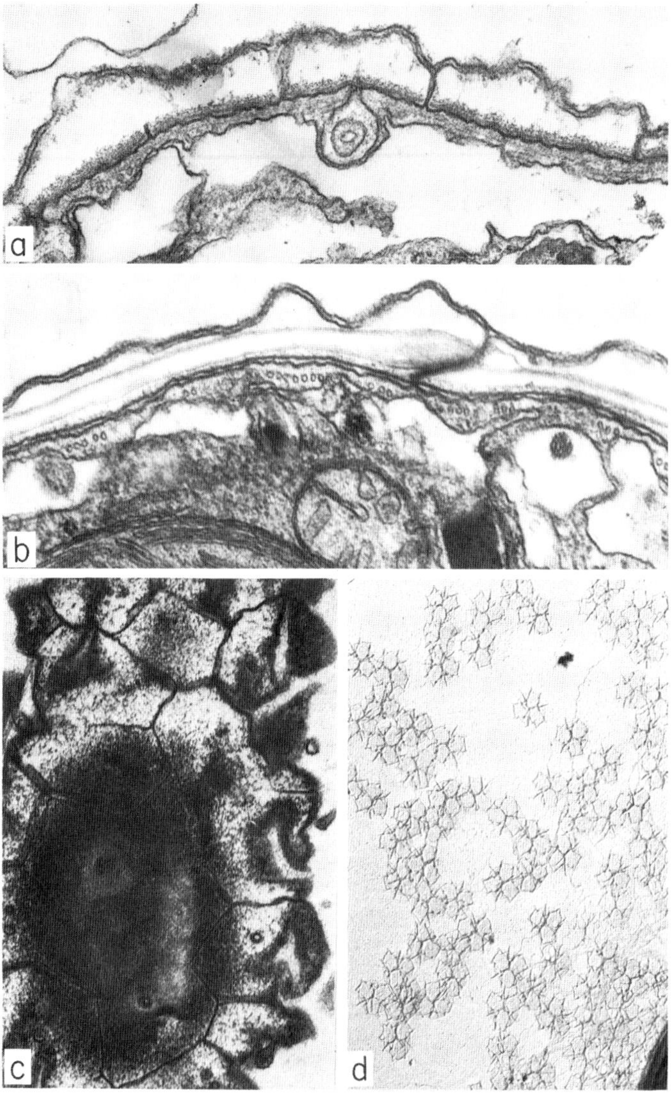

Fig. 3A.2. The theca or amphiesma. (a) Vertical section of the theca of *Amphidinium* in which the thecal vesicles are empty apart from some granular material at the proximal side (× 39 000). (b) Vertical section of the theca of *Glenodinium* showing the overlap of two plates and the presence of various membranes and microtubules (× 39 000). (c) A glancing section of the theca in *Amphidinium* showing the polygonal form of the thecal vesicles (× 14 000). (d) Scales detached from the outer thecal membrane of *Glenodinium* sp. (shadowed) (× 11 250).

studies but more recently, SEM has been successfully applied to these fragile organisms and the patterns of the mainly hexagonal plates have been described for *Scrippsiella gregaria* (Loeblich *et al.* 1979), *Symbiodinium* (= *Zooxanthella*) *microadriaticum* (Loeblich & Sherley 1979) and *Heterocapsa pygmaea* (Loeblich *et al.* 1981).

Numerous studies of the shape and ornamentation of thecal plates have been carried out, mainly with SEM for the purposes of taxonomy (e.g. Dodge & Hermes 1981). However, there has also been work on the arrangement of the overlap at the edges of the plates (Dürr & Netzel 1974; Dürr 1979b) which is consistent within a species (see Chapter 2).

Both Wetherbee (1975c) and Dürr (1979a) have studied the development of plates, following cell division, in organisms which do not shed the parental plates. Firstly, the membrane system is organized and thecal vesicles are formed. At an early stage there is a dark-staining layer and then plate precursor material is deposited in the vesicles. This appears to come from elongated cisternae which are formed by internally adjacent golgi bodies. Later, vesicles originating within the cytoplasm, and possibly part of the ER system, flatten out and fuse with the thecal vesicles. In genera such as *Ceratium* and *Gonyaulax*, which have thick thecal plates, it has been shown that microfibrils form a major component of the plates. There is no clear information as to the composition of the fibrils, although cytochemical tests indicate the presence of cellulose. There is no information on the methods used by the cell to arrange the ordered synthesis of the microfibrils and the particular placement of ridges and pores, etc. However, it is of interest that at the early stage of development numerous microtubules lie beneath the theca (Dodge & Crawford 1970b; Wetherbee 1975c) and these become reduced in number once the theca is complete.

Coccoid vegetative cells and resistant stages are normally bounded by a continuous wall which lies outside the cell membrane. This wall also contains numerous microfibrils which in *Pyrocystis* are arranged in many layers with each one orientated at right angles to the preceding one (Swift & Remsen 1970). In the cyst wall of *Peridinium cinctum* the microfibrils are more randomly arranged (Dürr 1979c). Sporopollenin is commonly found in resistant cyst walls.

3 FLAGELLA

Motile dinoflagellate cells possess two flagella; one, termed the transverse flagellum, is orientated around the cell and the other, the longitudinal flagellum, is orientated posteriorly (see Chapter 2). This latter is fairly conventional in structure, consisting of a wider portion which contains a '9 + 2' axoneme plus a complex spiral paraxial rod (Fig 3A. 3b) (Dodge & Crawford 1972) and a narrower tip region which may bear fine hairs (Leadbeater & Dodge 1967b). In *Oxyrrhis* both flagella may be of this type with hairs along most of their length and, additionally, the outer membrane is covered with tightly packed scales (Clarke & Pennick 1972).

The transverse flagellum has a construction which is unique to dinoflagellates, but its interpretation has proved somewhat problematical. The first ultrastruc-

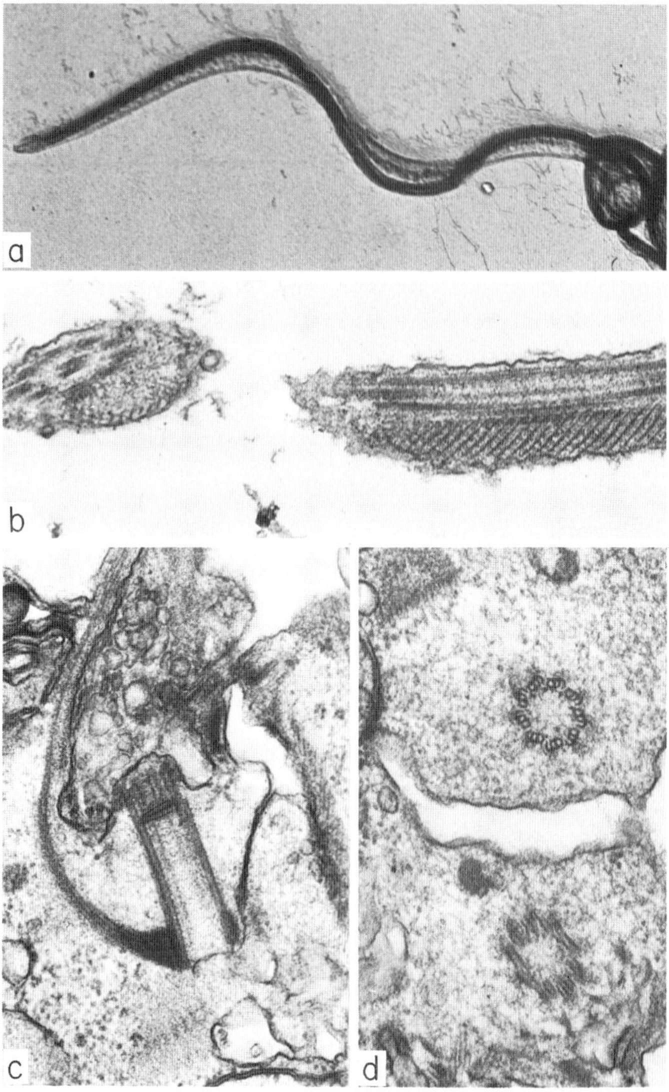

Fig. 3A.3. Flagella. (a) A shadowed preparation of the tip region of the transverse flagellum of *Prorocentrum* sp. Note the thick axoneme and the thinner striated strand (× 5300). (b) Sections through a flagellum of *Oxyrrhis* showing both the axoneme and a complex paraflagellar structure (× 30 500). (c) Longitudinal section of the basal body in *Prorocentrum triestinum* showing a root (× 39 000). (d) Transverse section of the two flagella bases in *P. triestinum* (× 32 000).

tural studies (Leadbeater & Dodge 1967a, b) which utilized TEM observations of shadowed, sectioned and replicated flagella, suggested that it had a helical structure with the '9 + 2' axoneme thrown into a spiral around a tight, striated, proteinaceous strand (Fig. 3A. 3a). Fine non-tubular hairs or mastigonemes were found to be attached to one side. Subsequently, a number of workers examined the flagella with scanning electron microscopy and some (F. Taylor 1975; Leblond & Taylor 1976; Herman & Sweeney 1977) reported that the striated strand formed a tight-draw-string at one side of an undulating axoneme and that the flagellum is possibly anchored in the girdle by a number of fine filaments. However, other workers using SEM on *Peridinium cinctum* (Berdach 1977; Berman & Roth 1979) and *Ceratium* (R. Wetherbee, person. comm.) and serial sectioning of *in situ* flagella (Rees & Leedale 1980) differ in finding that the axoneme is helically coiled but agree that it lies external to the striated strand. These latter workers did not see any anchoring threads and those reported by others may in fact represent aggregations of flagellar hairs. One of the problems here is undoubtedly the inevitable change brought about by fixing an extremely mobile and no doubt sensitive structure. There may also be differences of construction within the group. F. Taylor (1975, and Chapter 2) has suggested that one might represent the propulsive, and the other the relaxed state.

Several structures associated with the flagella are found inside the cells. As is usual, each has a basal body consisting of microtubules arranged in triplets (Fig. 3A. 3d) (Leadbeater & Dodge 1967b). The earlier work reported a rather unusual arrangement of basal discs and diaphragms in *Gymnodinium*. Wedermayer & Wilcox (1984) observed a chrysomonad-like transitional helix in the base of the longitudinal flagellum of *Peridiniopsis berolinense*. The basal bodies are anchored by a system of roots which generally consist of a microtubular root running from the inner end up to and then beneath the cell surface (Fig. 3A. 3c) and a striated root running more deeply into the cell. The exact arrangement varies from species to species (e.g. *Amphidinium*, Dodge & Crawford 1968; *Ceratium*, Dodge & Crawford 1970a; *Oxyrrhis*, Dodge & Crawford 1972) and is probably dependent on the various angles of insertion of the flagella and the shape of the cell.

Formation of flagella hairs also takes place inside the cell. As with algae that have tubular mastigonemes, the dinoflagellates produce their delicate flagellar hairs, which measure 10 nm wide and up to 500 nm long, in large perinuclear vesicles (Leadbeater 1971).

4 THE NUCLEUS

4.1 Interphase nucleus

The striking appearance of dinoflagellate nuclei in TEM studies results from the fact that the chromosomes normally are permanently condensed and visible as circular, oval or elongated profiles (Fig. 3A. 4a). By way of contrast the nucleolus appears to be similar to that of higher organisms, having both granular and fibrillar zones. Chromosomes, presumably nucleolar organizing, are often seen penetrating the nucleolus. Sometimes the chromosome width remains the same in the nucleolus (e.g. *Protog gonyaulax tamarensis*) and at other times it

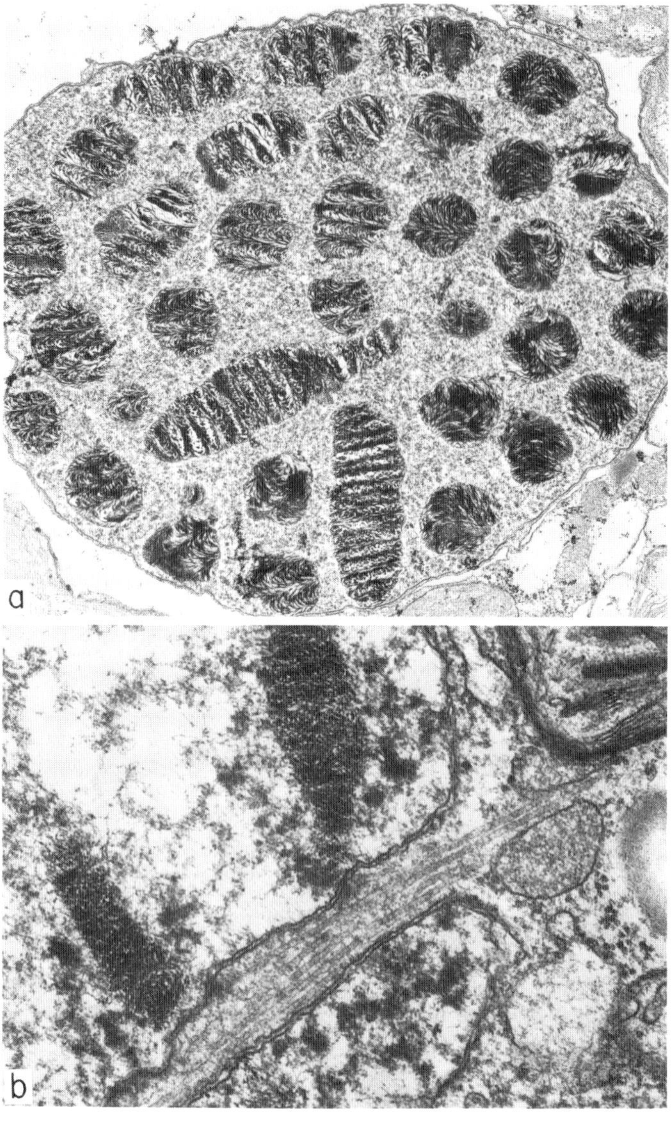

Fig. 3A.4. (a) Section through the interphase nucleus of *Amphidinium herdmanii*. Note the chromosomes sectioned both transversely and longitudinally (× 14 000). (b) Part of a dividing nucleus of *A. carterae* showing a spindle tunnel containing microtubules and two chromosomes which are attached to this tunnel (× 39 000).

becomes much thinner (e.g. *Glenodinium foliaceum*). The nucleolus probably remains intact during mitosis.

In general, the nuclear envelope consists of a typical two-membrane structure with numerous nuclear pores which freeze-etching has shown to be sometimes arranged in rows, to be close-packed or to be randomly arranged (Wecke &

Giesbrecht 1970; Soyer & Escaig 1980). In a few organisms much more elaborate, vesiculated nuclear envelopes have been found. In *Noctiluca* (Soyer 1969; Zingmark 1970) annulated vesicles are built into the envelope whilst in *Gymnodinium fuscum* (Dodge & Crawford 1969a) separate porous two-membrane depressions are found on the nuclear side. The vesicles are thought to provide reserve membrane material for nuclear division.

4.2 Chromosome structure

It has been known for some time that the dinoflagellate chromosome contains very little protein and consists almost entirely of approximately 2·5 nm diameter DNA fibrils. Its biochemistry is described in Chapter 4 (see also Spector 1984). Recently, it has been shown that nucleosomes are absent (Bodansky *et al.* 1979; Herzog & Soyer 1981). Numerous attempts have been made to interpret the intriguing arrangement of the DNA (for the earlier attempts see Dodge 1971b). The initial studies were based on random thin sections which gave a varying pattern of fibrils according to the plane of section (Fig. 3A. 4a). Some workers could see a similarity between sections of chromosomes and those obtained by cutting through other ordered structures and thus Bouligand *et al.* (1968) and later Livolant & Bouligand (1978) attempted to explain the observed arrangement by means of geometry. They concluded that the chromosome is a liquid crystal and that the fibrils are an illusion or an artefact (Livolant & Bouligand 1980). This, however, does not accord with the work of Haapala & Soyer (1973) who used spreading techniques to study the long chromosomes of *Prorocentrum micans*. They found that numerous fibrils were present and because the ends of the chromosomes were always rounded and intact they concluded that each chromosome was made up of a large number of circular chromatids which are normally twisted together. Oakley & Dodge (1979) used a combination of serial sectioning and spreading (Fig. 3A. 5a) to study the short chromosomes of *Amphidinium carterae*. They concluded that each chromosome consists of a toroidal bundle of DNA strands which is coiled into a tight double helix. It was thought that this represented a single chromonema. More recently, Spector *et al.* (1981a, b) have studied a brief 'uncoiled' phase during the cell cycle of *Peridinium cinctum*. Here, by means of serial sections, it was shown that the chromonema consists of a central core composed of many 9 nm fibrils. Around this is wrapped a right-handed helix composed of a fibre 2·5 nm in diameter. This is in turn surrounded by a second helix which is connected to the first by a series of fibres and which also has many 9 nm granules attached to it at fairly regular intervals. It is very difficult to correlate this model with the condensed-phase pictures of other workers.

Thus, the picture available is far from clear and whilst some of the accounts can now be discounted the true arrangement of the numerous DNA fibrils, which are so much thinner than any ultra-thin section, awaits to be revealed.

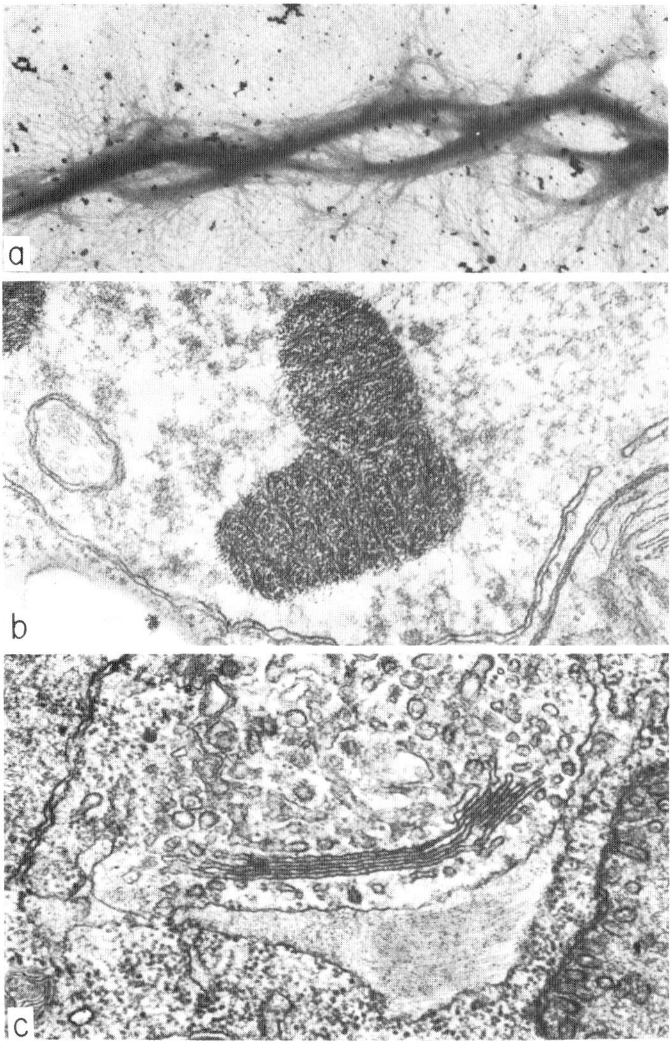

Fig. 3A.5. (a) A spread chromosome of *Amphidinium carterae* (uranyl acetate stained) showing the separation of the two major chromatin bundles (× 11 250). (b) A dividing chromosome in *A. carterae* (× 47 200). (c) Part of the Golgi region adjacent to the nucleus. The fibrous vesicle in the lower part of the micrograph represents the formation of flagellar hairs (× 39 000).

4.3 Nuclear division

The mitotic process in dinoflagellates is very unusual in that the nuclear envelope remains intact and the spindle is in the cytoplasm rather than in the nucleus as it is in others with a 'closed mitosis' (e.g. euglenoids). The only other organisms with a somewhat similar type of division are the hypermastigid flagellates (see Kubai (1975) and Dodge & Vickerman (1980) for refs).

The first obvious sign of division is the formation of bundles of microtubules along one side of the nucleus. These sink into the nucleus and are partially surrounded by nuclear envelope to form the mitotic spindle tunnels (Fig. 3A. 4b). About this time it becomes apparent that the chromosomes are attached to the nuclear envelope where it surrounds the tunnels (they may be attached at interphase but this is not clear). Soyer & Escaig (1980) have shown that in *Prorocentrum micans* there are no nuclear pores at these points but numerous pores are present between the contact areas. The chromosomes now split into pairs of chromatids, a process which starts at the distal end and works towards the attached end, giving Y and V shapes as it proceeds (Fig. 3A. 5b). It was thought that DNA synthesis must take place during this stage but recent work (Spector *et al.* 1981b and review by Triemer & Fritz 1984) shows an 'S' phase involving uncoiling of the chromosomes which probably takes place some time before the start of mitosis.

Metaphase consists of an ovoid nucleus perforated by one to several spindle tunnels, none of which has any centriole or spindle pole body attached to it. The pairs of chromatids are attached to the tunnel-lining at kinetochore-like structures to which spindle microtubules are also attached (Oakley & Dodge 1974; Cachon & Cachon 1979). Now the nucleus elongates and sister chromatids are separated whilst retaining contact with the nuclear envelope and indirectly with part of the spindle. Anaphase would seem to be a long drawn-out, process as the nucleus constricts and then the daughter nuclei become 'pushed' apart by the extension of the interzonal spindle (Oakley & Dodge 1977). Cytokinesis or cell cleavage now follows and in armoured species this commences with the development of cleavage vesicles beneath the theca (Dürr 1979a; Wetherbee 1975b) and also involves numerous sub-thecal microtubules.

In the phagotrophic dinoflagellate *Oxyrrhis* no microtubules were found in the single cytoplasmic tunnel by Cachon *et al.* (1979). Subsequently Triemer (1982) found that this species is unique among dinoflagellates in possessing an intranuclear spindle resembling that of euglenoids. Microtubules develop from dense plaques in the nuclear envelope. The nucleolus persists, as does the nuclear envelope. This organism is also unusual in that histone appears to be present in the chromosomes (Hollande 1974; see Chapter 4) as it also is in some parasitic dinoflagellates which have a variant mitosis involving centriolar structures (Cachon & Cachon 1973; Ris & Kubai 1974).

5 PLASTIDS AND PYRENOIDS

5.1 Plastid structure

Approximately half of the dinoflagellate species possess plastids (or chloroplasts) and they are found in most orders and families. The typical chloroplast envelope in this group consists of three parallel membranes having no obvious connections with the nuclear envelope (Dodge 1968). This structure appears similar to that found in the Euglenophyceae.

Within the plastid the general arrangement is for the lamellae to lie parallel to each other and each to be composed of three thylakoids which are fairly closely appressed together (Fig. 3A. 6a) (Dodge 1975). However, in some species

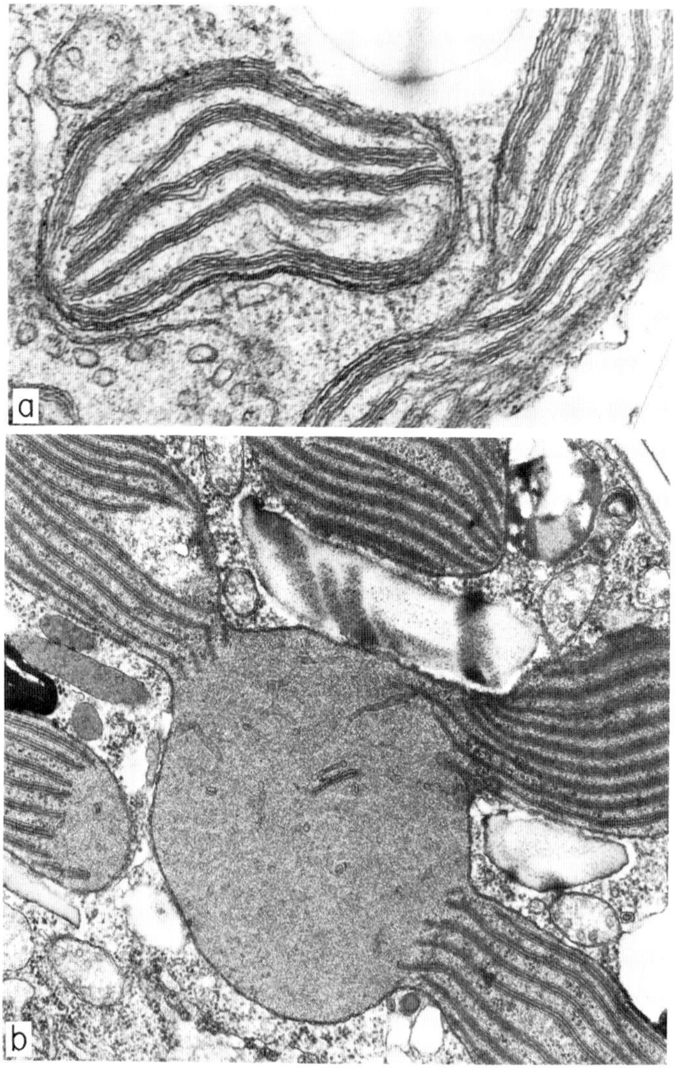

Fig. 3A.6. (a) Part of the chloroplast in *Amphidinium carterae* showing the three-membrane envelope and triple-thylakoid lamellae. A large starch grain is situated at the top of the picture (× 47 200). (b) The multiple-stalked pyrenoid of *Scrippsiella trochoidea* (26 500).

such as *Gymnodinium simplex* (Dodge 1974) the lamellae tend to be radial in arrangement and often consist only of a pair of thylakoids. In this species lamellar branching may also be found. In some species a single thylakoid lies parallel to the plastid envelope, but true girdle lamellae are not found except in *Glenodinium* (*Kryptoperidinium*) *foliaceum* and *Peridinium balticum* where the chloroplasts are part of an endosymbiont within the dinoflagellate cell (see Section 9).

Early studies were made on extracted thylakoids of *Gymnodinium* (Dodge 1968) where it was shown that they were round in profile and showed some granularity of the membrane. Recently, freeze-fracture has been applied to the chloroplast membranes of *Gonyaulax* (Sweeney 1981). As is common in other plants the external fracture face of the stacked thylakoid membranes (EFs) was found to carry far more particles than the external fracture face in unstacked regions (EFu). The protoplasmic face of the thylakoids showed twice as many particles as the EFs face. The chloroplast envelope membranes were quite distinct and the centre one of the three carried far fewer particles than either of the peripheral membranes.

In general, the chloroplast stroma is rather insignificant in amount and mainly granular. In some cases the DNA of the plastid genophore is scattered, as in *Prorocentrum micans* (Kowallik & Haberkorn 1971), but in *Scrippsiella sweeneyae* it is in the form of a distinct nucleus-like body (Bibby & Dodge 1974).

Rather little work has been carried out relating ultrastructure to physiology in dinoflagellate plastids. However, a study made into the effects of the 24-hour day/night cycle showed that in *Gonyaulax polyedra* the lamellae are more widely separated in the day in those parts of the plastid nearer the centre of the cell (Herman & Sweeney 1975).

5.2 Pyrenoids

Many photosynthetic dinoflagellates have distinctive regions of the chloroplast which are termed pyrenoids. To date, some five distinct types have been described (Dodge & Crawford 1971a) and other organisms such as *Gonyaulax polyedra* (Schmitter 1971) and *Ceratium hirundinella* have large areas of granular chloroplast stroma which may represent rather simple pyrenoids. In some species there is only one pyrenoid per cell but in others there may be more than one pyrenoid per chloroplast.

Basically, pyrenoids can be divided into two categories, those within the chloroplast and those protruding from it, although still within the chloroplast envelope. The most simple 'internal' pyrenoids consists of fusiform structures which lie between chloroplast lamellae. These are found in several fucoxanthin-containing *Gymnodinium* species and also in endosymbionts such as those in *Glenodinium* (*Kryptoperidinium*) *foliaceum* (Dodge 1975). Much more complex types are found in some members of the Prorocentrales (Kowallik 1969; Dodge

& Crawford 1971a) and here the pyrenoid matrix, which is regularly close-packed, occupies the spaces between many lamallae. Stalked pyrenoids are of several types. The simplest type consists of a projection from the chloroplast, bounded by the plastid envelope but containing little but granular material. Here, there may be a polysaccharide cap over the outside of the pyrenoid, external to the chloroplast. In *Heterocapsa* these stalked pyrenoids are perforated by narrow cytoplasmic channels (Dodge & Crawford 1971a). The more complex type consists of multiple-stalked pyrenoids (Fig. 3A. 6b) which are suspended from several arms of the chloroplast. In *Amphidinium carterae*, for example, several lamellae run into or through this type of pyrenoid (Dodge & Crawford 1968).

6 TRICHOCYSTS

The ejectile organelles or trichocysts which are found in the majority of dinoflagellates are in general more simple than those found in a number of other protistans. In the undischarged state the trichocyst consists of two distinct regions, the neck which is attached to the thecal membranes, and the body (Bouck & Sweeney 1966; Dodge 1973), the whole being surrounded by a single membrane. The neck usually contains a number of slightly twisted fibres and the bounding membrane may have a corrugated appearance. The main body of the trichocyst is some 2–4 μm long and consists mainly of a proteinaceous rod which is square or rhombic in section. The protein granules are packed into a regular paracrystalline array (Fig. 3A. 7a). The discharged trichocyst consists of a long thread, square in section, and of variable thickness. Negative staining or shadowing of whole mounts reveals a regular transverse banding which has been measured from many organisms and shows a major repeating period of 65–85 nm with some five minor bands between (Bouck & Sweeney 1966; Messer & Ben-Shaul 1971; Hausmann 1973). Longitudinally arranged filaments have a diameter of about 1 nm.

The trichocysts form within Golgi vesicles and stages of 'crystallization' of the protein material can be observed (Bouck & Sweeney 1966; Leadbeater & Dodge 1966). When mature they must migrate to the cell periphery and become located adjacent to a trichocyst pore. Discharge is interpreted as an unfolding process from one paracrystalline state to another (Hausmann 1973).

Some dinoflagellates have another type of ejectile organelle termed the mucocyst. These are flask-shaped vesicles associated with the theca (Cachon *et al.* 1976), and they contain an amorphous, finely granular material which appears to be secreted onto the outer surface of the cell. They are probably the source of material for the mucoid cysts produced by many naked dinoflagellates. The elaborate ejectile organelles termed nematocysts are considered in Chapter 3B.

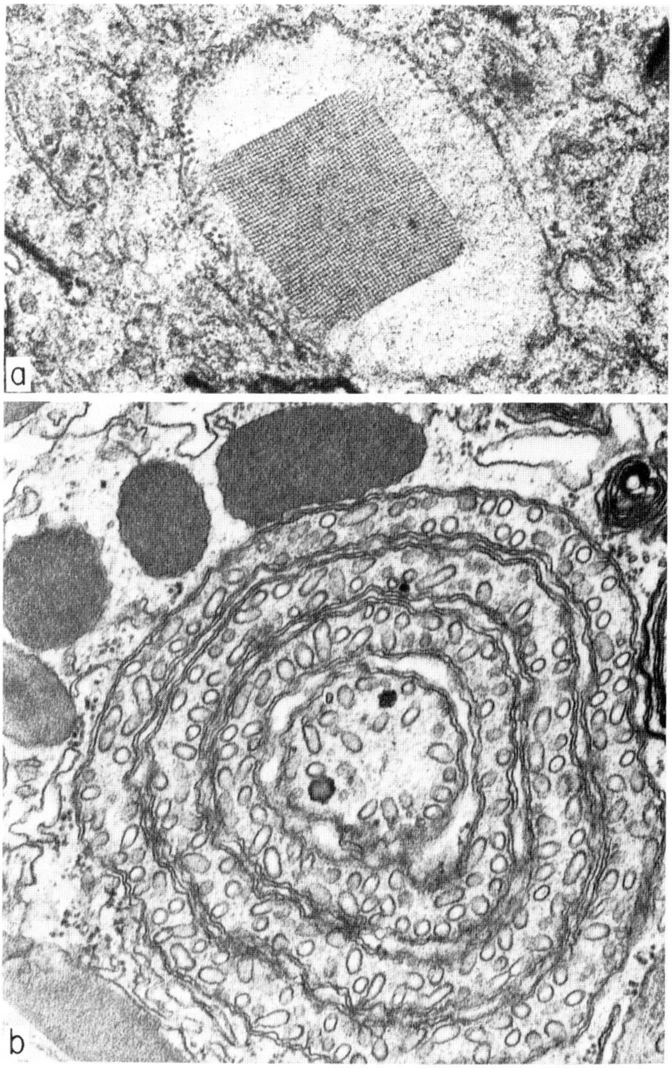

Fig. 3A.7. (a) A transverse section of an almost mature trichocyst in *Prorocentrum* sp. The trichocyst vesicle is bounded by one membrane and contains paracrystalline protein in addition to various fibrils (× 48 000). (b) Microbodies (dense structures, top left) and a rather complex mitochondrion in *Ceratium hirundinella*. Note the tubular cristae (× 39 000).

7 VARIOUS ORGANELLES

7.1 Pusule

This organelle is unique to dinoflagellates and, although its presence in larger cells can be seen by light microscopy, there was no idea of its complexity and variability until TEM was possible. Pusules are probably present in all freshwater dinoflagellates and are perhaps found in all marine species except the symbionts.

The pusule basically consists of a tubule or chamber, lined with an invagination of plasma membrane, which opens to the outside via a flagellar entry canal (Dodge 1972). What distinguishes the pusule from a contractile vacuole or a simple membrane invagination is that it is a permanent structure which, in part at least, is composed of two membranes in a region where the general cell vacuolar membrane is closely adpressed to an extension of the external membrane covering the cell.

A relatively simple pusule is found in the marine *Amphidinium herdmanii* (Dodge 1972). Here a chamber branches off from the flagellar canal (Fig. 3A. 8a) and around this are arranged a large number of subspherical vesicles with characteristic two-membrane walls. The vacuolar reticulum surrounds these vesicles. Far more elaborate pusules have been found in the freshwater dinoflagellates *Gymnodinium* sp. (Mignot 1970; Schnepf & Deichgräber 1972), *G. fuscum* (Dodge & Crawford 1969a), *Woloszynskia coronata* (Crawford & Dodge 1974) (Fig. 3A. 8b) and in the marine parasitic dinoflagellate *Oodinium* (Cachon *et al.* 1970). In non-photosynthetic marine *Protoperidinium* species the chambers are very large. In general, these more complex pusules are situated near the centre of the cell and involve considerable elaboration of the tubule system and vacuolar reticulum.

Although pusules are clearly involved in exchange, their exact function is at present uncertain.

7.2 Mitochondria

The mitochondria of dinoflagellates have tubular cristae with a conspicuously narrowed neck (Dodge 1973). Normally, cristae are not very numerous and there is a considerable amount of space, only some of which is occupied by DNA and occasionally by crystalline bodies. The shape of the mitochondria is rather varied. For example, extremely long thin ones have been found in *Prorocentrum* and concentrically wound mitochondria (Fig. 3A. 7b) in *Ceratium*.

7.3 Golgi bodies

These are generally situated in a hemispherical zone above the nucleus. The individual Golgi body may appear to be interlocked or might lie singly. Unlike many other organisms it is usually impossible to differentiate between the forming face and the maturing face of dinoflagellate Golgi bodies, each of which consists of a stack of around ten flattish cisternae. Trichocysts are formed within Golgi vesicles but there are no other known products of the Golgi stacks.

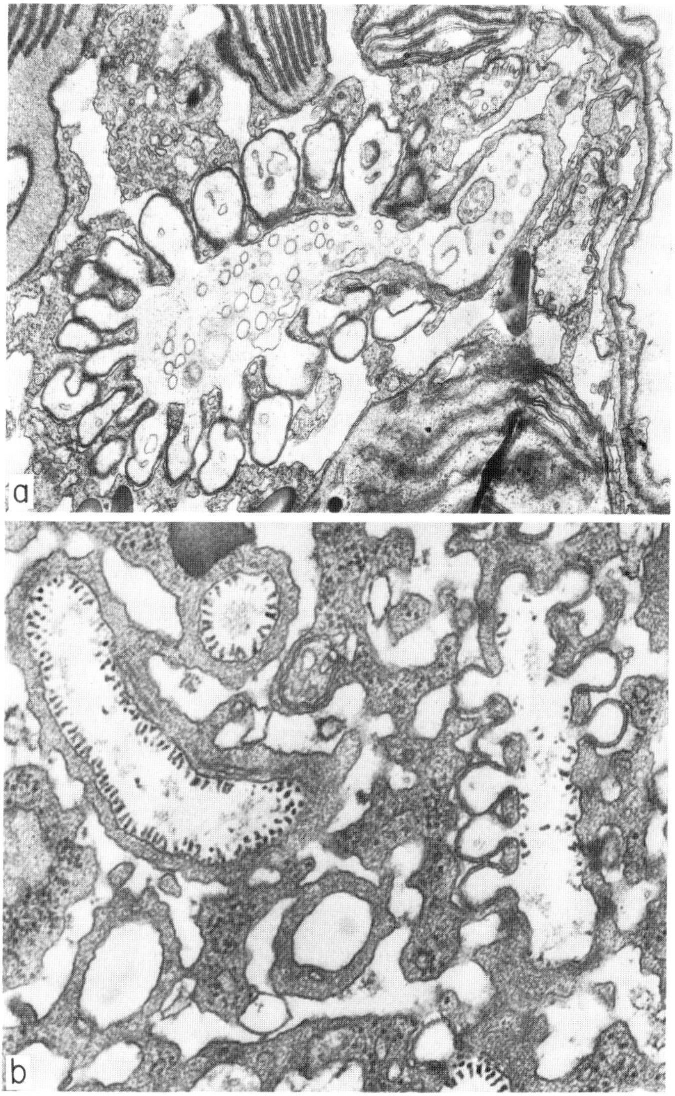

Fig. 3A.8. The pusule. (a) A rather simple pusule, consisting of a chamber opening from the flagellar canal and a large number of pusule-vesicles, in *Amphidinium herdmanii* (× 18 500). (b) Part of the extremely complex pusule of *Woloszynskia coronata*. Note the pusule-vesicles, tubules and vacuolar reticulum (× 47 200).

7.4 Microbodies

Simple microbodies have been observed in many dinoflagellates. They consist of a single membrane which encloses a granular material (Fig. 3A. 7b), the whole organelle being about 0·5 μm in diameter. No crystals or other structures have been found in dinoflagellate microbodies, neither do they stain with the DAB reaction. In *Scrippsiella sweeneyae*, Bibby & Dodge (1973) found microbodies to be associated with a complex reticulate structure apparently composed of endoplasmic reticulum. This was interpreted as the site of microbody assembly.

7.5 Simple eyespot (stigma)

Unlike a number of other algal groups, several distinct types of eyespot are found in dinoflagellates. Only a few species possess these structures which can be seen by light microscopy as a red-pigmented spot situated adjacent to the longitudinal groove or sulcus beneath the longitudinal flagellar base. The larger and much more complex ocelloids are the subject of a separate account (Chapter 3B).

The most simple eyespot known from this group has been found in *Wolozynskia coronata* (Crawford & Dodge 1974). This consists of a cluster of carotenoid-containing lipid globules situated behind the sulcus. It is not membrane-bound and does not seem to be associated with any particular organelle. In another member of the same genus, *W. tenuissima* (Crawford *et al.* 1971), in a *Gymnodinium* sp. (Schnepf & Deichgräber 1972) and in *Peridinium westii* (Messer & Ben Shaul 1969) the eyespot consists of a single layer of globules between the envelope and lamellae of a chloroplast situated just behind the sulcus. A flagellar root has been seen to pass in front of this eyespot which is rather like that found in the Chrysophyta (Dodge 1973).

A far more complex eyespot has been found in the endosymbiont-containing *Glenodinium* (*Kryptoperidinium*) *foliaceum* (Dodge & Crawford, 1969), and *Peridinium balticum* (Tomas & Cox 1973). Here a triple-layered envelope encloses two layers of carotenoidal globules which are separated by granular material. The eyespot is large and contains hundreds of globules which may be packed in an almost regular array. It is situated near to the flagellar bases and to an elaborate lamellar structure, although no direct connection has been observed.

7.6 Food reserves

Dinoflagellates are able to store both polysaccharides and fat as food reserves. Both types lie free in the cytoplasm and when they are present together there is a tendency for the lipid droplets to be situated at the anterior end of the cell and the starch grains towards the posterior (Crawford & Dodge 1974). This is also true for the young excysted cell of *Ceratium hirundinella* (Chapman *et al.* 1981). Starch grains may also be located adjacent to stalked pyrenoids. Recent studies on *C. hirundinella* have shown that there is a considerable build-up of starch prior to encystment (Chapman *et al.* 1982).

8 STUDIES OF ASPECTS OF THE LIFE CYCLE

Fertilization has been studied in *Peridinium cinctum* (Spector *et al.* 1981). Here, a fertilization tube forms between the gametes and one nucleus moves into this and fuses with the other before the zygote forms. The gametes possess few chloroplasts as compared to vegetative cells.

The formation of cysts has been followed in several species. In *Woloszynskia tylota* (Bibby & Dodge 1972) there was considerable reorganization of the cell structure as the characteristically shaped cyst wall developed. The mature cyst contained very few organelles and was bounded by a layer of lipid globules under a thick wall. In *Peridinium cinctum* (Dürr 1979c) it was found that the cyst wall was a multilayered structure with microfibrils in the endospore layer. In *Ceratium hirundinella* Chapman *et al.* (1982) found that the cyst wall was commenced before the theca was shed and at this stage numerous granule-containing vesicles were to be seen in the cell (Fig. 3A. 9a). The 60 nm silica-containing granules were discharged so that they formed a thin layer on the outside of the cyst (Fig. 3A. 9b, c) which has a distinctly granular appearance in the SEM. On excystment (Chapman *et al.* 1981) the motile gymnoceratium is at first naked but, as it develops a triangular shape, a complex of membranes is organized around the cell and here the characteristic plates are laid down within thecal vesicles. Other changes are described by Chapman *et al.* (1985). Further information on life cycles is given in Chapter 14.

9 ENDOSYMBIOSIS AND THE ORIGIN OF CHLOROPLASTS

There have been a number of reports of the presence of presumably symbiotic bacteria within the cytoplasm (Dodge 1973, D. V. Chapman, unpublished observation) or the nucleus (Silva 1978) of dinoflagellates. In none of these cases has the relationship between the two organisms been investigated. This is also the case in the large phagotrophic dinoflagellate *Noctiluca* where a strain was isolated which contained a green symbiotic alga (Sweeney 1978).

Much more interest and speculation has surrounded the discovery of endosymbiotic chrysophytic algae within certain dinoflagellates as this appears to provide a model which can explain the great diversity of chloroplast structure and pigment composition within the Dinophyceae. The first ultrastructural indication of this unusual condition was the discovery that *Glenodinium* (*Kryptoperidinium*) *foliaceum* had two nuclei, one being of the type normal in dinoflagellates ('dinokaryotic') and the other a more normal eukaryotic nucleus (Dodge 1971a) (Fig. 3A. 10). At this stage the status of the 'eukaryotic' nucleus was not understood although it was noted that it is surrounded by a distinct cytoplasmic band containing evenly dispersed granules. Studies were then made on a closely related organism, *Peridinium balticum*, which showed that this also had the two types of nucleus (Tomas *et al.* 1973) with a single membrane

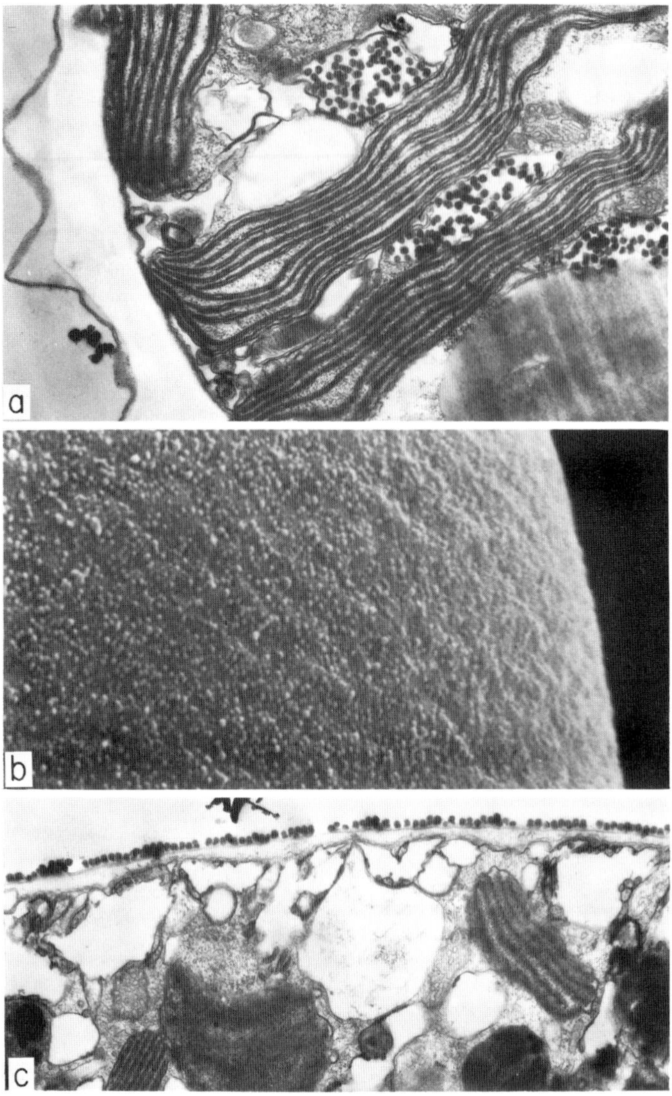

Fig. 3A.9. Cyst of *Ceratium hirundinella* (courtesy of D. V. Chapman). (a) A vegetative (thecate) cell preparing to encyst. Note the vesicles containing dark granules (× 14 400). (b) An SEM of part of the granular walled cyst (× 8800). (c) Vertical section of part of a young granular walled cyst showing a row of silica-containing granules affixed to the outside of the developing cyst wall (× 12 000).

separating the eukaryotic nucleus, all the chloroplasts, some of the mitochondria and ribosomes, etc. from the cytoplasm of the dinoflagellate host (Tomas & Cox 1973). Thus, it was concluded that there is a chrysophytic endosymbiont (possibly a wall-less diatom) within the dinoflagellate. Subsequent studies have

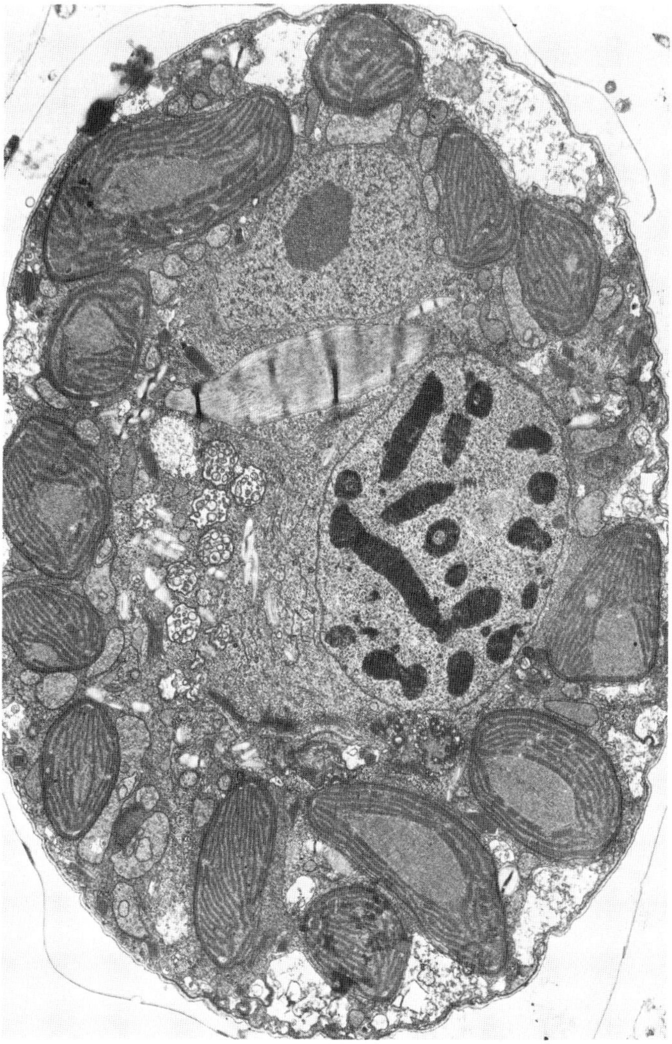

Fig. 3A.10. A longitudinal section of the endosymbiont-containing *Glenodinium foliaceum*. Note the normal dinoflagellate nucleus (centre) and the eukaryotic nucleus (top). The chloroplasts are of the chrysophytic type and belong to the endosymbiont (× 6800).

confirmed this state of affairs in *G. foliaceum* (Jeffrey & Vesk 1976), have shown that the endosymbiont divides amitotically (Tippit & Pickett-Heaps 1976), and that these organisms have pigments in which fucoxanthin is the dominant carotenoid (Withers *et al.* 1977; Withers & Haxo 1975). These two organisms also possess eyespots (see above) which, like most dinoflagellate chloroplasts, are bounded by three membranes (Dodge & Crawford 1969b; Tomas & Cox 1973), suggesting that they may be relicts of an earlier chloroplast system present in the cell before the host acquired the present chloroplast-containing endosymbionts (Gibbs 1978; Dodge 1979; F. Taylor 1980).

Recently, ultrastructural studies have been carried out on the blue-pigmented dinoflagellates *Gymnodinium acidotum, G. eucyaneum* and *Amphidinium wigrense* which contain cryptomonad endosymbionts. (Wilcox & Wedemayer 1984, 1985; H. Hu & J. West, pers. comm.)

This information, taken also with the fact that other dinoflagellates have been reported which appear to contain red or green plastids (although not examined by TEM) is strongly suggestive that dinoflagellate chloroplasts have derived from various endosymbiotic events, and have not necessarily evolved in parallel with the evolution of the morphology and general structure of the cells. The diversity of chloroplast types and the fact that in some organisms 'chloroplasts' appear to have been acquired twice suggests that the process has happened several times (Loeblich 1976; Dodge 1979; Steidinger & Cox 1980) in a group which can be regarded as basically non-photosynthetic and heterotrophic. Further comparative studies of chloroplast structure and pigmentation will undoubtedly reveal that these features show very little correlation with the present day classification of the group (e.g. see Steidinger & Cox 1980) and that an evolutionary scheme for plastids will look quite different from one developed on the more traditional characters of cell morphology and thecal pattern. It should be emphasized, however, that there are other views on the evolution of dinoflagellates and on the evidence of the chloroplasts. Readers are referred to F. Taylor (1980), Loeblich (1976) and Bujak & Williams (1981) for alternative hypotheses.

10 REFERENCES

Berdach J. T. (1977) *In situ* preservation of the transverse flagellum of *Peridinium cinctum* (Dinophyceae) for scanning electron microscopy. *J. Phycol.* **13**, 243–51.

Berman T. & Roth I. L. (1979) The flagella of *Peridinium cinctum* f. *westii*: *in situ* fixation and observation by scanning electron microscopy. *Phycologia* **18**, 307–11.

Bibby B. T. & Dodge J. D. (1972) The encystment of a freshwater dinoflagellate: a light and electron-microscopical study. *Br. phycol. J.* **7**, 85–100.

Bibby B. T. & Dodge J. D. (1973) The ultrastructure and cytochemistry of microbodies in dinoflagellates. *Planta* **112**, 7–16.

Bibby B. T. & Dodge J. D. (1974 The fine structure of the chloroplast nucleoid in *Scrippsiella sweeneyae* (Dinophyceae). *J. ultrastruct. Res.* **48**, 153–61.

Bodansky S., Mintz L. B. & Holmes D. S. (1979) The mesokaryote *Gyrodinium cohnii* lacks nucleosomes. *Biochem. biophys. Res. Commun.* **88**, 1329–36.

Bouck G. B. & Sweeney B. M. (1966) The fine structure and ontogeny of trichocysts in marine dinoflagellates. *Protoplasma* **61**, 205–23.

Bouligand Y., Soyer M.-O. & Puiseux-Dao, S. (1968) La structure fibrillaire et l'orientation des chromosomes chez les dinoflagellés. *Chromosoma* (*Berl.*) **24**, 251–87.

Bujak J. P. & Williams G. L. (1981) The evolution of dinoflagellates. *Can. J. Bot.* **59**, 2077–87.

Cachon J. & Cachon M. (1971) Ultrastructures du genre *Oodinium* Chatton. Différenciations cellulaires en rapport avec la vie parasitaire. *Protistologica* **7**, 153–69.

CACHON J. & CACHON M. (1973) Les Apodinidae Chatton. Revision Systèmatique rapports hote-parasite et métabolisme. *Protistologica* **9**, 17–33.

CACHON J. & CACHON M. (1979) Singular kinetochore structure in a parasitic dinoflagellate. *Arch. Protistenk.* **122**, 267–74.

CACHON J., CACHON M. & GREUET C. (1970) Le système pusulaire de quelques Péridiniens libres ou parasites. *Protistologica* **6**, 467–76.

CACHON J., CACHON M. & GREUET C. (1976) Les 'mucocystes' des Péridiniens. Constitution, évolution des structures et comparison avec celles des trichocystes. *Ann. Stn. Biol. Besse-en-Chandresse 1974–75* No. 9, 177–99.

CACHON J. CACHON M. & SALVANO P. (1979) The nuclear division of *Oxyrrhis marina*: an example of the rôle played by the nuclear envelope in chromosome segregation. *Arch. Protistenk.* **122,** 43–54.

CHAPMAN D. V., DODGE J. D. & HEANEY S. I. (1982) Cyst formation in the freshwater dinoflagellate *Ceratium hirundinella*. *J. Phycol.* **18**, 121–29.

CHAPMAN D. V., DODGE J. D. & HEANEY S. I. (1985) Seasonal and diel changes in ultrastructure in the dinoflagellate *Ceratium hirundinella*. *J. Plankt. Res.* **7**, 263–278.

CHAPMAN D. V., LIVINGSTONE D. & DODGE J. D. (1981) An electron microscope study of the excystment and early development of the dinoflagellate *Ceratium hirundinella*. *Br. Phycol. J.* **16**, 183–84.

CLARKE K. J. & PENNICK N. C. (1972) Flagellar scales in *Oxyrrhis marina* Dujardin. *Br. phycol. J.* **7**, 357–60.

CLARKE K. J. & PENNICK N. C. (1976) The occurrence of body scales in *Oxyrrhis marina* Dujardin. *Br. phycol. J.* **11**, 345–48.

CRAWFORD R. M. & DODGE J. D. (1974) The dinoflagellate genus *Woloszynskia*. II. The fine structure of *W. coronata*. *Nova Hedwigia* **22**, 699–719.

CRAWFORD R. M., DODGE J. D. & HAPPEY C. M. (1971) The dinoflagellate genus *Woloszynskia*. I . Fine structure and ecology of *W. tenuissimum* from Abbot's Pool, Somerset. *Nova Hedwigia* **19**, 825–40.

DODGE J. D. (1967) Fine structure of the dinoflagellate, *Aureodinium pigmentosum* gen. et sp. nov. *Br. phycol. Bull.* **3**, 327–36.

DODGE J. D. (1968) The fine structure of chloroplasts and pyrenoids in some marine dinoflagellates. *J. Cell Sci.* **3**, 41–48.

DODGE J. D. (1971a) A dinoflagellate with both a mesocaryotic and a eucaryotic nucleus. I. Fine structure of the nuclei. *Protoplasma* **73**, 145–57.

DODGE J. D. (1971b) Fine structure of the Pyrrophyta. *Bot. Rev.* **37**, 481–508.

DODGE J. D. (1972) The ultrastructure of the dinoflagellate pusule: A unique osmo-regulatory organelle. *Protoplasma* **75**, 285–302.

DODGE J. D. (1973) *The Fine Structure of Algal Cells*, pp. 261. Academic Press, London.

DODGE J. D. (1974) A redescription of the dinoflagellate *Gymnodinium simplex* with the aid of electron microscopy. *J. mar. biol. Ass. UK* **54**, 171–7.

DODGE J. D. (1975) A survey of chloroplast ultrastructure in the Dinophyceae. *Phycologia* **14**, 253–63.

DODGE J. D. (1976) Ultrastructural characteristics of dinoflagellates—the red-tide algae. *Symp. Soc. Appl. Bact.* **10**, 295–304.

DODGE J. D. (1979) The phytoflagellates, fine structure and phylogeny. In *Biochemistry and Physiology of the Protozoa*, eds S. H. Hutner & M. Levandowski, Vol. 1, pp. 7–57. Academic Press, London.

DODGE J. D. & BIBBY B. T. (1973) The Prorocentrales (Dinophyceae). I. A comparative account of fine structure in the genera *Prorocentrum* and *Exuviaella*. *Bot. J. Linn. Soc.* **67**, 175–87.

DODGE J. D. & CRAWFORD R. M. (1968) Fine structure of the dinoflagellate *Amphidinium carteri* Hulbert. *Protistologica* **4**, 231–42.

DODGE J. D. & CRAWFORD R. M. (1969a) The fine structure of *Gymnodinium fuscum*. *New Phytol.* **68**, 613–8.

DODGE J. D. & CRAWFORD R. M. (1969b) Observations on the fine structure of the eyespot and associated organelles in the dinoflagellate *Glenodinium foliaceum*. *J. Cell Sci.* **5**, 479–93.

DODGE J. D. & CRAWFORD R. M. (1970a) The morphology and fine structure of *Ceratium hirundinella*. *J. Phycol.* **6**, 137–49.

DODGE J. D. & CRAWFORD R. M. (1970b) A survey of thecal fine structure in the Dinophyceae. *Bot. J. Linn. Soc.* **63**, 53–67.

DODGE J. D. & CRAWFORD R. M. (1971a) A fine-structural survey of dinoflagellate pyrenoids and food reserves. *Bot. J. Linn. Soc.* **64**, 105–15.

DODGE J. D. & CRAWFORD R. M. (1971b) Fine structure of the dinoflagellate *Oxyrrhis marina*. I. The general structure of the cell. *Protistologica* **7**, 295–303.

DODGE J. D. & CRAWFORD R. M. (1972) Fine structure of the dinoflagellate *Oxyrrhis marina*. II. The flagellar system. *Protistologica* **8**, 399–409.

DODGE J. D. & CRAWFORD R. M. (1974) Fine structure of the dinoflagellate *Oxyrrhis marina*. III. Phagotrophy. *Protistologica* **10**, 239–44.

DODGE J. D. & HERMES H. (1981) A revision of the *Diplopsalis* group of dinoflagellates (Dinophyceae) based on material from the British Isles. *Bot. J. Linn. Soc.*, **82**, 15–26.

DODGE J. D. & VICKERMANN K. (1980) Mitosis and meiosis: nuclear division mechanisms. In *The Eukaryotic Microbial Cell*, eds G. W. Gooday, D. Lloyd & A. P. J. Trinci, pp. 77–102. Cambridge University Press, Cambridge.

DÜRR G. (1979a) Elektronenmikroskopische Untersuchungen am Panzer von Dinoflagellaten. I. *Gonyaulax polyedra*. *Arch. Protistenk.* **122**, 55–87.

DÜRR G. (1979b) Elektronenmikroskopische Untersuchungen am Panzer von Dinoflagellaten. II. *Peridinium cinctum*. *Arch. Protistenk.* **122**, 88–120.

DÜRR G. (1979c) Elektronenmikroskopische Untersuchungen am Panzer von Dinoflagellaten. III. Die Zyste von *Peridinium cinctum*. *Arch. Protistenk.* **122**, 121–39.

DÜRR G. & NETZEL, H. (1974) The fine structure of the cell surface in *Gonyaulax polyedra* (Dinoflagellata). *Cell Tiss. Res.* **150**, 21–41.

FAUST M. A. (1974) Micromorphology of a small dinoflagellate *Prorocentrum mariae-lebouriae* (Parke & Ballantine) Comb. Nov. *J. Phycol.* **10**, 315–22.

GIBBS S. P. (1978) The chloroplasts of *Euglena* may have evolved from symbiotic green algae. *Can. J. Bot.* **56**, 2883–9.

GIBBS S. P. (1981) The chloroplasts of some algal groups may have evolved from endosymbiotic eukaryotic algae. *Ann. N. Y. Acad. Sci.* **361**, 193–208.

HAAPALA O. K. & SOYER M. O. (1973) Structure of dinoflagellate chromosomes. *Nature (Lond.) New Biology* **244**, 195–7.

HAUSMANN K. (1973) Cytologishe Studen an Trichocysten. VI. Feinstruktur und Funktionsmodus der Trichocysten des Flagellaten *Oxyrrhis marina* und des Ciliaten *Pleuronema marinum*. *Helgoland wiss. Meersuntsers.* **25**, 39–62.

HERMAN E. M. & SWEENEY B. M. (1975) Circadian rhythm of chloroplast ultrastructure in *Gonyaulax polyedra*. Concentric organisation around a central cluster of ribosomes. *J. ultrastruct. Res.* **50**, 347–54.

HERMAN E. M. & SWEENEY B. M. (1977) Scanning electron microscope observation of the flagellar structure of *Gymnodinium splendens*. *Phycologia* **16**, 115–8.

HERZOG M. & SOYER M.-O. (1981) Distinctive features of dinoflagellate chromatin. Absence of nucleosomes in a primitive species *Prorocentrum micans* E. *European J. Cell Biol.* **23**, 295–302.

HOLLANDE A. Etude comparée de la mitose syndinienne et de celle des Péridiniens libres et des Hypermastigines. Infrastructure et cycle évolutif des Syndinides parasites de Radiolaires. *Protistologica* **10**, 413–51.

JEFFREY S. W. & VESK M. (1976) Further evidence for a membrane-bound endosymbiont within the dinoflagellate *Peridinium foliaceum*. *J. Phycol.* **12**, 450–5.

KALLEY J. P. & BISALPUTRA T. (1971) *Peridinium trochoideum*: the fine structure of the thecal plates and associated membranes. *J. ultrastruct. Res.* **37**, 521–31.

KEVIN M. J., HALL W. T., MCLAUGHLIN, J. J. A. & ZAHL P. A. (1969) *Symbiodinium microadriaticum* Freudenthal, a revised taxonomic description, ultrastructure. *J. Phycol.* **5**, 341–50.

KOWALLIK K. V. (1969) The crystal lattice of the pyrenoid matrix of *Prorocentrum micans*. *J. Cell Sci.* **5**, 251–69.

KOWALLIK K. V. & HABERKORN G. (1971) The DNA-structure of the chloroplast of *Prorocentrum micans* (Dinophyceae). *Arch. Mikrobiol.* **80**, 252–61.

KUBAI D. F. (1975) The evolution of the mitotic spindle. *Int. Rev. Cytol.* **43**, 167–227.

LEADBEATER B. S. C. (1971) The intracellular origin of flagellar hairs in the dinoflagellate *Woloszynskia micra* Leadbeater & Dodge. *J. Cell Sci.* **9**, 443–51.

LEADBEATER B. & DODGE J. D. (1966) The fine structure of *Woloszynskia micra* sp. nov., a new marine dinoflagellate. *Br. phycol. Bull.* **3**, 1–17.

LEADBEATER B. & DODGE J. D. (1967a) Fine structure of the dinoflagellate transverse flagellum. *Nature* (*Lond.*) **213**, 431–2.

LEADBEATER B. & DODGE J. D. (1967b) An electron microscope study of dinoflagellate flagella. *J. gen. Microbiol.* **46**, 305–14.

LEBLOND P. H. & TAYLOR F. J. R. (1976) The propulsive mechanism of the dinoflagellate transverse flagellum reconsidered. *BioSystems* **8**, 33–9.

LEE R. E. (1977) Saprophytic and phagocytic isolates of the colourless heterotrophic dinoflagellate *Gyrodinium lebourae* Herdman. *J. mar. biol. Ass. UK* **57**, 303–15.

LIVOLANT F. & BOULIGAND Y. (1978) New observations on the twisted arrangement of dinoflagellate chromosomes. *Chromosoma* (*Berl.*) **68**, 21–44.

LIVOLANT F. & BOULIGAND Y. (1980) Double helical arrangement of spread dinoflagellate chromosomes. *Chromosoma* (*Berl.*) **80**, 97–118.

LOEBLICH, A. R. III (1970) The amphiesma or dinoflagellate cell covering. pp. 867–929. *North Am. Paleont. Conv. Symp. Pt. G.*

LOEBLICH, A. R. (1976) Dinoflagellate evolution: speculation and evidence. *J. Protozool.* **23**, 13–28.

LOEBLICH A. R., SCHMIDT R. J. & SHERLEY J. L. (1981) Scanning electron microscopy of *Heterocapsa pygmaea* sp. nov., and evidence for polyploidy as a speciation mechanism in dinoflagellates. *J. plankton Res.* **3**, 67–79.

LOEBLICH A. R. & SHERLEY J. L. (1979) Observations on the theca of the motile phase of free-living and symbiotic isolates of *Zooxanthella microadriatica* (Freudenthal) comb. nov. *J. mar. biol. Ass. UK* **59**, 195–205.

LOEBLICH A. R., SHERLEY J. L. & SCHMIDT R. J. (1979) Redescription of the thecal tabulation of *Scrippsiella gregaria* (Lombard & Capon) comb. nov. (Pyrrhophyta) with light and electron microscopy. *Proc. Biol. Soc. Wash.* **92**, 45–50.

MESSER, G. & BEN-SHAUL Y. (1969) Fine structure of *Peridinium westii* Lemm a freshwater dinoflagellate. *J. Protozool.* **16**, 272–80.

MESSER G. & BEN-SHAUL Y. (1971) Fine structure of trichocyst fibrils of the Dinoflagellate *Peridinium westii*. *J. ultrastruct. Res.* **37**, 94–104.

MIGNOT J. P. (1970) Remarques sur le dévelopement du reticulum endoplasmique et du système vacuolaire chez les Gymnodiniens. *Protistologica* **6**, 267–81.

MORNIN L. & FRANCIS D. (1967) The fine structure of *Nematodinium armatum* a naked dinoflagellate. *J. Microscopie* **6**, 759–72.

MORRILL L. C. & LOEBLICH A. R. (1981a) A survey for body scales in dinoflagellates and a revision of *Cachonina* and *Heterocapsa* (Pyrrhophyta). *J. Plankton Res.* **3**, 53–65.

MORRILL L. C. & LOEBLICH A. R. (1981b) The dinoflagellate pellicular wall layer and its occurrence in the division Pyrrhophyta. *J. Phycol.* **17**, 315–23.

OAKLEY B. R. & DODGE J. D. (1974) Kinetochores associated with the nuclear envelope in the mitosis of a dinoflagellate. *J. Cell Biol.* **63**, 322–5.

OAKLEY B. R. & DODGE J. D. (1977) Mitosis and cytokinesis in the dinoflagellate *Amphidinium carterae*. *Cytobios* **17**, 35–46.

OAKLEY B. R. & DODGE J. D. (1979) Evidence for a double-helically coiled toroidal chromonema in the dinoflagellate chromosome. *Chromosoma* (*Berl.*) **70**, 277–91.

PENNICK N. C. & CLARKE K. J. (1977) The occurrence of scales in the Peridinian dinoflagellate *Heterocapsa triquetra* (Ehrenb.) Stein. *Br. phycol. J.* **12**, 63–6.

REES A. J. J. & LEEDALE G. F. (1980) The dinoflagellate transverse flagellum: three-dimensional reconstructions from serial sections. *J. Phycol.* **16**, 73–80.

RIS H. & KUBAI D. F. (1974) An unusual mitotic mechanism in the parasitic protozoan *Syndinium* sp. *J. Cell Biol.* **60**, 702–20.

SCHMITTER R. E. (1971) The fine structure of *Gonyaulax polyedra*, a bioluminiscent marine dinoflagellate. *J. Cell Sci.* **9**, 147–73.

SCHNEPF E. & DEICHGRÄBER G. (1972) Über den Feinbau von Theka, Pusule und Golgi-Apparat bein dem Dinoflagellaten *Gymnodinium* spec. *Protoplasma* **74**, 411–25.

SIEBERT A. E. & WEST J. A. (1974) The fine structure of the parasitic dinoflagellate *Haplozoon axiothellae*. *Protoplasma* **81**, 17–35.

SILVA E. DA S. (1978) Endonuclear bacteria in two species of dinoflagellates. *Protistologica* **14**, 113–9.

SOYER M. O. (1969) L'enveloppe nucléaire chez *Noctiluca miliaris* S. (Dinoflagellata). 1. Quelques données sur son ultrastructure et son évolution au cours de la sporogenèse. *J. Microscopie* **8**, 569–80.

SOYER M.-O. & ESCAIG J. (1980) Les structures nucléaires et leurs modification au cours de la division chez le dinoflagellé libre *Prorocentrum micans* E. Étude en cryofracture. *Protistologica* **16**, 485–95.

SPECTOR D. L. (1984). Dinoflagellate nuclei. Chapt. 4. In *Dinoflagellates*, ed. D. L. Spector, pp. 107–147. Academic Press, New York.

SPECTOR D. L., PFEISTER L. A. & TRIEMER R. E. (1981) Ultrastructure of the dinoflagellate *Peridinium cinctum* f. *ovoplanum*. II. Light and electron microscopic observations on fertilization. *Am. J. Bot.* **68**, 34–43.

SPECTOR D. L. & TRIEMER R. E. (1979) Ultrastructure of the Dinoflagellate *Peridinium cinctum* f. *ovoplanum*. I. Vegetative cell ultrastructure. *Am. J. Bot.* **66**, 845–50.

SPECTOR D. L. & TRIEMER R. E. (1981) Chromosome structure and mitosis in the dinoflagellates: an ultrastructural approach to an evolutionary problem. *BioSystems* **14**, 289–98.

SPECTOR D. L., VASCONCELOS A. C. & TRIEMER R. E. (1981) DNA Duplication and chromosome structure in the dinoflagellates. *Protoplasma* **105**, 185–94.

SPERO H. J. & MORÉE M. D. (1981) Phagotrophic feeding and its importance to the life cycle of the holozoic dinoflagellate *Gymnodinium fungiforme*. *J. Phycol.* **17**, 43–51.

STEIDINGER K. & COX E. R. (1980) Free-living dinoflagellates. In *Phytoflagellates*, ed. E. R. Cox, pp. 407–32. Elsevier/North Holland, New York.

STEIDINGER K. A., TRUBY E. W. & DAWES C. J. (1978) Ultrastructure of the red-tide dinoflagellate *Gymnodinium breve*. I. General description. *J. Phycol.* **14**, 72–79.

SWEENEY B. M. (1976) Freeze-fracture studies of the thecal membranes of *Gonyaulax polyedra*: Circadian changes in the particles of one membrane face. *J. Cell Biol.* **68**, 451–61.

SWEENEY B. M. (1978) Ultrastructure of *Noctiluca miliaris* (Pyrrhophyta) with green flagellate symbionts. *J. Phycol.* **14**, 116–20.

SWEENEY B. M. (1981) Freeze-fractured chloroplast membranes of *Gonyaulax polyedra* (Pyrrhophyta). *J. Phycol.* **17**, 95–101.

SWIFT E. & REMSEN C. C. (1970) The cell wall of *Pyrocystis* spp. (Dinococcales). *J. Phycol.* **6**, 79–86.

TAYLOR D. L. (1968) *In situ* studies on the cytochemistry and ultrastructure of a symbiotic marine dinoflagellate. *J. mar. biol. Ass. UK* **48**, 349–66.
TAYLOR D. L. (1971a) Ultrastructure of the 'Zooxanthella' *Endodinium chattonii in situ*. *J. mar. biol. Ass. UK* **51**, 227–34.
TAYLOR D. L. (1971b) On the symbiosis between *Amphidinium klebsii* and *Amphiscolops langerhansi* (Turbellaria: Acoela). *J. mar. biol. Ass. UK* **51**, 301–14.
TAYLOR F. J. R. (1975) Non-helical transverse flagella in dinoflagellates. *Phycologia* **14**, 45–7.
TAYLOR F. J. R. (1980) On dinoflagellate evolution. *BioSystems* **13**, 65–108.
TIPPIT D. H. & PICKETT-HEAPS J. D. (1976) Apparent amitosis in the binucleate dinoflagellate *Peridinium balticum*. *J. Cell Sci.* **21**, 273–89.
TOMAS R. N. & COX E. R. (1973) Observations on the symbiosis of *Peridinium balticum* and its intracellular alga. I. Ultrastructure. *J. Phycol.* **9**, 304–23.
TOMAS R. N., COX E. R. & STEIDINGER K. A. (1973) *Peridinium balticum* (Levander) Lemmermann, an unusual dinoflagellate with a mesocaryotic and an eucaryotic nucleus. *J. Phycol.* **9**, 91–8.
TRIEMER R. E. (1982) A unique mitotic variation in the marine dinoflagellate *Oxyrrhis marina* (Pyrrhophyta). *J. Phycol.* **18**, 394–411.
TRIEMER R. E. & FRITZ L. (1984) Cell cycle and mitosis. Chpt. 5. In *Dinoflagellates*, ed. D. L. Spector; pp. 149–179. Academic Press, New York.
WECKE J. & GIESBRECHT P. (1970) Freeze-etching of the nuclear membrane of dinoflagellates. *7th Int. Cong. EM*, Grenoble, 233–4.
WEDERMAYER G. J. & WILCOX L. W. (1984) The ultrastructure of the freshwater colorless dinoflagellate *Peridinlopsis berolinense* (Lemm.) Bourrelly. *J. Protozool.* **31**, 444–453.
WETHERBEE R. (1975a) The fine structure of *Ceratium tripos*, a marine armored dinoflagellate. I. The cell covering (theca). *J. Ultrastruct. Res.* **50**, 58–64.
WETHERBEE R. (1975b) The fine structure of *Ceratium tripos*, a marine armored dinoflagellate. II. Cytokinesis and development of the characteristic cell shape. *J. ultrastruct. Res.* **50**, 65–76.
WETHERBEE R. (1975c) The fine structure of *Ceratium tripos*, a marine armored dinoflagellate. III. Thecal plate formation. *J. Ultrastruct. Res.* **50**, 77–87.
WILCOX L. W. & WEDEMAYER G. J. (1984) *Gymnodinium acidotum* Nygaard (Pyrrhophyta), a dinoflagellate with an endosymbiotic cryptomonad. *J. Phycol.* **20**, 236–242.
WILCOX L. W. & WEDEMAYER G. J. (1985) Dinoflagellate with blue-green chloroplasts derived from an endosymbiotic eukaryote. *Science* **227**, 192–194.
WITHERS N. W., COX E. R., TOMAS R. W. & HAXO F. T. (1977) Pigments of the dinoflagellate *Peridinium balticum* and its photosynthetic endosymbiont. *J. Phycol.* **13**, 354–8.
WITHERS N. W. & HAXO F. T. (1975) Chlorophyll C_1 and C_2 and extraplastidic carotenoids in the dinoflagellate *Peridinium foliaceum Stein*. *Plant Sci. Lett.* **5**, 7–15.
ZINGMARK R. G. (1970) Ultrastructural studies on two kinds of mesocaryotic dinoflagellate nuclei. *Am. J. Bot.* **57**, 586–92.

B COMPLEX ORGANELLES

C. GREUET

1 STIGMAS AND OCELLOIDS

1.1 Introduction

It is first necessary to define the terminology that seems best. By stigma or 'eye-spot', we mean a cluster of osmiophilic droplets showing a red or orange colour due to the presence of carotenoid pigments, believed to be involved in

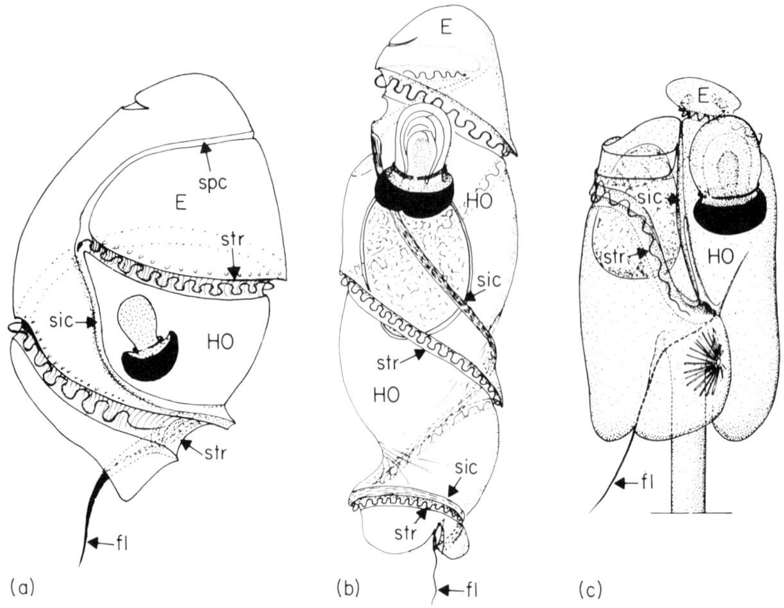

Fig. 3B.1. Ventral view of three trophonts of *Warnowiidae*: (a) *Nematodinium*; (b) *Warnowia*; (c) *Erythropsidinium*. Note the migration of the ocelloids towards the anterior part of the cell (Greuet 1978). E, episome; fl, posterior flagellum; HO, hyposome; sic, intercingular sulcus (part of ventral sulcus); spc, precingular sulcus; str, cingular sulcus with transverse flagellum.

photoreception. A more elaborate type of photoreceptor is the *ocelloid*, a complex organelle which, in addition to the stigma, possesses a refractile structure capable of acting as a focusing lens. In our opinion, the earlier term *ocellus* should be reserved for the multicellular photoreceptor structure of Metazoa. Dinoflagellates are the only organisms known to possess both stigmas and ocelloids and they show the same localization on the left side of the ventral surface (Dodge 1969).

1.2 The ocelloid of the Warnowiaceae (= *Warnowiidae*)

The first study of an organism possessing an ocelloid was by Kofoid & Swezy (1921) who examined the Pouchetidae later changed to the zoological family Warnowiidae by Lindemann (1928). The ultrastructure of the ocelloid has since been studied by Francis (1967), Greuet (1965, 1968, 1969, 1970, 1977, 1978) and Mornin & Francis (1967) in *Nematodinium*, *Warnowia* and *Erythropsidinium*. The ocelloid is made of three principal parts: the *hyalosome* and the *melanosome*, separated by an *ocelloid chamber*.

HYALOSOME

This is a focusing system corresponding to the lens of the eye which protrudes from the cell in more developed genera. The hyalosome is composed of a peripheral corneal zone and a central crystalloid body, the whole lying on a basal plate. It is constricted by rings in its proximal third (Fig. 3B.2).

The crystalline body is a pyriform lens formed by the superimposition of flat endoplasmic vesicles containing a hyaline, refractive substance secreted by the corneal layer. The crystalloid body lies in a local, differential region of the epiplasmic layer of the amphiesma: 'the basal plate'. The structure between plasmalemma and the basal plate, the 'core', depends on the relative abundance of the reticulum (*Nematodinium*, *Warnowia* and young *Erythropsidinium*) and also on the epiplasmic layer (mature *Erythropsidinium*). In the latter, the basal plate looks corrugated: the 'scalariform plate' (Fig. 3B.2). Striated fibres are located along the posterior margin of the basal plate and their contraction makes the hyalosome turn in a ventral direction, orienting the ocelloid with assistance from the ocelloid chamber.

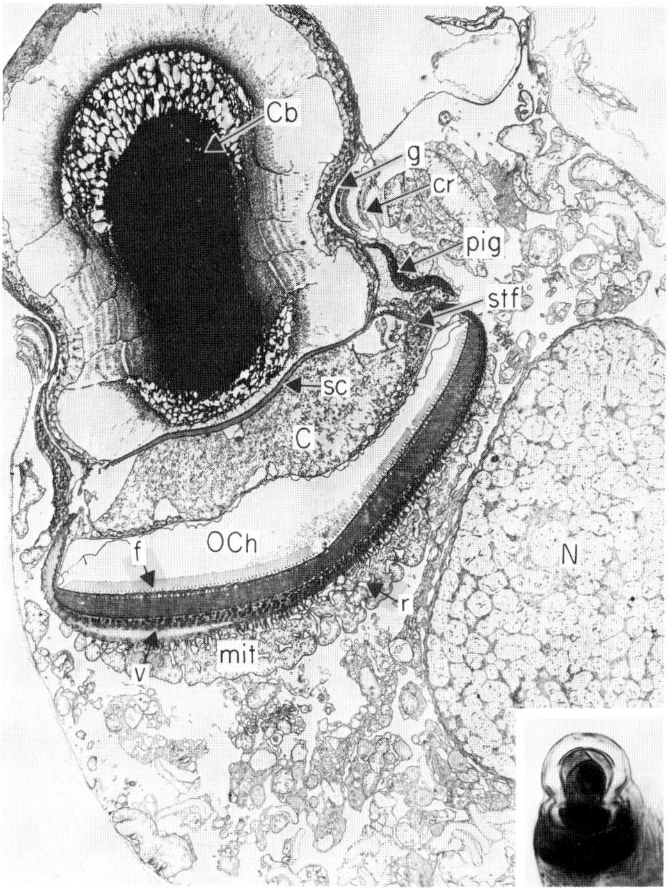

Fig. 3B.2. Longitudinal section of *Erythropsidinium* ocelloid (× 3900). C, core; N, nucleus. For other symbols see legend to Fig. 3B.3.

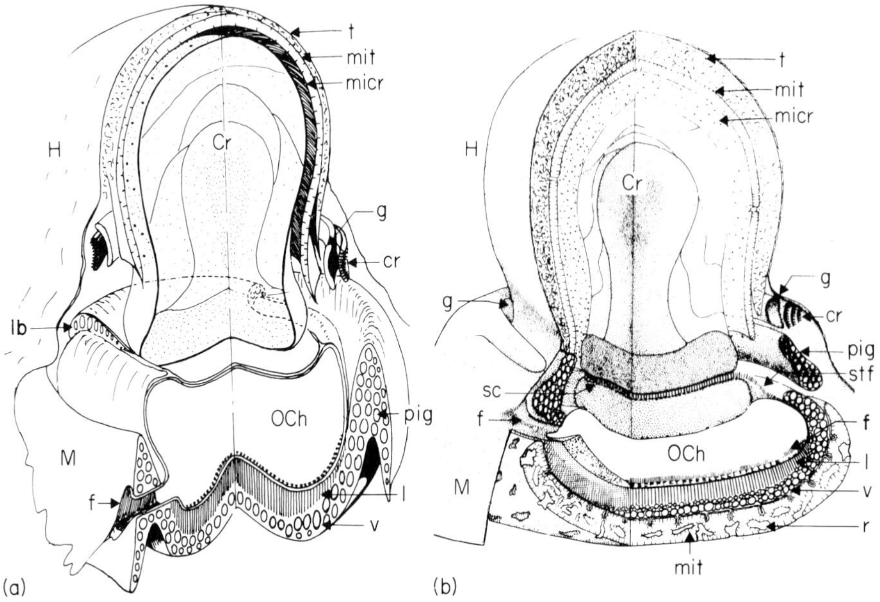

Fig. 3B.3. Ocelloid diagrams of (a) *Nematodinium* and (b) *Erythropsidinium*. Cr, crystalline body; f, ocelloid channel; cr, constricting ring; f, fibrillar formations on the floor of the ocelloid chamber; g, periocelloid gallery; H, hyalosome; l, lamellae of the retinal body (paired thylakoids); lb, basal plate (=sc. scalariforme plate of *Erythropsidinium*); M, melanosome; micr. microcrystalline layer; mit, mitochondrion; OCh, ocelloid chamber; pig, pigmentary ring; r, reticulum; sc, scalariforme plate (=lb, basal plate of *Nematodinium*); stf, striated fibre; t, microtubular layer; v, vesicular layer of the retinal body. (Greuet 1978.)

There are three constriction rings in *Erythropsidinium* (Fig. 3B.2), two in *Warnowia* and one in *Nematodinium*. Each ring is formed by a flat endoplasmic sac, containing crystals identical to those of the microcrystalline layer. The outermost ring is in contact with an invagination forming a belt around the hyalosome, the 'periocelloid gallery'.

OCELLOID CHAMBER

This separates the hyalosome and melanosome (Figs 3B.2, 3B.3). Fundamentally, it represents an invagination which communicates with the outside by means of an ocelloid channel opening into the sulcus. Lined by the amphiesma, its structure is much modified by its close association with the double membrane limiting the retinal body underneath.

The floor of the ocelloid chamber is coated by a fibrillar layer which can assume a paracrystalline aspect (Figs 3B.2, 3B.3).

MELANOSOME

Shaped like a cup, it is of plastid origin and divided into two parts: the 'retinal body' and the 'pigment ring'. The retinal body is in close connection with a mitochondrion, fenestrated by diverticula of endosplasmic reticulum, and with an ocelloid fibre which emerges at the level of the ocelloid channel.

The plastid origin of the retinal body and the pigment ring can be recognized only during cell division (Fig. 3B.5). The thylakoids, which do not apparently contain chlorophyll, are then dedifferentiated, having their usual appearance, and their reorganization can be followed. The same lamellar structure of the

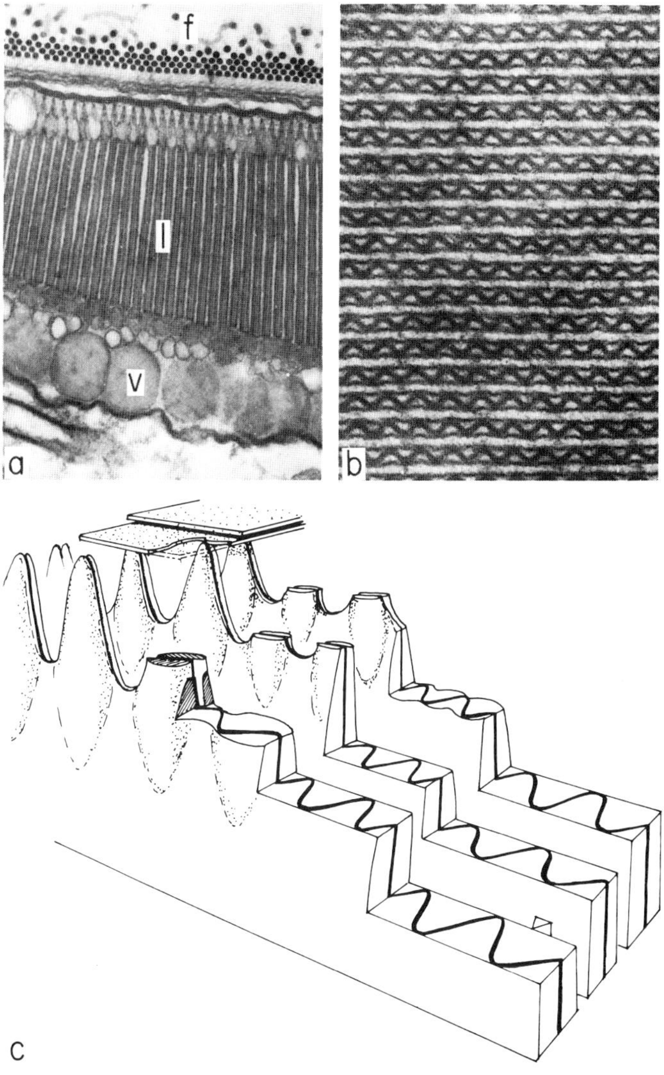

Fig. 3B.4. Retinal body organization. (a) Longitudinal section of *Warnowia* retinal body: f, fibrillar formations on the floor of the ocelloid chamber; l, lamellae (paired thylakoids); v, vesicular layer (× 30 000). (b) Transverse section the paired thylakaoids with the medium sinusoidal wall (× 62 000). (c) Diagram of the retinal body (Greuet 1978).

retinal body can be found in the three genera (Fig. 3B.4). Each lamella, formed by coalescence of two thylakoids, is made of three membranes, of which the central (median) thicker one, is sinuous with a period of 70 nm. The sinusoids of all the coupled thylakoids are in phase.

Towards the ocelloid chamber, each lamella is capped by a line of conical formations whose apices come in contact with the internal plastidial membrane of the retinal body (Fig. 3B.4a). To the exterior, the thylakoids hold back a matrix rich in vesicles and principally lipidic droplets of two kinds. The smallest contain carotenoid pigments and the largest, melanoid pigments.

In *Nematodinium*, the upper margin of the retinal body expands to form a band containing only droplets with melanoid pigments. This expansion, the 'pigment ring', also surrounds the base of the hyalosome. In *Warnowia* this ring tends to be isolated from the retinal body and in *Erythropsidinium* it seems to be totally independent (Fig. 3B.2, 3B.3). It seems that this ring is able to block oblique light.

The ocelloid fibre consists of microtubules which are connected to the retinal body through the ocelloid channel. This fibre is linked with the posterior basal body and in *Erythropsidinium* it is also bound with the posterior tentacle or 'piston'.

DIVISION OF THE OCELLOID

Cytokinesis of the protist is preceded by division of the ocelloid. The two fundamental constituents, hyalosome and melanosome, develop independently and are continuously visible. The layers which make up the crystalloid body of the hyalosome, develop as small spheres forming two equal groups of hyaline droplets; each of them divides into two. With the flattening and application of the layers onto each other, a new crystalloid body is constructed. This is completed by the addition of the different corneal layers. The melanosome migrates to the division plane and produces digitations. At this stage of differentiation its plastid origin can be recognized (Fig. 3B.5a). Its division is of the same type as that found in algal chloroplasts. The thylakoids are arranged in pairs and the common wall resulting from the joining of the two membranes, the 'central membrane', acquires its characteristic undulating appearance (Fig. 3B.5). Later, the melanosome retracts its extensions, while the pigment droplets migrate towards the periphery of the retina and towards the pigment ring. Complete reorganization of the ocelloid takes place after the connection of the hyalosome with the melanosome, between which the ocelloid chamber becomes differentiated as a result of coalescence of endoplasmic vesicles.

1.3 Phyletic origin of the ocelloid of Warnowiaceae

I have looked for the possible existence, in other dinoflagellates, of a photoreceptor whose simple organization would help us to understand the origin

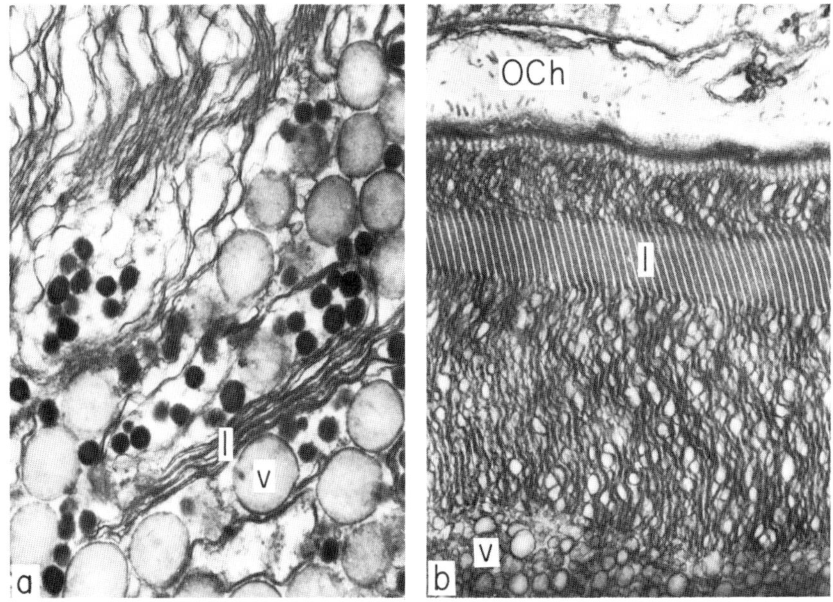

Fig. 3B.5. Two stages of the reorganizing retinal body in *Erythropsidinium*. (a) Pairing of the thylakoids and pleating of the median boundary: v, vesicular layer; l, lamellae (× 24 000). (b) Normalizing of the lamellar structure: l, lamellae; OCh, ocelloid chamber; v, vesicular layer.

of the ocelloid of Warnowiaceae (Greuet 1978). When compared to the stigma of *Kryptoperidinium* (*Glenodinium*) *foliaceum* (Dodge & Crawford 1969) one fact immediately appears significant: the stigma is positioned in the left interior corner of the hyposome, as is the ocelloid of Warnowiace. It is bounded by the proximal parts of the cingulum and the sulcus. There appears to be an opening in the stigma through which the posterior flagellum passes.

At first sight, the stigma of *Kryptoperidinium* is equivalent to a melanosome. But it has a cup shape in which a very large vacuole is present, identical to that found in *Erythropsidinium* during cytokinesis. This could be a simple variant of the ocelloid chamber in the Warnowiaceae. Furthermore, a turning of the lamellar body so that it blocks the concavity of the stigma, would be sufficient for the lamellar body to occupy the position of the basal plate on top of the future ocelloid chamber. Of course, in *Kryptoperidinium* the two basal bodies are in contact with the lamellar body, whereas in the 'trophonts' (feeding cells of *Erythropsidinium* and *Warnowia*), they are distant from one another and from the scalariform plate. But if one considers the trophont of *Nematodinium* and the young cells of *Erythropsidinium*, in which the two basal bodies lie very close, it can be concluded that their position is not fundamentally different from that of the basal bodies in *Kryptoperidinium*. In *Erythropsidinium*, following the later lengthening of the post-ocelloid region, the two basal bodies move slightly apart and their connection with the ocelloid stretches out. This stretching may be followed by the establishment of the ocelloid fibre which, in the trophont *Erythropsidinium*, binds the posterior basal body and ocelloid.

It appears then, that one can consider the stigma of *K. foliaceum* as a developmental stage which could be a precursor of the melanosome in the Warnowiaceae.

2 NEMATOCYSTS

2.1 Introduction

In the Gymnodiniales, certain members of the families Warnowiaceae and Polykrikaceae possess complex ejectile organelles besides trichocysts and mucocysts. Their shape and organization have caused some authors to compare them with the stinging cells of metazoan Cnidaria and to name them 'cnidocysts'. We prefer to keep the term cnidocyst solely for stinging cells, and, in agreement with Fauré-Fremiet (1913a, b) we use 'nematocyst' for these dinoflagellate organelles which may contain one or several ejectile filaments. In the Polykrikaceae *Polykrikos schwartzii* and *P. kofoidii* have nematocysts with one filament, straight or generally winding into a single or double spiral, while in the Warnowiaceae the nematocysts of *Nematodinium* contains several (up to ten) straight filaments.

2.2 Nematocyst of *Polykrikos schwartzii*

A number of authors, Fauré-Fremiet (1913a, b), Chatton (1914, 1952), Chatton & Grassé (1929), Chatton & Hovasse (1944), Hovasse (1951, 1962, 1965, 1969a, b), Greuet (1972a, b) and Greuet & Hovasse (1977) have studied the structure, function and development of the nematocyst in this organism and Westfall *et al.* (1983) have studied its ultrastructure in the closely similar species *P. Kofoidii*. This nematocyst is usually capped by an adjacent organelle which Chatton and Hovasse named the 'cnidoplastid'. According to these authors (Chatton & Hovasse, 1944), the nematocyst and cnidoplastid are two stages of development of one organelle, and partition of cnidoplastid should produce a nematocyst and a new cnidoplastid, this occurring over and over again.

This theory of cyclic 'autonematogenesis' was later completely discarded by Greuet (1972a) and Greuet & Hovasse (1977). The cnidoplastid is now known to be an autonomous organelle which can be compared to a trichocyst and has been renamed a 'taeniocyst' due to its ultrastructural organization (Greuet, 1972b). The nematocyst–taeniocyst complex includes the posterior nematocyst and anterior taeniocyst (Fig. 3B.6), which are separated by an intermediary piece.

The mature nematocyst ($15 \times 5\ \mu m$) contains an ovoid 'posterior body' crowned by an anterior 'operculum'. The posterior body has two cavities: an inner 'anterior chamber' (the 'bulb') which contains an axial stylet (Figs 3B.6a, 3B.7a) and an external posterior chamber.

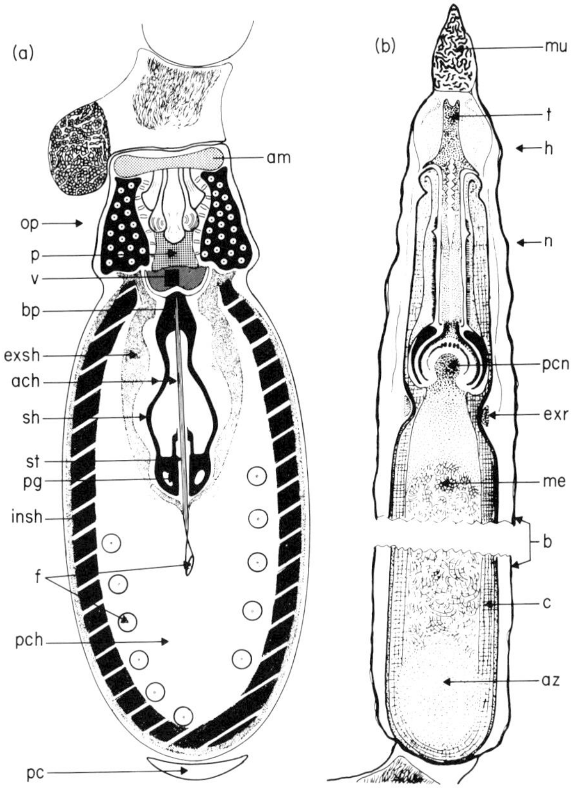

Fig. 3B.6. Schematic organization of the two components of the nematocyst-taeniocyst complex of *Polykrikos schwartzii*. (a) Nematocyst: ach, anterior chamber or bulb; am, apical meniscus; bp, basal plate; exsh, external sheath; f, filament; insh, internal striated shell; op, operculum; p, plug; pc, posterior cap; pch, posterior chamber; pg, posterior guide; sh, stylet sheath; st, stylet; v, valve. (b) Taeniocyst: az, amorphous zone (=Hovasse's posterior cap); b, body; c, cortex; exr, external ring; h, head; me, medulla; mu, mucron (=Hovasse's external or anterior cnidosome); n, neck; pcn, Hovasse's internal or posterior cnidosome; t, tigella.

The body consists of four parts. (a) The *posterior chamber* which is ovoid in shape and limited by a double layered 'shell'. (b) The *filament*, which has a more osmiophilic axis and wall. It is continuous with the stylet of the anterior chamber. The junction, where the bulb opens around the striker is a zone of weak resistance and it is here, according to Chatton (1914), that a break should occur, permitting the filament to extrude. (c) The anterior chamber, or *bulb*, has a double wall or sheath. The external sheath is connected with the internal sheath of the posterior chamber very precisely. (d) The *stylet* is a thin axial rod; it is formed of thin lamellae varying in width. The stylet is kept in place at both ends by dense, annular thickenings: the anterior and posterior guides.

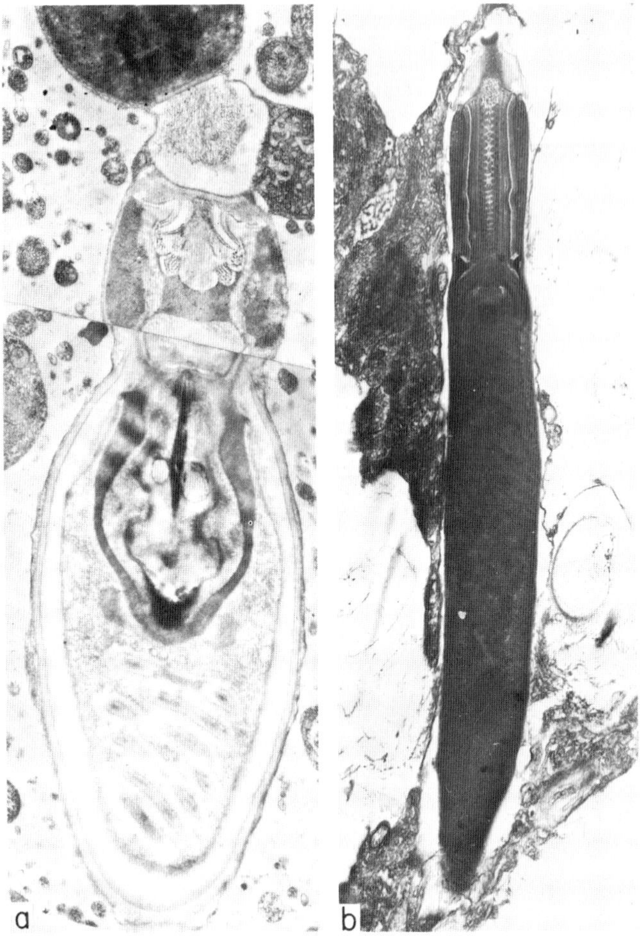

Fig. 3B.7. Micrographs of the two components of the nematocyst–taeniocyst complex of *Polykrikos schwartzii.* (a) Nematocyst (× 11 200), (b) taeniocyst (× 12 000). Compare with Fig. 3B.6a and b.

OPERCULUM

Its rigid peripheral structure consists of a ring surrounding central structures, with an 'apical meniscus' covering the whole. The ring is somewhat conical, its posterior surface having two concentric grooves. The inner groove supports the 'valve' capping the stylet, while the outer part supports the walls of the bulb and the posterior chamber. The axial cavity is filled by a fibrillar packing whose posterior part forms the 'plug.'

2.3 Development of the nematocyst of *Polykrikos*

It is differentiated from a specialized vesicle: the 'nematogene' which initially contains fine, homogeneous granules, which are only slightly osmiophilic. In most cases described by Chatton & Hovasse (1944), eight nematogenes occur, located in the ventral region of the organism and apparently connected with the eight kinetids (roots and basal bodies) and Golgi bodies which accompany them (Chatton & Grassé 1929).

ORGANIZATION OF THE NEMATOGENE

At the anterior pole of the nematogene, the stroma becomes denser and finally occupies two-thirds of the vacuole. From that stage on, the three essential parts of the future nematocyst are recognizable. The anterior hemisphere of the osmiophilic sphere will produce the operculum; the posterior hemisphere will form the bulb and its components; the rest of the nematogene will produce the posterior chamber and its components. The formation of these three components takes place in accord with an antero-posterior polarity. The operculum derives from a hemisphere which shows a concave and very osmiophilic anterior cap (the 'dome' of Hovasse). It lies on two discs: the future plug (anterior) and the future valve (posterior). The anterior guide and the stylet (Chatton & Hovasse's 'axe cnidogene') develop on either side of the valve disc. The internal sheath of the bulb condenses from granules and the remaining stroma in the posterior hemisphere forms the chamber and its appendages.

The nematogene is thus distinguished by: continuity of the apical dome; absence of the posterior chamber wall (= shell); and absence of clear separation between the valve and the nematocyst (Fig. 3B.9a).

PRE-NEMATOCYST STAGES

These can be divided into three developmental stages. The first is the differentiation of the operculum from a ring-like constriction. Then the differentiation of the bulb commences. It is characterized by lengthening of the bulb, the first signs of stylet development and internal sheath formation. The external shell is formed by the stroma. Finally, differentiation of the posterior chamber takes place. The stylet elongates and the filament, which extends from it stretches out usually describing a spiral. The residual stroma is then deposited on the existing walls, building the external sheath and the internal striated shell. Certain prenematocysts and young nematocysts display unusual features of development, such as those described by Hovasse (1951). He observed a considerable elongation of the body with the straight filament touching the bottom of the posterior chamber and this has also been seen by the present author. The rest of the organelle is quite normal (Fig. 3B.8b).

YOUNG NEMATOCYST STAGE

Development of the nematogene ends with the hollowing out of the bulb, the fusion of inner and outer shells and fusion with the interior surface of the ring.

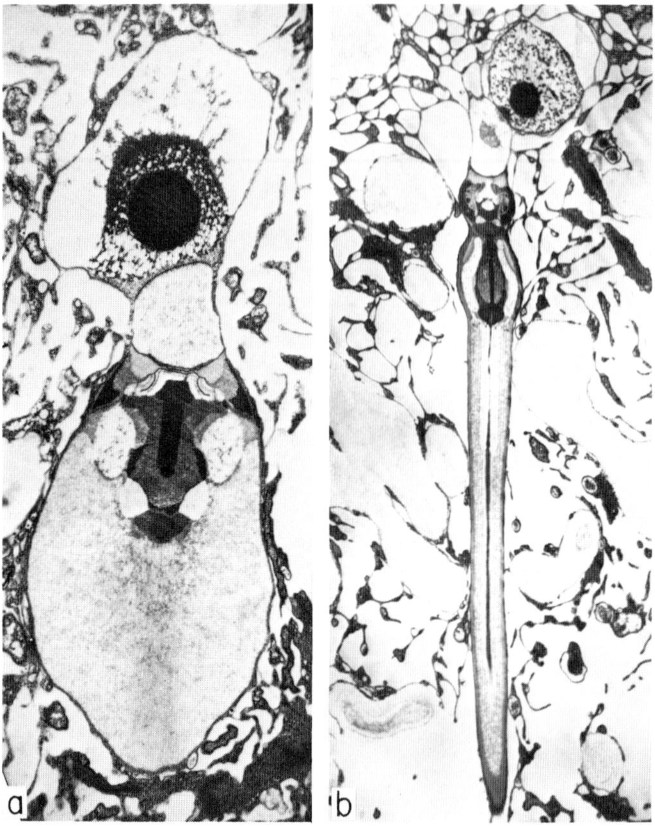

Fig. 3B.8. Two micrographs of the nematocyst–taeniocyst complex of *Polykrikos schwartzii,* (a) Prenematocyst stage topped by a taeniogene (× 6200). (b) Particular evolution of a nematocyst: lengthening of the posterior chamber containing a straight filament. The taeniogene above is normal (× 3600).

Striation of the inner shell is accomplished. Turgidity increases and widening of the nematocyst body can be observed (Fig. 3B.9a).

2.4 The mature taeniocyst

This is a very osmiophilic organelle (12 × 1 μm), generally connected to a nematocyst by means of an intermediary body containing a mucous plug which detaches itself at the end of its organogenesis. Rounded at its posterior extremity, the taeniocyst is made up of three parts: the posterior 'body' extending into a collar region and ending in a 'head' (Fig. 3B.6b).

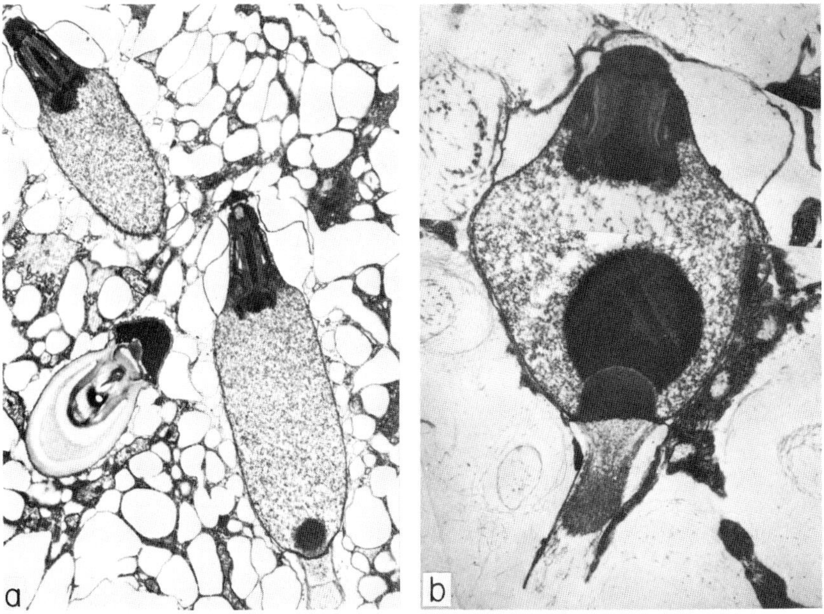

Fig. 3B.9. Development of the taeniocyst: (a) pretaeniocyst stage (× 5800); (b) advanced taeniogene (× 11 500).

THE BODY

It is divided into a cortex and a medullary zone. The cortex has a regularly layered structure which, in section, resembles that of the ribbon trichocysts ('toeniobolocysts') of cryptomonads. The medulla, on the other hand, is made up of interlinked concentric structures that resemble fingerprints. Towards the intermediary body end there is a part which is less osmiophilic and amorphous (the 'posterior crown' of Hovasse).

THE NECK AND HEAD

The connection of the neck with the body involves a complex zone, just posterior to an annular constriction holding a thick external ring. The medulla, which is amorphous at this level, forms a small, claviform mass surrounded by two or three thick internal rings. The cortex becomes thinner at this point before continuing into the tubular neck which contains the axial 'tigella' which emerges from the head. The taeniocyst is crowned at the free extremity of the tigella by a fibrillar vesicle which constitutes a 'mucron'.

2.5 Development of the taeniocyst

This takes place in the vesicle above the operculum of the nematocyst called a 'cnidosphere' by Hovasse, but which should be called a taeniosphere, and which begins with a taeniogene (Fig. 3B.8a).

ORGANIZATION OF THE TAENIOGENE

It is a vesicle with an irregular outline. Its granular content condenses in the posterior hemisphere into a pyriform mass, reduced at the anterior apex. At its first pole, this sphere, representing the body of the taeniocyst, is inserted into an ampoule-like 'posterior crown' whilst at the apex, another conical condensation forms the outline of the neck and head (Fig. 3B.9b)

DEVELOPMENT OF THE TAENIOGENE

In outline, the latter is characterized by (1) a condensation of the posterior crown region and at the opposite pole of the neck and head (ii) a regression in the material surrounding the head and (iii) the beginning of the organization of the head (Fig. 3B.9a). The next stage is marked by the termination of the organization of the head and the lengthening of the taeniocyst, which becomes

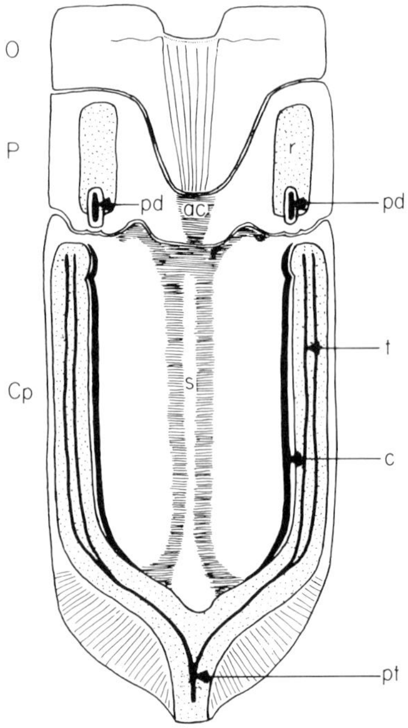

Fig. 3B.10. Schematic representation of a young nematocyst of *Nematodinium*: ac, axial cylinder; Cp, cupule of proximal part; c, coat; O, operculum of terminal part; P, intermediary plug; pd, pedicules of rods; r, rods; s, shaft; t, tigella; pt, pivot.

cylindrical (Fig. 3B.9a). Finally, the contents of the body lose their granular aspect, becoming increasingly compact and acquires fingerprint-like marks.

Westfall *et al.* (1983) have described how the nematocyst-taeniocyst pair in *Polykrikos kafoidii* come to lie in a special channel, the chute, projecting out of the sulcus, with the taeniocyst outermost.

2.6 The nematocyst of *Nematodinium*

The structure has been studied by Hovasse (1952) using light microscopy; then by Mornin & Francis (1967) and Greuet (1971) with electron microscopy. The nematocyst (10 × 5 μm) is composed of three superimposed elements (Fig. 3B.10): (i) a proximal basal part, the cupule, in which can be distinguished lengthened formations which look like the framework of an upside-down umbrella—tigellas at the periphery correspond to the ribs, and the axial shaft corresponds to the handle; (ii) an intermediary part, the plug, which contains small cylindrical bodies; (iii) a terminal distal part, the operculum, which is reinforced by a fibrillary axial funnel.

ORGANIZATION OF THE YOUNG NEMATOCYST

The shaft rests on the base of the tigellas and is joined at the top to the partition separating cupule and plug. It is crossed by a narrow canal. The wall of the cupule invaginates between the cortical tigellas.

The intermediary plug is a vesicle whose wall is depressed in the centre. It has two types of differentiation; the peripheral rods with basal pedicules and the axial cylinder which is part of the extension of the shaft (Figs 3B.10–3B.13).

The terminal operculum forms from a third vesicle, hollowed into a funnel, adapting to the shape of the underlying plug. At an early stage, it contains fibrils which stretch across it in the axis of the nematocyst (Fig. 3B.10).

2.7 Nematocysts: conclusion

The nematocyst of *Nematodinium* is characterized by the presence of multiple rectilinear filaments whose structure is far more complicated than that of the single filament of the nematocyst of *Polykrikos schwartzii*. The fairly constant orientation of the organelles, with their opercula in a centrifugal position and directed towards the sulcus, should be noted. This could not facilitate the extrusion of all or part of the organelle in the sulcal region. In *Nematodinium*, no one has seen the functioning of the nematocysts.

In *Polykrikos schwartzii*, however, Chatton managed to follow the extrusion of the nematocyst, but not while acting on prey and this leaves room for doubt as to the real functioning of the organelle. In this species the stylet goes through

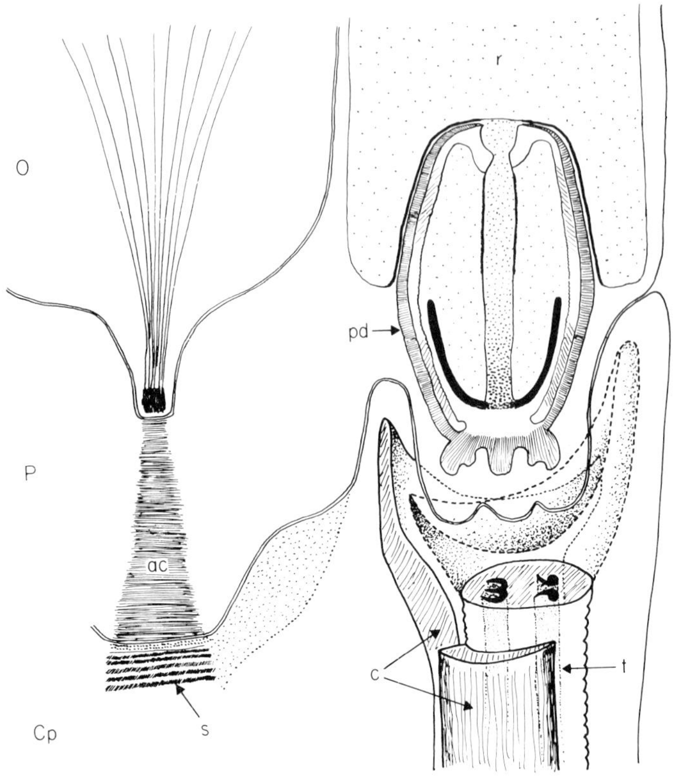

Fig. 3B.11. Schematic representation of the top of a tigella and its coadaptation with the pedicule of a rod. For symbols see Fig. 3B.10.

the operculum, bringing the filament behind it. Then the junction between the stylet and the filament breaks and the filament is expelled. Chatton also illustrated the return of the ampulla. Given the organization of this part of the nematocyst, it seems logical to think that only the internal sheath evaginates.

We are forced to admit that we know virtually nothing concerning the function, and especially the exact role of these nematocysts, although the species which possess them are phagotrophic and we assume they are used in food capture (Chapter 6).

3 THE PISTON

The genus *Erythropsidinium* is characterized by the possession of a posterior tentacle (Figs 3B.1, 6.16), which is covered by papillae. Because of its insertion in the post-cingular part of the ventral furrow, the piston cannot be considered homologous with the tentacles of other organisms (see Chapter 6). The latter

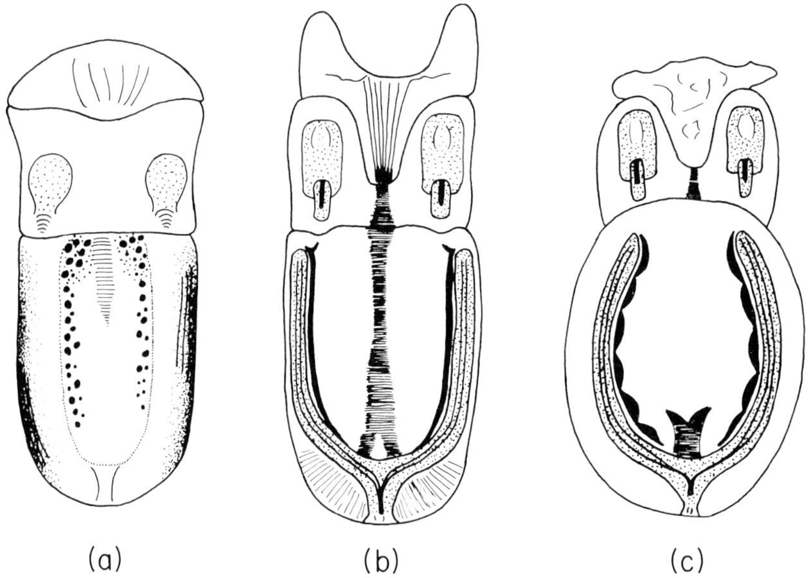

Fig. 3B.12. Evolution of the nematocyst of *Nematodinium*. (a) Prenematocyst stage; compare with Fig. 3B.10. (b) Young nematocyst; compare with Fig. 3B.6. (c) Old nematocyst.

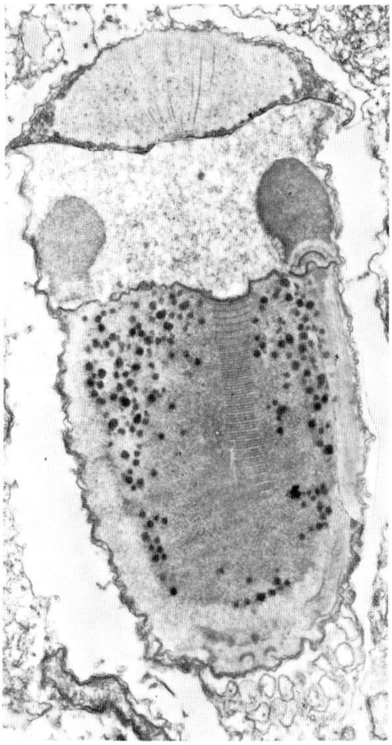

Fig. 3B.13. Micrograph of a prenematocyst of *Nematodinium* (× 10 400).

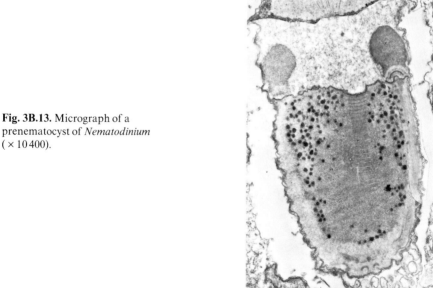

originate in the inter-cingular region of the sulcus, e.g. the tentaculiform pseudopodia of some gymnodinioids and peridinioids, or the tentacle of *Noctiluca* or the stomopod of *Erythropsidinium*. These non-actin contractile organelles and others in Protozoa have been recently reviewed by Cachon & Cachon (1981).

Erythropsidinium moves due to the tentacle: it travels extremely quickly and jerkily in a straight line (5 cm/s) thanks to the rapid backward and forward motion of this organelle. Thus, this particular tentacle has been termed the 'piston'. It has a remarkable capacity for contraction. The frequency of the backwards and forwards motion can be up to 10s and the proportional contraction is 1:20 or even up to 1:40.

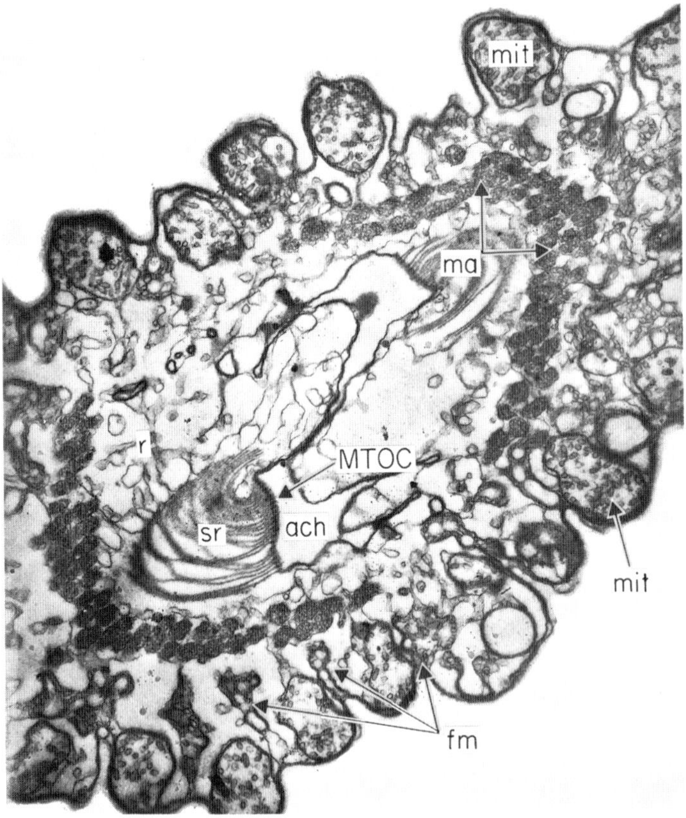

Fig. 3B.14. Transverse section of the piston (×9500): ach, axial canal; fm, contact area of myonemes on amphiesma invaginations; ma, myonematic arc; MTOC, microtubule organizing centre, basal plate; MTS, microtubules; mit, mitochondrion; r, reticulum; sr, skeletal rod.

3.1 Structural and ultrastructural organization

The piston has many mitochondrial papillae (Figs 3B.14–3B.16). In cross-section, the organization has a bilateral symmetry, with a flattened axial canal whose wall is considerably thickened. This canal is extended at both ends, the whole having a Z or S shape (Figs 3B.14, 3B.16).

Several cytoplasmic differentiations can be seen in the piston. Working inwards from the periphery these are: the mitochondrial cuff; two myonematic arcs; two skeletal bands; the axial canal and its extension in the cellular body; the 'compensatory sac'.

MITOCHONDRIAL CUFF

This is formed by a covering of spherical mitochondria, each in a papilla. At the top of the piston, between the mitochondria, invaginations are used for the attachment of elements which make up the myonematic arc (Figs 3B.14, 3B.18).

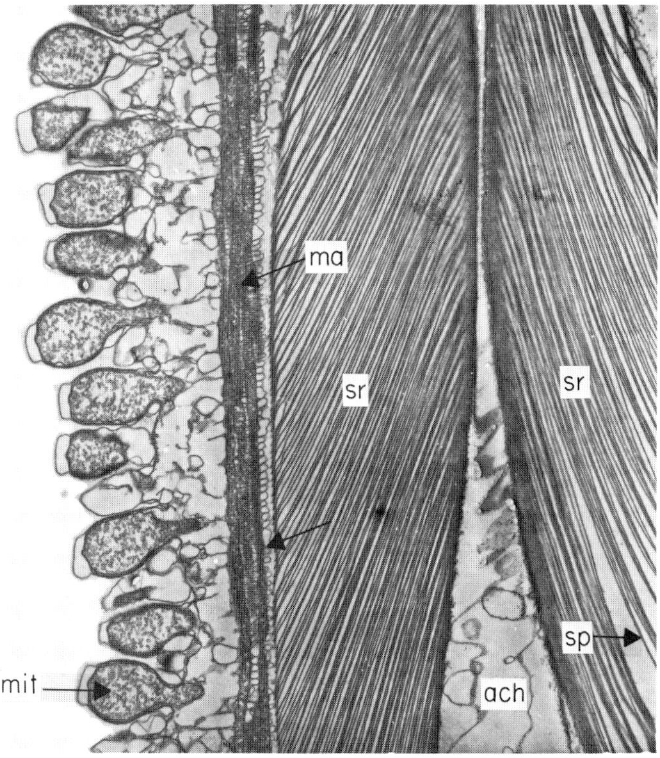

Fig. 3B.15. Longitudinal section of the piston (× 9500): ach, axial canal; ma, myonematic arc; mit, mitochondrion; sr, skeletal rod; sp, skeletal plate.

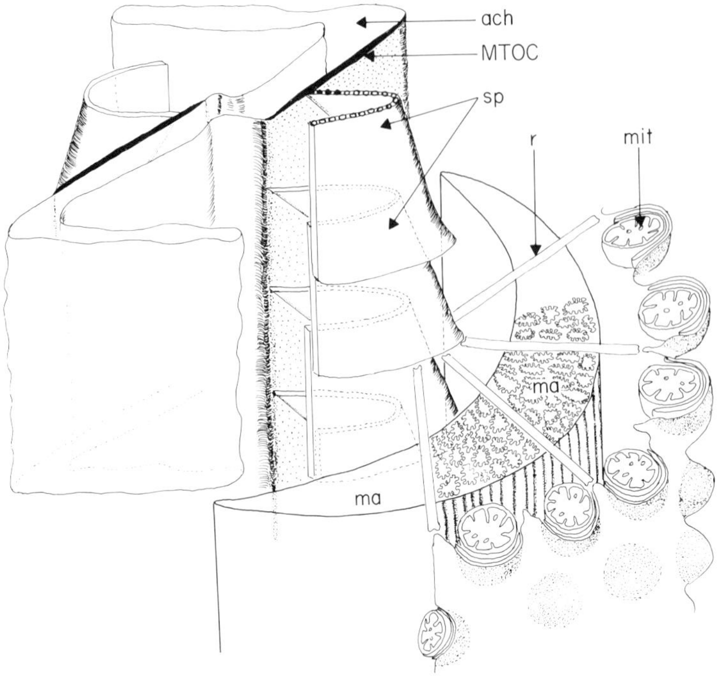

Fig. 3B.16. Diagram of the piston: ach, axial channel; ma, myonematic arc; MTOC, microtubule organizing centre; mit, mitochondrion; r, reticulum; sp, skeletal plate.

MYONEMATIC ARCS

A myonematic arc is a longitudinal strip formed by the association of fifty or so cylindrical elements, the myonematic bundles, which are found in a maximum of four layers. These bundles are slightly oblique, slanting from the exterior towards the interior, penetrating the myonematic arc and ending more posteriorly on the inner side of the myonematic arc. There they connect, either directly or via transverse endoplasmic reticulum vesicles with the posterior extremity of the microtubules of the skeletal plates. (Figs 3B.15–3B.17).

Each myonematic unit is made up of a reticulum bundle of 250 nm in diameter and several μm long. Its wall consists of a single membrane with irregular invaginations. Although we have not yet been able to see the microfilaments, we think that they are longitudinal between the bundles, and that they bridge the invaginations. These microfilaments would therefore be very short. The appearance of the myonematic bundles in section differs according to whether the piston is contracted or relaxed (Fig. 3B.17). In the first case the bundles appear dense and it is difficult to follow the membrane profiles. In the second case there are no membrane profiles to be seen in the lumen. It is

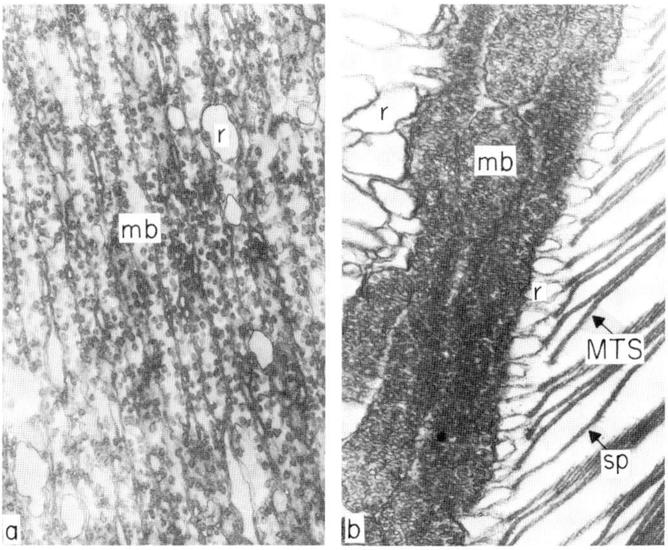

Fig. 3B.17. Two aspects of myonemes: (a) relaxed, (b) contracted (× 28 500). In (b), observe the connection between reticulum and the end of the microtubules of skeletal plates. mb, myonematic bundle; MTS, microtubules; r, reticulum; sp, skeletal plate.

in this lengthened state that the invaginations are most visible. The myonemes develop from a layer of endoplasmic vesicles arranged under the amphiesmal vesicles, initially arranged in irregular bundles (Fig. 3B.18). At the end of its differentiation, each myoneme is moulded into its proximal portion and hooked on to the amphiesma. The myonemes overlap each other along the tentacle, on the inside of each myonematic band.

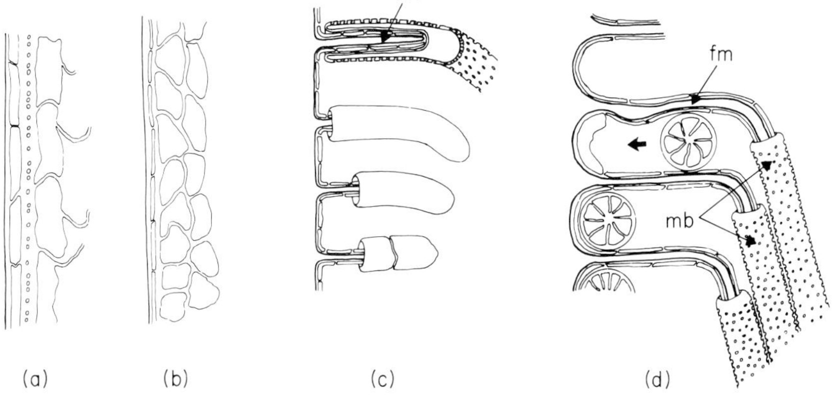

Fig. 3B.18. Diagram of organogenesis of myonematic bundles: (a) general amphiesma; (b) amphiesma of the piston before organogenesis; (c) and (d) two stages of organogenesis. fm, contact area of myonemes on amphiesma invaginations; mb, myonematic bundle.

SKELETAL RODS

Each skeletal rod is fixed in the concavity of the corresponding myonematic arc and has 30–150 skeletal plates (Figs 3B.14–3B.16). Successive plates overlap, the degree of telescoping being a function of the lengthening of the 'piston'. The incurved shape of the plates, and therefore the skeletal rods provide an excellent rigidity. Each plate is made up of microtubules; the microtubules are 80–100 nm apart and connected to each other by arms. One microtubule connects with the wall of the axial canal and forms the 'insertion plate' which joins the skeletal plate. This wall must be considered as a microtubule organizing centre (MTOC).

AXIAL CANAL AND COMPENSATORY SAC

During development of the piston, the axial canal is a simple flattened cavity, but then it becomes complicated by an expansion at each lateral extremity. Closed at its distal end, it extends into the cellular body by a trabecular structure which we think plays the role of a compensatory sac during the shortening of the piston.

3.2 Localization of motor functions

Observations lead us to intrepret the different structures of the piston as forming three systems, of which two are antagonistic. One, active and contractile, is most certainly located at the level of the tubular myonematic bundles. Their contiguity with the mitochondrial cuff adds to this theory. The second, passive, is skeletal and responsible for rigidity. It serves as a guide during the backwards and forwards motion of the piston. The third is also passive, but it is elastic and antagonistic to the first. It is the trabecular dilation which serves as a pressure reservoir to store the axial canal liquid during shortening of the piston. When the myonemes relax, the hydrostatic pressure of the released liquid is exerted on the wall of the canal thus causing lengthening of the piston.

3.3 Development of the piston

This has been observed during its formation in a daughter cell. The piston may undergo longitudinal duplication before binary fission of the organism. At first (Fig. 3B.18) there is a flattened and simple axial canal, cytoplasm containing a few sparse mitochondria, some vesicles of the reticulum, an amphiesma deprived of its layers of microtubules, and the mitochondria which migrate towards the amphiesma. Invaginations of the plasmalemma come into contact with the deeper sacs. This contact seems to induce their lengthening and the formation of bundles. The latter come into contact with the skeletal plates whilst the mitochondria form the papilla of the cuff. The myonematic bundles acquire their multiple indentations at the same time as the differentiation of the microfilaments which bridge them.

4 REFERENCES

CACHON J. & CACHON M. (1981) Movement by non-actin filament mechanisms. *BioSystems,* **14**, 313–326.

CHATTON E. (1914) Les cnidocystes du Péridinien *Polykrikos schwartzi. Arch. Zool. exp. et gén.* **54**, 157–94.

CHATTON E. (1952) La classe des Dinoflagellés ou Péridiniens. In *Traité de Zoologie* (ed. P-P. Grasse), vol. I, pp. 309–90. Masson, Paris.

CHATTON E. & GRASSE P-P. (1929) Le chondriome, le vacuome, les vésicules osmiophiles, le parabasal, les trichocystes et les cnidocystes du Dinoflagellé *Polykrikos schwartzi* Bütschli. *C. R. Soc. Biol. C.*, 281–4.

CHATTON E. & HOVASSE R. (1944) Sur les premiers stades de la cnidogenèse chez le Péridinien *Polykrikos schwartzi.* Leurs rapports avec les dictyosomes. *C. R. Acad. Sci. Paris* **218**, 60–2.

DODGE, J. D. (1969) A review of the fine structure of algal eyespots. *Br. phycol. J.* **4**, 199–210.

DODGE J. D. & CRAWFORD R. M. (1969) Observations on the fine structure of the eye-spots and associated organelles in the dinoflagellate *Glenodinium foliaceum. J. cell Sci.* **5**, 479–93.

FAURE-FREMIET E. (1913a) Sur les 'nématocystes' de *Polykrikos* et de *Campanella. C. R. Soc. biol.* **75**, 366–8.

FAURE-FREMIET E. (1913b) Sur les nématocystes et les trichocystes des *Polykrikos. Bull. Soc. zool., France* **38**, 289–90.

FRANCIS D. (1967) On the eyespot of the dinoflagellate *Nematodinium. J. exp. Biol.* **47**, 495–501.

GREUET C. (1965) Structure fine de l'ocelle d'*Erythropsis pavillardi* Hertwig. Péridineen Warnowiidae Lindemann. *C. R. Acad. Sci.* **261**, 1904–7.

GREUET C. (1968) Organisation ultrastructurale de l'orelle de deux Péridiniens Warnowiidae, *Erythropsis pavillardi* Kof. et Swezy et *Warnowia pulchra* Schiller. *Protistologica* **4**, 209–30.

GREUET C. (1969) Morphological and ultrastructural study of the stomopharyngian complex in *Erythropsis pavillardi* Kof. et Swezy. *Protistologica* **5**, 481–503.

GREUET C. (1970 Ultrastructure de l'ocelle du dinoflagellate *Nematodinium* comparées a celles d'autres representants de la famille des Warnowiidae. *Proc. 7th Int. Cong. EM. Grenoble*, pp. 385–6.

GREUET C. (1971) Etude ultrastructurale et évolution des cnidocystes de *Nematodinium,* Péridinien *Warnowiidae* Lindemann. *Protistologica* **7**, 345–55.

GREUET C. (1972a) Organisation ultrastructurale et organitogenèse du complexe cnidocyste-cnidoplaste chez le Péridinien *Polykrikos schwartzi* Bütshli. *J. Protozool* **19**, 199.

GREUET C. (1972b) La nature trichocystaire du cnidoplaste dans le complexe cnidoplaste-nématocyste de *Polykrikos schwartzi* Bütschli. *C. R. Acad. Sci. Paris* **275**, 1239–42.

GREUET C. (1977) Evolution structurale et ultrastructurale de l'ocelloide d'*Erythropsidinium pavillardi* Kof. et Swez. au cours des divisions binaire et palintomiques. *Protistologica* **23**, 127–43.

GREUET C. (1978) Organisation ultrastructurale de l'ocelloide de *Nematodinium.* Aspect phylogenetique du photorecepteur des Péridiniens Warnowiidae Lindemann. *Cytobiol.* **17**, 114–36.

GREUET C. & HOVASSE R. (1977) A propos de la genèse des nématocystes de *Polykrikos schwartzi* Bütschli. *Protistologica* **13**, 145–9.

Hovasse R. (1951) Contribution à l'étude de la cnidogenèse chez les Péridiniens. I. Cnidogenèse cyclique chez *Polykrikos schwartzi* Bütschli. *Arch. Zool. exp. et gen.* **87**, 299–334.

Hovasse R. (1952) Contribution à l'étude de la cnidogenèse chez les Péridiniens. II. Cnidogenèsne cyclique chez *Nematodinium armatum. Arch. Zool. exp. et gen.* **88**, 149–58.

Hovasse R. (1962) Quelques faits nouveaux concernant les trichocystes et nématocystes des *Polykrikos* (Dinoflagellés). *Arch. Zool. exp. et gen.* **102**, 189–98.

Hovasse R. (1965) Trichocystes, corps trichocystoides, cnidocystes et colloblastes. *Protoplasmatologia* **3**, 1–57.

Hovasse R. (1969a) Trichocystes ou corps trichocystoides et nematocystes chez les Protistes. *Ann Station biol. Besse en Chandesse* **4**, 245–69.

Hovasse R. (1969b) Les organites à detente des Protistes: trichocystes, corps trichocystoides et cnidocystes. *Progress in Protozool.* (*IIIrd Int. Protozool. Congress Leningrad*) p. 15.

Kofoid C. A. & Swezy O. (1921) The free-living unarmoured Dinoflagellata. *Mem. Univ. Calif.* **5**, 1–562.

Lindemann E. (1928) Peridineae. Die Natürlichen Pflanyenfamilien. Leipzig.

Mornin L. & Francis D. (1967) The fine structure of *Nematodinium armatum*, a naked Dinoflagellate. *J. Microsc.* **6**, 759–72.

Westfall J. A., Bradbury P. C. & Townsend J. W. (1983) Ultrastructure of the dinoflagellate *Polykrikos*. 1. Development of the nematocyst-taeniocyst complex and morphology of the site for extrusion. *J. Cell Sci.* **63**, 245–261.

CHAPTER 4
BIOCHEMISTRY OF THE DINOFLAGELLATE NUCLEUS

PETER J. RIZZO
Department of Biology, Texas A & M University, College Station, Texas 77843 USA

1 INTRODUCTION

While the cytoplasm of dinoflagellates is rather typical when compared to that of other eukaryotes (Loeblich 1976; Taylor 1980), the dinoflagellate nucleus is unique in several respects, both ultrastructurally (see Chapter 2) and biochemically. Numerous, identical-appearing chromosomes (Dodge 1966; Loeblich 1976) are attached to the nuclear envelope during division (Kubai & Ris 1969; Soyer 1969; Oakley & Dodge 1974), and remain condensed throughout the cell cycle (Dodge 1966; Soyer & Haapala 1947c). These chromosomes are devoid of typical histones (Rizzo 1981), contain substantial amounts of hydroxymethyluracil (Rae 1976), lack the differential heterochromatic cross-banding that is found in metaphase chromosomes (Haapala & Soyer 1974b), and undergo an atypical meiosis (Himes & Beam 1975). Although there are some exceptions to many of the above properties (some parasitic species in particular), and although some properties are shared with other protists, no other group of organisms is characterized by all of these unique nuclear features.

The presence of these unusual features has led many investigators to suggest that dinoflagellates are intermediate between prokaryotes and eukaryotes

(Dodge 1966; Soyer & Haapala 1974c; Allen *et al.* 1975; Hamkalo & Rattner 1977), although this is a subject of debate (see Conclusions, p. 168).

In spite of the interesting properties of dinoflagellate nuclei, few biochemical studies have been done. This is probably not due to a lack of interest, but rather due to the fact that many nuclear biochemical studies require isolated nuclei, and in relatively large amounts. The presence of a rigid cell wall in most dinoflagellates represents a major barrier to the former condition and a slow division rate complicates the latter. However, the development of methods for isolating nuclei and improvements in the culturing of dinoflagellates, have made possible some studies that could not have been done otherwise. The following is a review of the literature on the biochemical properties of the dinoflagellate nucleus, and does not include chromosome ultrastructure, chromosome models, or genetics.

2 CHEMICAL COMPOSITION

2.1 DNA content

One of the unique features of dinoflagellates as compared with other unicellular algae is the enormous amount of DNA in the nucleus. This can be seen from the data given in Table 4.1. The values range from 3 pg/cell in *Amphidinium* to 200 pg/cell in *Gonyaulax*. A comparison with DNA contents of various other algae shows that only *Euglena* and *Olisthodiscus* have greater than 1 pg DNA per cell, while 0·1–0·2 pg/cell is more common (Table 4.1). The high DNA content of dinoflagellates is, of course, due to the large numbers of chromosomes found in these algae (Dodge 1963; Loeblich 1976).

There are several factors that are noteworthy with regard to the DNA content of dinoflagellate nuclei. The first concerns a report by Allen *et al.* (1975), in which the DNA content per cell was found to depend on whether the cells were in the logarithmic or the stationary phase of growth. Specifically, in each of the three species tested, stationary phase cells had approximately half as much DNA as did cells in logarithmic growth, although this was not observed for cultures of *Amphidinium carterae* (Galleron & Durrand 1978). Furthermore, in the case of *Crypthecodinium cohnii* DNA content per nucleus was found to vary widely, and cysts may contain two, four or eight nuclei (Beam & Himes 1974). Another report suggests that autodiploidy may result from extended culturing of dinoflagellates. Loper *et al.* (1980) have found that field and recent culture samples of *Ptychodiscus brevis* (*Gymnodinium breve*) had chromosome counts of 121, while a 23-year old culture had 240 chromosomes per nucleus. However, regardless of these factors which introduce variability into estimates

Table 4.1. DNA content of dinoflagellate nuclei and whole cells

Organism (name used by source)	Preparation	Haploid DNA (pg)	Reference
Dinoflagellates			
Amphidinium carterae	Whole cells	3·2	Holm-Hansen 1969
Amphidinium carterae	Whole cells	2·7	Galleron & Durrand 1978
Cachonina niei	Whole cells	10·0	Holm-Hansen 1969
Cachonina niei	Whole cells	7·6	Loeblich 1976
Crypthecodinium cohnii	Nuclei	6·9	Rizzo & Nooden 1973
Crypthecodinium cohnii	Whole cells	6·9	Roberts *et al.* 1974
Crypthecodinium cohnii	Whole cells	7·3	Allen *et al.* 1975
Gonyaulax polyedra	Whole cells	200·0	Holm-Hansen 1969
Gymnodinium sp.	Whole cells	15·4	Allen *et al.* 1975
Gymnodinium breve	Whole cells	101·6	Kim & Martin 1974
Gymnodinium breve	Nuclei	112·5	Rizzo *et al.* 1982
Gymnodinium nelsoni	Nuclei	143·0	Rizzo & Nooden 1973
Gyrodinium resplendens	Whole cells	66·0	Allen *et al.* 1975
Peridinium trochoideum	Nuclei	34·0	Rizzo & Nooden 1973
Prorocentrum micans	Whole cells	42·0	Haapala & Soyer 1974b
Other algae			
Chlamydomonas reinhardtii	Whole cells	0·19	Cattolico & Gibbs 1975
Chlorella sorokinana	Whole cells	0·11	Cattolico & Gibbs 1975
Euglena gracilis	Whole cells	2·10	Charles 1977
Eudorina californica	Whole cells	0·17	Tautvydas 1976
Ochromonas danica	Whole cells	0·20	Charles 1977
Olisthodiscus luteus	Whole cells	1·7	Ersland & Cattolico 1981
Polyedriella helvetica	Whole cells	0·25	Charles 1977
Polytoma obtusum	Whole cells	0·19	Siu & Swift 1974
Porphyridium cruentum	Whole cells	0·10	Charles 1977
Scenedesmus obliquus	Whole cells	0·40	Charles 1977

of DNA content in dinoflagellates, it is clear that as a group, dinoflagellates contain much more DNA than do other algae.

2.2 Relative amounts of DNA, RNA and protein

The relative properties of DNA, RNA, basic, and non-basic protein are given in Table 4.2. If one compares the ratios of RNA, basic, and non-basic protein to DNA for isolated nuclei of various species, several features become evident. The RNA content ranges from 0·21 to 0·39, which is typical of other eukaryotic nuclei, providing there is insignificant contribution from contaminating ribosomal RNA. With the exception of *Oxyrrhis marina*, there is roughly ten times as much DNA as there is basic protein, which is merely a reflection of the absence of histones. *O. marina* appears to be a special case with regard to the presence of basic proteins. Unlike most dinoflagellates, nuclei of *O. marina* contain enough basic protein to stain with alkaline fast green (Cachon *et al.*

1979). Although this might be interpreted as a demonstration of the presence of histones, it should be stressed that the alkaline fast green test detects basic proteins and not histones *per se*. An interesting paper by Sun *et al.* (1978) reported that *O. marina* chromatin contains one major basic protein, which they call protein H (see also Table 4.3). The ratio of total protein to DNA was reported as 1·07, with the basic protein constituting more than 50% of the total protein (Sun *et al.* 1978). It thus appears that *O. marina* chromosomes might contain more histone-like protein per unit DNA when compared with other dinoflagellates. This may or may not be correlated with the observation that *O. marina* chromosomes do not have the characteristic dinoflagellate ultrastructure (Cachon *et al.* 1979; Triemer 1982). The amount of non-basic protein is in general equal to or greater than the amount of DNA. A lower value was obtained for *Pt. brevis*, but this is probably because these nuclei were washed with a dilute salt solution (Rizzo *et al.* 1982) which removes soluble nuclear proteins.

When chromatin is prepared from isolated nuclei (*Peridinium trochoideum* and *C. cohnii*), the relative proportions of RNA and non-basic protein are greatly reduced, as well as some of the basic protein. The reduction in non-basic protein can probably be accounted for by removal of the nuclear envelope, the nucleolus and soluble (nuclear sap) proteins. It is likely that the reduction in RNA and basic protein is due to removal of the nucleolus and soluble material ('ground substance') between the chromosomes. Although electron microscopy studies involving protease digestion confirm the presence of protein in dinoflagellate chromosomes (Soyer & Haapala 1974a), it is not known at present

Table 4.2. Chemical composition of dinoflagellate nuclei and chromatin

Organism (name used by source)	Preparation	RNA/DNA	Basic protein/DNA	Non-basic protein/DNA	Reference
C. cohnii	Nuclei	0·32	0·13	0·99	Rizzo & Nooden 1973
C. cohnii	Chromatin (log phase)	0·08	0·05	0·47	Rizzo & Nooden 1974a
C. cohnii	Chromatin (stationary phase)	0·04	0·04	0·25	Rizzo & Nooden 1974a
P. trochoideum	Nuclei	0·22	0·08	1·22	Rizzo & Nooden 1973
P. trochoideum	Chromatin (log phase)	0·06	0·06	0·50	Rizzo & Nooden 1974a
P. trochoideum	Chromatin (stationary phase)	0·03	0·02	0·21	Rizzo & Nooden 1974a
G. nelsoni	Nuclei	0·21	0·10	1·09	Rizzo & Nooden 1973
Pr. micans	Nuclei	—	<0·1	1·00	Herzog & Soyer 1981
G. breve	Nuclei	0·39	0·12	0·67	Rizzo *et al.* 1982
Oxyrrhis marina	Nuclei	0·27	>0·54	<0·54	Sun *et al.* 1978

what percentage of the protein associated with the DNA as isolated chromatin was actually bound to the chromosomes *in vivo*. This is because in the preparation of chromatin from nuclei in all eukaryotes, there is always the possibility that non-chromosomal nuclear material can absorb onto the DNA during the isolation of nuclei or chromatin. However, it is not likely that the DNA-associated protein in isolated chromatin is a result of cytoplasmic contamination. First of all, clean nuclei with no visible cytoplasmic contamination were used for chromatin isolation in these studies (Rizzo & Nooden 1973). Secondly, the low RNA/DNA ratio of the isolated chromatin (Table 4.2), is a good indication of negligible cytoplasmic contamination (Rizzo 1976).

From Table 4.2 it is evident that chromatin prepared from stationary phase cells contains less RNA and protein than does chromatin prepared from log phase cells. Since most of the RNA in chromatin from other eukaryotes is probably nascent or newly synthesized RNA, it would be expected that non-growing cells contain less RNA than do growing cells. Similarly, non-growing eukaryotic cells usually contain less non-histone chromosomal protein than growing cells.

2.3 Metal content of dinoflagellate chromosomes

Several recent papers have opened up a new aspect of dinoflagellate chromosome chemistry. Using X-ray microanalysis, Kearns & Sigee (1979) reported the presence of high levels of the transition metals iron, nickel, copper and zinc in chromosomes of *G. foliaceum*, *Pr. micans* and *A. carterae*. They noted that nickel was exceptional in being the only transition metal that was strictly localized in the chromosomes. Large amounts of calcium were also detected, which may be of importance in regard to the permanently condensed nature of dinoflagellate chromosomes (Kearns & Sigee 1979). In a subsequent report Kearns & Sigee (1980) discussed the presence of metal elements in unfixed chromosomes and the quantitation of Period IV elements in fixed chromatin, for several species of dinoflagellates. Among other metals, manganese was detected in unfixed cryosections. High levels of manganese in dinoflagellate chromosomes might explain the unexpected inhibition of RNA synthesis by low levels of manganese in isolated nuclei (Rizzo 1979). In all other eukaryotes, RNA polymerase II has a requirement for manganese and this metal is only inhibitory after the optimum concentration is reached. It is possible that the endogenous levels of manganese are already optimum and any additional manganese is inhibitory (see RNA Biosynthesis, p. 165). Quantitation of the Period IV elements on a nucleotide basis has revealed the presence of one Period IV atom per base pair of DNA. In comparison with other eukaryote chromatins, the levels of bound metal elements in dinoflagellate chromosomes are unusually high. Furthermore, the presence of these metals does not appear to be merely the result of deposition due to high levels in the growth medium (Sigee & Kearns 1981c). The authors therefore suggest that metals may play an important structural role in dinoflagellate chromosomes. This suggestion has received additional support from a study involving the extraction of chromosome-bound metals with ribonuclease and deoxyribonuclease (Sigee & Kearns 1981a).

In another study using X-ray microanalysis, Sigee & Kearns (1981b) compared the chromatin-bound Period IV metals in the dinokaryotic and eukaryotic (supernumerary) nuclei of the binucleate dinoflagellate *Glenodinium foliaceum*. They found that the two types of nuclei had some differences and some similarities in regard to their associated metal elements. The dinokaryotic nucleus differed from the eukaryotic nucleus in that it contained substantially higher levels of phosphorus, calcium, and the transition metals iron, nickel, copper and zinc. It is interesting that iron and nickel were undetectable in the eukaryotic nucleus. The X-ray microanalytical results were recently supported by an autoradiographic study using Ni^{63} (Sigee 1982). This lends support to the contention that the endosymbiont is of different phylogenetic origin and not a dinoflagellate (Tomas & Cox 1973; Tomas *et al.* 1973). Also noteworthy is the observation that the eukaryotic nucleus and the dinokaryotic nucleoplasm contained roughly the same levels of sulphur but the dinokaryotic chromatin contained nine times this amount (Sigee & Kearns 1981b). Although the occurrence of sulphur in chromosomes is in keeping with the data of Franker (1972) on the DNA binding of ^{35}S-labelled proteins (see Nuclear Proteins, p. 157), one would expect to find *less* sulphur in the chromosomes than in either the eukaryotic nucleus or the dinokaryotic nucleoplasm. Since the eukaryotic nucleus of *G. foliaceum* has cytochemically detectable basic proteins while the dinokaryotic nucleus does not (Rizzo 1981), one must conclude that at least the basic proteins are much more abundant in the eukaryotic nucleus. In regard to the nucleoplasm of the dinokaryotic nucleus, cytochemical tests indicated that the nucleoplasm is at least in part composed of protein (Dodge 1964). While the higher levels of sulphur in the dinokaryotic chromatin could be explained by large differences in the sulphur content of the proteins in question, the possibility that the bulk of the sulphur detected in dinokaryotic chromosomes is not a component of protein cannot be excluded at present.

3 NUCLEAR PROTEINS

3.1 Absence of histones and nucleosomes in dinoflagellates

There is now an overwhelming body of evidence indicating that in the nuclei of eukaryotic cells, the chromatin is arranged into regularly repeating subunits called nucleosomes (Lilley & Pardon 1979; McGhee & Felsenfeld 1980). These nucleosomes are formed by interactions between DNA and histones such that about 140 base pairs of DNA are wrapped around an octamer of histones (two each of histones H2A, H2B, H3 and H4), to form a nucleosome core, resulting in a five-fold compaction of the DNA. Nucleosome cores are connected by about 60 base pairs of 'linker DNA', which is continuous with the 140 base pairs of core DNA. A fifth histone, H1, is bound to this linker DNA and induces further compaction of the DNA by binding adjacent nucleosomes (Kornberg 1977). While this subunit structure of chromatin is not found in prokaryotes (or cellular organelles), it has been demonstrated in nuclei of all eukaryotes, with the exception of the dinoflagellates.

The standard methods for demonstrating the presence of nucleosomes in eukaryotes are electron microscopical observation of chromatin spreads, digestion of nuclei or chromatin with nucleases and *in vivo* photochemical cross-linking. The chromatin spreading technique has been used to demonstrate the absence of nucleosomes in *Prorocentrum micans* (Hamkalo & Rattner 1977; Livolant & Bouligand 1980; Herzog & Soyer 1981), *Crypthecodinium cohnii* (Rizzo & Burghardt 1980), the dinokaryotic nucleus of *Peridinium balticum* (Rizzo & Burghardt 1980), *Gymnodinium nelsoni* (=*sanguineum*) (Rizzo & Burghardt 1982) and *Ptychodiscus brevis* (Rizzo *et al.* 1982). Thus, while chromatin spreads reveal a subunit structure composed of 100 Å particles (nucleosomes) connected by short stretches of DNA in all other eukaryotes, dinokaryotic chromatin fibres appear as smooth strands. When nuclei or chromatin from other eukaryotes is digested with micrococcal nuclease and the deproteinized DNA fragments are subjected to electrophoresis, a 200 base pair repeat is seen. However, when this technique was applied to dinoflagellate nuclei no discrete size-classes were noted for *Crypthecodinium cohnii* (Bodansky *et al.* 1979; Shupe & Rizzo 1983) or *Prorocentrum micans* (Herzog & Soyer 1981). An interesting observation in all three of these studies was that only about 10–20% of the total DNA was resistant to nuclease attack. Whether or not this inaccessibility to micrococcal nuclease is due to DNA-protein complexes is not known at present. *In vivo* photochemical cross-linking of DNA with 4,5′8-trimethylpsoralen and ultra-violet light has also shown that eukaryotic DNA is protected from cross-linking (by histones) in lengths of about 200 base pairs. When this technique was applied to *C. cohnii* protected regions of DNA were found, but their lengths and distributions were very different from those of other eukaryotes (Yen *et al.* 1978). The results indicate that a basic structure involving 375 base pairs and multiples thereof, exists in the *C. cohnii* genome. However, only about 20% of the total genome is involved in these protected regions, which are clustered in stretches of 10–15 kilobase pairs that are spaced by unprotected regions of 30–60 kilobase pairs (Yen *et al.* 1978). It is not known if these protected regions are a result of DNA–protein interactions, as they are in other eukaryotes, or if they arise from some other mechanism.

The absence of nucleosomes in dinoflagellates is obviously due to the absence of typical histones. The first evidence that histones were lacking in dinoflagellates came from the cytochemical studies of Ris (1962) and Dodge (1964, 1966).

However, using an indirect immunofluorescence technique, Stewart & Beck (1967) detected histones in several species of dinoflagellates. This apparent contradiction was resolved by using a biochemical approach which showed that although typical histones were absent, chromatin from isolated nuclei of log phase *Crypthecodinium cohnii* (Rizzo & Nooden 1972; 1974a, b) and *Peridinium trochoideum* (Rizzo & Nooden 1974a, b) did contain a small amount of basic, 'histone-like' protein. It is probable that the amount of basic protein is too low to be detected by cytochemical staining. However, the validity of the immunofluorescent tests depends on the purity of the DNA–histone complexes used as antigens. Since the 'DNA–histone' complex used was whole chromatin rather than just DNA and histone, the test was not specific for histones and would be expected to detect non-histones as well.

Although proteolytic degradation is a major concern in studies on histone occurrence in micro-organisms (Rizzo 1976), proteolysis is not responsible for the absence of histones in dinoflagellates. First of all, if typical histones were present they should be detectable cytochemically, as in other micro-organisms. Secondly, the inclusion of proteolytic enzyme inhibitors in some (Rizzo & Nooden 1974b), or all (Bodansky *et al.* 1979) of the sample preparation buffers did not change the electrophoretic pattern of basic protein. Finally, attempts to demonstrate substantial protease activity in isolated nuclei of *Crypthecodinium cohnii* (P. J. Rizzo, unpublished observations) and *Prorocentrum micans* (Herzog & Soyer 1981) were unsuccessful. While typical histones are not present in dinoflagellate nuclei, basic 'histone-like' proteins have now been reported in several uninucleate dinoflagellates (Rizzo 1981). These electrophoretic studies are summarized below, along with the reports on histones in the endosymbiont nuclei of binucleate dinoflagellates.

3.2 Histone-like proteins in uninucleate dinoflagellates

Crypthecodinium cohnii

Initial electrophoretic studies on basic proteins in dinoflagellate nuclei and chromatin employed low pH urea-containing gels which were designed for histones (Panyim & Chalkley 1969). The majority of the basic protein extracted from *C. cohnii* chromatin migrated as a single band with a mobility slightly lower than that of histone H4 (Rizzo & Nooden 1972). Partial characterization of this histone-like protein showed that it was rich in lysine, but that it differed from the known histones in several respects (Rizzo & Nooden 1974b). In addition, it did not stabilize the DNA against melting (Rizzo & Nooden 1974a). While the molecular weight was estimated at 16 000 by electrophoresis in sodium dodecyl sulphate (SDS) gels, it was noted that this may be an overestimate due to the anomalous migration of low molecular weight basic proteins in these gels (Rizzo & Nooden 1974b). The molecular weight estimate has subsequently been revised to 13 000 due to the use of an improved gel system and by using histones as molecular weight standards (Rizzo 1981). This protein was not detectable in chromatin prepared for stationary phase cells (Rizzo & Nooden 1974b). However, we have recently shown that this protein is present in chromatin from stationary phase cells as well as from log phase cells (Rizzo *et al.* 1984). This is attributed to an improved nuclear isolation procedure, which allows much higher yields of nuclei due to a more efficient cell breakage step.

Studies on *C. cohnii* must now take into consideration the fact that different sibling species are known to exist for this organism (Beam & Himes 1977; Himes & Beam 1978). The existence of these sibling species (established from breeding experiments) was recently confirmed by restriction endonuclease cleavage and molecular hybridization of ribosomal RNA genes (Steele & Rae 1980b). The organism used for the earlier studies (Rizzo & Nooden 1972; 1974a, b), was the original Woods Hole strain isolated by Dr Luigi Provasoli, termed Whd or Gcd. More recently, a comparison was made of the basic chromatin proteins of four *C. cohnii*-like strains (Rizzo 1981). These were Gcd, WHA (a clonal isolate from Gcd), WHB (a mutant isolated from Gcd which does not mate with Gcd derivatives; Dr C. A. Beam, pers. comm.) and G (a Puerto Rican strain isolated by Dr Kenneth Gold). Strains Gcd, WHA and WHB all contained one major histone-like protein with a mobility in SDS gels slightly greater than that of calf H2A. Strain G, however, contained the above band plus a second one with a mobility the same as that of H2A (Fig. 4.1a). It appears that although Gcd is heterogeneous and contains at least two sibling species, they were derived from the same isolate and are more closely related to each other than to strain G.

PERIDINIUM TROCHOIDEUM

Chromatin from isolated nuclei of *P. trochoideum* was examined for the presence of basic proteins (Rizzo & Nooden 1974b). Log phase cells contained a protein with the same mobility in urea gels as that of *C. cohnii*. However, no comparison was made in SDS gels.

PROROCENTRUM MICANS

P. micans is considered to be a rather primitive dinoflagellate (Taylor 1980). A biochemical investigation on isolated nuclei of this organism revealed the presence of two major histone-like proteins with molecular weights of 12 000 and 13 000 (Herzog & Soyer 1981). As in other dinoflagellates the amount of basic nuclear protein was small, and there was no evidence of nucleosomes, as tested by nuclease digestion or electron microscopy studies.

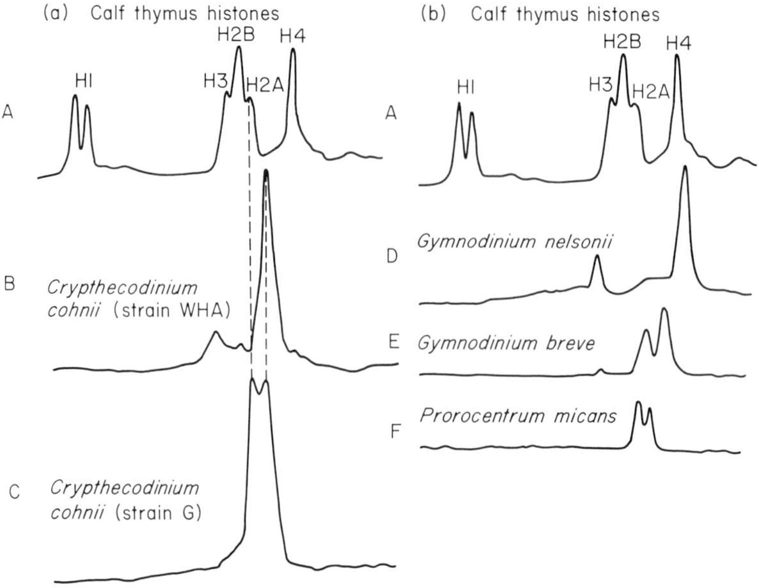

Fig. 4.1. Comparison between gel scans of calf thymus histones and histone-like proteins from uninucleate dinoflagellates.

Gymnodinium nelsoni (*G. Sanguineum*)

G. nelsoni would appear to be an ideal dinoflagellate for nuclear studies because it lacks a rigid cell wall and the nuclei are large and relatively stable (Rizzo & Nooden 1973). However, the organism has a slow growth rate and does not reach high cell densities, at least in the various media employed so far. Nevertheless, nuclear basic proteins were subjected to electrophoresis in SDS gels, which revealed one major band with a mobility slightly greater than that of calf H4, corresponding to a molecular weight of about 10 000 (Rizzo & Burghardt 1982). In addition, two minor basic proteins of about 13 000 and 17 000 were also present (Fig. 4.2b), and both of these were found to vary in amount, relative to the major basic protein. It is believed that the observed changes in the relative amounts of these three components are not due to proteolytic degradation, but are instead due to actual differences in these proteins during growth (Rizzo & Burghardt 1982). It would be interesting, therefore, to monitor the relative amounts of the three histone-like proteins during the various stages of growth.

Gymnodinium breve (*Ptychodiscus brevis*)

This Florida red tide dinoflagellate also lacks a rigid cell wall (although there is a thin, continuous wall layer beneath the vesicles), and it grows better than *G.*

nelsoni in culture. When isolated nuclei from log phase cells were examined, two histone-like proteins were detected, with molecular weights of 12 000 and 13 000. Although Herzog & Soyer (1981) have shown that *Prorocentrum micans* also contains two histone-like proteins with molecular weights of 12 000 and 13 000, preliminary results indicate that these proteins do not co-migrate exactly with the proteins from *G. breve* (Fig. 4.1b). It would be interesting to compare these two proteins with those of *P. micans*, using several different gel systems and peptide mapping.

Gyrodinium dorsum

Another wall-less dinoflagellate with a large nucleus is *Gyrodinium dorsum*. When acid extracts of isolated nuclei were subjected to electrophoresis in the presence of SDS, one major band was obtained, with a molecular weight of about 12 000 daltons (Rizzo & Morris 1984). Although this is close to the 13 000 dalton histone-like protein from *C. cohnii*, the peptide patterns produced from limited digestion with V8 protease (peptide maps) were significantly different (Rizzo & Morris 1984). These preliminary results suggest that the histone-like proteins from various dinoflagellates may differ significantly in primary structure.

Oxyrrhis marina

O. marina is a wall-less, non-photosynthetic, phagotrophic dinoflagellate (Droop 1959). The chromatin basic protein migrates as one major component in low pH urea gels (Sun *et al.* 1978). It would be extremely interesting to compare this protein (protein H) with the major histone-like protein from *C. cohnii* (HCc) in terms of molecular weight, behaviour in other gel systems, amino acid composition and peptide mapping.

A summary of work to date on basic proteins in uninucleate dinoflagellates is given in Table 4.3 and a comparison of the mobilities in SDS gels for *C. cohnii*, *G. nelsoni*, *P. micans*, *G. breve*, and *G. dorsum* is shown in Fig. 4.1.

3.3 Histones in binucleate dinoflagellates

The binucleate dinoflagellates *Peridinium balticum* (Tomas *et al.* 1973) and *Glenodinium foliaceum* (Dodge 1971) contain two dissimilar nuclei, usually referred to as dinokaryotic and eukaryotic. The dinokaryotic nucleus is ultrastructurally similar to that of uninucleate dinoflagellates and contains permanently condensed chromosomes. The chromatin of the eukaryotic nucleus, however, is not organized into chromosomes (Tomas *et al.* 1973; Tomas & Cox 1973; Tippit & Pickett-Heaps 1976). It has been suggested from ultrastructural

Table 4.3. Electrophoretic analysis of dinoflagellate histone-like proteins

Organism	Preparation			Reference
C. cohnii (Strain Gcd)	Nuclei and Chromatin	One	13 000	Rizzo & Nooden 1974b; Rizzo 1981
C. cohnii (Strain G)	Chromatin	Two	13 000; 14 000	Rizzo 1981
P. trochoideum	Chromatin	One	?	Rizzo & Nooden 1974b
Pr. micans	Nuclei	Two	12 000; 13 000	Herzog & Soyer 1981
G. nelsoni	Nuclei and Chromatin	One	10 000	Rizzo & Burghardt 1982
G. breve	Nuclei and Chromatin	Two	12 000; 13 000	Rizzo *et al.* 1982
Gyrodinium dorsum	Nuclei	One	12 000	Rizzo *et al.* 1984
Oxyrrhis marina	Nuclei	One	?	Sun *et al.* 1978

data, that the eukaryotic nucleus of *P. balticum* belongs to a chrysophyte-like endosymbiont, and this suggestion has been confirmed by pigment analysis (Withers *et al.* 1977). Furthermore, the chloroplast pigments are essentially the same in *P. balticum* and *G. foliaceum* but are not typical of dinoflagellates, which suggests that the two endosymbiotic organisms are closely related (Withers *et al.* 1977).

It was recently shown that when chromatin released from mixtures of dinokaryotic and eukaryotic nuclei was spread on electron microscope grids by the Miller technique (Miller & Bakken 1972), chromatin strands composed of nucleosomes and strands lacking nucleosomes were revealed; however, chromatin spreads of only eukaryotic nuclei revealed only beaded (nucleosome) strands (Rizzo & Burghardt 1980). Since histones are responsible for nucleosome formation, the results suggest that this class of proteins is present in the eukaryotic (endosymbiont) nucleus. Support for this suggestion comes from cytochemical tests using alkaline fast green, in which a positive test was obtained for isolated nuclei of both *P. balticum* and *G. foliaceum*, but a negative test was obtained for both dinokaryotic nuclei (Rizzo 1981). A typical dinokaryotic nucleus (*C. cohnii*) and a typical eukaryotic nucleus (*Olisthodiscus luteus*) were used as controls. As expected, *O. luteus* nuclei gave a positive staining reaction due to the presence of histones (Rizzo 1980), while the test for *C. cohnii* nuclei was negative.

Peridinium balticum

From the evidence presented above it is clear that the histones from *P. balticum* are located in the eukaryotic (endosymbiont) nucleus and not in the dinokaryotic nucleus. Preliminary electrophoretic analysis of *P. balticum* histones in low pH urea gels revealed the presence of only four histones (Rizzo & Cox 1977). More recent work using several different gel systems showed that while urea gels resolved four components, SDS gels resolved five components (Rizzo 1982). Two-dimensional electrophoresis consisting of urea gels in the first dimension and SDS gels in the second dimension showed that two proteins had the same mobility in the urea gels. Using calf thymus histones as standards, molecular weight estimates of 23 000, 20 000, 15 000, 13 000 and 11 000 were obtained for the five bands resolved in SDS gels (Fig. 4.2). It was suggested that the two high molecular weight proteins represent an H1-like doublet, that the 15 000 and 11 000 dalton components are analogous to histones H3 and H4 respectively, and that the 13 000 dalton protein may replace H2A and H2B in the nucleosome (Rizzo 1982). It was subsequently shown that the two H1-like components are located in the linker DNA and not in the nucleosome (Shupe & Rizzo 1983). It is interesting that there are bands with gel mobilities identical to the highly conserved histones H3 and H4, but no bands corresponding to the less conserved histones H2A and H2B. Furthermore, amino acid analysis and peptide mapping studies (Cleveland *et al.* 1977) support the suggestion that H3 and H4 are present in *P. balticum* but H2A and H2B are not (Rizzo *et al.* in preparation).

GLENODINIUM FOLIACEUM

Electrophoretic analysis in SDS gels of *G. foliaceum* basic proteins from mixtures of dinokaryotic and eukaryotic nuclei revealed the presence of six major components rather than five. A comparison of the gel patterns obtained from *P. balticum* and *G. foliaceum* is shown in Fig. 4.2. Note the co-migration of bands corresponding to histones H3 and H4. Due to the abundance of dinokaryotic nuclei in *G. foliaceum* nuclear preparations (they outnumber the eukaryotic nuclei), it is possible that the additional band is from the dinokaryotic nucleus.

It is noteworthy that while dinoflagellates are the only eukaryotes known to lack histones completely, other protists have been shown to contain an incomplete set. The occurrence of histones in various protists has been reviewed previously (Rizzo 1976; Horgen & Silver 1978; Ragan & Chapman 1978), and some of the earlier examples of missing histones in these organisms were later shown to be due to proteolytic degradation (*Neurospora* being a case in point). However, more recent work, in which proteolytic degradation was probably not significant, has shown that several protists are missing at least one of the five major histones found in all higher plants and animals. Among these are the yeast *Saccharomyces cerevisiae* (Mardian & Isenberg 1978), the slime mold *Dictyostelium discoideum* (Bakke & Bonner 1979), the aquatic fungus *Achyla ambisexualis* (Silver 1979), the marine chromophytic alga *Olisthodiscus luteus* (Rizzo 1980), and the hemoflagellate *Trypanosoma cruzi* (Rubio *et al.* 1980).

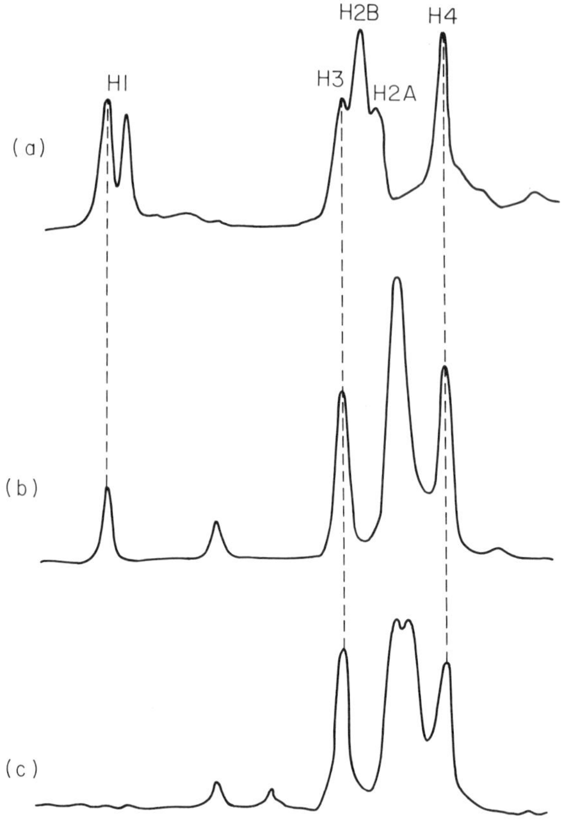

Fig. 4.2. Comparison of gel scans of basic chromatin proteins from binucleate dinoflagellates: (a) calf thymus histones; (b) *Peridinium balticum*; (c) *Glenodinium foliaceum*.

3.4 Non-basic nuclear proteins

While most of the work done on dinoflagellate chromosomal proteins has been concerned with basic proteins, non-basic proteins have also been found associated with the nuclear DNA. These non-basic DNA-associated proteins, or non-histones, are present in much greater amounts (relative to the DNA) than are the basic proteins (Rizzo & Nooden 1972, 1974a; Herzog & Soyer 1981), as discussed above (see Table 4.2).

When non-histones from *C. cohnii* chromatin were compared electrophoretically to those from a higher plant, the dinoflagellate banding pattern was less heterogeneous and the majority of the bands were restricted to a higher molecular weight range (>43 000). However, using a gel system designed to resolve low molecular weight proteins, analysis of total chromatin proteins of *C. cohnii* revealed the presence of many low molecular weight components (Rizzo 1981). Electrophoretic analysis of total nuclear proteins from *G. breve* also showed that the non-basic proteins contain a variety of molecular weight species (Rizzo *et al* 1982). Thus, while the non-basic proteins of dinoflagellate chromatin are less heterogeneous than the non-histones of higher organisms, they are much more heterogeneous than dinoflagellate basic proteins, and they span a wide range of molecular weights.

While non-histone chromosomal proteins have received much attention in other eukaryotes, very little work has been done on the non-histones of dinoflagellates. DNA-binding proteins from *C. cohnii* were studied in relation to the cell cycle (Franker 1972; Franker *et al.* 1973). Analysis of ^{35}S-labelled proteins that were absorbed to and step-eluted from DNA-cellulose columns revealed a variety of DNA-binding proteins in regard to molecular weight. While most proteins were in the molecular weight range of 80 000–140 000, two prominent components of 22 000 and 35 000 daltons were resolved (Franker 1972). Using synchronous cultures of *C. cohnii* Franker *et al.* (1973) detected two pronounced periods for the synthesis of DNA binding proteins during the cell generation cycle. Soluble ^{35}S-labelled proteins obtained at various intervals in the cell cycle were absorbed to DNA-cellulose columns and eluted with a linear salt gradient. During the two pronounced periods (right after excystment, and right before flagellar abcission) when the synthesis of cyst wall material begins, thirty-five well resolved peaks were eluted, while seventeen other phase-specific peaks were resolved during the remaining intervals in the cell cycle.

Further studies on dinoflagellate DNA-binding proteins during various phases of the cell cycle are warranted, especially at transitions such as encystment, excystment and the onset of DNA synthesis. Since nuclear isolation procedures are now available for several species (including *C. cohnii*), the use of isolated nuclei rather than whole cells can be expected to eliminate the 'high background' of DNA-binding proteins. Results from such studies will be helpful in determining the functions of non-histones in dinoflagellates and possibly in other eukaryotes. Studies aimed at elucidating the roles of certain non-histone proteins should be less complex with dinoflagellates than with other eukaryotes because there is less protein associated with the nuclear DNA. This is due not only to the absence of histones but also to the absence of other classes of DNA-associated proteins, such as those responsible for the transition of interphase chromatin to metaphase chromosomes. Therefore, one might expect a higher proportion of the total DNA-binding protein to have a regulatory function.

4 NUCLEAR DNA

4.1 Properties and sequence organization

Early work on the properties of dinoflagellate nuclear DNA focused on the determination of buoyant densities and thermal denaturation temperatures, and has been summarized by Loeblich (1976), and by Rae (1976). In general, dinoflagellate DNA exhibits considerable discrepancies when the percentage of G + C is determined from buoyant density and from the Tm (Rae 1976). This has been attributed to the presence of hydroxymethyluracil (see Modified Bases, p. 160).

Analytical caesium chloride gradient centrifugation can also be useful in comparing the DNAs from the two nuclei of binucleate dinoflagellates. If these nuclei are unrelated as suggested, the respective DNAs should have different compositions and thus different buoyant densities. This experiment was performed for *P. balticum* (Steele 1980), and the results are shown in Fig. 4.3. Note the presence of two distinct peaks in the buoyant density profile. Analysis of the buoyant density and thermal denaturation profiles has led to the tentative conclusion that the lighter peak (1·706) is from the endosymbiont nucleus and the heavier peak (1·730) is from the dinoflagellate nucleus (Steele 1980). The presence of two distinct peaks of DNA from *P. balticum* supports the conclusion that this organism is a dinoflagellate with an endosymbiont (Tomas & Cox 1973; Tomas *et al.* 1973).

Also, the presence of satellite DNA has been observed in the DNA of *C. cohnii* (Rae 1973) and *A. carterae* (Galleron & Durrand 1978). An interesting situation in regard to satellite DNA was revealed during a recent study of *Prorocentrum cassubicum* (*Exuviaella cassubica*) DNA (Steele 1980). About 50%

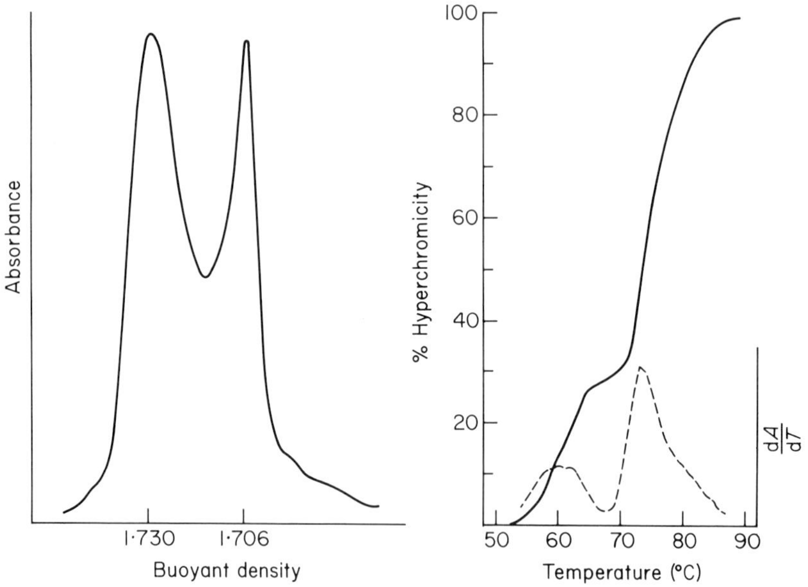

Fig. 4.3. Analytical CsCl gradient centrifugation (left) and thermal denaturation (right) profiles of *P. balticum* DNA. For the thermal denaturation profile the broken line is the first derivative of the melt curve. (From Steele 1980.)

of the DNA of *P. cassubicum* is present in a highly repeated fraction. This abundant DNA fraction might be expected to give rise to considerable amounts of satellite DNA, as is often the case in higher eukaryotes. However, when total DNA is subjected to analytical caesium chloride gradient centrifugation, no satellite peaks are seen (Fig. 4.4a), although a shoulder is seen on the heavy side of the DNA profile. When the DNA is centrifuged through a gradient containing the dye Hoechst 33258, a partial separation into two fractions is seen (Fig. 4.4b). Finally, in the presence of actinomycin D, a distinct satellite peak can be seen on the heavy side of the main DNA peak (Fig. 4.4c). The demonstration of satellite DNA in *P. cassubicum* is important in an evolutionary context in view of the conclusion that the Prorocentrales are primitive dinoflagellates. Highly repeated and satellite DNAs may therefore be an evolutionary old feature of dinoflagellates (Steele 1980).

The first evidence that dinoflagellates contain repeated DNA sequences was reported in a review article (Britten & Kohne 1968) and later in a report concerned primarily with dinoflagellate chromosome structure (Roberts *et al.* 1974). In a study involving DNA renaturation kinetics, Allen *et al.* (1975) reported that the DNA of *C. cohnii* is composed of two major kinetic classes; one repeated and the other highly complex. The repeated class comprised 55–60% of the DNA, and was interspersed in a manner similar to that of other eukaryotes. The highly complex class had properties consistent with unique DNA.

In a subsequent and more detailed study using hydroxylapatite binding, S1 nuclease digestion and electron microscopy of reassociated DNA, Hinnebusch *et al.* (1980) confirmed the finding by Allen *et al.* (1975) that a large fraction of the *C. cohnii* genome was composed of repeated, interspersed DNA. The study by Hinnebusch *et al.* (1980) revealed that about half of the genome is composed of interspersed unique and repeated sequences. About 95% of the total number

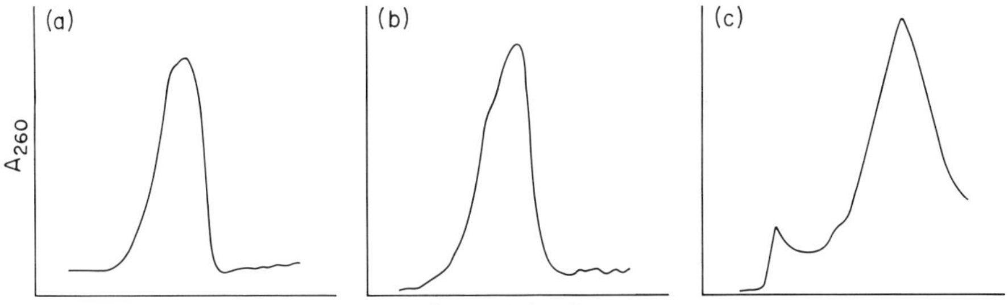

Fig. 4.4. Effect of DNA-binding dyes on the analytical CsCl gradient profile of *P. cassubicum* DNA; (a) no dye; (b) Hoechst 33258 at a dye to DNA ratio of 2·5 to 1; (c) actinomycin D at a dye to DNA ratio of 4 to 1. (from Steele 1980.)

of interspersed unique sequences is in this class, while the remaining 5% are uninterrupted by repeated sequences for 4000 or more nucleotides. About half of the repeated DNA is not interspersed among unique sequences. Evidence was also obtained for the presence of inverted repeats, and a novel arrangement of repeated sequences was observed with the electron microscope. Despite the various novel features observed in this latter study, the general sequence organization of *C. cohnii* was quite typical of higher organisms, and therefore eukaryotic in nature rather than prokaryotic. This eukaryotic nature of *C. cohnii* DNA organization is in contrast to the prokaryotic affinities of the chromosomes (see Introduction, p. 143). The conclusion that the dinoflagellate genome is comparable in structural complexity to higher eukaryote genomes was also reached by Steele (1980).

4.2 Modified bases

While eukaryotic DNA is known to contain modified bases such as 5-methylcytosine and the less common N^6-methyladenine, the only eukaryotes thus far known to contain 5-hydroxymethyluracil (HOMeU), are the dinoflagellates (see Rae & Steel (1978) for a review). The discovery of HOMeU in dinoflagellate DNA emerged from a discrepancy in the relationship between the buoyant density and thermal stability of *C. cohnii* DNA (Rae 1973, 1976). Since buoyant density and melting temperature (Tm) are both functions of the relative contributions of A+T and G+C nucleotide pairs, an expected relationship exists between the two. The DNA of *C. cohnii* had a high buoyant density but a low melting temperature due to the presence of considerable amounts of HOMeU, which effectively raises the buoyant density and lowers the thermal stability of DNA (Rae 1973). While the presence of substantial amounts of HOMeU is a common feature of dinoflagellate DNA, the relative amount varies considerably between species (Rae 1976). Specifically, HOMeU constitutes 4–19% of the total bases and replaces 12–70% of the thymine (Rae 1976). A high level of HOMeU has also been reported in the nuclear DNA of *Prorocentrum micans*, one of the most primitive of the living dinoflagellates (Herzog *et al.* 1982). The possible roles and the significance of HOMeU in dinoflagellate DNA have been discussed in detail, with the conclusion that the presence of HOMeU represents a neutral vestige of some mechanism that is no longer in operation, and therefore has no present functional significance (Rae & Steele 1978). In addition to HOMeU, some dinoflagellates also contain methylcytosine (Franker 1970; Rae 1976), and N^6-methyladenine (Rae 1976).

In regard to *C. cohnii*, the nuclear DNA of this organism has 38% of the thymine replaced by HOMeU and 3% of the cytosine is replaced by methylcytosine. Both of these modified bases are non-randomly distributed in the DNA (Steele & Rae 1980a). An indication that HOMeU is non-randomly distributed in *C. cohnii* DNA was provided earlier by Rae (1973) who showed that ratios of thymine to HOMeU varied in different buoyant density fractions. This was recently confirmed by nearest neighbour analysis and pyrimidine distribution analysis (Steele & Rae 1980a), which also revealed that HOMeU is preferentially located in the dinucleotides HOMeUpA and HOMeUpC. Steele & Rae (1980a) have shown that the amount and distribution of methylcytosine in *C. cohnii* are similar to what is found in eukaryotes. These authors also noted that cytosine in rDNA of *C. cohnii* is extremely methylated, which is unusual for rDNA of most animals with the exception of fish and amphibia (Steele & Rae 1980a).

While the levels of methylcytosine are similar in the DNA of dinoflagellates and other eukaryotes, the presence of large amounts of HOMeU has still not been detected in eukaryotes other than dinoflagellates. Even if HOMeU does not have any present functional significance, its abundance and non-random distribution constitute one of the most interesting biochemical features of dinoflagellate chromosomes.

4.3 Biosynthesis

Virtually all the available information on DNA biosynthesis in dinoflagellates is concerned with the timing of DNA synthesis in relation to the cell cycle. The interest in whether or not a discrete S-phase exists in dinoflagellates was stimulated by Dodge (1966), who suggested that DNA synthesis might be continuous, as it is in prokaryotes. Several subsequent reports have appeared, using different organisms and different methods. A distinct S-phase was reported by Franker (1971), Franker *et al.* (1974), Loeblich (1976), Galleron & Durrand (1979) and Spector *et al.* (1981). On the other hand, continuous DNA synthesis was suggested by Filfilan & Sigee (1977). Although different dinoflagellates and different methodologies were used in the above studies, it appears that the reason for the discrepancy as to whether or not a discrete S-phase exists depends heavily on the interpretation of the results, and possibly on the degree of cell synchrony in the culture used. The trend is that a maximum uptake of label is observed during part of the cell cycle, but sometimes residual uptake may occur during the entire cycle. This can be seen by comparing the pattern of DNA synthesis observed in *P. micans* (Filfilan & Sigee 1977), with that in *A. carterae* (Galleron & Durrand 1979). The patterns are similar in that a maximum period of uptake is observed, in addition to a residual uptake; however, Filfilan & Sigee (1977) interpret their results as continuous DNA synthesis while Galleron & Durrand (1979) interpret their results in favour of a discrete S-phase. From the available evidence, it appears that there is a discrete S-phase for DNA synthesis in dinoflagellates but this is often complicated by a residual uptake of label.

5 NUCLEAR RNA

5.1 Ribosomal RNA

Although ribosomal RNA (rRNA) is ultimately a component of the cytoplasm, it is included here because it is the product of transcription which occurs in the nucleus. The first study on the characterization of rRNA from a dinoflagellate was reported by Rae (1970). As judged by sedimentation velocities in isokinetic sucrose gradients, *C. cohnii* rRNA species were found to be 16 and 25S. The processing of precursor rRNA species to mature rRNA species was also studied, and the overall sequence was directly comparable with the processing sequence described for mammalian cells. Rae (1970) concluded that dinoflagellates have a eukaryotic form of rRNA and processing sequence.

Molecular weights of dinoflagellate rRNAs have been determined for *Peridinium cinctum* (Gressel *et al.* 1975) and *C. cohnii* (Werner-Schlenzka *et al.* 1978), by gel electrophoresis under native conditions. The *P. cinctum* rRNAs had molecular weights of $1{\cdot}23 \times 10^6$ and $0{\cdot}70 \times 16^6$ daltons for the heavy and light rRNAs respectively. The corresponding molecular weights for *C. cohnii* were $1{\cdot}25 \times 10^6$ and $0{\cdot}69 \times 10^6$ daltons. These molecular weights are of the same size range as that reported for other unicellular algae, as determined by electrophoresis under native conditions. A more accurate determination of rRNA molecular weights can be obtained by electrophoresis under denaturing conditions. This would prevent possible effects of secondary structure of the RNA during gel migration. This has recently been done by Steele (1980) and a typical gel is shown in Fig. 4.5a. The 24S rRNA of *C. cohnii* is smaller than the corresponding heavy rRNAs of *Dictyostelium* and *Tetrahymena*. The 17S rRNA of *C. cohnii* is also slightly smaller than the corresponding light rRNA of *Dictyostelium*. Using the sizes of known standards to plot migration distance versus molecular weight, values of the 17S and 24S rRNAs of *C. cohnii* were found to be $1{\cdot}16 \times 10^6$ and $0{\cdot}63 \times 10^6$ daltons respectively (Fig. 4.5b). Using these values for molecular weight in conjunction with previously determined data from other sources, Steele (1980) has calculated that there are about 380 copies of rDNA in each *C. cohnii* nucleus.

As a first step towards a detailed study of the ribosomal genes of *C. cohnii*, Steele (1980) has used recombinant DNA techniques to show that the rDNA in *C. cohnii* exists as a complex gene family. His studies indicate that a large fraction of the rDNA in *C. cohnii* consists of tandemly repeated, 15 kilobase units. While these tandem repeats are typical of most eukaryotes, *C. cohnii* differs from some other lower eukaryotes in which the rDNA is arranged in palindromes. In addition, his data suggests the possibility that *C. cohnii* rDNA may contain intervening sequences, although these sequences may be located at sites other than the usual site in the 3′ half of the rDNA coding for the large rRNA.

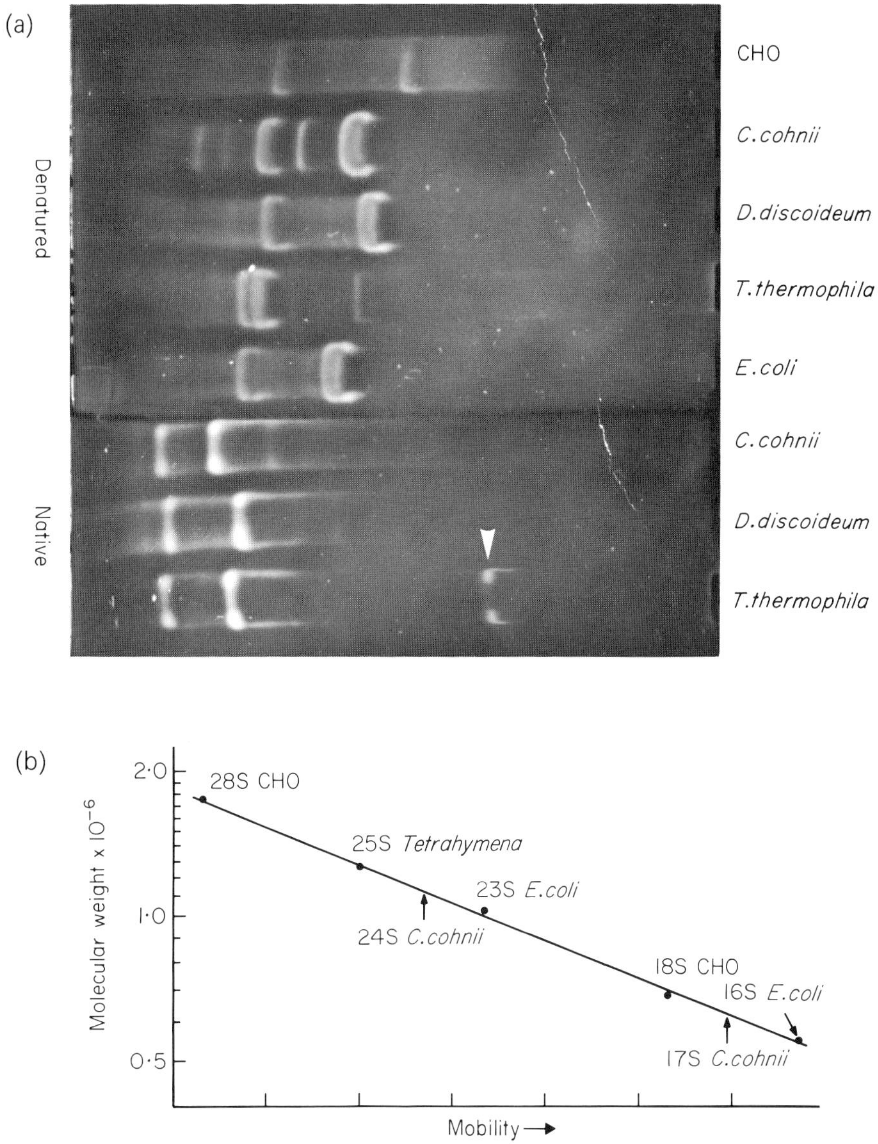

Fig. 4.5. Determination of molecular weight for the 17S and 24S rRNAs of *C. cohnii*. (a) Electrophoresis of total polysomal RNA from *C. cohnii* under native and denaturing conditions in 1% agarose. Denaturation was the glyoxal and dimethylsulphoxide. rRNAs from *Dictyostelium discoideum* and *Tetrahymena thermophila* are also present in the native gel. The arrowhead indicates the presence of contaminating DNA in the *T. thermophila* rRNA. *D. discoideum*, *T. thermophila*, *E. coli* and Chinese hamster ovary (CHO) rRNAs are also shown in the denatured gel. (From Steele 1980.) (b) Plot of molecular weight versus gel migration for denatured rRNAs shown in (a). The mobilities of *C. cohnii* 17S and 24S are indicated on the line, which was constructed from known molecular weights. (From Steele 1980.)

The size of the large rRNA subunit is considered to be important in an evolutionary context, because while the small rRNA subunit remains unchanged in eukaryotic evolution, the large rRNA subunit increases in size (Loening 1968). The data thus far would therefore place dinoflagellate rRNA at a rather early stage in eukaryotic evolution, as expected.

It has also been suggested that the ratio of ribose to base methylation in rRNA changes in the course of evolution, with a trend toward increased ribose to base methyl groups (Werner-Schlenzka *et al.* 1978). Although methylation studies in *C. cohnii* are complicated by the presence of an unspecified RNA (adenine-1-) methylase (Werner *et al.* 1975), Werner-Schlenzka *et al.* (1978) found a low ratio of ribose to base methyl groups in the rRNA of this organism, which also suggests that dinoflagellate rRNA represents an early form of eukaryotic rRNA.

While it is true that dinoflagellate rRNA indicates that these organisms are rather primitive in an evolutionary context, just *how* primitive they are is a matter of debate. Recently, Hinnebusch *et al.* (1981) have determined the primary structure of the 5S and 5·8S RNA from *C. cohnii*, for the purpose of studying dinoflagellate phylogeny. These authors found that the 5S RNA sequence shows 75% homology with the 5S sequences of higher eukaryotes but less than 60% homology with the prokaryotic 5S RNA sequences. This evidence, plus certain other features concerned with highly conserved regions and secondary structure of the molecule, suggests a distant relationship between dinoflagellates and prokaryotes (Hinnebusch *et al.* 1981). On the other hand Sigee (1985) concludes that the weight of evidence (mainly chromosomal) favours the view that dinoflagellates are the most primitive eukaryotes.

5.2 Chromosomal RNA

As shown in Table 4.2, dinoflagellate chromatin contains a small amount of RNA. While the nature of this RNA is unknown, it seems likely that at least a substantial part of it could be accounted for as nascent (newly synthesized) RNA. Although it is possible that this RNA was absorbed from other parts of the nucleus during isolation (the nuclei contain much more RNA than chromatin does), this possibility is not supported by electron microscopy studies which indicate the presence of RNA in dinoflagellate chromosomes (Soyer & Haapala 1974a, b).

An interesting possibility for the function of RNA in dinoflagellate chromosomes is that it might play a structural role (Soyer & Haapala 1974a, b). Early studies indicated that actinomycin D could affect chromosome structure at the light (Chunosoff & Hirshfield 1968) and electron (Babillot 1970b) microscope levels. It was shown later (Soyer & Haapala 1974a) that treatment with ribonuclease can loosen the structure of the dinoflagellate chromosomes and can increase their length threefold. Also of interest is the report that a

DNA–RNA hybrid was isolated from *Amphidinium carterae* chromatin (Galleron & Durand 1978).

The idea that RNA might be involved in the structure of dinoflagellate chromosomes is interesting, especially in light of the reports showing that when *E. coli* nucleoids are incubated with RNAse the DNA unfolds (Worcel & Burgi 1972; Stonington & Pettijohn 1981). It is noteworthy, however, that recent work has caused some reconsiderations concerning the role of RNA in bacterial nucleoid structure (see Sinden & Pettijohn, 1981, for references).

5.3 RNA biosynthesis

Very little information exists on the synthesis of RNA in dinoflagellate nuclei. An autoradiographic study showed that tritiated uridine was incorporated primarily in the nucleolus and the interchromosomal fibres of *A. carterae*, but not in the chromosomes themselves (Babillot 1970a). This suggests that the chromosomes are not active in RNA synthesis unless the DNA is unwound or looped out, which is understandable when one considers the difficulty RNA polymerase would have in penetrating the interior of a tightly coiled dinoflagellate chromosome. It is surprising that this type of study has not been pursued further in more detail. For example, one could use a synchronized culture and follow the incorporation of uridine into RNA at various stages in the cell cycle, and under various lengths of exposure to the isotope. This same approach could also be used to localize the activities of RNA polymerases that are sensitive and insensitive to a α-amanitin (see below). Brändle & Zetsche (1973) tested the effect of α-amanitin on *Acetabularia* nucleolar and nucleoplasmic RNA synthesis autoradiographically, and showed that the toxin inhibits incorporation into nucleoplasmic RNA but not into nucleolar RNA. This demonstrated that, as in other eukaryotes, mRNA synthesis occurs in the nucleoplasm and rRNA synthesis is localized in the nucleolus (see Roeder 1976). Interestingly, the earlier work of Babillot has recently been confirmed by a high resolution electron microscope autoradiographical study (Sigee 1983). A clear distinction between genetically inactive DNA, termed structural DNA, and genetically active DNA was provided for *Prorocentrum micans*. Since the bulk of the dinoflagellate chromosome was shown to be inactive (Babillot 1970a; Sigee 1983), real problems arise concerning questions of genome size and mutation frequency, etc.

Endogenous RNA polymerase activity was partially characterized in isolated nuclei of *C. cohnii* (Rizzo 1979). Under optimum assay conditions these nuclei incorporated tritiated UMP into RNA at a level that was similar per unit DNA to that of typical eukaryotes. It is difficult to relate these results to those just discussed above, although the absence of histones in dinoflagellates may be important in this regard. In addition, a unique situation was found in regard to the metal requirements of these nuclei. As in all other eukaryotes (but not in prokaryotes), α-amanitin caused a partial inhibition of RNA polymerase activity in *C. cohnii* nuclei. This suggests that as in other eukaryotes, α-amanitin sensitive and insensitive forms of RNA polymerase are present (Roeder 1976). It is interesting that while the α-amanitin sensitive enzyme (RNA polymerase II), prefers manganese over magnesium as a divalent cation cofactor in all other eukaryotes, manganese always inhibited RNA polymerase activity in *C. cohnii* nuclei (Rizzo 1979). It is possible that *C. cohnii* chromosomes already contain optimum amounts of endogenous manganese (see discussion in Section 2.3, p. 147), and any additional manganese results in inhibition (which is always noted in other eukaryotes at high concentrations). An alternative explanation is that the α-amanitin sensitive RNA polymerase in *C. cohnii* has unusual metal ion requirements.

A test for the above alternatives would be to solubilize the RNA polymerase (remove it from the DNA), and subject the enzyme to anion exchange chromatography. This would also prove or disprove the existence of multiple forms of the enzyme. Once the α-amanitin sensitive polymerase is eluted from the column, it could be dialysed and then tested for manganese effects. In this regard it is of interest that a DNA-dependent RNA polymerase has been extracted from *C. cohnii*, and partially characterized (Werner *et al.* 1978). Although the apparent molecular weight of this enzyme was much lower than expected for a eukaryotic nuclear RNA polymerase, several other properties of the enzyme were similar to those of RNA polymerase I (which synthesizes rRNA). Of particular interest is that this enzyme was not sensitive to α-amanitin, and no other types of RNA polymerase activity were detectable. This is in contrast to results obtained with isolated nuclei which showed that about 80% of the total activity was inhibited by α-amanitin (Rizzo 1979). It is probable that at least two types of RNA polymerase are present in *C. cohnii* nuclei, which differ in their sensitivity to α-amanitin. Preliminary attempts to remove total RNA polymerase from the DNA and still retain activity have been unsuccessful (P. J. Rizzo & R. L. Morris, unpublished observations). Partial success was obtained using methods developed for higher plants and animals (Polymin-P precipitation), and bacteria (polyethylene glycol precipitation under high or low salt conditions). It is possible that the enzyme activity is very unstable and a novel method will have to be developed. In all instances, α-amanitin sensitive RNA polymerase activity was lost after disruption of nuclei, or whole cells. Thus, as shown previously by Werner *et al.* (1978), only one peak (insensitive to α-amanitin) was obtained following chromatography on DEAE Sephadex (P. J. Rizzo & R. L. Morris, unpublished observations).

6 CONCLUSIONS

Biochemical studies have confirmed previous cytological studies on dinoflagellate nuclei, and have also provided some new information. The presence of large numbers of chromosomes suggested that high amounts of DNA could be found and this was confirmed by measurements on whole cells and isolated nuclei. The cytochemical studies indicating an absence of conventional histones were also confirmed, although a small amount of basic protein was found associated with these chromosomes. The amount of RNA in dinoflagellate nuclei and chromatin appears to be typical of eukaryotes in general, while the non-basic protein population is less heterogeneous than that of higher organisms. Although some of the RNA in dinoflagellate chromosomes may serve a structural role, it is likely that as in other eukaryotes, at least some of it is nascent RNA. Due to the absence of histones, heterochromatic regions, and a condensation–decondensation cycle, it is felt that a higher proportion of the non-histones in dinoflagellate chromatin may be of the regulatory type than in other eukaryotes. Dinoflagellates may therefore be good candidates for the study of transcriptional control.

Several comments are necessary in regard to the basic nuclear proteins. The first concerns nomenclature. Until these proteins are characterized somewhat, it is premature to decide whether or not they should be classified as unique histones. Such characterization would include amino acid analysis, N-terminal sequence data, peptide mapping in comparison with known histones and histone-like proteins, and ability to substitute for nucleosomal histones in chromatin reconstitution experiments. Some of these studies are now in progress. This would allow comparisons of these proteins with conventional histones, (Isenberg 1979), high mobility group (HMG) proteins (Johns 1982) or the histone-like proteins of prokaryotes (Cukier-Kahn *et al.* 1972; Rouviere-Yaniv & Gros 1975; Haselkorn & Rouviere-Yaniv 1976; Hubscher *et al.* 1980; Lathe *et al.* 1980; DeLange *et al.* 1981). Until such information is available it is best to use a term such as 'histone-like protein'. Thus, the major acid-soluble protein from *C. cohnii* nuclei has been designated HCc, for histone-like protein from *Crypthecodinium cohnii* (Rizzo & Morris 1984). Also, it is clear that the histones detected in the binucleate dinoflagellate *Peridinium balticum* and *Glenodinium folicaeum*, are associated with an endosymbiont nucleus and are therefore not dinoflagellate histones. The histone-like proteins detected in several uninucleate dinoflagellates are similar to each other but not identical, and in some species several are present. In view of the great diversity of living dinoflagellates (Taylor 1980), further differences will probably be revealed as different organisms are examined, especially for the dinoflagellates that have cytochemically detectable basic proteins (Hollande 1974; Ris & Kubai 1974; Soyer 1974; Cachon & Cachon 1977).

Studies on the nuclear DNA of dinoflagellates have shown that it differs from all other eukaryotic DNAs in that it contains substantial amounts of hydroxymethyluracil, but is similar in that it is made up of repeated sequences interspersed with unique sequences, and the rDNA is composed of tandem repeats. While the function of the modified base is unknown, it is probably not responsible for the characteristic structure of dinoflagellate chromosomes. The biosynthesis of DNA appears to involve a discrete S-phase, but with considerable residual synthesis.

The unusual features of dinoflagellate nuclei formed the basis for suggesting that dinoflagellates are very primitive and are possible evolutionary intermediates between prokaryotes and eukaryotes. Although there is no question that dinoflagellates are primitive eukaryotes, there is a controversy as to just how primitive they are. In addition to providing information to help elucidate dinoflagellate phylogeny, biochemical studies may also be of help in gaining an understanding of dinoflagellate chromosome structure. The biochemical studies have shown that there are only two truly unique chemical properties of these chromosomes: the presence of large amounts of hydroxymethyluracil, and the absence of histones. A major question is how all of this negatively charged DNA can be condensed into a chromosome without the presence of histones to exert a neutralization of charge. As discussed above, HOMeU probably does not serve a structural role in dinoflagellate chromosomes (Rae & Steele 1978). Although there are evidently ways in which DNA can be condensed *in vitro* without neutralization (Lerman 1974), it is also possible that metal ions (Kearns & Sigee 1980) or polyamines might be important in charge neutralization. Future biochemical studies may also help determine whether or not protein (Rizzo & Nooden 1974b) or RNA (Soyer & Haapala 1974a) is involved in dinoflagellate chromosome structure.

ACKNOWLEDGEMENTS

I wish to collectively thank all of the investigators who kindly provided preprints of papers prior to publication, and especially Drs Steele and Rae for providing figures of unpublished work.

7 REFERENCES

ALLEN J. R., ROBERTS T. M., LOEBLICH A. R. III & KLOTZ L. C. (1975) Characterization of the DNA from the dinoflagellate *Crypthecodinium cohnii* and implications for nuclear organization. *Cell* **6**, 161–169.

BABILLOT C. (1970a) Etude de l'incorporation d'uridine- ^{3}H dans le noyau chez l'*Amphidinium carteri*, dinoflagelle. *C. R. Acad. Sci. Paris* **271**, 828–831.

BABILLOT C. (1970b) Etude des effects de l'actionmycine D sur le noyau du dinoflagelle *Amphidinium carteri*. *J. Microscopie* **9**, 485–502.

BAKKE A. C. & BONNER J. (1979) Purification and the histones of *Dictyostelium discoideum* chromatin. *Biochemistry* **18**, 4556–4562.

BEAM C. A. & HIMES M. (1974) Evidence for sexual fusion and recombination in the dinoflagellate *Crypthecodinium* (*Gyrodinium*) *cohnii*. *Nature* **250**, 425–426.

BEAM C. A. & HIMES M. (1977) Sexual isolation and genetic diversification among strains of *Crypthecodinium cohnii*-like dinoflagellates: evidence of speciation. *J. Protozool.* **24**, 532–539.

BODANSKY S., MINTZ L. B. & HOLMES D. S. (1979) The mesokaryote *Gyrodinium cohnii* lacks nucleosomes. *Biochem. biophys. Res. Commun.* **88**, 1329–1336.

BRÄNDLE E. & ZETSCHE K. (1973) Zur lokalisation der α-amanitin sensitiven RNA-polymerase in Zellkernen von *Acetabulari. Planta (Berl)* **III**, 209–217.

BRITTEN R. J. & KOHNE D. E. (1968) Repeated sequences in DNA. *Science* **161**, 529–540.

CACHON J. & CACHON M. (1977) Observations on the mitosis and on the chromosome evolution during the life cycle of *Oodinium*, a parasitic dinoflagellate. *Chromosoma (Berlin)* **60**, 237–251.

CACHON J., CACHON M. & SALVANO P. (1979) The nuclear division of *Oxyrrhis marina*: an example of the role played by the nuclear envelope in chromosome segregation. *Arch. Protistenk.* **122**, 43–54.

CATTOLICO R. A. & GIBBS S. P. (1975) Rapid filter method for the microfluorometric analysis of DNA. *Anal. Biochem.* **69**, 572–582.

CHARLES D. (1977) Isolation and characterization of DNA from unicellular algae. *Plant Sci. Lett.* **8**, 35–44.

CHUNOSOFF L. & HIRSHFIELD H. I. (1968) The effects of chloramphenicol and actinomycin D on the nucleus of the dinoflagellate *Gonyaulax monilata. J. Gen. Microbiol.* **50**, 281–283.

CLEVELAND D. W., FISCHER S. G., KIRSCHNER M. W. & LAEMMLI U. K. (1977) Peptide mapping by limited proteolysis in sodium dodecyl sulfate and analysis by gel electrophoresis. *J. Biol. Chem.* **252**, 1102–1106.

CUKIER-KHAN R., JACQUET M. & GROS F. (1972) Two heat-resistant, low molecular weight proteins from *Escherichia coli* that stimulate DNA-directed RNA synthesis. *Proc. natl. Acad. Sci. USA* **69**, 3643–3647.

DELANGE R. J., GREEN G. R. & SEARCY D. G. (1981) A histone-like protein (HTa) from *Thermoplasma acidophilum*. I. Purification and properties. *J. Biol. Chem.* **256**, 900–904.

DODGE J. D. (1963) Chromosome numbers in some marine dinoflagellates. *Bot. Mar.* **5**, 121–127.

DODGE J. D. (1964) Chromosome structure in the Dinophyceae. II. Cytochemical studies. *Arch. Microbiol.* **48**, 66–80.

DODGE J. D. (1966) The Dinophyceae. In *The Chromosomes of the Algae* (ed. Godward M.B.E.), pp. 96–115. St Martin's Press, New York.

DODGE J. D. (1971) A dinoflagellate with both a mesokaryotic and a eukaryotic nucleus. I. Fine structure of the nuclei. *Protoplasma* **4**, 231–242.

DROOP M. R. (1959) Water-soluble factors in the nutrition of *Oxyrrhis marina. J. mar. Biol. Ass. UK* **38**, 605–620.

ERSLAND D. R. & CATTOLICO R. A. (1981) Nucleur deoxyribonucleic acid characterization of the marine chromophyte *Olisthodiscus luteus. Biochemistry* **20**, 6886–6893.

FILFILAN S. A. & SIGEE D. C. (1977) Continuous DNA replication in the nucleus of the dinoflagellate *Prorocentrum micans* (Ehrenberg). *J. Cell Sci.* **27**, 81–90.

FRANKER C. K. (1970) Some properties of DNA from zooxanthellae harbored by an anemone *Anthopleura elegantissima. J. Phycol.* **6**, 299–305.

FRANKER C. K. (1971) Division synchrony in primary cultures of an endozoic dinoflagellate. *J. Phycol* **7**, 165–169.

FRANKER C. K. (1972) DNA-binding protein and the cell cycle in *Crypthecodinium cohnii*. I. On the resolution of metabolically stable components. *J. Phycol.* **8**, 264–268.

FRANKER C. K., PRICHARD C. D. & LAMDEN C. A. (1973) DNA-binding protein and the cell cycle in *Crypthecodinium cohnii*. II. Short-lived protein with affinity for double stranded bulk DNA. *Differentiation* **1**, 383–391.

FRANKER C. K., SAKHRANI L. M., PRICHARD C. D. & LAMDEN C. A. (1974) DNA synthesis in *Crypthecodinium cohnii. J. Phycol.* **10**, 91–94.

GALLERON G. & DURRAND A. M. (1978) Characterization of a dinoflagellate (*Amphidinium carterae*) DNA. *Biochemie* **60**, 1235–1242.

GALLERON C. & DURRAND A. M. (1979) Cell cycle and DNA synthesis in a marine dinoflagellate *Amphidinium carterae. Protoplasma* **100**, 155–165.

GRESSEL J., BERMAN T. & COHEN N. (1975) Dinoflagellate ribosomal RNA; an evolutionary relic? *J. Mol. Evol.* **5**, 307–313.

HAAPALA O. K. & SOYER M.-O. (1974a) Size of circular chromatids and amounts of haploid DNA in the dinoflagellates *Gyrodinium cohnii* and *Prorocentrum micans*. *Hereditas* **76**, 83–90.

HAAPALA O. K. & SOYER M.-O. (1974b) Absence of longitudinal differentiation of a dinoflagellate (*Prorocentrum micans*) chromosome. *Hereditas* **78**, 141–145.

HALLER G. DE, KELLENBERGER E. & ROUILLER C. (1964) Etude au microscope electronique des plasmas contenant de l'acide desoxyribonuclique III. Variations ultra-structurales des chromosomes d'*Amphidinium*. *J. Microsc.* (*Paris*) **3**, 627–642.

HAMKALO B. A. & RATTNER J. B. (1977) The structure of a mesokaryote chromosome. *Chromosoma* (*Berl.*) **60**, 39–47.

HASELKORN R. & ROUVIERE-YANIV J. (1976) Cyanobacterial DNA-binding protein related to *Escherichia coli* HU. *Proc natl. Acad. Sci. USA* **73**, 1917–1920.

HERZOG M. & SOYER M.-O. (1981) Distinctive features of dinoflagellate chromatin. Absence of nucleosomes in a primitive species *Prorocentrum micans* E. *Eur. J. Cell Biol.* **23**, 295–302.

HERZOG M., SOYER M.-O. & DEMARCILLAC G. D. (1982) A high level of thymine replacement of 5-hydroxymethyluracil in the nuclear DNA of the primitive dinoflagellate *Prorocentrum micans*. E. *Eur. J. Cell Biol.* **27**, 151–155.

HIMES M. & BEAM C. A. (1975) Genetic analysis in the dinoflagellate *Crypthecodinium* (*Gyrodinium*) *cohnii*: evidence for unusual meiosis. *Proc. natl. Acad. Sci. USA* **72**, 4546–4549.

HIMES M. & BEAM C. A. (1978) Further studies of breeding restrictions among *Crypthecodinium cohnii*-like dinoflagellates. Evidence of a new interbreeding group. *J. Protozool.* **25**, 378–380.

HINNEBUSCH A. G., KLOTZ L. C., IMMERGUT E. & LOEBLICH A. R. (1980) Deoxyribonucleic acid sequence organization in the genome of the dinoflagellate *Crypthecodinium cohnii*. *Biochemistry* **19**, 1744–1755.

HINNEBUSCH A. G., KLOTZ L. C., BLANKEN R. L. & LOEBLICH A. R. III (1981) The primary structure and phylogeny of 5S and 5·8S RNA from the dinoflagellate *Crypthecodinium cohnii*. *J. mol. Evol.* **17**, 334–347.

HOLM-HANSEN O. (1969) Algae: amounts of DNA and organic carbon in single cells. *Science* **163**, 87–88.

HOLLANDE A. (1974) Etude comparée de la mitose Syndinienne et de celle des Peridiniens libres et des Hypermastigines. Infrastructure et cycles evolutifs des Syndinides parasites de Radiolaires. *Protistologica* **10**, 413–451.

HORGEN P. A. & SILVER J. C. (1978) Chromatin in eukaryotic microbes. *Ann. Rev. Microbiol.* **32**, 249–284.

HUBSCHER U., LUTZ H. & KORNBERG A. (1980) Novel histone H2A-like protein of *Escherichia coli*. *Proc. nat. Acad. Sci.* **77**, 5097–5101.

ISENBERG I. (1979) *Histones*. *Ann. Rev. Biochem.* **48**, 159–191.

JOHNS E. W. (1982) *The HMG Chromosomal Proteins*. Academic Press, London.

KEARNS L. P. & SIGEE D. C. (1979) High levels of transition metals in dinoflagellate chromosomes. *Experientia* **35**, 1332–1334.

KEARNS L. P. & SIGEE D. C. (1980) The occurrence of period IV elements in dinoflagellate chromatin: an X-ray microanalytical study. *J. Cell Sci.* **46**, 113–127.

KIM Y. S. & MARTIN D. F. (1974) Effects of salinity on synthesis of DNA, acidic polysaccharide, and ichthyotoxin in *Gymnodinium breve Phytochem.* **13**, 533–538.

KORNBERG R. D. (1977) Structure of chromatin. *Ann. Rev. Biochem.* **46**, 931–954.

KUBAI D. F. & RIS H. (1969) Division in the dinoflagellate *Gyrodinium cohnii* (Schiller). A new type of nuclear reproduction. *J. Cell Biol.* **40**, 508–528.

LATHE R., BUC H., LECOCQ J. P. & BAUTZ E. K. F. (1980) Prokaryotic histone-like protein interacting with RNA polymerase. *Proc. natl. Acad. Sci.* **77**, 3548–3552.

LERMAN L. S. (1974) Chromosomal analogues: long-range order in ψ-condensed DNA. *Cold Spring Harbor Symp. Quant. Biol.* **38**, 91–96.
LILLEY D. M. J. & PARDON J. F. (1979) Structure and function of chromatin. *Ann. Rev. Genet.* **13**, 197–233.
LIVOLANT R. & BOULIGAND Y. (1980) Double helical arrangement of spread dinoflagellate chromosomes. *Chromosoma* (Berl.) **80**, 97–118.
LOEBLICH A. R. III (1976) Dinoflagellate evolution: speculation and evidence. *J. Protozool.* **23**, 13–28.
LOENING U. E. (1968) Molecular weights of ribosonal RNA in relation to evolution. *J. Molec. Biol.* **38**, 355–365.
LOPER C. L., STEIDINGER K. A. & WALKER L. M. (1980) A simple chromosome spread technique for unarmored dinoflagellates and implication of polyploidy in algal cultures. *Trans. Amer. Micros. Soc.* **99**, 343–346.
MARDIAN J. K. W. & ISENBERG I. (1978) Yeast inner histones and the evolutionary conservation of histone–histone interactions. *Biochemistry* **17**, 3825–3832.
MCGHEE J. D. & FELSENFELD G. (1980) Nucleosome structure. *Ann. Rev. Biochem.* **49**, 1115–1156.
MILLER O. L. Jr. & BAKKEN A. H. (1972) Morphological studies of transcription. *Acta Endocrinol.* (*Suppl*) **168**, 155–177.
OAKLEY B. & DODGE J. D. (1974) Kinetochores associated with the nuclear envelope in the mitosis of a dinoflagellate. *J. Cell Biol.* **63**, 322–325.
PANYIM S. & CHALKLEY R. (1969) High resolution acrylamide gel electrophoresis of histones. *Arch. Biochem. Biophys.* **130**, 337–346.
RAE P. M. M. (1970) The nature and processing of ribosomal ribonucleic acid in a dinoflagellate. *J. Cell Biol.* **46**, 106–113.
RAE P. M. M. (1973) 5-Hydroxymethyluracil in the DNA of a dinoflagellate. *Proc. natl. Acad. Sci.* (USA) **70**, 1141–1145.
RAE P. M. M. (1976) Hydroxymethyluracil in eukaryotic DNA: a natural feature of the *Pyrrophyta* (dinoflagellates). *Science* **194**, 1062–1064.
RAE P. M. M. & STEELE R. E. (1978) Modified bases in the DNAs of unicellular eukaryotes: an examination of distributions and possible roles, with emphasis on hydroxymethyluracil in dinoflagellates. *BioSystems* **10**, 37–53.
RAGAN M. A. & CHAPMAN D. J. (1978) *A Biochemical Phylogeny of the Protists*. Academic Press. New York.
RIS H. (1962) Interpretation of ultrastructure in the cell nucleus. In *The Interpretation of Ultrastructure* (ed. Harris, R. J. C.), pp. 69–88. Academic Press, New York.
RIS H. & KUBAI D. F. (1974) An unusual mitotic mechanism in the parasitic protozoan *Syndinium sp. J. Cell Biol.* **60**, 702–720.
RIZZO P. J. (1976) Basic chromosomal proteins in lower eukaryotes: relevance to the evolution and function of histones. *J. Mol. Evol.* **8**, 79–94.
RIZZO P. J. (1979) RNA synthesis in isolated nuclei of the dinoflagellate *Crypthecodinium cohnii J. Protozool.* **26**, 290–294.
RIZZO P. J. (1980) Electrophoretic study of histones in the unicellular alga *Olisthodiscus luteus. Biochim. Biophys. Acta* **642**, 66–77.
RIZZO P. J. (1981) Comparative aspects of basic chromatin proteins in dinoflagellates. *BioSystems* **14**, 433–443.
RIZZO P. J. (1982) Analysis of histones from the endosymbiont nucleus of a binucleate dinoflagellate. *J. Protozool.* **29**, 98–103.
RIZZO P. J. & BURGHARDT R. C. (1980) Chromatin structure in the unicellular algae *Olisthodiscus luteus*, *Crypthecodinium cohnii* and *Peridinium balticum* Chromosoma (Berl.) **76**, 91–99.
RIZZO P. J. & BURGHARDT R. C. (1982) Histone-like protein and chromatin structure in the wall-less dinoflagellate *Gymnodinium nelsoni. BioSystems.* **15**, 27–34.

RIZZO P. J., CHOI J. & MORRIS R. M. (1984) The major histone-like protein from the nonphotosynthetic dinoflagellate *Crypthecodinium cohnii* (Pyrrophyta) is present in stationary phase cultures. *J. Phycol.* **20**, 95–100.

RIZZO P. J. & COX E. R. (1977) Histone occurrence in chromatin from *Peridinium balticum*, a binucleate dinoflagellate. *Science* **198**, 1258–1260.

RIZZO P. J., JONES M. & RAY S. (1982) Some properties of isolated nuclei of the Florida red tide dinoflagellate *Gymnodinium breve* (Davis). *J. Protozool.* **29**, 217–222.

RIZZO P. J. & MORRIS R. L. (1984) Some properties of the histone-like protein from *Crypthecodinium cohnii* (HCc). *BioSystems*, **16**, 211–216.

RIZZO P. J. & NOODEN L. D. (1972) Chromosomal proteins in the dinoflagellate *Gyrodinium cohnii*. *Science* **176**, 796–797.

RIZZO P. J. & NOODEN L. D. (1973) Isolation and chemical composition of dinoflagellate nuclei. *J. Protozool.* **20**, 666–673.

RIZZO P. J. & NOODEN L. D. (1974a) Isolation and partial characterization of dinoflagellate chromatin. *Biochim. Biophys. Acta* **349**, 402–414.

RIZZO P. J. & NOODEN L. D. (1974b) Partial characterization of dinoflagellate chromosomal proteins. *Biochim. Biophys. Acta* **349**, 415–427.

ROBERTS T. M., TUTTLE R. C., ALLEN J. R., LOEBLICH A. R. III & KLOTZ L. C. (1974) New genetic and physiochemical data on the structure of dinoflagellate chromosomes. *Nature* **248**, 446–447.

ROEDER R. G. (1976) Eukaryotic nuclear RNA polymerases. In *RNA Polymerase* (ed. Losick, R. & Chamberlin, M.), pp. 285–329. Cold Spring Harbor Press, New York.

ROUVIERE-YANIV J. & GROS F. (1975) Characterization of a novel low-molecular weight DNA-binding protein from *Escherichia coli*. *Proc. natl. Acad. Sci.* (*USA*) **72**, 3428–3432.

RUBIO J., ROSADO Y. & CASTANDEA M. (1980) Subunit structure of *Trypanosoma cruzi* chromatin. *Can. J. Biochem.* **58**, 1247–1251.

SHUPE K. A. & RIZZO P. J. (1983) Nuclease digestion of chromatin from the eukaryotic algae *Olisthodiscus luteus*, *Peridinium balticum*, and *Crypthecodinium cohnii*. *J. Protozool.* **30**, 599–606.

SIGEE D. C. (1983) Structural DNA and genetically active DNA in dinoflagellate chromosomes. *BioSystems*, **16**, 203–210.

SIGEE D. C. (1985) The dinoflagellate chromosome. In *Advances in Botanical Research* (ed. J. Callow), pp. 204–264. Academic Press, New York.

SIGEE D. C. & KEARNS L. P. (1981a) Nuclease extraction of chromosome-bound metals in the dinoflagellate *Glenodinium foliaceum*: an X-ray microanalytical study. *Cytobios* **31**, 49–65.

SIGEE D. C. & KEARNS L. P. (1981b) X-ray microanalysis of chromatin-bound period IV metals in *Glenodinium foliaceum*: a binucleate dinoflagellate. *Protoplasma* **105**, 213–223.

SIGEE D. C. & KEARNS L. P. (1981c) Levels of dinoflagellate chromosome-bound metals in conditions of low external ion availability: an X-ray microanalytical study. *Tissue & Cell* **13**, 441–451.

SIGEE D. C. (1982) Localized uptake of nickel^{63} into dinoflagellate chromosomes: an autoradiographic study. *Protoplasma* **110**, 112–120.

SILVER J. C. (1979) Basic nuclear proteins in the aquatic fungus *Achlya ambisexualis*. *Biochim. Biophys. Acta* **564**, 507–516.

SINDEN R. R. & PETTIJOHN D. E. (1981) Chromosomes in living *Escherichia coli* cells are segregated into domains of supercoiling. *Proc. natl. Acad. Sci* (USA) **78**, 224–228.

SIU C.-H. & SWIFT H. (1974) Characterization of cytoplasmic and nuclear genomes in the colorless alga *Polytoma* IV. Heterogeneity and complexity of the nuclear genome. *Chromosoma* (*Berl.*) **48**, 19–40.

SOYER M.-O. (1969) Rapports existant entre chromosomes et membrane nucleaire chez un Dinoflagelle parasite du genre *Blastodinium* Chatton. *C. R. Acad. Sci. Paris* **1268**, 2082–2084.

SOYER M.-O. (1974) Etude ultrastructurale de *Syndinium sp.* Chatton, parasite coelomique de

Copepodes pelogiques. *Vie Milieu* **24**, 191–212.

SOYER M.-O. & HAAPALA O. K. (1974a) Structural changes of dinoflagellate chromosomes by pronase and ribonuclease. *Chromosoma* (*Berl.*) **47**, 179–192.

SOYER M.-O. & HAAPALA O. K. (1974b) Electron microscopy of RNA in dinoflagellate chromosomes. *Histochem.* **42**, 239–246.

SOYER M.-O. & HAAPALA O. K. (1974c) Division and function of dinoflagellate chromosomes. *J. Microscopie* **19**, 137–146.

SPECTOR D. L., VASCONCELOS A. C. & TRIEMER R. E. (1981) DNA duplication and chromosome structure in the dinoflagellates. *Protoplasma* **105**, 185–194.

STEELE R. E. (1980) *Aspects of the composition and organization of dinoflagellate DNA*. Ph.D. Thesis, Yale University.

STEELE R. E. & RAE P. M. M. (1980a) Ordered distribution of modified bases in the DNA of a dinoflagellate. *Nucl. Acids Res.* **8**, 4709–4725.

STEELE R. E. & RAE P. M. M. (1980b) Comparison of DNAs of *Crypthecodinium cohnii*-like dinoflagellates from widespread geographic locations. *J. Protozool.* **27**, 479–483.

STEWART J. M. & BECK J. S. (1967) Distribution of the DNA and the DNA-histone antigens in the nuclei of free-living and parasitic *Sarcomastigophora*. *J. Protozool.* **14**, 225–231.

STONINGTON O. G. & PETTIJOHN D. E. (1981) The folded genome of *Escherichia coli* isolated in a protein-DNA-RNA complex. *Proc. natl. Acad. Sci.* (*USA*) **68**, 6–9.

SUN Y.-L., FAN P.-F. & SHANG W.-M. (1978) Characterization of acid-soluble proteins from the dinoflagellate *Oxyrrhis marina*. *Acta. Biol. Exp. Sinica* **11**, 297–302.

TAUTVYDAS K. J. (1976) Evidence for chromosome endoreduplication in *Eudrina californica*, a colonial alga. *Differentiation* **5**, 35–42.

TAYLOR F. J. R. (1980) On dinoflagellate evolution. *BioSystems* **13**, 65–108.

TIPPIT D. H. & PICKETT-HEAPS J. D. (1976) Apparent amitosis in the binucleate dinoflagellate *Peridinium balticum*. *J. Cell Sci.* **21**, 273–289.

TOMAS R. N. & COX E. R. (1973) Observations on the symbiosis of *Peridinium balticum* and its intracellular alga. I. Ultrastructure. *J. Phycol.* **9**, 304–323.

TOMAS R. N., COX E. R. & STEIDINGER K. A. (1973) *Peridinium balticum* (Levander) Lemmerman, an unusual dinoflagellate with a mesocaryotic and an eucaryotic nucleus. *J. Phycol.* **9**, 91–98.

TRIEMER R. E. (1982) A unique mitotic variation in the marine dinoflagellate *Oxyrrhis marina* (Pyrrophyta). *J. Phycol.* **18**, 399–411.

WERNER E., HOHOFF M. & KROGER H. (1978) A low molecular weight, DNA-dependent RNA polymerase from the dinoflagellate *Crypthecodinium cohnii*. *Hoppe-Seyler's Zeit. Physiol. Chem.* **359**, 1163–1164.

WERNER E., KAHLE P., LANGE U. & KROGER H. (1975) An unspecific RNA (adenine-1-) methylase from the dinoflagellate *Crypthecodinium cohnii*. *FEBS. Lett.* **55**, 245–248.

WERNER-SCHLENZKA H., WERNER E. & KROGER H. (1978) Dinoflagellate ribosomal RNA. An early stage in the evolution of eucaryotic ribosomal RNA. *Comp. Biochem. Physiol.* **61B**, 587–591.

WITHERS N. W., COX E. R., TOMAS R. & HAXO F. T. (1977) Pigments of the dinoflagellate *Peridinium balticum* and its photosynthetic endosymbiont. *J. Phycol.* **13**, 354–348.

WORCEL A. &. BURGI E. (1972) On the structure of the folded chromosome of *Escherichia coli*. *J. Molec. Biol.* **71**, 127–147.

YEN C. S., STEELE R. E. & RAE P. M. M. (1978) Chromatin structure in a dinoflagellate as revealed by electron microscopy of DNA cross-linked *in vivo* with trimethylpsoralen. *J. Cell Biol.* **79**, 120a.

CHAPTER 5
PHOTOSYNTHETIC PHYSIOLOGY OF DINOFLAGELLATES

BARBARA B. PRÉZELIN
Oceanic Biology Group, Marine Science Institute,
Department of Biological Sciences, University of California,
Santa Barbara, California 93106, USA

1 PHOTOSYNTHETIC APPARATUS

1.1 Pigmentation, chromoproteins and photosystems

The photosynthetic pigment system of dinoflagellates is unique. Among the chlorophyll *a*–chlorophyll *c*–carotenoid systems characteristic of the major marine algal classes, including brown algae, diatoms and chrysophytes, only the dinoflagellates contain one type of chlorophyll *c* (chl c_2) and they usually have peridinin instead of fucoxanthin as their major light-harvesting carotenoid (Table 5.1). Two exceptions are the dinoflagellate species, *Kryptoperidinium foliaceum* and *Peridinium balticum*, which contain fucoxanthin instead of peridinin and have both types of chl *c* (Riley & Wilson 1967; Mandelli 1968). However, the photosynthetic pigmentation of these algae represents an endosymbiont of possible chrysophyte origin and not the apochlorotic dinoflagellate host (Tomas & Cox 1973; Withers & Haxo 1975). But there are two *Woloszynskia* species and an *Exuviaella* species which contain fucoxanthin instead of peridinin, but where the presence of an endosymbiont has not been established (Whittle & Casselton 1968; Jeffrey *et al.* 1975). The dinoflagellates, *Prorocentrum (Exuviaella) cassubicum*, and *Gambierdiscus toxicus* which contains peridinin but has both chl c_1 and c_2 in its chloroplasts (Jeffrey *et al.* 1975; Durand & Berkaloff 1985).

Peridinin is structurally related to fucoxanthin and, together, they differ

Table 5.1. Major light-harvesting pigments in algae (Jeffrey 1980)

Taxon	Common name	Chlorophylls a	b	c_1	c_2	Bili-proteins	Carotenoids
Prokaryotes							
Cyanophyta	Blue-green algae (cyanobacteria)	+				+	
Prochlorophyta	Prochloron	+	+				
Eukaryotes							
Rhodophyta	Red algae	+				+	
Cryptophyta	Cryptomonads	+			+	+	
Brown algal line							
Dinophyta	Dinoflagellates	+		‡	+		Peridinin*
Chrysophyta							
Chrysophyceae	Chrysomonads, Silicoflagellates	+		+	+		Fucoxanthin
Rhaphidophyceae	Chloromonads	+		+	+		Fucoxanthin‡
Prymnesiophyta	Golden-brown flagellates (with haptonema), coccolithophorids	+		+	+		19′ hexanoyloxy-fucoxanthin
Eustigmatophyta	Eustigmatophytes	+					
Bacillariophyta	Diatoms	+		+	+		Fucoxanthin
Phaeophyta	Brown algae	+		+	+		Fucoxanthin
Xanthophyta	Yellow-brown algae	+		tr	tr†		
Green algal line							
Euglenophyta	Euglenoids	+	+				
Chlorophyta							
Chlorophyceae	Green algae	+	+				Siphonaxanthin‡
Prasinophyceae	'Scaly' green flagellates	+	+				Siphonaxanthin‡
Charophyceae	Stoneworts	+	+				

*Dinoflagellates with endosymbionts contain fucoxanthin and chlorophylls c_1 and c_2.
†Trace only.
‡Some species only.

from most other carotenoids because of their high oxygen content (Fig. 5.1). But where fucoxanthin is a bi-cyclic carotenoid, peridinin is the only carotenoid known to contain three closed cycles in its structure. Peridinin has a molecular weight of 630 daltons, having two less in-chain carbons and two less branching methyl groups than found in the usual carotenoid skeleton (Strain *et al.* 1971a). These structural modifications may be important to the strong blue-light absorbing characteristics of peridinin (λ max of 475 nm in ethanol), its abilities to attach to membrane proteins, and/or the very high efficiency with which light absorbed by peridinin is transferred to chl a within the photosynthetic membranes of dinoflagellates (cf. Prézelin & Boczar 1986).

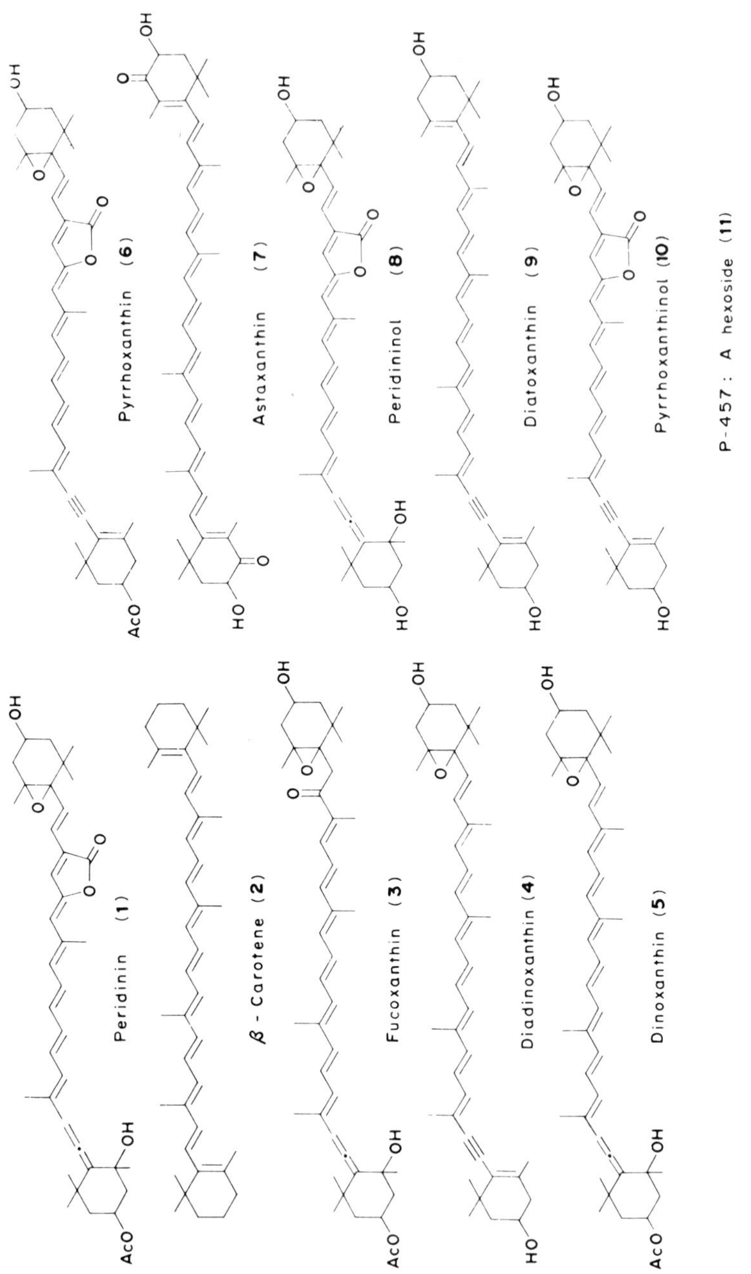

Fig. 5.1. Dinoflagellate carotenoids (from Johansen *et al.* 1974).

Dinoflagellates also contain small amounts of the yellow xanthophylls diadinoxanthin and dinoxanthin, as well as trace amounts of other carotenoids (Johansen *et al.* 1974; Jeffrey *et al.* 1975) (Fig. 5.1, Table 5.2). Unlike peridinin,

Table 5.2. Carotenoid components in order of polarity of six dinoflagellates (Johansen *et al.* 1974)

Order: Peridiniales Family: Gymnodiniaceae Species:												Glenodiniaceae	
		Amphidinium carterae		*Amphidinium carterae* Plymouth 450		*Gymnodinium nelsoni*		*Gymnodinium splendens*		*Gyrodinium dorsum*		*Glenodinium* sp.	
Pigment	$E^{1\%}_{1\,cm}$ used in acetone	(mg)	(% of total)	(mg)	(% of total)	(mg)	(% of total)	(mg)	(% of total)	(mg)	(% of total)	(mg)	(% of total)
β-Carotene	2500	0·30	0·4	0·14	9·0	0·06	3·7	0·12	0·9	0·08	0·4	0·20	0·4
Unknown	2500	0·18	0·2	0·04	0·3	0·01	0·6	0·03	0·2	0·40	1·9	0·14	0·3
Diatoxanthin								—	—				
Pyrrhoxanthin	2100	3·28	4·6	1·09§	7·3	0·05‖	3·1	—	—	2·02	9·4	2·25	1·9
Dinoxanthin				0·08	0·5			0·51	4·0				
Astaxanthin	2500	—	—	—	—	—	—	—	—	—	—	0·15	0·3
Diadinoxanthin	2250†	18·42	25·8	0·56	3·8	0·22	13·4	2·87	22·2	4·87	22·6	4·34	9·4
Peridinin	1350*	48·50	67·9	12·75	85·5	1·30	79·2	9·18	70·9	13·67	63·6	38·13	82·9
Pyrrhoxanthinol	1350	—	—	—	—	—	—	—	—	0·10	0·5	—	—
Peridininol	1350	0·09	0·1	0·25	1·7	tr.	—	0·12	0·9	0·12	0·6	0·28	0·6
P-457	2500	0·70	1·0	—	—	—	—	0·11	0·9	0·21	1·0	0·53	1·2
Total (mg)		71·47		14·91		1·64		12·94		21·47		46·02	
Total sample dry weight (g)		19·90		25·00¶		—		12·65		12·67		11·70	
Dry weight (mg/g)		3·59		2·91		—		1·02		1·70		3·93	

* Reference 6.
† Present study.
‡ The $E^{1\%}_{1\,cm}$ of dinoxanthin (= neoxanthin-3-acetate) is used.
§ Includes some diadinochrome (rearrange diadinoxanthin).
‖ Presence of diatoxanthin and pyrrhoxanthin are uncertain.
¶ Only $\frac{1}{5}$ of the extract was investigated.
tr. = trace.

these carotenoids are not major light-gathering pigments for photosynthesis. Some may be involved in stabilizing photochemical environments within the thylakoid, or they may be important to photoprotective and phototactic responses occurring inside and outside the chloroplasts (cf. Halldal 1970; Smith 1977). The minor carotenoids have also been used as biochemical determinants both of taxonomic affinities and for the presence of specific biosynthetic pathways (Isler 1971).

All chlorophylls contain a planar porphyrin ring structure with magnesium chelated in the centre and liganded at four sites to pyrole nitrogen atoms (Fig. 5.2). Differences in absorption properties between the different chlorophyll types arise from small changes in the side groups attached to the porphyrin ring (Figs 5.2, 5.3). For instance, chl c_1 and c_2 differ from each other only in whether they respectively have an ethyl or vinyl group at the R2 position on ring II (Doughtery *et al.* 1966; Strain *et al.* 1971b). This minor change gives rise to subtle absorption differences, i.e. chl c_1 absorption maxima are at about 445, 579, and 628 nm in ether with band ratios of 10·1:0·68:1·00, respectively, whereas absorption maxima of chl c_2 are at about 449, 582, and 629 nm in ether with band ratios of 14·5:1·13:1·00, respectively (Jeffrey 1969) (Fig. 5.3). Both chlorophyll c_1 and c_2 differ from chl *a* and chl *b* in that they have no phytol chain attached to their ring structure, but instead have an acrylic side group replacing the propionic acid group in ring IV. (Phytol is a hydrophobic terpenoid ($COO_{20}H_{39}$) attached via an ester linkage to the side group of ring IV.) As a result, chl *c* is smaller than chl *a* (610 vs. 893 daltons) and is much more hydrophilic.

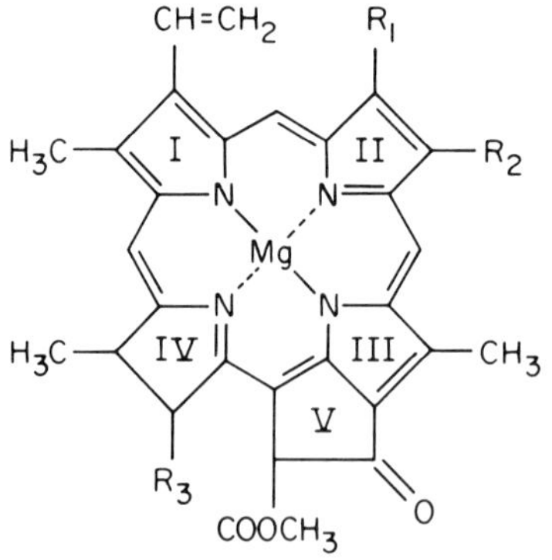

	R_1	R_2	R_3
Chl *a*	CH_3	CH_2CH_3	$(H_2C)_2$ — PHYTOL
Chl *b*	CHO	CH_2CH_3	$(H_2C)_2$ — PHYTOL
Chl c_1	CH_3	CH_2CH_3	$HC{=}CH_2$ — COOH
Chl c_2	CH_3	$CH{=}CH_2$	$HC{=}CH_2$ — COOH

Fig. 5.2. Structure of chlorophylls (redrawn from Seely 1966; Dougherty *et al.* 1966).

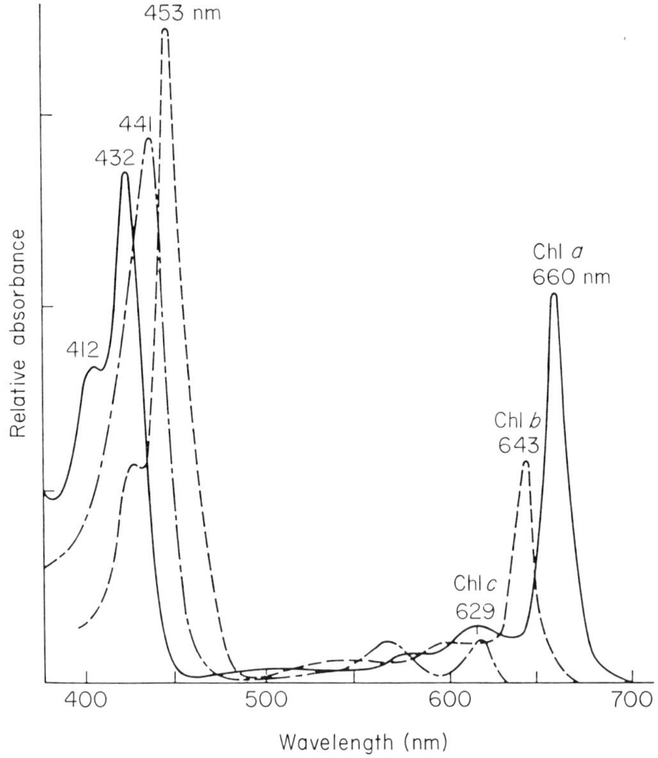

Fig. 5.3. Absorption spectra of chlorophylls extracted in acetone (redrawn and adapted from Hall & Rao 1977).

Proteins attach to chlorophylls through side groups of the porphyrin ring, thereby further modifying the absorption properties of the chlorophyll within chl–protein complexes. All photosynthetic pigments are protein-bound before becoming functional as light-harvesting components for photosynthesis. More than one kind of pigment can attach to the same protein, and different pigment–protein complexes can combine in aggregate structures in the thylakoid membranes. In such supramolecular aggregates light energy is both absorbed and passed on from one pigment type to another until it reaches the photochemical reaction centres buried deep in the thylakoid membrane. The photochemical reaction centres consist almost entirely of chl *a*–protein complexes, with all other accessory pigments situated in more peripheral light-harvesting components (Thornber *et al.* 1977). The structure/function of the two photochemical reaction centres, photosystem I and II, are believed to be conserved in all chl *a*-containing plants. However, the nature of the light-harvesting components differs with each major plant group, i.e. biliproteins in red algae and chl *a*–chl *b*–proteins in green algae and higher plants (cf. Thornber *et al.* 1977; Prézelin 1981; Prézelin & Boczar 1986).

Extensive investigation of the organization of chlorophylls and carotenoids of dinoflagellates has led to the characterization of distinct chl *a*-containing protein complexes (cf. Prézelin & Boczar 1981 (Boczar 1985; Boczar & Prézelin 1986)). The first is a brick-red, water-soluble peridinin–chl *a*–protein complex (PCP) which can be extracted from ruptured cells of most photosynthetic dinoflagellates (Haidak *et al.* 1966; Haxo *et al.* 1972; Prézelin & Haxo 1976; Haxo *et al.* 1976; Siegelman *et al.* 1977; Meeson 1981; Chang & Trench 1982) (Table 5.3). The percentage of total peridinin extracted as PCP varies considerably among different dinoflagellate species and appears related to the ease with which cells are ruptured. For instance, when two similarly pigmented strains of *Amphidinium carterae* are compared, large amounts of PCP can be isolated and characterized from the Plymouth 450 strain (Haxo *et al.* 1976) while the SIO-PY-1 strain will release little or no PCP (Prézelin & Haxo 1976). In easily ruptured dinoflagellates such as *Gonyaulax polyedra* and *Ceratium furca*, 80–95% of the total cellular peridinin is released as PCP (Prézelin & Sweeney 1978a; Meeson 1981).

The molecular topology of PCP has been characterized in detail for '*Glenodinium*' (see taxonomic appendix) sp. and *G. polyedra*, and to a lesser extent in *Amphidinium rhyncocephalum* and *A. carterae* (Song *et al.* 1976; Koka & Song 1977). With the first three species, the PCP chromophores contain peridinin with chl *a* in a 4 : 1 molar ratio and are devoid of all other photosynthetic pigments found in the chloroplast, notably chl *c*. A single chromophore is attached non-covalently to either one or two apoproteins, giving a total complex molecular weight of around 35 000 daltons. Depending upon the species, the PCP apoproteins are either single polypeptides with a molecular weight 30–35 KD or two polypeptides with molecular weights between 15–17 KD. The first 24 amino acids from the amino terminal of PCP apoproteins from *Glenodinium* sp. and *A. carterae* Plymouth 450 are identical (Fenna *et al.* in prep.). The amino acid sequence recently has been used to generate a PCP gene(s) probe to study the genetic base for photoregulation of PCP turnover in dinoflagellates (Prézelin & Triplett in prep.). The PCP chromophore is buried in a hydrophobic crevice of the protein, with externally exposed hydrophilic amino acids presumably conferring the water-soluble characteristic of the total complex (Song *et al.* 1976; Prézelin & Haxo 1976; Haxo *et al.* 1976; Siegelman *et al.* 1977).

The four peridinin molecules within the chromophore of *Glenodinium* sp., *G. polyedra* and *A. rhyncocephalum* appear to occur as two dimers around a chl *a* monomer (Fig. 5.4). In this configuration, the fluorescence lifetime of the peridinin is significantly lengthened and presumably accounts for the 100% efficient energy transfer from peridinin to chl *a* within this light-harvesting component (Prézelin & Haxo 1976) (Fig. 5.5) within 10 psec of excitation (Song *et al.* 1980). In addition to functioning as a light-gathering pigment for photosynthesis, peridinin in PCP also is capable of protecting the chl *a* monomer from photodestruction (Koka & Song 1977).

By contrast, *A. carterae* PCP chromophore is reported to be comprised of nine molecules of peridinin in association with two molecules of chl *a* attached to a single apoprotein (MW 31 800) (Haxo *et al.* 1976; Siegelman *et al.* 1977). This 9:2:1 ratio is not consistent with subsequent circular dichroism analyses of purified PCP from the same species, which suggested only one chromophore containing monomeric chl *a* and 4 peridinins of equal symmetry could be accommodated per apoprotein (Koka & Song 1977). The small size of the chromophore and the inherent variability in the analytical methods employed are technical factors which complicate determinations of percent chromophore composition of PCP complexes. Further studies are needed to resolve these differences.

The PCP fraction isolated from dinoflagellates can be resolved into several conformers, which are distinguishable from each other only on the basis of isoelectric point and small differences in the amino acid composition of the apoproteins (Haxo *et al.* 1976; Prézelin & Haxo 1976; Siegelman *et al.* 1977; Chang & Trench 1982). The PCP apoproteins are generally characterized by isoelectric points above pH7 and a larger percentage (45%) of polar amino acids (Prézelin & Haxo 1976; Haxo *et al.* 1976) than reported for membrane-bound chl *a*-protein complexes (cf. Thornber 1975). This may account for the ease with which PCP can be extracted from the chloroplast and implies a more peripheral placement on the photosynthetic membrane.

The isoelectrofocusing conformer (IEC) pattern appears to be species-specific, involving anywhere from four to twelve PCP conformers per species (Siegelman *et al.* 1977). Such observations have been used to distinguish

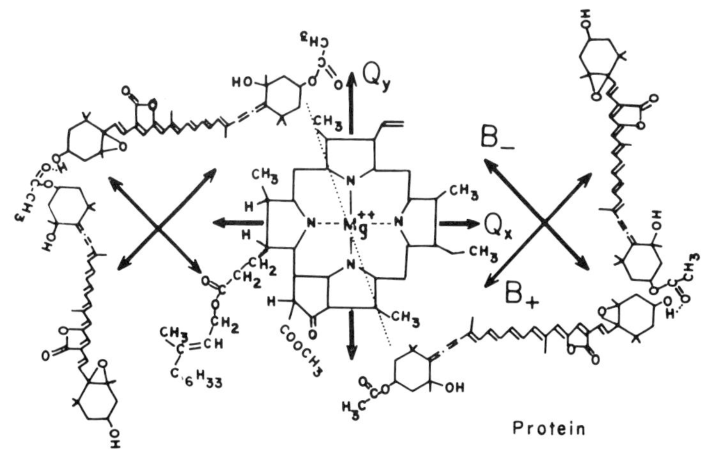

Fig. 5.4. A probable molecular arrangement of chlorophyll *a* and peridinins in the peridinin–chlorophyll *a*–protein complexes of dinoflagellates. Arrangement is based on relative orientations of transition moments (double arrows) of Q_y fluorescence and B^+ excitation transitions. (By permission, Song *et al.* 1976.)

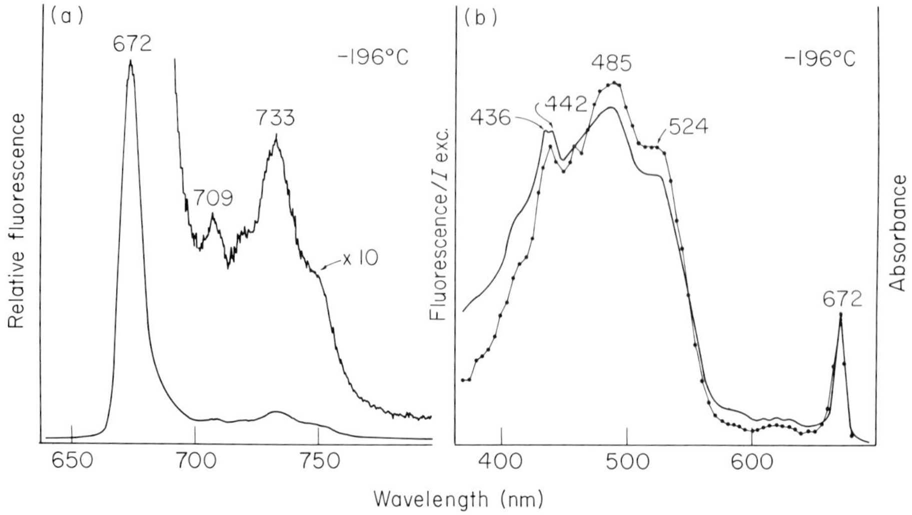

Fig. 5.5. *Glenodinium* sp. (a) Low temperature fluorescence emission spectra of purified pI 7·3 PCP. Excitation wavelength at 635 nm. Concentration of chlorophyll *a* was 1·5 μg ml^{-1}. (b) Low temperature absorption and fluorescence excitation (●) spectra of purified pI 7·3 PCP. Emission wavelength at 733 nm. Concentration of 0·03 μg ml^{-1} chlorophyll *a* for fluorescence excitation and 1·5 μg ml^{-1} chlorophyll *a* for absorption measurements. (By permission, Prezelin & Haxo 1976.)

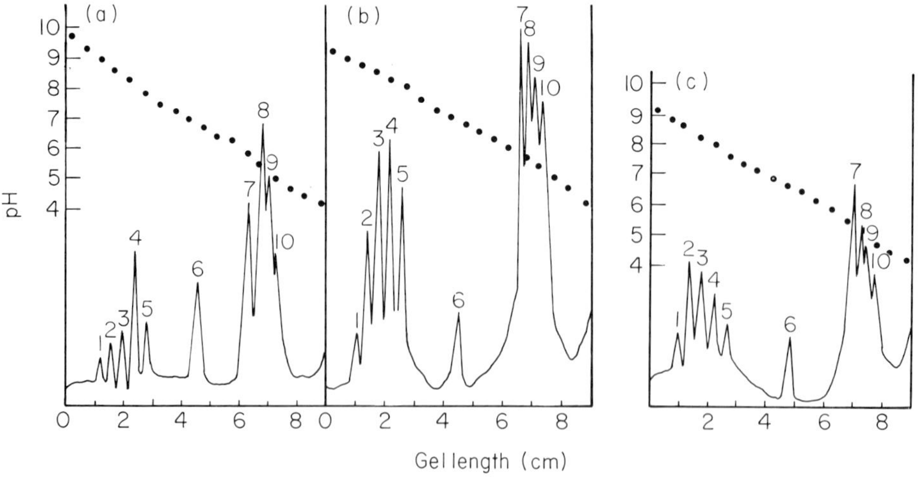

Fig. 5.6. *Symbiodinium* (= *Gymnodinium*) *microadriacticum Freudenthal*. Polyacrylamide gel isoelectric conformer patterns of peridinin–chlorophyll *a*–protein complexes: (a) freshly isolated from host; (b) from cultured isolates grown at 35 μE m^{-2} s^{-1}; (c) from cloned population of cultured isolates. (By permission, Chang *et al.* 1984.)

different algal populations of zooxanthellae isolated from distinct hosts. Several isolates from different hosts display unique PCP conformer patterns. These conformer patterns generally are stable when zooxanthellae freshly isolated from the host, cultured and subsequently cloned are compared (Fig. 5.6). With isolates from *Tridacna maxima*, the conformer IEC pattern does not alter when growth illumination for the algae is lowered (Chang & Trench 1982).

The second major light-harvesting complex to be isolated from dinoflagellates is an apple-green coloured chromoprotein, containing almost all the chl *c* in the cell and up to 30% of the total chl *a* (Boczar *et al.* 1980; Boczar 1985; Boczar & Prézelin 1986). This chl *c* can be released from the thylakoid membranes and kept intact only after gentle SDS detergent solubilization and combined SDS/Deriphat acrylamide gel electrophoresis. The band patterns are shown in Fig. 5.7, and the chl *a*–chl *c*–protein complex is localized in band II. The complex has an apparent molecular weight equivalence of 71 000 daltons, containing one

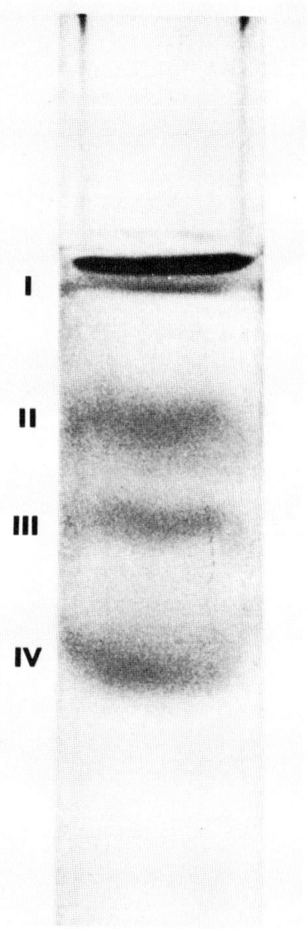

Fig. 5.7. *Glenodinium* sp. Deriphat electrophoretic pattern of SDS-solubilized thylakoid membranes. (By permission, Boczar *et al.* 1980.)

major (19 KD) and three minor (22, 24 and 37 KD) apoproteins (Boczar & Prézelin 1986). The chromophore is heavily enriched in chl *c*, with a chl *c*:chl *a* molar ratio of 4·2:1. Whole cell chl *c*:chl *a* ratios are about 1:1. The spectral characteristics of the complex are shown in Fig. 5.8, illustrating how the predominance of chl *c* (A453) contributes to the blue light-absorbing capabilities of dinoflagellates. Significant excitation energy absorbed by chl c_2 is transferred to chl *a* within the chromophore, which has a room temperature fluorescence emission maximum at 675 nm (Boczar & Prézelin 1986). The optical and

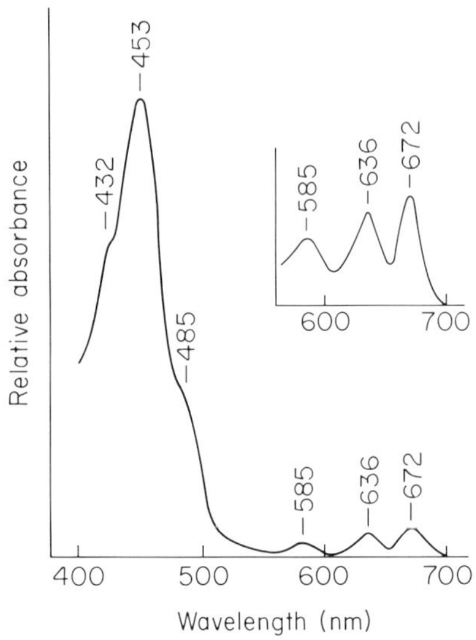

Fig. 5.8. *Glenodinium* sp. Room temperature absorption spectrum of the chlorophyll *a*–chlorophyll *c*–protein complex eluted from Band II of SDS/Deriphat gel system. (By permission, Boczar *et al.* 1980.)

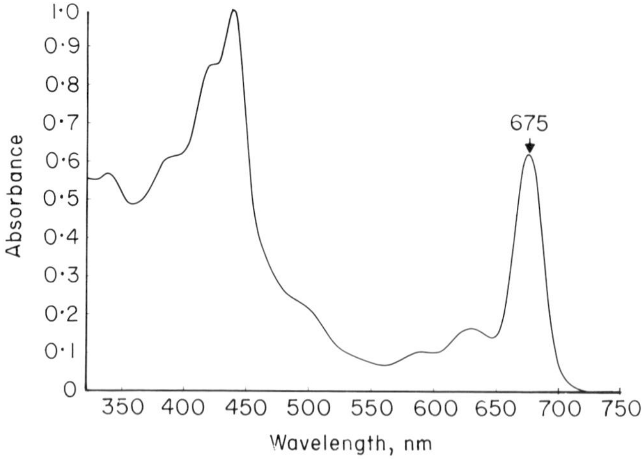

Fig. 5.9. *Glenodinium* sp. Room temperature absorption spectrum of the P_{700}–chlorophyll *a*–protein complex isolated on an SDS gel system. (By permission, Prézelin & Alberte 1978.)

biochemical characteristics of the light-harvesting chl c_2-containing component of dinoflagellates closely resemble those of chl c_2 light-harvesting complex found in cryptophytes (Ingraham & Hiller 1983).

Third, a P_{700}–chl *a*–protein complex representing the phototrap of photosystem I (Ps I) has been isolated from *Glenodinium* sp. and *G. polyedra* by SDS gel electrophoresis (Prézelin & Alberte 1978) (Fig. 5.9). It is identical in composition and photobleaching characteristics to similarly isolated reaction centre I chl *a*–protein complexes, now assumed to be ubiquitous throughout the plant kingdom (Brown *et al.* 1974). The P_{700} complex has a characteristic red absorption maximum around 675 nm, due to the presence of 20–40 light-harvesting chl *a* molecules specifically associated with the Ps I reaction centre. At the heart of the complex is P_{700}, the chl dimer whose photobleaching signal around 700 nm can be resolved and quantitated to provide an estimate of the number of Ps I reaction centres present in a sample. There is also a small amount of β-carotene present, contributing to the absorption shoulder around 480 nm (Thornber *et al.* 1977). Band I off the SDS/Deriphat gel system (Fig. 5.7) is spectrally equivalent to the P_{700} complex, although Band I contains more beta-carotene (A485). Minor amounts of chl c_2 are variably associated with Band I, dependent upon growth illumination or ionic strength of preparation buffers (Boczar & Prézelin 1986). This is consistent with analogous studies in higher plants, where minor amounts of chl *b* can be integrally associated with Ps I reaction centres (Argyroudi-Akoyunoglou & Thomou 1981; Anderson *et al.* 1983).

Based on the described distribution of chl *a* in discrete chromoproteins and a study of the fluorescence characteristics of whole cells (Govindjee *et al.* 1979), a working model for the organization of these components into a functional photosynthetic unit (PSU) has been proposed (Prézelin & Alberte 1978; Govindjee *et al.* 1979) (Fig. 5.10). Overall organization is identical to the model proposed by Thornber *et al.* (1977) to describe the distribution of chl in several other plant groups. The model has a chl *a* core comprised of Ps I (represented by the small interior square) and Ps II (represented by the small interior hexagon) coupled to receive light energy via an association with light-harvesting pigment–protein complexes (LHCs). The LHC is divided structurally into two or more components. A matrix of chl *a*–protein complexes (LH chl *a*, the rectangle) represents integral thylakoid components linking Ps I and Ps II. This component is assigned functions both in light harvesting and the distribution of absorbed light energy from extrinsic LHCs to and between intrinsic photosystems (i.e. energy transduction and spillover). The detergent-soluble chl *a*–chl *c*–protein is placed near the chl *a* core, to the interior side of the water-soluble PCP complexes. Together, these two LHCs account for the broad band light absorption by dinoflagellates and appear to be the functional analogues to the chl *a*/*b*–protein complexes of green plants and the phycobiliproteins of red and blue-green algae. The precise relationship between the LHCs of dinoflagellates

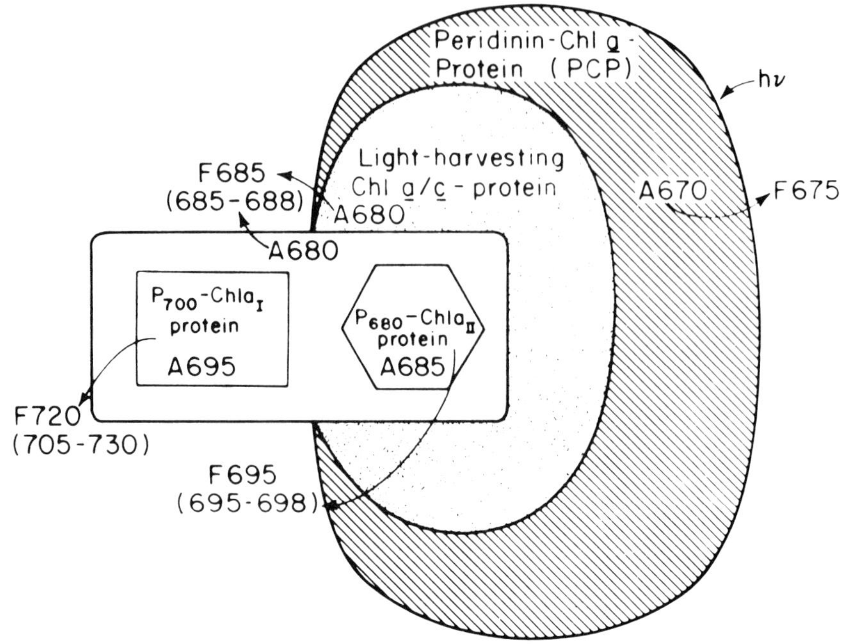

Fig. 5.10. *Gonyaulax polyedra*. A working model for the possible assignment of absorption (A) and emission (F) maxima, in nm, to the various chlorophyll *a* complexes in dinoflagellates. (By permission, Govindjee *et al*. 1979.)

and the chl *a*/*c*/fucoxanthin–protein particles isolated from brown algae is not yet clear (Barrett & Anderson 1977; 1980).

1.2 Chloroplast ultrastructure

The photosynthetic membranes of dinoflagellates occur as sets of closely appressed lamellae (see Chapter 3). They can be fused to the chloroplast envelope and can be differentially arranged within the chloroplast, depending upon the species (Dodge 1973, 1975, Chapter 3; Kirk & Tilney-Basset 1978). Within radially orientated chloroplasts, the lamellae lie parallel to each other along the long axis of the organelle and there are no connections between appressed membranes. In some species, the chloroplasts are arranged around the outline of the cell and are collectively termed the peripheral chloroplast reticulum. The lamellae of these chloroplasts tend to lie parallel to the cell boundary (Dodge 1975). The impact that these and other genetic variations in chloroplast form and lamellar arrangements might have on the physiological plasticity of photosynthesis in dinoflagellates is unknown.

Very little detailed information is available on the molecular substructure of thylakoids in dinoflagellates. The recent freeze-fracture study of *G. polyedra* does suggest that the membrane particle size-classes and density are consistent with those described for photosynthetic components of other algal and higher plant groups (Sweeney 1981a) (Fig. 5.11). The only exception is the absence of the well characterized 160 Å particle seen in higher plants and green algae, indicative of the LHC chl *a*/*b*–protein complex found in their pigment system. As of yet, it has not been possible to identify the physical location of the PCP and chl *a*–chl *c*-protein complexes predicted to be associated with the outer surfaces of the thylakoid membranes.

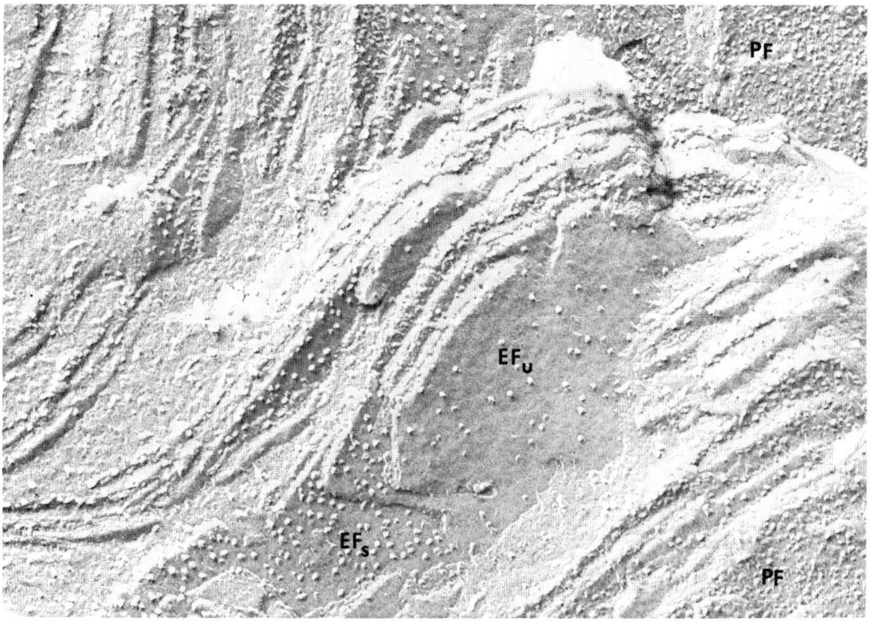

Fig. 5.11. *Gonyaulax polyedra.* Thylakoid membrane faces of the chloroplast exposed by freeze-fracture. All four faces of the thylakoid are exposed. × 84 000. (Courtesy of Sweeney 1982.)

2 REGULATION OF PHOTOSYNTHESIS BY ENVIRONMENTAL VARIABLES

As knowledge of the organization and function of photosynthetic components increased rapidly during the last decade (cf. Prézelin 1982), a number of endogenously-regulated and environmentally-induced photosynthetic responses in dinoflagellates were distinguished and described by simple mechanistic schemes. What follows here is a review of some of the major biochemical adaptations believed to underlie the photosynthetic responses of dinoflagellate to environmental and endogenous changes. There is an extensive discussion of the modifying effects of light intensity on photosynthesis and growth because it is presently an especially active area of dinoflagellate research. However, the single and combined effects of many other important variables still remain to be identified. As more sites of photosynthetic regulation are characterized and the environmental/endogenous variables acting at those sites are identified, our abilities to predict the regulating effects of multivariable environmental flux on photosynthesis, metabolism and growth at the whole cell level should improve greatly.

2.1 Light intensity

Numerous reports document that daily light intensities influence the organization and function of the photosynthetic apparatus of dinoflagellates and thereby often provide for their favourable accommodation to changing light fields (cf. Prézelin 1981; Richardson *et al.* 1983). All these studies indicate that increased cell pigmentation accompanies growth at low light levels (Fig. 5.12), i.e. the *Chlorella* type of light adaptation first described by Steeman-Nielsen & Jorgensen (1962, 1968a, b). Now it appears that increased pigmentation is achieved by one of two distinct photoadaptive strategies expressed by the 12 or more different species of dinoflagellates now characterized. An exception is *Prorocentrum marie-lebouriae*, whose pigment-dependent responses were described as a combination of both photo-adaptive strategies. However, the result in both strategies is to enhance photosynthesis and sometimes even maintain maximal cell division rates in low light environments (cf. Prézelin 1981; Richardson *et al.* 1983).

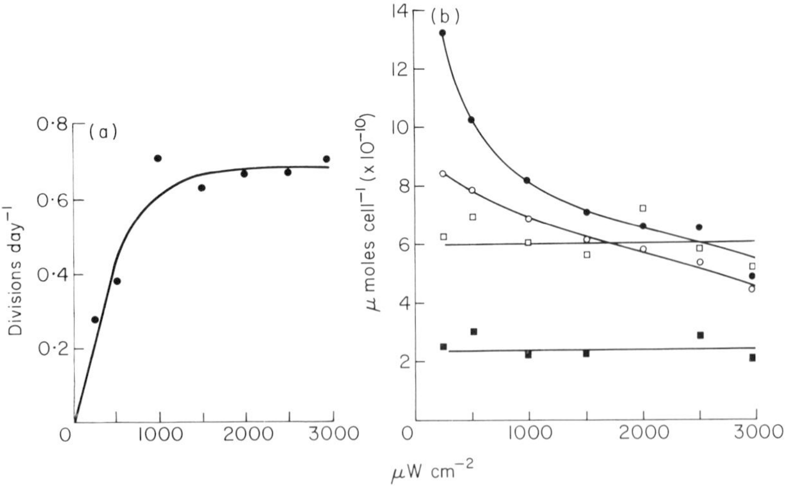

Fig. 5.12. *Glenodinium* sp. Effect of continuous light growth irradiances on (a) cell division rates and (b) pigment content. Concentrations of chlorophyll *a* (○), peridinin (●), chlorophyll *c* (□) are expressed as $\mu M\ cell^{-1}$ (10^{-1}). (Prézelin 1976.)

In either strategy, increased pigmentation characterizes growth at irradiance levels below 300 and down to about $15\,\mu E\,m^{-2}\,s^{-1}$ for cultured dinoflagellates grown on light:dark (LD) cycles. At the upper light limit for pigment change, both photosynthesis and cell division rates are light saturated. As growth illumination is lowered to about $100\,\mu E\,m^{-2}\,s^{-1}$, photosynthesis becomes increasingly light-limited while division rates remain high. Maximum rates of cell division are maintained down to light levels able to support photosynthetic rates equal to about one-third light-saturated rates of photosynthesis. It is over this particular light range that the mechanisms of photoadaptation are best elucidated. At light levels below $100\,\mu E\,m^{-2}\,s^{-1}$, cells are increasingly photo-stressed, division rates slow, and photoadaptive responses may or may not continue to be induced. Even when the photoadaptive response is still evident at the molecular level, the breakdown of other cell processes (i.e. light-dependent enzyme activities) can complicate interpretation of photosynthetic physiological responses measured in the whole cell.

(a) PSU SIZE

One photoadaptive response is to accumulate light-harvesting pigment–protein complexes (LHC) under low light conditions, thereby increasing the light-gathering capabilities of the cell (Fig. 5.13). Referred to as an increase in relative photosynthetic unit (PSU) size or optical cross-section for Photosynthesis, the LHC content of the cell increases relative to a fixed number of photosynthetic reaction centres when light levels fall and decreases again when light levels rise. An increase in PSU size is not meant to imply that the optical cross-section for each photosystem is identical, but rather that the *mean* optical cross-section for a summed population of photosystems has been enhanced. Often the increase in PSU size is characterized by a change in pigment molar ratios, because the increase in LHC can be in just one component, such as PCP in *Glenodinium* sp. (Fig. 5.12b; Prézelin 1976). If there is a coordinated increase in both LHC components, such as in the PCP and chl *a*–chl *c*–protein content of low-light *G. polyedra*, then cellular increases in pigmentation are inversely dependent on growth irradiance but occur without major changes in pigment molar ratios (Prézelin & Sweeney 1978a).

The increase in PSU size can be verified if an increase in the chl a/P_{700} ratio also is measured. This approach assumes the ratio of Ps I to II reaction centres is 1 : 1, which is not always true (cf. Prézelin 1981; Prézelin & Boczar 1986). The P_{700} technique has not been used widely with dinoflagellates, perhaps because relatively large samples are required and because extraction procedures require some modification if photochemical signals are to be reproduced accurately (cf. Prézelin 1981; Prézelin & Boczar 1986). However, low light ($100\,\mu E\,m^{-2}\,s^{-1}$) laboratory populations of *G. polyedra* and *Glenodinium* sp. have relatively large PSU sizes, i.e. around 600 chl a/P_{700} (Prézelin & Alberte 1978) as compared to

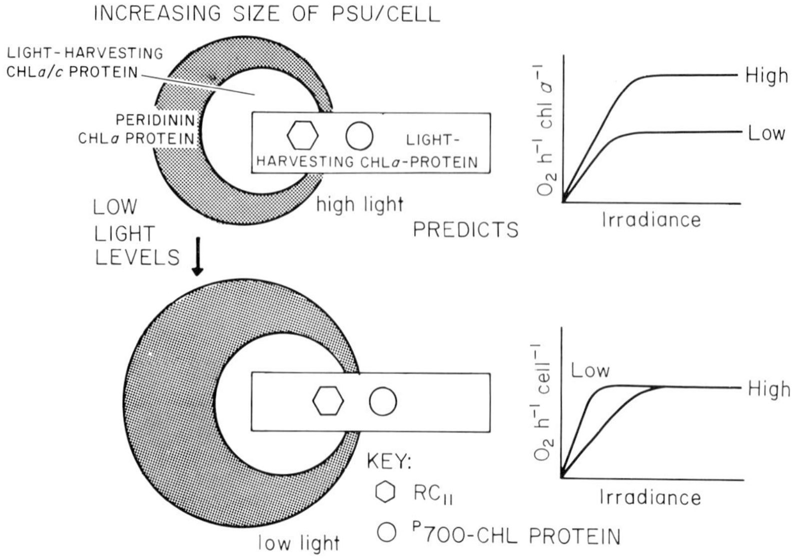

Fig. 5.13. Photoadaptation of photosynthesis by increasing the size of the light-harvesting component of the photosynthetic unit of dinoflagellates. (Prézelin & Sweeney 1978b.)

the typical green plant PSU size of about 300 chl a/P_{700} (Thornber *et al.* 1977). Perry *et al.* (1981) measured a smaller PSU size of 445 chl a/P_{700} for *G. polyedra* when grown under brighter light conditions (i.e. $300\,\mu E\,m^{-2}\,s^{-1}$), consistent with the photoadaptive strategies assigned to these dinoflagellates (Prézelin 1976; Prézelin & Sweeney 1978a). *Gymnodinium sanguineum* (= *splendens*), sampled from a red tide with midday light intensities within the bloom around $150\,\mu E\,m^{-2}\,s^{-1}$, had chl a/P_{700} values between 400 and 600:1 (unpublished data). Extremely high chl a/P_{700} ratios have been measured for coral zooxanthellae, i.e. shade populations had chl a/P_{700} ratios up to 1800:1 while sun populations of the same species had ratios four times smaller (Falkowski & Dubrinsky 1981). These very large chl a/P_{700} ratios have been also measured in other coral zooxanthellae, using the Ps II oxygen yield measurements instead of P_{700} (Zvalinskii *et al.* 1980). The large size of the light-harvesting component is exaggerated further when the accessory pigments are considered. In dinoflagellates the chl *c*:*a* molar ratios can approach 2:1, while *Glenodinium* sp. cells in transition to low light have been observed to have peridinin:chl *a* ratios reaching as high as 6:1 (Jeffrey *et al.* 1975; Prézelin 1976; Prézelin & Matlick 1980).

The effect of altered PSU size on photosynthesis–irradiance (*P–I*) relationship changes are exemplified by Fig. 5.13. On a cellular basis, maximum rates of photosynthesis (P_{max}) remain the same when PSU size increases because the total number of reaction centres involved in photochemistry is unchanged.

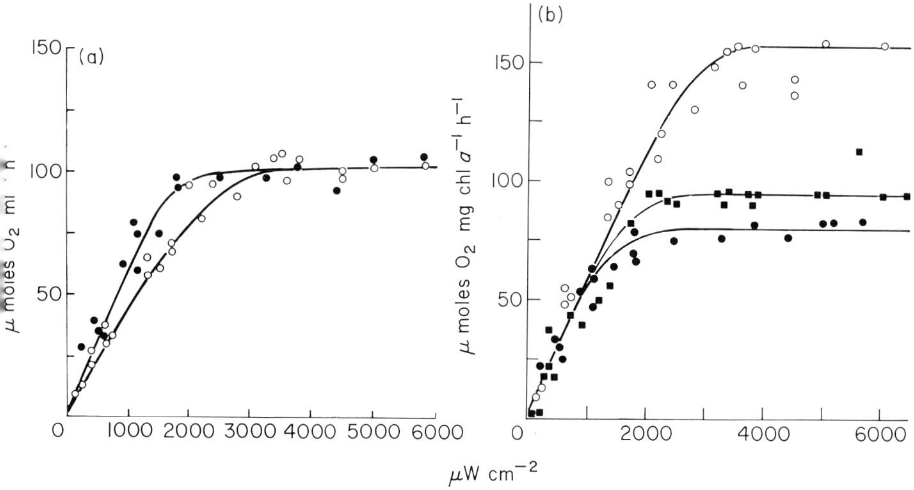

Fig. 5.14. *Glenodinium* sp. Photosynthesis–irradiance curves for continuous light cultures grown at 250 (●), 650 (■), and 2500 (○) μW cm^{-2}. Rates are expressed as (a) μM O_2 ml^{-1} h^{-1} and (b) μM O_2 mg chlorophyll a^{-1} h^{-1}. (Prézelin 1976.)

However, the irradiance required to saturate photosynthesis decreases in inverse proportion to the size of the PSU. Therefore, there is an increase in the light-limited slope, alpha, and a decrease in the light-saturation constant, $I = \frac{1}{2}P_{max}$, of the cellular *P–I* relationship (Fig. 5.13a). These changes cause the light-limited rates of photosynthesis to improve, presumably explaining why maximal growth rates are supported under some light-limited conditions and suggesting that suboptimal division rates might be lower still if these photoadaptive responses were inhibited. On a chl *a* basis, P_{max} declines in proportion to the increase in chl *a*. Therefore, the drop in assimilation number (P_{max}/chl *a*) characteristically associated with increasing PSU size does not reflect a true drop in photosynthetic performance, P_i.*

Light-induced changes in PSU size were first suggested in laboratory studies on *Glenodinium* sp. (Fig. 5.14a, b; Prézelin 1976), *Gonyaulax polyedra* (Prézelin & Sweeney 1978a) and most recently *Gyrodinium dorsum* (B. B. Prézelin, unpubl. data). *Glenodinium* sp. and *G. dorsum* have an increased pigment content and other related photoadaptive responses that continue well into the lowest light regime where cell division rates decline. The results with *G. polyedra* vary. We

* In Prézelin (1981), these photosynthesis–irradiance relationships were incorrectly redrawn from an earlier publication by Prézelin & Sweeney (1978b), i.e. the chl *a*-based light-limited rates of photosynthesis are predicted to fall in low light populations as the size of the light-harvesting antenna increases. However, the decline in chl *a*-based alpha is rarely seen in successfully photoadapting low light dinoflagellate populations, apparently due to accompanying compensations in the relative quantum yield of photosynthesis.)

found that long-term exposure to low light resulted in a progressive reduction in pigmentation and photosynthetic abilities in populations dividing at less than maximal rates (Prézelin & Sweeney 1978a). Meeson (1981), working with the same strain of dinoflagellate, and Rivkin *et al.* (1982b), working with a smaller-sized strain of *G. polyedra*, saw persistence of high pigmentation and photosynthetic abilities down to growth irradiances of about $25\,\mu E\,m^{-2}\,s^{-1}$. Differences between the studies may reflect difference in the length of light acclimation (i.e. light history) and/or in the nutrient status of the cultures. Employing single cell isolation techniques, Rivkin & Voytex (1985) have been able to document in the field that changes in photosynthetic characteristics of *Ceratium lineatum* and *Heterocapsa triguetra* isolated from high and low light environments are consistent with photoadaptive strategies based on increased PSU size.

Field studies suggest that increasing PSU size may operate as a successful photoadaptive response of zooxanthellae found in several species of corals (Titlyanov *et al.* 1980; Zvalinskii *et al.* 1980; Falkowski & Dubrinsky 1981), in cultured zooxanthellae isolated from the giant clam *Tridacna maxima* (Chang *et al.* 1984), and perhaps in *in situ* zooxanthellae of the anemone *Aiptasia pulchella* (Muller-Parker 1984a). Using oxygen flash yields to calculate PSU number and size in freshly isolated zooxanthellae from the coral *Pocillopora verrucosa*, individual zooxanthellae sampled from all light environments were shown to contain equal numbers of PSU per cell ($2{\cdot}7 \times 10^6$) (Zvalinskii *et al.* 1980). Pigmentation showed increases of 30% and near 80% respectively in chl *a* and peridinin when 15 m sun and shade populations were compared (Titlyanov *et al.* 1980). Also consistent with an increased PSU size, there was a proportional drop in assimilation number in shade populations, while no change in P_{max} per cell was observed and low light photosynthetic rates of the cells were significantly enhanced (Fig. 5.15). Very similar low light pigment increases and *P–I* changes were observed in *Porites nigrescens* zooxanthellae. Furthermore, there was no increase in chl *c* in either species, which is reminiscent of the photoadaptive responses of *Glenodinium* sp., and suggestive that major increases in PCP alone may account for the increases in PSU size. While less data are available, it appears that low light pigment and photosynthetic responses of the corals *Echinopora lamellosa* and *Acropora hyacinthus* also are consistent with enhanced PSU size, accompanied by an increase in both light-harvesting components.

Falkowski & Dubrinsky (1981) examined intact zooxanthellae in the coral *Stylophora pistillata* and reported P_{700} content to be the same in both light and shade samples, while the chl a/P_{700} ratio increased fourfold in shade populations. At first glance, the *P–I* relationship changes appear quite different from those predicted for a photoadaptive response based solely on an increase in PSU size. Most notably, P_{max} per cm surface area coral is much higher in low light cells while P_{max} per chl *a* is about the same (Fig. 5.16). However, if the *P–I* curves are corrected to approximate gross production by compensating for animal and

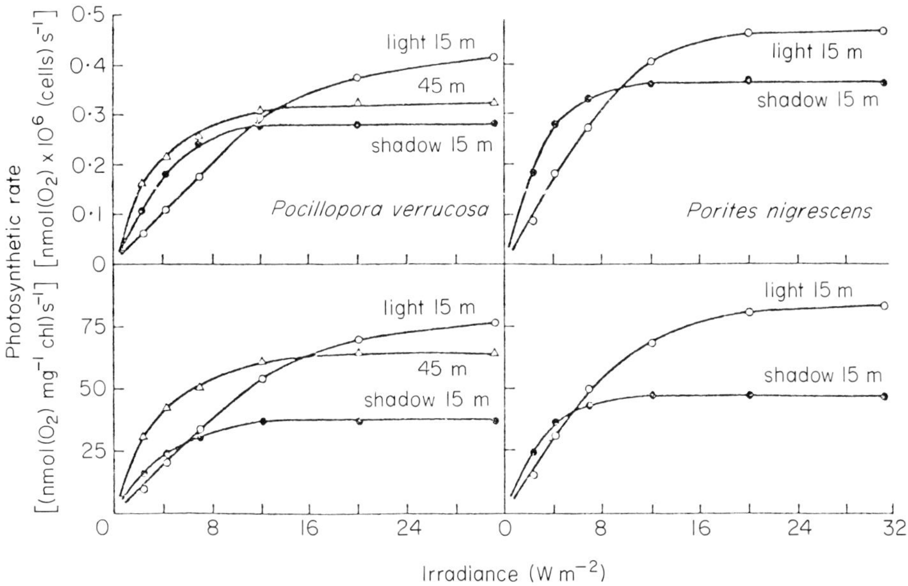

Fig. 5.15. Dependence of the rate of steady-state photosynthesis on irradiance in zooxanthellae from corals *Pocillopora verrucosa* and *Porites nigrescens* adapted to different irradiances. (Zvalinskii *et al.* 1980.)

plant respiration, assumed to be unaffected by brief exposure to various irradiance levels, then the differences in *P–I* relationship would more closely resemble those reported for other dinoflagellates which photoadapt by increasing PSU size under light-limited conditions.

The problems in interpreting photosynthesis data derived from intact symbiotic associations are discussed in detail by Trench (this volume) and a relevant example is found in studies of the zooxanthellae found in the anemone *Aiptasia pulchella*. Muller-Parker (1984a) compared the photosynthetic characteristics of zooxanthellae in their intact association with *Aiptasia pulchella*, sampled from a shaded mangrove lagoon and a sunlit reef flat. Lagoon anemones were chocolate-brown in colour whereas reef anemones were golden-brown, even though both contained equal numbers of zooxanthellae per mg anemone animal protein. Zooxanthellae removed from lagoon anemones contained 1·75 times more chl *a* per cell than did zooxanthellae from reef anemones (i.e. 2·97 pg chl *a* cell^{-1} vs. 1·70 pg chl *a* cell^{-1}). When photosynthesis as a function of irradiance was measured in anemones from both habitats, results differed according to whether they were based on weight of chl *a* or zooxanthellae cell number (Fig. 5.17). The curves most closely approximate a photoadaptive strategy based on increased PSU size. However, these results contrast to a laboratory study of cloned zooxanthellae from *A. pulchella*, where comparisons of cultures growing at different irradiances indicated a photoadaptive strategy

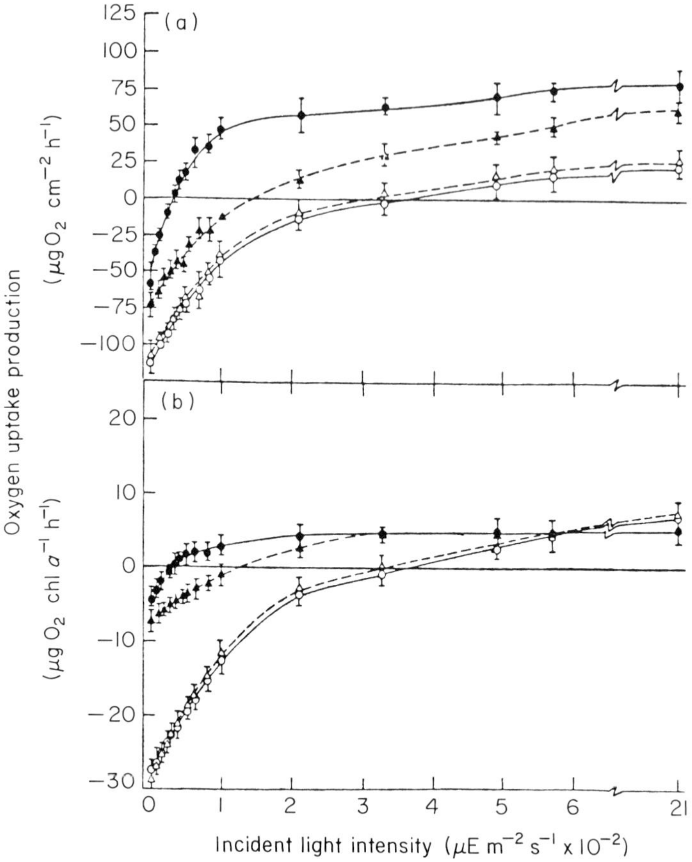

Fig. 5.16. Photosynthesis-irradiance profiles for light-adapted, shade-adapted, and transplanted *Stylophora pistillata*. (a) Light curves expressed on a surface area basis: light-adapted (○); shade-adapted (●); light-to-dark transplant after adaptation for 30 days (▲); dark-to-light transplant after adaptation for 30 days (△). (b) Light curves expressed on a chlorophyll *a* basis. The symbols used are the same as in (a). Data points represent means and standard deviations ($n = 3$ for light adapted, $n = 3$ for shade adapted, $n = 2$ for dark-to-light transplants, $n = 3$ for light-to-dark transplants). All measurements were made at 28 ± 1°C (Falkowski & Dubrinsky 1981).

based upon a major increase in PSU number, augmented by a secondary reorganization of pigmentation to optimize the overall quantum yield of photosynthesis (Chang *et al.* 1983).

(b) PSU NUMBER

A change in the number or density of PSUs usually induces a parallel change in the amount of thylakoid membranes present in an algal cell. As PSU number increases under lowered light conditions, pigment molar ratios and chl a/P_{700}

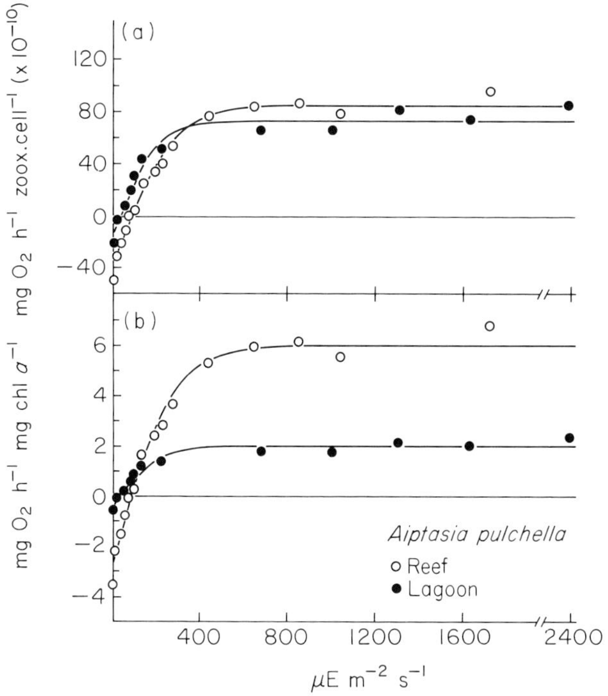

Fig. 5.17. Photosynthesis–irradiance curves for sunlit reef (○) and shaded lagoon (●) *Aiptasia pulchella* anemones containing symbiotic zooxanthellae. Light curves are expressed as (a) mg O_2 h^{-1} zoox. $cell^{-1}$ (10^{-10}) and (b) mg O_2 h^{-2} mg chlorophyll a^{-1}. (Muller-Parker 1984a).

stay the same but P_{700}/cell rises (Fig. 5.18). The cellular rates of light-limited and light-saturated photosynthesis increase in proportion to chl, while chl *a*-based *P–I* relationships stay the same, another example where assimilation numbers are a misleading index of primary productivity. (*Note.* This figure represents an updated interpretation of *P–I* changes associated with changes in PSU number since it was first presented by Prézelin & Sweeney (1978b) and is identical to the mechanistic scheme presented by Prézelin (1981).) This strategy represents a second low light induced pigment response in phytoplankton, such as that displayed by *Peridinium cinctum* between 400 and 65 $\mu E\,m^{-2}\,s^{-1}$ (Fig. 5.19, Prézelin & Sweeney, 1978b; Prézelin *et al.* 1984) and *Ceratium furca* between 2500 and 25 $\mu W\,cm^{-2}$ (Meeson 1981). A change in PSU density is also the most reasonable interpretation of the low light responses of zooxanthellae cultured from *Aiptasia pulchella* (Chang *et al.* 1984) and field populations of the free-living dinoflagellates *Ceratium tripos* and *C. fusus* (Rivkin & Voytek 1985).

It is not clear why two light adaptive strategies dependent on increased pigmentation should have evolved to increase light-absorptive capabilities of dinoflagellates in low light environments. Both strategies are displayed in a

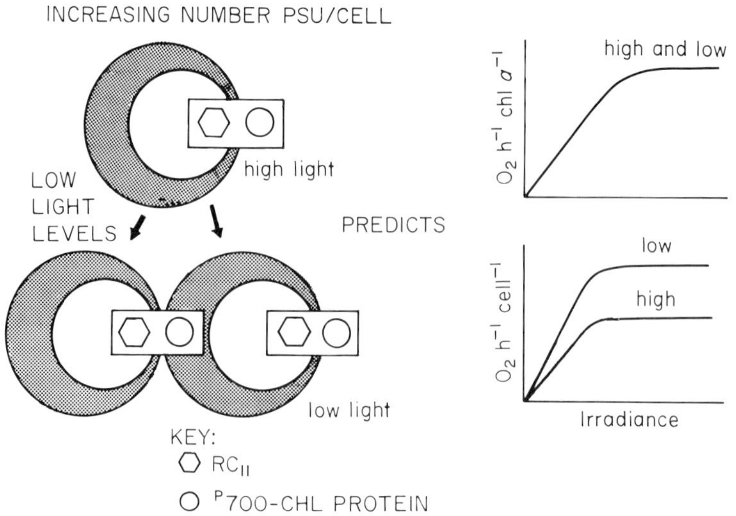

Fig. 5.18. Photoadaptation of photosynthesis by increasing the number of photosynthetic units per cell of dinoflagellate. (Modified from Prézelin & Sweeney 1978b.)

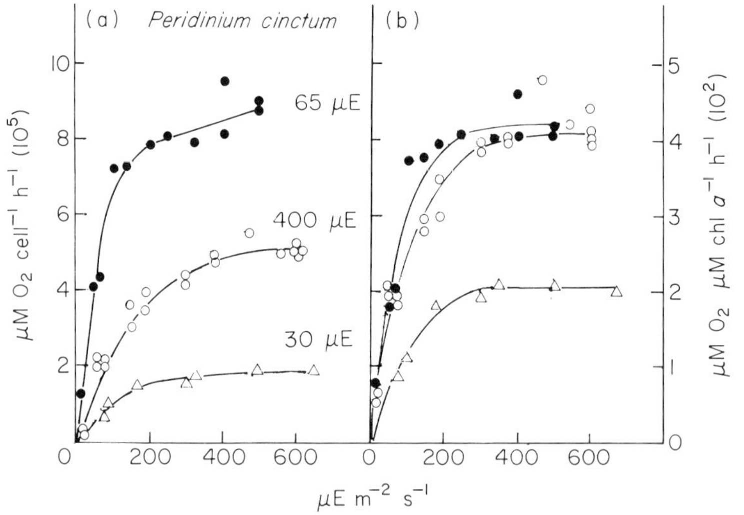

Fig. 5.19. *Peridinium cinctum*. Photosynthesis–irradiance curves for cultures grown on 12:12 hour light: dark cycles with light phase irradiances of 400 (○), 65 (●) and 30 (△) μE/m^2 s^{-1}. Light curves are expressed as (a) μM O_2 cell^{-1} h^{-1} (10^5) and (b) μM O_2 μM chlorophyll a^{-1} h^{-1} (10^2) (Prézelin unpubl.).

variety of other plant groups and therefore do not seem distributed on taxonomic bases. In both strategies, the light-limited rates of photosynthesis are enhanced and are apparently important to the survival of dinoflagellates in low light habitats. However, increasing the numbers of PSUs may have a greater synthetic demand on the cell than increasing PSU size, as a greater number of thylakoid components would have to be made and assembled. But, with increased PSU number, the algal population has the potential for a sustained burst of photosynthetic activity when light levels are increased once again. Such a response is not predicted solely on the basis of increased PSU size. It appears that the usefulness and distribution of photoadaptive strategies among different dinoflagellate species should be considered and compared in terms of the total autecology of the populations.

(c) DARK REACTIONS

The enzymatic reaction rates of photosynthesis also can slow in response to a number of environmental variables, especially changes in light and nutrient regimes. While this decline usually suggests a change in Calvin cycle enzyme activity (i.e. RUBP carboxylase), it also can mean a slowing of electron flow along the electron transport chains linking Ps I and II. It is possible to know that one or both of these dark reactions rates is the driving influence behind changes in primary production by first looking at pigmentation and *P–I* relationships (Fig. 5.20). If there is a slowing of dark reaction rates, there should

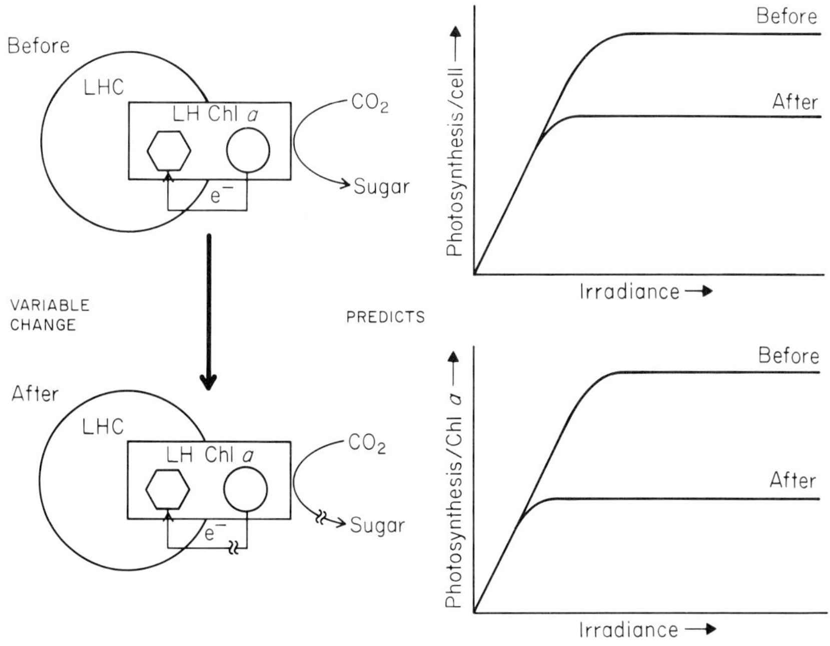

Fig. 5.20. Schematic representation of altered dark reaction rates and the resultant changes in photosynthesis–irradiance curves, expressed on a cellular basis (upper curve) and on a chlorophyll *a* basis (lower curve) (Prézelin 1982).

be a proportional drop in both light-saturated rates of cell- and chl-based photosynthesis, but there should be no change in light-limited rates of photosynthesis or in pigment content. If these photosynthetic characteristics are observed, then one can go on to distinguish between changes in enzymatic activity of Calvin enzymes and/or partial reaction rates of Ps I and II by direct biochemical assays. This pigment-independent photoresponse seems rare in dinoflagellates, but does appear as a mechanism of photoadaptation in zooxanthellae cultured from the scleractinian coral *Montipora verrucosa* (Chang *et al.* 1984).

The *P–I* relationship changes described above are also observed in laboratory cultured *Pyrocystis noctiluca*, but only when grown at very low light levels, i.e. below about 125 $\mu E\, m^{-2}\, s^{-1}$ (Rivkin *et al.* 1982a). A fourfold drop in photosynthetic capabilities is observed and is directly correlated to a similar decline in RUBP carboxylase activity. However, chl *a* content per ml rises more than sevenfold while carbon content and growth rate fall. It appears that the photosynthetic membranes remain intact and pigment accumulation remains high when cells are exposed to prolonged (4 months) low-light nutrient-rich environments, but that photosynthetic activity is severely impaired due to reduced enzyme levels. Similar physiological changes have been seen in the early stages of nutrient-dependent ageing in both higher plants (Thimann 1980) and *G. polyedra* (Prézelin 1982; Sterman 1986; see Section 2.5, this chapter).

(d) LOW LIGHT PHOTOSTRESS

Below about 50–100 $\mu E\, m^{-2}\, s^{-1}$, growth rates of cultured nutrient-sufficient dinoflagellates decline and many cell functions fall apart (Prézelin & Sweeney 1978a). The physiological changes associated with these lowest light responses seem to involve one or more cell processes and vary among species. For instance, in *Glenodinium* sp. only respiration and cell division rates seem to decline while pigment synthesis and photosynthetic capabilities remain optimal (Prézelin 1976). By contrast, there appears to be a general breakdown in most cell functions in *G. polyedra* over the same light range, including large drops in cell volume, cell division, pigment synthesis, photosynthesis, and respiration rates (Prézelin & Sweeney 1978a). Some, but not all, of the pigment decline in very low light cultures can be explained by the much smaller volume of photostressed cells. However, biomass-based rates of light-saturated photosynthesis fall while the assimilation ratio P_{max}/chl *a* remains the same. These photosynthetic changes are explained most simply by a reduction in thylakoid membranes, i.e. a loss of PSU number. Very similar changes also characterize the later stages of progressive nutrient stress in *G. polyedra* (Prézelin 1982; see Section 2.5, this chapter).

In *P. cinctum* grown at light levels below about 65 $\mu E\, m^{-2}\, s^{-1}$ (Fig. 5.19), there are large drops in both light-limited and light-saturated rates of photosynthesis unrelated to persistent high pigmentation (Prézelin *et al.* 1984). These changes cannot be attributed to any single mechanism yet described, although the changes are reminiscent of membrane state changes leading to uncoupling of energy transduction within photosynthetic light reactions and normally associated with photosynthetic diurnal periodicity in dinoflagellates (Section 2.3, this chapter). Examined together, these studies emphasize that photostress responses vary widely and if accurate interpretations of photoadaptive responses of dinoflagellates are to be ensured, it will be necessary that the biochemical and physiological characteristics should be compared over a broad range of light intensities.

(e) BRIGHT LIGHT RESPONSES

Only recently have studies been conducted which document the responses of dinoflagellates at light levels high enough to photoinhibit photosynthesis. Mandelli (1972) did document a proportional increase of yellow xanthophylls, especially diadinoxanthin, in *Amphidinium klebsii* when cultured at increasing irradiances up to about $900\,\mu E\,m^{-2}\,s^{-1}$. This increase may suggest a photoprotective role for these carotenoids in bright light environments, as well as photoinhibition of photosynthesis. *P–I* characteristics of the steady-state cultures of *Glenodinium* sp., *G. polyedra*, *Peridinium cinctum*, *Ceratium furca*, and an *Amphidinium* sp. grown at light levels between 50 and $400\,\mu E\,m^{-2}\,s^{-1}$ show no photoinhibition when illuminated briefly with light levels as high as $800\,\mu E\,m^{-2}\,s^{-1}$ (Prézelin 1976; Prézelin *et al.* 1977; Prézelin & Sweeney 1978; Humphrey 1979; Prézelin & Matlick 1980; Meeson 1981). In the dinoflagellate, *A. carterae*, photoinhibition can occur within 2 hours of high light exposure of low light cultures and is reversible if light levels are lowered within 4 hours (Samuelsson & Richardson 1982). This short-term photoinhibition is accompanied by chl photobleaching and a loss of variable fluorescence, indicative of an impairment of Ps II activity. Likewise, low-to-high light shifts of *Glenodinium* sp. and *G. polyedra* sometimes indicate photoinhibitory effects, especially when cultures are nutrient-limited, and are related to a pigment-independent uncoupling of electron flow within photosystems II (Prézelin *et al.* 1986). Only prolonged exposure (i.e. >3 to 6 h) induces chl photobleaching, accompanied by a variety of physiological responses dependent upon population age or nutrient status (Prézelin *et al.* 1986).

Photorespiration is another light-dependent process stimulated when illumination and oxygen production rates are high and further enhanced when carbon dioxide levels are low. Under such conditions, ribulose bisphosphate (RUBP) is oxygenated instead of carboxylated by the enzyme RUBP carboxylase/oxygenase and results in the production of phosphoglycolate which enters an enzymatic pathway and ends in the production of glycolate. Photorespiration can lead to a reduction in the amount of carbon fixed, and the subsequent release of carbon dioxide from glycolate as it is further metabolized outside the chloroplast (Tolber 1974). There are conflicting views regarding photorespiration in algae, but Burris (1981) presents data clearly indicating the negative effect of increasing oxygen content on photosynthetic rates in *Glenodinium* sp. However *Protogonyaulax* (=*Gonyaulax*) *tamarensis* shows oxygen inhibition of photosynthesis only when cells are exposed to 100% oxygen (Beardall *et al.* 1976). Little or no oxygen inhibition of photosynthesis is observed in zooxanthellae from *Tridacna maxima* or *Pocillopora capitata* (Burris 1981). But Phipps (1980) reports that algal symbionts from hydra exhibit marked inhibition of photosynthesis by oxygen, while symbionts from sea anemones appear relatively insensitive to environmental oxygen concentrations.

(f) TIME COURSE OF ADAPTATION

'High-to-low' light photoadaptation occurs quickly in *Glenodinium* sp. (Prézelin & Matlick 1980). Within 3 hours, the pigment content of the cell begins to increase rapidly, as expressed by the increase in a single type of light-harvesting component, PCP (Prézelin *et al.* 1976). PCP content can rise elevenfold in 5 days, only to fall to a steady-state level that is about fourfold greater than bright light concentrations. The overshoot of pigmentation in transition responses to changing light fields appears to be a common characteristic of cells growing in a nutrient-sufficient environment. By comparison to peridinin, the increases in chl *a* and chl *c* are slower, more modest in amount, and fluctuate over a 30-day period before reaching a steady-state level about 1·5 2·0 times greater than found in high light cells (Fig. 5.21). These fluctuations suggests that chl *a* is not always the best indicator of photoadaptive rates.

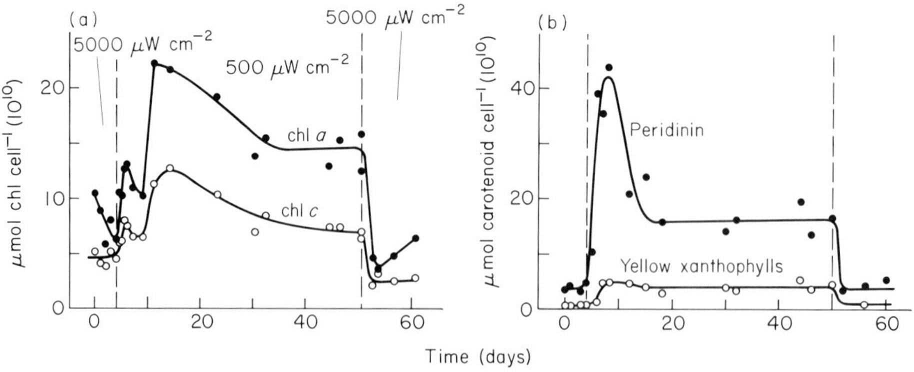

Fig. 5.21. *Glenodinium* sp. Whole cell pigmentation changes following transfer from high to low light levels and vice versa. Cells were cultured on a 12:12 LD cycle at 18°C; vertical dashed lines indicate times of transfer: concentrations of (a) chlorophyll *a* (●) and chlorophyll *c* (○); (b) peridinin (●) and yellow xanthophylls; (○) all expressed as μM pigment cell⁻¹ (10^{10}) (Prézelin & Matlick 1980).

An increase in PSU size can be monitored best by following the decrease in the light intensity which half-saturates photosynthesis, i.e. $I = \frac{1}{2}P_{max}$ (Fig. 5.22) or Talling's constant I_k (P_{max}/alpha). This photosynthetic parameter is preferable to following P_{max} or alpha changes, which oscillate in response to LD cycles (Prézelin & Matlick 1980; Section 2.2). In the small-sized, rapidly growing *Glenodinium* sp. populations, changes in the half-saturation constant for photosynthesis indicate the *P–I* relationships are fully low light adjusted within 12 hours of transfer, even though pigment synthesis is still occurring. The change in $I = \frac{1}{2}P_{max}$ is completed in *Glenodinium* sp. in less than one-quarter the generation time of the bright light culture. The comparable photoadaptive rates for log-phase populations of *G. polyedra* are 4 days (Prézelin & Matlick 1983), or about equal to the generation time of the population (Figs 5.23–5.25). In *P. noctiluca* populations with 9-day generation times, low light-induced chl *a* increases take 2 days to induce and 6 additional days to reach new steady-state levels (Rivkin *et al.* 1982b).

Since photoadaptation requires extensive synthesis of new molecules, it has been suggested that the nutrient status of phytoplankton exerts a strong influence on the rate and extent of photoadaptation (Prézelin 1982). Prézelin & Matlick (1983) tested this idea by measuring rates of photoadaptation in *Gonyaulax polyedra* as a function of increasing cell density in increasingly nutrient-limited batch cultures. As older bright-light cultures were shifted to low-light conditions, the onset of pigment synthesis was delayed, the rates of new chlorophyll synthesis slowed, and final yields of cell chlorophyll were sharply reduced (Fig. 5.23). Moreover, photoadaptive capabilities of photosynthesis observed in log-phase cultures were lost in nutrient-limited stationary cultures (Figs 5.24, 5.25). Addition of inorganic nutrients to stationary cultures at the time of transfer to low-light levels induced pigment synthesis at a rate and magnitude comparable to log-phase cultures and led to short-term increases in the rates of cell- and carbon-based photosynthesis. The nutrient status of the

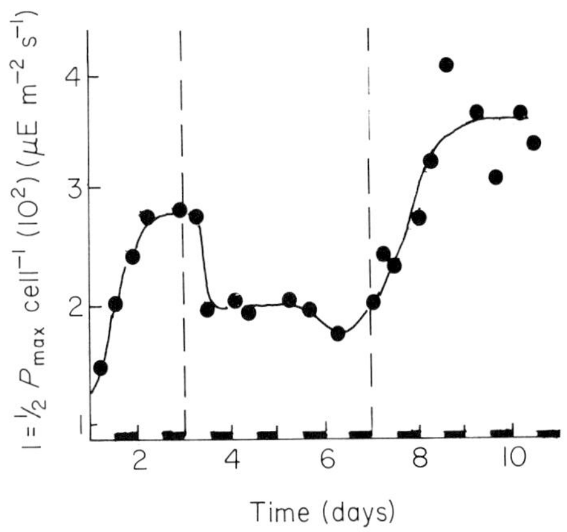

Fig. 5.22. *Glenodinium* sp. Changes in half-saturation constant for photosynthesis, $I = \frac{1}{2} P_{max}$, following transfer from high (2500 μ cm^{-2}) to low (250 μW cm^{-2}) light levels (indicated by first vertical dashed line) and vice versa (indicated by second vertical dashed line). (Adapted from Prézelin & Matlick 1980.)

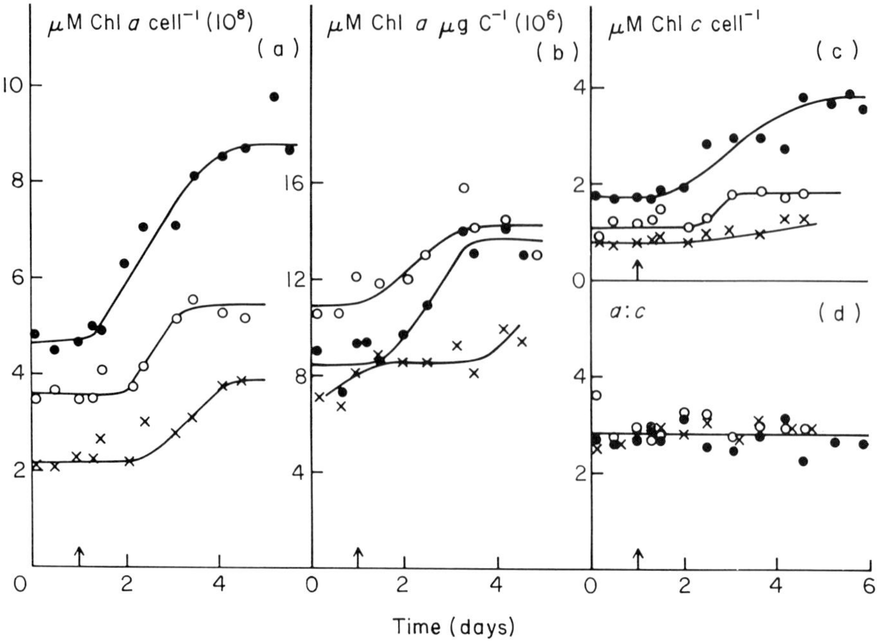

Fig. 5.23. *Gonyaulax polyedra*. Chlorophyll changes associated with a shift of growth illumination (indicated by arrow) from 330 to 80 μE m^{-2} s^{-1}: mid-log populations (●), late log populations (○) and stationary populations (×) (Prézelin & Matlick 1983).

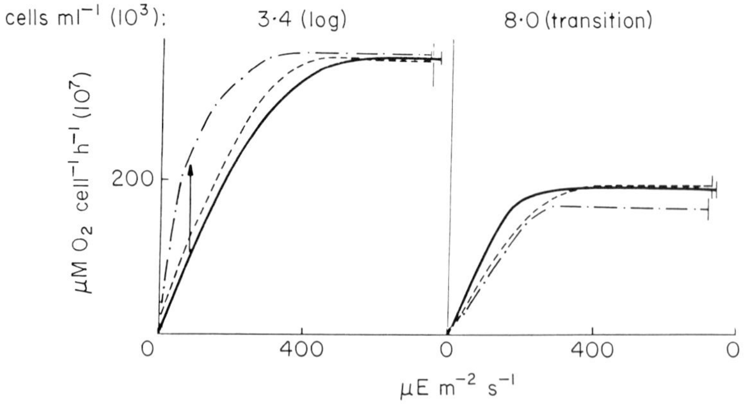

Fig. 5.24. *Gonyaulax polyedra*. Changes in photosynthesis–irradiance curves of early and late log cultures after growth illumination was shifted from 330 to 80 μ^{-2} s^{-1}. Comparisons were made of light curves from 18 hours before (——), 6 hours after (– – –) and 4 days after (–·–·–) high-to-low light shift. Vertical arrow indicates enhanced photosynthetic performance in photoadapted low light cultures. All measurements were made in the middle of the light period. (Prézelin & Matlick 1983).

population also increased, as witnessed by increased carbon and nitrogen content and a large, but transient, increase in nitrate reductase activity. However, improved photosynthetic status did not increase cell division rates over control populations.

'Low-to-high' photoresponses are not a simple or fast reversal of mechanisms outlined above. In nutrient-sufficient cultures of *Glenodinium* sp., pigment content of low light populations falls more rapidly when shifted to light levels bright enough to light-saturate photosynthesis (Prézelin & Matlick 1980) (Fig. 5.21). If increased light levels are not sufficient to light-saturate photosynthesis, then the pigment content falls much more slowly even though rates of cell division are increasing. It appears the return of low-light cultures to light-saturating intensities promotes an immediate return to higher growth rates, while a return to brighter but non-saturating intensities stimulates a slower

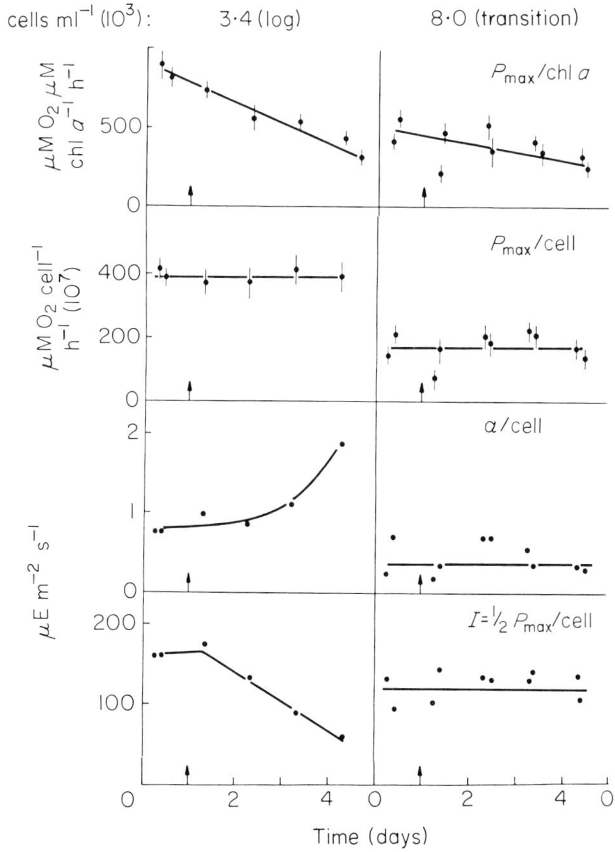

Fig. 5.25. *Gonyaulax polyedra.* Photosynthesis–irradiance parameters plotted as a function of time following transfer of early and late log cultures from 330 to 80 μE m^{-2} s^{-1}. Vertical arrows indicate time of transfer and bars indicate standard deviation of samples where more than four measurements were made. (Prézelin & Matlick 1983).

return to maximal growth rates which can take almost 2 weeks. Similar results have been observed in nutrient-sufficient cultures of *G. polyedra* (Rivkin *et al.* 1982b; Prézelin *et al.* 1986).

Again, the nutrient status of the culture affects the biochemical adaptability of photosynthesis to increased irradiances. If nutrient-deficient, low-light cultures of *G. polyedra* are shifted to bright light, cells die within 24–48 hours. Prior to cell death there is an accelerated drop in pigmentation and overall photosynthetic rates, as well as marked photoinhibition of Ps II electron transport rates (Prézelin *et al.* 1986). Such physiological observations are consistent with the behavioural observations showing that nutrient-limited *G. polyedra* will avoid the brightest light regions of the water column and will begin their downward vertical migration earlier in the day than nutrient-sufficient populations (Heany & Eppley 1981).

2.2 Total daily irradiance

Humphrey (1979) compared the growth and photosynthetic characteristics of *Amphidinium carterae* in day:night cycles, where the length of the light period varied from 4 to 24 hours but the incident irradiance was kept constant at $80\,\mu E\,m^{-2}\,s^{-1}$. The growth rate of cultures grown on 12:12 light:dark (LD) cycles had approximately the same growth rate as cultures grown under constant illuminations (i.e. 0·9 vs. 0·7 divisions per day respectively, with no standard deviations given). The 12:12 LD cells contained twice as much pigmentation and performed light-saturating rates of photosynthesis twice the magnitude of cells growing under constant illumination. Samuelsson & O. Richardson (1982) recently showed that a drop in irradiance levels within a fixed LD cycle will induce similar photosynthetic changes in *Amphidinium carterae*. These observations indicate that photoadaptive responses of dinoflagellates can be induced by changes in total daily irradiance, irrespective of whether the drop in illumination is brought about by lower light levels or shorter days. At photoperiods of less than 12 hours, the pigmentation of *A. carterae* begins to fall as does photosynthetic potential. It is likely that the total daily irradiance was below a threshold necessary to induce continued net synthesis of photosynthetic components, resulting in a light-dependent decline in photosynthesis at very low light levels (see Section 2.1(d)).

The respective growth rates of *Ceratium furca* (Weiler & Eppley 1979) and *Glenodinium* sp. (B. B. Prézelin & B. M. Sweeney, unpubl. data) are unchanged when cultured under different photoperiods, if incident irradiances are greater than $100\,\mu E\,m^{-2}\,s^{-1}$ and the light phase of the LD cycle is longer than 8 hours. The pigment content of *Glenodinium* sp. is twice as high in 12:12 hour LD cultures as compared to LL cultures of equal irradiance. Unlike *A. carterae*, the photosynthetic potential of *Glenodinium* sp. does not change when the photoperiod is shortened. If cultures of *Glenodinium* sp. are transferred from a

12:12 hour LD cycle to LL with only half the incident irradiance (i.e. the total daily illumination remains the same), then the pigment content of the population does not change. Under the same experimental manipulation, pigment synthesis in *G. polyedra* and *C. furca* is stimulated within 6 hours of transfer to constant illumination in transient response to a change in incident light intensity (Fig. 5.26). This trend is reversed within 24 hours and original pigmentation levels are re-established within a few days. Taken together, these observations suggest that integrated daily irradiance may have an overriding influence on incident irradiance in determining photosynthetic physiology of dinoflagellates.

Rivkin *et al.* (1982a) demonstrated the effect that fluctuating total daily irradiance within a fixed photoperiod has on the photosynthetic responses of dinoflagellates. Semi-continuous batch cultures of *P. noctiluca* were exposed to increments of increasing and decreasing light levels every few days and changes in pigmentation were followed (Fig. 5.27). The chl *a* content initially varies inversely to incident light intensity and then damps out to steady-state values within four constant light cycles, where each cycle lasts about 35 days and oscillates between 13 and 460 μE m^{-2} s^{-1} (first two cycles seen in Fig. 5.27) (Rivkin *et al.* 1981; R. B. Rivkin, pers. comm.). During the experiment, the cell division rates remain surprisingly constant and the chl content approaches a constant value by the beginning of the third cycle (R. B. Rivkin, pers. comm.). The authors suggest that in a changing photic environment, both cell division rates and chl *a* concentrations may adapt to the integral of the light experienced over one or more cell generations and physiologically resemble steady-state low light cultures (Rivkin *et al.* 1982a; R. B. Rivkin, pers. comm.).

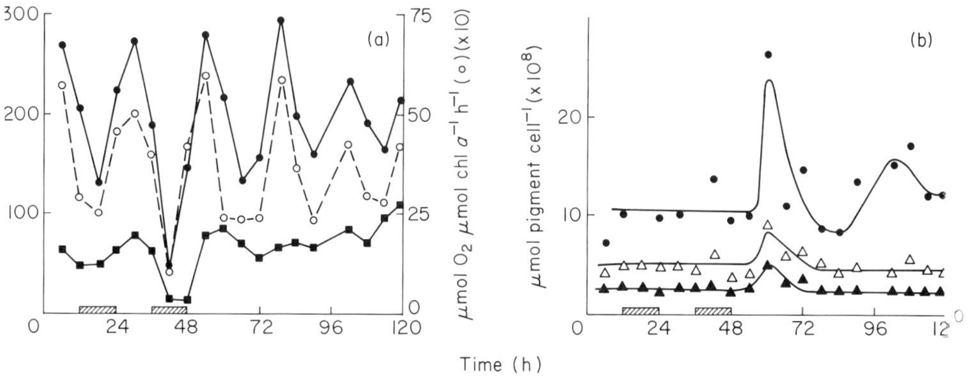

Fig. 5.26. *Ceratium furca.* (a) Rhythms in photosynthetic capacity expressed as μM O_2 cell^{-1} h^{-1} (●) and as μM O_2 μM chlorophyll a^{-1} h^{-1} (○). Respiration (■) is expressed as $-$μM O_2 cell^{-1} h^{-1}. Irradiance of light phase of 12:12 hour LD cycle was 1000 μW cm^{-2}. Dark periods are indicated by hatched bars on abscissa. Mid-log cultures were transferred to continuous dim illumination (500 μW cm^{-2}) at the beginning of the light phase. (b) Whole cell pigmentation with concentrations of peridinin (●), chlorophyll *a* (△) and chlorophyll *c* (▲) expressed as μM cell^{-1} (10^8) (Prézelin *et al.* 1977).

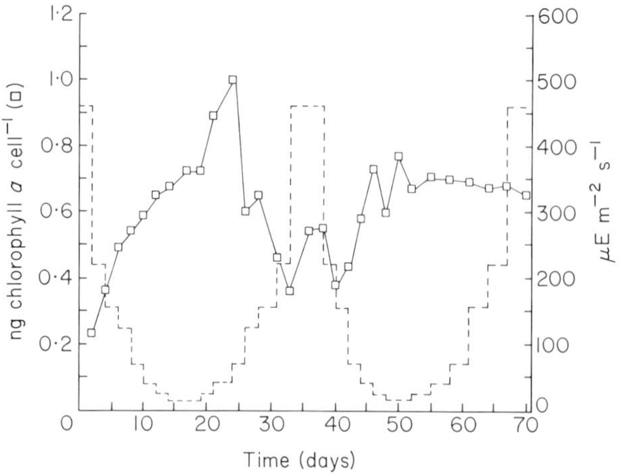

Fig. 5.27. Stepwise change in light intensity (----) and variations in chlorophyll *a* cell^{-1} (□) of *Pyrocystis noctiluca* after remaining at each light intensity for 2–4 days. Chl *a* was the mean of triplicate determinations (Rivkin *et al.* 1982a).

2.3 Vertical migration

There appears to be a relationship between photosynthesis and the vertical migration patterns of dinoflagellates. First, both photosynthesis and vertical migration are endogenous rhythms tied to each other through a biological clock (Section 2.4). Second, the photoreceptors for phototaxis include carotenoproteins (Forward 1976) whose absorption characteristics are reminiscent of the photosynthetic light-harvesting component PCP (Song *et al.* 1980). Third, the *in situ* migration phase and amplitude of dinoflagellates appears to be influenced strongly by the photosynthetic capabilities of the cells. For instance, *Ceratium hirundinella* follows *in situ* light gradients over the day which would maximize daily rates of photosynthesis (Harris *et al.* 1979). Kamykowski (1981), using published field observations on *Gymnodinium splendens* (Kiefer & Lasker 1975), related the diurnally integrated light experience of the dinoflagellate to their observed migration pattern and photosynthetic physiology. Using different *P–I* relationships and light histories to describe migrating populations in the upper and lower parts of the water column, he demonstrates that the differences in migration phases observed in the field (i.e. the onset of the ascent is delayed further into the day in lower light populations) places each population at their shallowest depth at midday. Furthermore, these light-regulated behavioural responses optimize the photosynthesis-dependent growth rate estimates for both shallow and deep populations. Other studies also demonstrate that high light and nutrient depletion can modify behavioural patterns of dinoflagellates (Eppley *et al.* 1968; Tyler & Seliger 1981; Heaney & Eppley 1981; Cullen &

Horrigan 1981). In the case of *Prorocentrum mariae-lebouriae*, which predominates in Chesapeake Bay in the summer, altered migrational patterns can be related to changes in the physiological state of the population over an annual cycle (Tyler & Seliger 1981).

2.4 Diurnal periodicity and biological clocks

All dinoflagellates examined so far have light-saturated rates of photosynthesis which oscillate over the day, i.e. they exhibit a diurnal periodicity in photosynthesis. While the LD cycle can entrain the periodicity of photosynthesis, it does not drive the oscillations. The diurnal periodicity in photosynthesis continues for many cycles after the LD cycles has been removed and the population placed in constant conditions, i.e. usually low continuous light (LL) (Fig. 5.26). Under such conditions, photosynthesis exhibits a circadian rhythmicity with a period close to, but not exactly, 24 hours, the rhythm is temperature-compensated and its phase can be reset by short pulses of visible light (cf. Sweeney 1981b, c).

The first dinoflagellate rhythm in photosynthetic capacity was documented in *G. polyedra* almost 25 years ago (Sweeney & Hastings 1958; Sweeney 1960; Hastings *et al.* 1961), but it is only recently that the impact such dinoflagellate rhythmicity might have on primary productivity estimates has been studied in some detail (Prézelin *et al.* 1977; Prézelin & Sweeney 1977; Prézelin & Matlick 1980; Harding *et al.* 1981a, b, 1982a, b, 1983). Now photosynthesis rhythms have been documented in continuous light in *Glenodinium* sp., *C. furca* (Prézelin *et al.* 1977) and *P. fusiformis* (Sweeney 1981a–c). Depending upon the species and its nutrient status (Section 2.6), day:night amplitudes in photosynthetic rate are between 2 and 10, and the peak activity occurs between late morning and mid-afternoon. Cell division rhythms also may be common in dinoflagellates, unlike diatoms (cf. Chisholm 1981). Where they do exist, i.e. in *Gonyaulax polyedra* (cf. Sweeney 1981c), all rhythms displayed in the organisms are closely coupled. Therefore, any agent which shifts the phase of the cell division rhythm in *G. polyedra* will equally shift the rhythms in photosynthesis, bioluminescence (cf. Sweeney 1981c) and possibly also the rhythm in vertical migration.

It is probable that circadian regulation of photosynthesis is a common property of all dinoflagellates (Prézelin & Sweeney 1977). As a result, there is a probable diurnal fluctuation in the availability of photosynthetic end-products, reducing power and ATP (Weiler & Karl 1979). If generally true, then there should be a predictable influence of photosynthesis rhythms on observed diel changes in cell composition (Weiler & Karl 1979), the uptake of inorganic nutrients (Eppley 1974; Mac Isaacs 1978; Rivkin & Swift 1982), and ATP-dependent enzymes like nitrate reductase (Hersey & Swift 1976).

Pigmentation does not oscillate over LD or LL cycles and the daily periodicity in respiration seen in LD cycles quickly dampens out in LL cycles

(Prézelin *et al.* 1977). However, the possibility that day/night differences in whole cell absorption might cause the photosynthetic rhythms was raised when Herman & Sweeney (1975) showed that the thylakoids of *G. polyedra* change their orientation over the LD cycle and that the movement is under the control of a biological clock. In the parts of the chloroplast that lie closest to the interior of the cell, thylakoid stacks are separated more during the day than at night and no ribosomes are present in the stroma when this region is expanded. But a comparison of absolute cellular absorption spectra of day and night cultures showed thylakoid movement has no effect on whole cell photosynthetic absorption capabilities (Prézelin & Sweeney 1977). At the organelle level, rhythmicity in whole chloroplast movement has been documented for *P. lunula* (Swift & Taylor 1967) and *P. fusiformis* (Sweeney 1981a–c). During the day, the chloroplasts are distributed evenly throughout the cytoplasm but at night they all collect around the nucleus in the centre of the cell. In this case, the effect of

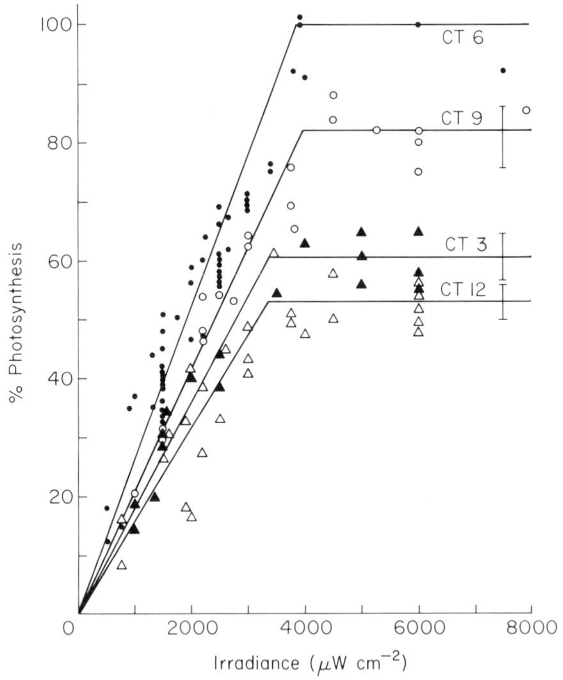

Fig. 5.28. *Gonyaulax polyedra*. Photosynthesis–irradiance curves determined 3 hours into the light phase (CT 3, ▲), in the middle of the light phase (CT 6, ●), 9 hours into the light phase (CT 9, ○) and at the beginning of the dark phase (CT 12, △). Cultures growing on a 12:12 LD cycle, with light phase irradiances of 1200 μW cm^{-2}, were harvested in log-phase growth. Photosynthesis was plotted as percentages of mid-day light-saturated rates. Standard deviations of P_{max} values are represented by vertical bars. Light-limited slopes were determined by simplified linear regression analyses, which assumed a zero intercept and utilized photosynthetic values measured between 0 and 3000 μW cm^{-2} (Prézelin & Sweeney 1977).

chloroplast migration on whole cell absorption capabilities over the day is not known.

In a detailed study of the photosynthesis rhythm of *G. polyedra*, the light-saturated and light-limited rates of photosynthesis change in parallel over the day, giving rise to the family of *P–I* relationships shown in Fig. 5.28. Photosynthetic rate changes with these characteristics occur with no change in pigmentation and can be explained in two ways. First, there could be a rhythmicity in the percentage of cells in the population capable of carrying out photosynthesis over the day. This possibility was eliminated when Sweeney (1960) demonstrated photosynthetic rhythms in single cells of *G. polyedra*. Second, there could be an uncoupling of energy transduction within some reaction centres, with more photosynthetic units uncoupled during the night than during the day. Such an arrangement would explain why the composition of the PSUs does not change over the day, why the shape of the *P–I* curves remains the same, and why both cell- and chl-based photosynthetic rates always fall in proportion to each other (Fig. 5.29) (Prézelin & Sweeney 1977).

Consistent with the above interpretation, there are observations that RUBP carboxylase activity does not change over the LD cycle (Bush & Sweeney 1972) but that there are chl fluorescence changes over the day (Prézelin & Sweeney

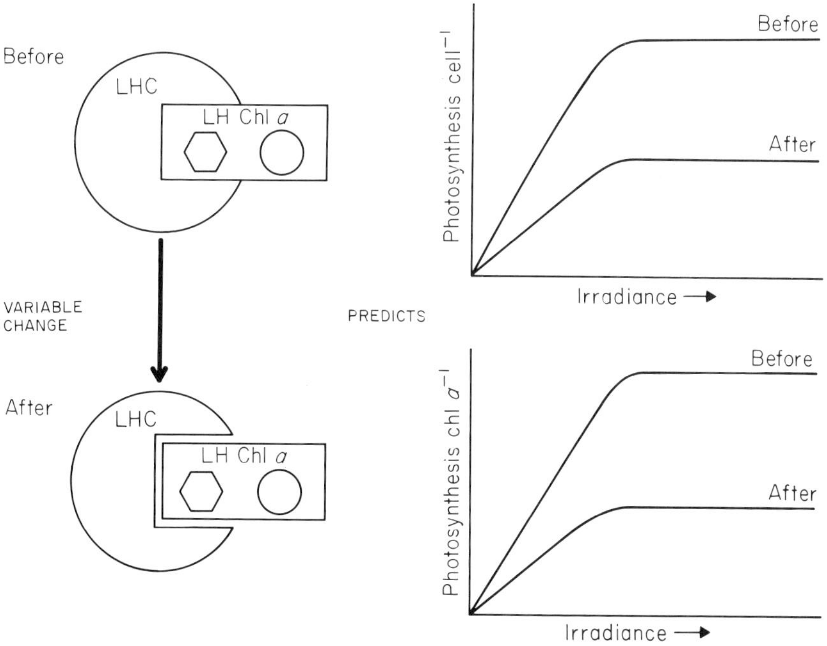

Fig. 5.29. Schematic representation of relationship between altered photosynthetic energy transduction and changes in the photosynthesis–irradiance curves, expressed either on a cellular basis (upper curves) or a chlorophyll *a* basis (lower curves) (Prézelin 1982).

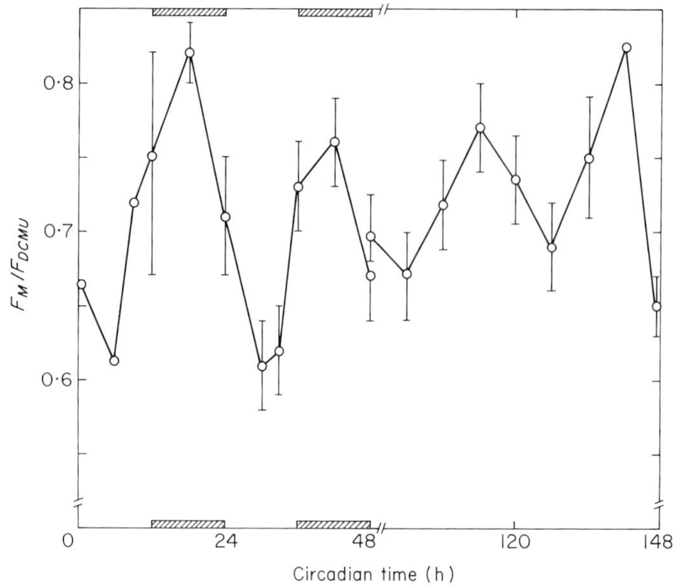

Fig. 5.30. *Gonyaulax polyedra.* Rhythm of room temperature *in vivo* chlorophyll *a* fluorescence in the absence and presence of *DCMU*, expressed as the ratio F_M/F_{DCMU}. Points represent averaged data from five experiments and percentage error is indicated by vertical bars. LD cycle was 12:12 hours and irradiance of light period was 1200 μW cm^{-2}. Dark periods are indicated by hatched bars on abscissa. Cultures were transferred to constant dim illumination (500 μW cm^{-2}) at 48 hours and measurements resumed at 96 hours. Cells were in mid-log phase (Prézelin & Sweeney 1977).

1977; Sweeney *et al.* 1979). The fluorescence ratio *F*/*F* + *DCMU* fluctuates in inverse proportion to the photosynthesis rhythm (Fig. 5.30), with *F*/cell and *F* + *DCMU*/cell fluctuating in direct proportion to photosynthesis. Since the daily amplitudes of *F* + *DCMU* changes are much greater than *F* − *DCMU*, it gives rise to the mid-day depression in the fluorescence ratio. These results indicate that the relative fluorescence yield of chl is not constant over the day and, in fact, is highest during the middle of the day when photosynthesis is greatest.

At first glance it appears contradictory that there could be parallel changes in photosynthesis and fluorescence over the day, given that the amount of light absorbed by the cell is constant. But detailed analyses of the fluorescence transients in *G. polyedra* at different times over the day, i.e. when the cells are in a LD cycle and after they have been transferred to continuous light, verify that photosynthetic capacity and relative quantum yield changes over the day are positively correlated (Govindjee *et al.* 1979; Sweeney *et al.* 1979). While the shape of these fluorescence transients show typical structure but no rhythmic changes in shape over the day, the intensity of fluorescence at all points on the transient was about twice as high during the day phase as during the night phase. This result can be interpreted as an indication that the non-radiative decay of chl excitation is less during the day than at night. Thus, it has been suggested that a circadian rhythm exists in the thylakoid membrane state, which in turn alters the arrangements of photosynthetic components so as to couple/uncouple energy transduction within the existing photosynthetic apparatus over the day (Prézelin & Sweeney 1977; Sweeney & Prézelin 1979). This interpretation is supported by the observation that the fluorescence oscillations are abolished by low temperatures, which should inhibit any membrane conformational changes (Sweeney *et al.* 1979). Likewise, thylakoid membranes have been examined by freeze-fracture but no circadian rhythmicity in particle size or number could be detected. This result would be predicted if the numbers of photosystems remains the same over the day (Sweeney 1981a). More recently, techniques were developed to measure photosynthetic electron transport rates in dinoflagellates (Samuelsson & Prézelin 1985) and employed to show that changes in Ps II activity accounted for all circadian rhythmicity in photosynthesis observed in *G. polyedra* (Samuelsson *et al.* 1983).

2.5 Light colour

The effect of light colour on the biology of dinoflagellates has been directly linked to their phototactic behaviour (Forward 1976), but our knowledge of the direct effect of light colour on photosynthetic processes is quite limited. A comparison of pigmentation of very low intensity white and blue-green light ($400\,\mu W\,cm^{-2}$) cultures indicated chlorophyll content generally was greater in blue light conditions, i.e. a 98% increase for *Gymnodinium* sp., a 35% increase for *Amphidinium carterae*, and a 19% increase for *Prorocentrum micans* (Vesk & Jeffrey 1977; Jeffrey 1980). Studies on *Prorocentrum mariae-lebouriae* indicate that at low light intensities, blue light cultures contain about half the chlorophyll *a*, chlorophyll *c*, and peridinin of white light cultures but are dividing four times as fast! However, when cells are grown under green light, pigmentation and growth rate are equivalent to white light cultures at all light levels (Figs 5.31, 5.32) (Faust *et al.* 1982). Such observations may suggest that metabolic accommodations to blue light occur in cell processes other than photosynthesis, i.e. changes in relative rates of carbohydrate and protein synthesis or excretion (Kowallik 1970).

While laboratory studies show a favourable accommodation of dinoflagellates to blue light, field light measurements of dinoflagellate blooms not only indicate strongly attenuate light penetration but also a spectral shift in available light colour toward the red end of the visible spectrum (Dubrinsky & Berman 1979; Faust *et al.* 1982). Given low intensity blue to red spectral shifts, the laboratory study on *P. mariae-lebouriae* could be taken to predict a doubling in cell pigmentation but a fourfold reduction in relative growth rate (Figs 5.31, 5.32). Increase in pigmentation, accompanied by decreases in assimilation

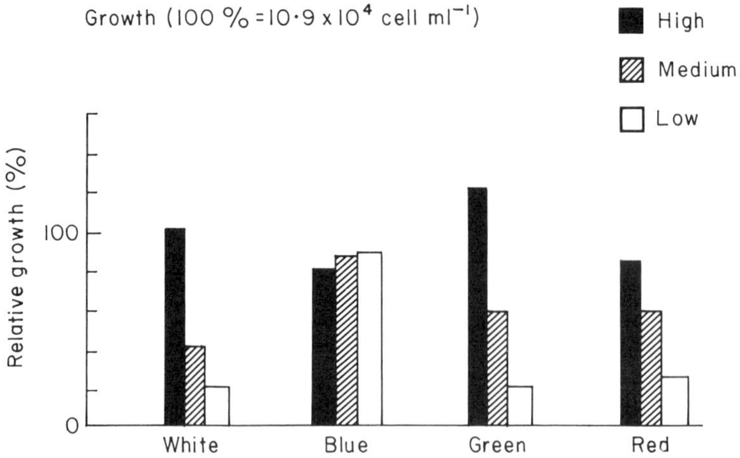

Fig. 5.31. Relative growth of *Prorocentrum mariae-lebouriae* irradiated at three different photon flux densities and four spectral qualities, i.e. white, blue, green and red light. The results were referenced to growth rates determined in cultures grown under bright white light. The photon flux densities were about 59 (■), 17 (□), and 8 (□) μM quanta m^{-2} s^{-1} (Faust *et al.* 1982).

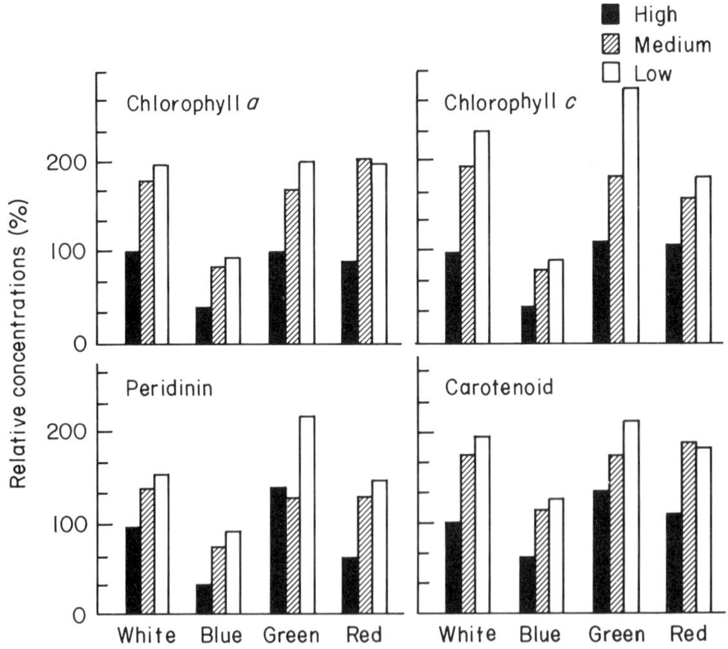

Fig. 5.32. Relative photosynthetic pigment concentrations for cells of *Prorocentrum mariae-lebouriae* after 12 days of growth when cultured at three different photon flux densities and four spectral qualities, i.e. white, blue, green and red light. Results were referenced to chlorophyll *a* content of bright white light grown cells. The photon flux densities were about 59 (■), 17 (□), and 8 (□) μM quanta m^{-2} s^{-1} (Faust *et al.* 1982).

numbers, have been associated with changing spectral characteristics of bloom waters of *Peridinium cinctum* (Dubrinsky & Polna 1976; Dubrinsky & Berman 1976). Enhanced red radiation also appears to increase the relative proportion of blue-greens, diatoms and green algae over dinoflagellates in some natural populations (Wall & Briand 1979). Together, these observations may suggest the demise of some red tide populations is influenced by increasingly inefficient use of the red-enhanced radiation by dinoflagellates in increasing low light environments.

2.6 Temperature

Temperature–light interactions regulating growth and photosynthesis in dinoflagellates have been studied by Meeson (1981). The two species she compared were chosen in part because they are known to photoadapt to low light levels by different strategies, i.e. *Gonyaulax polyedra* increases photosynthetic unit size while *Ceratium furca* appears to increase photosynthetic unit number. Meeson's work illustrates that these photoresponse mechanisms do not change with decreased growth temperature, even though the PSU size of *C. furca* appears larger when cells are cultured at lower temperatures. Pigment synthesis is induced at low light levels in both species at all temperatures tested (10, 15, and 20°C). Interestingly, cells grown at 15°C accumulate more chl *a* than cells grown at 20°C, and was especially pronounced in *C. furca* cultures (Fig. 5.33a, d). This large accumulation of pigmentation at lower temperature is unexplained. It may be that light-induced pigment synthesis is less susceptible to temperature regulation than are cell division rates, which fall as temperature is lowered from 20°C. In any event, a low-temperature induction of pigment accumulation may prove to be an advantageous anticipatory response of phytoplankton to the lower light environments generally associated with cooler seasons and/or water at the base of the photic zone.

Biomass-based gross photosynthetic rates of *G. polyedra* are also lowered with a decrease in growth and temperature, but the same productivity index is unaffected by lowering the growth temperature of cultures of *C. furca* (Fig. 5.33b, e). Surprisingly, while lower temperatures result in decreased respiratory rates for *G. polyedra* by about the same degree as lowered photosynthetic rates, the respiratory rates of *C. furca* increase almost twofold as growth temperature is lowered from 20 to 15°C (Fig. 5.33c, f). There is no biochemical explanation for this large respiratory increase, although it correlates well with the increased pigmentation and may simply reflect the higher metabolic rates required to synthesize large numbers of new PSUs, i.e. thylakoid membranes (Meeson 1981). If true, it would be the first indication of the differential energy demands of photoadaption by one of the two pigment-dependent strategies. The effect of these different photosynthetic and respiratory responses to lower growth temperature (20 vs. 15°C) is to lower significantly the light-saturated P : R ratios

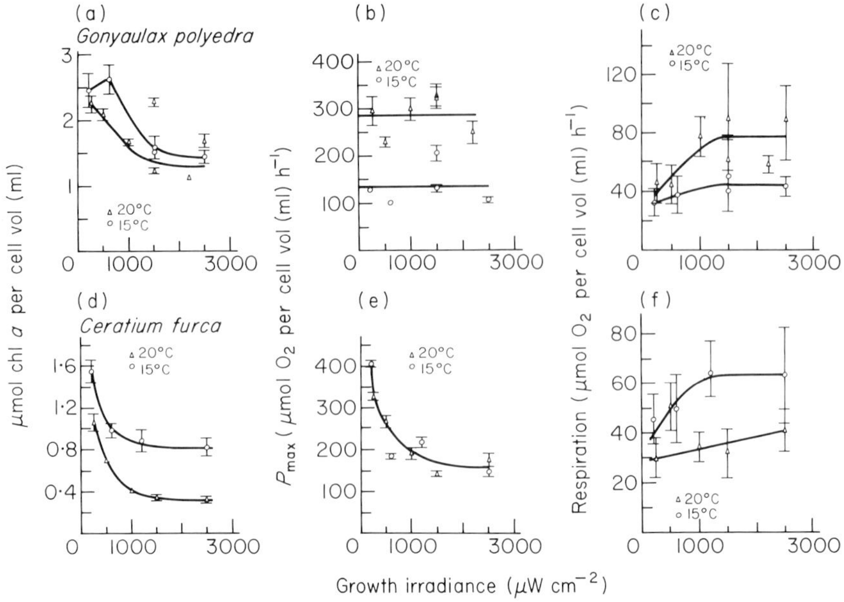

Fig. 5.33. Effect of growth temperature on pigmentation, photosynthesis and respiration for (a) *Gonyaulax polyedra* and (b) *Ceratium furca* grown on 12:12 LD cycles, where growth illumination during the photophase ranged from 2500 to 250 μW cm^{-2} (Meeson 1981).

of *C. furca* while those of *G. polyedra* remain about the same. This observation might suggest a competitive advantage to *G. polyedra* over *C. furca* at 15°C. Analyses of temperature-dependent division rates appear to bear out this suggestion in unialgal, nutrient-sufficient cultures. Maximal division rates of *Ceratium furca* fall twofold between 20 and 15°C, while comparable rates of *G. polyedra* cell division rates fall less than 50%. However, mixed culture experiments do not support this general conclusion and it may be that differential abilities of these two species to compete for limiting nutrients in batch culture reduces chances of predicting the eventual dominance of one species, based solely on steady-state photosynthesis and respiration abilities (Meeson 1981). The light–temperature interactions regulating dinoflagellate primary productivity are obviously complex, rarely studied, and yet important to understanding the *in situ* regulation of photosynthesis and growth. The biochemical bases of temperature regulation are not yet elucidated for dinoflagellates, although comparable studies of other phytoplankton groups suggest a variable temperature dependency of many enzyme processes within all major cell activities, including nutrient uptake, macromolecular synthesis, respiration, photosynthesis, cell division and motility (cf. Li 1980).

2.7 Light-dependent ageing

In *Gonyaulax polyedra*, a variety of cell functions alter as batch culture populations age (Prézelin 1982). In bright light cultures, the signs of ageing appear while cells are still dividing rapidly. First, there is a rapid depletion of cellular carbon and nitrogen during early exponential growth accompanied by a declining nitrate reductase activity, presumably reflecting the decline in external inorganic nutrient flux. By mid-log phase, photosynthetic rates peak and begin a steady decline through late log growth that is unrelated to pigmentation or respiration, but can be associated with an early drop in RUBPCase activity followed by increases in the percentage photosynthate lost from the cell due to excretion (Figs 5.34, 5.35; Prézelin 1982; Sterman & Prézelin 1983; Sterman 1986). When bright light cells are actively dividing in an environment with a declining amount of nutrients, it appears that cell reserves are expended so rapidly in the production of new daughter cells that inorganic nitrogen uptake or carbon fixation rates cannot be sustained, and a decline in productivity results. Thus, it is not surprising that the amplitude of light-saturated and light-limited photosynthesis rhythms decline with increasing population age (Fig. 5.36).

The declining primary productivity of bright light, log-phase cells continues as cell reserves of these nutrients reach a minimum and cells stop dividing.

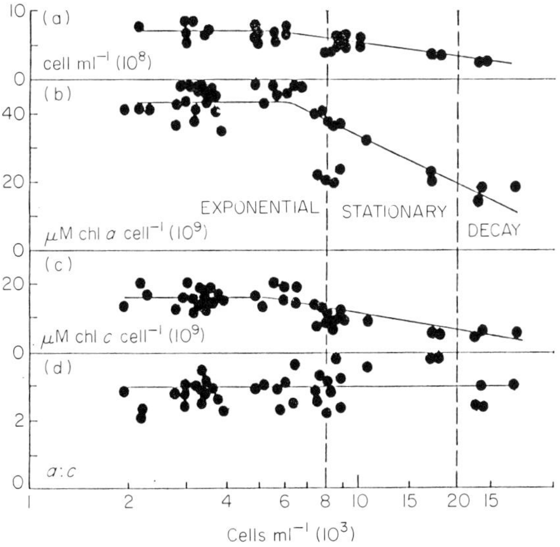

Fig. 5.34. *Gonyaulax polyedra.* Cell volume and chlorophyll content as a function of cell density in seven batch cultures grown at 330 μE m^{-2} s^{-1}. Density-dependent changes are shown for: (a) packed cell volume, expressed as cells ml^{-1} (10^8); (b) chlorophyll *a*; (c) chlorophyll *c*, expressed as μM $cell^{-1}$ (10^9); (d) molar ratios of chlorophyll *a*:chlorophyll *c*.

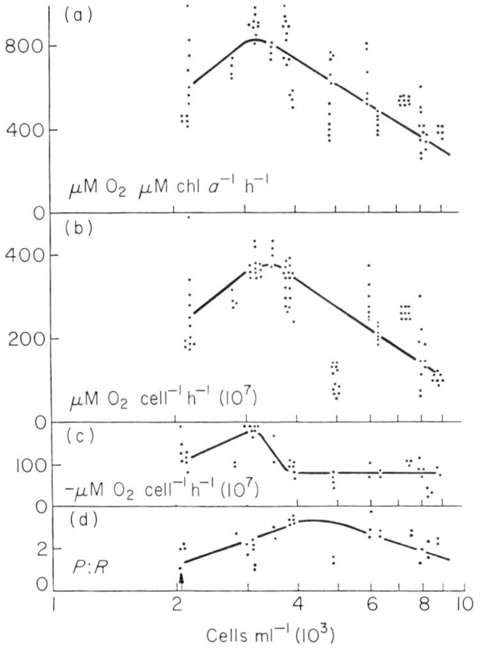

Fig. 5.35. *Gonyaulax polyedra.* Photosynthetic changes are shown as a function of cell density in three batch cultures grown at 330 $\mu E\ m^{-2}\ s^{-1}$. Density-dependent changes are shown for: (a) P_{max} chlorophyll a^{-1}, expressed as μM O_2 μM chlorophyll $a^{-1}\ h^{-1}$; (b) P_{max} cell^{-1}, expressed as μM O_2 cell^{-1} h^{-1} (10^7); (c) respiration rates, expressed as $-\mu M$ O_2 cell^{-1} h^{-1} (10^7); (d) $P:R$ ratios. (Prézelin 1982).

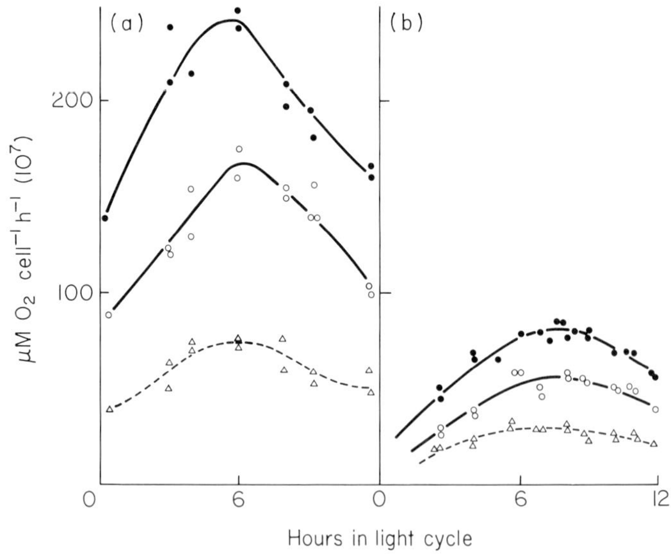

Fig. 5.36. *Gonyaulax polyedra.* Comparison of differences in photosynthetic rhythms in (a) mid-log and (b) stationary phase cultures. Cells were cultured on 12:12 LD cycle with photophase irradiance of 90 $\mu E\ m^{-2}\ s^{-1}$. Samples were briefly illuminated with 360 (●) and 180 (○) $\mu E\ m^{-2}\ s^{-1}$ at different times in the photophase to elucidate periodicity in light-saturating and light-limiting rates of photosynthesis. Respiration rates are seen as open triangles. (From Prézelin & Sweeney 1977, and unpubl. data.)

However, respiration rates do not fall until late stationary phase, suggesting that the basal metabolic demands of the population still are being met. A breakdown of pigment–protein complexes is initiated at the beginning of stationary phase, suggesting that their mobilization provides some of the nutrient requirements for non-dividing cells in a nutrient-depleted environment.

These comparative studies suggest ageing events in *G. polyedra* are similar at high and low growth illuminations, but that they occur earlier and are more widely spaced in time in bright light cultures. At high light intensities, the log-phase decline in photosynthesis appears to be brought on by declining enzyme activity induced by nutrient depletion within the cells. Support comes from the work of Beardall *et al.* (1976), which showed RUBP carboxylase activity dropping markedly during the log-phase growth of *Gonyaulax tamarensis* and our own work shows this to be true also for *G. polyedra* (Sterman & Prézelin 1983; Sterman 1986). The drop in enzyme activity approximated a drop in photosynthesis also measured over the growth curve of both dinoflagellates.

The second ageing stage of *G. polyedra* is seen in stationary phase, where nutrient-limited cells begin to break down photosynthetic membranes in order to sustain cell metabolic requirements. Structural evidence exists to document the breakdown of chloroplast membranes with cell ageing in dinoflagellates. In old cultures of zooxanthellae, Taylor (1968) notes many cell changes including a decrease in chloroplast lamellae and a reduction in stroma. Decreases in chloroplast area also occur in *Ptychodiscus brevis* with increasing population age (Steidinger 1979). Chloroplast structure changes of ageing cultures of *P. cinctum* are associated with large changes in cell volume and pigmentation (Messer & Ben-Shaul 1972). Electron micrographs of the cells show a marked depletion of both chloroplast number as well as thylakoid number, width, and especially length with age (Messer & Ben-Shaul 1972). However, unlike *G. polyedra*, the chlorophyll content of *P. cinctum* first increases and then declines steadily throughout log-phase growth. This early breakdown of pigments might reflect the inability of the stationary phase inoculum to acquire a significant nutrient reserve outside the photosynthetic apparatus when first exposed to new nutrients. At least three generations of nutrient-sufficient growth (accomplished by serial dilution) is required to restore the nutrient content of stationary phase *G. polyedra* cells to that of the log-phase populations studied here.

ACKNOWLEDGEMENTS

Financial support was provided by NSF research grant OCE 78–13919. Special thanks are given to Nanette Sterman for her fine editorial assistance.

3 REFERENCES

ANDERSON J. M., BROWN J. S., LAM E. & MALKIN R. (1983) Chlorophyll *b*: an integral component of photosystem I of higher plant chloroplasts. *Photochem. Photobiol.* **38**, 205–210.

ARGYROUDI-AKOYUNOGLOU J. H. & THOMOU H. (1981) Separation of thylakoid pigment-protein complexes by SDS-sucrose density gradient centrifugation. *FEBS Lett.* **135**, 177–181.

BARRETT J. & ANDERSON J. M. (1977) Thylakoid membrane fragments with different chlorophyll *a*, chlorophyll *c*, and fucoxanthin compositions isolated from the brown seaweed, *Ecklonia radiata*. *Plant Sci. Lett.* **9**, 275–283.

BARRETT J. & ANDERSON J. M. (1980) The P_{700} chlorophyll *a*-protein complex and two major light-harvesting complexes of *Acrocarpia paniculata* and other brown seaweeds. *Biochim. Biophys. Acta* **49**, 309–323.

BEARDALL J., MUKERJI D., GLOVER H. E. & MORRIS I. (1976) The path of carbon in photosynthesis by marine phytoplankton. *J. Phycol.* **12**, 409–417.

BOCZAR B. A. (1985) The functional organization of chlorophyll in chlorophyll *c*-containing marine phytoplankton. Thesis, University of California, Santa Barbara 205 pp.

BOCZAR B. A., PRÉZELIN B. B., MARKWELL J. & THORNBER J. P. (1980) A chlorophyll *c*-containing pigment-protein complex from the marine dinoflagellate, *Glenodinium* sp. *FEBS Letters* **120**, 243–247.

BOCZAR B. A. & PRÉZELIN B. B. (1986) Light and $MgCl_2$-dependent characteristics of four chlorophyll–protein complexes isolated from the marine dinoflagellate, *Glenodinium* sp. *Biochem. Biophys. Acta* (in press).

BROWN J. S., ALBERTE R. S. & THORNBER J. P. (1974) Comparative studies on the occurrence and spectral composition of chlorophyll–protein complexes in a wide variety of plant material. pp. 1951–1962 in *Proc. 3rd Internat. Congress on Photosynthesis.*

BURRIS J. E. (1981) Respiration and photorespiration in marine algae. In *Primary Productivity in the Sea* (Ed. P. G. Falkowski), pp. 411–432. Plenum Press, New York.

BUSH K. J. & SWEENEY B. M. (1972) The activity of ribulose diphosphate carboxylase in extracts of *Gonyaulax polyedra* in the day and night phases of the circadian rhythm of photosynthesis. *Plant Physiol.* **50**, 446–451.

CHANG S. S. & TRENCH R. R. (1982) Peridinin-chlorophyll *a*-proteins from the symbiotic dinoflagellate *Symbiodinium* (=*Gymnodinium*) *microadriaticum*, Freudenthal. *Proc. Royal Soc. London, series B* **215**, 191–210.

CHANG S. S., PRÉZELIN B. B. & TRENCH R. R. (1983) Mechanisms of photoadaptation in 3 strains of the symbiotic dinoflagellate, *Symbiodinium microadiatium Mar. Bio.* **76**, 219–229.

CHISHOLM S. W. (1981) Temporal patterns of cell division in unicellular algae. In *Physiological Bases of Phytoplankton Ecology* (Ed. T. Platt), pp. 150–181. Canadian Bulletin of Fisheries and Aquatic Sciences, No. 210.

CULLEN J. J. & HORRIGAN S. G. (1981) Effects of nitrate on the diurnal vertical migration, carbon to nitrogen ratio, and the photosynthetic capacity of the dinoflagellate *Gymnodinium splendens*. *Mar. Bio.* **62**, 81–89.

DODGE J. D. (1973) *The Fine Structure of Algal Cells.* Academic Press, New York.

DODGE J. D. (1975) A survey of chloroplast ultrastructure in the *Dinophyceae*. *Phycologia* **14**, 253–263.

DOUGHERTY R. C., STRAIN H. H., SVEC W. A., UPHAUS R. A. & KATZ J. J. (1966) Structure of chlorophyll *c*. *J. Am. Chem. Soc.* **88**, 5037–8.

DUBRINSKY Z. & BERMAN T. (1976) Light utilization efficiencies of phytoplankton in Lake Kinneret (Sea of Galilee). *Limnol. Oceanog.* **21**, 226–230.

DUBRINSKY Z. & BERMAN T. (1979) Seasonal changes in the spectral composition of downwelling irradiance in Lake Kinneret, Israel. *Limnol. Oceanog.* **24**, 652–663.

DUBRINSKY Z. & POLNA M. (1976) Pigment composition during a *Peridinium* bloom in Lake Kinneret (Israel) *Hydrobiologia* **51**, 239–243.

DURAND M. & BERKALOFF C. (1985) Pigment composition and chloroplast organization of *Gambierdiscus toxicus* Adachi & Fukuyo (Dinophyceae). *Phycologia* **24**, 217–223.

EPPLEY R. W., HOLM-HANSEN O. & STRICKLAND J. D. H. (1968) Some observations on the vertical migration of dinoflagellates. *J. Phycol.* **4**, 333–340.

EPPLEY R. W. & HARRISON W. G. (1975) Physiological ecology of Gonyaulax polyedra, a red water dinoflagellate off southern California. In *Toxic Dinoflagellate Blooms*, pp. 11–21. Mass. Sci. Technol. Found.

FALKOWSKI P. G. & DUBRINSKY Z. (1981) Light–shade adaptation of *Stylophora pistillata*, a hermatypic coral from the Gulf of Eilat. *Nature (Lond).* **289**, 172–174.

FAUST M. A., SAGER J. C. & MEESON B. W. (1982) Response of *Prorocentrum mariae-leboriae* to different spectral quality and irradiances: growth and pigmentation. *J. Phycol.* **18**, 349–356.

FORWARD R. B. JR (1976) Light and diurnal migration: photobehavior and photophysiology in plankton. *Photochem. Photobio. Rev.* **1**, 157–209.

GOVINDJEE O., WONG D., PRÉZELIN B. B. & SWEENEY B. M. (1979) Chlorophyll *a* fluorescence of *Gonyaulax polyedra* grown on a light–dark cycle after transfer to constant light. *Photochem. Photobio.* **30**, 405–411.

HAIDAK D., MATHEWS C. & SWEENEY B. M. (1966) Pigment protein complex from *Gonyaulax*. *Science* **152**, 212–214.

HALL D. O. & RAO K. K. (1977) *Photosynthesis*. Edward Arnold Publications, London.

HALLDALL P. (Ed.) (1970) *Photobiology of Microorganisms*. Wiley-Interscience, London.

HARDING L. W. JR, MEESON B. W., PRÉZELIN B. B. & SWEENEY B. M. (1981a) Diel periodicity of photosynthesis in marine phytoplankton. *Mar. Biol.* **61**, 95–105.

HARDING L. W. JR, MEESON B. W. & TYLER M. A. (1983) Photoadaptation and diel periodicity of photosynthesis in the dinoflagellate *Prorocentrum marie-lebouriae*. *Mar. Ecol. Prog. Ser.* **13**, 73–85.

HARDING L. W. JR, PRÉZELIN B. B., SWEENEY B. M. & COX J. L. (1981b) Diel oscillations of the photosynthesis–irradiance (*P–I*) relationship in natural assemblages of phytoplankton. *Mar. Biol.* **67**, 167–178.

HARDING L. W. JR, PRÉZELIN B. B., SWEENEY B. M. & COX J. L. (1982a) Diel oscillations of the photosynthesis–irradiance (*P–I*) relationship in natural assemblages of phytoplankton. *Mar. Biol.* **67**, 167–178.

HARDING L. W. JR, PRÉZELIN B. B., SWEENEY B. M. & COX J. L. (1982b) Primary productivity as influenced by diel periodicity of phytoplankton photosynthesis. *Mar. Biol.* **67**, 179–186.

HARRIS G. P., HEANEY S. I. & TALLING J. F. (1979) Physiological and environmental constraints in the ecology of the planktonic dinoflagellate *Ceratium hirundinella*. *Fresh. Bio.* **9**, 413–428.

HASTINGS J. W., ASTRACHAN L. & SWEENEY B. M. (1961) A persistent daily rhythm in photosynthesis. *J. Cell Physiol.* **45**, 69–76.

HAXO F. T., KYCIA J. H., SIEGELMAN H. W. & SOMERS G. F. (1972) *Carotenochlorophyll proteins from dinoflagellates*. Abst. 3rd Internat. Symp. on Carotenoids.

HAXO F. T., KYCIA J. H., SOMERS G. F., BENNETT A. & SIEGELMAN H. W. (1976) Peridinin–chlorophyll *a* proteins of the dinoflagellate *Amphidinium carterae* (Plymouth 450). *Plant Physiol.* **57**, 297–303.

HEANY S. I. & EPPLEY R. W. (1981) Light, temperature and nitrogen as interacting factors affecting diel vertical migrations of dinoflagellates in culture. *J. Plankton Research* **3**, 331–344.

HERMAN E. M. & SWEENEY B. M. (1975) Circadian rhythm of chloroplast ultrastructure in *Gonyaulax polyedra*, concentric organization around a central cluster of ribosomes. *J. Ultrastructure Res.* **50**, 347–354.

HERSEY R. L. & SWIFT E. (1976) Nitrate reductase activity of *Amphidinium carterae* and *Cachonina niei* (*Dinophyceae*) in batch culture: diel periodicity and effects of light intensity and ammonia. *J. Phycol.* **12**, 36–44.

HUMPHREY G. F. (1979) Photosynthetic characteristics of algae grown under constant illumination and light: dark regimes. *J. Exp. Mar. Biol. Ecol.* **40**, 63–70.

INGRAHAM K. & HILLER R. G. (1983) Isolation and characterization of a major chlorophyll a/c_2 light-harvesting protein from *Chroomonas* species (Cryptophyceae). *Biochem. Biophys. Acta* **772**, 310–319.

ISLER I. (Ed.) (1971) *Carotenoids*. Halsted Press, New York.

JEFFREY S. W. (1969) Properties of two spectrally different components in chlorophyll *c* preparations. *Biochim. Biophys. Acta* **177**, 456–467.

JEFFREY S. W. (1980) Algal pigment systems. In *Primary Production in the Sea* (Ed. P. Falkowski), pp. 33–58. Plenum Press, New York.

JEFFREY S. W., SIELICKI M. & HAXO F. T. (1975) Chloroplast pigment patterns in dinoflagellates *J. Phycol.* **11**, 374–384.

JOHANSEN J. F., SVEC W. A., LIAAEN-JENSEN S. & HAXO F. T. (1974) Carotenoids of the Dinophyceae. *Phytochem.* **13**, 2261–2271.

KAMYKOWSKI D. (1981) Dinoflagellate growth rate in water columns of varying turbidity as a function of migration phase with daylight. *J. Plankton Res.* **3**, 357–367.

KIEFER D. A. & LASKER R. (1975) Two blooms of *Gymnodinium splendens*, an unarmored dinoflagellate. *Fish. Bull. US.* **73**, 675–678.

KIRK J. T. O. & TILNEY-BASSETT R. A. E. (1978) *The Plastids*. Elsevier N. Holland, Amsterdam.

KOKA P. & SONG P. S. (1977) The chromophore topology and binding environment of peridinin–chlorophyll *a* protein complexes from marine dinoflagellate algae. *Biochim. Biophys. Acta* **495**, 220–231.

KOWALLIK W. (1970) Light effects on carbohydrate and protein metabolism. In *Photobiology of Microorganisms* (Ed. P. Halldal), pp. 165–186. Wiley-Interscience, New York.

LI W. K. W. (1980) Temperature adaptation in phytoplankton: cellular and photosynthetic characteristics. In *Primary Productivity in the Sea* (Ed. by P. G. Falkowski), pp. 259–280. Plenum Press, New York.

MAC ISAACS J. J. (1978) Diel cycles of inorganic nitrogen uptake in a natural phytoplankton population dominated by *Gonyaulax polyedra*. *Limnol. Oceanogr.* **23**, 1–9.

MANDELLI E. F. (1968) Carotenoid pigments of the dinoflagellate, *Glenodinium foliaceum* Stein. *J. Phycol.* **4**, 347–8.

MEESON B. (1981) Comparative physiology of a typical red tide dinoflagellate, *Gonyaulax polyedra*, with a cosmopolitan species, *Ceratium furca*. PhD. thesis University of California, Santa Barbara.

MEESON B. W. & SWEENEY B. M. (1982) Adaptation of *Ceratium furca* and *Gonyaulax polyedra* (Dinophyceae) to different temperatures and irradiances: growth rates and cell volumes. *J. Phycol.* **18**, 241–245.

MESSER G. & BEN-SHAUL Y. (1972) Changes in chloroplast structure during culture growth of *Peridinium cintum*. *Fa westii* (Dinophyceae). *Phycologia* **11**, 291–299.

MULLER-PARKER G. (1984a) Photophysiology of the symbiotic sea anemone *Aiptasia pulchella*. PhD thesis. University of California, Los Angeles 167 pp.

MULLER-PARKER G. (1984b) Photosynthesis-irradiance responses and photosynthetic periodicity in the sea anemone *Aiptasia pulchella* and its zooplankton. *Mar. Biol.* **82**, 225–232.

PERRY M. J., TALBOT M. C. & ALBERTE R. S. (1981) Photoadaptation in marine phytoplankton: response of the photosynthetic unit. *Mar. Biol.* **62**, 91–101.

PHIPPS D. W. JR (1980) Oxygen inhibition and photorespiration in 2 algal symbionts. *J. Phycol.* **16** (suppl.) 33.

PRÉZELIN B. B. (1976) The role of peridinin–chlorophyll *a*-proteins in the photosynthetic light adaptations of the marine dinoflagellate, *Glenodinium* sp. *Planta* **130**, 225–233.

PRÉZELIN B. B. (1981) Light reactions in photosynthesis. In *Physiological Bases of Phytoplankton Ecology* (Ed. by T. Platt), pp. 1–43. Canadian Bulletin of Fisheries and Aquatic Sciences, No. 210.

PRÉZELIN B. B. (1982) Effects of light intensity on ageing of a dinoflagellate, *Gonyaulax polyedra*. *Mar. Biol.* **69**, 129–135.

PRÉZELIN B. B. & ALBERTE R. S. (1978) Photosynthetic characteristics and organization of chlorophyll in marine dinoflagellates. *Proc. Nat. Acad. Sci. USA* **75**, 1801–1804.

PRÉZELIN B. B. & BOCZAR B. A. (1981) Chlorophyll–protein complexes from the photosynthetic apparatus of dinoflagellates. In *Photosynthesis* III. *Structure and Molecular Organisation of the Photosynthetic Apparatus* (Ed. by G. A. Akoyunglou), pp. 417–426. Balaban Internat. Sci. Serv., Philadelphia.

PRÉZELIN B. B. & BOCZAR B. A. (1986) Molecular bases of cell absorption and fluorescence in phytoplankton: potential applications to studies in optical oceanography. In: *Progress in Phycological Research* Vol. 4 (F. Round and D. Chapman, eds) Elsevier Science Publ. B.V. (in press).

PRÉZELIN B. B. & HAXO F. T. (1976) Purification and characterization of peridinin–chlorophyll *a*-proteins from the marine dinoflagellate *Glenodinium* sp. and *Gonyaulax* polyedra. *Planta* **128**, 133–141.

PRÉZELIN B. B., LEY A. C. & HAXO F. T. (1976) Effect of growth irradiance on the photosynthetic action spectra of the marine dinoflagellate *Glenodinium* sp. *Planta* **130**, 251–256.

PRÉZELIN B. B. & MATLICK H. A. (1980) Time-course of photoadaptation in the photosynthesis–irradiance relationship of a dinoflagellate exhibiting photosynthetic periodicity. *Mar. Biol.* **58**, 85–96.

PRÉZELIN B. B. & MATLICK H. A. (1983) Nutrient-dependent low light adaptation in the dinoflagellate *Gonyaulax polyedra*. *Mar. Bio.* **74**, 141–150.

PRÉZELIN B. B., SAMUELSSON G. & MATLICK H. A. (1986) Photosystem II photoinhibition and altered kinetics of photosynthesis during nutrient-dependent high light photoadaptation in *Gonyaulax polyedra*. *Mar. Biol.* (in press).

PRÉZELIN B. B., MEESON B. W. & SWEENEY B. M. (1977) Characterization of photosynthetic rhythms in marine dinoflagellates. I. Pigmentation, photosynthetic capacity and respiration. *Plant Physiol.* **60**, 384–387.

PRÉZELIN B. B. & SWEENEY B. M. (1977) Characterization of photosynthetic rhythms in marine dinoflagellates. II. Photosynthesis–irradiance curves and *in vivo* chlorophyll *a* fluorescence. *Plant Physiol.* **60**, 388–392.

PRÉZELIN B. B. & SWEENEY B. M. (1978a) Photoadaptation of photosynthesis in *Gonyaulax polyedra*. *Mar. Biol.* **48**, 27–35.

PRÉZELIN B. B. & SWEENEY B. M. (1978b) Photoadaptation of photosynthesis in two bloom-forming dinoflagellates. In *Toxic Dinoflagellate Blooms* (Ed. D. L. Taylor & H. H. Seliger), pp. 101–106. Elsevier North Holland, New York.

RICHARDSON K., BEARDALL J. & RAVEN J. A. (1983) Adaptation of unicellular algae to irradiance: an analysis of strategies. *New Phytol.* **93**, 157–191.

RILEY J. P. & WILSON T. R. S. (1967) The pigments of some marine phytoplankton species. *J. Mar. Biol. Assoc. UK.* **47**, 351–62.

RIVKIN R. B., SELIGER H. H., SWIFT E. & BIGGLY W. H. (1982a) Strategies of light–shade adaptation by the oceanic dinoflagellates *Pyrocystis noctiluca* and *Pyrocystis fusiformis*. *Mar. Biol.* **68**, 181–191.

RIVKIN R. B. & SWIFT E. (1982) Phosphate uptake by the oceanic dinoflagellate *Pyrocystis noctiluca*. *J. Phycol.* **18**, 113–120.

RIVKIN R. B. & VOYTEK M. A. (1985) Photoadaptations of photosynthesis by dinoflagellates from natural populations: a species approach. In: *3rd Internat. Cong. Toxic Dinoflagellates*.

RIVKIN R. B., VOYTEK M. A. & SELIGER H. H. (1982b) Strategies of phytoplankton division rates in light-limited environments. *Science* **215**, 1123–1125.

SAMUELSSON G. & PRÉZELIN B. B. (1985) Photosynthetic electron transport in cell-free extracts of diverse phytoplankton. *J. Phycol.* **21**, 453–457.

SAMUELSSON G., SWEENEY B. M., MATLICK H. A. & PRÉZELIN B. B. (1983) Changes in photosystem II account for the circadian rhythm in photosynthesis in *Gonyaulax polyedra*. *Plant Physiol.* **73**, 329–331.

SEELEY G. R. (1966) The structure and chemistry of functional groups. In *The Chlorophylls, Physical, Chemical and Biological Properties* (Ed. L. P. Vernon & G. R. Seely), pp. 67–110, Academic Press, New York.

SIEGELMAN H. W., KYCIA J. H. & HAXO F. T. (1977) Peridinin–chlorophyll *a* proteins of dinoflagellate algae. *Brookhaven Nat. Symp.* **28**, 162–169.

SMITH K. C. (1977) *The Science of Photobiology.* Plenum, New York.

SONG P. S., KOKA P., PRÉZELIN B. B. & HAXO F. T. (1976) Molecular topology of the photosynthetic light-harvesting pigment complex, peridinin–chlorophyll *a*-protein, from marine dinoflagellates. *Biochemistry* **15**, 4422–4427.

SONG P. S., WALKER E. B., JUNG J., AUERBACH R. A., ROBINSON G. W. & PRÉZELIN B. B. (1980) Primary processes of photobiological receptors. In *New Horizons in Biological Chemistry* (Ed. by M. Koike), pp. 79–93.

STEEMAN NIELSON E. & JORGENSEN E. G. (1962) The adaptation to different light intensities in *Chlorella vulgaris* and the time dependence on transfer to a new light intensity. *Physiol. Plant.* **15**, 505–513.

STEEMAN NIELSON E. & JORGENSEN E. G. (1968a) The adaptation of plankton algae. I. General part. *Physiol. Plant.* **21**, 401–408.

STEEMAN NIELSON E. & JORGENSEN E. G. (1968b) The adaptation of plankton algae. III. With special consideration of the importance in nature. *Physiol. Plant.* **21**, 647–657.

STEIDINGER K. A. (1979) Quantitative ultrastructure variation between culture and field specimens of the dinoflagellate *Ptychodiscus breve*. PhD. dissertation, University of South Florida.

STERMAN N. (1986) Light-dependent aging phenomena in the marine dinoflagellate *Gonyaulax polyedra*. Ms thesis Univ. California, Santa Barbara.

STERMAN N. T. & PRÉZELIN B. B. (1983) Light-dependent aging patterns in *Gonyaulax polyedra*. *Abst. Eos*, p. 953.

STRAIN H. H., SVEC W. A., AITZETMULLER K., GRANDOLFO M. C., KATZ J. J., KJOSEN H., NORGARD S., LIAAEN-JENSEN S., HAXO F. T., WEGFARHT P. & RAPOPORT H. (1971a) The structure of peridinin, the characteristic dinoflagellate carotenoid. *J. Am. Chem. Soc.* **93**, 1823–5.

STRAIN H. H., COPE B. T., MCDONALD G. N., SVEC W. A. & KATZ J. J. (1971b) Chlorophylls c_1 and c_2. *Phytochem.* **10**, 1109–14.

SWEENEY B. M. (1960) The photosynthetic rhythm in single cells of *G. polyedra*. *Cold Spring Harbor Symp. Quant. Biol.* **25**, 145–148.

SWEENEY B. M. (1981a) Chloroplast membranes of *Gonyaulax polyedra* (Pyrrophyta), a photosynthetic dinoflagellate with peridinin–chlorophyll-protein, studied by the freeze-fracture technique. *J. Phycol.* **17**, 95–101.

SWEENEY B. M. (1981b) The circadian rhythms in bioluminescence, photosynthesis and organellar movements in the large dinoflagellate, Pyrocystis fusiformis. In *International Cell Biology 1980–1981* (Ed. H. G. Schweiger), pp. 807–814. Springer-Verlag, Berlin.

SWEENEY B. M. (1981c) Circadian timing in the unicellular autotrophic dinoflagellate, *Gonyaulax polyedra*. *Ber. Dtsch. Bot. Ges. Bd.* **94**, S, 335–345.

SWEENEY B. M. & HASTINGS J. W. (1958) Rhythmic cell division in populations of *Gonyaulax polyedra*. *J. Protozool.* **5**, 217–221.

SWEENEY B. M. & PRÉZELIN B. B. (1979) Yearly review: circadian rhythms. *Photochem. Photobio.* **27**, 841–847.

SWEENEY B. M., PRÉZELIN B. B., WONG D. & GOVINDJEE O. (1979) *In vivo* chlorophyll *a* fluorescence transients and the circadian rhythm of photosynthesis in *Gonyaulax polyedra*. *Photochem.* and *Photobiol.* **30**, 309–311.

SWIFT E. & TAYLOR W. R. (1967) Bioluminescence and chloroplast movement in the dinoflagellate *Pyrocystis lunula*. *J. Phycol.* **3**, 77–81.

TAYLOR D. L. (1968) *In situ* studies of the cytochemistry and ultrastructure of a symbiont marine dinoflagellate. *J. Mar. Biol. Assn. UK.* **48**, 349–366.

THIMANN K. V. (1980) The senescence of leaves. In *Senescence in Plants* (Ed. K. V. Thimann), pp. 85–116. CRC Press, Boca Raton.

THORNBER J. P. (1975) Chlorophyll-proteins: Light-harvesting and reaction center components of plants. *Ann. Rev. Plant Physiol.* **26**, 127–158.

THORNBER J. P., ALBERTE R. S., HUNTER F. A., SHIOZAWA J. A. & KAN K. S. (1977) The organization of chlorophyll in the plant photosynthetic unit. In *Chlorophyll-Proteins, Reaction Centers, and Photosynthetic Membranes* (Ed. J. M. Olson & G. Hind), pp. 132–148. Brookhaven Symp. Biol. 28.

TITLYANOV E. A., SHAPOSHNIKOVA M. G. & ZVALINSKII V. I. (1980) Photosynthesis and adaptation of corals to irradiance. 1. Content and native state of photosynthetic pigments in symbiotic microalga. *Photosynthetica* **14**, 413–421.

TOLBERT N. E. (1974) Photo respiration. In: *Algal Physiology and Biochemistry* (Ed. W. D. P. Stewart), pp. 474–504. Univ. California Press, Berkeley.

TOMAS R. N. & COX E. R. (1973) Observations on the symbiosis of *Peridinium balticum* and its intracellular alga. I. Ultrastructure. *J. Phycol.* **9**, 304–23.

TYLER M. A. & SELIGER H. H. (1978) Annual subsurface transport of a red tide dinoflagellate to its bloom area: water circulation patterns and organism distributions in the Chesapeake Bay. *Limnol. Oceanog.* **23**, 227–246.

TYLER M. A. & SELIGER H. H. (1981) Selection for a red tide organism: physiological responses to the physical environment. *Limnol. Oceanog.* **26**, 310–324.

VESK M. & JEFFREY S. W. (1977) Effect of blue-green light on photosynthetic pigments and chloroplast structure in unicellular marine algae from six classes. *J. Phycol.* **13**, 280–288.

WALL, D. & BRIAND F. (1979) Response of lake phytoplankton communities to *in situ* manipulations of light intensity and colour. *J. Plankton Res.* **1**, 102–112.

WEILER C. S. & EPPLEY R. W. (1979) Temporal pattern of division in the dinoflagellate genus *Ceratium* and its application to the determination of growth rate. *J. exp. mar. Biol. Ecol.* **39**, 1–24.

WEILER C. S. & KARL D. M. (1979) Diel changes in phased dividing cultures of *Ceratium furca* dinophycease nucleotide triphosphates, adenylate energy charge, cell carbon and patterns of vertical migration. *J. Phycol.* **15**, 384–391.

WHITTLE S. J. & CASSELTON J. P. (1968) Peridinin as the major xanthophyll of the *Dinophyceae*. *Br. Phycol. Bull.* **3**, 602–3.

WITHERS N. & HAXO F. T. (1975) Chlorophyll c_1 and c_2 and extraplastidic carotenoids in the dinoflagellate, *Peridinium foliaceum* Stein. *Plant Science Letters* **5**, 7–15.

YENTSCH C. M., CUCCI T. L., PHINNEY D. A., SELVIN R. & GLOVER H. E. (1985) Adaptation to low photon flux densities in *Protogonyaulax tamarensis* var. *excavata*, with reference to chloroplast photomorphogenesis. *Mar. Biol.* **89**, 9–20.

ZVALINSKII V. I., LELETKIN V. A., TITLYANOV E. A. & SHAPOSHNIKOVA M. G. (1980) Photosynthesis and adaptation of corals to irradiance. 2. Oxygen exchange. *Photosynthetica* **14**, 422–430.

CHAPTER 6
HETEROTROPHIC NUTRITION

GREGORY GAINES
Department of Oceanography, University of British Columbia, 6270 University Boulevard, Vancouver, BC, Canada V6T 1W5

MALTE ELBRÄCHTER
Biologische Anstalt Helgoland, Litoralstation List, 2282 List auf Sylt, Hafenstrasse 3, West Germany (FRD)

1 INTRODUCTION

Dinoflagellates have been regarded both as plants and as animals because some are photosynthetic and others not. Approximately half of the original species descriptions did not specify the presence or absence of chloroplasts. Only about half of the rest have chloroplasts. Thus, it is estimated that approximately half of the known species of living dinoflagellates are obligate heterotrophs. Dinoflagellate heterotrophy (and particularly phagotrophy) was shown great interest in the nineteenth century, this being the main argument for classifying them as animals. Despite this, little is known of their nutrition. Most (of the few) modern studies have been concerned with the use of dissolved substances; very few of these have examined particle feeding. The fragility and apparently specific food requirements of these forms have prevented laboratory culture of all but a few neritic species, which may not be representative of the rest. A majority of open ocean dinoflagellates are colourless heterotrophic forms, yet they remain almost completely unknown. Detailed observations of phagotrophic feeding have been recorded, but the ecological implications of these events have only recently been shown some interest (Spero 1979; Morey-Gaines & Ruse

1980; Kimor 1981). Studies of predatory dinoflagellates show that they can have a substantial impact on higher trophic levels (Prasad 1953, 1958; Bursa 1961; Kimor 1981; Lessard 1984; Lessard & Swift 1985).

2 TERMINOLOGY

Heterotrophy is the use of organic compounds for metabolism, growth and reproduction. *Autotrophy* is the use only of inorganic compounds (for photosynthesis). While these terms leave a great deal unsaid (e.g. the sources of cellular energy and reducing power), they provide the most useful basic distinction between nutritional types of dinoflagellates (Lwoff *et al.* 1946; Pringsheim 1963).

It is important to stress that heterotrophy involves *growth or increased survival* of an organism by using organic substances. *Uptake* alone is not sufficient (Richardson & Fogg 1982). Many organic substances (e.g. fatty acids) can penetrate the plasmalemma by osmosis and yet be of no benefit to the organism (Wright & Hobbie 1966; Droop 1974).

3 TYPES OF HETEROTROPHY

3.1 Auxotrophy

Although phycologists commonly think of dinoflagellates as photosynthetic, only six species are presently known to have no need of organic substances for growth in the light (i.e. to be strict autotrophs): '*Exuviaella* sp.' (McLaughlin & Zahl 1959), *Peridinium cinctum* f. *ovoplanum* Lindemann (Carefoot 1968), *P. inconspicuum* Lemmermann, *P. volzii* Lemmermann, *P. willei* Huitfeld-Kass (Holt & Pfiester 1981) and *Symbiodinium microadriaticum* Freudenthal (= *Zooxanthella microadriatica* (Freudenthal) Loeblich & Sherley) (McLaughlin & Zahl 1959). Most of these (the *Peridinium* species) are freshwater forms (Holt & Pfiester 1981).

Most photosynthetic dinoflagellates are *auxotrophic*. That is, although their cell functions are supported mainly by photosynthesis, some specific external organic compounds (usually the three vitamins: B_{12} (cyanocobalamin), biotin, thiamine) are required in small amounts, presumably as catalysts (Table 6.1). Vitamin B_{12} is the most often required, and many species respond to analogues as well (Droop 1957; Loeblich 1966; Provasoli & Carlucci 1974). Very high concentrations can also enhance growth significantly (Provasoli & Carlucci 1974), but the mechanism for this is not known. Vitamins may limit growth in oligotrophic waters, especially in the tropics, because greater amounts are needed at higher temperatures (Gold & Baren 1966). Non-photosynthetic dinoflagellates also require some or all of the same three vitamins needed by photosynthetic forms.

Table 6.1. Auxotrophic requirements of dinoflagellates: +, required; −, not required; S, stimulatory but not required; I, inhibitory; ?, unclear

Species	B12	Biotin	Thiamine	Other	References
Amphidinium carterae					
Hulburt	+	+	+	−	2,4,5,6
A. rhynchocephalum					
Anissimowa	+	+	+	−	2,4,5,6
Ceratium hirundinella					
(O. F. Müller) Bergh	+	−	−	−	1
Coolia monotis					
Meunier	+	−	−	−	2
Crypthecodinium cohnii					
(Seligo) Chatton in Grassé	−	+	S	−	2,4,5
'*Exuviaella* sp'	−	−	−	−	4
Glenodinium hallii					
Freudenthal & Lee	+	−	S	−	2,4
Gonyaulax polyedra					
Stein	+	−	−	−	2,4,5
Gymnodinium breve					
Davis	+	+	+	?	2,4,5
G. sanguineum					
Hirasaka	+	−	−	−	2,4,5
Gyrodinium californicum					
Bursa	+	−	−	−	2,4,5,6
G. resplendens					
Hulburt	+	−	−	−	2,4,5,6
G. uncatenum					
Hulburt	+	−	−	−	2,4,5,6
Heterocapsa (=*Cachonina*)					
niei (Loeblich) Morrill & Loeblich	+	−	+	−	2
Oxyrrhis marina					
Dujardin	+	+	+	+*	2,4
Peridinium balticum					
(Levander) Lemmermann	+	−	−	−	2,4,5,6
P. chattoni					
Biecheler	+	−	−	−	2,4,5
P. cinctum f. *ovoplanum*					
Lindemann	−	−	−	−	2,4,5
P. foliaceum					
(Stein) Biecheler	+	−	−	−	2,4,5
P. hangoei					
Schiller	+	−	−	−	2,6
P. inconspicuum					
Lemmermann	−	−	−	−	3
P. limbatum					
(Stokes) Lemmermann	−	−	+	−	3
P. volzii					
Lemmermann	−	I	−	−	3
P. willei					
Huitfeld-Kass	−	−	−	−	3

Species	B12	Biotin	Thiamine	Other	References
Peridiniopsis polonicum (Woloszynska) Bourrelly	+	−	−	−	3
'*Polykrikos schwartzii*'†	+	+	+	−	2,6
Prorocentrum cassubicum (Woloszynska) Dodge	+	−	−	−	2,4,5,6
P. micans Ehrenberg	+	+	−	+‡	2,4
P. sp.	+	−	−	−	2,6
Scrippsiella trochoidea (Stein) Loeblich	+	−	−	−	2,4,5
Symbiodinium microadriaticum Freudenthal (= *Zooxanthella microadriatica* (Freudenthal) Loeblich & Sherley)	−	−	−	−	2,4
Woloszynskia limnetica Bursa	+	?	−	−	2,4,5,6

*Requires a quinone, either plastoquinone or ubiquinone, and a steroid, though this may not be absolute.
†This identification is incorrect. The organism is apparently a chain-forming, photosynthetic *Gymnodinium*.
‡Soil extract—isolate would not grow in synthetic medium.
References:
1 Bruno & McLaughlin (1977)
2 Hastings & Thomas (1977)
3 Holt & Pfiester (1981)
4 Loeblich (1966)
5 Provasoli (1958)
6 Provasoli & Carlucci (1974)

3.2 Mixotrophy

Mixotrophy is heterotrophy by an organism bearing chloroplasts (Pfeffer 1881). If either mechanism (heterotrophy or autotrophy) alone can support cell functions, the nutrition is termed *amphitrophy*. If both are required, it is called *mixotrophy s. str*.

AMPHITROPHY

Evidence of some degree of facultative heterotrophic activity has been found in several photosynthetic species (Table 6.2). Of twelve species tested for growth on dissolved organic compounds, nine have shown growth enhancement. In the light, with photosynthesis inhibited by DCMU (Gromet-Elhanan 1965), glycerol enhanced the growth or survival of *Amphidinium carterae* Hulburt, *Fragilidium heterolobum* Balech, *Heterocapsa* (=*Cachonina*) *niei* (Loeblich) Morrill & Loeblich, *Peridinium balticum* (Levander) Lemmerman, *P. foliaceum* (Stein) Biecheler and a free-living strain (but not a symbiotic strain) of

Table 6.2. Concentrations of various organic carbon sources found to support growth of dinoflagellates in heterotrophic culture or to enhance growth in tests for heterotrophic capabilities. Concentrations expressed in mM (—, not tested or not utilized)

Compound	*Amphidinium*[1,4] *carterae* Hulburt	*Amphidinium*[2] *hœfleri* Schiller & Diskus	*Crypthecodinium*[3] *cohnii* (Seligo) Chatton in Grassé
Carbohydrates:			
Ethanol	—	—	22–43
Fructose	—	—	—
Galactose	—	—	11–33
Glucose	—	—	11–220
Glycerol	50	—	22–65
Mannitol	—	—	—
Sucrose	—	—	3–18
Amino acids & peptides:			
Alanine	—	—	—
Asparagine	—	—	—
Glutamic Acid	—	—	—
Glycine	—	0·67	—
Glycylglycine	—	0·38	—
Proline	—	—	—
Valine	—	0·43	—
Organic acids:			
Acetic	—	—	15–67
Butyric	—	—	11–34
Caproic	—	—	2·6
Caprylic	—	—	0·07–0·21
Fumaric	—	—	4·3–22
Glycolic	—	—	—
Heptylic	—	—	2·3
Lactic	—	—	—
Malic	—	—	3·7–19
Malonic	—	—	—
Oleic	—	—	0·04–0·11
Propionic	—	—	13–40
Pyruvic	—	—	5·7–28
Succinic	—	—	4·2–21
Valeric	—	—	2·9–10
Miscellaneous:			
Soil extract	15 ml l^{-1}*	—	—
Growth conditions	Light and Light + DCMU[4]	Light	Dark
Response	Enhanced growth	Increased cell size	Growth

Compound	*Fragilidium*[4] *heterolobum* Balech *ex* Loeblich	*Gloeodinium*[5] *montanum* Klebs	*Gymnodinium*[6] *inversum* Nygaard
Carbohydrates:			
Ethanol	—	—	—
Fructose	—	—	—
Galactose	—	—	—
Glucose	—	—	—
Glycerol	54	—	—
Mannitol	—	—	—
Sucrose	—	—	—
Amino acids & peptides:			
Alanine	—	—	—
Asparagine	—	—	—
Glutamic Acid	—	—	—
Glycine	—	—	—
Glycylglycine	—	—	—
Proline	—	—	—
Valine	—	—	—
Organic acids:			
Acetic	—	—	1·7
Butyric	—	—	—
Caproic	—	—	—
Caprylic	—	—	—
Fumaric	—	—	—
Glycolic	—	—	—
Heptylic	—	—	—
Lactic	—	—	—
Malic	—	—	—
Malonic	—	—	—
Oleic	—	—	—
Propionic	—	—	—
Pyruvic	—	—	—
Succinic	—	—	—
Valeric	—	—	—
Miscellaneous:			
Soil extract	—	+†	—
Growth conditions	Light + DCMU	Dark at −8 to +49°C	Dark
Response	Enhanced growth	Growth	Growth

Table 6.2 (continued)

Compound	*Gyrodinium*[7] *californicum* Bursa	*Gyrodinium*[8] *lebouriae* Herdmann	*Gyrodinium*[7] *resplendens* Hulburt
Carbohydrates:			
Ethanol	—	—	—
Fructose	—	—	—
Galactose	—	0·06	—
Glucose	—	0·06	—
Glycerol	—	—	—
Mannitol	—	—	—
Sucrose	—	—	—
Amino acids & peptides:			
Alanine	1·1	—	1·1
Asparagine	0·76	—	0·76
Glutamic Acid	0·68	—	0·68
Glycine	1·3	—	1·3
Glycylglycine	—	—	—
Proline	—	—	—
Valine	—	—	—
Organic acids:			
Acetic	—	0·12	—
Butyric	—	—	—
Caproic	—	—	—
Caprylic	—	—	—
Fumaric	—	—	—
Glycolic	—	—	—
Heptylic	—	—	—
Lactic	—	—	—
Malic	—	—	—
Malonic	—	—	—
Oleic	—	—	—
Propionic	—	—	—
Pyruvic	—	—	—
Succinic	—	—	—
Valeric	—	—	—
Miscellaneous:			
Soil extract	—	—	—
Growth conditions	Light	Dark	Light
Response	Enhanced growth	Growth	Enhanced growth

Compound	*Gyrodinium*[7] *uncatenum* Hulburt	*Heterocapsa*[4] (=*Cachonina*) *niei* (Loeblich) Morrill & Loeblich	*Noctiluca*[9] *scintillans* Macartney
Carbohydrates:			
Ethanol	—	—	—
Fructose	—	—	—
Galactose	—	—	—
Glucose	—	—	0·28–11
Glycerol	—	54	0·54–22
Mannitol	—	—	—
Sucrose	—	—	—
Amino acids & peptides:			
Alanine	1·1	—	—
Asparagine	0·76	—	—
Glutamic Acid	0·68	—	—
Glycine	1·3	—	—
Glycylglycine	—	—	—
Proline	—	—	—
Valine	—	—	—
Organic acids:			
Acetic	—	—	0·61–24
Butyric	—	—	—
Caproic	—	—	—
Caprylic	—	—	—
Fumaric	—	—	—
Glycolic	—	66	—
Heptylic	—	—	—
Lactic	—	—	1
Malic	—	—	—
Malonic	—	—	—
Oleic	—	—	—
Propionic	—	—	—
Pyruvic	—	—	—
Succinic	—	18	—
Valeric	—	—	—
Miscellaneous:			
Soil extract	—	15 ml l^{-1}*	—
Growth conditions	Light	Light + DCMU	Dark
Response	Enhanced growth	Enhanced growth	Growth

Table 6.2 (continued)

Compound	*Oxyrrhis*[10] *marina* Dujardin	*Peridinium*[4] *balticum* (Levander) Lemmerman	*Peridinium*[11] *cinctum* f. *ovoplanum* Lindemann
Carbohydrates:			
Ethanol	0·43–43	—	—
Fructose	—	—	2·5
Galactose	—	—	2·5
Glucose	—	—	2·5
Glycerol	—	54	—
Mannitol	—	—	—
Sucrose	—	—	—
Amino acids & peptides:			
Alanine	—	—	—
Asparagine	—	—	—
Glutamic Acid	—	—	—
Glycine	—	—	—
Glycylglycine	—	—	—
Proline	—	—	—
Valine	—	—	—
Organic acids:			
Acetic	0·12–122	—	—
Butyric	—	—	—
Caproic	—	—	—
Caprylic	—	—	—
Fumaric	—	—	—
Glycolic	—	—	—
Heptylic	—	—	—
Lactic	—	—	—
Malic	—	—	2·5
Malonic	—	—	2·5
Oleic	—	—	—
Propionic	—	—	—
Pyruvic	—	—	2·5
Succinic	—	—	—
Valeric	—	—	—
Miscellaneous:			
Soil extract	—	—	—
Growth conditions	Dark	Light + DCMU	Light
Response	Growth	Enhanced growth	Enhanced growth

Compound	*Peridinium*[4] *foliaceum* (Stein) Biecheler	*Symbiodinium*[4] (= *Zooxanthella*) *microadriaticum* Freudenthal	Seawater[6,12]
Carbohydrates:			
Ethanol	—	—	—
Fructose	—	—	—
Galactose	—	—	—
Glucose	—	—	$(2{\cdot}2–1082) \times 10^{-6}$
Glycerol	54	54	—
Mannitol	28	—	—
Sucrose	—	—	—
Amino acids & peptides:			
Alanine	—	—	$(9{\cdot}0–415) \times 10^{-6}$
Asparagine	—	—	—
Glutamic Acid	—	—	$(4{\cdot}8–177) \times 10^{-6}$
Glycine	—	—	$(44–812) \times 10^{-6}$
Glycylglycine	—	—	—
Proline	—	—	$(40–174) \times 10^{-6}$
Valine	—	—	$(3{\cdot}4–196) \times 10^{-6}$
Organic acids:			
Acetic	—	—	$(0{\cdot}83–17) \times 10^{-3}$
Butyric	—	—	$(0–295) \times 10^{-6}$
Caproic	—	—	$(0–224) \times 10^{-6}$
Caprylic	—	—	$(0–180) \times 10^{-6}$
Fumaric	—	—	$(0–224) \times 10^{-6}$
Glycolic	—	—	$(0–342) \times 10^{-6}$
Heptylic	—	—	$(0–200) \times 10^{-6}$
Lactic	—	—	$(0–289) \times 10^{-6}$
Malic	—	—	$2{\cdot}2 \times 10^{-6}$
Malonic	—	—	$(0–250) \times 10^{-6}$
Oleic	—	—	$(0–92) \times 10^{-6}$
Propionic	—	—	$(0–351) \times 10^{-6}$
Pyruvic	—	—	$(0–295) \times 10^{-6}$
Succinic	—	—	$(0–220) \times 10^{-6}$
Valeric	—	—	$(0–255) \times 10^{-6}$
Miscellaneous:			
Soil extract	15 & 60 ml l^{-1}*	15 ml l^{-1}*	—
Growth conditions	Light + DCMU	Light + DCMU	—
Response	Enhanced growth	Enhanced growth	—

*Growth stimulation by soil extract was not due to its use as a carbon source.
†Cultured in bog water.

References:
1 Cheng & Antia (1970)
2 Elbrächter (1972)
3 Provasoli & Gold (1962), Gold & Baren (1966), Keller, Hutner & Keller (1968), Tuttle & Loeblich (1975)
4 Morrill & Loeblich (1979)
5 Hosiaisluoma (1975)
6 Wright & Hobbie (1966)
7 Provasoli & McLaughlin (1963)
8 Lee (1977)
9 McGinn & Gold (1969); McGinn (1971)
10 Droop (1959b)
11 Carefoot (1968)
12 Andrews & Williams (1971), Bohling (1970), Hanson & Snyder (1980), Lee & Bada (1977), Meyers (1976), Riley & Chester (1971), Wright & Hobbie (1966)

Symbiodinium microadriaticum (Morrill & Loeblich 1979). Soil extract stimulated *A. carterae*, *H. niei*, *P. foliaceum* and *S. microadriaticum*, as glycolate, lactate and succinate did *H. niei* (Morrill & Loeblich 1979). Without DCMU (i.e. with photosynthesis permitted), glycerol (at concentrations less than 50 mM) also enhanced the growth of *A. carterae* (Cheng & Antia 1970). *Gymnodinium inversum* Nygaard grew on acetate and glucose in the light and in the dark (Wright & Hobbie 1966). *Gloeodinium montanum* Klebs was cultivated in bog water in the dark for 1–2 years (Hosiaisluoma 1975). Growth of *Amphidinium hœfleri* Schiller & Diskus in the light was not enhanced in the presence of amino acids, but the mean cell size increased significantly (Elbrächter 1972).

Several species have failed to grow on organic substances in the dark. *Gymnodinium breve* Davis did not grow in the dark on any of a large variety of organic compounds (Aldrich 1962). *Prorocentrum micans* Ehrenberg did not grow on glucose (Droop 1957, 1974; Kain & Fogg 1960).

The importance of amphitrophy in the natural ecosystem is questionable. All of these studies employed unnaturally high concentrations of organic enrichments. Nearly natural concentrations of glucose, glycerol and acetate, did not enhance growth or increase dark survival of axenic cultures of *Amphidinium carterae*, *A. hœfleri*, *Gymnodinium incertum* Herdman, *Gymnodinium* sp. (a coral symbiont), *Prorocentrum balticum* (Lohmann) Loeblich or *Scrippsiella trochoidea* (Stein) Loeblich (Richardson & Fogg 1982). In addition, organic carbon incorporation was usually very small (less than 2% of inorganic CO_2 fixation). Only cultures of *A. carterae* with bacteria were growth-enhanced by organic additives, suggesting that this effect may have been caused by decreased CO_2 limitation of the algae due to increased bacterial respiration. These results for *A. carterae* (Richardson & Fogg 1982) directly contrast with those of other workers (Cheng & Antia 1970; Morrill & Loeblich 1979). In low light, *S. trochoidea* incorporated significant amounts of organic carbon (compared to inorganic carbon fixation), but not enough to support growth (Richardson & Fogg 1982). In contrast, Hellebust (1970) found that *S. trochoidea* took up glucose, lysine and alanine in amounts up to 110% of the light-saturated CO_2 photoassimilation rate. These discrepancies may be ascribable to differences among strains or in substrate concentrations.

Amino acids enhanced the growth of *Gyrodinium californicum* Bursa, *G. resplendens* Hulburt and *G. uncatenum* Hulburt in the light (Provasoli & McLaughlin 1961, 1963). However, these compounds were probably used as photosynthetic nitrogen sources (Morrill & Loeblich 1979).

Several photosynthetic dinoflagellates have been found to take up or incorporate dissolved organic substances, without showing any growth enhancement. *Gymnodinium sanguineum* (=*nelsoni*=*splendens*) Hirasaka and *Scrippsiella trochoidea* took up glucose, lysine and alanine in significant amounts (6–110% of the light-saturated CO_2 photoassimilation rate), although growth was not measured (Hellebust 1970). Leucine and phenylalanine were incorporated into *Gonyaulax polyedra*, *Gymnodinium sanguineum*, and *Scrippsiella trochoidea* (Ross & Abbott 1980). *Gymnodinium breve* used glucose in the light (but not in the dark) for the synthesis of cellular components, although the rate of uptake was too low to be a significant auxiliary source of cell carbon (Baden & Mende 1978). No growth enhancement was shown.

MIXOTROPHY *SENSU STRICTO*

Mixotrophy *sensu stricto* is difficult to separate in practice from auxotrophy, as the required organic compounds must be shown to be used directly for nutrition and not as trace substances (i.e. catalysts). Although unproven, the most likely candidates for this are the parasites *Dissodinium pseudolunula* Swift *ex* Elbrächter & Drebes, *Protoodinium chattonii* Hovasse (Fig. 6.1) and species of *Blastodinium* (Chatton 1920; Cachon & Cachon 1971; Drebes 1978; Elbrächter & Drebes 1978).

MIXOTROPHIC PHAGOTROPHY

Mixotrophic phagotrophy is the ingestion of food particles by dinoflagellates with chloroplasts. We treat it here under mixotrophy, but these food items may be used as sources of concentrated vitamins rather than as carbon sources.

The first record of this was *Gyrodinium fissum* Levander from the Baltic Sea

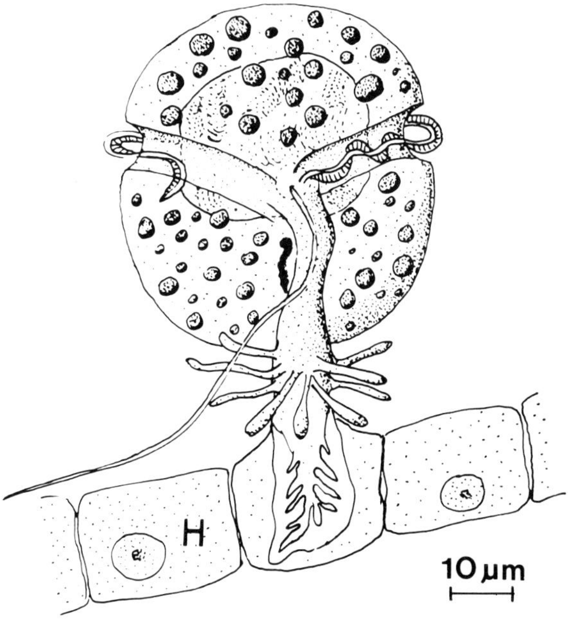

Fig. 6.1. Ventral view of *Protoodinium chattonii* Hovasse, showing its attachment or holdfast organelle (H). From Cachon & Cachon (1971).

(Levander 1894), in which foreign inclusions were seen: a diatom and an unidentified object. The process of ingestion was not observed. This record is complicated by taxonomic confusion (Elbrächter 1979), and has not been subsequently verified.

Biecheler (1936b) described the capture and digestion of a planktonic ciliate (*Strombidium* Claparède & Lachmann) by *Gyrodinium pavillardi* Biecheler, a marine species of the same size as the ciliate. *Strombidium* was immobilized and drawn into *G. pavillardi* through the posterior end of the sulcus (Fig. 6.2). *G. pavillardi* also ingested smaller dinoflagellates. These events have not been subsequently verified, either.

Gyrodinium pavillardi is itself the prey of another dinoflagellate with 'regressed chromatophores': *Gyrodinium vorax* Biecheler. This species displays another kind of phagotrophic nutrition, which we call myzocytosis (Schnepf & Deichgräber 1984): piercing the prey and sucking out its contents, including both dissolved and particulate organic substances. This is found in many ectoparasitic forms, e.g. *Chytriodinium* Chatton, *Dissodinium* Klebs *ex* Pascher and *Paulsenella* Chatton, and may be widespread. As Biecheler (1952) described, *G. vorax* immobilized its prey and adhered to its hypocone or girdle. The cell contents of *G. pavillardi* were then drawn into the body of *G. vorax*, after which the predator abandoned what was left, usually became spherical, discarded its flagella, and remained inactive for several hours.

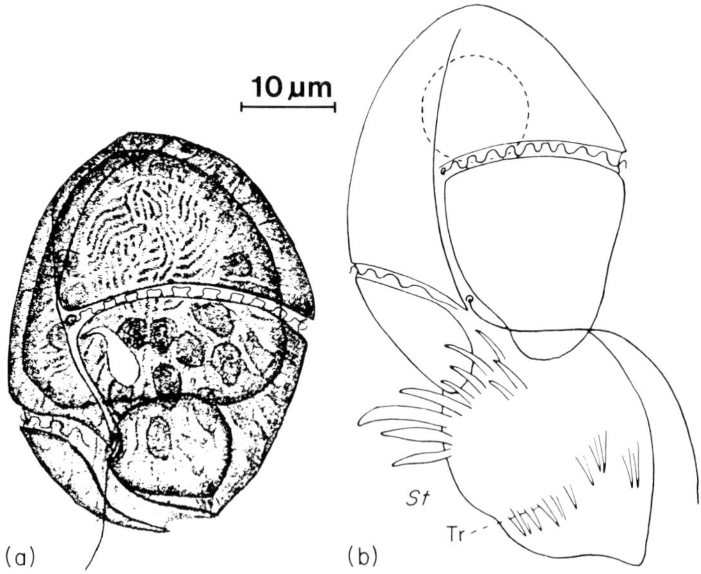

Fig. 6.2. Phagotrophy in *Gyrodinium pavillardi* Herdman. (a) Cell containing a large prey item and the remains of another, smaller one. The nucleus is displaced against the right wall of the cell. (b) *G. pavillardi* seizing the ciliate *Strombidium* (*st*) by its oral end. Tr, trichocysts. From Biecheler (1952).

Hofeneder (1930) described in detail two separate mechanisms of feeding in apparently colourless specimens of the common freshwater species, *Ceratium hirundinella* (O. F. Müller) Bergh. Chloroplasts were probably present but weakly pigmented and therefore overlooked. The more common feeding mechanism was extracellular digestion of food with subsequent uptake of the dissolved products. Odum (1971) has called this *'saprotrophy'*: delicate masses of cytoplasm emerged from the thecal pores; cytoplasmic threads reached outward, anastomosing and intertwining to form a delicate network (Fig. 6.3a,b) which entangled diatoms and other algae, even tiny nanoplankton; cells were digested and the dissolved products presumably taken up. Eventually, the plasma net was retracted, releasing the remains of the prey. Less detailed but similar observations had apparently been made previously by Krause (1910) of *C. hirundinella* and by Schütt (1899) of *Podolampas bipes*. Allmann (1855) depicted an unflagellated dinoflagellate with 'vibratile cilia' (perhaps a delicate pseudopodial network) covering its surface. This mode of nutrition has not been observed since Hofeneder (1930), despite recent attempts to duplicate these observations with cultured *C. hirundinella* cells (C. Happach-Kasan, pers. comm.). Recently, however, a similar feeding mechanism has been observed in the non-photosynthetic genus *Protoperidinium* Bergh (Gaines & Taylor 1984) which produces a 'feeding veil' for extracellular digestion.

The second feeding mechanism recorded in *C. hirundinella* was true phagotrophy (Hofeneder 1930). A large pseudopod emerged from the ventral area of the cell, engulfed a prey object (blue-green algae, diatoms or ciliates) and dragged it back through the ventral area (Fig. 6.3c). Hofeneder believed the ventral plates to be flexible, stretching to accommodate the entry of large (up to 75 μm diameter) prey. Similar events were reported in *Ceratium cornutum* (Ehrenberg) Claparéde & Lachmann, *C. furca* (Ehrenberg) Claparéde & Lachmann and *C. tripos* (O. F. Müller) Nitzsch. Stein (1883) depicted *C. cornutum* and *C. furca* with pseudopodia extending from their ventral areas. However, these cells may have been in distress during microscopical observation. Norris (1969) reported a specimen of *Ceratium lunula* (Schimper *ex* Karsten) Jörgensen containing a *Protoperidinium* Bergh cell found in (presumably) fixed material collected from the Gulf of Mexico. However, the figures are rather unclear about the spatial relationship of the two cells (Fig. 6.3d). This also applies to the observation of *Ceratium massilense* Gourret with an 'ingested' *Gymnodinium* sp. cell (Kimor 1981). Dodge & Crawford (1970), in their electron microscope study of photosynthetic *C. hirundinella*, reported finding remains of bacteria, blue-green algae, and diatoms in sectioned material, and detailed a cytostome (cell mouth) in the ventral area. However, Chapman *et al.* (1982) interpreted the apparent food vacuoles to be accumulation bodies. Culture experiments have shown no evidence of phagotrophy in *C. hirundinella* (C. Hapach-Kasan, pers. comm.) or in marine *Ceratium* Schrank species (M. Elbrächter, unpubl.).

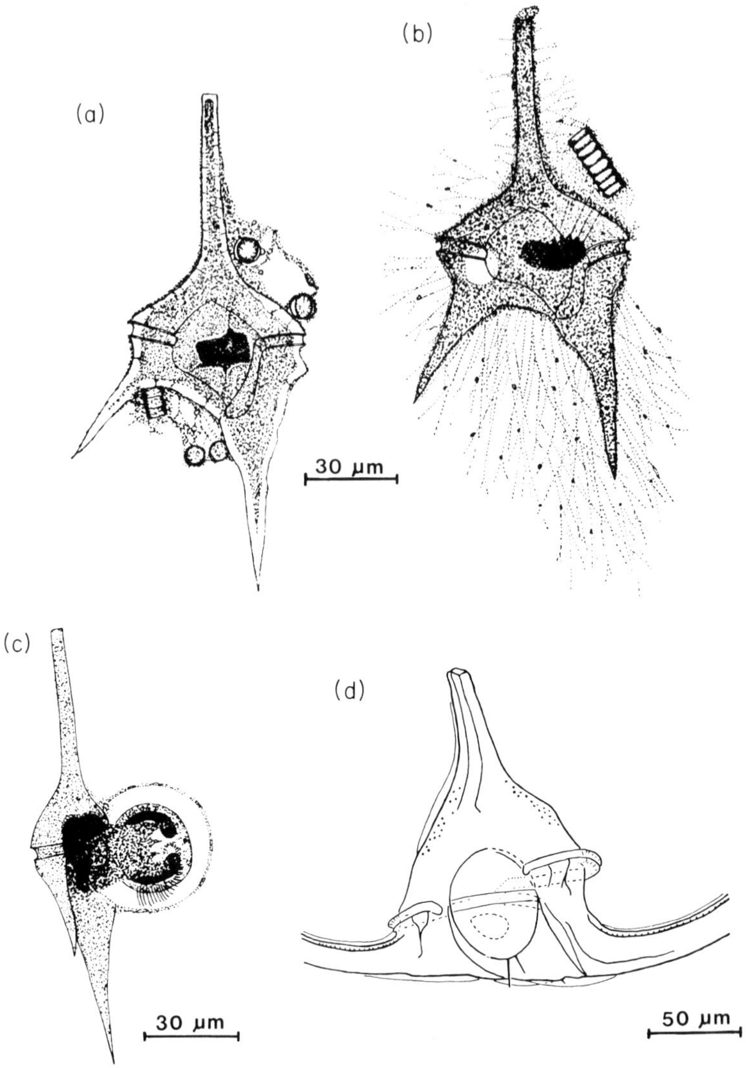

Fig. 6.3. Heterotrophy in the genus *Ceratium*. (a) and (b) saprotrophy in *Ceratium hirundinella* (O. F. Müller) Bergh. Fixed (a) and living (b) specimens, showing the plasma net trapping diatoms and small flagellates. (c) Pseudopodial feeding in *C. hirundinella*. (d) *Ceratium lunula* (Schimper *ex* Karsten) Jörgensen apparently ingesting a *Protoperidinium* cell through its ventral region. (a), (b), (c) from Hofeneder (1930); (d) from Norris (1969).

Of eleven athecate forms having chloroplasts, Hulburt (1957) described one species, *Gyrodinium resplendens*, as also containing ingested food bodies. Kofoid & Swezy (1921) listed five of fifty species clearly stated to have chloroplasts, which contained ingested food bodies as well: *Amphidinium steinii* Lemmerman,

Gymnodinium coeruleum Dogiel, *G. fulgens* Kofoid & Swezy, *G. ravenescens* Kofoid & Swezy and *Gyrodinium melo* Kofoid & Swezy. Kimor (1981) also reported chloroplasts in *G. coeruleum*, but chlorophyll fluorescence was absent in specimens examined by M. Elbrächter (unpubl.).

3.3 Organotrophy

Approximately half of all dinoflagellates, including the great majority of heterotrophic forms, have no chloroplasts (i.e. they are *apochlorotic*) and so live exclusively on organic compounds. This is referred to as *organotrophy* (Lwoff *et al.* 1946). More than 60% of the athecate species listed by Kofoid & Swezy (1921) were described as lacking chloroplasts. Several species have been described both with and without chloroplasts. This may be due to misidentification (e.g. *Gyrodinium fissum*) or to the difficulty of distinguishing between predator/host and prey/endosymbiont chloroplasts. Authors of taxonomic monographs (e.g. Schiller, 1931–37), have been liberal in inferring the presence of chloroplasts from statements that a species has a yellowish or greenish colour. *Protoperidinium pallidum* (Ostenfeld) Balech and *Oodinium poucheti* (Lemmermann) Chatton are examples recently shown by fluorescence microscopy to lack chlorophyll (M. Elbrächter, unpubl.). Many species have coloured cytoplasm (e.g. the '*Diplopsalis* group,' *Gymnodinium* spp., *Gyrodinium* spp., and *Protoperidinium*). Others have colour sequestered in distinct organelles (e.g. stigma, pyrenoid, reserve material bodies). *Gyrodinium britannicum* Kofoid & Swezy and *Protoperidinium depressum* (Bailey) Balech, have strikingly red oil storage globules. The ocelli of the Warnowiaceae are intensely coloured red to black.

Specimens of normally photosynthetic species are occasionally reported to lack chloroplasts (e.g. *Ceratium hirundinella*, *Gymnodinium fuscum* (Ehrenberg) Stein, *Peridinium cinctum* (Müller) Ehrenberg, *Polykrikos lebouriae*) (Dangeard 1892; Herdman 1924; Hofeneder 1930). The 'regressed chromatophores' of *Gyrodinium vorax* are variable and may appear to be absent altogether, leaving the cytoplasm with only vague greenish-yellow spots (Biecheler 1952). The loss of chloroplasts has been demonstrated with *Euglena gracilis* Klebs (see Pringsheim 1963), but not yet in a dinoflagellate. Leucoplasts, but no chloroplasts, have been described in *Amphidinium pellucidum* Herdman (Dragesco 1965).

Organotrophic dinoflagellates can be conveniently divided into several subcategories by the form of the compounds utilized, but the mechanisms involved are mostly unknown.

OSMOTROPHY

Despite the misleading prefix 'osmo-,' *osmotrophy* is the active uptake of dissolved organic substances (Pringsheim 1963). Four non-photosynthetic species have been successfully cultured on dissolved compounds (Table 6.2). *Crypthecodinium cohnii* (Seligo) Chatton *in* Grassé, a marine species frequently found amidst rotting *Fucus* (Linnaeus) Decaisne & Thuret (Küster 1908; Pringsheim 1956), was cultured axenically in a medium containing peptone, yeast autolysate and acetate (Pringsheim 1956). Provasoli & Gold (1957, 1962) replaced yeast extract with biotin and thiamine, and peptone with histidine. *C. cohnii* was able to utilize a variety of sugars, organic acids and ethanol, but glucose and glycerol were the best single carbon sources. A combination of glucose or glycerol with organic acids produced the best growth. Recently, extremely sensitive chemosensory responses have been found in *C. cohnii* (Hauser *et al.* 1973, 1975a,b, 1978).

Axenic cultures of *Oxyrrhis marina* Dujardin were first obtained in a medium containing liver infusion, tryptone, tryptophane, glucose and 'neutralized, strained but unfiltered, lemon juice' (Droop 1959a,b). The lemon juice provided a necessary quinone and a sterol (Droop & Doyle 1966; Droop 1970; Droop & Pennock 1971). Only ethanol and acetate could be used as carbon sources. The culture medium was eventually simplified to include acetate, vitamins (including proline), and the 'lemon factor.'

Because the necessary lipid-soluble substances are light-sensitive and insoluble in water, they exist in water only as particles, and would be quickly photodegraded if dispersed (Droop & Doyle 1966). Thus, only particle feeding can satisfy these auxotrophic requirements in nature (Provasoli 1977). Furthermore, the quinone is a common synthetic product of chloroplasts, easily obtained from a diet of microalgae (Dodge & Crawford 1971).

Noctiluca scintillans (Macartney) Ehrenb. was first brought into osmotrophic culture using casein and liver concentrate (McGinn & Gold 1969). These were later replaced by glucose, amino acids and vitamins (Gold 1970; McGinn 1971).

Gyrodinium lebourae Herdman was cultured in a natural seawater medium containing glucose, galactose and acetate (Lee 1977).

The osmotrophy exhibited by these four species may be a laboratory artefact. There have been no attempts to culture *C. cohnii*, characterized as an osmotroph (Kinne 1977), on particulate food except by providing rotting seaweed (Pringsheim 1956; Provasoli & Gold 1957). Yet Biecheler (1952) reported that in culture it catches small flagellates and sucks out their contents. *Oxyrrhis* is a well known phagotroph, having been grown for many years on yeast or flagellates (Droop 1953). *Noctiluca* is perhaps the most widely known, voracious and indiscriminant particle feeding dinoflagellate. In liquid culture, *Noctiluca* requires placebo particles (diatomaceous earth), probably to elicit vacuole formation (McGinn 1971). *G. lebourae* occurs in both phagotrophic and non-phagotrophic isolates (Lee 1977). However, the non-phagotrophic isolates contain endosymbiotic bacteria, indicating that the bacteria may provide nutrients the dinoflagellate cannot obtain by osmotrophy. Finally, the concentrations of dissolved organic substances naturally found in seawater

(Table 6.2) are from two to seven orders of magnitude lower than have ever been shown to affect dinoflagellate growth in culture (Richardson & Fogg 1982).

Histioneis, *Ornithocercus*, *Protoperidinium* and *Triposolenia* Kofoid have been thought to be osmotrophic (Droop 1974), as they have never been found to contain either chloroplasts or food bodies. Recently, however, *Protoperidinium* has been shown to perform extracellular digestion (Gaines & Taylor 1984). In addition, there are good theoretical grounds to doubt that these dinoflagellates can grow on dissolved organic compounds in nature. Dinoflagellates cannot compete with bacteria for nutrients because of the large numbers and large surface/volume ratios of bacteria (Hobbie & Wright 1965; Koch 1971; Azam & Hodson 1977; Sieburth 1978; Johnson & Sieburth 1979). Osmotrophy may be important for species that inhabit microenvironments with enormously increased levels of dissolved organic compounds (e.g. *C. cohnii*). Endoparasitic dinoflagellates, which inhabit microenvironments rich in dissolved organic substances, probably feed partly or exclusively by osmotrophy.

PHAGOTROPHY

The term *phagotrophy*, as it is used here, refers to the ingestion of discrete particles (by any of several mechanisms), with digestion occurring in phagocytic vacuoles (Table 6.3). It is by far the most common form of organotrophy, evidence of it having been recognized for more than 200 years (Slabber 1778); but the process has rarely been observed. Uhlig (1972) used microcinematography to document food uptake in *Noctiluca*, but there are no detailed electron microscopic studies of this process (Dodge & Crawford 1974, notwithstanding). It is reported to occur in thecate as well as in athecate forms. Some forms apparently specialize on certain foods. *Gymnodinium granii* Schiller feeds on a distinct species of blue-green algae (Schiller 1955). *Peridinium gargantua* Biecheler has a diurnal activity and migration pattern which parallels that of its photosynthetic prey (usually *Gonyaulax*) (Biecheler 1936a, 1952). Dinospores of *Dissodinium pseudolunula* are attracted to and infest eggs of *Acartia*, *Temora* and *Pseudocalanus* but not *Artemia* (G. Drebes, pers. comm.). Although not experimentally verified, specialization is obvious in several host-specific parasitic species. The mechanism of prey recognition is as yet obscure, but may involve sensitivity to dissolved organic compounds, shown for *Katodinium fungiforme* and *C. cohnii* (Hauser *et al.* 1978; Spero 1979).

Phagotrophy was apparently first recognized in *Noctiluca scintillans* (Slabber 1778), and first described in this species by Hofker (1930). In calm conditions, *Noctiluca* concentrates at the sea surface, reaching concentrations as high as 2000 cells l^{-1} (G. Uhlig, pers. comm.). *Noctiluca* is omnivorous, ingesting phytoplankton, zooplankton, detritus, cysts, and the eggs of fish and copepods, including items larger than itself and even pieces of glass (Gross 1934; Mironov 1954; Enomoto 1956; Prasad 1958; Uhlig 1972; Kimor 1979, 1981). It uses its

Table 6.3. Phagotrophic, free-living dinoflagellates that have been depicted as having food bodies. Only one reference, usually the original, is given. For additional references, see text

Amphidinium aeschrum Harris[15]
A. corpulentum Kof. & Sw.[22]
A. crassum Lohmann[23]
A. cucurbitella Kof. & Sw.[22]
A. fastigium Kof. & Sw.[22]
A. gyrinum Harris[15]
A. inconstans Schiller[35]
A. latum[23]
A. pellucidum Herdman[16]
A. phaeocysticola Lebour[23]
A. scissum Kof. & Sw.[22]
A. semilunatum Herdman[17]
A. cf. *sphenoides*[27]
A. steinii Kof. & Sw.[40]
A. vasculum Kof. & Sw.[22]
A. vorax Schiller[35].
A. wigrense Wol.[42]
Ceratium hirundinella (O.F.M.) Bergh[18]
C. lunula (Schimper *ex* Karsten) Jörg.[28]
Cochlodinium brandtii Wulff (sub *C. augustum*)[22]
C. clarissimum Kof. & Sw.[22]
C. conspiratum Kof. & Sw.[22]
C. faurei Kof. & Sw.[22]
C. helix (Pouchet) Lemm.[22]
C. lebourae Kof. & Sw.[22]
C. miniatum Kof. & Sw.[22]
C. radiatum Kof. & Sw.[22]
C. schüttii Kof. & Sw.[22]
C. scintillans Kof. & Sw.[22]
C. turbineum Kof. & Sw.[22]
C. vinctum Kof. & Sw.[22]
Crypthecodinium cohnii (Seligo) Chatton[14]
Dinamoebidium varians Pascher[29]
Entomosigma peridinioides Schiller[32]
Glenodinium bieblii Schiller[35]
G. edax Schilling[36]
G. eurystomum Harris[15]
G. leptodermum Harris[15]
G. pulvisculus (Ehrenb.) Stein[40]
Gymnodinium agile Kof. & Sw.[22]
G. albulum Lindemann[26]
G. amphora Kof. & Sw.[22]
G. arcuatum Kofoid[21]
G. aureum Kof. & Sw.[22]
G. austriacum Schiller[33]
G. blax Harris[15]
G. cnecnoides Harris[15]
G. coeruleum Dogiel[11]
G. colymbeticum Harris[15]
G. contractum Kof. & Sw.[22]
G. costatum Kof. & Sw.[22]
G. devorans Schiller[35]
G. doma Kof. & Sw.[22]
G. eufrigidum Schiller[35]
G. filum Lebour[23]
G. flavum Kof. & Sw.[22]
G. fuscum (Ehrenb.) Stein[10]
G. gracile Bergh (sub *G. abbreviatum* Kof. & Sw.)[22]
G. granii Schiller[35]
G. helveticum Penard[30]
G. heterostriatum Kof. & Sw.[22]
G. incisum Kof. & Sw.[22]
G. knollii Schiller[35]
G. lazulum Hulburt[20]
G. legiconveniens Schiller[35]
G. lineopunicum Kof. & Sw.[22]
G. marinum Saville Kent[31]
G. minor Lebour[23]
G. multistriatum Kof. & Sw.[22]
G. neapolitanum Schiller[33]
G. ovulum Kof. & Sw.[22]
G. pachydermatum Kof. & Sw.[22]
G. puniceum Kof. & Sw.[22]
G. ravenescens Kof. & Sw.[22]
G. roseolum (Schmarda) Stein[37]
G. rubricauda Kof. & Sw.[22]
G. rubrum Kof. & Sw.[4]
G. situla Kof. & Sw.[22]
G. sphaericum Calkins[7]
G. striatissimum Hulburt (sub *G. heterostriatum*)[22]
G. sulcatum Kof. & Sw.[22]
G. translucens Kof. & Sw.[22]
G. violescens Kof. & Sw.[22]
Gyrodinium biconicum Kof. & Sw.[22]
G. calyptoglyphe Lebour[23]
G. caudatum Kof. & Sw.[22]
G. dorsum Kof. & Sw.[22]
G. fissum (Levander) Kof. & Sw.[25]
G. flavescens Kof. & Sw.[22]
G. glaebum Hulburt[20]
G. herbaceum Kof. & Sw.[22]
G. hyalinum (Schilling) Kof. & Sw.[36]
G. intortum Kof. & Sw.[22]
G. lebourae Herdman[24]
G. maculatum Kof. & Sw.[22]
G. melo Kof. & Sw.[22]

G. ovoideum Kof. & Sw.[22]
G. pavillardi Biecheler[3]
G. pingue (Schütt) Kof. & Sw.[22]
G. postmaculatum Kof. & Sw.[22]
G. spirale (Bergh) Kof. & Sw.[22]
G. submarinum Kof. & Sw.[22]
G. truncatum Kof. & Sw.[22]
G. truncus Kof. & Sw.[22]
G. vorax Biecheler[3]
Katodinium astigmatum Christen[9]
K. asymmetricum (Massart) Loebl.[20]
K. austriacum (Schiller) Loebl.[35]
K. campylos (Harris) Loebl.[15]
K. edax (Schiller) Loebl.[34]
K. fungiforme (Anissimova) Loebl.[3]
K. hiemalis (Schiller) Loebl.[35]
K. hyperxanthum (Harris) Loebl.[15]
K. intermedium Christen[8]
K. molopica (Harris) Loebl.[15]
K. montanum (Schiller) Loebl.[34]
K. notatum (Skuja) Christen[38]
K. pratensis (Baumeister) Loebl.[2]
K. ptyrticum (Harris) Loebl.[15]
K. tetragonops (Harris) Loebl.[15]
K. vernale Christen[9]
K. vorticellum (Stein) Loebl.[39]
K. woloszynskaae (Schiller) Loebl.[41]
Kofoidinium pavillardi Cachon & Cachon[5]
Nematodinium partitum Kof. & Sw.[22]
N. radiatum Kof. & Sw.[22]
N. torpedo Kof. & Sw.[22]
Noctiluca scintillans (Macartney) Ehrenb.[19]
Oxyrrhis marina Dujardin[1]
Peridinium aciculiferum (Lemm.) Lemm.[13]
P. gargantua Biecheler[3]
Polykrikos kofoidi Chatton[22]
Pronoctiluca pelagica Pavillard[12]
P. spinifera sub *Protodinifer tentaculatum*[22]
Pratjetella (Leptodiscus) medusoides (Hertwig) Loebl. & Loebl.[6]
Proterythropsis crassicaudata Kof. & Sw.[22]
Protoperidinium globulus (Stein) Loebl.[4]
Protopsis neapolitana Kof. & Sw.[22]
Warnowia alba (Kof. & Sw.) Lindemann[22]
W. hatai (Kofoid) Schiller[21]
W. maculata (Kof. & Sw.) Lindemann[22]
W. maxima (Kof. & Sw.) Lindemann[22]
W. mutsui (Kofoid) Schiller[21]
W. pouchetii (Kof. & Sw.) Schiller[22]
W. purpurata (Kof. & Sw.) Lindemann[22]
W. purpurescens (Kof. & Sw.) Lindemann[22]
W. reticulata (Kofoid) Schiller[21]
W. rubescens (Kof. & Sw.) Lindemann[22]
W. violescens (Kof. & Sw.) Lindemann[22]
W. voracis (Kof. & Sw.) Schiller[22]

References:
1 Barker (1935)
2 Baumeister (1943)
3 Biecheler (1952)
4 Bursa (1961)
5 Cachon & Cachon (1968)
6 Cachon & Cachon (1969)
7 Calkins (1902)
8 Christen (1959)
9 Christen (1961)
10 Dangeard (1892)
11 Dogiel (1906)
12 Elbrächter (1979)
13 Entz & Sebestyn (1936)
14 Griessmann (1913)
15 Harris (1940)
16 Herdman (1922)
17 Herdman (1924)
18 Hofeneder (1930)
19 Hofker (1930)
20 Hulburt (1957)
21 Kofoid (1931)
22 Kofoid & Swezy (1921)
23 Lebour (1925)
24 Lee (1977)
25 Levander (1894)
26 Lindemann (1928)
27 Gaines & Elbrächter (this work)
28 Norris (1969)
29 Pascher (1916)
30 Penard (1891)
31 Saville Kent (1881)
32 Schiller (1925)
33 Schiller (1931–37)
34 Schiller (1954)
35 Schiller (1955)
36 Schilling (1891)
37 Schmarda (1854)
38 Skuja (1956)
39 Stein (1878)
40 Stein (1883)
41 Woloszynska (1917)
42 Woloszynska (1925)

tentacle to catch and 'glue' together small prey organisms to form a large clump. It may also use its tentacle to collect bacteria in the surface film, as has been observed in culture (M. Elbrächter, unpubl.; F. J. R. Taylor, pers. comm.).

In most cases, evidence of phagotrophy is limited to the presence of ingested food bodies (e.g. *Glenodinium edax* Schilling, *Gymnodinium gracile* Bergh, *G. helveticum* Penard, *G. roseolum* (Schmarda) Stein, *Gyrodinium spirale* (Bergh)

Kofoid & Swezy, *Polykrikos* spp.) (Schmarda 1854; Bergh 1881; Bovier-Lapierre 1888; Penard 1891; Kofoid & Swezy 1921; Harris 1940; Nauwerk 1963; Irish 1979; Frey & Stoermer 1980; Morey-Gaines & Ruse 1980; Popovský 1982). In cases where ingestion has been observed, it typically occurs either at the junction of the flagellar grooves (e.g. *Glenodinium pulvisculus* (Ehrenberg) Stein, *Gymnodinium marinum* Saville Kent) (Saville Kent 1881; Entz 1883; Stein 1883) or at the posterior of the cell (e.g. *Gymnodinium rubrum* Kofoid & Swezy, *Protoperidinium globulus* (Stein) (Loeblich) (Bursa 1961) (Figs 6.4, 6.5). Other species emit cytoplasmic extensions from the junction of the flagellar grooves (e.g. *Gyrodinium lebouriae*, *Katodinium vorticellum* (Stein) Loeblich (=*Gymnodinium vorticella*), *Oxyrrhis marina*, *Peridinium bicorne* Schmarda, *P. gargantua* Biecheler) (Schmarda 1854; Dangeard 1892; Barker 1935; Biecheler 1936a, 1952; Lee 1977) (Fig. 6.6). Most often, small green flagellates are the apparent prey of phagotrophic dinoflagellates (e.g. of *Glenodinium pulvisculus*, *Gymnodi-*

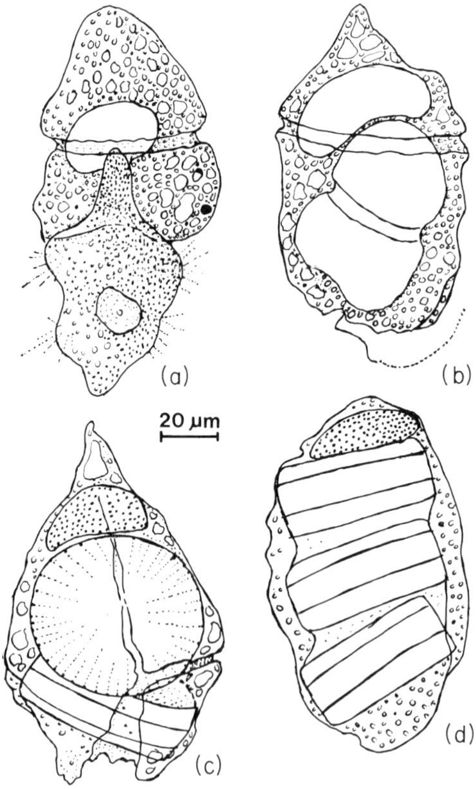

Fig. 6.4. Phagotrophy in *Gymnodinium rubrum* Kofoid & Swezy. (a) A cell ingesting the ciliate *Strombidium*. (b)–(d) Cells containing (b) *Gymnodinium*, (c) *Thalassiosira* and (d) three *Coscinosira* cells. Nuclei of (c) and (d) are displaced to the apices of the cells. From Bursa (1961).

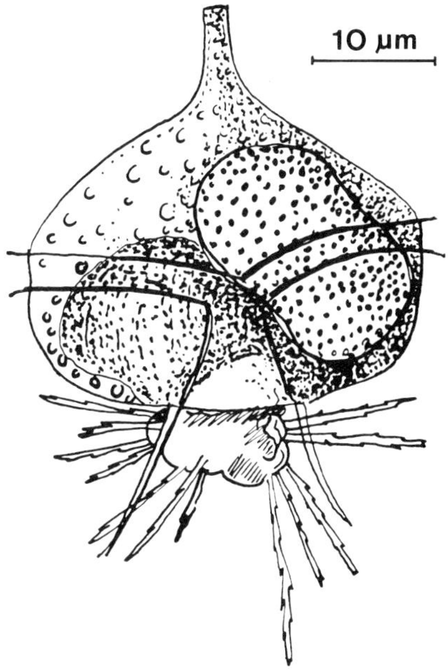

Fig. 6.5. The only report of a species of *Protoperidinium* (*P. globulus* (Stein) Loeblich) ingesting food: an unidentified flagellate resembling *Meringosphaera*. From Bursa (1961).

nium roseolum, *G. marinum*, *G. hyalinum* Schilling, *Katodinium vorticellum*) (Schmarda 1854; Stein 1878, 1883; Saville Kent 1881; Entz 1883; Schilling 1891), although *Gymnodinium helveticum* and *Oxyrrhis marina* feed on a wide variety of organisms (Penard 1891; Barker 1935; Droop 1963, 1966; Nauwerk 1963; Dodge & Crawford 1974; Irish 1979; Frey & Stoermer 1980; Popovský 1982). The report of green food bodies in *Hemidinium nasutum* Stein (1883) was probably a misinterpretation of its own chloroplasts.

These examples of phagotrophy involve ingestion of entire organisms. Other dinoflagellates ingest only parts of prey, usually by piercing cells and sucking out their cytoplasmic contents (myzocytosis). The best known example is *Katodinium fungiforme* (Anissomowa) Loeblich (=*Gymnodinium fungiforme*), whose feeding, chemosensory behaviour and life cycle have been studied in

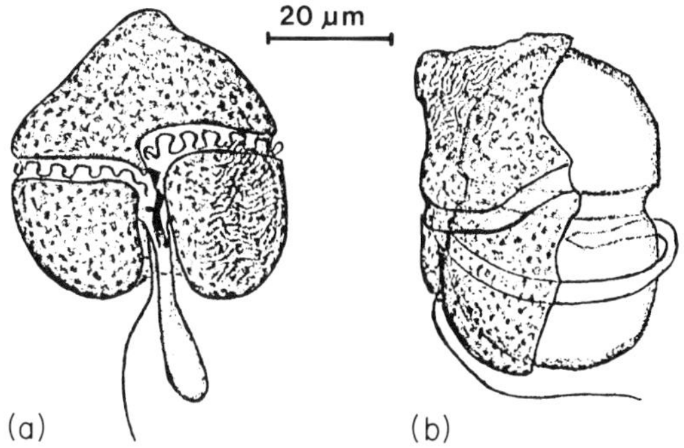

Fig. 6.6. Phagotrophic feeding by *Peridinium gargantua* Biecheler. (a) An individual with a pseudopod formed spontaneously upon contact with prey. (b) An individual enveloping its prey, which is held by the pseudopod. From Biecheler (1952).

culture (Biecheler 1952; Spero 1979, 1982; Spero & Morée 1981). It feeds by means of an extensible, tube-like 'peduncle' by which it attaches to unicellular algae, ciliates and even small metazoans (Biecheler 1952; Spero 1979, 1982; Spero & Morée 1981; G. Gaines & F. J. R. Taylor, unpubl.). Feeding is accompanied by 'swarming,' proportional to the size of the prey, triggered by chemosensation of senescent or injured individuals (Spero 1979). After feeding, the cell forms a 'zoosporangium' in which cell division takes place (Spero & Morée 1981). *Oxyrrhis marina* has been observed to attack amphipods during or shortly after moulting (M. Elbrächter, unpubl.). *Peridiniopsis berolinensis* (Lemmerman) Bourrelly (=*Glenodinium berolinense*) uses a peduncle to feed on dead algal cells (e.g. *Ceratium cornutum*), taking in chloroplasts whole, and remaining motile throughout feeding and digestion (Wedemayer & Wilcox 1980; Wedemayer 1982; H. A. von Stosch & C. Kollmann, pers. comm.).

At least five species of dinoflagellates have been observed to adhere temporarily to the spines of planktonic foraminifera, picking up food particles through the sulcal region while being transported along the spines (M. Elbrächter & H. J. Spero, unpubl.) (Fig. 6.7). *Amphidinium phaeocysticola* Lebour and *Gymnodinium striatissimum* Hulburt feed on single cells of the colony-forming haptophyte, *Phaeocystis pouchetii* (Harriot) Lagerheim, and the colonial diatom, *Thalassiosira partheneia* Schrader (Lebour 1925; Elbrächter 1979). *Cystodinedria inermis* Geitler and *Stylodinium sphaera* Pascher feed on *Oedogonium* Link or *Spirogyra* Link (Pfiester & Popovský 1979). *Amphidinium* cf. *sphenoides* Wulff, maintained for 2 weeks in the laboratory on *Chaetoceros borealis*, appeared to feed in the dark phase, as empty diatom frustules appeared at the beginning of the light phase (M. Elbrächter, unpubl.). Unusually, *A.* cf. *sphenoides* attached near its apex, rather than posteriorly or in the sulcal area.

Spero (1979) has speculated that myzocytosis may be an advantage in neritic

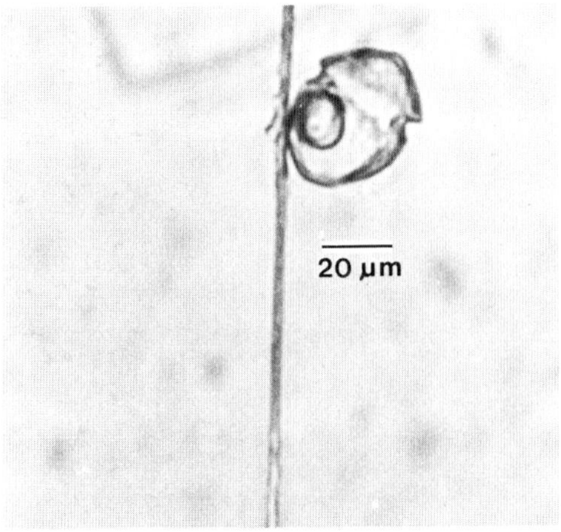

Fig. 6.7. *Amphidinium* cf. *phaeocysticola* Lebour feeding on spines of the foraminiferan, *Globigerinoides sacculifer* (M. Elbrächter, unpubl.).

habitats, where dying organisms, detritus, benthic algae, protozoa and soft-bodied meiofauna all may be usable food items. Because whole organisms need not be ingested, the array of possible food sources can be expanded to include extremely large items. In addition, food vacuoles can be efficiently packed and the contents digested without interference by a prey cell wall. A disadvantage is that the predator may not be able to puncture heavy cell walls (e.g. of diatoms), although they are punctured by the ectoparasite *Paulsenella*.

Erythropsidinium agile (Hertwig) P. C. Silva (=*Erythropsis pavillardi*) possesses an appendage (the 'stomopod') which may be used to inject lysing substances into its prey (Greuet 1969). Although feeding has never actually been observed, it is speculated that the cell contents of prey are sucked out (Greuet 1969).

Blepharocysta splendor maris Ehrenberg, *Protoperidinium depressum*, and *P. crassipes* (Kofoid) Balech have been found attached to other cells by a sort of connecting tube (Fig. 6.8), but ingested food has never been seen (Steidinger *et al.* 1967; K. A. Steidinger, pers. comm.). *Diplopsalis lenticula* Bergh and *Zygabikodinium lenticulatum* Loeblich & Loeblich have been observed adhering to colonial diatoms like *Ditylum brightwellii* (West) Grunow and *Cerataulina bergonii* Pergallo (G. Drebes & M. Elbrächter, unpubl.). Feeding was not observed, and the attachment mechanism could not be determined. Similarly, a colourless *Gyrodinium* species has been found attached to the umbrella of *Pratjetella* Lohmann (=*Leptodiscus* Hertwig) (Cachon 1964).

Gaines & Taylor (1984) have documented extracellular feeding in *Protoperidinium* , involving a 'feeding veil' extruded through the flagellor aperture and used to surround regions of the prey (Fig. 6.9). This mechanism appears to be the same as those seen in the species above, and allows them to eat large

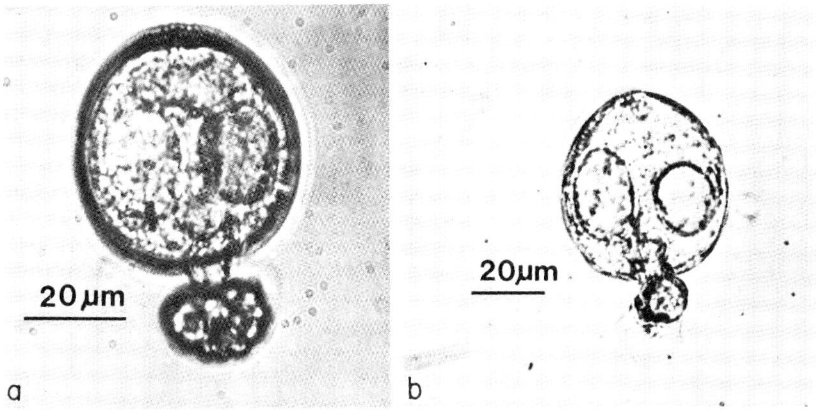

Fig. 6.8. Possible extracellular digestion by dinoflagellates. (a) Pigment mass and (b) *Scrippsiella trochoidea* (Stein) Loeblich cell attached to *Blepharocysta splendor maris* Ehrenberg. From Steidinger *et al.* (1967).

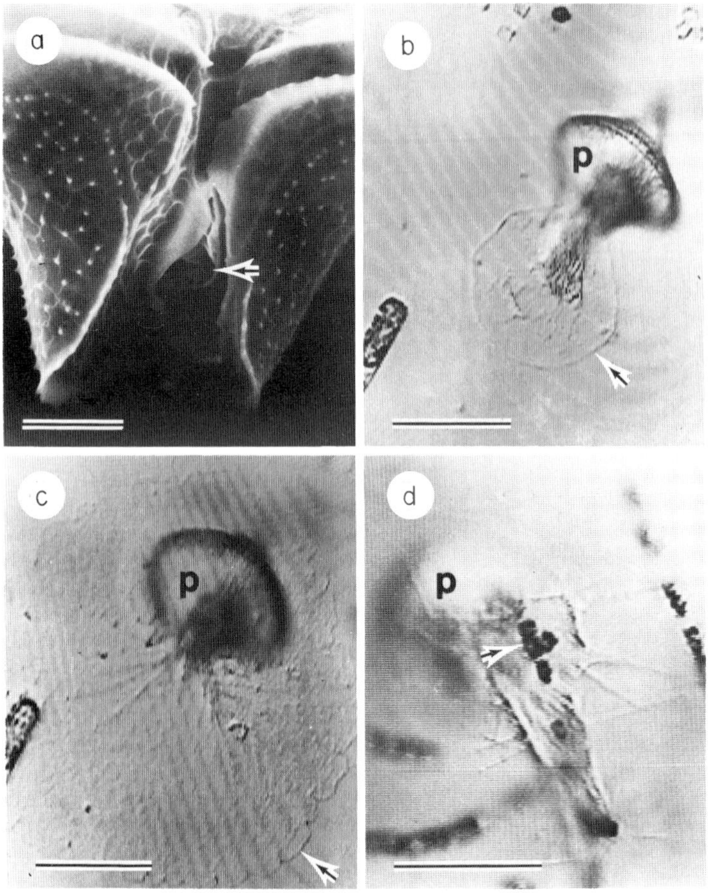

Fig. 6.9. Extracellular digestion in *Protoperidinium*. (a) Scanning electron micrograph of the ventral area of *P. conicum* (p), showing the flagellar aperture (arrow) from which the 'feeding veil' appears to extend. Scale = 10 μm. (b)–(d) Feeding veil (arrow) (b) in the initial stages of extension; (c) fully extended; (d) with engulfed cells of the diatom *Thalassiosira nordenskjöldii* (arrow). Scales = 50 μm. From Gaines & Taylor 1984.

diatoms, including parts of large chains, without having to physically penetrate their outer cell covering.

PARASITISM

Parasitism is difficult to define stringently (e.g. ectoparasitism or myzocytosis?). For the present purpose, parasites are distinguished as having morphologically different feeding and reproductive stages and as producing numerous progeny after only one feeding act. The biology and cytology of these forms are discussed in detail in Chapter 13. A few examples of the most important parasitic feeding characteristics are summarized in Table 6.4.

Table 6.4. Partial descriptions of feeding characteristics of parasitic dinoflagellate genera. The arrangement of this list does not imply any phylogenetic relationships. For complete discussion and references, refer to Chapter 13.

1. Ectoparasites with chloroplasts, whose function is unknown.
   Dissodinium pseudolunula: sucks out the contents of copepod eggs.
   Oodinium cyprinodontum Lawler: ectoparasitic on fish gills.
   Protoodinium chattonii: ectoparasitic on *Podocoryne minima*.
2. Ectoparasites without chloroplasts.
   Amyloodinium Brown & Hovasse: ectoparasitic on fishes, feeds by phagotrophy.
   Apodinium Chatton: ectoparasitic on Appendicularia, feeds by osmotrophy.
   Chytriodinium Chatton: sucks out the contents of copepod eggs, but in contrast to *Dissodinium*, these forms perform digestion extracellularly.
   Cystodinedria Pascher: recently reported to be ectoparasitic on the filamentous alga *Spirogyra* Link (Pfiester & Popovský 1979).
   Dissodinium pseudocalani (Gönnert) Drebes: sucks out the contents of copepod eggs.
   Myxodinium Cachon, Cachon & Bouquaheux: sucks out the contents of *Halosphaera* (Prasinophyceae).
   Oodinium pouchetii (Lemmermann) Chatton: ectoparasitic on Tunicata, feeds by osmotrophy.
   Paulsenella Chatton: sucks out the contents of diatom cells.
   Stylodinium Klebs: recently reported to be ectoparasitic on the filamentous alga *Oedogonium* Link (Pfiester & Popovský 1979).
3. Endoparasites with chloroplasts, whose function is unknown.
   Blastodinium Chatton: endoparasitic in copepods.
   Schizodinium Chatton: endoparasitic in copepods.
4. Endoparasites without chloroplasts. These forms feed first by osmotrophy, then later begin to ingest the cytoplasm of the host cells (Cachon 1964).
   Amoebophrya Koeppen: endoparasitic in dinoflagellates and siphonophores.
   Duboscquella Chatton: endoparasitic in ciliates and other protozoa.
   Syndinium Chatton: endoparasitic in the coeloms of copepods or the cytoplasm of Acantharia.

4 SYMBIOSIS

4.1 Dinoflagellate symbionts

Symbiosis is discussed in depth in Chapter 12. At least five dinoflagellate species are recognized as symbionts inhabiting various species of Ciliata, Coelenterata, Mollusca, Radiolaria, Foraminifera, and Turbellaria. Though many studies have been made of the contribution these symbionts make to the carbon nutrition of their hosts, very few have been directed toward the nutrition of the symbiont (D. Taylor 1974; Muscatine 1980; Trench 1980). However, those isolates that have been cultured outside their hosts have grown phototrophically on standard media (D. Taylor 1973, 1974); organic carbon sources have not been required.

Despite this, Taylor (1973) cited unpublished data showing acetate metabolism by *Amphidinium klebsi* Kofoid & Swezy. Morrill & Loeblich (1979) found growth enhancement by glycerol in a free-living isolate of *Symbiodinium microadriaticum*, but none in a symbiotic isolate of the same species. It was speculated that the symbiont had host-induced inhibition of glycerol permeability, similar to that found in a symbiotic species of the prasinomonad *Platymonas* (Gooday 1970). Other symbionts require vitamins (Table 6.1), which may be supplied by the host (D. Taylor 1973).

4.2 Dinoflagellates as symbiont hosts

EXTRACELLULAR SYMBIONTS

The most complex dinophysoid dinoflagellates (the apochlorotic genera *Citharistes* Stein, *Histioneis* Stein, *Ornithocercus* Stein and *Parahistioneis* Kofoid & Swezy), harbour symbiotic blue-green algae ('phaeosomes'), usually between the upper and lower girdle lists. They have been assigned to *Synechococcus carcerius* Norris and *Synechocystis consortia* Norris (see reviews by F. Taylor 1980, 1982). The host may absorb dissolved organic substances excreted by its guests (e.g. metazoan hosts with dinoflagellate symbionts) or it may periodically digest them. Although there have been no reports of food bodies resembling symbionts inside these hosts, individuals without symbionts are sometimes found (Norris 1967; Balech 1971).

INTRACELLULAR SYMBIONTS ('CYTOBIONTS')

Amphisolenia Stein has been reported to harbour intracellular blue-green algae (*Synechococcus* Nägeli) and golden-coloured algae (Norris 1967), but there is no information on the role of these symbionts in the nutrition of their hosts. *Kofoidinium splendens* Cachon & Cachon also harbours complete cytobionts (Cachon & Cachon 1968). There is evidence that these can be digested as well, since host specimens are found without symbionts but with residual symbiont chloroplasts. The nutritional importance of another intracellular flagellate to a population of *Noctiluca* has been demonstrated (Sweeney 1971). The symbiont, *Pedinomonas noctilucae* Sweeney, is motile but imprisoned in the vacuolar spaces of its host (Sweeney 1971, 1976). When reared in the dark, the symbionts disappeared and the host died in a few days, unless food organisms (*Dunaliella*) were supplied. Symbionts were only rarely found inside small cytoplasmic vacuoles, suggesting that although *Noctiluca* can digest its symbionts, that is not the primary relationship between the two.

Peridinium balticum and *P. foliaceum* apparently possess well integrated photosynthetic algal cytobionts (Tomas & Cox 1973; Jeffrey & Vesk 1976), but the dynamics of translocation between the two entities are not known.

The unusually-coloured chloroplasts of some species may be ingested food bodies or photosynthetic endosymbionts. Blue chloroplasts are found in *Amphidinium celestinium* Conrad & Kufferath, *A. coeruleum* Conrad & Kufferath, *A. conradi* (Conrad) Schiller, *A. cyanoturbo* Conrad & Kufferath, *A. dubium* Conrad & Kufferath, *A. geitleri* Huber-Pestalozzi, *A. truncatum* Kofoid & Swezy, *A. wigrense* Woloszynska, *Gymnodinium acidotum* Nygaard, *G. aeruginosum* Stein, *G. eucyaneum* Hu, *G. cyanofungiforme* Conrad & Kufferath, *G. hamulus* Kofoid & Swezy and *Gyrodinium louisae* Conrad & Kufferath. The blue pigments in *Gymnodinium eucyaneum* are phycobilin from cryptomonad endosymbionts (Hu *et al.* 1980; Wilcox & Wedemayer 1984). *Amphidinium carbunculus* Conrad & Kufferath and *A. coralinum* Conrad & Kufferath have red chromatophores.

Intracellular bacteria and viruses have been found in dinoflagellates (Lee 1977; Silva 1978; Soyer 1978). *Gyrodinium lebourae* cells possessing bacterial symbionts did not feed phagotrophically, whereas those lacking symbionts did (Lee 1977).

5 FEEDING ORGANELLES

5.1 Cytostomes

Some dinoflagellates, including members of the Noctilucales and Warnowiaceae, possess a well developed, permanent cytostome, or 'cell mouth.' In *Noctiluca* Suriray, this is near the base of the tentacle, at the bottom of a highly distensible, cone-shaped depression (buccal cone), surrounded by trichocysts, mucocysts, and a system of contractile, striated fibres. Prey items are carried to the buccal cone by the single flagellum or are pushed with the prehensile tentacle (Gross 1934; Soyer 1968, 1970; Lucas 1982). Prey are then forced into the cytostome, where they are enclosed in food vacuoles drawn into the interior of the cell, and digested (Soyer 1970).

The cytostome of *Kofoidinium* and *Pratjetella* (= *Leptodiscus*) is located on the anterior surface of the cell, connected to a thick-walled, fibrous 'cytopharynx,' which in *Kofoidinium* is surrounded by sheets of microtubules (Cachon & Cachon 1968, 1969, 1974). These structures together are termed the 'stomopharyngeal complex.' In *Kofoidinium*, the transverse flagellum conveys prey to the cytostome along a groove originating at the sulcus. The cytopharynx of *Pratjetella* is non-contractile, but is connected to a permanent digestive vacuole, from which a tubule leads to a 'cytoproct' (cell anus) (Fig. 6.10). It is, in short, a tubular digestive system. *Craspedotella* Kofoid also has a cytoproct, but its

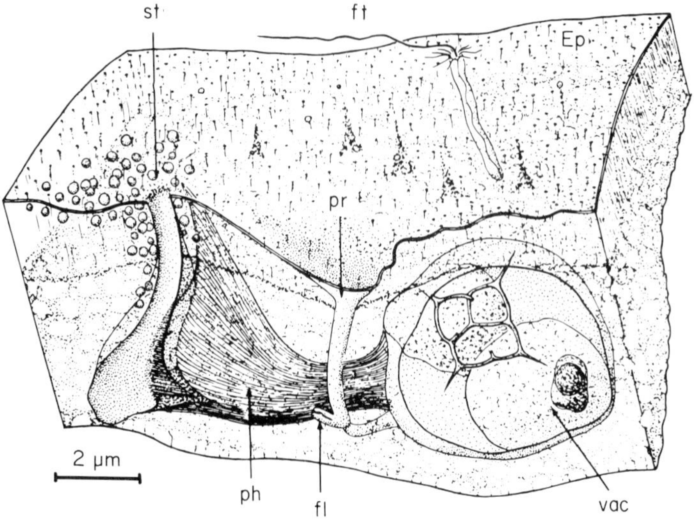

Fig. 6.10. The unique tubular digestive system of *Pratjetella medusoides* (Hertwig) Loeblich & Loeblich. st, cytostome; ph, cytopharynx; pr, cytoproct; vac, digestive vacuole; ft, transverse flagellum; fl, longitudinal flagellum; Ep, episome. From Cachon & Cachon (1969).

relationship to the cytostome and cytopharynx is not known, for this organism does not have a tubular digestive system (Cachon & Cachon 1969).

Recently, Greuet *et al.* (1980) reported finding a cytostome in *Oxyrrhis marina*, near the base of the residual hypocone, where the cell is bounded only by a single membrane (Dodge & Crawford 1971).

Peridinoid and gymnodinoid forms are not known to possess a stomopharyngeal complex. However, Wilcox *et al.* (1982) have discovered fine transverse fibres just beneath the thecal vesicles in the sulcus of *Amphidinium cryophilum* Wedermayer, Wilcox & Graham. The authors speculate that they may be supportive, but they may instead be contractile, consistent with interpretations of fibrillar systems in appendages of other dinoflagellates, and may be involved in distension of the sulcus during phagotrophy.

The slot-shaped sulcal opening of *C. hirundinella* leads to a cavity (the 'ventral chamber') lined by a single membrane and flattened, empty amphiesmal vesicles (Dodge & Crawford 1971). Beneath the surface of the right side of this chamber is a sheet of microtubules (the 'microtubular strand') (Fig. 6.11), which seems to be homologous to what Kubai & Ris (1969) called a 'microtubular basket' in *Crypthecodinium cohnii*, also observed in *Katodinium glandulum* (Lohmann) Loeblich (Dodge 1971) and in *Oxyrrhis marina* (Dodge & Crawford 1974; Greuet *et al.* 1980). These structures appear in cross-section as fan-like overlapping layers of microtubules arranged radially as if connected to a central point (Fig. 6.12d).

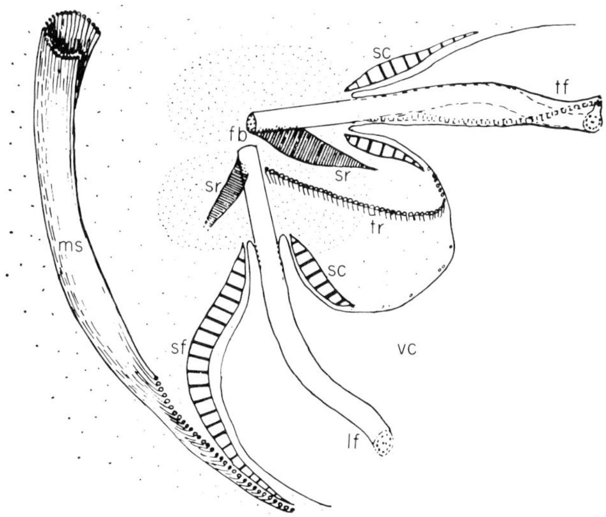

Fig. 6.11. Diagram of a longitudinal section through the ventral area of *Ceratium hirundinella* (O. F. Müller) Bergh, showing the relationship of the microtubular strand to other structures. ms, microtubular strand; fb, flagellar bases; vc, ventral chamber; sc, striated collar; sf, striated fibrils; sr, striated root; tr, tubular root. From Dodge & Crawford (1970). (Not to scale.)

5.2 Appendages

PEDUNCLE

The non-photosynthetic species, *Gyrodinium lebourae, Katodinium fungiforme* and *Peridiniopsis berolinense*, possess a clearly defined column of cytoplasm (peduncle) extending from between the flagellar insertions through an electron-dense ring (Figs 6.12, 6.13) (Lee 1977; Spero 1979, 1982; Wedemayer & Wilcox 1980, 1984; Wedemayer 1982). The peduncle is filled with longitudinal microtubules that are thought to be an extension of the internal microtubular basket (Lee 1977) (Fig. 6.13). It also contains many rod-shaped bodies similar to those found in the 'suction apparatus' of the parasite *Haplozoon* Dogiel (Siebert & West 1974), electron-translucent vacuoles, and sometimes endosymbiotic bacteria. Many parasitic forms, including those with both osmotrophic and phagotrophic feeding phases, attach to their hosts by a similar stalk (Cachon & Cachon 1974).

Peduncles have also been found among photosynthetic forms. *Amphidinium cryophilum* possesses a peduncle similar to that of *G. lebourae*, but the microtubular system is rather different, being composed of only a single, nearly straight row of about forty microtubules (Wilcox *et al.* 1982). The coral symbiont *Symbiodinium microadriaticum* has an apparently extensible peduncle that may

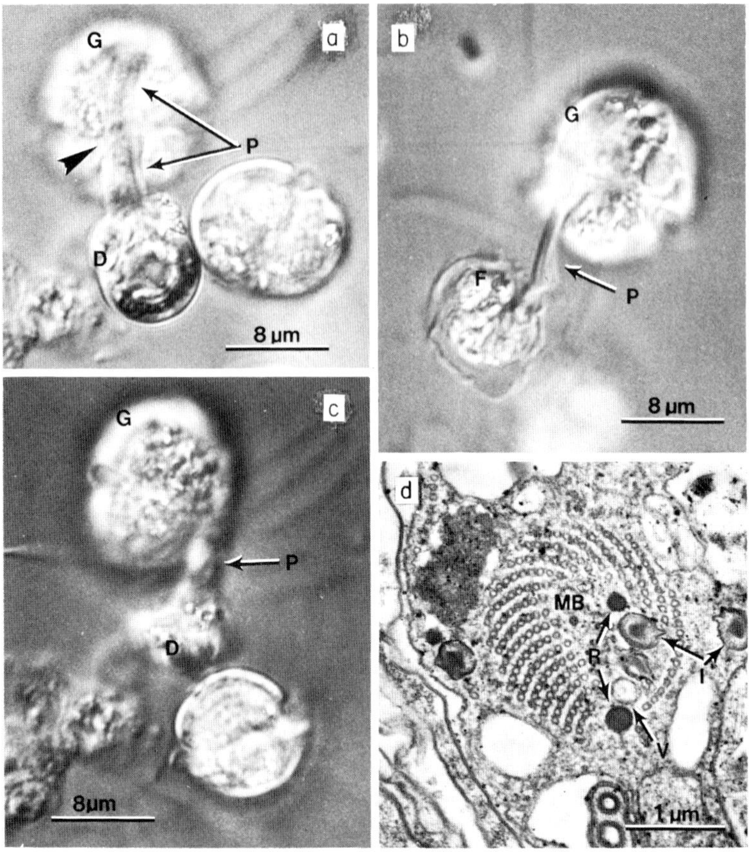

Fig. 6.12. Phagotrophy in *Katodinium fungiforme* (Anissomowa) Loeblich. (a) & (b) Successive stages of a cell (G) ingesting *Dunaliella salina* Teodoresco (D) through its peduncle (P). The prey cytoplasm collects in a vacuole in the hypocone, but an internal portion of the peduncle is visible in the epicone as well. The large arrow indicates the site of emergence of the peduncle. (c) Cell (G) about to detach itself from an unidentified food particle (P). Note the markedly thinner peduncle when no cytoplasm is flowing through it. (d) TEM section through the microtubular basket (MB) showing the fan-like arrangement of microtubules. R, dark osmiophilic bodies; V, electron translucent vacuoles; I, irregularly shaped structures. From Spero (1982).

be used to attach to substrates (Loeblich & Sherley 1979). *Ensiculifera* sp., *Gymnodinium sanguineum*, and *Gyrodinium instriatum* Freudenthal & Lee also possess peduncles (G. Gaines, unpubl.) (Fig. 6.14).

Peduncles of different species differ in certain structural details. Cell wall material is absent in the peduncle of *Katodinium fungiforme*, but present in *Gyrodinium lebourae*. In the parasites *Paulsenella* and *Dissodinium*, the peduncle is composed of different structures with one part that persists after feeding (Drebes & Schnepf 1982; Drebes 1984). The attachment organelles of other

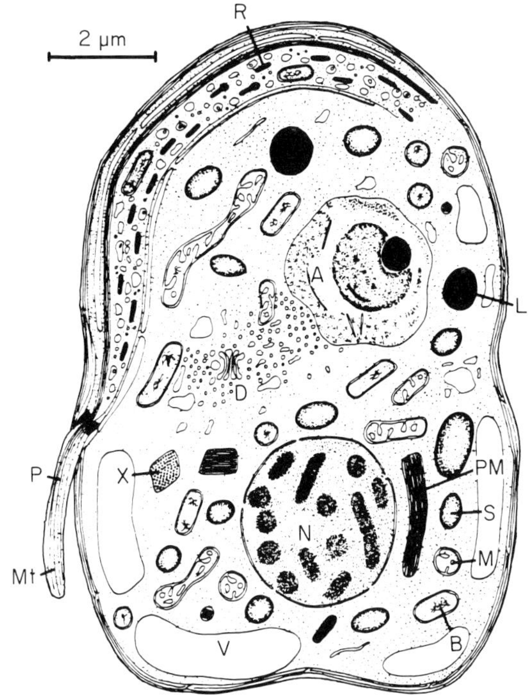

Fig. 6.13. Diagram of a longitudinal section of *Gyrodinium lebourae* Herdman, showing the peduncle and microtubules associated with it. A, accumulation body; B, endosymbiotic bacteria; D, dictyosome; L, lipid; M, mitochondrion; Mt, microtubules; N, nucleus; P, peduncle; PM, promastigonemes; R, rod-shaped body; S, starch; X, crystalline body. From Lee (1977).

parasitic forms have been described as 'modified hyposomes' (Lom & Lawler 1973). These differences indicate that inferences about homologies should be made cautiously at present. The mechanism of transport through the peduncle is not known, despite studies of *Katodinium* and *Paulsenella* (Spero 1979, 1982; Schnepf & Deichgräber 1983, 1985).

The peduncle may be a versatile organelle, adaptable to other functions in addition to feeding, e.g. sensation, attachment. In addition, the peduncle might form the 'copulation globule,' the cytoplasmic bridge connecting cells during sexual pairing (Stosch 1973) as the connection between the peduncle and the nucleus of *Ceratium hirundinella* (Dodge & Crawford 1974) suggests.

STOMOPOD

In the sulcal area, between the flagellar pits, *Erythropsidinium agile* possesses a complex appendage, the 'stomopod.' It is a slender, sharp rod that appears to be used to inject lytic substances through a thin tube (the 'sialoduct') into prey

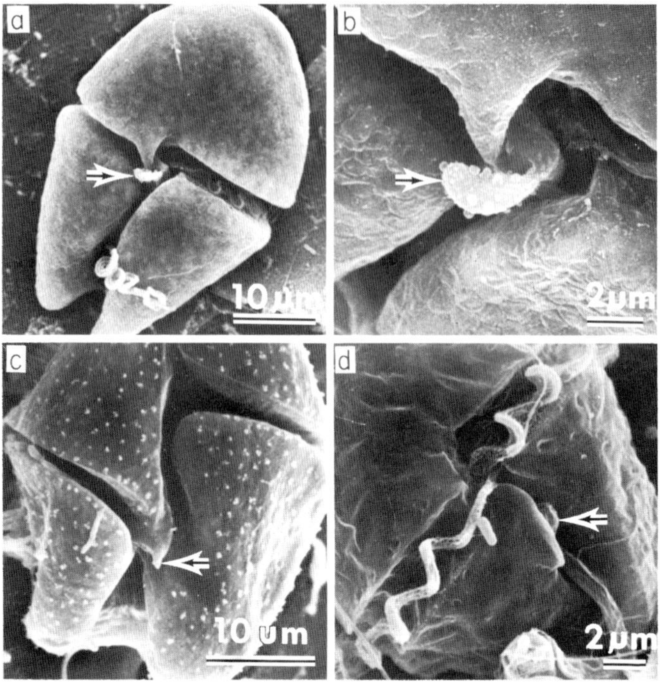

Fig. 6.14. Peduncles (arrows) of photosynthetic dinoflagellates. (a) and (b) *Gymnodinium sanguineum* Hirasaka. (c) *Gyrodinium instriatum* Fruedenthal & Lee. (d) *Ensiculifera* sp. (G. Gaines, unpubl.)

(Greuet 1969). A similar organelle is found in *Amyloodinium* Brown & Hovasse (Lom & Lawler 1973).

The stomopods of *Erythropsidinium* and *Amyloodinium* do not seem to be homologous with the peduncles or holdfast organelles of other free-living and parasitic forms, since both a stomopod and holdfast organelle are present in *Amyloodinium*. The stomopod does not seem to be homologous with the noctilucoid tentacle because *Erythropsidinium* possesses both a stomopod and another kind of tentacle (the 'piston').

TENTACLE AND PISTON

The tentacle of *Noctiluca scintillans* is directly involved in feeding (Soyer 1968, 1970). Its structure includes microtubular and fibrous systems, apparently in antagonism. The periphery of the tentacle contains microtubules and mucus vesicles, while the central region holds criss-crossed fibres (as well as mitochondria, endoplasmic reticulum and many ribosomes). The tentacle moves slowly but continuously, and has a flattened tip which it uses to convey food to the buccal cone near its base. Other genera in the Noctilucales (e.g. *Pomatodinium*

Cachon & Cachon-Enjumet, *Spatulodinium* Cachon & Cachon) also possess apparently homologous distensible masses or tentacles that have not yet been examined at the ultrastructural level (Cachon & Cachon-Enjumet 1966; Cachon & Cachon 1968).

Erythropsidinium agile possesses a kind of tentacle (the 'piston'), that extends posteriorly from the sulcal area below the flagellar pits (Greuet 1967) (Fig. 6.15). Inside, around the periphery, are sets of longitudinal fibres (myonemes). In the centre, stacked like inverted conical cups, are sheets of microtubules. From the lips of these cups, microtubules extend laterally out to the edges of the piston (see Chapter 3B for details). The microtubules and myonemes appear to be antagonistic systems, with the microtubules providing rigidity and tension and the myonemes providing swift contraction. Contraction can occur at great speed, providing bursts of rapid locomotion possibly related to feeding.

5.3 Food vacuoles

In most dinoflagellates, food vacuoles are temporary structures produced to enclose specific food items. At first, the vacuoles of *Oxyrrhis marina* consist of a single membrane surrounding ingested food (Dodge & Crawford 1974). Later,

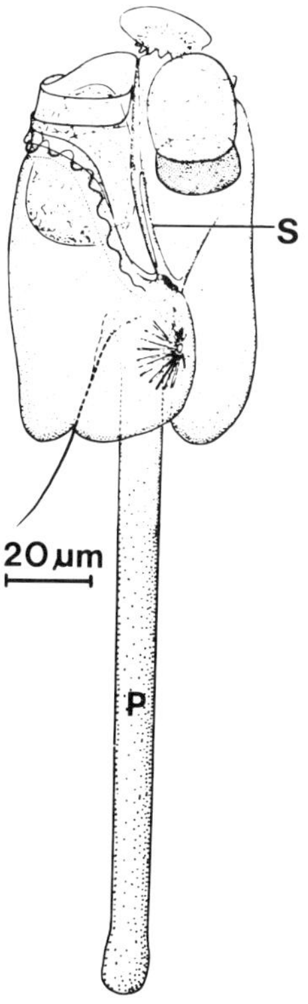

Fig. 6.15. *Erythropsidinium agile* (Hertwig) P. C. Silva, showing the stomopod (S) and 'piston' (P). from Greuet (1969) (as *Erythropsis pavillardi* Kofoid).

a thin layer of cytoplasm, containing microbodies and ribosomes, forms around it and grows progressively more complex. In the end, vacuoles move to the cell surface and their contents are released. Uhlig (1972) has documented the behaviour of food vacuoles in *Noctiluca*, using microcinematography.

It has been suggested (Chapman *et al.* 1981) that the food vacuoles of *C. hirundinella* reported by Dodge & Crawford (1974) were in fact accumulation bodies filled with the products of autolysis of cellular constituents. This is supported by observations of such bodies in what appear to be completely non-heterotrophic species grown axenically (Schmitter 1971). In addition, Chapman *et al.* (1981) report finding similar bodies in cysts of *C. hirundinella* preparing for excystment—where phagotrophy would be impossible through the thick cyst wall. However, unmistakable remains of blue-green algae, diatoms and bacteria have been seen in thin sections of *Oxyrrhis marina* and *Gyrodinium lebourae*, leaving little doubt about these being food vacuoles (Dodge & Crawford 1974; Lee 1977).

5.4 Pusules

Pusules are widely considered to be osmoregulatory or buoyancy-regulating organelles (Dodge 1972; Norris 1966; Abé 1981). However, Kofoid (1909) and Kofoid & Swezy (1921) thought that quantities of water could be taken into the pusule and the particles contained therein digested.

Schütt (1895) distinguished two kinds of pusules: the sack pusule ('Sackpusule') and the collecting pusule ('Sammelpusule'). However, most subsequent authors have failed to differentiate between the two. Abé (1981) is a recent and notable exception. Dodge (1972, see also Chapter 3A), described seven different kinds of pusules, four with vesicles and three without. Sack pusules, only known to occur in marine species, were examined in *Prorocentrum*. Unfortunately, no representatives of *Protoperidinium* or of the '*Diplopsalis* group' (which have one of each kind) were included, so it is not known whether their sack pusules are similarly constructed. Presumably then, the other six kinds of vesiculated and non-vesiculated pusules are all collecting pusules, and are found in both marine and freshwater forms. All pusules communicate with the surrounding medium through the posterior flagellar canal and pore. Abé (1981) reported observing the collapse and refilling of the sack pusule of *Protoperidinium spinulosum* (Schiller) Balech in a 'moment,' and concluded that its contents were nothing but the surrounding seawater. However, particles have never been seen inside pusules, except in *Oodinium*, where flakes of material (interpreted either as excretion products or as fixation artefacts) have been seen lining the internal surface of the collecting pusule (Cachon *et al.* 1970; Dodge 1972). Because the greatest relative development of the sack pusule is found in the non-photosynthetic genus *Protoperidinium*, and in the '*Diplopsalis* group,' and because its surface area could be very large, it has been thought to be involved in osmotrophic feeding (Taylor 1980). However, the principles opposed to osmotrophy as a way of life for large cells apply (see Section 3.3). In addition, extracellular digestion observed in *Protoperidinium* does not appear to involve the pusule (Gaines & Taylor 1984).

5.5 Water flow relative to feeding

The flow of water past a dinoflagellate cell must play a role in determining the potential food that comes within ingestion range. The importance of this may be inferred from the usual location of the feeding organelles: near the flagellar insertions, where water flowing past the cell meets water flowing around it (LeBlond & Taylor 1976).

Colourless peridinoids (e.g. *Diplopsalis*, *Protoperidinium*, *Zygabikodinium* Loeblich & Loeblich) possess wide flanges in the sulcal area that may hold a tunnel of water close to the cell body, where food particles may more easily be detected and taken in (LeBlond & Taylor 1976). Unfortunately, studies of water flow during swimming (Metzner 1929; Jahn *et al.* 1963; Gaines & Taylor 1985) have involved too few species and too gross a scale to decide.

Like other elaborate, colourless dinophysoids, *Ornithocercus* seems designed to resist normal dinoflagellate motion, yet it does turn and move slowly forward (Taylor 1971, 1976). Water flow appears to be greatest in the ventral area (Taylor 1971).

Noctiluca does not propel itself but rather floats freely, with the cytostome hidden by the monstrously swollen float. In those related forms that propel themselves through the water, the cytostome is located either on top of the cell (e.g. *Craspedotella*, *Pratjetella*) or at the junction of the cingulum and sulcus (e.g. *Cymbodinium* Cachon & Cachon) (Cachon & Cachon 1967). *Kofoidinium* and *Pomatodinium* carry hemispherical, extracellular 'domes' tipped to one side over the anterior part of the cell (Cachon & Cachon-Enjumet 1966; Cachon & Cachon 1974, see Fig. 2.7d, e). The cytostome is located on the epicone, toward the side least covered by the shell. The dome may thus serve to partly isolate the flow set up by the transverse flagellum from the longitudinal flow as the cell moves through the water.

6 APOCHLOROSIS

Table 6.5 presents a list of genera in which only colourless species have so far been described or reports of chloroplasts have proven false. All reportedly colourless *Ceratium* species examined with fluorescence microscopy (e.g. *Ceratium teres* Kofoid, *C. tripos*, *C. fusus* (Ehrenberg) Dujardin, *C. furca*, *C. hirundinella*) have been found to contain chromatophores (M. Elbrächter, unpubl.; F. J. R. Taylor, pers. comm.). A majority of athecate forms are probably apochlorotic.

Table 6.5. Families and genera of free-living dinoflagellates in which only apochlorotic species are known

Group	Remarks
Balechina Loeblich & Loeblich	
Citharistes Stein	Phaeosomes present
Clipeodinium Pascher	
Crypthecodinium Biecheler	
Diplopsalis Bergh	
Diplopsalopsis Meunier	
Ebria Borgert	
Entomosigma Schiller	
Entzia Lebour	
Erythropsidinium P. C. Silva	
Histioneis Stein	Phaeosomes present
Noctilucales	Species with photosynthetic endosymbionts reported
Oblea Balech	
Ornithocercus Stein	Phaeosomes present
Oxyphysis Kofoid	
Oxyrrhis Dujardin	
Parahistioneis Kofoid & Skogsberg	Phaeosomes present
Pronoctiluca Fabre-Domergue	
Protaspis Skuja	
Protoperidinium Bergh	
Warnowiaceae Lindemann	Reports of species with chloroplasts yet to be confirmed
Zygabikodinium Loeblich & Loeblich	

Colourless species seem to be more abundant in marine and especially in oceanic habitats, (Lessard & Swift 1986) whereas photosynthetic forms appear to predominate in freshwater and in coastal marine areas. Exclusively photosynthetic genera seem to be equally common in freshwater and marine environments. Despite the predominance of photosynthetic forms, freshwater environments have many colourless benthic forms (e.g., *Apsteinia acuta* (Apstein) Abé, *Gymnodinium helveticum*, *Peridiniopsis berolinense*, most *Katodinium* Fott).

ACKNOWLEDGEMENTS

We would like to thank H. von Stosch for information on pusules and photoheterotrophy, he and G. Drebes for their critical reading of the manuscript and E. Schnepf for information of terminology, ultrastructure and parasitic species.

7 REFERENCES

ABÉ T. H. (1981) Studies on the family Peridinidae. An unfinished monograph of the armoured dinoflagellata. *Publ. Seto Mar. Biol. Lab., Spec. Publ.* **6**, 1–413.

ALDRICH D. V. (1962) Photoautotrophy in *Gymnodinium breve* Davis. *Science, NY* **137**, 988–90.

ALLMANN G. J. (1855) Observations on *Aphanizomenon Flosaquae* and a species of Peridinea. *Q. Jl. microsc. Sci.* **3**, 21–25.

ANDREWS P. & WILLIAMS P. J. LEB. (1971) Heterotrophic utilization of dissolved organic compounds in the sea. III. Measurement of the oxidation rates and concentrations of glucose and amino acids in sea water. *J. mar. biol. Ass. UK* **51**, 111–125.

AZAM F. & HODSON R. E. (1977) Size distribution and activity of marine microheterotrophs. *Limnol. Oceanogr.* **22**(3), 492–501.

BADEN D. G. & MENDE T. J. (1978) Glucose transport and metabolism in *Gymnodinium breve*. *Phytochemistry* **17**(9), 1553–8.

BALECH E. (1971) Microplancton del Atlantico Ecuatorial Oeste (Equalant I). *Publnes. Serv. Hidrogr. Naval, Repub. Argent* **654**, 1–103.

BARKER H. A. (1935) The culture and physiology of the marine dinoflagellates. *Arch. Mikrobiol.* **6**, 157–81.

BAUMEISTER W. (1943) Die Dinoflagellaten der Kreise Pfarrkirchen und Eggenfelden (Gau Bayreuth). 2. Das Sumpfgebiet in Walde Südlich Altersham. *Arch. Protistenk.* **96**(3), 344–64.

BERGH R. S. (1881) Der Organismus der Cilioflagellaten. *Morph. Jb.* **7**, 177–288.

BIECHELER B. (1936a) Des conditions et du mécanisme de la prédation chez un dinoflagellé a enveloppe tabulée, *Peridinium gargantua n. sp. C. r. Séanc. Soc. Biol.* **121**, 1054–7.

BIECHELER B. (1936b) Observation de la capture et de la digestion des proies chez un Peridinien vert. *C. r. Séanc. Soc. Biol.* **122**, 1173–5.

BIECHELER B. (1952) Recherches sur les Péridiniens. *Bull. biol. Fr. Belg., suppl.* **36**, 1–149.

BOHLING H. (1970) Untersuchungen über freie gelöste Aminosäuren in Meerwasser. *Mar. Biol.* **6**, 213–25.

BOVIER-LAPIERRE M. E. (1888) Observations sur les Noctilucques. *C. r. Séanc. Soc. Biol.* **40**, 579–81.

BRUNO S. F. & MCLAUGHLIN J. J. A. (1977) The nutrition of the freshwater dinoflagellate *Ceratium hirundinella*. *J. Protozool.* **24**, 548–53.

BURSA A. S. (1961) The annual oceanographic cycle at Igloolik in the Canadian Arctic. II. The phytoplankton. *J. Fish. Res. Bd. Can.* **18**, 563–615.

CACHON J. (1964) Contribution à l'étude des Péridiniens parasites. Cytologie. Cycles évolutifs. *Annls Sci. nat., Zool.* **6**, 1–158.

CACHON J. & CACHON M. (1967) *Cymbodinium elegans* nov. gen. nov. sp., Péridinien Noctilucidae Saville-Kent. *Protistologica* **3**(3), 313–8.

CACHON J. & CACHON M. (1968) Contribution a l'étude des Noctilucidae Saville-Kent. I. Les Kofoidininae Cachon. Evolution, morphologique et systematique. *Protistologica* **3**(3), 427–44.

CACHON J. & CACHON M. (1969) Contribution a l'étude des *Noctilucidae* Saville-Kent. Evolution morphologique, cytologie, systematique. II. Les *Leptodiscinae* Cachon J. et M. *Protistologica* **5**(1), 11–33.

CACHON J. & CACHON M. (1971) *Protoodinium chattoni* Hovasse manifestations ultrastructurales des rapports entre le Péridinien et la meduse-hôte: fixation, phagocytose. *Arch. Protistenk.* **113**, 293–305.

CACHON J. & CACHON M. (1974) Le système stomopharyngien de *Kofoidinium* Pavillard. Comparisons avec celui de divers Péridiniens libres et parasites. *Protistologica* **10**(2), 217–22.

CACHON J., CACHON M. & BOUQUAHEUX F. (1969) *Myxodinium pipiens* gen. nov., Péridinien parasite d'*Halosphaera*. *Phycologia* **8**(3/4), 157–64.

CACHON J. & CACHON-ENJUMET M. (1966) *Pomatodinium impatiens,* nov. gen. nov. sp., Péridinien Noctilucidae Kent. *Protistologica* **2**(1), 23–30.

CACHON J., CACHON M. & GREUET C. (1970) Le système pusulaire de quelques Péridiniens libres ou parasites. *Protistologica* **6**, 467–76.

CALKINS G. (1902) Marine protozoa from Woods Hole. *Bull. U.S. Bureau Fish.* **21**, 413–68.

CAREFOOT J. R. (1968) Culture and heterotrophy of the freshwater dinoflagellate *Peridinium cinctum* fa. *ovoplanum* Lindemann. *J. Phycol.* **4**, 129–31.

CHAPMAN D. V., DODGE J. D. & HEANEY S. I. (1982) Cyst formation in the freshwater dinoflagellate *Ceratium hirundinella* (Dinophyceae). *J. Phycol.* **18**(1), 121–9.

CHAPMAN D. V., LIVINGSTONE D. & DODGE J. D. (1981) An electron microscope study of the excystment and early development of the dinoflagellate *Ceratium hirundinella*. *Br. phycol. J.* **16**(2), 183–94.

CHATTON E. (1920) Les péridiniens parasites. Morphologie, reproduction, éthologie. *Archs Zool. exp. gén.* **59**, 1–475.

CHENG J. Y. & ANTIA N. J. (1970) Enhancement by glycerol of photoheterotrophic growth of marine planktonic algae and its significance to the ecology of glycerol pollution. *J. Fish. Res. Bd. Can.* **27**, 335–46.

CHRISTEN H. R. (1959) Flagellaten aus dem Schützenweiher bei Veltheim. *Mitt. Naturw. Ges. Winterthur 1959* **29**, 167–89.

CHRISTEN H. R. (1961) Uber die Gattung *Katodinium* Fott (= *Massartia* Conrad). *Schweiz. Zeitschr. Hydrol.* **23**(2), 309–47.

DANGEARD P. A. (1892) La nutrition animale des Péridiniens. *Botaniste* **3**, 7–27.

DODGE J. D. (1971) Fine structure of the Pyrrhophyta. *Bot. Rev.* **37**, 481–508.

DODGE J. D. (1972) The ultrastructure of the dinoflagellate pusule: a unique osmo-regulatory organelle. *Protoplasma* **75**, 285–302.

DODGE J. D. & CRAWFORD R. M. (1970) The morphology and fine structure of *Ceratium hirundinella* (Dinophyceae). *J. Phycol.* **6**, 137–49.

DODGE J. D. & CRAWFORD R. M. (1971) Fine structure of the dinoflagellate *Oxyrrhis marina*. I. The general structure of the cell. *Protistologica* **7**, 295–304.

DODGE J. D. & CRAWFORD R. M. (1974) Fine structure of the dinoflagellate *Oxyrrhis marina*. III. Phagotrophy. *Protistologica* **10**, 239–44.

DOGIEL V. A. (1906) Beiträge zur Kenntnis der Peridineen. *Mitt. Zool. St. Neapel* **18**(1), 1–45.

DRAGESCO J. (1965) Étude cytologique de quelques flagellés mésopsammiques. *Cah. Biol. mar.* **6**, 83–115.

DREBES G. (1974) Marines Phytoplankton—eine Auswahl der Helgoländer Planktonalgen (Diatomeen, Peridineen). Thieme, Stuttgart. 186 pp.

DREBES G. (1978) *Dissodinium pseudolunula* (Dinophyta), a parasite on copepod eggs. *Br. phycol. J.* **13**(4), 319–28.

DREBES G. (1986) Life cycle and host specificity of marine parasitic dinophytes. *Helgoländer Meeresunters* **37**, 607–622.

DREBES G. & SCHNEPF E. (1982) Phagotrophy and development of *Paulsenella* cf. *Chaetoceratis* (Dinophyta), an ectoparasite of the diatom *Streptotheca thamesis, Helgoländer Meeresunters* **35**, 501–514.

DROOP M. R. (1953) Phagotrophy in *Oxyrrhis marina*. *Nature, Lond.* **172**, 250.

DROOP M. R. (1957) Auxotrophy and organic compounds in the nutrition of marine phytoplankton. *J. gen. Microbiol.* **16**, 286–93.

DROOP M. R. (1959a) A note on some physical conditions for cultivating *Oxyrrhis marina*. *J. mar. biol. ass. UK* **38**, 599–604.

DROOP M. R. (1959b) Water-soluble factors in the nutrition of *Oxyrrhis marina*. *J. mar. biol. ass. UK* **38**, 605–20.

DROOP M. R. (1963) A feeding experiment. *Br. phycol. Bull.* **2**, 278.

DROOP M. R. (1966) The role of algae in the nutrition of *Heteramoeba clara* Droop, with notes on *Oxyrrhis marina* Dujardin and *Philodina roseola* Ehrenberg. In *Some Contemporary Studies in Marine Science* (ed. H. Barnes), pp. 269–82. G. Allen & Unwin Ltd, London.

DROOP M. R. (1970) Nutritional investigation of phagotrophic protozoa under axenic conditions. *Helgoländer wiss. Meeresunters.* **20**, 272–7.

DROOP M. R. (1974) Heterotrophy of carbon. In *Algal Physiology and Biochemistry* (ed. W. D. P. Stewart), pp. 530–99. U. of Calif. Press, Berkeley.

DROOP M. R. & DOYLE J. (1966) Ubiquinone as a protozoan growth factor. *Nature* (*Lond.*) **212**(5069), 1474–5.

DROOP M. R. & PENNOCK J. F. (1971) Terpenoid quinones and steroids in the nutrition of *Oxyrrhis marina*. *J. mar. biol. Ass. UK* **51**, 455–70.

ELBRÄCHTER M. (1972) Begrenzte Heterotrophie bei *Amphidinium* (Dinoflagellata). *Kieler Meeresforsch.* **28**, 84–91.

ELBRÄCHTER M. (1979) On the taxonomy of unarmored dinophytes (Dinophyta) from the Northwest African upwelling region. *'Meteor' Forsch.-Ergebn.*, Ser. D, no. 30, pp. 1–22.

ELBRÄCHTER M. & DREBES G. (1978) Life cycles, phylogeny and taxonomy of *Dissodinium* and *Pyrocystis* (Dinophyta). *Helgoländer wiss. Meeresunters.* **31**, 347–66.

ENOMOTO Y. (1956) On the occurrence and the food of *Noctiluca scintillans* (Macartney) in the waters adjacent to the west coast of Kyushu, with special reference to the possibility of the damage caused to the fish eggs by that plankton. *Bull. Jap. Soc. scient. Fish.* **22**, 82–8.

ENTZ G. (1883) Beiträge zur Kenntniss der Infusorien. *Z. wiss. Zool.* **38**, 167–89.

ENTZ G. & SEBESTYN O. (1936) Morphologische, biologische und physico-chemische Untersuchungen an *Peridinium aciculifera* Lemm. mit besonderes Berücksichtigung der Gymnodinium-Form. *Arb. I. Abt. ung. Biol. Forsch.* **8**, 1–73.

FREY L. C. & STOERMER E. F. (1980) Dinoflagellate phagotrophy in the Upper Great Lakes. *Trans. Am. microsc. Soc.* **99**(4), 439–44.

GAINES G. & TAYLOR F. J. R. (1984) Extracellular digestion in marine dinoflagellates. *J. Plank. Res.* **6**(6), 1057–61.

GAINES G. & TAYLOR F. J. R. (1985) Form and function of the dinoflagellate transverse flagellum. *J. Protozool.* **32**, (2) 290–96.

GOLD K. (1970) Cultivation of marine ciliates (Tintinnida) and heterotrophic flagellates. *Helgoländer wiss. Meeresunters.* **20**, 264–71.

GOLD K. & BAREN C. F. (1966) Growth requirements of *Gyrodinium cohnii*. *J. Protozool.* **1**, 255–7.

GOODAY G. W. (1970) A physiological comparison of the symbiotic alga *Platymonas convolutae* and its free-living relatives. *J. mar. biol. Ass. UK* **50**, 199–208.

GREUET C. (1967) Organisation ultrastructurale du tentacule d'*Erythropsis pavillardi* Kofoid et Swezy, Péridinien *Warnowiidae* Lindemann. *Protistologica* **3**(3), 335–45.

GREUET C. (1969) Étude morphologique et ultrastructurale du trophonte d'*Erythropsis pavillardi* Kofoid et Swezy. *Protistologica* **5**(4), 481–503.

GREUET C., SALVANO P. & GAYOL P. (1980) Organisation du système stomopharyngien du Dinoflagellé *Oxyrrhis marina* Dujardin. *J. Protozool.* **27**(3), A80–1.

GRIESSMANN K. (1913) Über marine Flagellaten. *Arch. Protistenk.* **32**, 1–78.

GROMET-ELHANAN Z. & AVRON M. (1965) Effects of inhibitors and uncouplers on the separate light and dark reactions in photophosphorylation. *Pl. Physiol., Lancaster* **40**, 1053–9.

GROSS F. (1934) Zur Biologie und Entwicklungsgeschichte von *Noctiluca miliaris*. *Arch. Protistenk.* **83**, 178–96.

HANSON R. B. & SNYDER J. (1980) Glucose exchanges in a salt marsh-estuary: biological activity and chemical measurements. *Limnol. Oceanogr.* **25**(4), 633–42.

HARRIS T. M. (1940) A contribution to the knowledge of the British freshwater Dinoflagellata. *Proc. Linn. Soc. Lond.* **152**, 4–33.

HASTINGS J. L. & THOMAS W. H. (1977) Qualitative requirements and utilization of nutrients: algae. In *Handbook Series in Nutrition and Food. Section D: Nutritional Requirements* (ed. M. Rechcigl, Jr), Vol. 1, pp. 87–163. CRC Press, Cleveland.

HAUSER D. C. R., LEVANDOWSKY M. & GLASSGOLD J. M. (1975a) Ultrasensitive chemosensory responses by a protozoan to epinephrine and other neurochemicals. *Science, NY* **190**, 285–6.

HAUSER D. C. R., LEVANDOWSKY M., HUTNER S. H. & CHUSANOFF L. (1973) Chemoreception in *Crypthecodinium cohnii*. *J. Protozool.* **20**(suppl.), 496–7.

HAUSER D. C. R., LEVANDOWSKY M., HUTNER S. H., CHUSANOFF L. & HOLLWITZ J. S. (1975b) Chemosensory responses by the heterotrophic marine dinoflagellate *Crypthecodinium cohnii*. *Microb. Ecol.* **1**, 246–54.

HAUSER D. C. R., PETRYLAK D., SINGER G. & LEVANDOWSKY M. (1978) Calcium-dependent sensory-motor response of a marine dinoflagellate to CO_2. *Nature (Lond.)* **273**, 230–1.

HELLEBUST J. A. (1970) The uptake and utilization of organic substances by marine phytoplankters. In *Organic Matter in Natural Waters* (ed. D. W. Hood), pp. 225–56. Univ. Alaska Instn mar. Sci. occ. Publ. No. 1, Univ. of Alaska, Fairbanks.

HERDMAN E. C. (1922) Notes on dinoflagellates and other organisms causing discolouration of the sand at Port Erin. II. (1921) *Proc. Trans. Lpool biol. Soc.* **36**, 15–30.

HERDMAN E. C. (1924) Notes on dinoflagellates and other organisms causing discolouration of the sand at Port Erin. III. (1923) *Proc. Trans. Lpool biol. Soc.* **37**, 58–63.

HOBBIE J. E. & WRIGHT R.T. (1965) Competition between planktonic bacteria and algae for organic solutes. *Memorie Ist. ital. Idrobiol.* **18**(suppl.), 175–85.

HOFENEDER H. (1930) Über die animalische Ernährung von *Ceratium hirundinella* O. F. Müller und über die Rolle des Kernes bei dieser Zellfunktion. *Arch. Protistenk.* **71**, 1–32.

HOFKER F. (1930) Über *Noctiluca scintillans* Macartney. *Arch. Protistenk.* **71**, 57–78.

HOLT J. R. & PFIESTER L. A. (1981) A survey of auxotrophy in five freshwater dinoflagellates (Pyrrhophyta). *J. Phycol.* **17**, 415–6.

HOSIAISLUOMA V. (1975) *Gloeodinium montanum* (Pyrrhophyta, Algae) grown in darkness. *Annls bot. fennici* **12**, 55–6.

HU H., YU M. & ZHANG X. (1980) Discovery of phycobilin in *Gymnodinium cyaneum* Hu sp. nov. and its phylogenetic significance. *Kexue Tongbao* **25**(10), 882–4.

HULBURT E. M. (1957) The taxonomy of unarmored dinophyceae of shallow embayments on Cape Cod, Massachusetts. *Biol. Bull. mar. biol. Lab., Woods Hole* **112**, 196–219.

IRISH A. E. (1979) *Gymnodinium helveticum* Penard F. *achroum* Skuja. A case of phagotrophy. *Br. phycol. J.* **14**, 11–5.

JAHN T. L., HARMON W. M. & LANDMAN M. (1963) Mechanisms of locomotion in flagellates. I. *Ceratium*. *J. Protozool.* **10**(3), 358–63.

JEFFREY S. W. & VESK M. (1976) Further evidence for a membrane-bound endosymbiont within the dinoflagellate *Peridinium foliaceum*. *J. Phycol.* **12**, 450–5.

JOHNSON P. W. & SIEBURTH J. MCN. (1979) Chroococcoid cyanobacteria in the sea: a ubiquitous and diverse phototrophic biomass. *Limnol. & Oceanogr.* **24**(5), 928–34.

KAIN J. M. & FOGG G. E. (1960) Studies on the growth of marine phytoplankton. III. *Prorocentrum micans* Ehrenberg. *J. mar. biol. Ass. UK* **39**, 33–50.

KELLER S. E., HUTNER S. H. & KELLER D. E. (1968) Rearing the colorless marine dinoflagellate *Crypthecodinium cohnii* for use as a biochemical tool. *J. Protozool.* **15**, 792–5.

KIMOR B. (1979) Predation by *Noctiluca miliaris* Souriray on *Acartia tonsa* Dana eggs in the inshore waters of Southern California. *Limnol. Oceanogr.* **24**(3), 568–72.

KIMOR B. (1981) The role of phagotrophic dinoflagellates in marine ecosystems. *Kieler Meeresforsch. Sonderh.* **5**, 164–73.

KINNE O. (1977) In *Marine Ecology* (ed. O. Kinne), Vol. 3, pt. 2, pp. 587–591. John Wiley & Sons, New York.

KOCH A. L. (1971) The adaptive responses of *Escherichia coli* to a feast and famine existence. *Adv. Microb. Physiol.* **6**, 147–217.

KOFOID C. A. (1909) On *Peridinium steinii* Jörgensen, with a note on the nomenclature of the skeleton of the Peridinida. *Arch. Protistenk.* **16**, 25–47.

KOFOID C. A. (1931) Report on the biological survey of Mutsu Bay. 18. Protozoan fauna of

Mutsu Bay. Subclass Dinoflagellata; Tribe Gymnodinoidae. *Sci. Rept. Tôhoku Imp. Univ., 4th Ser., Biol.* **6**(1), 1–43.

Kofoid C. A. & Swezy O. (1921) The free-living unarmored Dinoflagellata. *Mem. Univ. Calif.* **5**, 1–562.

Krause F. (1910) Über das Auftreten von extramembranösem Plasma und Gallerthüllen bei *Ceratium hirundinella* O. F. Müll. *Int. Revue ges. Hydrobiol. Hydrogr.* **3**, 181–6.

Kubai D. F. & Ris H. (1969) Division in the Dinoflagellate *Gyrodinium cohnii* (Schiller). *J. Cell Biol.* **40**, 508–28.

Küster E. (1908) Eine kultivierbare Peridinee. *Arch. Protistenk.* **11**, 351–62.

LeBlond P. H. & Taylor F. J. R. (1976) The propulsive mechanism of the dinoflagellate transverse flagellum reconsidered. *BioSystems* **8**, 33–9.

Lebour M. V. (1925) *The Dinoflagellates of Northern Seas*, Mar. Biol. Ass. UK, Plymouth. 250p.

Lee C. & Bada J. L. (1977) Dissolved amino acids in the equatorial Pacific, the Sargasso Sea and Biscayne Bay. *Limnol. Oceanogr.* **22**(3), 502–10.

Lee R. E. (1977) Saprophytic and phagocytic isolates of the colorless heterotrophic dinoflagellate *Gyrodinium lebouriae* Herdman. *J. mar. biol. Ass. UK* **57**, 303–15.

Lessard E. J. (1984) Oceanic heterotrophic dinoflagellates: distribution, abundance, and role as microzooplankton. Ph.D. Dissertation, Univ. of Rhode Island, Kingston, Rhode Island.

Lessard, E. J. & Swift E. (1985) Species-specific grazing rates of heterotrophic dinoflagellates in oceanic waters, measured with a dual-label radioisotope technique. *Mar. Biol.* **87**, 289–96.

Lessard E. J. & Swift E. (1986) Dinoflagellates from the North Atlantic classified as phototrophic or heterotrophic with epifluorescence microscopy. *J. Plank. Res.* (in press).

Levander K. (1894) Materialien zur Kenntniss der Wasserfauna in der Umgebung von Helsnigfors mit besonderer Berücksichtigung der Meeresfauna. I. Protozoa. *Acta Soc. Fauna Flora fennica* **12**(2), 43–50.

Lindemann E. (1928) Neue Peridineen. *Hedwigia* **68**, 291–6.

Loeblich A. R. III (1966) Aspects of the physiology and biochemistry of the Pyrrhophyta. *Phykos* (Prof. Iyengar Mem. Vol.) **5**(1 & 2), 216–55.

Loeblich A. R. III & Sherley J. R. (1979) Observations on the theca of the motile phase of free-living and symbiotic isolates of *Zooxanthella microadriatica* (Freudenthal) comb. nov. *J. mar. biol. Ass. UK* **59**, 195–205.

Lom J. & Lawler A. R. (1973) An ultrastructural study on the mode of attachment in dinoflagellates invading gills of Cyprinodontidae. *Protistologica* **9**(2), 293–309.

Lucas I. A. N. (1982) Observations on *Noctiluca scintillans* (Macartney) Ehrenb. (Dinophyceae) with notes on an intracellular bacterium. *J. plankt. Res.* **4**, 401–9.

Lwoff A., van Niel C. B., Ryan F. J. & Tatum E. L. (1946) Nomenclature of nutritional types of microorganisms. *Cold Spring Harbor Symp. Quant. Biol.* **11**, 302–3.

McGinn M. P. (1971) *Axenic cultivation of* Noctiluca scintillans. Ph.D. Dissertation. Fordham University, New York.

McGinn M. P. & Gold K. (1969) Axenic cultivation of *Noctiluca scintillans. J. Protozool.* **16**(suppl.), 33.

McLaughlin J. J. A. & Zahl P. A. (1959) Vitamin requirements in symbiotic algae. In *International Oceanographic Congress Preprints*, pp. 930–1. AAAS, Washington, D.C.

Metzner P. (1929) Bewegungstudien an Peridineen. *Z. Bot.* **22**, 225–65.

Meyers P. A. (1976) Dissolved fatty acids in seawater from a fringing reef and a barrier reef at Grand Cayman. *Limnol. Oceanogr.* **21**(2), 315–8.

Mironov G. N. (1954) Nutrition of planktonic carnivores. I. *Noctiluca* nutrition. (In Russian) *Trud sevastopol'. biol. Sta.* **8**, 320–40.

Morey-Gaines G. & Ruse R. H. (1980) Encystment and reproduction of the predatory dinoflagellate, *Polykrikos kofoidi* Chatton (Gymnodiniales). *Phycologia* **19**(3), 230–2.

MORRILL L. C. & LOEBLICH A. R. III (1979) An investigation of heterotrophic and photoheterotrophic capabilities in marine Pyrrhophyta. *Phycologia* **18**(4), 394–404.

MUSCATINE L. (1980) Uptake, retention and release of dissolved inorganic nutrients by marine algae–invertebrate associations. In *Cellular Interactions in Symbiosis and Parasitism* (ed C. B. Cook, P. W. Pappas & E. D. Rudolf), pp. 229–44. Ohio State University Press, Columbus.

NAUWERCK A. (1963) Die Beziehungen zwischen Zooplankton und Phytoplankton im See Erken. *Symb. Bot. Ups.* **17**, 1–163.

NORRIS D. R. (1969) Possible phagotrophic feeding in *Ceratium lunula Schimper*. *Limnol. Oceanogr.* **14**(3), 448–9.

NORRIS R. E. (1966) Unarmoured marine dinoflagellates. *Endeavour* **35**, 124–8.

NORRIS R. E. (1967) Algal consortiums in marine plankton. In *Proceedings of the Seminar on Sea, Salt and Plants* (ed V. Krishnamurthy), pp. 178–89. Central Salt and Marine Chemicals Research Institute, Bhavnagar, India.

ODUM E. P. (1971) *Fundamentals of Ecology*, 3rd edn, W. B. Saunders Co., Philadelphia.

PASCHER A. (1916) Über eine neue Amöbe—*Dinamoeba* (*varians*)—mit dinoflagellatenartigen Schwärmern. (Der Studien über die rhizopodiale Entwicklung der Flagellaten. II. Teil.) *Arch. Protistenk.* **36**(2), 118–36.

PENARD E. (1891) Les Péridiniacées du Leman. *Bull. Soc. bot. Genève* **6**, 1–63.

PFEFFER W. (1881) *Handbuch der Pflanzenphysiologie*. Wilhelm Engelmann, Leipzig.

PFIESTER L. A. & POPOVSKÝ J. (1979) Parasitic, amoeboid dinoflagellates. *Nature* (*Lond.*) **279**, 421–4.

POPOVSKÝ J. (1982) Another case of phagotrophy by *Gymnodinium helveticum* Penard f. *achroum* Skuja. *Arch. Protistenk.* **125**, 73–8.

PRASAD R. R. (1953) Swarming of *Noctiluca* in the Palk Bay and its effect on the Choodai fishery, with a note on the possible use of *Noctiluca* as an indicator species. *Proc. Indian Acad. Sci., Sect. B* **38**, 40–7.

PRASAD R. R. (1958) A note on the occurrence and feeding habits of *Noctiluca* and their effects on the plankton community and fisheries. *Proc. Ind. Acad. Sci., Sec. B* **47**, 331–7.

PRINGSHEIM E. G. (1956) Micro-organisms from decaying seaweed. *Nature* (*Lond.*) **178**, 480–1.

PRINGSHEIM E. G. (1963) *Farblose Algen*. G. Fischer, Stuttgart. 471 pp.

PROVASOLI L. (1958) Nutrition and ecology of protozoa and algae. *A. Rev. Microbiol.* **12**, 279–308.

PROVASOLI L. (1977) Axenic cultivation. In *Marine Ecology* (ed O. Kinne), Vol. 3, pt. 3, pp. 1295–320. John Wiley & Sons, New York.

PROVASOLI L. & CARLUCCI A. F. (1974) Vitamins and growth regulators. In *Algal Physiology and Biochemistry* (ed. W. D. P. Stewart), pp. 741–87. Blackwell Scientific Publications, Oxford.

PROVASOLI L. & GOLD K. (1957) Some nutritional characteristics of *Gyrodinium cohnii*, a colorless marine dinoflagellate. *J. Protozool.* **4**(suppl.), 7.

PROVASOLI L. & GOLD K. (1962) Nutrition of the American strain of *Gyrodinium cohnii*. *Arch. Mikrobiol.* **42**, 196–203.

PROVASOLI L. & MCLAUGHLIN J. J. A. (1961) Limited heterotrophy of some photosynthetic dinoflagellates. *Bact. Proc.* 1961, 35.

PROVASOLI L. & MCLAUGHLIN J. J. A. (1963) Limited heterotrophy of some photosynthetic dinoflagellates. In *Symposium of Marine Biology* (ed. C. H. Oppenheimer), pp. 105–13. Charles C. Thomas, Springfield, Illinois.

RICHARDSON K. & FOGG G. E. (1982) The role of dissolved organic material in the nutrition and survival of marine dinoflagellates. *Phycologia* **21**(1), 17–26.

RILEY J. P. & CHESTER R. (1971) *Introduction to Marine Chemistry*. Academic Press, London.

ROSS M. R. & ABBOTT B. C. (1980) Heterotrophic metabolism of the marine red tide dinoflagellates. *Am. Soc. Limnol. Oceanogr., Abstr. 2nd Winter Meet.*

SAVILLE KENT W. (1881) *A Manual of the Infusoria*, Vol. 1. David Bogue, London.

SCHILLER J. (1925) Über die Besiedlung europäischer Meere mit Cryptomonaden und über einen Flagellaten peridineenähnlicher Organisation (*Entomosigma peridinioides*). *Öst. Bot. Z.* **74**(7–9), 194–8.

SCHILLER J. (1931–37) Dinoflagellatae (Peridineae). In *Kryptogamen-Flora von Deutschland, Österreich und der Schweiz* (Ed. by L. Rabenhorst), Vol. 10, Sec. 3. Akademisches Verlagsgesellschaft, Leipzig.

SCHILLER J. (1954) Über winterliche pflanzliche Bewohner des Wassers, Sees und des daraufregenden Schneebreies. I. *Öst. Bot. Zeitschr.* **101**, 236–84.

SCHILLER J. (1955) Untersuchen an Planktischen Protophyten des Neusidlersees 1950–1954, Teil I. *Wiss. Arb. Burgenland* **9**, 1–66.

SCHILLING A. J. (1891) Untersuchungen über die thierische Lebensweise einiger Perideen. *Ber. dt. bot. Ges.* **9**, 199–208.

SCHMARDA L. K. (1854) Zur Naturgeschichte Ägyptens. *Denkschr. Akad. Wiss., Wien* **7**(2), 1–28.

SCHMITTER R. F. (1971) The fine structure of *Gonyaulax polyedra*, a bioluminescent marine dinoflagellate. *J. Cell Sci.* **9**, 147–73.

SCHNEPFE E. & DEICHGRÄBER G. (1983) "Myzocytosis", a kind of endocytosis with implications to compartmentation in endosymbiosis. Obervations in *Paulsenella* (Dinophyta). *Naturwiss.* **71**, 218–9.

SCHNEPF E., DEICHGRÄBER G. & DREBES G. (1985) Food uptake and the fine structure of the dinophyte *Paulsenella* sp., an ectoparasite of marine diatoms. *Protoplasma* **124**, 188–204.

SCHÜTT F. (1895) Die, Peridineen der Plankton-Expedition. I Teil. Studien über die Zellen der Peridineen 170. In *Ergebn. Plankton-Expedition Humboldt-Stiftung* (ed. V. Hensen) **4**, pp. 1–170.

SCHÜTT F. (1899) Zentrifugales Dickenwachstum der Membran und extramembranöses Plasma. *Jahrb. wiss. Bot.* **33**, 594–690.

SIEBERT A. E. & WEST J. A. (1974) The fine structure of the parasitic dinoflagellate *Haplozoon axiothellae*. *Protoplasma* **81**, 17–35.

SIEBURTH J. MCN. (1978) About bacterioplankton. In *Phytoplankton Manual* (ed. A. Sournia), pp. 283–7. UNESCO, Paris.

SIEBURTH J. MCN. (1979) *Sea Microbes*. Oxford University Press, New York.

SILVA E. DE S. (1978) Endonuclear bacteria in two species of dinoflagellates. *Protistologica* **14**(2), 113–9.

SKUJA H. (1956) Taxonomische und biologische Studien über das Phytoplankton Schwedischer Binnengewässer. *Nova Acta Reg. Soc. Sci. Upsal. IV* **16**(3), 1–404.

SLABBER M. (1778) *Natuurkundige Verlustigengen, behelzende microscopice Waarneemingen van In- en Uitlandse, Waterend Land-Dieren*, 1st Edn, Dated 1771. Basch, Haarlem.

SOYER M.-O. (1968) Présence de formations fibrillaires complexes chez *Noctiluca miliaris* S. et discussion de leur rôle dans la motilité de ce Dinoflagellé. *C. r. hebd. Séanc. Acad. Sci., Paris* **266**, 2428–30.

SOYER M.-O. (1970) Les ultrastructures lièes aux fonctions de relation chez *Noctiluca miliaris* S. (Dinoflagella). *Z. Zellforsch.* **104**, 29–55.

SOYER M.-O. (1978) Particules de type viral et filaments trichocystoides chez les dinoflagellés. *Protistologica* **14**(1), 53–8.

SPERO H. J. (1979) *A study of the life-cycle, feeding and chemosensory behavior in the holozoic dinoflagellate* Gymnodinium fungiforme. M.Sc. Thesis, Texas A & M University, College Station, Texas.

SPERO H. J. (1982) Phagotrophy in *Gymnodinium fungiforme* (Pyrrhophyta): the peduncle as an organelle of ingestion. *J. Phycol.* **18**(3), 356–60.

SPERO H. J. & MORÉE M. D. (1981) Phagotrophic feeding and its importance to the life cycle of the holozoic dinoflagellate, *Gymnodinium fungiforme*. *J. Phycol.* **17**(1), 43–51.

STEIDINGER K. A., DAVIS J. T. & WILLIAMS J. (1967) A key to the marine dinoflagellate genera of the west coast of Florida. *Fla. Board Conserv. Mar. Lab. Publs., Tech. Ser.* **52**.

STEIN F. (1878, 1883) *Der Organismus der Flagellaten nach eigenen Forschungen in systematischer Reihenfolge bearbeitet. Der Organismus der Infusionsthiere.* III Abt. *Die Naturgeschichte der Flagellaten oder Geisselinfusorien.* Wilhelm Engelman, Leipzig.

STOSCH H. A. VON (1973) Observations on vegatative reproduction and sexual life cycles of two freshwater dinoflagellates, *Gymnodinium pseudopalustre* Schiller and *Woloszynskia apiculata* sp. nov. *Br. phycol. J.* **8**, 105–34.

SWEENEY B. M. (1971) Laboratory studies of a green *Noctiluca* from New Guinea. *J. Phycol.* **7**, 53–8.

SWEENEY B. M. (1976) *Pedinomonas noctilucae* (Prasinophyceae), the flagellate symbiotic in *Noctiluca* (Dinophyceae) in southeast Asia. *J. Phycol.* **12**, 460–4.

TAYLOR D. L. (1973) The cellular interactions of algal–invertebrate symbiosis. *Adv. Mar. Biol.* **11**, 1–56.

TAYLOR D. L. (1974) Symbiotic marine algae: taxonomy and biological fitness. In *Symbiosis in the Sea* (ed. W. B. Vernberg), pp. 245–62. University of South Carolina Press, Columbia.

TAYLOR F. J. R. (1971) Scanning electron microscopy of thecae of the dinoflagellate Genus *Ornithocercus*. *J. Phycol.* **7**, 249–58.

TAYLOR F. J. R. (1976) Dinoflagellates from the International Indian Ocean Expedition. *Biblthca bot.* **132**, 1–234.

TAYLOR F. J. R. (1979) Symbionticism revisited: a discussion of the evolutionary impact of intracellular symbioses. *Proc. Roy. Soc., Ser. B* **204**, 267–86.

TAYLOR F. J. R. (1980) On dinoflagellate evolution. *BioSystems* **13**, 65–108.

TAYLOR F. J. R. (1982) Symbioses in marine microplankton. *Ann. Inst. Océanogr. N.S.* **58**(suppl.). 61–90.

THOMAS W. H. (1955) Heterotrophic nutrition and respiration of *Gonyaulax polyedra*. *J. Protozool.* **2**(suppl.), 2–3.

TOMAS R. N. & COX E. R. (1973) Observations on the symbiosis of *Peridinium balticum* and its intracellular alga. I. Ultrastructure. *J. Phycol.* **9**, 304–23.

TRENCH R. K. (1980) Integrative mechanisms in mutualistic symbioses. In *Cellular Interactions in Symbiosis and Parasitism* (ed. C. B. Cook, P. W. Pappas & E. D. Rudolf), pp. 275–97. Ohio State University Press, Columbus.

TUTTLE R. C. & LOEBLICH A. R. III (1975) An optimal growth medium for the dinoflagellate *Crypthecodinium cohnii*. *Phycologia* **14**, 1–8.

UHLIG G. (1972) Entwicklung von *Noctiluca miliaris*. *Inst. Wiss. Film* C879, 1–15.

WEDEMAYER G. J. (1982) The peduncle of the dinoflagellate *Glenodinium berolinense*. *First International Phycological Congress, Scientific Program and Abstracts* a53.

WEDEMAYER G. J. & WILCOX L. W.(1984) An ultrastructural study of two freshwater, colorless dinoflagellates. *J. Phycol.* **16**(suppl.), 45.

WEDEMAYER G. J. & WILCOX L. W. (1984) The ultrastructure of the freshwater, colorless dinoflagellate *Peridiniopsis berolinense* (Lemm.) Bourrelly (Mastigophora, Dinoflagellida). *J. Protozool.* **31** (3), 444–53.

WILCOX, L. W. & WEDEMAYER G. J. (1984) *Gymnodinium acidotum* Nygaard (Pyrrhophyta), a dinoflagellate with an endosymbiotic cryptomonad. *J. Phycol.* **20**, 236–42.

WILCOX L. W, WEDEMAYER G. J. & GRAHAM L. E. (1982) *Amphidinium cryophilum* sp. nov. (Dinophyceae) a new freshwater dinoflagellate. II. Ultrastructure. *J. Phycol.* **18**(1), 18–30.

WOLOSZYNSKA J. (1917) Neue Peridineen-Arten nebst Bemerkungen über den Bau der Hülle bei *Gymno-* und *Glenodinium*. *Bull. Acad. Sci. Cracowie, Cl. Sc. Math. Nat., Sér. B.*

WOLOSZYNSKA J. (1925) Algologische Notizen 3(8). *Comptes Rendus St. Hydrobiol. Wigry* **1**(4), 1–9.

WRIGHT R. T. & HOBBIE J. E. (1966) Use of glucose and acetate by bacteria and algae in aquatic ecosystems. *Ecology* **47**, 447–64.

CHAPTER 7
BIOLUMINESCENCE AND CIRCADIAN RHYTHMS

BEATRICE M. SWEENEY
Department of Biological Sciences, University of California, Santa Barbara, CA 93106, USA

1 BIOLUMINESCENCE

1.1 Bioluminescence *in vivo*

The only photosynthetic organisms capable of bioluminescence are to be found among the dinoflagellates. A number of species of *Gonyaulax*, *Protogonyaulax*, *Pyrodinium* and *Pyrocystis*, all photoautotrophic, can emit flashes of light on either mechanical or chemical stimulation. Heterotrophic dinoflagellates such as *Noctiluca* are also bioluminescent. Isolated cells, both from cultures and collected directly from the sea, have been tested for bioluminescence (Table 7). From such studies, it is clear that not all dinoflagellates are able to emit light. For example, there are no known bioluminescent species in the genera *Gyrodinium*, *Prorocentrum* and *Dinophysis*, while only a few species of *Ceratium* have been reported to emit light. As far as I know, no freshwater dinoflagellates are bioluminescent. However, bioluminescent species are common in the marine phytoplankton and account for many of the beautiful displays recorded from all the seas of the world.

The light emitted from bioluminescent dinoflagellates is blue-green in colour, the maximum emission being at 474–476 nm. There is also one short report that a small amount of light in the red between 630 and 690 nm may be emitted from *Gonyaulax* (Reynolds 1976). In all this group, the natural stimulus is mechanical, cells responding to a deformation of the cell membrane produced by shear forces (Hamman & Seliger 1972), generated by strong stirring of the water, as in breaking waves or from the rapid swimming of fish or invertebrates. In the laboratory, it is sufficient to 'poke' the cell surface once with a piezoelectrically-driven probe to elicit a flash. At night, the threshold shear force required to elicit a bioluminescent flash in *Gonyaulax polyedra* is about 2 dynes cm^{-3} (Christianson & Sweeney 1972). In the two cells that have been

Table 7.1. Dinoflagellates from which bioluminescence has been recorded with photomultiplier tubes. Many other species are known *not* to be bioluminescent and still others have never been tested

Bioluminescent species	Reference
Ceratium fusus	Sweeney 1963
Fragilidium heterolobum	Sweeney 1963
Gonyaulax catenata	Tett 1971
G. digitale	Kelly & Katona 1966
G. hyalina	Sweeney 1963
G. polyedra	Haxo & Sweeney 1955
G. polygramma	Sweeney, unpubl.
G. sphaeroidea	Sweeney 1963
G. spinifera	Kelly & Katona 1966
Noctiluca scintillans	Nicol 1958
N. scintillans, green	Sweeney 1971
Protoperidinium brochi	Sweeney 1963
P. claudicans	Kelly & Katona 1966
P. conicum	Sweeney 1963
P. depressum	Sweeney 1963
P. divergens	Tett 1971
P. elegans	Sweeney, unpubl.
P. granii	Kelly & Katona 1966
P. leonis	Kelly & Katona 1966
P. oceanicum	Kelly & Katona 1966
P. ovatum	Tett 1971
P. pentagonum	Sweeney 1963
P. steinii	Tett 1971
P. subinerme	Kelly & Katona 1966
Polykrikos schwartzii	Tett 1971
Protogonyaulax acatenella	Esaias *et al.* 1973
P. catenella	Sweeney 1963
P. excavata (some strains)	Schmidt *et al.* 1978
P. monilata	Sweeney 1963
P. tamarensis	Esaias *et al.* 1973
Pyrocystis fusiformis	Swift & Meunier 1976
P. lunula	Swift & Taylor 1967
P. noctiluca	Swift & Meunier 1976
Pyrodinium bahamensis	Seliger *et al.* 1962

investigated with intracellular electrode recording, *Noctiluca* (Eckert & Sibaoka 1968) and *Pyrocystis fusiformis* (Widder & Case 1981a), the flash itself is known to be preceeded by an action potential. It is interesting that this action potential is negative with respect to an external reference (Fig. 7.1). This has been interpreted to mean that the potential arises on the vacuolar membrane rather than the plasmalemma (Eckert & Sibaoka 1968). In these large dinoflagellates, light can be observed to come from 'microsources' distributed throughout the cell in the cytoplasm at night (Fig. 7.2). The identity of these microsources is

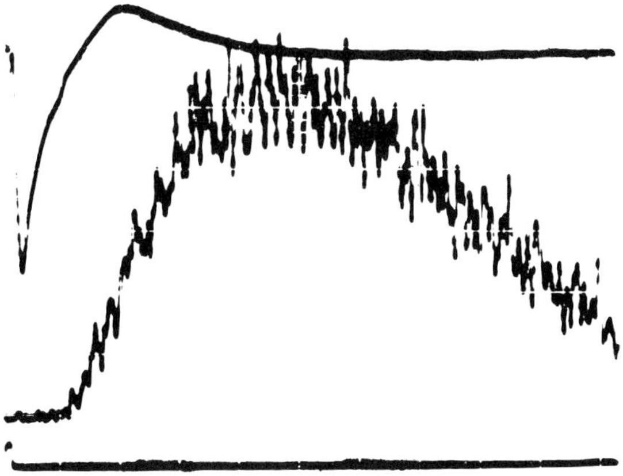

Fig. 7.1. Negative-going action potential (upper trace) and bioluminescent flash (lower trace) in *Pyrocystis fusiformis*. The calibration marker at the lower left corner indicates 20ms on the *x* axis and 20 mV on the *y* axis. (From Widder & Case 1981.)

not yet certain, although there is evidence that they may be membrane-bound particles in *Pyrocystis* (Sweeney 1979b) and in *Gonyaulax* (Nicolas *et al.* 1985).

The bioluminescence from a single isolated cell has been recorded in *Gonyaulax polyedra* (Christianson & Sweeney 1972; Krasnow *et al.* 1981), *Noctiluca scintillans*=(*N. miliaris*; Eckert & Reynolds 1967) and *Pyrocystis fusiformis* (Widder & Case 1981a, b).

The flash stimulated mechanically is of short duration (Fig. 7.3), with a rise time of 0·01 s in *Gonyaulax polyedra* (Christianson & Sweeney 1972) and about the same time for the first flash after a long dark interval in *Pyrocystis fusiformis* (Widder & Case 1981b). The maximum number of photons emitted by a single cell differs widely with the species, from 10^8 in *Gonyaulax polyedra* to more than 10^{11} in *P. fusiformis* (Sweeney 1981a).

The amount of light emitted in a flash appears to be highly conserved. It remains the same in stationary phase cultures as in the log phase of growth and is not altered when cultures of *Gonyaulax* are grown in media deficient in phosphorus or iron (Sweeney 1981a).

In addition to a flash, *Gonyaulax polyedra* is known to emit a dim glow without stimulation at certain times of night (see below). If a small amount of acid is added to the external medium containing *P. fusiformis* during the day, a continuous glow is emitted from a sphere in the centre of the cell (Sweeney 1981b). Whether or not other bioluminescent dinoflagellates can emit either a spontaneous or a stimulated localized glow is not known at present. Injured and dying cells often give off dim bioluminescence which persists for some time, but this is very different from the glows mentioned above. It spreads to the surrounding medium if the cells are broken and is not localized to any one region of the cell.

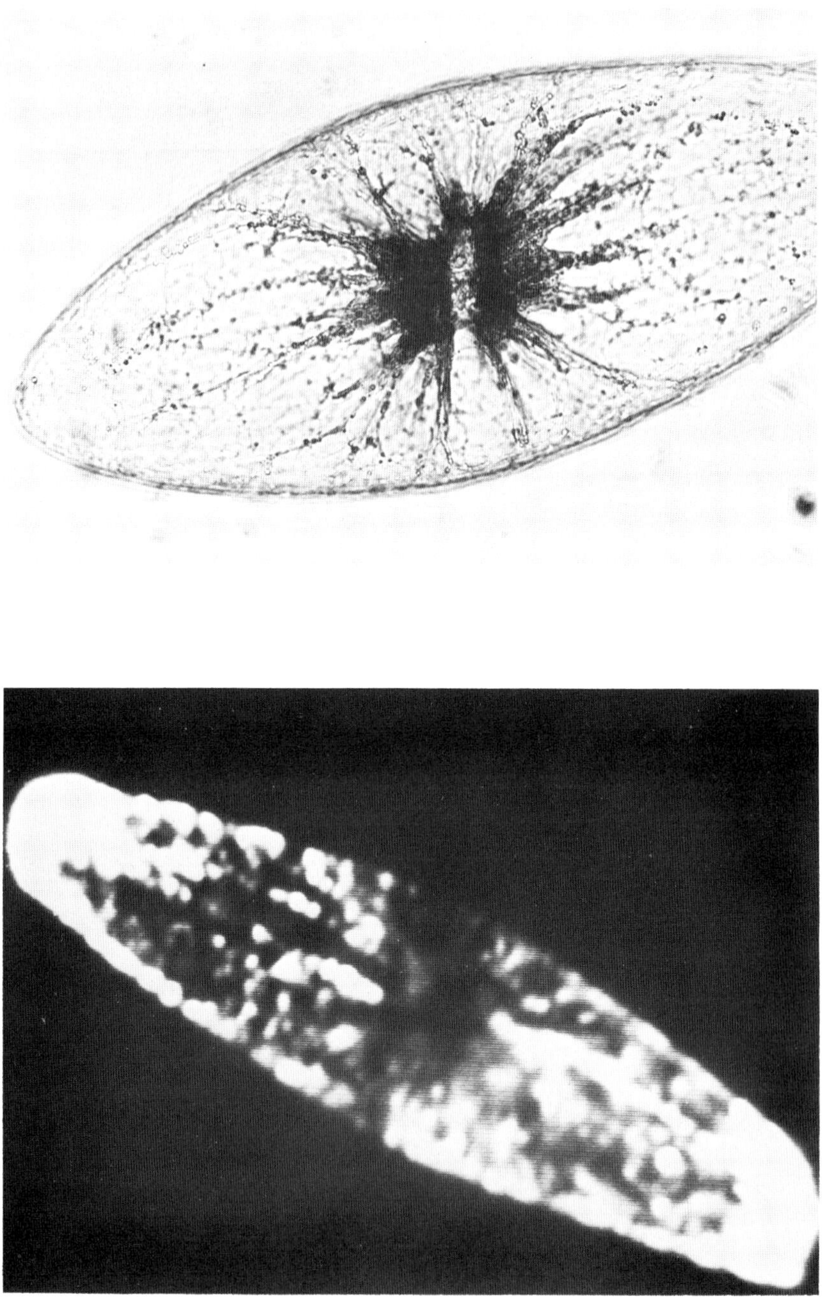

Fig. 7.2. *Pyrocystis fusiformis*, first night after cell division, light micrograph (above), and image intensified image of a smaller cell, 852 × 226 μm, stimulated to flash at night (below). Note microsources of bioluminescence. Light micrograph from Sweeney (unpubl.); photograph by image intensification from E. A. Widder (unpubl.).

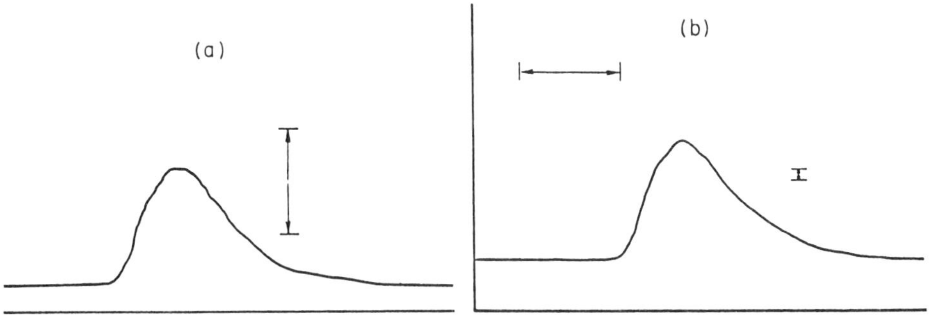

Fig. 7.3. Bioluminescent flash of a single *Gonyaulax polyedra* in the day (a) and the night (b) phases of the circadian rhythm in continuous darkness. Vertical bars represent 10^8 photons in each case; horizontal bar represents 40 ms in both graphs. (From Christianson & Sweeney 1972.)

In *Gonyaulax polyedra*, bioluminescence is inversely proportional to temperature over the range 13–27°C. When the temperature of a culture of *Gonyaulax* which has been grown at 20°C is lowered to about 10°C, bioluminescence begins to be emitted without mechanical or chemical stimulation (Fig. 7.4; Sweeney 1981a). Cultures grown at 11°C do not spontaneously emit light until the temperature is decreased to 4°C. *Pyrocystis fusiformis* also emits light spontaneously at low temperatures. High temperatures above 27°C elicit a similar spontaneous light emission. Other species of dinoflagellates have not yet been tested for these interesting responses to change in temperature.

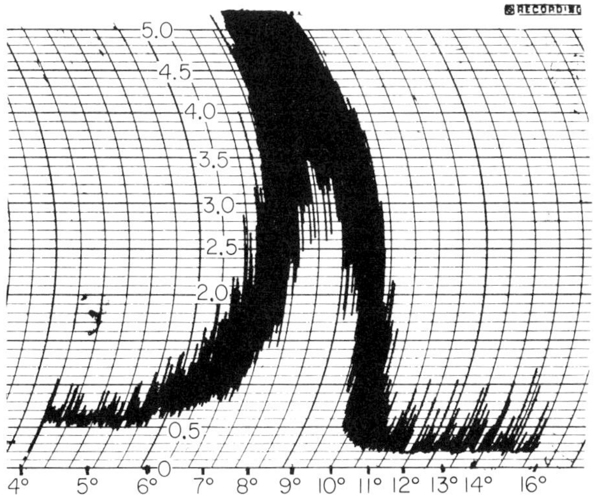

Fig. 7.4. Emission of bioluminescence in response to lowering the temperature in *Gonyaulax polyedra* grown at 20°C. Note that cooling started at the right side of the abscissa and proceeded at about 0·3°C per min. (From Sweeney, unpubl.). Three vertical bars = 1 minute.

Visible light inhibits bioluminescence in some dinoflagellates. The inhibition is proportional to the logarithm of the irradiance and different species show very different sensitivities, as shown by the thresholds of the response (Hamman *et al.* 1981a, b). The effect of light in *Gonyaulax polyedra* and in three species of *Protogonyaulax** is on the reception of a mechanical stimulus rather than on the biochemistry (Esaias *et al.* 1973). The same amount of light is finally emitted in light-inhibited cells as in cells remaining in darkness, if mechanical stimulation is continued long enough or luminescence is stimulated by the addition of acid. Light reduces the amount of luminescence emitted by *Pyrodinium bahamense* no matter how long stimulation is continued, so that the site affected may be different in this dinoflagellate (Hamman *et al.* 1981b). Light-inhibited *Pyrocystis* (*P. acuta, P. fusiformis, P. lunula* and *P. noctiluca*) do not bioluminesce over a long period on continued mechanical stimulation, but can be stimulated chemically to emit about the same amount of light as dark controls.

Action spectra for the photoinhibition of bioluminescence have been measured in several dinoflagellates. The first such action spectrum published was that for *Gonyaulax polyedra* (Sweeney *et al.* 1959) which showed both a blue and a red peak. Subsequently, only green light (maximum effectiveness at 567 nm) has been implicated as active in *Protogonyaulax catenella, P. acatenella* and *P. tamarensis* (Esaias *et al.* 1973). In *Pyrodinium bahamense, Pyrocystis acuta* and *P. lunula* (Hamman *et al.* 1981b), blue light is the most effective in light inhibition of bioluminescence.

Low bioluminescence during the day is not merely the result of light inhibition. This is clear from the measurements made either in constant light or constant darkness, where both *Gonyaulax* and *Pyrocystis* can continue to emit considerably less light during the subjective day than during the night (Fig. 7.3). Thus, the presence of a circadian rhythm must also be considered when interpreting the effects of light on bioluminescence (see below).

1.2 Biochemistry

It is not difficult to prepare extracts of bioluminescent dinoflagellates which are able to emit light. The manner of stimulating this bioluminescent emission is different in extracts than *in vivo* and depends on whether one is examining the soluble or particulate fraction of the extract. Light emission from the soluble components requires the presence of 1–1·5 M salt (Hastings & Sweeney 1957), a number of different salts being effective including $CaCl_2$, KCl and even $(NH_4)_2SO_4$. The pH optimum for this system is 6·6. The particulate fraction from *G. polyedra* emits light on acidification alone and the required pH is 5·5 or

**P. catenella, P. tamarensis* and *P. acatenella* were formerly attributed to *Gonyaulax*. They were transferred to *Protogonyaulax* by Taylor (1979) and to *Gessnerium* by Loeblich & Loeblich (1979).

lower (Fogel *et al.* 1972). Spent particles can be regenerated or 'recharged' in the presence of soluble luciferin and luciferase (Fogel & Hastings 1971; Fuller *et al.* 1972).

Both the luciferin and the luciferase of *G. polyedra* and the luciferin from *P. lunula* have been partially purified from extracts. Cross reactivity between luciferase from one species and luciferin from another show that these components are probably much the same in all luminescent dinoflagellates (see Sweeney 1979b). All luciferases examined so far from dinoflagellates are large proteins but differ somewhat in molecular weight in different species (Schmitter *et al.* 1976). The luciferin is a small molecule, 500–600 d molecular weight and is very unstable, even in the cold under argon. Recent work has disclosed that it is a tetrapyrrole (Dunlap *et al.* 1980a; Shimomura 1980; Dunlap & Hastings 1981a) whose structure suggests that it may be related to chlorophyll (Dunlap *et al.* 1981). Some dinoflagellates, particularly *Gonyaulax* and *Protogonyaulax*, also contain a luciferin-binding protein which binds the the luciferin at neutral pH, inactivating the light reaction. When the pH is lowered to 5·5 *in vitro*, the luciferin is freed from the luciferin-binding protein and the light reaction can take place. For this reason, it has been suggested (Hastings 1978) that release of protons may be the natural stimulating mechanism. However, several luminescent dinoflagellates, including *Pyrocystis fusiformis, P. lunula* and *P. noctiluca*, do not contain luciferin-binding protein (Schmitter *et al.* 1976).

Extracts of *Pyrocystis fusiformis* emit light, all of it in the soluble fraction. No particulate bioluminescence is observed.

Recently a very interesting cross-reaction between the luciferin of *Pyrocystis lunula* and the photoprotein of the euphausiid crustacean, *Meganyctiphanes norvegica*, has been reported (Dunlap *et al.* 1980a). Furthermore, compound 'F', the light-emitting molecule from euphausiids, has been characterized as a bile pigment which is very unstable and gives coloured products on oxidation. A tentative structure for it has been proposed (Shimomura 1980). It may be that the bioluminescence of the marine euphausiid shrimps is 'borrowed' from the dinoflagellates which they eat. This would be another example of transmission of components of bioluminescent systems from one organism to another up the food chain, similar in many respects to the transfer of bioluminescence from *Cypridina* to the fishes, *Apogon* and *Porichthys* (Johnson *et al.* 1960; Cormier *et al.* 1967).

Further details of the biochemistry of the dinoflagellate system have been reviewed by Sweeney (1979b).

1.3 Selective advantage of bioluminescence

The question of the selective advantage of bioluminescence in dinoflagellates has been a subject of much speculation and little experiment. However, two feeding experiments where luminous and non-luminous strains of the same species of dinoflagellates were fed to copepods have been conducted (Esaias & Curl 1972; White 1979). In both, predation was less in the presence of bioluminescence. Thus, one selective advantage of bioluminescence may be the ability to frighten away predators. The fact that the emission wavelength of dinoflagellate bioluminescence is close to the maximum transmission wavelength of sea water strongly suggests that its visibility is a major factor in its function.

2 CIRCADIAN RHYTHMS

When the bioluminescence of *Gonyaulax* was first examined (Haxo & Sweeney 1955), it was discovered that the amount of light emitted was very strongly dependent on the time of day when the cells were stimulated, being almost 100 times greater at night than during the day. This fluctuating behaviour continued when the cells were placed in constant darkness (Sweeney & Hastings 1957) or in constant light (Hastings & Sweeney 1958) provided that the light was not enough to inhibit mechanical stimulation or that acid stimulation was employed (Fig. 7.5). Subsequent studies showed that this rhythm in bioluminescence was similar to rhythms observed in *Drosophilla* emergence, *Euglena* phototaxis and other periodic physiology and behaviour, in being temperature-compensated and resettable by visible and ultraviolet light (Bruce & Pittendrigh 1957; Hastings & Schweiger 1976). All these phenomena are considered to be manifestations of the ability of cells to tell time over the interval of 24 hours.

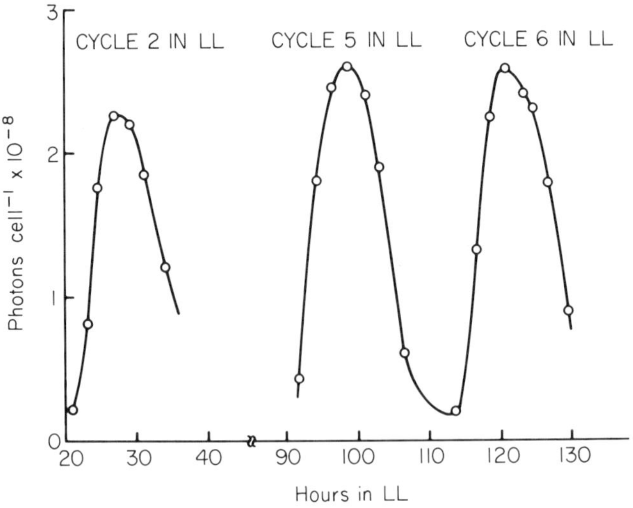

Fig. 7.5. Circadian rhythm in bioluminescence in *Gonyaulax polyedra* stimulated by the addition of acetic acid. The second, fifth and sixth cycles in continuous light (300 μW cm^{-2}) at 21°C are shown. (From Sweeney, unpubl.)

Such rhythms are now known as circadian, because the period is about a day. *Noctiluca*, which is non-photosynthetic, does not show this rhythm of luminescence (Nicol 1958).

The photosynthetic dinoflagellates, as a group, are especially well endowed with circadian rhythms. In addition to stimulated bioluminescence, spontaneous flashing and glow, *Gonyaulax* shows rhythms in photosynthetic capacity (Hastings *et al.* 1961), cell division (Sweeney & Hastings 1958) and relative membrane potential (Adamich *et al.* 1976). Circadian rhythmicity has been observed in the photosynthesis of *Ceratium furca* and in *Glenodinium* sp. (Prezelin *et al.* 1977). The photosynthetic capacity of natural phytoplankton populations varies over the day (Harding *et al.* 1981). However, most phytoplankton assemblages have never been shown to be rhythmic under constant conditions and consequently we do not know whether or not this apparently periodic behaviour is a true circadian rhythm. Such a phenomenon is called a 'diel rhythm', a term which does not imply time-keeping ability. Cell division in many dinoflagellates has been observed to occur in nature only at particular times of day (Weiler & Chisholm 1976; Pfeister & Anderson, this volume). In some species, the circadian nature of this phenomenon has been demonstrated.

In *Pyrocystis fusiformis*, bioluminescence continues to be rhythmic in both constant light and continuous darkness (Sweeney 1981b). When dividing pairs are isolated and their bioluminescence is measured in continuous darkness over several days as they develop, a circadian rhythm in luminescence is evident (Fig. 7.6). As these daughter cells proceed through the cell cycle the luminescence of each stage is still greater during the night phase than during the day. Rhythmicity in the bioluminescence of *P. lunula* in continuous darkness has also been documented (Swift & Meunier 1976). In both *P. fusiformis* and *P. lunula*, chloroplasts change position rhythmically. In the former, photosynthesis and cell division are also characteristically circadian and all stages in the cell cycle seem to be timed by the circadian system (Sweeney 1982).

The mechanism for generating circadian time-keeping is intracellular, but little is known regarding the details. Protein synthesis on 80S ribosomes is probably required since circadian rhythms in *Gonyaulax* (Walz & Sweeney 1979; Dunlap *et al.* 1980b), the green alga *Acetabularia* (Karakashian & Schweiger 1976), the optic nerve of the mollusc *Aplysia* (Rothman & Strumwasser 1976) and the fungus *Neurospora* (Nakashima *et al.* 1981) are all phase-shifted by the inhibitor of protein synthesis on 80S ribosomes, cycloheximide. Circadian rhythmicity in both total RNAa and specific RNA species has been demonstrated (Walz, Walz & Sweeney 1983). Furthermore, luciferase activity shows a circadian variation, probably a manifestation of clock-regulated synthesis or degradation (Dunlap & Hastings 1981b).

Ion transport through some membrane or membranes may also be important because substances which change membrane behaviour, e.g. valinomycin

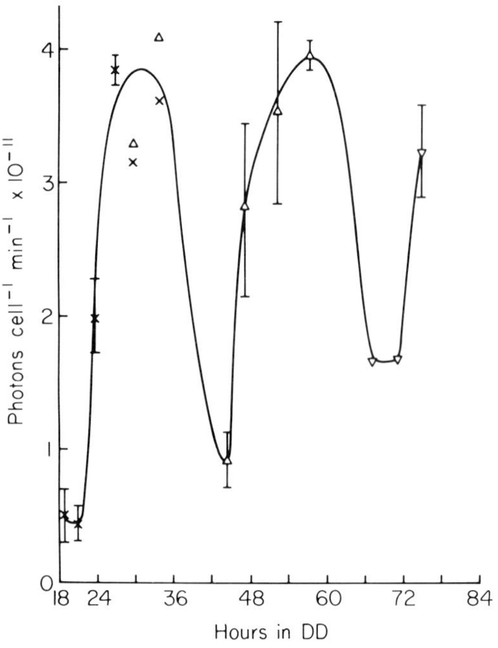

Fig. 7.6. Circadian rhythm in bioluminescence in *Pyrocystis fusiformis* stimulated by the addition of acetic acid. The second and third cycles in continuous darkness are shown. Temperature, 20°C. (From Sweeney, unpubl.) Δ, expt. 1; ×, expt. 2.

(Sweeney 1974), vanillic acid (Kiessig *et al.* 1979), alcohols and aldehydes (Sweeney & Herz 1977; Taylor *et al.* 1979; Taylor & Hastings 1979), alter timing in *Gonyaulax* and other rhythmic organisms. The nature of circadian time-keeping is being investigated vigorously at present, and dinoflagellates will no doubt play an important role here due to: (i) their diverse circadian rhythms, (ii) the finding that a single species can display several different rhythmic functions, and (iii) the possibility of growing large cultures.

3 REFERENCES

ADAMICH M., LARIS P. C. & SWEENEY B. M. (1976) *In vivo* evidence for a circadian rhythm in membranes of *Gonyaulax*. *Nature* (Lond.) **261**, 583–585.

BIGGLEY W. H., SWIFT E., BUCHANAN R. J. & SEILGER H. H. (1969) Stimulable and spontaneous bioluminescence in the marine dinoflagellates, *Pyrodinium bahamense, Gonyaulax polyedra* and *Pyrocystis lunula*. *J. gen. Physiol.* **54**, 96–122.

BRUCE V. G. & PITTENDRIGH C. S. (1957) Endogenous rhythms in insects and microorganisms. *Amer. Nat.* **91**, 179–195.

CHRISTIANSON R. & SWEENEY, B. M. (1972) Sensitivity to stimulation, a component of the circadian rhythm in luminescence in *Gonyaulax*. *Plant Physiol.* **49**, 994–997.

CORMIER M. J., CRANE J. M. JR & KAKANO Y. (1967) Evidence for the identity of the luminescent systems of *Porichthys porosissimus* (fish) and *Cypridina hilgendorfii* (Crustacean). *Biochem. biophys. Res. Commun.* **29**, 747–752.

DUNLAP J. C. & HASTINGS J. W. (1981a). Biochemistry of dinoflagellate bioluminescence: purification and characterization of dinoflagellate luciferin from *Pyrocystis lunula*. *Biochemistry* **20**, 983–989.

DUNLAP J. C. & HASTINGS J. W. (1981b) The biological clock in *Gonyaulax* controls luciferase activity by regulating turnover. *J. biol. Chem.* **256**, 10509–10518.

DUNLAP J. C., HASTINGS J. W. & SHIMOMURA O. (1980a). Cross reactivity between the light-emitting system of distantly related organisms: novel type of light-emitting compound.*Proc. natl. Acad. Sci. USA* **77**, 1394–1397.

DUNLAP J. C., HASTINGS J. W. & SHIMOMURA O. (1981) Dinoflagellate luciferin is structurally related to chlorophyll. *FEBS Lett.* **135**, 273–276.

DUNLAP J. C., TAYLOR W. & HASTINGS J. W. (1980b) The effects of protein synthesis inhibitors on the *Gonyaulax* clock. *J. comp. Physiol.* **138**, 1–8.

ECKERT R. & REYNOLDS G. T. (1967) The subcellular origin of bioluminescence in *Notiluca miliaris. J. gen. Physiol.* **50**, 1429–1458.

ECKERT R. & SIBOAKA T. (1968) The flash-triggering action potential of the luminescent dinoflagellate *Noctiluca. J. gen. Physiol.* **52**, 258–282.

ESAIAS W. E. & CURL H. C. JR. (1972) Effect of dinoflagellate bioluminescence on copepod ingestion rates. *Limnol Oceanogr.* **17**, 901–906.

ESAIAS W. E., CURL H. C. JR & SELIGER H. H. (1973) Action spectrum for a low intensity, rapid photoinhibition of mechanically stimulable bioluminescence in the marine dinoflagellates *Gonyaulax catenella, G. acatenella,* and *G. tamarensis. J. cell Physiol.* **82**, 363–372.

FOGEL M. & HASTINGS J. W. (1971) A substrate-binding protein in the *Gonyaulax* bioluminescent reaction. *Arch. Biochem. Biophys.* **142**, 310–321.

FOGEL M., SCHMITTER R. E. & HASTINGS J. W. (1972) On the physical identity of scintillons: bioluminescent particles in *Gonyaulax polyedra. J. cell Sci.* **11**, *305–317.*

FULLER C. W., KREISS P. & SELIGER H. H. (1972) Particulate bioluminescence in dinoflagellates: dissociation and partial reconstitution. *Science* (N. Y.) **177**, 884–885.

HAMMAN J. P. & SELIGER H. H. (1972) The mechanical triggering of bioluminescence in marine dinoflagellates: chemical basis. *J. cell Physiol.* **80**, 397–408.

HAMMAN J. P., BIGGLEY W. H. & SELIGER H. H. (1981a) Action spectrum for the photoinhibition of bioluminescence in the marine dinoflagellate *Dissodinium lunula. Photochem. Photobiol.* **33**, 741–747.

HAMMAN J. P., BIGGLEY W. H. & SELIGER H. H. (1981b) Photoinhibition of stimulable bioluminescence in marine dinoflagellates. *Photochem. Photobiol.* **33**, 909–914.

HAMMAN J. P. & SELIGER H. H. (1982) The chemical basis for the mechanical stimulation, photoinhibition and recovery from photoinhibition in the marine dinoflagellate, *Gonyaulax polyedra. J. cell. Physiol.* (*submitted*).

HARDING L. W. JR, MEESON B. W., SWEENEY B. M. & PREZELIN B. B. (1981) Diel periodicity in marine phytoplankton photosynthesis. *Mar. Biol.* **61**, 95–105.

HASTINGS J. W. (1978) Bacterial and dinoflagellate luminescent systems. In *Bioluminescence in Action* (ed. P. J. Herring), pp. 129–170. Academic Press, New York.

HASTINGS J. W. & SCHWEIGER H. G. (Eds) (1976) *The Molecular Basis of Circadian Rhythms.* Dahlem Koferenzen, Berlin.

HASTINGS J. W. & SWEENEY B. M. (1957) The luminescent reaction in extracts of the marine dinoflagellate, *Gonyaulax polyedra. J. cell. comp. Physiol.* **49**, 209–226.

HASTINGS J. W. & SWEENEY B. M. (1958) A persistent diurnal rhythm of luminescence in *Gonyaulax polyedra. Biol. Bull.* **115**, 440–458.

HASTINGS J. W., ASTRACHAN L. & SWEENEY B. M. (1961) A persistent daily rhythm in photosynthesis. *J. gen. Physiol.* **45**, 69–76.

HAXO F. T. & SWEENEY B. M. (1955) Bioluminescence in *Gonyaulax polyedra.* In *The Luminescence of Biological Systems* (ed. F. H. Johnson), pp. 415–420. AAAS Washington, D.C.

JOHNSON F. H., HANEDA Y. & SIE E. H.-C. (1960) An interphylum luciferin–luciferase reaction. *Science* (N. Y.) **132**, 422–423.

KARAKASHIAN M. W. & SCHWEIGER H. G. (1976) Temparture-dependence of cycloheximide-sensitive phase of the circadian cycle in *Acetabularia mediterranea*. *Proc. nat. Acad. Sci. USA* **73**, 3216–3219.

KELLY M. G. & KATONA S. (1966) An endogenous rhythm of bioluminescence in a natural population of dinoflagellates. *Biol. Bull.* **131**, 115–126.

KIESSIG R. S., HERZ J. M. & SWEENEY B. M. (1979) Shifting the phase of the circadian rhythm in bioluminescence in *Gonyaulax* with vanillic acid. *Plant Physiol.* **63**, 324–327.

KRASNOW R., DUNLAP J., TAYLOR W., HASTINGS J. W., VETTERLING W. & HAAS E. (1981) Measurements of *Gonyaulax* bioluminescence, including that of single cells. In *Bioluminescence Current Perspectives* (ed. K. H. Nealson), pp. 52–63. Burgess Publishing Co. Minneapolis.

LOEBLICH A. R. III & LOEBLICH L. A. (1979) The systematics of *Gonyaulax* with special reference to the toxic species. In *Toxic Dinoflagellate Blooms, Proceedings of the Second International Conference on Toxic Dinoflagellate Blooms* (eds D.L. Taylor & H. H. Seliger), pp. 41–46. Elsevier-North Holland, Amsterdam.

NAKASHIMA H., PERLMAN J. & FELDMAN J. F. (1981) Genetic evidence that protein synthesis is required for the circadian clock of *Neurospora*. *Science* **212**, 361–362.

NAWATA T. & SIBOAKA T. (1979) Coupling between action potential and bioluminescence in *Noctiluca*: effects of inorganic ions and pH in vacuolar sap. *J. comp. Physiol.* A **134**, 137–149.

NICOL J. A. C. (1958) Observations on luminescence in *Noctiluca*. *J. mar. biol. Assn. UK* **37**, 535–549.

PREZELIN B. B., MEESON B. W. & SWEENEY B. M. (1977) Characterization of photosynthetic rhythms in marine dinoflagellates. 1. Pigmentation, photosynthetic capacity and respiration. *Plant Physiol.* **60**, 384–387.

REYNOLDS G. T. (1976) Evidence for *in vivo* bioradiation at wavelengths greater than 6400 Å in *Gonyaulax polyedra*. *Biol. Bull.* **151**, 427 abstract.

ROTHMAN B. S. & STRUMWASSER F. (1976) Phase shifting the circadian rhythm of neuronal activity in the isolated *Aplysia* eye with puromycin and cycloheximide: electrophysiological and biochemical studies. *J. gen. Physiol.* **68**, 359–384.

SCHMIDT R. J., GOOCH V. D., LOEBLICH A. R. III & HASTINGS J. W. (1978) Comparitive study of luminescent strains of *Gonyaulax excavata* (Pyrrophyta). *J. Phycol.* **14**, 5–9.

SCHMITTER R. E., NJUS D., SULZMAN F. M., GOOCH V. D. & HASTINGS J. W. (1976) Dinoflagellate bioluminescence: a comparative study of *in vitro* components. *J. cell Physiol.* **87**, 123–134.

SELIGER H. H., FASTIE W. G., TAYLOR W. R. & MCELROY W. D. (1972) Bioluminescence of marine dinoflagellates. i. An underwater photometer for day and night measurements. *J. gen. Physiol.* **45**, 1003–1017.

SHIMOMURA O. (1980) Chlorophyll-derived bile pigment in bioluminescent Euphausiids. *FEBS Letters* **116**, 203–206.

SULZMAN F. M., KREIGER N. R., GOOCH V. D. & HASTINGS J. W. (1978) A circadian rhythm of the luciferin binding protein from *Gonyaulax polyedra*. *J. comp. Physiol.* **128**, 251–257.

SWEENEY B. M. (1963) Bioluminescent dinoflagellates. *Biol. Bull.* **125**, 177–181.

SWEENEY B. M. (1971) Laboratory studies of the green *Noctiluca* from New Guinea. *J. Phycol.* **7**, 53–58.

SWEENEY B. M. (1974) The potassium content of *Gonyaulax polyedra* and phase changes in the circadian rhythm of stimulated bioluminescence by exposures to ethanol and valinomycin. *Plant Physiol.* **53**, 337–342.

SWEENEY B. M. (1979a) Bright light does not immediately stop the circadian clock in *Gonyaulax*. *Plant Physiol.* **64**, 314–334.

SWEENEY B. M. (1979b) The bioluminescence of dinoflagellates. In *Biochemistry and Physiology of Protozoa,* 2nd edn (eds M. Levandowsky & S. H. Hunter), pp. 287–306. Academic Press, New York.

SWEENEY B. M. (1980) Intracelluar sources of bioluminescence. *Int. Rev. Cytol.* **68**. 173–195.

Sweeney B. M. (1981a) Variations in the bioluminescence per cell in dinoflagellates. In *Bioluminescence: Current Prospectives* (ed. K. H. Nealson), pp. 90–94. Burgess Publishing Co. Minneapolis.

Sweeney B. M. (1981b) The circadian rhythms in bioluminescence, photosynthesis and organellar movements in the large dinoflagellate, *Pyrocystis fusiformis*. In *International Cell Biology* 1980–1981 (ed. H. G. Schweiger), pp. 807–814. Springer Verlag, Berlin.

Sweeney B. M. (1982) The interaction of the circadian cycle with the cell cycle in *Pyrocystis fusiforms*. *Plant Physiol.* (in press).

Sweeney B. M. & Hastings J. W. (1957) Characteristics of the diurnal rhythm of luminescence in *Gonyaulax polyedra*. *J. cellular comp. Physiol.* **49**, 115–128.

Sweeney B. M. & Hastings J. W. (1958) Rhythmic cell division in populations of *Gonyaulax polyedra*. *J. Protozool.* **5**, 217–224.

Sweeney B. M., Haxo F. T. & Hastings J. W. (1959) Action spectra for two effects of light on luminescence in *Gonyaulax polyedra*. *J. gen. Physiol.* **43**, 285–299.

Sweeney B. M. & Herz J. M. (1977) Evidence that membranes play an important role in circadian rhythms. In *Proc. int. Soc. Chronobiol.*, pp. 751–761. El Ponte, Milano.

Swift E. & Taylor W. R. (1967) Bioluminescence and chloroplast movement in the dinoflagellate *Pyrocystis lunula*. *J. Phycol.* **31**, 77–81.

Swift E. & Meunier V. (1976) Effects of light intensity on division rate, stimulable bioluminescence and cell size of the oceanic dinoflagellate *Dissodium lunula, Pyrocystis fusiformis* and *P. noctiluca*. *J. Phycol.* **12**, 14–22.

Taylor F. J. R. (1979) The toxigenic gonyaulacoid dinoflagellates. In *Toxic Dinoflagellate Blooms, Developments in Marine Biology* 1 (ed. D. L. Taylor & H. H. Seliger), pp. 47–56. Elsevier/North Holland, Amsterdam.

Taylor W., Gooch V. D. & Hastings J. W. (1979) Period shortening and phase shifting effects of ethanol on the *Gonyaulax* glow rhythm. J. comp. Physiol. **130**, 355–358.

Taylor W. & Hastings J. W. (1979) Aldehydes phase shift the *Gonyaulax* clock. *J. comp. Physiol.* **130**, 359–362.

Tett P. B. (1971) The relation between dinoflagellates and the bioluminescence of sea water. *J. mar. biol. Assn. UK* **51**, 183–206.

Walz B. & Sweeney B. M. (1979) Kinetics of the cycloheximide-induced phase changes in the biological clock in *Gonyaulax*. *Proc. natl. Acad. Sci. USA* **76**, 6443–6447.

Walz B., Walz A. & Sweeney B. M. (1983) A circadian rhythm in RNA in the dinoflagellate, *Gonyaulax polyedra*. *J. comp. Physiol.* **151**, 207–213.

Weiler C. S. & Chisholm S. W. (1976) Phased cell division in natural populations of marine dinoflagellates from shipboard cultures. *J. exp. Mar. Biol. Ecol.* **25**, 239–247.

White H. H. (1979) Effects of dinoflagellate bioluminescence on the ingestion rates of herbivorous zooplankton. *J. exp. Mar. Biol. Ecol.* **36**, 217–224.

Widder E. A. & Case J. F. (1981a) Bioluminescence excitation in a dinoflagellate. In *Bioluminescence, Current Perspectives*. (ed. K. H. Nealson), pp. 125–132. Burgess Publishing Co., Minneapolis.

Widder E. A. & Case J. F. (1981b) Two flash forms in the bioluminescent dinoflagellate, *Pyrocystis fusiforms*. *J. comp. Physiol. A* **143**, 43–52.

CHAPTER 8
DINOFLAGELLATE TOXINS

YUZURU SHIMIZU
Department of Pharmacognosy and Environmental Health Sciences, College of Pharmacy, University of Rhode Island, Kingston, Rhode Island 02881, USA

1 INTRODUCTION

Despite the general impression of the toxic nature of dinoflagellates, only a few species are known to have confirmed toxicity. Nevertheless, the widespread notoriety reflects the serious problems associated with a few truly deleterious dinoflagellates. Several review articles on limited topics (e.g. Schantz 1972; Shimizu 1978) have appeared in the literature. This chapter is intended to provide an overview of dinoflagellate toxins in general.

Table 8.1 summarizes some of the dinoflagellates reported to be toxic in the past. However, older literature must be interpreted with extreme care. It is important to note that the identity of the cited organisms is questionable in many cases. It should also be noted that some of the generic and species names have recently been revised as the result of reinvestigation. Thus, the same organism may have been reported under two or three different names. The most

Table 8.1. Dinoflagellates reported to be toxic

Organism	Major distribution	Type of toxicity	Reference
Gonyaulax (= *Protogonyaulax*) *catenella*	North Pacific–California, British Columbia, Japan	Paralytic shellfish poison	Sommer & Meyer 1937
Gonyaulax tamarensis including *G. tamarensis* var. *excavata* or *G. excavata*	North Atlantic–New England, Canada, U.K., North Sea coasts, Japan, Argentina, etc.	Paralytic shellfish poison	Needler 1949; Prakash 1963; Ingham *et al.* 1968
Gonyaulax acatenella	British Columbia	Paralytic shellfish poison	Prakash & Taylor 1966
Gonyaulax polyedra	Southern California	Paralytic shellfish poison?	Schradie & Bliss 1962
Gonyaulax monilata	Florida coasts	Ichthyotoxic	Sievers 1969
Pyrodinium bahamense (var. *compressa*)	South Pacific	Paralytic shellfish poison	Maclean 1975
Pyrodinium phoneus	North Sea coast	Paralytic shellfish poison	Koch 1939
Gymnodinium veneficum	English Channel	Water-soluble neurotoxin	Abbot & Ballantine 1957
Gymnodinium (= *Ptychodiscus*) *breve*	Gulf of Mexico	Lipid-soluble neurotoxin Ichthyotoxic	Sievers 1969; Martin & Chatterjee 1970
Prorocentrum minimum var. *mariae-lebouriae* (= *Exuviaella mariae-lebouriae*)	Lake Hamana, Japan	Hepatotoxic	Nakajima 1968; Okaichi & Imatomi 1979
Prorocentrum lima	Tropical waters	Mouse toxicity, cytotoxic	Nakajima *et al.* 1981
Prorocentrum micans and/or *Prorocentrum minimum*	North Sea—Netherlands	Diarrhetic shellfish toxin	Kat 1979
Prorocentrum concavum	Tropical waters	Haemolytic, ichthytoxic	Nakajima *et al.* 1981
Dinophysis fortii	Northern Japan	Diarrhetic shellfish poison	Yasumoto *et al.* 1980
Gambierdiscus toxicus	Tropical waters	Ciguatera toxin	Yasumoto *et al.* 1981
Ostreopsis siamensis	Tropical waters	Water-soluble toxin, haemolytic	Nakajima *et al.* 1981
Ostreopsis ovata	Tropical waters	Haemolytic lipid-soluble toxin	Nakajima *et al.* 1981
Amphidinium carterae	Tropical waters	Haemolytic, ichthyotoxic	Ikawa & Sasner 1975 Nakajima *et al.* 1981
Amphidinium klebsii	Tropical waters	Haemolytic, ichthyotoxic	Nakajima *et al.* 1981
Noctiluca scintillans	South China Sea, etc.	Shellfish poison? Ichthyotoxic?	Grindley & Haydorn 1970
Peridinium polonicum	Lake Sagami (freshwater), Japan	Ichthyotoxic	Hashimoto *et al.* 1968

conspicuous examples are changes from *Gonyaulax* spp. to *Protogonyaulax* spp. or *Gessnerium* spp. and *Gymnodinium breve* to *Ptychodiscus brevis* (see Taylor 1984, 1985 and taxonomic notes in the appendix to this volume). These changes in names have caused great confusion in chemistry literature indexing. In this article, the author has used the traditional names and refers to the synonyms where appropriate in the text.

Another problem associated with the reported toxicity of certain organisms is the possibility of strain variations in toxin productivity. Schradie & Bliss (1962), for example, reported the toxicity of cultured *Gonyaulax polyedra*. Since then, nobody has confirmed this. However, the existence of a toxic strain of *G. polyedra* cannot be ruled out since many strains of normally toxic *G. tamarensis* have been shown to be non-toxic (see below). The toxin-producing ability of dinoflagellates seems to be rather whimsical, and the function of toxin production is still mysterious.

Important aspects of dinoflagellate toxins are their appearance in other organisms in the food chain and the possibility of symbiosis. In 1937 Sommer *et al.* reported that shellfish toxicity is related to the population in the water of the dinoflagellate, *Gonyaulax catenella*. Since then, it has been demonstrated that dinoflagellate toxins accumulate in shellfish or other animal bodies in either an intact or partially transformed state. The origin of ciguatera fish toxin, for long a mystery, is now believed to be due to dinoflagellates (Yasumoto *et al.* 1977). In view of the extensive participation of dinoflagellates in food chains and symbioses, it is quite possible that other marine toxins and secondary metabolites in general will be found to be associated with dinoflagellates.

2 *GONYAULAX* (= *PROTOGONYAULAX*) TOXINS

2.1 Occurrence and isolation of *Gonyaulax* toxins

This class of toxins causes paralytic shellfish poisoning (PSP). The occurrence of the poisoning is widespread over both northern and southern hemispheres including tropical waters. Incidents were reported at such places as the north eastern Atlantic coasts of the US and Canada (Needler 1949), the Pacific coasts of the US, Canada and Alaska (Meyer *et al.* 1928; Schantz *et al.* 1957; Prakash & Taylor 1966), the Japanese archipelagos from Hokkaido to the Inland Sea (Hashimoto *et al.* 1976; Oshima *et al.* 1978; Oshima & Yasumoto 1979; Satoh *et al.* 1977), the British Isles (Ingham *et al.* 1968), the North Sea shores of Europe, the Spanish west coast (Lüthy *et al.* 1978), South Africa (Grindley & Nel 1970), Peru (Avaria 1979), Chile (Guzman & Campodonico 1978) Venezuela (Reyes-Vasquez *et al.* 1979), Argentina (Carreto *et al.* 1985), Thailand (Tamayavanich *et al.* 1985) and India (Subramanian 1985). Tropical Pacific islands such as New Guinea (Worth *et al.* 1975), Palau (Kamiya & Hashimoto

1978) and the Philippines (Hermes & Villoso 1983) also have PSP problems caused by *Pyrodinium bahamense* var. *compressum* (Steidinger *et al.* 1980).

Saxitoxin was the first toxin isolated from toxic Alaska butter clams, *Saxidomus giganteus* (Schantz *et al.* 1957; Schantz 1960). Later it was reported as the toxin in *Gonyaulax catenella* (Schantz *et al.* 1966) and found in mussels contaminated with this organism on the Pacific west coast of North America. In 1975, Shimizu *et al.* (1975b) isolated a group of toxins from soft-shell clams, *Mya arenaria* which were exposed to a bloom of *Gonyaulax* (*Protogonyaulax*) *tamarensis* in Massachusetts in 1972 and named them gonyautoxins. Buckley *et al.* (1976) also isolated from the clams from New Hampshire two toxins, which were later identified as gonyautoxin-II and gonyautoxin-III. Subsequently the gonyautoxins were isolated from cultured *G. tamarensis* cells (Shimizu *et al.* 1975b). So far, eight gonyautoxins (I–VIII) have been isolated from various organisms (Shimizu 1980). In addition, a minor toxin in *S. giganteus*, was isolated and named neosaxitoxin. It later turned out to be one of the major toxins in *Gonyaulax* (Oshima *et al.* 1977; Shimizu 1978). Hall *et al.* (1980) noticed the presence of a series of new toxins in the cultured cells of *Protogonyaulax* sp. from Alaska, which can be easily converted to known toxins by mild acid treatment and were designated as B1, B2, C1 and C2. C2 was later identified as gonyautoxin-VIII and C1 as its epimer (Kobayashi & Shimizu 1981). It was unexpected to discover that saxitoxin and its analogs are synthesized via arginine or ornithine (Shimizu *et al.* 1984).

Contrary to the original contention, saxitoxin is a rather minor component in most *Gonyaulax* strains. The only instances in which saxitoxin was reported to be the major or exclusive toxin component are the very first report on the isolation of saxitoxin from *Gonyaulax catenella* from northern California (Schantz *et al.* 1966) and an Alaskan strain described by Hall *et al.* (pers. comm.). Regarding the northern California strains, mussels freshly exposed to the *Gonyaulax* bloom in 1980 were found to contain neosaxitoxin as the major component, accompanied by saxitoxin as a minor component and only trace amounts of gonyautoxins (K. F. Meyer & Y. Shimizu, unpubl.). Although the toxin content of the primary organism was not analysed, the toxin profile in the mussels is expected to closely reflect the toxin profile of the dinoflagellate in that particular bloom. The tropical dinoflagellate, *Pyrodinium bahamense* var. *compressum* (Steidinger *et al.* 1980) is responsible for the occurrence of toxic shellfish in tropical waters (Worth *et al.* 1975). Kamiya & Hashimoto (1978) recognized gonyautoxin and saxitoxin-like substance in contaminated shellfish from Palau. Most recently, Oshima *et al.* analysed the cells from a natural bloom on Palau and found that gonyautoxin-V and neosaxitoxin are the major constituents in the organism (Y. Oshima, pers. comm.). This, and *Gymnodinium catenatum* (H. Rapaport *et al.* in Morey-Gaines 1982) are the only dinoflagellates found to contain saxitoxin analogs outside of the genus *Gonyaulax* (= *Protogonyaulax*), although saxitoxin analogs were reported in the toxic strain of blue-green alga, *Aphanizomenon flos-aquae* (Jackim & Gentile 1968; Alam *et al.* 1978). Taylor (1979a) has noted that *Protogonyaulax*, *Pyrodinium* and *Gessnerium* are very closely related, judging from their tabulation patterns.

The purification method for saxitoxin was described by Schantz *et al.* (1957). It depends on the strongly basic character of the toxin. The toxin was tightly adsorbed on the carboxylate resin Amberlite C-50 (Sommer *et al.* 1948a, b). Subsequent washing with a buffer solution of pH 4·0 removed most of the contaminants. The toxin was then eluted with acid solutions. The final purification was done by chromatography on alumina. The procedure is summarized in Fig. 8.1. On the other hand the purification of gonyautoxins is not so simple because they only weakly bind with carboxylate resins (Schantz 1960; Ghazarossian *et al.* 1974). The breakthrough came when it was discovered that the toxins bind with Sephadex G-15 or Bio-Gel P-2 gels. Thus, a crude extract was passed through a gel column and subsequently the bound toxins were eluted with a dilute acetic acid solution (Shimizu *et al.* 1975a, b; Oshima *et al.* 1977). Chromatography with Bio-Gel P-2 can separate some of the toxins (Table 8.2). Fine fractionations of the toxins were done by Bio-Rex 70 resin chromatography using acetic acid solution gradients. The stability of individual toxins in this isolation process varies greatly (Fig. 8.2). Saxitoxin is the most stable toxin and in acidic pH ranges it is almost indefinitely unchanged. On the other hand, gonyautoxins lose their toxicity during handling, probably due to the solvolysis of the sulphate group and subsequent degradation (Shimizu 1980). Also, in high pHs, the 11-OSO_3^- group undergoes epimerization to form equilibrium mixtures of epimeric toxins (Shimizu *et al.* 1976b). The stability of the toxins is approximately in the order of saxitoxins > neosaxitoxin > gonyautoxin-II, -III > gonyautoxin-I, -IV. This difference in stability and the existence of interconversions between toxins can cause discrepancies in the toxin analysis. There is a general tendency for gonyautoxins, especially

Organisms
1. pH 2, 15% EtOH
2. Conc. *in vacuo*
Extract
Amberlite IRC-50 NA^+ column
H_2O and pH 4·0 buffer eluates
0·5 N AcOH eluate
Amberlite CG-50 H^+ column
0·1 N AcOH (repeat)
Almost pure toxin
Acidic alumina column
EtOH elution
Pure saxitoxin ($[\alpha]_D + 130°$)

Fig. 8.1. Isolation scheme of saxitoxin (Schantz *et al.* 1957, 1966).

Table 8.2. Chromatographic and electrophoretic behaviour of *Gonyaulax* toxins (Shimizu 1979a, b; 1980)

Toxin	R_f*	R_f†	Electrophoresis R_m‡	Order of elution§
Saxitoxin	0·62	0·51	1·00	10
Neosaxitoxin	0·70	0·54	0·50	8
Gonyautoxin-I	0·90	0·70	0·15	5
Gonyautoxin-II	0·81	0·65	0·56	7
Gonyautoxin-III	0·69	0·61	0·28	6
Gonyautoxin-IV	0·81	0·65	0	4
Gonyautoxin-V	0·61	0·52	0·28	3
Gonyautoxin-VI	0·57		0·08	2
Gonyautoxin-VII	0·44		0·97	9
Gonyautoxin-VIII¶	0·88		−0·40	1

* Silica gel 60 (pyridine : ethyl acetate : water : acetic acid = 75 : 25 : 30 : 15).
† Silica gel GF (*tert* butanol : acetic acid : water = 2 : 1 : 1).
‡ Tris buffer (pH 8·7) 200 V, 0·2 mA cm^{-1} cellulose acetate strip. Relative to saxitoxin mobility.
§ Bio Rex 70 < 400 mesh column, acid acetic gradient.
¶ Identical with C2 by Hall *et al.* (1980). The 11α-epimer was designated as C1.

gonyautoxin-I, to be more difficult to recover than the rest of the toxins. However, careful maintenance of the proper pH (3·0–6·0), avoidance of higher temperature and quick processing can keep the conversions to a minimum.

Table 8.3 shows the toxin content in some selected strains of organisms as analysed by the author's group and others. Although the toxin profiles in various

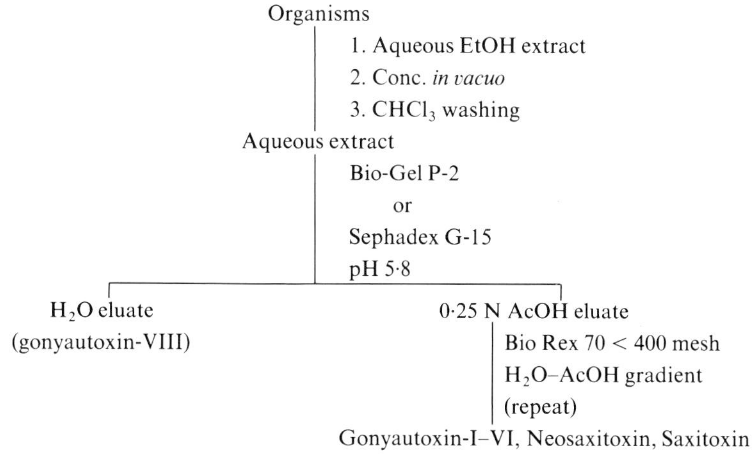

Fig. 8.2. Isolation scheme of *Gonyaulax* toxins or paralytic shellfish toxins in general (Shimizu *et al.* 1975; Shimizu 1977, 1978).

Table 8.3. Toxin profile of some *Gonyaulax* organisms of different origins (Shimizu 1979, 1979a, b)

	Toxin profile								
	STX	GTX-I	GTX-II	GTX-III	GTX-IV	GTX-V	GTX-VIII	NeoSTX	Toxicity
G. tamarensis[1]	++	++	+++	+	+	+++	++	+++	1 MU/10^4 cells
G. tamarensis[2]	+	++	++	++	+	+		+	1 MU/5 × 10^4 cells
G. tamarensis[3]	±	++	++	++	+	+		+	1 MU/10^5 cells
G. tamarensis[4]	+	++	+++	++	+	+	++	++	1 MU/10^4 cells
G. catenella[5]	+	+++	++	+	+	++			1 MU/4 × 10^4 cells
G. catenella[6]	++	++	++	++	+	+		++	1 MU/10^5 cells
G. catenella[7]					(±)				1 MU/10^6 cells

[1] Ipswich, MA; [2] Mill Pond, Cape Cod, MA; [3] Perch Pond, Cape Cod, MA; [4] Funkawan, Japan; [5] Owase Bay, Japan; [6] Sequim Bay, WA; [7] LaJolla, CA.
STX, saxitoxin; neoSTX, neosaxitoxin; GTX, gonyautoxin.

organisms are approximately the same except for a few strains, one fundamental problem to be considered is the species or strain difference in the production of toxin. It was already known that a strain of *G. tamarensis* from the Tamar river estuary, Plymouth, England, is non-toxic although it is morphologically very similar to the toxic strains (Loeblich & Loeblich 1975; Taylor 1975; Schmidt & Loeblich 1979). Attempts to link toxicity to a minor morphological character (presence of a ventral pore on the epitheca (Loeblich & Loeblich 1975) were abandoned when this was found to be variable (Alam *et al.* 1979; Turpin *et al.* 1978). It was demonstrated that three different isolates from proximate locations (Massachusetts) ranged from extremely toxic to moderately toxic to weakly toxic (Anderson & Wall 1978; Alam *et al.* 1979). Subsequent investigation of many strains from various parts of the world further confirmed the above finding. The toxicity ranges from non-toxic to toxic. Although toxic and non-toxic strains even co-exist sometimes (Yentsch *et al.* 1978), generally there is a great likelihood of finding more virulent strains the farther north one goes (Maranda & Shimizu 1981). There is also a report that differences in salinity influence the toxicity (White 1978). This difference in toxin productivity, which seems to be inherent in the strains, is an important determining factor in the toxification of shellfish. For example, the existence of only a few dinoflagellate cells per litre of sea water is enough to keep scallops at high toxicity levels at Funkawan Bay, Hokkaido, Japan, where the local strain is extremely toxic (Satoh *et al.* 1977). On the other hand, at Owase Bay, Mie, Japan, a bloom of very high density made the shellfish only moderately toxic (Hashimoto *et al.* 1976; Oshima *et al.* 1976).

2.2 Chemistry of *Gonyaulax* toxins

SAXITOXIN

The chemistry of saxitoxin is the best studied among the dinoflagellate toxins. It was isolated in large quantities from Alaska butter clams in south-east Alaska, where the shellfish maintain the toxin in their siphons all year-round (Schantz & Magnusson 1964). The chemistry of saxitoxin was elucidated primarily by Rapoport and his co-workers (for the details of chemistry, see Shimizu 1978). However, X-ray diffraction analyses were necessary for the final determination of the conspicuous tricycle structure (Schantz *et al.* 1975a, b; Bordner *et al.* 1975). Subsequently, the structure was also synthesized (Tanino *et al.* 1977).

Saxitoxin has two values of pKa, 11·6 and 8·24. This was a subject of discussion for a number of years (Wong *et al.* 1971; Bordner *et al.* 1975): 11·6 is undoubtedly attributable to one of the guanidium moieties, but 8·24 was considered to be too low for a guanidium moiety and various speculations were made for its origin. Recently, carbon and proton nuclear magnetic resonance (NMR) studies proved that both pKa values are derived from the guanidium groups (Rogers & Rapoport 1980; Shimizu *et al.* 1981). Another important feature of the crystal structure is the presence of a hydrated ketone at C-12, which is stabilized by the two electron withdrawing guanidium groups (Schantz *et al.* 1975a). In the NMR studies, it was shown that the saxitoxin molecule can exist in a keto form at pH 7 or higher at which the imidazole guanidine moiety is no longer protonated. Thus, saxitoxin exists in an equilibrium, shown in Fig. 8.3, around the physiological pH.

Saxitoxin can be oxidatively degraded to a highly fluorescent purine derivative (Wong *et al.* 1971). The reaction is very useful for detection and quantitation of the toxin (Bates & Rapoport 1975; Buckley *et al.* 1976, 1978). Typically, spraying developed thin-layer chromatography plates with 1% hydrogen peroxide, followed by heating and UV detection at 304 nm, reveals saxitoxin as a whitish blue spot. Oxidation of saxitoxin-containing solutions by H_2O_2 in the presence of NaOH gives excellent yields of fluorescence. This method is applicable to other *Gonyaulax* derived toxins.

Hydrogenation of saxitoxin with Adams' catalyst in acid (Schantz *et al.* 1961) produces mostly the 11α-isomer of dihydrosaxitoxin (Shimizu *et al.* 1981) which is almost completely devoid of toxicity (Schantz *et al.* 1961). Reduction with sodium cyanoborohydride or sodium borohydride gives a mixture of the 11α- and 11β-dihydrosaxitoxin (Rogers & Rapoport 1980; Shimizu *et al.* 1981). The latter also does not exhibit toxicity in mice. This observation suggests that the hydrated ketone at 11 position plays an important role in the manifestation of the toxicity.

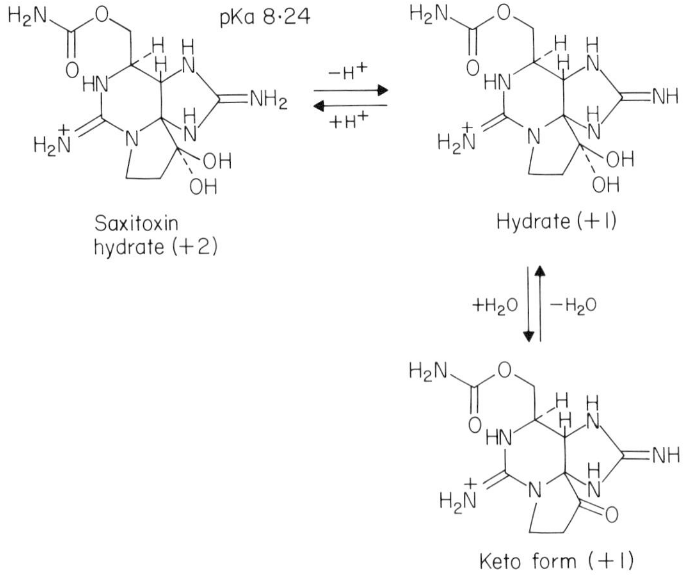

Fig. 8.3. Structures of saxitoxin in solutions.

NEOSAXITOXIN

As mentioned earlier, neosaxitoxin was first isolated from Alaska butter clams as a minor toxin (Oshima *et al.* 1977), but later it was found to be a major toxin in a number of other organisms (Shimizu 1979a, b). The compound resembles saxitoxin and is eluted almost simultaneously from carboxylate-type resin chromatography. However, in electrophoresis at pH 8·0, this toxin migrates only half as far as saxitoxin, suggesting the molecule carries only one positive charge at the particular pH (Fallon & Shimizu 1977). Titration did, indeed, indicate the presence of an additional dissociable functional group with pKa 6·7. On the basis of both spectroscopic and chemical evidence, the structure of 1(N)-hydroxysaxitoxin (Fig 8.4) was proposed for neosaxitoxin (Shimizu *et al.* 1978b). However, the proposed site for the functional group is based primarily on NMR data, and structures such as 2-hydroxy-amino saxitoxin cannot be totally excluded.

Reduction of neosaxitoxin with various agents leads to the formation of saxitoxin, and there is evidence that a similar transformation takes place in the biological system (Shimizu & Hsu 1981; Shimizu & Yoshioka 1981).

GONYAUTOXIN-II AND GONYAUTOXIN-III

These two compounds are the major toxins in most *Gonyaulax* cells. They are in an epimeric relationship and the epimerization easily takes place during the isolation process at neutral or slightly higher pH. Thus in most cases, they are isolated as an equilibrium mixture. However, in several cases, careful processing resulted in the isolation of mostly gonyautoxin-III (Y. Shimizu *et al.* unpubl.). Thus, it is quite possible that this thermodynamically less favoured isomer is the predominant toxin in organisms.

The structures of gonyautoxin-II and -III were incorrectly proposed as free 11α- and 11β-hydroxysaxitoxin respectively on the basis of degradation and spectroscopic studies (Shimizu *et al.* 1976b). They are, however, in the sulphate ester forms (Fig. 8.5) which are identical with the compounds reported by Boyer *et al.* (Boyer *et al.* 1978; Alam *et al.* 1981). The sulphate forms explain well their weak basicity compared to saxitoxin. Gonyautoxin-II and -III can both be oxidized to highly fluorescent compounds, which were the key compounds in

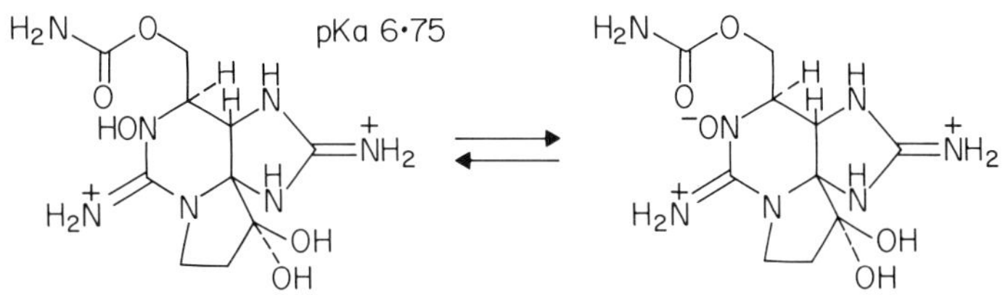

Fig. 8.4. Structure of neosaxitoxin.

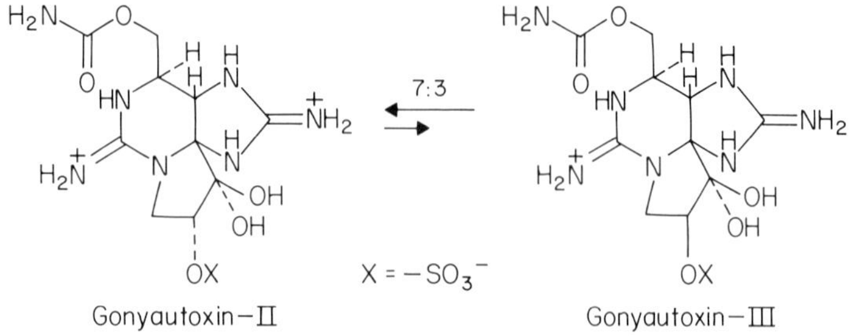

Fig. 8.5. Structures of gonyautoxin-II and -III.

the structure elucidation (Shimizu *et al.* 1976b; Shimizu 1978). This reaction is also very useful for the detection or quantitation of the toxins (Buckley *et al.* 1978). Gonyautoxin-II and -III were ultimately correlated with saxitoxin by reductive cleavage of the O–sulphate group with Zn–HCl (Shimizu & Hsu 1981).

GONYAUTOXIN-I AND -IV

Gonyautoxin-I and -IV are another pair of epimeric toxins. Compared to the other toxins, they are considerably less stable and tend to disappear during the isolation process. Nevertheless, they have been found as major toxins in a number of organisms, such as Alaskan mussels (Shimizu *et al.* 1978a). Their structures (Fig. 8.6), which are composites of gonyautoxin-II and -III and neosaxitoxin structures, have been proposed on the basis of spectroscopic and chemical evidence (Shimizu 1979c, 1980; Wichmann *et al.* 1981a). Gonyautoxin-I and -IV were correlated to gonyautoxin-II and -III respectively by zinc–acid reduction (Shimizu & Hsu 1981). The zinc–acid reduction also effects the reductive elimination of the O–sulphate group to afford neosaxitoxin. However, oxidation of gonyautoxin-I and -IV with H_2O_2–NaOH does not provide good yields of fluorescent compounds. Thus, the toxins cannot be assayed by fluorometry.

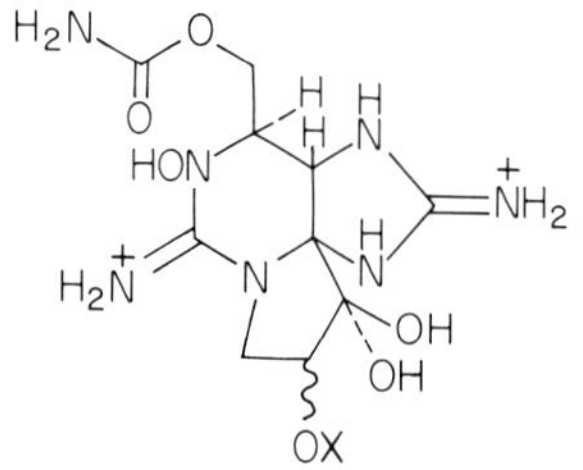

Fig. 8.6. Structures of gonyautoxin-I (11α) and gonyautoxin-IV (11β).

GONYAUTOXIN-V, -VI, -VII

The structures of these toxins have not been elucidated. Gonyautoxin-V was found in some *Gonyaulax* spp. and *Pyrodinium bahamense* var. *compressum* in fairly large quantities. This compound is probably a conjugated form of a toxin, and characteristically gives a high yield of fluorescence by treatment with H_2O_2–NaOH, indicating a low specific toxicity. Gonyautoxin-VI has a low toxicity and a low yield of fluorescence. At present, gonyautoxin-VII has been found only in shellfish and may possibly be a metabolic product.

GONYAUTOXIN-VIII (= C2 TOXIN)

Gonyautoxin-VIII is the first toxin which has a negative net charge in the molecule at neutral or higher pH. For this reason and because of its very low toxicity, this toxin was overlooked for some time. During the biosynthetic studies using ^{14}C-labelled precursors, the toxin appeared in *Gonyaulax tamarensis* cells as a weakly toxic but highly radioactive fraction (M. Kobayashi, unpubl. data). The compound was found to migrate to the anode at pH 8·0 (Shimizu 1980). Independently, Hall *et al.* (1980) recognized a set of compounds, which could be converted to known gonyautoxins by mild acid treatment. One of them, coded C2, was found to be identical with gonyautoxin-VIII (Kobayashi & Shimizu 1981). Gonyautoxin-VIII can be hydrolysed very easily to gonyautoxin-III, liberating 1 mol of sulphuric acid. This easily hydrolysable sulphate group was speculated to be in the form of N-sulphate or sulphamic acid (Kobayashi & Shimizu 1981). The site of conjugation was determined to be the carbamoyl amino group, based on spectroscopic evidence, and was confirmed by X-ray crystallography of the hydrate crystal (Wichmann *et al.* 1981b).

The toxin is easily isomerized to an epimer with the 11α-configuration (*epi*-gonyautoxin-VIII or C1; Fig. 8.7). In spite of this ease of isomerization, in the intact cells of *G. tamarensis* Ipswich, and *G. tamarensis* Funkawan strains, only gonyautoxin-VIII has been discovered (Y. Shimizu, unpubl.). On the other

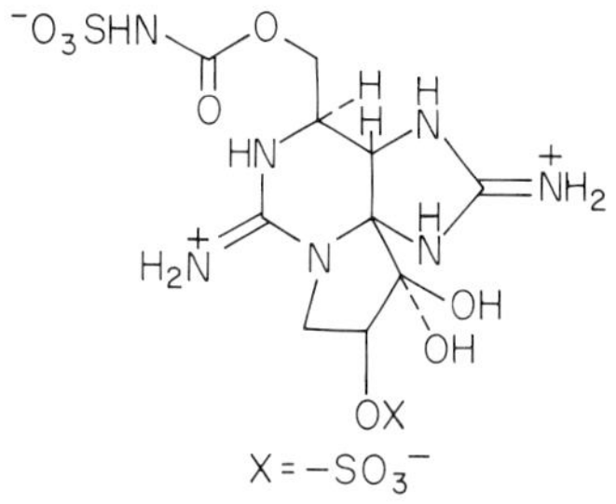

Fig. 8.7. Structure of gonyautoxin-VIII (=C1).

hand, the N-sulphated derivatives of all the existing toxins are likely to be found in some other strains as found in *Protogonyaulax* sp. clone P107 (Hall *et al.* 1980).

2.3 Pharmacology of *Gonyaulax* toxins

The action of *Gonyaulax* toxins is unique in that they selectively block the influx of sodium ions through the excitable membranes, effectively interrupting the formation of an action potential (Narahashi 1972). A similar action is observed only with another potent marine toxin, tetrodotoxin, found in the puffer fish and some other animals (see Kao 1966). This unique mode of action makes these toxins important tools in neurophysiology research. The ingestion of shellfish contaminated with toxic *Gonyaulax* results in the paralysis of the neuromuscular system. Death may occur, usually as a result of respiratory paralysis. The action is very fast and no antidote is known at the moment. Interestingly, the nerve cells of marine shellfish have tremendous tolerance against the toxins (Twarog *et al.* 1972). Others may be sensitive to them, and *Gonyaulax* blooms in South Africa are reported to cause the mortality of shellfish (Horstman 1981).

The toxicity is usually measured by the mouse assay, as established by the Association of Official Analytical Chemists (AOAC 1975). One mouse unit (MU) is defined as the amount needed to kill a 20 g mouse in 15 min, but in reality it is an asymptote point of the death–dose response curve. The assay is normally carried out using doses which cause death in several minutes. There are large fluctuations in the susceptibility of mice to the toxins, depending upon the colony. These variations are corrected using a standard solution of pure saxitoxin and the toxicity is generally expressed in μg equivalents of saxitoxin dihydrochloride of known toxicity (5500 MU/mg^{-1}). In the early days, saxitoxin was the only known toxin and everything was discussed in terms of saxitoxin. However, the reality is more complex, with many other toxins now confirmed, and rarely is saxitoxin the sole toxic component. An additional complication is the fact that with structural variations, the specific toxicity of each toxin is different. This may be partly due to variations in the affinity of toxins for the receptors and partly because the bio-availability of the toxins is expected to vary a great deal according to the physico-chemical properties. In fact, some of the toxins, such as gonyautoxin-V and -VII, are noticeably less toxic. Different data have been seen in the literature concerning the specific toxicity of the gonyautoxins. These differences may be due to difficulties in obtaining an absolutely pure specimen and to accurately weighing the highly hygroscopic compounds, especially since only limited quantities were available. An attempt was made to more accurately define the specific toxicity by concomitantly assaying the toxicity and the nitrogen content of the test solutions of highly purified toxins freshly eluted from column chromatography. The specific toxicity obtained by this method is listed in Table 8.4 together with the value obtained by other methods (Genenah & Shimizu 1981).

Generally, substitution at N-1 and/or C-11 results in a reduction of toxicity, yet most compounds retain the same order of toxicity. Gonyautoxin-III shows toxicity comparable to that of saxitoxin, whereas gonyautoxin-V and -VIII show remarkably lower toxicity. It is thought that the 12-keto group which is in equilibrium with the hydrated form may play an important role in the manifestation of toxicity. Thus, the reduction of the keto group of saxitoxin to a hydroxyl group results in a loss of toxicity.

2.4 Fate of *Gonyaulax* toxins in the food chain

The significance of the *Gonyaulax* toxins is primarily due to paralytic shellfish poisoning, which is a serious form of food poisoning. It is not only a serious health hazard in itself, there is also the enormous economic impact of the preventive measures. The fate of the toxins after being ingested by the shellfish, mostly bivalves, is not fully understood. According to an older theory, the toxic dinoflagellates in blooms are filter-fed by shellfish and the toxin, chiefly saxitoxin, accumulates in the shellfish bodies, primarily in the hepatopancreas (Sommer & Meyer 1937; Sommer *et al.* 1937). Recently it has become obvious that there are more factors involved.

First of all, the sources of toxins in shellfish are not limited to the motile forms of dinoflagellates. The existence of benthic *Gonyaulax* cysts prompts the possibility that shellfish may feed on them (Anderson & Wall 1978; Yentsch & Mague 1979; Anderson 1980). Dale *et al.* (1978) reported that the cysts of *Gonyaulax tamarensis* (= *excavata*) collected in Maine were far more toxic than the motile cells. The exact toxicity and the toxin profile of the cyst form have yet to be established, but the toxicity found in the sea scallop, *Placopecten magellanicus* from the deep sea (100 m) is considered to be derived from feeding on the cyst form and its toxin profile may reflect that of the cyst cells (Hsu *et al.* 1979).

Secondly, the major source of saxitoxin, the Alaska butter clam, in certain areas in Alaska mysteriously maintains the toxin year-round in its siphon despite the apparent absence of the causative organism in the surrounding water (Schantz & Magnusson 1964). Mussels from the same area, on the other hand, show completely different toxin profiles; gonyautoxins and neosaxitoxin are the dominant toxins (Shimizu *et al.* 1978a). A study with mussel and Alaska butter clam samples collected in the midst of a *Gonyaulax* bloom at Puget Sound in the State of Washington showed an interesting difference in the toxic contents, which probably indicates a conversion of certain toxins to saxitoxin (M. Yoshioka & Y. Shimizu, unpubl.). As in Alaska, in mussels the major toxins are gonyautoxins and neosaxitoxin, which may possibly reflect the toxin contents of the *Gonyaulax* sp. in the bloom. On the other hand, butter clams had

Table 8.4. Specific and relative toxicity of *Gonyaulax* toxins (Genenah & Shimizu 1981)

	STX	NeoSTX	GTX-I	GTX-II	GTX-III	GTX-IV	GTX-V	GTX-VIII
MU μmol^{-1}	2045 ± 126	1038 ± 44	1638 ± 128	793 ± 83	2234 ± 137	673 ± 38	354 ± 19	280
Specific toxicity (MU mg^{-1})	5494 ± 339	2363 ± 101	3976 ± 312	2003 ± 211*	5641 ± 346	1634 ± 92	—	589†
Relative toxicity	100	43	72	36	103	30	—	14

* A value of 2500 MU mg^{-1} was reported by Boyer *et al.* (1978).
† A value, 500 MU mg^{-1}, was given by Wichmann *et al.* (1981).
STX, saxitoxin; neoSTX, neosaxitoxin; GTX, gonyautoxin.

a significant amount of saxitoxin localized in the siphons. A plausible explanation for this discrepancy is the reductive conversion of neosaxitoxin, and possibly also gonyautoxins, to saxitoxin in the butter clam. A similar observation was made in an experiment using scallop, *Placopecten magellanicus*, homogenates (Shimizu & Yoshioka 1981). Incubation of a toxin mixture in the scallop homogenate resulted in a remarkable increase of saxitoxin in proportion to the other toxins. The conversion was most evident in the homogenates of locomotive organs. Very few data are available on the fate of saxitoxin and gonyautoxins in animals other than bivalves. There is a report that the filter-feeding crustacean *Emerita analoga*, can become toxic in a *Gonyaulax* bloom (Sommer 1932). The origin of saxitoxin and neosaxitoxin found in certain crabs in tropical waters is a mystery (Hashimoto *et al.* 1967; Konosu *et al.* 1968; Noguchi *et al.* 1969; Yasumoto *et al.* 1981). Since a toxic *Gonyaulax* sp. was not observed in the area, one possibility is that toxins derived from *Pyrodinium bahamense* var. *compressum* ended up in the carnivorous animals via filter-feeders. A similar origin is possible for the saxitoxin and neosaxitoxin found in carnivorous marine snails (Kotaki *et al.* 1981). Foxall *et al.* (1979) demonstrated under experimental conditions that the crab, *Cancer irroratus* can become toxic as a result of the secondary food chain. Massive filter-feeding fish-kills in the Bay of Fundy in Canada are also suspected to be due to *Gonyaulax* blooms (White 1977). White (1979) found that some zooplankton became toxic in the Bay of Fundy. Analysis showed that the toxin profile in the zooplankton was identical to that in *Gonyaulax tamarensis* (Hayashi *et al.*, unpubl.). The injection of the toxins was known to be toxic to fish (Down 1970) but recently White (1981) has shown the toxins are indeed lethal to fish by oral administration. Oral LD_{50} values are 400–750 μg saxitoxin equivalent per kg body weight.

3 *GYMNODINIUM BREVE* TOXINS

3.1 Isolation of *Gymnodinium breve* toxins

The toxins in *Gymnodinium breve* (=*Ptychodiscus brevis*), which cause massive fish-kills in the Gulf of Mexico, have been investigated for a number of years. Isolation work has been carried out mostly with the cultured cells. Several toxins of varied purity have been isolated by different groups. Martin & Chatterjee (1969, 1970) isolated an amorphous light yellow compound with a molecular weight 1545, containing phosphorus. Cummins & Stevens (1970) reported an amorphous powder containing a nitrogen atom. Trieff *et al.* (1970a, b, 1972) also reported phosphorus-containing toxins. Alam *et al.* (1975) chromatographed the ether extract of the cells on silica gel and isolated three separate toxins. T_1, T_2 and T_3. The toxin T_2 was the major toxin which showed UV absorption maxima at 260, 267 and 270 nm and was assigned a molecular formula of $C_{41}H_{59}NO_{10}$. Later this toxin was further purified to a specimen *G. breve* toxin-2

(GB-2) with a single UV absorption maximum, λ_{max} 213 nm (Shimizu *et al.* 1976a). Baden *et al.* (1979) isolated two toxins designated T_{17} and T_{34} from the cultured cells. T_{34} is the most abundant toxin (yield 5·7 pg/cell^{-1}) but less ichthyotoxic than T_{17}. Trieff's group at Galveston, Texas (Risk *et al.* 1979a) reinvestigated the toxin contents of the cells and was not able to observe T_2 toxin which they had previously isolated, but they obtained a fraction corresponding to T_4 which was further separated into two components, T_{46} and T_{47}, by high pressure liquid chromatography (HPLC). Independently, Padilla *et al.* (1979) and Kim & Padilla (1976) reported an ichthyotoxin (GBTXα) and hemolysin (GBTXβ), but their identity with other toxins was not established. The isolation of the first crystalline toxin was reported by Lin *et al.* (1981). They used flash chromatography to obtain three toxins, brevetoxin (BTX), -A, -B and -C. The major toxin BTX-B (5 mg from 5×10^8 cells) was crystallized from acetonitrile, and subsequently its structure was established by X-ray crystallography (Fig. 8.8). Brevetoxin-C differs from BTX-B only in the side chain from the terminal ring K (Golik *et al.* 1982).

The published high resolution PMR and IR spectral data of GB-2 (= T_2) indicated its identity with BTX-B, but the earlier report (Risk *et al.* 1979a) on the non-identity of T_{47} and T_2 conflicted with this observation. However, Baden *et al.* (1981) confirmed that the crystalline T_{34} toxin is identical with GB-2. Furthermore, the identity of GB-2 and BTX-B was unequivocally established by comparison of the 500 MHz NMR spectrum of crystalline GB-2 with that of BTX-B (Y. Shimizu, unpubl.). Spikes *et al.* (1968) noted toxicity variation in *G. breve* cultures. Recently, Alam observed that his present culture, which is the same strain used by Martin *et al.* in Alam's earlier work and Shimizu's work, is no longer producing the polyether-type *G. breve* toxins (M. I. Alam, pers. comm.). Instead, he now finds phosphorus-containing toxins chemically unrelated to brevetoxins. The structure of one of these, PB-1, **1** has been determined to be O,O-diphenyl-N-cyclooctyl phosphoramidate (DiNovi *et al.* 1983).

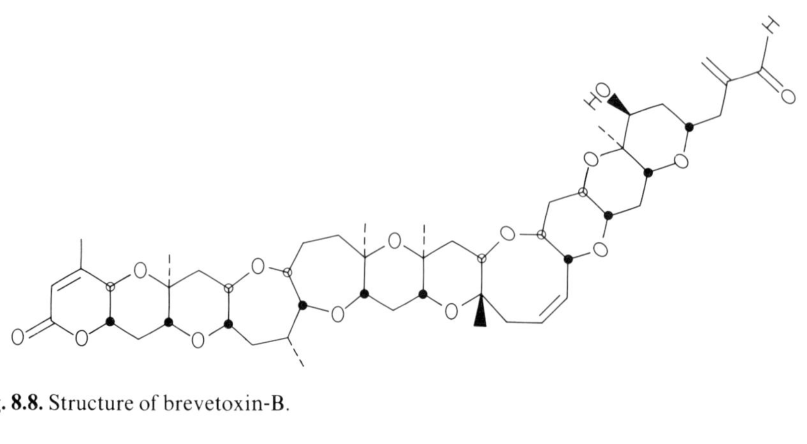

Fig. 8.8. Structure of brevetoxin-B.

3.2 Chemistry of the *Gymnodinium breve* toxins

Earlier reports on the chemical characteristics of *G. breve* toxins are filled with contradicting data. Martin & Chatterjee (1970) gave a formula, $C_{91}H_{166}O_{57}P$, for their toxin. A similar phosphorus-containing formula was also proposed by Trieff *et al.* (1971, 1972). Cummins & Stevens (1970) reported a formula containing one nitrogen atom, $C_{102}H_{157}N$. Sasner *et al.* (1972), Alam (1972) and Alam *et al.* (1973) assigned a molecular weight of 279 in their earlier work. Padilla *et al.* (1974) also reported low molecular weight toxins, which they suggested had substituted aromatic structures. T_2 toxin of Alam *et al.* (1975) was given a molecular weight of 725 on the basis of mass spectroscopic data. It had a UV absorption of 260, 267 and 270 nm, but its purified form, GB-2, showed a single maximum at 213 nm, probably for a conjugated ester, and a highest ion peak at m/e 850 in the EI mass spectrum (Shimizu *et al.* 1976a). The infra-red spectrum showed the presence of a conjugated carbonyl group, a hydroxyl group and a polyether structure. The 220 MHz NMR suggested the presence of an aldehydic function, several methyl groups and vinylic groups, probably on a terminal methylene group (Shimizu *et al.* 1976a). On the other hand, Baden *et al.* (1979) suggested the absence of an OH group and conjugation in their T_{17} and T_{34} toxins. Otherwise, T_{34} closely resembled GB-2 toxin in its possession of a terminal methylene and an aldehyde group. T_{17} gave the highest ion peak at m/e 726. The T_{47} toxin of Risk *et al.* (1979a, b) gave an IR spectrum very similar to GB-2 or T_{34}. In plasma ionization mass spectroscopy, T_{47} gave the highest mass peak at m/e 893. On the other hand, brevetoxin-B (BTX-B) studied by Lin *et al.* (1981) gave a $(M+H)^+$ ion at m/e 895 by desorption/chemical ionization mass spectrometry in accordance with its final molecular formula, $C_{50}H_{70}O_{14}$ (mol. wt. 894). The toxin was crystallized from acetonitrile to needles which were used for X-ray crystallography. The structure obtained is an unprecedented cyclic polyether derived from a partially methylated straight carbon chain with an enal group and a conjugated lactone on the terminals (Lin *et al.* 1981). The structure seems to coincide with various spectroscopic observations reported for GB-2 (= T_2), T_{34} and T_{47} and Nakanishi (1985) has concluded that they are identical with BTX-B. BTX-A also has an enal group but has a different skeleton, and BTX-C has a similar skeleton to BTX-B but has a CH_2Cl end group (Nakanishi 1985).

3.3 Pharmacology of *Gymnodinium breve* toxins

The pharmacological activities of *G. breve* toxins are quite diversified (Paster & Abbott 1969; Sievers 1969; Martin & Chatterjee 1969, 1970; Trieff *et al.* 1970, 1973, 1975; Sasner *et al.* 1972; Spiegelstein *et al.* 1973; Padilla *et al.* 1974; Abbott *et al.* 1975). The diversity is probably due to the fact that samples used for the pharmacological studies are not all uniform. Haemolytic activity seems to be the most frequently reported action of *G. breve* toxins. However, the T_1 and T_2 toxins reported by Alam *et al.* (1975) (identical with brevetoxin-B) lack haemolytic activity. These toxins were reported to have both ichthyotoxicity and mouse toxicity. The haemolytic activity seems to be mostly associated with the phosphorus-containing toxins reported by several researchers. In recent work, Padilla *et al.* (1979) obtained two toxins, GBTXα and GBTXβ, and observed haemolytic activity associated with the latter. GBTXβ was further separated into three toxic (β_1, β_2, β_3) and three non-toxic components. The anti-acetylcholinesterase activity claimed by Sasner *et al.* (1972) was reported to be absent in T_1 and T_2 toxins by Alam *et al.* (1975). Intraperitoneal injection of T_1, T_2 or GB-2 toxins causes rather quick deaths in mice. An LD_{50} of 0·25 mg kg^{-1} was reported for T_2. The mechanism for this animal toxicity has not been fully elucidated. However, it is conceivable that the toxins act on the neuromuscular system, as do other endotoxins. Sasner *et al.* (1972) reported that their toxin preparation blocked neuromuscular transmission in frog sartorious preparations without altering the membrane potential. Thus, they concluded that the blockage occurs at the neuromuscular junction. Padilla *et al.* (1979) reported that their GBTXα can induce repetitive firings in the squid giant axon at a concentration of 0·005 μg ml^{-1}. It also induced contractions in guinea-pig ileum at a concentration of 0·6–0·9 μg ml^{-1}. It was suggested that multivalent binding to the membrane at, or near, the excitable sodium channel is responsible for the toxin effect (Risk *et al.* 1979b).

Another important aspect of *G. breve* toxins is their cytotoxicity. Baden *et al.* (1979) reported that T_{34} is toxic against KB tumour cells *in vitro* at a concentration of 2–2·6 μg ml^{-1}. This cytotoxicity apparently does not parallel the ichthytoxicity. The more potent ichthyotoxin, T_{17}, did not show cytotoxicity at a level of 100 μg ml^{-1}.

The human toxicity of *G. breve* toxins is not well understood. Shellfish are known to become toxic by filter-feeding on the organism (Ray & Aldrich 1965). However, the intoxication symptoms (caused by ingestion of the shellfish) are not as drastic as those caused by toxins produced by *Gonyaulax*. Although the *G. breve* toxins are lipid soluble, the toxicity is not considered to be cumulative, judging by experimental data from mice (Baden *et al.* 1979). Generally, the poisonings are accompanied by mild neurotoxic symptoms, such as sensory abnormalities and mild diarrhoea (McFarren *et al.* 1965; Hughes 1979). The organism is also said to cause respiratory irritation when inhaled in an aerosol form. Ellis *et al.* (1979) investigated the action of *G. breve* toxin on the dog respiratory and cardiac system and found that it induces cardiac arrhythmias, fibrillation, and hypertension. At higher doses, it causes acute apnoea. The hypertension was attributed to an increase of catecholamine release.

3.4 Fate of *G. breve* toxins in food chain

As mentioned earlier, some *G. breve* toxins can accumulate in filter-feeding shellfish (Ray & Aldrich 1965; McFarren *et al.* 1965). Roberts *et al.* (1979) reported that molluscs were adversely affected by direct exposure to GB toxins. However, unlike paralytic shellfish poisons, little is known about the rate of accumulation and depuration, the biochemical fate of the toxins, etc. (See Baden 1983 for summary).

There were some earlier speculations that *G. breve* toxins could accumulate in fish, especially in view of the resemblance of *G. breve* toxin poisonings to ciguatera poisonings (McFarren *et al.* 1965). Although this connection with ciguatera has now been completely severed, it is still possible that the toxins may find a way into other forms of marine life as transformed molecules.

4 *GAMBIERDISCUS TOXICUS* TOXINS

4.1 Occurrence and isolation of *Gambierdiscus toxicus* toxins

Gambierdiscus toxicus Adachi and Fukuyo (Taylor 1979b) is a newly-discovered, large, epiphytic dinoflagellate species in tropical waters, found to be the long-sought causative organism of ciguatera poisonings. Studies by Yasumoto *et al.* (1977, 1979a, b) indicate a clear correlation between the occurrence of ciguatoxic fish and the population of *G. toxicus* at Tahiti and the Gambier Islands. Yasumoto and his co-workers (1979a, b) investigated the toxin content of wild and cultured cells from French Polynesia and determined that there are two different types of toxins in the cells.

One of them, an acetone-soluble toxin, was indistinguishable from the ciguatoxin isolated from Moray eels. The other, a water-soluble toxin, was determined to be maitotoxin, previously discovered in the viscera of surgeon fish from reefs (Yasumoto *et al.* 1976). The amounts and proportions of the two toxins in both wild and cultured cells can vary substantially from one source population to another (Table 8.5). In cultured cells, the amounts of ciguatoxin diminishes to almost nil in some cases.

Table 8.5. Toxin content in various *Gambierdiscus toxicus* samples (Yasumoto *et al.* 1979)

Sample	Number of cells	Ciguatoxin (total mu cells mu^{-1})		Maitotoxin (total mu cells mu^{-1})	
Wild specimen ('75)	$3{\cdot}0 \times 10^7$	2373	12 640	$7{\cdot}1 \times 10^3$	4210
Wild specimen ('78)	$3{\cdot}94 \times 10^8$	21 000	18 760	$6{\cdot}4 \times 10^5$	616
Unailgal culture 1	$2{\cdot}0 \times 10^6$	52	38 460	$6{\cdot}7 \times 10^4$	30
Unialgal culture 2	$7{\cdot}5 \times 10^6$	<30		$1{\cdot}3 \times 10^7$	5·8
Axenic culture	$4{\cdot}0 \times 10^5$	<30		$3{\cdot}2 \times 10^3$	130

G. toxicus has now been widely found in tropical waters such as the Caribbean, the Cape Verde Islands and the Hawaiian Islands in the northern hemisphere (Taylor 1979b) as well as the earlier reports from the South Pacific. This also seems to coincide with the wide distribution of ciguatoxic fish in tropical waters and their occurrence in the Indian Ocean suggests that it will be found there too. Wild cells of *G. toxicus* collected at Oahu, Hawaii, were also found to contain both ciguatoxin and maitotoxin fractions and their total toxicity was comparable to that found in Tahitian material (Shimizu *et al.* 1982).

4.2 Chemistry of *Gambierdiscus toxicus* toxins

Since the amounts of toxins obtainable from wild or cultured *Gambierdiscus toxicus* are very minute, most of the chemical work on ciguatoxin has been done on toxin isolated from Moray eels collected in ciguatera endemic areas.

Scheuer's group in Hawaii isolated what was thought to be pure ciguatoxin from the liver and viscera of Moray eels from Johnston Island, using a combination of the methods developed by Yasumoto & Scheuer (1969) and high pressure liquid chromatography (Tachibana 1980). The yield of the pure toxin was 1·31 mg from 122 g of the extract, corresponding to 60 kg of the viscera.

The pure toxin was obtained as an amorphous white powder of high toxicity (LD_{50}, 0·45 μg kg^{-1}). The infra-red spectrum showed a close resemblance to the spectra of other marine toxins such as okadaic acid (see below). The high resolution proton NMR indicated that the molecule is a medium-sized (mol. wt. 1120), highly oxygenated, alkyl compound. Further investigation has been hindered by the lack of sufficient material. In the course of the eel liver isolation work, the presence of two different toxins, one alkali-soluble and the other insoluble, was suspected. T. Yasumoto (pers. commun.) was able to separate the toxins into two fractions by chromatography on alumina. The nature of this second toxin has not been investigated.

Yasumoto *et al.* (1979a) reported that the lipid-soluble toxin from wild specimens of *G. toxicus* behaved just like the eel toxin on the silicic acid columns, DEAE-cellulose columns and Sephadex LH-20 columns. The other water-soluble toxin is acetone-insoluble and can be extracted with l-butanol from the aqueous solution. The toxin was eluted from silicic acid columns with chloroform–methanol and Bio-Gel P-10 and Sephadex G-50 columns in the same manner that maitotoxin is isolated from surgeon fish (Yasumoto *et al.* 1971, 1976). Maitotoxin is soluble in polar solvents such as methanol, ethanol, and l-butanol, but insoluble in ether, acetone, and chloroform. It loses toxicity upon heating, especially in strongly acidic or basic solutions. Purified maitotoxin is negative to Dragendorff, ninhydrin, 2,3-dinitrophenylhydrazine and *p*-anisidine phthalate. The IR spectrum suggests a polyhydroxyl nature of the molecule and the presence of an amide group. Earlier findings indicating the presence of an amino acid, sugar and fatty acid components in the toxin from surgeon fish may be due to contaminants in the sample used for analysis.

4.3 Pharmacology of *Gambierdiscus toxicus* toxins

Again, most of the pharmacological studies of ciguatera toxin were done with specimens from Moray eels and other ciguatoxic fish. Past studies on ciguatera toxins have been reviewed by several authors (Halstead 1978; Hashimoto 1979; Withers 1982). However, Yasumoto *et al.* (1979a) reported that the ciguatoxin preparation from the dinoflagellate increased the sodium ion influx through the excitable membrane, which is one of the reported properties of fish ciguatoxin. Similarly, the ciguatoxin fraction from the Hawaiian wild cells evoked a prolonged positive inotropic effect on the isolated guinea-pig atria, which is characteristic of ciguatoxin, while the response caused by the water-soluble maitotoxin was biphasic, enhancing the atrial contractility at low concentrations and depressing it at higher concentrations (Miyahara *et al.* 1979; Shimizu *et al.* 1982).

Intraperitoneal injection of ciguatoxin in mice usually causes increased excretion at first. Death occurs in about 30 min or longer depending upon the dose. The lethal dose for a purified sample from Moray eels was 0·45 $\mu g\ kg^{-1}$ in mice (Tachibana 1980). Maitotoxin causes a different response. There is generally no increased excretion and death usually ensues slowly (Yasumoto *et al.* 1976). Doses as low as 5 $\mu g/kg$ were reported to kill mice within 48 hours. The effects of fish ciguatera toxin on humans are said to be primarily neurological and gastrointestinal but numerous symptoms, some of them conflicting, have been reported (see Withers 1982), presumably because of the involvement of more than one toxin. The exact symptoms caused by maitotoxin are not known. It is interesting to note that although maitotoxin was isolated from fish, it can show ichthyotoxicity to guppies (Yasumoto *et al.* 1976).

4.4 Fate of *Gambierdiscus toxicus* toxins in the food chain

The geographical correlation between the distribution of *G. toxicus* and the occurrence of toxic fish is well documented (Yasumoto *et al.* 1979, 1980a; Bagnis *et al.* 1980). It has been speculated that herbivorous fish, grazing on macroalgae such as *Jania* sp., *Turbinaria ornata*, *Spyridia filamentosa*, *Acanthophora spicifera*, and *Ceramium* spp. ingest the epiphytic dinoflagellate. Then the toxin is transmitted to carnivores through the food chain. It has been reported that the larger the fish, the greater the chance that it is more toxic, e.g. barracuda; this indicates that the toxin is cumulative, although large fish may also be more effective predators. The toxin is initially concentrated in the liver and viscera but it can spread throughout the body. There is a good deal of evidence that the toxins from Moray eels and the dinoflagellate are the same. Such evidence includes the facts that both toxins behave similarly when subjected to chromatography and both react with the same antibody (Hokama *et al.* 1977; Shimizu *et al.* 1981). Nevertheless, the ultimate confirmation that they are one and the same has yet to be made. The question is whether all ciguatera toxins found in fish are derived from *G. toxicus*, and if the dinoflagellate toxin is transmitted and accumulated intact, without change in the molecular form. There are some indications that other species may also contribute to ciguatera poisonings (see below).

5 MISCELLANEOUS DINOFLAGELLATE TOXINS

5.1 *Prorocentrum lima* toxin

Prorocentrum lima is a benthic dinoflagellate suspected to be one of the secondary sources of ciguatera toxins (Yasumoto *et al.* 1980b; Nakajima *et al.* 1981). Its ether extract exhibits a toxicity of $1{\cdot}4 \times 10^{-6}$ cell and the butanol fraction $7{\cdot}1 \times 10^{-7}$ $cell^{-1}$ in mice intraperitoneal (ip) injection. The toxic component has been said to be identical with okadaic acid (Yasumoto *et al.* as quoted in Tachibana *et al.* 1981). Okadaic acid was originally isolated from the black sponge *Halichondria okadai*, commonly found along the Pacific coast of Japan (Tachibana *et al.* 1981) and *H. melanodocia*, a Caribbean sponge collected in the Florida keys (Schmitz *et al.* 1981). The structure (Fig. 8.9) was established by X-ray crystallography of the *o*-bromobenzyl ester (Tachibana *et al.* 1981). Okadaic acid is toxic to mice (LD_{50} 195 μg kg^{-1} ip) and to KB cells at concentrations of 2·5–5·0 ng ml^{-1}. The occurrence of this compound in the sponges prompts the speculation that the compound is actually produced by the dinoflagellate in symbiosis or that it may be taken in as a part of the diet.

5.2 *Prorocentrum minimum* var. *mariae-lebouriae* (= *Exuviaella mariae-lebouriae*) toxin

Prorocentrum minimum var. *mariae-lebouriae* is the alleged organism which caused massive poisoning following ingestion of the small clam, *Tapes semidecussata*, from Lake Hamana in Japan in 1942 (Akiba & Hattori 1949). In the incident, 114 out of 324 patients died. The toxin, named venerupin, was isolated from the clam and found to be hepatotoxic and caused symptoms in

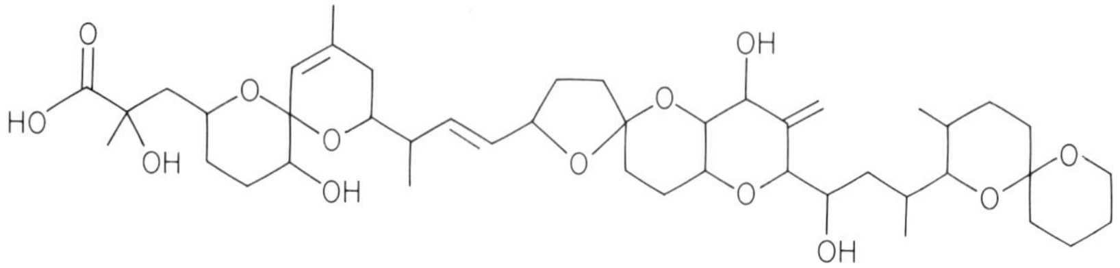

Fig. 8.9. Structure of okadaic acid.

mice similar to those observed in poisoned humans (Akiba & Hattori 1949). Nakajima (1965a, b, c, 1968) cultured the alleged causative organism and found toxic constituents similar to venerupin. Okaichi & Imatomi (1979) isolated two toxic components from the cultured cells of a strain isolated from Inland Sea. The lethality of both of the purified toxins was approximately 125 mg kg^{-1} (ip mice). The purified toxin was positive to Dragendorff reagent but negative to ninhydrin, Sakaguchi and Liebermann reagents. The yield of the toxins and their toxicity do not seem to warrant this particular organism (at least this strain) being the source of the virulent poison responsible for 114 deaths in 1942.

5.3 *Prorocentrum micans* and *P. minimum* toxins

A correlation was observed between the populations of *Prorocentrum* species and the occurrence of shellfish poisonings along the Dutch coast of the North Sea (Kat 1979). The poisonings were accompanied by gastrointestinal complaints: vomiting, abdominal pain and diarrhoea. In some cases the patients were seriously ill. Although shellfish fed with cultured *P. micans* and *P. minimum* did not cause ill effects in animals, it is still possible that these organisms produce toxins under natural conditions. F. Taylor & R. Waters (pers. comm.) have repeatedly tested cultures of *P. micans* for acid-extractable toxins with negative results.

5.4 *Dinophysis fortii* toxin

Shellfish poisonings very similar to those observed in the Netherlands occurred in the Pacific Tohoku District in 1976 and 1977 (Yasumoto *et al.* 1978). Typically the patients suffered from vomiting and diarrhoea. Yasumoto *et al.* (1980c) found a good correlation between the population of *Dinophysis fortii* and mussel toxicity. The lipid-soluble toxins extracted from both the toxic mussels and the organism cells corresponded well in chromatography. It has been found that this lipid-soluble toxin, dinophysistoxin-1, is chemically related to the okadaic acid produced by *Prorocentrum lima*. Murata *et al.* (1982) have identified it as 35S-methyl okadaic acid.

5.5 *Peridinium polonicum* toxin

This freshwater dinoflagellate was responsible for a massive fish kill in a lake in Japan. The toxic substance, glenodinine, was obtained as a crystalline reineckate containing nitrogen and sulphur (Hashimoto *et al.* 1968). The compound is an amine, positive to ninhydrin and Dragendorff reagent. The mass spectrum of the toxin resembles that of 12-methoxyibogamine, an indole alkaloid. Glenodinine is toxic to both fresh water and salt water fish. Acute mouse toxicity was observed at a dose around 200 μg g^{-1}.

5.6 *Gonyaulax monilata* toxin

Gonyaulax monilata is associated with massive fish kills in the Gulf of Mexico along with *Gymnodinium breve* (Connell & Cross 1950; Gates & Wilson 1960; Aldrich *et al.* 1967). Ray & Aldrich (1965) first suggested that the two organisms do not produce the same toxin based on the observation of the different reactions oysters and killifish displayed. Sievers (1969) further compared the toxicity of *G. monilata* and *Gym. breve* using the cultures and reached the same conclusion. *G. monilata* was strongly ichthyotoxic, while crustaceans were resistant to the organism. They also observed that annelids and molluscs were more sensitive to *G. monilata* than to *G. breve*. The effects of *G. monilata* toxin on humans is not known, although Ray & Aldrich (1965) reported that it is non-toxic to chickens and cats. Clemons *et al.* (1980) have recently tried to purify the toxin, following the toxicity against German cockroaches and haemolytic activity. In their experiment, the toxicity has been found in the water-soluble fraction of a molecular range over 100 000. Beside this report, little is known about the chemical nature of the toxin.

5.7 Other dinoflagellates reported to be toxic

Many other dinoflagellates are reported to be toxic, but the exact nature of the toxicity is not known. For example, Steidinger (1979) listed fourteen marine species which had been reported as toxic in laboratory and/or field tests. Among them, those which have not been discussed previously here are *Amphidinium carterae*, *A. rhynchocephalum*, *Gymnodinium veneficum*, and *Prorocentrum balticum*. The toxin of *Gymnodinium veneficum*, which caused a bloom in the English channel, was reported to cause depolarization of membranes (Abbott & Ballantine 1957) but the report is limited to this single incident and no harm was associated with the natural bloom. Yasumoto and his co-workers investigated the toxicity of benthic dinoflagellates as sources of toxins found in the coral reef environments (Yasumoto *et al.* 1980b; Nakajima *et al.* 1981). They found that *Prorocentrum concavum*, *Amphidinium klebsii*, *Ostreopsis siamensis*, *O. ovata* were moderately or slightly toxic in addition to the already mentioned *P. lima*, *A. carterae* and *Gambierdiscus toxicus*. All demonstrated haemolytic activity to mouse erythrocytes, and *A. carterae*, *A. klebsii* and *P. concavum* showed ichthyotoxicity to killifish. *A. carterae* has been previously reported to have high choline derivative content and shows toxicity to fish and mice (Ikawa & Sasner 1975). However, the toxicity reported by Nakajima *et al.* (1981) was found in the lipid-soluble fraction. *O. ovata* and *O. siamensis* are morphologically similar species (Yasumoto *et al.* 1980b) but the toxin contents are distinctly different; the latter contains a water-soluble toxin which is absent in the former (Nakajima *et al.* 1981).

Blooms of *Gyrodinium aureolum* (possibly synonymous with *Gymnodinium flavum* and *Gym. nagasakiense*: Taylor 1985) are reported to cause deaths of fish and littoral organisms (Lackey & Clendenning 1963; Tangen 1977). It was suggested that the kills were due to oxygen depletion rather than ichthyotoxins (Tangen 1977, 1979). *Gymnodinium nagasakiense* blooms also accompanied deaths of marine organisms in Omura Bay, Japan. Iizuka (1979) reported that the organism is toxic to rotifers but has only weak ichthyotoxicity.

6 REFERENCES

ABBOTT B. C. & BALLANTINE D. (1957) The toxin from *Gymnodinium veneficum* Ballantine. *J. mar. Biol. Assoc. UK* **37**, 169–89.

ABBOTT B. C., SIGER A. & SPIEGELSTEIN M. (1975) Toxins from the blooms of *Gymnodinium breve*. In *Proceedings of the First International Conference on Toxic Dinoflagellate Blooms* (ed. V. R. LoCicero), pp. 355–366. Mass. Sci. Tech. Found., Wakefield, MA.

AKIBA T. & HATTORI Y. (1949) Food poisoning caused by eating asari and oyster-toxic substance, Venerupin, *Jap. J. exp. Med.* **20**, 271–284.

ALAM M. (1972) *Chemical studies on toxins from* Gymnodinium breve *and* Aphanizomenon flos-aquae. Ph.D. thesis, Univ. of New Hampshire, Durham, USA.

ALAM M. I., HSU C. P. & SHIMIZU Y. (1979) Comparison of toxins in three isolates of *Gonyaulax tamarensis* (Dinophyceae) *J. Phycol.* **15**, 106–110.

ALAM M., OSHIMA Y. & SHIMIZU Y. (1981) About gonyautoxin-I, -II, -III and -IV. *Tetrahedron Letters,* **23**, 321–322.

ALAM M., SASNER J. J., Jr. & IKAWA M. (1973) Isolation of *Gymnodinium breve* toxin from Florida red tide water. *Toxicon* **11**, 201–202.

ALAM M., SHIMIZU Y., IKAWA M. & SASNER J. J. (1978) Reinvestigation of the toxin from *Aphanizomenon flos-aquae* by high performance chromatographic method. *J. environ. Sci. Health* **A13**, 493–499.

ALAM M., TRIEFF N. M., RAY S. M. & HUDSON J. E. (1975) Isolation and partial characterization of toxins from the dinoflagellate *Gymnodinium breve* Davis. *J. pharm. Sci.* **65**, 865–867.

ALDRICH D. V., RAY S. M. & WILSON W. B. (1967) *Gonyaulax monilata*: population growth and development of toxicity in cultures. *J. Protozool.* **14**, 636–639.

ANDERSON D. M. (1980) Effects of temperature conditioning on development and germination of *Gonyalulax tamerensis* (Dinophyceae) hypnozygotes. *J. Phycol.* **16**, 166–172.

ANDERSON D. M. & WALL D. (1978) Potential importance of benthic cysts of *Gonyaulax tamarensis* and *G. excavata* in initiating toxic dinoflagellate blooms. *J. Phycol.* **14**, 224–234.

ASSOCIATION OF OFFICIAL ANALYTICAL CHEMISTS (1975) Paralytic shellfish poison, biological method. In *Official Methods of Analysis, 12th edn* (rev.) Washington, DC. *Assoc. Offic. Anal. Chem.* **28**, 319–321.

AVARIA S. (1979) Red tides off the coast of Chile. In *Toxic Dinoflagellate Blooms* (ed. D. L. Taylor & H. H. Seliger), pp. 161–164. Elsevier/North Holland, New York.

BADEN D. G. (1983) Marine food-borne dinoflagellate toxins. *Int. Rev. Cytol.* **82**, 99–150.

BADEN D. G., MENDE T. J. & BLOCK R. E. (1979) Two similar toxins isolated from *Gymnodinium breve*. In *Toxic Dinoflagellate Blooms* (ed. D. L. Taylor & H. H. Seliger), pp. 327–334. Elsevier/North Holland, New York.

BADEN D. G., MENDE T. J., LICHTER W. & WELLHAM L. (1981) Crystallization and toxicology of T_{34}: a major toxin from Florida's red tide organism (*Ptychodiscus brevis*). *Toxicon* **19**, 455–462.

BAGNIS R., CHANTEAU S., CHUNGUE E., HURTEL J. M., YASUMOTO T. & INOUE A. (1980) Origins of ciguatera fish poisoning: a new dinoflagellate *Gambierdiscus toxicus* Adachi and Fukuyo, definitely involved as a causal agent. *Toxicon* **18**, 199–208.

BATES H. A. & RAPOPORT H. (1975) A chemical assay for saxitoxin, the paralytic shellfish poison. *J. agr. food Chem.* **23**, 237–239.

BORDNER J., THIESSEN W. E., BATES H. A. & RAPOPORT H. (1975) The structure of a crystalline derivative of saxitoxin. The structure of saxitoxin. *J. Am. Chem. Soc.* **97**, 6008–6012.

BOYER G. L., SCHANTZ E. J. & SCHNOES H. K. (1978) Characterization of 11-hydroxysaxitoxin sulphate, a major toxin in scallops exposed to blooms of the poisonous dinoflagellate *Gonyaulax tamarensis*, *J. C. S. Chem. Comm.* 1978, 889–890.

BUCKLEY L. J., IKAWA M. & SASNER J. J. Jr (1976) Isolation of *Gonyaulax tamarensis* toxins from softshell clams (*Mya arenaria*), and a thin-layer chromatographic-fluorometric method for their detection. *J. agr. food Chem.* **24**, 107–111.

BUCKLEY L. J., OSHIMA Y. & SHIMIZU Y. (1978) Construction of a paralytic shellfish toxin analyzer and its application. *Anal. Biochem.* **85**, 157–164.

CARRETO J. I., NEGRI R. M., BENAVIDES H. R. & AKSELMAN R. (1985). Toxic dinoflagellate blooms in the Argentine Sea. In *Toxic Dinoflagellates* (eds. D. M. Anderson, A. W. White & D. G. Baden), pp. 147–152. Elsevier, New York.

CLEMONS G. P., PINION J. P., BASS E., PHAM D. V., SHARIEF M. & WUTCH J. G. (1980) A hemolytic principle associated with the red-tide dinoflagellate *Gonyaulax monilata*. *Toxicon* **18**, 323–326.

CONNELL C. H. & CROSS J. B. (1950) Mass mortality of fish associated with the protozoan *Gonyaulax* in the Gulf of Mexico. *Science* (NY) **112**, 359–363.

CUMMINS J. J. & STEVENS A. A. (1970) In *Investigation of* Gymnodinium breve *toxins in shellfish*. Public Health Service Publ. US Department of Health, Education and Welfare, Washington, DC.

DALE B., YENTSCH C. M. & HURST J. W. (1978) Toxicity in resting cysts of the red tide dinoflagellate *Gonyaulax excavata* from deeper water coastal sediments. *Science* (NY) **201**, 1223–1225.

DINOVI M., TRAINOR D. A., NAKANISHIK, SANDUJA R. & ALAM M. (1983) The structure of PB-1, an unusual toxin isolated from the red tide dinoflagellate *Ptychodiscus brevis*. *Tetrahedron Lett.* **29**, 855–858.

DOWN R. J. (1970) The medical significance of shellfish and blowfish neurotoxins (saxitoxin and tetrodotoxin) as suggested by tests in killifish (Fundulus heteroclitus). *Proceedings 1969. Food-Drugs from the Sea* (ed. H. W. Youngken, Jr), pp. 327–343. Marine Technology Society, Washington, DC.

ELLIS S., SPIKES J. J. & JOHNSON G. L. (1979) Respiratory and cardiovascular effects of *G. breve* toxin in dogs. In *Toxic Dinoflagellate Blooms* (ed. D. L. Taylor & H. H. Seliger), pp. 431–434. Elsevier/North Holland, New York.

FALLON W. E. & SHIMIZU Y. (1977) Electrophoretic analysis of paralytic shellfish toxins. *J. environ. Sci. Health* **A12**, 455–464.

FOXALL T. L., SHOPTAUGH N. H., IKAWA M. & SASNER J. J., Jr (1979) Secondary intoxication with PSP in *Cancer irroratus*. In *Toxic Dinoflagellate Blooms* (ed. D. L. Taylor & H. H. Seliger), pp. 413–418. Elsevier/North Holland, New York.

GATES J. A. & WILSON W. B. (1960) The toxicity of *Gonyaulax monilata* Howell to *Mugil cephalus*. *Limnol. Oceanogr.* **5**, 171–174.

GENENAH A. A. & SHIMIZU Y. (1981) Specific toxicity of paralytic shellfish poisons. *J. ag. food Chem.* **29**, 1289–1291.

GHAZAROSSIAN V. E., SCHANTZ E. J., SCHNOES H. K. & STRONG F. M. (1974) Identification of a poison in toxic scallops from a *Gonyaulax tamarensis* red tide. *Biochem. biophys. Res. Comm.* **59**, 1219–1225.

GOLIK J., JAMES J. C. & NAKANISHI K. (1982) The structure of Brevetoxin C. *Tetrahedr. Lett.* **23**, 2535–2538.

GRINDLEY J. R. & NEL E. A. (1970) Red water and mussel poisoning at Elands Bay, December, 1966. *Fish. Bull. S. Afr.* **6**, 36–55.

GUZMAN L. & CAMPODONICO E. I. (1978) Mareas rojas en Chile. *Interciencia* **3**, 144–150.

HALL S., REICHARDT P. B. & NEVE R. A. (1980) Toxins extracted from an Alaskan isolate of *Protogonyaulax* sp. *Biochem. biophys. Res. Comm.* **97**, 649–653.

HALSTEAD B. W. (1978) *Poisonous and Venomous Marine Animals of the World*, pp. 328–402, Darwin, Princeton, NJ.

HASHIMOTO Y. (1979) Marine toxins and other bioactive marine metabolites. *Jap. sci. Soc.*, Tokyo, 360 p.

HASHIMOTO Y., KONOSU S., YASUMOTO T., INOUE A. & NOGUCHI T. (1967) Occurrence of toxic crabs in Ryukyu and Amami Islands. *Toxicon* **5**, 85–90.

HASHIMOTO Y., NOGUCHI T. & ADACHI R. (1976) Occurrence of toxic bivalves in association with the bloom of *Gonyaulax* sp. in Owase Bay. *Bull. Jap. Soc. sci. Fish.* **42**, 671–676.

HASHIMOTO Y., OKAICHI T., DANG L. D. & NOGUCHI T. (1968) Glenodinine, an ichthyotoxic substance produced by a dinoflagellate, *Peridinium polonicum*. *Bull. Jap. Soc. sci. Fish.* **34**, 528–534.

HERMES R. & VILLOSO E. P. (1983). A recent bloom of the toxic dinoflagellate *Pyrodinium bahamense* var. *compressa* in the central Philippine waters. *Fish. Res. J. Philippines* **8**(2), 1–8.

HOKAMA Y., BANNER A. H. & BOYLAN D. B. (1977) A radioimmunoassay for the detection of ciguatoxin. *Toxicon* **15**, 317–325.

HORSTMAN D. A. (1981) Reported red water outbreaks and their effects on fauna of the west and south coasts of South Africa, 1951–1980. *Fish. Bull. S. Afr.* **15**, 71–88.

HSU C. P., MARCHAND A., SHIMIZU Y. & SIMS G. G. (1979) Paralytic shellfish toxins in the sea scallop, *Placopecten magellanicus* in the Bay of Fundy. *J. fish. Res. Board Can.* **36**, 32–36.

HUGHES J. M. (1979) Epidemiology of shellfish poisoning in the United States, 1971–1977. In *Toxic Dinoflagellate Blooms* (ed. D. L. Taylor & H. H. Seliger), pp. 23–28. Elesevier/North Holland, New York.

IIZUKA A. (1979) Maximum growth rate of natural population of a *Gymnodinium* red tide. In *Toxic Dinoflagellate Blooms* (ed. D. L. Taylor & H. H. Seliger), pp. 111–114. Elsevier/North Holland, New York.

IKAWA M. & SASNER J. J., Jr (1975) Chemical and physiological studies on the marine dinoflagellate *Amphidinium carterae*. *Proceedings of the First International Conference on Toxic Dinoflagellate Blooms* (ed. V. R. LoCicero), pp. 323–332. Mass. Sci. Tech. Found., Wakefield, MA.

INGHAM H. R., MASON J. & WOOD P. C. (1968) Dinoflagellate crop in the North Sea. *Nature (Lond.)* **220**, 21–27.

JACKIM E. & GENTILE J. (1968) Toxins of a blue-green alga: similarity to saxitoxin. *Science* (N.Y.), **162**, 915–916.

KAMIYA H. & HASHIMOTO Y. (1978) Occurrence of saxitoxin and related toxins in Palauan bivalves. *Toxicon* **16**, 303–306.

KAO C. Y. (1966) Tetrodotoxin, saxitoxin and their significance in the study of excitation phenomena. *Pharm. Rev.* **18**, 997–1049.

KAT M. (1979) The occurrence of *Prorocentrum* species and coincidental gastrointestinal illness of mussel consumers. In *Toxic Dinoflagellate Blooms* (ed. D. L. Taylor & H. H. Seliger), pp. 215–220. Elsevier/North Holland, New York.

KIM Y. S. & PADILLA G. M. (1976) Purification of the ichthytoxic component of *Gymnodinium breve* (red tide dinoflagellate) toxin by high pressure liquid chromatography. *Toxicon* **14**, 379–387.

KOBAYASHI M. & SHIMIZU Y. (1981) Gonyautoxin VIII, a cryptic precursor of paralytic shellfish poisons. *J. C. S. Chem. Comm.* 1981, 827–828.

KOCH H. J. (1939) La cause des empoisonnements paralytiques par les moules. *Assoc. Fr. Adva. Sci.*, Paris, 63rd Session, 654–657.

KONOSU S., INOUE A., NOGUCHI T. & HASHIMOTO Y. (1968) Comparison of crab toxin with saxitoxin and tetrodotoxin. *Toxicon* **6**, 113–117.

KOTAKI H., OSHIMA Y. & YASUMOTO T. (1981) Analysis of paralytic shellfish toxins of marine snails. *Bull. Jap. Soc. sci. Fish.* **47**, 943–946.

LACKEY J. B. & CLENDENNING K. A. (1963) A possible fishkilling yellow tide in California waters. *Quart. J. Fla. Acad. Sci.* **26**, 263–268.

LIN Y. Y., RISK M., RAY S. M., VAN EUGEN D., CLARDY J., GOLICK J., JAMES J. C. & NAKANISHI K. (1981) Isolation and structure of brevetoxin-B from the 'red tide' dinoflagellate, *Ptychodiscus brevis* (*Gymnodinium breve*), *J. Am. chem. Soc.* **103**, 6773–6775.

LOEBLICH L. A. & LOEBLICH A. R. (1975) The organism causing New England red tides: *Gonyaulax excavata*. In *Proceedings of the First International Conference on Toxic Dinoflagellate Blooms* (ed. V. R. LoCicero), pp. 207–224. Mass, Sci. Tech. Found., Wakefield, MA.

LÜTHY J., ZWEIFEL U., SCHLATTER C., HUNYADY G., HASLER S., HSU C. P. & SHIMIZU Y. (1978) Vergiftungsfälle durch Miesmuscheln in der Schweiz 1976 Sanitätspolizeiliche Mapnahmen und Analyse der PSP-Toxine. *Mitt. Gebiete lebensen. Hyg.* **69**, 467–476.

MCFARREN E. F., TANABE H., SILVA F. J., WILSON W. B., CAMPBELL J. E. & LEWIS K. H. (1965) The occurrence of a ciguatera-like poison in oysters, clams and *Gymnodinium breve* cultures. *Toxicon* **3**, 111–123.

MARTIN D. F. & CHATTERJEE A. B. (1969) Isolation and characterization of a toxin from Florida red tide organism. *Nature* (Lond.) **221**, 59–60.

MARTIN D. J. & CHATTERJEE A. B. (1970) Some chemical and physical properties of two toxins from the red tide organism, *Gymnodinium breve*. *Fish. Bull.* **68**, 433–443.

MEYER K. F., SOMMER H. & SCHOENHOLZ P. (1928) Mussel poisoning. *J. prev. Med.* **2**, 365–394.

MARANDA L. & SHIMIZU Y. (1981) Distribution mapping and typing of toxic dinoflagellates. In *Sea Grant Report*, University of Rhode Island, pp. 160–172.

MIYAHARA J. T., AKAU C. K. & YASUMOTO T. (1979) Effects of ciguatoxin and maitotoxin on the isolated guinea-pig atria. *Res. Commun. chem. Pathol. Pharmacol.* **25**, 177–180.

MOREY-GAINES G. (1982) *Gymnodinium catenatum* Graham: morphology and affinities with armored forms. *Phycologia* **21**, 154–163.

MURATA M., SHIMATANI M., SUGITANI H., OSHIMA Y. & YASUMOTO T. (1982) Isolation and structural elucidation of the causative toxin of the diarrhetic shellfish poisoning. *Bull. Jap. Soc. sci. Fish.* **48**, 549–552.

NAKAJIMA I., OSHIMA Y. & YASUMOTO T. (1981) Toxicity of benthic dinoflagellates in Okinawa. *Bull. Jap. Soc. sci. Fish.* **47**, 1029–1033.

NAKAJIMA M. (1965a) Studies on the source of shellfish poison on Lake Hamana. I. Relation of the abundance of a species of dinoflagellate, *Prorocentrum* sp. to shellfish toxicity. *Bull. Jap. Soc. sci. Fish.* **31**, 198–203.

NAKAJIMA M. (1965b) Studies on the source of shellfish poison in Lake Hamana. II. Shellfish toxicity during the 'red-tide'. *Bull. Jap. Soc. sci. Fish.* **31**, 204–207.

NAKAJIMA M. (1965c) Studies on the source of shellfish poisoning in Lake Hamana. III. Poisonous effects of shellfish feeding on *Prorocentrum* sp. *Bull. Jap. Soc. sci. Fish.* **31**, 281–285.

NAKAJIMA M. (1968) Studies on the source of shellfish poison in Lake Hamana. IV. Identification and collection of the noxious dinoflagellates. *Bull. Jap. Soc. sci. Fish.* **34**, 130–131.

NAKANISHI K. (1985) The chemistry of brevetoxins: a review. *Toxicon* **23**, 473–479.

NARAHASHI T. (1972) Mechanism of action of tetrodotoxin and saxitoxin on excitable membranes. *Fedn. Proc. Fedn. Am. Soc. exp. Biol.* **31**, 1124–1132.

NEEDLER A. B. (1949) Paralytic shellfish poisoning and *Gonyaulax tamarensis*. *J. fish. Res. Bd. Can.* **7**, 490–504.

NOGUCHI T., KONOSU S. & HASHIMOTO Y. (1969) Identity of the crab toxin with saxitoxin. *Toxicon* **7**, 325–326.

OKAICHI T. & IMATOMI Y. (1979) Toxicity of *Prorocentrum minimum* var. *mariae-lebouriae* assumed to be a causative agent of short-neck clam poisoning. In *Toxic Dinoflagellate Blooms* (ed. D. L. Taylor & H. H. Seliger), pp. 385–388. Elsevier/North Holland, New York.

OSHIMA Y., BUCKLEY L. J., ALAM M. & SHIMIZU Y. (1977) Heterogeneity of paralytic shellfish poisons. Three new toxins from cultured *Gonyaulax tamarensis* cells, *Mya arenaria* and *Saxidomis giganteus*. *Comp. Biochem. Physiol.* **57c**, 31–34.

OSHIMA Y., FALLON W. E., SHIMIZU Y., NOGUCHI T. & HASHIMOTO Y. (1976) Toxins of the *Gonyaulax* sp. and infested bivalves on Owase Bay, *Bull. Jap. Soc. sci. Fish.* **42**, 851–856.

OSHIMA Y., SHIMIZU Y., NISHIO S. & OKAICHI T. (1978) Identification of paralytic shellfish toxins in shellfish from Inland Sea. *Bull. Jap. Soc. sci. Fish.* **44**, 395.

OSHIMA Y. & YASUMOTO T. (1979) Analysis of toxins in cultured *Gonyaulax excavata* cells originating in Ofunato Bay, Japan. In *Toxic Dinoflagellate Blooms* (ed. D. L. Taylor & H. H. Seliger), pp. 377–380. Elsevier/North Holland, New York.

PADILLA G. M., KIM Y. S. & MARTIN D. F. (1974) Separation and analysis of toxins isolated from a red tide sample of *Gymnodinium breve*. In *Proc. First International Conference on Toxic Dinoflagellate Blooms* (ed. V. R. LoCicero), pp. 299–308. Mass. Sci. and Tech. Found., Wakefield, MA.

PADILLA G. M., KIM Y. S., RAUCKMAN E. J. & ROSEN G. M. (1979) Physiological activities of toxins from *Gymnodinium breve* isolated by high performance liquid chromatography. In *Toxic Dinoflagellate Blooms* (ed. D. L. Taylor & H. H. Seliger), pp. 351–354. Elsevier/ North Holland, New York.

PASTER Z. & ABBOTT B. C. (1969) Hemolysis of rabbit erythrocytes by *Gymnodinium breve* toxin, *Toxicon* **7**, 245.

PRAKASH A. (1963) Source of paralytic shellfish toxin in the Bay of Fundy. *J. fish. Res. Bd. Can.* **20**, 983–996.

PRAKASH A. & TAYLOR F. J. R. (1966) A 'red water' bloom of *Gonyaulax acatenella* in the Strait of Georgia and its relation to paralytic shellfish toxicity. *J. fish. Res. Bd. Canada* **23**, 1265–1270.

RAY S. M. & ALDRICH D. V. (1965) *Gymnodinium breve*: induction of shellfish poisoning in chicks. *Science (NY)* **148**, 1748–1749.

REYES-VASQUEZ G., FERRAZ-REYES E. & VASQUEZ E. (1979) Toxic dinoflagellate blooms in Northeastern Venezuela during 1977. In *Toxic Dinoflagellate Blooms* (ed. D. L. Taylor & H. H. Seliger), pp. 191–194. Elsevier/North Holland, New York.

RISK M., LIN Y. Y., MACFARLEN R. D., SADAGOPA-RAMANUJAN V. M., SMITH L. L. & TRIEFF N. M. (1979a) Purification and chemical studies on a major toxin from *Gymnodinium breve*. In *Toxic Dinoflagellate Blooms* (ed. D. L. Taylor & H. H. Seliger), pp. 335–344. Elsevier/North Holland, New York.

RISK M., WERRBACH-PEREZ K., PEREZ-POLO J. R., BUNCE H., RAY S. M. & PARMENTIER J. L. (1979b) Mechanism of action of the major toxin from *Gymnodinium breve* Davis. In *Toxic Dinoflagellate Blooms* (ed. D. L. Taylor & H. H. Seliger), pp. 367–372. Elsevier/North Holland, New York.

ROBERTS B. S., HENDERSON G. E. & MEDLYN R. A. (1979) The effects of *Gymnodinium breve* toxin(s) on selected mollusks and crustaceans. In *Toxic Dinoflagellate Blooms* (ed. D. L. Taylor & H. H. Seliger), pp. 419–424. Elsevier/North Holland, New York.

ROGERS R. S. & RAPOPORT H. (1980) The pKa's of saxitoxin. *J. Am. chem. Soc.* **102**, 7335–7339.

SASNER J. J., IKAWA M., THURBERG F. & ALAM M. (1972) Physiological and chemical studies on *Gymnodinium breve* Davis toxin. *Toxicon* **10**, 163–172.

SATOH N., ICHIHARA N., KAWASE S., SATO H. & ISHIGE M. (1977) Poisonous bivalves in Hokkaido. *Rep. Hokkaido Inst. pub. Health* **27**, 66–68.

SCHANTZ E. J. (1960) Biochemical studies on paralytic shellfish poisons. *Ann. N.Y. Acad. Sci.* **89**, 843–855.

SCHANTZ E. J. (1972) The dinoflagellate poisons. In *Microbial Toxins* Vol. VII: *Algal and Fungal Toxins* (eds. S. Kadis, A. Ciegler & S. J. Ajil). Academic Press, New York.

SCHANTZ E. J., GHAZAROSSIAN V. E., SCHNOES H. K., STRONG F. M., SPRINGER J. P., PEZZANITE J. O. & CLARDY J. (1975a) The structure of saxitoxin. *J. Am. chem. Soc.* **97**, 1238–1239.

SCHANTZ E. J., GHAZAROSSIAN V. E., SCHNOES H. K., STRONG F. M., SPRINGER J. P., PEZZANITE J. O. & CLARDY J. (1975b) Paralytic poisons from marine dinoflagellates. In *Proceedings of the First International Conference on Toxic Dinoflagellate Blooms* (ed. V. R. LoCicero), pp. 267–277, Mass. Sci. Tech. Found., Wakefield, MA.

SCHANTZ E. J., LYNCH J. M., VAYVADA G., MATSUMOTO K. & RAPOPORT H. (1966) The purification and characterization of the poison produced by *Gonyaulax catenella* in axenic culture. *Biochem.* **5**, 1191–1195.

SCHANTZ E. J. & MAGNUSSON H. W. (1964) Observations on the origin of the paralytic poison in Alaska butter clams, *J. Protozool.* **11**, 238–242.

SCHANTZ E. J., MOLD J. D., HOWARD W. L., BOWDEN J. P., STANGER D. W., LYNCH J. M., WINTERSTEINER O. P., DUTCHER J. D., WALTERS D. R. & REIGEL B. (1961) Paralytic shellfish poison, VIII. Some chemical and physical properties of purified clam and mussel poisons. *Can. J. Chem.* **39**, 2117–2123.

SCHANTZ E. J., MOLD J. D., STANGER D. W., SHAVEL J., RIEL F. J., BOWDEN J. P., LYNCH J. M., WYLER R. S., REIGEL B. & SOMMER H. (1957) Paralytic shellfish poison. VI. A procedure for the isolation and purification of the poison from toxic clams and mussel tissues. *J. Am. chem. Soc.* **78**, 5230–5235.

SCHMIDT R. J. & LOEBLICH A. R. III. (1979) Distribution of paralytic shellfish poison among Pyrrhophyta. *J. mar. biol. Assoc. UK* **59**, 479–487.

SCHMITZ F. J., PRASAD R. S., GOPICHAND Y., HOSSAIN M. B., VAN DER HELM D. Z. & SCHMIDT P. (1981) Acanthifolicin, a new episulfide-containing polyether carboxylic acid from extract of the marine sponge, *Pandaros acanthifolium. J. Am. Chem. Soc.* **103**, 2467–2469.

SCHRADIE J. & BLISS C. A. (1962) The cultivation and toxicity of *Gonyaulax polyedra. Lloydia* **25**, 212–221.

SHIMIZU Y. (1977) Red tide toxins: assay and isolation of the toxic components. In *Handling of Natural Products—Extraction and Isolation of Bioactive Compounds* (ed. S. Natori, N. Ikekawa & M. Suzuki), pp. 151–170, Kodansha Scientific.

SHIMIZU Y. (1978) Dinoflagellate toxins. In *Marine Natural Products. Chemical and Biological Perspectives*, Vol. I. (ed. P. J. Scheuer), pp. 1–42, Academic Press, New York.

SHIMIZU Y. (1979a) Compounds from microalgae—their influence on the field of marine natural products. In *Recent Advances in Phytochemistry*, Vol. 13. Topics in the Biochemistry of Natural Products (ed. T. Swain & G. Waller), pp. 199–217. Plenum, New York.

SHIMIZU Y. (1979b) Developments in the study of paralytic shellfish toxins. In *Toxic Dinoflagellate Blooms* (ed. D. L. Taylor & H. H. Seliger), pp. 321–326. Elsevier/North Holland, New York.

SHIMIZU Y. (1979c) *Red Tide Toxins.* Symposium lecture No. 337. The American Chemical Society–Chemical Society of Japan Chemistry Congress, Honolulu, Hawaii.

SHIMIZU Y. (1979d) *Shellfish toxins and their origins.* Plenary section abstract of the 99th Annual Meeting of Pharmaceutical Society of Japan, pp. 87–89.

SHIMIZU Y. (1980) Red tide toxins—paralytic shellfish poisons produced by *Gonyaulax* toxins. *Kagaku to Seibutsu* **18** (12) 792–799.

SHIMIZU Y., ALAM M. & FALLON W. E. (1975a) Purification and partial characterization of toxins from poisonous clams. In *Proceedings of the First International Conference on Toxic Dinoflagellate Blooms* (ed. V. R. LoCicero), pp. 275–285. Mass. Sci. Tech. Found., Wakefield, MA.

SHIMIZU Y., ALAM M. & FALLON W. E. (1976a) Red tide toxins. *Proceedings 1974 Food Drugs from the Sea* (ed. H. H. Webber & G. D. Ruggieri), pp. 238–251. Marine Technology Society, Washington, DC.

SHIMIZU Y., ALAM M., OSHIMA Y., BUCKLEY L. J., FALLON W. E., KASAI H., MUIRA I., GULLO V. P. & NAKANISHI K. (1977) Chemistry and distribution of deleterious dinoflagellate toxins. In *Marine Natural Product Chemistry* (ed. D. J. Faulkner & W. H. Fenical), pp. 261–269. Plenum, New York.

SHIMIZU Y., ALAM M., OSHIMA Y. & FALLON W. E. (1975b) Presence of four toxins in red tide infested clams and cultured *Gonyaulax tamarensis* cells. *Biochem. biophys. Res. Commun.* **66**, 731–737.

SHIMIZU Y., BUCKLEY L. J., ALAM M., OSHIMA Y., FALLON W. E., KASAI H., MIURA I., GULLO V. P. & NAKANISHI K. (1976b) Structure of gonyautoxin II and III from the east coast toxic dinoflagellate, *Gonyaulax tamarensis, J. Am. chem. Soc.* **98**, 5414–5416.

SHIMIZU Y., FALLON W. E., WEKELL J. C., GERBER D., Jr. & GAUGLITZ E. J., Jr (1978a) Analysis of toxic mussels (*Mytilus* sp.) from the Alaskan Inside Passage, *J. agric. food Chem.* **26**, 878–881.

SHIMIZU Y., HSU C. P., FALLON W. E., OSHIMA Y., MIURA I. & NAKANISHI K. (1978b) Structure of neosaxitoxin, *J. Am. chem. Soc.* **100**, 6791–6793.

SHIMIZU Y. & HSU C. P. (1981) Confirmation of the structures of gonyautoxin I–VI by correlation with saxitoxin. *J. chem. Soc. chem. Commun.* 1981, 314–315.

SHIMIZU Y., HSU C. P. & GENENAH A. (1981) Structure of saxitoxin in solution and stereochemistry of dihydrosaxitoxin. *J. Am. chem. Soc.* **103**, 605–609.

SHIMIZU Y., NORTE M., HORI A., GENENAH A. & KOBAYASHI M. (1984) Biosynthesis of saxitoxin analogs: the unexpected pathway. *J. Amer. chem. Soc.* **106**, 6433–6434.

SHIMIZU Y., SHIMIZU H., SCHEUER P. J., HOKAMA Y., OYAMA M. & MIYAHARA J. (1982) *Gambierdiscus toxicus*, a ciguatera-causing dinoflagellate from Hawaii. *Bull. Jap. Soc. sci. Fish.* **48**, 811–813.

SHIMIZU Y. & YOSHIOKA M. (1981) Transformation of paralytic shellfish toxins as demonstrated in scallop homogenates. *Science* **212**, 546–547.

SIEVERS A. M. (1969) Comparative toxicity of *Gonyaulax monilata* and *Gymnodinium breve* to annelids, crustaceans and molluscs and a fish. *J. Protozoology* **16**, 401–404.

SOMMER H. (1932) The occurrence of the paralytic shellfish poison in the common sand crab. *Science* (NY) **76**, 574–575.

SOMMER H. & MEYER K. F. (1937) Paralytic shellfish poisoning. *Arch. Pathol.* **24**, 560–598.

SOMMER H., MONNIER R. P., RIEGEL B., STANGER D. W., MOLD J. D., WIKHOLM D. W. & KIRALIS E. S. (1948a) Paralytic shellfish poison. I. Occurrence and concentration by ion exchange. *J. Am. chem. Soc.* **70**, 1015–1018.

SOMMER H., RIEGEL B., STANGER D. W., MOLD J. D., WIKHOLM D. M. & MCCAUGHEY M. B. (1948b) Paralytic shellfish poison II. Purification by chromatography. *J. Amer. chem. Soc.* **70**, 1019–1021.

SOMMER H., WHEDON W. H., KOFOID C. A. & STOHLER R. (1937) Relation of paralytic shellfish poison to certain plankton organisms of the genus *Gonyaulax*. *Arch. Pathol.* **24**, 537–559.

SPIEGELSTEIN M. Y., PASTER Z. & ABBOTT B. C. (1973) Purification and biological activity of *Gymnodinium breve* toxin. *Toxicon* **11**, 85.

SPIKES J. J., RAY S. M., ALDRICH D. V. & NASH J. B. (1968) Toxicity variations of *Gymnodinium breve* cultures. *Toxicon* **5**, 171–174.

STEIDINGER K. A. (1979) Collection, enumeration and identification of free-living marine dinoflagellate. In *Toxic Dinoflagellate Blooms* (ed. D. L. Taylor & H. H. Seliger), pp. 435–442. Elsevier/North Holland, New York.

STEIDINGER K. A., TESTER L. S. & TAYLOR F. J. R. (1980) A description of *Pyrodinium bahamense* var. *compressa* (Bohm) stat. nov. from Pacific red tides. *Phycologia,* **19**, 329–337.

SUBRAMANIAN A. (1985) Noxious dinoflagellates in Indian waters. In *Toxic Dinoflagellates* (eds. D. M. Anderson, A. W. White & D. G. Baden), pp. 525–528. Elsevier, New York.

TACHIBANA K. (1980) *Structural studies on marine toxins*. Ph.D. dissertation Univ. of Hawaii.

TACHIBANA K., SCHEUER P. J., TSUKITANI Y., KIKUCHI H., VAN EUGEN D., CLARDY J., COPICHAND Y. & SCHMITZ F. J. (1981) Okadaic acid, a cytotoxic polyether from two marine sponges of the genus *Halichondria*. *J. Am. chem. Soc.* **103**, 2469–2471.

TAMAYAVANICH S., KODAMA M. & FUKUYO Y. (1985) The occurrence of paralytic shellfish poisoning in Thailand. In *Toxic Dinoflagellates* (eds. D. M. Anderson, A. W. White & D. G. Baden) pp. 521–524. Elsevier, New York.

TANGEN K. (1977) Blooms of *Gyrodinium aureolum* (Dinophyceae) in North European waters, accompanied by mortality in marine organisms. *Sarsia* **63**, 123–133.

TANGEN K. (1979) Dinoflagellate blooms in Norwegian waters. In *Toxic Dinoflagellate Blooms* (ed. D. L. Taylor & H. H. Seliger), pp. 111–114. Elsevier/North Holland, New York.

TANINO H., NAKATA T., KANEKO T. & KISHI Y. (1977) A stereospecific total synthesis of d,l-saxitoxin. *J. Am. chem. Soc.* **99**, 2818–2819.

TAYLOR F. J. R. (1975) Taxonomic difficulties in red tide and paralytic shellfish poison studies: The 'tamarensis complex' of *Gonyaulax*. *Environ. Letters* **9**, 103–119.

TAYLOR F. J. R. (1979a) Toxigenic Gonyaulacoid dinoflagellates. In *Toxic Dinoflagellate Blooms* (ed. D. J. Taylor & H. H. Seliger), pp. 47–56. Elsevier/North Holland, New York.

TAYLOR F. J. R. (1979b) A description of the benthic dinoflagellate associated with maitotoxin and ciguatoxin including observations on Hawaiian materials. In *Toxic Dinoflagellate Blooms* (ed. D. L. Taylor & H. H. Seliger), pp. 71–76. Elsevier/North Holland, New York.

TAYLOR F. J. R. (1984) Toxic dinoflagellates: taxonomic and biogeographic aspects with emphasis on *Protogonyaulax*. In *Seafood Toxins* (ed. E. P. Ragelis), pp. 77–97. *Amer. Chem. Soc., Symp. Ser.* **262**, Wash. D.C.

TAYLOR F. J. R. (1985) The taxonomy and relationships of red tide flagellates. In *Toxic Dinoflagellates* (eds. D. M. Anderson, A. W. White & D. G. Baden), pp. 11–26. Elsevier, New York.

TRIEFF N. M., MISHAN M., GRAJCER D. & ALAM M. (1973) Biological assay of *Gymnodinium breve* toxin using brine shrimp (*Artemia salina*). *Texas Rep. Biol. Med.* **31** (3) 409–422.

TRIEFF N. M., RAMANUJAM V. M. S., ALAM M., RAY S. M. & HUDSON J. E. (1975) Isolation, physico-chemical, and toxicologic characterization of toxins from *Gymnodinium breve* Davis. In *Proceedings of the First International Conference on Toxic Dinoflagellate Blooms* (ed. V. R. LoCicero), pp. 309–322. Mass. Sci. Tech. Found., Wakefield, MA.

TRIEFF N. M., SPIKES J. J., RAY S. M. & NASH J. B. (1970a) Isolation of *Gymnodinium breve* toxin. *Toxicon* **8**, 157–158.

TRIEFF N. M., SPIKES J. J., RAY S. M. & NASH J. B. (1970b) Isolation and purification of *Gymnodinium breve* toxin. In *Toxins of Plant and Animal Origin* (ed. A. DeVries & E. Kochva), pp. 557–577. Gordon & Breach, London.

TRIEFF N. M., VENKATASUBRAMANIAN V. & RAY S. M. (1972) Purification of *Gymnodinium breve* toxin—dry column chromatography technique. *Texas Rep. Biol. Med.* **30**, 97–104.

TURPIN D. H., DOBELL P. E. R. & TAYLOR F. J. R. (1978) Sexuality and cyst formation in Pacific strains of the toxic dinoflagellate *Gonyaulax tamarensis*. *J. Phycol.* **14**, 235–238.

TWAROG B. M., HIDAKA T. & YAMAGUCHI H. (1972) Resistance to tetrodotoxin and saxitoxin in nerves of bivalve molluscs. *Toxicon* **10**, 273–278.

WHITE A. W. (1977) Dinoflagellate toxins as probable cause of an Atlantic herring (*Clupea harengus harengus*) kill and pteropods as apparent vector. *J. fish. Res. Bd. Can.* **34**, 2421–2424.

WHITE A. W. (1978) Salinity effects on growth and toxin contents of *Gonyaulax excavata*, a marine dinoflagellate causing paralytic shellfish poisoning. *J. Phycol.* **14**, 475–479.

WHITE A. W. (1979) Dinoflagellate toxins in phytoplankton and zooplankton fractions during a bloom of *Gonyaulax excavata*. In *Toxic Dinoflagellate Blooms* (ed. D. L. Taylor & H. H. Seliger), pp. 381–384. Elsevier/North Holland, New York.

WHITE A. W. (1981) Sensitivity of marine fishes to toxins from the red-tide dinoflagellate *Gonyaulax excavata* and implications for fish kills. *Marine Biology* **65**, 255–260.

WICHMANN C. F., BOYER G. L., DIVAN C. L., SCHANTZ E. J. & SCHNOES H. K. (1981a) Neurotoxins of *Gonyaulax excavata* and Bay of Fundy scallops. *Tetrahedron Letters* **22**, 1941–1944.

WICHMANN C. F., NIEMCZURA W. P., SCHNOES H. K., HALL S., REICHARDT P. B. & DARLING S. D. (1981b) Structures of two novel toxins from *Protogonyaulax*. *J. Am. chem. Soc.* **103**, 6977–6978.

WITHERS N. W. (1982) Ciguatera fish poisoning. *Ann. Rev. Med.* **83**, 97–111.

WONG J. L., BROWN M. S., MATSUMOTO K., OSTERLIN R. & RAPOPORT H. (1971) Degradation of saxitoxin to a pyrimido (2,1-b) purine. *J. Am. chem. Soc.* **93**, 4633–4634.

WORTH G. K., MACLEAN J. L. & PRICE M. J. (1975) Paralytic shellfish poisoning in Papua New Guinea, 1972, *Pac. Sci.* **29**, 1–5.

YASUMOTO T., BAGNIS R. & VERNOUX J. P. (1976) Toxicity of the surgeonfishes. II. Properties of the principal water-soluble toxins. *Bull. Jap. Soc. sci. Fish.* **42**, 359–365.

YASUMOTO T., HASHIMOTO Y., BAGNIS R., RANDALL J. E. & BANNER A. H. (1971) Toxicity of the surgeonfishes. *Bull. Jap. Soc. sci. Fish.* **37**, 723–734.

YASUMOTO T., INOUE A., BAGNIS R. & GARCON M. (1979) Ecological survey on a dinoflagellate responsible for the induction of ciguatera. *Bull. Jap. Soc. sci. Fish.* **45**, 395–399.

YASUMOTO T., INOUE A., OCHI T., FUJIMOTO K., OSHIMA Y., FUKUYO Y., ADACHI R. & BAGNIS R. (1980a) Environmental studies on a toxic dinoflagellate responsible for ciguatera. *Bull. Jap. Soc. sci. Fish.* **46**, 1397–1404.

YASUMOTO T., NAKAJIMA I., OSHIMA Y. & BAGNIS R. (1979c) A new toxic dinoflagellate found in association with ciguatera. In *Toxic Dinoflagellate Blooms* (ed. D. L. Taylor & H. H. Seliger), pp. 65–70. Elsevier/North Holland, New York.

YASUMOTO T., NAKAJIMA I., BAGNIS R. & ADACHI R. (1977) Finding of a dinoflagellate as a likely culprit of ciguatera. *Bull. Jap. Soc. sci. Fish.* **43**, 1021–1026.

YASUMOTO T., OSHIMA Y. & KONTA T. (1981) Analysis of paralytic shellfish toxins of xanthid crabs in Okinawa. *Bull. Jap. Soc. sci. Fish.* **47**, 957–959.

YASUMOTO T., OSHIMA Y., MURAKAMI Y., NAKAJIMA I., BAGNIS R. & FUKUYO Y. (1980b) Toxicity of benthic dinoflagellates found in coral reef. *Bull. Jap. Soc. sci. Fish.* **46**, 327–331.

YASUMOTO T., OSHIMA Y., SUGAWARA W., FUKUYO Y., OGURI H., IGARASHI T. & FUJITA N. (1980c) Identification of *Dinophysis fortii* as the causative organism of diarrhetic shellfish poisoning. *Bull. Jap. Soc. sci. Fish.* **46**, 1405–1411.

YASUMOTO T., OSHIMA Y. & YAMAGUCHI M. (1978) Occurrence of a new type of shellfish poisoning in the Tohoku district. *Bull. Jap. Soc. sci. Fish.* **44**, 1249–1255.

YASUMOTO T. & SCHEUER P. J. (1969) Marine toxins of the Pacific VIII. Ciguatoxin from Moray eel livers. *Toxicon* **7**, 273–276.

YENTSCH C. M., DALE B. & HURST J. W. (1978) Coexistence of toxic and nontoxic dinoflagellates resembling *Gonyaulax tamarensis* in New England coastal water (NW Atlantic). *J. Phycol.* **14**, 330–332.

YENTSCH C. M. & MAGUE F. C. (1979) Motile cells and cysts: two probable mechanisms of intoxication of shellfish in New England waters. In *Toxic Dinoflagellate Blooms* (ed. D. L. Taylor & H. H. Seliger), pp. 127–130. Elsevier/North Holland, New York.

CHAPTER 9
DINOFLAGELLATE STEROLS

NANCY WITHERS
Department of Basic and Clinical Immunology and Microbiology, Medical University of South Carolina, Charleston, South Carolina 29425, USA

1 INTRODUCTION

Like other eukaryotes and unlike some prokaryotes, dinoflagellates contain sterols which presumably function as structural components of their membranes (Vanderheuvel 1965; Bloch 1976). However, some ultrastructural and biochemical peculiarities of this group reinforce the concept that the dinoflagellates occupy a position close to the base of eukaryotic radiation and are also a highly specialized group (Loeblich 1976; Taylor 1980). This specialization is evident biochemically in two terpenoid classes of dinoflagellate components: both in the C_{40} carotenoids and in the C_{30} sterols (dinosterol and others) (Shimizu *et al.*

1976; Liaaen-Jensen 1978; Withers *et al.* 1979c; Swenson *et al.* 1980). The dominent dinoflagellate carotenoid, peridinin, has a unique C-37 (i.e. nor-carotenoid) skeleton with an allenic lactone and its biosynthesis remains a mystery (Strain *et al.* 1971). Dinosterol has an unprecedented 4α-monomethyl saturated ring system as well as a side chain containing an extra C-23 methyl group on a Δ22 double bond in addition to the usual C-24 methyl group (Shimizu *et al.* 1976).

The persistence of the 4α-methyl group in the major sterols, dinosterol, amphisterol, peridinosterol and indeed in many minor sterols, particularly may be considered a primitive feature of dinoflagellates from a biosynthetic viewpoint. It is interesting to note that Bloch (1976) had previously postulated that there must have existed a group of primitive organisms with sterols which retained the methyl(s) at the 4-position as end-products of sterol biosynthesis, but he presumed that this group had become extinct. In most dinoflagellate species, 4α-methylsterols predominate over 4-demethyls (Shimizu *et al.* 1976; Withers *et al.* 1979a; Kokke *et al.* 1980). Further, 4-methyl- and 4,4-dimethylsterols, but no 'regular' 4-demethylsterols occur in the bacterium *Methylococcus capsulatus*; this species is one of the several bacteria known to synthesize sterols (Bird *et al.* 1971; Bouvier *et al.* 1976; Bloch 1976). The dinoflagellate sterols (particularly of the genus *Amphidinium*) are strikingly similar to these procaryotic sterols in two respects: the 4α-monomethyl group and the 8,14 double bond. Both structural features of the *Amphidinium* sp. and *Methylococcus* major sterols are presumably the result of a 'block' in 'normal' sterol biosynthesis (Bird *et al.* 1971; Bouvier *et al.* 1976; Withers *et al.* 1979a; Kokke *et al.* 1980).

Biochemical studies of dinoflagellates have been made possible only recently by mass culturing methods which eliminated previous dependency upon naturally-occurring 'red-tide' blooms, such as that of *Gonyaulax polyedra* (Patton *et al.* 1966), or the availability of invertebrates with zooxanthellae (Steudler *et al.* 1977). Second, the great technical advances in analytical equipment and methods including high performance liquid chromatography (HPLC), high resolution mass spectroscopy, computer based GC-MS analytical systems for structure analysis (Carlson 1977) and high field nuclear magnetic resonance spectroscopy (NMR) have enabled separation, quantitative recovery and structure elucidation of complex mixtures of sterols. In particular, for separating sterol homologues, the use of HPLC with a microparticle C_{18} reversed phase system as a supplement to $AgNO_3$-impregnated silica gel chromatography has been invaluable (Rees *et al.* 1976; Kokke *et al.* 1981). Reversed phase HPLC has been especially useful for several recent analyses of dinoflagellates which, like many marine invertebrates, contain a large number of trace sterols (Withers *et al.* 1982). Structural elucidation of the many new dinoflagellate sterols (about thirty-five since 1976) has been facilitated by the synthesis of many reference sterols, including several whose isolation from natural sources had been computer predicted (Varkony *et al.* 1978). The structural work has been further aided by the availability of refined NMR data, particularly the compilation of Zürcher values for the angular methyls C-18 and C-19 (Zürcher 1963; Cohen & Rock 1964) and measurements (NMR decoupling experiments) and of X-ray diffraction analysis for the determination of absolute configuration of crystals of novel sterols, e.g. dinosterol (Finer *et al.* 1977, 1978) and peridinosterol (Swenson *et al.* 1980).

Although the lipid chemists in early studies on dinoflagellates characterized the hydrocarbons, phospholipids, tri-(mono and di-) glycerides, and wax esters, they overlooked the dinoflagellate sterols, which are mainly saturated and therefore cannot be visualized with typical spray reagents (such as Lieberman–Burchard reagent) for Δ^5 sterols like cholesterol (Tonks 1967). The hydrocarbons of photosynthetic dinoflagellate species are unusually rich in 21:6 (heinacosahexaene) (Lee & Loeblich 1971). This compound appears to be bio-derived from the polyunsaturated fatty acid, 22:6 (*n*-3) which accumulates in especially high concentrations in the heterotrophic species *Crypthecodinium cohnii* (Beach & Holz 1973). Other long-chain polyunsaturated fatty acids are also common; these include fatty acids 20:5 and 18:5. The latter fatty acid is unique to dinoflagellates (Joseph 1975). Wax esters comprised of a series of long-chain polyunsaturated fatty acids esterified to phytol (= phytyl esters) have been found together with unspecified sterol esters in the extraplastidic lipid globules of two species (*Kryptoperidinium foliaceum* and *Peridinium balticum*) (Withers & Nevenzel 1977).

Bergmann (1962), in his extensive early survey of marine sterols, did not analyse free-living dinoflagellates but he did investigate symbiotic forms (zooxanthellae) in various host invertebrates in which he found gorgosterol (**2k**) (Bergmann *et al.* 1973). However, it was not clear if gorgosterol was synthesized by the invertebrates or the dinoflagellates or was derived directly from the diet. Except for a single report of cholesterol and 24-methylenecholesterol in the culture medium of *Pyrocystis lunula* (Ando & Barbier 1975), and a steroid requirement for the nutrition of the phagotroph, *Oxyrrhis marina* (Droop & Pennock 1971) the sterols of cultured free-living dinoflagellates were not studied until the exciting structural elucidation of dinosterol from the toxic North Atlantic-blooming species *Protogonyaulax (Gonyaulax) tamarensis* by Shimizu *et al.* (1976). Since that time, this has been an area of intensive research investigations, and more than thirty-five new sterol structures have been isolated from dinoflagellates. This quantity is phenomenal in consideration of the fact that only ninety-seven (total) marine sterol structures were known at the beginning of 1977 (Carlson 1977). Biosynthetic studies, focusing on side-chain alkylation schemes, have also been completed in recent years (Withers *et al.* 1979b; W. C. M. C. Kokke *et al.* unpubl.). This chapter summarizes information of dinoflagellate sterols with emphasis on the newly reported and unusual sterol structures.

2 UNIQUE STRUCTURAL FEATURES OF DINOFLAGELLATE STEROLS

A number of unusual sterols have been isolated from dinoflagellates. Some of these compounds had been previously reported and structurally characterized from other marine sources (invertebrates or sediments), and only recently has it become apparent that the dinoflagellates are the biosynthetic origin of these novel sterols.

There are several unique structural features in both the ring system and the side chain of dinoflagellate sterols which account for the plethora of new structures found in this group (Fig. 9.1). First, in the ring system, unusual characteristics include (i) the presence (or retention of) the 4α-methyl group, (ii) a saturated ring system (vs. the usual Δ^5 in ring B), or (iii) the presence of a double bond at 8,14 (ring C) or even at 14,15 (ring D). Dinoflagellate sterol side chains are heavily alkylated with two or even three methyls added to the side chain (one methyl or ethyl at C_{24} is typical for plant sterols). The alkylation at C_{23} is found almost exclusively in dinoflagellate sterols (Finer *et al.* 1978). One exception is acanthasterol isolated from an echinoderm (Gupta & Scheuer 1968; Sheikh *et al.* 1971). This sterol is apparently a Δ^7 metabolite of gorgosterol obtained from a zooxanthellae-rich coral diet (Steudler *et al.* 1977). Other unusual side chain features in dinoflagellate sterols include the presence of the cyclopropyl ring bridging carbons 22 and 23 in the gorgosterol series (Finer *et al.* 1978), and in peridinosterol a double bond at 17,20, the point of attachment of the side chain to the ring (Swenson *et al.* 1980).

In other algae, 4α-monomethyl sterols with 8(9) unsaturation have been identified as minor constituents in *Porphyridium cruentum* (unicellular red alga) (Beastall *et al.* 1974; Minale & Sodano 1976) and from the freshwater *Euglena*

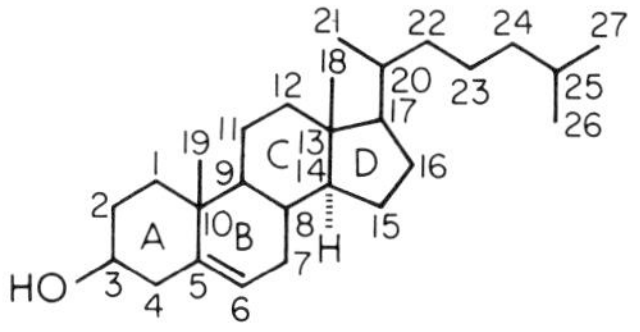

(2a)

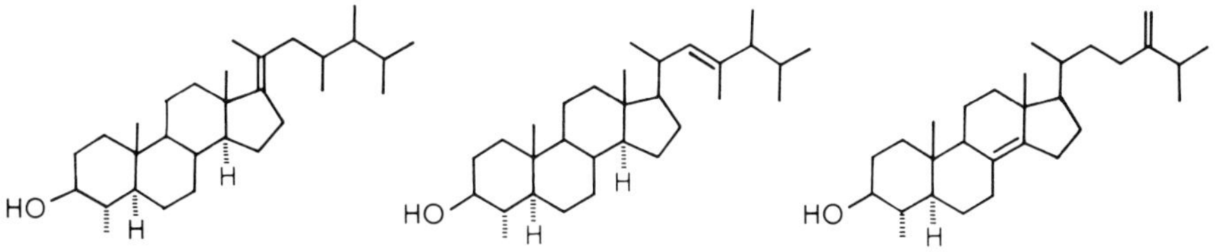

(1n) (7i) (10f)

Fig. 9.1. Cholesterol (**2a**) and three dinoflagellate sterols: peridinosterol (**1n**), dinosterol (**7i**), and amphisterol (**10f**).

gracilis (Anding *et al.* 1971). They are rare in terrestrial organisms, but have been reported from the skin of citrus fruits (Mazur *et al.* 1958). Saturated ring systems of sterols are common in immature sediments, where the absence of unsaturation is normally attributed to microbial and/or geochemical reductive processes (Ikan *et al.* 1975). Unsaturation at 8,14 has been reported in sterols of methanobacteria and hopanoids of bacteria (Bird *et al.* 1971; Bouvier *et al.* 1976; Ourisson *et al.* 1979). Because of the high degree of alkylation in the side chain and the retention of the methyl group at C_4, many dinoflagellate sterols have 30 or even 31 carbons (as opposed to 27 in cholesterol which has only 19 carbons in the ring system and 8 carbons in the side chain (Fig. 9.1). These high molecular weight dinoflagellate sterols are readily apparent during analysis: the GC traces show sterols with long retention times (RR_t of dinosterol = 1·5, 4α methyl gorgostanol = 2·67 vs. standard cholesterol = 1·0); the 4-methylsterols separate easily from regular sterols upon chromatography, and the presence of the 4-methyl group is apparent in proton NMR examination of crude sterol fractions (Kokke *et al.* 1980). A list of unique dinoflagellate sterols (C_{31} to C_{26}) reported recently from dinoflagellates (mostly cultured) are described below: 4α-methylsterols, 4α-demethylsterols, norsterols, and steroidal ketones, in decreasing order of carbon number. The structure (Fig. 9.2), molecular weight (mol. wt) and relative retention time (RR_t) from gas chromatographic analysis (using a 3% SP 2250 column at 260°C with the standard cholesterol RR_t of 1·0 unless otherwise indicated) are given for each sterol.

3 4α-METHYLSTEROLS

3.1 C_{31} 4α-methylsterols

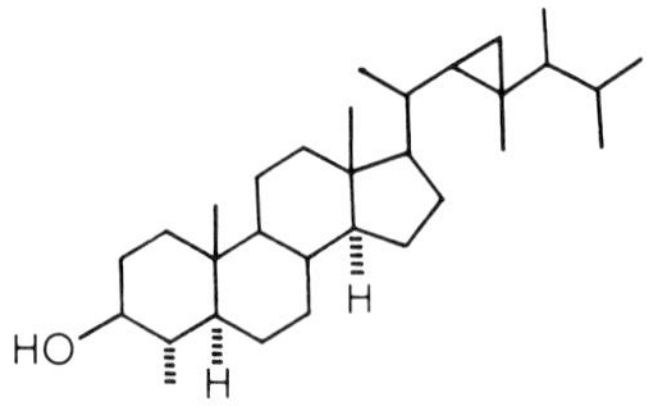

7k

1 22,23-Methylene-4α, 28, 24-trimethyl-5α-cholestan-3β-ol (4α-methylgorgostanol) (7k): mol. wt = 442, RR_t = 2·87

4α-methylgorgostanol (**7k**), was originally detected in zooxanthellae of *Briareum asbestinum* and tentatively identified by its mass spectrum, which was similar to that of 5α-gorgostanol except for a +14 AMU shift for the C-4 carbon (Steudler *et al.* 1977). The same sterol was isolated in quantities sufficient for proton NMR analysis from the cultured dinoflagellate, *Kryptoperidinium*

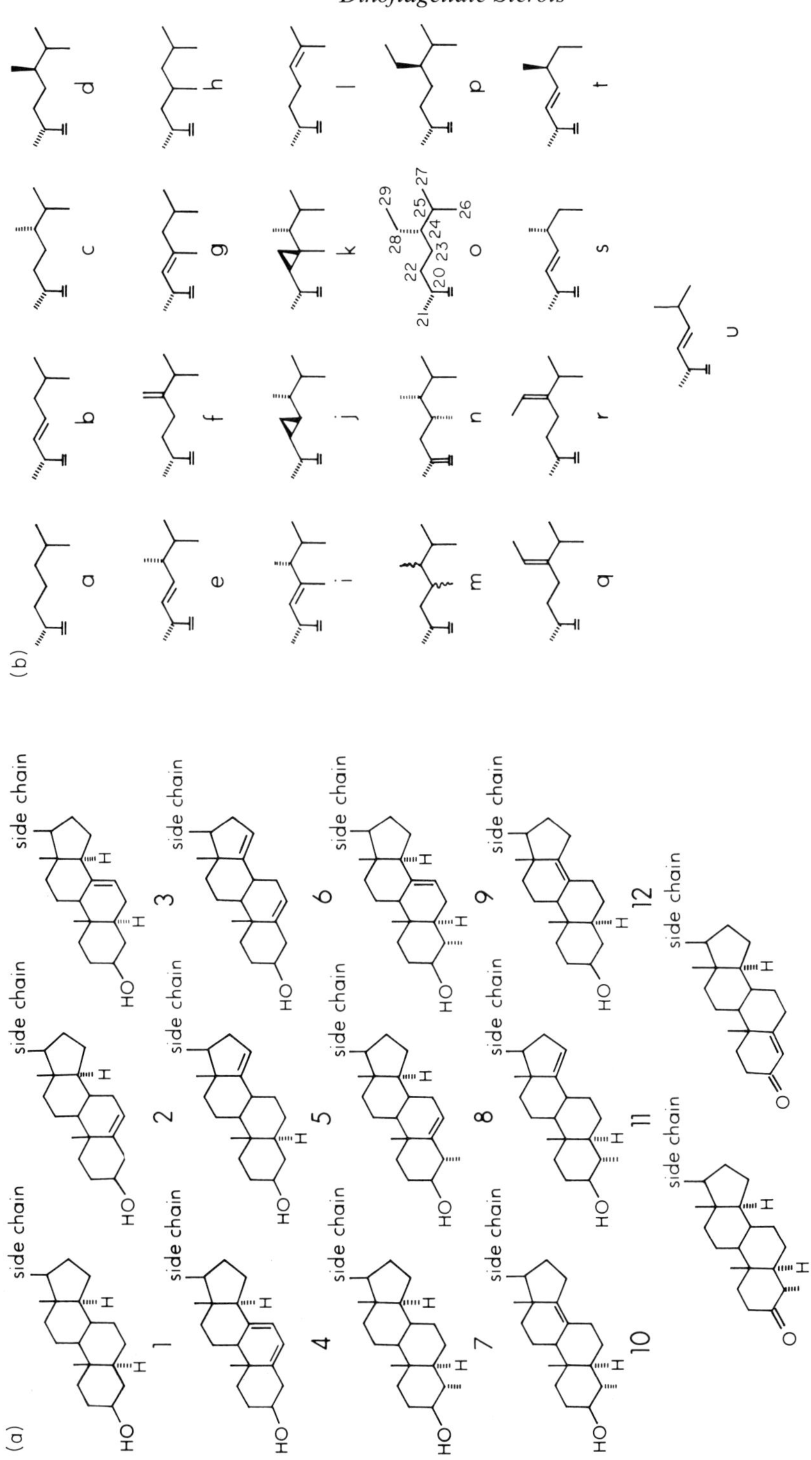

Fig. 9.2. Structures of dinoflagellate sterols: (a) rings; (b) side chains.

(= *Glenodinium*) *foliaceum* (Alam *et al.* 1979a; Withers *et al.* 1979b). The stereochemistry of the algal sterol was established as 4α-methyl-5α-gorgostanol (**7k**) by comparison with a partially synthetic reference prepared by Oppenauer oxidation of gorgosterol (**2k**). This yielded Δ^4-gorgostenone which was methylated to 4α-methyl-Δ^4-gorgostenone and reduced to **7k**, which gave the same 360 MHz NMR spectrum as the *K. foliaceum* compound, as well as the *B. asbestinum* sterol which was also isolated (Withers *et al.* 1979b; Kokke *et al.* 1982).

The isolation of **7k** from *K. foliaceum* was the first demonstration of a sterol with a cyclopropane-containing side chain from a free-living alga (Alam *et al.* 1979a). It was subsequently isolated from the dinoflagellates *Peridinium balticum* (Withers *et al.* 1979b), *Ptychodiscus brevis* (M. Alam *et al.* unpubl.) and *Gessnerium* (*Gonyaulax*) *monilatum* (Wengrovitz *et al.* 1981). It also has been detected in the foraminifer, *Marginopora vertebralis* which contains zooxanthellae (W. C. M. C. Kokke & G. Wefer, unpubl.), in the mollusc *Cyphoma gibbosa*, which is a predator of *B. asbestinum*, in both *C. gibbosa* and *Xenia elongata* faeces (Steuder *et al.* 1977) and in marine sediments (Brassell *et al.* 1980). The isolation of 4α-methylgorgostanol (**7k**), dinosterol (**7i**), gorgostanol (**ik**), and gorgosterol (**2k**) from a dinoflagellate supports the proposed mechanism for the formation of gorgosterol (**2k**) from dinosterol (**7i**) (Withers *et al.* 1979b).

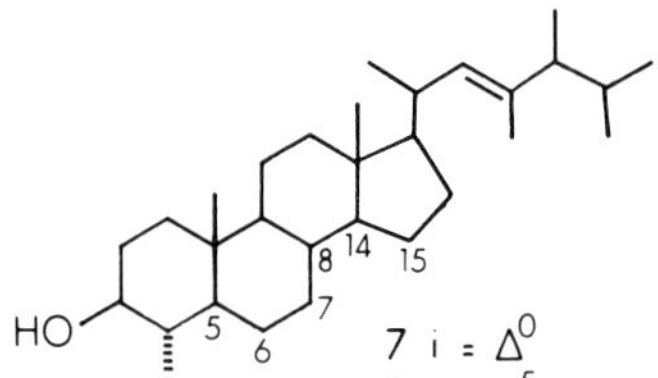

7 i = Δ^0
8 i = Δ^5
9 i = Δ^7
10 i = $\Delta^{8(14)}$
11 i = Δ^{14}

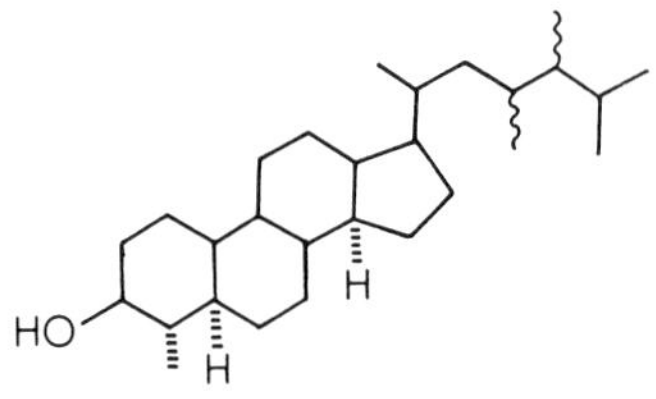

7m

3.2 C-30 4α-methylsterols

1 4α-23,24(R)-Trimethyl-5α-cholest-22-en-3β-ol (Dinosterol) (7i): mol. wt = 428, RR_t = 1·54

Dinosterol (**7i**) was isolated from the toxic dinoflagellate, *P. tamarensis* by Shimizu *et al.* (1976). The mass spectrum indicated an extra nuclear methyl group at C_4 and a C_{22} double bond in the side chain. Jones oxidation gave a ketone with a CD curve and mass spectrum identical to those of a model compound, 4α-methyl-5α-stigmasta-22-en-3-one. Both ketones gave the same diol upon ozonolysis followed by $NaBH_4$ reduction. The complete structure,

(22*E*, 24*R*)-4α,23,24-trimethyl-5α-cholesta-22-en-3*β*-ol (**7i**) was determined by X-ray analysis (Finer *et al.* 1978). This novel side chain (**i**), which had only been found previously in a 4-demethylsterol (**2i**) from the soft coral, *Sarcophyton elegans* (Kanazawa *et al.* 1974) is a postulated biosynthetic precursor for the gorgosterol-type side chain (**k**) (Shimizu *et al.* 1976; Kobayashi *et al.* 1979a, b). Dinosterol (**8i**) is a major sterol of many of the dinoflagellate species examined to date (approximately eleven of eighteen species) (Table 9.1, p. 341). A proposed scheme for the biosynthetic alkylation of the dinosterol side chain via s-adenosylmethionine is shown in Fig. 9.4 (Withers *et al.* 1979c). A stereospecific synthesis using a Claisen ortho-ester rearrangement for dinosterol and its C-24 epimer has recently been accomplished (Shu & Djerassi 1981).

2 5-Dehydrodinosterol (8i) : mol. wt = 426, RR_t = 1·61
4α, 23,24(R)-Trimethyl-cholesta-5,22-dien-3β-ol

Dehydrodinosterol (**8i**) was isolated as a second major sterol along with dinosterol (**7i**) from the heterotrophic dinoflagellate, *Crypthecodinium cohnii* (Withers *et al.* 1978). The mass spectrum of the acetate indicated a structure similar to dinosterol (**7i**) but with a double bond in the ring system. The TMS ether of the sterol showed fragmentation at m/e 369 (M^+-129, 41%) characteristic of a Δ^5 sterol (Knapp & Schroepfer 1976). Oxidation of the free sterol with Collin's reagent yielded a ketone which rearranged after the addition of HCl to the 4-ene-3-one. The ^1H NMR spectrum of dehydrodinosterol acetate revealed an olefinic proton which, together with characteristic shifts for the C-18 and C-19 methyl protons, located the double bond at the C_5, C_6 position (Withers *et al.* 1978). This sterol was subsequently detected as a trace sterol in several other species including the zooxanthellae isolated from *B. asbestinum* (7% free 4α-methylsterols) (Kokke *et al.* 1980; Bohlin *et al.* 1981).

3 8(14)-Dehydrodinosterol (10i) : mol. wt = 426, RR_t = 1·56
4α, 23,24-(R)-Trimethyl-5α-cholesta-8(14)-22-dien-3β-ol

This sterol was isolated from the free sterols of two *Amphidinium species: A. carterae* (12%) and *A. corpulentum* (5%) (Kokke *et al.* 1980) and from a gorgonian (Kokke *et al.* 1982). Structure **10i** was tentatively assigned and was based on its ^1H NMR and mass spectral features. The dinosterol side chain (**i**) was evident from spin-decoupling experiments and the mass spectral fragmentation pattern (Alam *et al.* 1980). The 8(14) double bond was confirmed by the ^1H NMR spectrum which showed no olefines in the ring and characteristic shifts of the angular methyls (C_{18} and C_{19}) (Zürcher 1963; Cohen & Rock 1964).

4 14-Dehydrodinosterol (11i) : mol. wt = 426, RR_t = 1·61

The trace sterol, 14-dehydrodinosterol (**11i**) was isolated from the free sterols (0·7%) and steryl esters (6%) of *Glenodinium* sp. (Kokke *et al.* 1980). The structure **11i** was assigned from the MS and proton NMR spectrum (clearest in deuterobenzene) and from its chromatographic behaviour on $AgNO_3$–Si gel where it gave a low R_f value like the sterically unhindered methylene sterols (Kokke *et al.* 1980).

5 7-Dehydrodinosterol (9i): mol. wt = 426, RR_t = 1·86

7-Dehydrodinosterol, 4α, 23,24(R) Trimethyl-5α-cholesta-7, 22*E*-dien-3*β*-ol (**9i**) was not isolated from a cultured dinoflagellate like most of the sterols reported here. Instead, this novel sterol was found in the 4α-methylsterol fraction (and is thus assumed to be of dinoflagellate origin) of the Caribbean gorgonians, *Muriceopsis flavida* (11·1%) and *Pseudoplexaura wagenaari* (1·2%). The structural assignment was mainly based on ^{1}H NMR, particularly of the C_{18} methyl protons which were at a very high field (about δ 0·57), and the mass spectral fragmentation (m/z 260) characteristic of 4α-methyl Δ^7 sterols (Kokke *et al.* 1982).

6 Dinostanol (7m): mol. wt = 430, RR_t = 1·81

The completely saturated dinostanol (**7m**) (of unknown configuration at C-23) was isolated from the dinoflagellates *Gonyaulax diegensis* (Alam *et al.* 1978), *Peridinium lomnickii* (Robinson *et al.* 1984), *Protoceratium reticulatum* (M. Eggersdorfer, unpubl.) and by L. Bohlin *et al.* (unpubl.) from zooxanthellae of *Velella velella*. The structure (**7m**) was later determined by comparison of NMR spectra of these sterols with a partially synthetic dinostanol (produced by hydrogenation of dinosterol) (Zielinski *et al.* 1983).

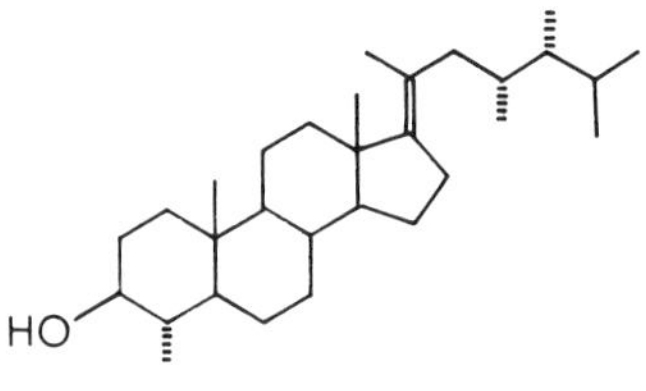

7n

7 Peridinosterol (7n): mol. wt = 428, RR_t = 1·77

Peridinosterol, (**7n**), 4α, 23*R*, 24*R*-trimethylcholest-17(20)-en-3*β*-ol is a rare $\Delta^{17,20}$ sterol isolated from the two binucleate species *Kryptoperidinium foliaceum* (6% free sterols) (Withers *et al.* 1979b; Swenson *et al.* 1980) and *Prorocentrum* (*Exuviaella*) *mariae-lebouriae* (W. C. M. C. Kokke, unpubl.) and *C. cohnii* (W. C. M. C. Kokke, unpubl.). Preliminary mass spectral and NMR analytical studies of this C_{30} sterol defined the typical saturated ring and 4-methyl group and indicated the $\Delta^{17,20}$ bond (Withers *et al.* 1979b). X-ray analysis of peridinosterol—*p*-bromobenzoate confirmed an assignment of the methyl groups at C_{23} (*R*) and C_{24} (*R*) (Swenson *et al.* 1980).

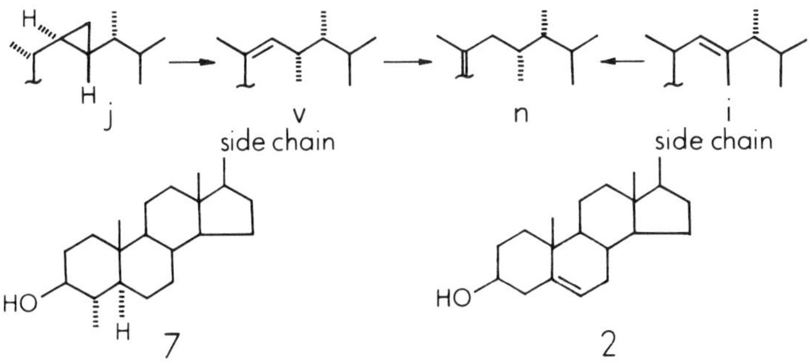

It is interesting to note that another $\Delta^{17,20}$ sterol, sarcosterol (**2n**), (stereochemistry unknown) was detected in the soft coral, *Sarcophyton glaucum* (Kobayashi *et al.* 1979a, b) (**2k**) and 23,24-dimethylcholesta-5,22-dien-3β-ol (**2i**). This side-chain pattern is coincidentally paralleled by the 4α-methyl analogs: peridinosterol, 4α-methyl gorgostanol (**7k**), and dinosterol (**7i**) in *P. balticum* and *K. foliaceum* (Swenson *et al.* 1980; Withers *et al.* 1979b). This suggests a biosynthetic relationship between side chains, **n**, **i** and **k**. Two speculative pathways for the formation of the 17,20 double bond (**n**) are (**i**) double bond migration from a Δ^{22} precursor (e.g. dinosterol) (**i**) or (**ii**) isomerization of a 23-demethylgorgosterol precursor (**j**) to a $\Delta^{20(22)\ 23,24}$ dimethyl intermediate (**v**) followed by double bond migration to the Δ^{17} isomer (**n**) (Swenson *et al.* 1980).

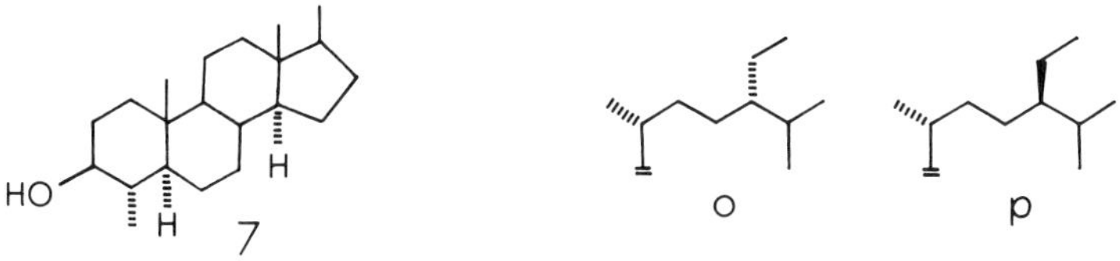

8 4α-Methyl-24R-ethyl-5β-cholestan-3β-ol (7p): mol. wt = 430, RR_t *= 1·83*
This saturated sterol was found in the steryl esters (4%) of the dinoflagellate *Glenodinium* sp. The ethyl group was initially assigned to the C_{24} (and not C_{23}) carbon only on biosynthetic grounds (Kokke *et al.* 1980). Comparison with a partially synthetic reference **7p**, confirmed the C_{24} assignment. The configuration was later found to be *R* from NMR comparisons, which showed an almost perfect correlation between the *Glenodinium* sterol and the reference sterol. Surprisingly, the C_{24} *S* epimer of this sterol, **7o**, was isolated from another dinoflagellate—the zooxanthellae of *B. asbestinum* (No. 9, see below) (Bohlin *et al.* 1981).

9 4α-Methyl-24S-ethyl-5α-cholestan-3β-ol (7o): mol. wt = 430, RR_t *= 1·83*
This new sterol was recently isolated from the cultured zooxanthellae of the host gorgonians *B. asbestinum* and *M. flavida* (Bohlin *et al.* 1981). The configuration

at C_{24} was found to be *S*, and thus structure **7o** was assigned since the NMR spectrum, particularly the shifts of the C_{29} protons, differed from that of **7p**, which had been isolated from the dinoflagellate *Glenodinium* sp. (No. 8) (Kokke *et al.* 1980; Kokke *et al.* 1982).

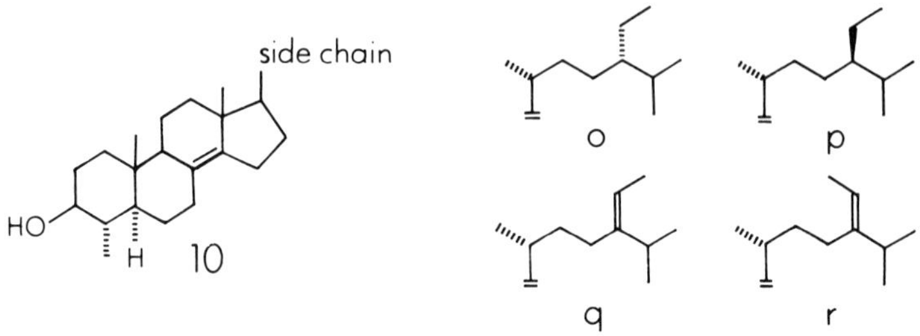

10 4α-Methyl-24R-ethyl-5α-cholest-8(14)-en-3β-ol (10p):
mol. wt = 428, RR_t = 1·83

This novel sterol occurs as 15% of the steryl esters (but not in the free sterols) of *Glenodinium* sp. and was identified by ^{1}H NMR (Kokke *et al.* 1982). The stereochemistry at C_{24} was resolved as 24-*S*, since the NMR spectrum had the same pattern of chemical shifts as that of **8o** (No. 9) (Bohlin *et al.* 1981).

11 4α-Methyl-24S-ethyl-5α-cholesta-8(14)-en-3β-ol (10o):
mol. wt = 428, RR_t = 1·83

This sterol (11.1% free 4α-methylsterols) was isolated along with its saturated analog **7o**, from the cultured zooxanthellae of *B. asbestinum*. The 24-*S* configuration was determined by NMR spectra, which showed the expected deviation from that of **10p**, parallel to the spectral differences between **7o** and **7p**. This assignment was supported by the isolation of clionasterol (**2o**) (along with **7o** and **10o**) from the same dinoflagellate (the *B. asbestinum* zooxanthellae) (Bohlin *et al.* 1981).

12 4α-Methyl-24(Z)-ethylidene-5α-cholest-8(14)-en-3β-ol (10q):
mol. wt = 428, RR_t = 2·03

This sterol (**10q**) was isolated from *Amphidinium carterae*, where it comprised 1·3% of the free sterols. It is interesting to note that the configuration (24*Z*) of the ethylidene at C-24 is like that of the 24(*Z*) isofucosterol (**q**) side chain and not 24(*E*) fucosterol (which has side chain **r**) (Kokke *et al.* 1980).

3.3 C_{29} 4α-methylsterols

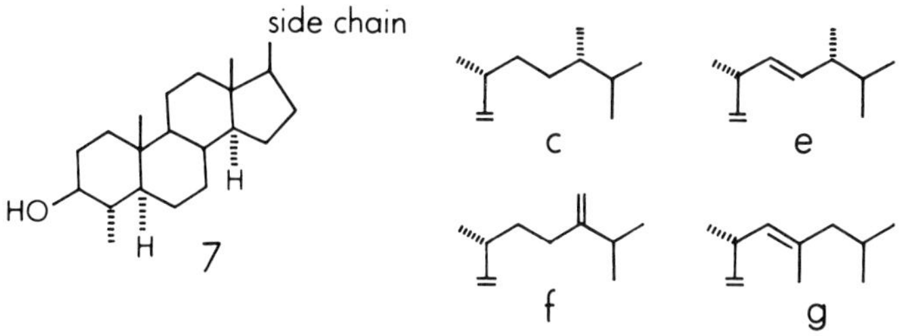

1 4α, 24S-Dimethyl-5α-cholestan-3β-ol (7c) : mol. wt = 416, RR_t = 1·49

This common saturated dinoflagellate sterol (**7c**) was isolated as the major 4α-methylsterol (95%) from the dinoflagellate, *Gymnodinium simplex* (Goad & Withers 1982), and from numerous zooxanthellae of hosts such as *Briareum asbestinum* (Bohlin *et al.* 1981), *Tridacna gigas*, *Melibe pilosa*, *Aiptasia pulchella* and many others (Kokke *et al.* 1981; Withers *et al.* 1982). In fact, in the genus *Gymnodinium* and in many of these zooxanthellae strains which are considered by some taxonomists to be gymnodinoid and to belong to the same genus, *Gymnodinium*, **7c** is a major sterol (Table 9.2, p. 348) (Withers *et al.* 1982). The biosynthesis of the side chain of **7c** has been studied along with that of dinosterol (**7i**) in the heterotroph, *C. cohnii* (Withers *et al.* 1979c) (Fig. 9.3).

2 4α, 24R-Dimethyl-5α-cholest-22-en-3β-ol (7e) : mol. wt = 414, RR_t = 1·27

The Δ^{22} analogue of **7c**, **7e**, occurs as a major constituent (10–37% total free sterols) in several zooxanthellae strains (Withers *et al.* 1981) in the free-living dinoflagellates *G. simplex* (Goad & Withers 1982), and *Glenodinium* sp. (4% of the steryl esters, 1% of the free sterols) (Kokke *et al.* 1980). Structure elucidation was made by mass spectral analysis (double bond in the side chain, methyl at C_{24}), and NMR which located the double bond at C_{22} (Kokke *et al.* 1981). This sterol, **7e**, was also found in *Plexaura homomalla* and several other gorgonians (Carlson 1977).

3 4α, 23-Dimethyl-5α-cholest-22-en-3β-ol (7g) (24-demethyldinosterol) : mol. wt = 414, RR_t = 1·23

4α, 23-Dimethyl-5α-cholest-22-en-3β-ol (**7g**) (24-demethyldinosterol) was isolated as one of the three major sterols (12·5%) in *Gonyaulax diegensis* (Alam *et al.* 1978). It was identified from the mass spectrum of the parent compound, and of the $NaBH_4$-treated ozonolysis product which were similar to that of dinosterol. It has been assumed that in the dinosterol type side chain (**i**), addition of the 24 methyl group precedes the introduction of Δ^{22} double bond and 23 methylation (Withers *et al.* 1979a, b; Djerassi *et al.* 1979; Kokke *et al.* 1979). The occurrence of sterol **7a** in *G. diagensis* (Alam *et al.* 1978) and the 4-demethylsterol, 23-methyl-cholesta-5,22-dien-3β-ol (**2g**), in the zooxanthellae of

Zoanthus sociatus (Kokke *et al.* 1979) suggests that either alkylation at C_{23} can precede that at C_{24}, or the less likely possibility that there may be a pathway for biomethylation of side chain **i** at C_{24} resulting in side chain **g** (Djerassi *et al.* 1979).

4 4α, 24-Methylene-5α-cholestan-3β-ol (7f) : mol. wt = 414, RR_t = 1·51

4α, 24-Methylene-5α-cholestan-3β-ol (**7f**) was isolated from the 4α-methyl fraction (= dinoflagellate-synthesized sterols) of four Caribbean gorgonians: in abundance of 1·4–1·7% in *Gorgonia mariae*, *Muriceopsis flavida* and *Pseudoplexaura wagenaari*, and 8% in *B. asbestinum*. This new sterol had a low R_f value on argentic silica gel indicating a methylene in the side chain, which was confirmed by NMR. The mass spectrum was consistent with the assignment of structure **7f** to the new sterol (Kokke *et al.* 1982).

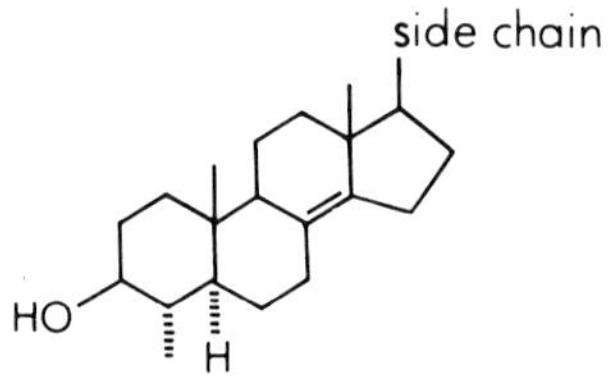

10

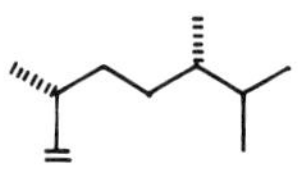

c

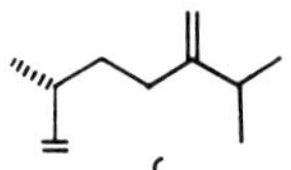

f

5 4α-Methyl-24-methylenecholest-8(14)-en-3β-ol (10f) :
mol. wt = 412, RR_t = 1·53

Amphisterol, the trivial name for 4α-methyl-24-methylene-cholesta-8(14)-en-3β-ol, **10f**, is the major sterol (43–58% free sterols) of all species of the dinoflagellate genus *Amphidinium* that have been studied for sterol content (Withers *et al.* 1979a). These include *A. carterae*, *A. corpulentum*, *A. klebsii*, *A. rhyncocephalum*, and *A. hoefleri* (Withers *et al.* 1979a; Kokke *et al.* 1980). The presence of the 24 methylene group was indicated by NMR measurements and by the low R_f value on $AgNO_3$-silica gel TlC. The 8,14 double bond was determined by NMR (Withers *et al.* 1979a). Sterols with this $\Delta^{8,14}$ unsaturation are rare; there are only a few terrestrial examples known (Bouvier *et al.* 1976). This 24-methylene group and the 8,14 double bond, together with the low sterol content (about one-seventh the yield of other genera) and the absence of dinosterol in all *Amphidinium* species suggest that, in members of this genus, 'normal' sterol synthesis is impeded (Withers *et al.* 1979a; Kokke *et al.* 1980). Amphisterol, **10f**, is probably an intermediate in the biosynthesis of dinosterol since the 24-methylene is apparently a precursor in dinosterol side chain formation (Withers *et al.* 1979c) and $\Delta^{8,14}$ unsaturated sterols are postulated intermediates for demethylation at C_{14} (Schroepfer *et al.* 1972) (Fig. 9.4).

6 *4α, 24R-Dimethyl-5α-cholesta-8(14),22-dien-3β-ol (10e): mol. wt = 412, RR_t = 1·28*

This new $\Delta^{8,14}$ unsaturated sterol, (**10e**), comprised 0·6% of the 4-methyl-sterols isolated by Kokke *et al.* (1982) from the Caribbean gorgonian, *Muriceopsis flavida*, which contains zooxanthellae. Structure **10** with an **e** side chain was assigned on the basis of MS and NMR spectral data, although the configuration at C_{24} (*R*) was only an assumption based on the consistent configuration (24*β*) of all known dinoflagellate sterols with a methyl substituent at C_{24}.

7 *4α, 24S-Dimethylcholest-8(14)-en-3β-ol (10c): mol. wt = 414, RR_t = 1·46*

This sterol has been detected in one species, *Glenodinium* sp., where it comprises 18% of the steryl esters and in the *B. asbestinum* zooxanthellae (2% free 4α-methylsterols) (Bohlin *et al.* 1981). The methyl group in the side chain was located at C_{24} from comparison of the NMR spectrum with that of a reference sample, hydrogenated amphisterol (**10c** and **10d**, a mixture of steroisomers at C_{24}). The 24*S* assignment for the *Glenodinium* sp. sterol was based on NMR measurements and the consistency of this stereochemistry in the other dinoflagellate sterols, dinosterol **7i** and 4,24-dimethylcholestanol (**7c**) (Kokke *et al.* 1980).

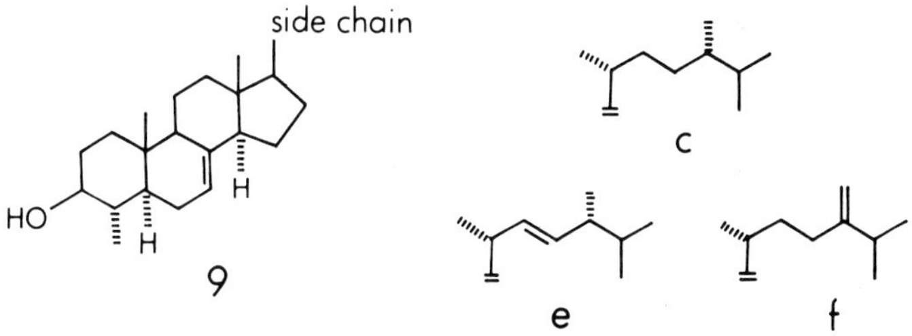

8 *4α, 24R-Dimethyl-5α-cholesta-7,22-dien-3β-ol (9e): mol. wt = 412, RR_t = 1·57*

A new Δ^7 unsaturated sterol, **9e**, was isolated from three zooxanthellae-rich Caribbean gorgonians, *B. asbestinum*, *G. mariae* and *M. flavida*, where it comprised 1–5% of the 4α-methylsterol fraction. Structure **9e** was determined by Kokke *et al.* (1982) on the basis of MS and NMR data. Hydrogenation of the Δ^{22} double bond yielded a known sterol, 4α, 24*S*-dimethyl-5α-cholest-8(14)-en-3*β*-ol (**10c**) due to migration of the Δ^7 bond, which confirmed the absolute configuration at C_{24}.

9 *4α, 24S-Dimethyl-5α-cholest-7en-3β-ol (9c) (or 9h): mol. wt = 414, RR_t = 1·77*

Another new Δ^7 unsaturated sterol, **9c**, was detected as a trace component (less than 1%) in the 4α-methylsterols of the gorgonians *B. asbestinum* and *Pseudoplexaura wagenaari*, and 4% of the same fraction from *M. flavida*.

Structure elucidation was made by both NMR and MS data although a C_{23} assignment of the methyl group could not be eliminated (but is less likely than 24_S) (Kokke *et al.* 1982).

10 4α-Methyl-24-methylene-5α-cholest-7-en-3β-ol (9f): mol. wt = 412, RR_t = 1·89

Although 24-methylene lophenol (**9f**) had been previously isolated from terrestrial sources, it was only recently found in a marine organism. This zooxanthellae-derived sterol (**4f**) comprised 1% of the 4-methyl sterols of the Caribbean gorgonian, *Briareum asbestinum* (Kokke *et al.* 1982). Like the previous Δ^7 unsaturated 4α-methyl sterols (**9c**, **9e**), the mass spectrum of **9f** from *B. asbestinum* revealed a diagnostic peak at m/z 260 (Partridge *et al.* 1977) and the ^{1}H NMR spectrum showed a very high C_{18} methyl signal (δ 0·536).

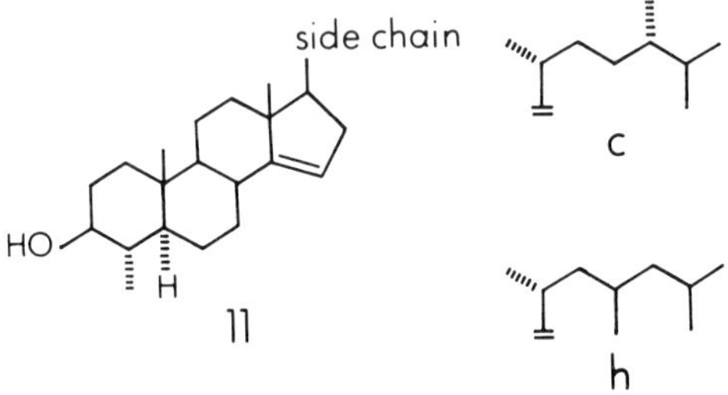

11

11 4α, 24S (or 23S)-Dimethyl-5α-cholest-14-en-3β-ol (11c) (or 11h): mol. wt = 414, RR_t = 1·48

This sterol was isolated along with 14 dehydrodinosterol (**11i**) as a minor sterol (0·3%) from the free sterols of *Glenodinium* sp. (Kokke *et al.* 1980). The mass spectrum was identical to that of 4α-methyl-5α-cholest-14-en-3β-ol (**11a**) except for the molecular ion; both showed an intense fragment at m/z 287 from loss of side chain, which is apparently enhanced in Δ^{14} unsaturated sterols (Djerassi 1970). The NMR shifts for the angular methyls (C_{18} and C_{19}) of the two *Glenodinium* sterols were also in good agreement with (**11**) (Zürcher 1963; Cohen & Rock 1964). The location of the methyl in the side chain is uncertain, but it is most likely at 24_S, so **11c** is the probable assignment of this new sterol (Kokke *et al.* 1980). Sterols with Δ^{14} unsaturation are extremely rare.

3.4 C_{28} 4α-methylsterols

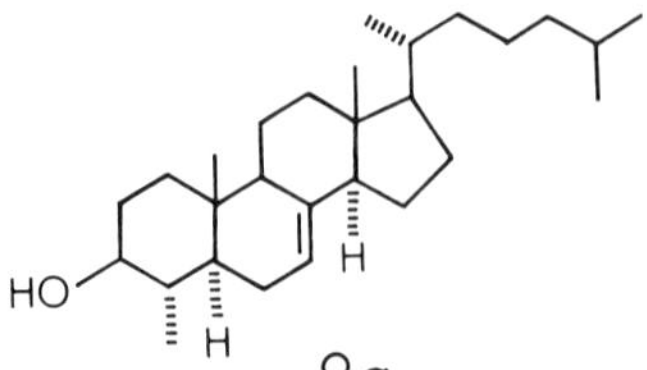

9a

1 4α-Methyl-5α-cholest-7-en-3β-ol (lophenol) (9a): mol. wt = 400, RR_t = 1·33
Lophenol (**9g**) is a trace constituent (0·9%) of the free sterols from *Amphidinium carterae* (Kokke *et al.* 1980). This sterol, originally isolated from cactus (Djerassi *et al.* 1958) is a major biosynthetic intermediate (prior to demethylation at C_4) in the mammalian route for cholesterol biosynthesis (as found in rats) (Nes & McKean 1977).

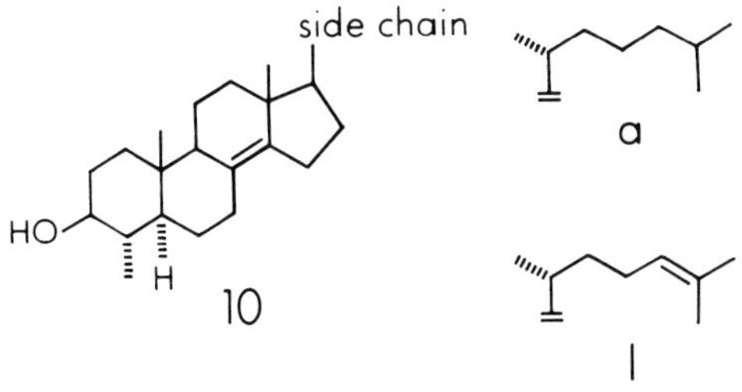

2 4α-Methyl-5α-cholesta-8(14),24-dien-3β-ol (10l): mol. wt = 398, RR_t = 1·46
This new sterol was found only in the sterol ester fraction (0·5%) of the zooxanthellae from the Hawaiian anemone *Aiptasia pulchella*. The mass spectrum showed a strong molecular ion, typical of 8(14) sterols (Withers *et al.* 1979a) and the NMR spectrum supported the presence of the 8(14) double bond and an ℓ (desmosterol) side chain having two olefinic methyl groups and only one methyl doublet (Withers *et al.* 1982).

3 4α-Methyl-cholest-8,(14)-en-3β-ol (10a): mol. wt = 400, RR_t = 1·14
4α-methyl-5α-cholesta-8(14)-en-3β-ol (**10a**) was isolated as a relatively abundant component (20–27%) of the free sterols from *A. carterae* and *A. corpulentum* (Kokke *et al.* 1980) and as a trace sterol from several cultured zooxanthellae (Withers *et al.* 1982). It was identified from NMR and mass spectral information. This sterol was first reported as one of the major sterols of the bacterium *Methylococcus capsulatus* (Bird *et al.* 1971; Bouvier *et al.* 1976).

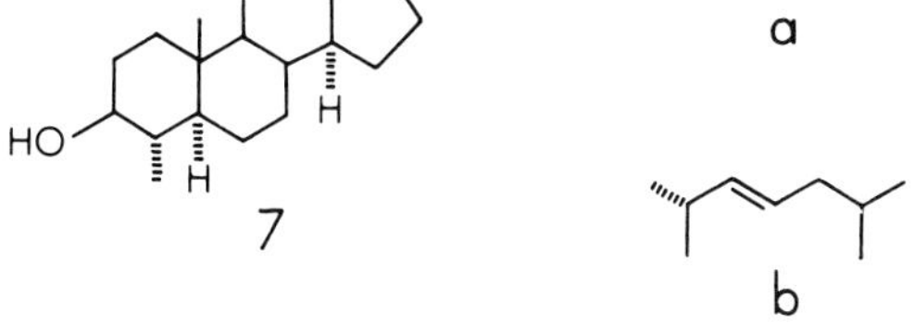

4 4α-Methyldehydrocholesta-22E-en-3β-ol (7b): mol. wt = 400, RR_t = 1·06
This C_{28} sterol was isolated with **7a** as another trace sterol (1% free sterols) from the *Z. sociatus* zooxanthellae (Withers *et al.* 1982; Kokke *et al.* 1979). It

was identified by mass spectral fragmentation and 360 MHz NMR spectra (especially of the angular methyls which were consistent with Zürcher values for the normal androstane skeleton) (Zürcher 1963). The double bond was found to be 22*E* by comparison of NMR chemical shifts of olefinic protons with 22-dehydrocholesterol (**2b**) which occurs in the same alga (Kokke *et al.* 1979).

5 4α-Methyl-5α-cholestanol (7a) : mol. wt = 402, RR_t = 1·12
4α-Methyl cholestanol (**7a**) occurs as a minor constituent (0·2–1·6% of the free sterols; 0·3–3·4% of the steryl esters) in all but one zooxanthellae (from the six hosts, *T. gigas*, *O. diffusa*, *M. pilosa*, *Z. sociatus* and *B. asbestinum*) which were recently mass-cultured and examined for sterol content (Withers *et al.* 1982). It was identified by GC-MS and NMR data (Kokke *et al.* 1979). The same sterol was also found (as 0·6% of the free sterols) in the 'free-living' species, *Glenodinium* sp. (Kokke *et al.* 1980).

4 4α-DEMETHYLSTEROLS

4.1 C_{30} 4α-demethylsterols

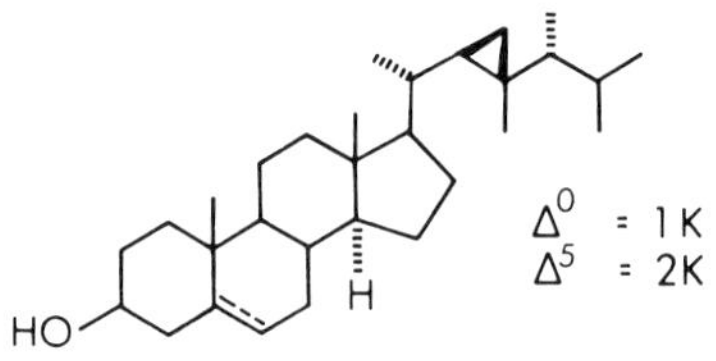

1 Gorgosterol (2k) : mol. wt = 428, RR_t = 2·36
This sterol has a radically different side chain pattern (cyclopropyl bridging C_{22}, C_{23} and methylation at C_{23} and C_{24}) and its discovery stimulated an intense search in marine organisms for novel sterols. It was originally isolated from a gorgonian *Plexaura flexuosa* in 1943 by Bergmann (Bergmann 1962; Bergmann *et al.* 1973), who found that it had an unusually high melting point (180–185°C). The sterol, named gorgosterol, was reisolated 25 years later and analysed by GC-MS by Ciereszko *et al.* (1968) from several coelenterates. Gupta & Scheuer (1969) isolated the same sterol from the zoanthid *Palythoa tuberculosa*. Further structural studies by NMR spectroscopy by several groups indicated the presence of a cyclopropane ring at either $C_{20,22}$ or $C_{22,23}$ (Hale *et al.* 1970). The $C_{20,22}$ position was eliminated by X-ray analysis of 3-bromogorgostene which defined the absolute configuration of gorgosterol as (22*R*, 23*R*, 24*R*)-22, 23-methylene-23,24-dimethylcholest-5-en-3β-ol (Ling *et al.* 1970). The presence of the C_{23} methyl group and the cyclopropane ring at $C_{22,23}$ was unprecedented at that time in either plant or animal sterols. Hale *et al.* (1970) suggested that a Δ^{22}, C_{23} methylsterol (e.g. one with a dinosterol side chain, (**i**) would be a likely biosynthetic precursor to gorgosterol (**2k**).

Gorgosterol and related compounds have been detected in many invertebrates containing zooxanthellae, but only recently have they been found and identified by NMR spectroscopy in cultured dinoflagellates. The relevant species, *K. foliaceum* and *P. balticum*, where **2k** is a trace component, are both binucleate (Dodge 1971; Withers *et al.* 1979b). In addition, the zooxanthellae from the Hawaiian cnidarian *Aiptasia pulchella* were cultured and found to contain gorgosterol: 10% in free sterols and 2% in steryl esters (Withers *et al.* 1982). Therefore, it is now clear that the dinoflagellates have a comprehensive capability of synthesizing gorgosterol.

2 5α Gorgostanol (1k) : mol. wt = 430, RR_t = 2·36
The saturated analogue of gorgosterol, 5α-gorgostanol (**1k**) has been isolated and identified by NMR spectroscopy from *K. foliaceum* (0·4% free sterols), *P. balticum*, (Withers *et al.* 1979b) and *Gessnerium monilatum* (Wengrovitz *et al.* 1981).

4.2 C_{29} 4α-demethylsterols

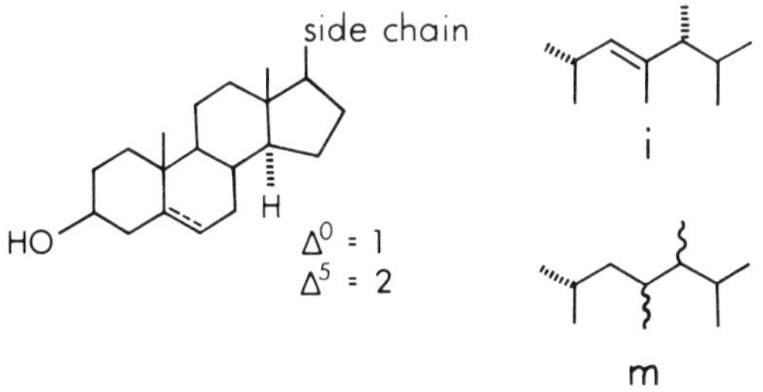

1 23,24 R-Dimethyl-5α-cholest-22E-en-3β-ol (1i) (4-demethyldinosterol) : mol. wt = 414, RR_t = 1·37
This sterol, **1i** was recently reported as a new sterol from the free sterols of the zooxanthellae from the scleractinian coral *Oculina diffusa* (Withers *et al.* 1982). It had been reported as a constituent of a jellyfish, with identification based on mass spectral evidence (Milkova *et al.* 1980). The *O. diffusa* sterol was isolated from other demethylsterols by argentic chromatography and showed a mass spectrum with characteristic side-chain fragmentation of dinosterol (Shimizu 1976; Alam *et al.* 1978). The NMR spectrum verified the dinosterol side-chain stereochemistry (22*E*, 23*R*, 24*R*). This sterol (**1i**) has also been identified, by GC and mass spectral data only, as a very minor component of the *B. asbestinum* zooxanthellae sterols (Bohlin *et al.* 1981).

2 23, 24R-Dimethyl-cholesta-5,22E-dien-3β-ol (2i) : mol. wt = 412, RR_t = 1·37
This sterol was first isolated from the soft coral *Sarcophyton glaucum* by Kanazawa *et al.* (1974) and identified by comparison of NMR shifts with those

of (22*E*)-23-methylcholesta-5,22-dien-3*β*-ol isolated as a trace sterol (0·1–1·6%) from several zooxanthellae from hosts *O. diffusa*, *T. gigas*, and *B. asbestinum* (Withers *et al.* 1982). The 360 MHz NMR data, when compared with those of dinosterol, confirmed the presence of the dinosterol side chain (**i**). This sterol or its C_{24} epimer (unspecified stereochemistry at C_{24}) has also been reported as a constituent of *Gymnodinium simplex* (14% total sterols) (Goad & Withers 1982), of several coelenterates (Kanazawa *et al.* 1977), of a coccolithophorid, *Hymenomonas carterae* (Volkman *et al.* 1981), and several other marine haptophytes (W. Kokke, N. Withers, W. Fenical & C. Djerassi, unpubl.).

3 23,24ζ-Dimethylcholest-5-en-3β-ol (2m) : mol. wt = 414, (RR_t = not reported)
In *G. simplex*, a trace sterol was tentatively identified as 23,24ζ-dimethylcholest-5-en-3*β*-ol (**2m**) on the basis of the mass spectrum (Goad & Withers 1982).

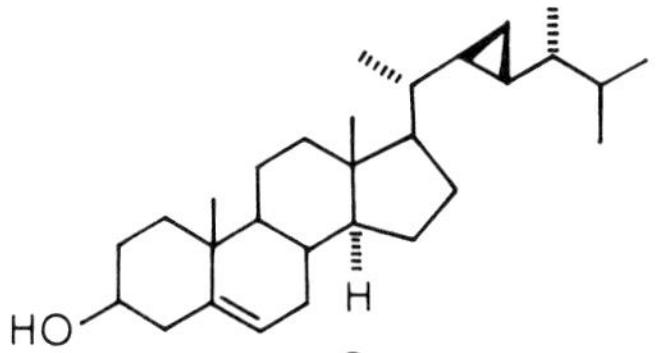

2j

4 23-Demethylgorgosterol (2j) : mol. wt = 414, RR_t = 1·78
23-Demethyl-gorgosterol was isolated and identified from the zooxanthellae derived from the Hawaiian anemone, *Aiptasia pulchella* (Withers *et al.* 1982). Mass cultures of the zooxanthellae grown apart from the host were found to contain **2j**: 4% in the free sterols, and 5% in the steryl esters. It was identified by both mass spectral and NMR analyses. The sterol was originally isolated and structurally elucidated from two gorgonians, *Gorgonia flabellum* and *G. ventalina*, both of which contain symbiotic dinoflagellates (Schmitz & Pattabhiraman 1970). Steudler *et al.* (1977) detected **2j** also in numerous reef coelenterates and their predators.

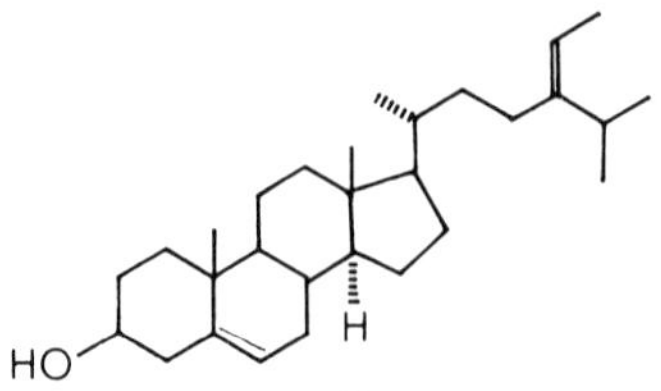

2q

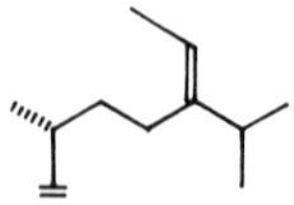

(r)

5 (24Z)-Stigmasta-5,24(28)-dien-3β-ol
Isofucosterol (2q): mol. wt = 414, RR_t = 1·84

Isofucosterol (**2q**), was reported to comprise 28·7% of the sterols from *Gonyaulax diagensis* (Alam *et al.* 1978). Both fucosterol (**2r**) and isofucosterol (**2q**) were isolated (together 6·0% of demethylsterols) from *A. carterae* (Kokke *et al.* 1980).

4.3 C_{28} 4α-demethylsterols

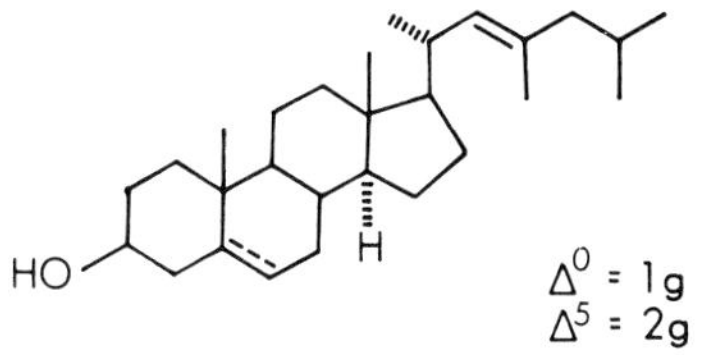

1 23-Methyl-cholesta-5,22-dien-3β-ol (2g): mol. wt = 398, RR_t = 1·07

This new sterol was isolated in 1979 from two sources: in trace quantities (1·1% free sterols) from the cultured zooxanthellae whose host was the Bermuda zoanthid, *Zoanthus sociatus* (Kokke *et al.* 1979; Withers *et al.* 1982), and from the soft coral, *Sarcophyton glaucum*, which contains zooxanthellae (Kobayashi *et al.* 1979a, b). The structure, except for the configuration of the C_{22} double bond, was verified by comparison of both NMR and mass spectra with a synthetic reference sample of **2g** (Kokke *et al.* 1979). The isolation of this sterol is significant because it is most likely the product of direct bioalkylation of a Δ^{22} double bond, which may be a new biosynthetic route to novel sterol side chains (Kobayashi *et al.* 1979; Kokke *et al.* 1979; Djerassi 1981).

2 23-Methyl-5α-cholesta-22E-en-3β-ol (1g): mol. wt = 400, RR_t = 1·07

From the free sterols of the cultured zooxanthellae from the scleractinian coral, *Oculina diffusa*, a new sterol, **1g** (1%) was isolated and identified by mass spectra and NMR comparison with published data on its 5 dehydro and 4α-methyl analogues (**2g** and **7g**) (Withers *et al.* 1982).

4.4 C_{27} 4α-demethylsterols

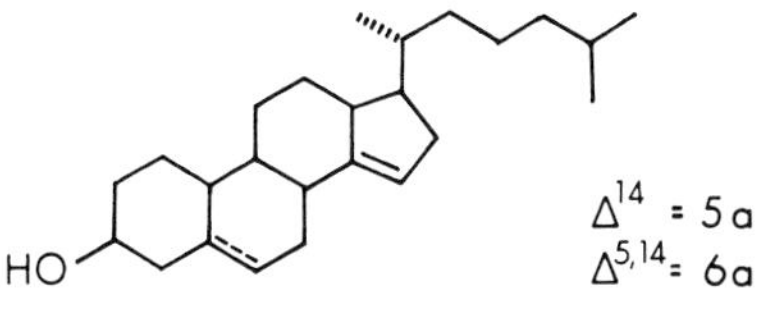

1 14 Dehydrocholestanol (5a): mol. wt = 386, RR_t = 1·08

This sterol, 14 dehydrocholestanol, **5a**, a known synthetic compound (Anastasia *et al.* 1978), was recently reported from the steryl esters (0·6%) of the zooxanthellae from *Aiptasia pulchella* (Withers *et al.* 1982). It was identified by mass spectral and NMR data. The mass spectra of 5a and 6a each showed a strong peak for loss of side chain (100%), characteristic of Δ^{14} sterols (Djerassi 1970).

2 Cholesta-5,14-dien-3β-ol (6a): mol. wt = 384, RR_t = 1·08

Isolation of this sterol **6a** from the steryl esters (0·7%) of *A. pulchella* zooxanthellae represents the first report of a naturally-occurring sterol with double bonds in both the Δ^5 and Δ^{14} positions (Withers *et al.* 1982).

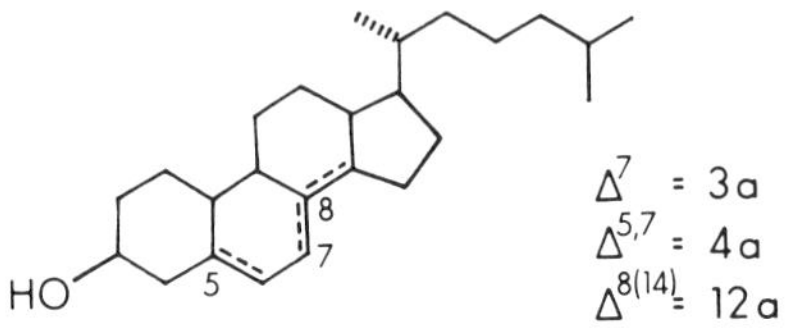

3 5α-Cholest-7en-3β-ol (3a): mol. wt = 386, RR_t = 1·16

7-Dehydrocholesterol **3a** occurs as 1·2% of the free 4-demethylsterol fraction of the *B. asbestinum* zooxanthellae (Bohlin *et al.* 1981). It is interesting to note that the abundance of Δ^7 and $\Delta^{5,7}$ sterols is much greater in the demethylsterol fraction than in the 4α-methylsterols of dinoflagellates (which are mainly saturated or 8,14-unsaturated).

4 Cholest-5,7-dien-3β-ol (4a) (7-dehydrocholesterol): mol. wt = 384, RR_t = 1·17

The major demethylsterols of the heterotrophic species *Crypthecodinium cohnii* Woods Hole strain d are 7-dehydrocholesterol (**4a**) and cholesterol (**2a**) in a ratio of about 7:1. The identity of the diene (**3a**) was confirmed by TLC, GLC, UV (λ_{max} 295, 282, 272 and 262 nm), ms (m/z 366, 351, 253) and HMR (δ 5·36, m, 3–6; 5·52, m, C_7) of the acetate. The preponderance of this sterol in a dinoflagellate is especially biologically relevant to the origin of vitamin D in the ocean, since in mammals, **4a** is the biosynthetic precursor of vitamin D (Holick *et al.* 1981b). This sterol **3a** has also been isolated, as 35·9% of the free 4-demethyl sterols from the zooxanthellae of *B. asbestinum* (Kokke *et al.* 1980; Bohlin *et al.* 1981).

5 5α Cholest-8(14) en-3βol (12a): mol. wt = 384, RR_t = 0·00

This sterol, 5α-cholest-8(14)-en-3βol, **12a** was isolated as a trace constituent (0·2%) of the sterylesters from both batches of mass-cultured zooxanthellae taken from host *Aiptasia pulchella* (Withers *et al.* 1982).

6 5α-Cholestan-3β-ol (1a) : mol. wt = 388, RR_t = 1·0

5α-cholestan-3β-ol comprised 5·6% and 7·5% respectively, of the total sterols of cultured dinoflagellates *Noctiluca scintillans (=miliaris)* (Teshima *et al.* 1980) and *Pyrocystis lunula* (Kokke *et al.* 1982). In a natural sample of the freshwater dinoflagellate *Peridinium lomnickii*, cholestanol was the most abundant component (100% vs. 90% for dinosterol, **7i**) and dominated over cholesterol **2a** (1%) which typically is the major 4-demethylsterol in dinoflagellates (Table 9.1). The abundance of cholestanol vs. cholesterol in overlying sediment samples of the same lake where *P. lomnickii* was collected was 15·7 vs. 7·1 μg per g dry sediment (0–6 cm) suggesting that an important source for 5α stanols in this and other sediments is dinoflagellate blooms (Robinson *et al.* 1984).

5 NORSTEROLS

5.1 C_{27} norsterols

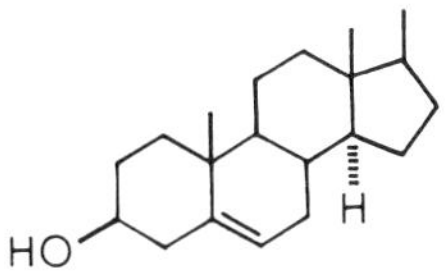

2

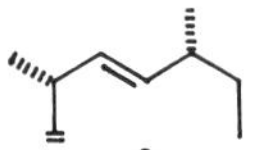

s

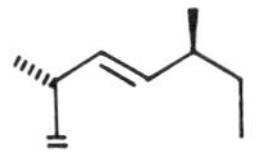

t

1 27-Nor-(24R) = 24-methylcholesta-5, 22-dien-3β-ol (2s) : mol. wt = 384, RR_t = 0·97

This novel norsterol, **2s**, was isolated and identified by Goad & Withers (1982) from the cultured dinoflagellated *Gymnodinium simplex*, where it comprised 32% of the total sterols. The biosynthetic route for this sterol is not known, i.e. whether it proceeds via demethylation of a parent 'regular' sterol or by *de novo* synthesis of the norsterol. A pathway for the latter (*de novo* synthesis of **2s**) in *G. simplex* is suggested (Goad & Withers 1982).

The presence of this unique norsterol in such quantities (one-third of the total sterols) in an alga indicates that the dinoflagellates are important primary producers of the 27-nor-sterols which frequently are found in marine invertebrates and seawater samples (Gagosian 1976; Goad 1978). This recent finding indicates that bacterial degradation may not be the major mechanism for the introduction of this sterol into the food web as had been previously postulated (Goad 1978; Goad & Withers 1982). The concept of bacterial origin of norsterols was reinforced by recent incubation experiments of C_{27}, C_{28} and C_{29} sterols with marine bacteria which yielded the short chain sterols (Boutry & Barbier 1981).

5.2 C_{24} norsterols

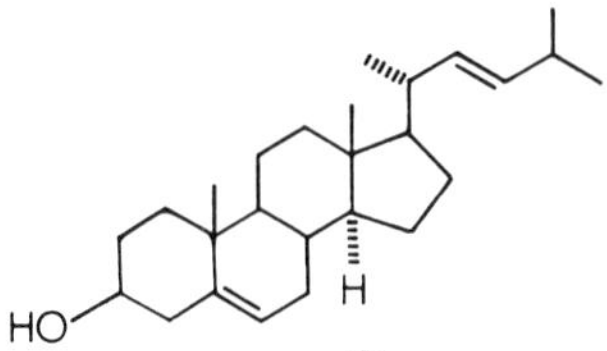

2u

1 24-Norcholest-5,22-dien-3β-ol (2u): mol. wt = 370, RR_t = 0·72
24-Norcholest-5,22-dien-3β-ol (**2u**) was isolated as a trace sterol (0·7%) with the major 27-norsterol (**2s**) from the dinoflagellate *G. simplex* (Goad & Withers 1982). It was identified by mass spectral data only. This sterol was first isolated from a scallop *Placopecten magellanicus* (Idler *et al.* 1970), and has since been recognized, together with its Δ^7 analogue, as a ubiquitous component of many marine invertebrates (Djerassi *et al.* 1979, Goad 1978; Schmitz 1978). A 24-norsterol accounts for 30% of the sterol mixture of a ctenophore (Djerassi 1981). Boutry *et al.* (1971) found a small amount of 24-norcholest-5,22-dien-3β-ol (**2u**) in the plankton, and suggested that a planktonic organism may be an important primary source of this C_{26} sterol. Teshima *et al.* (1980) reported small amounts (0·7%) of **2u** as well as occelasterol (**2t**) or possibly **2s** together with 24*S*-methylcholest-5,22-dien-3β-ol (72·5%) as sterols (both free and esterified) of the holozoic dinoflagellate *Noctiluca scintillans* (= *miliaris*) collected from a red tide near the coast of Japan. However, since (i) *Noctiluca* is a phagocytic form which ingests a variety of other planktonic organisms and (ii) other plankton might have been collected with the dinoflagellate, it was not clear from this study whether this sterol is actually produced by the dinoflagellate. The demonstration of both 24- and 27-norsterols from a cultured dinoflagellate (*G. simplex*) indicate that these algae represent at least one source of the short side chain sterols (Withers & Goad 1982). Further investigation is now necessary to determine the relative importance of the two processes: (i) the bacterial degradative pathway (Boutry & Barbier 1981), and (ii) dinoflagellate production, either by demethylation or *de novo* synthesis (Teshima *et al.* 1980; Goad & Withers 1982) and their contribution to the norsterols which are so commonly found in the marine environment (Kobayashi & Mitsuhashi 1974; Goad 1978; Schmitz 1978).

6 STEROID KETONES

6.1 4α-Methylsteroid ketones

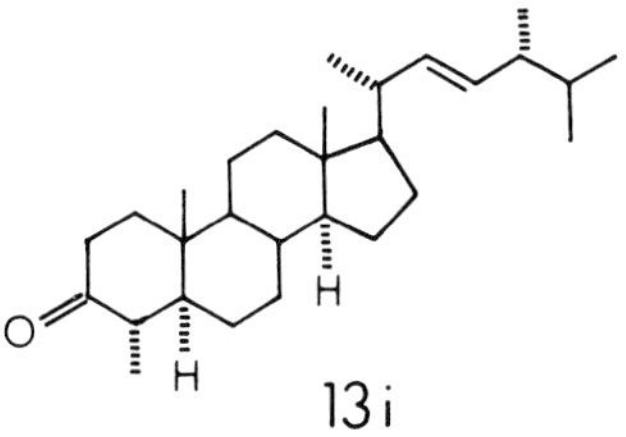

13i

1 (22E) 24R 4α23,24-Trimethylcholest-22en-3-one (dinosterone) (13i): mol. wt = 426, RR_t = 1·66

From axenic cultures of a heterotrophic dinoflagellate, *Crypthecodinium cohnii* (grown on a synthetic defined medium) a 3-oxosteroid fraction (14% total steroids) was isolated (Withers *et al.* 1978). The major component gave a mass spectrum similar to that reported (Shimizu *et al.* 1976) for the Jones oxidation product of dinosterol (**7i**) (m/z at 285, 100%).

The ^{1}H NMR spectra of the parent compound and its reduction product (with KBH_4) confirmed the structure of dinosterone, **13i**, based on the dinosterol skeleton (dinosterol is a major sterol of this species) (Withers *et al.* 1978). Mass spectral analysis of the sterone fraction from *C. cohnii* also revealed several minor sterones with M^+ ions at m/z 412, 414, and 440 with corresponding RR_ts of 1·37, 1·37 and 1·88. These sterones amount to 10% of the total sterone fraction and have not yet been identified. Dinosterone, **13i**, has also been identified in a natural population of *P. lomnickii* (major sterone, 100%), and from its underlying lacustrine sediment (5·9 μg per g dry sediment) (Robinson *et al.* 1984); as well as in marine sediments (Smith *et al.* 1983). These 3-oxo compounds may be intermediates in the demethylation reactions at C_4 and in production of the saturated ring system of dinosterol (Mulheirn & Ramm 1972) (Fig. 9.6).

2 4α-Methyl-5α-cholestan-3-one (13a): mol. wt = 398
3 4α,24-Dimethyl-5α-cholest-22en-3-one (13b): mol. wt = 396
4 4α,24-Dimethyl-5α-cholestan-3-one (13c): mol. wt = 398
5 4α,23,24-Trimethyl-5α-cholestan-3-one (13m): mol. wt = 428.

In a natural population of the freshwater dinoflagellate, *Peridinium lomnickii*, and in a bottom sediment sample from the same eutrophic lake (Priest Pot in the English Lake District), a series of 4α-methylsterones were isolated (Robinson *et al.* 1984). The major component in each sample was dinosterone (**13i**). In addition, 4α-Methyl-5α cholestan-3-one (**13a**), comprised 70% of the dinoflagellate sample and 1·8 μg per g dry sediment; 4α,24-Dimethyl-5α-cholest-22-en-3-one (**13b**) accounted for 30% and 2·5 μg per g sediment sample respectively;

4,24-Dimethyl-5α cholestan-3-one (**13c**) was present at 35% and 0·5 μg per g sediment; and finally 4α,23,24-Trimethyl-5α cholestan-3-one amounted to 40% of the dinoflagellate sample and 2·2 μg per g dry weight sediment sample. There were no 4-demethylsteroid ketones detected. These data suggest that the 4α-methylstanols and 4α-methylstanoles detected in ancient freshwater sediments, such as the lacustrine Messel shale, may also represent compounds of dinoflagellate origin (Mattern *et al.* 1970).

6.2 4α-Demethylsteroid ketones

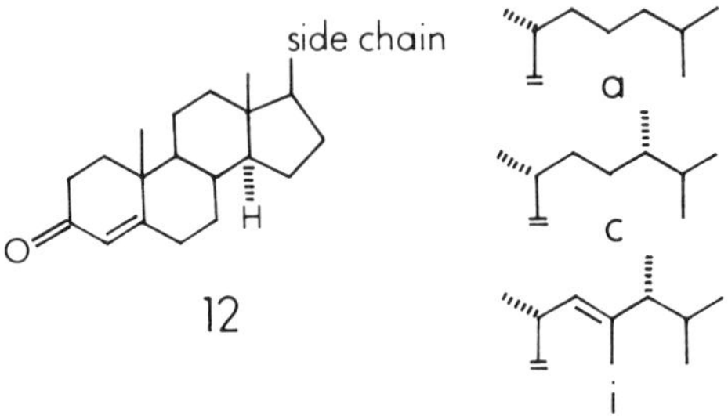

12

1 23,24R-Dimethylcholest-4, 22-dien-3-one (14i) : mol. wt = 412, RR_t *= 2·03*
2 24S-Methylcholest-4-en-3-one (14c) : mol. wt = 396, RR_t *= 1·94*
3 Cholest-4-en-3-one (14a) : mol. wt = 382, RR_t *= 1·47*

In *Pyrocystis lunula*, three Δ^4-3 ketones were recently isolated and identified. One ketone contains a dinosterol side chain (i), (**14i**); another has a 24-methylated side chain (c), (**14c**); and the third, (**14a**) is cholest-4-en-3-one (Kokke *et al.* 1982).

7 STRUCTURE AND TAXONOMY

7.1 Dinoflagellates as a group

By virtue of their unique structural features dinoflagellate sterols can be distinguished from the sterols of all other algal groups. This is consistent with other biochemical, morphological and structural characteristics unique to dinoflagellates which have led taxonomists to separate the group from the other 'Chromophytes' (unicellular chlorophyll-*c*-containing algae) as designated by Christensen (1962) and Hibberd (1979). None of these chromophyte algae contain 4-methylsterols. The sterols of marine algae have been reviewed by Goad (1978) and Nes & McKean (1977).

Table 9.1. Major sterols on dinoflagellates (cultured, except *Peridinium lomnicki* which was collected as a natural population)

Taxon	Major sterols*	References
Division Pyrrhophyta		
Class Dinophyceae		
Order Gymnodiniales		
Family Gymnodiniaceae		
Amphidinium carterae	**10f**, **10a**	Kokke *et al.* (1980), Withers *et al.* (1979c)
Amphidinium corpulentum	**10f, 10a**	Kokke *et al.* (1980), Withers *et al.* (1979c)
Amphidinium hoefleri (= *carterae*)	**10f**	Withers *et al.* (1979c)
Amphidinium klebsii	**10f**	Withers *et al.* (1979c)
Amphidinium rhyncocephalum	**10f**	Withers *et al.* (1979c)
Gymnodinium simplex	**7c, 2e, 2s**	Goad & Withers (1982)
Order Noctilucales		
Family Noctilucaceae		
Noctiluca scintillans	**2e**	Teshima *et al.* (1980)
Order Peridiniales		
Family Crypthecodiniaceae		
Crypthecodinium cohnii (Woods Hole d)	**7i, 7c**	Withers *et al.* (1978)
Family Glenodiniopsidaceae		
Glenodinium sp.	**7i, 7c**	Kokke *et al.* (1980)
Family Ostreopsidaceae		
Ostreopsis sp.	**2a, 2b, 7i**	L. Bohlin & N. W. Withers (unpubl.)
Family Peridiniaceae		
Kryptoperidinium foliaceum	**2a, 7i, 7c, 7k**	Withers *et al.* (1979a)
Glenodinium (= *Kryptoperidinium*) *foliaceum*	**7i, 2a, 7k, 7g**	Alam *et al.* (1979a)
Peridinium balticum	**2a, 7i, 7c, 7k**	Withers *et al.* (1979a)
Peridinium sociale	**7i, 7c**	W. C. M. C. Kokke & N. W. Withers (unpubl.)
Peridinium lomnicki	**1a, 7i**	Robinson *et al.* (1984)
Family Gonyaulacaceae		
Protoceratium reticulatum	**2a, 7i**	Eggersdorfer *et al.* (unpubl.)
Gonyaulax tamarensis	**2a, 7i**	Shimizu *et al.* (1976)
Gonyaulax acatenella	**2a, 7i**	Alam *et al.* (1978)
Gonyaulax catenella	**7i, 2a**	Alam *et al.* (1978)
Gonyaulax washingtonensis	**2a, 7i**	Alam *et al.* (1978)
Gonyaulax polyedra	**2a, 7i**	Alam *et al.* (1978)
Gonyaulax diegensis	**2a, 2a, 7q**	Alam *et al.* (1978)
Gonyaulax monilata	**2a, 7i, 11c, 7k**	Wengrovitz *et al.* (1981)
Order Pyrocystales		
Family Pyrocystaceae		
Pyrocystis lunula	**7c, 2a**	Kokke *et al.* (1982)
Order Zooxanthellales	**7c, 2a**	Withers *et al.* (1982)
	7c, 2a	
	7c, 7i	

* Listed in order of abundance.

7.2 Sterol structure patterns within the Dinophyceae

Peridinin is the major accessory carotenoid in almost all photosynthetic dinoflagellate species that have been examined (Jeffrey *et al.* 1975). However, although dinosterol predominates in representatives of both heterotrophic and autotrophic species, this compound does not have such a wide distribution as peridinin in the Dinophyceae. This was verified by a GC-MS survey of twenty-one dinoflagellate species by N. W. Withers & L. J. Goad (unpubl.), which showed that although dinosterol was a major sterol in eleven species, it was only a minor component in several species and was even lacking in others, including five *Amphidinium* species (Withers *et al.* 1979c). Subsequent detailed investigations of numerous species in several laboratories have confirmed that dinosterol (**7i**) is a common constituent among the dinoflagellates, but it is not the predominant sterol of the group (Table 9.1) (Kokke *et al.* 1980, 1981; Alam *et al.* 1979a, 1979b; Wengrovitz *et al.* 1981; Withers *et al.* 1982).

Careful investigation has revealed a complex mixture of sterols (both 4-methyl and demethyl) in most dinoflagellate species. For example, in a recent study of zooxanthellae sterols, which resolved compounds present at an 0·2% abundance level, one culture yielded seventeen different sterols in the steryl esters, and twelve sterols in the free sterol fraction (Withers *et al.* 1982). However, this complexity is mostly due to the presence of many trace sterols. When the 'minor' sterol components (those present in an abundance of less than 5% of each fraction, free sterols or sterol esters) are ignored, the sterol mixtures from the seven zooxanthellae examined are reduced to only 2–6 ($\bar{X} = 4$) sterols per organism. The two major sterols in each dinoflagellate comprised more than 50% of the total sterols (Table 9.2). There were more 4-monomethylsterols than demethylsterols in the free sterols from all zooxanthellae cultures analysed, although in the sterol esters demethyls sometimes prevailed. This preponderance of 4-methylsterols over demethylsterols seems to be a characteristic dinoflagellate trait (Kokke *et al.* 1981; Withers *et al.* 1982). However, *Gonyaulax* species and *Protoceratium reticulatum* are notable exceptions as they contain mainly cholesterol (Table 9.2).

Sterol composition is a useful chemotaxonomic criterion for distinguishing groupings within the Dinophyceae. For example, all members of the *Amphidinium* genus synthesize amphisterol and no dinosterol; members of the genus *Gymnodinium* have 4α,24 dimethylcholestanol as a major sterol (with little if any dinosterol); and those of the families Gonyaulacaceae, Peridiniaceae, and Ostreopsidaceae have a large amount of dinosterol and cholesterol (Table 9.1). Relative abundances of individual components in the free sterol fraction of each dinoflagellate species have not shown much (if any) variation due to alterations in culture conditions.

8 BIOSYNTHESIS

The intriguing aspects of biosynthesis of dinosterol and related sterols which indicate divergences from the well studied cholesterol biosynthesis include: (i) synthesis of the saturated ring system with retention of the 4α-methyl group, (ii) alkylation of the side chain, including formation of the cyclopropyl group, and (iii), coordination of these two processes (Fig. 9.3).

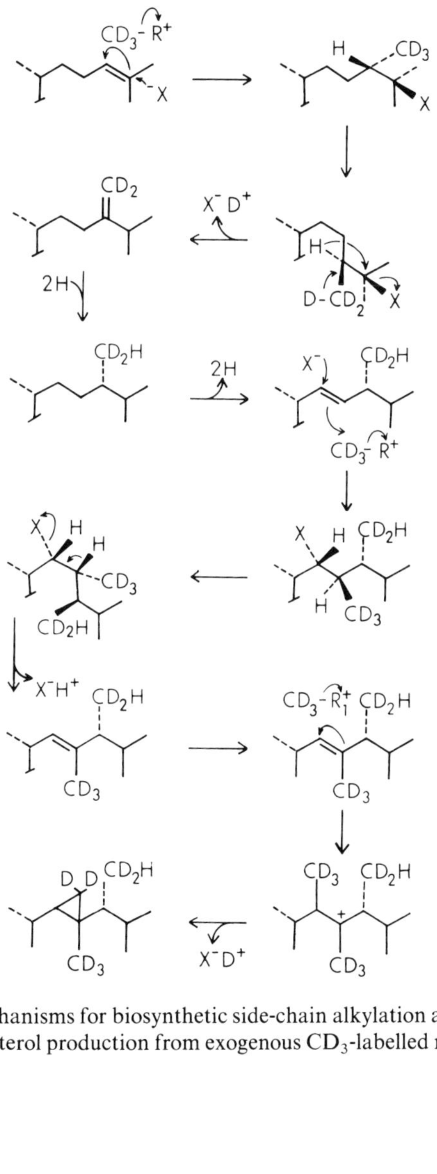

Fig. 9.3. Postulated mechanisms for biosynthetic side-chain alkylation at C_{22}, C_{23}, and C_{24} in dinosterol and gorgosterol production from exogenous CD_3-labelled methionine.

8.1 Ring-system synthesis

The presence of amphisterol as an end-product of sterol synthesis in the genus *Amphidinium* (Withers *et al.* 1979; Kokke *et al.* 1980) suggests the scheme in Fig. 9.4 for the two demethylation steps at C_{14} and C_4, based on cholesterol biosynthesis (Nes & McKean 1977), although no 4,4-dimethyl precursors have been isolated from dinoflagellates. The isolation of Δ^5-dinosterol and dinosterone from *C. cohnii* led to the subsequent pathway designated in Fig. 9.6 as postulated by Goad for the saturation from a Δ^7, 4α-methyl precursor for both dinosterol and 4,24 dimethylcholestanol production (Withers *et al.* 1978). This is analogous to the route producing 5α-cholestanol in animal tissues (Smith *et al.* 1975). However, 3-oxo-steroids are also intermediates in the demethylation reactions responsible for loss of the 4α- and 4β-methyl groups in the production of 4-demethylsterols in both plants and animals (Goad 1981) and a similar role in dinoflagellates would account for the presence of dinosterone in *C. cohnii* and the Δ^4, 3 ketones found in *P. lunula* (Withers *et al.* 1978; Kokke *et al.* 1982).

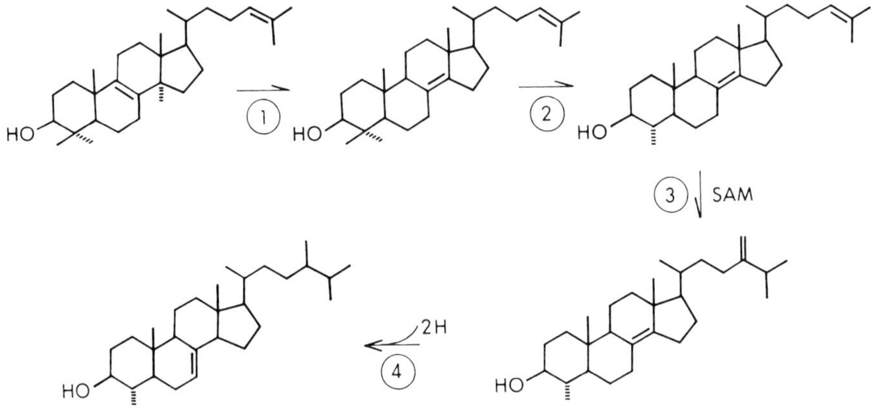

Fig. 9.4. Postulated scheme for the partial biosynthetic sequence of dinoflagellate sterols showing amphisterol (**10f**) as an intermediate. Steps 1 and 2 are the postulated demethylation steps at C_{14} and C_4, respectively. Step 3 indicates side-chain alkylation via SAM (S-adenosylmethionine) followed by hydrogenation and migration of the 8,14 double bond to the 7,8 position in step 4. Continuation of ring saturation from a $\Delta^{7,8}$ precursor is shown in Fig. 9.6.

8.2 Side-chain alkylation

Biosynthetic studies by the Liverpool group involved axenic cultivation of the heterotrophic species, *C. cohnii*, in a nutrient medium supplemented with CD_3-labelled methionine. The dinoflagellate incorporated deuterium into over 60% of the newly synthesized 4α-monomethylsterols (dinosterol and 4,24 dimethylcholestanol). The mass spectral fragmentation pattern indicated that the C_{23} methyl group contained three deuterium atoms and was introduced intact by transmethylation from methionine via S-adenosylmethionine. The C_{24} methyl group contained only two deuterium atoms, which is consistent with the production of a 24-methylenesterol intermediate that is subsequently reduced to give the 24-methyl side chain (Withers *et al.* 1979c). A proposed mechanism is shown in Fig. 9.3 (steps 1–8).

Indirect evidence suggests that the sequence of methylation (via SAM) of gorgosterol occurs in the order of C_{24}, C_{23}, and then C_{22} (Djerassi *et al.* 1979). The Stanford group recently induced uptake and incorporation of extracellular CD_3-methionine and $^{13}CH_3$-methionine into sterols of the photosynthetic binucleate species, *K. foliaceum* and *P. balticum*, and found a similar labelling pattern (of deuterium) at C_{24} and C_{23}. Analysis of the ^{13}C and ^{2}H NMR spectra revealed incorporation at the cyclopropyl bridge of a ^{13}C atom and two deuterium atoms at the cyclopropyl bridge in 4α-methylgorgostanol (W. C. M. C. Kokke *et al.* unpubl.).

A postulated mechanism for the formation of the gorgosterol-type side chain (from a dinosterol-type side chain) is shown in Fig. 9.3. The presence of 4α-methylgorgostanol indicates that perhaps cyclopropanation supersedes demethylation at C_4 for gorgosterol/gorgostanol formation, but this sequence is as yet unknown (Fig. 9.5).

8.3 Demethyls and 4α-monomethyls: separate pathways?

It is striking that the major sterols of dinoflagellates are 4α-methyl ring saturated compounds with additional methyl groups at the C_{23} and C_{24} of the side chain (i.e. dinosterol), but the abundant demethylsterols are rarely saturated (normally Δ^5 or $\Delta^{5,7}$) and have unsubstituted side chains (i.e. cholesterol). This seems to indicate that the dinoflagellates have a dichotomy in the sterol biosynthetic pathways, one branch apparently leading to the 4-methylsterols, (dinosterol, 4,24-dimethylcholestanol or amphisterol) and the other route, in which side-chain alkylation is relatively unimportant, leading to the 4-demethylsterols (Withers *et al.* 1978).

8.4 Steryl esters vs. free sterols

Almost all dinoflagellate species which have been examined for sterol ester content, by extraction without initial saponification followed by column chromatography, showed a sterol ester fraction, at a level of 2–20% of the weight of the free sterol fraction after saponification of the sterol esters (Kokke *et al.* 1981; Withers *et al.* 1982). Cholestanyl and 4α-methylstanyl tetradecanoates and hexadecanoates were identified in *P. lomnicki* (Robinson *et al.* 1984). In Black Sea sediments, esterified 4α-methylsterols were abundant (De Leeuw *et al.* 1983). The sterol esters in *K. foliaceum* were analysed for fatty acid content: these consisted of long-chain polyunsaturated fatty acids, mainly 16:3, 18:5,

20:5 and 22:6 similar to the fatty acids of the wax esters from that species (Withers & Nevenzel 1977). The sterol composition of the steryl esters generally shows a greater complexity (measured as total number of sterol structures) and a greater variation, both quantitative and qualitative, when compared with the free sterol fraction. The free sterols apparently function as structural components of the membrane, which explains the conservative nature of their composition, and the steryl esters are presumably the metabolic storage form for sterols in the algal cell (Nes & McKean 1977).

9 STEROL SYNTHESIS IN ZOOXANTHELLAE

9.1 Zooxanthella-contributed sterols

It is uncertain whether or not cnidarians (coelenterates) are capable of sterol synthesis, although the majority of evidence indicates they are not (Goad 1981). This presumed inability to complete sterol biosynthesis may explain the abundance of non-steroid terpenoids in cnidarians (and perhaps in sponges). The carbon isotope ($\delta^{13}C$) value indicating the $^{13}C/^{12}C$ ratio of the sterols from invertebrate/zooxanthellae associations suggests algal origin, whereas the $\delta^{13}C$ value for the terpenes is much higher, indicating 'animal' origin. The zooxanthellae sterols contain less ^{13}C (e.g. $\delta^{13}C$ of $-24{\cdot}7$ for gorgosterol vs. $-18{\cdot}2$ for briarein B, both from *Briareum asbestinum*) (Kokke *et al.* 1984).

The complex mixtures of sterols found in many corals and anemones (and other non-cnidarian invertebrates) probably reflect both a dietary accumulation and a contribution from their endosymbiotic dinoflagellates (zooxanthellae) (Ciereszko *et al.* 1968; Kokke *et al.* 1982). A recent survey of the sterols of three aposymbiotic Pacific gorgonians revealed that these invertebrates lack the typical 4-methylsterols and gorgosterol found in the zooxanthellae-containing gorgonians of the Caribbean (Kokke *et al.* 1981). The sterol content of Pacific gorgonians showed a direct dependence on dietary sterols or possibly gorgonian-synthesized sterols; cholesterol was the major component in all three species examined. The Caribbean gorgonians, which harbour zooxanthellae, also had cholesterol, together with some 4-methylsterols and an abundance (13–34%) of the cyclopropyl sterols, gorgosterol and 23-demethylgorgosterol, indicating dinoflagellate-derived sterols (Kokke *et al.* 1981). Also 4α-methylgorgostanol and gorgostanol were recently isolated by Wengrovitz *et al.* (1981) from *Gonyaulax monilata*, a chain-forming species (which does not contain an endosymbiont).

9.2 Gorgosterol in symbiotic associations

When the zooxanthellae from these Caribbean gorgonians were isolated and cultured free from their hosts, the algae synthesized mostly 4-methylsterols (dinosterol, 4,24-dimethylcholestanol, and 4,24 dimethylcholest-22-en-3β-ol), but none of the 4-demethylcyclopropyl sterols found in the algal/host consortia (Kokke *et al.* 1981). Also, the levels of 4α-methylsterols in the algal/coelenterate association are always much lower than in the host-free algae, which suggests coelenterate demethylation at C_4. The evidence suggests that formation (i.e. demethylation at C_4 and cyclopropanation at C_{20}, C_{22}) of gorgosterol from dinosterol (and 23-demethyl gorgosterol from 4,24 dimethylcholestanol) is enhanced when the algae live in symbiotic association with the host (Fig. 9.5). This conclusion is strengthened by the isolation of 4α-methylgorgostanol and gorgostanol from the dinoflagellates *K. foliaceum* and *P. balticum* (Alam *et al.* 1979a; Withers *et al.* 1979a). These species are algal/algal symbiotic associations of a heterotrophic dinoflagellate harbouring a photosynthetic haptophyte alga as an endosymbiont (Dodge 1971; Tomas & Cox 1973). *Ptychodiscus brevis* is an enigma because it apparently does not represent a symbiotic algal/algal association and is not binucleate yet is not a 'typical' dinoflagellate. Its major photosynthetic pigments do not include peridinin (or fucoxanthin) and ultrastructural features of its chloroplasts are not typical of dinoflagellates (Jeffrey *et al.* 1975; Steidinger & Cox 1980).

Careful screening of numerous strains of cultured zooxanthellae from various invertebrate hosts finally revealed one strain, the dinoflagellate isolated from the Hawaiian anemone, *Aiptasia pulchella*, which produces gorgosterol and 23-demethylgorgosterol in host-free culture. These data confirm the dinoflagellate capacity for synthesis of these unique cyclopropyl sterols (Withers *et al.* 1982). However, it should be emphasized that all other strains of zooxanthellae and all but one (*G. monilata*) of the 'typical' uninucleate free-living species examined to date do not produce cyclopropyl sterols in culture, although these sterols are found when the algae occur as symbionts of numerous invertebrates (including foraminifera and one sponge) (Delseth *et al.* 1979). Why the symbiotic existence (either algal/algal or algal/invertebrate) should enhance the dinoflagellate production of the cyclopropyl sterols remains an intriguing biological question.

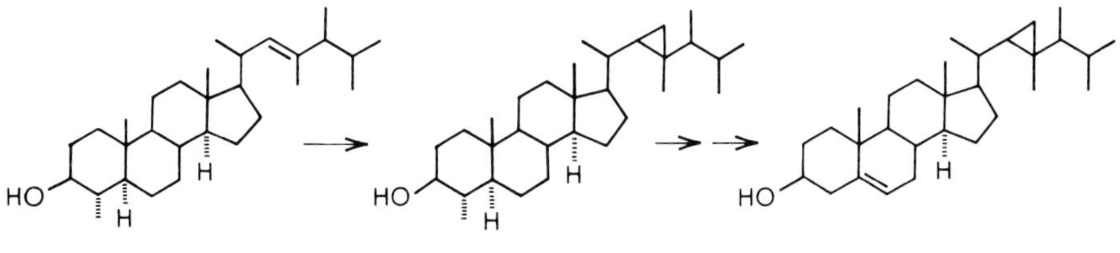

(7i) (7k) (2k)

Fig. 9.5. Conversion of gorgosterol (**2k**) from dinosterol (**7i**) via 4α-methylgorgostanol (**7k**) as postulated. The process is enhanced when the dinoflagellates exist as symbionts of invertebrates or with other algae.

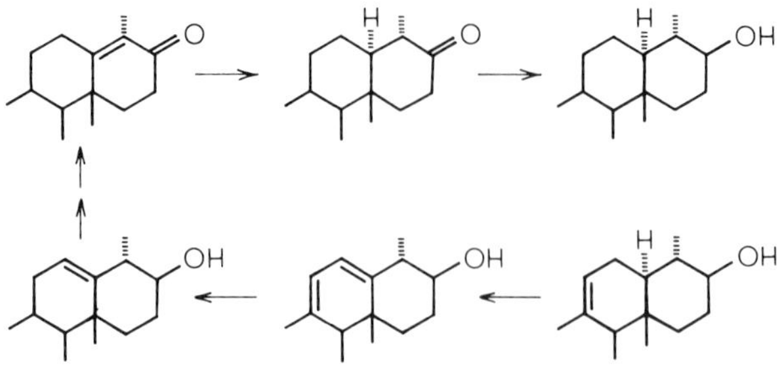

Fig. 9.6. Possible route for the saturation of the sterol ring system in the production of dinosterol and 4α,24-dimethylcholestanol.

9.3 Use of sterols as taxonomic markers for zooxanthellae strains

The sterol compositions of the symbiotic dinoflagellates isolated from these gorgonians and from a variety of other invertebrates clearly demonstrate that there is no special relationship between the taxonomic affiliation of the host and the sterol pattern of the isolated symbiont. For example, the sterols from the zooxanthellae of *Oculina diffusa* (Cnidaria) and *Tridacna gigas* (Mollusca) are the same. Furthermore, there are great differences in sterol patterns of various zooxanthellae strains (Table 9.2). These differences between strains, which are maintained by the zooxanthellae in the symbiotic condition (disregarding

Table 9.2. Major free sterols of zooxanthellae from various invertebrate hosts: expressed as percentages of total free sterols; values for the two most abundant sterols are underlined. From Withers *et al.* (1982) and Kokke *et al.* (1981).

Host invertebrate species	4α,24S-Dimethyl-cholesta-3β-ol **(7d)**	4α,24R-Dimethyl-cholesta-22-en-3β-ol **(7e)**	Dinos-terol **(7i)**	Choles-terol **(2a)**	Sum of two most abundant species (%)
Group 1					
Tridacna gigas	39	(10)	(9)	20	59
Melibe pilosa	45	(11)	(5)	17	62
Oculina diffusa	32	(14)	19	(17)	51
Group 2					
Aiptasia pulchella (average of 2 cultures)	(10)	24	(16)	22	46
Zoanthus sociatus	(24)	28	(4)	35	64
Gorgonia mariae	(11)	37	(4)	27	64
Group 3					
Briareum asbestinum	14	(1)	43	(2)	57
Muriceopsis flavida	31	—	38	—	69

cyclopropanation and demethylation at C_4) as well as under various culture conditions, suggest that the 'strains' of zooxanthellae are probably different species of dinoflagellates (see Chapter 12). The sterol patterns are useful as markers for identification and perhaps later assignment of taxonomic affiliation of the different zooxanthellae strains (Ciereszko *et al.* 1968; Kokke *et al.* 1981; Kokke *et al.* 1982; Bohlin *et al.* 1981; Withers *et al.* 1982).

10 DINOFLAGELLATE STEROLS AS TRACERS IN THE MARINE ENVIRONMENT

10.1 Marine sediments: dinosterol as a 'molecular fossil'

Huang & Menschen (1976) have emphasized the value of sterols as source indicators of organic materials in sediments because of their relative stability in the sedimentary environment and their characteristic distribution in different organisms. Sterol analysis offers an alternative to the carbon isotope method for a determination of marine vs. terrestrial contributions in marine sediments. Dinosterol and other dinoflagellate sterols have been isolated from Neogene and Holocene marine sediments. In fact, dinosterol has been termed a 'molecular fossil' for determining past dinoflagellate blooms (Boon *et al.* 1979). The 'Black Sea sterol' (Peake *et al.* 1974), the dominant component among the extractable sterols from the Black Sea sapropel layer (7000–3000 years BP) was identified as dinosterol by Boon *et al.* (1979). This sediment was deposited during the rise of the oxic–anoxic interface after the last glaciation in the Black Sea. The abundance of dinosterol as well as of dinoflagellate cysts found in this sediment indicate the existence of periodic blooms of dinoflagellates which generally have a wide salinity tolerance, at that time in the surface waters of the Black Sea. A salinity transition from freshwater to marine occurred during this period (Wall & Dale 1968). Ascertaining the earliest presence of gorgosterol and/or dinosterol in fossil coral samples should allow determination of the period when symbiosis was established between dinoflagellates and coral (and other invertebrates).

Lee *et al.* (1980) found dinosterol, dinostanol and 4,24-dimethylcholestanol in the Soxhlet-extractable free (SEF) sterols from sediment cores of the deepest part of the Black Sea. The anaerobic conditions of this sediment and the presence of the 4-methyl group initially suggested a bacterial origin for the sterols, but since these sterols are components of many dinoflagellate species, a phytoplanktonic source is much more probable. Dinoflagellate sterols were also prominent in free sterols of surface sediment samples collected at Walvis Bay and from the shelf and continental slope of the NW African coast. They were present but much less abundant in sediment samples from Buzzards Bay, Massachusetts and from the continental rise area of the western North Atlantic

(Lee *et al.* 1977, 1979, 1980). A number of 4α-methylstanols, 4α-methylstanones and corresponding esters of dinoflagellate origin (similar in composition to those of the overlying 'bloom' species *P. lomnicki*) were identified in the surface sediment of a freshwater lake. The prominence of 5α-cholestan-3β-ol suggested that dinoflagellates could be the 'long sought, direct biological source of sedimentary stanols' (Robinson *et al.* 1984). Lee *et al.* (1980) mentioned that in past studies of open ocean marine sediments, dinosterol might have been missed because it co-elutes by GC with 24-ethylcholesterol and 23,24-dimethylcholestanol. However, if the sterols are first separated into 4-methyl and demethyl fractions as suggested by Wardroper *et al.* (1978), this problem is eliminated.

In addition, 4α-methylgorgostanol was identified among the sixty-nine different sterols characterized from the extractable lipids from Neogene deep sea sediments from the Japan Trench (Wardroper *et al.* 1978). Gorgosterol has also been detected in marine sediments (Steudler *et al.* 1977; Wardroper *et al.* 1978). Because these sterols are unique to dinoflagellates, their presence is informative in geochemical evaluations of diagenetic pathways in marine sediments. The widespread occurrence of 4α-methyl steroid hydrocarbons in ancient sediments and petroleums may be products of diagenetic reactions on sterols and sterones of dinoflagellate origin (Robinson *et al.* 1984; Ensminger *et al.* 1978).

Sterols are particularly useful because they are stable molecules in the sedimentary environment and their relative concentrations show little quantitative or qualitative biological variation when compared with other 'marker' molecules such as the dinoflagellate carotenoid, peridinin (Huang & Menschen 1976). Dinoflagellate sterols with the combined traits of organism specificity, molecular stability, consistent concentration and composition in biological material are therefore superior to other organic compounds as useful tracers for both biological and geological processes in marine systems.

10.2 Use as tracers in the marine food web: coral reefs

Since coral reef cnidarians can be characterized by the presence of the C_{30} sterol gorgosterol, Steudler *et al.* (1977) decided to use gorgosterol as a tracer to identify animals which are cnidarian predators. The results of the sterol analyses showed a very close relationship between the content of each predator and its suspected prey organism: *Cyphoma gibbosa* (flamingo tongue snail) and gorgonian *Briareum asbestinum*; *Caphyra laevis* (crab) and soft coral *Xenia elongata*; *Mithraculus sculptus* (crab) and scleractinian *Porites porites*; and *Domicea acanthophora* (crab) and stony coral *Acropora palmata*. Gorgosterol has even been detected in the eggs of one gorgonian *Pseudopterogorgia americana* where it comprised 5% of total sterols (Kung & Ciereszko 1977), and in the sponge, *Biemna fortis* (2·5% of total sterols) (Delseth *et al.* 1979). Generally, gorgosterol and related cyclopropyl sterols are abundant in the tissues of most of the reef animals as well as in their faeces, which indicates a mode of entry of gorgosterol into marine sediments.

4α-Methylsterols, including 4α-methylcholest-8(14)-en-3β-ol, have been detected in the oyster, *Crassiostrea virginica*. These were considered by Teshima & Patterson (1981) to be of dietary (dinoflagellate?) origin.

10.3 Metabolism

Metabolism of dinoflagellate sterols has not been studied, although acanthasterol, the Δ^7 analogue of gorgosterol originally isolated by Gupta & Scheuer (1968) from *Acanthaster planci*, is undoubtedly a metabolite of gorgosterol since starfish are capable of converting Δ^5 sterols into their characteristic Δ^7-sterols (Smith & Goad 1975; Schmitz 1978; Goad 1981). Also, as discussed in Section 9.9, there is abundant circumstantial evidence which indicates that the sterols synthesized by zooxanthellae are further metabolized by the host invertebrate. Although the dinoflagellates synthesize the cyclopropyl bridge, it is probably the host which executes demethylation at C_4 and introduction of the Δ^5 double bond in gorgosterol and 23-demethylgorgosterol. A metabolite of gorgosterol, 3ζ-hydroxymethyl-A-nor-5α-gorgostanol was isolated by Bohlin *et al.* (1980) from a sponge, *Stylotella agminata*.

There are also a number of dinoflagellate sterols with 5,7 unsaturation, including 7-dehydrocholesterol which is the major demethyl sterol in *C. cochnii* (Withers *et al.* 1978), and *B. asbestinum* zooxanthellae (Bohlin *et al.* 1980). These provitamins D may be an important source of vitamin D in the ocean. Although the natural conversion (sunlight-mediated) of the dinoflagellate 5,7-diene sterols to vitamin D has not yet been demonstrated, M. F. Holick (pers. comm.) showed *in vivo* formation of pre-vitamin D_2 from the 5,7-diene sterol fraction in cultures of *Skeletonema menziesii* by exposing the diatoms to sunlight D_2 (Holick *et al.* 1981a).

Dinoflagellate (and other phytoplankton) sterols may be important dietary precursors of testosterone, oestradiol, and related steroid hormones in herbivorous zooplankton (O'Hara *et al.* 1979).

DIAGENETIC CONVERSION

The dinoflagellate sterols and sterones presumably undergo the same complex sequence of diagenetic reactions to hydrocarbons in the sediment, with an additional isomerization at C_4. The preponderance of the saturated ring system in dinoflagellate sterols (as in dinosterol) offers evidence in the 'stenol/stanol controversy.' As in the case of the norsterols in the plankton, the stanols presence in sediments may in part be due to dinoflagellate contribution, and thus are not necessarily the product of *in situ* microbial reduction of stenols to stanols (Ensminger *et al.* 1978; Robinson *et al.* 1984).

11 FUNCTION

The function and intracellular localization of these unique dinoflagellate sterols are unknown. Presumably dinosterol and its relatives function as membrane constituents in dinoflagellates, a role similar to that of cholesterol in most eukaryotes. However, in a fluorescence polarization study, dinosterol was found to be less soluble in model membranes and as effective in increasing microviscosity as cholesterol. This study thus did not show the expected special complementary structural interaction between dinosterol and other dinoflagellate membrane lipids (synthetic phosphatidyl choline, 16:0, 22:6, and the total phospholipid mixture of *C. cohnii*) (Harel & Djerassi 1980). Further studies involving isolation and purification of dinoflagellate membranes are necessary to determine whether or not dinosterol occurs as an integral part of this structure or if it perhaps serves a different biological role in these algal cells. Dinosterol esters have been isolated along with wax esters (phytyl esters) from extraplastidic lipid globules of the eyespot of *K. foliaceum* (Withers & Nevenzel 1977; Withers & Haxo 1978). Both wax esters and dinosterol esters presumably function as a storage reserve of lipids in the algal cell. Metabolism of dinoflagellate sterols to steroid hormones and/or vitamin D may be important in the marine environment. Dinoflagellate contribution of nutritionally vital sterols to host invertebrates, especially corals, has been well documented in recent years.

12 CONCLUSIONS

More than thirty-five new sterols have been isolated and identified from dinoflagellate species since the discovery of dinosterol. Dinoflagellates synthesize 24-norsterols, 26-norsterols, gorgosterol and other C_{22}, C_{23} cyclopropyl sterols, many new C_4 methylated sterols, and sterols with unusual unsaturation: at 17(20), and in the ring at 8(14), and 14(15) as well as completely saturated sterols. Most dinoflagellate species contain a complex mixture of sterols, but usually two sterols comprise over 50% of the total sterol fraction. There seems to be a taxonomic pattern for the distribution of the major sterols among the dinoflagellate families. The dominant sterols are dinosterol, 4,24-dimethylcholestanol, amphisterol and cholesterol.

The high molecular weight of dinoflagellate sterols is due to the 'extra' methyls at C_4 (retained from precursors in synthesis) and the three extra methyls in the side chain from bio-transmethylation via S-adenosyl-methionine. The structural features of dinoflagellate sterols are reminiscent of bacterial sterols and reinforce the 'lower eucaryotic' status of the class Dinophyceae. These unique molecular characteristics enable the use of dinosterol, gorgosterol, amphisterol and other dinoflagellate sterols as stable markers and tracers in freshwater and marine food-web systems, in sea-water and planktonic samples and in lacustrine and marine sediments. Gorgosterol and the related cyclopropylsterols are excellent indicators of symbiotic dinoflagellate associations. Synthesis of the cyclopropyl group in gorgosterol (probably from a dinosterol side chain) is performed by zooxanthellae and this process is enhanced when the algae co-exist with host invertebrates or another alga. The sterol patterns of zooxanthellae strains and other dinoflagellate species are useful chemotaxonomic markers.

Most of the research to date on dinoflagellate sterols has involved isolation and identification of new sterol structures from cultured algae. The biosynthesis, function, intracellular localization, and metabolism of these new dinoflagellate sterols are largely unknown and offer many intriguing areas of research.

13 REFERENCES

ALAM M., MARTIN G. E. & RAY S. M. (1979a) Dinoflagellate sterols. 2. Isolation and structure of 4-methyl gorgostanol from the dinoflagellate *Glenodinium foliaceum*. *J. Org. Chem.* **44**, 4466–4467.

ALAM M., SANSING T. B., BUSBY E. L., MARTINIZ D. R. & RAY S. M. (1979b) Dinoflagellate sterols. I. Sterol composition of dinoflagellates of *Gonyaulax* species. *Steroids* **33**, 197–203.

ALAM M., SCHRAM K. H. & RAY S. M. (1978) 24-demethyl dinosterol: an unusual sterol from the dinoflagellate *Gonyaulax diagensis*. *Tetrahedron Lett.* 3517–3518.

ANASTASIA M., FIECCHI A. & SCALA A. (1978) Side chain inversion of olefins promoted by hydrogen chloride. *J. Org. Chem.* **43**, 3505–3508.

ANDING C., BRANDT P. D. & OURISSON G. (1971) Sterol biosynthesis in *Euglena gracilis* Z. Sterol precursors in light-grown and dark-grown *Euglena gracilis* Z. *Eur. J. Biochem.* **24**, 259–263.

ANDO T. & BARBIER M. (1975) Sterols from the culture medium of the luminescent dinoflagellate *Pyrocystis lunula*. *Biochem. Syst. Ecol.* **3**, 245–248.

BEACH D. H. & HOLZ G. G. Jr (1973) Environmental influence on the docosahexaenoate content of triacyl glycerols and phosphatidyl choline of a heterotrophic marine dinoflagellate, *Crypthecodinium cohnii*. *Biochim. Biophys. Acta* **316**, 56–65.

BEASTALL G. H., TYNDALL A. M., REES H. H. & GOODWIN T. W. (1974) Sterols in *Porphyridium* species. *Eur. J. Biochem.* **41**, 301–309.

BERGMANN W. (1962) In *Comparative Biochemistry* (eds M. Florkin & H. Mason), Vol. III, pp. 103–162. Academic Press, New York.

BERGMANN W., MCLEAN M. J. & LESTER O. I. (1973) Contribution to the study of marine products XIII. Sterols from various marine invertebrates. *J. Org. Chem.* **8**, 271–282.

BIRD C. W., LYNCH T. M., PIRT F. T., REID W. W., BROOKS C. T. W. & MIDDLEDITCH B. S. (1971) Steroids and squalene in *Methylococcus capsulatus* grown on methane. *Nature (Lond.)* **230**, 473–474.

BLOCH K. (1976) On the evolution of a biosynthetic pathway. In *Reflections on Biochemistry* (eds A. Kornberg, B. L. Honecker, L. Cornudella & L. Oro), pp. 143–150. Pergamon Press, New York.

BOHLIN L., KOKKE W. C. M. C., FENICAL W. & DJERASSI C. (1981) 4α-methyl-24S-ethyl-5α-cholestan-3β-ol and 4α-methyl-24S-ethyl-5α-cholest-8(14)-en-3β-ol, two new sterols from a cultured marine dinoflagellate. *Phytochem.* **21**, 2397–2401.

BOHLIN L., GEHRKEN H. P., SCHEUER P. J. & DJERASSI C. (1980) Minor and trace sterols in marine invertebrates XVI. 3ζ-Hydroxymethyl-A-nor-5α-gorgostane, a novel sponge sterol. *Steroids* **35**, 295–304.

BOON J. J., RIPSTRA W. I. C., DELANGE F., DELEEUW J. W., YOSHIOKA M. & SHIMIZU Y. (1979) Black Sea sterol—a molecular fossil for dinoflagellate blooms. *Nature (Lond.)* **277**, 125–127.

BOUTRY J. L., ALCAIDE A. & BARBIER M. (1971) Sur le presence d'un sterolen $^{C}26$ dans un plancton marin vegetal. *C. R. Acad. Sci. Paris* **272**, 1022–1024.

BOUTRY J. L. & BARBIER M. (1981) Abstracts of the 72nd Annual Meeting of the American Oil Chemists Society, New Orleans, USA.

BOUVIER P., ROHMER M., BENVUENISTE P. & OURISSON E. (1976) Δ8(14) steroids in the bacterium *Methylococcus capsulatus*. *Biochem. J.* **159**, 267–271.

BRASSELL S. C., COMET P. A., EGLINTON G., ISAACSON P. J., MCEVOY J., MAXWELL J. R., THOMSON I. D., TIBBETTS P. J. C. & VOLKMAN J. K. (1980) The origin and fate of lipids in the Japan Trench. In *Advances in Organic Geochemistry*. (eds A. G. Douglas & J. R. Maxwell), pp. 375–392. Pergamon, Oxford.

BRASSELL S. C. & EGLINTON G. (1981) Biogeochemical significance of a novel sedimentary C_{27} sterol. *Nature (Lond.)* **290**, 579–582.

CARLSON R. M. K. (1977) *Rapid analysis of marine sterol mixtures through computer-assisted GC-MS*. Part II, pp. 99–252. Ph.D. Dissertation, Stanford Univ.

CHRISTENSEN T. (1962) In *Botanik 2 (Systematic Botanik)* (eds W. Bocher, M. Lange & T. Sorensen), pp. 1–178. Munksgaard, Copenhagen.

CIERESZKO L. S., JOHNSON M. A., SCHMIDT R. W. & KOONS C. B. (1968) Chemistry of coelenterates. VI. Occurrence of gorgosterol, a C-30 sterol in coelenterates and their zooxanthellae. *Comp. Biochem. Phys.* **24**, 899–904.

COHEN A. I. & ROCK S. Jr (1964) Anisotropy of substituents on the proton resonance of C-18 and C-19 methyl groups. Evaluation by computer regression. *Steroids* **3**, 243–257.

DE LEEUW J. W., RIPSTRA W. I. C., SCHENK P. A. & VOLKMAN J. K. (1983) Free, esterified and residual bound sterols in Black Sea Unit 1 sediments. *Geochim. Cosmochim. Acta* **47**, 455–465.

DELSETH C., KASHMAN Y. & DJERASSI C. (1979) Ergosta-5,7,9(11),22 tetraen-3βol and its 24 ξ-ethyl homologue, 2 new marine sterols from the Red Sea sponge, *Biemna fortis*. *Helv. Chim. Acta* **62**, 2037–2045.

DJERASSI C. (1970) Applications of mass spectrometry in the steroid field. *Pure Appl. Chem.* **21**, 205–225.

DJERASSI C. (1981) Recent studies in the marine sterol field. *Pure Appl. Chem.* **53**, 873–890.

DJERASSI C., KRAKOWER G. W., LEMIN A. J., LIV L. H., MILLS J. S. & VILOTTI R. (1958) The neutral constituents of the cactus *Lophocereus schotti*. The structure of lophenol-4α-methyl-Δ7-cholesten-3β-ol: a link in sterol biogenesis. *J. Am. Chem. Soc.* **80**, 6284–6292.

DJERASSI C., THEOBALD N., KOKKE W. C. M. C., YAK C. S. & CARLSON R. M. K. (1979) Recent progress in the marine sterol field. *Pure Appl. Chem.* **51**, 1815–1828.

DODGE J. D. (1965) Chromosome structure in the dinoflagellates and the problem of the mesocaryotic cell. *Excerpta Med. Int. Congr. Ser.* **91**, 339–375.

DODGE J. D. (1971) A dinoflagellate with both a mesokaryotic and a eukaryotic nucleus. I. Fine structure of the nuclei. *Protosplasma* **4**, 231–242.

DOW W. C., GEBREYESUS T., POPOV S., CARLSON R. M. K. & DJERASSI C. (1983) Marine 4-methyl sterols: synthesis of C-24 epimers of 4α,24-dimethyl-5α-cholestan-3β-ol and 360 MHz 'H NMR comparisons to the natural product from *Plexaura homomalla*.

DROOP M. R. & PENNOCK J. F. (1971) Terpenoid quinones and steroids in the nutrition of *Oxyrrhis marina*. *J. Mar. Biol. Assoc. UK.* **51**, 455–470.

ENSMINGER A., JOLY G. & ALBRECHT P. (1978) Rearranged steranes in sediments and crude oils. *Tetrahedron Letts* **18**, 1575–1578.

FINER J., CLARDY J., KOBAYASHI A., ALAM M. & SHIMIZU Y. (1978) Identity of the stereochemistry of dinosterol and gorgosterol side chain. *J. Org. Chem.* **43**, 1990–1992.

FINER T., HIROTZA K. & CLARDY J. (1977) X-ray diffraction and the structure of marine natural products. In *Marine Natural Products* (eds D. L. Faulkner & W. H. Fenical), pp. 147–158. Plenum Press, New York.

GAGOSIAN R. (1976) A detailed vertical profile of sterols in the Sargasso Sea. *Limnol. Oceanogr.* **21**, 702–710.

GAGOSIAN R. B., SMITH S. O. & NIGRELLI G. E. (1982) Vertical transport of steroid alcohols and ketones measured in a sediment trap experiment in the Equatorial Atlantic Ocean. *Geochim. Cosmochim. Acta* **46**, 1463–1472.

GIESBRECHT P. (1962) Vergleichende Untersuchunger an den Chromosomen des Dinoflagellaten *Amphidinium elegans* und denen der Bakterien. *Zentra Bakteriol. Parasitenkl., Infektionskr. Hyg. L. Orig.* **187**, 452–498.

GOAD L. J. (1978) Sterols of marine invertebrates. In *Marine Natural Products* (ed. P. J. Scheuer), Vol. II, pp. 76–172. Academic Press, New York.

GOAD L. J. (1981) Sterol biosynthesis and metabolism in marine invertebrates. *Pure Appl. Chem.* **51**, 837–852.

GOAD L. J. & WITHERS N. W. (1982) The identification of 27-nor (24*R*)-24-methyl cholesta-5,22-dien-3β-ol and brassicasterol as the major sterols of the marine dinoflagellate *Gymnodinium simplex*. *Lipids* **17**, 853–858.

GUPTA K. C. & SCHEUER P. J. (1968) Echinoderm sterols. *Tetrahedron* **24**, 5831–5837.

GUPTA K. C. & SCHEUER P. J. (1969) Zoanthid sterols. *Steroids* **13**, 343–356.

HALE R. L., LECLERCQ J., TURSCH B., DJERASSI C., GROSS R. A., WEINHEIMER A. J., GUPTA K. & SCHEUER P. J. (1970) Demonstration of a biogenetically unprecedented side chain in the marine sterol, gorgosterol. *J. Am. Chem. Soc.* **92**, 2179–2180.

HAREL & DJERASSI (1980) Dinosterol in model membranes: fluorescence polarization studies. *Lipids* **15**, 694–696.

HIBBERD D. J. (1979) The structure and phylogenetic significance of the fugellar transition in the chlorophyll *c*-containing algae. *Biosystems* **11**, 243–261.

HOLICK M. F., HOLICK S. A. & GUILLIARD R. L. (1981a) Photosynthesis of previtamin D in phytoplankton. In *Proceedings of the Ninth International Symposium on Comparative Endocrinology, Hong Kong, Dec 1981* (ed. B. Lofts). Hong Kong University Press (in press).

HOLICK M. F., MACLAUGHLIN T. A. & DOPPLET S. H. (1981b) Regulation of cutaneous previtamin D_3 photosynthesis in man: skin pigment is not an essential regulator. *Science* **211**, 590–593.

HUANG W. Y. & MENSCHEN W. G. (1976) Sterols as source indicators of organic materials in sediments. *Geochim. Cosmochim. Acta.* **40**, 323–330.

IDLER D. R., WISEMAN P. M. & SAFE L. M. (1970) A new marine sterol, 22-trans-24-norcholesta-5,22-dien-3β-ol. *Steroids* **16**, 451–461.

IIDA T., TAMURA T., WAINAI T., MASHIMO K. & MATSUMO T. (1977) Substituent effects on the PMR signals of methyl groups in methyl steroids. *Chem. Phys. Lipids* **19**, 169–178.

IKAN R., BAEDECKER M. J. & KAPLAN I. R. (1975) Thermal alteration experiments on organic matter in recent marine sediment. III. and steroidal alcohols. *Geochim. Cosmochim. Acta* **39**, 195–203.

JEFFREY S. W., SIELICKI M. & HAXO F. T. (1975) Chloroplast pigment patterns in dinoflagellates. *J. Phycol.* **11**, 374–384.

JOHNSON R., REES H. H. & GOODWIN T. W. (1974) The conversion of cholest-5-en-3β-ol into cholesta-5,7-dien-3β-ol by *Calliphora erythrocephala* under aseptic conditions. *Biochem. Soc. Trans.* **2**, 1062–1065.

JOSEPH J. D. (1975) Identification of 3,6,9,12,15-octadecapentaenoic acid in laboratory cultured photosynthetic dinoflagellates. *Lipids* **10**, 395–403.

KANAZAWA A., TESHIMA S. & ANDO T. (1977) Sterols of coelenterates. *Comp. Biochem. Physiol.* **57b**, 317–323.

KANAZAWA A., TESHIMA S., ANDO T. & TOMITA S. (1974) Occurrence of 23,24-dimethylcholesta-5,22-dien-3β-ol in a soft coral, *Sarcophyton elegans*. *Bull. Jap. Soc. Sci. Fish.* **40**, 729–731.

KANAZAWA A., TESHIMA S., ANDO T. & TOMITA S. (1976) Sterols in coral reef animals. *Mar. Biol.* **34**, 53–57.

KNAPP F. & SCHROEPFER G. J. (1976) Mass spectrometry of sterols. Electron ionization induced fragmentation of C-4 alkylated cholesterols. *Chem. and Physics of Lipids*, **17**, 466–500.

KOBAYASHI M. & MITSUHASHI H. (1974) Marine sterols. V. Isolation and structure of occelasterol, a new 27-nor-ergostane-type sterol from an annelida, *Pseudopotamilla occelata*. *Steroids* **24**, 399–410.

KOBAYASHI M., TOMIOKA A., HAYASHI T. & MITSUHASHI H. (1979a) Isolation of 23-methylcholesta-5,22-dien-3β-ol from the soft coral *Sarcophyton glaucum*. *Chem. Pharm. Bull.* **27**, 1951–1953.

KOBAYASHI M., TOMIOKA A. & MITSUHASHI H. (1979b) Marine sterols. VIII. Isolation and structure of sarcosterol, a new sterol with a Δ17,20 double bond from the soft coral, *Sarcophyton glaucum*. *Steroids* **34**, 273–284.

KOKKE W. C. M. C., BOHLIN L., FENICAL W. & DJERASSI C. (1982) Novel dinoflagellate 4α methylated sterols from four Caribbean gorgonians. *Phytochem.* **21**, 881–887.

KOKKE W. C. M. C., EPSTEIN S., LOOK S. A., RAU G. H., FENICAL W. & DJERASSI C. (1984) On the origin of terpenes in symbiotic associations between marine invertebrates and algae (zooxanthellae): culture studies and an application of $^{13}C/^{12}C$ isotope ratio mass spectrometry. *J. Bio. Chem.* **259**, 8168–8173.

KOKKE W. C. M. C., FENICAL W. & DJERASSI C. (1980) Sterols with unusual nuclear unsaturation from three cultured marine dinoflagellates. *Phytochem.* **20**, 127–134.

KOKKE W. C. M. C., FENICAL W. & DJERASSI C. (1982) Sterols of the cultured dinoflagellate *Pyrocystis lunula*. *Steroids* **40**, 307–308.

KOKKE W. C. M. C., FENICAL W., BOHLIN L. & DJERASSI C. (1981) Sterol synthesis by cultured zooxanthellae; implications concerning sterol metabolism in the host–symbiont association in Caribbean gorgonians. *Comp. Biochem. Physiol.* **68B**, 281–287.

KOKKE W. C. M. C., WITHERS N. W., MASSEY B. J., FENICAL W. & DJERASSI C. (1979) Isolation and synthesis of 23-methyl-22-dehydrocholesterol—a marine sterol of biosynthetic significance. *Tet. Letters* **38**, 3601–3636.

KUNG S. S. & CIERESZKO L. S. (1977) Lipids in the eggs of the gorgonian, *Pseudopterogorgia americana* (Ginelin). *Proc. Third Int. Coral Reef Symp., Miami, Fla.*, pp. 525–527.

LEE C., FARRINGTON J. W. & GAGOSIAN R. B. (1979) Sterol geochemistry of sediments from the western North Atlantic Ocean and adjacent coastal areas. *Geochim. Cosmochim. Acta* **43**, 35–46.

LEE C., GAGOSIAN R. B. & FARRINGTON T. W. (1980) Geochemistry of sterols in sediments from Black Sea and the southwest African shelf and slope. *Organic Geochem.* **2**, 103–113.

LEE C., GAGOSIAN R. B. & FARRINGTON J. W. (1977) Sterol diagenesis in recent sediments from Buzzards Bay, Massachusetts. *Geochim. Cosmochim. Acta* **41**, 985–992.

LEE R. F. & LOEBLICH A. R. III. (1971) Distribution of 21:6 hydrocarbon and its relationship to 22:6 fatty acid in algae. *Phytochem* **10**, 593–602.

LIAAEN-JENSEN S. (1978) Marine carotenoids. In *Marine Natural Products, Chemical and Biological Perspectives* (ed. P. J. Scheuer), Vol. II, pp. 1–73. Academic Press, New York.

LING N. C., HALE R. L. & DJERASSI C. (1970) The structure and absolute configuration of the marine sterol gorgosterol. *J. Amer. Chem. Soc.* **92**, 5281–5282.

LOEBLICH A. R. III. (1976) Dinoflagellate evolution: speculation and evidence. *J. Protozool* **23**, 13–28.

MATTERN G., ALBRECHT P. & OURISSON G. (1970) 4-methyl sterols and sterols in Messel Shale (Eocene). *Chem. Comm.* 1570–1571.

MAZUR V., WEIZMANN A. & SONDHEIMER F. (1958) Steroids and triterpenoids of citrus fruit.

III. The structure of citrastadienol, a natural 4α-methyl sterol. *J. Am. Chem. Soc.* **80**, 6293–6296.

MILKOYA T. S., POPOV S. S., MAREKOV N. L. & ANDREEV S. N. (1980) Sterols from Black Sea invertebrates. I. Sterols from Scyphozoa and Anthozoa (Coelenterata). *Comp. Biochem. Physiol.* **67B**, 633–638.

MINALE L. & SODANA G. (1976) Non-conventional sterols of marine origin. In *Marine Natural Products Chemistry* (eds D. J. Faulkner & W. A. Fenical), pp. 87–109. Plenum Press, New York.

MORILL L. C. & LOEBLICH A. R. III. (1981) The dinoflagellate pellicular wall layer and its occurrence in the division Pyrrhophyta. *J. Phycol.* **17**, 315–323.

MULHEIRN L. J. & RAMM P. J. (1972) The biosynthesis of sterols. *Chem. Soc. Rev.* **1**, 259–291.

NES W. K. & MCKEAN M. L. (1977) *Biochemistry of Steroids and Other Isopentenoids.* Univ. Park Press, Baltimore. pp. 375–690.

O'HARA S. C. M., GASKELL S. T. & CORNER E. D. S. (1979) On the nutrition and metabolism of zooplankton. XIII. Further studies of steroids in *Calanus. J. mar. biol. Assoc., UK* **59**, 331–340.

OURISSON G., ALBRECHT P. & ROHMER M. (1979) The hopanoids: palaeochemistry and biochemistry of a group of natural products. *Pure Appl. Chem.* **51**, 709–729.

PARTRIDGE L. G., MIDGLEY I. & DJERASSI C. (1977) Mass spectrometry in structural and stereochemical problems. 249. Elucidation of the course of the characteristic ring D fragmentation of unsaturated steroids. *J. Am. Chem. Soc.* **99**, 7686–7695.

PATTON S., FULLER G., LOEBLICH A. R. III & BENSON A. A. (1966) Fatty acids of the 'red tide' organism, *Gonyaulax polyedra. Biochem. Biophys. Acta* **116**, 577–579.

PEAKE E., CASAGRANDE J. & HODGSON G. W. (1974) The Black Sea—geology, chemistry and biology. *Am. Ass. petrol. Geol. Mem.* **20**, 505–523.

REES H. H., DONNAHEY P. L. & GOODWIN T. W. (1976) Separation of C27, C28, and C29 sterols in reversed-phase high-performance liquid chromatography of small particles. *J. Chromatogr.* **116**, 281–292.

RIZZO P. J. & NOODEN L. D. (1974) Isolation and partial characterization of dinoflagellate chromatin. *Biochim. Biophys. Acta* **349**, 415–427.

ROBINSON N., EGLINTON G., BRASSELL S. C. & CRANWELL P. A. (1984) Dinoflagellate origin for sedimentary 4α-methylsteroids and 5α(H)-stanols. *Nature* **308**, 439–442.

SCHMITZ F. J. (1978) Uncommon marine sterols. In *Marine Natural Products* (ed. P. J. Scheuer), Vol. I, pp. 291–297. Academic Press, New York.

SCHMITZ F. J. & PATTABHIRAMAN T. (1970) A new marine sterol possessing a side chain cyclopropyl group: 23 demethyl gorgosterol. *J. Am. Chem. Soc.* **92**, 6073–6074.

SCHROEPFER G. J., LUTZKY B. N., MARTIN J. A., HUNTOON S., FOURCANS B., LEE W. H. & VERMILION J. (1972) Recent investigations on the nature of sterol intermediates in the biosynthesis of cholesterol. *Proc. R. Soc. Lond. Ser. B.* **180**, 125–146.

SHEIKH Y., DJERASSI C. & TURSCH B. M. (1971) Acansterol: A cyclopropane-containing marine sterol from *Acanthaster planci. Chem. Commun. (J. Chem. Soc., Sect. D)* **5**, 217–218.

SHIMIZU Y., ALAM M. & KOBAYASHI A. (1976) Dinosterol, the major sterol with a unique side chain in the toxic dinoflagellate *Gonyaulax tamarensis. J. Am. Chem. Soc.* **98**, 1059–1060.

SHU A. Y. L. & DJERASSI C. (1981) Stereospecific synthesis of dinosterol. *Tetrahedron Letters,* **23**, 4627–4630.

SMITH A. G. & GOAD L. J. (1975) The conversion of cholest-5-en-3β-ol into cholest-7-en-3β-ol by the echinoderms *Asterias rubens* and *Solaster papposus. Biochem, J.* **146**, 35–40.

SMITH D. J., EGLINTON G., MORRIS R. J. & POUTANEN E. L. (1983) Aspects of the steroid geochemistry of an interfacial sediment from the Peruvian upwelling. *Oceanol. Acta,* **6**, 211–219.

SOYER M. O. (1969) Rapports existant entre chromosomes et membrane nuclaire chez un

Dinoflagelle parasite du genre *Blastodinium chatton*. *C. R. Acad. Sci. Paris*, **268**, 2082–2084.

STEIDINGER K. A. & COX E. R. (1980) Free-living dinoflagellates. In *Phytoflagellates* (ed. E. Cox), pp. 407–432. Elsevier North Holland, New York.

STEUDLER P. A., SCHMITZ F. J. & CIERESZKO L. S. (1977) Chemistry of coelenterates. Sterol composition of some predatory–prey pairs on coral reefs. *Comp. Biochem. Physiol.* **56B**, 385–392.

STRAIN H. H., SVEC W. A., AIZETMULLER K., GRANDOLFO M. C., KATZ T. T., KJOSEN H., NORGARD S., LIAAEN-JENSEN S., HAXO F. T., WEGFAHRT P. & RAPOPORT H. (1971) The structure of peridinin, the characteristic dinoflagellate carotenoid. *J. Amer. Chem. Soc.* **93**, 1823–1824.

SWENSEN W., TAGLE B., CLARDY L., WITHERS N. W., KOKKE W. C. M. C., FENICAL W. H. & DJERASSI C. (1980) Peridinosterol, a new Δ17-unsaturated sterol from two cultured marine algae. *Tetrahedron Lett.* **21**, 4663–4666.

TAYLOR F. J. R. (1980) On dinoflagellate evolution. *BioSystems* **13**, 65–108.

TESHIMA S. & PATTERSON G. W. (1981) Sterol biosynthesis in the oyster *Crassostrea virginica*. *Lipids* **16**, 234–239.

TESHIMA S., KANAZAWA A. & TAGO A. (1980) Sterols of the dinoflagellate, *Noctiluca miliaris*. *Mem. Fac. Fish.* **29**, 319–326.

TOMAS R. N. & COX E. R. (1973) Observations on the symbiosis of *Peridinium balticum* and its intracellular alga. I. Ultrastructure. *J. Phycol.* **9**, 304–323.

TONKS D. B. (1967) Δ5 sterols-Lieberman–Burchard. *Clin. Biochem.* **1**, 12-15.

VANDENHEUVEL F. A. (1965) Structural studies of biological membranes: the structure of myelin. *Ann. N. Y. Acad. Sci.* **122**, 57–76.

VARKONY T. H., SMITH D. H. & DJERASSI C. (1978) Computer-assisted structure manipulation. Studies in the biosynthesis of natural products. *Tetrahedron* **34**, 841–852.

VOLKMAN J. K., SMITH D. J., EGLINTON G., FORSBERG T. E. V. & CORNER F. D. S. (1981) Sterol and fatty acid composition of four marine haptophycean algae. *J. Mar. Biol. Assoc. UK* **61**, 509–529.

WAKEHAM S. G., FARRINGTON J. W., GAGOSIAN R. B., LEE C., DEBAAR H., NIGRELLI G. E., TRIPP B. N., SMITH S. O. & FREW N. M. (1980) Organic matter fluxes from sediment traps in the equatorial Atlantic Ocean. *Nature* **286**, 798–800.

WALL D. & DALE B. (1968) Modern dinoflagellate cysts and evolution of the Peridiniales. *Micropaleontology* **14**, 265–304.

WARDROPER A. M. K., MAXWELL T. R. & MORRIS T. R. (1978) Sterols of a diatomaceous ooze from Walvis Bay. *Steroids* **32**, 203–221.

WENGROVITZ P. S., SANDUJA R. & ALAM M. (1981) Dinoflagellate sterols. 3. Sterol composition of the dinoflagellate *Gonyaulax monilata*. *Comp. Biochem. Physiol.* **69B**, 535–539.

WITHERS N. W. (1983) Dinoflagellate sterols. In Marine Natural Products (ed. P. J. Shever) Vol. V, pp. 88–127, Academic Press, New York.

WITHERS N. W., GOAD L. J. & GOODWIN T. W. (1979a) A new sterol, 4α-methyl-5α-Ergosta-8(14),24(28)-dien-3β-ol from the marine dinoflagellate *Amphidinium carterae*. *Phytochem.* **18**, 879–901.

WITHERS N. W., KOKKE W. C. M. C., FENICAL W. & DJERASSI C. (1982) Sterol patterns of cultured zooxanthellae isolated from marine invertebrates. Synthesis of gorgosterol and 23-desmethylgorgosterol by aposymbiotic algae. *Proc. Nat. Acad. Sci. USA*, **79**, 3764–3768.

WITHERS N. W. & HAXO F. T. (1978) Isolation and characterization of carotenoid-rich lipid globules from *Peridinium foliaceum*. *Plant Physiol*, **62**, 36–39.

WITHERS N. W., KOKKE W. C. M. C., ROHMER M., FENICAL W. H. & DJERASSI C. (1979b) Isolation of sterols with cyclopropyl-containing side chains from the cultured marine alga *Peridinium foliaceum*. *Tetrahedron Lett.* **385**, 3605–3608.

WITHERS N. W. & NEVENZEL J. C. (1977) Phytyl esters in a marine dinoflagellate. *Lipids* **12**, 989–993.

WITHERS N. W., TUTTLE R. C., GOAD L. J. & GOODWIN T. W. (1979c) Dinosterol side chain biosynthesis in a marine dinoflagellate, *Crypthecodinium cohnii. Phytochem.* **18**, 71–73.

WITHERS N. W., TUTTLE R. C., HOLZ G. G., BEACH D. H., GOAD L. J. & GOODWIN T. W. (1978) Dehydrodinosterol, dinosterone, and related sterols of a non-photosynthetic dinoflagellate *Crypthecodinium cohnii. Phytochem.* **17**, 1987–1989.

ZIELINSK J., KOKKE W. C. M. C., TAM HA T. B., SHU A. Y. L., DUAX W. L. & DJERASSI C. (1983) Isolation, partial synthesis, and structure determination of sterols with the four possible 23,24-dimethyl-substituted side chains. *J. Org, Chem.* **48**, 3471–3477.

ZURCHER R. F. (1963) Protonenresonen-spektroskopie und sterodstruktur. II. Die Lage der—1% methyl Signale. In *Abhangigkeit von den Substituenten. Helv. Chim. Acta,* **46**, 2054–2088.

CHAPTER 10
BEHAVIOUR IN DINOFLAGELLATES

M. LEVANDOWSKY and PAMELA J. KANETA
Haskins Laboratories of Pace University, New York, NY 10038, USA

1 INTRODUCTION

By behaviour we mean those activities at the level of the whole organism that suggest animality. Though the term has occasionally been used rather loosely to include biochemical or metabolic events, we shall focus here mainly on those involving motility, and responses to sensory stimuli. For example, phototaxis will be viewed as a behaviour but not photosynthesis, though the initial events in both processes may have much in common at the molecular level. Similarly, we shall include phagotrophy as a behaviour, but not assimilation, growth or metabolism.

2 MOTILITY AT LOW REYNOLDS NUMBER

Since behaviour often involves motility, it will be worthwhile to review briefly some unusual features of movement on the microbial scale.

No naval architect would design a submarine with the shape of *Ceratium tripos* or *Dinophysis acuta*, and by the same token the macroscopic dynamic principles responsible for the shapes of boats and fishes are quite inappropriate at the microbial level. The dividing line between the familiar hydrodynamics of large objects and the microscopic regime is given by a dimensionless quantity, the *Reynolds number*, which expresses the ratio of inertial to viscous forces affecting an object moving in a fluid. Formally, it is defined as

$$R_e = \frac{Ul}{\nu}$$

where U is velocity, l is an appropriate length dimension of the moving object, and ν is the kinematic viscosity (viscosity of the medium divided by the density of the moving object). If R_e is much greater than 1, we are in the familiar *Eulerian* dynamic regime of macroscopic organisms, in which inertial effects dominate. But if R_e is much less than 1, we must deal with an unfamiliar *Stokesian* regime, in which viscosity dominates. (For R_e close to 1, both inertial and viscous forces are important and this situation is the least understood from both a physical and a biological point of view.) Table 10.1 shows some typical biological values.

Table 10.1. Typical values of length (l), velocity (U) and Reynolds number (R_e) for a range of organisms

Organism	l	U	R_e
Large fish	50 cm	1 m s^{-1}	5×10^4
Paramecium	0·1 mm	2 mm s^{-1}	0·2
Fast dinoflagellate (e.g. *Dinophysis acuta*)	<50 μm	0·5 mm s^{-1}	<0·025
Bacterium	1 μm	10 μm s^{-1}	10^{-5}

The general theory of flow in a viscous fluid is given by a nonlinear differential equation, the Navier-Stokes equation, which is the hydrodynamic version of Newton's second law of mechanics. When R_e is much larger than one (inertial forces are much larger than viscous forces), this equation can be simplified to a form known as Euler's equation, which is used in aerodynamics and naval architecture, and in the theory of bird flight and fish swimming. At the other extreme, when R_e is much smaller than one (viscous forces are much greater than inertial forces), the Navier-Stokes equation can be approximated by Stokes equation, and we are dealing with the microbial world. Stokes equation is linear, and therefore much easier to work with mathematically than the full Navier-Stokes equation. The biofluid dynamics of low Reynolds number has been reviewed by Lighthill (1975), Roberts (1981) and Childress (1981). A more extensive, non-biological treatment is given by Happel & Brenner (1973: See also Prandtl (1923) for an early comment on the importance of proper scaling in physical models).

In this framework, dinoflagellates, with swimming speeds up to 0·3–0·6 mm s^{-1} (see Table 10.7) for cells with linear dimensions in the range 10–

50 μm, have a swimming Reynolds number less than 0·05 and thus should be unambiguously Stokesian in their hydrodynamics. Many larger species, such as *Noctiluca*, swim more slowly). We shall now examine some consequences of this.

To begin with, since inertia is not significant at low Reynolds number, a micro-organism swimming at top speed stops moving abruptly when it stops swimming, and similarly the distance needed to accelerate from the rest to full speed is negligible on the scale of the organism. Roberts (1981) estimates that *Paramecium* has a stopping distance of 2 μm, and the theoretical distance for the bacterium *Escherichia coli* is less than the diameter of a hydrogen atom! Dinoflagellates, depending on size and swimming speed, fall between these two values with, in any case, stopping distances much smaller than their body lengths. Virtually all of the energy expended for locomotion is used to overcome the viscous resistance of the medium, and a given swimming speed is maintained only by continuous application of a proportional force against the medium. As soon as this force decreases or ceases, the organism *immediately* slows or halts. This is precisely analogous to the well known Stokes' law of settling for a particle at low Reynolds number, in which the terminal velocity is directly proportional to the force (weight of the particle) as well as inversely proportional to the viscosity of the medium.

An example where this is important is the stop response of dinoflagellate phototaxis. In many species, it is reported (Metzner 1929; Hand & Schmidt 1975) that, in response to an appropriate light stimulus, a spiral-swimming cell stops at the precise moment when its photoreceptor points to the light source, then reorients itself toward the stimulus before swimming again (see p. 366). In this case the ability to stop suddenly without any inertial drift has a clear physiological significance.

Another property of Stokesian dynamics can be seen in the external architecture of dinoflagellates. At low Reynolds number the hydrodynamic equations are symmetric in velocity, and the effect of reversing the sense of the motion is merely to reverse the direction of viscous drag without changing its magnitude. Thus, as Roberts (1981) notes, the drag on a hemispherical cup moving along its axis of symmetry is the same in either direction, in marked contrast to the more familiar high Reynolds number (Eulerian) situation, in which the difference in drag for the two directions is actually used in the design of a typical cup anemometer to measure wind speed. Thus, a meteorological anemometer built to dinoflagellate scale simply would not work, since forces acting on opposite facing cups would be the same. More generally, we have the somewhat surprising and counter-intuitive result that fore-aft asymmetry of cell form, a prominent feature of many dinoflagellates, is not ascribable to hydrodynamic streamline constraints as is often the case in large organisms such as fishes. Instead, its functional significance must be sought elsewhere in the organism's behaviour and natural history.

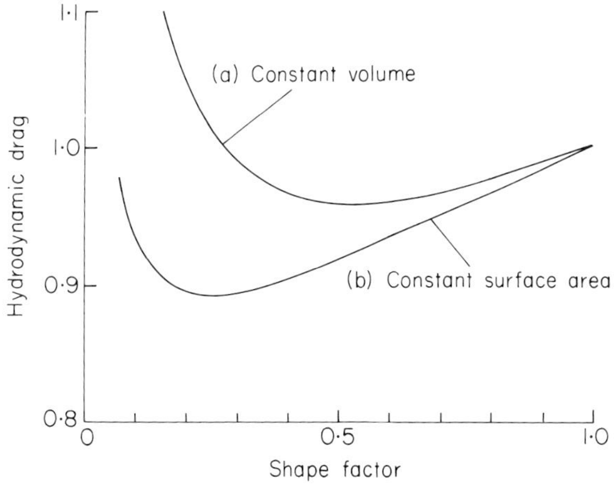

Fig. 10.1. Variation with shape of the viscous drag on prolate spheroids. The shape factor is the ratio of semimajor to semiminor axis lengths (being near zero for a needle-shaped object and unity for a sphere). The drag variation is given (a) at constant volume and (b) at constant surface area. (From Roberts 1981.)

Though the viscous drag forces are not influenced by the sense of the motion, they are a function of the overall shape. However, relative differences in drag associated with different shapes are not in general as great as they would be at high Reynolds numbers. Among various shapes, a sphere has relatively low drag, though a prolate spheroid, or sausage shape with the same surface area, has slightly less (Fig. 10.1). Many flagellates, including some dinoflagellates, are essentially spheres or flattened spheres. However the slow-moving forms do tend to have outlandish shapes (see Chapter 2), as in various *Ceratium* species. Such shapes have a higher viscous drag than a sphere of comparable size, but the relative difference in hydrodynamic resistance is much less than it would be at high Reynolds number, where the inertia of displaced fluid would also have to be considered. Thus, the shape of a dinoflagellate is less constrained by hydrodynamics than that of a fish, and may therefore display adaptive responses to many other aspects of its natural history. In fact, some unusually shaped organisms, such as *Dinophysis acuta*, are among the fastest swimmers (see Table 10.7).

3 SWIMMING BEHAVIOUR

3.1 General behaviour

Flagellar movement and dinoflagellate swimming were examined by several early investigators (Bütschli 1885; Metzner 1929; Peters 1929; Jahn *et al.* 1963).

Most studies dealt with the *dinokont* dinoflagellates, in which the two flagella are laterally inserted, with one lying in a transverse groove, or *cingulum* (girdle) and the other lying in a longitudinal groove or *sulcus* and projecting posteriorly (see Chapter 2).

The transverse flagellum is ribbon-shaped, with the axoneme lying along the outer (distal) edge of the ribbon and a *striated strand* lying on the inner (proximal) edge. The fact that the striated strand is shorter than the axoneme no doubt accounts for the curved shape that keeps the whole ribbon in the cingulum, and also for the ruffled or spiral shape of the outer (axonemal) edge, which in turn has important bearing on the mechanism of swimming (Fig. 10.6).

The longitudinal flagellum is ribbon-shaped in some species (e.g. *Ceratium* spp.), with the axoneme on one edge and a contractile paraxial rod or *retractile fibre* on the other: the latter may be homologous to a similar structure in the euglenids (F. J. R. Taylor, pers. comm.). In other species the longitudinal flagellum is not ribbon-shaped but appears to be a typical eukaryotic flagellum.

In desmokonts, where the swimming behaviour has been much less studied, the two flagella insert anteriorly. One is ribbon-shaped and usually curves back around the body during swimming, though there is no cingulum (see also Chapter 2). We shall now focus on the more numerous, and perhaps more evolutionarily-specialized dinokonts.

Most dinokonts swim in a spiral path, with the same side—usually the ventral (with the sulcus)—always facing the axis of the spiral.

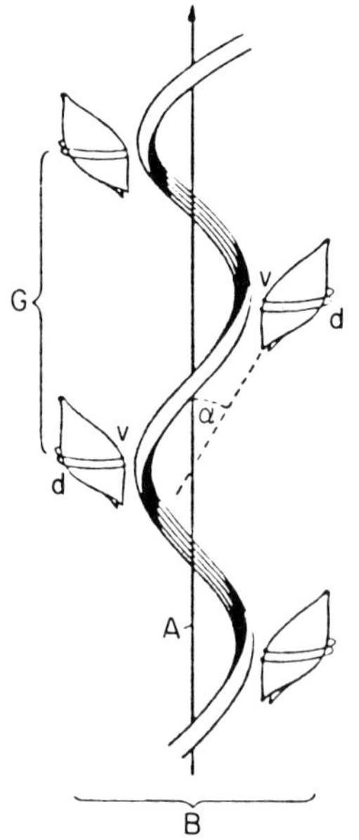

Fig. 10.2. Parameters of spiral swimming *Protoeridinium claudicans* (from Peters 1929). A axis of spiral, B width or amplitude of spiral, α pitch angle, G wavelength of spiral, v ventral side, d dorsal side.

Since most flagellates and ciliates spiral as they swim, a clear terminology is required for the sense of spiralling. Cells that rotate in a clockwise sense, looking in the direction of their motion, are said to be left-turning, or *laeotropic*. The opposite is right-turning or *dexiotropic*.

It appears that the dinoflagellates are laeotropic. Metzner (1929) observed this with four freshwater species, and recently Gaines & Taylor (1985) have analysed fifty two marine and brackish species using video-recordings and came to the same conclusion. Earlier writers, especially Peters (1929; see also Kofoid & Swezy 1921), reported reversals in the sense of spiralling, but Metzner and Gaines & Taylor attribute this to an optical illusion produced as the cell moves back and forth through the focal plane of the low-power microscope lens.

Some dinokonts swim without spiralling, including several large, slow-swimming species of *Ceratium*, *C. tripos* has been well studied because of its size and slow movement. Some *Ceratium* species however do usually spiral. Metzner (1929) concluded, from experiments with a model, that the shape of *C. cornutum* would cause it to spiral as it moved forward, even if its flagellar movement did not. One of the functions of spiral swimming is probably to stabilize the direction of overall motion, compensating to any curvilinear bias during the spiral. Non-spiralling species, such as *C. tripos*, have a tendency to rock back and forth as they swim.

Some of the smaller species have been seen to spin in one place for long periods (many authors, e.g. Kofoid & Swezy 1921). The reason for this is not known. An exotic form of swimming is seen in *Pratjetella* (*Leptodiscus*) *medusoides* and *Craspedotella pileolus*, which look very much like microscopic medusae, and move by medusa-like contractions (Chatton 1952). If the water is disturbed they contract suddenly. Contractions are caused by a layer of parallel non-actin filaments (Cachon & Cachon 1984).

We now turn to a more detailed consideration of the roles of the transverse and longitudinal flagella in dinokonts. Although most authors have assigned different roles to these two organelles, so different in structure and arrangement, it should be noted that Lindemann (1928) found that the swimming of the freshwater *Hemidinium nasutum* was virtually unchanged if either of the two were missing, indicating that the system can be quite flexible.

3.2 The longitudinal flagellum

The principal function of the longitudinal flagellum is thought to be orientation, but there is evidence that it can also contribute to forward movement. As noted above, Lindemann (1928) found that *H. nasutum* cells without longitudinal flagella swam with almost undiminished speed, the main difference being that the anterior of such cells had an added circular motion during swimming. More recently however, Hand & Schmidt (1975) measured swimming speed in *Gyrodinium dorsum* cells with normal, shortened or no longitudinal flagella, and

Table 10.2. Linear velocity of *Gyrodinium dorsum* cells with normal length, short or no longitudinal flagella. (From Hand & Schmidt 1975)

		Velocity ($\mu m\ s^{-1}$)		
Type of flagellum	No. of cells	Range	Mean	Standard Deviation
Normal length	28	254–454	324	43·8
Short	13	120–316	240	47·0
None	25	93–224	147	28·5

found a positive correlation between flagellar length and swimming speeds (Table 10.2), indicating some direct or indirect propulsive contribution from the longitudinal flagellum. The contribution to rotation however appears to be insignificant in *G. dorsum*, since cells with or without a longitudinal flagellum rotated at the same rate. Metzner (1929) and also Jahn *et al.* (1963) observed that the longitudinal flagellum generates backward currents which can be observed with suspensions of ink or polystyrene particles (Fig. 10.3).

In *Ceratium cornutum* (Metzner 1929) and in *C. tripos* and *C. furca* (Jahn *et al.* 1963), stroboscopic observations revealed that the longitudinal flagellum beats with a plane wave very close in shape to a sine wave. Brokaw & Wright

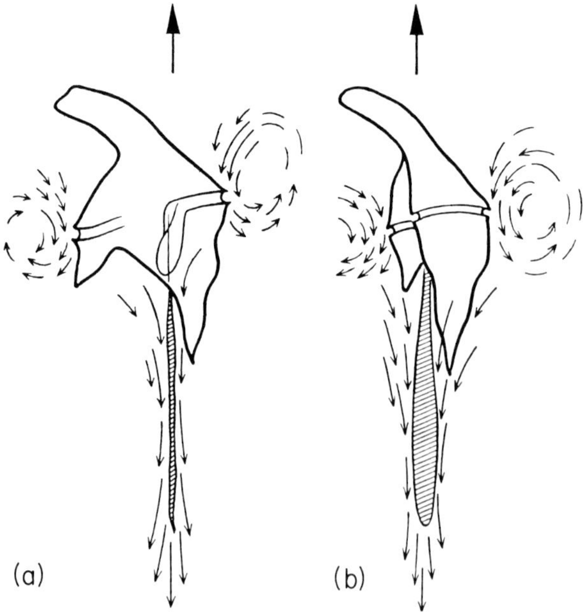

Fig. 10.3. Currents produced by the flagella of *Ceratium cornutum*: (a) ventral view; (b) side view. (From Metzner 1929.)

(1963) showed that the plane waves of the *Ceratium* longitudinal flagellum are actually semicircles joined by straight lines, rather than sine waves. This may also fit somewhat better with the sliding microtubule model of axonemal bending, although the ribbon-shaped longitudinal flagellum of *Ceratium* is not a very typical flagellum. Peters (1929), Butschli (1885) and others had reported cylindrical spiral waves of various shapes. Metzner, however, commenting on Peters' report, notes that it would be very easy to mistake the planar beat for a cylindrical spiral one in spiral swimmers, since the plane of beating is spiralling around with the organism. The species studied most extensively by Peters, *C. tripos* (not a spiral swimmer) was later shown by Jahn *et al.* (1963) to have a planar beat. It is of course always conceivable that the longitudinal flagellum may produce a cylindrical spiral wave form in other strains of *C. tripos*, or in other species.

The role of the longitudinal flagellum in cell orientation during the stop response preceding phototaxis has been demonstrated in *Ceratium cornutum* (Metzner 1929) and *Gyrodinium dorsum* (Hand & Schmidt 1975). In both species, if a light beam of suitable wavelength (see p. 371) is directed normal to the spiral swimming path, the cell keeps turning until the eyespot (a light antenna, p. 373) on the ventral face points to the light source, whereupon the cell stops, pauses, then swings the longitudinal flagellum out from its groove until it points in a latero-anterior direction. Then it begins again to beat, forcing the cell to turn until it points toward the light (Fig. 10.4) The longitudinal flagellum then straightens out and returns to the ventral groove, and the cell begins swimming in a spiral course toward the light source.

With low intensity stimuli, the stop response does not occur and the cell reorients during normal spiral swimming. This response is thought to be accomplished also by the longitudinal flagellum, though this has not been demonstrated.

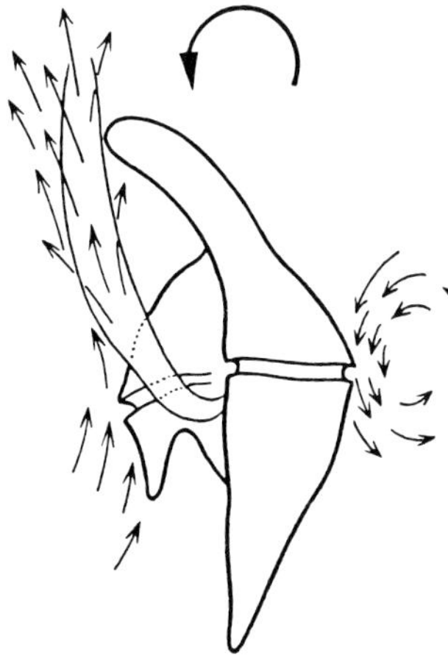

Fig. 10.4. Change in orientation of *Ceratium cornutum* by altering the position of the longitudinal flagellum. (From Metzner 1929.)

In addition to propulsion and orientation, the longitudinal flagellum is also responsible for a type of avoidance reaction ('Schreckbewegung') to mechanical stimuli. This was described for various species by Schütt (1895), Peters (1929), Afzelius (1969), and in great detail by Maruyama (1981) for *Ceratium tripos*. When the cell body is stimulated mechanically (either artificially, or by swimming into debris, another organism, or the surface of the medium), the longitudinal flagellum rapidly retracts into a series of folds which are then twisted secondarily in a right-handed helix (Fig. 10.5). The effect of this rapid retraction is to stop the cell and pull it backwards for a short distance. Peters (1929) likens this to the avoidance response of *Paramecium* (Jennings 1906). Maruyama (1981) suggests that the rapid retraction may be due to a *retractile fibre* (R-fibre) which runs parallel to the axoneme in the flagellum. Since the flagellum has a somewhat flattened elliptical cross-section with the axoneme running along one edge, and since the R-fibre in the retracted state appears clearly shorter than the axoneme, this is plausible. Standard fixation methods also cause retraction in *Ceratium*, and Maruyama suggests that a similar phenomenon may account for electron microscopic observations of folds or loops in longitudinal flagella of other species (Herman & Sweeney 1977; Berman & Roth 1979).

Re-extension usually occurs a few seconds after retraction but may take longer. The unfolding usually starts at the base, whereas during retraction folding usually starts at the tip.

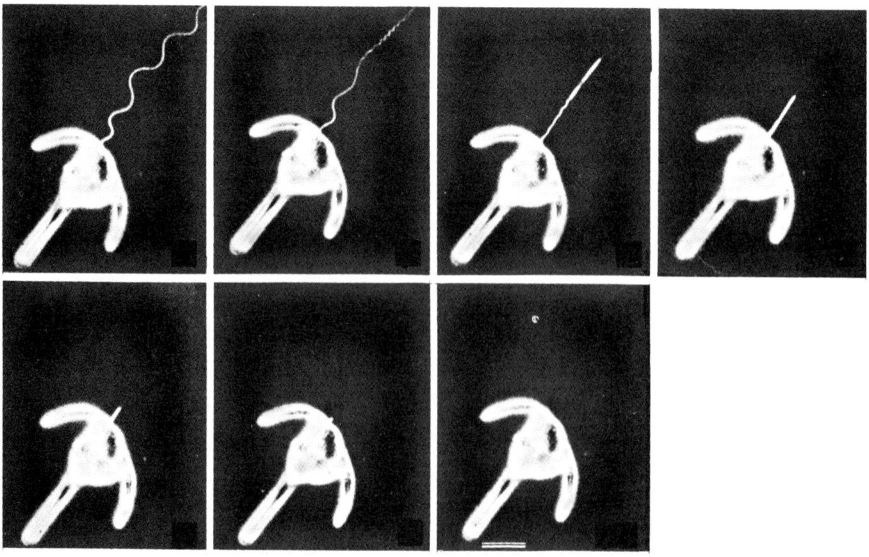

Fig. 10.5. Successive microcinematographs of a retraction of the *Ceratium tripos* longitudinal flagellum, 72 frames s^{-1}. Bar = 50 μm (from Maruyama 1981).

In addition to these rapid retraction movements, Peters also observed cells swimming backwards for long periods, by a method similar to that used during turning in the orientation response in phototaxis. This was accomplished by lifting the longitudinal flagellum out of its groove, as in the orientation response, and pointing it forward. Flagellar beating then causes the cell to move backward in a spiral path. How the cell lifts and controls the direction of beating of the longitudinal flagellum is not understood, but the retractile fibre is probably involved.

3.3 The transverse flagellum

This unique apparatus is thought to account for much of the forward motion as well as the spiral turning in dinokonts. Support for this view comes from the studies cited above, by Lindemann (1928) and Hand & Schmidt (1975), of swimming by cells that lack a longitudinal flagellum.

In most early treatments the transverse flagellum was described as a spiral appendage producing a spiral, rotating undulation in the cingulum. Since observations of currents produced in suspensions of polystyrene spheres or ink particles indicated that a forward thrust was produced all along the cingulum (Figs 10.3, 10.4), theories were proposed to derive this from a spiral undulation. Jahn *et al.* (1963) built a macroscopic spiral wire model which produced appropriate currents when rotating in a groove. However, there was no attempt to scale hydrodynamically by using a fluid of high viscosity. Both Metzner and Jahn *et al.* (the latter evidently unaware of Metzner's prior work) noted that a spiral rotating in a groove, with only one side exposed to the medium, should generate a thrust of the kind seen experimentally (Fig. 10.6a). Recent studies, however, show that the flagellum itself is ribbon-shaped, with a *striated strand* running along the edge opposite the axoneme.

Taylor (1975) noted that the axoneme is longer than the striated strand. In certain scanning electron micrographs (Fig. 2.3d, e) the axoneme had a 'ruffled' configuration (Taylor 1975; Herman & Sweeney 1977). Gaines and Taylor

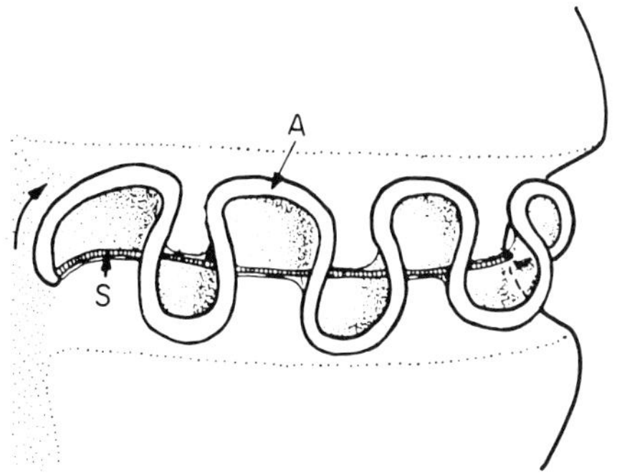

Fig. 10.6. Lateral view of model of transverse flagellum from Gaines & Taylor (1985). In this view, the axonemal wave travels from left to right, and the organism rotates in the same direction. Scale bar = 1 μm.

(1985) have concluded that this 'ruffled configuration' is really a fixation artifact which appears in certain species, and that in living material the axoneme always has a spiral or semi-spiral configuration, distal to the striated strand, as seen in SEM photographs of other species (Fig. 2.3c) (Berdach 1977; Berman & Roth 1979; Rees & Leedale 1980). (The axoneme is coiled within the flagellum, but not the propulsive surface of the flagellum, i.e. the axoneme does not spiral around the striated strand.)

Leblond & Taylor (1976) considered in some detail the hydrodynamics of propulsion by the transverse flagellum, since the ribbon-shaped organelle does not rotate within the cingulum, they suggested that the axonemal edge may oscillate in an asymmetric path (Fig. 10.7c). A symmetric beat as in Fig. 10.7b would not give a net propulsive force, but a helical or hemihelical beat would; it is possible for the axoneme itself (but not the propulsive surface) to beat in a helical wave, to yield a net propulsive force.

Leblond & Taylor (1976) thought also that the transverse flagellum was attached to the cell by a series of strands along the inner (proximal) edge of the ribbon. Subsequent studies, however, suggest that the strands were only flagellar hairs which, through a fixation artefact, pointed in toward the cingulum in some preparations, instead of outward as they do usually.

Gaines & Taylor (1985) have determined, using video-recordings, that: (i) the transverse flagellum invariably beats from the flagellar base to the tip, moving around the cell in the same direction in which the cell is turning; (ii) a circumferential current is generated in the medium in the reverse sense. They attribute this direction of flow to outward-projecting mastigonemes that are present but not held in alignment after fixation. These would operate as in such flagellates as *Ochromonas* and *Crithidia* (see, e.g. Holwill & Sleigh 1967) to pull the cell around in a laeotropic turn. A puzzling aspect of this very interesting model is that in most species the girdle spirals around the cell toward the rear, so that the circumferential current generated in this way would have a forward component that would counter the forward propulsion of the cell. If the model is correct, then this is presumably of little importance, since the cells do swim forward.

Most authors have supposed that the spiral turning component of swimming seen in many dinokonts arises from a torque generated by the transverse flagellum. If there are species with a spiral-beating longitudinal flagellum this might also contribute a torque in the opposing sense to the cell. On the other hand, Lindemann (1928) observed that spiralling in *H. nasutum* occurred virtually unchanged with either flagellum missing! (In *H. nasutum* the transverse flagellum only goes about half-way around the cell, and is apparently easily detached. It would be of great interest to repeat Lindemann's observations with other species). The spiralling may come simply from a combination of the pitch of the cell's axis with regard to the axis of the spiral, plus a continual bias toward the ventral direction.

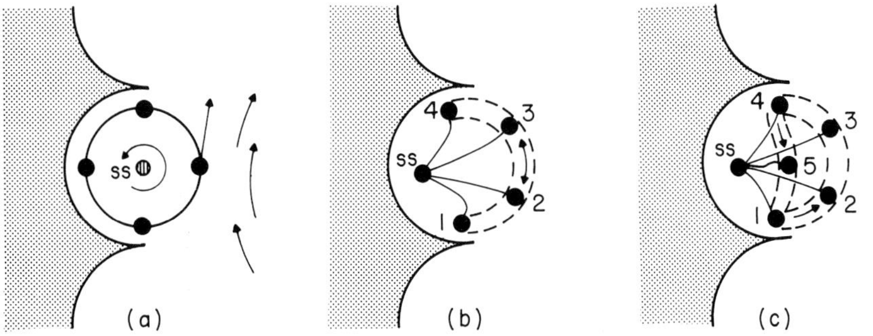

Fig. 10.7. (a) Cross-sectional view of moving transverse flagellum, according to the model of Metzner (1929) and Jahn & Bovee (1964). (b) Cross-sectional view of axonemal path of a ribbon-shaped transverse flagellum with symmetrical beat and; (c) with pseudo-helical beat SS, striated strand. (From Leblond & Taylor 1976.)

Clearly, much remains to be done before we will understand the mechanisms of swimming in dinokonts. Plausible qualitative models have been proposed for various aspects, but as yet there is no detailed quantitative theory for this unique and highly successful type of locomotion.

4 SENSORY RESPONSES

4.1 Light

Many dinoflagellates, probably all the photosynthetic forms, display a positive phototaxis. This has been studied mainly in dinokonts with a spiral swimming motion: in particular, *Ceratium cornutum* (Metzner 1929), *Gyrodinium dorsum* (Hand 1970; Hand & Schmidt 1975; Forward & Davenport 1968, 1970; Forward 1970, 1973, 1975), and *Gymnodinium sanguineum* (= *splendens*) (Forward 1974, 1976).

For these, a typical response to sudden illumination from the side has several parts.

1 The stop response: spiral movement ceases when the ventral surface faces the light. It has been suggested by several authors that this may be homologous to 'photophobic' responses in other organisms, such as *Euglena*.

2 Orientation: this was described in an earlier section. The longitudinal flagellum rises out of the sulcus, points anteroventrally, then beats and pushes the cell around until it points toward the light. Hand & Schmidt (1975) thought that in *Gyrodinium dorsum* the transverse flagellum does not move during the orientation, noting the absence of currents around the cingulum. Metzner, however, did detect currents from the transverse flagellum when *Ceratium cornutum* was turning (Fig. 10.4).

3 After orientation, the cell proceeds in the usual manner toward the light. In addition, the cell maintains its light-orientation during swimming by fine adjustments in the path. The mechanism for this is unknown but probably involves slight changes in the orientation of the longitudinal flagellum.

We shall now consider (i) the mechanism of light reception, and (ii) the difficult problem of the transduction mechanism(s) for generating the behavioural response described above.

RECEPTOR PIGMENTS

We begin with the pigments involved in reception. Since a large amount of the various pigments obtained by extraction methods are used for photosynthesis, the usual way to identify a photoreceptor pigment is by measuring the *action spectrum* of behaviour versus wavelength and correlating this with the absorption spectrum of a known pigment. (This approach requires caution, however: Song and Walker (1981) discuss in detail various pitfalls in arguing from *in vivo* response to *in vitro* absorption.)

Action spectra of the stop response and/or phototaxis have been obtained for *Gyrodinium dorsum* (Forward 1970, 1973), *Gymnodinium sanguineum* (Forward 1974), *Peridinium trochoideum, Protogonyaulax* (=*Gonyaulax) catenella* and the desmokont *Prorocentrum micans* (Halldal 1958). In the first four of these the spectra are similar to each other and different from other algae, suggesting that a distinctive pigment is involved. There is a peak at about 280 nm and in the region of 450–475 nm, with little absorption around 370 nm, where flavoproteins usually absorb. Foster & Smyth (1980) note that the visible peak is at a shorter wavelength than rhodopsin, but that the spectrum suggests a carotenoid. A possibility might be peridinin, a carotenoid restricted to, and abundant in the dinoflagellates, where it functions in a protein complex as an accessory pigment in photosynthesis (Song *et al.*, 1976 and Prezelin, Chapter 5, this volume). Another possibility suggested by Foster & Smyth is a rhodopsin with the peak shifted to the blue, as in some higher organisms.

The action spectrum for the desmokont *Prorocentrum micans*, on the other hand is quite different, with a peak at 570 nm. Foster & Smyth compare this to the cryptophyte *Cryptomonas* (Watanabe & Furuya 1974), suggesting that in both cases accessory pigments are involved.

In the case of *G. dorsum*, it was found that the sensitivity of the stop response to blue light varies with the exposure of the organism to red or far-red light (Forward & Davenport 1968). This has given rise to the hypothesis that an accessory system involving a phytochrome modulates the sensitivity of the receptor for the primary, photophobic response (Forward 1973). Song *et al.* (1979) have presented a detailed biochemical model for such a possibility. A similar red–far red effect was not found with *Gymnodinium sanguineum* however, so the involvement of a second, phytochrome system may not be general.

The tidepool organism, *Peridinium gregarium* (Lombard & Capon 1971) was 'activated' (i.e. stimulated to swim away from its plaque, see p. 390) by blue light (457 nm) but did not swim toward such light. A yellow (579 nm) light, on the other hand, induced phototaxis but only activated moderately. A complete action spectrum for both effects in this curious organism would obviously be of much interest.

LOCATION AND STRUCTURE OF THE RECEPTOR

Behavioural studies indicate that the dinoflagellate photoreceptor is in many—perhaps most—cases on the ventral side, just under the sulcus, and frequently near the junction of sulcus and cingulum below the base of the longitudinal flagellum. As regards the physical structure of the apparatus, however, there is a truly extraordinary variation. Eyespots have not been found in the desmokonts or in some of the dinokonts referred to above, e.g. *Gyrodinium dorsum, Ceratium* spp. (Foster & Smyth 1980; Piccinni & Omodeo 1975).

Hand (1970) conducted ingenious behavioural experiments to elucidate the nature and position of light receptors in *G. dorsum*. When illuminated simultaneously by two light sources at right angles to each other, most cells orientated in a direction 45° from each light beam; when illuminated by opposing sources, 180° apart, however, the cells exhibited no preferred orientation. Hand interpreted these data as favouring a model involving two light receptors and a shading body. The arrangement of these would be such that, in the first experiment, both receptors are shaded from the two beams when orientated at 45° between them. When the light sources are 180° apart, however there is no orientation in which both can be shaded. Thus, Hand postulated that a *Gyrodinium* cell exhibiting phototaxis has its own receptors fully shaded by the anterior of the cell. But there is no direct microscopic evidence for the putative receptor organelles and the shading body. Other flagellates (e.g. *Chlamydomonas*) exhibit a similar behaviour but do not fit this ultrastructural model.

Other dinokonts, however, such as *Gymnodinium* sp. (Schnepf & Deichgräber 1972) and *Peridinium cinctum* var *westii* (Messer & Ben-Shaul 1969), have simple eyespots, resembling those in some green flagellates, such as *Platymonas*; somewhat more complex eyespots have been studied in *Woloszynskia tenuissimum* (Crawford *et al.* 1971) and *Peridinium balticum* (Thomas & Cox 1973) (see Chapter 3). Foster & Smyth (1980) have shown that such eyespots probably act as light antennae: they greatly magnify light signals of a given wavelength, from a given direction. The light passes through a series of layers of reflecting pigment granules spaced at quarter wavelength intervals (for a given wavelength). Normally, incident light of the selected wavelength is reflected back in phase with the incoming radiation, leading to reinforcement of light intensity just in front of the eyespot, where the receptor pigment is located, probably in the plasma membrane or internal membranes. Interference effects minimize the contribution of light from other directions, or of other wavelengths, thus giving both directional and wavelength selectivity. The discovery that eyespots in dinoflagellates and many other algae are light antennae of the reflection-interference type is a major advance in the unravelling of receptor mechanisms.

Several non-photosynthetic, phagotrophic marine species in the family Warnowiaceae have developed antennae with a lens or ocelloid (Chapter 3B) sometimes backed by a quarter-wave reflecting layer (Francis 1967; Mornin & Francis 1967; Crawford *et al.* 1971; Greuet 1965, 1968a,b, 1977). Unfortunately, little is known of the natural history or physiology of these forms, and we have only the anatomical information so far. It seems most likely that the ocelloid, with a field of view in *Nematodinium* of about 30° (Francis 1967), but much wider in *Erythropsidinium* (Taylor 1980) acts also as a directional antenna. In some species it points forward. *Leucopsis* has about fifteen of these forward-directed organelles. Since these species are non-photosynthetic predators, the ocelloids may be used for catching prey rather than for phototaxis, and pattern-recognition might therefore be useful. Taylor (1980) has suggested that they act as 'range-finders,' leading to attack when contrast (form) is greatest on the retinoid. It is hoped that culture methods for these organisms will be developed so that their behaviour can be studied in detail.

TRANSDUCTION

Essentially nothing is known yet of photosensory transduction in dinoflagellates: the physiological processes coupling stimulus reception to phototactic response. The only glimpse that we have is a study by Forward (1977) in which exposure to the catecholamines dihydroxyphenylalanine (DOPA) and dopamine decreased light sensitivity in *Gymnodinium sanguineum*. Catecholamine-blockers (propranolol, dichloroisoproterenol, dibenzyline) or a DOPA synthesis inhibitor (α-methyl-p-tyrosine) increased sensitivity. 0·01 mM norepinephrine, epinephrine or isoproterenol had no discernible effect. Furthermore, both acetylcholine and the anticholinesterase eserine increased sensitivity, but the cholinergic blocker atropine decreased it. These intriguing data suggest that biogenic amines may be involved in the transduction chain. It would be of great interest to know if this had something to do with the putative phytochrome effect (reported, however, in a different species, *Gyrodinium dorsum*), since acetylcholine has been implicated in the regulation of phytochrome-mediated processes in the roots of higher plants (Jaffe 1980)

TEMPORAL VARIATION OF PHOTOTAXIS

Forward & Davenport (1970) observed a circadian rhythm in the stop response of *Gyrodinium dorsum*. Cells grown in a 12:12 LD cycle had maximum responsiveness 1 hour before the extended light period. Pre-irradiation with red light was required to observe a rhythmic response to blue light, which probably explains earlier failures to find a rhythm in this photoresponse. In *Gymnodinium sanguineum* a circadian rhythm was also present. Cells grown in a 12:12 LD cycle for a week and then kept under low illumination demonstrated greatest reponsivity during the first 4 hours of the original cycle for 3 days. It has been suggested that biogenic amines may function in single cells as transducers of the biological clock (Levandowsky 1981) and so it would be of interest to see if Forward's results with catecholamines and quaternary amines involved the cell's rhythm in some way. Hardeland (1980) reports blocking of some rhythmic effects in a dinoflagellate by propranolol, but attributes this to a non-specific membrane effect.

The probable ecological value of such rhythmic responsivities in the diurnal migration of many dinoflagellates is discussed below (p. 389).

4.2 Temperature

Relatively little is known about behavioural responses of dinoflagellates to temperature gradients, though these are no doubt important in the organism's adaptation to thermoclines in natural waters. Metzner (1929) observed avoidance reactions to heat by *Peridinium cinctum* and *Ceratium cornutum*. Unfortunately, he did not measure the temperatures, but observed that cells swimming toward a heated wire stopped suddenly 'as though there were an invisible wall,' and either remained motionless or swam off in another direction. By putting a light on the other side of the heat source he caused an accumulation at the 'invisible wall.' In some cases he had the impression that the heat may have reversed the phototaxic response, but could not be certain of this. Reimers (1927) had observed aggregation of *P. cinctum* in the colder region of a range from 6°C to 14°C. Metzner attributed Reimer's failure to observe phobic avoidance reactions to the relatively shallow gradients (2° mm^{-1}) in the latter's experiments.

There have been several studies of the effects of temperature on swimming speed (Hand *et al.* 1967; Kamykowski 1986a, b), but these experiments were not designed to detect thermotaxis, and may have simply measured the effects of temperature on metabolism, or of heat stress.

4.3 Electric fields

Verworn (1889a,b) reported that *Peridinium tabulatum* responded to an electric field by swimming to the negative pole. Metzner (1929) confirmed this galvanotaxis with careful studies of *Ceratium cornutum*, which appeared to be much more sensitive to low fields than other flagellates or ciliates in the same preparation. When the field was activated, cells swimming toward the negative pole continued, sometimes with a slight pause. Other cells stopped, reoriented themselves much as in the photic response, then swam toward the negative pole. In other protist groups, particularly ciliates, such responses have been explained electrophysiologically, in terms of the effects on membrane potential (Naitoh & Eckert 1974).

4.4 Gravity

Little detailed work has been done with geotaxis in dinoflagellates, despite its probable importance in vertical migrations. Peters (1929) noted that the marine species which studied all swam upward (negative geotaxis) if permitted. Metzner (1929) reported a positive geotaxis in *Peridinium umbonatum*, but Haupt (1962) considered that this may have been an indirect effect of light, working through photosynthesis to change the O_2/CO_2 tension in the medium. The bioconvective patterns formed by populations of the marine heterotrophic *Crypthecodinium cohnii* were considered to be due to a negative geotaxis, though this has not been rigorously shown experimentally (Levandowsky *et al.* 1975; Childress *et al.* 1975a, b).

Since many planktonic and other species have either diurnal or tidal rhythms of vertical migration (p. 383, 390, 391), and the downward movement is much faster than expected from passive settling by Stokes' law, it is clear that cells must swim downward at times. Even when this occurs during times of surface illumination, it is probably not a negative phototaxis, since that has never been seen experimentally, even in circadian rhythm studies. This implies that the downward migration is probably a positive geotaxis. Since the cells appear to swim upward by geotaxis at other times (including in continual darkness), there is a strong implication that they can change the sense of geotaxis with time.

Nothing is known of the mechanism(s) of geotaxis, negative or positive, in dinoflagellates. In other micro-organisms, such as various spermatozoa and the ciliate *Paramecium*, Roberts (1970, 1975, 1981) suggested that geotaxis is probably caused by a purely mechanical effect of cell shape, due to a torque generated by the centre of gravity and the centre of resistive (viscous) forces. This torque orients the cell to point upward (*Paramecium*) or downward (spermatozoa) as it swims. This is a plausible theory, and perhaps may apply to dinoflagellates; one might also be able to explain a shift from negative to positive geotaxis as due to something like a change in the relative contributions of longitudinal and transverse flagella to the forward thrust.

Kessler (1985a, b) reports an interaction between geotaxis and downwelling currents in chlorophycean flagellates, due to the torque generated by the position of the centre of mass, posterior to the geometric centre in these cells. It would be of interest to know if a similar 'hydrodynamic focusing' could occur in dinoflagellates.

A true gravitational sensory apparatus, analogous to the statoliths found in metazoa, has been reported in the ciliate *Loxodes* (Fenchel & Finlay 1984, 1985). It may be that crystalline inclusions found in some species of dinoflagellates could have a similar function.

4.5 Pressure

It is possible that the presumed gravity response is actually a pressure response (a *barotaxis*), since vertical swimming automatically confronts the organism with a pressure gradient. There are no published studies on pressure effects in dinoflagellates however, nor have potential baroreceptor organelles been identified. In fact, it would be difficult to distinguish experimentally between a geotaxis and a barotaxis on the basis of behavioural observations alone

4.6 Magnetism

Magnetotaxis is well-known in bacteria, where it is thought to play a role in orienting cells towards sediments (Blakemore 1975). Recent reports of magnetotaxis in other algal groups (de Barros *et al.* 1981; Araujo *et al.* 1986) suggest the possibility that a similar mechanism might occur in dinoflagellates.

4.7 Chemical stimuli

Given the widespread occurence of phagotrophy and parasitism, and the (presumably) universal presence of sexual interactions in dinoflagellates, it is likely that most, if not all of them have a chemical sense. So far, however, chemosensory responses have been detected and studied in only a few heterotrophic species.

Spero (1985) observed that cells of the voracious phagotroph *Gymnodinium fungiforme* accumulated in capillaries containing shrimp extract or various amino acids or other compounds (Table 10.3). He also saw responses to shrimp extract by three heterotrophs, *Crypthecodinium cohnii*, a *Gyrodinium* sp., and *Oxyrrhis marina*. The latter only responded after being precultured with the flagellate *Dunaliella salina* as a food organism. In addition, the following species were tested and showed no reaction to shrimp extract: *Peridinium balticum, P. cinctum, Kryptoperidinium (= Peridinium) foliaceum, Gonyaulax diegenesis, G. sphaeroidea, Scrippsiella sweeneyae, Ptychodiscus (= Gymnodinium) brevis, G. sanguineum* and *Gyrodinium resplendens*.

Fitt (1985) studied chemosensory responses in *Symbiodinium microadriaticum*, a symbiont of various invertebrates, using a capillary assay similar to that of Spero. Cells freshly isolated from the tissues of starved *Cassiopeia xamachana*, a scyphozoan, were attracted to inorganic and organic nitrogen sources: NH_4Cl, $NaNO_3$, urea, alanine, glycine and arginine, at levels as low as 10^{-8} M. The response was absent from cultured cells, unless they had been starved two days

Table 10.3. Threshold detection levels of *Gymnodinium fungiforme* to organic compounds. All amino acids are L-amino acids. (From Spero 1979)

Compound	Molar concentration	Compound	Molar concentration
Amino acids		Amino acids (*cont.*)	
Alanine	1×10^{-6}	Lysine	No response @ 1×10^{-3}
Arginine	1×10^{-4}	Methionine	1×10^{-4}
Asparagine	1×10^{-5}	Phenylalanine	No response @ 1×10^{-3}
Aspartic acid	1×10^{-5}	Proline	1×10^{-6}
Cysteine	1×10^{-5}	Serine	1×10^{-8}
Glutamine	1×10^{-4}	Taurine	1×10^{-8}
Glutamic acid	1×10^{-3}	Threonine	1×10^{-6}
Glycine	1×10^{-8}	Tryptophan	1×10^{-5}
Hydroxy proline	1×10^{-5}	Valine	1×10^{-5}
Histidine	1×10^{-5}	Other organic compounds	
Isoleucine	1×10^{-4}	Dextrose	1×10^{-7}
Leucine	1×10^{-3}	Glycolic acid	No response @ 1×10^{-3}
		Trimethylamine-HCl	2×10^{-4}

or more. In addition, normal, symbiont-containing hosts were not attractive unless they had been fed to repletion. Aposymbiotic hosts were attractive. Fitt suggests that in the normal, unfed host, ammonia or other nitrogenous wastes are taken up by the symbionts as fast as they are excreted.

Chemosensory responses by *Crypthecodinium cohnii*, a heterotroph often found in samples of decomposing seaweed (Pringsheim 1963), were also studied with an assay based on the tendency of this organism to embed itself in agar gels containing active substances (Hauser *et al.* 1975a). Responses were detected to aqueous extracts of seaweeds, and to pure compounds that might be produced by decomposing seaweeds (Table 10.4), such as the sulphur compound dimethyl-β-propiothetin, produced by seaweeds and many other algae, including *C. cohnii* itself. The quaternary amine, betaine, a common nitrogenous waste product, is produced by *C. cohnii*, and is also a growth stimulant (Provasoli & Gold 1962). Since it is also a neurohumour in metazoa, other pharmacologically interesting compounds were examined (Table 10.5) (Hauser *et al.* 1975b).

The strikingly low threshold of the negative response (fewer cells embedding in the test agar) to epinephrine is comparable to the sensitivity of postsynaptic membranes in vertebrate neurones. The functional significance of such responses is not known. There is a short, early report that *Noctiluca* produces a catecholamine chromatographically distinct but pharmacologically indistinguishable from epinephrine (Ostland 1954) and, as noted earlier, *Amphidinium carterae* produces three choline esters (Taylor *et al.* 1974). These scattered observations, and those of Forward (1977) on adrenergic and cholinergic effects in phototaxis already discussed, suggest the involvement of biogenic amines in the sensory-motor system of dinoflagellates.

Table 10.4. Chemosensory responses of *Crypthecodinium cohnii* to ecologically interesting compounds. (From Hauser *et al.* 1975a)

Chemical	Active range (g 100 ml^{-1})
Positive response	
Dimethyl-β-propiothetin	10^{-5}–10^{-1}
α-L-fucose	10^{-4}–10^{-2}
Betaine	10^{-4}–10^{-3}
Fucus spp. infusions	
Negative response	
Formalin	10^{-6}–10^{-1}
Glutathione*	10^{-2}–10^{-1}
Agar hydrolyzate†	10^{-5}–10^{-1}
Protamine sulphate	10^{-5}–10^{-1}
L-Glutamic acid	10^{-6}–10^{-2}
L-Aspartic acid	10^{-5}–10^{-2}
Inert (no response)	
Dimethyl acetothetin chloride	—
Glucose	—
cAMP	—
α-D-fucose	—
Trimethylamine hydrochloride	—
Mannose	—
Rhamnose	—
Macrocystis spp. infusions	—
Sargassum spp. infusions	—

*The reduced form was used, but may have become oxidized under experimental conditions (solution in hot agar).
†Supernatant from agar, autoclaved at pH 2·0.

Of various blocking agents for these responses, antitubulins proved effective. Colchicine had no effect, but the *Vinca* alkaloids, vincristine and vinblastine, at 10^{-7}M, abolished negative but not positive responses (Table 10.6) (Levandowsky *et al.* 1975). In the case of vinblastine, some negative responses were transformed to positive ones. The possibility that microtubules, or at any rate tubulin, may be involved in chemoreception or chemosensory transduction in these organisms is intriguing from a comparative viewpoint, since many metazoan chemoreceptor cells have either modified axonemes or other prominent microtubular structures (Barber 1974; Atema 1975).

C. cohnii has also been shown to respond chemically to CO_2 gradients. When dense populations are observed with the microscope they often form dynamic aggregations near the coverslip edge, or around entrapped bubbles, resembling those produced by 'aerotaxis' in other flagellates such as *Bodo* (Fox 1921) and *Euglena* (Colombetti & Diehn 1978). However, in the case of *C. cohnii* the

Table 10.5. Chemosensory responses of *C. cohnii* to neurologically active compounds*. (From Hauser *et al.* 1975b)

Compound	Active concentration range (M)
Positive response	
Dihydroxyphenylalanine	5×10^{-7}–5×10^{-4}
Betaine	$8{\cdot}5 \times 10^{-6}$–$8{\cdot}5 \times 10^{-5}$
Glycine	10^{-6}–10^{-5}
Carbamylcholine (carbachol)	5×10^{-7}–5×10^{-4}
Negative response	
Epinephrine	5×10^{-15}–5×10^{-8}
Norepinephrine	4×10^{-9}–5×10^{-6}
Choline citrate	3×10^{-7}–3×10^{-3}
Choline bitartrate	2×10^{-7}–2×10^{-3}
Choline chloride	7×10^{-7}–7×10^{-3}
L-Glutamic acid	7×10^{-8}–7×10^{-4}
Tryptophan	5×10^{-7}–5×10^{-4}
Putrescine dihydrochloride	6×10^{-7}–6×10^{-3}
Taurine	8×10^{-7}–8×10^{-3}
5-Hydroxytryptophan	5×10^{-4}
Phenylalanine	6×10^{-3}
Ergotamine	$1{\cdot}7 \times 10^{-11}$–$1{\cdot}7 \times 10^{-5}$
Guanethidine sulphate	5×10^{-10}–5×10^{-4}
Nicotine	6×10^{-5}–6×10^{-3}

*The following chemicals were inert: serotonin, reserpine, amphetamine, eserine salicylate, atropine, decamethonium bromide, dibenamine hydrochloride.

aggregations persist in the absence of oxygen, and proved to be a response to a gradient in dissolved CO_2 instead (Hauser *et al.* 1978). Cells swimming down the CO_2 gradient were inhibited from turning and had greater mean free path than cells swimming up the gradient, or in the absence of a gradient. This inhibition was calcium-dependent, and the aggregation disappeared in the

Table 10.6. Effects of antitubulins on chemosensory responses of *C. cohnii.** (From Levandowsky *et al.* 1975)

Antitubulin	Response to choline (10^{-4}M)	Response to DOPA (10^{-4}M)
Control	−	+
Vincristine sulphate (10^{-7}–10^{-4}M)	0	+
Vinblastine sulphate (10^{-7}–10^{-4}M)	+	+
Trifluralin (3×10^{-7}M)	±	+
Griseofulvin (10^{-8}–10^{-6}M)	0	+
Colchicine (10^{-7}–10^{-4}M)	−	+

*Responses classified as negative (−), positive (+), random (no response detected) (0) or variable (±). Antitubulins were added to a suspension of organisms at the levels indicated.

presence of calcium antagonists such as barium salts, or calcium-binders such as oxalic acid; addition of more calcium then restored the response.

Other kinds of dynamic aggregation or swarming are observed during mating in many dinoflagellate species, including *C. cohnii*. Von Stosch (1973) gave these the apt name of 'dancing figures.' A chemosensory response to some secreted pheromone seems likely, but this has not been isolated.

The chemoreceptors in dinoflagellates are as yet unidentified, and very little is known about the sensory transduction mechanisms (Levandowsky & Hauser 1978). Two hints available from the work with *C. cohnii* are: (i) the elimination of negative, but not positive responses in the agar plate assay by low levels of antitubulins; (ii) the calcium requirement for turning inhibition in the response to a CO_2-gradient.

4.8 Mechanical stimuli

Little is known in detail about mechanoreception in dinoflagellates, but it is clearly important in the natural history of the cell. As noted already (p. 367), cells that collide with objects retract the longitudinal flagellum rapidly, jerking themselves back from the obstruction.

In bioluminescent species, mechanical stimuli (shear in the medium) elicit luminescence, which may be of adaptive value in preventing predation by zooplankters (Esaias & Curl 1972; Buskey *et al.* 1983; Buskey & Swift 1983, 1985). In some species at least, this response is inhibited by low intensity light (Esaias *et al.* 1973).

5 AMOEBOID PHENOMENA, PHAGOTROPHY AND PARASITISM

These topics are thoroughly covered in Chapters 6 and 13, and we shall not review them here, but merely note their importance in dinoflagellate behaviour and natural history. Indeed, amoeboid movements are widespread, probably universal in dinoflagellates, a fact underappreciated by many experimentalists. They occur during phagotrophy and parasitism, as well as in mating.

Phagotrophy is also much underestimated as a feature of dinoflagellate nutrition. It has been observed in a large number of species, including many photosynthetic forms. Predation may vary from unselective to highly selective, and prey range from bacteria to small metazoa and seaweed tissues. There are many histophagic and/or parasitic species. Though many accounts of such behaviour exist, little work has been done on the underlying mechanisms (see Spero 1982; Spero & Moreé 1981).

A number of species have trichocysts which may sometimes aid in immobilizing and attacking prey. Some warnowiacean species have well developed nematocysts (Greuet & Havasse 1977), similar in structure to those of coelenterates with presumably a similar function though this has not been experimentally verified.

For examples and detailed discussion of this rich and understudied behavioural area, we refer the reader to Chapters 6 and 13.

6 MATING BEHAVIOUR

Until recent years sexuality in dinoflagellates was controversial, many workers being sceptical of early scattered reports (see review by Beam & Himes 1980 and Chapter 14). However recent studies of Von Stosch (1965, 1973), Zingmark (1970), Beam & Himes (1974, 1980), Tuttle & Loeblich (1974), Pfiester (1975, 1976), Spector *et al.* (1981) and others have firmly established the phenomenon. In many cases there is a behavioural prelude, in which gametes swarm in small groups, swimming around each other for long periods (von Stosch's 'dancing figures'). Following these manoeuvres, cells fuse (Fig. 10.8) or attach by protoplasmic strands through which genetic exchange can occur. It seems likely that chemosensory and perhaps also mechanosensory cues are involved in all this, but these have not been studied yet. (See Chapter 14 for more details of mating behaviour).

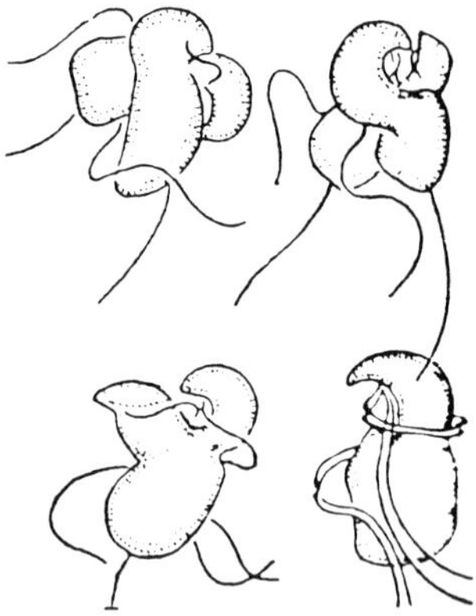

Fig. 10.8. Four stages in the mating fusion of *Coccidinium mesnili* gametes (from Chatton 1952).

7 ECOLOGICAL ASPECTS OF BEHAVIOUR

The general ecology of dinoflagellates is treated in Chapter 11, but it would not be possible to deal adequately with dinoflagellate behaviour without some discussion of its ecological implications. Indeed, behaviour is crucial in the ecology of this group. Vertical migrations interact with water movements, pycnoclines, wind, light and tide conditions to determine distribution on a time scale much faster than that of growth and grazing by predators. Table 10.7 gives swimming speeds for different species, measured under various conditions in the field and the laboratory. Kamykowski (1986a) found a positive correlation between swimming speed and body length.

Table 10.7. Swimming speeds of dinoflagellates. For ease of comparison, these have all been converted to metres per hour. However, many come from microscopic observations, and it is therefore not clear how long or under what conditions these rates would occur in the field. Data correct to two decimal places.

Organism	Speed (m h^{-1})	Reference
Katodinium rotundatum	1·1–2·0	Throndsen 1973
Gyrodinium sp.	0·72–0·88	
Prorocentrum micans	0·17–0·48	Kamykowski 1981
Gonyaulax polyedra	0·36–1·08	Kamykowski 1986a
Cachonina niei	0·18–0·36	
Gyrodinium dorsum	0·18–1·26	
Prorocentrum micans	0·36–0·90	
Peridinium gregarium	2·8–6·5	Lombard & Capon 1971
Gyrodinium sp.	1·0	Hand *et al.* 1965
Ceratium furca & *C. fusus*	0·5–1·0	Hasle 1954
Peridinium trochoideum	0·13–0·25	
Various spp.	0·4–0·7	Blasco 1978
Protoperidinium cf. *quinquecorne*	5·4	Horstmann 1980
Gonyaulax polygramma	1·8	Kamyowski 1980
Prorocentrum mariae-lebouriae	0·3	Tyler & Seliger 1981
Gymnodinium sanguineum	1·1	Cullen & Horigan 1981
Amphidinium carterae	0·05–0·44	Gittleson *et al.* 1974
Dinophysis acuta	1·8	Peters 1929
Protoperidinium pentagonum	0·72–1·1	
P. subinerme	1·0	
P. ovatum	0·45–0·90	
P. claudicans	0·45–1·1	
P. crassipes	0·36	
Ceratium furca	0·8	
Ceratium furca (chain)	0·6	
C. tripos	0·2–0·9	
C. fusus	0·23–0·9	
C. longipes	0·06	
C. horridum	0·03–0·12	
C. macroceros	0·055	
Gonyaulax polyedra & *Cachonina niei*	1·0	Eppley *et al.* 1968
Ceratium hirundinella	0·7–1·0	Heaney & Eppley 1981
Hemidinium nasutum	0·38	Metzner 1929
Peridinium umbonatum	0·9	
P. cinctum	0·72	
Ceratium cornutum	0·45–0·83	

7.1 Vertical migration, periodicity and physicochemical barriers

Many, perhaps most freshwater and marine free-living planktonic dinoflagellates tend to migrate up and down in the water column with a circadian rhythm. Since the populations typically swim upward in the morning and downward in the evening, many early workers considered this to be simply a phototaxis (Baldi 1941; Hasle 1950, 1954; Forward 1974, 1975) but, although phototaxis may play a role in some cases, it cannot be the entire or even the main mechanism. As noted above (p. 376), it is observed in both field and laboratory populations that downward migrations reach greater depths than could be obtained by passive, Stokesian settling, so that active downward swimming is implied, i.e. a negative phototaxis. But there are no reports of negative phototaxis in all the studies of dinoflagellate phototaxis, including studies of circadian changes in phototaxis (Weiler & Karl 1979; Forward & Davenport 1970). Dinoflagellate phototaxis does not change in sense, only in sensitivity, during the diel cycle. Second, in various laboratory studies, upward migration began before the light phase. After entrainment, *Cachonina niei* continued rhythmic migrations up and down for 24 hours in complete darkness (Eppley *et al.* 1968). Finally, Weiler & Karl (1979) observed rhythmic up and down migrations of a laboratory culture of *Ceratium furca* for 6 days in complete darkness. When cells moving upward were given only lateral illumination, they collected on the illuminated side of the container but continued to move up. Furthermore it is difficult to imagine that a response to chemical cues could be involved in the well oxygenated water columns where these phenomena often occur.

Thus, we seem to be left with geotaxis (or perhaps barotaxis; see p. 377) as the principal mechanism of vertical migrations—not a simple positive or negative geotaxis, but one which changes its sign rhythmically. As noted above, this poses an interesting problem: a plausible model for geotaxis in a single cell is that proposed by Roberts (1981), in which vertical orientation during swimming is determined by the cell's shape and centre of gravity (see p. 376). To change the sense of geotaxis with this model would seem to require some architectural changes in the cell, and it would be of great interest to know whether there are such changes in upward and downward swimming populations. Other possibilities include an as yet undiscovered geotactic organelle, or a rhythmically varying magnetotaxis (see p. 377).

Even if phototaxis is not the main driving mechanism in upward migration, light does have an auxiliary role in the phenomenon. Thus, many field observations indicate that on cloudy days, in the absence of direct sunlight, blooms may fail to appear at the surface, or are much more diffuse (e.g. Baldi 1941). On the other hand, strong illumination under certain nutritional conditions may keep the cells from rising to the very surface (see p. 386).

PHYSICOCHEMICAL BARRIERS TO VERTICAL MIGRATION

These are of great importance. Heaney & Talling (1980; see also Reynolds 1976) observed that the freshwater species *Ceratium hirundinella* did not enter an anoxic hypolimnion, and this was attributed to its inability to cross the thermocline. Blasco (1978) noted that even a weak pycnocline served as a barrier to vertical migration of marine dinoflagellates off Baja California. Kamykowski & Zentara (1977) noted that a thermocline was an effective barrier to *Cachonina niei* downward migration but not for *Amphidinium carterae*. In subsequent experiments with four marine species, it appeared that the size of the temperature gradient was less important than the absolute temperatures involved, and the effect was greatly influenced by other environmental conditions (Kamykowski 1981). Tyler & Seliger (1981) showed that *Prorocentrum mariae-lebouriae*, although strongly phototactic and illuminated from above, would not cross a halocline from 15‰ to 10‰. This was correlated with observations of reduction in swimming rate after rapid salinity changes: a 5‰ change caused circular swimming. In Chesapeake Bay, *P. mariae-lebouriae* remains trapped under a halocline at certain seasons see p. 389). Taylor *et al.* (1966) studied a similar case in which a *Pyrodinium bahamense* bloom could not cross a pycnocline caused by a lens of fresh water in a small Jamaican bay. Kamykowski (1980) observed populations of *Gonyaulax polygramma* and *Gymnodinium simplex* which migrated diurnally between depths of 15 m and 32 m, underneath both a halocline and a thermocline.

An intriguing phenomenon is the observed tendency of dinoflagellates to avoid the top 0·5 m of the water column. Hasle (1950) and the others have attributed this to excess light at the surface (see also Talling 1971). Another possibility is that the surface water might experience more small-scale wind generated turbulence, which could serve as a mechanical cue: some dinoflagellates are mechanically delicate and tend to immobilize when agitated (White 1976). However, Eppley *et al.* (1968) report a similar phenomenon with *Cachonina niei* populations migrating under low illumination in a 10 m deep tank (with, presumably, no particular turbulence at the surface). Perhaps the cells, after reaching the meniscus, swim down for a distance and then drift for a bit before swimming up again (in several marine species observed by Peters (1929) the longitudinal flagellum retracted when the cell reached the meniscus). This could cause a net steady-state accumulation just below the surface. It is also conceivable, though perhaps not very plausible, that slight thermal or chemical gradients (e.g. O_2 or CO_2) could provide a sensory cue near the surface in some cases. Most recently, Heaney & Eppley (1981; also Heaney & Furnas 1980) have shown that nitrogen depletion causes dinoflagellates to migrate downward from high light intensity, and this may account for some cases of stable subsurface accumulations, particularly when thermocline barriers are also present.

7.2 Open waters

We shall be largely concerned here with interactions between the organism's sensory-motor system, operating on a microscopic scale, and the medium, whose dynamic properties are best defined on a larger scale. To express the latter conveniently, there is another dimensionless parameter, the *Richardson number*, which expresses the ratio of the density gradient to the square of the shearing motion:

$$R_i = G\frac{d\rho}{dz}\bigg/\bar{\rho}\left(\frac{du}{dz}\right)^2$$

Here G is the gravitational constant, ρ is the density of the water (a function of salinity and temperature), $\bar{\rho}$ is the average density over the relevant interval, u is mean water velocity and z is depth. For small R_i (below about 0·25), turbulence and vertical mixing are more important than horizontal flow; for high R_i, vertical stability tends to prevent vertical mixing.

The Richardson number is potentially an extremely useful index of water stability. The main problem with it is that the data needed to calculate it are not available from synoptic surveys of the usual sort.

Some dinoflagellates do not tolerate the high mechanical shearing of high turbulence, and prefer stable waters (White 1976). Autotrophic dinoflagellates and other flagellates in stable waters have the behavioural advantage over diatoms and blue-green algae of being able to swim to the appropriate light levels, shading their competitors. On the other hand *extremely* stable waters tend to be nutrient-depleted in the photic zone, through lack of advective mixing with nutrient-rich deep waters. This will select for a specialized oligotrophic biota, including some of the flattened, slower-swimming dinoflagellate species. Thus, the fast-swimming dinoflagellates occupy a middle region of the Richardson number scale, in which turbulence is low, but not so low as to cause undue depletion of nutrients. Margalef *et al.* (1979) have refined this concept, dividing open water autotrophic species into two major groups: (i) typical fast-swimming species, with rounded, chlorophyll-rich cells, which live in high-nutrient, moderate to high stability waters. (ii) oligotrophic species, with flattened, chlorophyll-poor cells, often associated with nutritional symbionts, with a tendency to complex morphology which maximizes surface area and water flow, and disrupts gradients at the cell surface (see Chapter 11 for further discussion).

Recent progress in understanding open water phytoplankton ecology has come largely from specialized efforts focusing on upwelling systems and on red tides, in both of which dinoflagellate behaviour is prominent, as well as the Richardson number of the medium. Wyatt (1974, 1975) provides a very useful, clear review of physical mechanisms by which sorting and concentration of migrating organisms might occur through convergences, convection cells, and other movements of the medium (see also Chapter 11A).

Particular attention has been given to the system off the coast of Peru. In a one week study (Huntsman *et al.* 1980) an initial diatom-dominated population was transformed into one with 80% *Gymnodinium sanguineum* (= *splendens*). The authors considered it likely that the latter species was originally advected from elsewhere and then persisted during a period of stability (low winds and solar heating of surface waters, hence greater stability and no upwelling). After high winds resumed, the dinoflagellates persisted at about 50% of the population. The authors speculated that a poleward current at a depth of 10 m may have prevented them from being carried northward out of the system. By migrating down at night they might have been carried by this current, compensating more or less for the daytime equatorward transport at the surface. Subsequently the dinoflagellates declined in the face of vigorous upwelling and mixing, which favoured the diatoms (see also Barber & Smith 1981).

Wind-driven convection cells (Langmuir cells) are a widespread collecting mechanism which, in some cases, may favour motile organisms such as dinoflagellates. Where the mixing layer is shallow due to a thermocline or shallow bottom, long cylindrical convection cells form parallel bands of upwelling and downwelling, at an angle to the direction of the wind (Fig. 11A.13). Buoyant organisms, such as *Noctiluca*, and upward-swimming organisms such as many smaller dinoflagellates, might be expected to automatically collect in the downwelling bands, providing the downwelling current is not too strong—a function of the wind strength. Actually, as pointed out by Stommel (1949), they will tend to move toward the centre of the cell (a centripetal motion) as well, and this in combination with the upward drift would produce a spiral trajectory along the cell. Evans & Taylor (1980) consider these possibilities in more detail, including the effects of variation in swimming speed.

On a larger, oceanic scale, where two water masses of different density meet, a *front* is formed of downwelling water that can also concentrate buoyant or upward-swimming phytoplankton (Fig. 11A.13). Such fronts appear to be responsible for concentrations of *Gyrodinium aureolum* and other dinoflagellates in the English channel and nearby waters (Pingree *et al.* 1975; Holligan 1979). This mechanism is known to occur in open waters, but may be of particular importance in coastal regions and especially bays, where bay water and ocean water often come from fronts at the point of contact (see p. 389).

Kamykowski (1974, 1978) suggested that interactions of internal waves or (in lakes) seiches of the thermocline with vertical migration by phytoplankton can have an important sorting and concentrating effect, particularly during periods of light winds or calm. More field work is required to determine the importance of these mechanisms.

Finally, for very high Richardson numbers (low winds or calm, plus thermal stratification), the medium becomes still and behaviour is a major determinant of dinoflagellate distribution. Wyatt & Horwood (1973) note the possibility that swimming dinoflagellates might collect at the level of optimum light intensity for photosynthesis, giving them a great advantage over non-motile forms such as diatoms, and possibly leading to bloom formation. While it is true that such calm periods are often associated with red tides, this behavioural mechanism, though plausible, has not yet been demonstrated to occur.

A very early red tide model that requires high Richardson numbers was that of Kierstead & Slobodkin (1953), originally proposed to account for *Ptychodiscus* (= *Gymnodinium*) *brevis* blooms in the Gulf of Mexico. In this, it is assumed that suitable water for growth of the bloom organisms occurs in patches of finite size, and that the organisms diffuse laterally by random swimming. There is a critical patch size at which growth can dominate diffusion and a bloom will occur. Refinements, in the form of predation, density-dependent growth rates, and recruitment from a 'seed population' (e.g. cysts) were considered by Wroblewski *et al.* (1975) and Levandowsky (1979; Levandowsky & White 1977). Kierstead & Slobodkin supposed that the patches of suitable water could be lenses of bay water washed out to sea (there is evidence of correlation of rainfall with red tides off the Florida west coast). As for the random diffusion of organisms, very little is known about the horizontal component of swimming in natural dinoflagellate populations (or, for that matter, in laboratory populations!), most research having addressed the problem of vertical migration.

In the absence of wind and current, the depletion of nutrients might limit the growth of dinoflagellates in such still waters. Winet (1975) suggested that bioconvection of dense dinoflagellate populations could perhaps provide some mixing with nutrient-rich deeper layers. This is a phenomenon in which negatively geotactic, or positively phototactic populations collect in a dense surface layer, over the less dense medium below. This hydrodynamic instability leads to convection, since organisms fall down in plumes, entraining the water with them, and then swim up again individually (Levandowsky *et al.* 1975a). The phenomenon is often seen in laboratory cultures, and sometimes in natural populations during complete calm.

Much current thinking about the Florida coast *Pt. brevis* red tides however, favours a primarily physical mechanism for bloom initiation, involving the intrusion of an oceanic loop current. This colder water would resuspend sedimented cysts, concentrate them and foster germination (Steidinger & Haddad 1981; Haddad & Carter 1979). Behaviour of the organisms is not a feature here, except for the subsequent vertical migration.

7.3 Bays

Much of the work and thinking discussed above involved events in open coastal waters, where red tides are common. From these to a consideration of semi-enclosed bays is almost a continuum, the main differences being in the reduction of overall scale, and the importance of haloclines as well as thermoclines. Seliger and his associates have made a major contribution with their studies of *Prorocentrum mariae-lebouriae* red tides in Chesapeake Bay (Tyler & Seliger

1978, 1981; Seliger *et al.* 1979, 1975). This population overwinters in the southern bay by the York River estuary, where a convergence between saline water and York River water, plus phototaxis or negative geotaxis lead to vertical concentrations. In the spring, with changes in the flow rate of the York River, a large population is transported in a subsurface saline current to the north bay where it collects in an upwelling near the Susquehanna River estuary. During transport the organisms form a horizontal subsurface layer, being phototactic and/or geotactic but unable to cross the halocline (see Fig. 11A.12).

Earlier studies of *Pyrodinium bahamense* blooms in much smaller tropical bays, showed that a combination of upward taxis of the organism, and tidal and wind-driven variation in convergences of oceanic, bay and fresh waters combined to maintain the population in the bay (Seliger *et al.* 1970, 1971). Such mechanisms may be widespread (e.g. Anderson & Stolzenbach 1985; Cohen 1985). Thus, vertical migration is important in many red tides and other blooms.

In all their studies, Seliger and his associates refer to the upward swimming, as a phototaxis, but as noted earlier (p. 384), a negative geotaxis is perhaps more likely, and has not been ruled out in these cases.

7.4 Tide pools and tidal flats

Decreasing the physical scale yet further, we now consider two rather remarkable cases. The first is *Peridinium gregarium*, studied by Lombard & Capon (1971) in tide pools of Southern California, where it forms aggregations in a secreted mucoid matrix. During periods of low temperatures and heavy tides the population stays on the bottom in small plaques, with thousands of cells in each. During other seasons it leaves the bottom in the morning and goes to new mucoid masses on the upper surface of rocks. Such masses often float to the surface. Whenever an incoming tide reaches the pool, the surface cells migrate to the bottom with the first incoming wave. This behaviour was also followed in the laboratory. In one aquarium, a submerged bottle contained rocks with plaques. Every evening the populations swam 'home' into the bottle with the rocks, and every morning they streamed out on the way to the surface. How did they find their way to the rocks in the bottle?

Another case is a population of *Peridinium* cf. *quinquecorne* in a shallow, eutrophic tidal flat in the Philippines, studied by Horstman (1980). This population moves to a very distinct near-surface layer during bright sunlight, but only during the incoming tide. At reduced light, or just before high tide regardless of the light, the organisms seems to disappear. In the laboratory it was found that they swim down and attach themselves to the dark underside of solid objects. In this case then, a combination of tidal rhythm and photic response appears to maintain the dense population in the tidal area. Such phenomena may be widespread in shallow tidal waters.

7.5 Sharp interfaces: epineustonic and sand-dwelling forms

Timpano and Pfiester (1985) observed a peculiar behaviour in the freshwater species, *Cystodinium bataviense*. Mature cells are found in the epineuston, resting on top of the water's surface film. As the cell ages it becomes larger and possibly denser, and eventually falls through the surface film into the water, where it liberates swimming zoospores. These are strongly phototactic, and swim up to the surface. They then swim horizontally just under the surface, changing their swimming from a strongly apical direction to a 'sidelong gliding motion'. Eventually the zoospore places its ventral side against the under side of the surface film and stops moving. It then splits open its theca, producing an elongate, asymmetric cell, and ruptures through the surface film to become epineustonic. One can see small depressions in the surface film at points where the mature cell rests upon it.

Finally, mention should be made of the rhythmic migrations of sand-dwelling dinoflagellates (Herdman 1923; Eaton & Simpson 1979). *Amphidinium herdmanii* migrates to the surface of tidal sandflats with a tidal rhythm. Presumably, there is limited space for free-swimming in the interstitial water; this might pose a problem for a theory of geotaxis involving orientation during swimming, since the mean free path might be too constricted. It is possible that chemical gradients at the sand surface provide an orienting signal.

ACKNOWLEDGEMENTS

We thank the following colleagues for helpful conversation or correspondence (names arranged alphabetically): R. Barber, M. Estrada, R. Forward, K. Foster, W. Hand, S. Heaney, D. Kamykowski, L. Provasoli, H. Spero, F. J. R. Taylor, G. Vargo.

8 REFERENCES

AFZELIUS B. A. (1969) Ultrastructure of cilia and flagella. In *Handbook of Molecular Cytology* (Ed. A. Lima-de-Faria), pp. 1219–1241. North-Holland, Amsterdam.

ANDERSON D. M. & STOLZENBACH K. D. (1985) Selective retention of two dinoflagellates in a well-mixed estuarine embayment: the importance of diel vertical migration and surface avoidance. *Mar. Ecol. Prog. Ser.* **25**, 39–50.

ATEMA J. (1975) Stimulus signal transmission along microtubules in sensory cells: an hypothesis. In *Microtubules and Microtubule Inhibitors* (Eds. M. Borgers & M. de Brabander), pp. 247–260. North-Holland, Amsterdam.

BALDI E. (1941) Mechanismus der Rotfärbung im Tovel-see. *Arch. Hydrobiol.* **38**, 299–302.

BARBER R. T. & SMITH W. O. (1981b) The role of circulation, sinking and vertical migration in physical sorting of phytoplankton in the upwelling centre at 15°S. *Coastal and Marine Science*, I, pp. 366–371. Amer. Geophys. Union, Washington DC.

BARBER V. (1974) Cilia in sense organs. In *Cilia and Flagella* (Ed. M. A. Sleigh), pp. 403–441. Academic Press, New York.

BEAM C. A. & HIMES M. (1974) Evidence for sexual fusion and recombination in the dinoflagellate *Crypthecodinium* (*Gyrodinium*) *cohnii*. *Nature* (*Lond.*) **250**, 435–436.

BEAM C. A & HIMES M. (1980) Sexuality and meiosis in dinoflagellates. In *Biochemistry and Physiology of Protozoa*, 2nd edn, Vol. 3 (Eds M. Levandowsky & S. H. Hutner), pp. 171–207. Academic Press, New York.

BERDACH J. T. (1977) *In situ* preservation of the transverse flagellum of *Peridinium cinctum* (Dinophyceae) for scanning electron microscopy. *J. Phycol.* **13**, 243–251.

BERMAN T. & ROTH I. L. (1979) The flagella of *Peridinium cinctum* fa. *westii*: *in situ* fixation and observation by scanning electron microscopy. *Phycologia* **18**, 307–311.

BLAKEMORE R. (1975) Magnetotactic bacteria. *Science* **190**, 377–379.

BLASCO D. (1978) Observations on the diel migration of marine dinoflagellates off the Baja California coast. *Mar. Biol.* **46**, 41–47.

BROKAW C. J. & WRIGHT L. (1963) Bending waves of the posterior flagellum of *Ceratium*. *Science* (*N.Y.*) **142**, 1169–1170.

BUSKEY E. J., MILL L. & SWIFT E. (1983) The effects of dinoflagellate bioluminescence on the swimming behaviour of a marine copepod. *Limnol. Oceanogr.* **28**, 575–578.

BUSKEY E.J. & SWIFT E. (1983) Behavioural responses of the coastal copepod *Acartia hudsonica* to simulated dinoflagellate bioluminescence. *J. Exp. Mar. Biol. Ecol.* **72**, 43–58.

BUSKEY E. J. & SWIFT E. (1985) Behavioural responses of oceanic zooplankton to simulated bioluminescence. *Biol. Bull.* **168**, 263–275.

BUTSCHLI O. (1885) Protozooen. In *Klassen und Ordnungen der Thierreich*, Bd. I (Ed. Bronn).

CACHON J. & CACHON M. (1984) An unusual mechanism of cell contraction: Leptodiscinae dinoflagellates. *Cell Motility* **4**, 41–55.

CHATTON E. (1952) Classe des Dinoflagelles ou Peridiniens. In *Traite de Zoologie I* (Ed. P.-P. Grassé), pp. 309–406. Masson, Paris.

CHILDRESS S. (1981) *Mechanics of Swimming and Flying*. Cambridge University Press.

CHILDRESS W. S., LEVANDOWSKY M. & SPIEGEL E. A. (1975a) A mathematical model of bioconvection. J. Fluid Mech. **63**, 591–613.

CHILDRESS W. S., LEVANDOWSKY M. & SPIEGEL E. A. (1975b) Solutions of equations describing bioconvection. In *Swimming and Flying in Nature*, vol. I (Eds T. Wu, C. Brokaw & C. Brenner), pp 361–375. Plenum, New York.

COHEN R. R. H. (1985) Physical processes and the ecology of a winter dinoflagellate bloom of *Katodinium rotundatum*. *Mar. Ecol. Prog. Ser.* **26**, 135–144.

COLOMBETTI G. & DIEHN B. (1978) Chemosensory responses toward oxygen in *Euglena gracilis*. *J. Protozool.* **25**, 211–217.

CRAWFORD R. M., DODGE J. D. & HAPPEY C. M. (1971) The dinoflagellate genus *Woloszynskia*: I. Fine structure and ecology of *W. tenuissimum* from Abbot's Pool, Somerset. *Nova Hedw.* **19**, 825–840.

CULLEN J. J. & HORRIGAN S. G. (1981) Effects of nitrate on the diurnal vertical migration, carbon to nitrogen ratio, and the photosynthetic capacity of the dinoflagellate *Gymnodinium splendens*. *Mar. Biol.* **62**, 81–89.

DE ARAUJO F. F. T., PIRES M. A., FRANKEL R. B. & BICUDO C. E. M. (1986) Magnetite and magnetotaxis in algae. *Biophys. J.* (In press).

DE BARROS L., ESQUIVEL D. M. S. & OLIVEIRA L. P. H. (1981) Magnetotactic algae. *An. Acad. Bras. Cienc.* **54**, 258–259.

EATON J. W. & SIMPSON P. (1979) Vertical migrations of the intertidal dinoflagellate *Amphidinium herdmanae* Kofoid and Swezy. In *Cyclic Phenomena in Marine Plants and Animals* (Ed. E. Naylor & E. G. Hartnoll), pp. 339–345. Pergamon, Oxford.

EPPLEY R. W., HOLM-HANSEN O. & STRICKLAND J. D. H. (1968) Some observations on the vertical migrations of dinoflagellates. *J. Phycol.* **4**, 333–340.

ESAIAS W. E. & CURL H. H. (1972) Effect of dinoflagellate bioluminescence on copepod ingestion rates. *Limnol. Oceanogr.* **17**, 901–906.

ESAIAS W. E., CURL H. C. & SELIGER H. H. (1973) Action spectrum of a low intensity, rapid photoinhibition of mechanically stimulable bioluminescence in the marine dinoflagellates *Gonyaulax catenella, G. acatenella* and *G. Tamarensis. J. cell Physiol.* **32**, 363–372.

EVANS G. T. & TAYLOR F. J. R. (1980) Phytoplankton accumulation in Langmuir cells. *Limnol. Oceanogr.* **25**, 840–845.

FENCHEL T. & FINLAY B. J. (1984) Geotaxis in the ciliated protozoon *Loxodes. J. Exp. Biol.* **110**, 17–33.

FENCHEL T. & FINLAY B. J. (1985) The structure and function of Muller vesicles in loxodid ciliates. *J. Protozool.* **33**, 69–76.

FITT W. K. (1985) Chemosensory responses of the symbiotic dinoflagellate *Symbiodinium microadriaticum* (Dinophyceae). *J. Phycol.* **21**, 62–67.

FORWARD R. B. (1970) Change in the photoresponse of the dinoflagellate *Gyrodinium dorsum* Kofoid by red and far-red light. *Planta* (*Berl.*) **92**, 248–258.

FORWARD R. B. (1973) Phototaxis in a dinoflagellate: action spectra as evidence for a two-pigment system. *Planta* (*Berl.*) **111**, 167–178.

FORWARD R. B. (1974) Phototaxis by the dinoflagellate *Gymnodinium splenden* lebour. *J. Protozool.* **21**, 312–315.

FORWARD R. B. (1975) Dinoflagellate phototaxis: pigment systems and circadian rhythm as related to diurnal migration. In *Physiological Ecology of Estuarine Organisms* (Ed F. J. Vernberg), pp. 367–381. Univ. S. Carolina, Columbia, USA.

FORWARD R. B. (1976) Light and diurnal vertical migration: photobehaviour and photophysiology of Plankton. *Photochemical and Photobiological Reviews* **1**, 157–209.

FORWARD R. B. (1977) Effects of neurochemicals upon a dinoflagellate photoresponse. *J. Protozool.* **24**, 401–405.

FORWARD R. B. & DAVENPORT D. (1968) Red and far-red light effects on a short-term behavioural response of a dinoflagellate. *Science* (N.Y.) **161**, 1028–1029.

FORWARD R. B. & DAVENPORT D. (1970) The circadian rhythm of a behavioral photoresponse in the dinoflagellate *Gyrodinium dorsum. Planta* (*Berl.*) **92**, 259–266.

FOSTER K. W. & SMYTH R. D. (1980) Light antennas in phototactic algae. *Microbiol. Rev.* **44**, 572–630.

FOX H. (1921) An investigation into the cause of the spontaneous aggregation of flagellates and into the reactions of flagellates to dissolved oxygen. I. *J. Gen. Physiol.* **3**, 483–499.

FRANCIS D. W. (1967) On the eyespot of the dinoflagellate, *Nematodinium. J. exp. Biol.* **47**, 495–501.

GAINES G. & TAYLOR F. J. R. (1985) Form and function of the dinoflagellate transverse flagellum. *J. Protozool.* **32**, 290–296.

GITTLESON S. M., HOTCHKISS S. K. & VALENCIA F. G. (1974) Locomotion in the marine dinoflagellate *Amphidinium carterae* (Hulburt). *Trans. am microscop. Soc.* **93**, 101–105.

GREUET C. (1965) Structure fine de l'ocelle d'*Erythropsis pavillardi* Kofoid et Swezy. Peridinien Warnowiidae Lindemann. C. R. Acad. Sci. Paris **261**, 1904–1907.

GREUET C. (1968a) *Leucopsis cylindrica* nov. gen., nov. sp. Peridinien Warnowiidae Lindemann: considerations phylogenetiques sur les Warnowiidae. *Protistol.* **4**, 419–422.

GREUET C. (1968b) Organization ultrastructurale de deux Peridiniens Warnowiidae, *Erythropis pavillardi* Kofoid et Swezy et de *Warnowia pulchra* Schiller. *Protistol.* **4**, 209–230.

GREUET C. (1977) Evolution structurale et ultrastructurale de l'ocelloide d'*Erythropsidinium pavillardi* Kofoid et Swezy (Peridinien Warnowiidae Lindemann) au cours des divisions binaire et palintomiques. *Protistol.* **13**, 127–143.

GREUET C. & HOVASSE R. (1977) A propos de la genese des nematocystes de *Polykrikos schwartzi* Butschli. *Protistol.* **13**, 145–149.

HADDAD K. & CARTER K. (1979) Oceanic intrusion: one possible initiation mechanism of red tide blooms on the west coast of Florida. In *Toxic Dinoflagellate Blooms* (Eds. D. L. Taylor & H. H. Seliger), pp. 269–274. Elsevier North-Holland, Amsterdam.

HALLDAL P. (1985) Action spectra of phototaxis and related problems in Volvocales, *Ulva* gametes and Dinophyceae. *Physiol. plant.* **14**, 133—139.

HAND W. G. (1970) Phototactic orientation by the marine dinoflagellate *Gyrodinium dorsum* Kofoid. I. A mechanism model. *J. exp. Zool.* **174**, 33–38.

HAND W. G., COLLARD P. & DAVENPORT D. (1965) The effects of temperature and salinity change on the swimming rate in the dinoflagellates *Gonyaulax* and *Gyrodinium*. *Biol. Bull.* **128**, 90–101.

HAND W. G., FORWARD R. B. & DAVENPORT D. (1967) Short-term photic regulation of a receptor mechanism in a dinoflagellate. *Biol. Bull.* **133**, 150–165.

HAND W. G. & SCHMIDT J. A. (1975) Phototactic orientation by the marine dinoflagellate *Gyrodinium dorsum* Kofoid. II. Flagellar activity and overall response mechanism. *J. Protozool.* **22**, 494–498.

HAPPEL J. & BRENNER H. (1973) *Low Reynolds Number Hydrodynamics*. Groningen, Leyden.

HARDELAND R. (1980) Effects of catecholamines on bioluminescence in *Gonyaulax polyedra*. *Comp. Bioch. Physiol.* **66**, 53–58.

HASLE G. R. (1950) Phototactic migration in marine dinoflagellates. *Oikos* **2**, 162–175.

HASLE G. R. (1964) More on phototactic diurnal migration in marine dinoflagellates. *Nytt Mag. Bot.* **2**, 139–147.

HAUPT W. (1962) Geotaxis. In *Handbuch der Pflanzenphysiologie,* Vol. 17, part 2 (Ed. E. Bunning), pp. 390–395. Springer-Verlag, Heidelberg.

HAUSER D., LEVANDOWSKY M., HUTNER S., CHUNOSOFF L. & HOLLWITZ J. (1975a) Chemosensory responses by the heterotrophic marine dinoflagellate *Crypthecodinium cohnii*. *Microb. Ecol.* **1**, 246.

HAUSER D., LEVANDOWSKY M. & GLASSGOLD J. (1975b) Ultrasensitive chemosensory responses by a protozoan to epinephrine and other neurochemicals. *Science* **190**, 285–186.

HAUSER D., PETRYLAK D., SINGER G. & LEVANDOWSKY M. (1978) Calcium-dependent sensory-motor response of a marine dinoflagellate to CO_2. *Nature* (*Lond.*) **273**, 230–231.

HEANEY S. I. & EPPLEY R. W. (1981) Light, temperature and nitrogen as interacting factors affecting diel vertical migration of dinoflagellates in culture. *J. plankton Res.* **3**, 331–334.

HEANEY S. I. & FURNASS T. I. (1980) Laboratory models of diel vertical migration in the dinoflagellate *Ceratium hirundinella*. *Freshwater Biol.* **10**, 163–170.

HEANEY S. I. & TALLING J. F. (1980) Dynamic aspects of dinoflagellate distribution in a small productive lake. *J. Ecol.* **68**, 75–94.

HERDMAN E. C. (1923) Notes on dinoflagellates and other organisms causing discolouration of the sand at Port Erin III. 37th *Ann. Rep. oceanogr. Dept. Univ. Liverpool,* pp. 32–37.

HERMAN E. M. & SWEENEY B. M. (1977) Scanning electron microscope observations of the flagellar structure of *Gymnodinium splendens*. *Phycologia* **16**, 115–118.

HOLLIGAN P. M. (1979) Dinoflagellate blooms associated with tidal fronts around the British Isles. In *Toxic Dinoflagellate Blooms* (Eds. D. L. Taylor & H. H. Seliger), pp. 249–250. Elsevier/North-Holland, Amsterdam.

HOLLWILL M. & SLEIGH M. (1967) Propulsion by hispid flagella. *J. Exp. Biol.* **47**, 267–276.

HORSTMANN U. (1980) Observations on the peculiar diurnal migration of a red tide dinoflagellate in tropical shallow waters. *J. Phycol.* **16**, 481–485.

HUNTSMAN S. A., BRINK K. H., BARBER R. T. & BLASCO D. (1981) The role of circulation and stability in controlling the relative abundance of dinoflagellates and diatoms over the Peru shelf. *Coastal and Marine Science,* I, pp. 357–365. Amer. Geophys. Union, Washington DC.

JAFFE M. (1980) Evidence for the regulation of phytochrome-mediated processes in bean roots by the neurohumor, acetylcholine. *Plant Physiol.* **46**, 768.

JAHN T. L. & BOVEE E. C. (1964) Protoplasmic movements and locomotion of protozoa. In *Biochemistry and Physiology of Protozoa,* 1st Edn, Vol. 3 (Ed S. H. Hutner), pp. 62–130. Academic Press, New York.

JAHN T. L., HARMON W. M. & LANDMAN M. (1963). Mechanism of locomotion in flagellates. I. *Ceratium*. *J. Protozool.* **10**, 358–363.

JENNINGS H. S. (1906) *Behaviour of the Lower Organisms*. Columbia, New York. (Reprinted

1976 by Indiana University Press.)
KAMYKOWSKI D. (1974) Possible interactions between phytoplankton and semidiurnal tides. *J. mar. Res.* **32**, 67–89.
KAMYKOWSKI D. (1978) Organism patchiness in lakes resulting from the interaction between the internal seiche and plankton diurnal migration. *Ecol. Mod.* **4**, 197–210.
KAMYKOWSKI D. (1980) Subthermocline maximums of the dinoflagellate *Gymnodinium simplex* (Lohman) Kofoid and Swezy and *Gonyaulax polygramma* Stein. *Northeast Gulf Science* **4**, 39–43.
KAMYKOWSKI D. (1981) Laboratory experiments on the diurnal vertical migration of marine dinoflagellates through temperature gradients. *Mar. Biol.* **62**, 57–64.
KAMYKOWSKI D. (1986) A survey of laboratory temperature studies of dinoflagellate behavior from the viewpoint of ecological implications. In *Migration: Mechanism and Adaptive Significance*. (Ed. M. A. Rankin) U. Texas Press, Austin (In press).
KAMYKOWSKI D. & MCCOLLUM S. A. (1986) The temperature-acclimated swimming speed of selected marine dinoflagellates. *J. Plankton Res.* (In press).
KAMYKOWSKI D. & ZENTARA S.-J. (1977) The diurnal vertical migration of motile phytoplankton through temperature gradients. *Limnol. Oceanogr.* **22**, 148–151.
KIERSTEAD H. & SLOBODKIN L. B. (1953) The size of water masses containing plankton blooms. *J. mar. Res.* **12**, 141–146.
KOFOID C. A. & SWEZY O. (1921) The free-living unarmored dinoflagellates. *Mem. Univ. Calif.* **5**, 1–526.
LEBLOND P. H. & TAYLOR F. J. R. (1976) The propulsive mechanism of the dinoflagellate transverse flagellum reconsidered. *BioSystems* **8**, 33–39.
LEVANDOWSKY M. (1979) On a class of mathematical models for *Gymnodinium breve* red tides (Appendix to Chapter 12). In *Biochemistry and Physiology of Protozoa*, 2nd edn, Vol. 1 (Eds M. Levandowsky & S. H. Hutner), pp. 394–402. Academic Press, New York.
LEVANDOWSKY M., HAUSER D. & GLASSGOLD J. (1975) Chemosensory responses of a protozoan are modified by antitubulins. *J. Bacteriol.* **124**, 1037.
LEVANDOWSKY M. (1981) Endosymbionts, biogenic amines and a heterodyne hypothesis for circadian rhythms. *Ann. N.Y. Acad. Sci.* **361**, 369–374.
LEVANDOWSKY M & WHITE B. S. (1977) Randomness, time scales and the evolution of biological communities. *Evolutionary Theory* **10**, 69–161.
LEVANDOWSKY M. & HAUSER D. C. R. (1978) Chemosensory responses of swimming algae and protozoa. *Int. Rev. Cytol.* **53**, 145–210.
LIGHTHILL M. J. (1975) *Mathematical Biofluiddynamics*. Soc. Ind. Appl. Math., Philadelphia.
LINDEMANN E. (1928) Uber die Schwimmbewegung einer experimentell eingeisselig gemachten Dinoflagellaten. *Arch. Protistenk.* **64**, 507–510.
LOEBLICH A. R. III (1966) Aspects of the physiology and biochemistry of the Pyrrophyta. *Phykos* **5**, 216–255.
LOMBARD E. H. & CAPON B. (1971) Observations on the tidepool ecology and behaviour of *Peridinium gregarium*. *J. Phycol.* **7**, 188–194.
MARGALEF R., ESTRADA M. & BLASCO D. (1979) Functional morphology of organisms involved in red tides, as adapted to decaying turbulence. In *Toxic Dinoflagellate Blooms* (Eds D. L. Taylor & H. H. Seliger). pp. 89–94. Elsevier, Amsterdam.
MARUYAMA T. (1981) Motion of the longitudinal flagellum in *Ceratium tripos* (Dinoflagellida): a retractile flagellar motion. *J. Protozool.* **28**, 328–336.
MESSER G. & BEN-SHAUL Y. (1969) Fine structure of *Peridinium westii* Lemm., a freshwater dinoflagellate. *J. Protozool.* **16**, 272–280.
METZNER P. (1929) Bewegungstudien an Peridineen. *Z. Bot.* **22**, 225–265.
MITCHELL R. (1983) Ca and protein kinase C: two synergistic cellular signals. *Trends in Bioch. Sci.* **8**, 263–264.
MORNIN L. & FRANCIS D. (1967) The fine structure of *Nematodinium armatum*, a naked dinoflagellate. *J. Microscopie* **6**, 759–772.

NAITOH Y. & ECKERT R. (1974) The control of ciliary activity in protozoa. In *Cilia and Flagella* (Ed. M. Sleigh) pp. 305–352.

NISHIZUKA Y. (1983) Protein kinases in signal transduction. *Trends in Biochem. Sci.* **8**, 13–16.

NORDLI E. (1957) Experimental studies on the ecology of Ceratia. *Oikos* **8**, 200–265.

OSTLUND E. (1954) The distribution of catecholamines in lower animals and their effect on the heart. *Acta Physiol Scand.* **31**, *Suppl.* 112, 5.

PETERS N. (1929) Orts-und Geisselbewegung bei marinen Dinoflagellaten. *Arch. Protistenk.* **67**, 291–321.

PICCINI E. & OMODEO P. (1975) Photoreceptors and phototactic programs in protista. *Boll. Zool.* **42**, 57–79.

PFIESTER L. A. (1975) Sexual reproduction of *Peridinium cinctum* f. *ovoplanum* (Dinophyceae) *J. Phycol.* **11**, 259–265.

PFIESTER L. (1976) Sexual reproduction of *Peridinium willei. J. Phycol.* **12**, 234–238.

PINGREE R. D., PUGH P. R., HOLLIGAN P. M. & FORSTER G. R. (1975) Summer phytoplankton blooms and red tides along the tidal fronts in the approaches to the English channel. *Nature* (*Lond.*) **258**, 672–677.

POHL R. (1962) Die äussere Mechanik der Geisselbewegung. In *Handbuch der Pflanzen-physiologie,* Vol. 17, part 2 (Ed. E. Bünning), pp. 843–875. Springer-Verlag, Heidelberg.

PRANDTL L. (1923) Zur Hydrodynamik der Infusorien. *Naturwiss.* **11**, 640.

PRINGSHEIM E. G. (1963) *Farblose Algen.* Gustav Fischer, Stuttgart.

PROVASOLI L. & GOLD K. (1962) Nutrition of the American strain of *Gyrodinium cohnii. Arch. Microbiol.* **42**, 196–206.

REES A. & LEEDALE G. (1980) The dinoflagellate transverse flagellum: three-dimensional reconstruction from serial sections. *J. Phycol.* **16**, 73–80.

REIMERS H. (1927) Uber die Thermotaxis niederer Organismen. *Jahrb. wiss. Bot.* **67**, 242–290.

REYNOLDS C. S. (1976) Sinking movements of phytoplankton indicated by a simple trapping method II. Vertical activity ranges in a stratified lake. *Br. J. Phycol.* **11**, 293–303.

ROBERTS A. M. (1970) Geotaxis in motile microorganisms. *J. exp. Biol.* **53**, 687–699.

ROBERTS A. M. (1975) The biassed random walk and the analysis of microorganism movement. In *Swimming and Flying in Nature,* Vol. 1 (Eds T. Y. Wu, C. J. Brokaw & C. Brennen), pp. 377–393. Plenum, New York.

ROBERTS A. M. (1981) Hydrodynamics of protozoan swimming. In *Biochemistry and Physiology of Protozoa*, 2nd edn. Vol. 4 (Eds M. Levandowsky & S. H. Hutner), pp. 5–66. Academic Press, New York.

SCHNEPF E. & DEICHGRÄBER G. (1972) Über die Feinbau von Theka, Pusule und Golgi Apparat bei dem Dinoflagellaten *Gymnodinium* sp. *Protoplasma* **74**, 411–425.

SELIGER H. H., CARPENTER J. H., LOFTUS M. & MCELROY W. D. (1970) Mechanisms for the accumulation of high concentrations of dinoflagellates in a bioluminescent bay.

SELIGER H. H., CARPENTER J. H., LOFTUS M., BIGGLEY W. H. & MCELROY W. D. (1971) Bioluminescence and phytoplankton successions in Bahia Fosforescente, Puerto Rico. *Limnol. Oceanogr.* **16**, 608–622.

SELIGER H. H., LOFTUS M. E. & SUBHA RAO D. V. (1975) Dinoflagellate accumulation in Chesapeake Bay. In *Proc. 1st Int. Congr. Toxic Dinoflagellate Blooms* (Ed. V. R. LoCicero), pp. 181–205. Massachussetts Sci. Technol. Found., Wakefield, Mass.

SELIGER H. H., TYLER M. A. & MCKINLEY K. R. (1979) Phytoplankton distributions and red tiles resulting from frontal circulation patterns. In *Toxic Dinoflagellate Blooms* (Eds D. L. Taylor & H. H. Seliger), pp. 239–248. Elsevier/North-Holland, Amsterdam.

SONG P. S., KOKA P., PREZELIN B. & HAXO E. T. (1976) Molecular topology of the photosynthetic light-harvesting pigment complex, peridinin–chlorophyll *a*–protein, from marine dinoflagellates. *Biochemistry* **15**, 4422–4427.

SONG P. S., CHAE Q. & GARDNER J. (1979) Spectroscopic properties and chromophore conformations of the photomorphogenic receptor, phytochrome. *Bioch. Bioph. Acta* **576**, 479–485.

SONG P. S. & WALKER E. B. (1981) Molecular aspects of photoreceptors in protozoa and other microorganisms. In *Biochemistry and Physiology of Protozoa*, 2nd edn, Vol. 4 (Eds M. Levandowsky & S. H. Hutner), pp. 199–233. Academic Press, New York.

SPECTOR D. L., PFIESTER L. A. & TRIEMER R. E. (1981) Ultrastructure of the dinoflagellate *Peridinium cinctum* f. *oviplanum* II. Light and electron microscope observations on fertilization. *Am. J. Bot.* **68**, 34–43.

SPERO H. (1979) *A study of the life-cycle, feeding and chemosensory behaviour of the holozoic dinoflagellate* Gymnodinium fungiforme. Master's thesis, Texas A & M University, USA.

SPERO H. (1982) Phagotrophy in *Gymnodinium fungiforme* (Pyrrhophyta): the peduncle as an organelle of ingestion. *J. Phycol.* **18**, 356–360.

SPERO H. (1985) Chemosensory capabilities in the phagotrophic dinoflagellate *Gymnodinium fungiforme*. *J. Phycol.* **21**, 181–184.

SPERO H. J. & MOREÉ M. D. (1981) Phagotrophic feeding and its importance to the life cycle of the holozoic dinoflagellate *Gymnodinium fungiforme*. *J. Phycol.* **17**, 43–51.

STEIDINGER K. A. & HADDAD K. (1981) Biologic and hydrographic aspects of red tides. *Bioscience* **31**, 814–819.

STOMMEL H. (1949) Trajectories of small bodies sinking slowly through convection cells. *J. mar. Res.* **8**, 25–29.

TALLING J. F. (1971) The underwater light climate as a controlling factor in the production ecology of freshwater phytoplankton. *Mitt. int. Verein. theor. angew. Limnol.* **19**, 214–243.

TAYLOR F. J. R. (1975) Non-helical transverse flagella in dinoflagellates. *Phycologia* **14**, 45–47.

TAYLOR F. J. R. (1980) On dinoflagellate evolution. *BioSystems* **13**, 65–108.

TAYLOR R., IKAWA M., SASNER J., THURBERG F. & ANDERSEN K. (1974) Occurrence of choline esters in the marine dinoflagellate *Amphidinium carterae*. *J. Phycol.* **10**, 279–282.

TAYLOR W. R., SELIGER H. H. FASTIE W. G. & MCELROY W. D. (1966) Biological and physical observations on a phosphorescent bay in Falmouth Harbour, Jamaica, W. I. *J. mar. Res.* **24**, 28–43.

THRONDSEN J. (1973) Motility in some marine nanoplankton flagellates. *Norwegian J. Zoology* **21**, 193–200.

TIMPANO P. & PFIESTER L. A. P. (1985) Colonization of the epineuston by *Cystodinium bataviense* (Dinophyceae): behavior of the zoospore. *J. Phycol.* **21**, 56–62.

TOMAS R. N. & COX E. R. (1973) Observations on the symbiosis of *Peridinium balticum* and its intracellular alga I. Ultrastructure. *J. Phycol.* **9**, 304–323.

TUTTLE R. C. & LOEBLICH III A. R. (1974) The discovery of genetic recombination in the dinoflagellate *Crypthecodinium cohnii*. *Science* **185**, 1061–1062.

TYLER M. A. & SELIGER H. H. (1978) Annual subsurface transport of a red tide dinoflagellate to its bloom area: water circulation patterns and organism distributions in the Chesapeake Bay. *Limnol. Oceanogr.* **23**, 227–246.

TYLER M. A. & SELIGER H. H. (1981) Selection for a red tide organism: physiological responses to the physical environment. *Limnol. Oceanogr.* **26**, 310–324.

VERWORN M. (1889a) Die polare Erregung der Protisten durch den galvanischen Strom. *Pflügers Arch.* **46**, 267–303.

VERWORN M. (1889b) *Psycho-physiologische Protistenstudien*. Jena.

VON STOSCH H. A. (1965) Sexualität bei *Ceratium cornutum* (Dinophyta). *Naturwiss.* **52**, 112.

VON STOSCH H. (1973) Observations on vegetative reproduction and sexual life cycles of two freshwater dinoflagellates, *Gymnodinium pseudopalustre* Schiller and *Woloszynskia apiculata* sp. nov. *Br. J. Phycol.* **8**, 105–134.

WATANABE M. & FURUYA M. (1974) Action spectrum of phototaxis in a cryptomonad alga *Cryptomonas* sp. *Plant cell. Physiol.* **15**, 413–420.

WEILER C. S. & KARL D. M. (1979) Diel changes in phased-dividing cultures of *Ceratium furca* (Dinophyceae): nucleotide triphosphates, adenylate energy charge, cell carbon, and patterns of vertical migration. *J. Phycol.* **15**, 384–391.

WHITE A. W. (1976) Growth inhibition caused by turbulence in the toxic marine dinoflagellate *Gonyaulax excavata*. *J. Fish. Res. Bd. Canada* **33**, 2598–2602.

WINET H. (1975) Does bioconvection play a role in red tide ecology? *J. Protozool.* **22**, 19a.

WROBLEWSKI J., O'BRIEN J. J. & PLATT T. (1975) On the physical and biological scales of phytoplankton patchiness in the ocean. *Mem. Soc. R. Liege, Collect.* **7**, 43–57.

WYATT T. (1974) Red tides and algal strategies. In *Ecological Stability* (Eds M. B. Usher & M. H. Williamson), pp. 35–40. Chapman and Hall, London.

WYATT T. (1975) The limitation of physical models for red tides. In *Proceedings of the First International conference on Toxic Dinoflagellate Blooms* (Ed. V. R. LoCicero), pp. 81–93. Mass. Science and Technol. Found., Wakefield, Massachusetts.

WYATT T. & HORWOOD J. (1973) A model which generates red tides. *Nature* (*Lond.*) **244**, 238–240.

ZINGMARK R. (1970) Sexual reproduction in the dinoflagellate *Noctiluca miliaris* Suriray. *J. Phycol.* **6**, 122–126.

CHAPTER 11
ECOLOGY OF DINOFLAGELLATES

F. J. R. TAYLOR
Departments of Oceanography and Botany, University of British Columbia, Vancouver, BC, Canada V6T 1W5

UTSA POLLINGHER
Israel Oceanographic and Limnological Research and Kinneret Limnological Laboratory, P.O. Box 8030, Haifa 31080, Israel

A. GENERAL AND MARINE ECOSYSTEMS

F. J. R. TAYLOR

1 GENERAL CONSIDERATIONS

Because of their physiological diversity it is difficult to generalize about the roles of dinoflagellates in ecosystems without being excessively simplistic. We have distinguished here between their roles in freshwater and marine ecosystems, although there is considerable parallelism between the two.

Photosynthetic dinoflagellates can be significant primary producers in aquatic ecosystems: less than diatoms under most marine conditions but periodically of much greater impact. The properties of their photosynthetic systems are described in detail in Chapter 5. Non-photosynthetic members acquire their nutrition by the uptake of dissolved organics, extracellular digestion or by phagotrophy (Chapter 6). Attention has been focused on their role as primary producers (e.g. in Round 1981), and there is a great need for investigations into their importance as secondary producers

The habitats in which dinoflagellates may be found are also very varied. The majority (nearly 90%) are marine planktonic or benthic forms, with greatest diversity in tropical waters. Dinoflagellates can be found in polar waters, in sea ice and even in snow but other groups, such as diatoms or green algae, are more successful in these cold environments. The photosynthetic members are restricted to illuminated water, although many can survive under very dim light conditions, whereas the heterotrophs can extend into non-illuminated depths, both in the water and in sediments. Much more is known about the distribution of the autotrophs than the heterotrophs, and those in sediments have received little study.

Although most dinoflagellates are free-living, there are many that are parasitic (Chapter 13), some having serious consequences for aquatic animals. A few photosynthetic species are beneficial symbionts, 'zooxanthellae', in a great variety of hosts (Chapter 12). Because of their abundance, widespread occurrence, particularly in tropical coral communities, and influence on their hosts, zooxanthellae have considerable ecological importance. The present chapter deals with the ecology of the free-living forms. After some general features are described, dinoflagellate physiological ecology and biogeography are considered, and the chapter concludes with an outline of the special factors involved in 'red tides'.

2 GROUP LEVEL ECOLOGICAL FEATURES

Before considering marine and freshwater dinoflagellate ecology one may ask whether there are some features of their ecology which characterize dinoflagellates as a whole or for the most part. For example, the majority are flagellated and their motility permits them to optimize their location to a limited extent and to avoid sinking out under very stable water conditions. These seem to be key factors in their competition with diatoms, their principal rivals in the plankton. The majority of dinoflagellates occur in warmer waters and tend to be most abundant in the warmer months of the year. Most photosynthetic forms have similar pigments and photosynthetic physiology (see Chapter 5), and require organic growth factors such as biotin and vitamin B_{12} in addition to the usual inorganic macronutrients (nitrate, ammonium, phosphate). Most of the phagotrophic dinoflagellates are athecate, although the pellicle of forms like *Noctiluca* and *Kofoidinium* provides a flexible peripheral strengthening. The wall is internal to the cell membrane in dinoflagellates but external in diatoms. Thus, all cell extensions are usually covered by potential uptake surface in the former, although no eco-physiological differences have been linked to this feature yet. Both wall types have pores.

2.1 Size

In size dinoflagellates span almost precisely the same size range as diatoms, from 2 μm (*Gymnodinium simplex*) to 2 mm (*Noctiluca*), with the majority between 20 and 200 mμ. These two groups seem to achieve a larger size than members of the other microalgal groups found in the plankton, most of which (cryptomonads, prymnesiomonads including coccolithophorids, etc.) fall into the nanoplanktonic range (less than 20 μm). Although similar in size, diatoms and dinoflagellates differ in their volume-specific biomass.

2.2 Cell composition

Planktonic diatoms tend to have a large, central vacuole, the cytoplasm forming a thin surrounding layer: the larger the diatom the greater the relative volume of vacuole to cytoplasm (Strathmann 1967). Flagellates, including dinoflagellates, on the other hand, are relatively 'solid', the bulk of the cell consisting of cytoplasm, although they may have large pusules or a substantial peripheral vacuole. The oceanic, tropical dinoflagellate *Pyrocystis*, which remains non-motile throughout most of its life-cycle, interestingly mimics diatoms in vacuole formation and *Noctiluca* is also greatly inflated by a vacuole, its single flagellum being ineffective in locomotion.

In the last century Brandt (in Gran 1912) realized that dinoflagellates should be of greater nutritive value to grazers than diatoms of equal volume because of their lower liquid content, their general chemical composition being similar. More recent studies have confirmed the higher carbon content of dinoflagellates than diatoms (Parsons *et al.* 1961; Chan 1978; Hitchcock 1982) and have also revealed higher protein, carbohydrate and sometimes lipid content in dinoflagellates. Hitchcock (1982) found that the calorific content of eight species of dinoflagellates gave a linear relationship between log calorific content and log cell volume indicated by the regression:

$$\log \text{mcal cell}^{-1} = 0{\cdot}80 \log \text{volume} + 0{\cdot}90 \ (r = 0{\cdot}93)$$

Diatoms have a similar slope but are consistently lower in calorific content (Fig. 11A.1). Finlay & Uhlig (1981) estimated the calorific content of *Ceratium hirudinella* as 45 J mg C^{-1} and *Noctiluca* as 43·4–44·6 J mg C^{-1}, figures which are not markedly different from various protozoa and bacteria.

Thomas *et al.* (1978) estimated that diatoms consistently contributed a greater percentage of the photosynthesis of natural communities (including dinoflagellates) off California than their percentage carbon would suggest, but the method used (inhibition by 5 mg l^{-1} germanium dioxide) to estimate the diatom contribution may also inhibit some dinoflagellate photosynthesis (see below). In a comparison between two diatoms and the dinoflagellates *Prorocentrum micans* and *Gymnodinium simplex*, Chan (1980) found that all had similar maximum photosynthetic rates per unit chlorophyll *a* but the diatoms had higher chlorophyll—and therefore a higher photosynthetic rate—per unit carbon. Respiration rates in dinoflagellates are high (see below) and some have very little chlorophyll (e.g. some *Dinophysis* and tropical, surface *Ceratium* spp.).

For ecological purposes it would be useful to have a readily measurable, biochemical index of the abundance of dinoflagellates in a mixed, natural community. In most biochemical respects members of this group are very similar or indistinguishable from other closely related protist groups. For example, their chlorophylls are shared by several other common phytoplankton groups, such as cryptomonads, chrysomonads, coccolithophorids and diatoms (see Chapter 5). Dinoflagellates resemble diatoms and several other groups in the predominance of highly unsaturated 3,6,9,12,15,18-heneicosahexaene (HEH) in their hydrocarbon composition (Blumer *et al.* 1971). Their principal storage products, starch and oils, are also shared by several groups, as are cellulose walls. There are two compounds which could be useful, however.

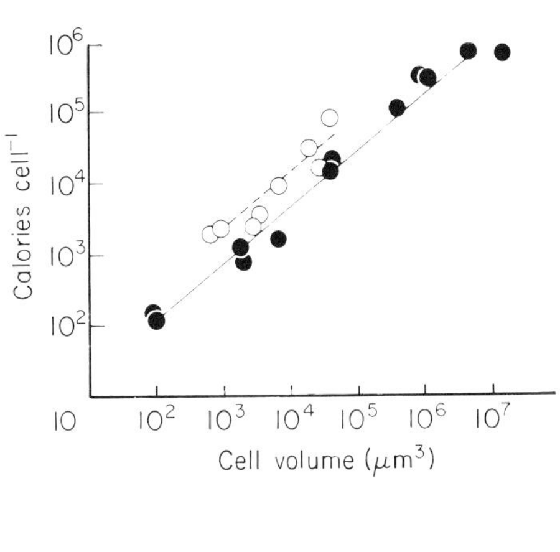

Fig. 11A.1. Calorific content of dinoflagellates (○) relative to diatoms (●) as a function of cell volume. From Hitchcock (1982), with permission.

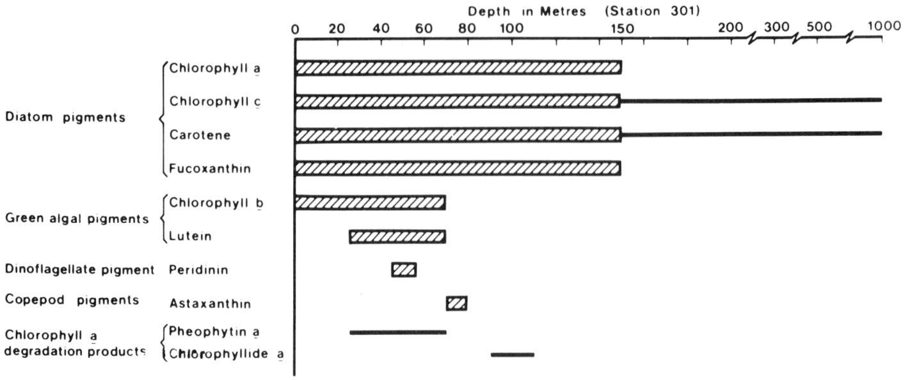

Fig. 11A.2. The depth distribution of various phytoplankton pigments extracted from populations at a station off the east coast of Australia. Peridinin was detectable only between 40 and 60 m. Bar extensions indicate detrital material. From Jeffrey & Hallegraeff (1980), with permission.

Peridinin is a xanthophyll which is unique to the group, although it is only present in the photosynthetic members and not all the latter possess it (Chapter 5). Jeffrey & Hallegraeff (1980) have examined the depth distribution of various pigments extracted from phytoplankton east of Australia (Fig. 11A.2) and compared them to the abundance of various groups present. Peridinin was found only between 40 and 60 m. The correlation with dinoflagellates was not very good, although the authors noted that non-photosynthetic members could have caused much of the discrepancy (see also Jeffrey 1980).

Dinosterol is one of several sterols unique to the group (Chapter 9). Its presence in Black Sea sediments has been used to argue for a dinoflagellate origin for much of this material (Boon *et al.* 1979).

In many ways the nutrient requirement spectrum (Figs 11A.3, 11A.4) of dinoflagellates is similar to diatoms (e.g. nitrogen, phosphorus, iron requirements). However, there are some key differences: dinoflagellates, with the exception of those rare species producing silicified internal structures, are not

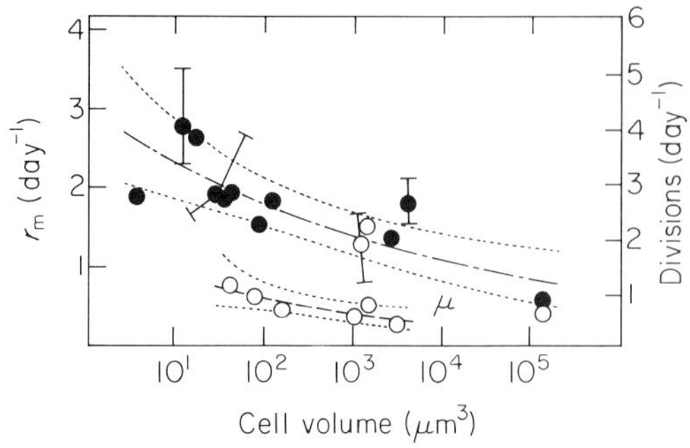

Fig. 11A.3. Maximum division rates from diatoms (●) and dinoflagellates (○) as a function of cell volume at 20°C. r_m = maximum intrinsic growth rate scaled to cell mass. Redrawn from Banse (1982).

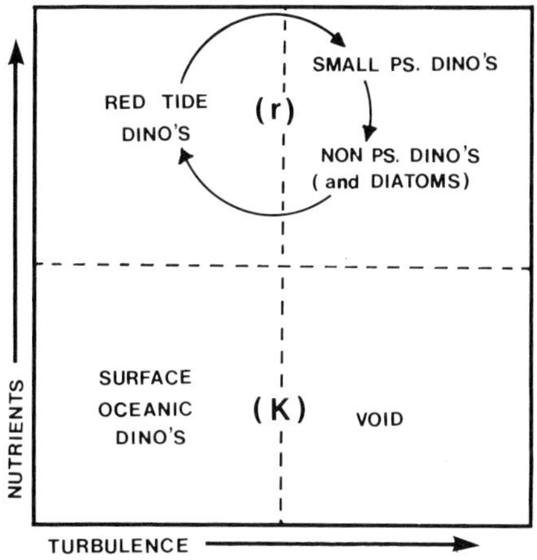

Fig. 11A.4. The relationship of various dinoflagellate types, and diatoms, to nutrient concentrations and turbulence. The upper circle indicates the annual cycle in temperate coastal waters and regions of greater '*r*' and '*K*' selection are also shown. Redrawn and modified from Margalef's *phytoplankton mandala* (Margalef 1978a, b; Margalef *et al.* 1979).

growth-limited by silicon in the sea as diatoms may be (e.g. Thomas 1959), silicon being needed not only for the wall but also for DNA synthesis. Diatom growth can be inhibited by the addition of 1–5 mg l^{-1} germanium dioxide, a competitive inhibitor of silicon uptake, but its use to favour dinoflagellates should be made with caution. Tatewaki & Mizuno (1979) found that *Protogonyaulax* (*Gonyaulax*) *catenella* and *Prorocentrum micans* were unaffected for 1–2 months in batch cultures containing up to 30 mg GeO_2 l^{-1}, but a *Peridinium* sp. was inhibited above 2·5 mg l^{-1}, and other groups lacking silicon walls are nevertheless inhibited by GeO_2, particularly the brown algae (McLaughlin *et al.* 1971). Virtually all dinoflagellates require vitamin B_{12} but many diatoms do not.

Dinoflagellates often contain very high quantities of DNA per cell (200–650 pg $cell^{-1}$ compared with 0·6–25 pg $cell^{-1}$ in diatoms: Allen *et al.* 1975; Holm-Hansen 1969: see also Chapter 4). Its synthesis relative to cell division has been studied by Karentz (1983). It is not always in phase.

An obvious difference between diatoms and dinoflagellates is the large number of non-photosynthetic members of the latter, whereas diatoms are nearly all photosynthetic. The exact proportion of heterotrophs is difficult to determine because of the bleaching or masking of pigments in preserved samples.

2.3 Rhythmicity

A major group-level characteristic is the strong rhythmicity found, not only in photosynthesis and luminescence (Chapter 7), but also in cell division. Because dinoflagellates divide predominantly in the dark period (many close to the dark/light transition: Chapter 14 and see Doyle & Poore 1974; Sournia 1974; Weiler & Chisholm 1976; Weiler & Eppley 1979; Nelson & Brand 1979 and the review by Chisholm 1981), they are effectively limited to the relative slow maximum division rate of 1 doubling per day. Thomas (1966) measured division rates of 1·25 and 1·17 divisions per day for a very small, tropical *Gymnodinium* species (*G. simplex?*). It is possible that some species may divide more rapidly in short bursts (e.g. *Amphidinium carterae*, Jitts *et al.* 1964) but probably cannot sustain this over longer periods of time (Karentz 1983). Yentsch & Mague (1980) observed an apparent increase in division rates in cultures of *Protogonyaulax tamarensis* (= *excavata*) to 2·0 doublings per day in summer, despite growth in artificial medium under constant conditions. These and other division rates of more than 1 per day (Zgyrovskaya; 1970) seem to be anomalous and are surprising. Diatoms are less regulated in this way, although some also exhibit dark division (Chisholm 1981), and consequently are capable of more rapid increases in numbers, many being capable of 2 or 3 doublings per day. This may be a key reason why diatoms can outcompete dinoflagellates in waters well supplied with inorganic nutrients and with sufficient small-scale turbulence to keep them up in the water column (Chan 1978).

2.4 Motility

The majority of dinoflagellates, both planktonic and benthic, spend most of their life cycle in the flagellated (mastigote) state. This permits them to exploit their environment to a greater extent than non-motile cells. For example, by undergoing diel vertical migrations (Chapter 10), they can exploit both surface waters and deeper, nutrient-rich waters. The costs versus the benefits of such behaviour have been evaluated by Raven & Richardson (1984).

Thus, dinoflagellates can be thought of essentially as exploiters of calm, stratified conditions. If mixing, usually due to wind, is too great, they are unable to maintain their optimum position in the water column and they become more evenly distributed within the mixed layer. George & Heaney (1978) found that wind strengths below 50 km day $^{-1}$ permitted *Ceratium hirundinella* to aggregate in patches, whereas 100 km day^{-1} broke down plankton patchiness (Harris, 1983, has provided an extensive review).

Turbulence also may be directly inhibitory to the cells. In the culture of dinoflagellates it has been found that shaking decreases growth and division rates (Tuttle & Loeblich 1975; Karentz, 1983). White (1976) found that shaking for as little as 15 minutes at 125 rpm once every 6 hours doubled the division time for *Protogonyaulax tamarensis* (= *Gonyaulax excavata*), with no growth if shaken every 2 hours. Galleron (1976) used shaking to synchronize cultures of *Amphidinium carterae*. Thus, it appears that in dinoflagellates, the most evident first effect of turbulence is on the division process.

It can be questioned whether such agitation can be related to natural small-scale turbulence. Pollingher & Zemel (1981) were able to demonstrate the effect of turbulence on the division of the freshwater species *Peridinium westii*, both in the field and under experimental conditions (see Chapter 11B.4 here). Eppley *et al.* (1978) found that dinoflagellates and other flagellates succeeded diatoms in large, unstirred plastic bags whereas diatoms persisted in stirred bags, but nutrient availability may also be a factor in this case.

Moderate turbulence is favourable to diatoms, returning them (and their resting spores) to the surface of the mixed layer after periods of quiescence. If they are mixed deeper than the euphotic zone (winter in higher latitudes) bloom development is impeded. In essence, then, diatoms can be viewed as possessing a more economical, conservative 'strategy', reducing their density as much as possible, but sinking out during relatively rare periods of calm, sunny weather when the water becomes stratified and the upper layer depleted in nutrients. Dinoflagellates (and other flagellates) are at least potentially able to optimize their depth throughout the day.

Margalef (1978 a, b; Margalef *et al.* 1979) devised a 'phytoplankton mandala' which illustrates the relationship between various forms of phytoplankton, turbulence and nutrients (Fig. 11A.3). The 'main axis' extends from diatoms, occupying the high nutrient/high turbulence range, to oceanic dinoflagellates in low turbulence/low nutrients. Red tide dinoflagellates occupy the low turbulence/high nutrient sector. His diagram has been added to here, to reflect the position of dinoflagellates other than 'typical' ones.

Another advantage of swimming is that it may flush the cell's uptake surfaces with fresh, nutrient-containing fluid, avoiding the formation of a nutrient-poor 'halo' around the cells, resulting in diffusion-limited uptake rates (Munk & Riley 1952). This was shown in a practical way by Pasciak & Gavis (1974), who found a lower nutrient uptake by non-motile cells of *Gymnodinium sanguineum* (= *splendens*) than by motile cells. On a very fine scale, a thin film of water may still be carried along with the cell, but how thin it is, relative to swimming speed, has not been calculated.

2.5 Respiration

In a comparison between two marine dinoflagellates, *Glenodinium* sp. and the zooxanthella *Symbiodinium microadriaticum,* and five other photosynthetic algae (including the diatom *Thalassiosira pseudonana*) Burris (1977) found that the dinoflagellates had relatively high dark respiration rates and an unusual photorespiration pathway. He estimated a net photosynthetic rate: steady-state dark respiration rate ratio of 1·3–2·4 for the dinoflagellates, compared with 8·3 for the diatom and 3·6–10·3 for members of the other groups, none of which were motile. Earlier Hochachka & Teal (1964) had discovered that *Gymnodinium sanguineum* (= *splendens*) had a lower respiration in the dark than in the light (1·5 versus 2·5 mm^3 0_2 mg^{-1} h^{-1} respectively). The high dark respiration of the dinoflagellates should result in a slower growth rate than the others.

Most respiration rates measured in the laboratory are within the range of 2–

20×10^{-15} MO_2s^{-1} $cell^{-1}$. High respiration rates can also be detected in field populations. For example, Tanaka & Sano (1980) estimated that respiration in a bloom of *Prorocentrum levantinoides* reached 45% of the gross photosynthesis, compared with 10–20% in blooms of other phytoplankters such as diatoms. Barker (1935) found that a species of *Peridinium* had a respiration rate only 10% of the maximum photosynthetic rate, but Moshkina (1961) recorded much higher rates in four dinoflagellates (up to 155×10^{-15} MO_2 s^{-1} $cell^{-1}$). A bloom of *G. sanguineum* off Peru appeared to lose 78% due to respiration (Smith 1977) compared with 14·8% for a diatom bloom off NW Africa.

At first it seems reasonable to assume that swimming is a major reason for high respiration rates in dinoflagellates (Moshkina 1961). In Burris' study (above) the dark respirations of zooxanthellae, relative to net photosynthetic rate, was roughly half that of *Glenodinium,* but he did not indicate what proportion of the former were 'zoospores' relative to coccoids and the photosynthetic rate of the zooxanthellae per unit chlorophyll was much less than the *Glenodinium.* Computations (courtesy of G. Gaines) based on the energy consumption of uniflagellar swimming by *Ciona* sperm ($0{\cdot}5 \times 10^{-19}$ moles ATP $beat^{-1}$ $flagellum^{-1}$: Brokaw & Benedict 1968), in which it is reasonable to assume that virtually all energy is expended on swimming, suggest that only a very low proportion of energy expenditure is used for swimming: 0·0042–0·192%! Another approach to the question is to consider the maximum possible ATP consumption based on the number of dynein/ATPase molecules per unit flagellar length. Using a total flagellar length of 250 μm for *Gonyaulax polyedra*, and a figure of 600 ATPase molecules per μm, Raven & Beardall (1981) arrived at a maximum of $2{\cdot}5 \times 10^{-17}$mol ATP s^{-1} and calculated that it should result in a specific respiration requirement of 0·000015 h^{-1}, compared with a maintenance respiration of at least 0·001 h^{-1}.

These extraordinarily low figures indicate that, unless there are significant factors which have been omitted from the calculations, the high respiration rates of dinoflagellates are due to processes other than swimming.

It is probably partly due to their high respiration rates that dinoflagellates do not have prolonged dark survival abilities unless they encyst, although many can survive 2–3 weeks of darkness (Richardson & Fogg 1982). *Amphidinium carterae* was one of the briefest dark survivors in a study of thirty-one algal species by Antia (1976). Wood (1966) and Kimor & Wood (1975) found apparently healthy dinoflagellates in closing bottle samples from 1000 m.

A few dinoflagellates seem to have adopted the economical diatom strategy of suppressed motility and size enlargement by means of vacuole development, such as *Pyrocystis* and *Noctiluca*, as noted earlier. The former is the most successful, widespread dinoflagellate in tropical oligotrophic waters (Section 4.2). The latter appears to be a highly successful phagotrophic predator in temperate and tropical coastal waters. However, both may also benefit from their large size by escaping the grazing pressure of ciliates.

It is difficult to suggest reasons why dinoflagellates are more prominent in the microplankton than other flagellate groups in the sea. Coccolithophorids are of major importance in oceanic waters and their coccolith-lacking prymnesiomonad relatives are more important in cold temperate and polar coastal waters. Other groups usually predominate in the nanoplankton (<20 μm) or picoplankton (<2 μm).

3 FACTORS GOVERNING GROWTH AND DISTRIBUTION

In this section an attempt will be made to summarize some of the principal features of marine dinoflagellate distributions (freshwater distributional features are reviewed in Chapter 11B). Most of the section summarizes influences on photosynthetic forms, although some factors (temperature, salinity) apply to both, and the ecology of heterotrophic forms is briefly discussed separately (Section 3.5; see also Chapter 6).

Before discussing the influence of individual factors in photosynthetic dinoflagellates it must be stressed that one-factor control of growth or photosynthetic rate may be extremely transient (light over-riding nutrient at night, for example) and within a population not all the individuals are in the same physiological state (some in a different cell- or life-cycle stage, for example). Nutrient uptake varies according to light, internal pools (past history) and temperature. Acclimation (short-term adaption) or change in genotype dominance (short-term, infraspecific selection) may alter the response of populations fairly rapidly.

3.1 Light

The influence of light on photosynthetic dinoflagellates is discussed extensively in Chapter 5 and so will only be briefly covered here. The response to light is variable, not only according to species, but also according to nutritional state, and generalizations are difficult to make (see also Section 3.4b). In an early study, Ryther (1956) compared the photosynthesis at various irradiances of four dinoflagellates (*Gymnodinium sanguineum* (= *splendens*), *Gyrodinium* sp., *Prorocentrum* sp. and *Amphidinium klebsii*) with seven green algae and four diatoms, and his results strongly suggested that dinoflagellates require, and are more tolerant of higher irradiances than the others. However, Dunstan (1973) and Chan (1978) found no consistent group differences in the amount of light required for saturated growth rate or photoinhibition. *Amphidinium carterae*, for example, was effectively light-saturated at irradiances as low as 50 μE m^{-2} s^{-1} and its growth rate remains at approximately 1 doubling per day up to more than 250 μE m^{-2} s^{-1} (Chan 1978). On the other hand, *Scrippsiella sweeneyae* is inhibited by light above 100 μE m^{-2} s^{-1}, which is lower than any of the diatoms tested. *Protogonyaulax* (*Gonyaulax*) *tamarensis* was not different from four diatoms in a study by Cole *et al.* (1975). Although the initial slope of the photosynthesis versus illumination curve was more gradual than three of the diatoms, maximum saturation was similar.

3.2 Temperature

Temperature is of importance in marine ecology due both to its direct effects on metabolic rates (generally positive until inhibiting temperatures are reached) and its indirect effect on the vertical stability of the water column, surface heating increasing the latter. Usually this stabilizing effect is thought to be more significant than the direct physiological action, although Eppley (1972) has indicated that *maximum* growth and photosynthetic rates follow a generally predictable relationship with temperature.

Although dinoflagellates can occur over a very wide range of temperatures, as a rule they are more predominant in warm water communities, i.e. in the tropics throughout the year, or during the warmest part of the temperate season.

From culture experiments the highest recorded temperature tolerance for a dinoflagellate is 35°C for *Crypthecodinium cohnii* (Loeblich 1966). Length of exposure is important. In the field it appears that *Peridinium quinquecorne* can tolerate 38°C for up to 5 hours in shallow waters in a Philippine bay (Horstmann 1980). In shallow tropical tidepools, where the temperature may reach 40°C Taylor (1983) found blooms of *Scrippsiella subsalsa* and a gymnodinoid resembling the motile cell of *Symbiodinium microadriaticum*. Wilson *et al.* (1984) acclimated *Amphidinium carterae* to 30°C in 45 days, after which it would not grow at 12°C.

A few dinoflagellates have been recorded from snow or ice, e.g. 'red snow' caused by *Gymnodinium pascheri* (Gerath & Nichols 1974) and the winter species *Amphidinium cryophilum* (Wedemayer *et al.* 1982). In a study of the former, Schiller (1954) found that optimal growth occurred between 0·2 and 3·0°C. *A. cryophilum* can reach concentrations of $3{\cdot}5 \times 10^6\ l^{-1}$ beneath the ice in pond water of 0 to 1°C. In the Baltic and polar sea water some species of *Ceratium* (which are also euryhaline) can occur at −1°C (Nordli 1957), *C. tripos, C. fusus* and *furca* also being able to tolerate 29·5°C.

As a general rule the temperature–response curve determined in the laboratory for most organisms, including dinoflagellates, is skewed towards the higher temperature range, maximum growth rates occurring close to the lethal temperature. In the field, most blooms occur in suboptimal temperature water (Barker 1935). For example, in the Pamlico River estuary *Heterocapsa triquetra* usually blooms in early spring in temperatures of 2–8°C, even though its optimum is between 10–27°C (Carpenter 1973).

3.3 Salinity

The great majority of living dinoflagellates are marine, only approximately 220 species living in freshwater (see Chapter 11B). Among the marine species one can find examples of both eury- and stenohaline forms although, again, the majority of species live in ocean waters with salinity greater than 20 or 30‰, rarely greater than 40‰. No dinoflagellates possess contractile vacuoles for osmoregulation and suggestions that pusules carry out this function (Dodge 1972) are not supported by experimental evidence.

Nearshore and intertidal forms can be remarkably tolerant. For example, the temperate non-photosynthetic, seaweed-associated species *Oxyrrhis marina* and *Crypthecodinium* (*Gyrodinium*) *cohnii* can tolerate 4–64‰ and 3–50‰ respectively (Droop 1959; Provasoli & Gold 1962). The more tropical/warm-temperate species *Coolia monotis* can survive 15–60‰ for 10 days or more, and shows reasonable growth between 20–50‰. An isolate from the Mediterranean had an optimum of 30–35‰ (Pincemin 1972). A few marine species, such as *Prorocentrum compressum* and *Amphidinium klebsii* have been recorded from freshwater environments (Thompson 1950).

As is often the case, species which can tolerate very low salinities can also tolerate very high ones, two species of *Ceratium*: *C. furca* and *C. fucus,* being striking in this regard. Although common in brackish water (the optimum for photosynthesis for *C. furca* from Cochin backwaters in India was given as 6–10‰ by Qasim 1973), they play a major role as primary producers in Bardawil Lagoon, northern Sinai, where the salinity exceeds 70‰ (Kimor 1972) and are dominants in the Persian Gulf.

Several species of *Prorocentrum* are common or abundant in estuarine water, e.g. *P. minimum, P. balticum, P. cassubicum, P. cordatum* and *P. compressum*, but this is not true for the whole genus. For example, *P. micans* cannot tolerate salinities less than 15‰; it has an optimum of 25‰ and an upper limit of more than 40‰ (Kain & Fogg 1960).

A preference for estuarine conditions may not be due to a lower salinity optimum (see Sections 3.6 and 4.5b). For example, *Protogonyaulax tamarensis* (=*Gonyaulax excavata*) strongly favours estuarine, or at least run-off influenced areas, growing well between 20 and 40‰ and tolerating 11–42‰ (Prakash 1967; White 1978), but it seems to need the protective effect of organic complexing agents which restricts its distribution more than its salinity tolerance suggests. In this species salinity also affects toxin production, increasing from 21‰ to 37‰ (White 1978).

High stenohaline distributions may be due to an intolerance of some constituents of land run-off other than low salinity. Aldrich & Wilson (1960) experimentally determined a growth tolerance of 27–37‰ (with survival of a few cells from 22·5 to 45‰) for *Ptychodiscus* (*Gymnodinium*) *brevis,* this well known 'red tide' species seems to be excluded from low salinity, lagoonal, or river-influenced areas of the Gulf of Mexico, being replaced by another, more euryhaline red tide species, *Gessnerium* (*Gonyaulax*) *monilatum*. The primary source organism for 'ciguatera' fish poison, *Gambierdiscus toxicus,* is a benthic species which appears to avoid areas with appreciable land run-off, occurring most abundantly near remote islands or off-shore reefs in low rainfall parts of the tropics (Taylor & Gustavson 1985).

Low stenohaline species, such as *Peridinium balticum, Peridinium chattonii* and *Prorocentrum* (*Exuviaella*) *cassubicum* will stay motile up to relatively high salinities (>30‰), but will only divide at salinities below 20‰, with optima near 10‰ (Provasoli & McLaughlin 1963). *Kryptoperidinium* (*Glenodinium*) *foliaceum* usually forms blooms in water below 10‰, although Silva (1962) and Prager (1963) both observed good growth at higher salinities, the latter recording an optimim of 24‰ NaCl. *Prorocentrum balticum* grows equally well over a wide range from 10 to 40‰ (Braarud 1951), which accords well with its distribution as a common, important species both in the Baltic Sea and open North Atlantic Ocean.

Because species can acclimate to various salinities, depending on length of time and the rate and direction of change, caution is needed in the interpretation of published data. Braarud (1951) found that varying light and temperature did not alter the optimum for *Scrippsiella trochoidea* (15–20‰), although the upper tolerance limit did change.

3.4 Nutrients

Six elements, known by the mnemonic 'CHONSP', make up more than 99% of a cell's mass. Heterotrophs obtain all their requirements from organic sources (Chapter 5). Evidently, for an aquatic photosynthetic organism there is not usually a shortage of H_2 and O_2. An analysis of the proportions of the others in the composition of apparently unstarved phytoplankton reveals 106 C:16 N:1 P (by atoms) (Redfield *et al.* 1963). The C:N:P 'Redfield' ratio, can be shown to vary, in cultured phytoplankters, depending on which element is in least supply relative to demand, but if the cells grow more slowly due to another limiting factor, such as light, the normal ratio can be maintained (Tett *et al.* 1985). Protein usually makes up about 50% of the cell composition (C:N $\simeq$ 5). Depletion of nitrogen may shift the composition towards a higher relative proportion of carbohydrate. Growth rate, in turn, is also a function of light and temperature, and degrees of adaptation to them. Much emphasis in recent phytoplankton physiological ecology has been on interactions between these parameters and transient responses, as well as steady states (Table 11A.1). In the following section some basic features of the nutrient requirements of photosynthetic dinoflagellates will be summarized. More details are provided in the reviews by Loeblich (1966), Iwasaki (1979) and those referred to in the individual sections below.

Table 11A.1. Examples of half saturation values (K_s, the substrate concentration that supports half the maximum growth rate) for nitrate, ammonium and phosphate. Note that, although often referred to as half-saturation constants (from enzyme kinetics), they can be influenced by previous history of the cells (starved, replete, etc.), temperature, light and timing of the measurement

Species	K_s-No_3- (μM)	K_s-NH_4^+ (μM)	K_s-PO_4^{-3} (μM)	Temp. (°C)	Culture Preconditioning	Sources
Amphidinium carterae	—	2·0		21	Ammonium limited	1,2
			0·01	21	Ammonium replete	1
Amphidinium klebsii	—	—	0·01	21	replete?	1
Gonyaulax polyedra	8·6, 10·3	5·7, 5·3	—	20	Nitrite deplete*	3
Gymnodinium sanguineum	3·8	1·1	—	20	,, ,, *	3
	1·0	—	—	18	replete	4
	6·5	—	—	25	,,	4
Prorocentrum minimum	—	—	1·9	17	limited	5
Protogonyaulax tamarensis	1·5, 2·8	1·9	—	18	limited	6
	—	—	0·4	17	limited	5
Pyrocystis noctiluca	—	—	2·0–26·0	22–24	replete†	7
	—	—	1·9–34·0	22–24	starved†	7
Symbiodinium microadriaticum	—	—	0·01	26	replete?	1

Sources: 1, Deane & O'Brien (1981a); 2, Hersey & Swift (1976); Eppley *et al.* (1969b); 4, Thomas & Dodson (1974); 5, A. Cembella (unpubl.); 6, MacIsaac *et al.* (1979); 7, Rivkin & Swift (1982).

*Preincubated with 1 μm nitrate for 3–4 hours to induce nitrate reductase.

†Light and dark effects studied as well (within the above ranges).

(A) CARBON

Although carbon as CO_2 or HCO_3^- may be rate-limiting under intense growth situations in culture or in fresh water, in the marine environment it is not usually thought to play an important ecological role. Caperon & Smith (1978) have studied the uptake kinetics for several species, including an axenic culture of *Amphidinium carterae*. They found the half-saturation constant for CO_2 uptake to be 2.8 + 0.3 μg CO_2 l^{-1}, less than in natural populations of phytoplankton (4·3–5·3 ± 0·7 μg l^{-1}). In dinoflagellates the principal carbon fixation pathway seems to be the usual carboxylation of ribulose-1, 5-bisphosphate (RUBP) in the Calvin–Benson cycle: the 'C_3' pathway (Morris 1980; Kremer 1981). However, like a number of algae and higher plants, the dinoflagellates *Amphidinium carterae* and a *Gymnodinium* sp. have been found to have an additional, β-carboxylation, 'C_4', pathway for carbon fixation (giving rise to aspartate, and malate as initial products) in addition to the classical pathway producing 3-phosphoglycerate as an initial product. These two species (see Chapter 12.4.1 for zooxanthellar pathways) have been shown to differ from those of the other groups studied in the use of the enzyme pyruvate carboxylase rather than phosphoenolpyruvate carboxykinase: PEPCKase (most diatoms, a prymnesiomonad) or phosphoenolpyruvate carboxylase: PEPCase (higher plants, green algae, euglenoids, some diatoms, red algae and a cyanobacterium) (Appleby *et al.* 1980). From an ecological point of view, the significance of β-carboxylation is that carbon fixation can continue in the dark. 'C_4' plants on land are capable of very high photosynthetic rates under high irradiances, but this depends upon the high affinity of PEPCase for carbon dioxide (Kremer 1981), and so it cannot be assumed that dinoflagellates share this property.

Several studies have examined the ability of photosynthetic dinoflagellates to take up dissolved organic carbon in low light or in the dark. This subject is of particular ecological interest because of the common occurrence of dinoflagellates in layers deep in the euphotic zone (Section 6.1c). Early studies by Aldrich (1962) on *Ptychodiscus* (*Gymnodinium*) *brevis* and Thomas (1955) on *Gonyaulax polyedra*, failed to detect any heterotrophic growth in these species. The freshwater species *Gymnodinium inversum* could grow in the dark on high levels of acetate (100 mg l^{-1}). Cheng & Antia (1970) and Morrill & Loeblich (1979) found growth stimulation in the light by the addition of high levels of glycerol (0·01–0·05 M: natural levels would not be expected to reach 10 μM) in axenic cultures. However, Richardson & Fogg (1982) could not detect any stimulation in several strains of marine dinoflagellates, including *Amphidinium carterae* in which it had been reported earlier, when the cells were axenically provided with 10 μM glycerol, 10 μM glucose or 10 μM acetate. A very small amount of uptake was noted, but it was not thought to be significant except in *Scrippsiella trochoidea* and in the latter there was no increase when changed from high to low light. A growth stimulation was observed in bacterized cultures of *A.*

carterae, which the authors suggested may have been due to an indirect effect.

Baden & Mende (1978) obtained some growth stimulation of lightly bacterized *Pt. brevis* cultures in the light using glucose. The addition of 1–5 mg l^{-1} glucose to a very large bag containing mixed phytoplankton did not cause an increase in the dinoflagellate component. Instead a strong increase in bacterial growth suppressed the phytoplankton (Parsons *et al.* 1981).

(B) NITROGEN

Nitrogen is essential for its role as a key component of all amino acids and hence proteins, the nitrogenous bases (purines, pyrimidines) of nucleotides (and some toxins), the porphyrin ring of chlorophyll, and cytochromes. Some phospholipids also contain nitrogen. Taking into account supply versus need, nitrogen has been considered to be the major limiting nutrient in the ocean (Dugdale & Goering 1967; Eppley *et al.*1969b). Like other plankters, dinoflagellates can use the oxidised forms, nitrate and nitrite, as well as reduced forms, chiefly ammonium and urea, but all must be converted to ammonium before incorporation into amino acids (see reviews by Collos & Slawyk 1980; McCarthy 1980; Syrett 1981; Parsons & Harrison 1983) and it may be for this reason that dinoflagellates and other eukaryotic phytoplankters preferentially take up ammonium (Eppley *et al.* 1969a). None can use ('fix') dissolved nitrogen. Nitrate is apparently actively taken up, but Falkowski (1975) was unable to demonstrate the same enzyme activity in *Amphidinium carterae* that he was able to observe in other phytoplankton. In most algae chlorate may be taken up as an analogue for nitrate, but not in dinoflagellates (Balch 1985a).

Dinoflagellates appear to be able to store nitrate internally, even though they lack a large cell vacuole like diatoms, but tropical, oceanic species of *Pyrocystis* are an exception (Bhovichitra & Swift 1977). Internal concentrations of inorganic nitrogen in *A. carterae* and *Protogonyaulax tamarensis* grown on sufficient nitrate (1.8 mM NO_3^-, 8.2 mM NH_4^+ and 49 mM NO_3^-, 3.7 mM NH_4^+ respectively) and ammonium (NO_3^- 0 mM, 92 mM in *A. carterae*) support the view that they can store inorganic nitrogen (Dortsch *et al.* 1984). Usually the nitrate is immediately reduced by the enzyme nitrate reductase (NR) to nitrite, and by nitrite reductase (NiR) to ammonium. NR is an inducible enzyme, its activity depending on the presence of nitrate (Dortch & Maske 1982), Harrison (1973, 1976) found appreciable NR activity in a bloom of *Heterocapsa triquetra* during which ammonium often remained between 5 and 20 μM. Hersey & Swift (1976) found that NR activity in *Amphidinium carterae* and *Heterocapsa* (*Cachonina*) *niei* at midday was double that at night. However, laboratory and field studies suggest that this difference is less in some dinoflagellates (Packard & Blasco 1974; Eppley & Harrison 1975; Harrison 1976).

Some dark nitrate uptake is important if vertically migrating dinoflagellates are to exploit oxidized nitrogen sources ('new' or 'renewed' nitrogen) below the nutricline in dim light or at night (see Section 6.1c). Their high dark respiration activity (Section 2.5) may account for this activity, as well as the use of NADH (predominantly respiratory) rather than NADPH (chiefly from photosynthesis) for reduction (Eppley *et al.* 1969a). NiR does require NADH or ferridoxin. MacIsaac (1978) noted that the observations showing little or no rhythmicity in NR activity were all made under relatively low light conditions and the activity could have been less than under high light conditions. Despite this, it can be concluded that some nitrate uptake and reduction can occur in low light or darkness in migrating species such as *Gymnodinium sanguineum* (Dortch & Maske 1982) or *Gonyaulax polyedra* (Eppley *et al.* 1969b). In the latter, dark uptake of ammonium is twice as fast as nitrate (MacIsaac 1978), and Balch (1985) found light requirement for methylamine uptake.

Like most nutrient excesses, ammonium can be inhibitory at higher levels. Thomas *et al.* (1980a, b) have shown that 150–200 μM ammonium values which can be encountered in major sewage outfalls, are inhibitory to the red tide organisms *Gonyaulax polyedra* and *Gymnodinium sanguineum* (*splendens*), but not to other non-dinoflagellate phytoplankters tested, including three diatoms. Earlier, Provasoli & McLaughlin (1963) found that ammonium sulphate was toxic to *Amphidinium carterae, A. rhynchocephalum, Gyrodinium californicum* and *Prorocentrum balticum* at levels ranging from 70 to 1400 μM, the two *Amphidinium* species being the least sensitive. The lower threshold for toxicity in the fresh water species *Ceratum hirundinella* is 500 μM (Bruno & McLaughlin 1977). An unusual reduction of chlorophyll *c* relative to *a*, was reported by Zgurovskaya & Kustenko (1968) in *P. micans* exposed to high ammonium levels.

Once ammonium has been assimilated in plants it may follow either a pathway involving glutamic dehydrogenase (GDH): the classical pathway in which glutamic acid is the first product, or another in which glutamine is first produced, involving the enzyme glutamine synthetase (GS) and later glutamine oxoglutarate aminotransferase, otherwise known as glutamate synthase (GOGAT): see Syrett (1981) for a review. In dinoflagellates there is evidence for the existence of both the GDH and GS/GOGAT pathways (see also Chapter 12) although Ahmed *et al.* (1977) could find virtually no GDH activity in *Amphidinium carterae* and ammonium-limited cells of *Gymnodinium simplex* were shown to have GS/GOGAT by Turpin & Harrison (1978). Ecologically this is significant because GS has a higher affinity for ammonium than GDH and should confer an advantage in an ammonia-limited environment (Falkowski & Rivkin 1976).

When phytoplankters are subjected to water depleted in nitrogen the ratio of cell C:N by atoms, can increase once the internal supply is exhausted (Eppley & Renger 1974), but this flexibility may not be the same in all dinoflagellates. The C:N ratio, by atoms, in *Gymnodinium sanguineum* can vary from 5·02 to

8·07, depending on the availability of nitrate, if other sources of nitrogen are not available (Cullen & Horrigan 1981), similar values (6·5–8·5) being found in *Gonyaulax polyedra* (Heaney & Eppley 1981).

Urea is a good organic nitrogen source for dinoflagellates (Antia *et al.* 1975), *Prorocentrum micans* growing equally well on nitrate, ammonium or urea (Eppley *et al.* 1971), as does *Protogonyaulax catenella* (Norris & Chew 1975). Other organic nitrogen sources can be used by members of the group under axenic conditions although they do not give as good growth. These include the amino acids arginine, asparagine, glycine, alanine, methionine, glutamic acid and the purine hypoxanthine (Provasoli & McLaughlin 1963; Antia *et al.* 1975, 1976, 1980; Berland *et al.* 1976). The spectrum of usable amino acids varies according to the species (Table 11A.2). Histidine is poor and some, such as D-glucosamine are not usable at all. In an autoradiographic study of natural samples *P. micans* showed no appreciable labelling with ^{14}C L-arginine or L-glutamic acid (Hollibaugh 1976).

The levels of total amino acids dissolved in seawater range from 10 to 400 μg l^{-1} and free amino acids from 5 to 110 μg l^{-1} (refs in Orlova *et al.* 1979). Phytoplankton, including dinoflagellates, can be involved in the release, as well as the uptake of amino acids. Some details of the amino acid content and release by *Scrippsiella* (*Peridinium*) *trochoidea* and *Gymnodinium lanskayae* have been provided by Orlova *et al.* (1979) and in *A. carterae* and *P. tamarensis* by Dortsch

Table 11A.2. Utilization of amino acids as organic nitrogen sources for growth by dinoflagellates. + positive growth; − not used when tested

	Alanine	Arginine	Asparagine	Cysteine	Glucosamine	Glutamic acid	Glycine	Histidine	Methionine	Phenylalanine	Proline	Serine	Source
Prorocentrum balticum	+		+			+	+						1
cassubicum	+		+			+	+						1
chattonii	+		+			+	+						1
micans		−				−							3
Amphidinium carterae	+	+	+		−	+	+		+				2,4
rhynchocephalum	+	+	+			+	+		+				2
Gyrodinium californicum	−	+	+			−	+		−				2
resplendens	−	+	+			−	−		−				2
uncatenum	−	+	+			−	−		−				2
Gymnodinium sanguineum	+	−	−			+	+		−				2
Ceratium hirundinella	−	+	+	−			+	−	−	−	+	+	5

Sources: 1, Provasoli & McLaughlin (1963); 2, Provasoli & McLaughlin (1974); 3, Eppley *et al.* (1971); 4, Antia *et al.* (1975); 5, Bruno & McLaughlin (1977).

et al. (1984). Antia *et al.* (1980) have suggested that hypoxanthine and guanine may be excreted by ciliates in sufficient quantities to be ecologically important in seawater.

(C) PHOSPHORUS

Phosphorus is an essential macronutrient, used in the manufacture of sugar phosphates, nucleotides, phospholipids and the adenosine phosphates (AMP, ADP, ATP) crucial for energy metabolism, plus other important phosphorus-containing metabolites. Phytoplankton can utilize both inorganic (Pi) and organic (Po) phosphorus. Phosphorus occurs in seawater chiefly in the form of inorganic orthophosphate (principally as HPO_4^-), pyrophosphate, polyphosphate and their organic esters: see Nalewajko & Lean (1980) for freshwater and the extensive review by Cembella *et al.* (1983, 1984). Although Pi is usually looked on as the chief source in seawater under some conditions, Po may be more important under some circumstances, e.g. in Japanese coastal waters, in which 87–98% of the total phosphorus may be Po (Koburi & Taga 1979). In oligotrophic, tropical waters Pi can be extremely low (less than 0·1 μM in the southern Sargasso Sea: Rivkin & Swift 1979) and phosphorus, usually considered more important as a limiting nutrient in fresh water than the marine environment, has been suggested as the principal limiting nutrient in the Mediterranean Sea (Bonin & Maestrini 1981) and off Barbados (Sander & Moore 1979).

Dinoflagellates have been shown to take up pyrophosphate as well as orthophosphate, over the range of 30–300 μM (Deane & O'Brien 1981a), including the symbiotic species *Symbiodinium microadriaticum* (see Chapter 12). Phosphate esters such as glycerophosphate, adenylic, guanylic and cytidilic acids can also serve as phosphorus sources for some dinoflagellates in culture (Provasoli & Pintner 1960: Kuenzler 1965; Prakash 1967; Provasoli & McLaughlin 1963; Loeblich 1975; Mahoney & McLaughlin 1977).

Because internal P levels are always likely to be higher than external levels, the nutrient requires active uptake, although the precise mechanism has not been established. For example DCMU [3-(3,4-dichlorophenyl) 1,1-dimethyl urea], an inhibitor of photosynthetic, non-cyclic photophosphorylation, reduced the uptake of Pi in the light by *S. microadriaticum,* but increased that by *Amphidinium carterae* and *A. klebsii* (Deane & O'Brien 1981a). Cyanide, which inhibits aerobic respiratory ATP production, totally blocked Pi uptake by P-starved *Pyrocystis noctiluca* after 48 hours of exposure (Rivkin & Swift 1982) and uptake could occur in the dark as well as the light for this species and *A. carterae*. Anderson & Lindquist (1985) have determined that the subsistence cell phosphorus quota for *Protogonyaulax tamarensis* is 27 pg $cell^{-1}$; declines below this may induce sexuality.

Although phosphate uptake usually follows Michaelis-Menten enzyme kinetics, dissappearance of phosphate from the medium is not necessarily linear. Rivkin & Swift (1982) found in Pi-depletion batch cultures, that it was triphasic between 0·1 and 100 μM and biphasic in a Pi-limited continuous culture between 0·2 and 25 μM. The internal phosphate pool may also influence the uptake kinetics. Some dinoflagellates, such as *Peridinium cinctum,* appear to be able to store phosphate better than diatoms during periods when it is abundant ('luxury consumption') drawing on this reserve when it is needed (see Chapter 11B). *Amphidinium carterae* can store polyphosphate in P-rich medium (Sakshaug *et al.* 1983). *Ceratium furca* was found to have a lower half saturation constant (K_s = concentration of nutrient producing half the maximum growth or uptake rate) than the larger diatom *Biddulphia sinensis* (Qasim & Bhattathiri 1973). Size can over-ride group efficiency, however. Correll *et al.* (1975) observed a dinoflagellate bloom in which other nanoplanktonic organisms accounted for most of the Pi uptake, even though they constituted less than 10% of the biomass. Nevertheless, it is striking that the three marine dinoflagellates tested by Deane & O'Brien (1981a) had extremely low K_s values for orthophosphate (0·01 μM), much lower than any diatoms or other algae tested (Cembella *et al.* 1983, 1984). *Pyrocystis* was somewhat higher (1·90 μM) and *Peridinium* sp. was high (6.30 μM).

In order to obtain the phosphorus from organic phosphates it is necessary for cells to separate it enzymatically from the organic part of the molecule, before taking up the phosphate generally. This is usually done by alkaline phosphatases. Intracellular utilization is usually associated with internal acid phosphates, while external Po uptake involves inducible alkaline phosphatases. Both can be formed by estuarine and freshwater dinoflagellates (Kuenzler & Perras 1965; Wynne 1977), external alkaline phosphatases being released by *P. noctiluca* (Rivkin & Swift 1982). Acid phosphatases are formed in greater quantity when ambient Pi levels fall (e.g. by *Peridinium cinctum* in Lake Kinneret: Berman 1970: Wynne 1977) and elevated Pi suppresses alkaline phosphatases. In *P. noctiluca* this suppression occurs above 0·05 μM (Rivkin & Swift 1982).

A relationship with light has also been found for alkaline phosphatases in *P. noctiluca*, being higher under high irradiances. This suggests that under field conditions such activity is low at the bottom of the euphotic zone relative to the waters above it.

(D) SULPHUR (AND ANOXIC CONDITIONS)

Although sulphur is known to be required as an inorganic macronutrient, being an integral part of some proteins, it is relatively abundant in seawater, 2·76 g l^{-1} at 35‰, so is unlikely to become a limiting nutrient. Lewin & Busby (1967) found that it is basically 100–200 times the requirement of most of the algae they tested. However, of the nine algae from five groups that they tested, *Amphidinium carterae* had the highest requirement, followed by the red alga *Porphyridium*. *A. carterae* showed no growth at 0·024 g l^{-1} of sulphate (added as magnesium sulphate), although this was sufficient to support the maximum yield of all others except *Porphyridium* (which had 40–50% growth at this concentration). Even at 0·045 g l^{-1}, growth of the dinoflagellate was only 20% of maximum. They attributed this poor growth to a possible requirement for the production of mucilage-containing sulphated polysaccharide. *A. carterae*, as a tide-pool/sand flat organism, lives in naturally high sulphur environments.

Collier *et al.* (1969) and Wilson (1975) found that the chelator disodium EDTA, which produced the greatest growth enhancement of *Ptychodiscus brevis* of several culture additives to natural seawater, could be replaced by hydrogen sulphide. The sulphides of trace metals are extremely insoluble in seawater, (Provasoli, 1979) and so a stimulatory effect may be due to a reduction of the latter from toxic levels. Okaichi (1975) found that one effect of sulphite pulpmill waste was the chelation of iron by lignosulphonic acid, which enhanced the growth of *Prorocentrum micans* and a euglenoid flagellate.

In Omura Bay in Japan *Gymnodinium nagasakiense* blooms after the bottom waters of the bay become anoxic (Iizuka & Irie 1969; Iizuka 1972), and even migrates into the anoxic layer. It is stimulated by addition of 1 mg l^{-1} or more of sodium sulphide (Iizuka & Nakashima 1975). Anaerobically decomposed mud extracts from the time of the bloom were also stimulatory (Hirayama & Numaguchi 1972). Other dinoflagellates, such as *Ceratium furca* and *Gymnodinium sanguineum* (= *nelsonii*) are less stimulated by additions of sodium sulphide.

In Walvis Bay, South West Africa, *Gymnodinium galatheanum* (which greatly resembles the species from Omura Bay) often blooms in the waters over anoxic coastal sediments (Pieterse & Van der Post 1967). Marine fauna mortalities add to the anoxia of the bottom waters and an autostimulatory cycle seems to exist in these and other similar coastal areas (e.g. Peru, the west coast of India, or the Black Sea). *Gyrodinium aureolum* and *Gymnodinium flavum* (which are also morphologically similar to the above species (Taylor 1985) and also kill marine fauna, including fish) appear to bloom when previously mixed water becomes stabilized, although Tangen (1977) could not find a correlation with anoxic bottom conditions in Norwegian blooms of the former.

Dinoflagellates can also use organic sources of nutritional sulphur. Deane & O'Brien (1981b) have shown that *A. carterae, A. klebsii* and *Symbiodinium microadriaticum* can use cysteine, methionine or taurine as sulphur sources, using an active transport mechanism. Interestingly, there was little difference between the usually symbiotic *S. microadriaticum* and the others in this regard, even though taurine, which is abundantly present in the haemolymph of its giant clam host, *Tridacna*, is not found freely in the ocean. There were some small differences in inhibitory interactions in the uptake of the organic sulphur sources.

(E) IRON

This is an essential trace element, used for numerous oxidation–reduction (redox) reactions in cells, including components of the electron transport chains in both photosynthesis and respiration and various enzymatic activities. Metalloproteins such as ferredoxin (an iron–sulphur protein), some cytochromes (a, a_3, b, c_1) and one type of superoxide dismutase, are well known examples. It is also involved in chlorophyll synthesis in an indirect way.

Ryther & Kramer (1961) studied the uptake of chelated ferric iron by several phytoplankters, including *Amphidinium carterae*, showing stimulation up to 65 μg l^{-1}, with possible suppression at higher concentrations. They found that coastal phytoplankters required higher amounts, but did not have an oceanic dinoflagellate to test. Complex interactions of iron and other metals with natural organic compounds, such as humic acids (Prakash & Rashid 1968) or hydroxamates (McKnight & Morel 1978) in the water, greatly complicate the interpretation of field observations (see Huntsman & Sunda 1980). Ingle & Martin (1971) found an apparent correlation between the iron load in run-off in the vicinity of Charlotte Harbour on the west coast of Florida, and blooms of *Ptychodiscus* (*Gymnodinium*) *brevis*. An 'iron index' was devised, being the amount of iron potentially delivered to the coastal waters in a 3-month period. A minimum load of 235 000 pounds of iron (measured in a nearby river), together with suitable water temperatures (16–27°C) and an absence of strong wind, were thought to be necessary for blooms in that vicinity. The authors acknowledged possible multiple effects of the organic load in the run-off, including the detoxifying of inhibitory metals such as copper (as appears to be the case for pond populations of *Protogonyaulax* (*Gonyaulax*) *tamarensis* around Massachusetts: Anderson & Morel 1978). Furthermore, the index has no correlation with blooms over wider areas. Steidinger (1973) noted that *Pt. brevis* also followed blooms of the filamentous cyanobacterium *Trichodesmium* (*Oscillatoria*), the latter also being a potential source of iron or other organic nutrients.

Glover (1978) was able to show, by nutrient enrichment of natural waters, that iron concentrations apparently limited a bloom of *Pr. tamarensis* in the Gulf of Maine in 1975, levels of total iron in the coastal water dropping to less than 1·5 μg atoms l^{-1} in August and September. Higher iron levels were correlated with lower salinities, implicating run-off, rather than upwelling, as the primary source in this area. Internal levels are once again important. Sergeyeva & Krupatkina-Atkinia (1971) found that *Prorocentrum micans* grew almost as well for more than 10 days in a medium lacking trace metals, as in a complete medium, but *Gymnodinium kovalevskii* did not. The background levels of metals in the unenriched seawater were not reported.

The importance of iron to oceanic dinoflagellates is unknown, with very few measurements of its level in open ocean waters and no correlations with oceanic dinoflagellate growth shown. However, there are preliminary indications that horn-length in some oceanic species of *Ceratium* may be partly influenced by the levels of trace metals (see Chapter 2, Section 1.3).

There has been considerable interest in the possibility of the production by phytoplankton of organic compounds, such as hydroxamic acids, with specific iron-binding abilities ('siderophores'). Trick *et al.* (1983) have shown that *Prorocentrum minimum* produces a water soluble compound in culture, 'prorocentrin', which has the characteristics of a hydroxamate-containing siderophore. The precise mechanism by which such binding compounds facilitate uptake is not known.

(F) OTHER METALS AND COMPLEXING AGENTS

Manganese and copper are other trace metals whose abundance in natural waters may limit the growth of algae due to insufficiency, but in seawater both are usually in adequate supply and copper has attracted attention more because of its acute toxicity to dinoflagellates at relatively low levels (see also Section 3.9). Concentrations of 0·215–0·25 mg l^{-1} have been used to kill dinoflagellates parasitic on fish (Dempster 1955; Loeblich 1967). The earliest attempt to control red tides involved the addition of copper sulphate to the water (Section 6.3).

The usual levels of copper in seawater are between 10^{-9} and 10^{-8} mol l^{-1}. Even this may be inhibitory to some dinoflagellates. Anderson & Morel (1978) found that although *Protogonyaulax tamarensis* could excyst in offshore water near Cape Cod, Massachusetts, subsequent development in this water was inhibited, whereas the dinoflagellate grows well in shallow, coastal marine ponds where it is presumed to be protected by the complexing of copper by natural organic matter in such a way as to detoxify the water. Schenk (1984) also found *P. tamarensis* to be highly sensitive to copper, but also that it had a relatively high requirement for it at lower levels, being limited at pCu 12.5. Clearly, other species cannot be as sensitive. Huntsman & Sunda (1980) class dinoflagellates and cyanobacteria as the algal groups most sensitive to cupric ions, although Saward *et al.* (1975) concluded that diatoms are more sensitive. *Kryptoperidinium* (*Glenodinium*) *foliaceum* was found to be as sensitive to copper as the cyanobacterium *Coccochloris elabans,* both being harmed by levels of 0·03 μg Cu l^{-1} (Mandelli 1969), but the former concentrated much more of the metal within its cells (45 μg mg^{-1} wet weight, a concentration factor of 1500 ×). Collier *et al.* (1969) suggested that *Ptychodiscus* (*Gymnodinium*) *brevis* was excluded from some water types because of copper sensitivity.

W. G. Harrison *et al.* (1977) found that even blooms of non-photosynthetic *Noctiluca* were damaged by concentrations of 10–20 μg l^{-1}, cell debris accumulating in the bottom of copper-containing chambers and vacuolar ammonium levels declining within the cells.

Copper is usually classed as a 'heavy metal'. Other heavy metals such as mercury, lead, cadmium, arsenic and zinc are also toxic to dinoflagellates at low levels, as they are to most organisms. Although experimental studies on the effects of these metals on dinoflagellates have been carried out (see Miller 1973; Berland *et al.* 1976; Andreae & Klumpp 1979; Prevot & Soyer 1979; Sanders & Windom 1980 and Section 3.9 for toxic levels), there are no reports of their probable actions on natural marine dinoflagellate communities.

Cobalt is a required element for all organisms. In dinoflagellates it is usually obtained in organically-bound form as vitamin B_{12} although cobalt chloride is also usually added to culture media.

Selenium has also been reported to be a micronutrient for some dinoflagellates. For example, Lindstrom & Rodhe (1978) found that selenium could be growth-limiting in fresh water to *Peridinium cinctum* at concentrations below $50\,\mathrm{ng\,l^{-1}}$.

Humic acids in land run-off are believed to be beneficial to dinoflagellates due to their metal-complexing activities and have been suggested to be significant in the promotion of red tide blooms (Prakash & Rashid 1968; Collier *et al.* 1969; Doig & Martin 1974). Meat extract from fish such as mackerel, improves the growth of *Gymnodinium nagasakiense*, especially if heated, but the reason(s) for this are not known (Nishimura 1982). Experimentally, gibberellic acid has been shown to stimulate the growth of *Gymnodinium sanguineum* and *Ptychodiscus brevis*, greatly reducing the lag period in culture (Paster & Abbott 1970; Bentley-Mowat & Reid 1969).

(G) VITAMINS

Organic compounds are required, not only by heterotrophic dinoflagellates but also, in very small amounts, by photosynthetic forms. Those of the latter which require vitamins are termed 'auxotrophs' (see reviews by Provasoli & Carlucci 1973; Swift 1980; and Chapter 6 here). Nearly all photosynthetic dinoflagellates tested have a requirement for vitamin B_{12} (cyanocobalamin), and some require, or are stimulated by, thiamin and/or biotin as well (Table 11A.3). *Crypthecodinium cohnii* (a heterotroph) requires biotin only, except at high temperatures. Vitamins may enter the marine environment in land run-off or may be produced *in situ* by bacteria and algae, including phytoplankton which do not require them (Swift 1980). A 'binding factor' has been recognized, which restricts the availability of B_{12} (Pintner & Altmyer 1979), at least temporarily before its breakdown by bacteria (Provasoli 1979). The vitamins also have analogues: compounds of differing structure but with the same effect, e.g. cobamides, and dinoflagellates can use many of these (details in Iwasaki 1979). *Amphidinium carterae* has been used to assay natural vitamin levels in seawater (Cattell 1969, 1973).

Where the supply of a nutrient is limited, an inverse correlation between the nutrient and the user population may be expected. Cattell (1969) found such a

Table 11A.3. Dinoflagellate vitamin requirements, from the compilation by Iwasaki (1979), with nomenclatural modifications (see footnotes): s = stimulated growth; / = differing results with different strains

Species	B_{12}	Thiamin	Biotin
Amphidinium carterae	+	+	+
Amphidinium rhynchocephalum	+	+	+
Crypthecodinium cohnii	−	s	+
'Glenodinium' hallii	+	s	−
Gonyaulax polyedra	+	−	−
Gymnodinium sanguineum[1]	+/s	+/s	−
Gymnodinium sp. 'SP4'	+	s	s
Gymnodinium californicum	+	−	−
Gymnodinium resplendens	+	−	−
Gyrodinium uncatenum	+	−	−
Heterocapsa niei[2]	+	+	−
Kryptoperidinium foliaceum[3]	+	−	−
Oxyrrhis marina	+	+	+
Peridinium balticum	+	−	−
Peridinium chattonii	+	−	−
Peridinium hangoei	+	−	−
Prorocentrum cassubicum	+	−	−
Prorocentrum micans	+	?	?
Prorocentrum sp.	−	−	−
Protogonyaulax catenella[4]	×/−	×/−	−/s
Ptychodiscus brevis[5]	+	+	+
Scrippsiella trochoidea[6]	+	−	−
Symbiodinium microadriaticum	−/s	−	−
Woloszynskia limnetica	+	+	?

[1]As *G. nelsonii* and *G. splendens*. [2]As *Cachonina*. [3]As *Glenodinium*. [4]As *Gonyaulax*. [5]As *Gymnodinium*. [6]As *Peridinium*.

correlation with dinoflagellate numbers in the Strait of Georgia, British Columbia in spring and, to a lesser extent, later in the year. Carlucci (1970) observed a strong drop in B_{12} levels during the development of a bloom of *Gonyaulax polyedra* off California. On the other hand, Prakash & Taylor (1966) found B_{12} levels as high as 13 ng l^{-1} were present after a bloom of *Protogonyaulax acatenella* (probably = *P. tamarensis*) and high levels were also present after a bloom of *Ceratium* sp. off California (Carlucci & Bowes 1970). The latter are possibly the result of cell lysis or release during stationary phase (Swift 1980). In the upper layer of the open ocean, B_{12} concentrations may be as low as 1 ng l^{-1} (Carlucci 1970).

Vitamin B_{12} generally increases with depth, and is highest in coastal waters. Depending on the region it may be highest in winter or summer (Bruno *et al.* 1981). Because biotin is an absolute requirement for some species (Loeblich 1966), this is also a potentially limiting micronutrient. Swift (1980) found that biotin levels in the upper 25 m in the Gulf of Maine in summer were 2–5 ng l^{-1}

and noted that the biotin half-saturation constant (K_s) for photosynthesis by *Amphidinium carterae* was 4 ng l^{-1} (Carlucci & Silbernagel 1969).

Unfortunately, vitamins are difficult to measure (requiring the use of a bacteria-free bioassay), because analogues sometimes may be in greater concentration than the primary forms (Bruno *et al.* 1981) and binding factor may be present, so they have only been measured in a few studies with conflicting results. Almost nothing is known of the vitamin requirements of dinoflagellates inhabiting open ocean water and yet they must cope with extremely low levels. Furthermore, the vertical distribution of vitamins in the central gyres of the ocean is relatively unknown.

3.5 Heterotrophs

Although various subtle distinctions can be made on a physiological basis (Chapter 6), from the ecological point of view there is interest in whether any can use dissolved organic carbon (DOC) sources instead of attacking prey organisms, ingesting them whole or performing extracellular digestion.

Sieburth *et al.* (1978) have maintained that it is unlikely that 'osmotrophs', living on DOC, could compete against bacterial uptake abilities (Wright & Hobbie 1966) and that they are therefore probably feeders on extremely small particles. Many heterotrophic marine dinoflagellates have never been seen to contain ingested material. This is true for nearly all species of the common thecate genera *Protoperidinium, Diplopsalis* (and closely related genera), *Podolampas, Blepharocysta, Ornithocercus, Histioneis,* some *Dinophysis* (those formerly assigned to *Phalacroma*) etc., as well as some individuals of athecate genera such as *Gymnodinium.* Recent observations on extracellular feeding, by means of a 'feeding veil', in some species of *Protoperidinium* (Gaines & Taylor 1984, see Fig. 6.10) suggest that this may be a common form of feeding in thecate, heterotrophic dinoflagellates. Using a dual-label radioisotope method, Lessard & Swift (1985) calculated clearance rates in oceanic heterotrophic dinoflagellates feeding on phytoplankton (1–28 μl indiv.$^{-1}$ h^{-1}) that are equivalent to those of ciliates in the same regions.

In contrast to the larger zooplankton, the 'protozooplankton' are more closely coupled to the timing of maximal food abundance because of their more rapid reproductive rates (Smetacek 1981). This is particularly so for ciliates, but is also true for flagellates. Species of *Protoperidinium* are most abundant in temperate coastal waters during the latter part of the spring diatom bloom. Presumably their excystment factors are similar to those for diatom resting spores. During diatom blooms they can reach abundances of 10^2–10^3 l^{-1}. Biogeographically they are most prevalent where diatoms are, i.e. coastal and high latitudes (Section 4).

Cases where dinoflagellates are more abundant than ciliates as grazers are rare. Swarms of *Oxyrrhis* may occur in tide pools or coastal ponds. *Noctiluca scintillans* is the most common, abundant dinoflagellate grazer in coastal waters, causing spectacular displays of orange 'red water' in many localities in summer, due to its buoyancy and aggregation at convergences (Section 6.1c). Uhlig & Sahling (1982) has documented the occurrence of *Noctiluca* blooms in the vicinity of Helgoland, and Kimor (1979, 1981) has described the feeding of *Noctiluca* on *Acartia tonsa* eggs off California. Its diet is extremely varied, including objects greater in diameter than the hungry *Noctiluca*, and it can be considered an opportunistic feeder.

The ecological impact of parasitic dinoflagellates has not been evaluated although it may be significant in regulating toxic blooms (Section 6.3).

3.6 Neriticism

The reasons for neritic preference, other than low salinity, may be many. Coastal waters are not only areas where land-derived nutrients are abundant, but are also where dissolved organics may serve complexing roles, where most nutrient-rich deeper water upwells, where run-off and physical sheltering from wind may add greater vertical stability, or where tidal mixing may occur. Other plankton groups, particularly diatoms, are most abundant in neritic environments; see Hasle (1976) and Guillard & Killam (1977) for extensive reviews of the distribution of the latter. Planktonic organisms which rely on a benthic cyst phase in order to survive poor conditions, e.g. 'overwintering' cysts, are limited to continental shelf (or fresh) waters because the length of time to settle, isolation from surface conditions and unlikelihood of transport to the surface again, mitigate against the use of benthic cysts by plankton organisms in deeper waters. A few oceanic dinoflagellates, notably *Pyrocystis* and *Thoracosphaera* seem to have modified the cyst phase to be their predominant, pelagic, photosynthetically and reproductively active (vegetative) state and, puzzlingly, a few cyst types are common in oceanic sediments (see Chapter 15).

Another factor leading to a near-shore, brackish distribution may be an inability to tolerate natural levels of metals without the protection afforded by the complexing action of organic compounds present in river waters or shallow coastal embayments (Section 3.4f). The restriction of *Protogonyaulax* (*Gonyaulax*) *tamarensis* to shallow ponds around the Massachusetts coast (Anderson & Morel 1978) or its occurrence in or near estuaries such as those of the Tamar, Forth, St Lawrence, Fraser and Plate Rivers (UK, UK, eastern, western Canada and Argentina, respectively) accords with this hypothesis. The distribution of *Gessnerium* (*Gonyaulax*) *monilatum* in the southern US suggests a similar autecology, as does that of *Pyrodinium bahamense* in the New Guinea region which blooms in months in which rainfall exceeds 10 cm (MacLean 1979).

3.7 Upwelling

Dinoflagellates as a whole seem to be less favoured by the upwelling of nutrient-rich water than diatoms, although they can also benefit from the nutrients, and so this distinction is probably more one of competition than suitability or unsuitability of the water. However, upwelling water can be high in copper, which may inhibit copper-sensitive species (Steemann Neilsen & Wium-Anderson 1970).

Between 1964 and 1966 Dodson & Thomas (1977) found that both diatoms and dinoflagellates, or diatoms alone, predominated in the upwelling region off Baja California, whereas dinoflagellates predominated (in low numbers) in the oligotrophic water further offshore. This pattern was also seen by Schnack & Elbrächter (1981) off NW Africa. Blooms of photosynthetic dinoflagellates can occur within this upwelling zone off Baja California, however (Blasco 1977), but are usually present in the upwelled water in lower numbers. A detailed picture of the relative abundance of dinoflagellates versus diatoms in the Peru Shelf zone has been provided by Huntsman *et al.* (1981) and the species composition by Rojas de Mendiola (1981). Off NW Africa, Blasco *et al.* (1981) could distinguish six cross-shelf phytoplankton groups corresponding to water of different upwelling characteristics (e.g. 'age'), each of which contained at least some dinoflagellates.

Reports of dinoflagellate blooms from other upwelling regions, such as Walvis Bay, Africa (Braarud 1957), usually concern near-shore, sheltered localities (Pieterse & Van der Post 1967). Where upwelling intensity fluctuates, dinoflagellates are more important in the interval periods (Barth & Gomez 1973; Wangersky 1977). Off Peru red tides ('aguajes'), caused principally by *Gymnodinium sanguineum* and the ciliate *Mesodinium rubrum,* occur commonly during periodic El Niño phenomena, when warm equatorial counter current water invades the area (Rojas de Mendiola 1979; Sorokin 1980). Conversely, Lasker (1978) observed that the disruption of flagellate abundance by upwelling, with replacement by diatoms, could be harmful to larval development.

Hulburt (1976) found that the commonest dinoflagellates in several upwelling regions off the western coast of Africa were the athecate species *Katodinium rotundatum* (which is also found in brackish waters) and *Gymnodinium punctatum,* neither being abundant and blooms of *G. sanguineum* occur off the Ivory Coast (Dandonneau 1970).

Non-photosynthetic dinoflagellates may be exploiters of the bacterial concentrations or organic matter which can occur in freshly-upwelled water. Sorokin & Kogelschatz (1979) found that the bacterial biomass in newly-upwelled water off Peru was greater than the phytoplankton by two orders of magnitude. Ciliates are usually the principal feeders on the bacteria in these regions.

In a number of 'red tides' the blooms occurred after wind-induced upwelling had died down and the water column became stabilized, e.g. blooms of *Gymnodinium sanguineum* (= *splendens*) off Norway (Tangen 1977), *Gym. flavum* off California (Cullen *et al.* 1982), *Gonyaulax polyedra* off California (Eppley *et al.* 1969a), *Gon. polygramma* (or probably *G. reticulata*) near Cape Town (Grindley & Taylor 1964, 1971) and *Protogonyaulax tamarensis* off Cape Ann, Massachusetts (Mulligan 1973).

3.8 Tidal influences

Where tides are of sufficient magnitude they can influence the blooming of phytoplankton, including dinoflagellates, on a shorter time scale than seasonal influences. For example, in the Gulf of Maine, Balch (1981) found that diatoms were most abundant during major spring tide periods whereas dinoflagellates bloomed principally during neaps or minor spring tides. He was unable to correlate this with nutrient changes, inorganic nutrients being generally low throughout the study period. Spring tides give rise to an upward mixing of colder subsurface water, but the reason(s) for the phenomenon have not yet been determined. In the Damariscotta River estuary in Maine, two surface bloom patches were correlated spatially and in degree or formation with the tidal cycle (Incze & Yentsch 1981). The patches were located in the least stratified parts of the estuary.

Tidal mixing in shallow estuaries, greatest during spring tides, may also resuspend sediment and cysts which have previously settled. This may not only influence the passive distribution of cysts but may be responsible for the exposure of cysts to excystment-inducing conditions at some seasons of the year, although diatoms should also be favoured by such resuspension phenomena.

Tidally mixed waters are less favourable for dinoflagellate development around the British Isles, higher concentrations occurring at the fronts between the well-mixed water and the more stable offshore water (Holligan (1978); see Section 6.1c).

3.9 Pollution

The question of the influence of pollutants on dinoflagellates is a difficult one to treat briefly. Although there is a growing body of literature on the toxic levels of various substances, particularly metals, it is very difficult to extrapolate these single-factor, acute toxicity studies to the natural environment in which many interacting factors (e.g. organic complexing and speciation of metals) may significantly alter the results. Adaptation can occur, often to a remarkable extent, provided that the increases are gradual enough (Stockner & Antia 1976). For example, Antia & Klut (1981) found that *Amphidinium carterae* could be very gradually acclimated (66 days) to 200 $\mu g\,ml^{-1}$ of fluoride, a lethal concentration if suddenly exposed to it in estuarine water.

There are several major areas subject to eutrophication which support numerous dinoflagellate blooms. Examples are: the inner Oslo Fjord, Norway; Chesapeake Bay, the New York Bight and Long Island Sound, USA; Tokyo Bay, and parts of the Seto Inland Sea, Japan. These regions are characterized by high dissolved organic carbon (DOC) levels and a greater importance of ammonium relative to nitrate (e.g. McCarthy *et al.* 1977; Paasche & Kristiansen 1982). Urea is also an important nitrogen source in such areas. In some sewage outfalls ammonium may reach inhibitory levels (Thomas *et al.* 1980a,b). Some of the earliest studies on these effects of eutrophication on phytoplankton blooms were carried out by Braarud (1945) on the inner portion of Oslo Fjord. Since 1935, thirty major dinoflagellate blooms have been documented in that area (Tangen 1979).

Okaichi (1975) has summarized the increased organic pollution in the Inland Sea of Japan in the late 1960s and early 1970s. The incidence of red tide blooms, 'akashio' (many caused not by dinoflagellates but by species of the chloromonad *Chattonella*), became a serious problem after 1965 and seemed to be directly related to eutrophication, both from domestic (3000 tons per day of untreated sewage) and industrial sources. The number of outbreaks increased from 44 y^{-1} in 1965 to 298 y^{-1} in 1974, most harm being caused to cultured fish (yellowtail) and oysters. A similar change has occurred in Tokyo Bay, which has altered from a diatom-dominated, to a flagellate-dominated community with increasing eutrophication and anoxia of the bottom waters (Seki *et al.* 1974).

Although sewage effluent may enrich the water with nutrients, the final effect depends also on the concentrations of other biostimulatory and bioinhibitory constituents. Nitrilotriacetic acid (NTA), a sometime substitute for phosphate in detergents, was found under experimental conditions to be stimulatory to *Protogonyaulax tamarensis* at concentrations of 10–80 ppm (Yentsch *et al.* 1974); the possibility of selective enhancement of the organism, leading to an aggravation of the paralytic shellfish poisoning problem associated with this organism, was raised but this was doubted by Martin (1974). The effect of surfactants depends on the ionic type involved (Kutt & Martin 1974), with anionic sulphonated aromatic hydrocarbons and sulphated ether, but not cationic or non-ionic surfactants, being lethal to *Ptychodiscus brevis*.

Industrial effluents vary according to the composition of the outflow. For example, Kayser (1969) observed inhibitory effects on *Prorocentrum micans* and *Ceratium furca* exposed to effluent from a titanium dioxide factory (principally H_2SO_4 and $FeSO_4$, with metal contaminants). Stockner & Costella (1976) found that *Amphidinium carterae,* in axenic culture, could tolerate relatively high levels of kraft and other pulpmill effluents provided that an adaptive pre-exposure at lower levels was given, newsprint effluent having very little effect except at very high concentrations (40%).

Hydrocarbons from oil spills are toxic to plankton in high concentrations, but at low levels may even be slightly stimulatory (Dunstan *et al.* 1975), although *Amphidinium carterae* was inhibited by low molecular weight hydrocarbons such as benzene and toluene, but not xylene, until concentrations exceeded $10^4 \mu g l^{-1}$ (the diatom *Skeletonema costatum* was not inhibited by benzene or toluene up to the same concentration, but it was inhibited by lower concentrations of xylene). The low molecular weight fractions of oil are the most active, stimulatory or neutral at low levels and inhibitory at higher levels. Following the wreck of the 'Torrey Canyon' in 1967 large blooms of *Noctiluca* appeared off Brittany and, in one instance, were seen to be accumulated with the chalky substance used to absorb the oil (Cooper 1968; Smith 1970), some having ingested the chalk particles. In an early study, Elmhurst (1922) found that *Oxyrrhis marina* became abundant in some containers to which oil tanker discharge had been added, other non-photosynthetic protists also flourishing. Numerous dinoflagellate blooms also occurred off the French coast in 1978 following the 'Amoco Cadiz' disaster, but LeFevre (1979) concluded that there was no significant connection between the two. The blooms occurred, as usual, at fronts between favourable and unfavourable water conditions.

DDT is known to inhibit photosynthesis in concentrations of parts per billion (Wurster 1968), but no cases of apparent inhibition under field conditions are known. DDE is even more toxic (Powers *et al.* 1975). Other insecticides such as chlordane and heptachlor (Magnani *et al.* 1978), lindane (Jeanne 1979) and herbicides (isopropyl N phenyl carbamate: Matthys-Rochon 1980; 2,4,5-T: DeBilly & Soyer 1979) are also inhibitory.

The influence of metals on dinoflagellates has been discussed in part earlier (Section 3.4f) and lethal and inhibitory levels of mercury, cadmium, copper and lead in cultures were reported by Berland *et al.* (1976). Under their experimental conditions mercury was most toxic to *Amphidinium carterae* and *Prorocentrum minimum* var. *mariae-lebouriae* (inhibitory at $5 \mu g l^{-1}$, lethal at 15–$20 \mu g l^{-1}$), then cadmium and copper (10–50, lethal 20–250), with lead less so (250–1000, lethal $>2000 \mu g l^{-1}$). Zingmark & Miller (1975) obtained similar results for mercury with *A. carterae*, with some inhibition detectable even at $1 \mu g l^{-1}$. Ultrastructural changes as a result of exposure to cadmium (Soyer & Provot 1981), fluoride (Klut *et al.* 1981), 2,4,5-T (DeBilly & Soyer 1979) and carbamate (Matthys-Rochon 1980) have been described. Kayser (1976) developed an assay for mercuric acetate using dinoflagellates in continuous culture.

Increasing temperature, due to the discharge from power plant cooling systems has a variable effect, depending not only on the degree and rate of heating, but also the temperature of the receiving water. Thus, a gradual rise of 5–10°C in cold water (less then 10 or 15°C) will cause a shift in species composition towards a warmer community structure (greater diversity, more dinoflagellate-dominated) in the receiving waters; if the water is already close to or above the optimum of warm communities (20–25°C) a similar rise will either cause little or no change in a subtropical population (Bienfang & Johnson 1980) or, if higher than 20–35°C, lead to domination by the most extremely thermotolerant species of cyanobacteria. Thus, *Heterocapsa triquetra* dominated pools raised 5·5°C to 10–27°C, even though its natural abundance in the Pamlico River estuary was in temperatures from 2–8°C (Carpenter 1973), with an abundance fifteen times that in unheated controls. Off California, where intake water of 17–20°C was raised 9–11°C, most diatoms were severely damaged and dominance by the local red tide species *Gonyaulax polyedra* and *Ceratium divaricatum* were reinforced (Briand 1975). A similar selective effect was observed by Domotor *et al.* (1982), with enhanced growth of *Prorocentrum minimum*.

Concern has been expressed that the construction of a tidal power generating facility in the Bay of Fundy, eastern Canada, will alter the hydrography such that paralytic shellfish poisoning, caused by *Protogonyaulax tamarensis,* may be aggravated (Reid 1980) but, as the area is usually closed to harvesting due to high PSP, the effects on other species may be of equal concern.

3.10 Grazing

Like other organisms of similar size, dinoflagellates are subject to grazing pressure from protozooplankton (ciliates, including tintinnids, foraminiferans, radiolarians, other flagellates or even dinoflagellates: *Noctiluca* readily feeds on dinoflagellates such as *Protogonyaulax catenella*; F. J. R. Taylor, unpubl. obs.) and metazooplankton, including larval or larger planktivorous fishes.

Apart from the direct impact on cell numbers, grazing has also presumably had a long-term influence on cell size and shape. Small cells (10–20 μm) are frequently too small for the filtering apparatus of copepods (e.g. *Acartia tonsa*: Ryther & Sanders 1980), but are readily fed upon by tintinnids and non-loricate ciliates (Smetacek 1981) or web-feeders. However, tintinnids do not seem to be able to feed on many species of *Ceratium*, the large, robust horns preventing entry into the oral aperture of the lorica; this may be a significant factor in the predominance of ceratia in warm oceanic waters in which tintinnids are diverse and common. Non-loricated ciliates, such as *Strombidium* spp., do eat smaller *Ceratium* spp. (Smetacek, 1981) and Elbrächter (1971) observed that the barrel-shaped ciliate, *Tiarina fusus*, could attack dividing cells of *Ceratium*. Prakash (1963) and Stoecker *et al.* (1981) observed that the tintinnid *Favella ehrenbergii* is a strong grazer on *Protogonyaulax tamarensis,* the former attributing a prolongation of toxic material in the water to the tintinnid. Stoecker *et al.* (1984) found that *Favella*, as well as other micrograzers such as *Balanion* sp. and *Strobilidium* sp. were behaviourally correlated with dinoflagellate concentrations. Similarly White (1981) showed that zooplankton, particularly pteropods, could accumulate toxins from *P. tamarensis* blooms and harm fish which fed upon them. In paralytic shellfish poisoning benthic, filter-feeding molluscs, such as mussels, clams and oysters, concentrate saxitoxin-like compounds from the ingested dinoflagellates. Some can be harmed by the toxin (see Section 6.2) and oysters may slow their filtering rate in the presence of some dinoflagellates (e.g.

Gymnodinium sanguineum; Nightingale 1936). Ukeles & Sweeney (1969) showed that a preparation of the trichocysts from *Prorocentrum micans* discouraged feeding by oysters on the prymnesiomonad *Pavlova* (*Monochrysis*) *lutheri* to which the trichocyst preparation was added, but it is not clear if trichocysts may have this effect under natural circumstances. *P. tamarensis* and *P. catenella* are equipped with numerous trichocysts but are evidently readily fed upon by mussels, clams and scallops. Their toxins do not discourage predation by zooplankton either, as noted above.

Dinoflagellates should be an excellent food source for the rearing of larvae in aquaculture, in view of their high protein, carbohydrate and calorific content per cell (Hitchcock 1982). Klein Breteler (1980) found that it was necessary to add heterotrophic dinoflagellates, notably *Oxyrrhis marina,* to other flagellates used in the rearing of several copepods, in order to get good growth. Despite an unconfirmed early report that *Gonyaulax polyedra* may produce toxin(s) (Schradie & Bliss 1962), it was used successfully as a food source for rats, its amino acid composition resembling casein (Patton *et al.* 1967). However neither *G. polyedra,* nor *Prorocentrum micans* was a particularly good food source for anchovy larvae (Scura & Jerde 1977). *Gymnodinium sanguineum* (= *splendens*) has proven to be a very useful food source for the culture of anchovy larvae (Hunter 1976), although zooplankton show a marked avoidance for dense layers of it (Fig. 11A.5) and *Gym. flavum* in natural water off California (Huntley 1982; Fiedler 1982).

In dense concentrations the release of metabolites may discourage zooplankton. Perhaps also, contact of the sensory appendages with too high a particle density may cause avoidance, with feeding from the edge of the bloom. In addition, bioluminescent species appear to discourage predation by their flashing in response to vibration, as demonstrated for *Protogonyaulax tamarensis* (= *excavata;* White 1979), *P. catenella, P. acatenella* and *Gonyaulax polyedra* (Esias & Curl 1972).

4 BIOGEOGRAPHICAL FEATURES

The essential continuity of the oceans has led some authors to reject the nineteenth century view of P. T. Cleve & F. Schütt that water masses are discrete bodies with unique indicator species (Williams *et al.* 1981). However, the 'single ocean' concept, while provocative and timely, disregards strong population barriers known to exist (e.g. temperature/salinity fronts, seasonal cycles, major convergences) and clear cases of regionally-unique restricted dinoflagellate distribution patterns ('endemicism'). An appreciation of the importance of quantitive community composition rather than the presence/absence of endemic 'indicator species' may provide the most useful synthesis of these views.

General aspects of phytoplankton distributions have been summarized by

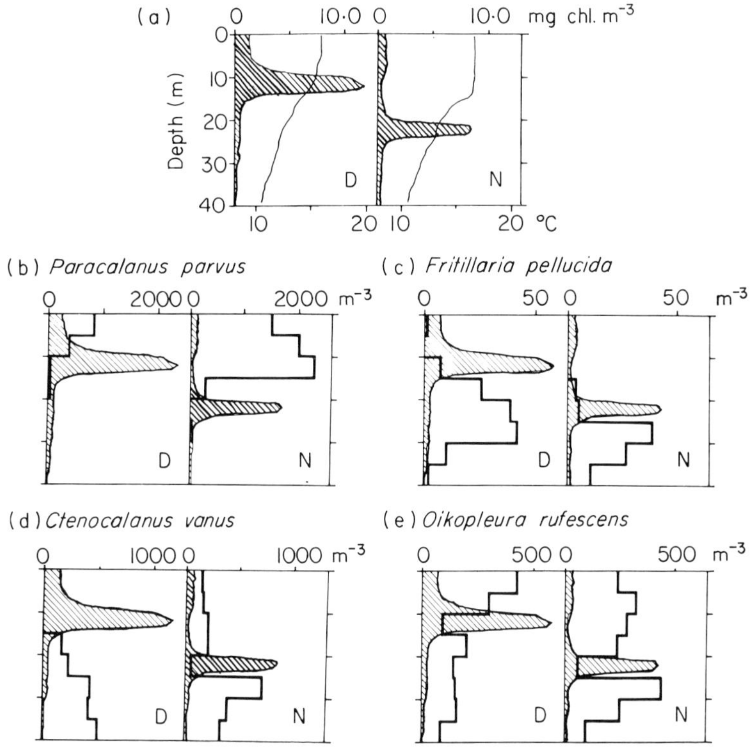

Fig. 11A.5. Chlorophyll, temperature and zooplankton with depth at a station off California in May 1980. The chlorophyll was produced predominantly by *Gymnodinium sanguineum* (=*splendens*). (a) day (D) and night (N) profiles for chlorophyll and temperature, illustrating the migration of the SCM (hatched area). (b)–(e) vertical distributions of various zooplankton species (heavy line). From Fiedler (1982), with permission.

Round (1981) and by Smayda (1958), Margalef (1961a), Braarud (1962) and Williams (1971a), toxic species by Taylor (1984), cysts in sediments by Wall (1971), Williams (1971b), Wall *et al.* (1977) and others summarized by Goodman in Chapter 15 here. Regional references are cited below where appropriate.

4.1 Modified latitudinal cosmopolitanism

The essence of marine dinoflagellate distribution, as with other protists, is modified latitudinal cosmopolitanism: the occurrence of the same (morpho-) species around the world within broad latitudinal limits, the boundaries of which approximate to particular upper layer temperatures.

Major ocean currents, such as the Gulf Stream, Kuroshio, Brazil or Agulhas Currents, can extend the zones north and south within their areas of flow. Upwelling zones of colder water can create seasonal hiatuses within the distribution range (see Section 3.7). Major fronts, such as the polar front, subarctic front, subtropical convergence region, or antarctic convergence, delimit the oceanic boundaries for many species because of their relatively rapid change in water conditions and so are used to recognize major biogeographic zones. Figure 11A.6 illustrates those which are used in the present text.

Within the latitudinal cosmopolitanism further limits are set according to the proximity of the coast and the continental shelf (<200 m), with *neritic* species (e.g. *Gonyaulax spinifera* (Fig. 11A.7) and most species with benthic cysts) occurring predominantly over the shelf and *oceanic* species beyond it (*panthalassic* species showing no preference of this type). Additionally, areas of dilution by run-off will contain estuarine species (generally in water ranging from 5–15 or 20‰). The reasons for neriticism, and the influence of salinity, are discussed in Sections 3.3 and 3.6. The seasonal succession of species (Section 5) further complicates the picture since low latitude species may appear briefly in high latitude areas during the summer.

Latitudinal cosmopolitans are not solely the oceanic planktonic species; neritic or benthic species are circumglobal as well, although the few endemics appear to be mostly neritic. Remote islands are surrounded by typical neritic species despite major, often unidirectional current flows. If ciguatera poisoning is an indicator, benthic *Gambierdiscus toxicus* appear to be present in all tropical

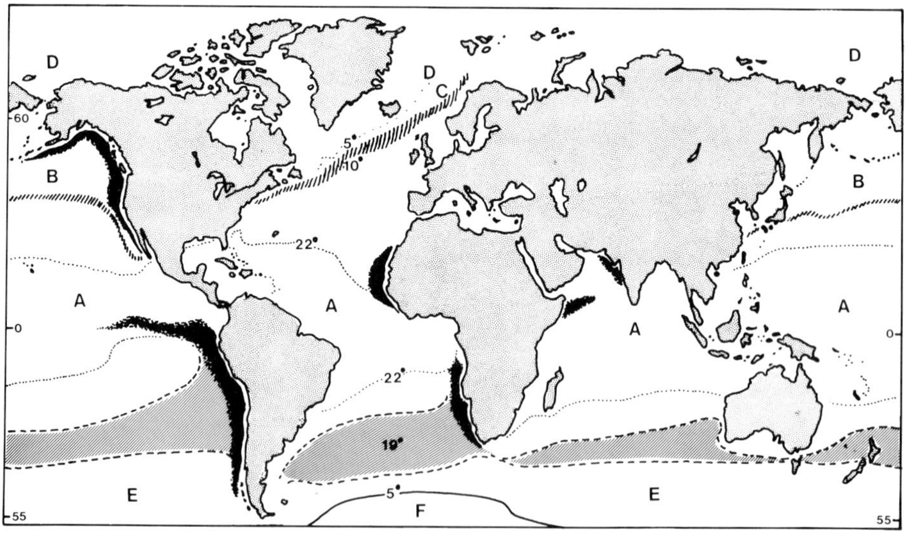

Fig. 11A.6. Major biogeographic zones used in this chapter (isotherms indicated in some regions): (A) tropical–temperate macrozone, the 22°C isotherm indicating strictly tropical water according to some authors although it does not coincide with any major oceanographic features; (B) subarctic North Pacific; (C) subarctic North Atlantic, poorly separated from (D) the Arctic zone; (E) the subantarctic zone; (F) Antarctic zone. Black areas are subjected to substantial seasonal upwelling. Hatched areas are regions of mixed character (in the south reflecting the seasonal shift of the 19°C isotherm).

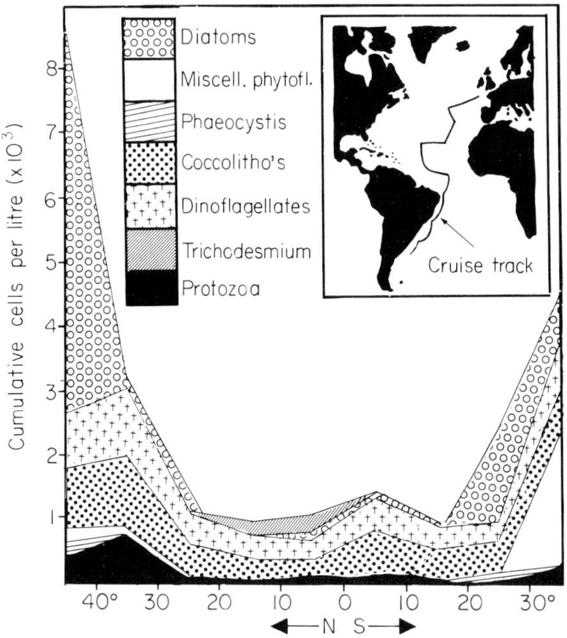

Fig. 11A.7. Cumulative cell counts and group composition of centrifuged samples (average for 0–200 m) collected by Lohmann (1912) on an Atlantic cruise of the 'Deutschland' from 50°N to 40°S (see inset). Redrawn from Lohmann (1912).

oceans between approximately 35°N and 35°S. Intertidal species are even more ubiquitous, presumably due to their wide tolerances.

4.2 General zonal characteristics

Here the principal features of the dinoflagellate communities of the primary latitudinal zones are reviewed, with examples of common species (see Figs. 11A.8–11A.10 for typical distribution maps). Focus has been placed on the general characteristics and some of the major anomalies, with selected references for the reader wanting more information about a particular region. Records of dinoflagellates are also included in many of the regional references provided in the more detailed review of diatom distributions by Guillard & Kilham (1977).

TROPICAL–TEMPERATE MACROZONE

This large region, extending from the transition zone of the subarctic front (in the North Pacific) and polar front (in the North Atlantic), between 40 and 50°N, to the broad, variable transition regions of the subtropical convergence in the southern Atlantic, Indian and Pacific Oceans at 30–50°S, contains few abrupt gradients within it, other than at the edges of upwelling zones or current

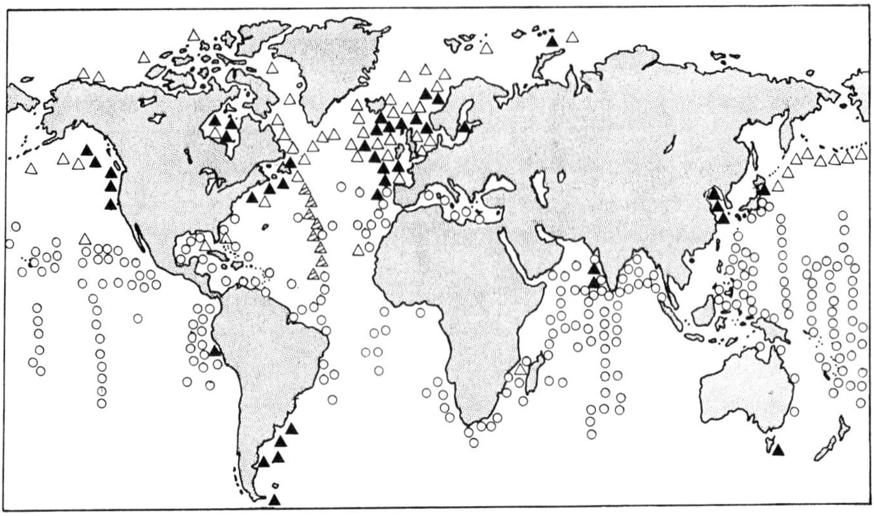

Fig. 11A.8. Distributions of *Pyrocystis noctiluca* (○), a tropical oceanic species, *Ceratium arcticum* var. *arcticum* (△), an Arctic variety (the hatched triangles down the central North Atlantic are subsurface records; see Graham & Bronikovsky, 1944 and the present text), and *C. arcticum* var. *longipes* (▲). Various sources including the above, Semina (1974) and Taylor (1976).

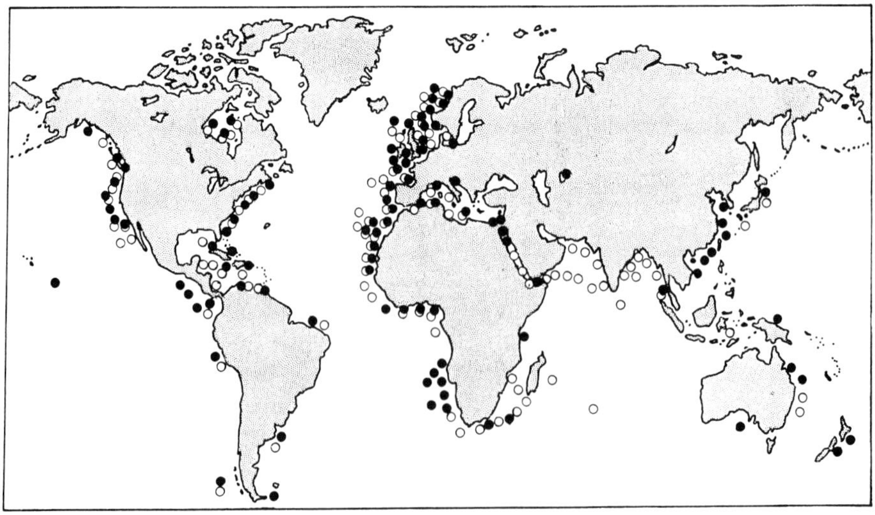

Fig. 11A.9. Distributions of two thermotolerant, neritic species: *Gonyaulax spinifera* (●) and *Protoperidinium divergens* (○). Various sources.

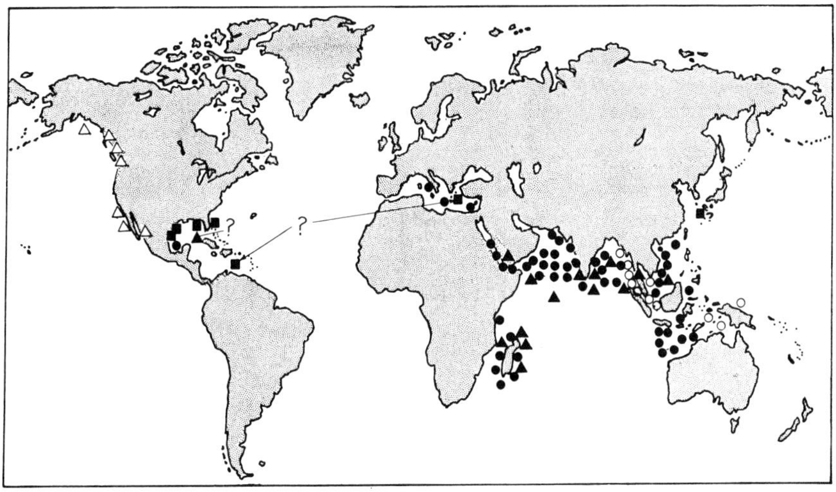

Fig. 11A.10. Distributions of some endemic species; *Dinophysis miles* (●); *D. miles* var. *schroeteri* (○) and *Ceratium dens* (▲), which are Indo-west Pacific taxa; *Oxyphysis oxytoxoides*, a Pacific endemic (△); and *Ptychodiscus brevis* (■).

cores. Surface temperatures are greater than 10°C, exceeding 25°C near the equator. Salinities are generally high, usually above 34‰ or 36‰ in high evaporation belts in the Atlantic and Pacific Oceans.

The euphotic zone (which can exceed 150 m due to the high transparency) is stratified for most, or all, of the year in the tropics, with a sharp pycnocline at or near its base. The upper layer is usually extremely low in nutrients (Parsons & Harrison 1983)—due to uptake by phytoplankton and transport to below the pycnocline by faecal pellets—with little upward renewal except in upwelling regions, including oceanic divergences. Thus, as Dugdale (1967) noted, it can be thought of as a two-layer pelagic system, with a bright sunlit, oligotrophic upper layer and a deeper, low-light or dark, high nutrient region, separated by the pycnocline/nutricline. Much of the phytoplankton (often more than half the total chlorophyll) including dinoflagellates, is aggregated in a deep, subsurface chlorophyll maximum (SCM; see Section 6.1c) associated with the pycnocline where, despite the very low light (less than 1% of the surface light), as much as 20% of the primary productivity may occur (Venrick *et al.* 1973). Because of the relative stability of the system there is the presence of a 'shade flora' (Section 4.5a), consisting of species which have become adapted to the low light levels in order to exploit the higher nutrient levels below the nutricline. Species habitually in the upper layer appear to have slow division rates. Doubling times from 2 days to 2 months have been made for several *Ceratium* species (Elbrächter, 1973; Weiler & Chisholm, 1976; Weiler 1980). For *Pyrocystis* spp.

in the surface layer of the Sargasso Sea the values were 7–19 days, with a low of 4 days in the Tyrrhenian Sea (Swift & Durbin 1972). Maximum doubling times (4–5 days) are found between 80 and 100 m (Sukhanova & Rudyakov 1973; Swift *et al.* 1976).

Goldman *et al.* (1979) found that the oligotrophic oceanic phytoplankton they analysed had C:H:N ratios (Section 3.4) close to those of nutritionally replete cultures and argued that the growth rates should therefore be close to maximal (0·5–1 doubling per day at those temperatures in saturating light). The contrast between their hypothesis and the few direct measurements has raised much interest (see, e.g. Sharp *et al.* 1980). Perhaps oceanic dinoflagellates are less flexible than most cultured species in their composition and grow slowly in order to maintain a normal ratio.

Sournia (1969) adopted a minimum value of 22°C as a convenient limit for tropical waters in a review of the productivity of tropical waters, but dinoflagellate species distributions do not usually show a marked change at this isotherm. The closest, perhaps, is *Pyrocystis noctiluca,* one of the most abundant and widespread tropical oceanic species, found in all oceans where surface water is above 20°C and not appreciably diluted (Sukhanova & Rudyakov 1973; Taylor 1973) (see Fig. 11A.6). Other members of the 'Basic Tropical Complex' (Sukhanova 1962, expanded by Taylor 1973, 1976) are more highly stenothermal, e.g. *Ceratium carriense* and *C. trichoceros,* or slightly more thermotolerant, e.g. *Triadinium* (*Heteraulacus*) *polyedricum, Ceratocorys horrida, Oxytoxum variabile, Ceratium euarcuatum* and *C. massiliense.* Quite often there may be a gap in tropical oceanic distributions, e.g. *Ceratium cephalotum* in the Pacific (McGowan 1971) or *Cladopyxis brachiolata* (Taylor 1976), with low numbers in the low-nutrient equatorial currents, but increases along their margins (Sukhanova 1969; Taylor 1976).

Dinoflagellates and coccolithophorids dominate the microplankton of tropical and temperate oceanic waters, and show greatest diversity in tropical waters. Taylor (1976) found 291 dinoflagellate taxa in 213 preserved net hauls (mostly 200–0 m) from the Indian Ocean. Subrahmanyan & Sarma (1970) found that approximately half the cells not retained by nets off the southwest coast of India were also dinoflagellates. In one net sample in the Bay of Bengal, Taylor (1976) identified 88 taxa of dinoflagellates. *Ceratium* (58 species in Taylor 1976) *Protoperidinium* (56 species in Taylor 1976), and dinophysoids (*Ornithocercus, Histioneis, Amphisolenia* and *Triposolenia*) are particularly prominent and diverse in tropical oceanic waters. Many of these elaborate dinophysoids have symbiotic cyanobacteria between their lists or intracellularly (Taylor 1982; Hallegraeff & Jeffrey 1984) and free living cyanobacteria are an important component of the picoplankton (Waterbury *et al.* 1979).

One reason for the predominance of dinoflagellates in tropical oceanic waters is the quantitive impoverishment of diatoms. Lohmann's (1912) north–south transect from the 'Deutschland' Atlantic cruise, illustrates this quite well

(Fig. 11A.7). However, in tropical and temperate neritic microplankton diatoms are usually much more abundant and diverse. In the Great Barrier reef region a 3-year study recorded 220 diatom and 176 dinoflagellate species (Revelante *et al.* 1982). In a 6-month study of the nearshore waters of Phuket, Thailand (Andaman Sea) Taylor recorded 136 diatoms and 84 planktonic dinoflagellate species (unpubl. ms., UNESCO). In the latter study the proportion of dinoflagellates increased sharply at the edge of the shelf, with *Protoperidinium* and *Gonyaulax* species among the commonest over the shelf.

There is relatively little difference between the oceans, although the Atlantic seems to have relatively fewer species. Sournia (1968a,b) lists seven stenothermal species: *Ceratium dens. C. filicorne, C. humile, C. aultii, C. deflexum, C. lanceolatum* and *C. petersii* from the Indo-Pacific which appeared to be absent from the Atlantic. Some of these have since been found in the Atlantic (M.C. Carbonell, pers. comm.).

In the Sargasso Sea dinoflagellates are sparse in the upper layer, represented chiefly by *Oxytoxum variabile* and, curiously, *Katodinium rotundatum* (usually abundant in estuaries), both small species. In the upper 30–50 m there may be more dinoflagellates present as zooxanthellae in acanthometrids than free in the water (Taylor 1982), with *Pyrocystis noctiluca* more abundant at greater depths (80–100 m). Khmeleva (1967) asserted that zooxanthellae in radiolarians are major primary producers in the Gulf of Aden.

The equatorial waters of the Atlantic range across a narrower distance than the other oceans and are bounded by an area of major river influence, the Amazon plume, to the west and upwelling coasts of West Africa to the east. Perhaps as a result of this proportionately greater eutrophication, diatoms are more dominant than they are in the other oceans, and coccolithophorids and cyanobacteria are more important in the oligotrophic waters (Zernova 1979). The communities of the upwelling zones off NW Africa have been summarized by Margalef (1978a,b: see also Section 3.7). Extensive blooms of *Gymnodinium sanguineum* can occur off the Ivory Coast, analogous to similar areas off Peru and California. Other athecate species from NW Africa have been described by Elbrächter (1979).

Wood (1966) has described the region of Amazon influence, in which dinoflagellates are sparse. The plume extends northwards, occasionally exerting some influence in the Windward Islands.

Off Venezuela there is further river input from the Orinoco, but increased productivity in the vicinity of the Cariaco Trench seems to be associated more with upwelling. Halim (1967) has described the dinoflagellate community from this region.

In the Caribbean Sea and Gulf of Mexico dinoflagellate communities are essentially tropical–subtropical (Graham 1954), with areas of strong neritic influence and major run-off. The phytoplankton in the vicinity of the Mississippi delta was characterized by Simmons & Thomas (1962). The brackish marine water is dominated by diatoms, the few dinoflagellates (*Ceratium fusus, C. furca* plus a mixture of some more open Gulf species such as *Pyrocystis* and a few fresh water forms, e.g. *Hypnodinium sphericum*) never becoming abundant.

Dinoflagellate diversity may exceed that of diatoms in some parts, e.g. in bottle samples taken in March in the southern Gulf of Mexico (Zernova 1969), but in November in the closely adjacent western Caribbean, with much more shelf influence, diatoms predominate (Hulburt 1968), dinoflagellates being almost negligible.

In lines of stations close to the lesser Antilles Hargraves *et al.* (1970) found more diatoms than dinoflagellates, but a number of the former were littoral species found 'downstream' (west) of the islands.

Steidinger's (1973) review of eastern Gulf of Mexico phytoplankton ecology is a useful introduction to the region, together with the floristic works of Wood (1968a) and Steidinger & Williams (1970). A general picture of the phytoplankton composition in the northern Bahamas was provided by Wood (1968b), who recorded a total of sixty-five species of dinoflagellates, all typical tropical or eurythermal species.

Between 6 and 5·5 million years ago (Late Miocene), the Mediterranean appears to have been almost completely dried out, presumably killing resident species. Then it is believed to have refilled from the Atlantic Ocean. Thus, the dinoflagellate community has been principally introduced from the Atlantic.

The Mediterranean is essentially an oligotrophic basin, depauperate in phytoplankton except near the three large river plumes (Rhone, Po and formerly the Nile, the latter not having flooded since the Aswan Dam came into effect in 1965, causing a drastic decline in the eastern Mediterranean sardine fishery: Aleem 1972) and some upwelling in the west.

The Golfe du Lion is one of the more productive regions of the Mediterranean, with some seasonal upwelling and more pronounced neritic influence due to the outflow of the Rhone river. Jacques (1968) has reviewed the phytoplankton ecology of the region. Diatoms are generally more abundant than dinoflagellates. The neriticism is more reduced nearer the Strait of Gibraltar along the Spanish coast (see Margalef 1969a,b). Localized eutrophication from cities has led to very localized red tides in many areas. Jacques & Sournia (1979) have reviewed the occurrence of red tides in the Mediterranean region.

The Black Sea is a less saline body of water (surface approximately 18‰), now connected to the Mediterranean, although in glacial times it was connected to the Caspian Sea, a feature reflected in its fauna (Ekman 1953). It is anoxic and sulphide-rich below approximately 100 m. In winter it reaches 3–4°C and so only euryhaline, eurythermal Mediterranean species survive within it.

Dinoflagellate remains are abundant in the sediments of the Black Sea (Wall & Dale 1974), this being consistent with the interpretation of Boon *et al.* (1979) that the dinosterol content of the sediments is due to past dinoflagellate blooms (see also Chapter 9). Sediments back to 7000 years BP (before present) are dominated by *Lingulodinium machaerophorum*, the cyst of *Gonyaulax polyedra*, and organic and calcitic cysts produced by species of *Protoperidinium* and *Scrippsiella trochoidea*, with smaller members of other gonyaulacoid cysts. Blooms of *Prorocentrum* (e.g. see *P. cordatum*: Nesteroffa 1979) occur in the north-western area, but are not known to produce cysts.

Euryhaline diatoms and dinoflagellates also occur in the land-locked Caspian and Aral Seas, with *Prorocentrum cordatum* a common and abundant, dominant species in both. In the Caspian, *P. micans, Gonyaulax polyedra* and *Diplopsalis* (*Glenodinium*) *caspica* are also important. In the Aral Sea diatoms dominate, but *P. obtusum, D. caspica* and *Gonyaulax spinifera* form occasional intense blooms, particularly in summer, accounting for up to 54% of the total phytoplankton biovolume (Pichkily 1970).

The dinoflagellates of the Mediterranean Sea received much early study, the communities described by authors such as Pavillard (1937) or Halim (1960) differing little from the adjacent tropical Atlantic. There do not seem to be any species unique to the Mediterranean. In the northern areas winter surface temperatures may be low (12°C or less). Léger (1972) has characterized the Ligurian Sea as permanently low in abundance, linked with the oligotrophy of the surface waters, with little seasonal variability in the dinoflagellates or the coccolithophorids. Summer stratification does lead to a subsurface chlorophyll maximum (SCM) containing dinoflagellates, even nearshore, e.g. in the bay of Villefranche. The study by Kimor & Wood (1975) provides a picture of the ecology of eastern Mediterranean dinoflagellates.

The cooler temperate waters of all three oceans, lying between the 22°C isotherm and subpolar zones are essentially diatom-dominated regions with sporadic, summer coastal red tide blooms (Section 6). In the oceanic regions a summer thermocline breaks down, providing winter mixing and a spring diatom bloom similar to, although smaller than the coastal spring bloom. Coccolithophorids are still important in the more oligotrophic waters, but dinoflagellates (other than some nanoplanktonic species) are less important. The north central Sargasso Sea is an example of these regions (Riley 1957). The temperate eastern coast of North America is highly diatom-dominated (Marshall 1976), although many dinoflagellate species (147) have been recorded, principally from the Gulf Stream core where dinoflagellates predominate (Doyle & Poore 1974).

The east coast of the US is characterized by numerous marshy, lagoonal habitats. Campbell (1973) has provided a thorough description of a warmer brackish locality in North Carolina. Such habitats extend to Cape Cod. In these environments flagellates of several groups other than dinoflagellates are often dominant, and the dinoflagellates consist chiefly of *Prorocentrum* spp., athecate gymnodinoids and *Scrippsiella trochoidea*. Larger estuarine embayments, such as Chesapeake Bay, are subject to periodic red tides (Section 6). Narragensett Bay, Rhode Island is well mixed through tidal action and is essentially diatom-dominated, with sporadic small-scale, summer flagellate blooms of dinoflagellates being relatively sparse (Smayda 1983). The dominant dinoflagellates of the open coastal waters are *Ceratium fusus, C. tripos, C. lineatum* and *Prorocentrum micans* (Marshall 1978).

The tropical–temperate macrozone extends over a greater latitudinal range in the Atlantic Ocean than in the Pacific or Indian Oceans. Extensions of the Gulf Stream, the North Atlantic and Norwegian Currents, carry temperate oceanic species north-eastward past the west coasts of Ireland, Scotland and Norway. This influence extends beyond the Norwegian Sea, mixing with polar water. Some temperate species can even be found in the Barents and Kara Seas above 70°N (Meunier 1910). In the southwestern Atlantic the Brazil Current flowing down the east coast of South America extends temperate species to 35–40°S or more on the Argentine shelf, where it encounters the subantarctic water of the Patagonian Coastal Current and Malvinas/Falklands Current.

Temperate species may extend even further, at subsurface depths (perhaps as much as 200 m or more), well into the 60°S subantarctic zone of Drake Passage (Balech 1971).

The distribution of forty-three species from the South Atlantic cruises of the 'Meteor' was described in a classic study by Peters (1932). He used them to recognize several warm water regions in which they were most diverse (10–30°S) and cold water zones (upwelling or subantarctic) in which the diversity was much less (less than 10 species). In this study he suggested that horn length was not so much a function of water density (salinity differences), but rather was related to nutrient content (phosphate in particular).

The tropical Indo-West Pacific region is particularly rich in species of dinoflagellates, as it is in most other plant and animal groups, the richest and most abundant populations occurring in the Andaman Sea (Taylor 1976) and South China Sea. Seasonal upwelling associated with the SW Monsoon and the Somali Current in the Arabian Sea does not seem to favour dinoflagellates.

INDO-PACIFIC

In the south-western Indian Ocean the Agulhas Current extends tropical species southwards, but diatoms generally predominate (Taylor 1967; Grindley 1979). As elsewhere in the tropics, diatoms are dominant in neritic waters (Subrahmanyan & Sarma 1970; Sournia 1968a,b).

The plankton characteristics of the Red Sea have been summarized by Halim (1969). Like the Persian Gulf, the region typically has high surface temperatures (21–32°C, highest in the south) and salinities (37–41‰, highest in the north). There is usually a subsurface chlorophyll maximum at 50–75 m. During some months dinoflagellates predominate over diatoms in species richness and they are nearly always more abundant, although not as diverse as in the Indian Ocean. All species can be thought of as tropical, some not occurring in the adjacent Mediterranean (see below.) Dinoflagellates also usually predominate in the Gulf of Elat/Aqaba (Kimor & Golandsky 1977). Appreciable numbers of dinoflagellates can occur to 500 m in the latter area.

In the open eastern Pacific Ocean upwelling occurs in a tropical divergence zone, contributed to by the advection of upwelled water from the Peru coast. Some dinoflagellates, such as *Gymnodinium simplex* (Thomas 1959) and *Pyrocystis noctiluca*, seem to be more abundant in this region (Semina 1974), although not in the most recently upwelled water. Further away from the divergence zone, north of the Marquesas Islands, Hasle (1959, 1960) found the phytoplankton concentrations and composition were similar to those of the Sargasso Sea, with maximum dinoflagellate populations at subsurface depths. The most abundant dinoflagellate was *Oxytoxum variabile* (max. $4 \times 10^3\,l^{-1}$) which occurred from the surface to 400 m.

In the south-eastern Pacific coccolithophorids appear to be dominant ($>10^3\,l^{-1}$), but nanoplanktonic dinoflagellates, again *Oxytoxum variabile,* as well as small gymnodinoids, are also relatively abundant. Larger members of *Ceratium, Podolampas* and other *Oxytoxum*s are present, but sparse (Desrosières & Wauthy 1972).

The extensive study by Smayda (1966) on the Gulf of Panama, provides a picture of eastern Pacific, tropical coastal phytoplankton. As with other tropical coastal areas it is diatom-dominated, both in species composition and abundance. *Prorocentrum balticum* was the most abundant dinoflagellate ($2 \times 10^4\,l^{-1}$). He did not observe the usual succession from diatoms to dinoflagellates following stabilization of the water column. The area is subject to periodic upwelling.

Semina's (1974) book is a valuable source of information on phytoplankton distributions in the Pacific, but is only available in Russian. However, distributions in the North Pacific are reviewed in English by Semina & Tarkhova (1972) and McGowan (1971) has provided a general review which contains dinoflagellate data, mostly drawn from the study of Graham & Bronikovsky (1944; see also Graham 1942). The seasonal cold water off the west coast of South America extends the distribution of cold water species to the equator and there is a similar extension to Baja California in the northern hemisphere (see the Boreal section below). In both coastal areas dinoflagellates such as *Gymnodinium sanguineum* can form extensive blooms in the interim or following upwelling, and this species is a dominant associated with warm water invasions of the Peru Coast during 'El Niño' phenomena (Rojas de Mendiola 1981).

In the north-western North Pacific the Kuroshio current conducts warm water northwards past the east coast of Japan, but strongly diverges eastward at approximately 36°N (east of Tokyo), becoming the North Pacific current. The east coast of the northern island of Hokkaido is therefore within the subpolar influence of the Oyashio, flowing south from the Bering Sea and Sea of

Okhotsk. Between the two is a transitional zone corresponding to the north-east (Sanriku) coast of Honshu.

The dinoflagellates of Japanese waters have received considerable attention, particularly due to the taxonomic studies of T. Abé and Y. Fukuyo and concern over red tides (Section 6). Yamaji (Yamazi 1972) has provided an illustrated compendium of the planktonic species of Japanese waters, both temperate and warm water, the latter being essentially similar to those of the East Indies. In the more oligotrophic, equatorial Pacific waters dinoflagellates are principally contributors to the nanoplankton, with species of *Oxytoxum* and *Prorocentrum* important (e.g. Philippine Sea; Taniguchi 1977). The rates of phytoplankton ecological processes in the open North Pacific have been described by Sharp *et al.* (1980).

A detailed review of the vertical distribution of phytoplankton in the North Central Gyre of the Pacific, north of Hawaii, has been provided by Venrick (1982), supplementing the study of *Ceratium* from the same region by Weiler (1980). The transition between the surface and deep communities is at approximately 100 m. Species such as *O. variabile* occur throughout the water column, from the surface to 150 m, whereas *C. boehmii* occurs only below 75 m.

In the south-western Pacific the monograph by Wood (1954) on Australian waters can serve as a taxonomic introduction, the region having a range, like Japan, from tropical (northern and north-eastern coasts) to temperate waters. A more recent summary of phytoplankton ecology in the eastern Australian region has been provided by Jeffrey (1981) and Hallegraeff (1981). Off the east coast dinoflagellates are a minor component and only increase after the spring bloom of diatoms, the dominants being species of *Prorocentrum, Ceratium, Dinophysis, Protoperidinium* and *Gymnodinium*.

Several particular aspects of dinoflagellate communities from the tropics: shade flora, coral reefs and other benthic community features, are dealt with separately later (Section 4.5). Hallegraeff & Jeffrey (1984) have described the tropical plankton community of the northern shelf of Australia.

SUBPOLAR (COLD TEMPERATE) WATERS

Subpolar water comprises a belt of water of intermediate characteristics between the polar/subarctic fronts and polar waters in the north (boreal/subarctic), and between the subtropical convergence regions and the Antarctic convergence in the south (antiboreal/subantarctic). It is bounded approximately by the 10–12°C summer isotherms in lower latitudes (roughly 40°N and S) and 5°C at the edge of the polar waters.

In the northern hemisphere this type of water constitutes a large part of the North Pacific, including the Bering Sea and Sea of Okhotsk. In the North Atlantic the zone is much narrower, subject to more mixing of polar and temperate water, and is more easily distinguished from Arctic water by its mixed properties than by isotherms.

In the southern hemisphere the zone forms a continuous, circumglobal belt in the Southern Ocean, of more uniform properties, sharply delimited from Antarctic surface water by the Antarctic convergence (50–60° S) and less well separated from the subtropical convergence zones (broad areas of lateral mixing ranging from 12 to 19°C, varying in position and width seasonally).

SUBANTARCTIC

This zone is generally oceanic, with the exception of the southern tip of South America, which supports neritic populations including *Protogonyaulax catenella* (west coast and Strait of Magellan: Guzman & Campodonico 1978) and *P. tamarensis* (east coast: Carreto *et al.* 1981). Other species also resemble those of the northern hemisphere at similar latitudes.

Diatoms are quantitatively and qualitatively more abundant than dinoflagellates. In the subantarctic Pacific 'Brategg' material studied by Hasle (1969) the number of diatoms beneath 1 m^2 of surface was approximately twenty-five times greater than the dinoflagellates. On each of the three transects dinoflagellates showed a slight increase near 60°S before declining sharply in the Antarctic water.

Among the larger cells dinoflagellates, principally belonging to *Ceratium* and *Protoperidinium* are important (Cassie 1963). Occasionally oceanic stations are encountered at which dinoflagellates are most abundant in summer, e.g. south of New Zealand (Holm-Hansen *et al.* 1977). *Ceratium pentagonum* is the most important large oceanic species (Jacques *et al.* 1979). *Prorocentrum balticum* is important in the nanoplankton, at least during December to February in the Pacific sector (Hasle 1969). This zone contains many species that are more abundant to the north, e.g. *Protoperidinium divergens, P. pellucidum, Scrippsiella trochoidea* and *Gonyaulax spinifera*. *Ceratium fusus,* which is common and occasionally abundant north of this zone, does not appear to penetrate it some times (Hart 1934; Jacques *et al.* 1979) but does at other times (Hasle 1969).

Essentially, then, the subantarctic zone involves persistence of the more eurythermal or cold-loving species of temperate waters, rather than a unique zonal assemblage. The *grande* form of *Ceratium pentagonum* is one of the few taxa which is most abundant in this zone (shown most clearly in the transect studied by Jacques *et al.* 1979), replacing the forma *tenerum* found most commonly in warmer waters.

BOREAL NORTH PACIFIC

This consists of the North Pacific, north of the subarctic (or polar) front, centred on 40°S, with a transitional belt which broadens near its eastern and western boundaries. Unlike the zone of similar temperature in the North Atlantic, it is not subjected to significant mixing with Arctic water, except within the Arctic Ocean, because flow over the shallow sill in the Bering Strait is almost entirely northward. Thus, there is a continuous export of Pacific cells into the Arctic Ocean, with no northern hemisphere input from the Atlantic. Bursa (1963a,b) has attributed the large number of dinoflagellates at Point Barrow, Alaska to an input from the Bering Sea.

The surface waters have a relatively lower salinity, due to land run-off. Typically, the range is from 32‰ to 33·5‰ except in close proximity to rivers near-shore. For example, within the Strait of Georgia surface salinities are generally 29‰, but much less near the Fraser River plume. In the Bering Sea and Sea of Okhotsk the temperatures are lowest and can resemble the Arctic in winter (to −2°C). Run-off is greatest in the Gulf of Alaska, leading to a region known as the Dilute Domain, with a surface salinity less than 32·6‰. Although this has little apparent influence on the dinoflagellates, it can result in the replacement of coccolithophorids by other prymnesiomonads for a long distance offshore (500 miles or more).

There are three principal subdivisions of the region: the Alaskan Gyre in the east, the Bering Sea, and Sea of Okhotsk. The Oyashio Current carries cold, Bering Sea water southwards to the coasts of the northern Japanese Island of Hokkaido. The region is significantly more productive than the temperate oceanic waters to the south (Semina & Tarkhova 1972).

Diatoms generally predominate over the whole region, particularly in neritic areas (reviewed by Karohji 1972), but the smaller nanoplanktonic components of the oceanic regions are much more varied in composition (e.g. Booth *et al.* 1981; Taylor & Waters 1982), at least in spring, with variable contributions by cryptomonads, chrysomonads, prymnesiomonads and athecate dinoflagellates (*Gymnodinium simplex*). In summer *Prorocentrum balticum* may be relatively abundant. Only a few species of *Ceratium* (*C. arcticum* var. *longipes, C. pentagonum*) are present, and are not abundant in the oceanic waters.

In coastal waters in summer, blooms of several dinoflagellate species, such as *Noctiluca scintillans, Gymnodinium sanguineum, Ceratium divaricatum* (commonly misidentified as *C. dens*), *Protogonyaulax catenella, P. tamarensis* and several species of *Gonyaulax* may occur (Harrison *et al.* 1983). A distinctive species is *Oxyphysis oxytoxoides,* whose range extends from Mexico to Japan in neritic waters. *Gonyaulax triacantha* also occurs in the Arctic.

As noted earlier, the coastal boundary between boreal and temperate is less readily defined than in oceanic waters. Cold surface water (not exceeding 16°C in summer, and dropping to less than 10°C in winter) extends from California to Alaska, the southern cold water being due to coastal upwelling. Periodically, warm oceanic water moves nearer the coast during periods with weaker upwelling, and then southern California and Baja California have a temperate marine environment, characterized by frequent summer blooms of *Gonyaulax polyedra* and *Gymnodinium sanguineum,* following upwelling events (Section 6).

North of Point Conception (34–35°N) there are fewer near-shore warming events, but in late summer, in sheltered waters such as San Fransisco Bay, Puget Sound or the Strait of Georgia, the temperature may exceed 17–20°C, with accompanying dinoflagellate blooms resembling those further south, e.g. by *Gonyaulax spinifera*, and this can also occur in the Gulf of Alaska in late summer. Species typical of the cool coastal waters are *Protogonyaulax catenella, Ceratium divaricatum, Prorocentrum gracile, P. micans* and *Oxyphysis oxytoxoides,* with other coastal species of *Ceratium, Protoperidinium* and *Dinophysis* common. *Ceratium arcticum* var. *longipes* reaches British Columbia. Species such as *Gym. sanguineum* appear to avoid low salinity, run-off water; others, such as *Protogonyaulax tamarensis*, prefer it and bloom in estuarine inlets such as those in the eastern Strait of Georgia (Taylor 1984). The report by Wailes (1939) provides a general picture of the coastal dinoflagellate flora/fauna. Although the coastal sediments of British Columbia are dominated by the cyst of *Gonyaulax grindleyi* (= *Protoceratium reticulatum*; cyst name *Operculodinium centrocarpum*) the species is only occasionally seen in summer, and not in bloom proportions (Dobell 1978).

BOREAL NORTH ATLANTIC

This zone is much narrower than its counterpart in the North Pacific and its recognition is complicated by mixing between the Arctic surface water to the north, the Gulf Stream and its extension to the south. In this sense the whole region is essentially a transition region, although its south-western edge, between Cape Cod (or perhaps, Cape Hatteras) and the Gulf of Maine, is particularly variable, with temperatures ranging from less than 3°C in winter to more than 19°C in summer.

Like the subantarctic zone it is essentially a region where less tolerant, warm water species disappear but, unlike that zone, there is also an input of species from the Arctic. The varieties of *Ceratium arcticum* (considered to be separate species by some authors) illustrate this well. *C. arcticum* var. *longipes* is less cold-stenothermal than the var. *arcticum*, extending throughout the neritic waters of the boreal zones in both the Atlantic and Pacific, particularly in spring. It is thus typical of the zone. The var. *arcticum,* on the other hand, which is common in the Arctic, extends only into the more northern mixed areas, e.g. those influenced by the Labrador Current and East Greenland Current. Frost (1938) described the distribution of these and other ceratia around Newfoundland, concluding that they were particularly useful as indicators of hydrographic conditions. *C. fusus* is relatively eurythermal, but dislikes Arctic water, and *C. tripos, C. macroceros* and *C. lineatum* indicate the presence of mixed Atlantic water.

Diatoms dominate throughout most of the year, but dinoflagellates exhibit summer blooms in the neritic zone.

In the central North Atlantic the island of Jan Mayen lies within the boreal zone, with *Prorocentrum balticum* most important in spring (Smayda 1958).

The northern part of the North Sea and the Norwegian Sea are also subpolar in character, whereas the southern region is more temperate. In summer the northern region is stratified, whereas the shelf zones between this and in the Irish Sea remain mixed due to tidal action (Pingree 1978; Holligan 1978). Holligan *et al.* (1980) have distinguished four principal groups of dinoflagellates around the British Isles: a northern *Ceratium lineatum* group (including *Dinophysis norvegica, D. dens, C. arcticum* var. *longipes* and *Helgolandinium subglobosum*); a *Scrippsiella* group (including *Heterocapsa triquetra, Mesoporos perforatus* and *Gonyaulax triacantha*) in the English Channel and northern Irish Sea; *Gyrodinium aureolum* forming its own category in the south-west English Channel and southern Irish Sea, associated with the stratified edge of fronts; and a *Prorocentrum micans* group (many species, including *C. tripos, C. fusus, C. furca, D. acuta* and *D. acuminata*). The distributions of individual species have been plotted by Dodge (1981), and cyst assemblages by Reid (1975). A taxonomic guide has been recently prepared by Dodge (1982).

Offshore Norwegian coastal waters are similar to those of the northern group and Jan Mayen. However, the coast is a strongly fjord-indented region, with shallow sills and varying degrees of run-off. The communities described in the latter in studies such as those by Braarud (1976), Tangen (1979) and in Oslo fjord (Heimdal *et al.* 1973), are extremely similar to those inhabiting British Columbia fjords, although the latter usually lack the shallow sills commonly present in Norway and some species found in the North Pacific are lacking in the North Atlantic. The estuarine circulation in the fjords (outflow at the surface, inflow below) tends to export diatoms, dinoflagellates being better able to exploit such environments except when turbulence breaks down the stratification. Larger populations are usually found within the fjords than in open coastal waters. Common species, most abundant in summer, include *Ceratium furca, C. fusus, Scrippsiella faeroense* (= *trochoidea*?), *Gonyaulax grindleyi* (whose cysts dominate in the sediment), *G. spinifera, Protogonyaulax tamarensis* and *Prorocentrum balticum. Gymnodinium sanguineum* forms blooms on the outer coast. *C. arcticum* var. *longipes* appears earlier in the season in the fjords than other *Ceratium* species, in accordance with its preference for colder waters.

The Baltic Sea, communicating through the shallow Kattegat, is basically a brackish, boreal region, although in the Gulfs of Bothnia and Finland, species also found in Arctic Ocean waters such as *Peridiniella* (*Gonyaulax*) *catenata*, indicate a biogeographic link. Surface salinities are low throughout the Sea (most less than 10‰, reaching 1‰ in the extreme north). In the southwest, dinoflagellates form a major component of the summer phytoplankton, with a succession from *Prorocentrum balticum*, blooming in May, to *Ceratium* spp. in summer (reviewed by Lenz 1981). Further north *Peridiniella* (*Gonyaulax*) *catenata* is one of the earliest bloom species (Hobro 1979). At the end of the diatom bloom *Protoperidinium* species are common, as well as *Dinophysis acuminata, Gonyaulax verior* (= *diacantha*) and *G. triacantha*. *Heterocapsa triquetra* may form blooms later in the season.

POLAR WATERS

These consist of the waters of the Arctic ocean north of 65°N, with southward extensions of arctic admixed, subpolar water in the north-western Atlantic as the East Greenland and Labrador Currents, reaching the Labrador Sea north and east of Newfoundland; and the Antarctic water south of the Antarctic Convergence (approximately 60°S but reaching 50°S south of the Atlantic Ocean). Much of the Arctic Ocean is permanently covered by ice 3–5 m thick, with seasonally open water occurring as a belt along the north coasts of Canada, Greenland, Norway and Russia. In summer the offshore Arctic ice does have extensive open 'leads'. Unlike the Antarctic it has relatively low salinity over some shallow areas due to the discharge of rivers, particularly north of Russia (Kara Sea, Laptev Sea, East Siberian Sea) but also in Canada (Beaufort Sea), and salinities do not exceed 32‰. The Baltic Sea and Hudson's Bay represent southern extensions of the Arctic community. Although the former is not in direct connection with the Arctic Ocean there is subsurface continuity (the 'Yoldia Sea' having a strong Arctic input c. 9600 BP; Holmes & Holmes 1978).

In the Antarctic the surface salinity is close to 34‰ and there are few areas of appreciable dilution. The 5°C isotherm is a convenient isotherm by which to limit both regions. Surface temperatures within them can reach below 0°C.

In both there may be a single or double summer phytoplankton bloom, heavily dominated by relatively large diatoms (Allen 1971). Dinoflagellates do not usually play a major role in the microplankton or the community inhabiting the underside of the ice, the bulk of which consists of pennate diatoms (Horner 1985), although ice discolourations produced by *Gonyaulax nivicola* and *Peridiniella* (*Gonyaulax*) *catenata* have been recorded in the Arctic. Phytoplankton cells are frozen into the ice as it freezes and dinoflagellates are present in this component (Usachev 1949), but this passively-preserved assemblage is readily distinguished from the active communities, consisting heavily of pennate diatoms in the underside of the ice or in pools upon it. Non-photosynthetic species and the cysts of planktonic species may also occur in the ice community.

The role of nanoplankton in polar waters has received relatively little attention. Much of it consists of prymnesiomonads such as *Phaeocystis*, both in the arctic and antarctic. However, Von Brockel (1981) has noted that dinoflagellates may be a significant component of nanoplankton in the Bellinghausen Sea, Antarctica, in summer, but the proportion was not indicated. In Disko Bay, western Greenland, the only dinoflagellate in nanoplankton samples examined by Thomsen (1982) was *Oxyrrhis marina*.

ARCTIC DINOFLAGELLATES

Since the classic study by Meunier (1910) of the Kara and Barents Seas there have been quite a few studies that include information on Arctic dinoflagellates. The Russian work has been reviewed by Usachev (1949, 1961), with floristic information in the compendium by Kisselev (1950) and in the Canadian and American Arctic by Bursa (1961a,b; 1963a,b).

There is an interesting, distinct gradient in diatom versus dinoflagellate dominance from north to south, as well as an increase in diversity (Table 11A.4), with dinoflagellates only reaching appreciable abundance near the coast and south, except in estuarine areas. The Russian Arctic has large regions of lower salinity due to the discharge of numerous, substantial rivers, and this may act as a barrier to the eastward spread of thermotolerant, Atlantic species. The Bering Strait is a region of significant input from the Pacific Ocean, influencing communities off Northern Alaska (Point Barrow) and in the Chukchi Sea. Bursa (1963a) attributed the high number of dinoflagellate species at the former, relative to the latitude, to this input. *Protogonyaulax* (*Gonyaulax*) *tamarensis* can be abundant in the inshore waters and leads at Point Barrow (maximum $28 \times 10^3\,l^{-1}$ in August), its range in the Pacific thus extending north from British Columbia, in reduced salinity waters. Further offshore, various species of *Ceratium* including *C. arcticum* var. *arcticum* and *C. arcticum* var. *longipes*, are abundant to a depth of 40 m. They do not usually enter low salinity water.

In the Canadian Arctic, river input is primarily in the Beaufort Sea (MacKenzie River delta) and Hudson's Bay. In the former a typically Arctic,

Table 11A.4. Diatom and dinoflagellate species composition data for selected Arctic localities

	No. of Species			
Locality	Diatoms (a)	Dinoflagellates (b)	Ratio of a to b	Source(s)
'North Pole' drift ice station (88°–84°N)	61	5	12·10	Usachev (1949)
Arctic Ocean shelf (N of 80°N)	44	13	3·38	Bursa (1963b)
Isachsen (78°N, brackish)	48	33(38?)	1·45	"
Igloolik (69°N)	78	30	2·60	"
Kara/Barents Sea (70–75°N)	36	29	1·24	Usachev (1961)
Beaufort Sea (69°N, brackish)	75	6	12·50	Hsiao *et al.* (1977)
Hudson's Bay (60°N) first survey	86	56	1·54	Bursa (1961a)
Hudson's Bay (60°N) with additions from recent survey	110	89	1·24	Anderson *et al.* (1981)

estuarine flora occurs, but with diatoms more abundant near-shore and flagellates more important offshore (Hsiao *et al.* 1977). Dinoflagellates form only a small part of this community, with *Peridiniella* (*Gonyaulax*) *catenata* a convenient indicator of such waters, and a few *Protoperidinium* species are present.

There is an apparent seasonal succession, despite the shortness of period suitable for blooms, resembling a compressed boreal cycle, with diatoms forming an early summer (or Arctic 'spring') bloom and dinoflagellates following with non-photosynthetic, thecate (principally *Protoperidinium*) and athecate (*Gyrodinium* spp.) forms appearing close to the peak of the diatom bloom. *Ceratium arcticum* appears early, species of *Gonyaulax* and *Protogonyaulax* reaching maximum numbers later. Slightly further south, small flagellates may also occur early in the season, e.g. in the Denmark Strait (Braarud 1935). In the Polar Sea, Usachev (1961) found that dinoflagellates were present, but in negligible quantities, until September, when they could make up to 41% of the net haul; *Protoperidinium islandicum* was the most dominant species. The phytoplankton flora of the open Arctic was 37% endemic, with 26% arctic-boreal and 37% boreal.

Hudson's Bay is considered here to represent a southern extension of the Arctic zone, although it has several water types within it, ranging from arctic water entering from the north, boreal water from the south, and brackish regions. The community has been described by Bursa (1961a) and Anderson *et al.* (1981). Dinoflagellates can become locally abundant ($125 \times 10^3 \, l^{-1}$ in the north-west) and are higher in the Bay than outside it. Common lower salinity species include *Katodinium rotundatum* and *Protogonyaulax tamarensis*. Fresh-water species are also present.

ANTARCTIC DINOFLAGELLATES

Repeatedly, during all periods studied (spring to autumn), authors have reported diatoms as predominant over all other groups, both quantitatively, from bottle samples, and qualitatively, from net and bottle samples (e.g. Zernova 1968; Holm-Hansen *et al.* 1977; Jacques *et al.* 1979). During the summer and autumn of 1956–58, diatoms comprised 99% of all cells in the samples (Zernova 1968). At other times they may comprise 80% of the microplankton. The peak phytoplankton abundance is close to the Antarctic ice coasts, or near and south of the Antarctic divergence (65°S). Hasle (1969) also observed an abrupt decline in dinoflagellate numbers south of 60°S on transects along 90°W and 150°W, principally due to a decline in the numbers of *Prorocentrum balticum*. Near the Antarctic continent (Graham Land) Hasle found dinoflagellates stocks to be 'extraordinarily small'.

There is a relatively abrupt change, south of the convergence, with a decline in all species of *Ceratium*. *C. pentagonum* forma *grande* and *C. lineatum* disappear close to the convergence. Other dinoflagellates important in this zone are *Prorocentrum antarcticum, Dinophysis tuberculata,* various non-photosynthetic species of *Protoperidinium* and non-photosynthetic athecate members of *Gymnodinium* and *Gyrodinium*. Holm-Hansen *et al.* (1977) considered *Protoperidinium* as the most important dinoflagellate genus south of the Antarctic convergence, this being consistent with its preference for regions in which diatoms predominate. Balech (1968) noted that two-thirds of the species recorded by early studies belonged to this genus (referred to as *Peridinium*), the remainder belonging to *Dinophysis* and *Gonyaulax*, but athecate or small species had not been thoroughly examined. Typical species are *P. adeliense, P. antarcticum, P. defectum. P. mediocre* and *P. applanatum,* the latter being particularly common. Balech recorded a total of fifty species of dinoflagellates from Antarctic waters of which one, *Diplopeltopsis minor,* was truly cosmopolitan and six also occurred in subantarctic waters. *D. granulosa* is common in the coastal waters of the Bellinghausen Sea. Balech & El-Sayed (1965) noted the presence of a few individuals resembling warmer water species of the genera *Podolampas Oxytoxum* and *Kofoidinium velleloides,* but a close examination revealed that they had distinct morphological features which warranted their recognition as separate species.

Unlike the Arctic region, there are little signs of major regional differences around Antarctica, other than areas of increased productivity. Although the antarctic convergence constitutes a strong barrier, some mixing may occur in the Antarctic Peninsula region due to deflection of water northward into Drake Passage and eddy formation.

Balech (1975) has provided an illustrated flora of forty-five dinoflagellates which are native to Antarctica. He noted that records of many northern species, such as *Protoperidinium cerasus, P. divergens, P. subinerme* and *Dinophysis ovum,* in the early literature, were probably based on inadequately close examination whereas some others (e.g. *Diplopeltopsis minor, P. thulense, P. saltans, P. obtusum, Ceratium pentagonum, C. lineatum*) are evidently bipolar. Many of the northern species reported as Antarctic by Wood (1954) seem to have been unsound identifications.

The presence of very similar, but slightly different species in the waters of the poles is interesting, suggesting divergence due to isolation, although when the latter occurred is not evident. Some other morphological parallelisms are also apparent. For example, the tubercular outgrowths usually present in the dorsal and antapical sagittal regions of the northern species *Dinophysis norvegica* are similar to those developed by *D. tuberculata* and *D. cornuta* in the Antarctic, but the latter are more extremely developed than any northern species. Other species from opposite polar or boreal waters, which superficially resemble each other are *P. defectum*, which resembles *P. minusculum* from the northern hemisphere, and *P. applanatum* (southern) which resembles *P. bulbosum*.

4.3 Endemicism

Despite the potential continuity of all ocean waters there are, nevertheless, cases of apparent endemicism (in the strict sense of the word, i.e. restricted to only one region). It is obvious that in attempting to recognize such restricted distributions, all species that have only been observed once or twice should be omitted, and also those likely to be easily misidentified. Also, it is evident that, as the study of natural phytoplankton communities continues and there is greater awareness of the need for closer time scale, or deeper, sampling, many species thought to be limited to only one region will be found in another. It is difficult to distinguish these increased-search results from genuine new introductions to a region (e.g. through the Suez Canal).

It might be expected that the north polar regions would have endemics because of the great barrier to exchange between the two and this is indeed the case, with roughly 80% endemicism in each. *Ceratium arcticum* var. *arcticum* is an example of an Arctic endemic whose presence may be used as an indication of Arctic water (the var. *longipes* being arcto-boreal). It has been recorded as far south as 20°N, but only at subsurface depths (Graham & Bronikowsky 1944). The presence of Arctic, brackish species, such as *Peridiniella* (*Gonyaulax*) *catenata* in the northern Baltic suggests some link, perhaps in the past.

As with other groups, the Indo-West Pacific region possesses a variety of endemic species, most of which are neritic. *Dinophysis miles* is a distinctively shaped, neritic dinophysoid which occurs from the western Indian Ocean and Red Sea to the Gulf of Tonkin. A record from the Mediterranean is dubious (possibly confused with *D. tripos*, but, if not, it could be good evidence for trans-Suez Canal migration; Kimor 1972). The species occurs in several varieties, one of which, the var. *scroeteri* (with an extremely elongated dorsal extension), is restricted to the East Indies region from the eastern Bay of Bengal to the Gulf of Tonkin (Taylor 1973, 1976), co-occurring with the more common form.

It is difficult to understand such restricted endemicism in a region in which there is not only a strong inter-ocean flow, but one that reverses seasonally with the monsoons. *Ceratium dens* also seems to resist eastward transport, although it extends to the western Indian Ocean like the principal variety of *D. miles* (records off California appear to be due to confusion with *C. divaricatum,* endemic to the latter region, which commonly forms blooms and has a body shape which is not triangular like that of *C. dens*). A record of *C. dens* in the Gulf of Mexico has not been confirmed. *Pyrodinium bahamense* var. *compressum* has a similarly limited distribution, extending from the Persian Gulf to the East Indies (Taylor 1984).

Another *Ceratium* species that resembles *C. dens*, but has a distinctively different left antapical horn, *C. egyptiacum,* occurs only in the Suez Canal, the Red Sea and the south-eastern Mediterranean (Dowidar 1971). It is perhaps the best example of the presence of a Red Sea dinoflagellate species in the latter region (Kimor 1972). As the high salinity barrier in the Great Bitter Lakes continues to diminish, further introductions into the eastern Mediterranean are likely.

The tropical Atlantic Ocean also has a few readily recognizable endemics, restricted chiefly to the Gulf of Mexico and adjacent waters. Both draw attention to their presence by producing fish-killing blooms. *Gessnerium* (*Gonyaulax*) *monilatum* is a chain-forming species known from coastal lagoons and bays within the Gulf of Mexico, but also can be found on the east coast of Florida and as far south as Venezula (Connell & Cross 1950; Howell 1953; Halim 1967; Perry *et al.* 1979). *Ptychodiscus* (*Gymnodinium*) *brevis* blooms principally along the south-west coast of Florida, but has also caused fish-kills in the western Gulf of Mexico and the east coast of Florida (Murphy *et al.* 1975). Reports of the latter species from Spain, the Mediterranean and Japan are based on an unusual form which may not be conspecific (Steidinger 1983). Those from northern Denmark (Hansen *et al.* 1969) appear to have been an erroneous confusion with *Gyrodinium aureolum,* a fish-killing species in the NE Atlantic as well as Japan (reviewed by Tangen 1977). The latter is very similar to *Gymnodinium flavum*, which has been implicated in fish kills or harmless blooms off southern California (Lackey & Clendenning 1963; Wilton & Barham 1968) and the two may have been confused at times (Taylor 1985).

Pyrodinium bahamense var. *bahamense* appears to be limited to the tropical Atlantic, and does not usually co-occur with the toxic Indo-West Pacific var. *compressum* (there is overlap off the west coast of Mexico).

The North Pacific is another area with readily recognizable endemics, of which *Oxyphysis oxytoxoides* is perhaps the best example, ranging from the west coast of Mexico to Japan, through the Gulf of Alaska, at neritic localities. The largely unidirectional flow through the Bering Strait (northwards), minimises exchange with Atlantic species, and other potential exchange routes are of considerable distance.

4.4 Special communities

4.4A SHADE FLORA

As noted elsewhere in this chapter, dinoflagellates are common constituents of subsurface chlorophyll maxima (SCM) found in the lower euphotic zone (10% to less than 1% of the subsurface light), although other nanoplankters may be more abundant. While some migrate to shallower depths on a daily cycle, others in low latitudes in which the surface is stratified for much of the year may remain at depth, being adapted to living at low light intensities.

The search for a shade flora ('Schattenflora') extends back to the studies by Schimper, using closing nets, during the German Deep Sea Expedition (in

Karsten 1907). Since then several others have attempted to recognize this community in their material, particularly in tropical, open ocean studies, where the lower euphotic zone may extend deeper than 100, or even 150 m. Sournia (1982) has reviewed much of the existing literature. Using very conservative criteria, he listed ten dinoflagellate species: *Ceratium gravidum*, *C. incisum*, *C. longissimum*, *C. platycorne*, *C. praelongum*, *C. ranipes*, *C. vultur*, *Heterodinium scrippsii*, *Pyrocystis noctiluca* and *Triposolenia truncata* (plus four diatoms, three coccolithophorids, one prasinophycean and a collective, coccoid cyanobacterial category) as shade species. He noted the existence of many other possible candidates, particularly within the genera *Heterodinium* and *Triposolenia* (nearly all!), whose records are too rare to justify their assignment at this time.

Balech's (1971) list of 'shade species' in tropical Atlantic material is more extensive, including sixteen species of *Ceratium*, several *Heterodinium* spp., *Dinophysis* (*Phalacroma* type) spp., one species each of *Triposolenia* and *Ornithocercus*, and *Gonyaulax pacifica*. Several species of *Histioneis* were too rare to assign to a depth category.

Part of the problem in the recognition of shade species is that their depth of occurrence is not absolute, but is probably governed by an interaction of light and nutritional factors (Section 6.1c) and, under unusual circumstances, they may even occur in surface samples. Indeed the name is probably a misnomer, the ocean environment being a far more dynamic habitat than a forest. Absolute light values are usually not given in the available data. It is perhaps more appropriate to think of the ocean consisting of surface, high-light, non-migrating, slow-growing species, tolerant of very low nutrient levels, adapted to the rapid uptake of regenerated nutrient, produced by animals (e.g. ammonium), and others capable of exploiting the deep nutricline at the bottom of the euphotic zone. Sournia (1982) was not able to point to morphological features in phytoplankton in general which might be adaptations to a deep existence. However, if one looks only at the dinoflagellates there do appear to be certain forms which are more common or elaborately developed among the species considered to be deep-living. In particular, cell flattening, either of the body (*Heterodinium* subgenus *Platydinium*) or horns (*Ceratium* subgenus *Archaeceratium* plus *C. platycorne*), or thin body extensions (*Ceratium* species such as *C. extensum* and *C. ranipes*) is so pronounced as to seem a common adaptive feature (Steemann Nielsen 1934, 1939; Graham 1941; Taylor 1973). There is also a predominance of small species (Holligan & Harbour 1977). Presumably all these features optimize the illumination of chloroplasts within the cells.

Halim (1967) and Taylor (1973) have noted that the occurrence of these so-called shade species nearer the surface may be indicative of the rise of nutrient-rich water in tropical waters. Thus, Halim listed several of those listed as shade species by Sournia as 'upwelling indicator species' associated with the Cariaco Trench. Off the Caribbean coast of Columbia and north of the Dominican Republic, several of the 'shade' *Ceratium* spp. such as *C. incisum*, *C. praelongum*,

C. longirostrum, C. ranipes and *C. reflexum* have been found in surface samples, as well as species of *Heterodinium* and *Histioneis* (M. C. Carbonell, pers. comm.).

4.4B ESTUARIES

Within estuaries there is considerable environmental variability, depending on the intensity of run-off, sediment load, depth and ocean tidal effects. They include river valleys, fjords and coastal lagoons. Smayda (1983) has reviewed the phytoplankton communities of estuaries.

The estuaries of high run-off rivers such as the Mississippi or Amazon Rivers are usually impoverished in phytoplankton, containing mostly freshwater species washed out of more productive tributaries, and there is little microbenthos because of the heavy sedimentation rate and small sediment size.

Medium to low run-off estuaries can support considerable phytoflagellate populations, diatoms finding it difficult to form blooms, except in the lowest run-off types, due to export by the estuarine circulation (Braarud 1976). The high stratification resulting from halocline development also tends to favour flagellates, provided that they can tolerate the upper salinities or, if staying below the halocline, the low light levels. Diatoms often dominate nearer the mouth of the estuaries.

Generally, species living in these regions readily tolerate salinities in the 10–20‰ range or lower. Dinoflagellates which are typically abundant in temperate estuaries, such as those of the eastern seaboard of the US, include *Prorocentrum minimum, Katodinium rotundatum, Heterocapsa triquetra, Gymnodinium danicum* and *Gyrodinium estuariale* (e.g. Patten *et al.* 1963; Campbell 1973). These can also occur in higher latitude estuaries in summer. The toxic species *Protogonyaulax* (*Gonyaulax*) *tamarensis* and *Gessnerium* (*Gonyaulax*) *monilatum* are essentially estuarine, the former in cold, and the latter in tropical Atlantic waters. Other species in the community found in tropical estuaries are euryhaline members of *Ceratium*, such as *C. fusus* and *C. furca* (Wood 1954; Qasim 1973).

Estuarine species occurring off the Belgian coast were the subject of many studies, of which the most important report is that by Conrad & Kufferath (1954). Many of the species occurring in coastal ponds and lagoons are also euryhaline, e.g. *Kryptoperidinium foliaceum* (see below). The entire Baltic Sea can also be considered to be estuarine, populated by dinoflagellates similar to those of Arctic, river-influenced coasts in the north and east (e.g. *Peridiniella catenata*) and *Prorocentrum* and *Ceratium* spp. in the south.

4.4C TROPICAL BAYS WITH PERSISTENT BLOOMS

In some bays in the tropical Atlantic ocean *Pyrodinium bahamense* has formed extraordinarily persistent, high concentrations with resultant luminosity of the bay waters, permitting their exploitation as tourist attractions. The best known of these—Fire Lake in the Bahamas, Oyster Bay in Jamaica, and the Bahia Fosforescente in Puerto Rico—have now ceased to produce persistent blooms. At the time of writing such blooms can be found only in the Isla de Vieques, Puerto Rico (W. Biggley, pers. comm.).

The ecology of the blooms was studied particularly in Puerto Rico (Margalef 1961b; Seliger *et al.* 1971) and Jamaica (Seliger *et al.* 1969) and has been summarized by Smayda (1980). One common characteristic in the undeveloped bays was a mangrove-covered (*Rhizophora*) shoreline, which has become reduced or eliminated by land development. It is not known if pollution has had a negative effect as well. The change in Bahia Fosforescente occurred between the studies of Margalef and Seliger *et al.*, the latter finding that other dinoflagellates had largely replaced *P. bahamense*. Smayda (1970) found that the bay water inhibited growth of some diatoms, but a primary reason for poor growth in these areas seemed to be related to the shallow, stable nature of the bay waters and a combination of estuarine circulation and dinoflagellate behaviour; light, daily onshore winds may also act (Seliger *et al.* 1969). Other dinoflagellates that were abundant in the bays included *Ceratium furca, C. fusus* and *Dinophysis caudata.*

4.4D TIDE POOLS AND COASTAL PONDS

In view of their ease of access it is remarkable how little study had been done until very recently on the dinoflagellates of these localities. Many are brackish and share features with the areas described in Section 4.5b above, but lacking in the estuarine circulation (outflow at the surface, inflow below). It appears to be their stability and shallowness which favours flagellates of many types (prymnesiomonads, prasinomonads, euglenoids and dinoflagellates: see the classic paper by Carter 1937).

Tide-pool species must cope with the environmental influences which occur while the relatively small body of water is isolated from the sea as well as flushing when the tide returns wave-actions to the pool. No dinoflagellates are known to be exclusively adapted to tide pools but some, such as *Scripsiella (Peridinium) gregaria* (Lombard & Capon 1971) are very common in them. Thermotolerance of many members may aid their survival in tropical areas with a significant tidal range (principally in SE Asia). In Phuket, Thailand, pools on black rock cut off from the sea for several days, became dominated by *Scrippsiella subsalsa* and a species resembling the gymnodinoid stage of *Symbiodinium microadriaticium* (Taylor 1983) (Section 3.2). Horiguchi (1983) observed blooms of several dinoflagellates, including *Alexandrium (Gessnerium) pseudogonyaulax*, *Scripsiella hexapraecingula*, *S. pseudosubsalsa* and *Gymnodinium pyrenodosum* (almost certainly the same species observed in Taylor 1983) in Japanese tide pools.

4.4E BENTHIC AND NEAR-BENTHIC COMMUNITIES

Benthic dinoflagellates have been much neglected for reasons which are not obvious, other groups such as diatoms and ciliates having received much attention.

PSAMMOPHILIC (SAND) DINOFLAGELLATES

Dinoflagellates are common and sometimes abundant inhabitants of marine sands, usually in the upper centimetre, but a few non-photosynthetic species can occur down to, or into, the sulphide-rich, anoxic 'black layer' sand underlying the light-coloured, oxygenated sand above (see Fenchel & Riedl (1970) for a review of the sulphide-sand community and its environment). Vertical migration is a characteristic feature of psammophilic protists; the cells are often flattened or elongate, apparently as an adaptation to swimming in the interstitial fluid between the sand grains. For example, *Polykrikos lebourae* is strongly laterally compressed in comparison with planktonic *Polykrikos* species.

Early studies were limited to brief notes by W. A. and E. C. Herdman on the Isle of Man. They saw that the cells migrated and could be abundant enough to discolour the sand (e.g. by *Amphidinium operculatum*) (W. A. Herdman 1914; E. C. Herdman 1922, 1924). Approximately forty species were described, most of them new. Further species were described by Balech (1956) and Dragesco (1965).

Little ecological information was provided by the above authors. The Danish study of Larson (1984), those of Baillie (1971), and observations by this author, indicate that the dinoflagellates of the beaches of British Columbia are largely the same as those from European coasts, except for new species, with very few individuals present intertidally in winter (some subtidal at this time). They become abundant in summer with blooms of *Amphidinium testudo, Thecadinium inclinatum* and *A. asymmetricum* occurring on 'clean' sand flats away from the shore, with little organic content and an average grain size of 125–250 μm. Silt clogs the interstitial spaces of finer sands, and coarser sands are too frequently disturbed. The photosynthetic species are most abundant on extensive sand flats in sheltered bays with little siltation from rivers, diatoms usually being abundant in moderately organic-rich sediments. Both groups migrate onto the upper edges of ripples during low tides in the day time (see Round 1981), migrating down the upper centimetre before the tide comes in. On the sand flats the cells must be extremely tolerant of strong fluctuations of temperature, and sometimes salinity for periods of several hours, as are intertidal sand protists in general. This may explain their very broad cosmopolitanism.

On sloping, open-surf beaches, non-photosynthetic species (mostly *Amphidinium* species) can be found by sampling the interstitial water which drains from the sand as the tide recedes.

Many dinoflagellates which do not usually extend into the intertidal zone are also found on subtidal sands. Species of *Prorocentrum* and *Amphidinium* and *Ostreopsis* are particularly common in warm water areas (Fukuyo 1981). They can appear in the plankton after storm turbulence.

The maximum depth to which benthic dinoflagellates have been observed is relatively shallow (diving depth, i.e. 50 m) and there has not been a deliberate effort to search for them at greater depths, although it seems possible that non-photosynthetic species could occur on the deep-ocean floor.

PHYCOPHILIC (ALGA-ASSOCIATED) DINOFLAGELLATES

Macroalgae are well known to harbour a variety of associated protists, including attached (epiphytic) diatoms and foraminiferans. Less well known is the occurrence of a distinctive community of flagellates, including dinoflagellates, which either swim in the still water between the branches (e.g. *Amphidinium* spp.) or are attached by threads or pads of mucilage.

In temperate and cold water *Oxyrrhis marina* is often found with the intertidal green alga *Ulva*, and *Crypthecodinium cohnii* seems to associate with many brown algae, particularly *Fucus* spp. (see Beam & Himes 1982).

Recently tropical, algal-associated dinoflagellates have received attention because it has become evident that this is the preferred habitat of the toxin-producing species *Gambierdiscus toxicus* (Adachi & Fukuyo 1979; Taylor & Seliger 1979; Besada *et al.* 1982; Taylor & Gustavson 1986) and other possible toxic species. Although not species-specific in habitat, the latter dinoflagellate strongly favours highly branched 'bushy' members of red, brown or green algae, sometimes blanketing the surface of the alga, the cells adhering by mucilage (Fukuyo 1981; Taylor & Gustavson 1986). Members of the genus *Ostreopsis* can also be found attached to algae by fine threads, but their association is not discriminate and they can also be found in large numbers on sand grains (see above) or even attached to benthic worms (F. J. R. Taylor, unpubl. obs., Villefranche-sur-Mer, France).

CORAL REEFS

In coral reef areas, including the lagoons surrounded by atolls, the bulk of the primary productivity is benthic, the crystal clear overlying waters being very low in inorganic nutrients. Within the reef most primary productivity results from the activities of calcareous red and green algae which act as reef-binding cement together with the dinoflagellate zooxanthellae (morphologically referable to *Symbiodinium microadriaticum*: see Chapter 12) within the coelenterates ('hermatypic', i.e. reef-building corals, various types of anemones and jellyfish) and molluscs (principally the tridacnid clams) of the reef.

In the plankton of atoll lagoons it appears that dinoflagellates predominate in cell number over diatoms, e.g. Fanning Island (Gordon *et al.* 1971) and

Takapoto Atoll (Tuamoto Islands; Sournia 1976, Sournia & Ricard 1976), which may be a function of the relative stability of the water column, other flagellates such as coccolithophorids also being more important than diatoms. In the bottom water of a shallow (15–25 m) bay on the south coast of St John, US Virgin Islands, Hickel (1974) found that dinoflagellates (unspecified, but mostly small or medium-sized) comprised 71·2% of the total particulate organic carbon (POC): 30·1 $\pm$ 20·4 of 42·2 $\pm$ 22·9 μg C l^{-1}, most of the remainder being diatoms. The dinoflagellates were most abundant 0·5 m over a sandy bottom. Unpublished observations by the present author, on similar layers in the Belau (Palau) Islands indicates that these layers rise off the bottom, consisting of species common on sand; Fukuyo (1981) has provided descriptions of epibenthic species from coral areas. The layers may only be a few centimetres thick and can be seen from the side while diving.

Other major components in the productivity of coral areas are the filamentous cyanobacteria ('blue-green algae') and the photosynthetic intracellular symbionts of large, benthic foraminifera living on and in the lagoon sands. Sournia (1976) has estimated, on the basis of studies at Takapoto Atoll, that the productivity of the lagoon sediments at 1 m is fifty times that of the water column above it, with 1000 times as much chlorophyll. Although the foraminifera contain a variety of photosynthetic cytobionts, including zoochlorellae (green algae of the genera *Chlamydomonas* and *Chlorella*: Lee *et al.* 1979; Saks 1981) and diatom zooxanthellae (Schmaljohann & Rottger 1976; Lee *et al.* 1980), many also contain dinoflagellate zooxanthellae resembling *Symbiodinium*, particularly in the genera *Sorites* and *Amphisorus* (Muller-Merz & Lee 1976; Leutenagger 1977). It is not known what the relative contributions of benthic dinoflagellate versus non-dinoflagellate symbionts are for a given region.

5 TEMPORAL DISTRIBUTION

5.1 Seasonal succession

It has been repeatedly observed that dinoflagellate blooms usually follow the diatom spring bloom in temperate coastal waters (e.g. Gran & Braarud 1935; Braarud 1976; and many authors cited by Smayda 1980). The illustration provided here (Fig. 11A.11) is fairly typical for a temperate region, although fluctuations are often more abrupt and striking and the spring bloom of diatoms is greater. Also, the summer increase of *Ceratium* species in the southern Baltic seems to be due largely to advection through the Kattegat (Smetacek 1975; Lenz 1981).

In fact, a close examination of the phytoplankton of temperate areas such as the Strait of Georgia, British Columbia or Narragansett Bay reveals that very small dinoflagellates, typically *Katodinium rotundatum* and *Gymnodinium* spp., may be an important component of the winter or very early spring nanoplankton

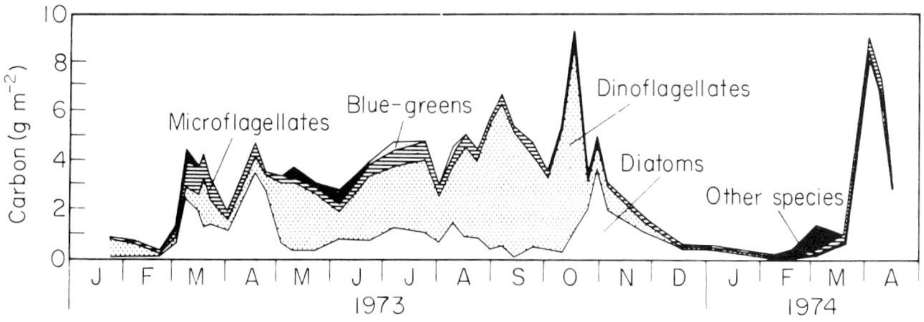

Fig. 11A.11. The seasonal cycle of phytoplankton biomass (as carbon) and the proportions contributed by various groups, at 'Boknis Eck' in the south western Baltic Sea. The spring diatom bloom in 1973 was unusually low. From Smetacek (1975; reproduced in Lenz 1981), with permission.

(Cattell 1969; Takahashi *et al.* 1978). In the Baltic Sea *Prorocentrum balticum* blooms early in the season (Lenz 1981). Despite this it is unquestionably true that the majority of intense 'red tide' blooms occur in mid-to-late summer in such regions (Smayda 1983) in warm, stratified water low in organic nutrients, although possibly high in inorganic nutrients.

The surface appearances of such summer blooms are often highly sporadic but, as Holligan (1978) has noted, subsurface layers dominated by dinoflagellates are often relatively prolonged summer features in stratified temperate waters (Section 6.1). Holligan & Harbour (1977) observed that shallower thermoclines in British coastal waters often have larger dinoflagellates, such as *Ceratium fusus* or *Gyrodinium aureolum* (*Gonyaulax polyedra* or *Gym. sanguineum* off California), whereas deeper, more offshore thermoclines have smaller dinoflagellates and other flagellates associated with them.

Margalef (1958) systematized the succession of temperate coastal phytoplankton as a series of stages, designated by numerals, linked to characteristic species diversity features, based on his studies on the Ria de Vigo, NW Spain. In his succession small-celled diatoms predominate early in the year (stage 1). This is succeeded by large diatoms (stage 2), with the appearance of dinoflagellates and dominance by the latter in stratified water during the summer (stage 3). Red tides are anomalies which develop from stage 3. Diversity increases from stage 1 to 2 and 3, dropping dramatically in red tide blooms. As noted above, it is now known that very small dinoflagellates can also be members of Margalef's stage 1.

Although the release of biologically-active compounds is frequently invoked as a significant factor in phytoplankton successions (additional to the depletion of essential nutrients), the demonstration of such compounds has been difficult and examples in dinoflagellates are few.

The toxins released by *Ptychodiscus brevis* and similar toxic species, harming higher trophic level organisms (Chapter 8), would only be significant in this regard if they acted on their immediate predators, which they usually do not. They may act on competitors, however. Freeberg *et al.* (1979) found that material(s) released into the medium by cultures of *Pt. brevis* inhibited the axenic growth of eighteen species of phytoplankton (including most diatoms tested, except at low concentrations). Some other dinoflagellates were lysed by the exudate. The toxins of other dinoflagellates have not yet been demonstrated to have competitive significance although *Protogonyaulax tamarensis* is known to produce antifungal compounds (Accorinti 1983; Hernandez & Almodovar 1983).

The spring bloom of diatoms may strip the surface waters of inorganic nutrients, but they may also add to the vitamin levels, particularly of B_{12} (Section 3·4 g). Production of 'binding factor' (Provasoli 1979) complicates the ecological interpretation of interactions among those species which require it.

Steidinger (1973) has noted that blooms of the cyanobacterium *Trichodesmium* (*Oscillatoria*) occur before those of *Pt. brevis,* and has suggested that it may thus provide vitamins or iron required by the latter.

Prorocentrum micans has been found frequently as a co-dominant with gonyaulacoid blooms, although in much smaller numbers. For example, it accompanied a 'red tide' caused by *Gonyaulax reticulatum* (identified as *G. polygramma*), which killed marine fauna near Cape Town in 1961 (Grindley & Taylor 1964); it was co-dominant with a toxic bloom of *Protogonyaulax* (*Gonyaulax*) *tamarensis* in Obidos Lagoon, Portugal in 1959 (Silva 1963) and accompanied a bloom of *Gonyaulax grindleyi* off the west coast of South Africa in 1974, which was associated with both shellfish toxicity and paralysis of the black and white mussels themselves (Horstman 1981). Elbrächter (1976) found that *Gymnodinium sanguineum* (= *splendens*) would grow in the filtrate from a culture of *P. micans,* but not *vice versa*, the former presumably releasing an inhibiting factor into the medium. On the other hand, similar evidence suggests that *P. micans* may inhibit the growth of the diatoms *Skeletonema costatum* and *Chaetoceros didymus* while growing on ammonium, but not when supplied with nitrate (Uchida 1981). In another series of experiments Elbrächter (1977) found no evidence of growth rate stimulation or inhibition when *Ceratium horridum* and *P. micans* were grown together with the diatoms *Coscinodiscus concinnus* or *Biddulphia regia,* reduced cell yields being attributable to competition for nutrients rather than the release of metabolites. A further multispecies study was carried out by Kayser (1979), using *Scrippsiella faeroense*, *Gymnodinium sanguineum* and *P. micans*. In batch cultures *S. faeroense* outgrew *G. sanguineum, P. micans* dominated *S. faeroense,* and *G. sanguineum* prevailed over *P. micans* when equal cell numbers were used as inocula, but if unequal numbers were used the species with highest initial concentration invariably prevailed. In continuous culture the species grew independently in log phase, with generation time (fastest in *S. faeroense*) the only determinant of cell numbers. Some inhibitory material seemed to be released only in plateau phase.

5.2 Long-term changes

In addition to seasonal cycles, plankton can also exhibit long-term changes which may fluctuate or be unidirectional in trend. Here we can recognize three common, often related types, although others undoubtedly exist: (i) fluctuations in the timing and abundance of blooms from year to year, correlated with climatic fluctuations; (ii) introduction and persistence, or extinction from an area; and (iii) a gradual, unidirectional change in water quality, often anthropogenic, causing a community change.

The first is evident to anyone who carries out a multi-year series of observations at one locality, although its long-term periodicity and the underlying mechanisms are usually not analysed. Cloudy, turbulent weather delays spring blooms; calm, sunny weather sometimes causing blooms to appear anomalously.

One of the best documented long-term changes is reflected in the Continuous Plankton Recorded survey of the North Sea and NE Atlantic from 1948 (at the species level from 1958) to the present. The resolution of this record is poor because it is only monthly, samples are from only one depth (10 m) and the mesh is relatively coarse, but it can indicate long-term relative changes in larger dinoflagellates such as *Ceratium* spp. For example, after 1971 the timing of seasonal abundance of *C. fusus, C. furca* and *C. tripos* in the eastern and southern North Sea showed a marked advance, these years also showing increased early diatom abundance and a decline in the abundance of a boreal, grazing copepod (*Calanus finmarchicus*), with the beginning of a reversion to the more usual timing in 1974 (Colebrook *et al.* 1978). *Ceratium arcticum* var. *longipes* showed no obvious change during this time. The changes could be coupled with a change in flow from the English Channel into the North Sea, which in turn might have been related to a change in the distribution of atmospheric pressure during this time. 1970–71 marked a period in which the weather quality in the north-eastern North Atlantic began to improve after a progressive deterioration since the 1940s (Dickson & Lamb, in Reid 1977), but this was not reflected in diatom or *Ceratium* abundance in more open North Atlantic areas, such as the region south of Iceland (Reid 1977).

Because of their undesirable effects, the incidence of red tide blooms is a matter of general concern; mortalities of marine fauna, sea birds or humans draw attention to blooms which might otherwise be overlooked. In particular, there is the possibility that toxic red tides may be increasing due to man's activities. At present, there is evidence both for and against such effects, depending on the organism and region concerned, but it is difficult to subtract the effect of increased awareness and sampling after particularly severe outbreaks and, while areas of increasing incidence are recorded, the converse situation is often not recorded.

In the British Isles *Protogonyaulax tamarensis* blooms have been recorded frequently and every year since a severe outbreak in 1968 (Armstrong *et al.*

1978), this being particularly manifest in the death of birds in the north-east (Farne Islands). However, their intensity varies considerably and it is doubtful if some would have been noticed without increased vigilance. Blooms of this, and similar paralytic shellfish poison (PSP)-producing organisms, have been known for many years. The earliest undoubted illness due to PSP, indicated by characteristic symptoms after eating mussels, appears to be that of a 20-year old French girl in 1689 (Chevallier & Duchesne 1851). After that there were similar occurrences, often fatal, in very uninhabited regions, such as that of a crew member of the explorer George Vancouver in a British Columbia inlet in 1793, or the death of more than 100 Aleut Indians accompanying Alexander Baranoff in Peril Strait, southern Alaska in 1799 (references in Halstead 1965). In British Columbia, the most seriously affected areas have generally been the least populated mainland north coast and west coast of Vancouver Island (Quayle 1969).

Off the east coast of North America *P. tamarensis* was first recorded from Woods Hole, Massachusetts (Lillick 1937), although it is usually more prevalent further north in the Gulf of Maine, Bay of Fundy and Gulf of St Lawrence (Prakash *et al.* 1971) and PSP-related fatalities in the latter region probably extend back earlier than European settlement of the coast, Indians in that region refusing to eat mussels, even when starving. Nevertheless, White (1982) has presented data which show an increasing intensity of toxicity and duration of toxin retention at several localities in the Bay of Fundy since the early to mid 1970s, for reasons that are not apparent. The severe outbreak in Massachusetts in 1972 seems to have been more a result of advection from further north than local shallow coastal ponds, although populations of *P. tamarensis* have persisted in them since then. In British Columbia, PSP severity appears to cycle on a 6-year (5–7) basis, in which case, severe toxicity would be expected in 1986 (Gaines & Taylor 1985). *Protogonyaulax*-produced PSP was not recognized in Japan prior to 1975 (possibly to 1961: Fukuyo 1979), but is now common in the north and east.

Gyrodinium aureolum announced its presence in waters off southern Norway by a major bloom in 1966 (Braarud & Heimdal 1970), this being the first record of its mass occurrence and related fish mortality in European waters. It was originally described from the north eastern US. As a 'dinoflagellate-aware' school of phytoplankton specialists has been active in Norway since the turn of the century it is unlikely that any major blooms were missed, although the presence of small numbers earlier cannot be ruled out. Braarud & Heimdal (1970) noted that the organism was probably identical to one recorded from Obidos Lagoon in the 1950s. Blooms of this species have also occurred frequently around the British Isles since the early 1970s, occasionally causing mortalities of lugworms and other marine fauna (refs. in Pybus 1980).

Blooms of *Ptychodiscus* (*Gymnodinium*) *brevis* are essentially an annual event off the west coast of southern Florida (Steidinger 1973; Steidinger & Haddad

1981). The first record of such blooms from this area was in 1844 and they have continued to be observed roughly every 2–10 years, with an apparent 30-year hiatus from 1916 to 1946, followed by a very severe outbreak (1946–1947) which was worse than more recent ones. It is tempting to attribute the greater frequency of recent blooms to human influence, but no satisfactory links have been established. Rounsefell & Dragovich (1966) found no correlation between the occurrence of major blooms in the 1954–61 period and fluctuations in the inshore levels of inorganic nutrients other than iron, and the initiation of the blooms offshore and transport nearer the coast also argues against human influence.

6 RED TIDES

The term 'red tide' has its origin in the periodic, reddish discoloration of sea-water due to intense blooms of plankton. However, it has come to be associated with specific, harmful consequences of such blooms in regions where one type is more common than another, e.g. fish kills in Florida, fish and shellfish kills in Japan, and paralytic shellfish poisoning of humans (PSP) on the Atlantic and Pacific coasts of North America and elsewhere. The colour can range from yellowish-brown or chocolate, orange, rusty brown ('blood-like') to maroon, depending on the causative organism, its concentration and physiological state.

Diatoms, the most common coastal, bloom-forming phytoplankters, do not usually cause red water. The most frequent producers of such discolorations are dinoflagellates, but other flagellates, (raphidophytes, chrysophytes), cyanobacteria 'blue-green algae' = purple bacteria and ciliates can also be causes.

Although they are considered unusual events, a close watch of almost any coastal area will reveal localized, periodic, brief duration (days, weeks), red water occurrences with no obvious harmful consequences. Because these are not reported, the literature gives a deceptive impression of the incidence of harmful effects associated with red tides. Also, because only their surface manifestations are usually noticed, they may seem of briefer duration and more abrupt development than is real. Harmful consequences are described later (see also Chapter 8).

Toxic dinoflagellate blooms have also been the subject of three meetings which led to published proceedings (Lo Cicero 1975; D. Taylor & Seliger 1979; Anderson *et al.* 1985). Earlier literature has been reviewed by Bainbridge (1957), Brongersma-Saunders (1957) and Rounsefell & Nelson (1966).

6.1 Special ecology of red tide blooms

Dinoflagellate red tides represent intense concentrations. *Prorocentrum minimum* reached concentrations of $1{\cdot}7 \times 10^9\,l^{-1}$ in Oslo Fjord in 1979 (Tangen 1980).

Blooms of 10–20 × 10^6 l^{-1} or more are common. In such blooms levels of 400–500 μm chl *a* l^{-1} can be achieved (Holmes *et al.* 1967). Because of their practical consequences they have been the subject of many field studies, although usually after the blooms have manifested themselves. A general picture has emerged, with variations depending on the species involved, location and transient environmental influences.

The principal regulating factors change as the bloom develops. Although the process is essentially continuous, it is helpful to recognize several distinct phases: (i) *initiation;* (ii) *development;* (iii) *aggregation* and (iv) *dissipation* (Steidinger 1973).

INITIATION

For a bloom to begin a 'seed population' of the species must be present to act as an inoculum. Most coastal red tide dinoflagellates are not present, even in very low numbers, in the water column throughout the year. Benthic cysts are produced by many, e.g. *Protogonyaulax* (*Gonyaulax*) *catenella, P. tamarensis, Gessnerium* (*Gonyaulax*) *monilatum, Gonyaulax polyedra,* but not, apparently, by others, e.g. *Ptychodiscus* (*Gymnodinium*) *brevis, Gymnodinium sanguineum, Gyrodinium aureolum, Prorocentrum* spp.

The initial vegetative cells may be introduced to the area by a mixing event, or lateral transport to a physically more suitable (stable) area for growth, or the cells may excyst from benthic cysts. In Chapter 14 the conditions leading to excystment are summarized. It is of interest to note that excystment may not occur under 'ideal' conditions for rapid growth, a lag occurring between excystment and maximal division. Furthermore, since many cysts may not 'hatch' in a given year (Anderson *et al.* 1983), the presence of abundant cysts in an area cannot be assumed to be a harbinger of massive blooms to come.

The question of the role of excystment in red tide bloom initiation has been specifically addressed by Steidinger (1975) and Anderson & Wall (1978): see also Chapter 14 here.

DEVELOPMENT

During this phase the population undergoes its principal growth, dividing at least part of the time at the maximum rate conditions permit. The rate of increase is a product of all the many factors summarized in earlier sections in this chapter: light, temperature, salinity, nutrient supply and grazing. However, it should be noted that laboratory-determined optima are only partly indicative of field success. Blooms often occur in conditions which appear to be sub-optimal, particularly in terms of temperature (often lower) and salinity (often higher).

Despite this complexity, some generalizations can be made. Red tides in temperate and high latitudes are predominantly summer, coastal phenomena. They follow the main diatom bloom, typically appearing at the surface during calm periods of relatively low surface, inorganic nutrient concentration (most commonly low nitrate), and high stability due to surface salinity and/or surface warming. In the tropics there may be a correlation with seasonal rainfall, immediately following heavy rains. In some instances they may follow an apparent input of nutrients from upwelling or storms (e.g. Morton & Twentyman 1971), but in other instances they may appear with no apparent enrichment of the surface water. In the latter cases the blooms may have developed in nutrient-rich, subsurface water, being advected to the surface later.

The development of a pycnocline (usually caused by heating, i.e. a thermocline) during the calm weather results in the relative physical isolation of the low nutrient surface water from the cooler, higher nutrient water below. This situation seems to favour dinoflagellates in several ways. Small-scale turbulence no longer keeps the diatoms in suspension and their sinking rates are believed to increase with nutrient depletion (Smayda 1970) causing them to sink out of the upper layer. The dinoflagellates, which are frequently able to develop blooms in low nutrient water (but perhaps, due to migration, exploiting deeper, high-nutrient water) can continue to grow and are able to regulate their position in the water column by swimming as long as the vertical stability persists.

In experiments with large experimental enclosures the diatoms may sink out quite rapidly from the stabilized water column, being succeeded by flagellates, including dinoflagellates (e.g. Parsons *et al.* 1978). Stirring, on the other hand, allows the diatoms to continue to bloom (Eppley *et al.* 1978).

As noted earlier, dinoflagellates have relatively low growth rates, rarely exceeding one doubling per day, although they may achieve bursts of higher division rates (Karentz 1983). Thus, the time to develop a bloom sufficiently concentrated to discolour the water is usually of the order of several weeks. It is striking how frequently there are reports of 2 weeks of sunshine and calm weather prior to visible red water, e.g. Grindley & Taylor (1964), but in several recent studies it is evident that the blooms develop at deeper levels over a longer period (see next section).

AGGREGATION

During red tide blooms the cells are often found at concentrations far higher than that predictable from measurements of ambient nutrient levels before or during the outbreak (Ryther 1955). The explanation for this lies in their ability to swim and therefore migrate, actively aggregating at a compromise position in the water column, or to be passively aggregated by physical factors due to their upward-swimming behaviour (Chapter 10).

Subsurface aggregation

In stratified water the upper water rapidly becomes stripped of nutrient(s) due to the activities of the phytoplankters within it. The upper zone is separated from the lower by a relatively sharp density discontinuity.

Phytoplankters living under these conditions face a dilemma: light increases upwards but nutrients increase downwards. Those that sink passively have little choice in the matter, although diatoms may lessen their sinking rate when they are nutritionally replete, sometimes being liable to aggregate at the nutricline. Flagellates can regulate their position in the water column over a relatively short time scale and some are capable of migrating 1 m h^{-1} or more (refs in Chapter 10 and Sournia 1982). It is now well established that some dinoflagellates carry out daily migrations, moving towards the surface during the day and down to the nutricline at night. Such vertical aggregations were first observed in the studies by Hasle (1950, 1954), Ragotzkie & Pomeroy (1957), and Wheeler (1966), but are now known from many locations.

The extent of migration under stratified conditions seems to vary, the dinoflagellates often remaining near the bottom of the euphotic zone during the day. Species that occur together may migrate to different depths, such as *Ceratium fusus* and *Gonyaulax polyedra* (Blasco 1978). The absence of a critical nutrient in the overlying water may discourage any movement into it (Heaney & Eppley 1981; Cullen & Eppley 1981). Under other circumstances the overlying water may be unfavourable in another aspect, such as temperature too high or salinity too low. In these situations motile photosynthetic organisms will rise as near to the surface as the environment, and their tolerances, will permit, leading to a strong aggregation within a layer which may be less than 1 m thick and easily missed by point sampling techniques (Fig. 11A.5). 'Self shading' by the cells probably contributes to the narrowness of the layers. Such vertical, upward aggregation, leads to the formation of a *subsurface chlorophyll maximum* (SCM), named for its usual means of its detection.

Although many studies on SCMs have not determined the component organisms it is evident that dinoflagellates are often contributors (Reid *et al.* 1970; Venrick 1982). Off California, Lasker (1975) observed the occurrence of *Gymnodinium sanguineum* (= *splendens*) in a subsurface layer that extended along the coast for at least 100 km, the cells being aggregated to a level above the threshold needed for a good feeding by anchovy larvae (greater than $3 \times 10^5\ l^{-1}$, depending on particle size). Kiefer & Lasker (1975) showed a coincidence between this species and SCM only 1–2 m thick which migrated from 6 and 15 m. A similar case is illustrated in Fig. 11A.5, although in connection with avoidance by zooplankton.

Because phytoplankton utilization is the cause of the nutrient depletion of the upper, euphotic layer, it is not surprising that the nutricline often occurs at or below the 1% surface light isolume. In the 'North Pacific Central Enviroment'

Venrick *et al.* (1973) found that the SCM was below the 1% isolume, containing more than half of the chlorophyll of the water column and responsible for 7–20% of the total primary productivity. Downward light is abruptly attenuated by the SCM due to shading by the cell aggregations.

A particularly striking example of a dinoflagellate SCM had been provided by Tyler & Seliger (1978, 1981) in their studies on *Prorocentrum minimum* var. *mariae-lebouriae* in Chesapeake Bay. They have shown that this species is transported as a subsurface layer from year-round populations at the mouth of the bay, moving 240 km in northward-flowing water below the pycnocline (sometimes deeper than 20 m) and being brought to the surface in mixed water at the head (Fig. 11A.12). The cells are aggregated at convergences in that region. This behaviour is due principally to an interaction of temperature and salinity effects on the organism, good growth in salinities less than 15‰ only occurring at high temperatures, but good over a wider range of temperature at higher salinities. This affects the species survival in surface waters. Also, chlorophyll concentration in the deeper cells appears to increase from approximately 2·3 pg to 3·2 pg cell^{-1} (65%), allowing survival at very low light levels ($\sim$4 μE m^{-2} s^{-1}), decreasing when the cells approach the surface once more.

This type of subsurface aggregation and transport may be significant in the development of blooms in other, less estuarine situations, Seliger *et al.* (1979) suggesting its involvement in the movement of *Ptychodiscus brevis* and *Protogonyaulax tamarensis* from offshore to near-shore in Florida and Maine respectively, although this has not been clearly established. In both situations some upwelling has been noted (Haddad & Carder 1979; Hartwell 1975), but it has been assumed to be the mechanism for bringing benthic resting cysts (not yet demonstrated for *Pt. brevis*) into surface water.

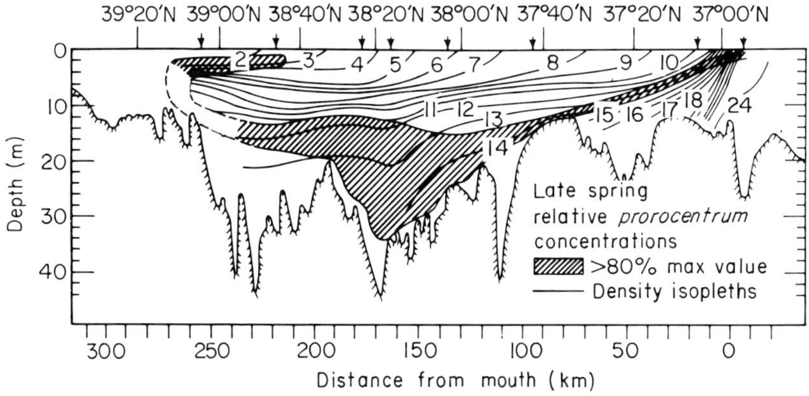

Fig. 11A.12. The vertical distribution of *Prorocentrum minimum* var. *mariae-lebouriae* relative to density in a north–south section of Chesapeake Bay in late spring. From Tyler & Seliger (1978, 1981), with permission.

Horizontal aggregation

When dinoflagellate blooms occur at the surface they usually occur in highly concentrated streaks and patches (see Bainbridge, 1957 for a review of the extent and shape of such aggregations). There are several causes for such horizontal aggregations (Fig. 11A.13), some of which were noted by Ryther (1955) and earlier authors, other receiving much recent attention. Essentially, they depend on the swimming abilities of the dinoflagellates (or buoyancy in the case of *Noctiluca*) which resist the slow downward movement of water. The water may sink due to a density difference or due to wind.

Thus, dinoflagellates may be aggregated onshore, particularly at the head of a bay, by moderate onshore winds, the cells resisting downwelling. A strongly buoyant organism such as *Noctiluca*, is particularly liable to this type of concentration.

Moderate winds (above 3 m s^{-1}) can produce a special type of effect: Langmuir circulation, if they blow for long enough. Essentially the wind

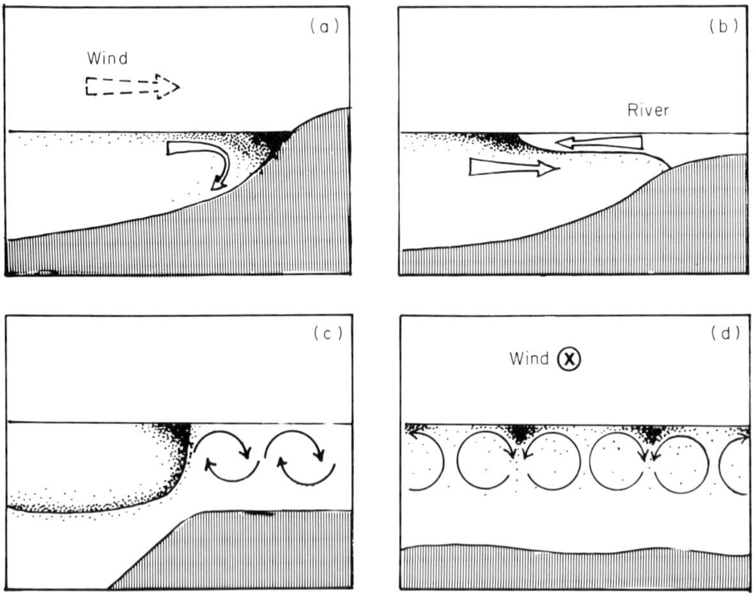

Fig. 11A.13. Horizontal concentration mechanisms. (a) an onshore wind with downwelling. (b) Aggregation at the boundary of a river, or other low density flow, the more dense water containing the dinoflagellates. (c) Aggregation at the boundary between stratified (left) and mixed water, such as that found at the edge of a shelf region with strong tidal mixing (similar conditions may occur between sheltered, stratified and open, turbulent coastal waters). (d) Langmuir cells produced by wind (perpendicular to the page) of moderate strength, with buoyant or upwardly-migrating cells aggregated at the convergences (there is spiral downwind transport within the rotating 'Langmuir cells'). Redrawn from various sources, including Ryther (1955) and Holligan (1979).

produces contra-rotating cells of water, whose size and rate of rotation is proportional to the wind strength. This results in linearly-arranged alternating convergences and divergences aligned slightly to the right of direction of the wind (see Wyatt 1975; Pollard 1977; Evans & Taylor 1980) with spiral downwind transport. The downwelling rate in the Langmuir cells is approximately $1\,cm\,s^{-1}$ per $2\,m\,s^{-1}$ of wind. Cells which can resist sinking will become aggregated into 'windrows' coinciding with the convergences. This phenomenon is essentially transient, since stronger wind force, above $25\,m\,s^{-1}$, will break the structure down into a more generally turbulent state. The structure is also lost if the wind dies down. It is unlikely that the cells can adapt significantly to this circulation because of its short duration. Rivkin *et al.* (1982) have shown that *Gonyaulax polyedra* can maintain a high division rate for at least two divisions when transferred to low light.

More commonly seen, and of longer duration, are the aggregations at 'fronts', horizontal boundaries of water types. Although not invariably, in most cases they also correspond to convergences, water of greater density sinking below that of less density. Thus, surface water is transported horizontally to the convergence. Any objects which have an upward buoyancy or upward mobility greater than the downwelling rate will accumulate in a belt along the convergence. Such density convergences are usually present around the plumes of rivers or within estuaries (e.g. Chesapeake Bay, Fig. 11A.12) but they also occur in more open waters, such as the accumulation of *Noctiluca* at fronts off Brittany studied by LeFèvre & Grall (1970; see Tyler & Seliger 1981, their fig. 11). Tyler *et al.* (1982) have also shown that cysts may be transported along such an interface, accumulating at a 'benthic front',

Fronts at the boundary between strong, tidally-mixed shelf-waters and more stratified, offshore water, have been the subject of much study around the British Isles (e.g. Pingree 1978; Holligan 1978, 1979), where they have been very clearly correlated with dinoflagellate blooms, particularly of *Gyrodinium aureolum*. Carreto *et al.* (1981) have suggested that a similar mechanism influences the distribution of *Protogonyaulax* (*Gonyaulax*) *tamarensis* off Argentina. In protected bays the reverse situation often occurs in summer, with the bay water more stratified and less dense than the shelf water.

Further offshore, larger-scale 'shelf-break' fronts can be formed (see review by Owen 1981), often associated with upwelling, but less information is available on their aggregative effects. Very large-scale convergences, such as the subtropical of Antarctic convergence zones, also aggregate phytoplankton, but have not been involved in red tide formation.

Off the west coast of Florida, *Pt. brevis* seems to be brought to the coast from an 'initiation zone' 18–74 km offshore by mechanisms accompanying intrusions of the southward-flowing Loop current (Haddad & Carder 1979), the surface blooms occurring on the landward side of fronts between Loop current and coastal water (Steidinger & Haddad 1981).

Internal waves can also act as producers of visible banding or fluctuations of concentration at a particular depth. Slight surface convergence effects accompany the troughs in such waves. Kamykowski (1979, 1981) has studied the aggregational potential of this mechanism using computer simulation and data from migrating blooms of *Gym. sanguineum* and *Gonyaulax polyedra* respectively. A subsurface layer of the latter very clearly illustrated the presence of internal waves at an observation tower off San Diego (Wilton & Barham 1968). Kamykowski concluded that lateral aggregational effects can also occur if the thermocline is relatively shallow and the wave amplitude is large.

DISSIPATION

Dinoflagellate blooms are terminated by different mechanisms, depending on the stability of the water column. Under stable conditions the bloom may exhaust a key nutrient, leading to a decline in population numbers. In cyst-forming species, the induction of sexuality (presumed to be associated with nutrient depletion, although this may be difficult to demonstrate: see Chapter 14), may lead to the eventual production of dormant cysts which sink out of the water column, settling in the sediments below. Cyst formation can be seen even in rapidly multiplying populations, for reasons which are not understood.

Grazing by zooplankton, particularly tintinnids such as *Favella* spp., may hasten the decline of blooms of *Protogonyaulax tamarensis* off eastern Canada and the US (Prakash 1963; Stoecker *et al.* 1981). Similar grazing by *Polykrikos*, *Noctiluca*, rotifers and ciliates, including tintinnids, is thought to hasten the end of red water blooms, e.g. by *Gonyaulax polyedra*, off the southern California (Eppley & Harrison 1975 and references therein: see also Section 3.10).

The dissappearance of surface blooms is most frequently the result of a meteorological change, from the calm, sunny conditions which permitted their aggregation, to turbulent dispersal, flushing of the area and general rapid change due to mixing. Diatoms are often re-introduced into the water column, with renewed growth permitted by the replenishment of nutrients from below.

In some exceptional circumstances a densely concentrated bloom may die overnight, the patches containing healthy, actively-photosynthesizing cells one day, and dead, decaying cells the next. Such mass plankton death can lead to the death of other marine fauna in the area (see below).

6.2 Unfortunate consequences of red tide blooms

Although most dinoflagellate blooms represent useful contributions to aquatic production, some periodically produce harmful results. Together with blooms produced by other phytoplankters, these can be subdivided into several basically different types.

GENERAL MARINE FAUNA MORTALITIES CAUSED BY O_2 DEPLETION

These are non-specific kills of marine fauna following intense blooms of dinoflagellates in which the strongly aggregated cells die in restricted waters, such as shallow bays.

Bacterial activity on the mass of decaying cells results in high biological oxygen demand (BOD), causing oxygen levels to become undetectable in the affected area. Any animals which are contained in such an area (sessile, territorial or trapped) soon die, adding to the load of decaying organic matter. The water often becomes greasy to the touch due to the release of less-rapidly decomposed fats (e.g. Grindley & Taylor 1964).

In essence, then, almost any species which forms intense blooms can lead to massive fauna kills under rare circumstances in which they can be 'overconcentrated' and die. In practice, this phenomenon has been associated more with some species than others. Many examples have been summarized by Brongersma-Sanders (1957).

An intense bloom of *Ceratium tripos* in New York Bight in 1976 is believed to have contributed to prolonged, anoxic bottom water over 13 000 km^2 in that region, with resulting fish and shellfish kills (Mahoney & Steimle 1979). Falkowski *et al.* (1980) concluded that large subpycnocline populations of *C. tripos* were the result of a warm winter with high run-off, little storm mixing, a deep summer pycnocline and little grazing on the *Ceratium*, the high respiration of the latter substantially contributing to the high BOD and subsequent anoxia. Seki *et al.* (1974) believed that the rain of the faecal pellets resulting from the grazing on red tide flagellates contributed to anoxic bottom water in Tokyo Bay. This type of phenomenon is not limited to the marine environment. For example, the death of intensive blooms of *Ceratium hirundinella* has caused fish-kills in lakes (Nicholls *et al.* 1980). However, in general, cyanobacteria are a more common cause of mortalities in fresh water than dinoflagellates.

The role of bacteria, particularly due to the high BOD produced by such kills (and those produced by toxins: see below), has been insufficiently evaluated, despite Evans' (1973) indication that such effects might be important.

'Clogging of the gills', with consequent suffocation, has been invoked as a cause of death by various authors, but such a mechanism has not been established as a cause of marine fauna mortality caused by dinoflagellates.

SELECTIVE MARINE FAUNA MORTALITIES RESULTING FROM TOXINS

Marine fauna are more commonly killed directly by toxins released from the blooms of certain species of dinoflagellates. In these cases oxygen levels are usually still high when the fish or shellfish begin dying, although they subsequently drop sharply in confined waters due to the BOD associated with the decay of marine fauna.

When concentrations of *Ptychodiscus* (*Gymnodinium*) *brevis* reach $1–2{\cdot}5 \times 10^5 \, l^{-1}$, fish in the vicinity begin dying, lower concentrations being insufficient to harm them. It is not clear whether the toxins from this species (see Chapter 8) are actively excreted into the surrounding water or whether cell lysis is necessary. Clearly the latter will increase the harmful effects.

Several *Gymnodinium* species have also been associated with fish kills, e.g. *G. flavum*, *G. nagasakiense* and *Gyrodinium aureolum*, but these can also kill benthic organisms such as the worm *Arenicola* (see refs in Chapter 8). Blooms of *Ceratium fusus* have been associated with larval oyster kills several times (e.g. Cardwell 1978; Cho 1979). A bloom tentatively identified as *Gymnodinium simplex*, led to a drastic reduction in the settlement of oyster larvae ('spatfall') in Hiroshima Bay, Japan in 1977 (Wisely *et al.* 1978), but this species has not been shown to produce a toxin. Some of these cases may be due to secondary effects, such as enhanced bacterial growth, rather than production of a specific toxin.

Although many molluscs and crustaceans are immune to saxitoxin and related toxins, thus allowing them to act as concentrators of the poisons, others are not and die during blooms of *Protogonyaulax* spp., e.g. the 'white mussel', *Donax serra* on the west coast of South Africa (Horstmann 1981). Oysters are sensitive, but seem to protect themselves by reducing filtration (Twarog & Yamaguchi 1975).

Flagellates of other groups such as *Chattonella* spp. or *Heterosigma* spp., are major fish killers in Pacific waters.

INDIRECT POISONING OF UPPER TROPHIC LEVEL PREDATORS

Certain dinoflagellate blooms can lead indirectly to death of the predators which consume herbivores feeding on the blooms. Generally, this involves the production of toxins by the dinoflagellates, concentration of the toxins by grazers which are unharmed by them ('food-chain amplification'), and illness or death of predators on the grazers. A classical example of this is paralytic shellfish poisoning (PSP) in which the grazers are filter-feeding shellfish such as clams, mussels and scallops and, to a lesser extent, oysters. The dinoflagellates responsible (*Protogonyaulax* spp.; *Pyrodinium bahamense* var. *compressum; Gymnodinium catenatum,* etc.) and their toxins are reviewed in Chapter 8. Other indirect toxic effects include 'ciguatera': the accumulation of toxins from benthic, sea-weed associated species, particularly *Gambierdiscus toxicus* (Section 4.5d) by fish; or severe or fatal liver damage caused by some *Prorocentrum* and *Dinophysis* species, and gastro-intestinal distress (diarrheic shellfish poisoning: DSP).

The degree to which filter-feeding shellfish become toxic depends on the dinoflagellate population concentration, rate of feeding, the duration of exposure while feeding and storage phenomena. Toxic *Protogonyaulax* species do not need to be abundant enough to produce visible 'red water' in order to cause the shellfish to become dangerously toxic. In a study of blooms of *P. tamarensis* in shallow ponds in Massachussetts, Anderson *et al.* (1983) found that 10^4–10^5 cells l^{-1} accompanied the closure of the areas to harvesting due to excessive toxin levels in 1980, whereas no closure was required in 1981 when the populations peaked between 10^3 and 10^4 cells l^{-1}.

Herbivorous zooplankton may also accumulate enough PSP toxins while grazing on toxic blooms (White 1979) to kill fish feeding on them. Although so far only the death of herring (White 1980) and sandlances (Adams *et al.* 1968) have been linked to this food-chain in field situations, White (1981) has shown that flounder, pollock and salmon are also killed by oral ingestion of PSP toxins.

AEROSOL OR GAS EFFECTS

When water from a *Pt. brevis* bloom becomes an aerosol as a result of wind-produced surf, local residents often experience stinging eyes and nostrils, a burning throat and severe coughing from the acrid, but odourless, spray (Woodcock 1948). Krzanowski *et al.* (1981) have concluded that this condition arises from the opening of sodium channels by the toxin(s), releasing acetylcholine and causing smooth tracheal muscle contraction. The effects are only temporary. No irritation of this type occurs under calm conditions, no matter how intense the bloom. From this distinct phenomenon, it can be argued strongly that the fish mortalities off Vera Cruz, Mexico in the 1800s (Nunez Ortega 1879) were due to *Pt. brevis*.

Hydrogen sulphide (H_2S) is a common, noxious product of the sea-bed in areas of major marine fauna mortalities, such as off SW Africa (Brongersma-Sanders 1957) or Peru (Rojas de Meniola 1979). Although it is usually rapidly oxidized there may be sufficient to escape as gas from the sea surface. At such times the air is not only unpleasant to breathe, but the H_2S causes discoloration of paint (in Peru the phenomenon was called the 'Callao painter' because of its effect on ships) or blackening of silver.

ECONOMIC COSTS

It is obvious that the death of marine fauna may incur high costs, not only in the loss of harvestable resources, but also in clean-up costs and cancellation of holiday reservations in tourist–recreational areas. Habas & Gilbert (1974) estimated that the 1971 fish kill by *Ptychodiscus brevis* cost seven Florida counties \$20 000 000 and noted that future costs would be much higher (35% or more) if of the same magnitude. The 1972 outbreak of shellfish poison in the vicinity of Massachusetts was estimated to have cost fisherman \$1 000 000, partly due to public over-reaction (National Marine Fisheries Service, in Jensen

1975). Anoxia in the New York Bight in 1976, contributed to by *Ceratium tripos*, was estimated to have caused a loss of \$60 000 000 to the shellfish industry (Falkowski *et al.* 1980).

In view of the massive kills of fish and other marine fauna off Florida's south-west coast—approximately 500 000 000 fish, estimated to weigh 25 000 tons in 1946–47 (Gunter *et al.* 1948; Rounsefell & Nelson 1966)—it might be expected that this would result in greatly reduced subsequent fish landings. However, most of the fish killed are usually menhaden and other low-value fish. Also, it appears that catches of even these fish are relatively unaffected and sport-fishing may even be enhanced by such red tides (Steidinger & Joyce (1973) and references therein).

Aquaculture projects are particularly vulnerable to harmful red tide effects and this is particularly evident in Japan. The pioneering cultured pearl experiments of Mikimoto in the late nineteenth/early twentieth century suffered from oyster-kills (one in 1933 costing 15 000 000 yen, equivalent to approximately \$7 000 000 at the time), but not sufficient to prevent a successful outcome of the project (Fàge 1953).

Fish culture involves the holding of the larger fish, such as red seabream (*Chrysophyrys major*), yellowtail (*Seriola quinqueradiata*), and salmon in mesh pens, usually in bay water adjacent to the aquaculture facility. These expensively-reared fish are vulnerable to red tide damage, being unable to escape the toxins or low oxygen water. Red tides in the Seto Inland Sea, Japan in 1972, were estimated to have killed \$35 000 000 worth of caged fish (Iwasaki 1979). Similar kills have occurred in caged fish in Norway (Braard & Heimdal 1970; Tangen 1977) and eels in Denmark.

Ironically, the excretion products from the cultured organisms can exacerbate the problem, enriching the water in a way which favours flagellates if the water is stable. This was demonstrated for fish farms and *Gymnodinium nagasakiense* by Nishimura (1982) and in the stimulation of *Peridinium hangoei* by pearl oyster faeces (Iwasaki 1969). The BOD from fish faeces can also cause anoxia if fish are too crowded.

In an unusual action, fisherman's co-operatives from Tokuyama Bay in the Seto Inland Sea successfully sued twelve firms for the sum of 200 million yen for dumping cadmium and mercury waste into the bay, causing the death of benthic fauna and increasing red tides (not all due to dinoflagellates) as a result of the release of nitrogen and phosphorus from the sediment (New Scientist, 7 August, 1975).

6.3 Prediction and control?

In view of the harmful potential of red tides there is an evident need for the development of reliable prediction and control measures to cope with, reduce or terminate them.

PREDICTION

The fact that red tides are virtually monospecific blooms, is both a hindrance and an advantage in modelling and prediction. On the one hand, they are situations in which autecology and synecology converge, i.e. the ecological pattern determined for the species should be representative of the whole local community, instead of some 'average' condition. On the other hand, since all species have different tolerances, optima and behaviour, it should be evident that models developed from data relating to one species, will probably not work well for other species, and areas differ in the nutrients and physical conditions which are most likely to limit growth.

Some of the earliest models were based on *Ptychodiscus brevis* blooms in the Gulf of Mexico. Kierstead & Slobodkin (1953) proposed a simple model for the early stage of such a bloom, involving patches of 'conditioned' waters, originating from coastal run-off, the dinoflagellates growing within the patches, with turbulent diffusion being a key factor. In their equation, $L_c = \pi D/K$ where L_c is the critical dimension of a water mass below which diffusion exceeds growth, K is the population, and D is the diffusion; see Levandowsky (1979), and Chapter 10 here for a review of their model. Although the premise and mathematics of the model are interesting, it did not take into account the more recently known fact that the blooms in question originate offshore and only move inshore under unusual circumstances, *Pt. brevis* usually avoiding low-salinity run-off (Steidinger 1973; Steiding & Haddad 1981). The Kierstead–Slobodkin model does not seem to be directly applicable to the prediction of most red tides, although there are situations in which land run-off is significant and it may aid in the modelling of these, e.g. in a *Gymnodinium* sp. bloom studied by Ragotzkie & Pomeroy (1957).

Another simple approach was that by Wyatt & Horwood (1973; also Wyatt 1974) the essence of which is that, by aggregating in a layer, the grazing pressure (assumed to be constant over the water column) is significantly reduced. There are certainly numerous examples of such aggregation (Section 6.1c), but the assumption of equal grazing pressure at all depths is highly arguable. Ironically, a recent study by Fiedler (1982) on zooplankton distributions relative to a layer of *Gymnodiniun sanguineum* (Fig. 11A.5 here) strongly indicates an *avoidance* of such a layer by zooplankton even though *G. sanguineum* is a good food source for larval fish. This should produce an even stronger release from grazing pressure than in Wyatt & Horwood's model.

Early authors tended to assume or suggest that a real or theoretical controlling factor, operating on one species in one location, could be extrapolated to all, but the limitations of this approach are readily apparent. Thus, the factors operating in Gulf of Mexico blooms of *Ptychodiscus brevis,* compared with *Gessnerium monilatum*, which is estuarine in the Gulf, are obviously going to be different in some critical aspects. In the same location, several different red tide blooms may succeed each other as the water conditions in the bay change (e.g. seasonal anoxia of the bottom water in Omura Bay, Japan: Iizuka 1976). Within the distribution range of the same species there is no reason to assume that the same factor(s) control its growth over the range: indeed, because of differing rates of supply and other environmental variability, the opposite should be expected. Blooms of *Protogonyaulax tamarensis* (= *excavata*) off the north-eastern coast of North America illustrate this clearly. In the Gulf of Maine the blooms appear to develop offshore, at subsurface levels (P. Holligan, pers. comm.), moving in towards the coast as the season progresses, whereas the species is restricted to shallow embayments in the southern part of its range (Cape Cod, Long Island, etc.: Anderson *et al.* 1982a,b). In the latter localities light and nutrients do not appear to be limiting and population densities can be predicted quite well from the interaction of temperature and salinity (Watras *et al.* 1982) but this is not likely to be the case further north.

Surveys of benthic cyst densities of red tide species have offered hope of the prediction of blooms, but this has not proved to be of sufficient precision. For example, not all the cysts may 'hatch' in a season (see Chapter 14), and even if excystment does occur, a bloom need not necessarily follow. Reid (1972) raised the possibility of red tides due to *Gonyaulax polyedra* around the British Isles due to the abundance and prevalence of its cysts (*Lingulodinium machaerophorum*), but such has not occurred since then. The commonest, most abundant cyst in many temperate areas such as Norway, the Grand Banks or British Columbia, is that of *Gonyaulax grindleyi* (*Operculodinium centrocarpum*) (Wall *et al.* 1977) and yet blooms by this species in these areas are relatively rare. Anderson *et al.* (1982a) have found that differential burial of cysts can lead to differing vertical profiles of viable cysts in sediments, and that subsurface, as well as those in the surface floc, can contribute to blooms if resuspended. Nevertheless, it is to be expected that such benthic cysts can cause blooms to re-occur in an area in which a toxic species has been introduced by advection from an adjacent region. Care should be taken, in the transplanting of shellfish from a red tide region to an unaffected area, to avoid the transport of cysts to the latter. Despite intensive searches, undoubted benthic cysts have not been discovered in some red tide species, such as *Ptychodiscus brevis* or *Gymnodinium sanguineum*.

On the west coast of Florida the 'iron index' of Ingle & Martin (1971) was reasonably predictive in the Charlotte Harbour region but not in other parts. Baldridge (1975) proposed a predictive model for the same organism using the temperature occurring during the period 19 January to 2 April of the preceding winter, known as the 'temperature departure' or TD:

$$\mathrm{TD} = \sum[(T - T_a)^2/(N - 1)]^{1/2}$$

where T is the actual daily mean temperature, T_a is the average temperature for the day and N is the number of days in the period studied. He noted that minima in TD seem to be correlated with succeeding red-tide outbreaks. Problems with this approach were noted by Morel & Anderson (1976), and a reply to their comments published by Baldridge (1976).

In order to make useful predictions, models are needed which take into account not only the autecology of the particular species, but also the characteristics of the locality. Multiple regression analysis is one form of analysis which can aid in ascertaining the most probably important controlling factors in some dinoflagellate blooms, e.g. as used by Cattell (1969) in British Columbia and Ouchi (1982) in Japan. This type of 'situational ecology' may seem excessively detailed and provincial to some, but it is necessary in order to produce a predictive model of practical value.

CONTROL

The earliest attempts to control red tides appear to have been the addition of copper sulphate to the water in which cultured pearl experiments were conducted in the 1930s (Mikimoto, in Fage 1957; Oda 1935). Copper sulphate was suspended in the water in 1 kg bags, resulting in a concentration of 1 ppm after dissolving for 20–30 mim. At the time this was considered acceptable, with little harm to the pearl oysters. However, because of the wide-spectrum effects of copper toxicity, this method would not now be considered suitable, despite the greater sensitivity of dinoflagellates to copper than most other organisms (Section 3.4f). Further information on copper toxicity to dinoflagellates has been provided by Saifullah (1978) and in Section 3.9.

The search for other chemical control agents has met with little success, none being exclusively lethal to dinoflagellates and few practical in terms of cost effectiveness. Marvin & Proctor (1966) tested 4306 compounds for possible control of *Pt. brevis* blooms, but only a few carbamates were effective at the required concentrations (0·01 ppm or less) and they are prohibitively expensive to use on the appropriate scale and not species specific. More recently this species has been found to be particularly sensitive to a surfactant, 'C^{13}LAS', a linear alkylbenzene sulphonate (Kutt & Martin 1974) and also 'aponin', a surface-active natural compound produced by the cyanobacterium *Gomphosphaeria aponina* (Kutt & Martin 1974; Hitchcock & Martin 1977). The latter organism occurs naturally in the southern Florida waters in which *Pt. brevis* blooms.

A biological control solution would be preferable to chemical control if one could be found, but those for red tides are in the realm of science fiction at present. The closest to such control is the possibility of using a parasitic dinoflagellate to attack the red tide organism.

Taylor (1968) observed that, although *Amoebophrya ceratii* is known to parasitize a variety of different dinoflagellates (see Chapter 13), it greatly prefers *Protogonyaulax* (*Gonyaulax*) *catenella*; in one bloom 30–40% of the cells were visibly infected by the parasite, with only two cells of other species having similar visible infections. This strongly suggests a possibility of control, growing seed populations of *A. ceratii* on a suitable maintenance organism for inoculating local waters during an early stage of bloom development. At present little is known of several key questions needed to assess the potential of such a mechanism: (i) What becomes of the PSP toxins of the infected host cells? (ii) What other effects would the artificial introduction of *A. ceratii* have on local plankton, even if it does occur there naturally? Observations by Nishitani *et al.* (1985) indicate that significant natural control of *P. catenella* occurs in Puget Sound.

From the point of view of PSP the control of toxic dinoflagellate populations may not be necessary if the toxic shellfish can be safely recognized by a good monitoring programme, but this still results in severe actual or potential economic losses due to closure of shellfish harvesting. Detoxification of the shellfish offers a pragmatic solution to the problem. Although the steaming and canning process reduces toxicity to a certain extent (Quayle 1969) and may allow the use of shellfish with low toxicity levels for this purpose, exposure to ozone gas in water for 2 weeks seems to be much more effective (Thurberg 1975; Blogoslawski & Stewart 1978).

ACKNOWLEDGEMENT

The author is indebted to Paul J. Harrison and Quay Dortsch for their valuable, critical reading of this chapter.

7 REFERENCES

ACCORINTI J. (1983) Antifungal product from algal origin (Review). *Rev. Int. D'Oceanogr. Med.* **72**, 45–53.

ADACHI R. & FUKUYO Y. (1979) The thecal structure of a marine toxic dinoflagellate *Gambierdiscus toxicus* gen. et. sp. nov. Collected in a ciguatera-endemic area. *Jap. Soc. Sci. Fish.* **45**(1), 67–71.

ADAMS J A., SEATON D. D., BUCHANAN J. B. & LONGBOTTOM M. R. (1986) Biological observations associated with the toxic phytoplankton bloom off the east coast. *Nature (Lond.)* **220**, 24–25.

AHMED S. E., KENNER K. A. & PACKARD T. T. (1977) A comparative study of the glutamate dehydrogenase activity in several species of marine phytoplankton. *Mar. Biol.* **39**, 93–101.

ALDRICH D. V. (1962) Photoautotrophy in *Gymnodinium breve* Davis. *Science N.Y.* **137**, 988–990.

ALDRICH D. V. & WILSON W. B. (1980) The effect of salinity on *Gymnodinium breve* Davis. *Biol. Bull.* **119**(1), 57–64.

ALEEM A. A. (1972) Effect of river outflow management on marine life. *Mar. Biol.* **14**, 200–208.

ALLEN M. B. (1971) High latitude phytoplankton. *Ann. Rev. Ecol. System.* **2**, 261–276.

ALLEN J. R., ROBERTS T. M., LOEBLICH A. R. III & KLOTZ L. C. (1975) Characterization of the DNA from the dinoflagellate *Crypthecodinium cohnii* and implications for nuclear organization. *Cell (Cambridge Mass.)* **6**, 161–169.

ANDERSON D. M., AUBREY D. G., TYLER M. A. & COATS D. W. (1982a) Vertical and horizontal distributions of dinoflagellate cysts in sediments. *Limnol. Oceanogr.* **27**(4), 757–765.

ANDERSON D. M., CHISHOLM S. W. & WATRAS C. J. (1983) The importance of life cycle events in the population dynamics of *Gonyaulax tamarensis*. *Mar. Biol.* **76**, 179–183.

ANDERSON D. M., KULIS D. M., ORPHANOS J. A. & CEURVELS A. R. (1982b) Distribution of the toxic dinoflagellate *Gonyaulax tamarensis* in the southern New England region. *Estuar. coast. shelf Sci.* **14**, 447–58.

ANDERSON D. M. & LINDQUIST N. L. (1985) Time-course measurements of phosphorus depletion and cyst formation in the dinoflagellate *Gonyaulax tamarensis* Lebour. *J. exp. mar. Biol. Ecol.* **86**, 1–13.

ANDERSON D. M. & MOREL F. M. M. (1978) Copper sensitivity of *Gonyaulax tamarensis*. *Limnol. Oceanogr.* **23**, 283–295.

ANDERSON D. M. & MOREL F. M. M. (1979) The seeding of two red tide blooms by the generation of benthic *Gonyaulax tamarensis* hypnocysts. *Estuar. coast. mar. Sci.* **9**. 279–294.

ANDERSON D. M. & WALL D. (1978) Potential importance of benthic cysts of *Gonyaulax tamarensis* and *Gonyaulax excavata* in initiating toxic dinoflagellate blooms. *J. Phycol.* **25**, 224–234.

ANDERSON J. T., ROFF J. C. & GERRATH J. (1981) The diatoms and dinoflagellates of Hudson Bay. *Can. J. Bot.* **59**, 1793–1810.

ANDREAC M. O. & KLUMPP D. (1979) Biosynthesis and release of organic arsenic compounds by marine algae. *Envir. Sci. Tech.* **13**, 738–741.

ANTIA N. J. (1976) Effects of temperature on the darkness survival of marine microplankton algae. *Microbiol. Ecol.* **3**, 41–54.

ANTIA N. J., BERLAND B. R. & BONIN D. J. (1980) Proposal for a red tide nitrogen turnover cycle in certain marine planktonic systems involving hypoxanthine-guanine excretion by ciliates and their reutilization by phytoplankton. *Mar. Ecol. prog. Ser.* **2**, 97–103.

ANTIA N. J., BERLAND B. R., BONIN D. J. & MAESTRINI S. Y. (1975) Comparative evaluation of certain organic and inorganic sources of nitrogen for phototrophic growth of marine microalgae. *J. mar. biol. Ass. UK* **55**, 519–539.

ANTIA N. J., BERLAND B. R., BONIN D. J. & MAESTRINI S. Y. (1976) Utilisation de la matière organique dissoute en tant que substrat par les algues unicellulaires marines. *Actual. Biochim. mar. Colloque Gabim CNRS*, 147–178.

ANTIA N. J. & KLUT E. M. (1981) Fluoride addition effects on euryhaline phytoplankter growth in nutrient-enriched seawater at an estuarine level of salinity. *Bot. Mar.* **24**, 147–152.

APPLEBY G., COLBECK J. & HOLDSWORTH E. S. (1980) β-Carboxylation enzymes in marine phytoplankton and isolation and purification of pyruvate carboxylase from *Amphidinium carterae* (Dinophyceae). *J. Phycol.* **16**, 290–295.

ARMSTRONG I. H., COULSON J. C., HAWKEY P. & HUDSON M. J. (1978) Further mass seabird deaths from paralytic shellfish poisoning. *Br. Birds* **71**, 58–68.

BADEN D. G. & MENDE T. J. (1978) Glucose transport and metabolism in *Gymnodinium breve*. *Phytochem.* **17**, 1553, 1558.

BAILLIE K. D. (1971) *A taxonomic and ecological study of the intertidal, sand-dwelling dinoflagellates of the north eastern Pacific Ocean.* M.Sc. thesis, Univ. British Columbia.

BAINBRIDGE R. (1957) The size shape and density of marine phytoplankton concentrations. *Biol. Rev.* **32**, 91–115.

BALCH W. MC. (1981) An apparent lunar tidal cycle of phytoplankton blooming and community succession in the Gulf of Maine. *J. exp. mar. biol. Ecol.* **55**, 65–77.

BALCH W. MC. (1985a) Differences between dinoflagellates and diatoms in the uptake of ^{36}Cl–ClO_3^-, an analogue of NO_3. In *Toxic Dinoflagellates* (eds D. M. Anderson, A. W. White & D. G. Baden) pp. 121–124. Elsevier, New York.

BALCH W. MC. (1985b) Lack of an effect of light on methylamine uptake by phytoplankton.

Limnol. Oceanogr. **30**, 665–674.

BALDRIDGE H. D. (1975). Temperature patterns in the long-range prediction of red tide in Florida waters. In *The First International Conference on Toxic Dinoflagellate Blooms* (ed. V. R. LoCicero), pp. 69–74. Mass. Sci. & Technol. Found., Mass. USA.

BALDRIDGE H. D. (1976) Reply to comment by F. M. Morel and D. M. Anderson. *Limnol. Oceanogr.* **21**, 627–632.

BALECH E. (1956) Etude des dinoflagellés du sable de Roscoff. *Rev. Algol.* NS **2**(1–2), 29–52.

BALECH E. (1968) The distribution and endemism of some Antarctic microplankters. In *Antarctic Biological Research in Prospect and Retrospect* (ed. M. W. Holdgate), pp. 143–146. Cambridge Univ. Press, Cambridge.

BALECH E. (1971) Microplankton del Atlantico Ecuatorial Oeste (Equalanti). *Serv. Hidrogr. Naval-Armada Argentina* **654**, 1–199.

BALECH E. (1975) Clave illustrada de Dinoflagelades Antarticos. *Inst. Antart. Argentino IAA* (Buenos Aires) 11, 1–99.

BALECH E. & EL-SAYED S.Z. (1965) Microplankton of the Weddell Sea. *Biol. Antarctic Seas II, Antarctic Res. Ser.* **5**, 107–124.

BANSE K. (1982) Cell volumes, maximal growth rates of unicellular algae and ciliates, and the role of ciliates in the marine pelagial. *Limnol. Oceanogr.* **24**, 1059–1071.

BARKER H. A. (1935) The culture and physiology of marine dinoflagellates. *Archiv. Mikrobiol.* **6** 156–181.

BARTH R. & GERALDO GOMES G. (1973). Observaceos en nanoplancton en una estacao fundeada. *Publ. Inst. Pesq. marin.* (Brasil) **70**, 1–7.

BENTLEY-MOWAT J. A. & REID S. M. (1969). Effect of gibberellins, kinetin and other factors on the growth of unicellular marine algae in culture. *Bot. Mar.* **12**, 185–199.

BERLAND B. R., BONIN D. J., KAPKOV V. I., MAESTRINI S. Y. & ARLHAC D. D. (1976) Action toxique de quatre metaux lourde sur la croissance d'algues unicellulaires marines. *C. R. Acad. sci. Paris* **282**, 633–636.

BERMAN T. (1970) Alkaline phosphates and phosphorous availability in Lake Kinneret. *Limnol. Oceanogr.* **15**, 663–674.

BESADA E. G., LOEBLICH L. A. & LOEBLICH A. R. III (1982). Observations on tropical benthic, dinoflagellates from ciguatera-endemic areas: *Coolia*, *Gambierdiscus* and *Ostreopsis*. *Bull. Mar. Sci.* **32**, 723–735.

BHOVICHITRA M. & SWIFT E. (1977) Light and dark uptake of nitrate and ammonium by large oceanic dinoflagellates: *Pyrocystis noctiluca, Pyrocystis fusiformis,* and *Dissodinium lunula*. *Limnol. Oceanogr.* **22**, 73–83.

BIENFANG P. JOHNSON W. (1980) Response of subtropical phytoplankton to power plant entrainment. *Environ. Poll.* (Ser.A) **22**, 165–178.

BLASCO D. (1970) Red tide in the upwelling region of Baja, California. *Limnol. Oceanogr.* **22**, 255–258.

BLASCO D. (1978) Observations on the diel migration of marine dinoflagellates off the Baja California coast. *Mar. Biol* **46**, 41–47.

BLASCO D., ESTRADA M. & JONES B. H. (1981) Short time variability of phytoplankton populations in upwelling regions—the example of Northwest Africa. In *Coastal Upwelling: Coastal and Estuarine Sciences* 1. (ed. F. A. Richards), pp. 339–347. Geophysical Monography Bd., Washington, D. C.

BLOGOSLAWSKI W. J. & STEWARD M. E. (1978) Paralytic shellfish poison in *Spisula solidissima*: Anatomical location and ozone detoxification. *Mar. Biol.* **45**, 261–264.

BLUMER M., GUILLARD R. R. L. & CHASE T. (1971) Hydrocarbons of marine phytoplankton. *Mar. Biol.* **8** (3), 183–189.

BONIN D. J. & MAESTRINI S. Y. (1981) Importance of organic nutrients for phytoplankton growth in natural environments: implications for algal species succession. *Can. Bull. Fish. aquat. Sci.* **210**, 279–291.

BOON J. J., RIJPSTRA W. I. C., DE LANGE F., DE LEEU J. W., YOSHIOKA M. & SHIMUZU Y. (1979) Black Sea sterol—a molecular fossil for dinoflagellate blooms. *Nature* (Lond.) **277**,

125–127.

BOOTH B. C., LEWIN J. & NORRIS R. E. (1981) Silicified cysts in North Pacific phytoplankton. *Biol. Oceanogr.* **1**, 57–80.

BRAARUD T. (1935). The Øst expedition to the Denmark Strait, 1929. II. The Phytoplankton and its conditions of growth. *Hvalråd Skr.* **10**, 1–173.

BRAARUD T. (1945). A phytoplankton survey of the polluted waters of inner Oslo Fjord. *Hvalråd. Skr.* **28**, 1–142.

BRAARUD T. (1951) Salinity as an ecological factor in marine phytoplankton. *Physiol. Plant.* **4**, 28–32.

BRAARUD T. (1957) A red water organism from Walvis Bay (*Gymnodinium galatheanum* n. sp.). *Galathea Rep.* **1**, 157–158.

BRAARUD, T. (1962). Species distribution in marine phytoplankton. *J. oceanogr. Soc. Japan,* 20th Ann. Vol., 628–649.

BRAARUD T. (1976) The natural history of the Hardanger fjord. (13) *Sarsia* **60**, 41–62.

BRAARUD T. & HEIMDAL B. R. (1970) Brown water on the Norwegian Coast in Autumn 1966. *Nytt. Mag. Bot.* **17** (2), 91–97.

BRIAND F. J. -P. (1975) Effects of power-plant cooling systems on marine phytoplankton. *Mar. Biol.* **32**, 135–146.

BROKAW C. J. & BENEDICT B. (1968) Mechanochemical coupling in flagella. II. Effects of viscosity and thiourea on metabolism and motility of *Ciona* spermatozoa. *J. gen. Physiol.* **52**, 283–299.

BRONGERSMA-SANDERS M. (1957) Mass mortality in the sea. Chapter 29 In *Treatise on Marine Ecology and Paleoecology,* Vol. 1. (ed. J. W. Hedgpeth), pp. 941–1010. Geol. Sci. Am., Mem. 67.

BRUNO S. F. & MCLAUGHLIN J. J. A. (1977) The nutrition of the freshwater dinoflagellate *Ceratium hirundella. J. Protozool.* **24**, 548–553.

BRUNO S. F., STAKER R. D. & CURTIS L. L. (1981). Seasonal variation of B_{12}, B_{12} analogs and phytoplankton in a Long Island estuary. *J. mar. Res.* **39**, 335–352.

BURRIS J. E. (1977) Photosynthesis, photorespiration and dark respiration in eight species of algae. *Mar. Biol.* **39**, 371–379.

BURSA A. (1961a) Phytoplankton of the 'Calanus' expeditions in the Hudson Bay, 1953–1954. *J. fish Res. Bd. Canada* **18**, 51–83.

BURSA A. S. (1961b) The annual oceanographic cycle in Igloolik in the Canadian Arctic. II. The phytoplankton. (Calanus Repts. No. 17), *J. fish. Res. Bd. Canada* **18**(4), 563–615.

BURSA A. (1963a) Phytoplankton in coastal waters of the Arctic Ocean at Point Barrow, Alaska. *Arctic* **16**, 239–262.

BURSA A. S. (1963b) Phytoplankton successions in the Canadian Arctic. In *Symposium on Marine Microbiology,* Chapter 58 (ed. C. H. Oppenheimer), pp. 625–628. Carl. C. Thomas Publ., Springfield, Ill., USA.

CAMPBELL P. H. (1973) Studies on brackish water phytoplankton. *Univ. North. Carolina Sea Grant Program, Publ.* UNC-SG-73-07. 407 pp. (also US Dept. Comm., nat. tech. Inf. Serv. Publ. COM-73-10672).

CAPERON J. & SMITH D. F. (1978) Photosynthetic rates of marine algae as a function of inorganic carbon concentration. *Limnol. Oceanogr.* **23**, 704–708.

CARDWELL R. D. (1978) Oyster larvae mortality in South Puget Sound, Washington, USA. *Proc. nat. Shellfish. Ass.* **68**, 88–89.

CARLUCCI A. F. (1970) The ecology of the plankton off La Jolla, California in the period April through September, 1967. II. Vitamin B_{12}, thiamin and biotin. *Bull. Scripps Inst. Oceanogr.* **17**, 23–30.

CARLUCCI A. F. & BOWES P. M. (1970) Production of vitamin B_{12}, thiamin and biotin by phytoplankton. *J. Phycol.* **6**, 351–357.

CARLUCCI A. F. & SILBERNAGEL S. B. (1969). Effect of vitamin concentrations on growth and development of vitamin-requiring algae. *J. Phycol.* **5**, 64–67.

CARPENTER E. J. (1973) Brackish-water phytoplankton response to temperature elevation.

Estuar. coast. mar. Sci. **1**, 37–44.

Carreto J. I., Lasta M. L., Negri R. & Benvanides H. (1981) Los fenomenos de Marea Roja y toxicidad de moluscos bivalvos en el mar Argentino. *Rep. INIDEP Argentina,* contrib. 399, 1–21 + appendix.

Carter N. (1937) New or interesting algae from brackish water. *Arch. Prtotistenk.* **90**, 1–68.

Cassie V. (1963) Distribution of surface phytoplankton between New Zealand and Antarctica, December 1957. *Trans-Antarctic Exped. 1955–58, Sci. Rep. No. 7*, 11 pp.

Cattell S. A. (1969) *Dinoflagellates and vitamin B_{12} in the Strait of Georgia, British Columbia.* Ph.D. thesis, Univ. British Columbia. pp. 132.

Cattell S. A. (1973) The seasonal cycle of vitamin B_{12} in the Strait of Georgia, British Columbia. *J. fish. Res. Bd. Can.* **30**, 215–222.

Cembella A. D., Harrison P. J. & Antia N. J. (1983) (1984). The utilization of inorganic and organic phosphorus compounds as nutrients by eukaryotic microalgae: a multidisciplinary perspective. *Crit. Rev. Microbiol.*, Part 1 (Vol. 10) 317–319, Part 2 (Vol. 11) 13–81.

Chan A. T. (1978) Comparative physiological study of marine diatoms and dinoflagellates in relation to irradiance and cell size. I. Growth under continuous light. *J. Phycol.* **14**, 396–402.

Chan A. T. (1980) Comparative physiological study of marine diatoms and dinoflagellates in relation to irradiance and cell size. II. Relationship between phytosynthesis, growth, and carbon/chlorophyll *a* ratio. *J. Phycol.* **16**, 428–32.

Cheng J. & Antia N. J. (1970) Enhancement by glycerol of phototrophic growth of marine planktonic algae and its significance to glycerol pollution. *J. fish. Res. Bd. Can.* **27**, 335–45.

Chevallier A. & Duchesne E. A. (1851) Memoire sur les empoisonnements par les huitres, les moules, les crabes, et par certains poissons de mer et de rivière. *Ann. Hyg. publ. (Paris)* **45**, 387–437; **46**, 108–47.

Chisholm W. S. (1981) Temporal patterns of cell division in unicellular algae. *Can. Bull. Fish. aquat. Sci.* **210**, 150–181.

Cho C. H. (1981) Mass mortalitys (sic) of oyster due to red tide in Jinhae Bay in 1978. *Bull. Korean Fish. Soc.* **12**, 27–33.

Cole E. J., Yentsch C. N., Yentsch C. S. & Salvaggio M. (1975) Some of the growth characteristics of *Gonyaulax tamarensis*. *Environ. Lett.* **9**, 153–166.

Colebrook J. M., Reid P. C. & Coombs S. M. (1978) Continuous plankton records: a change in the plankton of the southern North Sea between 1970 and 1972. *Mar. Biol.* **45**, 209–213.

Collier A., Wilson W. B. & Borkowski M. (1969) Responses of *Gymnodinium breve* Davis to natural waters of diverse origin. *J. Phycol.* **5**, 168–172.

Collos Y. & Slawyk G. (1980) Nitrogen uptake and assimilation by marine phytoplankton. In *Primary Productivity in the Sea* (ed. P. G. Falkowski), pp. 195–211. Plenum Press, New York.

Connell C. H. & Cross J. B. Mass mortality of fish associated with the protozoan *Gonyaulax* in the Gulf of Mexico. *Science. NY,* **112**, 359–363.

Conrad W. & Kufferath H. (1954). Recherches sur les eaux saumâtres des environs de Lilloo. II. *Inst. Roy. Sci. Nat. Belgique Mem.* **127**, 1–346.

Cooper L. H. N. (1968) Scientific consequences of the wreck of the 'Torrey Canyon.' *Helgoland wiss. Meeresunters.* **17**, 340–355.

Correll D. L., Faust M. A. & Severn D. J. (1975) Phosphorus flux and cycling in estuaries. In *Estuarine Research,* Vol. 1 (ed. L. E. Cronin), pp. 108–136. Academic Press, New York.

Cullen J. J. & Eppley R. W. (1981) Chlorophyll maximum layers of the southern California Bight and possible mechanisms of their formation and maintenance. *Oceanol. Acta.* **4**, 23–32.

Cullen J. J. & Horrigan S. G. (1981) Effects of nitrate on the diurnal vertical migration, carbon to nitrogen ratio, and the photosynthetic capacity of the dinoflagellate *Gymnodinium splendens*. *Mar. Biol.* **62**, 81–89.

CULLEN J. J., HORRIGAN S. G., HUNTLEY M. E. & REID F. M. H. (1982) Yellow water in La Jolla bay, California, July 1980. A bloom of the dinoflagellate, *Gymnodinium flavum*. *J. exp. mar. Biol. Ecol.* **63**, 67–80.

DANDONNEAU Y. (1970) Un phénomène d'eaux rouges au large de la Côte d'Ivoire causé par *Gymnodinium spendens* Lebour. *Doc. Scient. Centre Rech. Océanogr.* Abidjan. **1**, 11–19.

DEANE E. M. & O'BRIEN R. W. (1981a). Uptake of phosphate by symbiotic and free-living dinoflagellates. *Arch. Microbiol.* **128**, 307–310.

DEANE E. M. & O'BRIEN R. W. (1981b) Uptake of sulphate, taurine, cysteine and methionine by symbiotic and free-living dinoflagellates. *Arch. Microbiol.* **128**, 311–319.

DEBILLY F. & SOYER M. O. (1979) Toxic effects and cellular adaptions after the action of a defoliant 2,4,5-T of the marine dinoflagellate *Prorocentrum micans* E. Observations in light and electron microscopy. *J. Protozool.* **26**, (3pt.1); 53A.

DEMPSTER R. P. (1955) The use of copper sulphate as a cure for fish diseases caused by parasitic dinoflagellates of the genus *Oodinium*. *Zool.* (NY) **40**, 133–138.

DESROSIÈRES R. (1975) Some observations on the oceanic phytoplankton collected in the vicinity of New Caledonia (southwestern Pacific Ocean). *Norw. J. Bot.* **22**, 195–200.

DESROSIÈRES R. & WAUTHY B. (1972) Distribution du phytoplancton et structure hydrologique dans la région des Tuamotu (Océan Pacifique Central). *Cah. ORSTOM, sér. oceanogr.* **10**, 275–287.

DOBELL P. E. R. (1978) *A study of dinoflagellate cysts from recent marine sediments of British Columbia*. Masters Thesis, University of British Columbia, B.C. pp. 1–176.

DODGE J. D. (1972) The ultrastructure of the dinoflagellate pusule: a unique osmo-regulatory organelle. *Protoplasma* **75**, 285–302.

DODGE J. D. (1981) *Provisional Atlas of the Marine Dinoflagellates of the British Isles*. Biological Records Centre Monks Wood Exp. Station, England. pp. 1–130.

DODGE J. D. (1982) *Marine Dinoflagellates of the British Isles*. HMSO, London.

DODSON A. N. & THOMAS W. H. (1977) Marine phytoplankton growth and survival under simulated upwelling and oligotrophic conditions. *J. exp. mar. Biol. Ecol.* **26**, 153–161.

DOIG M. T. III & MARTIN D. F. (1974) The effect of naturally occurring organic substances on the growth of a red tide organism. *Water Res.* **8**, 601–606.

DOMOTOR S. L., MOUNTFORD K. & D'ELIA C. F. (1982) Autoradiographic detection of species-specific thermal stress effects on natural phtoplankton assemblages. *Mar. environ. Res.* **6**, 27–35.

DORTSCH Q. & MASKE H. (1982) Dark uptake of nitrate and nitrate reductase activity of a red tide population off Peru. *Mar. Ecol., prog. Ser.* **9**, 299–303.

DORTSCH Q., CLAYTON J. R. Jr., THORESEN S. S. & AHMED S. I. (1984). Species differences in accumulation of nitrogen pools in phytoplankton. *Mar. Biol.* **81**, 237–250.

DOWIDAR N. M. (1971) Distribution and ecology of *Ceratium egyptiacum* Halim and its validity as indicator of the current regime in the Suez Canal. Int. Rev. ges. Hydrobiol. **56**, 957–966.

DOYLE R. & POORE R. V. (1974) Nutrient competition and division synchrony in phytoplankton. *J. exp. mar. Biol. Ecol.* **14**, 201–210.

DRAGESCO J. (1965) Étude cytologique de quelques flagelles mesopsammiques. *Cahiers biol. mar.* **4**, 83–115.

DROOP M. R. (1959) A note on some physical conditions for cultivating *Oxyrrhis marina*. *J. mar. biol. Ass. UK* **38**, 599–604.

DUGDALE R. C. (1967) Nutrient limitation in the sea: dynamics, identification and significance. *Limnol. Oceanogr.* **12**, 685–695.

DUGDALE R. C & GOERING J. J. (1967) Uptake of new and regenerated forms of nitrogen in primary productivity. *Limnol. Oceanogr.* **12**, 196–206.

DUNSTAN W. M. (1973) A comparison of the photosynthesis–light intensity relationship in phylogenetically different marine microalgae. *J. exp. mar. Biol. Ecol.* **13**, 181–187.

DUNSTAN W. M., ATKINSON L. P. & NATOLI J. (1975) Stimulation and inhibition of phytoplankton growth by low molecular weight hydrocarbons. *Mar. Biol.* **31**, 305–310.

ECKMAN S. (1953) *Zoogeography of the Sea*. Sidgwick & Jackson, London.

ELBRÄCHTER M. (1971) *Untersuchungen uber die Populations-dynamik und Ernahrungsbiologie von Dinoflagellaten im Freiland und im Labor*. Ph.D. thesis, Kiel University.

ELBRÄCHTER M. (1973) Population dynamics of *Ceratium* in coastal waters. *Oikos* **15**, (suppl.) 43–48.

ERBRÄCHTER M. (1976) Population dynamic studies on phytoplankton cultures. *Mar. Biol.* **35**, 201–209.

ERBRÄCHTER M. (1977) On population dynamics in multi-species cultures of diatoms and dinoflagellates. *Helgoland wiss. Meeresunters.* **30**, 192–200.

ELBRÄCHTER M. (1979) On the taxonomy of unarmored dinophytes (Dinophyta) from the Northwest African upwelling region. *Meteor Forsch.-Ergebnisse.* **30**, 1–22.

ELMHIRST R. (1922) Investigation on the effects of oil tanker discharge. *Rep. Scottish mar. Biol. Ass.* **1922**, 8–9.

EPPLEY R. W. (1972) Temperature and phytoplankton growth in the sea. *Fish. Bull.* **70**, 1063–1085.

EPPLEY R. W., COATSWORTH J. L. & SOLORZANO L. (1969a) Studies of nitrate reductase in marine phytoplankton. *LImnol. Oceanogr.* **14**, 194–205.

EPPLEY R. W. & HARRISON W. G. (1975) Physiological ecology of *Gonyaulax polyedra*, a red water dinoflagellate off Southern California. In *Proc. 1st Int. Conf. on Toxic Dinoflagellate Blooms, Nov. 1974* (ed. V. R. LoCicero), pp. 11–22. MSTF, Wakefield, Mass, USA.

EPPLEY R. W., KOELLER P., & WALLACE G. T. JR (1978) Stirring influences the phytoplankton species composition within enclosed columns of coastal sea water. *J. exp. mar. Biol. Ecol.* **32**, 219–240.

EPPLEY R. W. & RENGER E. H. (1974) Nitrogen assimilation of an oceanic diatom in nitrogen-limited continuous culture. *J. Phycol.* **10**, 15–23.

EPPLEY R. W., RODGERS J. N., MCCARTHY J. J. (1969b) Half-saturation constants for uptake of nitrate and ammonium by marine phytoplankton. *Limnol. Oceanogr.* **14**, 912–920.

EPPLEY R. W., ROGERS J. N., MCCARTHY J. J. & SOURNIA A. (1971) Light/dark periodicity in nitrogen assimilation of the marine phytoplankton *Skeletonema costatum* and *Coccolithus huxleyi* in N-limited chemostat culture. *J. Phycol.* **7**, 150–154.

ESAIAS W. E. & CURL H. C. JR (1972) Effect of dinoflagellate bioluminescence on copepod ingestion rates. *Limnol. Oceanogr.* **17**, 901–906.

EVANS E. E. (1973) The role of bacteria in the Florida red tide. *Environ. Lett.* **5**, 37–44.

EVANS G. T. & TAYLOR F. J. R. (1980) Phytoplankton accumulation in Langmuir cells. *Limnol. Oceanogr.* **25**, 840–845.

FAGE L. (1953) Commentaires sur la première plaie d'Égypte; l'eau du fleuve changée en sang. *Conferences Palais Découverte* (Univ. Paris) ser. A, **184**, 1–20.

FALKOWSKI P. G. (1975) Nitrate uptake in marine phytoplankton: (nitrate, chloride) activated adenosine triphosphatase from *Skeletonema costatum*. *J. Phycol.* **11**, 323–326.

FALKOWSKI P. G., HOPKINS T. S. & WALSH J. J. (1980) An analysis of factors affecting oxygen depletion in the New York Bight. *J. mar. Res.* **38**, 479–506.

FALKOWSKI P. G. & RIVKIN R. B. (1976) The role of glutamine synthetase in the incorporation of ammonium in *Skeletonema* costatum (Bacillari & phyceae) *J. Phycol.* **12**, 448–450.

FENCHEL T. M. & RIEDL R. J. (1970) The sulfide system: a new biotic community underneath the oxidized layer of marine sand bottoms. *Mar. Biol.* **7**, 255–268.

FIEDLER P. C. (1982) Zooplankton avoidance and reduced grazing response to *Gymnodinium splendens* (Dinophyceae). *Limnol. Oceanogr.* **27**, 961–965.

FINLAY B. J. & UHLIG G. (1981) Calorific and carbon values of marine and freshwater Protozoa. *Helgol wiss. Meeresunters.* **34**, 401–412.

FREEBURG L. R., MARSHALL A. & HEYL M. (1979) Interrelationships of *Gymnodinium breve* (Florida red tide) within the phytoplankton community. In *Toxic Dinoflagellate Blooms* (eds D. L. Taylor & H. H. Seliger) pp. 139–144. Elsevier, North Holland, New York

FROST N. & WILSON A. M. (1938) The genus *Ceratium* and its use as in indicator of hydrographic conditions in the Newfoundland waters. *Newfoundland Res. Bull.* **5**, 1–15.

FUKUYO Y. (1979) Theca and cyst of *Gonyaulax excavata* (Braarud) Balech found at Ofunato Bay, Pacific Coast of Northern Japan. In *Toxic Dinoflagellate Blooms* Developments in Marine Biology, Vol. 1. (eds D. L. Taylor & H. H. Seliger), pp. 61–64. Elsevier, North-Holland. New York.

FUKUYO Y. (1981) Taxonomical study on benthic dinoflagellates collected in coral reefs. *Bull. Jap. Soc. Sci. Fish.* **47**, 967–978.

GAINES G. & TAYLOR F. J. R. (1985) An exploratory analysis of PSP patterns in British Columbia: 1942–1984. In *Toxic Dinoflagellates* (eds D. M. Anderson, A. W. White & D. G. Baden) pp. 439–444, Elsvier, New York.

GALLERON C. (1976) Synchronization of the marine dinoflagellate *Amphidinium carteri* in dense cultures. *J. Phycol.* **12**, 69–73.

GEORGE D. G. & HEANEY S. I. (1978) Factors influencing the spatial distribution of phytoplankton in a small productive lake. *J. Ecol.* **66**, 133–155.

GERRATH J. F. & NICHOLLS K. H. (1974) A red snow in Ontario caused by the dinoflagellate *Gymnodinium pascheri*. *Can. J. Bot.* **52**, 683–685.

GLOVER H. E. (1978) Iron in Maine coastal waters; seasonal variation and its apparent correlation with a dinoflagellate bloom. Limnol Oceanogr. **23**, 534–537.

GOLDMAN J. C., MCCARTHY J. J. & PEAVEY D. G. (1979) Growth rate influence on the chemical composition of phytoplankton in oceanic waters. *Nature (Lond.)* **279**, 210–215.

GORDON D. C. Jr, FOURNIER R. D. & KRASNICK G. J. (1971) Note on the planktonic primary production in Fanning Island lagoon. *Pac. Sci.* **25**, 228–233.

GRAHAM H. W. (1941) An oceanographic consideration of the dinoflagellate genus *Ceratium*. *Ecol. Monogr.* **11**, 99–116.

GRAHAM H. W. (1942) Studies in the morphology, taxonomy and ecology of the Peridiniales. In *Carnegie Dinoflagellate Reports*, pp. 1–127 Biology III. Carnegie Inst. Wash. Publ. 542, Washington, USA.

GRAHAM H. W. (1954) Dinoflagellates of the Gulf of Mexico. *US Dept. Int., Fish wild. Serv. Spec. Sci. Rept.* **52**, 223–226.

GRAHAM H. W. & BRONIKOWSKY N. (1944) The genus *Ceratium* in the Pacific and North Atlantic Oceans. Biology 5. *Carnegie Inst. Wash. DC. Pub.* **565**, 1–209.

GRAN H. H. (1912) Pelagic plant life. In *The Depths of the Ocean* (eds J. Murray & J. Hjort), pp. 307–386. Macmillan, London.

GRAN H. H. & BRAARUD T. (1935) A quantitive study of the phytoplankton in the Bay of Fundy and the Gulf of Maine (including observations on hydrography, chemistry and turbidity). *J. biol. bd Can.* **1**, 279–433.

GRINDLEY J. R. (1979) Factors determining the productivity of South African coastal waters. In *Marine Production Mechanisms* (ed M. J. Dunbar), pp. 89–132. Cambridge Univ. Press, Cambridge.

GRINDLEY J. R. & TAYLOR F. J. R. (1964) Red water and marine fauna mortality near Cape Town. *Trans. Roy. Soc. S. Afr.* **37**, 113–130.

GRINDLEY J. R. & TAYLOR F. J. R. (1971) Factors affecting plankton blooms in False Bay. *Trans. Roy. Soc. S. Afr.* **31**, 201–210.

GUILLARD R. R. L. & KILHAM P. (1977) The ecology of marine plankton diatoms. In *The Biology of Diatoms* (ed. D. Werner), pp. 372–469. Blackwell Scientific Publications, Oxford and University of California Press, Berkeley.

GUNTER G., WILLIAMS R. H., DAVIS C. C. & SMITH F. G. W. (1984) Catastrophic mass mortality of marine animals and coincident phytoplankton bloom on the west coast of Florida, November 1946 to May 1947. *Ecol. Monogr.* **18**, 310–324.

GUZMAN L. & CAMPODONICO I. (1978) Mareas rojas en Chile. *Interciencia* **3**, 144–151.

HABAS E. J. & GILBERT C. K. (1974) The economic effects of the 1971 Florida red tide and the damage it presages for future occurrences. *Environ. Lett.* **6**, 139–147.

HADDAD K. D. & CARDER K. L. (1979) Oceanic intrusion: one possible initiation mechanism of red tide blooms on the west coast of Florida. In *Toxic Dinoflagellate Blooms* (eds D. L. Taylor & H. H. Seliger), pp. 269–274. Elsevier/North Holland, New York.

HALIM Y. (1960) Étude qualitative et quantitative du cycle ecologique des Dinoflagèlles dans les eaux de Villefranche-sur-mer (1953–1955). *Ann. Inst. Océanogr.* **38**, 123–232.

HALIM Y. (1967) Dinoflagellates of the South-East Caribbean Sea (East Venezuela). *Int. Rev. ges. Hydrobiol.* **52**, 701–755.

HALIM Y. (1969) Plankton of the Red Sea. *Oceanogr. mar. Biol. Ann. Rev.* **7**, 231–275.

HALLEGRAEFF G. (1981) Seasonal study of phytoplankton pigments and species at a coastal station off Sydney: importance of diatoms and nanoplankton. *Mar. Biol.* **61**, 107–118.

HALLEGRAEFF G. & JEFFREY S. W. (1984) Tropical phytoplankton species and pigments of continental shelf waters of north and north-west Australia. *Mar. Ecol. Progr. Ser.* **20**, 59–74.

HALSTEAD B. W. (1965) *Poisonous and Venomous Marine Animals of the World,* Vols 1–3. US Gov't Printing Office, Washington, D.C. (Revised edn, 1 Vol. 1978, Darwin Press, Princeton, N. J.), pp. 1–1042.

HANSEN V., ALBRECHTSEN K. & FRANSDEN C. (1969) De døde fisk og planteplanktonet i Nordsøeni 1968. *Skr. Dansk fisk. Havunders.* **19**, 36–53.

HARGRAVES P. E., BRODY R. W. & BURKHOLDER P. R. (1970) A study of phytoplankton in the Lesser Antilles region. *Bull. mar. Sci.* **20** (2), 331–349.

HARRIS G. P. (1983) Mixed layer physics and phytoplankton populations: Studies in equilibrium and non-equilibrium ecology. In *Progress in Phycological Research* Vol. 2. (eds. F. E. Round & D. J. Chapman), pp. 1–52. Elsevier Press, New York, Oxford.

HARRISON P. J., FULTON D. J., TAYLOR F. J. R. & PARSONS T. R (1983) Review of the biological oceanography of the Strait of Georgia: pelagic enviroment. *Can. J. Fish. Aquat. Sci.* **40**, 1064–1094.

HARRISON W. G. (1973) Nitrate reductase activity during a dinoflagellate bloom. *Limnol. Oceanogr.* **18**, 457–465.

HARRISON W. G. (1976) Nitrate metabolism of the red tide dinoflagellate *Gonyaulax polyedra* Stein. *J. exp. mar. Biol. Ecol.* **21**, 199–209.

HARRISON W. G., EPPLEY R. W. & RENGER E. H. (1977) Phytoplankton nitrogen metabolism, nitrogen budgets, and observations on copper toxicity: controlled ecosystem pollution experiment. *Bull. mar. Sci.* **27**, 44–57.

HART T. J. (1934) On the phytoplankton of the south-west Atlantic and the Bellingshausen Sea 1929–1931. *'Discovery' Rep.* **8**, 1–268.

HARTWELL A. D. (1975) Hydrographic factors affecting the distribution and movement of toxic dinoflagellates in the western Gulf of Maine. In *Proc. of the First International Conference on Toxic Dinoflagellate Blooms* (ed. V. R. LoCicero), pp. 47–68. Mass. Sci. & Technol. Found., Waksfield, Mass., USA.

HASLE G. R. (1950) Phototactic vertical migration in marine dinoflagellates. *Oikos* **2**, 162–175.

HASLE G. R. (1954) More on phototactic diurnal migration in marine dinoflagellates. *Nytt. Mag. Bot.* **2**, 139–147.

HASLE G. R. (1959) A quantitative study of phytoplankton from the equatorial Pacific. *Deep-Sea Res.* **6**, 38–59.

HASLE G. R (1960) Phytoplankton and cilliate species from the tropical Pacific. *Norske. Vidensk. Akad. Skr. Mat. Nat. Kl.* 1960 (2), 5–49.

HASLE G. R. (1969) Scientific results of marine biological research. *Hvalrådets Skr.* **52**, 1–168.

HASLE G. R. (1976) The biogeography of some marine planktonic diatoms. *Deep-Sea Res.* **23**, 319–338.

HEANEY S. I. & EPPLEY R. W. (1981) Light, temperature and nitrogen as interacting factors affecting diel vertical migrations of dinoflagellates in culture *J. plankt. Res.* **3**, 331–343.

HEIMDAL B. R., HASLE G. R. & THRONDSEN J. (1973) An annotated check-list of plankton algae from the Oslofjord, Norway (1951–1972). *Norw. J. Bot.* **20**, 13–19.

HERDMAN E. C. (1922) Notes on dinoflagellates and other organisms causing discolouration of the sand of Port Erin II. (1921) *Proc. Trans. Liverpool. Biol. Soc.* **36**, 15–30. (Session 1921–1922.)

HERDMAN E. C. (1924) Notes on dinoflagellates and other organisms causing discolouration of the sand at Port Ering III. *Proc. Trans. Liverpool. Biol. Soc.* **38**, 58–63.

HERDMAN W. A. (1914) The minute life of the sea beach. *Proc. Trans. Liverpool. Biol. Soc.* **28**, 39–43.

HERNANDEZ V. & ALMODOVAR M. R. (1983) *Gonyaulax tamarensis* Lebour (Pyrrhophyta) and *Falkenbergia hillebrandii* (Barnet) Falkenberg (Rhodophyta), two organisms with fungistatic capabilities of the coasts of Puerto Rico. *Science* **10**, 24–26.

HERSEY R. L. & SWIFT E. (1976) Nitrate reductase activity of *Amphidinium carteri* and *Cachonina niei* (Dinophyceae) in batch culture: Diel periodicity and effects of light intensity and ammonia. *J. Phycol.* **12**, 36–44.

HICKEL W. (1974) Seston composition of the bottom waters of Great Lameshur Bay, St John, US Virgin Islands. *Mar. Biol.* **24**, 125–130.

HIRAYAMA K. & NAMAGUCHI K. (1972) Growth of *Gymnodinium* 'type 65', causative organism of red tide in Omura Bay, in medium supplied with bottom mud extract. *Bull. plankt. Soc. Japan* **19**, 13–21.

HITCHCOCK G. L. (1982) A comparative study of the size-dependent organic composition of marine diatoms and dinoflagellates. *J. plank. Res.* **4**, 363–377.

HITCHCOCK W. S. & MARTIN D. F. (1977) Effects and fate of a surfactant in cultures of red tide organisms, *Gymnodinium breve*. *Bull. environ. Contam. Toxicol.* **18**, 291.

HOBRO R. (1979) Stages of the annual phytoplankton succession in the Asko area (northern Baltic Sea). *Acta Bot. Fennici* **10**, 79–80.

HOCHACHKA P. W. &TEAL J. M. (1964) Respiratory metabolism in a marine dinoflagellate. *Biol. Bull mar. biol Lab. Woods Hole* **126**, 274–281.

HOLLIBAUGH J. T. (1976) The biological degradation of arginine and glutamic acid in seawater in relation to the growth of phytoplankton. *Mar. Biol.* **36**, 303–312.

HOLLIGAN P. M. (1978) Patchiness in subsurface phytoplankton populations on the northwest European continental shelf. In *'Spatial Patterns in Plankton Communities,* pp. 221–238. Plenum Press, New York.

HOLLIGAN P. M. (1979) Dinoflagellate blooms associated with tidal fronts around the British Isles. In *Toxic Dinoflagellate Blooms* (eds D. L. Taylor & H. H. Seliger), pp. 249–256. Elsevier/North Holland, New York.

HOLLIGAN P. M. & HARBOUR D. S. (1977) The vertical distribution and succession of phytoplankton in the western English Channel in 1975 and 1976. *J. mar. biol. Ass. UK* **57**, 1075–1093.

HOLLIGAN P. M., MADDOCK L. & DODGE J. D. (1980) The distribution of dinoflagellates around the British Isles in July 1977: a multivariate analysis. *J. mar. biol. Ass. UK* **60**, 851–867.

HOLM-HANSEN, O. (1969) Algae: amounts of DNA and organic carbon in single cells. *Science (N.Y.)* **163**, 76–88.

HOLM-HANSEN O., EL-SAYED, S. Z., FRANCESCHINI G. A. & CUHEL R. L. (1977) Primary production and factors controlling phytoplankton growth in the Southern Ocean. In *The Proceedings of the Third Scar Symposium on Antarctic Biology,* pp. 11–49. Gulf Publ. Co., Houston, Tex., USA.

HOLMES A. & HOLMES D. L. (1978). *Principles of Physical Geology.* Thomas Nelson & Sons, England.

HOLMES R. W., WILLIAMS P. M. & EPPLEY R. W. (1967) Red water in La Jolla Bay, 1964–1966. *Limnol. Oceanogr.* **12**, 503–512.

HORIGUCHI T. (1983) Life history and taxonomy of benthic dinoflagellates. PhD Thesis, Univ. of Tsukuba. 141 pp.

HORNER R. A. (ed.) (1985) *Sea Ice Biota.* 215 pp. CRC Press, Boca Raton, Florida.

HORSTMAN D. A. (1981) Reported red-water outbreaks and their effects on fauna of the west and south coasts of South Africa, 1959–1980. *Fish. Bull. S. Afr.* **15**, 71–88.

HORSTMANN U. (1980) Observations on the peculiar diurnal migration of a red tide Dinophyceae in tropical shallow waters. *J. Phycol.* **16**, 481–485.

HOVIS W. A., CLARK D. K., ANDERSON F., AUSTIN R. W., WILSON W. H., BAKER E. T., BALL D., GORDON H. R., MUELLER J. L., ELSAYED S. Z., STURM B., WRIGELY R. C. & YENTSCH C. S. (1980) Nimbus-7 coastal zone colour scanner: system description and initial imagery. *Science* (N.Y.) **210**, 60–63.

HOWELL J. F. (1953) *Gonyaulax monilata* sp. nov., the causative dinoflagellate of a red tide on the east coast of Florida in Aug./Sept. 1951. *Trans Amer. micr. Soc.* **72**, 153–156.

HSIAO S. I. C., FOY M. G. & KITTLE D. W. (1977). Standing stock, community structure, species composition, distribution, and primary production of natural populations of phytoplankton in the southern Beaufort Sea. *Can. J. Bot.* **55**, 685–694.

HULBURT E. M. (1968) Phytoplankton observations western Caribbean Sea. *Bull. mar. Sci.* **18**, 387–399.

HULBERT E. M. (1976) Limitation of phytoplankton species in the ocean off western Africa. *Limnol. Oceanogr.* **21**, 193–211.

HUNTER J. R. (1976) Culture and growth of northern anchovy, *Engraulis mordax*, larvae. *Fish. Bull. US* **74**, 81–88.

HUNTLEY M. E. (1982) Yellow water in the La Jolla Bay, California, July 1980. II. Suppression of zooplankton grazing. *J. exp. mar. Biol. Ecol.* **63**, 81–91.

HUNTSMAN S. A., BRINK K. H., BARBER R. T. & BLASCO D. (1981). The role of circulation and stability in controlling the relative abundance of dinoflagellates and diatoms over the Peru Shelf. In *Coastal Upwelling; Coastal and Estuarine Sciences,* 1. (ed. F. A. Richards), pp. 357–365. Geophysical Monography Bd., Washington, D.C.

HUNTSMAN S. A. & SUNDA W. G. (1980) The role of trace metals in regulating phytoplankton growth, with emphasis on Fe, Mn and Cu. In *The Physiological Ecology of Phytoplankton* (ed. I. Morris), pp. 285–328, Blackwell Scientific Press, Oxford.

IIZUKA S. (1972) *Gymnodinium* type '65 red tide in occurring anoxic environment of Omura Bay. *Bull. plankt. Soc. Japan* **19**, 22–23.

IIZUKA S. (1976) Succession of red tide organisms in Omura Bay, with relation to water pollution. *Bull. plankt. Soc. Jap.* **23**, 31–44.

IIZUKA S. & IRIE H. (1969) Anoxic status of bottom waters and occurrences of *Gymnodinium* red water in Omura Bay. *Bull. plankt. Soc. Japan* **16**, 99–115.

IIZUKA S. & NAKASHIMA T. (1975) Response of red tide organisms to sulphide. *Bull. plankt. Soc. Jap.* **22**(1–2), 27–32.

INCZE L. S. & YENTSCH L. M. (1981) Stable density fronts and dinoflagellate patches in a tidal estuary. *Estuar, coast shelf Sci.* **13**, 547–556.

INGLE R. M. & MARTIN D. F. (1971) Prediction of the Florida red tide by means of the iron index. *Environ. Lett.* **1**, 69–74.

IWASAKI H. (1969) Studies on the red tide dinoflagellates. III. On *Peridinium hangoei* Schiller appeared in Gokasho Bay, Shima Peninsula. *Bull. plankt. Soc. Japan* **16**, 132–139.

IWASAKI H. (1979) Physiological ecology of red tide flagellates. In *Biochemistry and Physiology of Protozoa,* 2nd Edn. Vol. 1 (eds M. Levandowsky & S. H. Hutner), pp. 356–393. Academic Press, New York.

JACQUES G. (1968) Aspects quantitatifs du phytoplancton de Banyuls-sur-Mer (Golfe du Lion). III. Diatomees et dinoflagèlles de Juin 1965 à Juin 1968. *Vie Milieu* (B)**20**, 91–126.

JACQUES G., DESCOLAS-GROS C., GRALL J-R. & SOURNIA A. (1979) Distribution du phytoplankton dans la partie Antarctique de l'Ocean Indièn en fin d'ete. *Int. Rev. ges. Hydrobiol.* **64**, 609–628.

JACQUES G. & SOURNIA A. (1978–79) Les 'Eaux Rouges' dues au phytoplancton en Méditerranée. *Vie Milieu* **28–29**, 175–187.

JEANNE N. (1979) Effects of Lindane on division cell cycle and biosynthesis in two uni-cellular algae. *Can. J. Bot.* **57**, 1464–1472.

JEFFREY S. W. (1980) Algal pigment systems. In *Primary Productivity in the Sea* (ed. P. G. Falkowski), pp. 33–58. Plenum Press, New York.

JEFFREY S. W. (1981) Phytoplankton ecology with particular reference to the Australasian region. In *Marine Botany: an Australasian Perspective* (eds M. N. Clayton & R. J. King), pp. 241–291. Longman Cheshire, Australia.

JEFFREY S. W. & HALLEGRAEFF G. M. (1980) Phytoplankton species and photosynthetic pigments in a warm core eddy of the East Australian Current 1. Summer Populations. *Mar. Ecol., Prog. Ser.* 3(4), 285–294.

JENSEN A. C. (1975) The economic halo of red tide. In *Proceedings of the First International Conference on Toxic Dinoflagellate Blooms* (ed. V. R. LoCicero), pp. 507–516. Mass. Sci. Technol. Found., Wakefield, Mass., USA.

JITTS H. R., MCALLISTER C. D., STEPHENS K. & STRICKLAND J. D. H. (1964) The cell division rates of some marine phytoplankters as a function of light and temperature. *J. fish. Res. Bd. Canada* **21**, 139–157.

KAIN J. M. & FOGG G. E. (1960) Studies on the growth of marine phytoplankton. III. *Prorocentrum micans* Ehrenberg. *J. mar. biol. Ass. UK* **39**, 33–50.

KAMYKOWSKI D. (1979) The growth response of a model *Gymnodinium splendens* in stationary and wavy water columns. *Mar. Biol.* **50**, 289–303.

KAMYKOWSKI D. (1981) Simulation of a southern California USA red tide using characteristics of a simultaneously measured internal wave field. *Ecological Modelling* **12**, 253–266.

KARENTZ D. (1983) Patterns of DNA synthesis and cell division in marine dinoflagellates *J. Protozool* **30**, 581–588.

KAROHJI K. (1972) Regional distribution of phytoplankton in the Bering Sea and western and northern subarctic regions of the North Pacific Ocean in summer. In *Biological Oceanography of the Northern North Pacific Ocean.* (ed. A/ Y. Takenouti) pp. 99–115. Idemitsu Shoten, Japan.

KARSTEN G. (1907) Das indische Phytoplankton nach dem Material der deutschen Tiefsee-Expedition 1898–1899. *Wiss. Ergebn. dt. Tiefsee-Exped. Valdivia* **2**(2), 221–548.

KAYSER H. (1969) Zuchtungs experimente an zwei marine Flagellaten (Dinophyta) und ihre Anwendung im toxikologischen Abwassertest. *Helgol. wiss. Meeresunters.* **19**, 21–44.

KAYSER H. (1976) Waste-water assay with continuous algal cultures: the effect of mecuric acetate on the growth of some marine dinoflagellates. *Mar. Biol.* **36**, 61–72.

KAYSER H. (1979) Growth interactions between marine dinoflagellates in multispecies culture experiments. *Mar. Biol.* **52**, 357–369.

KEARNS L. P. & SIGEE D. C. (1979) High levels of transition metals in dinoflagellate chromosomes. *Experientia* **35**, 1332–1334.

KHEMELEVA N. N. (1967) Rol'radiolyarii pri otsenke pervichnoi produktsii v Krasnom More i Adenskom zalive. *Dokl. Adad. Nauk SSSR* **172**, 1430–1433.

KIEFER D. A. & LASKER R. (1975) Two blooms of *Gymnodinium splendens,* an unarmored dinoflagellate. *Fish. Bull. US* **73**, 675–678.

KIERSTEAD H. & SLOBODKIN L. B. (1953) The size of water masses containing plankton blooms. *J. mar. Res.* **12**, 144–147.

KIMOR B. (1972) The Suez Canal as a link and a barrier in the migration of planktonic organisms. *Proc. XVII Congr. Int. Zool. 3,* (Les Consequences biologiques des canuax inter-oceans) 1–17.

KIMOR B. (1979) Predation of *Noctiluca miliaris* Suriray on *Acartia tonsa* Dana eggs in the inshore waters of southern California. *Limnol. Oceanogr.* **24**, 568–572.

KIMOR B. (1981) The role of phagotrophic dinoflagellates in marine ecosystems. *Kieler Meeresforsch. Sonderh.* **5**, 164–173.

KIMOR B. & GOLANDSKY B. (1977) Microplankton of the Gulf of Elat: aspects of seasonal and bathymetric distribution. *Mar. Biol.* **42**, 55–67.

KIMOR B. & WOOD E. J. F. (1975) A plankton study in the eastern Mediterranean Sea. *Mar. Biol.* **29**, 321–333.

KISSELEV N. A. (1950) Armoured Dinoflagellates of the seas of the SSSR. 1. Introd. to Genera. *Akad. Nauk SSSR* **33**, 1–279.

KLEIN BRETELER W. D. M. (1980) Continuous breeding of marine pelagic copepods in the presence of the heterotrophic dinoflagellates. *Mar. Ecol., prog. Ser.* **2**(3), 229–234.

KLUT M. E., BISALPUTRA T. & ANTIA N. J. (1981) Abnormal ultrastructural features of a marine dinoflagellate adapted to grow successfully in the presence of inhibitory fluoride concentration. *J. Protozool.* **28**, 406–414.

KOBORI H. & TAGA N. (1979) Occurrence and distribution of phosphatase in neritic and oceanic sediments. *Deep-Sea Res.* **26**A, 799–807.

KREMER B. P. (1981) Dark reactions of phytosynthesis. *Can. Bull. Fish. aquat. Sci.* **210**, 44–54.

KRZANOWSKI J., ASAI S., ANDERSON W., MARTIN D., LOCKEY R., BUKANTZ S. & SZENTIANYI A. (1981) Contractile effects of *Ptychodiscus brevis* toxin on canine airway smooth muscle. *Fed. Proceed. -Fed. Amer. Soc. exper. Biol.* **40**(3(1)), 721.

KUENZLER E. J. & PERRAS J. P. (1965) Phosphatase of marine algae. *Biol. Bull.* **128**, 271–284.

KUTT E. C. & MARTIN D. F. (1974) Effect of selected surfactants on the growth characteristics of *Gymnodinium breve. Mar. Biol.* **28**, 253–259.

LACKEY J. B. & CLENDENNING K. A. (1963) A possible fish-killing yellow tide in California waters. *Quart. J. Fla. Acad. Sci.* **26**, 263–268.

LARSEN J. (1984) Algal studies from the Wadden sea I. Dinoflagellates. *Nord. J. Bot.* (in press).

LASKER R. (1975) Field criteria for survival of anchovy larvae: the relation between inshore chlorophyll maximum layers and successful first feeding. *Fish. Bull. US* **73**, 453–462.

LASKER R. (1978) The relation between oceanographic conditions and larval anchovy food in the California Current: identification of factors contributing to recruitment failure. *Rapp. P.-v. Cons. int. Explor. Mer* **173**, 212–230.

LEE J. J., MCENERNY M. E. & GARRISON J. R. (1980) Experimental studies of larger foraminifera and their symbionts from the Gulf of Elaf on the Red Sea. *J. Foram. Res.* **10**, 31–47.

LEE J. J., MCENERNY M. E., KAHN E. G. & SCHUSTER F. L. (1979) Symbiosis and the evolution of larger foraminifera. *Micropal.* **25**, 118–140.

LEFÈVRE J. (1979) On the hypothesis of a relationship between, dinoflagellate blooms and the 'Amoco-Cadiz' oil spill. *J. mar. biol. Ass. UK* **59**, 524–528.

LEFÈVRE J. & GRALL J. R. (1970) On the relationships of *Noctiluca* swarming off the western coast of Brittany with hydrological features and plankton characteristics of the environment. *J. exp. mar. Biol. Ecol.* **4**, 287–306.

LÉGER G. (1972) Les populations phytoplanctoniques au point $\varphi = 42°47'$N, G $= 7°29'$E Greenwich. *Bull. Inst. océanogr. Monaco.* **70**(1417A), 1–56.

LENZ J. (1971) Phytoplankton standing stock and primary production in the Western Baltic. *Kieler Meeresforsch. Sonderh.* **5**, 29–40.

LESSARD E. J. & SWIFT E. (1985) Species-specific grazing rates of heterotrophic dinoflagellates in oceanic waters, measured with a dual-label radioisotope technique. *Mar. Biol.* **75**, 289–296.

LEUTENEGGER S. (1977) Symbiosis between larger foraminifera and the unicellular algae in the Gulf of Elat. *Utrecht Micropaleont. Bull.* **15**, 225–239.

LEVANDOWSKY M. (1979). On a class of mathematical models for *Gymnodinium breve* red tides. In *Biochemistry and Physiology of Protozoa*, 2nd Edn, Vol. 1. (eds M. Levandowsky & S. H. Hutner), pp. 394–402. Academic Press, New York.

LEWIN J. & BUSBY W. F. (1967) The sulfate requirements of some unicellular marine algae. *Phycologia* **6**, 211–217.

LILLICK L. C. (1937) Seasonal studies of the phytoplankton off Woods Hole, Massachusetts. *Biol. Bull.* (*Woods Hole*) **73**, 488–503.

LINDSTROM K. & RODHE W. (1978) Selenium as a micronutrient for the dinoflagellate *Peridinium cinctum* fa. *westii. Mitt. internat. Verein Limnol.* **21**, 168–173.

LO CICERO V. R. (Ed.) (1975) *Proceedings of the First International Conference on Toxic Dinoflagellate Blooms.* Mass. Sci. Tech. Found. Wakefield, Mass. pp. 1–541.

LOEBLICH A. R. (III) (1966, appeared 1967) Aspects of the physiology and biochemistry of the Pyrrhophyta. *Phykos* **5**, 216–255.

LOEBLICH A. R. III (1975) A sea water medium for dinoflagellates and the nutrition of

Cachonina niei. *J. Phycol.* **11**, 80–86.
LOHMAN H. (1912) Untersuchungen uber das Pflanzen und Tier-Leben der Hochsee im Atlantischen Ozean wahrend der Ausreise der 'Deutschland'. *Sitzber. Ges. naturforsch. Freunde,* Berlin. pp. 32.
LOMBARD E. H. & CAPON B. (1971) Observations on the tidepool ecology and behaviour of *Peridinium gregarium*. *J. Phycol.* **7**, 184–187.
MCCARTHY J. J. (1980) Nitrogen. In *The Physiological Ecology of Phytoplankton* (ed. I. Morris), pp. 191–233. Blackwell Scientific Publ., Oxford.
MCCARTHY J. J., TAYLOR W. R. & TAFT J. L. (1977) Nitrogenous nutrition of the plankton in Chesapeake Bay. I. Nutrient availability and phytoplankton preferences. *Limnol. Oceanogr.* **22**, 996–1011.
MCGOWAN J. A. (1971) Oceanic biogeography of the Pacific. In *The Micropaleontology of Oceans* (eds B. M. Funnel & W. R. Riedel), pp. 3–74. Cambridge University Press, Cambridge.
MACISAAC J. J. (1978) Diel cycles of inorganic nitrogen uptake in a natural phytoplankton population dominated by *Gonyaulax polyedra*. *Limnol. Oceanogr.* **23**, 1–7.
MACISAAC J. J., GRUNSEICH G. S., GLOVER H. E. & YENTSCH C. M. (1979) Light and nutrient limitation in *Gonyaulax excavata*: Nitrogen and carbon trace results. In *Toxic Dinoflagellate Blooms* (eds D. L. Taylor & H. H. Seliger), pp. 107–110. Elsevier/North Holland, New York.
MCKNIGHT D. M. & MOREL F. M. M. (1978) Release of weak and strong copper-complexing agents by algae. *Limnol. Oceanogr.* **24**, 823–836.
MCLAUGHLIN J., CHEN L. C.-M. & EDELSTEIN T. (1971) The culture of four species of *Fucus* under laboratory conditions. *Can. J. Bot.* **49**, 1463–1469.
MACLEAN J. L. (1979) Indo-Pacific red tides. In *Toxic Dinoflagellate Blooms* (eds D. L. Taylor & H. H. Seliger), pp. 173–178. Elsevier/North Holland, New York.
MAGNANI B., POWERS C. D., WURSTER C. F. & O'CONNORS H. B. JR (1978) Effects of chlordane and heptachlor on the marine dinoflagellate *Exuvialla baltica*. *Bull. environ. Contam. Toxicol.* **20**, 1–8.
MAHONEY J. B. & MCLAUGHLIN J. J. A. (1977) The association of phytoflagellate blooms in lower New York bay with hypertrophication. *J. exp. mar. Biol. Ecol.* **28**, 53–65.
MAHONEY J. B. & STEIMLE F. W. JR (1979) A mass mortality of marine animals associated with a bloom of *Ceratium tripos* in the New York Bight. In *Toxic Dinoflagellate Blooms* (eds D. L. Taylor & H. H. Seliger), pp. 225–230. Elsevier/North Holland, New York.
MANDELLI E. F. (1969) The inhibitory effects of copper on marine phytoplankton. *Contrib. mar. Sci., Univ. Texas* **24**, 47–57.
MARGALEF R. (1958) Temporal succession and spatial heterogeneity in phytoplankton. In *Perspectives in Marine Biology* (ed. A. A. Buzzati-Traverso), pp. 323–349. Univ. Calif. Press, Calif., USA.
MARGALEF R. (1961a) Distribution ecologica y geografica de las especies del fitoplankton marino. *Inv. Pesq.* **19**, 81–101.
MARGALEF R. (1961b) Hidrografia y fitoplancton de un area marina de la costa meridional de Puerto Rico. *Inv. Pesq.* **18**, 33–96.
MARGALEF R. (1969a) Comunidades planctonicas en lagunas litorales. Lagunas Costeras, un simposio. *Mem. Simp. Intern. Lagunas Costeras UNAM-UNESCO Nov. 28–30, Mexico DF,* 545–562.
MARGALEF R. (1969b) Composicion especifica del fitoplancton de la costa catalano-levantina (Mediterraneo occidental) in 1962–1967. *Inv. Pesq.* **33**, 345–380.
MARGALEF R., (1978) Phytoplankton communities in upwelling areas: the example of NW Africa. *Oecologia Aquatica* **3**, 97–132.
MARGALEF R. (1978) Life-forms of phytoplankton as survival alternatives in an unstable environment. *Oceanologia Acta* **1**(4) 493–509.
MARGALEF R., ESTRADA M. & BLASCO D. (1979) Fuctional morphology of organisms involved in red tides as adapted to decaying turbulence. In *Toxic Dinoflagellate Blooms* (eds D. L.

Taylor & H. H. Seliger), pp. 89–94. Elsevier/North Holland, Amsterdam.

MARSHALL G. H. (1976) Phytoplankton distribution along the eastern coast of the USA. Part I. Phytoplankton Composition. Mar. Biol. **38**, 81–89.

MARSHALL H. G. (1978) Phytoplankton distribution along the eastern coast of the USA. Part II. Seasonal assemblages north of Cape Hatteras, N. Carolina. *Mar. Biol.* **45**, 203–208.

MARTIN D. F. (1974) A comment on the conclusions of Yentsch and co-workers on the biostimulatory effect on nitrilotriacetic acid on growth and photosynthetic rate of the red tide dinoflagellate *Gonyaulax tamarensis*. *Environ. Lett.* **7**, 175–177.

MARVIN K. T. & PROCTOR R. R. JR (1966) Laboratory evaluation of red tide controls agents. *Fish. Bull. Wildl. Serv. US*, **66**, 163–164.

MATTHYS-ROCHON E. (1980) Effects of isopropyl N Phenyl carbamate on the growth and dino mitosis in a free living dinoflagellate *Amphidinium carterae*. *Cryptog. Algol.* **1**, 3–18.

MEUNIER A. (1910) *Duc d'orleans Campagne Arctique de 1907. Microplankton des mers de Barents et de Kara*, pp. 1–355. Charles Bulens, Bruxelles.

MILLER T. G. (1973) Studies on the effects of mercury on the marine dinoflagellate *Amphidinium carteri*. *Bull. S. Carolina Acad. Sci.* **35**, 131.

MOREL F. M. M. & ANDERSON D. M. (1976) On the subject of red tide predictions from temperature patterns. *Limnol. Oceanogr.* **21**, 625–627.

MORRILL L. C. & LOEBLICH A. R. III (1979) An investigation of heterotrophic and photoheterotrophic capabilities in marine Pyrrhophyta. *Phycologia* **18**, 394–404.

MORRIS I. (1980) Paths of carbon assimilation in marine phytoplankton. In *Primary Productivity in the Sea* (ed. P. G. Falkowski), pp. 139–161. Plenum Press, New York.

MORTON B. & TWENTYMAN P. R. (1971) The occurrence and toxicity of a red tide caused by *Noctiluca scintillans* (McCartney) Ehrenb., in the coastal waters off Hong Kong. *Environ. Res.* **4**, 544–557.

MOSHKINA L. V. (1961) Nekotorye dannye o fotosinteze chenomorskikh dinoflagellate. *Fiziologiya Rast.* **8**, 172–177.

MULLER-MERZ E. & LEE J. J. (1976) Symbiosis in the larger foraminiferan *Sorites marginalis* (with notes on *Archaias* spp.). *J. Protozool.* **23**, 390–396.

MULLIGAN H. F. (1973) Probable causes for the 1972 red tide in the Cape Ann region of the Gulf of Maine. *J. fish. Res. Bd. Canada* **30**, 1363–1366.

MUNK W. H. & RILEY G. A. (1952) Absorption of nutrients by aquatic plants. *J. mar. Res.* **11**, 215–240.

MURPHY E. B., STEIDINGER K. A., ROBERTS B. S., WILLIAMS J. & JOLLEY J. W. JR (1975) An explanation for the Florida coast *Gymnodinium breve* red tide of November 1972. *Florida Dept. nat. Res. mar. Res. lab.* **250**, 481–486.

NALEWAJKO C. & LEAN D. R. S. (1986) *Phosphorus*. In *The Physiological Ecology of Phytoplankton*. (ed I. Morris), pp. 235–257. University of California Press, Berkeley.

NELSON D. M. & BRAND L. E. (1979) Cell division periodicity in 13 species of marine phytoplankton on a light:dark cycle. *J. Phycol.* **15**, 65–75.

NESTEROFF D. A. (1979) The development of *Exuviaella cordata* Ostf. and the 'red tide' phenomenon in the north-western part of the Black Sea. *Biol. Morya* **5**, 24–29.

NICHOLLS K. H., KENNEDY W. & HAMMETT C. (1980) A fish kill in Heart Lake, Ontario, Canada associated with the collapse of a massive population of *Ceratium hirundinella*, Dinophyceae. *Freshwater Biol.* **10**, 553–562.

NIGHTINGALE H. W. (1936) *Red Water Organisms*. Angus Press, Seattle, Wash. 24 pp.

NISHIMURA A. (1982) Effects of organic matters produced in fish farms on the growth of red tide algae *Gymnodinium* type-65 and *Chattonella antiqua*. *Bull. plankt. Soc. Japan* **29**, 1–7.

NISHITANI L., ERICKSON G. & CHEN K. K. (1985) Role of the parasitic dinoflagellate *Amoebophyra ceratii* in control of *Gonyaulax catenella* populations. In *Toxic Dinoflagellates* (eds D. M. Anderson, A. W. White & D. G. Baden) pp. 225–230, Elsevier, New York.

NORDLI E. (1957) Experimental studies on the ecology of Ceratia *Oikos* **8**, 200–265.

NORRIS L. & CHEW K. K. (1975) Effect of environmental factors on growth of *Gonyaulax catenella*. In *Proc. 1st Int. Conf. Toxic Dinoflagellate Blooms* Nov. 1974 (ed. V. R.

LoCicero), pp. 41–46. Mass. Sci. Tech. Found., Wakefield, Mass., USA.

Nunez Ortega D. A. (1879) Ensayo de una explicacion del origin de las grandes mortandades de peces que ocurren en el golfo de Mexico: *La Naturaleza* **4**, 188–197 (1877–79).

Oda M. (1935) *Gymnodinium mikimotoi* Miyake et Kominami n. sp. (MS) no akashio to ryusando no kokwa (Red tide of *G. mikimotoi* M. & K. and the effect of copper sulphate on it: In Japanese). *Dobutsugaku Zasshi, zool. Soc. Japan* **47**, 35–48.

Okaichi T. (1975) Organic pollution and the outbreaks of red tides in Seto Inland Sea. *Science for Better Environment* Proc. int. Congr. on the Human Environ., Kyoto 1975, 455–460.

Orlova T. A., Voznya G. I. & Zlobin V. S. (1979) Metabolites in marine bacteria—and phytoplankton cultures. *Ecologiya* (Ekologiya) **4**, 45–50.

Ouchi A. (1982) Prediction of red tide occurrence by means of Multiple Linear Regression Model. *Bull. Jap. Soc. sci. fish.* **48**, 1245–1250.

Owen R. W. (1981) Fronts and eddies in the sea: mechanisms, interactions and biological effects. In *Analysis of Marine Ecosystems* (ed. A. R. Longhurst), pp. 197–233. Academic Press, London.

Paasche E. & Kristiansen S. (1982). Nitrogen nutrition of the phytoplankton in the Oslofjord. *Estuar. coast. shelf Sci.* **14**, 237–249.

Packard T. T. & Blasco D. (1974) Nitrate reductase activity in upwelling regions. 2. Ammonia and light dependence. *Tethys* **6**, (1–2), 269–280.

Parsons T. R., Albright L. J., Whitney F., Wong C. S. & Williams P. J. & LeBlond P. L.(1981) The effect of glucose on the productivity of seawater: an experimental approach using controlled aquatic ecosystems. *Mar. environ. Res.* **4**, 229–242.

Parsons T. R. & Harrison P. J. (1983) Nutrient cycling in marine ecosystems. In *Physiological Plant Ecology IV. Ecosystem Processes: Mineral Cycling, Productivity and Pollution* (eds B. Richards & J. L. Charley), pp. 85–115. Springer-Verlag, Heidelberg.

Parsons T. R., Harrison P. J. & Waters R. (1978) An experimental simulation of changes in diatom and flagellate blooms. *J. exp. mar. Biol. Ecol.* **32**, 285–294.

Parsons T. R., Stephens K. & Strickland J. D. H. (1961) On the chemical composition of eleven species of marine phytoplankton. *J. fish. Res. Bd. Canada* **18**, 1001–1016.

Pasciak W. J. & Gavis J. (1974) Transport limitation of nutrient uptake in phytoplankton. *Limnol. Oceanogr.* **19**, 881–888.

Paster Z. & Abbott B. C. (1970) Gibberellic acid: a growth factor in the unicellular alga *Gymnodinium breve Science* (N.Y.) **169**, 606–601.

Patten B. C., Mulford R. A. & Warinner J. E. (1963) An annual phytoplankton cycle in the lower Chesapeake Bay. *Chesapeake Sci.* **4**, 1–20.

Patton S., Chandler P. T., Kalan E. B., Loeblich A. R. III, Fuller G. & Benson A. A. (1967) Food value of red tide (*Gonyaulax polyedra*). *Science* (N. Y.) **158**, (3802), 789–790.

Pavillard J. (1937) Les péridiniens et diatomées pélagiques de la mer de Monaco de 1907 a 1913. Observations générales et conclusions. *Bull. Inst. océanogr. Monaco* **738**, 1–56.

Perry H. M., Stuck K. C. & Howse H. D. (1979) First record of a bloom of *Gonyaulax monilata* in coastal waters in Mississippi, USA. *Gulf Res. Rep.* **6**, 313–316.

Peters N. (1932) Die Bevolkerung des Sudatlantischen Ozeans mit Ceratien. *Wiss. Ergebn. Deutsh. Atlant. Exped. 'Meteor' 1925–1927.* **12**, 1–69.

Pichkily L. O. (1970) Dynamics of the abundance and biomass of phytoplankton in the Aral Sea. *Hydrobiol. J.* **6**, 22–26.

Pieterse F. & Van der Post D. C. (1967) The pilchard of South West Africa. Oceanographical conditions associated with red-tides and fish mortalities in the Walvis Bay region. *Admin. S. W. Africa mar. Res. Lab. invest. Report.* **14**, 1–125.

Pincemin J. M. (1972) Influence of salinity on the dinoflagellate *Glenodinium monotis. Rev. int. océanogr. Médicale* **25**, 71–87.

Pingree R. D. (1978) Mixing and stabilization of phytoplankton distributions on the northwest European continental shelf. In *Spatial pattern in Plankton Communities* (ed. J. H. Steele), pp. 1–470. Plenum Press, New York.

PINTNER I. J. & ALTMYER V. L. (1979) Vitamin B_{12} binder and other algal inhibitors. *J. Phycol.* **15**, 391–398.

POLLARD R. T. (1977) Observations and models of the structure of the upper ocean. In *Modelling and Prediction of the upper layers of the ocean* (ed. E. B. Kraus), pp. 102–117. Pergamon Press, Oxford.

POLLINGHER U. & ZEMEL E. (1981) *In situ* and experimental evidence of the influence of turbulence on cell division processes of *Peridinium cinctum* fa. *westii* (Lemm.) LeFèvre. *Br. Phycol. J.* **16**, 281–287.

POWERS C. D., ROWLAND R. G., MICHAELS R. R., FISHER N. S. & WURSTER C. F. (1975) The toxicity of DDE to a marine dinoflagellate. *Environ. Pollut.* **9**, 253–262.

PRAGER J. C. (1963) Fusion of the family Glenodiniaceae into the Peridiniaceae with notes on *Glenodinium foliaceum* Stein. *J. Protozool.* **10**, 195–204.

PRAKASH A. (1963) Source of paralytic shellfish poisoning research in Canada. In *Proceedings of the shellfish sanitation workshop* (*Nov. 1961*), pp. 248–251. App. V. US Dept. Health Educ. Welfare Public Health Service.

PRAKASH A. (1967) Growth and toxicity of a marine dinoflagellate, *Gonyaulax tamarensis*. *J. fish. Res. Bd. Canada* **24**, 1589–1606.

PRAKASH A., MEDCOF J. C. & TENNANT A. D. (1971) Paralytic shellfish poisoning in eastern Canada. *Bull. fish. Res. Bd. Canada* **177**, 1–88.

PRAKASH A. & RASHID M. A. (1968) Influence of humic substances on the growth of marine phytoplankton: Dinoflagellates. *Limnol. Oceanogr.* **13**, 598–606.

PRAKASH A. & TAYLOR F. J. R. (1966) A 'red water' bloom of *Gonyaulax acatenella* in the Strait of Georgia and its relation to paralytic shellfish toxicity. *J. fish. Res. Bd. Canada* **23**, 1265–1270.

PREVOT P. & SOYER M.-O. (1979) Action du cadmium sur un Dinoflagèlle libre: *Prorocentrum micans E.*: croissance, absorption due cadmium et modifications cellulaires. *C. R. Acad. Sci. Paris* **287**, 833–836.

PROVASOLI L. (1979) Recent progress, an overview. In *Toxic Dinoflagellate Blooms; Proc II Int. Conf. on Toxic Dinoflagellate Blooms* (eds D. L. Taylor & H. H. Seliger), pp. 1–14. Elsevier/North Holland, New York.

PROVASOLI L. & CARLUCCI A. F. (1973) Vitamins and growth regulators. In *Algal Physiology and Biochemistry*. Botanical Monographs Vol. 10 (ed. W. D. P. Stewart), pp. 741–787. Blackwell Scientific Publ., Oxford.

PROVASOLI L. & GOLD K. (1962) Nutrition of the American strain of *Gyrodinium cohnii*. *Arch. Mikrobiol.* **42**, 196–203.

PROVASOLI L. & MCLAUGHLIN J. J. A. (1963) Limited heterotrophy of some photosynthetic dinoflagellates. In *Symposium on Marine Microbiology* (ed. C. H. Oppenheimer), pp. 105–113. C. C. Thomas, Springfield, Illinios.

PROVASOLI L. & PRINTER I. J. (1960) Artificial media for freshwater algae: problems and suggestions. In *The Ecology of Algae* (Pymatuning Symposia in Ecology, Special Publ. No. 2) (eds C. A. Tryon Jr & R. T. Hartman), pp. 84–96. Univ. Pittsburg Press, Penn., USA.

PYBUS C. (1980) Observations on a *Gyrodinium aureolum* (Dinophyta) bloom off the south coast of Ireland. *J. mar. biol. Ass. UK* **60**, 661–674.

QASIM S. Z. (1973) Productivity of back waters and estuaries. In *The Biology of the Indian Ocean* (ed. B. Zeitzchel), pp. 1–549. Springer-Verlag, New York.

QASIM S. Z. & BHATTATHIRI P. M. A. (1972) The influence of salinity on the rate of photosynthesis and abundance of some tropical phytoplankton. *Mar. Biol.* **12**, 200–206.

QUAYLE D. B. (1969) Paralytic shellfish poisoning in British Columbia. *Fish. Res. Bd. Can. Supp.* **168**, 1–68.

RAGOTZKIE R. A. & POMEROY L. R. (1957) Life history of a dinoflagellate bloom. *Limnol. Oceanogr.* **2**, 62–69.

RAVEN J. A. & BEARDALL J. (1981) Respiration and photorespiration. *Can. Bull. Fish. aquat. Sci.* **210**, 55–82.

RAVEN J. A. & RICHARDSON K. (1984) Dinoflagellate flagella: a cost-benefit analysis. *New Phytol.* **98**, 259–276.

REDFIELD A. C., KETCHUM F. A. & RICHARDS F. A. (1963) The influence of organisms on the composition of seawater. In *The Sea* (ed. M. N. Hill), Vol. 2, pp. 26–77. Interscience Publ., New York.

REID P. C. (1972) Dinoflagellate cyst distribution around the British Isles. *J. mar. biol. Ass. UK* **52**, 939–944.

REID P. C. (1975) Regional sub-division of dinoflagellate cysts around the British Isles. *New Phytol.* **75**, 589–603.

REID P. C. (1977) Continuous plankton records: changes in the composition and abundance of the phytoplankton of the North-Eastern Atlantic Ocean and North Sea, 1958–1974. *Mar. Biol.* **40**, 337–339.

REID P. C. (1980) Toxic dinoflagellates and tidal power generation in the Bay of Fundy, Canada. *Mar. Poll. Bull.* **11**, 47–51.

REID F. M. H., FUGLISTER E. & JORDAN J. B. (1970) Phytoplankton taxonomy and standing crop. V. The ecology of the plankton off La Jolla, California, in the period April through September, 1967. *Bull. Scripps. Inst. Oceanogr.* **17**, 51–66.

REVELANTE N., WILLIAMS W. T. & BUNT J. S. (1982) Temporal and spatial distribution of diatoms, dinoflagellates and *Trichodesmium* in waters of the Great Barrier Reef. *J. exp. mar. Biol. Ecol.* **63**, 27–45.

RICHARDSON K. & FOGG G. E. (1982) The role of dissolved organic material in the nutrition and survival of marine dinoflagellates. *Phycologia* **21**, 17–26.

RILEY G. A. (1957) Phytoplankton of the North Central Sargasso Sea, 1950-1952. *Limnol. Oceanogr.* **2**(3), 252–270.

RIVKIN R. B. & SWIFT E. (1979) Diel and vertical patterns of alkaline phosphatase activity in the oceanic dinoflagellate *Pyrocystis noctiluca*. *Limnol. Oceanogr.* **27**, 107–116.

RIVKIN R. B. & SWIFT E. (1982) Phosphate uptake by the oceanic dinoflagellate *Pyrocystis noctiluca*. *J. Phycol.* **18**(1), 113–120.

RIVKIN R. B., VOYTEK M. A. & SELIGER H. H. (1982) Phytoplankton division rates in light-emitted environments—two adaptions. *Science* (N.Y.) **215**, 1123–1125.

ROJAS DE MENDIOLA B. (1979) Red tide along the Peruvian Coast. In *Toxic Dinoflagellate Blooms: Proc. II Int. Conf. Toxic Dinoflagellate Blooms.* (eds D. L. Taylor & H. H. Seliger), pp. 183–194. Elsevier/North Holland, New York.

ROJAS DE MENDIOLA B. (1981) Seasonal phytoplankton distribution along the Peruvian coast. In *Coastal Upwelling: Coastal and Estuarine Sciences* 1 (ed. F. A. Richards), pp. 348–356. Geophysical Monograph Bd., Washington, D.C.

ROUND R. E. (1981) *The Ecology of Algae.* Cambridge University Press, Cambridge.

ROUNSEFELL G. & NELSON W. R. (1966) Red tide research summarized to 1964 including annotated bibliography. *Sci. Rept. US Fish. Wildlife Serv.*, **535**, 1–85.

RYTHER J. H. (1955) Ecology of autotrophic marine dinoflagellates with reference to red water conditions. In *The Luminescence of Biological Systems* (ed. F. H. Johnson), pp. 387–413. Am. Assoc. Adv. Sci, New York.

RYTHER J. H. (1956) Photosynthesis in the ocean as a function of light intensity. *Limnol. Oceanogr.* 1, 61–70.

RYTHER J. H. & KRAMER D. D. (1961) Relative iron requirement of some coastal and offshore plankton algae. *Ecology* **42**, 444–446.

RYTHER J. G. & SANDERS J. G. (1980) Experimental evidence of zooplankton control of species composition and size distribution of marine phytoplankton. *Mar. Ecol., Prog. Ser.* **3**, 279–283.

SAKS N. M. (1981) Growth productivity and excretion of *Chlorella* spp. endosymbionts from the Red Sea: Implications for host foraminifera. *Bot. Mar.* **24**, 445–449.

SAKSHAUG E., ANDRESEN K., MYKLESTAD S. & OLSEN Y. (1983) Nutrient status of phytoplankton communities in Norwegian waters (marine, brackish and fresh) as revealed by their chemical composition. *J. plankt. Res.* **5**, 175–196.

SAIFULLAH S. M. (1978) Inhibitory effects of copper on marine dinoflagellates. *Mar. Biol.* **44**, 299–308.

SANDER F. & MOORE E. (1979) Significance of ammonia in determining the N:P ratio of the sea water off Barbados, West Indies. *Mar. Biol.* **55**, 17–21.

SANDERS J. G. & WINDOM H. L. (1980) The uptake and reduction of arsenic species by marine algae. *Estuar. coast. mar. Sci.* **10**, 555–568.

SAWARD D., STIRLING A. & TOPPING G. (1975) Experimental studies on the effects of copper on a marine food chain. *Mar. Biol.* **29**, 351–361.

SCHENK R. C. (1984) Copper and *Gonyaulax tamarensis*: physiology, toxicity and requirement. *Diss. Abstr. Int. P.T.B.—Sci. & Engin.* **45**, 1–176.

SCHILLER J. (1954) Uber winterliche pfanzliche Bewohner des Wassers, Eises und des daraufliegenden Schneebreies. *Oesterr. bot. Z.* **101**, 236–284.

SCHMALJOHANN R. & ROTTGER R. (1976) Die symbionten der Grossforaminifere *Heterostegina depressa* sind Diatomeen. *Naturwiss.* **10**, 486.

SCHNAK S. B. & ELBRÄCHTER M. (1981) On the food of calanoid copepods from the Northwest African upwelling region. In *Coastal Upwelling:* Coastal and Estuarine Sciences 1 (ed. F. A. Richards), pp. 433–439. Geophysical Monograph Bd., Washington, D.C.

SCHRADIE J. & BLISS C. A. (1962) The cultivation of *Gonyaulax polyedra*. *Lloydia* **25**, 214–221.

SCURA E. D. & JERDE C. W. (1977) Various species of phytoplankton as food for larval northern achovy, *Engraulis mordax,* and relative nutritional value of the dinoflagellates *Gymnodinium spendens* and *Gonyaulax polyedra*. *Fish. Bull. US* **75**, 577–583.

SEKI H., TSUJI T. & HATTORI A. (1974) Effect of zooplankton grazing on the formation of the anoxic layer in Tokyo Bay. *Estuar. coast. mar. Sci.* **2**, 145–151.

SELIGER H. H., CARPENTER J. H., LOFTUS M., BIGGLEY W. H. & MCELROY W. D. (1971). Bioluminescence and phytoplankton successions in Bahia Fosforescente, Puerto Rico. *Limnol. Oceanogr.* **16**, 608–622.

SELIGER H. H., CARPENTER J. H., LOFTUS M. & MCELROY W. D. (1969) Mechanisms for the accumulation of high concentrations of dinoflagellates in a biolumnescent bay. *Limnol. Oceanogr.* **15**, 234–245.

SELIGER H. H., TYLER M. A. & MCKINLEY K. R. (1979) Phytoplankton distributions and red tides resulting from frontal circulation patterns. In *Toxic Dinoflagellate Blooms; Proc. II. Int. Conf. on Toxic Dinoflagellate Blooms* (eds D. L. Taylor & H. H. Seliger), pp. 239–248. Elsevier/North Holland, New York.

SEMINA H. J. (1974) *Fitoplankton Tichogo Okeana* (Pacific Phytoplankton; in Russian), pp. 237. Nauka, Moscow.

SEMINA H. J. & TARKHOVA I. A. (1972) The ecology of phytoplankton in the North Pacific. In *Biological Oceanography of the Northern North Pacific Ocean* (ed. A. Y. Takenouti) pp. 117–124. Idetmitsu Shoten, Tokyo.

SERGEYEVA L. M. & KRUPATKINA-ATKINIA D. K. (1971) Growth of dinoflagellates in the absence of certain minerals. *Hydrobiol. J.* (English translation) **7**(5), 65–69.

SHARP J. H., PERRY M. J., RENGER E. H. & EPPLEY R. W. (1980) Phytoplankton rate processes in the oligotrophic waters of the central North Pacific Ocean. *J. plankt. Res.* **2**, 335–353.

SIEBURTH J. M., SMETACEK V. & LENZ J. (1978). Pelagic ecosystem structure: heterotrophic compartments of the plankton and their relationship to plankton size fractions. *Limnol. Oceanogr.* **23**, 1256–1263.

SILVA E. DE S. (1962) Some observations on marine dinoflagellate cultures. II. *Glenodinium foliaceum* Stein and *Gonyaulax* (sic) *diacantha* (Meunier) Schiller. *Notas Estudos* **24**, 75–100.

SILVA E. DE S. (1963) Les 'Red Waters' a la Lagune d'Obidos. Ses causes probables et ses rapports avec la toxicite des bivalves. *Notas Estud. Inst. Biol. Maritima* **27**, 265–275.

SIMMONS E. G. & THOMAS W. H. (1962) Phytoplankton of the Eastern Mississippi Delta. *Texas Univ. Inst. Mar. Sci. Publ.* **8**, 269–298.

SMAYDA T. J. (1958) Biogeographical studies of marine phytoplankton. *Oikos* **9**, 159–191.

SMAYDA T. J. (1958) Phytoplankton studies around Jan Mayen island. March-April 1955. *Nytt. Mag. Bot.* **6**, 75–96.

SMAYDA T. J. (1966) A quantitative analysis of the phytoplankton of the Gulf of Panama. *Inter-Amer. trop. Tuna Comm.* **2**, 355–612.

SMAYDA T. J. (1970) Growth potential bioassay of water masses using diatom cultures: phosphorescent Bay (Puerto Rico) and Caribbean waters. *Helgolander wiss. Meeresunters* **20**, 172–194.

SMAYDA T. J. (1980) Phytoplankton species succession. In *The Physiological Ecology of Phytoplankton*. Studies in Ecology, Vol. 7 (ed. I Morris), pp. 493–570. Blackwell Scientific Publ., Oxford.

SMAYDA T. J. (1983) The phytoplankton of estuaries. In *Estuaries and Enclosed Seas* (ed. B. H. Ketchum) pp. 65–102, Elsevier, Amsterdam.

SMETACEK V. (1975) *Die Sukzession des Phytoplanktons in der westlichen Kieler Bucht*. Ph.D. thesis, Kiel Univ.

SMETACEK V. (1981) The annual cycle of protozooplankton in the Kiel Bight of W. Germany. *Mar. Biol.* **63**, 1–12.

SMITH J. E. (1970) *'Torrey Canyon' Pollution and Marine Life*. Cambridge Univ. Press, Cambridge.

SMITH W. O. JR (1977) The respiration of photosynthetic carbon in eutrophic areas of the ocean. *J. mar. Res.* **35**, 557–565.

SOROKIN Y. I. (1980, transl.) The 'Red Tide' in the region of the Peruvian upwelling, *Doklady Akad. Nauk SSR* **249**(1), 253–256 (orig. Nov. 1979).

SOROKIN Y. I. & KOGELSCHATZ J. (1979) Analysis of heterotrophic microplankton in an upwelling area. *Hydrobiologia* **66**, 195–208.

SOURNIA A. (1968a). Le genre 'Ceratium' (Péridinien) planktonique dans le canal de Mozambique contribution a une révision mondiale. *Vie et Milieu Ser. A* (*Biol. Mar*). **18**, 375–500.

SOURNIA A. (1968b) Variations saisonnieres et nycthémérales du phytoplancton marin et de la production primaire dans une baie tropicale, à Nosy-Be (Madagascar). *Int. Rev. ges Hydrobiol.* **53**, 1–76.

SOURNIA A. (1969) Cycle annuel du phytoplankton et da la production primaire dans les mers tropicales. *Mar. Biol.* **3**, 287–303.

SOURNIA A. (1974) Circadian periodicities in natural populations of marine phytoplankton. *Adv. mar. Biol.* **12**, 325–389.

SOURNIA A. (1976) Primary production of sands in the lagoon of an atoll and the role of Foraminiferan symbionts. *Mar. Biol.* **37**, 29–32.

SOURNIA A. (1982) Is there a shade flora in the marine plankton? *J. plankt. Res.* **4**, 391–399.

SOURNIA A. & RICARD M. (1976) Données sur l'hydrologie et la productivite du lagon d'un atoll fermé (Takapoto, iles Tuamotu). *Vie Milieu* **26**(2), sér. B, 243–279.

SOYER M. O. & PREVOT P. (1981) Ultrastructural damage by cadmium in a marine dinoflagellate *Prorocentrum micans*. *J. Protozool.* **28**, 308–313.

STEEMANN NIELSEN E. (1934) Untersuchungen uber die Verbreitung, Biologie und Variation der Ceratien im sudlichen Stillen Ozean. *Dana Report* **4** (Carlsberg Found.), 1–66.

STEEMANN NIELSEN E. (1939) Uber die vertikale Verbreitung der Phytoplanktonten im Meere. *Int. Rev. ges. Hydrob. Hydrogr.* **38**, 421–440.

STEEMAN NIELSEN E. & WIUM-ANDERSON S. (1970) Copper ions as poison in the sea and in fresh water. *Mar. Biol.* **6**, 93–97.

STEIDINGER K. A. (1973) Phytoplankton ecology: a conceptual review based on eastern Gulf of Mexico research. CRC Rev. Microb. **3**(1), 49–68.

STEIDINGER K. A. (1975) Implications of dinoflagellate life cycles in initiation of *Gymnodinium breve* red tides. *Environ. Lett.* **9**(2), 129–136.

STEIDINGER K. A. (1983) A re-evaluation of toxic dinoflagellate biology and ecology. In *Progress in Phycological Research,* Vol. 2 (eds. F. E. Round & D. J. Chapman), pp 147–188. Elsevier Science Publ. N.Y.

STEIDINGER K. A. & HADDAD K. (1981) Biologic and hydrographic aspects of red tides. *BioSciences* **31**(11), 814–818.

STEIDINGER K. A. & JOYCE E. A. JR (1973) Florida red tides. *State Florida Dept. nat. Res. Educ. Series* **17**, 1–26.

STEIDINGER K. A. & WILLIAMS J. (1970) *Memoirs of the Hourglass Cruises.* Vol. II, Marine Res. Lab., Florida, pp. 1–251.

STOCKNER J. G. & ANTIA N. J. (1976) Phytoplankton adaptation to environmental stresses from toxicants, nutrients and pollutants—a warning. *J. fish. Res. Bd. Canada* **33**, 2089–2096.

STOCKNER J. G. & COSTELLA A. C. (1976) Marine phytoplankton growth in high concentrations of pulpmill effluent. *J. fish. Res. Bd. Canada* **33**, 2758–2765.

STOECKER D., GUILLARD R. R. L. & KAREE R. M. (1981) Selective predation by *Favella ehrenbergii* (Tintinnida) and on and among dinoflagellates. *Biol. Bull.* **160**, 136–145.

STRATHMAN R. R. (1967) Estimating the organic carbon content of phytoplankton from cell volume or plasma volume. *Limnol. Oceanogr.* **12**, 411–418.

SUBRAHMANYAN R. & SARMA A. H. V. (1970) Studies on the phytoplankton of the west coast of India. III. Seasonal variation of the phytoplankters and environmental factors. *Indian J. Fish.* **7**, 307–336.

SUKHANOVA I. N. (1962) The tropical phytoplankton of the Indian Ocean. *Dokl. Akad. Nauk. SSSR* **142**(5), 1162–1164 (transl.)

SUKHANOVA I. N. (1969) Some data on the phytoplankton of the Red Sea and western Gulf of Aden. *Oceanology* **9**, 243–248.

SUKHANOVA I. N. & RUDYAKOV A. (1973) Population composition and vertical distribution of *Pyrocystis pseudonoctiluca* (W. Thompson) in the western equatorial Pacific. In *Life Activity of Pelagic Communities in the Ocean Tropics* (ed. M. E. Vinogradov), pp. 218–228. Israel Program Sci. Transl., Jerusalem.

SWIFT D. G. (1980) Vitamins and phytoplankton growth. In *The Physiological Ecology of Phytoplankton* (ed. I. Morris), pp. 329–368. Studies in Ecology, Vol. 7. Blackwell Scientific, Oxford.

SWIFT E. & DURBIN E. G. (1972) The phased division and cytological characteristics of *Pyrocystis* spp. can be used to estimate doubling times of their populations in the sea. *Deep-Sea Res.* **19**, 189–198.

SWIFT E., STUART M. & MEUNIER V. (1976) The *in-situ* growth rates of some deep living oceanic dinoflagellates *Pyrocystis fusiformis* and *Pyrocystis noctiluca*. *Limnol. Oceanogr.* **21**, 418–426.

SYRETT P. J. (1981) Nitrogen metabolism of microalgae. *Can. Bull. Fish. aquat. Sci.* **210**, 182–210.

TAKAHASHI M., BARWELL-CLARKE J., WHITNEY F. & KOELLER P. (1978) Winter conditions of marine plankton populations in Saanich Inlet, B.C. Canada. I. Phytoplankton and its surrounding environment. *J. exp. mar. Biol. Ecol.* **31**, 283–301.

TANAKA T. & SANO M. (1980) Dynamic aspects of primary production by phytoplankton in eutrophic Mikawa Bay, Japan. *Bull. plankt. Soc. Japan* **27**, 75–86.

TANGEN K. (1877) Blooms of *Gyrodinium aureolum* (Dinophyceae) in north European waters, accompanied by mortality in marine organisms. *Sarsia* **63**, 123–133.

TANGEN K. (1979) Dinoflagellate blooms in Norwegian waters. In *Toxic Dinoflagellate Blooms* (eds D. L. Taylor & H. H. Seliger), pp. 179–182. Elsevier/North Holland, New York.

TANGEN K. (1980) Brown water in the Oslofjord, Norway, in Sept. 1979, caused by the toxic *Prorocentrum minimum* and other dinoflagellates. *Blyttia* **38**, 145–158.

TANIGUCHI A. (1977) Distribution of microzooplankton in the Philippine Sea and the Celebes Sea in summer, 1972. *J. oceanogr. Soc. Japan* **33**, 82–89.

TATEWAKI M. & MIZUNO M. (1979) Growth inhibition by germanium dioxide in various algae, especially brown algae. *Jap. J. Phycol.* **27**, 205–212. (in Japanese).

TAYLOR D. L. & SELIGER H. H. (Eds) (1979) *Toxic Dinoflagellate Blooms* Vol. 1. Elsevier/North Holland, Amsterdam.

TAYLOR F. J. R. (1979) Phytoplankton of the south western Indian Ocean. *Nova Hedw.* **12**, 432–476.

TAYLOR F. J. R. (1979) Parasitism of the toxin-producing dinoflagellate *Gonyaulax catenella* by the endoparasitic dinoflagellate *Amoebophyra ceratii. J. fish. Res. Bd. Can.* **25**, 2241–2245.

TAYLOR F. J. R. (1973) General features of dinoflagellate material collected by the 'Anton Bruun' during the International Indian Ocean Expedition. In *'The Biology of the Indian Ocean* (ed. B. Zeitzschel), pp. 153–169. Springer-Verlag, New York.

TAYLOR F. J. R. (1976) *Dinoflagellates from the International Indian Ocean Expedition—A report on material collected by the R. V. 'Anton Bruun' 1963–1964.* (Bibliotheca Botanica Vol. 132) E. Schweizerbart'sche Verlagsbuchhandlung, Stuttgart.

TAYLOR F. J. R. (1982) Symbioses in marine microplankton. *Ann. Inst. océanogr. Paris.* **58**(S) 61–90.

TAYLOR F. J. R. (1983) Possible free-living *Symbiodinium microadriaticum* (Dinophyceae) in tide pools in Southern Thailand. In *Endocytobiology* II. (ed. H. E. A. Schenk & W. Schwemmler), pp. 1009–1014. Walter de Gruyter & Co. Berlin.

TAYLOR F. J. R. (1984) Toxic dinoflagellates: taxonomic & biogeographic aspects with emphasis on *Protogonyaulax*. In *Seafood Toxins* ACS Symposition Series (ed. E. D. Ragelis) Amer. Chem. Soc. Washington.

TAYLOR F. J. R. (1985) The taxonomy and relationships of red tide flagellates. In *Toxic Dinoflagellates* (eds D. M. Anderson, A. W. White & D. G. Baden) pp. 11–26. Elsevier, New York.

TAYLOR F. J. R. & GUSTAVSON M. (1986) An underwater survey of the organism chiefly responsible for 'Ciquatera' fish poisoning in the Eastern Caribbean: the benthic dinoflagellate *Gambierdiscus toxicus*. In *Proc. 7th Int. Diving Science Symp.*, Padova, Sept. 1983, pp. 95–111.

TAYLOR F. J. R. & WATERS R. E. (1982) Spring phytoplankton in the Subarctic North Pacific Ocean. *Mar. Biol.* **67**, 323–335.

THOMAS W. H. (1955) Heterotrophic nutrition and respiration of *Gonyaulax polyedra. J. Protozool.* **2**(suppl.), 2–3.

THOMAS W. H. (1959) The culture of tropical oceanic phytoplankton. *Preprints Int. oceanogr. Congr. Amer. Ass. Adv. Sci.,* 207–208.

THOMAS W. H. (1966) Effects of temperature and illuminance on cell division rates of three species of tropical oceanic phytoplankton. *J. Phycol.* **2**, 17–22.

THOMAS W. H. & DODSON A. N. (1974) Effect of interactions between temperature and nitrate supply on the cell-division rates of two marine phytoflagellates. *Mar. Biol.* **24**, 213–217.

THOMAS W. H., DODSON A. N. & REID F. M. H. (1978) Diatom productivity compared to other algae in natural marine phytoplankton assemblages. *J. Phycol.* **14**, 250–253.

THOMAS W. H., HASTINGS J. & FUJITA M. (1980a) Ammonium input to the sea via large sewage outfalls. 2. Effects of ammonium on growth and photosynthesis of Southern California, USA on phytoplankton cultures. *Mar. environ. Res.* **3**, 291–296.

THOMAS W. H., POLLOCK M., SIEBERT D. C. R. (1980b). Effects of simulated upwelling and oligotrophy on chemostat grown natural marine phytoplankton assemblages. *J. exp. mar. Biol. Ecol.* **45**, 25–36.

THOMPSON R. H. (1950) A new genus and new records of fresh water Pyrrophyta in the Desmokontae and dinophyceae. *Lloydia* **13**, 277–299.

THOMSEN H. A. (1982) Planktonic chaonoflagellates from Disko Bugt, West Greenland, with a survey of the marine nanoplankton of the area. *Medd. orn Grønland. Biosci.* **8**, 1–36.

THURBERG F. P. (1975) Inactivation of red-tide toxins by ozone treatment. In *Aquatic Applications of Ozone,* (eds W. J. Blogoslawski & R. G. Rice), pp. 50–58. Int. Ozone Instit., Syracuse, New York.

TRICK C. G., ANDERSEN R. J., GILLAM A. & HARRISON P. J. (1983) Prorocentrin: An extra cellular siderophore produced by the marine dinoflagellate *Prorocentrum minimum. Science (N.Y.)* **21,** 306–308.

TURPIN D. H. & HARRISON P. J. (1978) Fluctations in free amino acid pools of *Gymnodinium simplex* (Dinophyceae) in response to ammonia perturbation. Evidence for glutamine synthetase pathway. *J. Phycol.* **14**, 461–464.

TUTTLE R. C. & LOEBLICH A. R. III (1975) An optimal growth medium for the dinoflagellate *Crypthecodinium cohnii*. *Phycologia* **14**, 1–18.

TWAROG B. M. & YAMAGUCHI H. (1975) Resistance to paralytic shellfish toxins in bivalve molluscs. In *Proceedings of the First International conference on Toxic Dinoflagellate Blooms* (ed. V. R. LoCicero), pp. 381–393. Mass. Sci. and Technol. Found., Wakefield, Mass.

TYLER M. A., COATS D. W. & ANDERSON D. M. (1982) Encystment in a dynamic environment: deposition of dinoflagellate cysts by a frontal convergence. *Mar. Ecol., Prog. Ser.* 7, 163–178.

TYLER M. A. & SELIGER H. H. (1978) Annual subsurface transport of a red tide dinoflagellate to its bloom area: water circulation patterns and organism distributions in the Chesapeake Bay. *Limnol. Oceanogr.* **23**, 227–246.

TYLER M. A. & SELIGER H. H. (1981) Selection for a red tide organism *Prorocentrum mariae-lebouriae*: physiological responses to the physical environment. *Limnol. Oceanogr.* **26**, 310–324.

UCHIDA T. (1981) The relationships between *Prorocentrum micans* growth and its ecological environment. *Sci. Pap. Inst. algol. Res. Fac. Sci. Hokkaido Univ.* **7**, 17.

UHLIG G. & SAHLING G. (1982) Rhythms and distributional phenomena in *Noctiluca miliaris Ann. Inst. océanogr. Paris* **58**(5) 277–284.

UKELES R. & SWEENEY B. M. (1969) Influence of dinoflagellate trichocysts factors on the feeding of *Crassostrea virginica* larvae on *Monochrysis lutheri*. *Limnol. Oceanogr.* **14**, 403–410.

USACHEV P. I. (1949) The microflora of Arctic ice. *Trudy Inst. Okeanologii* **3**, 216–259. (*Fish. Res. Bd. Can.* transl. ser. no. 1305.).

USACHEV P. I. (1961) Phytoplankton of the North Pole. *Trudy Vsesoyuznogo Gidrobiol. Obshchestra* **11**, 189–208. (Fish. Res. Bd Can. transl. ser. no. 1285).

VENRICK E. L. (1982) Phytoplankton in an oligotrophic ocean: observations and questions. *Ecol. Monogr.* **52**, 129–154.

VENRICK E. L., MCGOWAN J. A. & MANTYLA A. W. (1973) Deep maxima of photosynthetic chlorophyll in the Pacific Ocean. *Fishery Bull.* **71**, 41–52.

VON BROCKEL K. (1981) The importance of nanoplankton within the pelagic Antarctic ecosystem. *Kieler Meeresforsch., Sonderh.* **5**, 61–67.

WAILES G. H. (1939) *Canadian Pacific Fauna* I. Protozoa, pp. 1–45. U. Toronto Press, Toronto.

WALL D. (1971) The lateral and vertical distribution of dinoflagellates in Quaternary sediments. In *The Micropalaeontology of Oceans* (ed B. M. Funell & W. R. Riedel), pp. 399–405. Cambridge, University Press, Cambridge.

WALL D. & DALE B. (1974) Dinoflagellates in late quaternary deep water sediments of Black Sea. In *The Black Sea—Geology, Chemistry and Biology* (eds E. T. Degens & D. A. Ross), pp. 364–380 (Am. Assoc. petro. Geol., Mem. 20).

WALL D., DALE B., LOHMANN G. P. & SMITH W. K. (1977) The enviromental and climatic distribution of dinoflagellate cysts in modern marine sediments from regions in the North and South Atlantic Oceans and adjacent seas. *Mar. Micropal.* **2**, 121–200.

WANGERSKY, P. J. (1977) The role of particulate matter in the productivity of surface waters. *Helgol. wiss. Meeresunters.* **30**, 546–564.

WATERBURY J. B., WATSON S. W., GUILLARD R. R. L. & BRAND L. E. (1979) Widespread occurrence of a unicellular, marine, planktonic cyanobacterium. *Nature* (*Lond.*) **277**, 293–294.

WATRAS C. J., CHISHOLM S. W. & ANDERSON D. M. (1982) Regulation of growth in an estuarine clone of *Gonyaulax tamarensis* Lebour: salinity dependent temperature responses. *J. exp. mar. Biol. Ecol.* **62**, 25–37.

WEDEMAYER G. J., WILCOX L. W. & GRAHAM L. E. (1982) *Amphidinium cryophilum* sp. nov. (Dinophyceae) a new freshwater dinoflagellate 1. Species description using light and scanning electron microscopy. *J. Phycol.* **18**, 13–17.

WEILER C. S. (1980) Population structure and *in situ* division rates of *Ceratium* in oligotrophic waters of the North Pacific central gyre. *Limnol. Oceanogr.* **25**, 610–619.

WEILER C. S. & CHISHOLM S. W. (1976) Phased cell division in natural populations of marine dinoflagellates from shipboard cultures. *J. exp. mar. Biol. Ecol.* **25**, 239–247.

WEILER C. S & EPPLEY R. W. (1979) Temporal pattern of division in the dinoflagellate genus *Ceratium* and its application to the determination of growth rate. *J. exp. mar. Biol. Ecol.* **39**, 1–24.

WHEELER B. (1966) Phototactic vertical migration in *Exuviella* (sic) *baltica. Bot. Mar.* **9**, 15–17.

WHITE A. W. (1976) Growth inhibition caused by turbulence in the toxic marine dinoflagellate *Gonyaulax excavata. J. fish. Res. Bd. Canada* **33**, 2598–2602.

WHITE A. W. (1978) Salinity effects on growth and toxin content of *Gonyaulax excavata*, a marine dinoflagellate causing paralytic shellfish poisoning. *J. Phycol.* **14**, 475–479.

WHITE A. W. (1980) Recurrence of kills of Atlantic herring (*Clupea harengus harengus*) caused by dinoflagellate toxins transferred through herbivorous zooplankton. *Can. J. Fish. aquat. Sci.* **37**, 2262–2265.

WHITE A. W. (1981) Sensitivity of marine fishes to toxins from the red-tide dinoflagellate *Gonyaulax excavata* and implications for fish kills. *Mar. Biol.* **65**, 255–260.

WHITE A. W. (1982) Intensification of *Gonyaulax* blooms and shellfish toxicity in the Bay of Fundy. *Can. Tech. Rep. fish. Aquat. Sci.* **1064**, 1–12.

WHITE H. H. (1979) Effects of dinoflagellate bioluminescence on the ingestion rates of herbivorous zooplankton. *J. exp. mar. Biol. Ecol.* **36**, 217–224.

WILLIAMS D. B. (1971a) The distribution of marine dinoflagellates in relation to physical and chemical conditions. In *The Micropalaeontology of Oceans* (eds B. M. Funnell & W. R. Riedel), pp. 91–95. Cambridge University Press, Cambridge.

WILLIAMS D. B. (1971b) The occurrence of dinoflagellates in marine sediments. In *The Micropalaeontology of the Oceans* (eds B. M. Funnell & W. R. Riedel), pp. 231–243. Cambridge University Press, Cambridge.

WILLIAMS W. T., BUNT J. S., JOHN R. D. & ABEL D. J. (1981) The community concept and phytoplankton. *Mar. Ecol., Prog. Ser.* **6**, 115–121.

WILSON M. K., KELLEY G. J. & JEFFREY S. W. (1984) *Interplay between light and temperature in the growth of phytoplankton algae*. Abst. 2nd Div. Res. Seminar. CSIRO, Australia.

WILSON W. B. (1975) *Gymnodinium breve* growth requirements. In *Proceedings of the Florida Red Tide Conference* (ed. E. A. Joyce Jr) 10–12 Oct., 1974, *Fla. Mar.Res. Publ.* **8**, 1–18.

WILTON J. W. & BARHAM E. G. (1968) A yellow-water bloom of *Gymnodinium flavum. J. exp. mar. Biol. Ecol.* **2**, 167–173.

WISELY B. OKAMOTO R. & REID B. L. (1978) Pacific oyster (*Crassostrea gigas*) spatfall prediction at Hiroshima, Japan, 1977. *Aquaculture* **15**, 227–241.

WITHERS N. W. (1982) Ciguatera fish poisoning. *Ann. Rev. Med.* **33**, 97–111.

WOOD E. J. F. (1954) Dinoflagellates in the Australian Region. *Austral. J. mar. freshw. Res.* **5**, 171–351.

WOOD E. J. F. (1966) A phytoplankton study of the Amazon region. *Bull. mar. Sci.* **16**, 102–123.

WOOD E. J. F. (1968a) *Dinoflagellates of the Caribbean Sea and Adjacent Areas.* Univ. Miami Press, Florida.

WOOD E. J. F. (1968b) Studies of phytoplankton ecology in tropical and subtropical enviroments of the Atlantic ocean. 3 Phytoplankton communities in the Providence Channels and the Tongue of the Ocean. *Bull. mar. Sci.* **18**, 481–543.

WOODCOCK A. H. (1948) Note concerning human respiratory irritation associated with high concentrations of plankton and mass mortality of marine organisms. *J. mar. Res.* **7**, 56–62.

WRIGHT R. T. & HOBBIE J. E. (1966) Use of glucose and acetate by bacteria and algae in aquatic ecosystems. *Ecology* **47**, 447–464.

WURSTER C. F. JR (1968) DDT reduces photosynthesis by marine phytoplankton. *Science*, (N. Y.) **159**, 1474–1475.
WYATT T. (1974) Red tides amd algal strategies. In *Ecological Stability* (eds M. B. Usher & M. H. Williamson), pp. 35–40. Chapman & Hall, Great Britain.
WYATT T. (1975) The limitations of physical models for red tides. In *Proc. of the First Int. Conf. on Toxic Dinoflagellate Blooms* (ed V. R. LoCicero), pp. 81–93. Mass. Sci. & Technol. Found., Wakefield, Mass.
WYATT T. & HORWOOD J. (1973) Model which generates red tides. *Nature* (*Lond.*) **244**, 238–240.
WYNNE D. (1977) Alterations in activity of phosphatase during the *Peridinium* bloom in Lake Kinneret. *Physiol. Plant.* **40**, 219–224.
YAMAZI I. (1972) *Illustrations of the Marine Plankton of Japan.* Hoikasha Publ., Osaka.
YENTSCH C. M. & MAGUE F. C. (1980) Evidence of an apparent annual rhythm in the toxic red tide dinoflagellate *Gonyaulax excavata*. *Int. J. Chronobiol.* **7**, 77–84.
YENTSCH C. M., YENTSCH C. S., OWEN C. & SALVAGGIO M. (1974) Stimulatory effects on growth and photosynthesis of the toxic red tide dinoflagellate, *Gonyaulax tamarensis*, with the addition of nitrilotriacetic acid (NTA) $N(CH^2COOH)_3$. *Environ. Lett.* **6**, 231–238.
ZERNOVA V. V. (1968) Phytoplankton of the Southern Ocean. In *Antarctic Biological Research in Prospect and Retrospect* (ed. M. W. Holdgate), pp. 136–142. Cambridge Univ. Press, Cambridge.
ZERNOVA V. V. (1969) Horizontal distribution of phytoplankton in the Gulf of Mexico. *Okeanologia* **9**, 695–706.
ZERNOVA V. V. (1979) The distribution and temporal variability of phytoplankton biomass in the tropical waters of the Atlantic Ocean. *Mar. Sci. Comm.* **5**, 399–422.
ZGUROVSKAYA L. M. (1970) Chlorophyll content and the rate of cell division in some algae in relation to the phosphorous content of nutrient media. *Hydrobiol. J.* **6**, 39–45.
ZGUROVSKAYA L. M. & KUSTENKO N. G. (1968) The effect of ammonia nitrogen on cell division, photosynthesis and pigment accumulation in *Skeletonema costatum* (Grev.) Cl., *Chaetoceros* sp. and *Prorocentrum micans* Ehr. *Oceanology* **8**, 90–98.
ZINGMARK R. G. & MILLER T. G. (1975) The effects of mercury on the photosynthesis and growth of estuarine and oceanic phytoplankton. In *Physiological Ecology of Estuarine Organisms* (ed F. J. Vernberg), pp. 45–57. Univ. S. Carolina Press, Columbia, S. Carolina.

B. FRESHWATER ECOSYSTEMS

UTSA POLLINGHER

1 INTRODUCTION

The freshwater dinoflagellates are less common qualitatively and quantitatively than the marine forms. Only 220 species of dinoflagellates occur in freshwater bodies (Bourrelly 1970). Most of them are autotrophic motile forms. A group of non-motile forms (Dinococcales) represented by ten genera, populate very small water bodies: peat pools, bogs and reservoirs (Huber-Pestalozzi 1951; Baumeister 1963; Javornicky & Popovsky 1971).

2 WORLDWIDE DISTRIBUTION OF FRESHWATER DINOFLAGELLATES

Freshwater dinoflagellates are cosmopolitan. They occur in lakes near the equator: Lake Victoria (*Ceratium brachyceros, Peridinium cunningtonii, P. westii, P. africanum*, etc.; Woloszynska 1914), Lake Tanganyika (*Peridinium africanum, P. apiculatum, Gymnodinium mirabilis, G. profundum, G. varians, G.* cf. *albulum, Glenodinium pulvisculus, G. quadridenus*: Hecky *et al.* 1978), Lake Bangweulu (*Ceratium brachyceros, P. volzii, Stylodinium truncatum*; Thomasson 1960), in lakes in Iceland (64° N lat.) (*P. aciculiferum*; Ostenfeld & Wesenberg-Lund 1904) and in ponds in Vestspitsbergen (78° N lat.) (*Amphidinium* spp., *Glenodinium pulvisculus, Gyrodinium pascheri, Peridinium* spp.; Willen 1980). In Arctic lakes dinoflagellates do not play an important role but *P. willei, P. cinctum* and *P. umbonatum* have been recorded (Dakin & Latarche 1913).

The range of altitudinal distribution is considerable. *P. cinctum* (fa. *westii*) blooms in Lake Kinneret (Sea of Galilee, Lake Tiberias, Israel) 209 m below sea level (Komarovsky 1959; Pollingher 1968) and Sasuma reservoir (Kenya) 2474 m above sea level (Lind 1968). *P. willei* and *P. volzii* are present in high mountain Mexican lakes (2700–4150 m above sea level; Löffler 1972), *Gymnodinium uberrimum* is a perennial form in Vorderer Finstertaler Lake in Austria (2237 m above sea level; Tilzer 1973). *Ceratium hirundinella* is present in high Alpine lakes in Europe as high as 2000 m above sea level (Pesta 1929; Messikommer 1942), in Sasura reservoir (Kenya) accompanying *P. cinctum* (Lind 1968) and in lakes Jenny (alt. 2066 m) and Jackson (alt. 2052 m) (Grand Teton National Park, Wyoming) (Thomasson 1962).

The most common and bloom-forming genera are *Ceratium* and *Peridinium*. Their presence in different water bodies is well documented, although ecological and quantitative data on them are less abundant.

The worldwide distribution of *Ceratium* was described by Huber-Pestalozzi (1951). Since then many new records have become available, especially from water bodies located in less accessible regions.

3 DINOFLAGELLATES IN DIFFERENT TYPES OF FRESHWATER ECOSYSTEMS

3.1 Lakes and ponds

BLOOMS OF *CERATIUM HIRUNDINELLA*

Ceratium hirundinella is a very common planktonic form in temperate zone lakes. It is found in abundance generally in small, relatively shallow alkaline lakes with low concentrations of nutrients, especially available phosphorus.

The ecology of *C. hirundinella* was studied in Lake Erken, a flatland unpolluted lake of the Baltic type (Nauwerck 1963; Dottne-Lindgren & Ekbohm 1975). The surface area of the lake is ~ 25 km^2, its mean depth is 9 m and its pH is 8·2. During Nauwerck's survey, performed in 1957, the first *Ceratium* appeared in March, started increasing only in July and peaked in August (70 cells ml^{-1}). The population remained in a steady state for 3 weeks, then decreased suddenly.

Twelve years later, the *Ceratium* population was studied by Dottne-Lindgren & Ekbohm. The temporal development of the bloom was similar to that described by Nauwerck but the maximum number of cells recorded was 416 cells ml^{-1}, and the doubling time of the population varied between 70 and 107 hours. Cyst formation started in early August when the population was in its stationary phase and the water reached its highest temperature (20°C). The cysts constituted only 1–2% of the number of vegetative cells, and at the end of the bloom their number exceeded that of the motile cells (13 cysts ml^{-1} were recorded). The authors suggested that the difference in abundance of the *Ceratium* population in 1969 in comparison with that in 1957 may have been correlated with the occurrence of *Gloeotrichia echinulata*. In years when blue-green algae are abundant, *C. hirundinella* is relatively rare. A similar development of *Ceratium* has occurred for many years in a large shallow lake, Lake Balaton (Hungary) (Entz 1927; Sebestyen 1952, 1958; Padisak 1985).

Entz (1931) studied the development of *Ceratium* populations in large ponds located in Lagymanyos (near Budapest, Hungary). He found that the maximum population division rate (DR) of *Ceratium* was 33% and the minimum 3%. The optimal temperature for division was 20–22°C, and from late March until the beginning of October, twenty-four generations of *Ceratium* developed. Recently, Frempong (1984) and Hickel (1985) have described the population dynamics of *Ceratium* in small, more eutrophic lakes.

An abundant population of *C. hirundinella* develops during the summer months on some of the English meres (Reynolds 1973, 1976, 1978) and in small lakes of the English Lake District (Heaney & Talling 1980). A detailed study has been performed in one of these—Esthwaite Water, a small productive lake (surface area 1 km^2 and maximum depth 15 m). The average concentration of nutrients in the 0–5 m layer during the summer 1976 was: ammonium nitrogen 25–100 μg l^{-1}; nitrate nitrogen 350–400 μg l^{-1}; soluble phosphate-phosphorus from 0·5 μg l^{-1} to $<$1 μg l^{-1} in September–October (Harris *et al.* 1979). The seasonal growth of *C. hirundinella* based on weekly integral samples, was described by Heaney & Talling (1980) over a 3 year period (Fig. 11B.1). Winter is characterized by a very low concentration of *Ceratium* (less than one cell per 100 ml water). A small increase is observed in March, followed by an exponential increase until July and a stationary phase until September. The bloom ends with a rapid decline in late September or early October and is sometimes accompanied by cyst formation. In 1976 the maximum density of cells (632 cells ml^{-1}) was recorded in late September (Harris *et al.* 1979). The doubling time up to 6 July was *c.* 10 days and 34 days during July and August. Heaney (1976) has recorded in 1975, in the same lake, a population doubling time of 6·7 days from the middle of May to the beginning of August (temperature 11–22°C). In 1975 at the end of the bloom (early October), less than 1% of normal cells were transformed into cysts, but during a 1-month period the number of cysts increased and by November was equal to that of vegetative cells (Heaney 1976). In 1973 a large population of vegetative cells underwent a sudden lysis at the end of the bloom.

Phased cell division has been reported by Mueller-Elser & Smith (1985). High abundance of *C. hirundinella* (638 cells ml^{-1} was found during the summer months in 1971 in Rostherne Mere (area ~ 49 ha) (Reynolds 1978).

A different temporal development of *C. hirundinella* occurs in the warm, large (surface area ~ 170 km^2) monomictic Lake Kinneret (Israel) (U. Pollingher, unpubl.) *Ceratium* appears in the lake at the end of March (temperature 15·3–16·8°C), increases slowly in April, reaching a maximum in May (22–23°C). and disappears every year in mid-June (25°C). It is a characteristic feature of the *Ceratium* population in Lake Kinneret to have some discrete blooms for only a few days. The maximum number of cells recorded

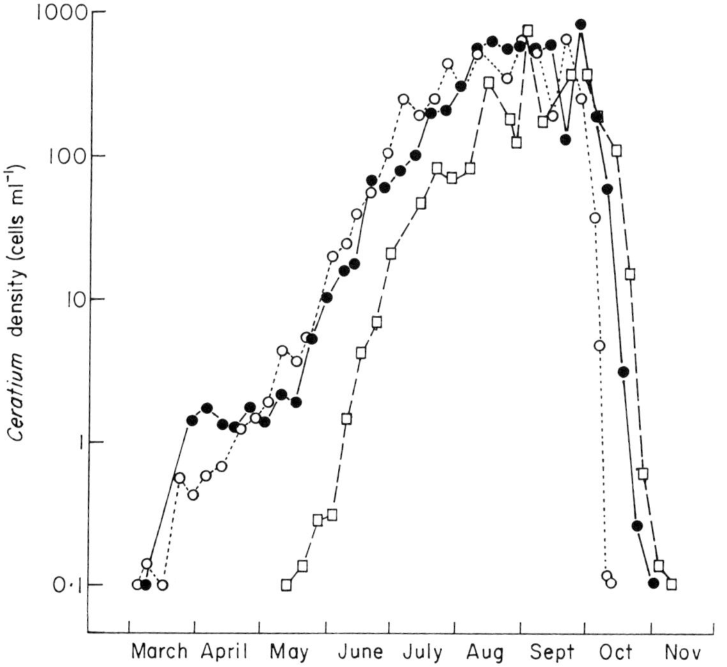

Fig. 11B.1. Seasonal growth and decline of *Ceratium hirundinella* during 1975–77, as measured by mean cell concentration in the water layer at 0–5 m depth (from Heaney & Talling 1980).

was 31 cells ml^{-1}. *Ceratium* disappears completely during the summer months. It seems that it has a narrow tolerance range for water temperature in Lake Kinneret.

The first cysts of *Ceratium*, which are resuspended from the sediments by turbulence, are recorded in the epilimnion at the beginning of March. These cysts generate the *Ceratium* population. At the decline of the *Ceratium* population in June, newly-formed cysts are recorded in the epi- and metalimnion. The cyst of *Ceratium* in Lake Kinneret is an 'oversummering' form and not an 'overwintering' one as in the water bodies in the temperate zones, but the temperature ranges of the vegetative cells' development are the same. *Ceratium* always disappears in Lake Kinneret when the temperature exceeds 25°C. The warm water form of *Ceratium*, which develops in summer in the temperate water bodies, develops during the spring months in the subtropical Lake Kinneret. In the case of Lake Kinneret, *Ceratium* accompanies the *Peridinium cinctum* fa. *westii* bloom, whereas in the water bodies of the temperate zones *Ceratium* appears in association with algae belonging to other taxonomic groups. *C. hirundinella* is dominant in the upland lakes on the Malmsjo Plateau (Sweden) (Florin 1957). These lakes are 7500–5500 years old and are characterized by very low concentrations of nutrients due to edaphic conditions and are not influenced by cultivation. *Ceratium* is also common in low inland lakes of the same area, rich in nutrients (Florin 1957; see also Hickel 1985 for a north German lake). In both groups of lakes, *Ceratium* appears in association with Chrysophyta species.

FORM VARIATION

Many authors have recorded the great morphological variability of *C. hirundinella*; some have studied the local variations, others the seasonal ones. Most of the investigations were based on qualitative net samples (Wesenberg-Lund 1908; Lemmermann 1910; Krause 1911–1912; Langhans 1925; Entz 1927; Pearsal 1929; Hubber-Pestalozzi 1950; Florin 1957; Komarovsky 1959; Daily 1960; etc.).

The relative distribution of form types combined with the abundance of *C. hirundinella* in Lake Erken, was studied by Dottne-Lindgren & Ekbohm (1957) using an integrated water sample and statistical analysis. The whole *Ceratium* population was divided into three different form types: (i) with two posterior horns; (ii) with three posterior horns although the third one is rudimentary; (iii) with three well developed posterior horns (Fig. 11B. 2). The authors have found that the total length of the cells was the most variable parameter and could give the most reliable indication of a possible variation in the enviroment. They concluded that long individuals appear in early summer, disappear in June (probably dying off) and reappear in mid-October. They are succeeded by a

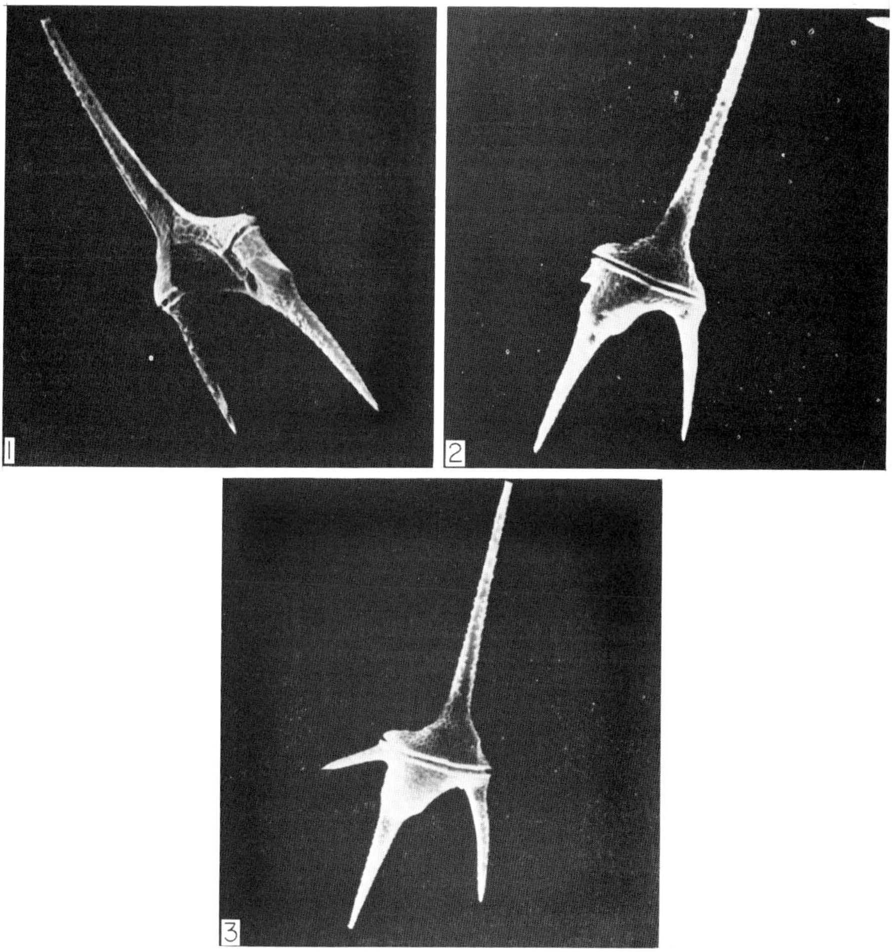

Fig. 11B.2. Form types 1, 2 and 3 of *Ceratium hirundinella* (From Dottne-Lindgren & Ekbohm 1975).

'summer form' population consisting of smaller individuals originating from resting cysts.

The decreasing length of organisms can also be explained by the increase of temperature which increases the division rate. A decrease of body size of *Peridinium cinctum* due to increase in division rate was observed in Lake Kinneret (Serruya & Pollingher 1977) and the theca may be shed altogether if nutrients are stressful (Crisculo *et al.* 1981). Huber & Nipkow (1923) determined experimentally that the length of the hatched *Ceratium* cells decreased with increased incubation temperature. At present, Dottne-Lindgren & Ekbohm suggest that the population of *C. hirundinella* in Lake Erken constitutes the sum of the populations of three form types which are independent of each other and originate from specific resting cysts.

In cultures of *C. hirundinella* isolated from different lakes, the cells usually have two posterior horns (Bruno 1975), even when the cultures were started with three posterior horned cells. The author has been able to bring about the production of a third posterior horn by varying temperature and/or trace metals. At 21–25°C all cells in his cultures had two posterior horns but at 15°C a large percentage of cells had three posterior horns. Different form types have also been observed in the small population of *C. hirundinella* in Lake Kinneret (Komarovsky 1959).

DISTRIBUTION

Records of *Ceratium* from nearly all over the world are available but quantitative data are scanty. A review of the literature gives the impression that *Ceratium* is present in low numbers in warm lakes and more abundant in small water bodies located in the warm belt. Hickel (1973) has found in a small lake–Rupa Tall (surface area 1·4 km^2, maximum depth 4 m; Western Nepal)—a mass development of *Ceratium* in winter (temperature 18·3–19·8°C). The density was 680 cells ml^{-1}, causing a brownish water colour. Ganapati (1957) described a heavy bloom of *C. hirundinella* (4112 org. ml^{-1}) in Ootacamund Lake, a small lake formed by the damming of a mountain stream. The lake is located at 2317 m altitude in the Nilgri mountains (Madras State, India). *Ceratium* is present in Lake Wabby (Australia, pH 5·9) (Bayly *et al.* 1975) and in lakes located in Middle America (Brezonik & Fox 1974; Brinson & Nordlie 1975; B. Hickel, pers. comm.). It is absent in Araucanian lakes (Löffler 1972) situated west of the Andes (Thomasson 1963); *C. hirundinella* and *C. cornutum* were recorded in lakes from the eastern part of the Andes (Thomasson 1963), but they are very scarce. *C. hirundinella* was not found in Venezuelan freshwater bodies (Lewis & Weibezahn 1976) and was not recorded in the Amazonian basin (Uherkovich 1976; Uherkovich & Rai 1979).

BLOOMS OF *PERIDINIUM* SPP.

The second genus, *Peridinium*, is represented by four very common species: *P. cinctum*, *P. willei*, *P. volzii* and *P. bipes*.

P. cinctum (reviewed by Berman & Dubinsky 1985) is a eutrophic cosmopolitan form which blooms frequently in ponds, reservoirs, small shallow lakes and rarely in large lakes. One of the few large freshwater bodies in the world where a yearly *Peridinium* bloom can be observed is Lake Kinneret

(Pollingher 1968; Pollingher & Berman 1975; Pollingher & Serruya 1976). Lake Kinneret is a warm monomictic lake, located in northern Israel at 209 m below m.s.l. It has surface area of ~ 170 km^2 and a mean depth of 24 m. The winter turnover plays a decisive role in the lake metabolism. Full homothermy occurs between the months of December and February (temperature 14–16°C; the concentrations of available nutrients are at their maximum) and a steady thermocline develops after April. During the stratification period (May–December) the epilimnion (temperature 15–30°C) is rich in oxygen and poor in nutrients. Concentrations of dissolved phosphorus rarely exceed 10 μg l^{-1} (C. Serruya 1978). In spite of this limitation of assimilable phosphorus, a bloom of *P. cinctum* fa. *westii* develops during the winter–spring months. Vegetative cells of *P. cinctum* have been absent, during the summer months in recent years. The bloom starts from the cysts which occur mostly in the shallow sediments. Turbulence, generated by winds, leads to the resuspension of the sediments and cysts. The cysts appear in the water in mid-November and are common in the littoral area until February (temperature 20°C and 14°C, respectively). During the same period juvenile cells with 1–2 red bodies are observed in the water (Pollingher 1978). Cells emerging from the cysts begin their motile life at the mud–water interface, where the nutrient concentrations are at their maximum, especially phosphorus. Their capacity for 'luxury consumption' (Serruya & Berman 1975) allows them to store large quantities of phosphorous intracellularly. This mechanism may explain why the early stage of the bloom does not depend entirely on nutrient supply. The quiet weather generally prevailing in March enables the cells to divide rapidly and to reach their maximum in March–April (Fig. 11B.3). At that time the concentration of nutrients decreases. The continuation of the bloom (after a certain number of divisions) depends on the

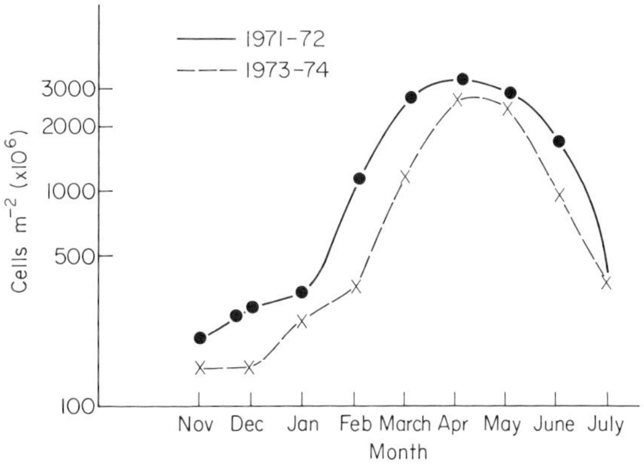

Fig. 11B.3. Growth curve of *Peridinium cinctum* fa. *westii* in Lake Kinneret (represented as × 10^6 cells m^{-2}, monthly averages) in the trophogenic layer.

nutrients supplied by the lake water. In 1973, when the concentration of nutrients in the winter–spring period was low (due to complete absence of floods; Serruya & Pollingher 1977), the bloom declined earlier than in other years. The disappearance of the bloom is generally very rapid, lasting 1–2 weeks, probably due to a combination of physical and chemical factors, e.g. high temperature, high irradiance, high wind intensity, current patterns, nutrient deficiency, crowding, etc. (Hertzig *et al.* 1981)

At the end of the bloom, the bulk of the *Peridinium* population dies off and only 1% of the population encysts. In Lake Kinneret the cysts formed during the bloom period (March–June, when the temperature increases) germinate in December–February (when the temperature decreases). The resting cysts of *Peridinium* seem to be resistant to attack by fungi and bacteria generally present in the sediments. The *Peridinium* cyst in Lake Kinneret is an 'oversummering' form like that of *Ceratium hirundinella* in the same lake, and not an 'overwintering' form as in lakes in the temperate zone.

Development of the bloom in Lake Kinneret is directly correlated with the time of lake turnover and with the duration and intensity of the mixing period (Pollingher & Zemel 1981). *Peridinium* appears in greater numbers at the end of the mixing period. In 1973–74 and 1974–75, when the mixing period extended until the beginning of March, higher numbers of *Peridinium* were recorded only in March (Table 11B.1). In the next year (1975–76), when the mixing period was unusually long (until the end of March) greater numbers of *Peridinium* were observed only in April. Conversely, the poor mixing during a short period in 1970 allowed the development of *Peridinium* early in January. In the years 1971–72, 1972–73 and 1976–77, when the mixing period ended at the beginning of February, high numbers of *Peridinium* were found during this month.

The abundance of the *Peridinium* population varies during the bloom period. The increase of the population depends on its division rate (DR). The division of *Peridinium* in Lake Kinneret takes place only at night between 01.00 and 07.00 hours, with a peak between 02.00 and 04.00 hours (Pollingher & Serruya 1976). The bloom cycle of 1973–74 showed a DR of 10% from November until February 1974, then it increased and remained between 30 and 40% until late March, dropping again to 10% by the end of May. In summer the population DR was no higher than 2·5% and similar values were recorded during three consecutive years. The multiplication of the cells occurs in the temperature range of 14–22°C (see Lindstrom 1984), with maximum values between 15–18°C. It is worth noting that during a drastic decrease of temperature (on very cold nights), no division was recorded, but the DR of a sample collected in the late afternoon of the same day and kept in the laboratory was 15%, showing that the cells had a high division potential which was repressed by unfavourable external factors. Although the division occurs in complete darkness, the dividing cells are not found in the whole water column. Cells in division are not found in water layers where the oxygen concentration is low (Pollingher & Serruya 1976).

Table 11B.1. The timing of the mixed period (shaded area) and the fluctuations of the number of *Peridinium* cells ($\times 10^6 m^{-2}$, monthly averages) in the trophogenic layer. The number of cells per surface area of the trophogenic column was obtained by planimetry. (From Pollingher & Zemel 1981.)

	1969	1970	1971	1972	1973	1974	1975	1976	1977
January	146	1320	460	356	140	175	33	25	101
February	417	2040	877	1210	1473	400	52	258	655
March	2040	4780	2345	3193	3053	1343	788	740	1761
April	4505	4870	3705	3430	4027	2793	2033	2683	2068
May	3700	2885	3850	2990	1270	3148	2912	2158	2542
June	1640	2017	2015	1787	1315	1087	1274	1362	1391

The DR is not homogeneous in the water column, and the highest rate of cells in division does not occur at the depth where the greatest population is concentrated. The intensity of the wind blowing throughout the whole day does not affect the DR of *Peridinium*. However, a strong inhibitory effect is observed when strong windy episodes occur between 18.00 and 02.00 hours, which corresponds to the premitotic and mitotic phase of cell division (Pollingher & Zemel 1981).

The success of *Peridinium* in Lake Kinneret is explained by its ability to build up high biomass per unit of phosphorus and nitrogen. In the case of *Peridinium* in Lake Kinneret, 1 mg of phosphorus leads to the synthesis of 250–650 mg algal biomass (dry weight; Serruya *et al.* 1974). *Peridinium* blooms in the lake at low concentrations of phosphorus, when other algae are not able to develop. Ammonium and nitrate uptake have been described by Berman *et al.* (1984). During the years 1973–76, major environmental changes occurred in Lake Kinneret (Serruya & Pollingher 1977), with a consequential increase in the concentration of dissolved phosphorus. The new enviromental conditions allowed the development of all other algal groups, which delayed the appearance of *Peridinium* and the development and intensity of the bloom. A drastic decrease of the *Peridinium* population occurred, from a semiannual average of 2160×10^6 cells m^{-2} (1972) to 1180×10^6 cells m^{-2} (1975), and from a semiannual average of dinoflagellate biomass of 131 g m^{-2} (1972) to 41 g m^{-2} (1975) (Fig. 11B.4). Since 1977, the situation has returned to normal. The maximum number of *Peridinium* recorded in a patch was ~ 5000 cells ml^{-1}.

Cell size of *Peridinium*, as in other dinoflagellates, varies considerably in

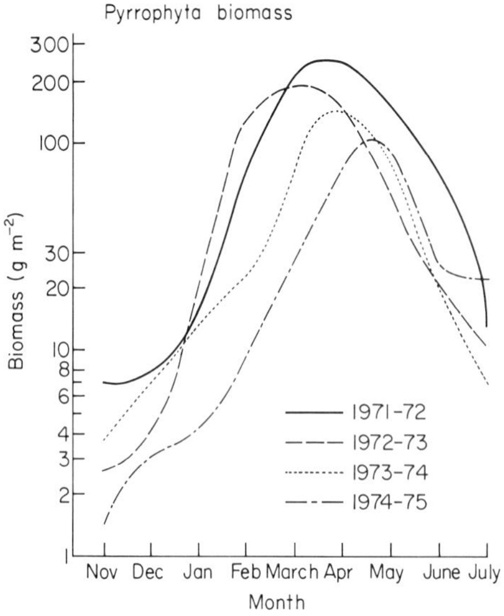

Fig. 11B.4. Annual fluctuations of the dinoflagellate biomass in Lake Kinneret (expressed as g m^{-2}, monthly averages) in the trophogenic layer.

Lake Kinneret. The same situation was observed in cultures grown from one single cell by W. Rodhe (pers. comm.). This broad spectrum of body size variation is a natural feature of the *Peridinium* population in the lake and in cultures and should be correlated with the relatively low division rate and long span of life. The body-size variation during the bloom period was described by Pollingher & Berman (1975).

A bloom of *Peridinium penardii* was recorded in Clear Lake (California) between late March and mid-May 1970 (Horne *et al.* 1971). The bloom was defined as a 'red tide' since it was concentrated in the shore area as narrow, coffee-brown strips. The maximum number of cells recorded was also 5000 cells ml^{-1}. In April the concentration of cells in the tide was 1240 cells ml^{-1}, whereas in the open water sites there were only 11–250 cells ml^{-1}. A bloom of the same species, *P. penardii* fa. *californicum*, occurred in four consecutive years (1970–74) in Lake Berreyessa, Napa County, California (Sibley *et al.* 1974); a maximum of 747 cells ml^{-1} was recorded.

Blooms of *Peridinium* species were recorded in small monomictic lakes in New Zealand. Burns & Mitchell (1974) described a bloom of *P. cinctum* in Lake Johnson (580 cells ml^{-1}) (South Island). Thomasson (1974) found *Ceratium furcoides* and *Peridinium sydneyense* dominant forms in Lake Okareka (North Island) and *C. furcoides* and *P. volzii* v. *meandricum* in Lake Rotoma. Cassie (1978) described *Ceratium* as a dominant form in the same lake from September 1973 to April 1974. The same author (Cassie 1968) recorded *P. striolatum* as the most common dinoflagellate in Lake Rotorua (North Island, New Zealand).

An unidentified species of *Peridinium* was a dominant form in Lake Mainit,

Philippines (Lewis 1973). *Peridinium* blooms were also recorded in two small crater lakes in Java (Ruttner 1952).

Water blooms of *Peridinium* species are common in ponds. A bloom of *Peridinium borgei* was observed by Entz (1926) in a pond near Budapest (Hortyteich). In 1907, 1000 cells ml^{-1} were found; 60 years later blooms of this species still occur there. Bourrelly (1968), on the basis of a sample from there, redescribed the organism as *Peridiniopsis borgei* due to the presence of 0–1 accessory plates on the epitheca. Cronberg (1981) has found the same species blooming in an abandoned chalk mine (St Kalkbrottsdammen, Klagshamn, Sweden).

3.2 Old lakes

Lake Baikal is the world's deepest lake and one of the largest (surface area 31 500 km^2 and depth 1741 m; its bottom lies at 1285 m below sea level) (Kozhov 1963). The waters of Baikal belong to the group of alkaline waters of low mineral content. The concentration of PO_4–P varies from 0·01 to 0·06 mg l^{-1} (Vereshiagin 1949; in Kozhov 1963). Dinoflagellates are represented by four species, three of which are endemic forms: *Gymnodinium baicalense* Ant, *G. coeruleum* Ant, and *Peridinium baicalense* Kissel & Zwetk. Large numbers of *Gymnodinium* spp. appear in open waters in the sub-ice period (March–April). *Ceratium hirundinella* occurs only in summer, being more common in shallow gulfs and bays (Kozhov 1963). *P. baicalense* was described by Kisselew & Zwetkow (1935) as a cold stenotherm form. Its optimal temperature ranges from 2–4·5°C. The maximum number of cells recorded by these authors was 698 cells m^{-3} on 25 August 1929, in a water layer of 25–50 m depth.

Lake Tanganyika (Africa) is the second largest mass of freshwater in the world (surface area 32 600 km^2, mean depth 700 m). Hecky *et al.* (1978) listed nine dinoflagellates from the lake: *Peridinium africanum, P. apiculatum, P. mirabilis, Gymnodinium profundum, G. varians, G.* cf. *albulum, Glenodinium pulvisculus, G. quadridens and Glenodinium* sp. Dinoflagellates do not play an important role in this lake: only in the Malagarazi Plume in April–May 1975 did a dinoflagellate, *G. pulvisculus,* dominate (Hecky & Kling 1981). The absence of *Ceratium* is worth noting.

Lake Biwa (Japan) (surface area 685 km^2, mean depth 41 m, maximum depth 104 m) contains many endemic green algae. The dinoflagellates recorded by Mori & Miura (1980) were *Gymnodinium* sp., *Glenodinium* sp. and *C. hirundinella*. In the paper *Symposium on some problems of water bloom or so-called freshwater 'red tide'* (Anonymous 1979), *P. africanum* was cited as a dominant species in Lake Biwa in June 1975, reaching 3000 cells ml^{-1} and giving a dark red colour to the water.

Lake Ohrid (Yugoslavia) is a typical karstic lake, located in a calcareous area and fed by under-ground water (surface area 348 km^2, maximum depth

286 m, mean depth 145 m). Dinoflagellates are represented by *Gymnodinium mirabile* var. *rufescens*. *P. cunningtonii, P. borgei, C. hirundinella* and the endemic form *Cystodinium dominii* Fott. *Ceratium, Peridinium* and *Cystodinium* develop during the summer months and spend the winter in an encysted state (Stankovic 1960).

Lake Kinneret (Israel) seems to be one of the few large old lakes with a true bloom of dinoflagellates. The dinoflagellates which accompany *P. cinctum* fa. *westii* during its bloom are: *Glenodinium gymnodinium, P. cunningtonii, P. inconspicuum, Diplopsalis acuta* and *C. hirundinella* (Pollingher 1978).

3.3 Reservoirs and man-made lakes

Dinoflagellates are very common in water reservoirs in Japan. A list of eighteen reservoirs located in western Japan, where water blooms or 'red tides' occurred, is given in an anonymous (1979) paper. In most cases the bloom is caused by *Peridinium* spp., and more rarely by *C. hirundinella*. In these reservoirs, the organisms reach high densities (7000 cells ml^{-1}) and cause a deep liver, reddish-brown, green or yellowish-brown water colour.

A heavy freshwater 'red tide' was described by Nakomoto (1975) in a new water reservoir near Tokyo—Lake Kanao-ko (surface area 323 km^2). The concentration of nutrients was extremely low. The blooms was caused by a *Peridinium* species (diameter 20–50 μm) with cell numbers varying between 10 000 and 93 000 cells ml^{-1}.

In 1962 an extensive water bloom of *P. polonicum* occurred for the first time in the man-made Lake Sagami, also near Tokyo. The bloom was accompanied by a mass mortality of fish (Adachi 1965; Hashimoto *et al.* 1968). The bloom developed in limited areas of tributaries and never reached the central part of the lake. The number of cells recorded varied between 4000 and 7000 ml^{-1}. This is one of the very rare cases of a toxic freshwater dinoflagellate bloom. Another case was described in Lake Austin (Texas) (Jurgens 1953), where a bloom of *Gymnodinium* sp. turned the water red and caused fish kills. Blooms in Sagami lake occurred also in 1964, 1965 and 1967, but they were not toxic. In 1964, *P. polonicum* reached 3820 cells ml^{-1} in June and was accompanied by *P. elpatiewsky* (420 cells ml^{-1}) and *P. penardii* (40 cells ml^{-1}) (Adachi 1965).

In some man-made lakes and reservoirs the dominant organism can be *C. hirundinella*. This is the case in Madden Lake (Panama) (Gliwicz 1976) and Bratzkii reservoir (Siberia) (Vorobieva *et al.* 1981), where *Ceratium* reached 345 cells ml^{-1} during the summer months of 1974. In the Govindsagar reservoir in India, *Microcystis* was replaced by *Ceratium* (Sarkar *et al.* 1977). A water bloom of *Ceratium* was described in the Pilloor artificial hydroelectric reservoir (India). Its surface area is 250 ha, pH 8·6, and nitrates and phosphate were undetectable at the time. A density of 1000 cells ml^{-1} was recorded in the central part of the lake; in the littoral the abundance was lower—only 480 cells ml^{-1} (Sreenivasan

1970). Ruttner (1952) found a bloom of *Ceratium* (136 cells ml^{-1}) in a small dam reservoir in Java, and in a second one a bloom of *Peridinium gutwinskii* (594 cells ml^{-1} at 3 m depth).

In the man-made Lake Kariba (Africa), Thomasson (1980) found *C. hirundinella* f. *brachyceros* in July and August, *P. cunningtonii* in January and April, and *P. willei* and *P. volzii* v. *cinctiforme* in April. In the largest man-made lake of Africa, the Volta Lake, the blue-green algae abundant in the pre-dam river were replaced by diatoms and dinoflagellates a few years after impoundment and then reappeared after conditions stabilized. The common dinoflagellate was *Peridinium africanum* (19 cells ml^{-1}) (Biswass 1975); due to its development the phytoplankton biomass nearly doubled from 1965 to 1966.

3.4 Natural acidic and acidified lakes

Most freshwater dinoflagellates are alkalinophile. A small number prefer acidic water bodies. Among the latter are *P. pusillum* and *P. willei*, described by Gessner & Hammer (1967) as dominant forms in Laguna Negra, located at 3470 m altitude in the Andes, with a pH of 5·8.

P. limbatum made up 50–60% of the total phytoplankton biomass from May to October 1974 in Carlyle Lake (Canada). The lake is a typical deep north temperate lake with a pH varying from 4·5 to 6·0. The lowest pH levels were recorded in spring and were attributable to an influx of acidic meltwater (Yan & Stokes 1978). Acidification from sulphur and nitrogen compounds affects the lakes and rivers in Scandinavia. On the Swedish west coast in 1970–72, 22–36% of the lakes had pH values lower than 5·0. The concentration of phosphorus in these lakes is very low. At a pH of about 4·0 the algal population consists mainly of dinoflagellates (generally *P. inconspicuum* and *Gymnodinium*) (Almer 1974). In some acidic lakes in western Sweden, with a pH of 4·5–6·8, Morling (1979) found *Ceratium carolinianum*, which is characterized by the author as preferring deep lakes with a high transparency, low conductivity, low degree of total hardness, low alkalinity and low concentrations of phosphorus (phosphate 5–10 $\mu g\ l^{-1}$).

3.5 Rivers

Dinoflagellates are very scarce in rapid flowing streams or rivers. Kofoid (1908) observed *C. hirundinella, C. cornutum, P. tabulatum* and *Glenodinium cinctum* in the potamoplankton of the Illinois River and its backwaters, and noted the minor significance of dinoflagellates here, relative to their importance in the phytoplankton of the Great Lakes.

Lindemann (1927a) described *Diplopsalis acuta* as a common form in the

Rivers Volga and Bolda. In the backwaters and in the delta of the Volga River the most common form is *C. hirundinella,* accompanied by *Gl. gymnodinium* and *P. cunningtonii.*

The genera *Ceratium, Peridinium* and *Gymnodinium* are represented in the potamoplankton of the Danube River (Brtek & Rothschein 1964). More species are found in the backwaters, especially in the delta of the Danube River (*C. cornutum, C. curvirostra, C. hirundinella, Gl. gymnodinium, Gl. pulvisculus, P. palatinum, P. tabulatum* and *P. willei*) (Olteanu 1967). *C. hirundinella* (vegetative cells and cysts), *P. cinctum* and two species of *Gymnodinium* were recorded in the River Drava by Uherkovich (1978). *C. hirundinella,* with a great variety of forms, is common in the majority of dead arms of the River Tisza (Hungary), (Kiss 1978). Moore (1977) recorded *C. hirundinella* near the mouth of the Yellowknife river (Canadian, subarctic region) and Evans (1962) found *Gymnodinium varians* in river Kafu (Central East Africa). Many species of *Peridinium* have been recorded from the Amazon basin, in the Rio Negro and its tributaries and in Rio Tapajos. Among them the most common forms are: *P. pygmaeum, P. cunningtonii, P. wisconsinense, P. baliense, P. cinctum, P. inconspicuum,* etc. (Uherkovich 1976; Uherkovich & Rai 1979). Thomasson (1955) provided a drawing of an unknown species of *Peridinium* found in the Rio Negro near Manaus.

3.6 Pools, bogs, peat and swamp waters

In permanent and ephemeral pools and in peat and swamp waters, motile and non-motile dinoflagellates are present. *C. hirundinella* occurred in 31 of 56 lowland pools and small lakes situated between Newcastle-on-Tyne and Durham (England), studied by Griffiths (1923). It did not occur in pools with a strong stream course and it was less abundant than *P. cinctum.*

P. pusillum was recorded from three pools in the southern part of the Black Forest, turning the water red (Klotter 1954). *Hemidinium ochraceum* lives on rocks in ephemeral rain pools and in sphagnum-filled hollows, giving them a yellow-rust colour (Droop 1953). In ephemeral and permanent pools, *Katodinium* (=*Massartia*) *vorticella, K. stigmatica* and *P. volzii* are also found (Droop 1953),

In peat pools, bogs and swamp waters the following dinoflagellates can be found: *P. cinctum, P. palustre, P. allorgei, P. umbonatum, P. inconspicuum, P. centenniale, P. pseudolaeve, Glenodinium uliginosum, G. steinii* and *Hemidinium nasutum* (Lefèvre 1932). In the same environment can also be found species belonging to the following motile genera: *Amphidinium, Gymnodinium, Woloszynskia, Gyrodinium, Sphaerodinium* (Thompson 1950; Javornicky & Popovsky 1971), as well as attached species belonging to the genera *Cystodinium, Dinococcus, Tetradinium, Stylodinium, Phytodinium,* etc. (Javornicky & Popovsky 1971).

3.7 Brackish water bodies

In brackish water, species of *Amphidinium, Glenodinium* (*G. foliaceum, G. armatum*), *Gymnodinium* (*G. albulatum*), *Gyrodinium, Katodinium* (Droop 1953) and *Heterocapsa triquetra* (Höll 1928) are found.

Freshwater dinoflagellates are usually stenohaline and most are halophobic (Höll 1928), but a bloom of *Gymnodinium aeruginosum* was described (Walker 1973) in Lake Werowrap (Western Victoria, Australia), a small crater lake with a sodium chloride and sodium bicarbonate type of water, pH 9·8 and salinity 23–56 g l^{-1}. The water bloom occurred during the summer–autumn months (1970) with a peak of 158 000 cells ml^{-1}; a second peak was recorded during mid-winter. *Gymnodinium* replaced *Anabaena spiroides* bloom when the salinity increased. Also in *Gymnodinium*, body size variations were observed during the bloom period.

The only record of a clearly freshwater dinoflagellate in a marine environment is *Diplopsalis acuta* fa. *halophyta* from the Caspian Sea (Lindemann 1927b). In another case, a marine form was found in a freshwater body: *Ceratium pentagonum* f. *turgidus* in Lago Cabacero (Tierra del Fuego, South America) (Thomasson 1955). Most freshwater dinoflagellates will not tolerate more than 100 mg chloride per litre, but *P. willei* and *P. cinctum* have been found in water bodies with 220 mg Cl l^{-1} (Höll 1928). *P. cinctum* fa. *westii* blooms in Lake Kinneret which has had a water chloride content of 240 mg Cl^{-1} in the last 13 years, although it was 345 mg Cl l^{-1} in 1967 (Oren, in C. Serruya 1978) before the diversion of the salt springs.

4 FACTORS AFFECTING DISTRIBUTION AND ABUNDANCE OF FRESHWATER DINOFLAGELLATES

4.1 Wind and currents

Wind plays an important role in the development of freshwater dinoflagellate blooms and in their vertical and horizontal distribution just as it does in the marine environment (Chapter 11A), because it causes turbulent mixing. One consequence is the resuspension of sediments holding cysts of dinoflagellates. The turbulence generated by the wind (intensity $>3{\cdot}5$ m s^{-1}) blowing during the premitotic and mitotic phase of *P. cinctum* fa. *westii* (in Lake Kinneret) negatively affects its division rate (Pollingher & Zemel 1981). Strong wind velocity (>5 m s^{-1}) eliminates the vertical stratification of *Peridinium* in Lake Kinneret and pushes down the population which then aggregates above the metalimnion. *Peridinium* and *Ceratium* never sink down to the thermocline in Lake Kinneret. Very high turbulence can sometimes destroy cells of *Peridinium* and *Ceratium*.

Wind-induced water motion leads to a heterogeneous, patchy distribution of *Peridinium* in Lake Kinneret (Berman & Rodhe 1971; Pollingher & Berman 1975; S. Serruya 1975; Berman 1978). The correlation between wind intensity and the spatial distribution of *Ceratium* was described in Esthwaite Water lake (George & Heaney 1978; Heaney & Talling 1980). Heaney & Talling have demonstrated that high densities of cells occur at the surface as a result of the combined effect of horizontal transport and upwelling of deep water rich in *Ceratium*.

The horizontal circular current pattern prevailing in Lake Kinneret causes patches of *Peridinium* to concentrate near the shore (Serruya & Serruya 1972). Strong upwelling near the shore, due to the oscillation of internal seiches in the late spring and summer months and to wind-driven currents throughout the year, forces the *Peridinium* cells into the surface layer at the end of the bloom. The conditions prevailing at the surface of the lake (maximum radiation and heat) at that time are unfavourable and damage the cells (S. Serruya 1975, 1978).

4.2 Light

Most photosynthetic dinoflagellates prefer abundant light and are located in the upper layers of the water bodies. They generally avoid red-brown peat waters (Höll 1928), but they are present in the black waters of the Amazon River (Uherkovich 1976; Uherkovich & Rai 1979). The positive phototrophic response of the *Peridinium* cells in Lake Kinneret leads to a phytoplankton stratification in February–March, earlier than thermal stratification of the water (Pollingher & Berman 1975). In *P. cinctum* fa. *westii* cultures, at three different temperatures between 15 and 25°C and a light/dark period of 12:12 hours, yields were found to increase with temperature at the two highest, but not at the lowest intensities of light (Dubinsky 1975 in Rodhe 1978). Harris *et al.* (1979) studied the vertical distribution of *Ceratium* in Esthwaite Water lake in relation to irradiance, temperature and dissolved oxygen, and found a maximum population density at an irradiance level of 140 μmE m^{-2} s^{-1}. Highest yields of *P. cinctum* fa. *westii* in cultures have been obtained by Dubinsky & Berman (1981) with an irradiance of 170 μmE m^{-2} s^{-1} and a temperature of 25°C. *P. cinctum* fa. *westii* is able to survive in complete darkness for 7–9 days (U. Pollingher, unpubl.). This means that if cells are carried down from the photic zone they can survive for some time and return to the upper layer. Self-regulation of the effective light climate, with reference to self-shading behaviour and active vertical migration of the *Ceratium* population, is discussed by Talling (1971).

4.3 Temperature

Temperature does not seem to play an important role in the distribution of freshwater dinoflagellates, since most species are not strong stenotherms (Höll

1928). Höll indicated seven cold and six warm stenotherm forms among the sixty species he studied. Those he classified as cold stenotherm forms are: *Gymnodinium tenuissimum, Peridinium quadridens, Gymnodinium leopoliense, Gymnodinium veris, Glenodinium aciculiferum, Gymnodinium helveticum* and *Peridinium apiculatum*. It is worth noting that *P. aciculiferum* lives at the surface below the ice cover in Lake Erken, accompanied by *Gymnodinium helveticum* and *P. euryceps* considered by Nauwerck (1963) to be extreme cold stenotherm forms. During the winter, *Amphidinium hialinum* can also be found. The endemic species *Peridinium baicalense* has an optimal temperature of 2–4·5°C, and *Gyrodinium pascheri* (syn. *G. nivale*) is known as an ice or snow form (Huber-Pestalozzi 1950). According to Höll the warm stenotherm forms are: *Kolkwitziella salebrosa, Gymnodinium mirabile, Diplopsalis acuta, Peridinium elpatiewskyi, Glenodinium gymnodinium* and *Ceratium hirundinella*. The following species were described as tropical forms, recorded only in warm regions: *Ceratium brachyceros, P. gutwinskii, P. baliense, P. wildemani, P. africanum, P. gatunense* and *Gymnodinium varians*.

The following species are not stenotherm but prefer the warmest time of the year: *P. volzii, P. cinctum, P. tabulatum* and *P. minusculum* (Höll 1928). Some species are relatively indifferent to temperature: *Glenodinium cinctum, Ceratium cornutum* (temperature range 5·4–21·5°C) (Findenegg 1943), *P. bipes* and *P. willei*; the latter was found to be perennial under the ice cover (Höll 1928). Temperature is also important in the process of excystation and encystation of *Peridinium cinctum* (Huber & Nipkow 1923; Eren 1969): see Chapter 14. Wesenberg-Lund (1908) and Dottne-Lindgren & Ekbohm (1975) concluded that the *Ceratium* population reacts in development only to an increase of temperature and not to a decrease.

4.4 Oxygen

Most of the dinoflagellates prefer well-oxygenated water bodies and avoid the polluted ones. They are absent in fishponds and ponds supplied with sewage water (Höll 1928). A lack of oxygen leads to sheddings of the flagella, immobility and ecdysis by *P. cinctum* fa. *westii* in Lake Kinneret. *Peridinium* from Lake Kinneret, sampled during the steady phase of the bloom and maintained for some hours in conditions with insufficient oxygen, undergo encystment.

4.5 pH

Dinoflagellates are present in water bodies with a pH ranging from 3·5 to 9·0. The number of species found in acidic and acidified lakes is low (see p. 515), but they can occur in very high numbers. The most common bloom-forming species thrive in alkaline water bodies. Höll (1928) found that among the sixty forms of dinoflagellates he studied, forty species were 'stenoion' and six 'euryion'. Among the stenoion, thirteen were oxyphile and twenty-four alkalinophile. *Ceratium hirundinella* can tolerate a wide range of pH under experimental conditions but the optimal levels are 7·0–7·5 (Bruno 1975). *P. cinctum* fa. *ovoplanum* preferred pH 7·5 in Carefoot's experiments (1968). *P. cinctum* fa. *westii* blooms in Lake Kinneret at a pH ranging from 8 to 9, and grows in cultures at a pH of 8·3 (Rodhe 1978).

4.6 Calcium amd magnesium

Dinoflagellates are generally associated with high calcium and magnesium concentrations. Calcium causes the precipitation of phosphate-phosphorus. Whereas the resulting low concentration of phosphorus in the water is detrimental to most algae, it has little effect on dinoflagellates as they are able to store intracellular phosphorus reserves. (Serruya & Berman 1975; Elgavish *et al.* 1982). Most freshwater dinoflagellates prefer hard waters with high levels of calcium. In calcium-poor waters ($<10\,mg\,l^{-1}$) with a pH below 6, *P. raciborski* dominates.

In waters with 10–18 mg Ca l^{-1} and pH 6–7, *Gymnodinium* species are present. *Gymnodinium neglectum* was found in water containing more than 100 mg Ca l^{-1} (Höll 1928). Most dinoflagellate species are found in waters containing more than mg Ca l^{-1} and with a pH of more than 7·0, with a low content of humic compounds (Höll 1928). Some species are eurytrophic, such as *P. cinctum, P. willei* and *C. hirundinella*. *P. cinctum* fa. *westii* blooms at 50 mg Ca l^{-1} in Lake Kinneret (C. Serruya 1978).

4.7 Nitrogen and phosphorus

Some dinoflagellates can utilize very low concentrations of nitrogen. Barker (1935; quoted by Ketchum 1954) found that growth rate of *Prorocentrum micans* and *Peridinium sp.* was not limited unless the sodium nitrate was less than 0·001 mg l^{-1} (Ketchum 1954). Carefoot (1968) obtained maximum growth of *P. cinctum* fa. *ovoplanum* with nitrate as compared with glycine, glycylglycine and glutamate. *Ceratium hirundinella* in cultures prefers $NaNO_3$ and urea (Bruno 1975); NH_4Cl and NH_4NO_3 permit growth of *Ceratium* but are less efficient than $NaNO_3$. It seems that NH_4 has a tendency to be toxic to *Ceratium* at high levels (Bruno 1975). McCarthy *et al.* (1982; and Sherr *et al.* 1982) found that *P. cinctum* fa. *westii* in Lake Kinneret and in cultures prefers ammonium and not nitrate.

In most freshwater bodies in which dinoflagellate blooms occur, the concentration of orthophosphate is low (1–10 $\mu g\,l^{-1}$). In cultures, *Ceratium hirundinella* is able to use both inorganic and organic sources of phosphate with better growth resulting from the organic sources (Bruno 1975). *P. cinctum* fa. *westii* cells grown in a culture medium containing 1500 mg P l^{-1}, when washed and transferred to a medium without phosphorus, continued to grow through three to four divisions (Serruya & Berman 1975). *P. cinctum* fa. *westii* will grow in culture on compounds such as glucose-6-phosphate, glycerophosphate and ATP as the only phosphorus source (Serruya & Berman 1975).

4.8 Trace metals and vitamins

Trace metals are used in cultures of *Ceratium* (Bruno 1975), *P. cinctum* fa. *ovoplanum* (Carefoot 1968) and *P. cinctum* fa. *westii* (Lindstrom & Rodhe 1978). Carefoot (1968) found that *P. cinctum* fa. *ovoplanum* does not require vitamins for growth. Thiamine, biotin and vitamin B_{12} did not stimulate its growth. *P. cinctum* fa. *westii* can grow in a synthetic medium (Lindstrom's medium) without any addition of vitamins, but with an addition of selenium. Maximum cell production was reached at a selenium concentration of 50 ng l^{-1} (Lindstrom & Rodhe 1978).

Ceratium hirundinella cultures need a vitamin mix of thiamine, nicotinic acid, biotin and B_{12} (Bruno 1975).

5 SPECIES INTERACTIONS

5.1 Species accompanying *Ceratium* and *Peridinium* water blooms

In many lakes, the *Ceratium* bloom does not reach high densities and occurs in association with other phytoplankters, such as *Aphanizomenon, Dinobryon, Asterionella* and *Melosira* (Stange-Bursche 1963), or in association with green-algae (Hickel 1978). In most cases, when one species of a dinoflagellate is dominant, other less abundant dinoflagellate species accompany it. The bloom of *P. cinctum* fa. *westii* in Lake Kinneret is accompanied by *P. cunningtonii, P. inconspicuum, C. hirundinella, Glenodinium gymnodinium* and *Diplopsalis acuta*. The last two species are more abundant in the littoral area. The maximum density of the *Ceratium* population always occurs in May, after the *Peridinium* peak which generally occurs in April. A similar situation was observed by Pfiester (1971) in a small pond in Calloway County (Kentucky), where a maximum population density of *Ceratium* was observed in April, *P. cinctum* achieving its maximum development in March. A different timing was observed in Crosemere (England) (Reynolds 1976), where *Peridinium* reaches its maximum at the beginning of July, whereas that of *Ceratium* was at the end of August.

Shading and self-shading effects are discussed in Chapter 10.

5.2 Parasitism

At the end of the bloom in Lake Kinneret, only a very small number of dead cells of *Peridinium* are infected by Chitridiales. Infected cysts of this species have not been recorded. Canter (1961) described the chytrid *Amphicypellus elegans* as infecting cells of *Ceratium* and *Peridinium* from Blelham Tarn, and Esthwaite Water (England). She also described the chytrid *Rhizophidium nobile* parasitizing the resting spores of *Ceratium hirundinella,* especially during the autumn months of September and October (Canter 1968). Canter & Heaney (1985) have discussed the impact of fungal infection on population dynamics in English lakes.

6 ROLE OF DINOFLAGELLATES IN THE FRESHWATER FOOD CHAIN

Due to their relatively large body size, most of the dinoflagellates cannot be used directly as food by freshwater zooplankters. However, at the end of the *Peridinium* bloom in Lake Kinneret, the ciliate *Bursariella truncatella* has been seen to contain one to six cells of *Peridinium.* The rotifers *Hydatina* and *Asplanchna* sp., *A. priodonta* and *A. herricki*, are known to feed on *Ceratium* and *Peridinium* (Barrois 1894; Entz 1927; Gophen 1973; Gilyarov 1977). *P. penardii* was found in the gut of *Daphnia* during the bloom period in Clear Lake (Horne *et al.* 1971). Entz (1927) described the presence of *Ceratium* in the stomach of fingerlings from Lake Balaton. *Glenodinium quadridens* was found in the gut contents of the fry of *Tilapia esculenta* as encysted cells or non-motile vegetative cells. Motile cells developed, in cultures derived from gut contents, indicating that the vegetative cells and/or the cysts were able to survive passage through the gut (Evans 1962).

The fish *Tilapia galilaea* (*Sarotherodon galilaeus*) feeds selectively on *P. cinctum* fa. *westii* in Lake Kinneret (Spataru 1976; Spataru & Zorn 1976). Most of the cells are found to be disintegrated in the digestive tract; only a small fraction is excreted undigested, but as non-viable protoplasts. It is likely that no more than 50 000 tons of *Peridinium* (fresh weight) per year are consumed by *T. galilaea* (Landau 1979). The second species of *Tilapia* present in the Lake, *T. aurea,* ingests *Peridinium,* but the dinoflagellate is expelled intact (Spataru & Zorn 1978). At the end of the bloom, most of the *Peridinium* population dies off.

The specific productivity of dinoflagellates *in situ* is at least tenfold lower than that of nannoplanktonic species (Gutelmacher 1975; Desortova 1976; Pollingher & Berman 1982). The values of carbon assimilated per carbon biomass by *Peridinium* cells in Lake Kinneret vary between 0·01 and 0·19 compared with those of *Chroococcus* which are 0·9–0·77 (Pollingher & Berman 1982).

Measurements made *in situ* showed that *Peridinium* cells at the beginning of the bloom are more active than those at the end of the bloom (Pollingher & Berman (1975, 1982). 'The dinoflagellates are able to build up high biomass, due to their long life span and low specific productivity they immobilize large amounts of nutrients and slow down the overall activity of the ecosystem' (Serruya *et al.* 1980).

7 CONCLUSIONS

Most freshwater dinoflagellates are eurytrophic forms. For no obvious reason, their blooms occur generally during the summer months, when the levels of nutrients are at their minimum in most water bodies. Some dinoflagellates replace other algae when drastic changes in environmental conditions occur, e.g. acidification, very high salinity, etc. Some are present in ephemeral water bodies and in newly built man-made lakes, when the conditions are not stabilized. The following features favour the development of dinoflagellates in freshwater.

1 Life cycle: the alternation of a motile vegetative cell and a resistant benthic resting cyst, allowing encystment to withstand unfavourable environmental conditions and hatching when conditions become optimal.

2 The capacity for 'luxury consumption' and storage of nutrients, especially phosphorus.

3 Motility, which enables the dinoflagellates to (i) screen nutrients from much of the water column, (ii) aggregate at an optimal or compromise depth, and (iii) reduce the development of other algae by shading.

4 A relatively long life span.

5 A relatively large body size (400 μm), which makes these algae inedible to most freshwater grazing organisms, thus preventing losses due to grazing.

All these facts suggest that the abundance of dinoflagellates is due to their ability to develop in conditions which are unfavourable to other algae. The abundance of dinoflagellates since the Jurassic (Tappan & Loeblich 1971) is additional evidence of their ability to adapt to drastic changes in the environmental conditions.

8 REFERENCES

Adachi R. (1965) Studies on a dinoflagellate *Peridinium polonicum* Wol. I. The structure of skeleton. *J. Fac. Fish. Pref. Univ. Mie* **6**, 317–326.

Almer B. (1974) Effects of acidification on Swedish lakes. *Ambio* **3**, 30–36.

Anonymous. (1979) Symposium on some problems of water bloom or so-called freshwater red tide. *Bull. Plankt. Soc. Jpn.* **26**, 110–122.

Barrois T. (1894) Contribution a l'étude de quelques lacs de Syrie. *Rev. biol. Nord. France* **6**, 224–314.

Baumeister W. (1963) Dinophyceen aus perennierenden Gewässern des Schwingrasenmoores bei Burgperg, sowie aus Alpsee, Freibergsee und dem Moorweicher in Oberstdorf (Algän). *Arch. Protistenk.* **106**, 535–552.

Bayly I. A. E., Ebsworth E. P. & Wan H. F. (1975) Studies on the lakes of Fraser Island, Queensland. *Aust. J. mar. freshw. Res.* **26**, 1–13.

BERMAN T. (1978) *Peridinium cinctum* fa. *westii* (Lemm.) Lef. Spatial distribution in the lake. In *Lake Kinneret* (ed. C. Serruya), pp. 291–292. W. Junk, The Hague.

BERMAN T. & DUBINSKY Z. (1985) The autoecology of *Peridinium cinctum* fa. *westii* from Lake Kinneret. *Verh. Int. Ver. Limnol.* **22**, 2850–4.

BERMAN T. & RODHE W. (1971) Distribution and migration of *Peridinium* in Lake Kinneret. *Mitt. Int. Verein. Limnol.* **19**, 266–276.

BERMAN T., SHERR B. F., SHERR E., WYNNE D. & MCCARTHY J. J. (1984) The characteristics of ammonium and nitrate uptake by phytoplankton in Lake Kinneret. *Limnol. Oceanogr.* **29**, 287–97.

BISWASS, S. (1975) Phytoplankton in Volta Lake, Ghana, during 1964–1973. *Verh. Int. Verein. Limnol.* **19**, 1928–1934.

BOURRELLY P. (1968) Note sur *Peridiniopsis borgei* Lemm. *Phykos* **7**, 1–2.

BOURRELLY P. (1970) *Les Algues d'eau douce. III. Algues bleues et rouges.* Editions N. Boubée & Cie., Paris.

BREZONIK P. L. & FOX L. (1974) The limnology of selected Guatemalan lakes. *Hydrobiology* **45**, 467–487.

BRINSON M. M. & NORDLIE F. G. (1975) Lake Izabel, Guatemala. *Verh. Int. Verein. Limnol.* **19**, 1468–1479.

BRTEK J. & ROTHSCHEIN J. (1964) Ein Beitrag zur Kenntnis der Hydrofauna und des Reinhes zustandes des Tschechoslowakischen Abschnittes der Donau. *Biol. Pr. Slov. Akad. Vied.* **10**, 1–61.

BRUNO S. F. (1975) *Cultural studies on the ecology of* Ceratium hirundinella *(OFM) Bergh., with notes on cyclomorphis.* Ph.D. Thesis, Fordham Univ.

BURNS C. W. & MITCHELL S. F. (1974) Seasonal succession and vertical distribution of phytoplankton in Lake Hayes and Lake Johnson, South Island, New Zealand. *N.Z. J. mar. freshw. Res.* **8**, 167–209.

CANTER H. M. (1961) Studies on British chytrids. XVIII. Further observations on species invading planktonic algae. *Nova Hedwigia* **3**, 73–78.

CANTER H. M. (1968) Studies on British chytrids. XXVIII. *Rhizophydium nobile* sp. nov. parasitic on the resting spore of *Ceratium hirundinella* O. F. Müll. from the plankton. *Proc. Linn. Soc. Lond.* **179**, 197–201.

CANTER H. M. & HEANEY S. I. (1984) Observations on zoosporic fungi of *Ceratium* spp. in lakes of the English Lake District: importance for phytoplankton population dynamics. *New Phytol.* **97**, 601–12.

CAREFOOT J. R. (1968) Culture and heterotrophy of the freshwater dinoflagellate *Peridinium cinctum* fa. *ovoplanum* Lindeman. *J. Phycol.* **4**, 129–131.

CASSIE V. (1968) Seasonal variation in phytoplankton from Lake Rotorua and other inland waters, New Zealand, 1966–67. *N.Z. J. mar. freshw. Res.* **3**, 98–123.

CASSIE V. (1978) Seasonal changes in phytoplankton densities in four North Island lakes, 1973–74. *N.Z. J. mar. freshw. Res.* **12**, 153–166.

CRISCULO C. M., DUBINSKY Z. & AARONSON S. (1981) Skeleton shedding in *Peridinium cinctum* from Lake Kinneret—a unique phytoplankton response to nutrient imbalance. In *Developments in Arid Zone Ecology and Environmental Quality.* (ed. H. Shuval), pp. 169–76. Balaban ISS, Philadelphia

CRONBERG G. (1981) SEM and ecological investigations of *Peridiniopsis borgei* Lemm., *P. polonicum* (Wolosz.) Bourr. from Sweden and *P. palustre* var. *raciborskii* (Wolosz.) Lef. from Brazil. Preprint. Hexrose Conf. on Modern & Fossil Dinoflagellates. Tübingen F. R. Germany

DAILY W. A. (1960) Forms of *Ceratium hirundinella* (O. F. Müller) Schrank in lakes and ponds of Indiana. *Proc. Indiana Acad. Sci.* **70**, 213–215.

DAKIN W. J. & LATARCHE M. (1913) The plankton of Lough Neagh: a study of the seasonal changes in the plankton by quantitative methods. *Proc. R. Irish. Acad.* **30B**, 20–96.

DESORTOVA B. (1976) Productivity of individual algal species in natural phytoplankton assemblage determined by means of autoradiography. *Arch. Hydrobiol. Suppl.* **49**, 415–449.

DOTTNE-LINDGREN A. & EKBOHM G. (1975) *Ceratium hirundinella* in Lake Erken: horizontal distribution and form variation. *Int. Rev. ges. Hydrobiol.* **60**, 115–144.

DROOP M. R. (1953) On the ecology of flagellates from the brackish and freshwater rockpools of Finland. *Acta Bot. Fenn.* **51**, 3–52.

DUBINSKY Z. & BERMAN T. (1981) Environmental parameters affecting growth and photosynthesis in *Peridinium cinctum*. Abstract ASLO 44th Ann. Mtg, Milwaukee, Wisconsin, 15–18 June 1981.

ELGAVISH A., HALMAN M. & BERMAN T. (1982) A comparative study of phosphorus utilization and storage in batch cultures of *Peridinium cinctum, Pediastrum duplex* and *Cosmarium* sp. from lake Kinneret (Israel). *Phycologia* **21**, 47–54.

ENTZ G. (1926) Beitrage zur Kenntnis der Peridineen. 1. Zur Morphologie und Biologie von *Peridinium borgei. Arch. Protistenk.* **56**, 397–446.

ENTZ G. (1927) Uber Peridineen der Balaton-Sees. *Arch. Balaton.* **1**, 275–342.

ENTZ G. (1931) Analyse des Wachstums und der Teilung einer Population sowie eines Individuums des Protisten *Ceratium hirundinella* unter den natürlichen Verhältnissen. *Arch. Protistenk.* **74**, 310–361

ENTZ G., KOTTASZ J. & SEBESYTEN O. (1937) Quantitive Untersuchungen am Bioseston des Balatons. *Arbeiten der I. Abteilung des Ungarischen Biologischen Forschungsinstitutes* **11** (Tihany), 1–144.

EREN J. (1969) Studies on development cycle of *Peridinium cinctum* fa. *westii. Verh. int. Verein. Limnol.* **17**, 1013–1016.

EVANS J. H. (1962) Some new records and forms of algae in Central East Africa. *Hydrobiologia* **20**, 59–86.

FINDENEGG I. (1943) Untersuchungen über die Ökologie und die Produktions-verhältnisse des Planktons im Kärntner Seengebiete. *Int. Rev. ges. Hydrobiol.* **43**, 368–429.

FLORIN M. B. (1957) Plankton of fresh and brackish waters in the Södertälje area. *Acta Phytogeogr. Suec.* **37**, 1–144.

FREMPONG E. (1984) A seasonal sequence of diel distribution patterns for the planktonic dinoflagellate *C. hirundinella* in a eutrophic lake. *Freshwat. Biol.* **14**, 401–21.

GANAPATI S. V. (1957) Limnological studies of two upland waters in the Madras State. *Arch. Hydrobiol.* **53**, 30–61.

GEORGE D. G. & HEANEY S. I. (1978) Factors influencing the spatial distribution of phytoplankton in a small productive lake. *J. Ecol.* **66**, 133–155.

GESSNER F. & HAMMER L. (1967) Limnologische Untersuchungen an Seen der venezolanischen Hochanden. *Int. Rev. ges. Hydrobiol.* **52**, 301–320.

GILYAROV A. (1977) Observations on food composition of rotifiers of the genus *Asplanchna. Zoologhiceskii J.* **56**, 1874–1876.

GLIWICZ Z. M. (1976) Plankton photosynthetic activity and its regulation in two neotropical man-made lakes. *Pol. Arch. Hydrobiol.* **23**, 61–93.

GOPHEN M. (1973) Zooplankton in Lake Kinneret. In *Lake Kinneret Data Record* (Ed. by T. Berman), pp. 61–67. Israel Scientific Research Conference.

GRIFFITHS B. M. (1923) The phytoplankton of bodies of freshwater and the factors determining its occurrence and composition. *J. Ecol.* **11**, 184–213.

GUTELMACHER B. L. (1975) Relative significance of some species of algae in plankton primary production. *Arch. Hydrobiol.* **75**, 318–328.

HARRIS G. P., HEANEY S. I. & TALLING J. F. (1979) Physiological and environmental constraints in the ecology of the planktonic dinoflagellate *Ceratium hirundinella. Freshw. Biol.* **9**, 413–428.

HASHIMOTO Y., OKAICHI T., DANG D. L. & NOGUCHI T. (1968) Glenodinine, an ichthyotoxic substance produced by a dinoflagellate, *Peridinium polonicum. Bull. Jap. Soc. Sci. Fish.* **34**, 528–533

HEANEY S. I. (1976) Temporal and spatial distribution of the dinoflagellate *Ceratium hirundinella* O. F. Müller within a small productive lake. *Freshw. Biol.* **6**, 531—542.
HEANEY S. I. & TALLING J. F. (1980) Dynamic aspects of dinoflagellate distribution patterns in a small productive lake. *J. Ecol.* **68**, 75–94.
HECKY R. E. & KLING H. J. (1981) The phytoplankton and protozooplankton of the euphotic zone of Lake Tanganyika: Species composition, biomass, chlorophyll content, and spatio-temporal distribution. *Limnol. Oceanogr.* **26**, 548–564.
HECKY R. E., FREE E. J., KLING H. & RUDD J. W. M. (1978) Studies on the planktonic ecology of Lake Tanganyika. *Fish. mar. Serv. Tech. Rep., Fish. Env. Can.* No. 816.
HERTZIG R., DUBINSKY Z. & BERMAN T. (1981) Breakdown of *Peridinium* biomass in Lake Kinneret. In *Developments in Arid Zone Ecology and Environmental Quality.* (ed. H. Shuval), pp. 179–85. Balaban ISS, Philadelphia.
HICKEL B. (1973) Limnological investigations in lakes of the Pokhara Valley, Nepal. *Int. Rev. ges. Hydrobiol.* **58**, 659–672.
HICKEL B. (1978) Phytoplankton succession in the Grebiner See during artificial aeration of the hypolimnion. *Arch. Hydrobiol.* **82**, 216–230.
HICKEL B. (1985) The population structure of *Ceratium* in a small eutrophic lake. *Verh. Int. Ver. Limnol.* **22**, 2845–9.
HÖLL K. (1928) Oekologie der Peridineen. *Pflantzen Forschung* **11**, 1–105.
HORNE A. J., JAVORNICKY P. & GOLDMAN C. R. (1971) A freshwater 'red tide' on Clear Lake, California. *Limnol. Oceanogr.* **16**, 684–688.
HUBER G. & NIPKOW F. (1923) Experimentelle Untersuchungen über Entwicklung und Formbildung von *Ceratium hirundinella* O. F. M. *Flora* **16**, 114–215.
HUBER-PESTALOZZI G. (1950) *Das Phytoplankton des Susswassers.* Die Binnengewasser 16(3). E. Schweizerbart'sche Verlagsbuch, Stuttgart.
HUBER-PESTALOZZI G. (1951) Die Verbreitung der Susswasserceratien auf der Erde. *Verh. int. Verein. Limnol.* **11**, 152–188.
JAVORNICKY P. & POPOVSKY J. (1971) *Pyrrhophyta common in Czechoslovakia.* Mimeograph.
JURGENS K. C. (1953) The red tide of Lake Austin, Texas. *Game and Fish* **2**, 8.
KETCHUM B. H. (1954) Mineral nutrition of phytoplankton. *Ann. Rev. Plant. Physiol.* **5**, 55–74.
KISS I. (1978) Algological investigations in the Dead-Tisza at Lakitelek-Töserdö. *Tiscia* **13**, 27–47.
KISSELEW J. A. & ZWETKOW W. N. (1935) Zur Morphologie und Okologie von *Peridinium baicalense* n.sp. *Beih. Bot. Centralbl.* **3**, 518–524.
KLOTTER H. R. (1954) Eine Wasserblüte von *Peridinium pusillum* (Penard) Lemmermann. *Arch. Hydrobiol. Suppl.* **20**, 144–156.
KOFOID C. A. (1908) The plankton of the Illinois River 1894–1899, with introductory notes upon the hydrography at the Illinois River and its basin. II. Constituent organisms and their seasonal distribution. *Bull. Illinois State Lab. nat. Hist.* **8**, 1–360.
KOMAROVSKY B. (1959) The plankton of Lake Tiberias. *Bull. Sea Fish. Res. Stn, Haifa* **25**, 65–96.
KOZHOV M. (1963) *Lake Baikal and its Life.* W. Junk, The Hague.
KRAUSE F. (1911–12) Studien über die Formveränderung von *Ceratium hirundinella* O. F. Müll. als Anpassungserscheinung an die Schwebefähigkeit. *Int. Rev. ges. Hydrobiol.* **3**, 1–32.
LANDAU R. (1979) Growth and population studies on *Tilapia galilaea* in Lake Kinneret. *Freshw. Biol.* **9**, 23–32.
LANGHANS V. H. (1925) Gemischte Population von *Ceratium hirundinella* (O. F. M.) Schrank und ihre Deutung. *Arch. Protistenk.* **52**, 585–602.
LEFÈVRE M. (1932) Monographie des espèces d'eau douce du genre *Peridinium* Ehrb. *Mem. No. 5* (Caen), 1–208.
LEMMERMAN E. (1910) *Kryptogamenflora der Mark Brandenburg.* Algen I. Bd. III. Gebrüden Borntraeger, Leipzig.

LEWIS W. M. (1973) A limnological survey of Lake Mainit, Philippines. *Int. Rev. ges. Hydrobiol.* **58**, 801–818.

LEWIS W. M. & WEIBEZAHN F. (1976) Chemistry, energy flow and community structure in some Venezuelan freshwaters. *Arch. Hydrobiol. Suppl.* **50**, 145–207.

LIND E. M. (1968) Notes on the distribution of phytoplankton in some Kenya waters. *Br. Phycol. Bull.* **3**, 481–493.

LINDEMANN E. (1927a) Dinoflagellaten aus der Wolga. *Arb. Biol. Wolga Stn* **9**, 1–3.

LINDEMANN E. (1927b) Über einige Dinoglafellaten des Kapischen Meeres. *Arch. Protistenk.* **59**, 417–422.

LINDSTROM K. (1984) Effect of temperature, light and pH on growth, photosynthesis and respiration of the dinoflagellate *Peridinium cinctum* fa. *westii* in laboratory cultures. *J. Phycol.* **20**, 212–20.

LINDSTROM K. & RODHE W. (1978) Selenium as a micronutrient for the dinoflagellate *Peridinium cinctum* fa. *westii. Mitt int. Verein. Limnol.* **21**, 168–173.

LÖFFLER H. (1962) Contribution to the limnology of high mountain lakes in Central America. *Int. Rev. ges. Hydrobiol.* **57**, 397–408.

MCCARTHY I. I., WYNNE D. & BERMAN T. (1982) The uptake of nitrogenous nutrients by microplankton in Lake Kinneret. *Limnol. Oceanogr.* **27**, 673–680.

MESSIKOMER E. (1942) Beitrag zur Kenntnis der Algenflora und Algenvegetation des Hochgebirges um Davos. *Beitr. Geobot. Landes. Schweiz* **24**, 305–380.

MOORE J. M. (1977) Some factors influencing the density of subarctic populations of *Bosmina longirostris, Holopedium gibberum, Codonella cratera* and *Ceratium hirundinella. Hydrobiologia.* **56**, 199–207.

MORI S. & MIURA T. (1980) List of plant and animal species living in Lake Biwa. *Mem. Fac. Sci. Kyoto Univ. Ser. Biol.* **8**, 1–33.

MORLING G. (1979) *Ceratium carolinianum* and the related *Ceratium carolinianum* var. *elongatum* and *Ceratium cornutum.* A study of the distribution and morphology. *Nova Hedwigia* **31**, 937–956.

MUELLER-ELSER M. & SMITH W. O. (1985) Phased cell division and growth rate of a planktonic dinoflagellate *Ceratium hirundinella* in relation to environmental variables. *Arch. Hydrobiol.* **104**, 477–91.

NAKAMOTO N. (1975) A freshwater red tide on a water reservoir. *Jap. J. Limnol.* **36**, 55–64.

NAUWERCK A. (1963) Die Beziehungen zwischen Zooplankton und Phytoplankton im See Erken. *Symb. bot. Upsal.* **17**, 1–163.

OLTEANU M. (1967) Fitoplanctonul cu o privire generala asuzpra florei algale. *In* Acad. Rep. Soc. Romania, *Limnologia Sectorului Roman Dunarii*, Stud. Monogr., Ed. Acad. Rep. Soc. Romania.

OSTENFELD C. H. & WESENBERG-LUND C. (1904–1905) A regular fortnightly exploration of the phytoplankton of the two Icelandic lakes, Thingvallavatn and Myvatn. *Proc. R. Soc. Edinburgh* **25**, 1092.

PADISAK J. (1985) Population dynamics of the freshwater dinoflagellate *Ceratium hirundinella* in the largest shallow lake of Central Europe, Lake Balaton, Hungary. *Freshwat. Biol.* **15**, 43–52.

PEARSAL W. H. (1929) Form variation in *Ceratium hirundinella* O. F. M. *Proc. Leeds Phil. Soc.* **1**, 432–439.

PESTA O. (1929) *Der Hochgebirgsee der Alpen.* Die Binnengewasser, Bd. VIII. E. Schweizerbart'sche Verlagsbuch, Stuttgart.

PFIESTER L. A. (1971) Periodicity of *Ceratium hirundinella* (O. F. M.) Dujardin and *Peridinium cinctum* (O. F. M.) Ehrenberg in relation to certain ecological factors. *Castanea* **36**, 246–257.

POLLINGHER U. (1968) Fluctuations de la biomasse du phytoplancton de Lac Tiberiade. *Verh. int. Verein. Limnol.* **17**, 352–357.

POLLINGHER U. (1978) *Peridinium cinctum* fa. *westii* (Lemm.) Lef. Life cycle. in *Lake Kinneret*

(Ed. by C. Serruya), pp. 271–274. W. Junk, The Hague.

POLLINGHER U. & BERMAN T. (1975) Temporal and spatial patterns of dinoflagellate blooms in Lake Kinneret, Israel (1969–1974). *Verh. Int. Verein. Limnol.* **19**, 1370–1382.

POLLINGHER U. & BERMAN T. (1982) Relative contributions of net and nannophytoplankton to primary production in Lake Kinneret (Israel). *Arch. Hydrobiol.* **96**, 33–46.

POLLINGHER U. & SERRUYA C. (1976) Phased division of *Peridinium cinctum* f. *westii* (Dinophyceae) and development of the Lake Kinneret (Israel) bloom. *J. Phycol.* **12**, 162–170.

POLLINGHER U. & ZEMEL E. (1981) *In situ* and experimental evidence of the influence of turbulence on cell division processes of *Peridinium cinctum* fa. *westii* (Lemm.) Lefèvre. *Br. Phycol. J.* **16**, 281–287.

REYNOLDS C. S. (1973) The phytoplankton of Crose Mere, Shropshire. *Br. Phycol. J.* **8**, 153–162.

REYNOLDS C. S. (1976) Succession and vertical distribution of phytoplankton in response to thermal stratification in a lowland mere, with special reference to nutrient availability. *J. Ecol.* **64**, 529–551.

REYNOLDS C. S. (1978) Notes on phytoplankton periodicity of Rostherne Mere, Cheshire, 1967–1977. *Br. Phycol. J.* **13**, 329–335.

RODHE W. (1978) *Peridinium cinctum* fa. *westii* (Lemm.) Lef. Growth characteristics. In: *Lake Kinneret* (Ed. by C. Serruya), pp. 275–283. W. Junk, The Hague.

RUTTNER F. (1952) Planktonstudien der Deutschen Limnologischen Sunda-Expedition. *Arch. Hydrobiol. Suppl.* **21**, 1–274.

SARKAR S. K., GOVIND B. V. & NATARAJAN A. V. (1977) A note on some distinctive limnological features of Govindsagar Reservoir, Himachal Pradesh, India. *Indian J. anim. Sci.* **47**, 435–439.

SEBESTYEN O. (1952) Quantitative Planktonstudien und das Problem der Produktion. *Acta biol. Acad. Sci. Hung.* **3**, 319–332.

SEBESTYEN O. (1959) The ecological niche of *Ceratium hirundinella* (O. F. Müller) Schrank in the plankton community and in lacustrine life in general. *Acta biol. Acad. Sci. Hung.* **10**, 235–244.

SERRUYA C. (1978) Water chemistry. In *Lake Kinneret* (Ed. by C. Serruya), pp. 185–204. W. Junk, The Hague.

SERRUYA C. & BERMAN T. (1975) Phosphorus, nitrogen and the growth of algae in Lake Kinneret. *J. Phycol.* **11**, 155–162.

SERRUYA C. & POLLINGHER U. (1977) Lowering of water level and algal biomass in Lake Kinneret. *Hydrobiologia* **54**, 73–80.

SERRUYA C. & SERRUYA S. (1972) Oxygen content in Lake Kinneret: physical and biological influences. *Verh. Int. Verein. Limnol.* **18**, 580–587.

SERRUYA C., EDELSTEIN M., POLLINGHER U. & SERRUYA S. (1974) Lake Kinneret sediments: nutrient composition of the pore water and mud–water exchanges. *Limnol. Oceanogr.* **19**, 489–508.

SERRUYA C., GOPHEN M. & POLLINGHER U. (1980) Lake Kinneret: carbon flow patterns and ecosystem management. *Arch. Hydrobiol.* **88**, 265–302.

SERRUYA S. (1975) Wind, water temperature and motions in Lake Kinneret: general pattern. *Verh. Int. Verein. Limnol.* **19**, 73–87.

SERRUYA S. (1978) Water motions. In *Lake Kinneret* (Ed. C. Serruya), pp. 167–171. W. Junk, The Hague.

SHERR E. B., SHERR B. F., BERMAN T. & MCCARTHY J. J. (1982) Differences in nitrate and ammonia uptake among components of a phytoplankton population. *J. Plankton Res.* **4**, 961–5.

SIBLEY T. H., HERRGESELL P. L. & KNIGHT A. W. (1974) Density-dependent vertical migration in the freshwater dinoflagellate *Peridinium penardii* (Lemm.) Lemm. fa. *californicum* Javorn. *J. Phycol.* **10**, 475–476.

SPATARU P. (1976) the feeding habits of *Tilapia galilaea* (Artedi) in Lake Kinneret (Israel). *Aquaculture* **9**, 47–59.

SPATARU P. & ZORN M. (1976) Some aspects of natural feed and feeding habits of *Tilapia galilaea* (Artedi) and *Tilapia aurea* Steindachner in Lake Kinneret. *Bamidgeh* **28**, 12–17.

SPATARU P. & ZORN M. (1978) Food and feeding habits of *Tilapia aurea* (Steindachner) (Cichlidae) in Lake Kinneret (Israel). *Aquaculture* **13**, 67–79.

SREENIVASAN A. (1970) A note on chemical stratification and blooming of *Ceratium* sp. in a man-made lake. *J. mar. biol. Ass. India* **12**, 217–218.

STANGE-BURSCHE E. M. (1963) Beobachtungen und Untersuchungen über die horizontale Phytoplanktonverteilung in Seen. *Arch. Hydrobiol.* **59**, 351–372.

STANKOVIC S. (1960) *The Balkan Lake Ohrid and its Living World.* W. Junk, The Hague.

TALLING F. (1971) The underwater light climate as a controlling factor in the production ecology of freshwater phytoplankton. *Mitt. Int. Verein. Limnol.* **19**, 214–243.

TAPPAN H. & LOEBLICH A. R. JR (1971) Geobiological implications of fossil phytoplankton evolution in time and space. In *Symp. Palynology of the Late Cretaceous and Early Tertiary* (ed R. M. Kosanke & A. T. Cross), pp. 247–340 Geol. Soc. Am. Spec. Pap. 127.

THOMASSON K. (1955) Studies on South American freshwater plankton. 3. Plankton from Tierra del Fuego and Valdiwia. *Acta horti Gotoburg.* **19**, 193–225.

THOMASSON K. (1960) Notes on the plankton of Lake Bangweulu. *Nova Acta Reg. Soc. Sci. Ups.* **17**, 43 pp.

THOMASSON K. (1962) Planktological notes from western North America. *Arkiv Bot.* **4**, 437–463.

THOMASSON K. (1963) Plankton studies in North Patagonia with notes on terrestrial vegetation. Araucanian lakes. *Acta Phytogeogr. Suec.* **47**, 1–139.

THOMASSON K. (1974) Rotorua phytoplankton reconsidered (North Island of New Zealand). *Int. Rev. ges. Hydrobiol.* **59**, 703–727.

THOMASSON K. (1980) Plankton of Lake Kariba re-examined. *Acta phytogeogr. Suec.* **68**, 157–162.

THOMPSON R. H. (1950) A new genus and new records of freshwater Pyrrhophyta in the Desmokontae and Dinophyceae. *Lloydia* **13**, 277–299.

TILZER M. M. (1973) Diurnal periodicity in the phytoplankton of a high mountain lake. *Limnol. Oceanogr.* **18**, 15–30.

UHERKOVICH G. (1976) Algen aus den Flüssen Rio Negro und Rio Tapajos. *Amazoniana* **5**, 465–515.

UHERKOVICH G. (1978) Uber die Algenvegetation der ungarlandischen Strecke der Drau. *Janus Pannonius Muzeum Evkonyve, Pecs* (*Hungaria*) **23**, 7–23.

UHERKOVICH G. & RAI H. (1979) Algen aus dem Rio Negro und seinen Nebenflüssen. *Amazoniana* **6**, 611–638.

VOROBIEVA S. S., ZEMSKAIA T. L., SKRIABIN A. G. & SPIGLAZOVA G. N. (1981) *Plankton Bratzkovo vodohraniliscia.* Izd. Acad. 'Nauka' Sibirskoe Otd., Novosibirsk.

WALKER K. F. (1973) Studies on a saline lake ecosystem. *Aust. J. mar. freshwat. Res.* **24**, 21–71.

WESENBERG-LUND C. (1908) *Plankton investigations of the Danish lakes.* General part. The Baltic freshwater plankton, its origin and variation, pp. 1–389. Nordish Forlag, Copenhagen.

WILLEN T. (1980) Phytoplankton from lakes and ponds on Vestspitsbergen. *Acta Phytogeogr. Suec.* **68**, 173–188.

WOLOSZYNSKA J. (1914) Studien über das Phytoplankton des Victoria Sees. *Hedwigia* **55**, 184–223.

YAN N. D. & STOKES P. (1978) Phytoplankton of an acidic lake, and its responses to experimental alterations of pH. *Environ. Conserv.* **5**, 93–100.

CHAPTER 12
DINOFLAGELLATES IN NON-PARASITIC SYMBIOSES

ROBERT K. TRENCH
Department of Biological Sciences and the Marine Science Institute, University of California at Santa Barbara, Santa Barbara, CA 93106

1 INTRODUCTION

Because of the lack of appropriately preserved material in the fossil record, it is not known when in their evolutionary past dinoflagellates began to explore a symbiotic way of life. The evolutionary history of dinoflagellates in general is somewhat obscure (Tappan 1980; F. J. R. Taylor 1980), and our knowledge of the origins of those dinoflagellates which are symbiotic is highly inadequate. Neither is it known whether those extant dinoflagellate groups which exploit the symbiotic habit are derived from a long line of symbiotic ancestors or represent a relatively recent shift from a free-living existence to symbiosis. Finally, it is not known whether the evolutionary initiation of symbioses involving dinoflagellates was mono- or polyspecific (Trench 1979). The evidence, as indicated by the relatively broad range of algal taxa now recognized as symbionts with an equally broad range of marine invertebrates, suggests that there were probably many attempts at initiating symbioses by different algae (Trench 1981). Although dinoflagellates are currently found as symbionts in a phyletically wide range of marine invertebrates (see McLaughlin & Zahl 1966; D. L. Taylor 1974) and are thought to be important, if not central elements in the ecological success of their respective host groups, only among the 'hermatypic' corals and perhaps the foraminifera (Lee *et al.* 1979) does the fossil record tend to support such conjecture. Among invertebrate paleontologists, it is a widely held view that the rapid ecological success of the scleractinian corals since their appearance in the Triassic was a direct result of their acquisition of dinoflagellate symbionts (Wells 1956; Heckel 1974). Indeed, some coral reef biologists believe that a modern coral reef ecosystem, known for its richness and diversity (Fig. 12.1a, b), would probably not exist but for the mutualistic symbiosis between corals and dinoflagellates. Whatever the selective forces might initially have been, the coexistence of phototrophic dinoflagellates with

Fig. 12.1. (a) Underwater photograph of a portion of the fore reef slope at Discovery Bay, Jamaica, at a depth of about 13 m. The two dominant corals shown are *Montastrea annularis* and *Acropora cervicornis*. (b) A 'soft coral', the gorgonian *Plexaura* is also a common inhabitant of this area.

heterotrophic invertebrates has resulted in ecologically significant phenomena that hardly need to be itemized; the exploitation of nutrient-poor environments being only one, frequently cited example (see Lewis 1973; Muscatine & Porter 1977).

Currently the evidence indicates that at least two different groups, amphidinioid and gymnodinioid dinoflagellates, are involved as non-parasitic symbionts in the marine environment. The initial view of a single alga (a 'zooxanthella') distributed among all the various host taxa is beginning to give way to a more selective and restrictive view (Trench 1981; Trench *et al.* 1981a; F. J. R. Taylor 1982).

It is important to recognize that dinoflagellate-invertebrate symbioses represent the coexistence of at least two genomes of evolutionarily divergent origins (Trench 1979, 1980; F. J. R. Taylor 1979). Hence, particularly in those cases where the algae reside within animal cells, mechanisms whereby the potential host normally eliminates a foreign invader would have had to be circumvented. Following the initiation of such associations, further adaptations on the part of both host and symbiont could potentially lead to a mutually beneficial coexistence and perpetuation of such consortia through subsequent generations.

Details of many aspects of the mechanisms involved in the initiation, maintenance, regulation and inheritance of symbioses involving dinoflagellates are poorly understood. In this article, I shall attempt a synthesis of the pertinent data available. Many of my own biases will inadvertently be expressed. These ideas and interpretations should be regarded as those that I currently advocate and not an expression of dogma.

2 TAXONOMY AND PHYLETIC DISTRIBUTION

Using the colloquial term 'zooxanthellae' to refer to symbiotic dinoflagellates (or any symbiotic 'brown' algae), it is possible to construe an extremely broad phyletic range for symbioses between 'zooxanthellae' and marine invertebrates. Comprehensive lists of the host taxa involved may be found in McLaughlin & Zahl (1966) and D. L. Taylor (1974). Recognizing that the term 'zooxanthella' was without taxonomic significance, D. L. Taylor (1971a) proposed the identification of dinoflagellate symbionts by virtue of their systematic affinities to free-living dinoflagellates. Based on his studies, it is now generally accepted that both gymnodinioid and amphidinioid dinoflagellates may be partners in symbioses with invertebrates. The organism called *Symbiodinium* is fundamentally different from other dinoflagellates in the genus *Gymnodinium* in its division in the non-motile state and so I shall continue to use the name *Symbiodinium* as introduced by Freudenthal (1962) without agreeing with him that the organisms be placed in the family Blastodiniidae. In fact, I would suggest that the algae be placed in the family Gymnodiniidae and the genus *Symbiodinium*. (See D.

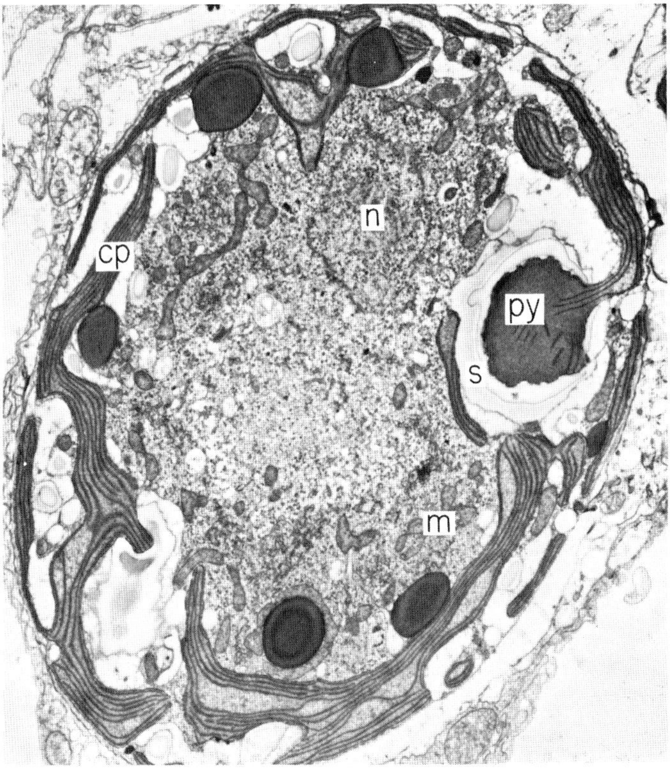

Fig. 12.2. Transmission electron micrograph of *Amphidinium* (= *Endodinium*) *chattonii* in the 'By-the-Wind-Sailor' *Velella velella*. Note the presence of two pyrenoids with chloroplast thylakoid membranes invading them. cp, chloroplast; py, pyrenoid; n, nucleus; m, mitochondrion; s, starch. Magnification approximately × 8000.

Taylor 1984; F. Taylor 1982; Blank & Trench 1986 and Appendix, this volume.)

D. L. Taylor (1974) recognized two (and possibly a third) species of *Amphidinium* as symbionts: *A. chattonii* in the 'By-the-Wind-Sailor' *Velella velella* (Fig. 12.2) (however, contrast Hollande & Carré 1974), *A. klebsii* in the flatworm *Amphiscolops* (= *Convoluta*) *langerhansi* (Fig. 12.3a, b) and a third in the radiolarians *Collozoum inerme* and *Sphaerozoum punctatum* (cf. Hollande & Carré 1974 and Anderson 1976). This third type is also known as *Zooxanthella nutricola*, the type of the genus *Zooxanthella*. Among the gymnodinioid dinoflagellates, only one species, *S. microadriaticum*, has so far been recognized. Based on the *apparent* morphological uniformity of this alga (Fig. 12.4) in the respective hosts (cf. Leutenegger 1977) and in culture (cf. Schoenberg & Trench 1980b), an equivocal position has been taken by some investigators (D. L. Taylor 1982) on the possibility of the existence of different species within the genus *Symbiodinium*, while others (Loeblich & Sherley 1979) tend towards a view of one algal species for each host species. The evidence suggests that the 'truth' probably lies somewhere between these two extremes.

Before mentioning various pieces of evidence indicating differences in *S. microadriaticum*, I would like to draw attention to the gymnodinioid dinoflagellate symbionts of the planktonic foraminiferans *Orbulina universa* and *Globigerinoides sacculifer*. The ultrastructure of the algae *in situ* (Fig. 12.5) has

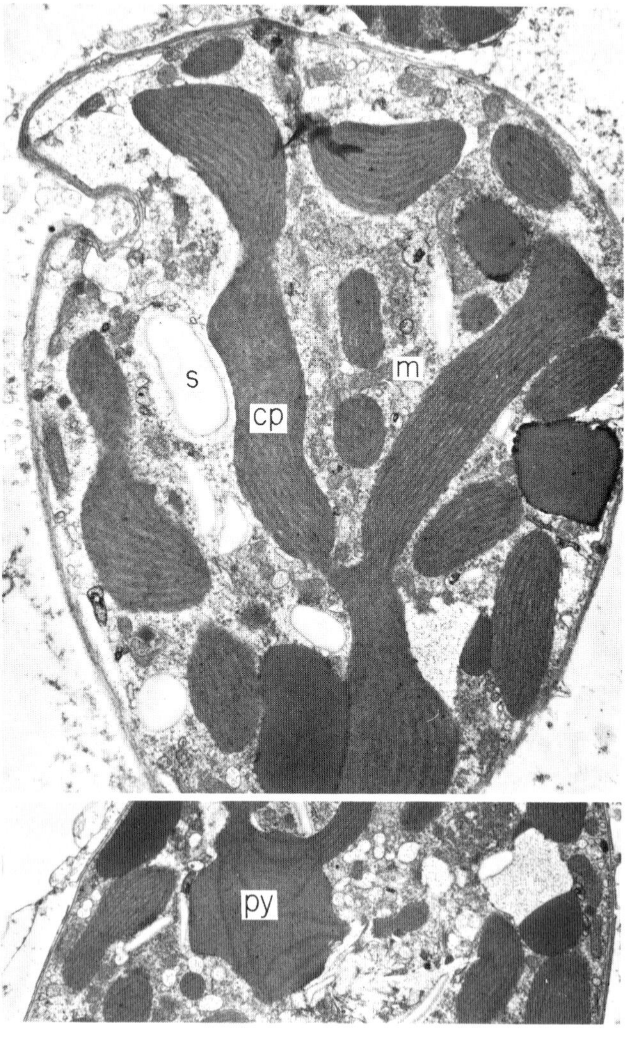

Fig. 12.3. (a) Transmission electron micrograph of *Amphidinium klebsii* (?) in the flatworm *Amphiscolops* (*Convoluta* ?) *langerhansi* (?) collected in Belau, Western Caroline Is. All abbreviations as in Fig. 12.2. Magnification approximately ×9000. (b) Same as above, but showing details of the pyrenoid with invasive chloroplast thylakoids. Abbreviations as before. Magnification approximately ×10 000.

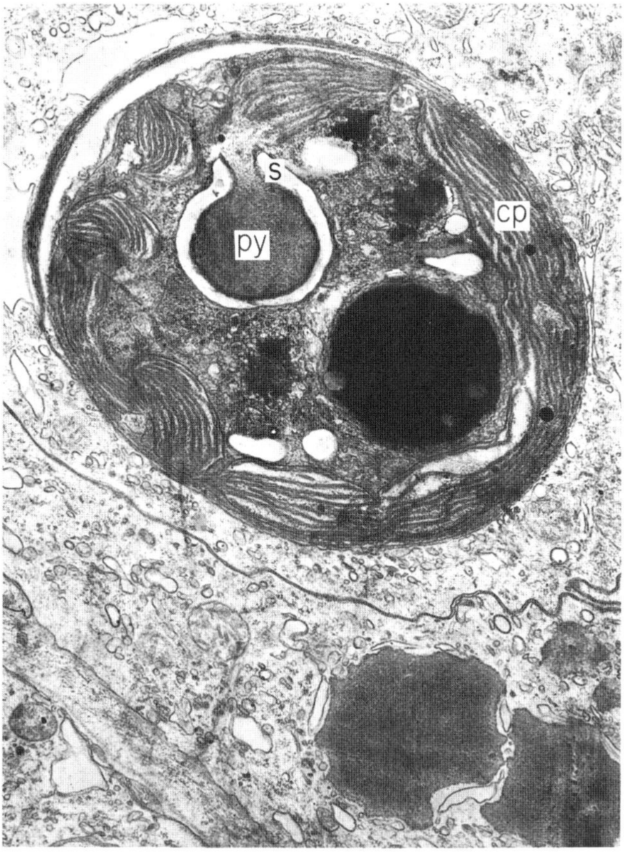

Fig. 12.4. Transmission electron micrograph of *Symbiodinium microadriaticum* in the hydroid *Myrionema amboinense*. Note the single, stalked pyrenoid without invasive chloroplast thylakoid membranes. Abbreviations as above. Magnification approximately × 13 000.

been described by Anderson & Bé (1976) and Spindler & Hemleben (1980). Recently, the symbionts of *O. universa* were successfully isolated and brought into unialgal culture. Based on their morphology in the hosts and their appearance in culture, the algae from these two foraminiferan hosts are thought to be the same organism, an opinion supported by successful cross infection experiments (Spero 1986). However, these algae show an invasion of the pyrenoid by chloroplast thylakoids, and are gymnodinioid in the hosts (except for the absence of flagella) and in their non-coccoid morphology (F. J. R. Taylor 1982). They remain permanently motile in culture and divide in the motile state (Spero 1986). In addition, their major sterol is 24-epioccelasterol (C. Djerassi, pers. comm.). All of these features are distinct from *S. microadriaticum* (cf. Kevin *et al.* 1969; Fitt *et al.* 1981; Müller-Merze & Lee 1976; Withers *et al.*

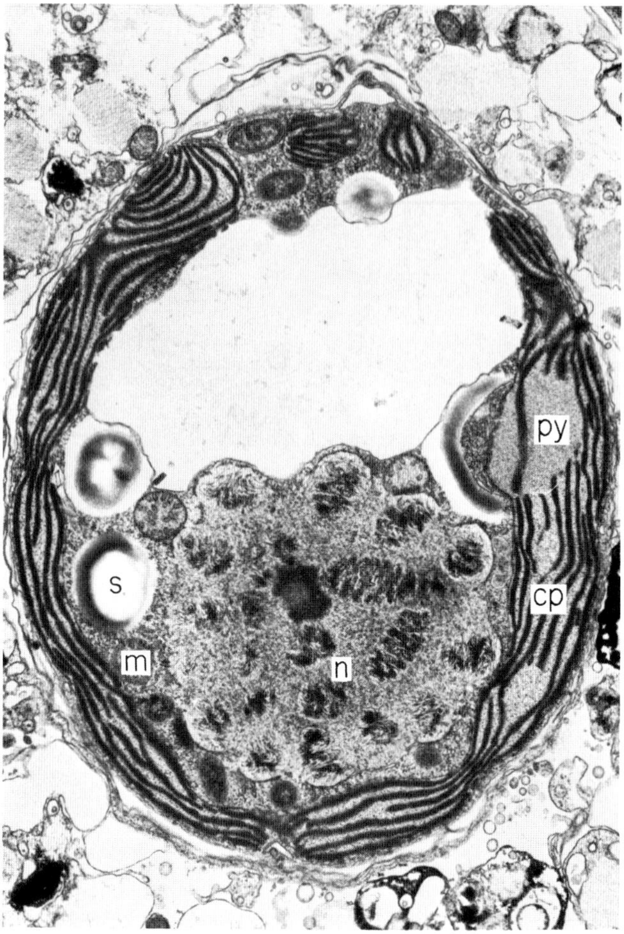

Fig. 12.5. Transmission electron micrograph of the symbiotic dinoflagellate in the planktonic foraminiferan *Globigerinoides sacculifer*. Note the unstalked pyrenoid invaded by chloroplast thylakoid membranes. Abbreviations as before. Magnification × 15 000. This photograph was kindly provided by Dr O. Roger Anderson, Biological Oceanography, Columbia University.

1982). These algae do not infect hosts of *S. microadriaticum* such as the scyphistomae of *Cassiopeia xamachana* (Colley & Trench 1983).

Since *S. microadriaticum* appears to be symbiotic with a phyletically broad range of hosts, to what extent does the host milieu interact with the algal genotype to result in the *apparent* phenotypic uniformity so often emphasized? In many symbiotic associations the algae are intracellular while in others they are intercellular. Is it reasonable to assume that the intracellular environment in a coelenterate or a foraminiferan is the same as the interstitial environment in the haemal sinuses of a tridacnid clam or the gills of *Corculum*?

Recognizing the inherent difficulties of analysing fundamental differences among algal symbionts because of potential host influences, Schoenberg & Trench (1976, 1980a, b, c) cultured the various algal isolates in the same growth medium under identical conditions of temperature, radiant flux and photoperiod, before attempting to analyse them. With the environmental parameters maintained constant, their observations of stable differences in isoenzymes, morphology and infective ability—the data from which corroborate each other—are consistent with the interpretation that there are genetic differences within the species called *S. microadriaticum*. Other investigators have reported stable differences in fatty acid composition (Bishop & Kenrick 1981), sterols (Kokke *et al.* 1981; Withers *et al.* 1982; Withers, ch. 9, this volume), diel rhythmicity of motility (Fitt *et al.* 1981), morphology *in situ* (Leutenegger 1977), and in the isoelectric forms of peridinin–chlorophyll *a*-proteins (PCP) (Chang & Trench 1982) in *Symbiodinium* isolated from different hosts.

An illustration can be made based on the algal isolates derived from *Zoanthus sociatus* (strain Z of Schoenberg & Trench 1980a): (i) analyses of their isoenzyme patterns in terms of 'genetic distance' showed that they were least similar to any of the other isolates studied; (ii) the isoelectric forms of PCP isolated from these algae (Trench *et al.* 1981a, b) is unique in having only alkaline forms; (iii) they are the only isolates which possess 'tufts of hairlike projections' on the surface of the cell wall of the coccoid cells in culture; (iv) they are among the largest strains of *Symbiodinium* recognized to date (Schoenberg & Trench 1980b). Such differences are probably sufficient to justify recognition of this isolate as a separate species from the type species *S. microadriaticum* isolated from *C. xamachana* (see Fig. 12.7), particularly since strain Z algae do not infect *C. xamachana*.

Based on any single criterion cited above, it is probably not reasonable to propose speciation within the genus *Symbiodinium*. However, considering the stability of these characteristics, when they are taken together, the case for speciation becomes much more plausible. The stumbling block to progress in this area remains the lack of any experimental evidence for genetic recombination in *Symbiodinium*, in spite of reports to the contrary (cf. Freudenthal 1962; D. L. Taylor 1973a). Approaches based on recombination analyses similar to those used with *Crypthecodinium cohnii* (Beam & Himes 1974, 1977, 1980, 1982; Steele & Rae 1980) would probably go a long way to resolve some of the problems of possible speciation in *S. microadriaticum*. However, recent reports of differences in chromosome number (Blank & Trench 1985a, b) and in details of chloroplast morphology are consistent with the interpretation that there are several species within the genus *Symbiodinium*.

3 HOST–SYMBIONT INTERACTIONS IN ESTABLISHING SYMBIOSES

3.1 Acquisition of symbionts by hosts

Among those associations in existence today, the mechanisms of acquisition of dinoflagellate symbionts by the progeny of hosts can be divided into two basic categories: (i) acquisition by maternal inheritance (a 'closed' system) and (ii) acquisition from the environment by either larval or adult stages (an 'open' system) (cf. Trench 1980, 1981; Trench *et al.* 1981a). Table 12.1 provides some examples of organisms which employ one or the other mechanism.

When a coral polyp or a sea anemone containing algal symbionts reproduces asexually by fission or budding, each 'daughter' polyp receives a portion of the algal population harboured by the parent. However, during sexual reproduction, more elaborate mechanisms are used to ensure the passage of algae from parent to offspring. An example of this can be found in the hydroid *Myrionema amboinense* where the algae are transmitted via the egg (cf. Fraser 1931; Trench 1981). A recent ultrastructural analysis of the process of passage of algae from parent hydroid to the developing egg indicates that cytoplasmic processes from the parent endodermal cells bearing algae pass through the mesoglea separating the egg from the endoderm tissue (Fig. 12.6a, b). The parental cell cytoplast, including organelles such as mitochondria, gradually disappears and the algae come to lie free in the egg plasm, until cleavage. The algae are eventually sequestered into vacuoles within presumptive endodermal cells (R. K. Trench, unpubl.).

It has been observed that corals and anemones may 'eliminate' their algal symbionts when they experience environmental 'stress' such as lowered salinity (Goreau 1964) or elevated temperatures (Pearse 1974; Fankboner & Reid 1981). Often such 'bleached' animals will reacquire a normal population of algae, but

Table 12.1. Mode of acquisition of symbiotic dinoflagellates by sexual progeny of selected invertebrate hosts

	References
(A) Direct transmission via the egg or larvae (closed systems)	
Hydrozoa	
Millepora sp.	Magnan 1909
Myrionema amboinense	Fraser 1931; Trench 1981
Velella velella	Kuskop 1921; Brinkmann 1964
Aglaophenia pluma	Faure 1960
Alcyonaria	
Eunicella stricta stricta	Theodor 1969
Actiniaria	
Anthopleura sp.	Atoda 1954

Scleractinia	
Six species	Durden 1902
Pocillopora damicornis	Atoda 1947a
*Stylophora pistillata**	Atoda 1947b
Porites spp. (4 species)	Kojis 1982
(B) From the ambient environment, by post-larval stages (open systems)	
Scyphozoa	
Mastigias papua	Sugiura 1963, 1964
Cassiopeia andromeda	Ludwig 1969
C. xamachana	Bigelow 1900; Trench *et al.* 1981a
Alcyonaria	
Eunicella stricta aphyta	Theodor 1969
Briarium asbestinum	Kinzie 1974
Pseudopterogorgia elizabethae	Kinzie 1974
P. bipinnata	Kinzie 1974
Actiniaria	
Anthopleura elegantissima	Siebert 1974
A. xanthogrammica	Siebert 1974
Aiptasia tagetes	R. D. Steele, pers. comm.
A. pulchella	G. Muller-Parker, pers. comm.
Scleractinia	
Favia doreyensis (= *F. pallida*)	Marshall & Stephenson 1933
Acropora bruggemanni	Atoda 1951
Pocillopora meandrina	J. Stimpson, pers. comm.
Fungia scutaria	Krupp 1981
Goniastrea australensis	Kojis & Quinn 1981
Astrangia danae	Szmant-Froelich *et al.* 1980
Turbinaria mesenterina	B. Willis, pers. comm.
Acropora formosa	J. Oliver, pers. comm.
Platyhelminthes	
Amphiscolops langerhansi	Taylor 1971b
Mollusca	
Corculum cardissa	Kawaguti 1950
Tridacna squamosa	LaBarbera 1975; Fitt & Trench 1981
T. derasa	Beckvar 1981
T. maxima	LaBarbera 1975; Jameson 1976
T. gigas	Beckvar 1981
T. crocea	Jameson 1976
Hippopus hippopus	Jameson 1976; Fitt *et al.* 1984

* Rinkevich & Loya (1979) report no algae in eggs of *S. pistillata*, but algae were present in the brooded planulae.

it is unclear whether this is achieved by the acquisition of new algal cells from the ambient environment or by the proliferation of those few algae that may have remained in the polyps.

The acquisition of algae by sexually produced offspring which do not inherit them is relatively unclear. Probably the most reasonable mechanism is that the free-swimming stages of the symbiotic algae which are occasionally released from some hosts (see Reimer 1971; Trench 1974; Steele 1975, 1977; Trench *et al.* 1981b) are dispersed by water currents and their own motility, and somehow possibly through chemotropic behaviour gain access to the host (Fitt 1984). This possibility is strengthened by laboratory observations of motile *Symbiodinium* successfully infecting aposymbiotic juveniles of the gorgonian *Pseudopterogorgia bipinnata* (Kinzie 1974), the scyphistomae of the jellyfish *Cassiopeia xamachana* (Trench 1980) and juveniles of the clams *Tridacna squamosa* (Fitt & Trench 1981) and *Hippopus hippopus* (Fitt *et al.* 1984). Other mechanisms of dispersal may include fishes, e.g. the puffer-fish *Arothron meleagris* which feeds on polyps, but passes the algal symbionts undigested and alive through the alimentary tract (G. Muller-Parker 1984a). This latter situation, where the algae pass intact through the digestive tract of an animal, is not unlike that described by Trench *et al.* (1981b) for tridacnid clams.

3.2 Recognition and specificity

Careful analysis of the phyletic distribution of symbiotic consortia involving dinoflagellates and their respective host groups demonstrates that such associations are randomly distributed, but the individual associations appear to be highly specific. First, symbioses do not appear to be readily created. Some 50% of all extant genera of scleractinian corals are non-symbiotic (cf. Trench 1979; Wells 1956) and non-symbiotic corals cannot be rendered symbiotic merely by contact with symbiotic dinoflagellates. Similarly, most bivalves found on coral reefs, with the exception of the tridacnids and *Corculum cardissa*, are non-symbiotic. Secondly, those invertebrates that are symbiotic with *Amphidinium* have never been observed under natural circumstances to be symbiotic with *Symbiodinium*. The reverse situation also appears to be true. In fact, although the juvenile stages of tridacnid clams will accept virtually any strain of *Symbiodinium* provided from culture (Fitt & Trench 1981), they do not accept the cultured *Amphidinium* symbionts from the flatworm *Amphiscolops*. Similarly, when larval stages of *Amphiscolops* (which do not inherit their symbionts) are exposed simultaneously to *Amphidinium klebsii*, *Amphidinium* sp., and several different strains of *Symbiodinium*, only *A. klebsii* becomes successfully established as symbionts (D. L. Taylor 1971b; R. K. Trench, unpubl.). Finally, Schoenberg & Trench (1980c) found that the anemone *Aiptasia tagetes* collected from Florida, Barbados, Bermuda and Jamaica all harboured a strain of *Symbiodinium* that was indistinguishable based on analyses of isoenzyme patterns, morphology and infective ability, and Trench (1981) found that the algae in the jellyfish *Cassiopeia mertensii* collected from different localities around Oahu, Hawaii all harboured isozymically identical symbionts. The same phenomenon is observed with regard to the algal symbionts of *Tridacna maxima* based on analyses of isoelectric forms of peridinin–chlorophyll *a*–protein complexes (Chang & Trench 1982).

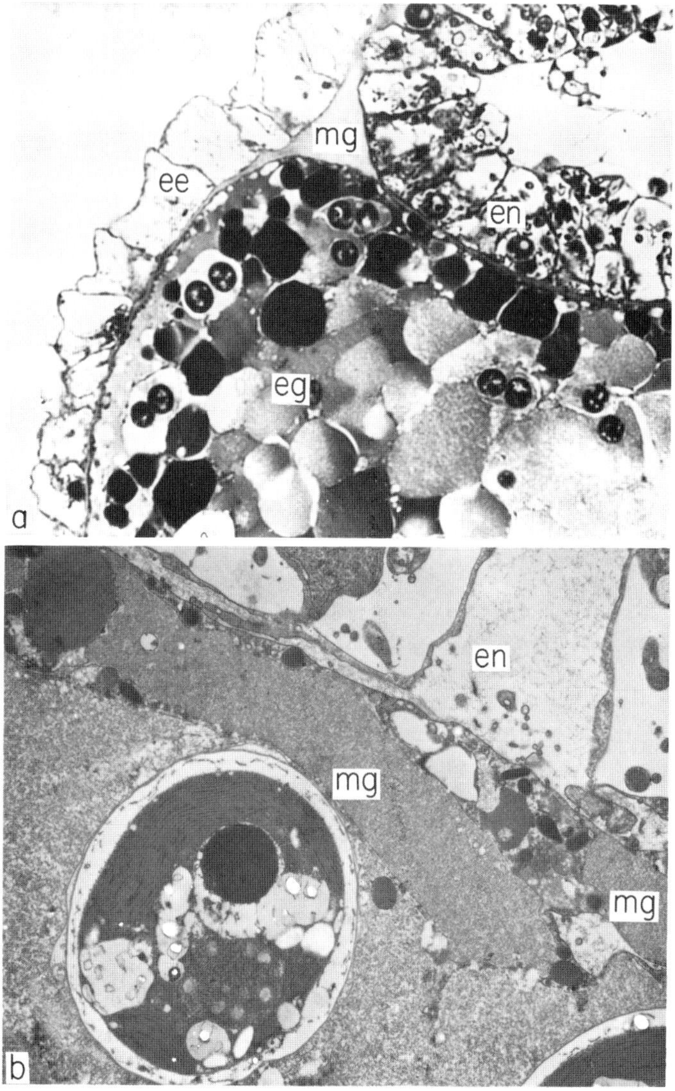

Fig. 12.6. (a) Light micrograph of a section through a portion of a developing egg attached to the hydroid *Myrionema amboinense*. ec, animal ectoderm; en, animal endoderm; eg, egg plasm; mg, mesogloea. Magnification approximately × 250. (b) Transmission electron micrograph showing the algae in the egg plasm juxtaposed to the mesogloea and endoderm. Abbreviations as above. Magnification approximately × 10 000.

The term specificity usually implies a selective process whereby one entity interacts exclusively with another. A major quest among students of biological specificity is the identification of membrane-bound binding sites on the surfaces of interacting cells which lead to intercellular recognition. The result of such ligand–receptor mediated phenomena could lead to specificity, but this is not the only mechanism whereby specificity could be established. Dubos & Kessler (1963) and Heslop-Harrison (1978) have suggested that biological specificity may not be based on a single recognition event occurring during intercellular contact, but might be the end result of a series of dynamic interactions between the two components. Consistent with these views, and pertinent to dinoflagellate–invertebrate symbioses, Trench *et al.* (1981a, b) proposed that the specificity expressed in these symbioses was the end result of a variety of spatial and temporal factors which independently or together influence the establishment and perpetuation of specific associations. Some of these factors are: (i) ecological, behavioural and physiological parameters of the algae and their potential hosts which regulate their distribution in time and space; (ii) possible cell surface recognition modulators which come into play during intercellular contact; (iii) post-uptake phenomena which in the short-term may lead to sequestration and maintenance, or elimination of symbionts; (iv) any force acting on the consortium that may select an optimum combination of associants. Evidence supporting the involvement of all four of these hypotheses are not uniformly strong.

Since the Caribbean and Indo-Pacific are discontinuous, and have been so since the last rise of the Panama Bridge 3–5 million years ago (Marshall 1981), free exchange of symbiotic dinoflagellates between the two oceans would not be expected, and hence the two populations can be regarded as having been in genetic isolation for at least this period. The symbiotic dinoflagellates, particularly *Symbiodinium*, have never been positively identified in the free-living state in the reef environment, although F. J. R. Taylor (1983) has noted that a tide-pool gymnodinoid observed in Thailand shows strong resemblances to them. Hence, other mechanisms of genetic isolation may exist. Few studies have been conducted that might be relevant to aspects of their ecology, although there are some scattered data on temperature tolerances (D. L. Taylor 1975) suggesting that they are stenothermal, and some studies on nutrient utilization (D. L. Taylor 1980; Deane & O'Brien 1978, 1981a, b) suggesting differences in kinetic aspects of nutrient uptake. Recent studies of Fitt *et al.* (1981) confirmed by Lerch & Cook (1982) have provided evidence in support of the concept that different strains of *Symbiodinium* show characteristic periodicities in their circadian rhythm of motility. Several potential hosts tested, e.g. the larvae of *Tridacna squamosa*, acquire their algal symbionts at significantly higher rates when the algae are motile than when non-motile (Fitt & Trench 1981). Such phenomena may also operate in the reef environment, but nonetheless, motile *Amphidinium* are not acquired by the juveniles of *Hippopus*.

Specific molecules on the surface of algal cells may interact with their counterparts on host endoderm cells during contact, regulating selective uptake and hence specificity, but this mechanism has not been confirmed experimentally as yet. It is known that in culture, the surface of the coccoid cells of *Symbiodinium* is composed of a substance with the histochemical characteristics of an acid polysaccharide or glycoprotein, staining positively with ruthenium red and alcian blue (Trench *et al.* 1981a, b). This substance, found external to the cell wall or pellicle (Fig. 12.7a), has not been analysed chemically, but has been observed on algae in the haemal sinuses of *Tridacna maxima* (Schoenberg & Trench 1980b). Similarly, the endodermal cell surface of the few marine invertebrates studied also possess a mucopolysaccharide or glycoprotein 'coat' indicating the presence of a glycocalyx (Fig. 12.8). It is conceivable that molecular adhesion or binding based on electrical charge (Frazier & Glaser 1979) occur at cell surfaces during contact, but whether these interactions represent the basis of molecular recognition remains unsubstantiated. In fact, experimental evidence indicates that endocytosis of freshly isolated algae and potential food particles by the endodermal cells of the scyphistomae of *C. xamachana* is a competitive process, when both particles are provided simultaneously (Colley & Trench 1983, 1985). This latter observation is consistent with the phagotrophic function of the endodermal cells.

In studies of the 'resynthesis' of symbiotic associations involving intracellular algae, it has become standard procedure to render an experimental host aposymbiotic and then expose it to algae from a variety of sources, either soon after isolation or after maintenance in culture (cf. Schoenberg & Trench 1980c; Trench 1980). In our laboratory, we have developed a system based on the naturally aposymbiotic scyphistomae of the jellyfish *C. xamachana*, which has been cloned. An additional advantage of this organism is that strobilation and the production of ephryae are arrested until the scyphistomae are infected with a 'compatible' strain of *Symbiodinium*, thus providing another criterion by which to assess the successful establishment of an association (Trench *et al.* 1981a).

The process of 'recognition' can be divided conceptually into contact recognition (phenomena occurring at the interface of interacting cells), and post-endocytotic selection (as indicated by persistence or loss of the algae over the short term). When algae, freshly isolated from *C. xamachana*, are provided to aposymbiotic scyphistomae of *C. xamachana* (homologous infections), the algae are endocytosed by a large number of endodermal cells, and many endodermal cells demonstrate multiple endocytotic events. When algae freshly isolated from other host species are similarly provided (heterologous infections), fewer endodermal cells endocytose algae, and fewer multiple endocytotic events occur (Table 12.2). By contrast, when cultured homologous or some heterologous algae are provided, rates of endocytosis are extremely low. Other cultured heterologous algae are never endocytosed in detectable numbers (cf. Trench *et al.* 1981a, b; Colley & Trench 1983). These observations suggested that some property of the freshly isolated algae was responsible for the enhanced endocytosis. Removal of host membranes from the freshly isolated algae, either with Triton X-100, or allowing the host membranes to be shed when the algal

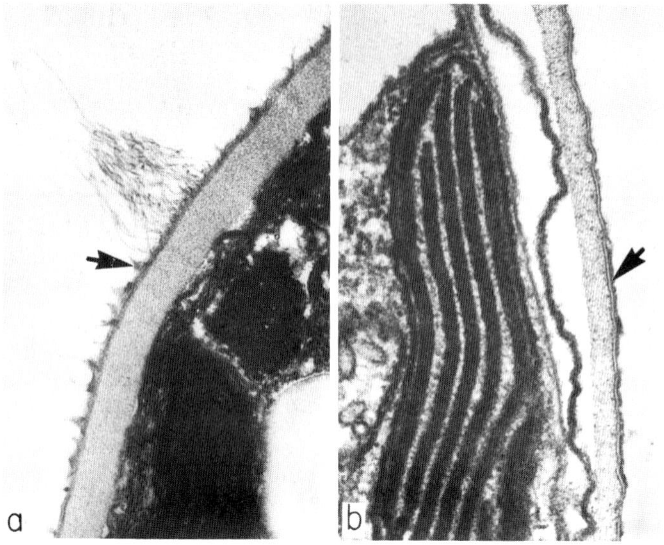

Fig. 12.7. (a) The cell wall of *Symbiodinium*. isolated from *Zoanthus sociatus* and cultured in ASP-8A. The fibrous material on the surface of the wall (arrow) is the polysaccharide. Magnification approximately × 56 000. (b) The cell wall of *S. microadriaticum* isolated from *Cassiopeia xamachana* and cultured in ASP-8A. Arrow indicates the surface polysaccharide. Magnification approximately × 60 000.

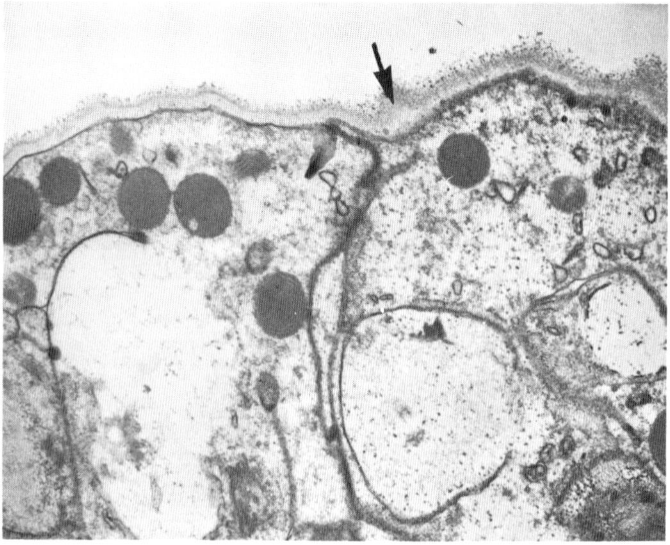

Fig. 12.8. The glycocalyx (arrow) on the surface of the endodermal cells of the hydroid *Myrionema amboinense*. Magnification approximately × 17 500.

Table 12.2. Endocytosis of freshly isolated homologous and heterologous *Symbiodinium* by endoderm cells of the scyphistomae of *C. xamachana*. All scyphistomae were 48 h starved, measured 1·0 mm oral disc diameter, and were exposed to algae for 2 h before maceration. Colley & Trench 1983; courtesy of the Royal Society of London.

Donor host	*n*	Number of endodermal cells infected with algae (per scyphistoma)	Number of algae phagocytosed (per scyphistoma)	Distribution of algae (per endodermal cell) 1	2	3	4	>4 algae
C. xamachana	5	6552 ± 575	20192 ± 1934	1628 ± 192	1600 ± 101	1140 ± 119	716 ± 97	1468 ± 208
A. pallida	5	2184 ± 221	3710 ± 453	1138 ± 145	718 ± 86	218 ± 24	76 ± 27	34 ± 20
A. pulchella	5	818 ± 145	1136 ± 237	602 ± 78	154 ± 57	48 ± 19	10 ± 5	4 ± 4
Zoanthus sp.	5	2084 ± 245	3576 ± 614	1119 ± 73	612 ± 111	227 ± 56	91 ± 35	35 ± 17
A. elegantissima	5	552 ± 24	592 ± 38	512 ± 15	40 ± 16	0	0	0

Numbers represent means ± standard error, *n* = number of scyphistomae tested.

cells divide in culture, resulted in reduced uptake, leading us to suspect that the use of freshly isolated algae to study aspects of contact recognition may lead to conclusions which have their foundation in artefact (Colley & Trench 1983). This suspicion is further strengthened by the following observation. Cultured algae from *Aiptasia pallida*, when provided to scyphistomae of *C. xamachana*, are endocytosed at low rates. Freshly isolated algae from *A. pallida* are endocytosed at higher rates than cultured algae, but at lower rates than homologous algae freshly isolated from *C. xamachana*. When cultured algae from *A. pallida* are allowed to proliferate in *C. xamachana* and are then freshly isolated and offered to the scyphistome, they are endocytosed at the same rate as freshly isolated homologous algae (Table 12.3) (cf. Schoenberg & Trench 1980c).

In further experiments to test the role of surface contact in recognition, scyphistomae of the non-symbiotic jellyfish *Aurelia aurita* were exposed to

Table 12.3. Endocytosis of homologous and heterologous algae after maintenance in culture, and in scyphistomae of *C. xamachana*. Colley & Trench 1983; courtesy of the Royal Society of London.

Original algal host	No. of endoderm cells with algae (per animal) (mean ± standard error)	*N*
C. xamachana (freshly isolated)	1693 ± 229	5
C. xamachana (cultured)	6 ± 2	27
A. pallida (freshly isolated)	688 ± 37	5
A. pallida (cultured)	8 ± 3	5
A. pallida (grown 1·5 months in scypthistomae of *C. xamachana*, freshly isolated)	1496 ± 354	5

symbiotic algae freshly isolated from a variety of hosts. The endodermal cells of *A. aurita* actively endocytosed the algae, regardless of source. However, all were exocytosed within 72 hours. When the same algae were provided after maintenance in culture, they were repeatedly rejected.

As stated before, cultured homologous algae and some cultured heterologous algae are endocytosed by the endoderm cells of the scyphistomae of *C. xamachana* when they are injected into the coelenteron, but many of the heterologous algae do not persist, and disappear within 49–72 hours after uptake, probably by exocytosis and regurgitation. Similarly, scyphistomae provided with strains of *Symbiodinium* carried live in *Artemia*, endocytosed *all* the different algal strains, but most disappeared from the animals within 72 hours (W. K. Fitt, unpubl.). Hence, initial uptake does not guarantee the successful establishment of a symbiosis. Algae that do not persist do not 'stimulate' the scyphistomae of *C. xamachana* to strobilate (Table 12.4).

Table 12.4. Infection of scyphistomae of *C. xamachana* by different strains of motile *Symbiodinium*, and its influence on strobilation.* (Data of W. K. Fitt)

Host source of algal isolate	Strain designation†	Percentage infections	No. of days to first strobilation‡	Percentage strobilation	*N*
Cassiopeia xamachana	C	100	23–30	100	87
Aiptasia tagetes	A	100	28–36	100	59
Gorgonia ventalina	A	0	∞	0	15
Condylactis gigantea	D	100	26–43	100	27
Lebrunia danae	U	100	40–47	100	27
Tridacna maxima	U	100	56–68	100	15
Tridacna gigas	T	0	∞	0	42
Bartholomea annulata	B	0	∞	0	23
Meandrina meandrites	M	0	∞	0	15
Fungia actiniformis	U	0	∞	0	5
Zoanthus sociatus	Z	0	∞	0	20

* The conditions under which the experiments were conducted are as follows: Illumination, 10 μE m^{-2} s^{-1}; temperature, 25 $\pm$ 1°C. All polyps used were of similar size (*c.* 1 mm oral disc diameter). The scyphistomae were fed to repletion daily. Approximately equal densities of algae were provided to all test scyphistomae.
† Strain designations after Schoenberg & Trench (1980a). 'U' denotes uncharacterized strains.
‡ Values refer to the number of days taken for all test individuals to produce the first ephyra.

A similar phenomenon is demonstrated by the flatworm *Amphiscolops langerhansi*. When aposymbiotic larvae were offered a range of cultured symbiotic algae, only *A. klebsii* and another large *Amphidinium* sp. were engulfed (D. L. Taylor 1971b; confirmed by R. K. Trench, unpubl.). When adult worms, rendered aposymbiotic with DCMU, were fed *Artemia nauplii* laden with either *A. klebsii* or different strains of *Symbiodinium*, although all the algae gained access to the worms, only *A. klebsii* moved from the central digestive region to the peripheral parenchyma and persisted. All the other algae were eliminated within 27 hours (R. K. Trench, unpubl.).

The observations cited above are supportive of the possibility of some selective process occurring after endocytosis, but the mechanisms by which an endocytosed alga is deemed compatible by a host cell is unknown. The idea that the release of photosynthetic products by the algae is central to their acceptance is unwarranted here, since all the algae tested were photosynthetically competent, and the freshly isolated heterologous algae, including *Amphidinium*, typically released large proportions of their photosynthate as glycerol, glucose and alanine (cf. Trench 1971b; Trench *et al.* 1981a).

From evidence obtained by experimental manipulations of symbiotic dinoflagellates and their respective hosts, it is clear that a given host may initially accept algae other than those with which they normally associate (see D. L. Taylor 1971b; Trench *et al.* 1981a). It is also clear that in several instances, associations of a given host with unnatural algal symbionts may be of selective disadvantage in the context of host growth rates (Kinzie & Chee 1979), fecundity (D. L. Taylor 1971b) and the ability to complete important morphogenetic events in the life cycle (Trench *et al.* 1981a, b). Another possibility is that the different algae which may initially occupy the same host have different demands for nutrients derived from the host (cf. D. L. Taylor 1980), or may have different intrinsic growth rates (Fitt & Trench 1983). Thus, competitive exclusion of one algal strain by another in a given host could also lead to a final expression of specificity (cf. Schoenberg & Trench 1980c).

4 METABOLIC INTERACTIONS IN SYMBIONTS AND HOSTS

4.1 Photosynthetic performance of symbionts

Studies on the photosynthetic performance of symbiotic dinoflagellates have been approached in two fundamentally different ways: (i) using intact associations and assaying light-dependent CO_2 fixation or O_2 evolution; (ii) using freshly isolated symbionts suspended in seawater or homogenates of the tissues of the hosts. For recent reviews covering much of this work, the reader is referred to Trench (1979) and Muscatine (1980a). Over recent years, much of the interest in the photobiology of symbiotic dinoflagellates has been placed on photoadaptation and mechanisms of light-dependent CO_2 fixation.

Like many of their non-symbiotic counterparts (cf. Prézelin, Ch. 5, this volume), symbiotic dinoflagellates possess chlorophylls *a* and c_2 (Jeffrey 1969; Jeffrey & Shibata 1969), with peridinin as the major antenna pigment, accompanied by β-carotene, neoperidinin, dinoxanthin, diadinoxanthin, and other minor xanthophyll pigments. One of the photosynthetic light-harvesting components, the peridinin–chlorophyll *a*–protein complex (PCP), derived from various strains of *Symbiodinium* has now been characterized extensively with respect to its isoelectric conformers (Chang & Trench 1982, 1984; Figs 12.9 & 12.10). The *in vivo* action spectrum, as measured with the coral *Favia* (Halldal 1968), and the *in vitro* action spectrum measure with *Symbiodinium* from the clam *T. maxima* (Scott & Jitts 1977), are consistent with the participation of these pigments in the light reactions of photosynthesis.

Since photon flux density is attenuated, and there are changes in the spectral quality of light as it passes through the water column, it might be anticipated, by analogy with 'sun'- and 'shade'-adapted higher plants (Bjorkman *et al.* 1972) that symbiotic dinoflagellates also photoadapt. From the point of view of the intact individual coral on the reef, particularly as photosynthesis influences nutrition (Muscatine 1980a) and calcification (Barnes & Taylor 1973; Goreau 1977), it would be of selective advantage to maintain high photosynthesis in the face of decreasing insolation. One way to achieve this would be to decrease the number of algae per unit of coral tissue, thereby reducing the potential for self-shading, and concomitantly increase the light-capturing efficiency of the algae.

Most studies which have been conducted on intact associations experiencing only differences in photon flux density suggest that in light- and shade-adapted situations, the number of symbiotic algae per unit host tissue remains the same (Redalje 1976; Svoboda & Porrmann 1980; Falkowski & Dubinsky 1981; G. Muller-Parker, 1984, 1985). However, Titlyanov *et al.* (1980) found increases in algal cell density in shade-adapted corals, and Houck (1978) found no change in algal density with decreasing light in the coral *Pocillopora damicornis*, but found an increase in algal density with decreasing light in the coral *Montipora verrucosa*.

In studies conducted on coral reefs where, with increasing depth both photon flux density and spectral quality change, the reported observations also vary. Drew (1972) interpreted his observations as indicating no correlation between algal density and increasing depth. By contrast, Dustan (1979), W. K. Fitt (unpubl.) and J. F. Battey (unpubl.) found decreasing algal densities with increasing depth in the coral *Montastrea annularis*, as did McCloskey and Muscatine (unpubl.) in *Stylophora pistillata*.

Alternative mechanisms to changing algal densities to optimize photosynthesis in symbiotic systems would be to modify the light-capturing apparatus of the chloroplasts or alter the rates of dark reactions in the algae. From studies of the mechanisms of photoadaptation of photosynthesis in non-symbiotic dinoflagellates (Prézelin & Sweeney 1979; Prézelin 1981; Perry *et al.* 1981; Rivkin *et al.* 1982; see also Prézelin, Ch. 5, this volume), four possibilities are thought to exist. The algae may (i) increase the size of the photosynthetic units (PSUs), (ii) keep the PSU size constant and increase the number of photosynthetic units, (iii) without varying the PSUs, alter the activities of CO_2 fixing enzymes or (iv) alter the activities of the photosynthetic electron transport system. A fundamental aspect of these studies is that measurements are made

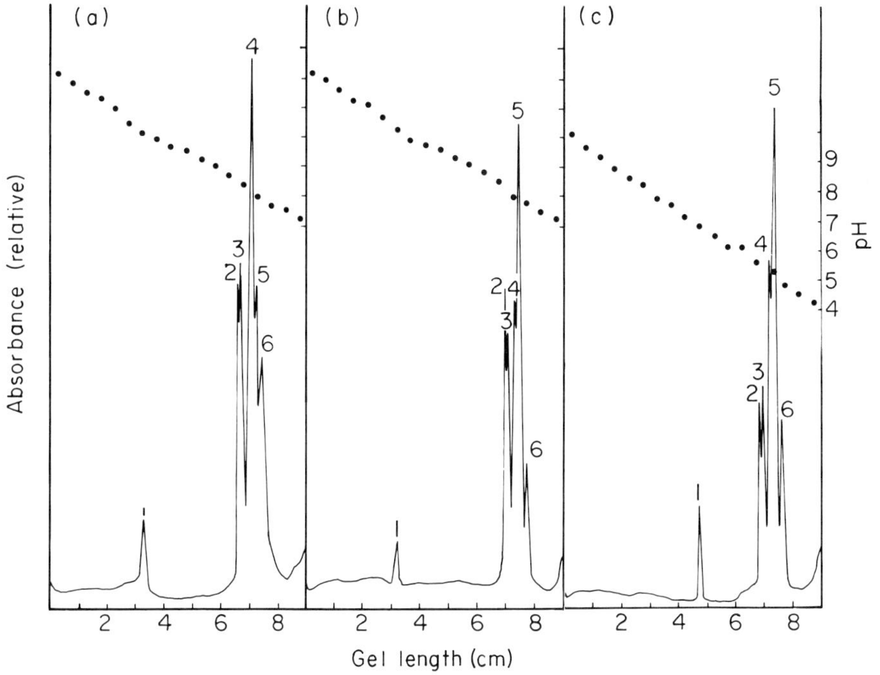

Fig. 12.9. Densitometer scans (478 nm) of isoelectric focused gels containing PCP from (a) algae freshly isolated from the host, *Rhodactis sancti-thomae*, (b) the same algae after culture in ASP-8A, and (c) a cloned population of the same algae. All tested clones produced the same pattern. (From Chang & Trench 1982; courtesy of the Royal Society of London.)

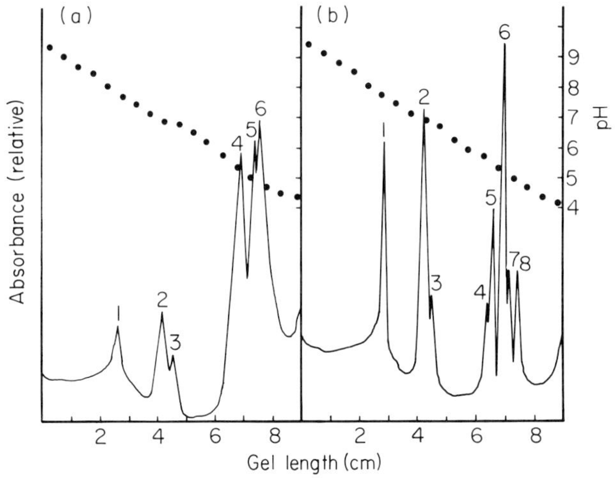

Fig. 12.10. Densitometer scans (478 nm) of i.e.f. gels containing PCP from cultured symbionts from (a) *Zoanthus solanderi* and (b) *Condylactis gigantea*. (From Chang & Trench, 1982; courtesy of the Royal Society of London.)

of the photosynthetic performance as a function of photon flux density, (*P* versus *I*) of algae grown under both light-saturating and light-limited conditions.

Wethey & Porter (1976a, b) were among the first to demonstrate that maximum gross photosynthesis remained almost the same in several corals which were adapted to light regimes found in shallow and deep habitats on coral reefs, and in addition, showed that photosynthesis in the deep corals saturated at lower values of photon flux density than their shallow counterparts. Since then, several studies have been initiated in an attempt to analyse the mechanism of photoadaptation demonstrated by symbiotic dinoflagellates and their respective hosts. Unfortunately, these studies are not uniform in the approaches taken by the respective investigators.

When changes in pigment composition as a function of light versus shade adaptation (at the same depth) was analysed, Redalje (1976) and Titlyanov *et al.* (1980) found increases in chl *a* and c_2 in the shade-adapted corals. The ratio of chl a:c_2 remained the same. However, a unique situation (as far as the available literature is concerned) was found in *Montipora verrucosa* by Houck (1978), where an increase in chl *a*, c_2 and peridinin was correlated with increasing light. Data on changes in pigment composition as a function of depth are more variable. R. Warnock (unpubl.), W. K. Fitt (unpubl.) and J. F. Battey (unpubl.) found increases in chl *a* and c_2 (on a cellular basis) with increasing depth in the algae of the coral *M. annularis*. By contrast, R. Warnock (unpubl.) found no change in the algae associated with the corals *Agaricia agaricites* and *Acropora palmata*. S. S. Chang (unpubl.) found an increase in PCP content/zooxanthella with increasing depth in *M. annularis*, but R. Warnock (unpubl.) found no change in peridinin content of the algae in *A. agaricites* and *A. palmata*. Titlyanov *et al.* (1980) found increases in chl *a* and c_2 with increasing depth in the algae in *Acropora hyacinthus* and *Echinopora lamellosa*, but also found decreases in these pigments in the algae in *Montipora foliosa* and *Pocillopora verrucosa*.

The results of the studies of Zvalinskii *et al.* (1980) and Falkowski & Dubinsky (1981) conducted on intact corals adapted to light and shade conditions at the same depth are consistent with the interpretation that the algae photoadapt by increasing the size of the PSUs. Consistent with this conclusion is the observation in both instances, that the P_{700} content per algal cell remained constant, and the chl a/P_{700} increased in the shade-adapted corals. The data presented by Zvalinskii *et al.* (1980) on adaptation as a function of depth vary somewhat from predictions, but in this instance also, the P_{700} content of the algae remained constant, suggesting photoadaptation by virtue of increase in PSU size.

The use of intact associations is made difficult because measurements of photosynthetic response are based on apparent net O_2 production by the intact association, but host respiration varies with depth, as does algal photosynthesis. (Spenser-Davies 1977, 1980; Wethey & Porter 1976b; L. McCloskey &

Muscatine 1984; Muscatine *et al.* 1981). Hence, gross production of the symbiotic algae in intact associations is difficult to estimate (Muscatine 1980a). There are no published reports on photoadaptation in *S. microadriaticum* under controlled conditions of laboratory culture. In the only study of the symbiont *Amphidinium klebsii** in culture that I am aware of, Mandelli (1972) found that at low illumination the cells increased their content of peridinin, but the chlorophylls remained unchanged at intensities ranging from 2 to 100 W m^{-2}.

To circumvent the problems involved with analyses of interassociations, the mechanisms of photoadaptation of photosynthesis have been compared in three strains of *Symbiodinium* isolated from the clam *Tridacna maxima*, the sea anemone *Aiptasia pulchella*, and the stony coral *Montipora verrucosa*. The algae were maintained in ASP-8A and were grown at 22, 57, 157, and 248 $\mu E\ m^{-2}\ s^{-1}$ on a 14:10 hours (light:dark) photoperiod at 26°C. All analyses were conducted during log phase of growth, and rates of growth and photosynthesis, as well as cell volumes, pigmentation and carbon and nitrogen contents determined.

Low light-induced alterations in pigment composition were evident to varying degrees in all three algal strains; in algae from *T. maxima* the molar ratio of chl a:c_2 remained essentially constant throughout, while in the other two strains, the molar ratios of chl a:c_2 increased with increasing irradiance.

The relations between photosynthetic performance and irradiance (*P* versus *I*) demonstrated by the three strains were different. The algae from *T. maxima* appeared to photoadapt by altering the size of the photosynthetic units (PSU), while the algae from *A. pulchella* appeared to alter PSU numbers and the algae from *M. verrucosa* altered the activities of CO_2-fixing enzymes or electron transport systems. Interestingly, the adaptive capabilities of the algae appear to correlate well with the ecological distribution of their respective hosts. For further details of this study, the reader is referred to Chang *et al.* (1983).

Published studies on productivity of symbiotic systems involving dinoflagellates are extremely difficult to compare because of the wide range of productivity and photon flux units employed (cf. Muscatine 1980a; Chalker 1981). Table 12.5 presents assimilation numbers (cf. Falkowski 1981) on some selected systems. The original data were in several instances converted from O_2 units to carbon equivalents as discussed by Muscatine (1980a). Saturating photon flux densities for intact systems are appreciably higher than those observed with the same algae after isolation, suggesting that the multilayered packing of the algae (resulting in self-shading), and/or the animal tissue itself may be responsible for absorption or scattering of light impinging on the surface. The self-shading phenomenon was clearly demonstrated in *Tridacna gigas* (Fisher *et al.* 1985) and *T. erocea* (Banaszak 1985) but was not observed in *Aiptasia pulchella* (Muller-Parker 1984b). Generally, *Symbiodinium* appears to saturate photosynthetically at relatively low photon flux densities. There is high variability in calculated

* Apparently there is some doubt on the identity of the organism used. It may have been *A. carterae* which is not known to be symbiotic.

assimilation numbers when comparisons are made based on data obtained with intact systems, but reasonable agreement when analyses are conducted on either freshly isolated or cultured symbionts.

The mechanism of photosynthetic CO_2 fixation in marine algae in general, and in symbiotic dinoflagellates in particular, remains an area of controversy (see Chapter 11A and cf. Trench 1979; Muscatine 1980a; Morris 1980; Burris 1981). The relative importance of CO_2 fixation via HCO_3^- and phosphoenol pyruvate (PEP) carboxylase versus CO_2 and ribulose bisphosphate (RuBP) carboxylase in the marine environment is currently a much disputed topic. General references to this subject include Beardall *et al.* (1976), suggesting significant C_4-like metabolism and Kremer & Küppers (1977) and Hofmann & Kremer (1981), emphasizing the C_3 pathway.

With respect to the symbiotic dinoflagellate *Symbiodinium*, Ting (1976) found equivocal evidence for the existence of PEP carboxylase (EC4.1.1.31) activity in algae isolated from *Tridacna maxima*. No distinction was made between PEP carboxylase and PEP carboxykinase. He also found high levels of NADH malate dehydrogenase (EC1.1.1.37) activity (confirmed by Schoenberg & Trench 1980a; and Trench *et al.* 1981a), and qualitative indications of the existence of NADP malate dehydrogenase (decarboxylating, EC1.1.1.40, malic enzyme). The presence of malic enzyme in *Symbiodium* has also been confirmed (N. J. Colley & R. K. Trench, unpubl.). Graham & Smillie (1976) found higher levels of carbonate hydrolyase (carbonic anhydrase, EC4.2.1.1) activity in coral and clam tissue than in symbiotic algae. We have found several isoenzymes of carbonate hydrolyase by polyacrylamide gel electrophoresis of extracts of algae freshly isolated from all six species of tridacnid bivalves, the medusa *Mastigias papua*, and in several cultured strains of *Symbiodinium*.

Burris (1977) found low levels of photosynthetic ^{14}C incorporation into glycine and serine (which did not vary when the algae were exposed to various O_2 tensions) by algae from *Pocillopora capitata*. G. Steen (pers. comm.) also detected low levels of ^{14}C incorporated into glycine and serine in algae from *Aiptasia pulchella*. Trench & Fisher (1983) have been unable to demonstrate inhibition of photosynthesis in cultured symbionts from *A. pallida*, freshly isolated symbionts from *A. pulchella* at 21% O_2 and 0·02–20 mM HCO_3^- or freshly isolated algae from tridacnid clams. At 100% O_2 saturation, inhibition of photosynthesis was about 20% in algae from *A. pulchella*, varying slightly as a function of the concentration of HCO_3^-. Crossland & Barnes (1977b) found no evidence of inhibition of photosynthetic O_2 production at saturating radiant fluxes and above in intact *Acropora acuminata*, and found very low rates of photorespiration in the isolated algae. These pieces of evidence tend to be supportive of C_4-like metabolism as do the recent observations of activities of pyruvate P_1-dikinase, PEP carboxylase, PEP carboxykinase and NAD and NADP-dependent malate dehydrogenases in cultured *Symbiodinium* derived from *Montipore verrucosa* (Tytler & Trench 1986).

Table 12.5. Selected data on photosynthesis and light relations in symbiotic dinoflagellates (*in vivo* and *in vitro*)

Species	Photon flux density at saturation (I_k) ($\mu E\ m^{-2}\ s^{-1}$)	P_{max} (gross) (mg C mg chl a^{-1} h^{-1})	Reference
Pocillopora damicornis (intact)	> 500	1·6	Scott & Jitts 1977
Stylophora pistillata (intact)	100–200	11·3*	Falkowski & Dubinsky 1981
Acropora acuminata (intact)	300		Crossland & Barnes 1977
A. acuminata (freshly isolated algae)	25		Crossland & Barnes 1977
A. cervicornis (intact)	236		Chalker 1981
A. formosa (intact)	54		Chalker 1981
Manicina areolata (intact)	342		Chalker 1981
Pavona praetorta (intact)			
10 m	175	4·2*	Wethey & Porter 1976a, b
25 m	75	3·4*	Wethey & Porter 1976a, b
Tridacna maxima (intact, 5 m)	> 400	13·6*	Trench *et al.* 1981b
T. maxima (freshly isolated algae)	~ 240	3·3	Scott & Jitts 1977
T. maxima (freshly isolated algae)	300†	3·6–4·8	Trench *et al.* 1981
T. maxima (freshly isolated algae)	100	2·6	C. R. Fisher (unpubl.)
T. maxima (cultured algae)	85–115	3·2–4·3	S. S. Chang (unpubl.)
T. crocea (freshly isolated algae)	120	3·6	C. R. Fisher (unpubl.)
T. derasa (freshly isolated algae)	155	6·0	C. R. Fisher (unpubl.)
T. gigas (freshly isolated algae)	160	7·4	C. R. Fisher (unpubl.)
T. squamosa (freshly isolated algae)	85	3·8	C. R. Fisher (unpubl.)
Hippopus hippopus (freshly isolated algae)	130	5·2	C. R. Fisher (unpubl.)
Aiptasia pulchella (intact, bright light adapted)	~ 600	3·2*	G. Muller-Parker (unpubl.)
A. pulchella (intact, shade adapted)	~ 200	1·4*	G. Muller-Parker (unpubl.)
A. pulchella (freshly isolated algae)	~ 100	7·0	C. R. Fisher (unpubl.)
Amphidinium klebsii (?) (freshly isolated algae)	~ 100	3·1	C. R. Fisher (unpubl.)

* These data were originally presented in terms of net O_2 production. Although not completely appropriate (cf. Muscatine 1980a), the respiration values were added to the net production values to approximate gross production. For conversion to carbon equivalents, the equation of Muscatine (1980a) was used, assuming a *PQ* value of 1·0.

† This was the photon flux density at which the assay was conducted, not an actual measurement of *P* versus *I*.

In contrast, and consistent with a C_3 fixation pathway, Downton *et al.* (1976) and Black *et al.* (1976) found evidence of O_2 inhibition of photosynthesis in algae freshly isolated from *Pocillopora*, *Tridacna* and *Hippopus*, but the values varied from one preparation to another. Similarly, D. Phipps and R. Pardy (pers. comm.) found O_2 inhibition of photosynthesis in algae from *Aiptasia*. The synthesis and release of glycolate, an indicator of C_3 metabolism, is well documented for *S. microadriaticum* (Trench 1971b; Muscatine 1973) and Burris (1977) indicates that such release was highest under conditions of high O_2 tensions. Black & Bender (1976) found $\delta^{13}C$ values of $-23‰$ in the algae from *T. maxima*, and interpreted these as indicating a C_3 pathway, but Land *et al.* (1975) found $\delta^{13}C$ values of -12 to $-18‰$ in algae from corals, and these values, according to Smith & Epstein (1970) or Benedict (1978) would indicate C_4 metabolism. However, according to Smith & Walker (1980), $\delta^{13}C$ values provide little information on the mechanism of CO_2 fixation in aquatic plants.

Resolution of these conflicting data must await further studies. It is, at this point, not justified to conclude that either the C_3 or C_4 pathway predominates in symbiotic dinoflagellates (cf. Muscatine 1980a; Hofmann & Kremer 1981; Kremer *et al.* 1980). The possibility of some form of a mixed C_3–C_4 system or some novel system with the characteristics of both, is still viable (cf. Raven & Glidewell 1978; Raven & Beardall 1981). Arguments based on assumptions of the P_{CO_2} and P_{O_2} existing in host tissues where the algae reside are not warranted, since we know virtually nothing about the concentrations of HCO_3^-, CO_2 or O_2, or their relative rates of diffusion across the tissues of animals harbouring symbiotic algae, particularly under conditions of intense photosynthesis (Crossland & Barnes 1977b; Dykens & Shick 1982). Sensitivity to O_2 in photorespiring systems is a function of CO_2 availability because of the relative affinity of RuBP carboxylase/oxygenase to CO_2 and O_2. Some of the evidence presented could be interpreted as suggesting that RuBP carboxylase in *Symbiodinium* may be relatively insensitive to O_2. Alternatively, the possibility of inducible enzyme systems, such as carbonate hydrolyase, influencing the availability of CO_2 or HCO_3^- and thus the pathway of carbon fixation (Reed & Graham 1977; Bird *et al.* 1980), particularly under conditions of high P_{O_2}, cannot now be ruled out (see Trench & Fisher 1983; Tytler & Trench 1986).

Some observations pertinent to the subject of O_2 diffusion in *Tridacna* are warranted at this point. Under conditions of high light, the amount of O_2 diffusing across the 'mantle' surface of *T. maxima* was undetectable, but O_2 leaving the excurrent siphon was measured at 600–760 mmHg (J. J. Childress, unpubl.). Since the algae are located in the haemal spaces of the 'mantle' tissue it would have to be assumed that O_2 produced by the algae is transported to and unloaded at the gills (Trench *et al.* 1981a). Therefore, depending on the efficiency of removal of O_2, there is the potential for high blood P_{O_2} and therefore a potential for oxygen inhibition of photosynthesis and for photorespiration (cf. Crossland & Barnes 1977b).

Similarly, Dykens & Shick (1982) found P_{O_2} values in excess of 300 mmHg in endodermal tissues of the sea anemone *Anthopleura elegantissima* after brief illumination, and emphasized the importance of animal superoxide dismutase in dealing with the potential deleterious effects of hyperbaric O_2. It should be pointed out in this context that all strains of *Symbiodinium* tested possess several isoenzymes of superoxide dismutase.

4.2 Nitrogen and phosphorus metabolism of the algae

In light of the fact that symbiotic dinoflagellates abound in what is usually regarded as nutrient-depauperate environments such as coral reef ecosystems, it is important to understand how these algae obtain nitrogen and phosphorus, potentially limiting nutrients. In intact associations, two sources of supply are readily identifiable: (i) the animal host may supply NH_4^+ as the end-product of amino acid catabolism, and may also provide organic phosphorus; (ii) the environment may supply nitrogen as NH_4^+ or NO_3^- and phosphorus as inorganic phosphate.

If the algae in their hosts utilize exogenous NO_3^-, then they must possess the necessary enzyme systems, nitrate and nitrite reductases, to reduce NO_3^- to NH_3. Evidence supporting the existence of these enzymes in freshly isolated *Symbiodinium* is equivocal (Franzisket 1975; Crossland & Barnes 1977a; Summons & Osmond 1981), even though intact corals show high affinity for NO_3^- in light (K_s values of 0·26–0·65 μM) (Muscatine 1980b). However, there appears to be no NO_3^- flux in *Aiptasia pulchella* (F. Wilkerson, pers. comm.), and Muscatine & Marian (1982) report no NO_3^- uptake by the intact symbiotic jellyfish *Mastigias papua*. That the nitrate reducing enzymes, which are inducible, exist in the algae in culture is self-evident, since a variety of strains of *Symbiodinium*, and *A. klebsii* may be successfully cultured in media containing NO_3^- in the absence of NH_4^+ (Schoenberg & Trench 1980a; D. L. Taylor 1980).

The pathway by which symbiotic dinoflagellates utilize NH_4^+ is unclear. Among plants, two predominant pathways exist (Miflin & Lea 1976): the incorporation of NH_3 via NAD(P)H-dependent glutamate dehydrogenase (GDH; EC.1.4.1.3),

$$\text{NAD(P)H} + \alpha\text{-ketoglutarate} + \text{NH}_3 \longrightarrow \text{glutamate} + \text{NAD(P)}^+,$$

and via glutamine synthetase (GS; EC.6.3.1.2),

$$\text{glutamic acid} + \text{ATP} + \text{NH}_3 \xrightarrow{\text{Mg}^{++}} \text{glutamine} + \text{ADP} + \text{P}_i.$$

This latter system is linked to the transaminase system involving glutamine and α-ketoglutaric acid (GOGAT; EC.2.6.1.53), resulting in the synthesis of glutamate.

Cloned cultures of a variety of strains of *Symbiodinium* grown in ASP-8A demonstrate GDH activity; indeed several isoenzymes of GDH have been resolved (N. J. Colley, unpubl.). Crossland & Barnes (1977a) demonstrated GDH activity in algae freshly isolated from the coral *A. acuminata*, but Summons & Osmond (1981), based on ^{15}N studies, concluded that freshly isolated algae from *Hippopus* used the GS/GOGAT system rather than GDH. Intact corals showed high affinity for NH_3 (K_s values 0·29–1·05 μM) (Muscatine 1980b). Similarly, K_s values for NH_3 in intact *Tridacna crocea* is 0·4 μM (R. K. Trench, unpubl.). Muscatine & Marian (1982) found a K_s for NH_4^+ uptake of 4·2 μM for the algal symbionts isolated from *M. papua*. Freshly isolated algae from the coral *Seriatopora* incorporated apparently increasing quantities of photosynthetically fixed ^{14}C into glutamine when incubated in the presence of increasing concentrations of NH_4^+ (Muscatine 1980b), suggesting an active GS system. It is possible that both the GDH and the GS/GOGAT systems exist in some *Symbiodinium*; D. J. Miller (pers. comm.) has found activities of both systems in algae from *Montipora verrucosa* maintained in culture from ASP-8A.

If it is assumed that symbiotic dinoflagellates in their respective hosts use NH_4^+ derived from host catabolism of protein, then by analogy with free-living phytoplankton, the activity of nitrate reductase should be repressed. However, the data of Franzisket (1975), D'Elia & Webb (1977) and Webb & Wiebe (1978) indicate that *Symbiodinium* in corals appear to simultaneously use NO_3^- and NH_4^+, while the algae in *M. papua* and *A. pulchella* do not use NO_3^-. Reconciliation of these apparently conflicting pieces of evidence must await further study. The possibility that other microbial flora (e.g. bacteria) in intact associations use NH_4^+ and/or NO_3^- needs clarification.

When in their respective hosts, symbiotic dinoflagellates probably also utilize organic nitrogen. McLaughlin & Zahl (1959) showed that *Symbiodinium* in culture could use a variety of amino acids. Interestingly, McLaughlin & Zahl (1959) could not demonstrate absolute requirements for vitamin B_{12} (cf. D. L. Taylor 1975), and we have found that some strains of *Symbiodinium* can grow in ASP-8A without added vitamin B_{12} while others do not.

It should also be noted that, under normal conditions of incubation (i.e. in seawater with host tissue homogenate), *Symbiodinium* readily demonstrates the synthesis of glutamate, glutamine, aspartate, leucine and alanine (Trench 1971b, c, 1974). In the case of alanine, its synthesis and translocation in response to increased levels of exogenous NH_4^+ is enhanced in intact corals (Lewis & Smith 1971).

Different strains of *S. microadriaticum* in culture in ASP-8A use phosphate provided as glycerophosphate. These algae may also use other sources of organic phosphorus such as nucleic acids (D. L. Taylor 1980), and there is an implication that they may use such sources when residing in their host tissues.

Deane and O'Brien (1981a) found that *Amphidinium klebsii* and *Symbiodinium* in culture demonstrate high affinities for inorganic phosphate, with K_s values of 0·005–0·016 μM. Phosphate uptake was found to be greater in the dark than in the light. The photosynthetic inhibitor DCMU stimulated phosphate uptake in the light by *A. klebsii*, but inhibited uptake by *Symbiodinium* under the same conditions. Carbonylcyanide 3-chlorophenyl hydrazone (CCCP) inhibited uptake in both light and dark in *A. klebsii*, but only inhibited uptake in *Symbiodinium* in the light.

D'Elia (1977) showed that symbiotic corals were able to remove inorganic phosphate from seawater at low ambient concentrations while Yonge (1936) demonstrated higher uptake of phosphate in the symbiotic clam *Tridacna crocea* than in the non-symbiotic clam *Spondylus*.

4.3 Bilateral metabolite exchange

The movement of photosynthetically-fixed carbon from symbiotic dinoflagellates to hosts has been repeatedly reviewed (Smith *et al.* 1969; Muscatine 1973; Trench 1979; Cook 1983) and hence will not be reiterated here. There is a concensus that small molecular weight metabolites such as glycerol, glucose, alanine and some organic acids are readily transported from the algae, and are metabolized by the hosts. Details of the chemical nature of the factor in homogenates of host tissues that enhance the release of photosynthetic products by the algae have still not been ascertained.

More recently, there has been growing interest in the possibility of other classes of metabolites being transported from symbiotic dinoflagellates to their respective hosts. Some indirect evidence suggests that acetate from the host may be incorporated by the algae, and there elaborated into fatty acids and triglycerides, which in turn are transported to the host and used, among other things, in the synthesis of wax esters such as cetyl palmitate (Benson *et al.* 1978; Kellogg & Patton 1983; Patton & Burris 1983; Patton *et al.* 1983). Again, there is indirect evidence for the transport of sterols from symbiotic dinoflagellates to their hosts (see Withers, Ch. 9, this volume). The biological function of these substances should prove an interesting area for future investigation.

The movement of metabolites in the opposite direction, i.e. from the hosts to the dinoflagellates, has not received the experimental attention it richly deserves. Based on studies of nutrient uptake by symbiotic dinoflagellates in culture (McLaughlin & Zahl 1966; D. L. Taylor 1975, 1980; Deane & O'Brien 1981a, b) it is clear that the algae are able to utilize organic nitrogen (e.g. urea and uric acid), organic phosphorus (glycerophosphate, adenylic, guanylic or cytidylic acids), and sulphur amino acids (e.g. cysteine and taurine). The *in vivo* analyses mentioned in Trench (1979) suggesting the incorporation of ^{35}S-amino acids from hosts by their algal symbionts, would tend to support the facultative auxotrophy observed in culture. A supply of organic compounds from the host may be important to the survival of the algae under conditions of light limitation, as well as a mechanism of nutrient recycling.

5 REGULATION OF SYMBIONT POPULATION

A characteristic feature of symbiotic consortia involving non-parasitic dinoflagellates is that genetically distinct cell types (i.e. animal and plant) proliferate in harmony (cf. Trench 1979). There are no reported examples of the algae, under normal circumstances, ever overgrowing their hosts, or of hosts outgrowing their algal symbionts. In situations where the algae reside intracellularly, their uncontrolled proliferation would eventually lead to their combined volume exceeding that of the host cell containing them. In other instances, e.g. the tridacnid bivalves or *Corculum*, where the algae are intercellular in the haemal sinuses (Trench *et al.* 1981b; Kawaguti 1950, 1968), unchecked growth of the algae might lead to their exceeding the carrying capacity of the animals' body spaces.

Generally, the number of algae within endodermal cells of coelenterates harbouring dinoflagellates is variable (Fig. 12.11), differing from one consortium to another, and from one anatomical aspect of the same individual host to another (Trench 1979). Unfortunately, the possible mechanisms by which algae and animal cells adjust their respective rates of proliferation to maintain constant relative biomass have not yet been analysed. It might nevertheless be useful to advance some testable hypotheses, and remark on some relevant circumstantial evidence.

Three possible mechanisms, which are not necessarily mutually exclusive, relating to regulation of algal numbers are: (i) in intracellular symbioses, the host cell divides in 'synchrony' with the algal population, partitioning equal numbers of algae to daughter cells, (ii) the host cell eliminates excess algae by exocytosis or digestion, and (iii) the host cell 'controls' increase in algal population by prohibiting algal division, withholding critical nutrients or promoting the flow of metabolites out of the algae. Although, as expressed here, the active component in the regulatory process appears to be the host, there is no good reason why regulation could not be effected by the algae.

Very little is known about the frequency of mitotic division of animal cells harbouring dinoflagellates. It is known that different symbiotic dinoflagellates *in culture* have different intrinsic growth rates (D. L. Taylor 1980; Fitt & Trench 1983), and while *Symbiodinium* has been observed to produce tetraspores in culture (Schoenberg & Trench 1980b), tetraspores have never been reported in intact associations. Hence, if one mitotic cycle in the algal population is followed by one mitotic cycle of the host cell, and equipartitioning of the algae, then constancy can be maintained. The entire algal population within a given host does not undergo synchronous division. In fact, depending on the association in question, only a small proportion (less than 10%) of the algae are usually observed dividing at any instant.

There are many examples in the literature of animal hosts 'expelling' their algal symbionts, and some investigators are of the opinion that this mechanism

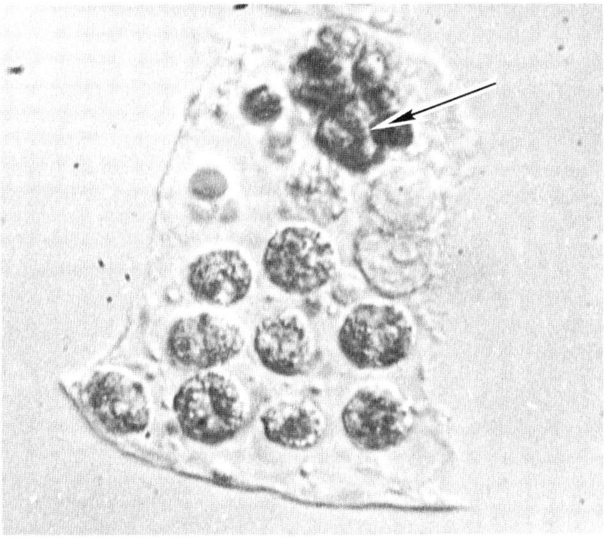

Fig. 12.11. Light micrograph of an isolated endoderm cell from the hydroid *M. amboinense* showing nine functional algae located towards the proximal end of the cell and several pycnotic algae at the distal end. Arrow indicates pycnotic algae. Magnification approximately 500 ×.

represents the regulation of the algal population (D. L. Taylor 1969; Steele 1977).

The question of digestion of algae by hosts remains controversial and problematic. Two current hypotheses prevail. First, the algae are digested intracellularly by enzymes liberated into the vacuoles containing the algae following fusion with host lysosomes. The evidence supporting this view is the histochemical demonstration of acid phosphatase activity near or in the algae (Fankboner 1971). This interpretation is rendered ambiguous in light of the presence of intrinsic acid phosphatases in the algae (Trench *et al.* 1981a). Second, the algae undergo autolysis (Trench 1974). This idea is supported by observations of degraded *Symbiodinium* in senescent axenic cultures. However, there is some preliminary evidence, based on ferritin labelling of lysosomes, that host lysosomes may fuse with phagosomes containing recently endocytosed heat-killed algae (Fitt & Trench 1983; Colley & Trench 1985). This phenomenon was not observed with recently endocytosed live algae, but was observed after algae in the hosts' cells were treated with the photosynthetic inhibitor DCMU. Thus, release of photosynthetic products by the algal symbionts may somehow inhibit phago–lysosome fusion.

6 DISCUSSION AND CONCLUDING REMARKS

In this article, I have attempted to highlight those aspects of the interactions between non-parasitic dinoflagellates and their invertebrate hosts which are currently being actively investigated. However, I have not exhausted the examples of dinoflagellates involved in symbiosis. Dinoflagellates may themselves be hosts to other organisms such as viruses (Franca 1976), bacteria (Gold & Pollingher 1971; Silva 1978) and other algae. *Kryptoperidinium foliaceum* and *P. balticum* (Tomas *et al.* 1973; Jeffrey & Vesk 1976) are dinoflagellates which harbour an intracellular chrysophyte. *Noctiluca scintillans* harbours a green prasinophyte (Sweeney 1971, 1976, 1978). With the exception of the biochemical studies conducted on *K. foliaceum* and *K. balticum* (Withers & Haxo 1978; Swenson *et al.* 1980; Withers *et al.* 1979) studies on most of the consortia mentioned above remain at the descriptive stage, but *K. foliaceum* and *P. balticum* have been central in speculations on the symbiotic origins of dinoflagellate chloroplasts (Gibbs 1980; Whatley 1980).

One area that I have consciously avoided pertains to the role of symbiotic dinoflagellates in the ecosystems in which they occur (see Cook 1985). What are the pathways whereby mass or primary carbon fixed by symbiotic dinoflagellates finds its way into components of the biosphere other than the respective hosts of the algae? Some of the possibilities are discussed by D. L. Taylor (1973b), but even now, many of the details remain, for the most part, in the realm of conjecture. Benson & Muscatine (1974) related how primary fixed carbon from symbiotic dinoflagellates, translocated to corals and metabolized to cetyl palmitate, is secreted in mucus and consumed by reef fishes as a major energy source (cf. Krupp 1982).

Again, to what extent do ecosystems such as coral reefs effectively conserve nutrients such as nitrogen and phosphorus by recycling between algal symbionts and hosts? High ambient levels of nitrogen and phosphorus have been shown to adversely affect reef coral growth (Kinsey & Davies 1979) and normally, coral reef systems are depauperate with respect to these nutrients. It is believed that the bulk of N and P is 'locked up' within the living system, where it is efficiently recycled between plant and animals (D'Elia 1977; Muscatine & D'Elia 1978; Muscatine 1980b; Muscatine & Porter 1977). Recently, Trench *at al.* (1981b) demonstrated the presence of phosphorite concretions in the kidney of *Tridacna*. The possibility that these concretions could serve as a 'reservoir' of phosphorus needs to be experimentally explored.

It is apparent that many aspects of the ecological significance of algal symbioses need to be placed on a firmer quantitative basis. It is also equally apparent that several parameters of the cell biology and biochemistry of symbiotic interactions involving dinoflagellates need to be further explored, both in the context of the intact consortium as well as the separated components.

ACKNOWLEDGEMENTS

I wish to thank my graduate students Suknan S. Chang, Nansi J. Colley, C. R. Fisher and W. K. Fitt for consenting to have unpublished data included in this manuscript. Others who were equally kind are A. Banaszak, G. Muller-Parker, G. Steen, R. Warnock, H. Spero, J. Battey and D. Krupp. My special thanks also go to Dr O. Roger Anderson for providing the micrograph in Fig. 12.5, to Dr D. Dunn for identification of the jellyfish-eating anemone from Palau, and to Dr C. J. Crossland for his comments on the manuscript. The reported studies from our laboratory have been supported by NSF, PCM 78-15209, BSR83-20450 and DEP-5542-G-SS-1085 from the US AID.

7 REFERENCES

ANDERSON O. R. (1976) Ultrastructure of a colonial radiolarian *Collozoum inerme* and a cytochemical determination of the role of its zooxanthellae. *Tissue Cell* **8**, 195–208.

ANDERSON O. R. & BÉ A. W. H. (1976) The ultrastructure of a planktonic foraminifer, *Globigerinoides sacculifer* (Brady), and its symbiotic dinoflagellates. *J. Foraminiferal Res.* **6**, 1–21.

ATODA K. (1947a) The larva and postlarval development of some reef-building corals. I. *Pocillopora damicornis cespitosa* (Dana). *Sci. Rep. Tohoku Univ.* **4**, 18, 24–47.

ATODA K. (1947b) The larva and postlarval development of some reef-building corals. II. *Stylophora pistillata* (Esper). *Sci. Rep. Tohoku Univ. Ser. 4*, **18**, 48–64.

ATODA K. (1951) The larval and postlarval development of some reef-building corals. III. *Acropora brüggemanni* (Brook). *J. Morphol.* **89**, 1–16.

ATODA K. (1954) Post-larval development of the sea anemone *Anthopleura* sp. *Sci. Rep. Tohoku Univ. Ser. 4*, **20**, 274–286.

BANASZAK A. (1985) The importance of zooxanthellae at various stages of development in the intertidal burrowing clan *Tridacna crocea* Lamark 1819. Unpublished Honours Thesis, James Cook University, Townsville, Australia.

BARNES D. J. & TAYLOR D. L. (1973) *In situ* studies of calcification and photosynthetic carbon fixation in the coral *Montastrea annularis*. *Helgolander. wiss. Meeresunters.* **24**, 284–291.

BEAM C. A. & HIMES M. (1974) Evidence for sexual fusion and recombination in the dinoflagellate *Crypthecodinium cohnii*. *Nature* **250**, 435–436.

BEAM C. A. & HIMES M. (1977) Sexual isolation and genetic diversification among some strains of *Crypthecodinium chonii*-like dinoflagellates: evidence of speciation. *J. Protozool.* **24**, 532–539.

BEAM C. A. & HIMES M. (1980) Sexuality and meiosis in dinoflagellates. In *Biochemistry and Physiology of Protozoa* (eds M. Levandowsky, S. H. Hunter & L. Provasoli), pp. 171–206. Academic Press, New York.

BEAM C. A. & HIMES M. (1982) Distribution of members of the *Crypthecodinium cohnii* (Dinophyceae) species complex. *J. Protozool.* **29**, 8–15.

BEARDALL J., MUKERJI D., GLOVER H. E. & MORRIS I. (1976) The path of carbon in photosynthesis by marine phytoplankton. *J. Phycol.* **12**, 409–417.

BECKVAR N. (1981) Cultivation, spawning and growth of the giant clams *Tridacna gigas*, *T. derasa* and *T. squamosa* in Palau, Caroline Islands. *Aquaculture* **24**, 21–30.

BENEDICT C. R. (1978) Nature of obligate photoautotrophy. *Ann. Rev. Plant Physiol.* **29**, 67–93.

BENSON A. A. & MUSCATINE L. (1974) Wax in coral mucus: energy transfer from corals to reef fishes. *Limnol. Oceanogr.* **19**, 810–814.

BENSON A. A., PATTON J. S. & ABRAHAM S. (1978) Energy exchange in coral reef ecosystems. *Atoll Res. Bull.* **220**, 33–53.

BIGELOW R. P. (1900) The anatomy and development of *Cassiopeia xamachana. Mem. Boston Soc. Natur. Hist.* **5**, 191–236.

BIRD I. F., CORNELIUS M. J. & KEYS A. J. (1980) Effect of carbonic anhydrase on the activity of ribulosebisphosphate carboxylase. *J. exp. Bot.* **31**, 365–369.

BISHOP C. A. & KENRICK J. R. (1981) Fatty acid composition of symbiotic zooxanthellae in relation to their hosts. *Lipids* **15**, 799–804.

BJORKMAN O., ANDERSON N. K., ANDERSON J. M., THORNE S. W., GOODCHILD D. J. & PYLIOTIS N. A. (1972) Effect of light intensity during growth of *Atriplex patula* on the capacity of photosynthetic reactions, chloroplast component and structure. *Carnegie Inst. Wash.* **107**, 115–135.

BLACK C. C. & BENDER M. M. (1976) $\delta^{13}C$ values in marine organisms from the Great Barrier Reef. *Aust. J. Plant Physiol.* **3**, 25–33.

BLACK C. C., BURRIS J. E. & EVERSON R. G. (1976) Influence of oxygen concentration on photosynthesis in marine plants. *Aust. J. Plant Physiol.* **3**, 81–86.

BLANK R. J. & TRENCH R. K. (1985a) Speciation and symbiotic dinoflagellates. *Science* **229**, 656–658.

BLANK R. J. & TRENCH R. K. (1985b) *Symbiodinium microadriaticum*: a single species? *Intl. Congr. Coral Reefs* **5**.

BLANK R. J. & TRENCH R. K. (1986) Nomenclature of endosymbiotic dinoflagellates. *Taxon* (in press).

BRINKMANN A. (1964) Observations on the structure and development of the medusa of *Velella velella* (Linné, 1758). *Vidensk. Medd. dansk. naturh. Foren.* **126**, 327–336.

BURRIS J. E. (1977) Photosynthesis, photorespiration and dark respiration in eight species of algae. *Mar. Biol.* **39**, 371–379.

BURRIS J. E. (1981) Effects of oxygen and inorganic carbon concentrations on the photosynthetic quotients of marine algae. *Mar. Biol.* **65**, 215–219.

CHALKER B. E. (1981) Simulating light-saturation curves for photosynthesis and calcification by reef-building corals. *Mar. Biol.* **63**, 135–141.

CHANG S. S., PRÉZELIN B. B. & TRENCH R. K. (1983) Mechanisms of photoadaptation in three strains of the symbiotic dinoflagellate *Symbiodinium microadriaticum. Mar. Biol.* **76**, 219–231.

CHANG S. S. & TRENCH R. K. (1982) Peridinin–chlorophyll *a*-proteins from the symbiotic dinoflagellate *Symbiodinium* (= *Gymnodinium*) *microadriaticum* Freudenthal. *Proc. R. Soc. Lond.* (*B*) **215**, 191–210.

CHANG S. S. & TRENCH R. K. (1984) The isoelectric forms, quaternary structure and amino acid composition of peridinin-chlorophyll *a*—proteins from the dinoflagellate *Symboidinium microadriaticum* Feudenthal. *Proc. R. Soc. Lond. (B)* **222**, 259–271.

COLLEY N. J. & TRENCH R. K. (1983) Selectivity in phagocytosis and persistence of symbiotic algae by the scyphistoma stage of the jellyfish *Cassiopeia xamachana. Proc. R. Soc. Lond.* (*B*) **219**, 61–82.

COLLEY N. J. & TRENCH R. K. (1985) Cellular events in the re-establishment of a symbiosis between a marine dinoflagellate and a coelenterate. *Cell Tissue Res.* **239**, 93–103.

COOK C. B. (1983) Metabolic interchange in algae–invertebrate symbiosis. *Int. Rev. Cytol.* **14**, 177–210.

COOK C. B. (1985) Equilibrium populations and long term stability of mutualistic algal and invertebrate hosts. In *The Biology of Mutualism: Ecology and Evolution* (ed. D. B. Boucher) pp. 171–191, Croom-Helm, Amsterdam.

CROSSLAND C. J. & BARNES D. J. (1977a) Nitrate assimilation enzymes from two hard corals, *Acropora acuminata* and *Goniastrea australensis. Comp. Biochem. Physiol.* **57B**, 151–157.

CROSSLAND C. J. & BARNES D. J. (1977b) Gas exchange studies with the staghorn coral *Acropora acuminata* and its zooxanthellae. *Mar. Biol.* **40**, 185–194.

DEANE E. M. & O'BRIEN R. W. (1978) Isolation and axenic culture of *Gymnodinium microadriaticum* from *Tridacna maxima*. *Br. Phycol. J.* **13**, 189–195.

DEANE E. M. & O'BRIEN R. W. (1981a) Uptake of phosphate by symbiotic and free-living dinoflagellates. *Arch. Microbiol.* **128**, 307–310.

DEANE E. M. & O'BRIEN R. W. (1981b) Uptake of sulphate, taurine, cystine and methionine by symbiotic and free-living dinoflagellates. *Arch. Microbiol.* **128**, 311–319.

D'ELIA C. (1977) The uptake and release of dissolved phosphorus by reef corals. *Limnol. Oceanogr.* **22**, 301–315.

D'ELIA, C. F. & WEBB K. L. (1977) The dissolved nitrogen flux of reef corals. *Proc. 3rd. Int. Coral Reef Symp.* vol. 1, pp. 325–330. Univ. Miami.

DOWNTON W. J. S., BISHOP D. G., LANKUM A. W. D. & OSMOND C. B. (1976) Oxygen inhibition of photosynthetic oxygen evolution in marine plants. *Aust. J. Plant Physiol.* **3**, 73–79.

DREW E. A. (1972) The biology and physiology of alga-invertebrate symbioses. I. The density of symbiotic algal cells in a number of hermatypic hard corals and alcyonarians from various depths. *J. exp. mar. Biol. Ecol.* **9**, 71–75.

DUBOS R. & KESSLER A. W. (1963) Integrative and disintegrative factors in symbiotic associations. *Symp. Soc. Gen. Microbiol.* **13**, 1–11.

DURDEN J. E. (1902) West Indian madreporarian polyps. *Mem. Nat. Acad. Sci. USA*. **7**, 403–597.

DUSTAN P. (1979) Distribution of zooxanthellae and photosynthetic chloroplast pigment of the reef-building coral *Montastrea annularis* Ellis and Solander in relation to depth on a West Indian coral reef. *Bull. mar. Sci.* **29**, 79–95.

DYKENS J. A. & SHICK J. M. (1982) Oxygen production by endosymbiotic algae controls superoxide dismutase activity in their animal host. *Nature* **297**, 579–580.

FALKOWSKI P. (1981) Light and shade adaptation and assimilation numbers. *J. Plankton Res.* **3**, 203–216.

FALKOWSKI P. G. & DUBINSKY Z. (1981) Light–shade adaptation of *Stylophora pistillata*, a hermatypic coral from the Gulf of Eilat. *Nature* **289**, 172–174.

FANKBONER P. V. (1971) Intracellular digestion of symbiotic zooxanthellae by host amoebocytes in giant clams (Bivalvia: Tridacnidae), with a note on the nutritional role of the hypertrophied siphonal epidermis. *Biol. Bull.* **141**, 222–234.

FANKBONER P. V. & REID R. G. (1981) Mass expulsion of zooxanthellae by heat-stressed reef-corals: a source of food for giant clams? *Experientia* **37**, 251–252.

FAURE C. (1960) Etude des phenomenes de reproduction chez *Aglaophemia pluma* (L.). *Cah. Biol. mar.* **1**, 185–204.

FISHER C. R., FITT W. K. & TRENCH R. K. (1985) Photosynthesis and respiration in *Tridacna gigas* as a function of irradiance and size. *Biol. Bull.* **169**, 230–245.

FITT W. K. (1984) The role of chemosensory behavior of *Symbiodinium microadraticum*, intermediate hosts and host behavior in the infection of coelenterates and molluscs with zooxanthellae. *Mar. Biol.* **81**, 9–17.

FITT W. K. (1985) The effect of different strains of the zooxanthella *Symbiodinium microadriaticum* on growth and survival of their coelenterate and molluscan hosts. *Intl. Congr. Coral Reefs* **5**.

FITT W. K., CHANG S. S. & TRENCH R. K. (1981) Motility patterns of different strains of the symbiotic dinoflagellate *Symbiodinium* (= *Gymnodinium*) *microadriaticum* (Freudenthal) in culture. *Bull. mar. Sci.* **31**, 436–443.

FITT W. K., FISHER C. R. & TRENCH R. K. (1984) Larval biology of tridacnid clams. *Aquaculture* **39**, 181–195.

FITT W. K. & TRENCH R. K. (1981) Spawning, development, and acquisition of zooxanthellae by *Tridacna squamosa* (Mollusca, Bivalvia). *Biol. Bull.* **161**, 213–235.

FITT W. K. & TRENCH R. K. (1983a) The relation of diel patterns of cell division to diel patterns of motility in the symbiotic dinoflagellate *Symbiodinium microadriaticum* Freudenthal. *New Phytol.* **94**, 421–432.

FITT W. K. & TRENCH R. K. (1983b) Endocytosis of the symbiotic dinoflagellate *Symbiodinium microadriaticum* Freudenthal by endodermal cells of the scyphistomae of Cassiopeia xanachana and resistance of the algae to host digestion. *J. cell Sci.* **64**, 195–212.

FRAIZER W. & GLASER L. (1979) Surface components and cell recognition. *Ann. Rev. Biochem.* **48**, 491–523.

FRANCA S. (1976) On the presence of virus-like particles in the dinoflagellate *Bymnodinium resplendens* (Hulbunt). *Protistologica* **12**, 425–430.

FRANZISKET L. (1975) Nitrate uptake by reef corals. *Int. Revue ges. Hydrobiol.* **59**, 1–7.

FRASER E. A. (1931) Observations on the life-history and development of the hydroid, *Myrionema ambionense. Sci. Rep. Great Barrier Reef. Exped.* **3**, 135–144.

FREUDENTHAL H. D. (1962) *Symbiodinium* gen. nov. and *S. microadriaticum* sp. nov., a zooxanthella: taxonomy, life cycle, and morphology. *J. Protozool.* **9**, 45–52.

GIBBS S. P. (1980) The chloroplasts of some algal groups may have evolved from endosymbiotic eukaryotic algae. *Ann. NY Acad. Sci.* **361**, 193–208.

GOLD K. & POLLINGHER U. (1971) Occurrence of endosymbiotic bacteria in marine dinoflagellates. *J. Phycol.* **7**, 264–265.

GOREAU T. F. (1964) Mass expulsion of zooxanthellae from Jamaican reef communities after Hurricane Flora. *Science* **145**, 383–386.

GOREAU T. J. (1977) Coral skeletal chemistry: physiological and environmental regulation of stable isotopes and trace metals in *Montastrea annularis. Proc. R. Soc. Lond.* (*B*) **196**, 291–315.

GRAHAM D. & SMILLIE R. M. (1976) Carbonate dehydratase in marine organisms of the Great Barrier Reef. *Aust. J. Plant. Physiol.* **3**, 113–119.

HALLDAL P. (1968) Photosynthetic capacities and photosynthetic action spectra of endozoic algae of the massive coral *Favia. Biol. Bull.* **134**, 411–424.

HAMNER W. M. & HAURI I. R. (1981) Long-distance horizontal migrations of zooplankton (Scyphomedusae: Mastigias). *Limol. Oceanogr.* **26**, 414–423.

HECKEL P. H. (1974) Carbonate buildup in the geologic record: a review. In *Reefs in Time and Space* (ed. L. F. Laporte), pp. 90–154, No. 18. Spec. Publs. Soc. Econ. Palont. Miner. Tulsa, Oklahoma.

HESLOP-HARRISON J. (1978) *Cellular Recognition Systems in Plants*, pp. 60. University Park Press, Baltimore, Maryland, USA.

HOFMANN D. K. & KREMER B. P. (1981) Carbon metabolism and strobilation in *Cassiopea andromeda* (Cnidaria: Scyphozoa): significance of endosymbiotic dinoflagellates. *Mar. Biol.* **65**, 25–33.

HOLLANDE A. & CARRÉ D. (1974) Les xanthelles des radiolaires sphaerocollides, des acanthaires et de *Velella velella*: Infrastructure–cytochimie–taxonomie. *Protistologica* **10**, 573–601.

HOUCK J. E. (1978) *The potential utilization of scleractinian corals in the study of marine environments.* Unpublished Ph.D. dissertation, University of Hawaii. pp. 178.

JAMESON S. C. (1976) Early life history of the giant clams *Tridacna crocea* Lamarck, *Tridacna maxima* (Röding) and *Hippopus hippopus* (Linnaens). *Pac. Sci.* **30**, 219–233.

JEFFREY S. W. (1969) Properties of two spectrally different components in chlorophyll *c* preparations. *Biochem. Biophys. Acta* **177**, 456–467.

JEFFREY S. W. & SHIBATA K. (1969) Some spectral characteristics of chlorophyll *c* from *Tridacna crocea* zooxanthellae. *Biol. Bull.* **136**, 54–62.

JEFFREY S. W. & VESK M. (1976) Further evidence for a membrane-bound endosymbiont within the dinoflagellate *Peridinium foliacium. J. Phycol.* **12**, 450–455.

KAWAGUTI S. (1950) Observations of the heart shell, *Corculum cardissa* (L.), and its associated zooxanthellae. *Pac. Sci.* **4**, 43–49.

KAWAGUTI S. (1968) Electron microscopy on zooxanthellae in the mantle and gill of the heart shell. *Biol. J. Okayama Univ.* **14**, 1–12.

KELLOGG R. B. & PATTON J. S. (1983) Lipid droplets, medium of energy exchange in the symbiotic anemone *Condylactis gigantea*: a model coral polyp. *Mar. Biol.* **75**, 137–149.

KEVIN M. J., HALL W. T., MCLAUGHLIN J. J. A. & ZAHL P. A. (1969) *Symbiodinium microadriaticum* Fruedenthal, a revised taxonomic description–ultrastructure. *J. Phycol.* **5**, 341–350.

KINSEY D. W. & DAVIES P. J. (1979) Effects of elevated nitrogen and phosphorus on coral reef growth. *Limnol. Oceanogr.* **24**, 935–939.

KINZIE R. A. III (1974) Experimental infection of aposymbiotic gorgonian polyps with zooxanthellae. *J. Exp. mar. Biol. Ecol.* **15**, 335–345.

KINZIE R. A. III & CHEE G. S. (1979) The effect of different zooxanthellae on the growth of experimentally reinfected hosts. *Biol. Bull.* **156**, 315–327.

KOJIS B. L. (1982) Reproductive strategies in four species of Porites (Seleractinia). *Proc. 4th Int. Coral Reef Symp.* Manila, Philippines.

KOJIS B. L. & QUINN N. J. (1981) Aspects of sexual reproduction and larval development in the shallow water hermatypic coral, *Goniastrea australensis* (Edwards & Haime, 1957). *Bull. Mar. Sci.* **31**, 448–573.

KOKKE W. C. M. C., FENICAL W., BOHLIN L. & DJERASSI C. (1981) Sterol synthesis by cultured zooxanthellae: implications concerning sterol metabolism in the host–symbiont association in Caribbean gorgonians. *Comp. Biochem. Physiol.* **68B**, 281–287.

KREMER B. P. & KÜPPERS U. (1977) Carboxylating enzymes and pathway of photosynthetic carbon assimilation in different marine algae—evidence for the C_4 pathway? *Planta* **133**, 191–196.

KREMER B. P., SCHMALJOHANN R. & RÖTTGER R. (1980) Features and nutritional significance of photosynthates produced by unicellular algae symbiotic with larger foraminifera. *Mar. Ecol. Prog. Ser.* **2**, 225–228.

KRUPP D. A. (1981) *Sexual reproduction in the solitary scleractinian coral* Fungia scutaria. Abstract. Western Soc. Nature. 62nd Ann. Meet. Santa Barbara.

KRUPP D. A. (1982) *The composition of the mucus from the mushroom coral* Fungia scutaria. Proc. 4th Int. Coral Reef Symp. Manila, May, 1981.

KUSKOP M. (1921) Über die symbiose von Siphonophoren und zooxanthellen. *Zool. Anz.* **52**, 258–266.

LABARBERA M. (1975) Larval and post-larval development of the giant clams *Tridacna maxima* and *Tridacna squamosa* (Bivalvia, Tridacnidae). *Malacologia* **15**, 69–79.

LAND L. S., LANG J. C. & SMITH B. N. (1975) Preliminary observations on the carbon isotopic composition of some reef coral tissues and symbiotic zooxanthellae. *Limnol. Oceanogr.* **20**, 283–297.

LEE J. J., MCENERY M. E., KAHN E. G. & SCHUSTER F. L. (1979) Symbiosis and the evolution of larger foraminifera. *Micropaleontology* **25**, 118–140.

LERCH K. & COOK C. B. (1982) The effect of photoperiod on motility in zooxanthellae. *Bull. mar. Sci.* (in press).

LEUTENEGGER S. (1977) Ultrastructure and motility of dinophyceans symbiotic with larger imporforated foraminifera. *Mar. Biol.* **44**, 157–164.

LEWIS D. H. (1973) The relevance of symbiosis to taxonomy and ecology, with particular reference to mutualistic symbioses and the exploitation of marginal habitats. In *Taxonomy and Ecology* (ed. V. H. Heywood), pp. 151–172. Academic Press, London.

LEWIS D. H. & SMITH D. C. (1971) The autotrophic nutrition of symbiotic marine coelenterates with special reference to hermatypic corals. I. Movement of photosynthetic products between the symbionts. *Proc. R. Soc. Lond. (B)* **178**, 111–129.

LOEBLICH A. R. & SHERLEY J. L. (1979) Observations on the theca of the motile phase of the free-living and symbiotic isolates of *Zooxanthella microadriatica* (Freudenthal) comb. nov. *J. mar. biol. Ass. UK* **59**, 195–205.

LUDWIG F. D. (1969) Die Zooxanthellen bei *Cassiopeia andromeda* Eschscholtz 1929 (Polyp-Stadium) und ihr Bedeuting fur die Strobilation. *Zool. Jb. Anat. Bd.* **86**, 238–277.

MCCLOSKEY L. & MUSCATINE L. (1984) Production and respiration in the Red Sea coral *Stycophora pistillata* as a function of depth. *Proc. R. Soc. Lond. (B)* **222**, 215–230.

McLaughlin, J. J. A. & Zahl P. A. (1959) Axenic zooxanthellae from various invertebrate hosts. *Ann. NY Acad. Sci.* **77**, 55–72.

McLaughlin J. J. A. & Zahl P. A. (1966) Endozoic algae. In *Symbiosis*, vol. I (ed. S. M. Henry), pp. 257–297. Academic Press, New York.

Magman J. (1909) The entry of zooxanthellae into the ovary of *Millepora* and some particulars concerning the medusa. *Quart. J. Microscop. Sci.* **33**, 697–709.

Mandelli E. F. (1972) The effect of growth illumination on the pigmentation of a marine dinoflagellate. *J. Phycol.* **8**, 367–369.

Marshall L. G. (1981) The great American interchange—an invasion-induced crisis for South American mammals. In *Biotic Crises in Ecological and Evolutionary Time* (ed. M. H. Niticki), pp. 113–229. Academic Press, New York.

Marshall S. & Stephenson T. A. (1933) The feeding of reef animals. I. The corals. *Sci. Rep. Great Barrier Reef Exped.* **3**, 219–245.

Miflin B. J. & Lea P. J. (1976) The pathway of nitrogen assimilation in plants. *Phytochemistry* **15**, 873–885.

Morris I. (1980) Paths of carbon assimilation in marine phytoplankton. In *Primary Productivity in the Sea* (ed. P. Falkowski), pp. 139–159. Plenum Press, New York.

Müller-Merz E. & Lee J. J. (1976) Symbiosis in the larger foraminiferan *Sorites marginalis* (with notes on *Archaias* spp.). *J. Protozool.* **23**, 390–396.

Muller-Parker G. (1984a) Dispersal of zooxanthellae on coral reefs by predators on cnidarians. *Biol. Bull.* **167**, 159–167.

Muller-Parker G. (1984b) Photosynthesis-irradiance responses and photosynthetic periodicity in the sea anemone *Aiptasia pulchella* and its zooxanthellae. *Mar. Biol.* **82**, 225–232.

Muller-Parker G. (1985) Effects of feeding regime and irradiance on the photophysiology of the symbiotic sea anemone *Aiptasia pulchella. Mar. Biol.* **90**, 65–74.

Muscatine L. (1973) Nutrition of corals. In *Biology and Geology of Coral Reefs*, Vol. II (eds. O. A. Jones & R. Endean), pp. 77–115. Academic Press, New York.

Muscatine L. (1980a) Productivity of zooxanthellae. In *Primary Productivity in the Sea* (ed. P. G. Falkowski), pp. 381–402. Plenum Press, New York.

Muscatine L. (1980b) Uptake, retention, and release of dissolved inorganic nutrients by marine alga–invertebrate associations. In *Cellular Interactions in Symbiosis and Parasitism* (eds. C. B. Cook, P. W. Pappas & E. D. Rudolph), pp. 229–244. Ohio State University Press, Columbus.

Muscatine L. & D'Elia C. (1978) The uptake, retention, and release of ammonium by reef corals. *Limnol. Oceanogr.* **23**, 725–734.

Muscatine L. & Marian R. E. (1982) Dissolved inorganic nitrogen flux in symbiotic and nonsymbiotic medusae. *Limnol. Oceanogr.* **27**, 910–917.

Muscatine L., McCloskey L. R. & Marian R. E. (1981) Estimating the daily contribution of carbon from zooxanthellae to coral animal respiration. *Limnol. Oceanogr.* **26**, 601–611.

Muscatine L. & Porter J. W. (1977) Reef corals: mutualistic symbioses adapted to nutrient-poor environments. *Bio. Sci.* **27**, 454–460.

Patton J. S. & Burris J. E. (1983) Lipid synthesis and extrusion by freshly isolated zooxanthellae (symbiotic algae). *Mar. Biol.* **75**, 131–136.

Patton J. S., Battey J. F., Rigler M. W., Porter J. W., Black C. C. & Burris J. E. (1983) A comparison of the metabolism of bicarbonate ^{14}C and acetate 1-^{14}C and the variability of species lipid compositions in reef corals. *Mar. Biol.* **75**, 121–130.

Pearse V. B. (1974) Modification of sea anemone behavior by symbiotic zooxanthellae: phototaxis. *Biol. Bull.* **147**, 630–640.

Perry M. J., Talbot M. C. & Alberte R. S. (1981) Photoadaptation in marine phytoplankton: Response of the photosynthetic unit. *Mar. Biol.* **62**, 91–101.

Prézelin B. B. & Sweeney B. M. (1979) Photoadaptation of photosynthesis in two bloom-forming dinoflagellates. In *Toxic Dinoflagellate Blooms* (eds. D. L. Taylor & H. H. Seliger), pp. 101–106. Elsevier North Holland, Inc.

PRÉZELIN B. B. (1981) Light reactions in photosynthesis. In *Physiological Bases of Phytoplankton Ecology* (ed. T. Plah). Canadian Bulletin of Fisheries and Aquatic Sciences, No. 201.

RAVEN J. A. & BEARDALL J. (1981) Respiration and photorespiration. In *Physiological Bases of Phytoplankton Ecology* (ed. T. Platt). *Can. Bull. Fish Aquat. Sci.* **210**, 55–82.

RAVEN J. A. & GLIDEWELL S. M. (1978) C_4 characteristics of photosynthesis in the C_3 alga *Hydrodictyon africanum*. *Plant, Cell Envir.* **1**, 185–197.

REDALJE R. (1976) Light adaptation strategies of hermatypic corals. *Pac. Sci.* **30**, 212.

REED M. L. & GRAHAM D. (1977) Carbon dioxide and the regulation of photosynthesis: activities of photosynthetic enzymes and carbonate dehydratase (carbonic anhydrase) in *Chlorella* after growth or adaptation in different carbon dioxide concentrations. *Aust. J. Plant Physiol.* **4**, 87–98.

REIMER A. A. (1971) Observations on the relationships between several species of tropical zoanthids (Zoanthidea, Coelenterata) and their zooxanthellae. *J. Exp. mar. Biol. Ecol.* **7**, 207–214.

RINKEVICH B. & LOYA Y. (1979) The reproduction of the Red Sea coral *Stylophora pistillata*. I. Gonads and planulae. *Mar. Ecol. Prog. Ser.* **1**, 133–144.

RIVKIN R. B., SELIGER H. H., SWIFT E. & BIGGLEY W. H. (1982) Light–shade adaptation by the oceanic dinoflagellates *Pyrocystis noctiluca* and *P. fusiformis*. *Mar. Biol.* **68**, 181–191.

SCHOENBERG D. A. & TRENCH R. K. (1976) Specificity of symbioses between marine cnidarians and zooxanthellae. In *Coelenterate Ecology and Behaviour* (ed. G. O. Mackie), pp. 423–432. Plenum Press, New York.

SCHOENBERG D. A. & TRENCH R. K. (1980a) Genetic variation in *Symbiodinium* (= *Gymnodinium*) *microadriaticum* Freudenthal, and specificity in its symbiosis with marine invertebrates. I. Isoenzyme and soluble protein patterns of axenic cultures of *S. microadriaticum*. *Proc. R. Soc. Lond.* (*B*) **207**, 405–427.

SCHOENBERG D. A. & TRENCH R. K. (1980b) Genetic variation in *Symbiodinium* (= *Gymnodinium*) *microadriaticum* Freudenthal, and specificity in its symbiosis with marine invertebrates. II. Morphological variation in *S. microadriaticum*. *Proc. R. Soc. Lond.* (*B*) **207**, 429–444.

SCHOENBERG D. A. & TRENCH R. K. (1980c) Genetic variation in *Symbiodinium* (= *Gymnodinium*) *microadriaticum* Freudenthal, and specificity in its symbiosis with marine invertebrates. III. Specificity and infectivity of *S. microadriaticum*. *Proc. R. Soc. Lond.* (*B*) **207**, 445–460.

SCOTT B. D. & JITTS J. R. (1977) Photosynthesis of phytoplankton and zooxanthellae on a coral reef. *Mar. Biol.* **41**, 307–315.

SIEBERT A. E. (1974) A description of the embryology, larval development, and feeding of the sea anemone *Anthopleura elegantissima* and *A. xanthogrammica*. *Can. J. Zool.* **52**, 1383–1388.

SILVA E. S. (1978) Endonuclear bacteria in two species of dinoflagellates. *Protistologica* **14**, 113–119.

SPERO H. (1986) Symbiosis chamber formation and stable isotope incorporation in the planktonic foraminifer *Orbulina universa*. PhD Thesis, University of California.

SMITH B. N. & EPSTEIN S. (1970) Biogeochemistry of the stable isotopes of hydrogen and carbon in salt marsh biota. *Plant Physiol.* **46**, 738–742.

SMITH D. C., MUSCATINE L. & LEWIS D. H. (1969) Carbohydrate movement from autotrophs to heterotrophs in parasitic and mutualistic symbiosis. *Biol. Rev.* **44**, 17–90.

SMITH F. A. & WALKER N. A. (1980) Photosynthesis by aquatic plants: effects of unstirred layers in relation to assimilation of CO_2 and HCO_3^- and to carbon isotopic discrimination. *New Phytol.* **86**, 245–259.

SPENSER-DAVIES P. (1977) Carbon budgets and vertical zonation of Atlantic Reef corals. *Proc. 3rd Int. Reef. Coral Symp.* **1**, 392–396.

SPENSER-DAVIES P. (1980) Respiration in some Atlantic reef corals in relation to vertical distribution and growth form. *Biol. Bull.* **158**, 187–194.

SPINDLER M. & HEMLEBEN Ch. (1980) Symbionts in planktonic foraminifera (Protozoa). In *Endocytobiology* (ed. W. Schwemmler & H. E. A. Schenk), pp. 133–140. Walter de Gruyter and Co., Berlin.

STEELE R. D. (1975) Stages in the life history of a symbiotic zooxanthella in pellets extruded by its host *Aiptasia tagetes* (Duch. and Mich.) (Coelenterata, Anthozoa). *Biol. Bull.* **149**, 590–600.

STEELE R. D. (1977) The significance of zooxanthella-containing pellets extruded by sea anemones. *Bull. Mar. Sci.* **27**, 591–594.

STEELE R. E. & RAE P. M. (1980) Comparisons of *Crypthecodinium cohnii*-like dinoflagellates from widespread geographic locations. *J. Protozool.* **27**, 479–483.

SUGIURA Y. (1963) On the life history of rhizostome medusae. I. *Mastigias papua* (L) Agassiz. *Annot. Zool. Japan.* **36**, 194–202.

SUGIURA Y. (1964) On the life history of rhizostome medusae. II. Indispensibility of zooxanthellae for strobilation in *Mastigias papua*. *Embriologica* **8**, 223–233.

SUMMONS R. E. & OSMOND C. B. (1981) Nitrogen assimilation in the symbiotic marine alga *Gymnodinium microadriaticum*: direct analysis of ^{15}N incorporation by GC-MS methods. *Phytochemistry* **20**, 575–578.

SVOBODA A. & PORRMANN T. (1980) Oxygen production and uptake by symbiotic *Aiptasiadiaphana* (Rapp), (Anthozoa: Coelenterata) adapted to different light intensities. In *Nutrition in the lower Metazoa* (eds. D. C. Smith & Y. Tiffon), pp. 87–99. Pergamon Press, Oxford.

SWEENEY B. M. (1971) Laboratory studies of a green *Noctiluca* from New Guinea. *J. Phycol.* **7**, 53–58.

SWEENEY B. M. (1976) *Pedinomonas noctilucae* (prasinophyceae), the flagellate symbiotic in *Noctiluca* (Dinophyceae) in Southeast Asia. *J. Phycol.* **12**, 460–464.

SWEENEY B. M. (1978) Ultrastructure of *Noctiluca miliaris* (Pyrrophyta) with green flagellate symbionts. *J. Phycol.* **14**, 116–120.

SWENSON W., TAGLE B., CLARDY J., WITHERS N. W., KOKKE W. C. M. C., FENICAL W. & DJERASSI C. (1980) Peridinosterol—a new Δ^{17}-unsaturated sterol from two cultured marine algae. *Tetrahedron Lett.* 4663–4666.

SZMANT-FROELICH A., YEUICH P. & PILSON M. E. Q. (1980) Gametogenesis and early development of the temperate coral *Astrangia danae* (Anthozoa: Scleractinia). *Biol. Bull.* **158**, 257–269.

TAPPAN H. (1980) *The Paleobiology of Plant Protists*, pp. 1082. W. H. Freeman and Co., San Francisco.

TAYLOR D. L. (1969) On the regulation and maintenance of algal numbers in zooxanthellae-coelenterate symbiosis, with a note on the nutritional relationship in *Anemonia sulcata*. *J. mar. biol. Ass. UK* **49**, 1057–1065.

TAYLOR D. L. (1971a) Ultrastructure of the 'zooxanthella' *Endodinium chattonii in situ*. *J. mar. Biol. Ass. UK* **51**, 227–234.

TAYLOR D. L. (1971b) On the symbiosis between *Amphidinium klebsii* (Dinophyceae) and *Amphiscolops langerhansi* (Turbellaria: Acoela). *J. mar. Biol. Ass. UK* **51**, 301–313.

TAYLOR D. L. (1973a) The cellular interactions of algal–invertebrate symbioses. *Adv. Mar. Biol.* **11**, 1–56.

TAYLOR D. L. (1973b) Symbiotic pathways of carbon in coral reef ecosystems. Present status and future prospects. *Helgolander wiss. Meeresunters.* **24**, 276–283.

TAYLOR D. L. (1974) Symbiotic marine algae: taxonomy and biological fitness. In *Symbiosis in the Sea* (ed. W. B. Vernberg), pp. 245–262. University of South Carolina Press, Columbia.

TAYLOR D. L. (1975) Symbiotic dinoflagellates. *Symp. Soc. Exp. Biol.* **29**, 267–277.

TAYLOR D. L. (1980) Nutrient competition as a basis for symbiont selection in associations involving *Convoluta roscoffensis* and *Amphiscolops langerhansi*. In *Endocytobiology* (eds. W. Schwemmler & H. E. A. Schenk), pp. 279–292. Walter de Gruyter and Co., Berlin.

TAYLOR D. L. (1982) Coral–algal symbioses. In *Algal Symbiosis: A Continuum of Interaction Strategies* (ed. L. Goff), pp. 19–35. Cambridge University Press, Cambridge.

TAYLOR D. L. (1984) Autotrophic eukaryotic marine symbionts. In *Encyclopedia of Plant Physiology*, Vol. 17 (eds A. Pirson & M. H. Zimmerman), pp. 75–90. Springer-Verlag, Berlin.

TAYLOR F. J. R. (1979) Symbiontism revisited: a discussion of the evolutionary impact of intracellular symbioses. *Proc. R. Soc. Lond.* (*B*) **204**, 267–286.

TAYLOR F. J. R. (1980) On dinoflagellate evolution. *BioSystems* **13**, 65–108.

TAYLOR F. J. R. (1982) Symbioses in marine microplankton. *Ann. Inst. oceanogr. Paris* **58**, 61–90.

TAYLOR F. J. R. (1983) Possible free-living *Symbiodinium microadriaticum* (Dinophyceae) in tide pools in southern Thailand. In *Endocytobiology III.* (eds H. E. A. Schenk & W. Schwemmler), pp 1009–1014. Walter de Gruyter, Berlin.

THEODOR J. (1969) Contribution a l'étude des gorgones (VIII): *Eunicella stricta aphyta* sousespèce nouvelle sans zooxanthelles, porche d'une espèce normalement infesteé par ces algues. *Vie et Milieu* **20**, 635–637.

TING I. (1976) Malate dehydrogenase and other enzymes of C_4 acid metabolism in marine plants. *Aust. J. Plant Physiol.* **3**, 121–127.

TITLYANOV E. A., SHAPOSHMIKOVA M. G. & ZVALINSKII V. I. (1980) Photosynthesis and adaptation of corals to irradiance. I. Contents and native state of photosynthetic pigments in symbiotic microalga. *Photosynthetica* **14**, 413–421.

TOMAS R. N., COX E. R. & STEIDINGER K. A. (1973) *Peridinium balticum* (Levander) Lemmermann, an unusual dinoflagellate with a mesokaryotic and a eukaryotic nucleus. *J. Phycol.* **9**, 91–98.

TRENCH R. K. (1971a) The physiology and biochemistry of zooxanthellae symbiotic with marine coelenterates. I. The assimilation of photosynthetic products of zooxanthellae by two marine coelenterates. *Proc. R. Soc. Lond.* (*B*) **177**, 225–235.

TRENCH R. K. (1971b) The physiology and biochemistry of zooxanthellae symbiotic with marine coelenterates. II. Liberation of fixed ^{14}C by zooxanthellae *in vitro*. *Proc. R. Soc. Lond.* (*B*) **177**, 237–250.

TRENCH R. K. (1971c) The physiology and biochemistry of zooxanthellae symbiotic with marine coelenterates. III. The effects of homogenates of host tissues on the excretion of photosynthetic products *in vitro* by zooxanthellae from two marine coelenterates. *Proc. R. Soc. Lond.* (*B*) **177**, 250–264.

TRENCH R. K. (1974) Nutritional potentials in *Zoanthus sociatus* (Coelenterata, Anthozoa). *Helgolander wiss. Meeresunters.* **26**, 174–216.

TRENCH R. K. (1979) The cell biology of plant–animal symbiosis. *Ann. Rev. Plant Physiol.* **30**, 485–531.

TRENCH R. K. (1980) Integrative mechanisms in mutualistic symbioses. In *Cellular Interactions in Symbiosis and Parasitism* (eds. C. B. Cook, P. W. Pappas & E. D. Rudolph), pp. 275–297. Ohio State University Press, Columbus.

TRENCH R. K. (1981) Cellular and molecular interactions in symbioses between dinoflagellates and marine invertebrates. *Pure & Appl. Chem.* **53**, 819–835.

TRENCH R. K., COLLEY N. J. & FITT W. K. (1981a) Recognition phenomena in symbioses between marine invertebrates and 'zooxanthellae'; uptake, sequestration and persistence. *Ber. Deutsch. Bot. Ges. Bd.* **94**, 529–545.

TRENCH R. K. & FISHER C. R. (1983) Carbon dioxide fixation in *Symbiodinium microadriaticum*: problems with mechanisms and pathways. In *Endocytobiology* II (eds. H. E. A. Schenk & W. Schwemmler), pp. 659–675. Walter de Gruyter & Co., Berlin.

TRENCH R. K., WETHEY D. S. & PORTER J. W. (1981b) Observations on the symbiosis with zooxanthellae in the Tridacnidae (Mollusca: Bivalvia). *Biol. Bull.* **161**, 180–198.

TYTLER E. M. & TRENCH R. K. (1986) Activities of enzymes in β-carboxylation reactions and of catalase in cell-free preparations from the symbiotic dinoflagellates *Symbiodinium* spp. from a coral, a clam and two sea anemones. *Proc. R. Soc. Lond. (B)* (in press).

WEBB K. L. & WIEBE W. J. (1978) The kinetics and possible significance of nitrate uptake by several alga invertebrate symbioses. *Mar. Biol.* **47**, 21–27.

WELLS J. W. (1956) Scleractinia. In *Treatise on Invertebrate Paleontology* (ed. R. C. Moore), pp. F328–F444. University of Kansas Press, Lawrence.

WETHEY D. S. & PORTER J. W. (1967a) Sun and shade differences in productivity of reef corals. *Nature* **262**, 281–282.

WETHEY D. S. & PORTER J. W. (1976b) Habitat-related patterns of productivity of the foliaceous reef coral, *Pavona praetorta* Dana. In *Coelenterate Ecology and Behavior* (ed. G. O. Mackie), pp. 59–66. Plenum Publishing Corp., New York.

WHATLEY J. M. (1980) Chloroplast evolution—ancient and modern. *Ann. NY Acad. Sci.* **361**, 154–165.

WITHERS N. W. & HAXO F. T. (1978) Isolation and characterization of carotenoid-rich lipid globules from *Peridinium foliaceum. Plant Physiol.* **62**, 36–39.

WITHERS N. W., KOKKE W. C. M. C., ROHMER M., FENICAL W. H. & DJERASSI C. (1979) Isolation of sterols with cyclopropyl-containing side chains from the cultured marine alga *Peridinium foliaceum. Tetrahedron Lett.* 3605–3608.

WITHERS N. W., KOKKE W. C. M. C., FENICAL W. & DJERASSI C. (1982) Sterol patterns of cultured zooxanthellae isolated from marine invertebrates: synthesis of gorgosterol and 23-desmethyl gorgosterol by aposymbiotic algae. *Proc. Natl. Acad. Sci. USA* **79**, 3764–3768.

YONGE C. M. (1936) Mode of life, feeding, digestion and symbiosis with zooxanthellae in the Tridacnidae. *Sci. Rep. Gr. Barrier Reef Exped. Brit. Mus.* (*Nat. Hist.*), **1**, 283–321.

ZVALINSKII V. I., LELETKIN V. A., TITLYANOV E. A. & SHAPOSHNIKOVA M. G. (1980) Photosynthesis and adaptation of corals to irradiance. 2. Oxygen exchange. *Photosynthetica* **14**, 422–430.

CHAPTER 13
PARASITIC DINOFLAGELLATES

JEAN & MONIQUE CACHON
Université de Nice, Laboratoire de Protistologie Marine, Station Zoologique, 06230 Villefranche sur Mer, France

1 INTRODUCTION

The dinoflagellate nature of many organisms parasitizing other protists, metazoans and even algae, was first recognized by Chatton (1910, 1912 . . . 1952). Despite great variety of morphology he pointed out particular features of the nuclear structures (dinokaryon) and mitosis (dinomitosis) which are common to both free-living and parasitic dinoflagellates. The parasites have special morphological features not present in free-living forms. However, life-history stages resembling free-living species (dinospores) occur and are responsible for the dispersal of the species. On this basis Chatton attributed several *incertae sedis* parasites to this class. He also described many new species and though he suspected that these were polyphyletic organisms, he grouped them into the single zoological tribe Blastodinida.

Chatton's original research on dinoflagellates concluded in 1937. In 1964, Jean Cachon gave more precise details of their cytology and life cycles and asserted their polyphyletic origin. He distinguished two tribes: Blastodinida and Duboscquodinida, differing in the morphology of their vegetative phase, their nuclear development and the structural and metabolic relations with their host. The first tribe is composed essentially of ectoparasites, the second of intracellular ones. The mechanisms of parasitism usually have major consequences for the morphology, physiology, reproduction and dispersal of the parasite. Loeblich (1976), using biochemical data as a basis, proposed that

Syndinium should be separated from the Duboscquodinida as an autonomous order (Syndiniales), to which he later (1983) transferred all those included in the Duboscquodinida here. He also recognized *Chytriodinium* and its relatives as a separate order (Chytriodiniales).

As a result of recent knowledge, including ultrastructural data, we shall distinguish primarily three zoological tribes, recognizing also the parasitic propensities of some of the Dinococcida (Dinococcales or Phytodiniales).

At present, the fungal-like Elliobiopsidae do not seem to belong among the parasitic dinoflagellates. Galt & Whistler (1970) and Hovasse (1974) have placed them there for the single reason that they possess flagella as dinospores, but other morphological and cytological features are unknown. Loeblich (1983) considers them to be a separate class.

This chapter concentrates on information recorded subsequent to the chapter by Chatton in Grassé's *Traité de Zoologie* (1952). The latter should be consulted for more general information.

2 DINOCOCCIDA (botanical order Dinococcales or Phytodiniales)

These organisms are characterized by the importance of a non-motile, vegetative, encysted stage. Some are planktonic, others are benthic, living epiphytically mainly in fresh water. Most possess chlorophyll, starch and a pyrenoid. They have long been considered to be strictly autotrophic. However, a great number are able to feed holozoically, perforating algal cell walls by means of a stalk and rapidly sucking out the cytoplasm. *Stylodinium sphaera*, *Cystodinedria inermis* (Pfiester & Popofsky 1979) and *Cystodinium bataviense* (Pfiester & Lynch 1980) form not only dinospores (naked or thecate, non-flagellated autospores), but also amoeboid forms bearing many thin straight pseudopods resembling axopodia. Autospores and amoebae are motile and attach themselves successively to cells of several algae (*Oedogonium*, *Spirogyra*, *Mougeotia*), finally emptying them. Using their pseudopodia these amoebae can also phagocytose small protists, e.g. algal zoospores, thus behaving as predators as well. Their apparent autotrophy needs re-examination. They possess chloroplasts, but the pigmentation varies from green to brown according to the life-history stage. The chloroplasts may result from feeding, disappearing when the cell contents are metabolized. These amoeboid forms can return to the typical dinococcal form by retracting their pseudopods and secreting an envelope. Dispersal is then by flagellated dinospores which behave either as zoospores with a direct development, or as gametes. The eventual development of zygotes is unknown.

Some previously described organisms could actually be stages of the complex life cycle of these parasites. For example, *Hypnodinium sphaericum* could be a part of the cycle of *Cystodinium*, and *Dinococcus* a part of the cycle of *Cystodinium bataviense* (Pfiester & Lynch 1980). *Dinamoebidium varians* of the monospecific tribe Rhizodinida, some *Rhizochrysis* (chrysomonads) and *Vampyrella* (pseudoheliozoans) might also be stages of such Dinococcida. The genus *Paulsenella*, epiphytic and parasitic on marine pelagic diatoms, was considered by Chatton (1952) to be in the Apodinidae; but it is far from well studied and could also be a Dinococcidian, possessing a permanent cyst-like wall (according to Paulsen's description). On the other hand, *Dissodinium*, which has been placed in the Dinococcida (and is often mistaken for *Pyrocystis*, a free-living form) has to be considered as a member of the Blastodinida according to Drebes (1969, 1978, 1981) and Elbrächter & Drebes (1978).

In the Blastodinida, Duboscquodinida and Syndinida, the life cycle of the protist has two periods, the first corresponding to the vegetative phase (trophont stage), the other to a reproductive phase (sporont stage), which is responsible for new infections. Once attached to their host, spores develop *in situ*. As is nearly always the rule for parasites, reproduction is greatly developed: increased fertility and dispersal counterbalance the improbability of finding and infecting a new host. Table 13.1 lists the species of parasitic dinoflagellates which are currently recognized.

3 THE VEGETATIVE PHASE OF THE NON-DINOCOCCID PARASITES

Infective processes are known in only a few cases. The dinospore is biflagellated but loses its flagella as soon as it comes into contact with the host. It is then either passively swallowed by the host together with food, or it attaches itself by a posterior tentacle-like projection and eventually perforates the host membrane. This projection seems to be homologous to the peduncle, developed from the sulcus in some free-living dinoflagellates and used in phagocytosis (Chapter 6). The development of the parasite is defined at the very outset by the mode of attachment: extracellular (Blastodinida), or intracellular (Duboscquodinida, Syndinida). Comparisons between free-living and parasitic dinoflagellates are made more easily with the first group; the others are more deeply modified by parasitism. The cytology of the Blastodinida, Duboscquodinida and Syndinida will be successively described here.

3.1 Extracellular parasites

Blastodinida (botanical order Blastodiniales)

These organisms mainly parasitize marine protists and metazoans (copepods, siphonophores, appendicularians, jelly-fish, thaliaceans, annelids, fishes) a few algae, and diatoms. They show very clearly gradual modifications of morphology, constitution and physiology from free-living to parasitic dinoflagellates.

Table 13.1. Species of parasitic dinoflagellates currently recognized and their hosts

Tribe/Family	Genus	Species	Host
Dinococcida (semi-parasitic, semi-free)			
		Cystodinium bataviense	Siphonophones
		Stylodinium sphaera	,,
		S. gastrophilium Cachon *et al.*	,,
		Cystodinedria inermis	,,
I Blastodinida (entirely extra-cellular)			
Blastodinidae	*Blastodinium*	*B. apsteini* Sewell	Copepods
	13 spp.	*B. chattoni* Sewell	,,
		B. contortum Chatton	,,
		B. hyalinium Chatton	,,
Protoodinidae	*Protoodinium*	*P. chattoni* Hovasse	Hydromedusae
		P. hovassei Cachon & Cachon	Siphonophores
Apodinidae	*Apodinium*	*A. mycetoides* Cachon & Cachon	Appendicularia
		A. chattoni Cachon & Cachon	,,
		A. zygorhizum Cachon & Cachon	,,
		A. rhizophorum Cachon & Cachon	,,
Haplozoonidae	*Haplozoon*	*H. inerme* Cachon	,,
		H. axiothellae Siebert	Annelida
		H. armatum Dogiel	,,
Oodinidae	*Oodinium*	*O. poucheti* Chatton	,,
		O. dogieli Cachon & Cachon	Annelida
		O. acanthometrae Cachon	Acantharia
		O. fritillariae Cachon	Appendicularia
	Oodinioides	*O. vastator* Reichenback	Fish
	Amyloodinium	*A. ocellatum* Brown	,,
	Crepidoodinium	*C. cyprinodontum* Lawler	,,
	Piscinoodinium	*P. pillulare* Schläperclaus	,,
		P. limneticum Jacobs	,,
Diplomorphidae	*Cachonella*	*C. paradoxa* Cachon	Siphonophores
		C. echinaria Cachon	,,

Table 13.1. (continued)

Tribe/Family	Genus	Species	Host
Chytriodinidae	*Chytriodinium*	*C. affine* Dogiel	Crustacean eggs
		C. roseum Dogiel	" "
		C. parasiticum Dogiel	" "
	Myxodinium	*M. pipiens* Cachon *et al.*	*Halosphaera*
II Duboscquodinida (partly intra, partly extra-cellular)			
Amoebophryidae	*Amoebophrya*	*A. ceratii* Cachon (in Gymnodinians, Adinides, nuclei of Peridinians)	
		A. acanthometrae Borgert (in Acantharia)	
		A. sticholonchae Koeppen (in *Sticholonche*)	
		A. grassei Cachon (in *Oodinium poucheti* etc. *Oodinium acanthometrae*)	
		A. leptodisci Cachon (in *Pratjetella*)	
		A. rosei Cachon (in Ciliates, living in Siphonophores and *Sagitta*)	
		A. tintinni Cachon (in *Xystonella*)	
Duboscquellidae	*Duboscquella*	*D. anisospora* Grasse	
		D. aspida Cachon (in Tintinnids)	
		D. caryophaga Cachon	
		D. cnemata Cachon (in Tintinnids)	
		D. melo Cachon (in *Noctiluca*)	
		D. nucleocola Cachon	
	Duboscquodinium	*D. collini* Grasse	
		D. kofoidi Grasse	
	Keppenodinium (*Hollandella*)	*K. mycetoides* Cachon (*Spongosphaera* in Sphaerellaria)	
		K. lobata Cachon (in Sphaerellaria, *Plegmosphaera*)	
		K. piriformis Cachon (in Sphaerellaria, *Actinosphaera*)	
Sphaeriparidae	*Sphaeripara* (*Neresheimeria*)	*S. catenata* Loeblich (in *Fritillaria*)	
		S. paradoxa Loeblich (in Appendicularia)	
	Atlanticellodinium	*A. tregouboffi* Cachon & Cachon (in Radiolaria *Planktonetta*)	

Table 13.1. (continued)

Tribe/Family	Genus	Species	Host
III Syndinida (entirely intra-cellular)			
Syndinidae	*Syndinium* Chatton	*S. brandti* Hovasse	(in *Collozoum*)
		S. insidiosum Chatton	(in *Collozoum*)
		S. vernale Hovasse	(in *Collozoum*)
		S. mendax Chatton	(in *Myxosphaera*)
		S. dolosum Chatton	(in *Sphaerozoum*)
		S. chattoni Hovasse & Brown	(in *Sphaerozoum*)
		S. astutum Chatton	(in *Sphaerozoum*)
		S. breve Hovasse & Brown	(in *Sphaerozoum*)
		S. globiforme Hollande & Enjumet	(in *Collozoum*)
		S. belari Hollande & Enjumet	(in *Collozoum*)
		S. borgerti Hollande *et al*	(in *Aulacantha*)
		S. sp.	(in *Thalassophysa*, *Spongosphaera*, *Eucyrtidium*, *Cochlodinium*)
	Solenodinium	*S. fallax* Chatton	(in *Thalassicolla*)
		S. leptotaemia Hovasse & Brown	(in *Thalassicolla*)
		S. denseum Hovasse & Brown	(in *Thalassicolla*)
		S. sp. Hollande & Enjumet	(in *Sphaerellaria*)
	Synhemidinium Chatton		
	Ichthyodinium	*I. chabelardi* Hollande & Cachon	(in eggs of *Sardinia*)
	Cochlosyndinium Chatton		*Coryceus*
	Merodinium Chatton		Radiolarians
	Haematodinium Chatton & Poisson		Crabs
	Trypanodinium Chatton		Copepod eggs

Blastodinium, the type genus of this tribe, is not typical but it is the least modified for parasitism. Commonly observed in the gut of copepods, it is the only one which does not have a direct physical attachment to the host; all other Blastodinida are fixed by a posterior stalk. Nevertheless, it produces castration in its host.

A wall is observed in all Blastodinida. In *Blastodinium* (Fig. 13.1) the left-handed helical girdle and short sulcus are conspicuously lined by a cellulosic wall, and a row of small spines decorates one of the girdle edges, aiding perhaps in the anchorage of the organism. Though fixed by a stalk to the membrane of the umbrella of a jelly-fish or siphonophore, *Protoodinium* (Hovasse 1935; Cachon & Cachon 1971a) has a typical theca, with visible girdle and sulcus in which the transverse and longitudinal flagella undulate normally. The tabulation of the theca is still discernible by silver staining in some species of *Oodinium* (Hovasse 1935) but the girdle and sulcus are not lined by the wall. In all other Blastodinida (*Oodinium* pro parte, *Apodinium*, *Haplozoon*, etc.) the wall consists

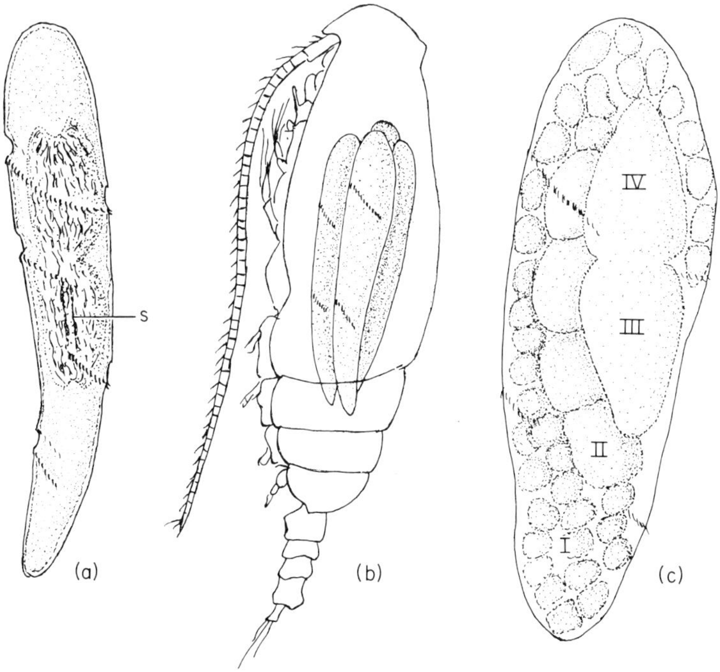

Fig. 13.1. *Blastodinium spinulosum* Chatton. (a) Young trophont. The nucleus is figured at telophase. Its wall bears two helical girdles which are decorated with small spines; they are connected by a short sulcus(s). (b) This species can reproduce vegetatively while it is inside the copepod gut. (c) The parasite is undergoing palisporogenetic divisions within its original wall. Three generations of sporocysts may be seen: (I) the oldest sporocysts (stages 32–64), (II) the four sporocysts of the following stages, (III) one gonocyte and one trophocyte (IV) which are not yet separated.

of a thin continuous, resistant pellicular envelope, bearing no decoration. Beneath the cell membrane there is the usual layer of flat amphiesmal vesicles found in free-living dinoflagellates (Dodge & Crawford 1970, and see Chapter 3). Many of these organisms are able to shed their wall and build a new one, both during growth and sporogenesis.

Chloroplasts are highly modified in parasites. *Blastodinium* still possesses well developed chloroplasts, but Chatton noticed that the pigmentation of the chloroplasts progressively disappears during the vegetative phase, appearing again during sporogenesis (*B. spinulosum*, *B. pruvoti*, *B. contortum*). This pigmentation may be more intense in warm water species which have a rapid development than in those living in colder waters. *B. hyalinum* appears to have no chloroplasts. Thus, there are various intermediary stages between autotrophy and complete heterotrophy in this genus.

In *Protoodinium*, *Piscinoodium* and *Crepidoodinium*, chloroplasts are well developed and appear functional, although partial heterogeneity is probable: certainly something is sucked from the host through the stalk. The host attachment must be made before the parasite can continue its development. All other Blastodinida and most other parasitic dinoflagellates are heterotrophic and lack chlorophyll.

'Chromoplasts' responsible for dark pigmentation, have been described in *Oodinium*, but they have not been observed with electron microscopy. Pigmented lipid droplets are found scattered throughout the cytoplasm of *Oodinium*; they are food reserves. In other parasitic Blastodinida (*Amyloodinium*, *Crepidoodinium*) there are starch granules.

Trichocysts and mucocysts, which are so commonly observed in free-swimming and all other parasitic dinoflagellates, occur in *Protoodinium*, *Amyloodinium* and *Crepidoodinium*. In *Cachonella* they are numerous and grouped in three large bundles beneath the cell membrane. Their 'snap action' may be involved in the strange transformation of the protist at the beginning of sporogenesis (see below). The mitochondria of these marine aerobes do not appear to differ from the norm.

The most remarkable feature of the blastodinid trophont (excluding *Blastodinium*) is the stalk. It may be homologous to temporary pseudopodia which issue from the bottom of the sulcus of some phagotrophic dinoflagellates, e.g. *Podolampas* or *Protoperidinium* (G. Gaines & F. Taylor, pers. comm.), and the peduncle through which many organisms feed (*Gymnodinium*, *Gyrodinium*) (Chapter 6). It may also be homologous to the movable tentacle of forms like *Noctiluca*. This homology is clear in *Protoodinium* but in other organisms (*Oodinium*, *Haplozoon*, *Apodinium*) the hyposome is part of the stalk; the cell body of the latter group consists almost exclusively of the episome. In *Chytrodinium*, which must penetrate the thick chorion before reaching the egg of a crustacean, the hyposome is extraordinarily active and acts as a drill (Fig. 13.6a, b).

The ultrastructure of the stalk makes it clear that it can be an anchoring organelle as well as a feeding apparatus. It may remain totally or partially on the host cuticle, developing a system of ramified rhizoids. Or, the stalk may perforate the cuticle and penetrate the host cell. The parasite is then directly in contact with the host cytoplasm through its rhizoids; the host does not form any membrane around the parasite (parasitophorous membrane). This direct contact seems to be of the same type as in Gregarines, which are also extracellular.

Oodinium's stalk (Cachon & Cachon 1971b) (Fig. 13.2) is the simplest. With the light microscope it appears to be fibrous. It originates in the vicinity of the nucleus. It is globular in the median part but its extremity is flattened into a wide, sucking disc on the host surface. Its double function becomes evident with the electron microscope: the cell membrane of the stalk has deep, cylindrical, closely-packed invaginations which open on the host side producing a 'brush border' appearance. The invaginations reach the perinuclear cytoplasm of the dinoflagellate, where they appear as large, flattened and contorted bags. They appear to be empty but are clearly for exchange. The cell membrane of

Fig. 13.2. The stalk of *Oodinium fritillariae* Chatton: the host (h), tubular invaginations of the cell membrane (c), pusule (pv), perinuclear mitochondria (mi), multivesicular bodies (mvb).

the stalk intimately adheres to the host surface between the invaginations; the areas of contact are not altered ultrastructurally.

In *Crepidoodinium* (Lom & Lawler 1973; Lom 1977, 1981), the anchoring disc has an irregular appearance. Its surface possesses a large number of small ramifications closely applied to the fish integument. The extremity of each ramification is conical and slightly inserted into the host cell. A bundle of tonofibrils similar to a desmose is developed. At the point of contact the host membrane appears fuzzy. Vesicles and microtubules perpendicular to the cell surface (transport?) are observed in the dinoflagellate but no particulate matter can be seen coming from the host.

The stalk apparatus is more complex in *Amyloodinium* (Fig. 13.3) (Lom 1973, 1981), and helps to explain that of *Protoodinium*. The stalk is essentially an anchoring organelle. A separate tentacle-like structure, the stomopod, is associated with a cytopharyngeal feeding apparatus. Rhizoids from the margin of the stalk disc radiate into long filiform projections which insert into the gill epithelial cells of fish and penetrate the cell membrane. The end of the rhizoids are ampulla-shaped and contain small vesicles which suggest an absorbtive function. A cylindrical, movable tentacle arises from the base of the stalk; it has an axial tube made of twenty concentrically overlapping microtubular sheets, thus resembling a suctorian tentacle. Small vesicles (membranous dark bodies) and organelles (clove-like bodies, spindle-shaped bodies) arise from the perinuclear cytoplasm of the parasite and are carried along the stalk. As no food vacuole has been observed here, lytic substances are thought to be injected by the stalk into the prey. If so, the stalk would resemble the stomopod or peduncle of some free-living gymnodinoids, e.g. *Erythropsidinium* (Greuet 1969). In the latter, a cytopharyngeal apparatus made of microtubular sheets exists at the base of the stomopod and many food vacuoles are observed above it.

The parasite that lives beneath the umbrella of jelly-fish and siphonophores, *Protoodinium* (Cachon & Cachon 1971a) (Fig. 6.1 and 13.4), shows similarities to *Amyloodinium*, although in this genus the anchoring and feeding systems are combined in a single organelle. The stalk, which arises from the sulcus at the girdle level, is tap-root shaped, bearing lateral radicles, but is hollow. Its wall is strengthened by microtubular sheets which overlap concentrically and penetrate as far as the perinuclear cytoplasm of the parasite, as in *Amyloodinium*. This stalk becomes deeply embedded in the host cytoplasm and acts as a large cytopharynx, filled with lumps of host cytoplasm. Numerous food vacuoles are seen in the parasite cytoplasm.

Piscinoodinium (Lom 1977, 1981), a fish parasite, possesses a short stalk, the extremity of which is spread out as a disc on the fish epithelium. This disc has a honeycombed sole, bearing numerous rod-like holdfast corpuscles (rhizocysts), which are made in the perinuclear cytoplasm before migrating to their final position in the disc. Microtubular sheets exist at the base of the stalk as in the previous organisms.

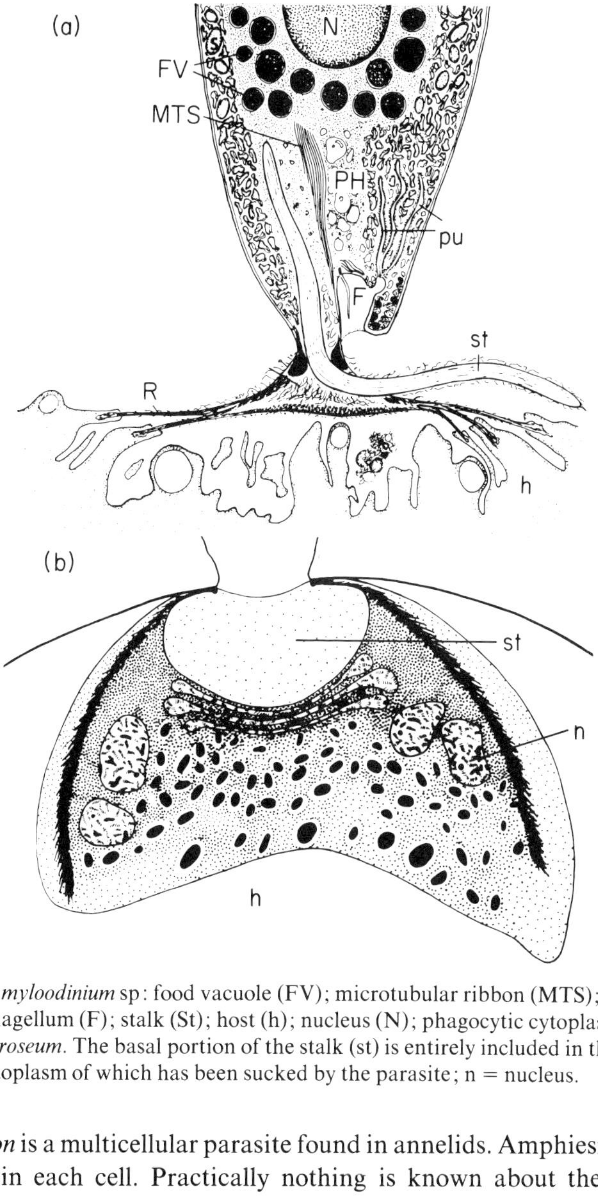

Fig. 13.3. (a) *Amyloodinium* sp: food vacuole (FV); microtubular ribbon (MTS); pusule (pu); rhizoids (R); flagellum (F); stalk (St); host (h); nucleus (N); phagocytic cytoplasm (PH). (b) *Chytriodinium roseum*. The basal portion of the stalk (st) is entirely included in the copepod egg (h), the cytoplasm of which has been sucked by the parasite; n = nucleus.

Haplozoon is a multicellular parasite found in annelids. Amphiesmal vesicles are present in each cell. Practically nothing is known about the *Haplozoon* anchoring and feeding system. An anterior individual has a suction disc in contact with the intestinal epithelium of the host and a movable 'stylet' which is thrust into the host cells (Shumway 1924; Siebert 1973). The 'stylet' may be similar to a stomopod. Nutrients must pass from the anterior individual to those behind it but this process has not been studied.

Apodinium (Fig. 13.5), a parasite of appendicularians, has a well developed stalk (Cachon & Cachon 1973). A long, cylindrical stem penetrates the cuticle

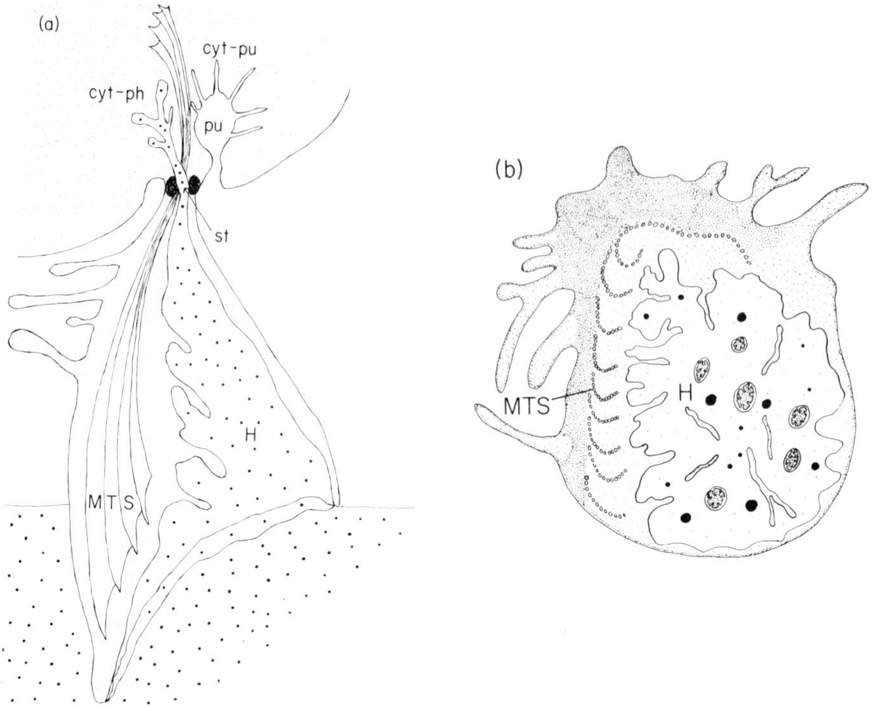

Fig. 13.4. *Protoodinium chattoni* Hovasse. Longitudinal (a) and transverse (b) sections of the stalk in which there are ribbons of microtubules (MTS), a stomopharynx (st), pharyngeal cytoplasm (cyt ph); the pusule (pu), its cytoplasm (cyt pu); and host cytoplasm (H).

of the appendicularian host and divides into two flattened roots. This system shows an astonishing duality: the stem is made of two hollow parts limited by a thick dense wall, the whole system being surrounded by the cell membrane. Between these two elements there is an axial cavity, often contorted and containing multimembranous vesicles. The elements continue as rhizoids running in opposite directions between the host muscles or along the urochord. These rhizoids are covered by a thick, tomentous wall, containing a thin granular material. Where the axial cavity ends at the base of the stalk, a great number of multimembranous vesicles can be seen. One of the elements is postero-dorsal, the other postero-ventral.

A well developed pusule opens into the axial cavity between the two stem elements. Differences in the stalk are used in the taxonomy of the genus (size, morphology, ultrastructural arrangements). *Amyloodinium*'s system is also made of two dorso-ventral elements (an attachment disc and a stomopod; Lom & Lawler 1973) which are associated with a pusule.

Chytriodinium (Cachon & Cachon 1968) (Fig. 13.3 and 13.6) uses its hyposome to drill through the thick chorion of the crustacean egg. It can

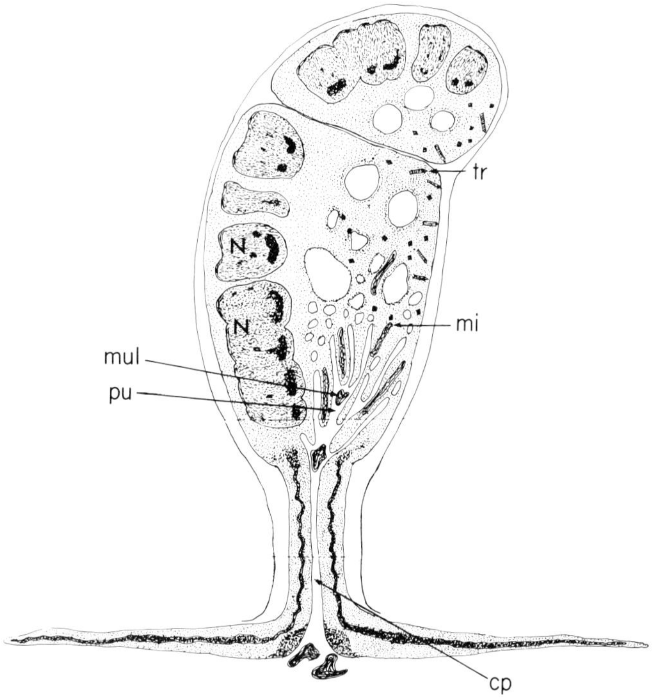

Fig. 13.5. *Apodinium* sp., showing the duality of the stalk structure and in its axis a canal (cp), related to the pusule (pu). Multilamellar bodies are observed (mul), as well as trichocysts (tr) and perinuclear mitochondria (mi).

lengthen considerably (up to approximately 50 μm) while becoming extremely thin. Unfortunately, no electron microscopical observations have been made. Once the egg cytoplasm is reached a crown of holdfast organelles develops. The hyposome then shortens so that the episome is brought in contact with the egg. The parasite seems to be rivetted onto the egg envelope. It feeds through the holdfast organelles, between which the ampulla (a vacuole, undoubtedly for absorbtion) appears. The host cytoplasm is progressively absorbed and drawn towards the ampulla where its nuclei, mitochondria and vitelligenous plates are gathered. The egg is emptied in 1–2 hours.

Dissodinium is also a parasite of crustacean eggs (copepods) and its behaviour is similar to the previous genus (Drebes 1978; Elbrächter & Drebes 1978). The infecting dinospores penetrate in a few minutes, using an anchoring and feeding stalk which issues from the hyposome.

In *Myxodinium* (Cachon *et al.* 1970), a pseudopodium issues from the sulcus of the spore. It comes in contact with the cellulosic wall of the phycoma of the prasinophycean alga, *Halosphaera*, and spreads out. In a few minutes the wall is perforated by two small holes (2 μm in diameter). Two small circular cups

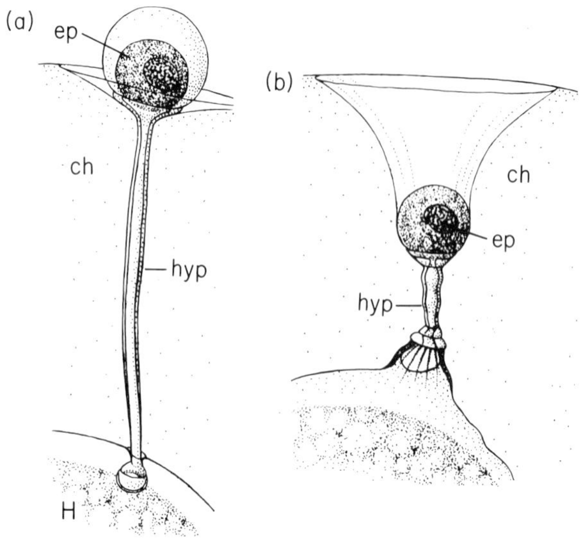

Fig. 13.6. *Chytriodinium parasiticum.* (a) The organism consists mainly of its episome (ep) while its hyposome (hyp) is reduced to a drill which goes through the chorion (ch) of the crustacean egg. (b) The stalk has been contracted so that the episome comes nearer to the egg.

develop in the stalk cytoplasm and the alga is rapidly sucked out through the two apertures until the shell is emptied. *Paulsenella*, an ectoparasite of diatoms (e.g. *Chaetoceros*) is imperfectly known: it also possesses a small peduncle used for sucking its host (Schnepf 1985).

Finally, the stalk of *Cachonella* seems to be similar to that of *Oodinium*, except that ramified rhizoids appear on the margin of the anchoring sole (Fig. 13.14a).

The pusule is generally well developed in members of the Blastodinida. For example, in *Oodinium* (Cachon *et al.* 1970; Cachon & Cachon 1971b) a system of lacunar vesicles makes up a 'spongiome' containing several types of vesicles. They are connected to a system of long collecting canals which finally flow into an ampulla (sometimes two). Around the ampulla are large ribbons of striated fibrils that may be responsible for its contractility. Multilamellar bodies are often observed within the vesicular system. They appear to be periodically ejected from the ampulla. The pusule is always positioned postero-ventrally and is associated with the stalk. In *Apodinium* the accumulation of multilamellar bodies in the vicinity of the rhizoids proves that these structures are excretory.

3.2 Intracellular parasites

These have been observed in a great variety of hosts, both protists (e.g. dinoflagellates, radiolarians, ciliates) and in the cytoplasm of tissues of multicellular organisms (e.g. crustaceans, coelenterates, appendicularians). The mechanisms of their infection (passive, active, phagocytic) are usually not known. If the parasite is intracellular, the host reacts by forming a membrane around it (the parasitophorous vacuole), of which two types may be distinguished. The Duboscquodinida grow considerably as trophonts and lose all the external morphological features typical of dinoflagellates; these features then reappear during late sporogenesis. The Syndinida, on the other hand, lose not only their morphological features, but also some of the cytological and biochemical features typical of dinoflagellates. As trophonts they divide actively or form large plasmodia.

Duboscquodinida (botanical order Coccidiniales)

Except for *Sphaeripara* these all parasitize protists. They are always intracellular, sometimes even intranuclear. Besides being intracellular they can easily be distinguished from the Blastodinida because they do not have normal amphiesmal organization (Chapter 3). Theca, chloroplasts, pusule and generally mitochondria are absent.

The dinospores of the Amoebophryidae, parasites of both unicellular and multicellular hosts, are highly modified (Cachon 1964). They are the only dinoflagellate endoparasites to keep their flagella during their trophic phase, and they can proliferate extraordinarily. The girdle (or a depression homologous to a girdle) is helical. In the spore, it extends only half a turn. As the parasite grows, it becomes considerably lengthened at both ends, making several helical turns. At the same time, the episome progressively sinks into the hyposome. The *Amoebophrya* trophont appears as a round body in the host parasitophorous vacuole. A cavity, 'the mastigocoel' (Cachon 1964), becomes visible with the light microscope. It has a double system of helical structures, one parietal involving the hyposome, the other axial on the part corresponding to the episome (termed the 'columelle' by Chatton 1922). The edges of the hyposome expand and finally completely surround the episome with a small aperture at the apex of the parasite. The girdle may become extremely long and twisted so that its helical coils are not regular. The number of basal bodies and nuclei increases enormously; they form a long chain lying just beneath the girdle. There are as many flagella as basal bodies; the flagella remain motionless during the trophic phase. In some hosts, e.g. the dinoflagellates *Triadinium polyedricum* (Cachon 1964) or *Protogonyaulax catenella* (Taylor 1967), the nucleus of the host is rapidly digested.

Microtubular bundles run beneath the helical girdle, with others beneath the limiting membrane of the hyposome where they are arranged like the petals of a corolla (Cachon & Cachon 1969, 1970). Microfilaments are associated with these bundles (non-actin filaments, Cachon & Cachon 1984). A great number of trichocysts are present, mainly in the girdle. The cell membrane of the protist makes contact with the parasite membranes through structures which resemble nuclear pores. No food vacuoles can be seen in the cytoplasm of the parasite.

The last intracellular phase involves great changes in organization and physiology. Suddenly, the flagella begin to beat. In less than 1 minute the protist goes through an eversion which is the inversion of its earlier development (Fig. 13.7). The episome pushes forwards through the aperture of the mastigocoel. The edges rapidly turn up and are thrown backwards so that they imprison a large portion of the host cytoplasm in a big trophic vacuole as the parasite

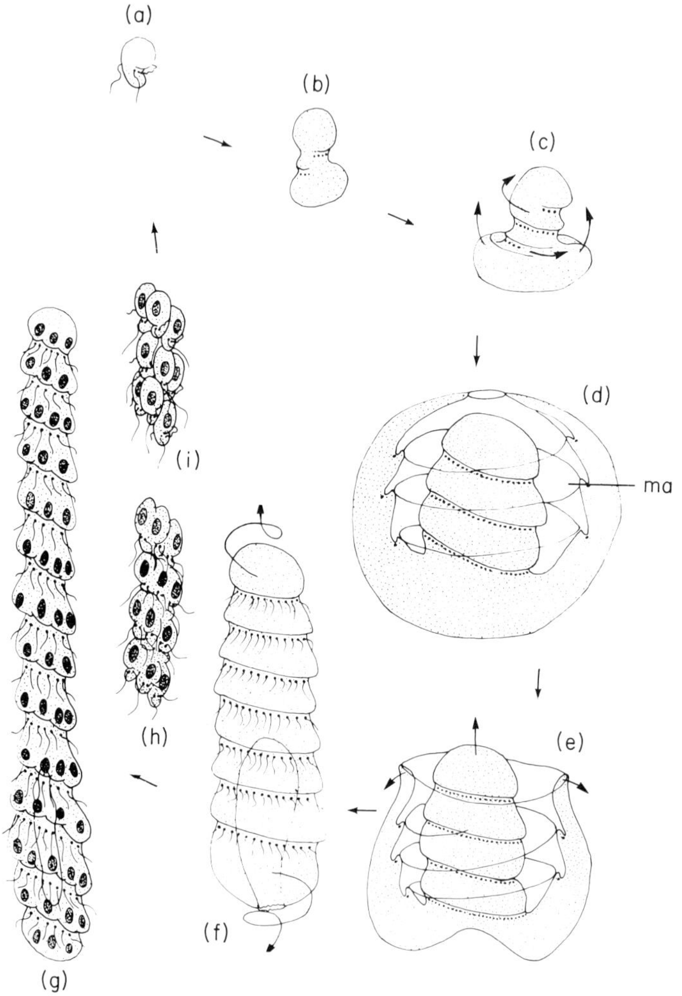

Fig. 13.7. Diagram of the life cycle of *Amoebophrya*: (a) dinospore; (b, c, d) invagination of the growing intracellular trophont (Ma = mastigocoel); (e, f) evagination of the trophont phagocytosis of the host and formation of a worm-shaped organism, the 'vermiform'; (g) lengthening of the vermiform; (i, h) formation of the swarmers.

leaves the host. The cell periphery is three layered, with a layer of flat vesicles and bears a great number of flagella. The parasite now appears worm-like (the vermiform stage) (Fig. 13.8b). It swims in a spiral manner with all its flagella

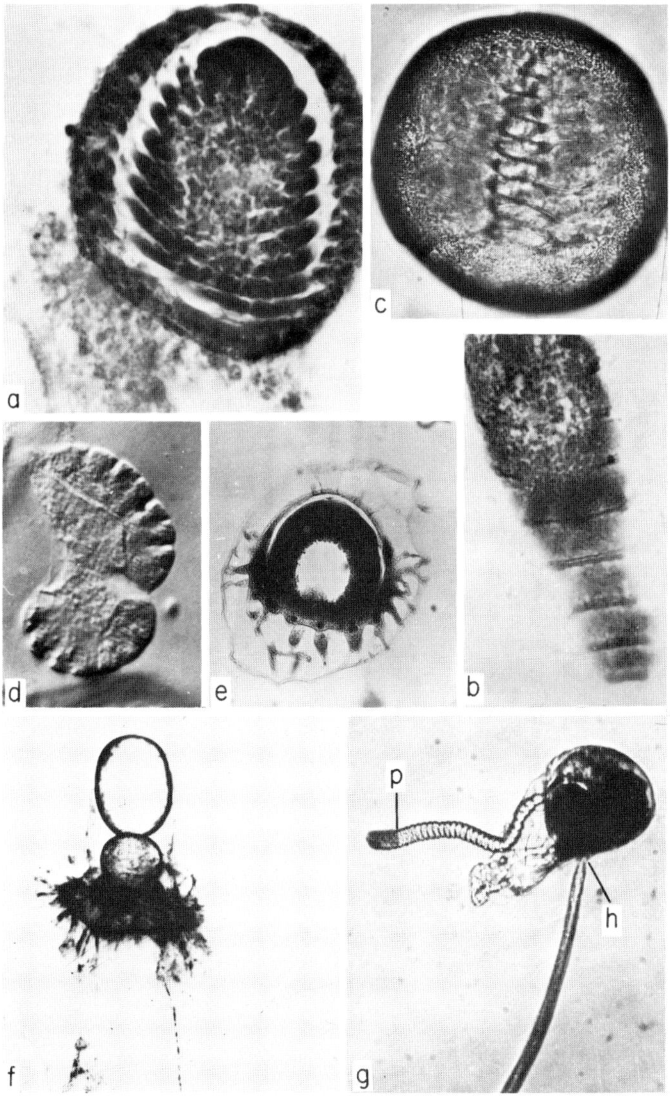

Fig. 13.8. All these micrographs are from living organisms. (a) *Amoebophrya sticholonchae* Koeppen, trophont. (b) *A. sticholonchae*, the vermiform. (c) *Amoebophrya grassei* Cachon which parasitizes *Oodinium*. (d) *Apodinium* sp. beginning its sporogenesis. (e, f) *Sphaeripara catenata*, a young trophont and an older one (g) *Haplozoon* sp. parasitizing *Appendicularia sicula* (h, host; p, parasite).

beating, its jerky movements resembling those of Orthonectids. The large, posterior digestive vacuole contents are progressively absorbed.

Duboscquella is also a parasite of protists (Cachon 1964). Although its trophont appears to be different from that of *Amoebophrya*, there are a number of common features in the development of the trophont and during sporogenesis. For example, *D. aspida* (Fig. 13.9) parasitizes tintinnids such as *Cyttarocylis*. The infection is caused by biflagellated spores with hyposomes smaller than their episomes. Inside the cytoplasm of the host the spore flagella, girdle and sulcus vanish (by phagocytosis?). The young trophont becomes spherical. Two distinct parts can be recognized in the nucleus: the anterior region is occupied by chromosomes, the other contains a cap-shaped nucleolus. A circular plate appears at the anterior pole of the cell; it is due to the development of fibrous elements beneath the cell membrane. This watch-glass shaped anterior plate is inserted into the cytoplasm and induces the formation of a circular groove around it. While growing (the trophont may reach 80 μm in diameter) the plate develops greatly and hides the anterior part (which corresponds to the episome). The girdle is recognizable only by its anterior (upper) edge which is lined with dense material. Beneath a small, weakly developed sulcus there is a dense helical ribbon made of a bundle of microtubules, which could be a non-functional

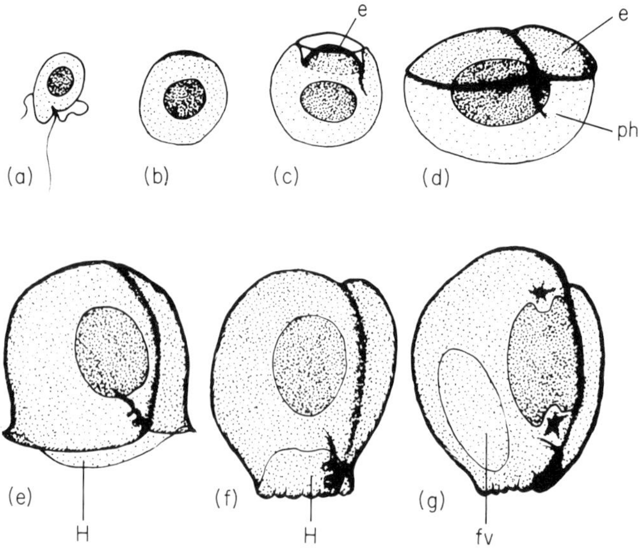

Fig. 13.9. Diagram of the life cycle of *Duboscquella aspida*: (a) spore; (b, c, d) growth of the trophont, formation of a fibrous cap at the anterior part (e = episome) in which there is a trace of a slight longitudinal groove; beneath this groove a ribbon of microtubules is observed (pharyngeal area, ph); (e, f) the hyposome (H) is pushed forward inside; (g) the episome which has incorporated a portion of the host in a large food vacuole (fv). The nucleus in (g) is beginning its first sporogenetic prophase.

cytopharynx and may indicate the ventral side. During its development the protist remains uninucleate (see Section 5 for further information of the development of synenergid nuclear structures). The cell membrane of the hyposome has considerable pinocytotic activity.

As in *Amoebophrya*, major structural rearrangements accompany the end of the feeding phase and the beginning of the reproductive one (Fig. 13.10). The

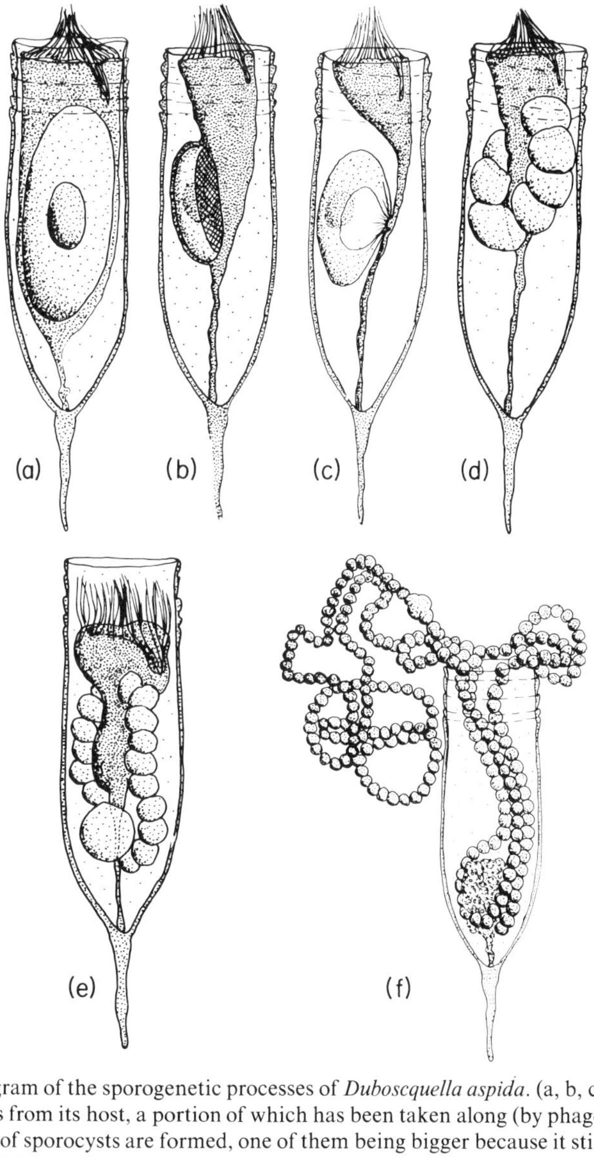

Fig. 13.10. Diagram of the sporogenetic processes of *Duboscquella aspida*. (a, b, c) The parasite escapes from its host, a portion of which has been taken along (by phagocytosis). (d, e) Large chains of sporocysts are formed, one of them being bigger because it still has the food vacuole. (f) The chain of sporocytes, moving with the aid of flagella, leaves the host lorica and soon will give rise to dinospores.

cap-shaped hyposome is suddenly pushed inwards, forming a large cavity at the posterior end of the cell. This cavity encloses a large part of the host cytoplasm in a food vacuole. The mechanisms are similar to those of *Amoebophrya*, although the microtubular cytoskeleton of *Duboscquella* is much less important than the microtubular ribbons of the corolla.

The differences between the various species of *Duboscquella* are solely in the construction and behaviour of the wall-plate of the episome. In *D. cnemata*, for example, the subpellicular fibrils are arranged like the spokes of a wheel and the episome, which is normally rather flat, has radial grooves. In *D. melo*, a parasite of *Noctiluca*, the episome is rather flat and appears to have longitudinal ribs.

Among the Sphaeriparidae, *Atlanticellodinium* (Cachon & Cachon 1965) parasitizes protists, particularly radiolarians. However, the type genus of this family, *Sphaeripara* (*Neresheimeria*) parasitizes Metazoans (Fig. 13.8e, f). At the beginning of its cycle it behaves like a parasite of unicellular hosts (Fig. 13.8e, f); it selectively infects the oikoplast, a giant glandular cell of *Fritillaria* (Cachon & Cachon 1964, 1966). In this cell, it forms in a parasitophorous vacuole. The beginning of its cycle is similar to *Duboscquella*. The young trophont, which has no girdle, sulcus or flagella, soon forms a dense-walled hyposome, separated from the episome by a circular groove. The hyposome flattens to form a disc which widens progressively; many rhizoids grow from the margin of the disc and sink into the host cytoplasm (Fig. 13.11). A small, axial cavity appears which extends to the hyposome wall by a fibrillar funnel. The hyposome wall forms highly ramified villi. The host cell becomes hypertrophied and contains many nuclei, most of these polyploid. Its mitochondria are attached to the hyposome of the parasite in invaginations of the cell membrane. It is difficult to know whether they are phagocytosed or externally digested (the latter seems more probable). The axial cavity is surrounded by a 'spongiome' made of small tubes, as in Ciliates; the whole system appears empty. All the host–parasite exchanges in *Sphaeripara* go through the hyposome wall. The episome does not act in feeding: it is distant from the host tissues and the parasitophorous vesicle membrane; it has no villi or invaginations and its surface consists only of a cell membrane with normal amphiesmal vesicles. In *Sphaeripara*, trophic and sporogenetic processes occur together and are intricate. *Coccidinium* (Chatton & Biecheler 1934, 1936), *Duboscquodinium* (Grasse 1952) and *Keppenodinium* (Cachon 1964) are insufficiently known. As they are plasmodial forms they may be closer to the Syndinida.

Syndinida (Botanical order Syndiniales)

These dinoflagellates can parasitize other protists, such as radiolarians and the curious organism *Sticholonche* (Fig. 13.13). In these they are intracellular (Fig. 13.12) and may even be intranuclear. Other syndinians parasitize Metazoans

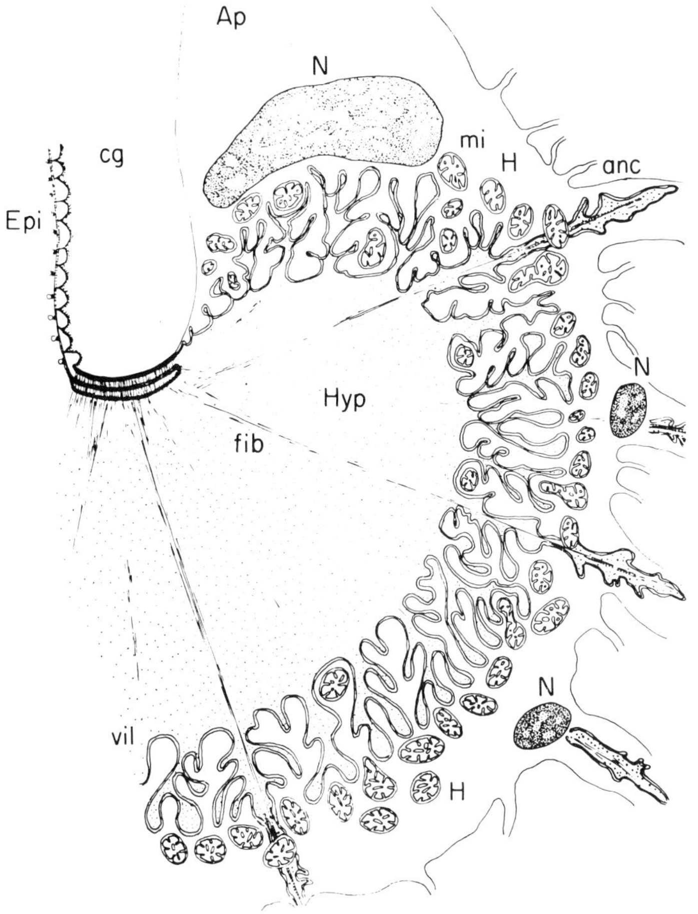

Fig. 13.11. Latero-basal portion of *Sphaeripara*. Episome (epi), hyposome (Hyp), cingulum (cg); host cell (H), nucleus (N) and mitochondria (mi) of the Appendicularium. Fibres (fib), anchoring rhizoids (anc) and villi (vil) of the membrane of the parasite hyposome.

and are mainly found in cavities (Soyer 1974; Chatton 1910; Manier *et al.* 1971), e.g. the gonocoele of hydrozoans, the hematocoele of crustaceans such as copepods and crabs, appendicularians and the yolk sac of fishes (Fig. 13.13). They occur as plasmodial forms which continue to grow and divide until they form swarmers. During the trophic phase they lose their dinoflagellate morphology: the girdle, sulcus and flagella disappear, although basal bodies remain and the amphiesma is normal. There is a thin polysaccharidic cell coat and amphiesmal vesicles without thecal plates. There are polysaccharidic and lipidic inclusions, mitochondria with poorly developed cristae and many

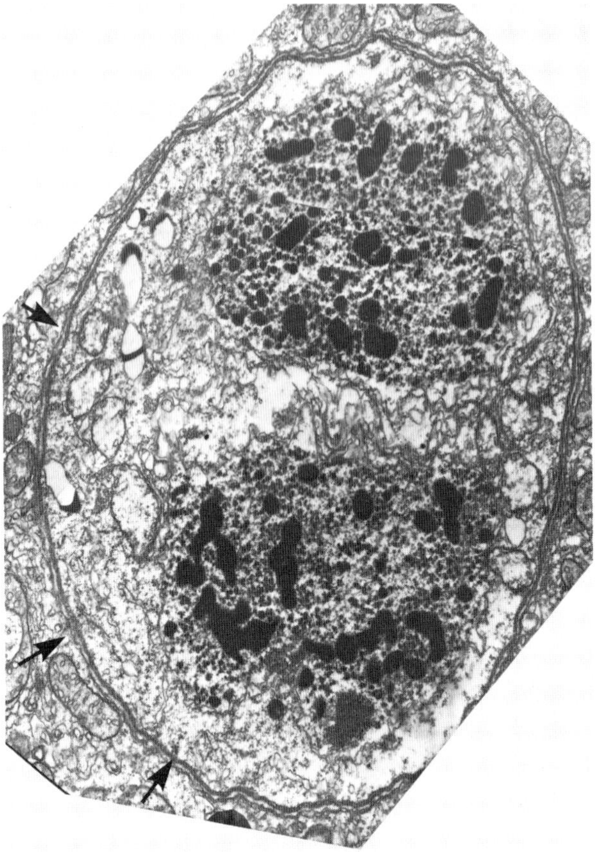

Fig. 13.12. Young trophont of *Syndinium borgerti* in the Phaeodarian *Aulacantha scolymantha*. This micrograph shows the parasitiphorous membrane of the host (arrows).

trichocysts. Chloroplasts are absent. Osmotrophy is the rule, although remnants of a cytopharyngeal funnel can sometimes be seen (*Syndinium* sp., Cachon 1964).

4 REPRODUCTIVE PROCESSES

Of all parasitic dinoflagellates, only some species of *Blastodinium* reproduce by simple binary fission. *B. spinulosum* divides in the same way as free-living forms, but in the intestine of a copepod (Chatton 1920); there may be as many as twenty in a single host. This endogenous multiplication may interfere with the

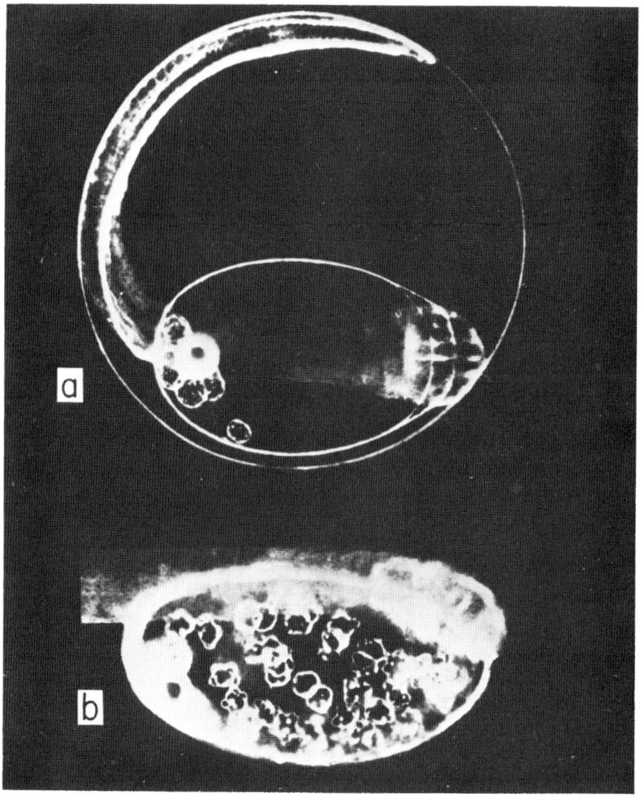

Fig. 13.13. *Ichthyodinium chabelardi* Hollande and Cachon. This syndinian parasitizes the vitelline vesicle of an embryo of the sardine. The vesicle finally bursts, the embryo dies and the dinospores are then free.

exogenous multiplication which produces swarmers and which is the only type of reproduction in most of the other parasites.

Sporogenesis may be by three different mechanisms.

1 The simplest is found in the Syndinida; the parasite multiplies its nuclei, plasmodia and cytoplasmic divisions until the last generation of sporocysts, when biflagellated dinospores are produced and liberated.

2 Alternatively, nuclear and cell divisions may not occur during the growth phase of the parasite, which remains uninucleate and may reach a large size. As soon as the parasitic stage is ended and maximum size is reached, nuclear and cytoplasmic divisions take place without interruption until swarmers are formed; Chatton (1938) termed this process 'palintomy'.

3 The parasite becomes multinucleate, after which one of two processes may occur: the cytokineses may occur (i) progressively throughout the parasitic phase, or (ii) only at the end of the parasitic phase with the organism producing a great number of spores simultaneously. A single trophont can thus produce, through repeated reproductive phases, several generations of dinospores; Chatton (1906) termed this type of mechanism 'iterative sporogenesis' or 'palisporogenesis'.

4.1 Palintomic sporogenesis

This is commonly observed in *Oodinium*, *Cachonella*, *Dissodinium*, *Duboscquella*, *Amoebophrya* and *Chytriodinium*. It is easy to detach a sufficiently old *Oodinium* from its host. This induces mitosis and cytokinesis after 7–8 hours which proceed without interruption until progressively smaller sporocysts form, each enveloped by a cyst wall, later becoming liberated as flagellated spores. This triggering of sporogenesis is a good demonstration of the relationship between host and parasite and also shows that the parasites are extracellular. Under natural conditions, the rupture of the stalk always precedes sporogenesis.

Cachonella has a peculiar behaviour (Rose & Cachon 1951, 1952; Cachon 1953) (Fig. 13.14). It grows like *Oodinium*, while fixed to the wall of the swimming bell of a siphonophore. In order to develop it has to be swallowed by a gastrozoid of the siphonophore. Once in the presence of gastric juices, the cuticle is rapidly ejected, perhaps by the simultaneous extrusion of all its trichocyst batteries. Immediately after this, long digitations develop from the base to the top of the cell. These involve the stretching of a pre-existing membranous system. The cytoplasm of the trophont is filled with flattened and contorted membranous formations. Under the influence of the gastric juices of the host, a sudden increased turgidity of the cytoplasm causes the evagination of the membranous tubes, like the fingers of a glove turning inside out. The parasite is then ejected from the gastric cavity, these processes having taken only a few minutes. The cytoplasm in the digitations retracts and the empty digitations collapse. The cell leaves its envelope and the first sporogenetic divisions begin.

Dissodinium behaves in a similar way (Drebes 1978; Elbrachter & Drebes 1978). The parasite begins to divide only after it becomes detached from the egg by rupture of its stalk, even if the egg has been reduced to an empty shell. The sporogenetic processes of *Dissodinium* occurs in two stages. First, a spherical cyst wall is secreted within which the parasite divides to form eight to sixteen lunate or oval secondary cysts, which are then released. Next, inside each secondary cyst, eight dinospores are formed then released when the secondary cyst wall bursts. These directly infest new eggs. The primary and secondary cysts of the parasite and those of *Pyrocystis* (Dinococcales) are morphologically similar.

The end of feeding in *Amoebophrya* and *Duboscquella* triggers sporogenesis immediately after cell inversion. In *Duboscquella* the sporocysts remain connected end to end in long chains by a polysaccharide coat. The sporocyst

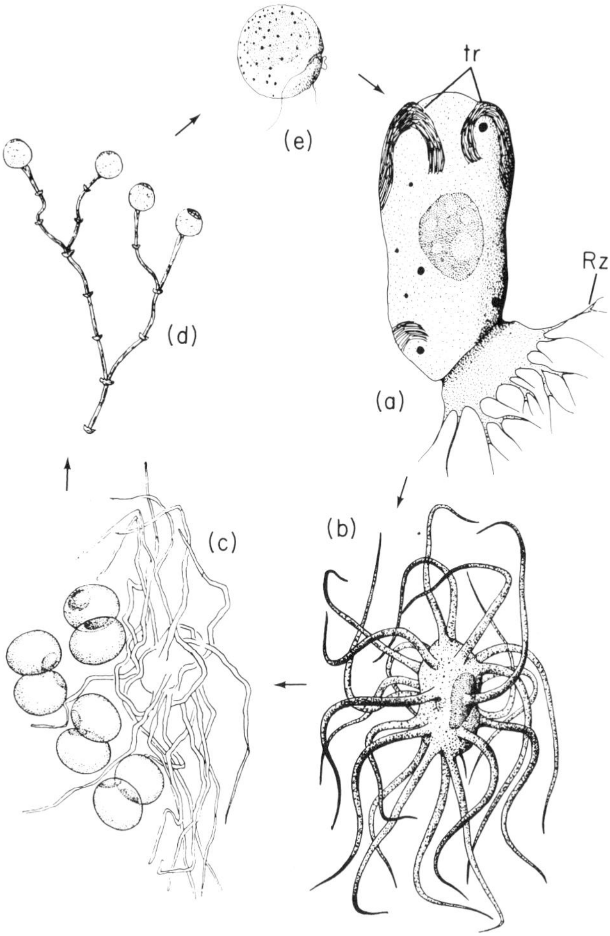

Fig. 13.14. *Cachonella paradoxa* Cachon and Rose. The trophont (a) is fixed on the cell membrane of the swimming bell of a siphonophore by a large stalk bearing many rhizoids (Rz). Three bundles of trichocysts (tr) are observed. At the end of the vegetative phase, the protist escapes (b) and once swallowed by the gastrozoid, long tubular digitations are developed. It is then ejected into the sea (c) and undergoes palintomic sporogenesis. Many cyst membranes are successively formed at each generation; they keep the sporocysts attached to each other (d). Finally, spherical biflagellated swarmers are formed (e).

inheriting the phagotrophic inclusion (which will be progressively absorbed) is larger than the others. In the final generation the sporocysts develop two flagella each and the chain moves. The individuals separate to become spores.

In *Amoebophrya*, the trophont either remains uninucleate during the first phase (e.g. *A. acanthometrae*), or it becomes multinucleate (e.g. *A. ceratii*), with the nuclei lying beneath the kinetosomes along the helical girdle. In the first case, the nuclear divisions begin only after the formation of the 'vermiform' organism. In both cases a great number of small nuclei are observed in the multinucleate vermiform, each related to a pair of basal bodies. The vermiform stretches like a spring, while the helical girdle twists. This lengthening is fantastic: in some species, e.g. *A. grassei* (Fig. 13.8c), which is a hyperparasite of *Oodinium*, the length of the vermiform increases from 250 μm to more than 4000 μm in a few hours. The helical coils multiply while the number of nuclei per coil decreases to three or four. Long chains of sporocysts are formed, as in the previous genus. Finally, the spores are liberated, each with a nucleus and a pair of flagella.

4.2 Palisporogenesis

This is found mainly in *Blastodinium*, *Apodinium*, *Haplozoon* and *Sphaeripara*. The first nuclear division of the trophont is differential (Cachon & Cachon 1965). One daughter nucleus continues dividing without interruption, first forming the nuclei of the sporocysts and finally those of the swarmers. The other daughter nucleus divides much later to produce new generations of sporocysts. The distinction between a *trophocyte*, which continues to grow (containing the non-dividing nucleus), and a *gonocyte*, which is involved in sporogenesis, may be due to starvation of the latter.

Chytriodinium (Fig. 13.15) is an interesting example (Cachon & Cachon 1968). Its sporogenesis is usually direct and palintomic, but, depending on the food reserves in the parasitized egg, there can be an alteration in the development time of the sporocysts arising from the two cells of the first division. This provides a rough model of palisporogenesis.

In *Apodinium* (Figs 13.8d, 13.16b) and *Haplozoon* (Fig. 13.8g) palisporogenesis is complex; at a very early phase, the trophont undergoes two nuclear divisions but without complete separation of the nuclei. The nuclei of the sporocysts are tetrapolar and in final divisions the swarmer tetrads separate before becoming spores. In *Sphaeripara* the trophocyte nuclei divide actively throughout the trophic phase (Fig. 13.17). Palisporogenesis occurs by strobilation of the trophocyte; each strobilus is multinucleate. This development is reminiscent of *Amoebophrya*: the parasite lengthens while its diameter decreases; basal bodies and flagella appear but they are arranged in successive circles. The final processes are similar to those described above.

5 THE SWARMERS

All parasitic dinoflagellates produce swarmers which ensure dispersal of the species and are responsible for new host infection. They always have two dissimilar, laterally inserted, flagella, one trailing and the other transverse and undulating. The girdle and sulcus are often poorly developed and are associated with a striated root and a few microtubules, as in free-living dinoflagellates.

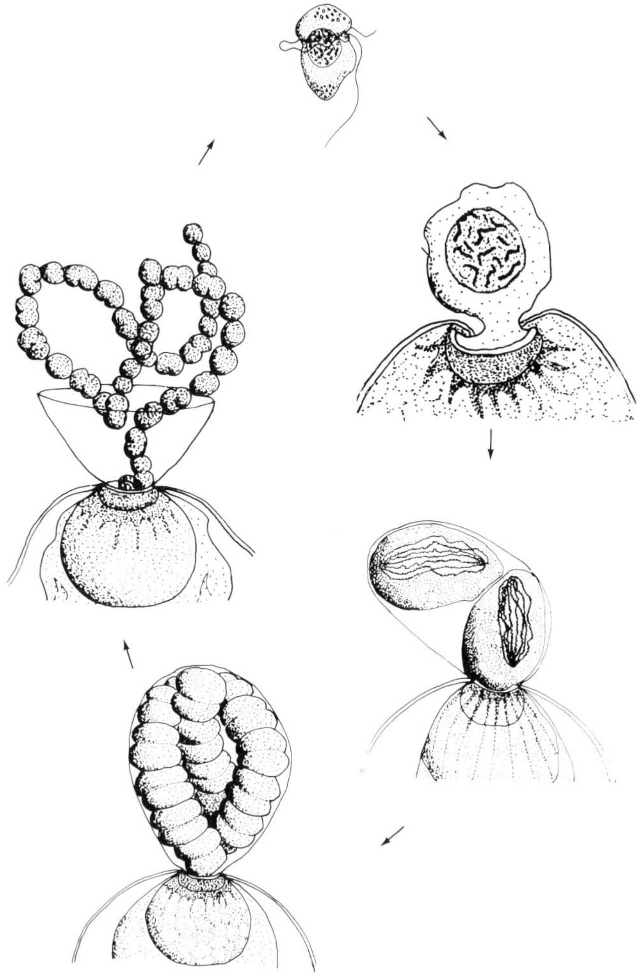

Fig. 13.15. Diagram of the life cycle of *Chytriodinium affine* Dogiel. The swarmer perforates the shell of the egg of a crustacean and a strong holdfast organelle develops. The palintomic sporogenetic divisions begin beneath the initial envelope of the trophont long before the egg is emptied. A chain of sporocysts is formed. The spores become free (64–128 cell stage) by the rupture of the trophont envelope.

Trichocysts and large refringent inclusions are often present. Chatton (1938) thought that swarmer morphology (gymnodiniform, gyrodiniform, cochlodiniform) could be important for the systematics of the parasitic taxa. However, our observations showed their morphology to be unstable. Thus, the spores do not seem to be sufficiently conservative to serve as the basis for the systematics of these species.

Some species may produce swarmers of two different sizes (macro- and microspores), which arise from different parent individuals. Sexual reproduction

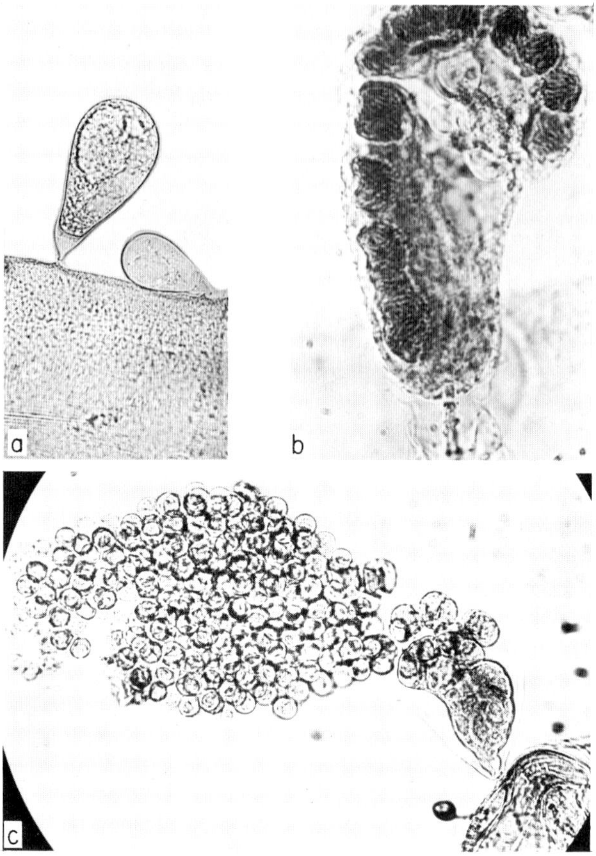

Fig. 13.16. *Apodinium* sp. (a, b) Young trophonts. (c) The parasite undergoes palisporogenesis.

comes to mind; syngamy has been observed only in *Duboscquella*, i.e. *D. tintinnicola* (isogamy) by Duboscq & Collin (1910), and *D. anisospora* (with micro- and macrogametes) by Grassé (1952), without knowledge of their further development. In *D. aspida*, infection of the tintinnid host can occur by macrospores without prior syngamy (Cachon 1964).

The spores of dinoflagellate fish parasites (*Amyloodinium*, *Crepidoodinium*, *Piscinoodinium*) and those of *Protoodinium* possess an attachment organelle, a small ventral pseudopod, by which they infect the host immediately.

6 PARASITE NUCLEAR STRUCTURE AND CYCLE

This aspect is difficult to resolve, even though it was the structure and development of the nucleus as well as the spore morphology which enabled

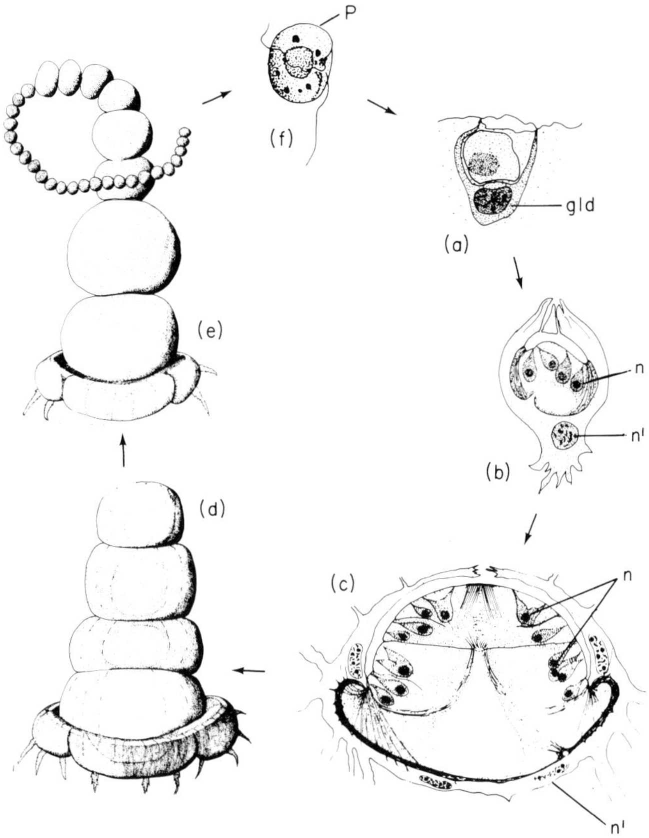

Fig. 13.17. Life cycle of *Sphaeripara* (= *Neresheimeria*) *catenata*. (a) The young parasite (p) is inserted in the glandula cell (gld) of an appendicularian which will eventually enclose it completely (b). Its hyposome develops an anchoring and feeding apparatus while there is much nuclear division in the episome (c). n = nucleus of the parasite, n′ = host nucleus. The episome becomes stabilized (d) and produces successive generations of multinucleated sporocysts by palisporogenesis (e). Each primary sporocyst produces a chain of secondary sporocysts which finally become dinospores (f).

Chatton to recognize the dinoflagellate nature of a great number of parasites. He thought that the simplest and most typical dinomitosis occurred in parasitic dinoflagellates (Fig. 13.18a, b) (syndinium mitosis), although they are now known to be anomalous in several ways. He also recognized the peculiar aspects of the nucleus of some trophonts which remain uninucleate during the growth phase. This large nucleus, which stops undergoing mitosis, progressively accumulates material which can later be used during the numerous nuclear divisions necessary to reach the final spore stage. He termed this type of nucleus ‘synenergide’, for it can store away a great potential for eventual activity.

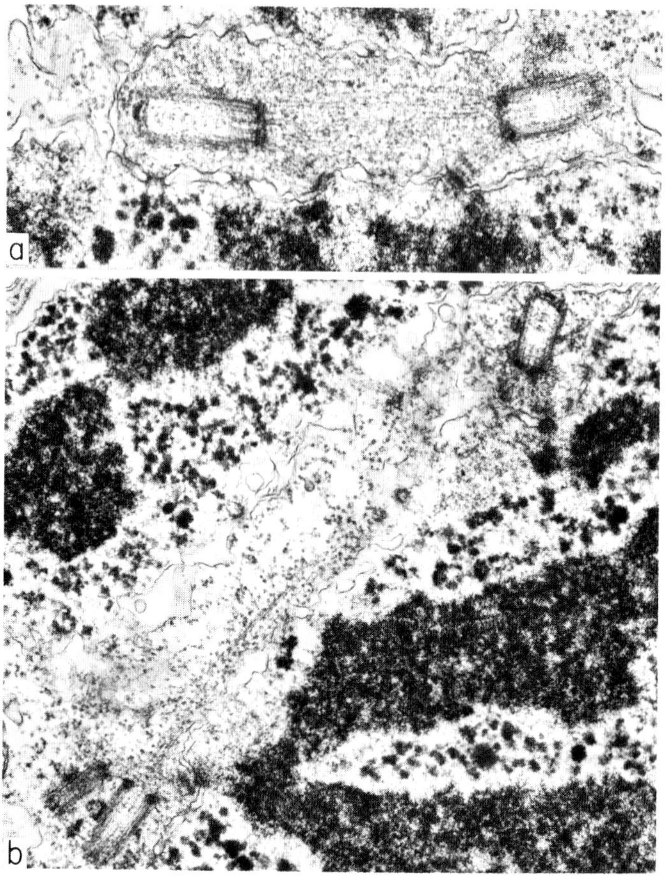

Fig. 13.18. Micrographs of syndinian mitosis seen by Ris & Kubai (a) and by Cachon & Cachon (b).

The data from electron microscopy and cytochemistry have confirmed the unusual structure and composition of the nucleus and mitosis in free-living dinoflagellates (see Chapter 3). However, most parasites do not have all these features. Most do not have condensed chromosomes throughout the whole cycle, and in some they do not condense at all. Furthermore, although the number of chromosomes is relatively high in Blastodinida, it is low in Duboscquodinida and Syndinida.

Oodinium (Cachon & Cachon 1974, 1977) is a typical example. The nucleus of the spore and of young trophonts (Fig. 13.19a–e) has typical rod-like dinoflagellate chromosomes. There are about thirty with arched fibrils, lacking histones. They stain strongly with Schiff reagent (DNA) but remain unstained with alkali fast green (for histones). As growth proceeds, the chromosomes lengthen and their first fibrils become irregularly arranged. The Schiff reaction disappears and nucleoli appear. Finally, in the adult trophont the chromosomes

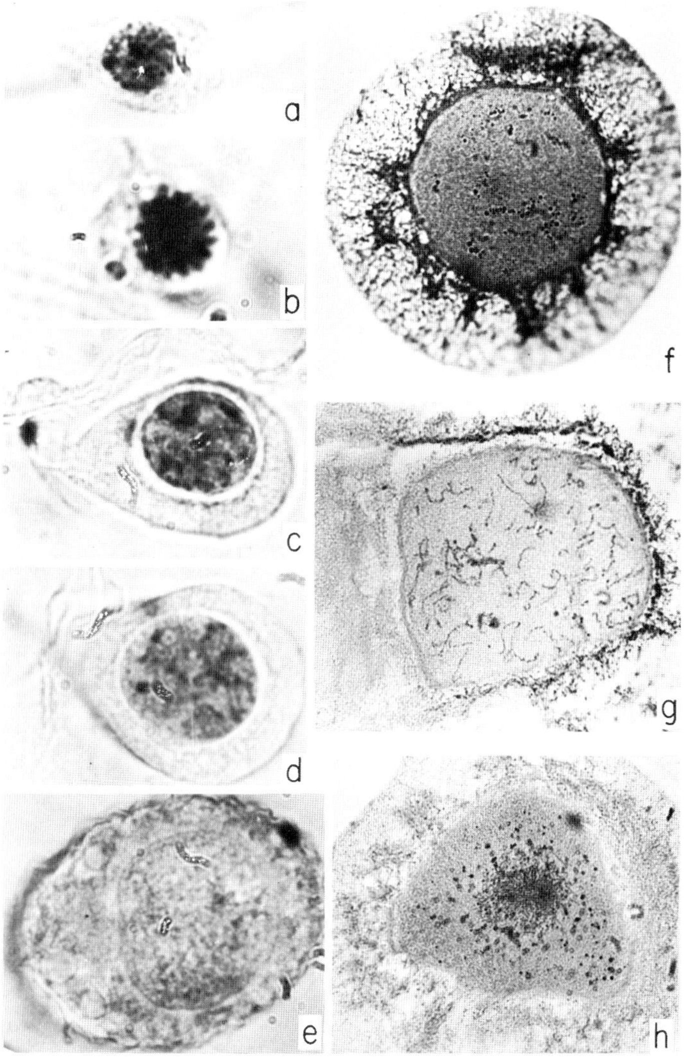

Fig. 13.19. Nuclear development in *Oodinium fritillariae* Chatton. (a, e) The Schiff reaction *in vivo* (× 800). The nucleus becomes progressively more homogeneous. The chromosomes which had normal dinoflagellate structure vanish. (f, g, h) (bi-acidic Mann staining). As soon as *Oodinium* becomes detached from the host, the nuclear substance disappears (f) while the chromosomes again become conspicuous (g) and are radially arranged (h). The first mitosis is then ready to begin.

become conspicuous; the Schiff reaction is negative and the fast green reaction is positive, diffuse except in the nuclei where it is intense. By the time sporogenesis begins, the protist is in a peculiar state, fundamentally different from that of a free-living dinoflagellate (Fig. 13.19f, g, h).

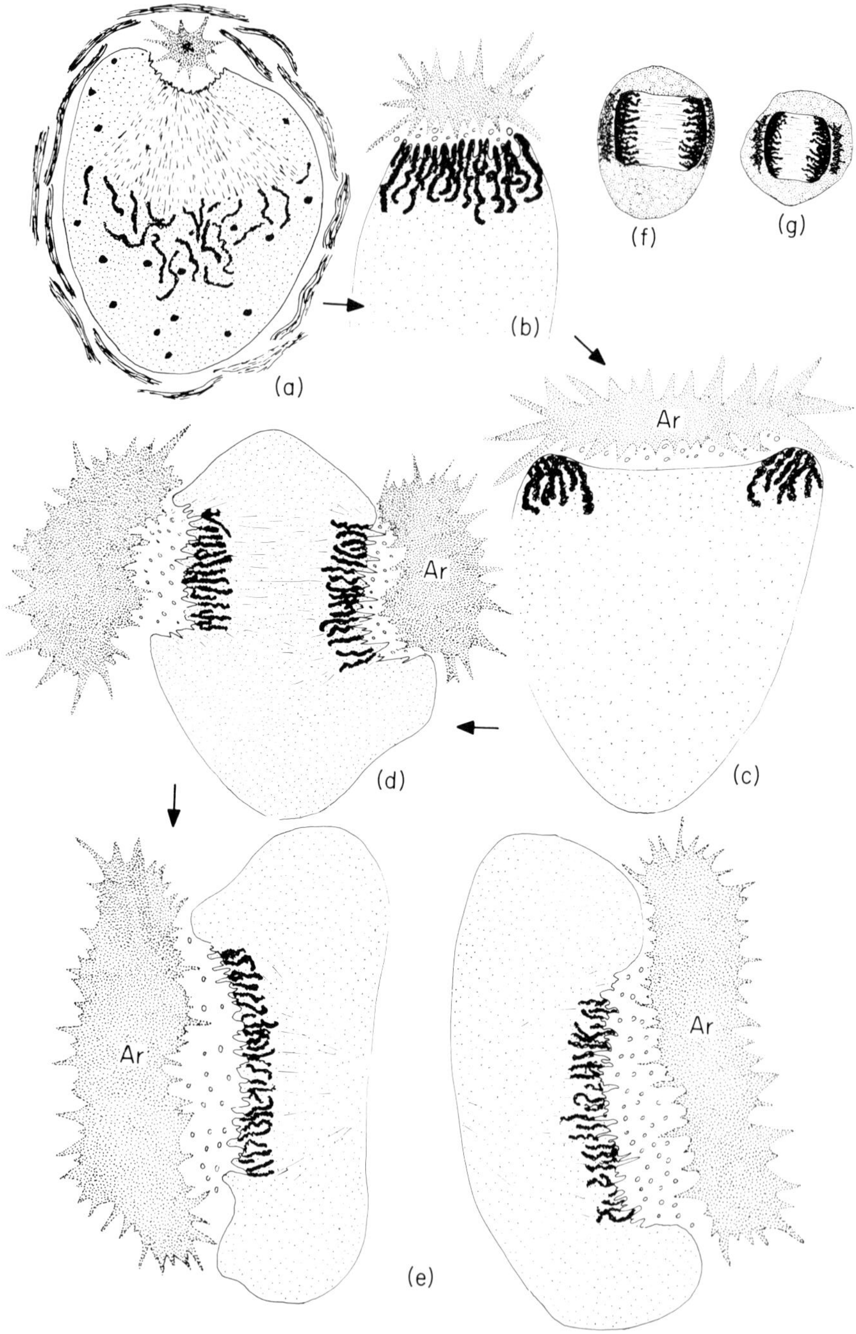

Fig. 13.20. *Oodinium fritillariae*. Diagram showing the first mitosis. (a) The chromosomes are pulled towards the archoplasmic area. (b, c) The beginning of the division. (d) Anaphase. (e) Telophase. (f, g) Late divisions of sporogenesis.

The beginning of sporogenesis (Fig. 13.20) is indicated by the development of an 'archoplasmic mass' (Boveri 1895), a dense cytoplasmic area consisting only of golgi vesicles and ribosomes and surrounded by dictyosomes. This 'archoplasm' is embedded in the anterior pole of the nucleus. Large chromosomes with dense axes and irregularly arranged fibrils reappear. They are stained only by fast green. They are pulled towards the archoplasmic area: one of their extremities becomes attached to the nuclear envelope (which does not break down during mitosis: 'closed' mitosis) (Jenkins 1967). The archoplasmic mass begins to stretch and soon after gives rise to two daughter archoplasmic masses (also known as 'archoplasmic spheres'). A spindle forms between them. The surface of the nucleus undergoes extensive changes: cytoplasmic strands containing spindle microtubules (MTS) penetrate the nucleus, some of which appear to originate from the spindle pole body and terminate at the attachment points of daughter chromosomes: each microtubule ends at a single nuclear pore in which an amorphous dense body (kinetochore?) can be observed.

This mitosis has several features which are typically dinoflagellate: closed mitosis, archoplasmic masses, pole-to-pole MTS, cytoplasmic channels, and the participation of the nuclear envelope in chromosome segregation. There are features peculiar to parasitic dinoflagellates, at least at the beginning of sporogenesis: a bundle of loosely arranged DNA fibrils with no arches, a weak Schiff reaction and a histone-like fast green reaction (see Chapter 4).

The sporogenic divisions proceed without pause. The chromosomes (Fig. 13.21, 13.22) become smaller and denser, and the arched fibrillar arrangement progressively reappears. At the same time, the Schiff reaction intensifies, while the Alfert and Gerschwind (fast green) test becomes negative. The kinetochore-like arrangements become less conspicuous. Finally, the chromosome attachments to the nuclear envelope become invisible. A pair of orthogonally arranged basal bodies appear *de novo* at the anterior pole of the nucleus. They act as centrioles during the final mitosis.

Oodinium-type chromosomes occur in all Blastodinida. However, in the Duboscquodinida and Syndinida the chromosomes are associated with basic proteins (judging by the fast green reaction) and the arched fibrillar arrangement never appears. They are sufficiently condensed to produce an intense, positive Schiff reaction.

Kinetochores have not been observed in Blastodinida or any of the Duboscquodinida. However, they are well developed in *Apodinium* (three-layered formations) (Cachon & Cachon 1979) and in the Syndinida (Hollande 1974; Ris & Kubai 1974).

Basal bodies, which appear in the final stages of sporogenesis in the Blastodinida and Duboscquodinida, seem to be superfluous in mitosis. This seems to be true in the Syndinida also, even though they are permanent in the latter.

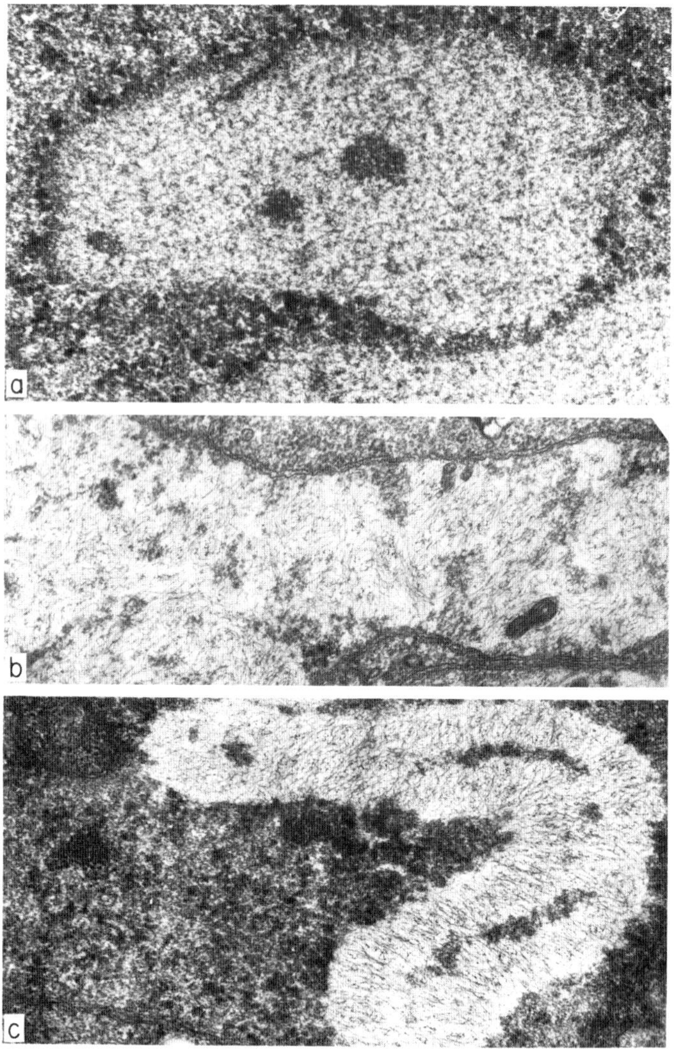

Fig. 13.21. *Oodinium fritillariae*. Development of the chromosome during the first sporogenetic prophase (× 40 000). The fibrous elements begin to appear in the axis of the chromosome (a, b); then they form a transverse array around this axis (c).

7 PRACTICAL ASPECTS

Parasitism in freshwater and marine organisms has both ecological and human economic repercussions. In marine plankton, for example, we often observe that there are sudden outbreaks of parasitism by particular dinoflagellates. In a very short time, a sudden fall in the density of the zooplankton population occurs. Syndinida causes rapid death of copepods, *Duboscquella* kills tintinnids,

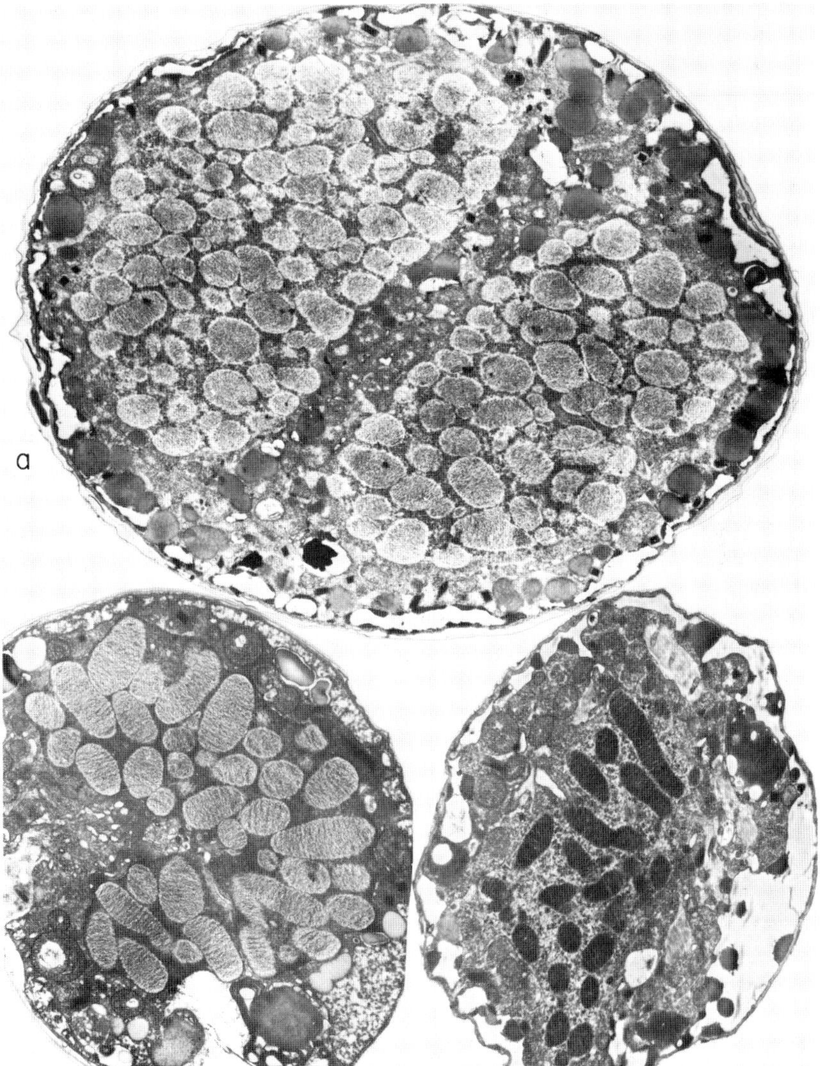

Fig. 13.22. *Oodinium fritillariae.* Development of chromosome ultrastructure during the last generation of sporocysts (a, b) and in the swarmer (c). The typical dinoflagellate structure has reappeared (× 15 000).

Chytriodinium empties the eggs of Euphausiacea, and *Sphaeripara* castrates *Fritillaria.* We have sometimes had great difficulty in collecting uninfected organisms for experimental studies.

In some cases there can be a more direct effect on resources. *Ichthyodinium chabelardi* (Hollande & Cachon 1952) infects and destroys the eggs of sardines in 24 hours. Under certain conditions, still unknown, it can destroy nearly 100% of the egg population.

The Oodinidae which parasitize fish (*Oodinioides, Amyloodinium, Crepido-odinium, Piscinoodinium*) (Brown 1934; Geus 1960; Lom 1981; Needham & Wotten 1979; Nigrelli 1936, 1943; Reichenbac-Klinke 1955, 1956, 1970; Schubert 1978) cause great damage in aquaculture and require prophylactic measures.

Other parasitic dinoflagellates could be indirectly useful for human economy. Taylor (1968) observed that *Gonyaulax catenella*, one of the main producers of toxic red tides, is parasitized by *Amoebophrya ceratii*. The author suggested that *A. cerattii* might eventually be used as a biological control agent for that host.

8 CONCLUSION

Little is known about a great many parasitic dinoflagellates, e.g. *Amoebophrya, Sphaeripara*. Our knowledge of the development of their chromosomes and their mitosis confirms that these protists are truly dinoflagellates, even the Syndinida which, if taken out of context, could be considered to be members of a distinct class (Ris & Kubai 1974).

These comparisons of trophont structure and spore morphology do not provide a clear idea of their phylogeny. Chatton (1938) placed some importance on the morphology of the spore but this is questionable. With some exceptions, the morphology of the trophont is too peculiar to be compared to free-living cells. *Protoodinium* and *Blastodinium* have a structure and tabulation which are typical. *Dissodinium* has sporogenetic stages that have long been misidentified as being those of *Pyrocystis* (Dinococcales).

The Blastodinida have a wall which may have markings, whereas the Duboscquodinida and Syndinida are naked. Are the former more closely related to Peridinida or, conversely, could the two others, which are intracellular and thus more modified by parasitic life, have lost these features more recently? Dinoflagellate parasites may even be polyphyletic and the three tribes recognized here may resemble each other due to convergent evolution.

9 REFERENCES

Boveri (1895) Über das Verhalten der Centrosomen bei der Befruchtung des Seeigeleies nebst allgemeinen Betrachtungen über Centrosomen und Verwandtes. *Verh. physik. med. Ges. Würzburg NF* **29**.

Brown E. M. (1934) On *Oodinium ocellatum* Brown, a parasitic Dinoflagellate causing epidemic disease in marine fishes. *Proc. zool. Soc. London*, **33**, 583–607.

Brown E. M. & Hovasse R. (1945) *Amyloodinium ocellatum* (Brown), a Peridinian parasitic on marine fishes. A complementary study. *Proc. zool. Soc. London* **116**, 33–46.

Cachon J. (1953) Morphologie et cycle évolutif de *Diplomorpha paradoxa* (Rose et Cachon), Péridinien parasite des Siphonophores. *Bull. Soc. zool. France* **78**, 408–414.

CACHON J. (1964) Contribution à l'étude des Péridiniens parasites. Cytologie, cycles évolutifs. *Ann. Sci. nat. zoologie*, 12ème série, **VI**, 1–158.

CACHON J. & CACHON-ENJUMET M. (1965) Cycle évolutif et cytologie de *Neresheimeria catenata* Neresheimer, Péridinien parasite d'Appendiculaires. Rapports de l'hôte et du parasite. *Ann. Sci. nat. Biol. Ser.* **12**, 6, 779–800.

CACHON J. & CACHON M. (1965) *Atlanticellodinium tregouboffi* nov. gen. nov. sp. Peridinien Blastuloidae Neresheimer, parasite de *Planktonetta atlantica* Borgert, Phaedarié Atlanticellide. Cytologie, cycle évolutif, évolution nucléaire au cours de la sporogenèse. *Arch. zool. exp. gen.* **105**, 369–379.

CACHON J., CACHON M. & BOUQUAHEUX F. (1965) *Stylodinium gastrophilum* Cachon, Péridinien Dinococcide parasite de Siphonophores. *Bull. Inst. Ocean.* **65**, 1359, 1–8.

CACHON J. &. CACHON M. (1968) Cytologie et cycle évolutif des *Chytriodinium* (Chatton). *Protistologica* **4**, 249–262.

CACHON J. & CACHON M. (1969) Ultrastructures des Amoebophryidae (Péridiniens: Duboscquodinida), I. Manifestations des rapports entre l'hôte et le parasite. *Protistologica* **5**, 535–547.

CACHON J. & CACHON M. (1970) Ultrastructures des Amoebophryidae (Péridiniens Duboscquodinida), II. Systèmes atractophoriens et microtubulaires, leur intervention dans la mitose. *Protistologica*, **6**, 57–70.

CACHON J., CACHON M. & BOUQUAHEUX F. (1970) *Myxodinium pipiens* gen. nov. sp. nov. péridinien parasite *d'Halosphaera*. *Phycologia* **8**, 157–164.

CACHON J., CACHON M. & GREUET C. (1970) Le système pusulaire de quelques Péridiniens libres ou parasites. *Protistologica* **6**, 467–476.

CACHON J. & CACHON M. (1971a) *Protoodinium chattoni* Hovasse. Manifestations ultrastructurales des rapports entre le péridinien et la Méduse-hôte: fixation, phagocytose. *Arch. Protistenk.* **113**, 293–305.

CACHON J. & CACHON M. (1971b) Ultrastructures du genre *Oodinium* Chatton. Différenciations cellulaires en rapport avec la vie parasitaire. *Protistologica* **7**, 153–169.

CACHON J. & CACHON M. (1973) Les Apodinidae Chatton. Révision systématique. Rapports hôte-parasite et métabolisme. *Protistologica* **9**, 17–33.

CACHON J. & CACHON M. (1974) Comparaison de la mitose des Péridiniens libres et parasites à propos de celle des *Oodinium. C. R. Acad. sci. Paris* **278**, 1735–1737.

CACHON J. & CACHON M. (1977) Observations on the mitosis and on the chromosome evolution during the life-cycle of *Oodinium.* A parasitic Dinoflagellate. *Chromosoma* **60**, 237–251.

CACHON J. & CACHON M. (1979) Singular kinetochore. Structure in a peculiar dinoflagellate. *Arch. f. Protist.* **122**, 267–274.

CACHON J. & CACHON M. (1984) A new Ca^{2+} dependent function of flagellar rootlets in Dinoflagellates, the releasing of a parasite from its host. *Biology of the Cell* **52**, 61–76.

CHATTON E. (1888). Les Bastodinides, ordre nouveau de Dinoflagellatés parasites. *C. R. Acad. Sci.* **142**, 981–983.

CHATTON E. (1910) Sur l'existence de Dinoflagellés parasites coelomiques. Les *Syndinium* chez les Copépodes parasites. *C. R. Acad. sci. Paris* **102**, 654–656.

CHATTON E. (1912) Diagnoses préliminaires de Péridiniens parasites nouveaux. *Bull. Soc. Zool. France*, **37**, 85–93.

CHATTON E. (1920) Les Péridiniens parasites. Morphologie, reproduction, éthologie. *Arch. Zool. exp. gén.* **59**, 1–475.

CHATTON E. (1921) Sur un mécanisme cinétique nouveau: la mitose syndinienne chez les Péridiniens parasites plasmodiaux. *C. R. Acad. sci. Paris* **173**, 859.

CHATTON E. (1922) Sur le polymorphisme et la maturation des spores de Syndinides (Péridiniens). *C. R. Acad. sci. Paris* **175**, 126.

CHATTON E. (1938) *Titres et travaux scientifiques* (1906–1937), 1–406. Sottano, Sète.

CHATTON E. (1952) Classe des Dinoflagellés ou Péridiniens. In *Traité de Zoologie.* P. P. Grassé **1** (1), pp. 309–390. Masson, Paris.

CHATTON E. & BIECHELER D. (1934) Les Coccidinides, Dinoflagellés coccidiomorphes parasites des Dinoflagellés, et le phylum des Phytodinozoa. *C. R. Acad. sci. Paris* **199**, 252–255.

CHATTON E. & BIECHELER B. (1935) Les *Amoebophrya* et les *Hyalosaccus* leur cycle évolutif. L'ordre nouveau des Coelomastigines dans les flagellés. *C. R. Acad. sci. Paris* **200**, 505–507.

CHATTON E. & BIECHELER B. (1936) Documents nouveaux relatifs aux Coccidinides (Dinoflagellés parasites). La sexualité de *Coccidinium mesnili. C. R. Acad. sci. Paris* **208**, 573–575.

DODGE J. D. & CRAWFORD R. M. (1970) A survey of thecal fine structure in the Dinophyceae. *Bot. J. Linn. Soc.* **63**, 53–67.

DREBES G. (1969) *Dissodinium pseudocallani* sp. nov. ein parasitischer Dinoflagellat auf Copepodeneiern. *Helgol wiss. Meersunters* **19**, 58–67.

DREBES G. (1978) *Dissodinium pseudolunula* (Dinophyta) a parasite on copepod eggs. *Br. phycol. J.* **13**, 319–327.

DREBES G. (1981) Possible resting spores of *Dissodinium pseudolunula* (Dinophyta) and their relation to other taxa. *Br. phycol. J.* **16**, 207–215.

DUBOSCQ O. & COLLIN B. (1910) Sur la reproduction sexuée d'un Protiste parasite des Tintinnides. *C. R. Acad. Sci. Paris* **151**, 340–341.

ELBRÄCHTER M. & DREBES G. (1978) Life cycles, phylogeny and taxonomy of *Dissodinium* and *Pyrocystis* (Dinophyta). *Helgol. wiss. Meeresunters* **31**, 347–366.

GALT J. H. & WHISTLER H. C. (1970) Differentiation of flagellated spores in *Thalassomyces*, Elloliopsid parasite of marine crustacea. *Arch. Mikrobiol.* **71**, 295–383.

GEUS A. (1960) Nachtrâgliche Bemerkungen zur Biologie des fisch pathogenen Dinoflagellaten *Oodinium pillularis* Schäperclaus. *Aquarien Terrarien Z.* **13**, 305–306.

GRASSÉ P. P. (1952) Les Coelomastigina (Chatton et Biecheler, 1935) et les Blastuloidae Neresheimer, 1904. In *Traité de Zoologie* **1**, (1), pp. 1020–1022. Masson, Paris.

GREUET C. (1969) Etude morphologique et ultrastructurale du trophonte d'*Erythropsis pavillardi* Kofoid et Swezy. *Protistologica*, **5**, 481–503.

HOFFMAN G. L. & MEYER F. P. (1974) *Parasites of freshwater fishes*. TFM Publications inc. New Jersey.

HOLLANDE A. & CACHON J. (1952) Un parasite des oeufs de Sardine: l'*Ichthyodinium chabelardi* nov. gen. nov. sp. (Péridiniens parasite). *C. R. Acad. sci., Paris* **235**, 976–977.

HOLLANDE A. & CACHON J. (1953) *Morphologie et évolution d'un Péridinien parasite des oeufs de Sardine* (Ichthyodinium chabelardi) Station d'Aquiculture et de Pêche de Castiglione (Alger), n° 4, pp. 7–17.

HOLLANDE A. & ENJUMET M. (1953) Contribution à l'étude biologique des Sphaerocollides (Radiolaires Collodaires et Polycyttaires) et de leurs parasites. *Ann. Sci. Nat. Zool.*, 11ème série, **15**, 99–183.

HOLLANDE A. & ENJUMET M. (1955) Parasites et cycles évolutifs des Radiolaires et des Acanthaires. *Bull. Stat. Aquic. et Pêche, Castiglione*, **7**, 51–176.

HOLLANDE A. (1974) Etude comparée de la mitose syndinienne et de celle des Péridiniens libres et des Hypermastigines. Infrastructure et cycle évolutif des Syndinides parasites de Radiolaires. *Protistologica*, **10**, 413–451.

HOVASSE R. (1923) Les Péridiniens intracellulaires. Zooxanthelles et *Syndinium* chez les Radiolaires coloniaux. Remarques sur la reproduction des Radiolaires. *Bull. Soc. Zool. de France*, **48**, 247–254.

HOVASSE R. (1923) Sur les Péridiniens parasites des Radiolaires coloniaux. *Bull. Soc. Zool. de France* **48**, 337–338.

HOVASSE R. (1935) *Oodinium poucheti* (Lemin) *Protoodinium chattoni* gen. nov. sp. nov. *Bull. Biol. France et Belgique*, 59–86.

HOVASSE R. (1935) Deux Péridiniens parasites convergents: *Oodinium poucheti* (Lemm.); *Protoodinium chattoni* gen. nov. sp. nov. *Bull. Biol.* **LXIX**, 59–85.

HOVASSE R. (1974) A propos des Ellobiopsidae. *Actualités protozooliques* **1**, 362.

JACOBS D. L. (1946) A new Dinoflagellate from freshwater fish. *Trans. Amer. microsc. Soc.* **65**, 1–17.

JENKINS R. A. (1967) Fine structure of division in Ciliates protozoa. I. Micronuclear mitosis in *Blepharisma. J. Cell Biol.* **34**, 463–481.

LAWLER A. R. (1967) *Oodinium cyprinodontum* n. sp., a parasitic Dinoflagellate on gills of Cyprinodontidae of Virginia. *Ches. Sci.* **8**, 67–68.

LAWLER A. R. (1968) Occurrence of the parasitic Dinoflagellate *Oodinium cyprinodontum* Lawler 1967, in North Carolina. *Va. J. Sci.* **19**, 240.

LAWLER A. R. (1979) North America fishes reported as hosts of *Amyloodinium ocellatum Blepharisma. J. Cell. Biol.* **34**, 463–481.

LOEBLICH A. R. III (1976) Dinoflagellate evolution: Speculation and evidence! *J. Protozool.* **23**, 13–28.

LOEBLICH A. R. III (1982) Dinophyceae. In *Synopsis and Classification of Living Organisms* (ed. S. P. Parker) Vol. 1, pp. 101–115. McGraw Hill, New York.

LOM J. & LAWLER A. R. (1973) An ultrastructural study on the mode of attachment in Dinoflagellates invading gills of Cyprinodontidae. *Protistologica* **9**, 293–309.

LOM J. (1977) Mode of attachment to the host in a Dinoflagellate parasite of freshwater fishes. *J. Protozool.* **24**, (suppl.) 16 A.

LOM J. (1981) Fish invading Dinoflagellates: a synopsis of exiting and newly proposed genera. *Folia parasitologica* (*Praha*) **28**, 3–11.

MANIER J. F., FIZE A. & GRIZEL H. (1971) *Syndinium gammari* nov. sp., Péridinien Duboscquodinidae syndinidae, parasite de *Gammarus locusta* (Lin). Crustacé Amphipode. *Protistologica* **7**, 213–219.

NEEDHAM T. & WOTTEN R. (1979) Parasitology of teleosts. In *Fish Pathology* (ed. R. Roberts), pp. 144–183. Baillière Tyndell, Aberdeen.

NIGRELLI R. F. (1936) The morphology, cytology and life history of *Oodinium ocellatum* Brown, a Dinoflagellate on marine fishes. *Zoologica* **21**, 129–164.

NIGRELLI R. F. (1943) Causes of diseases and death of fishes in captivity. *Zoologica*, **28**, 203–216.

PFIESTER L. A. & LYNCH R. A. (1980) Amoeboid stages and sexual reproduction of *Cystodinium bataviexx* and its similarity to *Dinococcus* (Dinophyceae). *Phycologia*, **19** (3), 178–183.

PFIESTER L. A. & POPOFSKY J. (1979) Parasitic, amoeboid Dinoflagellates. *Nature* (*London*), **279**, 421–424.

REICHENBACH-KLINKE H. H. (1955) Die Artzu behörigkeit in Mitteleuropa workommenden *Oodinium.* Art und Beobachtungen über ihr parasitäres stadium (Dinoflagellata, Gymnodiniidae). *J. gen. Microbiol.* **1**, 106–111.

REICHENBACH-KLINKE H. H. (1956) Die Dinoflagellatenart *Oodinium pilularis* Shäperclaus als Bindegewebsparasit von Süsswasser-Fischen. *Gior. Microbiol.* **1**, 263–265.

REICHENBACH-KLINKE H. H. (1970) Vorläufige Mitteilung und Neubeschreibung einer parasitären Blastodinidae (Dinoflagellate) bei Süsswasserfischen. *z. Fish. N.F.* **18**, 289–297.

RIS H. & KUBAI D. (1974) An unusual mitotic mechanism in the parasitic protozoan *Syndinium* sp. *J. cell Biol.* **60**, 702–720.

ROSE M. & CACHON J. (1951) *Diplomorpha paradoxa*, nov. gen. nov. sp. Protiste de l'ectoderme de Siphonophores. *C. R. Acad. sci., Paris* **233**, 451–452.

ROSE M. & CACHON J. (1952) Le mouvement chez *Diplomorpha paradoxa*, Parasite de Siphonophores. *C. R. Acad. sci., Paris* **234**, 669–671.

SCHNEPF E. (1985) Cytoskeleton and food uptake in the parasitic Dinoflagellate *Pausenella. Europ. J. Cell. Biol.* **7**, 36.

SCHUBERT G. (1978) *Krankheiten der Fishe.* Kosmos, Franckh'sche Verlagshandlung Stuttgart. 68 pp.

SHUMWAY W. (1924) The genus *Haplozoon* Dogiel. Observations on the life history and systematic position. *J. Parasitol.* **11**, 59–74.

SIEBERT A. Jr (1973) A description of *Haplozoon axiothellae* n. sp. an endosymbiont of the polychaete *Axiothella rubrocinta*. *J. Phycol.* **9**, 185–190.

SOURNIA A., CACHON J. & CACHON M. (1975) Catalogue des espèces et taxons intraspécifiques de Dinoflagellés marins actuels publiés depuis la révision de J. Schiller, II. Dinoflagellés parasites ou symbiotiques. *Arch. Protistenk.* **117**, 1–19.

SOYER M. O. (1969) Rapports existant entre chromosomes et membrane nucléaire chez un Dinoflagellé du genre *Blastodinium* Chatton. *C. R. Acad. sci., Paris* **268**, 2082–2084.

SOYER M. O. (1971) Structure du noyau des *Blastodinium* (Dinoflagellés parasites). Division et condensation chromatique. *Chromosoma*, **33**, 70.

SOYER M. O. (1974) Etude ultrastructurale de *Syndinium* sp. Chatton, parasite coelomique des Copépodes pélagiques. *Vie et Milieu* **24**, 191–212.

TAYLOR F. J. R. (1967) Phytoplankton of the South Western Indian Ocean. *Nora Hedwigia*, **12**, 433–476.

TAYLOR F. J. R. (1968) Parasitism of the toxin-producing dinoflagellate *Gonyaulax catenella* by the endoparasitic dinoflagellate *Amoebophyra ceratii*. *J. Fish Res. Bd.* Canada, **25**, 2241–2245.

CHAPTER 14
DINOFLAGELLATE REPRODUCTION

LOIS A. PFIESTER
Department of Botany and Microbiology, University of Oklahoma, Norman, OK 73019, USA

DONALD M. ANDERSON
Woods Hole Oceanographic Institution, Woods Hole, Mass. 02543, USA

1 INTRODUCTION

Early descriptions of both marine and freshwater dinoflagellates were based on living or preserved field samples. Only a small percentage of these organisms are even now available in culture. Thus, the older descriptions of their life cycles are only complete to the degree that the author was able to sample frequently and/or be fortuitous enough to have collected a population in all stages of its life history. Thus, researchers who are able to culture dinoflagellates and follow their life history in the laboratory are finding previously unknown vegetative and sexual stages not contained in original descriptions. In some instances these stages have been described as separate taxa.

2 VEGETATIVE REPRODUCTION

Bold & Wynne (1978) distinguish between vegetative or asexual cell division in which the products of cell division are either naked or surrounded completely by new cell walls that are not intimately related with the parental cell walls (*eleutheroschisis*) and that in which the development of the cell walls of the newly divided protoplast is initiated adjacent to, and continuous with, the parental cell wall; each of the young protoplasts forms a wall over its entire surface; the walls of the young protoplasts remain closely contiguous with the parental cell wall, for at least a short period; and immediate rupture or hydration of the parental cell wall does not occur liberating the contained division products (*desmoschisis*). Since there is confusion in the literature concerning the term 'vegetative cell division' (Smith 1950 and Fritsch 1935) the authors have used the terms desmoschisis and eleutheroschisis to distinguish between methods of non-sexual reproduction.

Dinoflagellates either have a cell covering consisting only of membranes (unarmoured forms) or have structural cellulose or other polysaccharides in vesicles (armoured forms) which form plates (see Chapters 2 and 3). In 1970, Loeblich reintroduced Schutt's term 'amphiesma' for the dinoflagellate peripheral complex to replace the term 'theca', which is still used by some authors to designate dinoflagellate cell coverings ranging from those with no plates to those with many (Dodge & Crawford 1970; Steidinger & Cox 1980).

Armoured taxa which shed both epitheca and hypotheca at cell division often have a pellicular layer internal to the plates and external to the plasma membrane called a pellicle.

2.1 Desmoschisis

Three types of desmoschitic reproduction are known to occur in dinoflagellates. The most common is a simple pinching and splitting of one cell into two (binary fission) while continuously synthesizing the cell covering in the course of daughter cell separation (Fig. 14.1). This occurs in unarmoured or so called 'naked forms' such as *Gymnodinium* (Ehrenberg 1838) where it is oblique. Oblique division always occurs from the upper right cell quadrant to the lower left, even in dinophysoids (Taylor 1973). There is evidence that in some species of *Gymnodinium* the nuclear division which occurs at right angles to the plane of mitosis is unequal, resulting in daughter cells having different numbers of chromosomes. There is also evidence of nuclear fragmentation or amitosis occurring within a cell (Shyam & Sarma 1978).

In *Ceratium* the division of the cell wall and protoplast is oblique. Each daughter cell was thought to reproduce the other half (Wetherbee 1975a,b). Plate separation occurs along predetermined sutures thus separating adjacent thecal plates (Fig. 14.2). Dürr & Netzel (1974) have shown that in *Ceratium* the cell eventually sheds the half of the parental theca it inherited; this produces an entire new wall. Nuclear division in *C. tripos* usually begins during the last 2 hours of the light period. The cytoplasm splits by binary fission, with development of the missing half of the two new daughter cells beginning immediately. Non-disjunction may occur following division. This results in chains of two or more cells which remain attached during and after maturity (Wetherbee 1975b and Chapter 2). *Prorocentrum* and *Dinophysis* divide

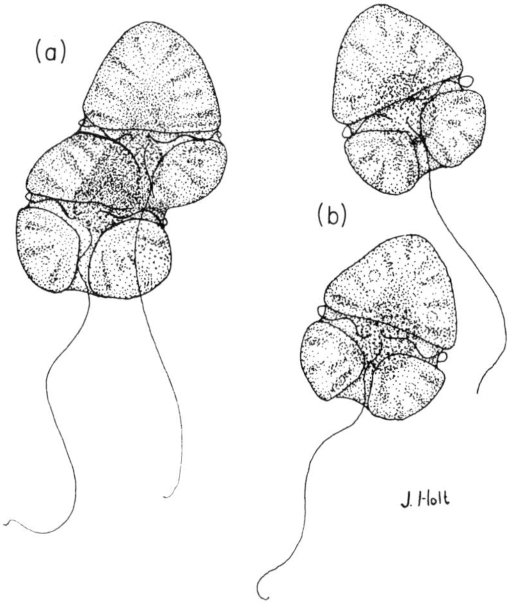

Fig. 14.1. Cell division in *Gymnodinium* sp. (a) Mitosis in oblique plane of motile parent cell. (b) Daughter cells.

longitudinally, separating the two large valves of the parent cell such that each daughter cell retains one valve. The new wall, however, is almost fully formed before the separation of the parent wall along the 'megacytic zone' as Taylor (1973) showed in *Ornithocerus*. Whether a dinoflagellate divides by longitudinal or oblique division, the plane of division passes through the region of the cell from which the flagella emerge.

Dinoflagellate cells increase in volume during the course of the division cycle. Armoured species then must have some mechanism for expansion. In the Dinophysiales this is accomplished by growth of the cell wall along the margins of the sagittal fission line that divides the cell wall into two portions. Cells exhibiting this secondary growth are called megacytic. Some secondary growth does occur in young cells (Taylor 1973) (Fig. 14.3). Secondary growth is said to dissolve during cytokinesis or just following cytokinesis (Pavillard 1915; Tai & Skogsberg 1934; Taylor 1973).

Dinothrix is one of the few dinoflagellate genera that has a filamentous organization. In cell division the protoplast contracts slightly and divides obliquely into two (Fig. 14.4). The parent cell wall remains but the new daughter cells become enveloped in their own cell wall. Filaments thus formed are immobile and may branch sparingly.

2.2 Eleutheroschisis

There are two basic forms of asexual reproduction that occur in the dinoflagellates. The theca may be shed prior to division (ecdysis) as in

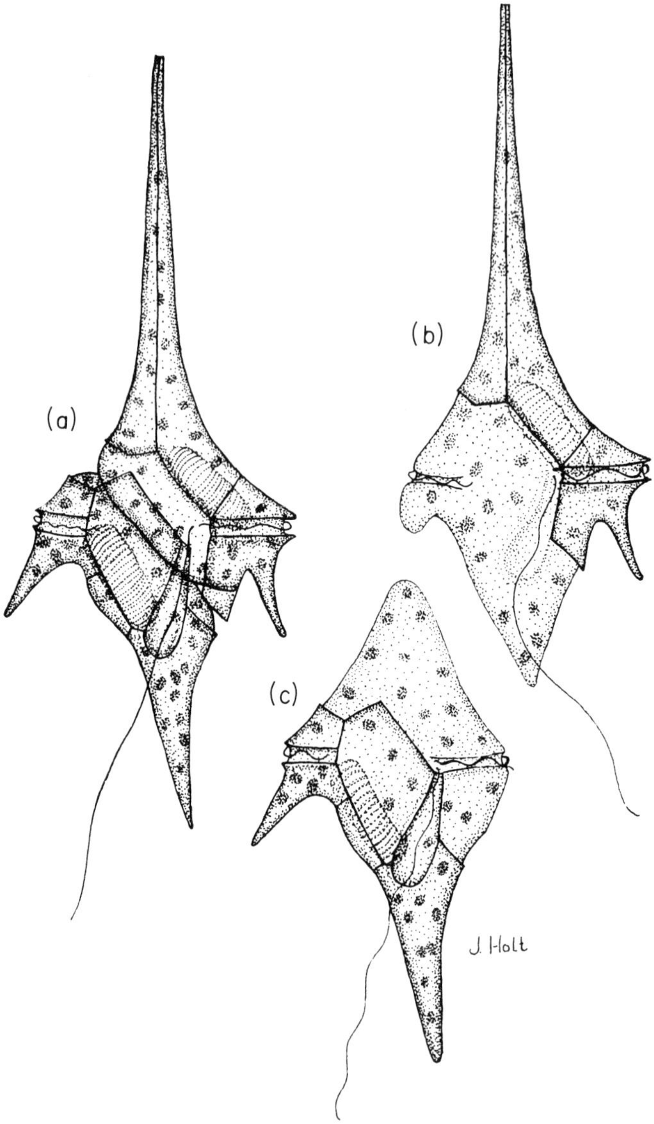

Fig. 14.2. Cell division in *Ceratium*. (a) Mitosis in oblique plane of motile parent cell. (b) Daughter cell with parent epitheca. (c) Daughter cell with parent hypotheca.

Symbiodinium and *Peridinium sanguineum* (Carter 1858) (Fig. 14.5) or it may undergo cytokinesis within the old wall which is shed by the daughter cells; these then form completely new walls (Fig. 14.6). Loeblich (1969) has provided tables (Tables 14.1, 14.2) listing the dinoflagellates that undergo each of the above-mentioned types of reproduction.

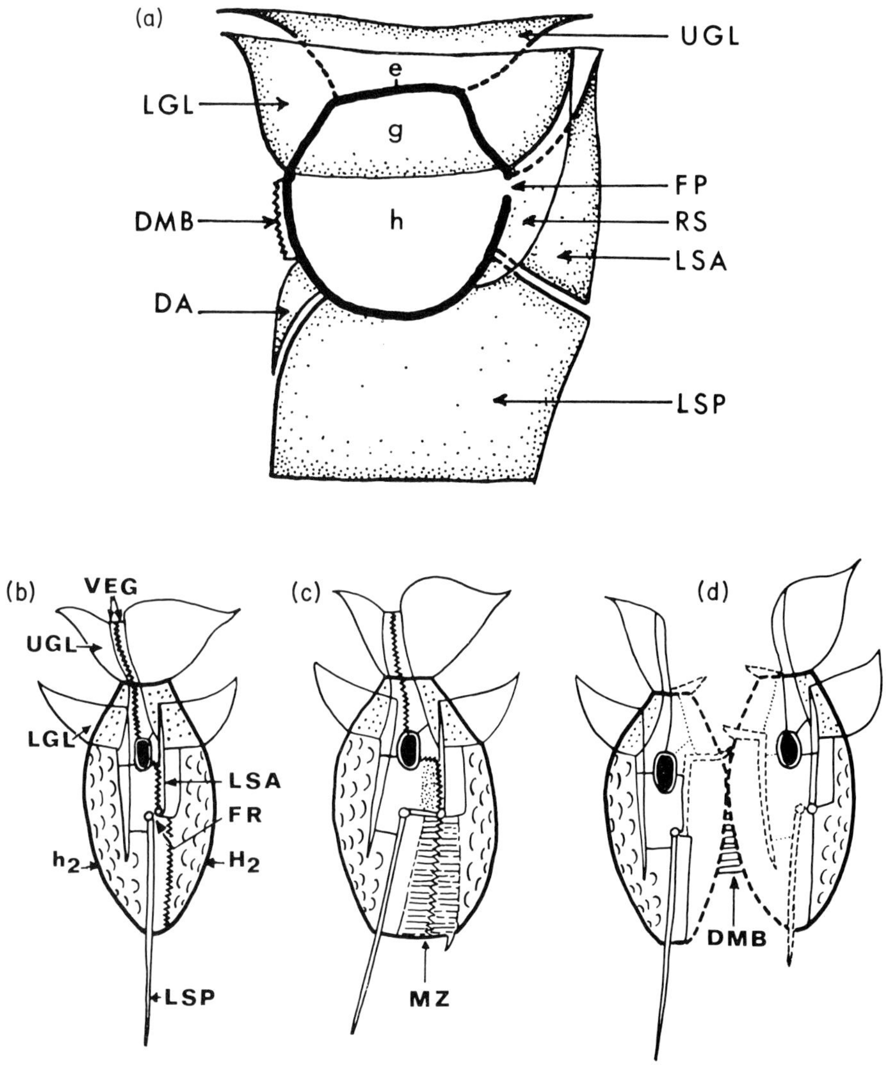

Fig. 14.3. (a) Diagrammatic illustration of the principal thecal components of *Ornithocerus*, seen from the right side. The cell body is shown in heavy outline; the lists are indicated: e, epitheca; g, girdle; h, hypotheca; DA, dorsal accessory list; DMB, dorsal megacytic bridge; FP, flagellar pore; LGL, lower girdle list; LSA, left sulcal list, anterior moiety; LSP, left sulcal list, posterior moiety; UGL, upper girdle list. (b–d) A diagrammatic representation of stages in division of a member of *Ornithocerus* as seen from the ventral (flagellar pore) side: (c) would be termed a megacytic cell. The incomplete lines in (d) represent newly formed material. The dotted part of the central region in (c) indicates a region of lateral growth hypothesized but not observed as yet. New list material forms between the two parts of the fission rib (FR) as they separate. DMB, dorsal megacytic bridge; LGL, lower girdle list; LSA, left sulcal list, anterior moiety; LSP, left sulcal list, posterior moiety; MZ, megacytic zone. (From F. J. R. Taylor (1973) Topography of cell division in the structurally complex genus *Ornithocerus*. *J. Phycol.* **9,** 1–10.)

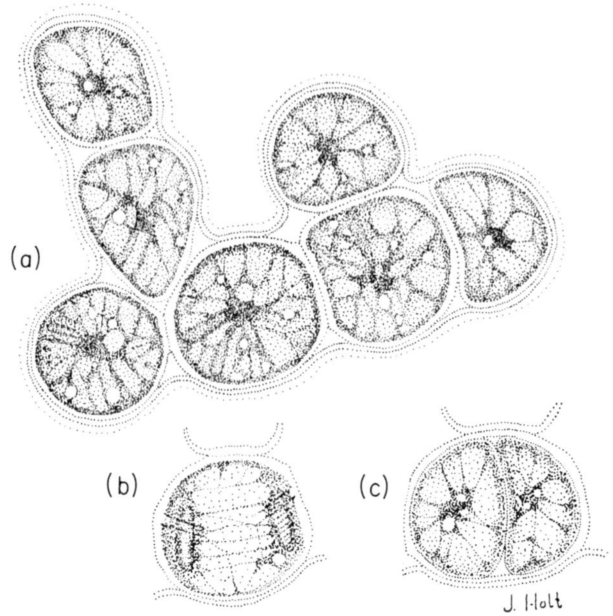

Fig. 14.4. Cell division in *Dinothrix*. (a) Filamentous growth habit. (b) Mitosis within parent cell wall. (c) Daughter cell formation and enlargement within old parent cell wall.

In the Peridiniales, the cell wall accommodates increased cell volume during the mitotic cycle by thecal growth in the sutural area between plates, producing intercalary bands between adjacent plates. Pfiester & Skvarla (1979, 1980) have shown that such growth is slight in vegetative thecae but great in the thecae of plano- and hypnozygotes.

Yet another type of asexual reproduction occurs in those dinoflagellates known to have amoeboid forms. For a detailed description of these forms see

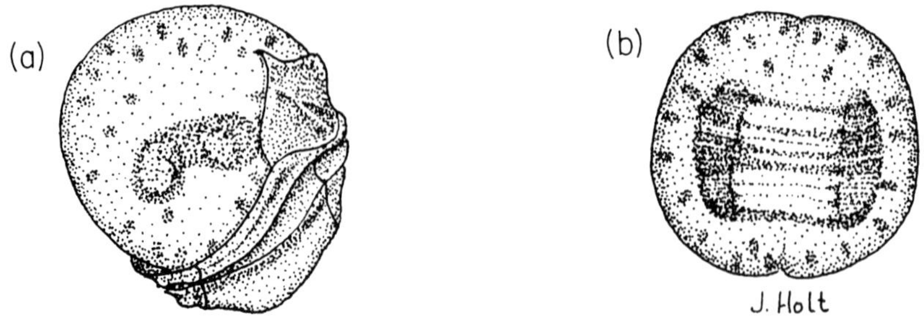

Fig. 14.5. Cell division in *Peridinium sanguineum*. (a) Ecdysis of parent theca. (b) Mitosis of protoplast.

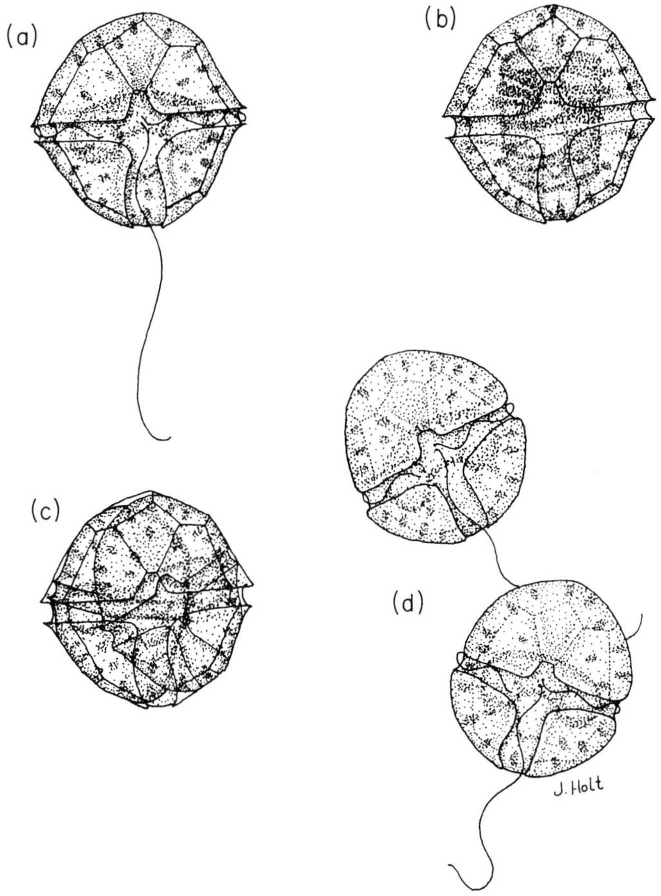

Fig. 14.6. Cell division in *Peridinium volzii.* (a) Vegetative cell. (b) Mitosis. (c) Daughter cells within parent theca. (d) Daughter cells forming armoured theca.

the chapter on parasitic dinoflagellates by Cachon & Cachon (Chapter 13). *Stylodinium* and *Cystodinedria* both have non-motile, walled stages that are epiphytic on freshwater filamentous algae. The protoplast of this non-motile stage at times divides into two to four amoebae which are released, phagocytize intact filamentous algal cells and eventually develop back into their non-motile form. At some stage in development the protoplast of the non-motile stage divides producing gymnodinoid-like cells which are probably gametes (Pfiester & Popovský 1979). Fusion of these cells has been observed in *Cystodinium* (Pfiester & Lynch 1980), a genus similar to the non-motile stage of *Dinamoebidium* (Pascher 1930), another amoeboid-producing dinoflagellate (Fig. 14.7).

Table 14.1. Dinoflagellate species retaining one-half of the parent wall and regenerating the missing half (type 2)*

Species	Reference
Order Prorocentrales	
Exuviaella ssp.	Lebour 1925
Prorocentrum spp.	Lebour 1925
Order Dinophysiales	
Amphisolenia spp.	Kofoid and Skogsberg 1928
Dinophysis spp.	Schütt 1895
Ornithocercus spp.	Kofoid and Skogsberg 1928
Oxphysis oxytoxoides Kofoid, 1926	Kofoid 1926
Order Peridiniales	
Cachonina niei Loeblich, 1968	This paper
Ceratium spp.	Stein 1883; Lauterborn 1895
Ceratocorys spp.	Kofoid & Rigden 1912
Gonyaulax catenata (Levander) Kofoid, 1911	Kofoid & Rigden 1912
G. fragilis (Schütt) Kofoid, 1911	Kofoid & Rigden 1912
G. polygramma Stein, 1883	Kofoid 1911
G. triacantha Jörgensen, 1899	Kofoid 1906
Heteraulacus polyedricus (Pouchet) Drugg and Loeblich, Jr, 1967	Schütt 1895; Nie & Wang 1942
Protoceratium reticulatum (Claparède and Lachmann) Butschli, 1885	Braarud 1945
Pyrodinium bahamense Plate, 1906	
Spiraulaxina kofoidii (Graham) comb. nov.	Kofoif 1911

*From Loeblich (1969).

2.3 Cellular processes involved in vegetative reproduction

Vegetative reproduction involves cytokinesis and karyokinesis. Nuclear organization in the free-living dinoflagellates is unusual in that the chromosomes lack regular histones (Chapter 4), maintain the same appearance throughout the cell cycle and are visible as rod-shaped bodies within the interphase nucleus (Kubai & Ris 1969; Ris & Kubai 1974). Chromosome numbers of only 77 of the more than 2000 dinoflagellate taxa have been reported (Holt & Pfiester 1982). Dinoflagellate chromosome numbers range from 5 for *Syndinium turbo* (Chatton 1920) to approximately 274 for *Ceratium hirundinella* (Entz 1921). Recent studies have shown that dinoflagellates undergo polyploidy and/or aneuploidy in culture. Loper *et al.* (1980) demonstrated polyploidy in the unarmoured dinoflagellate, *Ptychodiscus brevis* in culture. Shyam & Sarma (1978) have suggested a possible polyploid or aneuploid series within the genus *Peridinium* and Holt & Pfiester (1982) have shown a relationship between chromosome numbers and years in culture in six freshwater dinoflagellates.

Table 14.2. Dinoflagellate species forming an entire new wall after cytokinesis (type 3)*

Species	Reference
Order Peridiniales	
Crypthecodinium cohnii (Seligo) Chatton in Grasse, 1952	Kubai & Ris 1969
Dissodium assymetricum (Mangin) comb. nov.	Lebour 1925
D. lenticulum (Bergh) comb. nov.	Lebour 1925
Gonyaulax polyedra Stein, 1883	This paper
G. series Kofoid and Rigden, 1912	Kofoid & Rigden 1912
Helgolandinium subglobosum von Stosch, 1969	von Stosch 1969
Heteraulacus sp.	Sousa & Silva 1969
Peridinium bipes Stein, 1883	Lefèvre 1932
P. cinctum (Muller) Ehrenberg, 1830	Lefèvre 1932
P. foliaceum (Stein) Biecheler, 1952	Sousa & Silva 1962
P. nudum Meunier, 1919	Paredes 1962
P. raciborskii var. *palustre* (Lindemann) Lindemann, 1928	Lindemann 1929
'*P sanguineum*' Carter, 1858	Carter 1858
'*P. tabulatum*' Ehrenberg, 1832	Schilling 1913
P. trochoideum (Stein) Lemmermann, 1910	Braarud 1957
Protoperidinium minutum (Kofoid) comb. nov. (cited as *Peridinium monospinum* Paulsen, 1907)	Lebour 1925
P. ovatum Pouchet, 1883	Schütt 1883
Pyrophacus horologium Stein, 1883	Stein 1883
Woloszynskia coronatum (Woloszynska) Thompson, 1950	Woloszynska 1917

*From Loeblich (1969).

In dinoflagellates the onset of cell division is marked by the duplication of the two flagellar bases from which the new flagella emerge. At this same time the nucleus enlarges and some chromosomes are found with Y-shapes and some with Vs. Nuclear division is unusual in that the nuclear envelope to which the chromosomes are attached remains intact, but a few to many cytoplasmic

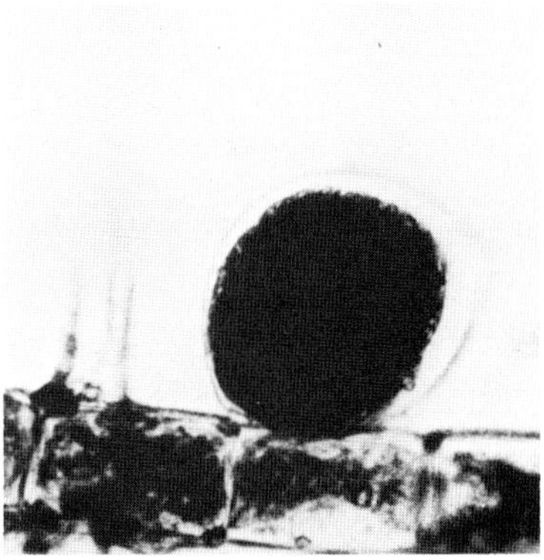

Fig. 14.7. *Cystodinium* vegetative cell.

channels containing microtubules run completely through the nucleus (Leadbeater & Dodge 1967). V-shaped chromosomes make contact at their apices with the nuclear membrane where it surrounds cytoplasmic channels. Thus, the nuclear membrane is probably involved in some way in daughter chromosome separation (Kubai & Ris 1969). Division of the chromosome into two chromatids starts at one end of the chromosome and works toward the other. Once the chromosomes are separate the two longitudinal flagella move slightly apart and the nucleus becomes laterally invaginated by cytoplasmic streams from both sides. These invaginations become continuous through the nucleus and contain many microtubules (Fig. 14.8). Free-living and parasitic dinoflagellates have been found to have kinetochores though they are not all structurally identical (Oakley & Dodge 1974). In *Amphidinium* the nuclear envelope remains intact (Oakley & Dodge 1974), while in *Syndinium* the kinetochores are found in perforations or pores in the envelope (Ris & Kubai 1974; Fig. 13.18). In *Amphidinium* the kinetochore is completely outside the nuclear membrane. Thus, dinoflagellates do have a type of spindle but it is topologically outside the nucleus. As the chromatids separate, the nucleus becomes dumbell shaped and eventually separates in the middle forming two daughter nuclei. The cytoplasmic invaginations eventually disappear in the daughter nuclei. The nucleolus persists throughout mitosis (Leadbeater & Dodge 1967). The cytoplasm cleaves, separating the two nuclei and thus forming two separate protoplasts. As described above, thecal division occurs at this time in some dinoflagellates and not in others.

2.4 Cysts

As previously mentioned the term 'cysts' has different meanings for the palynologists and the neontologists. Palynologists use the term dinoflagellate cyst to refer to the fossilized forms of dinoflagellates (Chapter 15) while

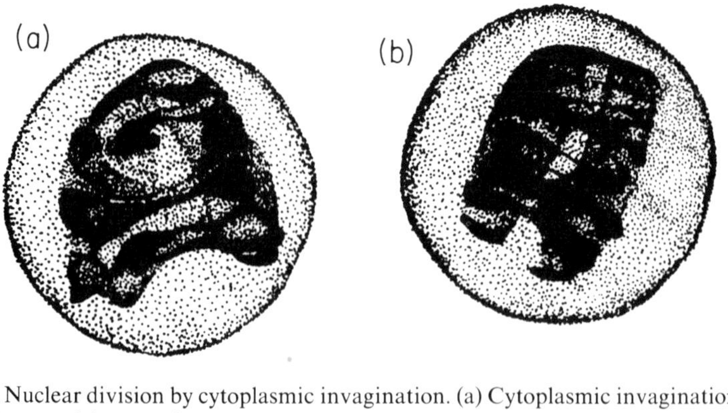

Fig. 14.8. Nuclear division by cytoplasmic invagination. (a) Cytoplasmic invagination. (b) Daughter nuclei separating.

neontologists have used it to refer to (i) a temporary resting state (temporary or digestive cysts), (ii) what is now known to be a dormant zygote (resting cyst), or (iii) a coccoid condition in which the cells are still photosynthetically active, e.g. *Pyrocystis*. Since we now know that some (probably many) of the fossilized cysts studied by palynologists are hypnozygotes their meaning of the term cysts will be discussed later under dinoflagellate sexual reproduction.

The occurrence of temporary cysts or spores has long been known and documented in numerous dinoflagellate studies. Fritsch (1935) referred to them as 'thin-walled' cysts. They are formed in *Peridinium cinctum* when cessation of movement is followed by a marked contraction of the protoplast. The contracted protoplast acquires a new wall before its release from the old theca, and the thin-walled spherical cysts thus formed may rest for a varying length of time. When placed in fresh media at 20°C the cyst will divide and germinate within 24 hours. It remains viable, however, for at least 5 months at 4°C in the dark. Thus, freshwater dinoflagellates may overwinter as thin-walled temporary cysts (Pfiester 1975). However, Anderson & Wall (1978) found that temporary cysts of *Gonyaulax tamarensis* which occur by ecdysis (Chapter 2) could not survive beyond a month at low temperatures. This may represent a basic difference between marine and freshwater dinoflagellate taxa. Evitt (pers. comm.) has found that the 'thin-walled' cyst of *P. limbatum* is fossilizable.

3 SEXUAL REPRODUCTION

Dinoflagellate sexual reproduction has long been disputed in the literature (Grell 1973) but has now been well documented and established by Von Stosch (1965, 1969, 1972, 1973), Cao Vien (1967a,b, 1968), Zingmark (1970), Tuttle & Loeblich (1975), Beam & Himes (1974, 1980), Pfiester (1975, 1976, 1977), Pfiester & Skvarla (1979, 1980), Pfiester & Lynch (1980), Walker & Steidinger (1979), Spero (1980), Anderson (1980), Turpin *et al.* (1978), and Yoshimatsui (1981). While a general pattern of sexual reproduction appears to be emerging it is important to remember that the sexual life histories of only twenty-two of the approximately 2000 extant species of dinoflagellates have been observed and described in the literature. Researchers are now only in the beginning stages of studying dinoflagellate sexual life histories. Developments leading to the culture of sensitive dinoflagellates have been of major importance in the discovery of and the ability to induce their sexual reproduction. Sexual reproduction appears to be of common occurrence in cultures but has not been recognized because (i) gametes usually look similar to regular cells, (ii) fusion has been confused with division, and (iii) 'warty' zygotes have been interpreted as aberrant cells. Given the previous studies on sexuality and further efforts to refine culture techniques, we should expect an increase in the number of sexual life histories known. Mechanisms of sexual reproduction in the algae are varied. Some of the life cycle types present in algae are: (i) haplontic (vegetative cells are haploid with the zygote being the only diploid cell in the life cycle), (ii) diplontic (all cells are diploid except the gametes), and (iii) diplohaplontic (an alternation of haploid and diploid generations). All dinoflagellates studied thus far exhibit the haplontic type of life cycle with the exception of *Noctiluca*. For that reason Zingmark (1970) maintains that it is not a true dinoflagellate.

Haplontic dinoflagellates differ in the types of gametes produced. That is, in many species the gametes are *hologamous*, (Beam & Himes 1974; Cao Vien 1967a,b), i.e. gametes do not differ morphologically from vegetative cells. Some species are *isogamous*. That is, while the two gametes which fuse may differ morphologically from vegetative cells by size and/or amount of pigmentation and presence of theca they are morphologically identical to each other (von Stosch 1964, 1972, 1973; Pfiester 1975, 1976, 1977; Pfiester & Skvarla 1980; Anderson 1980; Spero & Moree 1981; Walker & Steidinger 1979). When fusing gametes differ from each other morphologically a species is said to be *anisogamous*. *Ceratium cornutum*, *C. horridum*, *Helgolandinium subglobosum* (Stosch 1972), *Protogonyaulax* (=*Gonyaulax*) *tamarensis* (Turpin *et al.* 1978), and *Coccidinium mesnili* (Chatton & Biecheler 1936) produce anisogamous gametes. Algal species that produce isogamous or anisogamous gametes are considered to be monoecious if sexual reproduction may occur within a clone such as *Peridinium cinctum* (Pfiester 1975) and dioecious if two different mating strains (usually designated plus and minus) must be combined in order for sexual reproduction to occur. Most dinoflagellates studied are monoecious. Dioecism has been reported for *Glenodinium lubiniensiforme* (Diwald 1937), *Woloszynskia apiculata* (Stosch 1973), *C. cornutum* (Stosch 1972), *Peridinium volzii* (Pfiester & Skvarla 1979), *Protogonyaulax catenella* (Yoshimatsu 1981) and one population of *Noctiluca scintillans* (=*miliaris*) (Hoefker 1930).

3.1 Nuclear phenomena associated with the sexual process

As discussed elsewhere in this chapter and book the dinoflagellate nucleus is unique in that it has features in common with both prokaryotes and eukaryotes. With the exception of *Noctiluca* the vegetative cells of dinoflagellates studied thus far are haploid, although the chromosome numbers may be high. Nuclear fusion has been observed to occur during plasmogamy but prior to its completion in fusing gametes of most species studied (Spector *et al.* 1981; Pfiester 1976, 1977; Pfiester & Skvarla 1979). The resulting nucleus is large and occupies a considerable part of the cell volume. In a growing number of described dinophycean zygotes the nucleus has been seen to enlarge further and to rotate rapidly within the cell. This phenomenon termed 'cyclosis', was first described by G. Pouchet (1883). Later Biecheler (1952) described the same process in six species of *Peridinium*. She showed that some time after the nucleus had come to rest cell division followed. Biecheler postulated that the phenomenon might be associated with meiosis. Von Stosch (1972, 1973) has clearly shown this to be the case. The exact phase of the life cycle at which this occurs differs slightly for the organisms in which it has been noted. In *Ceratium horridum* nuclear cyclosis occurs during the planozygotic (motile zygote) stage. Von Stosch was able to show the double nature of the chromosomes during the late stages of cyclosis. Thus, nuclear cyclosis corresponds to the late zygotene or postzygotene of meiosis. In *C. cornutum* however, nuclear cyclosis occurred in the cell which emerged from the hypnozygote (non-motile zygote). In *Wolosynskia apiculata* it occurs for 9 hours before the onset of meiosis (von Stosch 1973). On excystment one cell or two may emerge.

Dinoflagellates have many characteristics in common with the ciliates (Phylum Protozoa). For this reason some (Taylor 1980) believe that they may be more closely related than previously supposed. While the ciliate macronucleus does not undergo a true cyclosis during division it enlarges greatly, nuclear reorganization occurs forming two zones with each zone moving toward the centre (Kudo 1966). Cyclosis does occur in ciliates' endoplasm. In *Paramecium* the endoplasm moves along the aboral side to the anterior region and down the other side, with a short cyclosis in the posterior half of the body (Kudo 1966).

Beam & Himes (Himes & Beam 1975; Beam & Himes 1980) have proposed an unusual one-division meiosis in dinoflagellates resulting from their work on the sexual life history of *Crypthecodinium cohnii* (Beam *et al.* 1977). Working with motility mutants that show complementation shortly after zygote formation they showed that segregations were always 1:1, i.e. in all tetrads showing recombination, only the two reciprocal recombinant genotypes were found; there were no tetratypes. They postulated that this could result from (i) centromere linkage, (ii) the absence of crossing over in an otherwise conventional meiosis, or (iii) an unusual one-division 'meiosis'. They concluded that reduction in *C. cohnii* does not employ a second meiotic division on the basis of two-celled zygotic cysts, many of which showed recombination, and the presence of eight-celled cysts. Von Stosch, Theil and Happach-Kasan (pers. comm.) recently found evidence for a two-step meiosis in *Ceratium cornutum*, a heterothallic freshwater dinoflagellate that reproduces anisogamously. Its planozygote increases in size for several weeks, eventually developing into a dormant hypnozygote. Under laboratory conditions one uninucleate swarmer escapes. Its size, shape and flagellation differs from vegetative cells. The two subsequent nuclear divisions of this cell are thought to be meiotic. Aberrant thecal halves of the meiocyte are transmitted to the offspring. These aberrant halves were used as markers. Von Stosch *et al.* were able to isolate ordered tetrads and to raise meiospore clones from them. Their sexual determination was then tested by subjecting them to conditions favourable to sexual reproduction in clonal cultures or in combination with each of the two standard female and male clones and determining the number of hypnozygotes formed. Of the 125 complete tetrads analysed, 69 had the sex factors segregated in the first meiotic division, 59 in the second meiotic division, while 7 tetrads were either 1–3 segregants or non-classifiable. Von Stosch *et al.* concluded from these data that meiosis is a two-step process in *C. cornutum* as opposed to the one-step division of Beam & Himes. Von Stosch *et al.* did observe parthenogenesis in male haploid and diploid clones. This phenomenon has also been observed in *Peridinium volzii* (L. A. Pfiester, pers. obs.) and in *Protogonyaulax tamarensis* (D. M. Anderson, pers. obs.).

3.2 Examples of sexual reproduction

NON-CYST FORMERS

There are dinoflagellates studied thus far that do not produce resting zygotes (cysts) in their sexual cycle: *Noctiluca scintillans* (Zingmark 1970), *Ceratium tripos* (von Stosch 1969), and *Peridinium gatunense* (Pfiester 1977).

Noctiluca (Fig. 14.9 and Chapter 2) is a marine dinoflagellate that may range up to 2 mm in size. It is a naked, spherical, non-photosynthetic colourless cell that is usually capable of luminescence. *Noctiluca* has a conspicuous nuclear mass near the oral groove, a cytostome and extended tentacle. Food vacuoles extend into the large cell vacuole but are separated from it.

Noctiluca's morphology changes during gametogenesis. Its tentacle disappears and the cell becomes more spherical. At this stage the nuclear mass is located at the periphery of the cell and food vacuoles have disappeared. The nucleus divides within a few hours, only to divide again in about 45 min. The four nuclei thus formed are arranged in a tetrad. This appears to signal the end

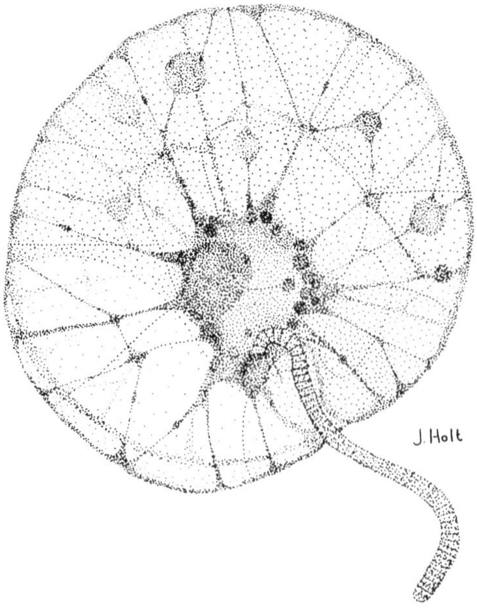

Fig. 14.9. *Noctiluca scintillans* vegetative cell.

of meiosis. The tetrad continues to divide every hour until it reaches the 256–1024 stage, depending on the original gametocyte size. Uniflagellated gametes are formed and released by a budding process. These gametes have a prominent constriction on one side of the flattened surface which has been interpreted as showing morphological affinity with dinoflagellates. Gametes fuse with the flattened side of one individual touching the narrow edge side of another. As fusion continues the gametes become non-motile as their flagella appear shortened. The nuclei have fused at this stage. Zygotes measure about 25 μm in diameter when first formed. All zygotes formed during Zingmark's (1970) research died shortly after formation except for one, which increased in size, became vacuolated and developed a tentacle and nuclear mass. It too died, after reaching about 200 μm in size.

Vegetative cells of *Noctiluca* are thus diploid, with meiosis occurring during gamete formation. Gametes are isogamous, Zingmark's strain was monoecious. *Noctiluca* thus has a diplontic life cycle (Fig. 14.10). The vegetative nucleus is the eukaryotic type (the chromosomes dispersing at interphase) and thus not typical of dinoflagellates, but the nucleus of the gametes is typically mesokaryotic. Thus, it appears that in the case of *Noctiluca* the two types of nuclei found in the dinoflagellates occur together in the same life history. Zingmark (1970) who named this type of nuclear condition 'noctikaryotic' did not believe that *Noctiluca* was a true dinoflagellate because: (i) while the gamete's body is partially constricted in the middle resembling a *Gymnodinium* cell, closer examination reveals that this is not a transverse girdle and a transverse flagellum is lacking; (ii) the vegetative nucleus is eukaryotic rather than mesokaryotic as in the dinoflagellates. However, as illustrated in Chapter 2, a ribbon-like flagellum is present in the feeding stage ('trophont').

CYST-FORMERS: HYPNOZYGOTES

Increasing numbers of dinoflagellate life cycles are being reported in the literature in which the zygote undergoes a long resting stage before germination similar to other algal groups such as the Volvocalan green algae. The life histories described to date have been cited elsewhere in this chapter, so that the serious student of dinoflagellate sexuality should be able to pursue this phenomenon in detail should he/she wish. Here we select two 'typical' taxa for which we describe the phenomenon in detail. The first is *Peridinium cinctum*.

Peridinium cinctum *life history*

The sexual life history of *P. cinctum* has been reported in detail at the light (Pfiester 1974, 1975), scanning (Pfiester & Skvarla 1980), and transmission electron microscopical levels (Spector *et al.* 1981). It is isogamous and monoecious (see Chapter 11B for its autecology).

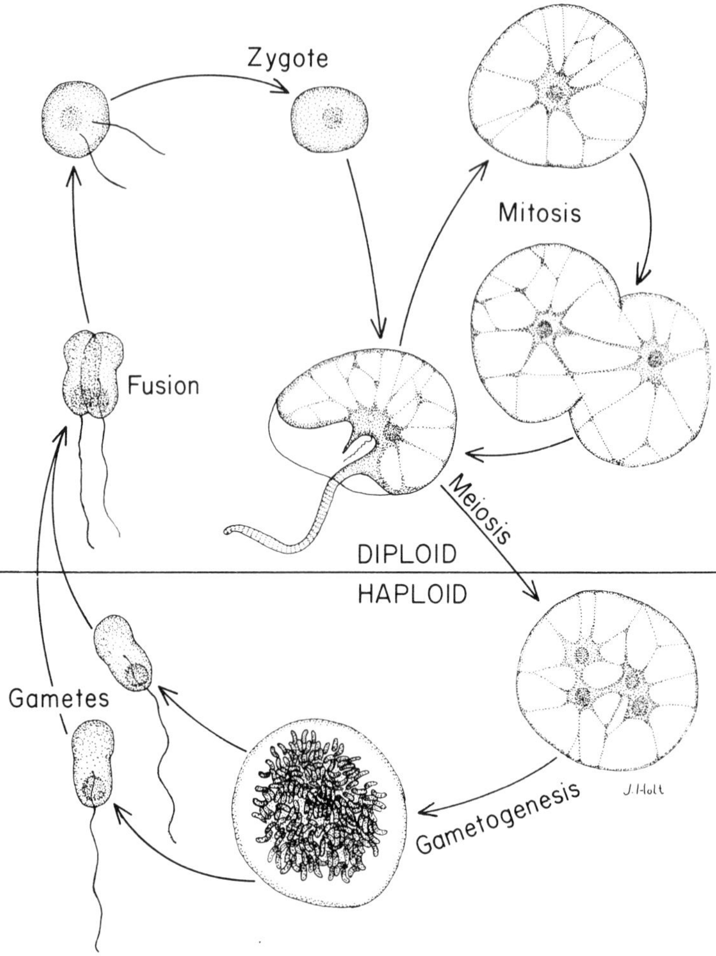

Fig. 14.10. Life cycle of *N. scintillans*.

Peridinium is a thecate genus, i.e. its cell wall has a number of plates joined by what have been called intercalary bands (Chapter 2). The girdle (cingulum) divides the cell into two halves, the epi- and hypo-theca. The number and arrangements of the plates in these two halves has thus far been the major taxonomic character used in classifying these organisms and is referred to as a tabulation or formula. The plate pattern formula for *Peridinium* is 4′, 3a, 5c, ?s, 7″, 5‴, 2⁗ (Bourrelly 1968). Members of the genus tend to be flattened dorsoventrally. Vegetative cells are approximately 35–38 μm in length and have discoid, parietal chloroplasts. Cells contain large amounts of a brownish pigment, peridinin, the name derived from the generic name, giving a light to dark brown appearance.

Gamete formation, structure and development. When a vegetative population is placed in a nitrogen deficient medium, gametes are produced by the longitudinal mitotic division of small vegetative cells, not involving the parent theca (Pfiester 1975). Gametes so formed turn before being released so that the plane of cell division appears to be horizontal. Maximum numbers of gametes are observed 5–7 days following inoculation into the N-deficient medium. Transmission electron microscopy shows that the parent protoplast is surrounded by three membranes internal to the thecal plates (Spector & Treimer 1979) and that mucilage is deposited between the outer two membranes during gamete formation (Spector *et al.* 1981). Gametes, approximately 25 μm in diameter, are released by the breakdown of the parent theca. Initially gametes are naked, i.e. they lack a theca. However, a theca begins forming shortly after release. Gametes may fuse immediately upon release from the parent cell or remain motile for 24–36 hours, during which time they develop a cell wall. The final thickness of this wall depends upon the length of time the gametes remain motile before fusing (L. A. Pfiester, pers. obs.). Fusing gametes with thecal plates measuring 20–200 nm in thickness have been observed with the TEM (Spector *et al.* 1981) (Fig. 14.11). Fusing thecate gametes have also been

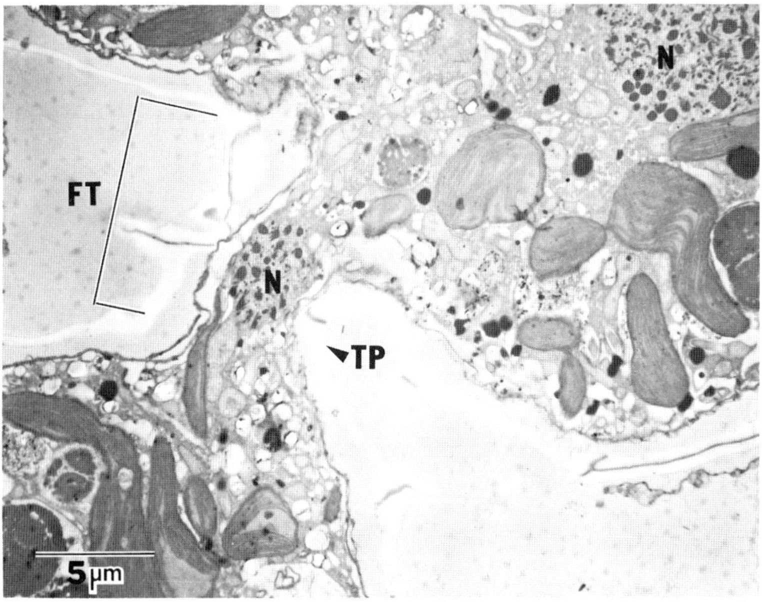

Fig. 14.11. Section through the fertilization tube (FT) of *Peridinium cinctum* connecting two gametes. Thecal plates (TP) are continuous between the gamete and appear to have fused. One nucleus (N), containing fibrous chromosomes, is present in the fertilization tube while the other nucleus has not yet migrated into the fertilization tube (FT) (× 3200). From Spector *et al.* (1981) *Amer. J. Bot.* **68,** 34–43.

observed with the light microscope (Pfiester 1976). However, at this observational level thecate gametes were never seen to complete fusion. Gametes have fewer chloroplasts than vegetative cells, many membrane-bound storage bodies and starch grains. The nucleus is large in proportion to the cell size and contains many slightly unwound chromosomes with 'arms' (loops) extending from them. As the gametes fuse laterally, a fertilization tube is formed which widens along the sulcus as fusion continues. At the level of the basal bodies the sulcus is devoid of thecal plates (in thecate gametes) and covered by a membrane. Initially the fertilization tube is located beneath the basal bodies. At that stage a fibrous chromosome is usually present in the tube. At later stages the two nuclei fuse in the fertilization tube. Fusion between naked gametes occurs as described for thecate gametes except that the fusing pair are surrounded by three membranes. Fusion takes approximately 45 minutes.

Planozygote formation, structure and development. The zygote is at first spherical but within 45 minutes resembles a *P. cinctum* vegetative cell 38 × 35 μm. A theca develops within 24 hours. Some time later a wall is laid down internal to the thecal plates. At this stage the planozygote, which has two trailing flagella, measures approximately 40 μm. It remains motile for approximately 2 weeks, enlarging during this time to a maximum size of 70 × 75 μm. The planozygote appears 'warty' as it enlarges, partly because of the protrusions present on the thecal plates (Spector *et al.* 1981) and partly because of an unequal widening of suture bands between plates (Pfiester & Skvarla 1980). At this stage the chromosomes are similar to those of vegetative cells. Once the maximum size is reached the planozygote loses motility, its protoplast contracts, one or more large red bodies develop and a third wall develops.

Hypnozygote structure, germination and meiosis. The non-motile zygote is referred to as a hypnozygote. It has been reported to contain chitin in its outer wall, which is extremely thick and is called the exospore (Pfiester 1975). The thin middle wall is referred to as the mesopore and the inner wall the endospore. The endospore has been reported to contain sporopollenin (Spector *et al.* 1981). The red body resists acetolysis and is fossilizable. Thus, it too may contain sporopollenin (W. Evitt, pers. comm.).

Meiosis has not been observed in *P. cinctum.* Nuclear stains will not penetrate the hypnozygote until just prior to germination. At that stage the cell has been shown to contain four nuclei (Pfiester 1975). On that basis meiosis is thought to have been completed before germination of the hypnozygote. It is highly possible that the binucleated planozygote observed by Pfiester (1975) was the result of the first meiotic division and that the second division was delayed until just prior to germination from the hypnozygotic resting stage. Thus, vegetative cells and gametes are said to be haploid. This is supported by the report (Pfiester 1975) of isolated gametes placed in N-enriched medium giving rise to vegetative populations which can later be induced sexually.

Ceratium cornutum *sexual life history*

Von Stosch (1964, 1965), in his work on the genus *Ceratium*, was the first to recognize that the 'Knauelstadium' (knotty stage) described by Borgert (1910) represented the postzygotene stage of meiosis in the *C. cornutum* zygote. Von Stosch, Theil and Happach-Kasan (pers. comm.) have recently completed further work on the sexual life history of *C. cornutum* wherein they have described meiosis in detail. *C. cornutum* is dioecious and anisogamous. Male gametes are considerably smaller than female gametes. They are engulfed by the female gamete through the sulcal area. Plates in this region appear 'hinged' and open for the male gamete and close again when syngamy is complete. The planozygote grows in size for several weeks until it develops into a dormant hypnozygote. If maintained at 4°C for at least 3 months the hypnozygotes will then germinate, producing a uninucleate swarmer. This meiocyte differs in size, shape and flagellation from vegetative cells. Von Stosch *et al.* (pers. comm.) have shown that its subsequent two nuclear divisions are meiotic. The thecal halves of the meiocyte are aberrant and are transmitted to the offspring of both steps of meiotic cell divisions. Thus, ordered tetrads are produced. From a detailed study of these tetrads von Stosch *et al.* have shown that meiosis is a two-division process in *C. cornutum* as opposed to the one-division process proposed for *Crypthecodinium*. They have shown further that some male clones may undergo haploid and diploid parthenogenesis.

A generalized dinoflagellate life cycle is shown in Fig. 14.12. Such a diagram may change in time as more dinoflagellates are cultured and their vegetative and sexual life histories elucidated.

4 SIGNIFICANCE OF SEXUAL REPRODUCTION TO DINOFLAGELLATE SYSTEMATICS

The earliest documented report of sexual reproduction in the Dinophyceae was Joseph's (1879) description of pairing and fusion of swimming cells of *Peridinium stygium*. The development of culture methods and techniques has enabled researchers to study and document sexual life histories in detail. While these studies are in themselves valuable and interesting, their effects on dinoflagellate systematics make them even more so. From these studies we now know that many (perhaps nearly all) of the fossil forms are remains of hypnozygotes rather than vegetative cells as previously thought. Their excellent preservation in the fossil record may indeed be due to the presence of sporopollenin in the cell wall as Spector *et al.* (1981) and earlier authors have postulated. Further, in the laboratory, researchers have been able to trigger the sexual process by drastically lowering the phosphorus and/or nitrogen concentrations in the culture media (see next section). While this may not be the only mechanism that induces sexuality in nature, it is not unreasonable to speculate that at least some of the fossil hypnozygotes were initially formed under conditions of nutrient limitation. Thus, the knowledge of the sexual process in dinoflagellates may lead to a further understanding of the environment where fossil or extant hypnozygotes are found.

A number of descriptions and figures of taxa at various levels discuss and show extremely widened intercalary bands. We now know that in the armoured dinoflagellates the planozygote theca accommodates protoplast enlargement by increased width of intercalary bands (Pfiester & Skvarla 1979, 1980) (Fig. 14.12). Thus, these descriptions are not of normal vegetative cells as the authors

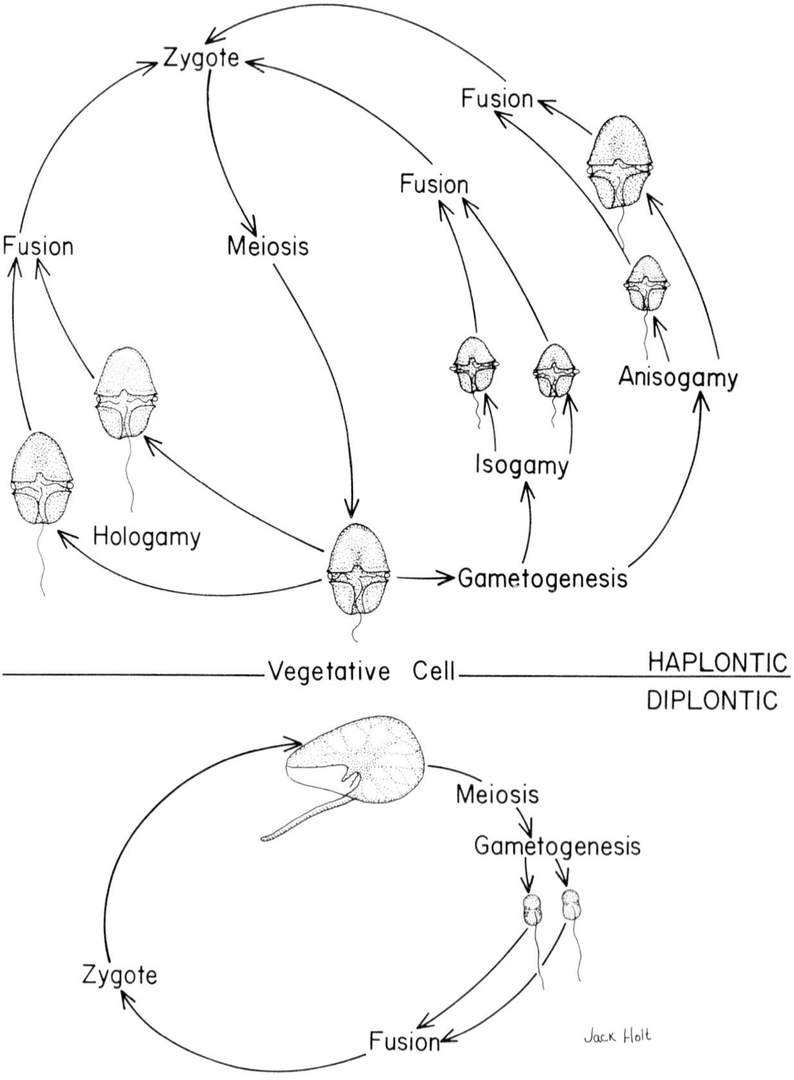

Fig. 14.12. Generalized dinoflagellate life cycle.

thought but are merely stages in a long-lived sexual cycle. Many described taxa may thus be invalid and simply represent a stage in the sexual life history of an already validly described taxon. Such is the case with *Pyrocystis margalefii* Leger and its connection with *Dissodinium pseudolunula* (Drebes 1981). Drebes (1981), in describing resting spores of *D. pseudolunula*, has presented strong evidence to show that *P. margalefii* (described in 1973) is but a description of the resting spores of *D. pseudolunula* and thus an invalid taxon. He further presents information that suggests that *Palaeocystodinium gozlowense* Alberti, might indeed be a fossil form of *D. pseudolunula*. To show further the effects of culturing and life-history studies on dinoflagellate systematics, photographs of *D. pseudolunula*, a marine species, appear similar to stages in the life history of *Cystodinium bataviense* (L. A. Pfiester, unpub. obs.).

Studies of life histories have shown that a number of dinoflagellates, e.g. *Symbiodinium microadriaticum* (Freudenthal 1962), have gymnodinoid stages in their life cycles. However, the motile stage appears to be transitory in the life history of *S. microadriaticum* while it is the dominant phase in the life history of *Gymnodinium* species, and division occurs in the cyst stage in the former and the motile phase of the latter. Thus, on the basis of life-history studies Freudenthal (1962) placed *Symbiodinium* in the family order Dinococcales, family Blastodiniaceae, recognizing the transient nature of the motile stages. D. Taylor (1971a,b) however, has renamed it *Gymnodinium microadriaticum* on the basis of the morphology of the motile phase. Thus, once again systematists must in some way weigh characters in classifying living organisms (see also the taxonomic appendix here).

The more characters a systematist can analyse in classifying an organism the more valid that classification should be. Thus, the knowledge of sexual life histories gives the systematist a vast array of new morphological information to help probe more accurately into phylogenetic relationships. There is no reason why the whole life cycle should not be taken into consideration for taxonomic purposes (F. Taylor 1979).

5 ENVIRONMENTAL CONTROL

Dinoflagellate life cycles are affected by a variety of environmental factors, with the major effects being changes in asexual growth rates and control of specific stages in sexual reproduction. Since the former topic is covered in another chapter (11), this section will focus primarily on environmental control of sexual reproduction.

5.1 Induction of sexuality

Environmental factors such as day length, temperature, nutrient depletion, light intensity, and dissolved gases have been suggested as possible causes of sexuality in dinoflagellates (von Stosch 1967). However, many studies conducted to date were designed to demonstrate sexuality for a particular species, with little or no emphasis on underlying mechanisms.

The most common culture manipulation that induces sexuality in autotrophic species is nutrient starvation, and more specifically, nitrogen depletion (Tables 14.3, 14.4). Typically, exponentially growing cells are removed from their growth medium and resuspended in similar medium with no added nitrogen (e.g. Pfiester 1975; Turpin *et al.* 1978). Deletion of this single nutrient has induced sexuality in seven of the fourteen species on which culture manipulations have been attempted. Two studies used simultaneous reductions in both N and P. Sexuality and cyst formation have also been observed in ageing cultures with no manipulations, presumably due to the depletion of one nutrient (Cao Vien 1967a,b; Von Stosch 1973; Pfiester 1976).

Table 14.3. Carbon and nitrogen pools during sexual induction of *G. tamarensis**

	Days after inoculation (batch culture)					
	0	6	7	8	9	12
Inoculum: from f/2 into f/2						
cell conc. (ml^{-1})	85	936		2840		4000 (no zygotes)
external N (μM)	883	686		592		
N/cell (pg/cell)	371	653		512		515
C/cell (pg/cell)	1505	3284		2510		1920
C/N (by weight)	4·1	5·1		4·9		3·7
Inoculum: from f/2 into f/2 with f/20 N						
cell conc. (ml^{-1})	85		1341		5010	4500 (zygotes present)
external N (μM)	70		15		—	0·2
N/cell (pg/cell)	371		519		189	188
C/cell (pg/cell)	1505		2692		1469	2584
C/N (by weight)	4·1		5·2		7·8	13·8

*Data from D. M. Anderson (unpubl.).

The time to gamete formation varies, but induction periods as short as 30 min or as long as 5–7 days have been reported, following the change in external nutrients (Pfiester 1975; Walker & Steidinger 1979). Once gametogenesis has occurred, addition of nutrients may have no effect on sexuality (Zingmark 1970; Von Stosch 1973; Pfiester 1976; Pfiester & Skvarla 1979), or it may cause the gametes to revert to vegetative cells (Von Stosch 1973; Pfiester 1975).

Since the objective of many studies was to document sexuality for a particular species, success with nitrogen starvation often precluded the testing of other nutrients. It would thus be incorrect to conclude from Table 14.4 that nitrogen is the most important nutrient in sexuality. For three species, (*Peridinium willei*, *Scrippsiella trochoidea*, and *Gonyaulax*/*Protogonyaulax tamarensis*), separate manipulations of nitrogen and phosphorus induced sexuality (Pfiester 1976; Watanabe 1981; Anderson *et al.* 1984; Anderson & Linquist 1985).

In most of the studies listed in Table 14.4, details of the nutrient depletion are lacking. External nutrient concentrations were not measured, nor were the internal storage pools of the inoculum cultures or the sexually active cells. Significant changes in cellular metabolism are associated with variations in nutrient supply—changes mediated in part by internal storage pools that enable cells to divide for several generations after external nutrients disappear (Fogg 1959). An example of the importance of stored reserves is seen in Table 14.3 for *G. tamarensis*. In cultures grown in nutrient rich f/2 medium (Guillard & Ryther 1962), sexuality does not occur, whereas it is readily induced in the same medium with 90% less nitrogen. The major changes in metabolism are: (i) a rapid increase in N/cell following inoculation (luxury uptake), with only the nitrogen-depleted culture showing a significant decrease thereafter; and (ii) a relatively constant C/N ratio in the presence of excess nutrients, but a sharp increase as external nitrogen disappears.

Similar changes in internal nutrient pools are possible with phosphorus as the limiting nutrient (Anderson *et al.* 1984; Anderson & Lindquist 1985), and also with changes in physical conditions such as temperature or light (Goldman 1977, 1979). In this context, it is noteworthy that Von Stosch (1964) induced sexuality in *Ceratium cornutum* without altering external nutrient levels but instead by decreasing temperature, day length and light intensity. Since three factors were varied concurrently, individual effects remain obscure. In a recent attempt to isolate the effects of low light levels on sexuality, Anderson *et al.* (1984) found that cyst formation of *G. tamarensis* was negligible in nutrient-replete medium, even with a 50% reduction in growth rate due to non-optimum lights. Recent field data suggest that day length is also not a factor in sexuality for this species since sexual stages have been observed during blooms in two estuaries 40 km apart but at times that differed by over 1 month (Anderson *et al.* 1983).

Temperature alone has never been shown to be a factor that induces sexuality directly, but it can alter the process once initiated by nutrient depletion. For example, Anderson *et al.* (1984) showed that encystment of *G. tamarensis* was more sensitive to temperature than was growth rate, with optimal cyst production occurring over a relatively narrow temperature range and no encystment at some temperatures that permitted growth. Apparently some metabolic process unique to gamete formation, fusion, or encystment requires higher temperatures than those that support division of this species.

Another factor recently linked to sexuality is dissolved CO_2. When *S. trochoidea* was stressed with separate N and P depletion, sexuality and cyst production were observed, but the addition of less than 1 μM bicarbonate enhanced the phenomenon significantly (Watanabe *et al.* 1982). Perhaps the change in total carbonate is important because this marine species forms calcitic

Table 14.4. Induction of sexuality

Organism	Method	Comments	References
Amphidinium carteræ (marine)	Ageing cultures		Cao Vien 1967
Ceratium cornutum (freshwater)	21–12°C LD 14/10–10/14 High light to low light	More recent data suggests reduced nutrients (N + P together) successful; also that reduced temperature and shorter day each work independently	von Stosch 1965; von Stosch, pers. comm.
Ceratium horridum (freshwater)	Ageing cultures	Produces gametes, but copulations relatively scarce	von Stosch 1972
Crypthecodinium cohnii (marine)	(a) None	(a) Heterotrophic nutrition; spontaneous induction of sexuality	Beam & Himes 1974, 1977
	(b) P + N depletion	(b) Nutrient depletion implicated	Tuttle & Loeblich 1974
Glenodinium lubiniensiforme (freshwater)	Nutrient depletion; ageing cultures	Best results from older, high nutrient cultures at high light intensities	Diwald 1937
Gonyaulax monilata (marine)	N-depletion	P-depletion not tested	Walker & Steidinger 1979
Gonyaulax tamarensis (marine)	(a) N-depletion	(a) Zygotes formed were irregular and warty; unsuccessful transition to resting cyst	Turpin *et al.* 1978
	(b) N-depletion; P-depletion	(b) Viable resting cysts formed	Anderson *et al.*, in press
Gymnodinium fungiforme (marine)	Decrease in concentration of algal food source	Phagotrophic nutrition	Spero & Moree 1981
Gymnodinium pseudopalustre (freshwater)	21–15°C LD 14/10–10/14	More recent data suggest N and P depletion successful; some sexuality in old cultures	von Stosch 1973; von Stosch, pers. comm.
Helgolandinium subglobosum (marine)	Ageing cultures		von Stosch 1972; von Stosch, pers. comm.
Noctiluca scintillans (marine)	None	Phagotrophic nutrition; spontaneous induction of sexuality	Zingmark 1970
Oxyrrhis marina (marine)	Change in food source from *Pyraminonas* to *Dunaliella*	Phagotrophic nutrition	von Stosch 1972; von Stosch, pers. comm.
Peridinium cinctum (freshwater)	N-depletion; P-depletion	Light necessary; other nutrients tested, but N depletion gave optimal response	Pfiester 1975
Peridinium gatunense (freshwater)	N-depletion	Other nutrients tested, but N depletion gave optimal response	Pfiester 1977

Peridinium volzii (freshwater)	N-depletion	Other nutrients tested, but N depletion gave optimal response	Pfiester & Skvarla 1979
Peridinium willei (freshwater)	N-depletion	Other nutrients tested, but N depletion gave optimal response	Pfiester 1976
Polykrikos kofoidi (marine)	None	Phagotrophic nutrition; spontaneous induction of sexuality	Morey-Gaines & Ruse 1980
Scrippsiella trochoidea (= *Peridinium trochoidium*) (marine)	(a) N-depletion; P-depletion; Vitamin and $NaHCO_2$ additions	(a) Encystment induced by either N or P starvation; addition of $NaHCO_3$ beneficial; vitamin additions inhibitory. Light and temperature optima were same for encystment as for maximum vegetative growth.	Watanabe *et al.* 1982
	(b) None	(b) Spontaneous induction of sexuality	Wall *et al.* 1970
Woloszynskia apiculata (freshwater)	Lower light (14 000–2000 lux) Reduced N + P		von Stosch 1973

cysts (Wall *et al.* 1970), but there are other possible effects, including those associated with changes in alkalinity that must be considered and evaluated for other organisms.

For some species, sexuality does not require direct environmental changes, although mostly heterotrophic or phagotrophic species are included in this category thus far. Zingmark (1970) observed gamete fusion and zygote formation in *Noctiluca scintillans* cultures fed on *Dunaliella*. Morey-Gaines & Ruse (1980) reported spontaneous cyst formation in phagotrophic *Polykrikos* cultures, while Beam & Himes (1974, 1980) describe sexuality in heterotrophic *Crypthecodinium cohnii* cultures without manipulation (Tuttle & Loeblich (1974) used N and P depletion to induce sexuality in the same species). Sexuality in a third phagotrophic species, *Gymnodinium fungiforme*, was recently observed following depletion of the algal food source (Spero & Moree 1981).

Experiments by Wall *et al.* (1970) demonstrated that the photosynthetic dinoflagellate *Peridinium trochoideum* (= *Scrippsiella trochoidea*) formed cysts in culture without any apparent stress. Although details of the induction experiments were not given, the authors suggested that for this species, encystment does not occur in response to an adverse environment but is a naturally occurring stage in the life history, favoured rather than inhibited by optimal conditions for vegetative growth. Watanabe *et al.* (1982) also demonstrated encystment in control (nutrient-rich) cultures of this same species, but increased the magnitude of the encysting fraction of the population by using nutrient limitation. Thus, an apparent spontaneous encystment process could be enhanced by nutrient depletion.

Field observations linking sexual stages to environmental parameters are limited, but there is evidence for a succession in encystment of different species throughout the year, especially in temperate regions. West (1909) found a strong correlation between temperature and encystment for four freshwater dinoflagellates, concluding that two species were 'winter forms' that encysted in response to the vernal rise of temperature while the other two were dominant during the summer and encysted during the autumnal temperature decrease. There may also be a difference between tropical and temperate species in this regard (Chapter 11B). Wall & Dale (1968) observed this successional phenomenon for marine species. They reported cysts in the water column during the declining stages of three dinoflagellate blooms, but did not attempt to explain the reasons for the declines. Anderson & Morel (1979), reported cyst formation for natural populations of *G. tamarensis* in the presence of relatively high levels of nitrate ($>5\,\mu M$) and with phosphate consistency above $0{\cdot}1\,\mu M$. It is not known whether such levels are limiting to this organism.

Despite the emphasis on nutrient depletion in experimental studies to date, it seems that a variety of factors could be important in the induction of sexuality in dinoflagellates. In a search for common mechanisms, it is difficult to reconcile reports of sexuality in some species under apparently optimal growth conditions with the need for nutrient stress in other species. In this respect it is noteworthy that Watanabe *et al.* (1982) observed cyst formation in nutrient-rich control cultures of *Scrippsiella trochoidea* but enhanced the process through nutrient limitation. It seems prudent to emphasize that: (i) all studies to date have been conducted in batch cultures where growth conditions may not be as 'optimal' as they appear; (ii) that no measurements of the physiological status of the cells has been provided; and (iii) that within any culture there may well be some cells growing slower and responding differently to their environment than the bulk of the population. Until these issues are addressed, spontaneous sexuality without an applied stress must remain an unproven hypothesis.

The bulk of the reports of sexuality in dinoflagellates involve changes in growth conditions. One possibility is that gametogenesis occurs when cellular nitrogen or phosphorus drops to a critical or threshold level, either through a change in uptake rate due to low ambient concentrations, or to a change in the relative rates of uptake, storage, and metabolism under environmental stress. In *G. tamarensis* there are indications that when levels fall below the subsistence cell quota of 27 pg $cell^{-1}$ phosphorus, gametes may be induced (Anderson & Lindquist 1985). Since several species can be forced into sexuality by at least two different stresses (N and P starvation) (Pfiester 1976; Watanabe *et al.* 1982; Anderson *et al.* 1984) such a direct link between sexuality and nutrient limitation requires that the pathways for uptake and utilization of nitrogen or phosphorus each include a mechanism for the shift to sexual metabolism.

Alternatively, it is tempting to hypothesize that general conditions resulting in slower growth induce sexuality, without specific emphasis on nutrients. Thus, the control of sexuality would lie at a stage more closely associated with cellular division. Although this has considerable appeal from an ecological standpoint, it is already evident from Table 14.3 that populations in nutrient-rich batch cultures have reduced growth rates as they reach maximum cell concentrations, but are not induced to sexuality. In this case, at least some of the non-nutritional factors that limit growth in the presence of excess nutrients (e.g. pH, self-shading, or excreted metabolites) do not trigger a sexual response.

One hypothesis advanced to explain the induction of gametogenesis in a green alga is consistent with the range of observations described above, but has yet to be tested for dinoflagellates. Cain & Trainor (1976) argued that sexual induction in *Scenedesmus obliquus* was inhibited by nitrogen concentrations above a specific threshold, but readily occurred at lower levels that were apparently not limiting. In other words, no gametes were formed at high nutrient concentrations, but they were plentiful as nutrients dropped below a critical level that seemed sufficient for continued vegetative division.

The key concept here is the 'limiting' concentration of nutrients. The uptake characteristics of cyst-forming dinoflagellates are not yet known, but it is probable that at the optimal temperatures typically used for batch culture studies, growth rate can exceed the rate of nutrient uptake at seemingly adequate extra-cellular concentrations (Anderson & Lindquist 1985). The cells could thus be dividing at the expense of internal nutrient pools until they reach a threshold cell quota that triggers sexuality. What at first glance appears to be sexuality without stress might in actuality be a response to nutrient limitation caused by the dinoflagellate's relatively inefficient nutrient uptake capability.

It is evident from the foregoing discussion that the links between nutrient uptake, internal storage pools and sexuality are important and complex. Despite the difficulties in growing dinoflagellates in continuous cultures, a detailed understanding of these processes awaits such studies in which the cells can be maintained under physiological and environmental steady state conditions.

In overview, the environmental control of sexuality in many dinoflagellate species is both intuitively appealing and consistent with most laboratory results. Studies on field populations are too few, however, and the laboratory experiments too limited in scope to define the detailed mechanisms. Since spontaneous encystment has been reported (i.e. without any apparent stimulus), internal control such as that from a metabolic clock may be important for some species. However, it is still likely that such a process would be mediated in part by light, temperature or other environmental factors. Resolution of the complex interactions between external nutrients, internal storage pools, and the mediating effects of physical variables such as temperature, salinity or light require more comprehensive studies than have been attempted in the past.

5.2 Encystment and dormancy

Once sexuality is induced and gametes have fused, the product is generally a swimming zygotic cell (planozygote). Cao Vien (1967a, 1968) reported a non-motile cell immediately after fusion in *Amphidinium carteriæ*. In most cases, dinoflagellate planozygotes are large and deeply pigmented, but little is known of their physiology or the factors affecting longevity. In most cases, this cell swims for a variable length of time before becoming a non-motile resting cyst. The swimming stage can be as short as 3–5 days when sexuality leads directly to either division and haploid vegetative cells (e.g. *C. horridum*; von Stosch 1972) or to a very short dormancy in hypnozygote form (*P. gatunense*; Pfiester 1977). When a long-lived resting cyst or hypnozygote is the result, the planozygote swims from 1 to 3 weeks (Von Stosch 1973; Pfiester 1975, 1976, 1977; Walker & Steidinger 1979; Anderson *et al.* 1984). The reasons for this prolonged swimming phase without division are unknown but presumably relate to the metabolic changes associated with transition from active photosynthesis and nutrient uptake to prolonged dormancy. No experiments have examined the effects of environmental variations on this life cycle stage.

It should be emphasized that not every resting cyst need be sexual. H. A. Von Stosch (pers. comm.) reports male haploid 'parthenozygotes' in *Ceratium cornutum* cultures, and Happach-Kasan (1980) describes vegetative cysts in *Ceratium hirundinella*.

It is important to define the terminology of dormancy carefully, recognizing that much of the dinoflagellate cyst literature is less rigorous. Borrowing from the terminology used for seeds of higher plants (Jann & Amen 1977), 'dormancy' is considered the suspension of growth by active endogenous inhibition, and 'quiescence' as the suspension of growth by unfavourable environmental conditions. Thus, it is not possible to germinate a dormant cyst until a mandatory resting period is completed. Similarly, a quiescent cyst cannot germinate until an applied external constraint (such as cold temperature) is removed.

Once a resting cyst is formed, the length of dormancy is highly variable (Table 14.5). In *P. gatunense*, the hypnozygote can excyst within 12 hours of formation (Pfiester 1977). In most other species, the mandatory dormancy period lasts from several weeks to 6 months, during which germination is either not possible under optimal conditions or occurs at a very low rate (Wall & Dale 1968; Dale *et al.* 1978; Anderson 1980). The effect of temperature on this process can be significant. For *G. tamarensis*, the first germination of new cysts was possible after 1·5 months of storage at 22°C or after 6 months at 5°C (Anderson 1980). In contrast, maturation of *S. trochoidea* cysts is equally long (3 weeks) at high and low temperatures (B. J. Binder, pers. comm.).

The factors that initiate excystment are obscure since previous studies have emphasized the process of sexuality rather than the underlying mechanisms. When cysts of several freshwater dinoflagellate species were stored continuously under normal culture conditions, spontaneous germination occurred after 2–4 months (Von Stosch 1973; Pfiester 1975, 1976, 1977; Pfiester & Skvarla 1979). This type of storage has not been tested for other species, so the generalization that excystment simply occurs without stimulus if suitable conditions are continuously maintained is not yet justified.

Several species have been stored at cold (4–5°C) or high (22°C) temperatures and tested for germination by raising or lowering the temperature thereafter (*P. cinctum* and *C. hirundinella* (Huber & Nipkow 1922, 1923); *C. cornutum* (Von Stosch 1965); *G. pseudopalustre* and *W. apiculata* (Von Stosch 1973); *G. tamarensis* (Anderson & Wall 1978; Anderson 1980). One generalization holds for all but one of the species tested: storage of cysts at low temperatures maintains quiescence until the temperature is increased. One exception is *C. cornutum* which, after a mandatory resting period, excysted spontaneously at 4°C in a laboratory refrigerator (H. A. von Stosch, pers. comm.).

Long-term survival is possible at low temperatures. Huber & Nipkow (1922, 1923) found that constant exposure to cold temperatures in deeper sections of Lake Zurich suppressed cyst germination of *P. cinctum* for as long as 16·5 years. The only species tested at high temperatures (22°C) was *G. tamarensis*, which remained quiescent for 1 year as storage conditions were held constant (Anderson & Morell 1979; Anderson 1980).

Table 14.5. Dormancy and excystment

Organism	Method	Comments	References
Ceratium cornutum (freshwater)	Cold storage (0–4°C)	Spontaneous excystment; even in cold; laboratory cultures	von Stosch 1965; von Stosch, pers. comm.
Ceratium hirundinella (freshwater)	Temperature change (0 to 4–7°C successful, but 0 to 20°C optimal)	Excystment of planktonic cysts retarded for 1½ months, optimal after 7·5 months; excystment rate of cysts from sediments accelerated by high temperature; older cysts took longer to excyst; no effect from water type, potassium, salinity or light; excystment retarded by 5% glucose or various qualities of monochromatic light; cysts killed by freezing or drying.	Huber & Nipkow 1922, 1923; Morey-Gaines & Ruse 1980
Gonyaulax tamarensis (marine)	Temperature increase (4–15°C) Temperature decrease	Cysts from sediments were sonicated prior to testing; no effect from nutrients, or light; constant low or high temperature maintains dormancy; intermediate temperature not tested; mandatory dormancy 1–6 months, depending on storage temperature.	Anderson 1980; Anderson & Wall 1978; Anderson & Morell 1979
Gymnodinium pseudopalustre (freshwater)	No change from normal culture conditions; or storage at 3°C followed by higher light, temperature	Some spontaneous excystment after several weeks; more complete and better synchronized encystment if cold conditioned; laboratory cultures.	von Stosch 1973
Noctiluca scintillans (marine)	Normal culture conditions	Zygotes remain non-motile for several days without visual development, then enlarge; diplontic life history.	Zingmark 1970
Peridinium cinctum (freshwater)	No change from normal culture conditions; also tested storage at 4°C then increase to 20°C	Spontaneous excystment after 7–8 weeks with constant 20°C; no germination at 4°C for 5 months; laboratory cultures.	Pfiester 1975
Peridinium gatunense (freshwater)	No change from normal culture conditions	Spontaneous excystment after 12 h; laboratory cultures.	Pfiester 1977

Peridinium volzii (freshwater)	No change from normal culture conditions	Spontaneous excystment after 4 months; laboratory cultures.	Pfiester & Skvarla 1979
Peridinium willei (freshwater)	No change from normal culture conditions	Spontaneous excystment after 7–8 weeks; laboratory cultures.	Pfiester 1976
Woloszynskia apiculata (freshwater)	No change from normal culture conditions; also tested cold storage at 6°C	Some spontaneous excystment, but more complete and better synchronized with cold storage	von Stosch 1973

As demonstrated by Von Stosch (1973), cysts of some species are capable of limiting germination several weeks after formation, without an external stimulus but storage in the cold results in more complete and better synchronized excystment when warmer temperatures are restored. This emphasizes another common characteristic: increasing temperatures above cold storage levels breaks quiescence. In a similar manner, *G. tamarensis* cysts stored at high temperatures excyst if the temperature is lowered into an optimum range (Anderson & Morell, 1979; Anderson 1980). This is consistent with the existence of a temperature 'window' for excystment. High or low temperatures maintain quiescence, but once the temperature drops or rises into the permissive range, germination occurs.

Temperature is thus very important throughout dormancy. It can maintain quiescence for extended periods, determine the duration of dormancy after cyst formation, synchronize or entrain cyst populations for more uniform germination, and initiate the excystment process.

The effects of other environmental factors on dormancy and excystment are poorly understood. The first and most comprehensive study was that of Huber & Nipkow (1922, 1923) on *C. hirundinella* cysts from lake sediments. They found that manipulation of growth medium (well water, distilled water, filtered lake water, dilutions of KNO_3, Knop's and Kleb's artificial media, weakly saline medium) and incubation in the dark had little effect on the germination of cysts. A 5% glucose solution (as a non-electrolyte) and various qualities of monochromatic light (green, blue and red) retarded germination. Freezing and drying killed the cysts. These researchers also noted the adverse effect of non-optimum temperatures on the viability and shape of the germinated cells.

Another study examined the effects of light, nutrients, and trace metals on germination of *G. tamarensis* cysts (Anderson & Wall 1978). Unlike Huber & Nipkow (1922, 1923) who placed small quantities of sediment directly into culture medium, experiments with *G. tamarensis* used cysts isolated by micropipette from sediment samples after sonication and size fractionation (Wall & Dale 1968). These methods may have introduced artefacts associated with sonication, bacterial regeneration of nutrients, and light exposure during isolation. Laboratory incubation at 16°C after storage for 6 months at 5°C initiated excystment with no appreciable effect from light regime, nutrient concentration, or even toxic metal concentrations. Motility and viability of the excysted germlings, however, required highly chelated medium and light. In general, once initiated by the temperature increase, excystment proceeded through emergence regardless of the suitability of the ambient environment.

In comparing these various laboratory studies with the few field observations that have been made on cyst populations, one discrepancy is noteworthy. For certain species, between 80 and 100% of the cysts isolated in the laboratory will germinate during incubation at higher temperature (Von Stosch 1973; Anderson & Wall 1978; Anderson & Morell 1979). Examination of sediments during blooms when temperatures have risen significantly reveals an abundance of viable cysts which will excyst rapidly when brought to the laboratory (Anderson & Morell 1979). It seems clear that there are factors that can override the temperature stimulus after overwintering and prevent excystment (e.g. anoxia), and/or that laboratory incubations are not adequate simulations of natural conditions. Here again the resolution of this important issue awaits further study.

There are also environmental factors that influence sexuality in dinoflagellates by operating indirectly on the behaviour or physical location of the organisms. One example is light, which induces either positive or negative photoactic movement (see Chapter 10) and thus acts to concentrate a population in dense accumulations at specific depths. On theoretical grounds, this could facilitate gamete pairing and may even contribute to the rapid decrease of external nutrients. Gametes of *W. apiculata* are negatively photoactic (Von Stosch 1973), and thus would accumulate away from a light source, while gametes of the other species that were examined exhibit the same positive photoaxis as vegetative cells and accumulate near the highest intensity (H. A. Von Stosch, pers. comm.).

A variety of hydrographic or wind-related features also accumulate dinoflagellates in surface or subsurface patches. An example of the importance of these mechanisms on a cyst-forming dinoflagellate is seen for *Gyrodinium uncatenum* in the Potomac River. In this system, a persistent estuarine front serves to concentrate and recirculate the motile population in downstream-flowing surface waters and upstream-flowing bottom waters (Tyler *et al.* 1982). Sexual stages also accumulate within this 'conveyor belt', and the resulting cysts accumulate in highest numbers along the subsurface transport pathway. Clearly these frontal regions and convergence zones are areas of increased sexual activity.

If we examine the accumulated information on environmental control of sexuality in dinoflagellates, it is clear that generalizations are not readily apparent. This is due in part to the paucity of specific experimental data and partly to the difference between species. Sexuality is not a simple, automatic response to adverse environmental conditions. It involves complex interaction between physical or chemical parameters and the metabolism of the organisms. Conceivably there are a variety of combinations of these factors that produce the physiological state that dictates the shift into sexual metabolism. Thus, low temperatures with high nutrient concentrations may have the same effect as the combination of high temperatures and low nutrients. Variations in these interactions at the species level can further explain the temporal distribution and succession of co-occurring organisms.

The environmental control of dormancy and excystment is equally complex. Temperature seems to be the most obvious controlling parameter, but it is clear that other factors contribute to the timing and magnitude of germination as well. The existence of living dinoflagellate species whose cysts can be found in early Tertiary sediments (Wall & Dale 1970) or earlier, is strong evidence of the effectiveness of sexuality and cyst formation in responding to short- and long-term environmental fluctuations.

ACKNOWLEDGEMENTS

We thank Dr Jack Holt for his original figures without which this chapter would have been incomplete. This article was prepared while the second author was supported by National Science Foundation grant OCE-8011039. This is contribution No. 5620 from the Woods Hole Oceanographic Institution.

6 REFERENCES

ANDERSON D. M. (1980) Effects of temperature conditioning on development and germination of *Gonyaulax tamarensis* (Dinophyceae) hypnozygotes. *J. Phycol.* **16**, 166–172.

ANDERSON D. M., CHISHOLM S. W. & WATRAS C. J. (1983) Importance of life cycle events in the population dynamics of *Gonyaulax tamarensis*. *Marine Biology* **76**, 179–189.

ANDERSON D. M., KULIS D. M. & BINDER B. J. (1984). Sexuality and cyst formation in the dinoflagellate *Gonyaulax tamarensis*: I. Cyst yield in batch cultures. *J. Phycol.* **20**, 418–425.

ANDERSON D. M. & LINDQUIST N. L. (1985). Sexuality and cyst formation in the dinoflagellate *Gonyaulax tamarensis*: II. Time course measurements of phosphorus limitation. *J. exp. mar. Biol. Ecol.* **86**, 1–13.

ANDERSON D. M. & MORELL F. M. M. (1979) The seeding of two red tide blooms by the germination of benthic *Gonyaulax tamarensis* hypnocysts. *Estuarine Coastal Mar. Sci.* **8**, 279–93.

ANDERSON D. M. & WALL, D. (1978) Potential importance of benthic cysts of *Gonyaulax tamarensis* and *G. excauata* in initiating toxic dinoflagellate blooms. *J. Phycol.* **14**, 224–234.

BEAM C. & HIMES M. (1974) Evidence for sexual fusion and recombination in the dinoflagellate *Crypthecodiium* (*Gyrodinium*) *cohnii*. *Nature* (*Lond.*) **250**, 435–436.

BEAM C. A. & HIMES M. (1980) Sexuality and meiosis in dinoflagellates. In *Biochemistry and Physiology of Protozoa* (eds. M. Levandowsky & S. H. Hutner), Vol. 3, pp. 171–206. Acad. Press, Inc., New York.

BEAM C. A., HIMES M., HIMELFARB J., LINK C. & SHAW K. (1977) Genetic evidence of unusual meiosis in the dinoflagellate *Crypthecodinium cohnii*. *Genetics* **87**, 19–32.

BIECHELER B. (1952) Recherches sur les Péridiniens. *Bull. Biol. France et Belg.*, Suppl. 36, 1–149.

BOLD H. C. & WYNNE M. J. (1978) *Introduction to the Algae*. Prentice-Hall, Inc., Englewood Cliffs, N.J.

BORGERT A. (1910) Kern und Zellteilung bei marinen *Ceratium*-Arten. *Arch. Protistenk.* **20**, 1–46.

BOURRELLY P. (1968) Note sur les Peridiniens d'eau douce. *Prostistologica* **4**, 5–14.

BRAARUD T. (1945) Morphological observations on marine dinoflagellate cultures (*Parella perfarta, Gonyaulax tamarensis, Protoceratium reticulatum*). *Avh. Norsk. Vidensk. Acad. Oslo, Nat-Natur. Klasse*, 1944. No. 11 p. 1–18.

BRAARUD T. (1957) Observations on *Peridinium trochoideum* (Stern) Linn. in culture. Cell division and size variation; encystment. *Nytl. Mag. Botanikk.* **6**, 39–42.

CAIN J. R. & TRAINOR F. R. (1976) Regulation of gametogenesis in *Scenedesmus obliquus* (Chlorophyceae). *J. Phycol.* **12**, 383–390.

CAO VIEN M. (1967a) Sur l'existence de phénomènes sexuels chez un peridinien libre, *l'Amphidinium carteri. C. R. Acad. Sci. Ser. D.* **264**, 1006–1008.

CAO VIEN M. (1967b) Un mode particular de multiplication vegetative chez un *Peridinien* libre, le *Prorocentrum micans* Ehrenberg. *C. R. Acad. Sci (Paris)* **265** D, 108–110.

CAO VIEN, M. (1968) Sur la germination du zygote et sur un mode particular de multiplication vegetative chez le peridinien libre, *l'Amphidinium carteri. C. R. Acad. Sci. Ser. D.* **267**, 701–703.

CARTER H. J. (1858) Note on the red colouring matter of the sea round the shores of the island of Bombay. *Ann. Mag. Nat. Hist. Ser 3*, **1**, 258–262.

CHAPMAN D. V., LIVINGSTONE D. & DODGE J. D. (1981) An electron microscope study of the excystment and early development of the dinoflagellate *Ceratium hirundinella Br. Phycol. J.* **16**, 183–194.

CHATTON E. (1920) Les Peridiniens parasites, morphology, reproduction, ethologie. *Arch. Zoo. exp. gen.* **59**, 1–475.

CHATTON M. E. & BIECHELER B. (1936) Documents nouveaux Relatifs aux Coccidinides (Dinoflagelles parasites). La sexualite du *Coccidinium Mesnili* n. sp. *C. R. Acad. Sci. Ser. D.* **203**, 573–576.

DALE B. (1977) Cysts of the toxic red tide dinoflagellate *Gonyaulax excavata* (Braarud) Balech from Olsofjorden, Norway. *Sarsia* **63**, 29–34.

DALE B., YENTSCH C. M. & HURST J. W. (1978) Toxicity in resting cysts of the red-tide dinoflagellate *Gonyaulax excavata* from deeper water coastal sediments. *Science* **201**, 1223–1225.

DIWALD K. (1937) Die ungeschlechtliche und geschlechtliche fortpflanzung von glenodiium lubinensiforme spec. nov. *Flora* **132**, 174–192.

DODGE J. D. & CRAWFORD R. M. (1970) A survey of thecal fine structure in the Dinophyceae. *Bot. J. Linn. Soc.* **63**, 53–67.

DURR G. & NETZEL H. (1974) The fine structure of the cell surface in *Gonyaulax polyedra* (Dinoflagellata). *Cell Tiss. res.* **150**, 21–41.

DREBES G. (1981) Possible resting spores of *Dissodinium pseudolunula* (Dinophyta) and their relation to other taxa. *Br. Phycol. J.* **16**, 207–215.

EHRENBERG C. G. (1838) *Die Infusionsthierchen als vollkommene Organismen. Ein Blick in das Tiefere organische Leber der Natur.* Leopold Voss, Leipzig.

ENTZ G. (1921) Uber diemitotische Teilung von *Ceratium hirundinella. Arch. Protistenk.* **43**, 415–430.

FOGG G. E. (1959) Nitrogen nutrition and metabolic patterns in algae. In *Symposia Soc. Exp. Biol.* (ed. H. K. Porter), **13**, 107–125.

FREUDENTHAL H. D. (1962) *Symbiodinium* gen. nov. and *Symbiodinium microadriaticum* sp. nov., a Zooxanthella, Taxonomy, Life Cycle, and Morphology. *J. Protozool.* **9**, 45–52.

FRITSCH F. E. (1935) *The Structure and Reproduction of the Algae*, Vol. 1. Cambridge University Press, Cambridge.

GOLDMAN J. C. (1977) Temperature effects on phytoplankton growth in continuous culture. *Limnol. Oceanogr.* **22**, 932–936.

GOLDMAN J. C. (1979) Temperature effects on steady state growth, phosphorus uptake, and the chemical composition of a marine phytoplankter. *Microbial Ecology* **5**, 153–166.

GRELL K. G. (1973) *Protozoology*. Springer-Verlag, Berlin.

GUILLARD R. R. L. & RYTHER J. H. (1962) Studies of marine planktonic diatoms. I. *Cyclotella nana* Huystedt and *Detonula confervacea* (Cleve). *Gran. Can. J. Microbiol.* **8**, 229–239.

HAPPACH-KASAM C. (1980) Beobachtungen zur Entwicklungsgeschichte der Dinophycee *Ceratium cornutum* Sexualitat. Dissertation, Philipps Univ., Marburg.

HIMES M. & BEAM C. A. (1975) Genetic analysis in the dinoflagellate *Crypthecodinium cohnii.* Evidence for unusual meiosis. *Proc. Nat. Acad. Sci. USA* **72**, 4546–4549.

HOEFKER I. (1930) Uber *Noctiluca scintillans* (Macartney). *Arch. Protistenk.* **71**, 57–78.

HOLT J. R. & PFIESTER L. A. (1982) A technique for counting chromosomes of armored dinoflagellates, and chromosome numbers of six freshwater dinoflagellate species. *Amer. J. Bot.* **69**, 1165–1168.

HUBER G. & NIPKOW F. (1922) Experimentelle untersuchungen uber die entwicklung von *Ceratium hirundinella* O.F.M. *Zeitschrift fur Botanik* **14**, 337–371.

HUBER G. & NIPKOW F. (1923) Experimentelle untersuchungen uber die entwicklung und formbildung von *Ceratium hirundinella* O.F.M. *Flora, Jena* **116**, 114–215.

JANN R. C. & AMEN R. D. (1977) What is germination? In *The Physiology and Biochemistry of Seed Dormancy and Germination* (ed. A. A. Kuhn), pp. 7–28. Elsevier, North Holland.

JOSEPH G. (1879) Ueber Grotten-Infusorien Zool. *Anz.* **2**, 114–118.

KOFOID C. A. (1906) On the structure of *Gonyaulax triacantha*. *Jörg. Zool. Anzeigen.* **30**, 102–105.

KOFOID C. A. (1911) The genus *Gonyaulax* with notes on its skeletal morphology and a discussion of its generic and specific characters. *Univ. Calif. Publ. Zool.* **8**, 187–249.

KOFOID C. A. (1926) On *Oxyphysis oxytoxoides* gen. nov., sp. nov. a dinoflagellate convergent toward the peridinioid type. *Univ. Calif. Publ. Zool.* **28**, 203–216.

KOFOID C. A. & RIGDEN E. J. (1912) A peculiar form of schizogeny in *Gonyaulax*. *Bull. Mus. Comp. Zool. Harvard.* **54**, 335–48.

KOFOID C. A. & SKOGSBERG T. (1928) Reports on the scientific results of the expedition to the Eastern Tropical Pacific. The Dinophysioideae. *Mem. Mus. Comp. Zool. Harvard Coll.* **51**.

KUBAI D. F. & RIS H. (1969) Division in the dinoflagellate *Gyrodinium cohnii* (Schiller). A new type of nuclear reproduction. *J. Cell. Biol.* **40**, 508–28.

KUDO R. R. (1966) *Protozoology*. 5th Edn. 1, 174 pp. Charles C. Thomas, City.

LAUTERBORN (1895).

LEADBEATER B. & DODGE J. D. (1967) An electron microscope study of nuclear and cell division in a dinoflagellate. *Archiv. fur Mikrobiol.* **57**, 239–254.

LEBOUR M. V. (1925) *The Dinoflagellates of the Northern Seas*. Marine Biol. Ass. U.K. Plymouth. 250pp.

LEFÈVRE M. (1932) Monographie des especes d'eau douce du genre *Peridinium*. *Arch. Bot.* **2**, 1–210.

LEGER G. (1973) Diatomees et Dinoflagelles de la mer Ligure. Systematique et distribution en juliet 1963. *Bull. Inst. oceanogr. Monaco*, **71** (1425), 1–36.

LINDEMANN (1929).

LOEBLICH A. R. III (1969) The amphiesma or Dinoflagellate Cell covering. *North American Palaeontological Convention*. Part G. pp. 867–929.

LOEBLICH A. R. III (1970) The amphiesma or dinoflagellate cell covering. *Proc. N. Amer. Paleontol. Conv.*, Part G, 867–929.

LOPER C. L., STEIDINGER K. A. & WALKER L. M. (1980) A simple chromosome spread technique for unarmored dinoflagellates and implications of polyploidy in Algal Cultures. *Trans. Amer. Micros. Soc.* **99**, 343–346.

MOREY-GAINES G. & RUSE R. H. (1980) Encystment and reproduction of the predatory dinoflagellate *Polykrikos kofoidii* Chatton (Gymnodiniales). *Phycologia* **19**, 230–236.

NIE D. & WANG C. C. (1942) Dinoflagellata of the Hainan region, V. On the thecal morphology of the genus *Goniodoma*, with description of species of the region. *Sinensia* **13**, 41–48.

OAKLEY B. R. & DODGE J. D. (1974) Kinetochores associated with the nuclear envelope in the mitosis of a dinoflagellate. *J. Cell Biol.* **63**, 322–325.

PAREDES J. F. (1962) A brief comment on *Peridinium nudum* Meunier. *Mem. Junta Invest. Ultramar. Estud. Biol. Marit. Lisboa, Ser. 2*, **33**, 117–120.

PASCHER A. (1930) Eine neue, stigmatisierte und phototakische Amoebe Biol. Centralbl. **50**, 1–7.

PAVILLARD J. (1915) Accroissement et scrissipartie chez les peridiniens. *C. R. Acad Sci. Paris* **160**, 372–375.

PFIESTER L. A. (1974) *Effects of nitrogen on asexual and sexual reproduction of* Peridinium cinctum *f.* ovoplanum *Lindemann*. PhD. Thesis, The Ohio State Univ. Columbus.

PFIESTER L. A. (1975) Sexual reproduction of *Peridinium cinctum* f. *ovoplanum* (Dinophyceae). *J. Phycol.* **11**, 259–265.

PFIESTER L. A. (1976) Sexual reproduction of *Peridinium willei* (Dinophyceae). *J. Phycol.* **12**, 234–238.

PFIESTER L. A. (1977) Sexual reproduction of *Peridinium gatunense* (Dinophyceae). *J. Phycol.* **13**, 92–95.

PFIESTER L. A. & POPOVSKY J. (1979) Parasitic, amoeboid dinoflagellates. *Nature* (*Lond.*) 421–424.

PFIESTER L. A. & SKVARLA J. J. (1979) Heterothallism and thecal development in the sexual life history of *Peridinium volzii* (Dinophyceae). *Phycologia* **18**, 13–18.

PFIESTER L. A. & SKVARLA J. J. (1980) Comparative ultrastructure of vegetative and sexual thecae of *Peridinium limbatum* and *Peridinium cinctum* (Dinophyceae). *Amer. J. Bot.* **67**, 955–958.

PFIESTER L. A. & LYNCH R. A. (1980) Amoeboid stages and sexual reproduction of *Cystodinium bataviense* and its similarity to *Dinococcus* (Dinophyceae).

POUCHET G. (1883) Contribution a l'histoire des Cilio-flagelles. *Journ. Anat. Physiol.* **19**, 399–455.

RIS H. & KUBAI D. F. (1974) An unusual mitotic mechanism in the parasitic protozoan *Syndinium* sp. *J. Cell Biol.* **60**, 702–720.

SCHILLING A. J. (1913) Dinoflagellatae. In *Die Süsswasserflora Deutschland, Österreichs und der Schweiz*. **3**, 1–66.

SCHÜTT (1895) *Die Peridineen der Plankton-Expedition. Ergebn. Plankton-Exped. Humboldt-Stiftung*, **4**, 1. Kiel & Leipzig.

SHYAM R. & SARMA Y. S. K. K. (1978) Cytology of Indian freshwater Dinophyceae. *Brit. J. Linn. Soc.* **76**, 145–159.

SMITH G. (1950) *The Freshwater Algae of the United States*. McGraw-Hill, N.Y.

SOUSA & SILVA E. DE (1962) Some observations on marine dinoflagellate cultures. *Bot. Mar.* **3**, 75–100.

SOUSA & SILVA E. DE (1969) Cytological aspects on multiplication of *Goniodoma* sp. *Bot. Mar.* **12**, 233–243.

SPECTOR D. & TRIEMER R. E. (1979) Ultrastructure of the dinoflagellate *Peridinium cinctum* f. ovoplanum. I. Vegetative cell ultrastructure. *Amer. J. Bot.* **66**, 845–50.

SPECTOR D. L., PFIESTER L. A. & TREIMER R. E. (1981) Ultrastructure of the dinoflagellate *Peridinium cinctum* f. *ovoplanum*. II. Light and electron microscope observations on fertilization. *Amer. J. Bot.* **68**, 34–43.

SPERO H. J. & MOREE M. D. (1981) Phagotrophic feeding and its importance to the life cycle of the holozoic dinoflagellate, *Gymnodinium fungiforme*. *J. Phycol.* **17**, 43–51.

STEIDINGER K. & COX E. R. (1980) Free-living dinoflagellates. In *Phytoflagellates* (ed. E. R. Cox) Elsevier, New York. pp. 407–432.

STEIN F. R. (1883) *Der Organismus der Infusionsthiere*. III Abt. II. Halfte, Leipzig. 30 pp.

STOSCH H. A. VON (1964) Zum Problem der sexuellen Fortpflanzung in der Peridineengattung *Ceratium. Helgolander Wiss. Meeresunters.* **10**, 140–152.

STOSCH H. A. VON (1965) Sexualitat bei *Ceratium cornutum* (Dinophyta). *Naturwiss.* 52, 112–113.

STOSCH H. A. VON (1969) Dinoflagellaten ausder Nordsee. II. *Helgolandium subglobosum* gen. et spec. nov. *Helgol. Wiss. Meeresunters.* **19**, 569–77.

STOSCH H. A. VON (1972) La signification cytologique de la 'cyclose nucléaire' dans le cycle de vie des Dinoflagallés. *Mem. Soc. Bot. Fr.* 201–212.

STOSCH H. A. VON (1973) Observations on vegetative reproduction and sexual life cycles of two freshwater dinoflagellates, *Gymnodinium pseudopalustre* Schiller and *Woloszynskia apiculata* sp. nov. *Br. Phycol. J.* **8**, 105–134.

STOSCH H. A. VON & DREBES G. (1964) Entwicklungsgeschichtliche Untersuchungen an zentrischen Diatomeen. IV. Die Planktondiatomee *Stephanopyxis turris*—ihre Behandlung und Entwicklungsgeschichte. *Helgol. Wiss. Meeresuntersuch.* 11, 209–257.

TAI L. S. & SKOSBERG T. (1934) Studies on the Dinophysoidae, marine armored dinoflagellates, of Monterey Bay, California. *Archiv. Protistenk.* **82**, 380–482.

TAYLOR D. L. (1971a) On the symbiosis between *Amphidinium klebsii* (Dinophyceae) and *Amphiscolops langerhansl* (Turbellaria: Acoela). *J. Mar. Biol. Ass. UK* **51**, 301–14.

TAYLOR D. L. (1971b) Ultrastructure of the 'zooxanthella' *Endodinium* Chattonii *in situ*. *J. Mar. Biol. Ass. UK* **51**, 227–34.

TAYLOR F. J. R. (1973) Topography of cell division in the structurally complex dinoflagellate genus *Ornithocerus*. *J. Phycol.* **9**, 1–10.

TAYLOR F. J. R. (1979) The toxigenic Gonyaulacoid dinoflagellates. In *Toxic Dinoflagellate Blooms*. Proc. 2nd Int. Conf. (eds. D. L. Taylor & H. H. Seliger), pp. 47–56. Elsevier North Holland.

TAYLOR F. J. R. (1980) On dinoflagellate evolution. BioSystems **13**, 65–108.

TURPIN D. H., DOBELL P. E. R. & TAYLOR F. J. R. (1978) Sexuality and cyst formation in Pacific strains of the toxic dinoflagellate *Gonyaulax tamarensis*. *J. Phycol.* **14**, 235–238.

TUTTLE R. C. & LOEBLICH A. R. III (1974) Genetic recombination in the dinoflagellate *Crypthecodinium cohnii*. *Science* **185**, 1061–1062.

TUTTLE R. C. & LOEBLICH A. R. III (1975) Sexual reproduction and segregation analysis in the dinoflagellate *Crypthecodinium cohnii*. *J. Phycol.* **11** (suppl.), **15**.

TYLER M. A., COATS D. W. & ANDERSON D. M. (1982) Encystment in a dynamic estuarine environment, selective deposition of dinoflagellate cysts by a frontal convergence. *Mar. Ecol. Proj. Series* **7**, 163–178.

WALKER L. M. & STEIDINGER K. A. (1979) Sexual reproduction in the toxic dinoflagellate *Gonyaulax monilata*. *J. Phycol.* **15**, 312–315.

WALL D. & DALE B. (1968) Modern dinoflagellate cysts and evolution of the Peridiniales. *Micropaleontology* **14**, 265–304.

WALL D. & DALE B. (1970) Living hystrichosphaerid dinoflagellate spores from Bermuda and Puerto Rico. *Micropaleontology* **16**, 47–58.

WALL D., GUILLARD R. R. L., DALE B., SWIFT E. & WATANABE N. (1970) Calcitic resting cysts in *Peridinium trochoideum* (Stein) Lemmermann, an autotrophic marine dinoflagellate. *Phycologia* **9**, 151–156.

WATANABE M. M., WATANABE M. & FUKUYO Y. (1982). *Scrippsiella trochoidea*: encystment and excystment of red tide flagellates I. Induction of encystment of *Scrippsiella trochoidea*. *Res. Rep. Natl. Inst. Environ. Stud.* **30**, 27–43.

WEST G. S. (1909) A biological investigation of the Peridineae of Sutton Park, Warwickshire, *New Phytologist* **8**, 181–196.

WETHERBEE R. (1975a) The fine structure of *Ceratium tripos*, a marine armored dinoflagellate. I. The cell covering. *J. Ultrastructural Res.* **50**, 58–64.

WETHERBEE R. (1975b) The fine structure of *Ceratium tripos*, a marine armored dinoflagellate. II. Cytokinesis and development of the characteristic cell shape. *J. Ultrastructural Res.* **50**, 65–76.

WOLOSZYNSKA J. (1917) Neue Perideen-arten, nebst Bemerkungen über den Bau der Hulle bei *Gymno-* und *Glenodinium*. *Bull. Int. Inst. Acad. Sci. Cracovie, B: Sci. Nat.* 114–122.

YOSHIMATSU S. (1981) Sexual reproduction of *Protogonyaulax catenella* in Culture. *Bull. Plankton Soc. Jap.* **28**, 131–139.

ZINGMARK R. (1970) Sexual reproduction in the dinoflagellate *Noctiluca milnaris* suriray. *J. Phycol.* **6**, 122–126.

CHAPTER 15
DINOFLAGELLATE CYSTS IN ANCIENT AND MODERN SEDIMENTS

D. K. GOODMAN
Exxon Production Research Company, P.O. Box 2189, Houston, Texas 77001, USA*

'Well! I've often seen a cat without a grin,' thought Alice; 'but a grin without a cat! It's the most curious thing I ever saw in my life!'

The Cat seemed to think that there was enough of it now in sight, and no more of it appeared.

LEWIS CARROLL

1 INTRODUCTION

The dinoflagellates have an extensive fossil record for the past 215–220 million years (Ma). The group is widely distributed in Mesozoic and Cenozoic sediments in many areas of the world, and at present more than 400 genera and 2200 species of fossils have been described. Most fossil dinoflagellates are considered to represent non-motile spores, or cysts, produced during a sexual phase of the life cycle (Chapter 14) and which function as hypnozygotes. Fossilized motile stage walls are rare and include siliceous internal skeletal elements and probable pellicular walls in the Gymnodiniales, and calcareous vegetative cells in the Thoracosphaerales. A small number of fossil cysts are siliceous or calcareous, but the vast majority are composed of an organic material which chemically appears similar to the wall material in spores and pollen grains (sporopollenin,

* Present address: ARCO Exploration and Technology Company, 2300 West Plano Parkway, Plano, Texas 75076, U.S.A.

a complex carotenoid polymer), although its composition is not known in detail (see Chapter 2).

Dinoflagellate cysts occur primarily in marine sediments, although there are several reported occurrences of freshwater cysts in latest Cretaceous to Quaternary deposits (see Tappan 1980, and Harland & Sharp 1980, for summaries; also see Evitt & Wall 1968; Poplawski & Norris 1979; Liengjarern *et al.* 1980). Among living marine species, meroplanktonic (cyst-producing) forms are, for the most part, neritic in habit (Chapter 11A); ancient and modern cysts are consequently most typically found in shelf and upper slope sediments (see Stover & Williams 1982, their Fig. 2), and they are most abundant in relatively fine-grained sedimentary rock such as shales, siltstones and sandstones. Cysts are distributed in broadly latitudinal bands as a function of marine climatic regimes and hydrographic systems, similar to other major plankton groups. Within these latitudinal belts, cyst assemblages also differ both in the number and relative abundance of species along an onshore–offshore gradient, depending on the local ocean bottom topography and current conditions.

The fossil record of the dinoflagellates is a highly selective one, perhaps more so than for other microfossil groups. Preservable cysts are produced by a relatively few living species, and at present only six of the eighteen orders (see Section 3) into which dinoflagellates are classified are known to have a fossil history, with some records being unusually sporadic with respect to time. Dinoflagellates are thought to be very primitive organisms (Taylor 1979, 1980); however, the earliest form with some recognizable dinophycean affinity is the Late Silurian genus *Arpylorus* (Sarjeant 1978a). Following that occurrence, there is a nearly 200 Ma year gap in the fossil record until the Middle Triassic, from which a single species is known. The first abundant and relatively diverse occurrence of dinoflagellates is in the Late Triassic, when the general trend of an increasing number of species began that continued through the Jurassic and into the mid-Cretaceous (Albian). A progressive decrease in the number of fossil cyst species has characterized the post-Albian history of the group, and today only about 100 species of cysts occur among living taxa.

The study of fossil dinoflagellates has progressed at an ever increasing pace during the past two decades. To a large extent, this growing attention to fossil dinoflagellates has been fuelled by the use of these organisms to date and correlate nearshore marine sediments in regions of hydrocarbon exploration. In industrial laboratories, dinoflagellate cysts (along with several other major groups of microfossils) are utilized routinely for biostratigraphic and biochronologic determinations in Mesozoic and Cenozoic sedimentary sequences. A relatively recent development in dinoflagellate micropalaeontology has been the use of cysts as palaeoenvironmental indicators. Data concerning Quaternary cysts are rare, but this situation is improving as an increasing number of neontologists study them in relation to life cycles and environment. In any case, the bulk of our knowledge for this group has been collected in the past 25 years, and the study of dinoflagellate cysts in the modern sense is quite a recent development. The history of the study of dinoflagellates is an interesting story of itself, and it has been discussed in various detail by Deflandre (1947a), Sarjeant (1974), Tappan (1980), Stover & Williams (1982), Williams & Bujak (1985) and Evitt (1985).

The most recent development in dinoflagellate micropalaeontology has been the comprehensive treatment of various aspects of the fossil record of the group. In recent years, Eisenack & Kjellström (1971a, b) and Lentin & Williams (1973, 1977, 1981) have produced a series of indexes for fossil dinoflagellate species with accompanying bibliographic data. Broad systematic generic revisions for selected groups have been made (Davey *et al.* 1966; Vozzhennikova 1967; Wiggins 1975; Lentin & Williams 1976; Sarjeant 1978b, 1982b) to synthesize information and to alleviate certain taxonomic problems in some groups. The generic catalogue of Stover & Evitt (1978) was the first truly comprehensive synthesis of morphologic and taxonomic criteria for distinguishing organic-walled cyst genera. Higher in the taxonomic hierarchy, the most recent suprageneric classification of fossil cysts of Norris (1978a, b) basically followed the approach of the earlier systems of Downie & Sarjeant (1966, 1974). Two illustrated guides to fossil cyst genera, which complement these systematic studies, have been published by Artzner *et al.* (1979), and by Wilson & Clowes (1980). Attempts to provide stability to the growing morphologic terminology for cysts have been outlined by Evitt *et al.* (1977) and Sarjeant (1982a). Prior to this volume, general treatments of dinoflagellates can be found in the books by Sarjeant (1974; predominantly palaeontological), Spector (1984) and Tappan (1980; predominantly biological), and in review articles by Williams (1977, 1978) and Taylor (1980). The recent book by Evitt (1985) summarizes our understanding to date of many aspects of dinoflagellate morphology and systematics. Stratigraphic distribution charts illustrating composite (worldwide) ranges for literally hundreds of species, as well as regional zonation schemes, are becoming more common as our data base rapidly expands. The most complete of these in past years have been compiled by Sarjeant (1967, 1979), Riley & Sarjeant (1972), Harker & Sarjeant (1975), Williams (1977), Williams & Bujak (1985), and in a group effort by Millioud, Sarjeant, Drugg, Stover, Lentin and Williams in Evitt (1975).

Predicting the future direction of research activity on dinoflagellate cysts is about as tenuous as inferring the evolutionary history of the group based on fossil evidence alone. It appears certain, however, that detailed lineage studies, the development of a widely accepted phylogeny and suprageneric classification for the group based on a clearer, more fundamental understanding of structural homology, an expanded data base for Quaternary cysts and thecae, and an increased appreciation and application of cyst palaeoecology and biogeography, are unquestionably among the unexplored 'new frontiers' of dinoflagellate micropalaeontology.

2 SELECTIVITY OF THE FOSSIL RECORD

The fossil record of dinoflagellates is fundamentally different from that of most other microfossil groups in terms of both the qualitative (species, or groups, represented) and quantitative (relative abundance of those forms) relationship between the living planktonic assemblage and the thanatocoenosis ultimately preserved in the sediments. This situation is due to the fact that in the past, fossilizable cysts with demonstrable dinophycean affinity seem to have been produced by only certain groups or species during certain intervals of time. Selective, or sporadic, cyst production in geologic time is suggested by some peculiarities of cyst production among living dinoflagellates (Evitt 1981), and it is supported by a number of observations from the fossil record consistent with those made on modern forms. Wall (1971) proposed four functions of cysts—reproduction, protection, propagation and dispersion—that might bear on the interpretation of cyst distribution in ancient sediments; the most recent and compelling discussion of the preserved record and of constraints on its interpretation is that of Evitt (1981), from which the following summary, with additions and modifications, is largely taken.

Basically, biological and environmental bias is introduced into the record in several ways, as observed in modern forms. The resistant resting cyst—which is probably a hypnozygote (Dale 1976, 1977a and Chapter 14)—is the principal phase of the life cycle capable of preservation under most conditions (thecal molds may exist in cherts, see Lejeune-Carpentier & Sarjeant 1981; and the pellicular layer of the motile stage may be preservable, see Taylor 1980, and Chapter 2); the zygotic theca, in contrast, disassociates and is destroyed, and thus from the start only half of the life cycle is usually available for preservation. Further biological selectivity occurs because relatively few modern species produce preservable resting cysts (Dale 1976); this ability to produce cysts may even vary among species of a single genus, notably in *Gonyaulax* and *Peridinium* (Evitt 1981; Dale 1976, in which the selective removal of cysts of certain *Peridinium* species due to processing techniques was also noted). Finally, some modern dinoflagellates, including *Peridinium* (Dale 1977a, 1978) and *Protogonyaulax* (Dale 1977b; Turpin *et al.* 1978; Anderson & Wall 1978), produce fossilizable acritarchous cysts which lack the morphological features indicative of dinophycean affinity.

Environmental factors combine to produce a cyst assemblage which does not necessarily reflect accurately the characteristics of the planktonic (thecate or athecate) community in the overlying water column. This is even true in areas of relatively simple hydrographic conditions and in which plankton records have been maintained for several years (Dale 1976). The majority of cyst-producing dinoflagellates in modern oceans are neritic in habit, living over the continental shelf or perhaps upper slope (Wall *et al.* 1977; Reid 1978). Most oceanic species, for example those of *Ceratium*, do not produce dormant cysts;

however, Wall *et al.* (1977) found several cyst-based genera (*Leptodinium, Nematosphaeropsis, Lingulodinium*, and others) occurring in oceanic environments. Relationships between the number of cyst-producing and non-cyst-producing motile species in particular environments, and correlation of these data to the bottom cyst assemblage, have not been documented. However, the studies above strongly indicate that both the absolute number and the taxonomic representation of cysts in the sediment are not related to that in the motile community in any constant or determinable manner other than the majority being neritic.

Evitt (1981) assumes a comparable degree of selectivity among ancient dinoflagellates, and there are several examples of significant gaps in the fossil record of certain groups which apparently support that assumption. For example, ceratioid cysts are characteristic elements of Jurassic and Cretaceous assemblages (Wall & Evitt 1975; R. Helby, in press), but there is no Tertiary record of this group and no modern species of *Ceratium* produces a resistant cyst. Likewise, the only fossil gymnodinioids, *Dinogymnium* and *Polykrikos* (see Section 3), are quite restricted in time (*Dinogymnium* to the Late Cretaceous, and *Polykrikos* to the post-Late Oligocene) despite the widespread occurrence and morphologic diversity of modern representatives of the group and known cyst production of the latter genus. Among living peridinioids, a considerable number of tabulation patterns have been documented; however, only two of these—the ortho-hexa and ortho-quadra (see Section 3)—have any significant fossil representation. Among fossil cysts, the ortho-quadra style is found only in the Palaeogene *Wetzeliella*-complex; there is no post-Oligocene fossil record of this style, but it occurs in living species of *Protoperidinium*. Finally, the major temporally disjunct distribution pattern for the dinoflagellates is that between their predicted evolutionary appearance in the Late Pre-Cambrian/Early Cambrian (see Taylor 1980 and Section 5) and the considerably younger earliest fossil record of a dinoflagellate (*Arpylorus antiquus*) in the Late Silurian, and the subsequent 200 Ma gap devoid of dinophycean fossils before the record of the group resumes in the Middle to Late Triassic. Some of these apparent gaps in the dinoflagellate record during the Palaeozoic may be 'filled' by acritarchous cysts whose affinities cannot be recognized, but few, or no resistant cysts may have been produced during this time.

The selective nature of the dinoflagellate fossil record imposes constraints on interpreting several critical aspects of that record; notably, the determination of phylogenetic or evolutionary trends, and the reconstruction of a biocoenosis from the cyst assemblage at a point in, or as a function of time. According to Evitt (1981), the occurrence of particular distinctive morphological features in cysts of certain ages cannot be accepted unequivocally as documenting either the time or sequence in which these features evolved. It merely sets the times during which resistant cysts were formed within the groups having the appropriate genetics for those characters. Thus, it is possible for a fossil from older strata to be either more primitive or more advanced than one from a younger deposit. A primitive ancestral form simply may not have developed cyst-producing capabilities (if it did so at all) until after that ability was acquired by a later, more advanced species which evolved directly from the primitive form. In this context, it is impossible to determine, for example, whether the temporal distribution of *Nannoceratopsis* and certain other genera suggests the dinophysoid lineage is ancestral to the peridinioid–gonyaulacoid lineages, or vice versa (Piel & Evitt 1980). However, the preserved record of fossil cysts can still be used to interpolate general patterns of evolutionary development, and to indicate the time and sequence in which certain fractions of the evolutionary record became 'fossilizable.' The development of a particular feature of thecal morphology or tabulation, or the evolution of a separate lineage in the thecate community, may significantly predate the time when the feature appears in the cyst record, if in fact it was ever recorded. However, fossil cysts can be used to indicate the most recent date at which an evolutionary event may have occurred.

Since we are unable to determine characteristics of the motile biocoenosis from the cyst thanatocoenosis it is also difficult to make accurate statements regarding fossil dinoflagellate assemblages. Details of species composition and population density for an ancient assemblage of living cells cannot be made by making taxonomically valid counts of fossil cysts, because there is no way to accurately estimate either the number of motile cells or the relative percentage of non-cyst-producing species using that method. Because of the sporadic distribution of cysts (both temporally and taxonomically), and our inability to reconstruct a living community from them, it is inappropriate to use fossil cysts to estimate the role of dinoflagellates in the environment, e.g. as primary producers. Furthermore, plotting species diversity or survivorship curves for dinoflagellates in geologic time is misleading in the sense that such curves show trends only for fossilizable cysts (Fig. 15.1), which are simply a part (and an unknown part at that) of the living assemblages. Comparing curves based on cysts to those for other microfossil groups is therefore apt to be misleading unless taken in context. Because the criteria for species and generic recognition differ substantially between living and fossil taxa (Dale 1978; Taylor 1980; Harland 1982a, b), comparisons between them are also hazardous.

Despite the fact that the dinoflagellate fossil record is unusual in several respects, informed statements about the geologic history of the group can be made if those peculiarities are taken into consideration. In short, the record can be used. In addition, the biased fossil record does not significantly affect routine application of dinoflagellate cysts to biochronology (age determination) and biostratigraphic correlation. These techniques rely upon definition and recognition of the temporal distribution of individual species, and the 'on-again, off-again' phenomenon of sporadic cyst production is not considered to operate at such a low taxonomic level. Dinoflagellate cysts can also be used to indicate general palaeoenvironmental trends within a depositional basin by correlating

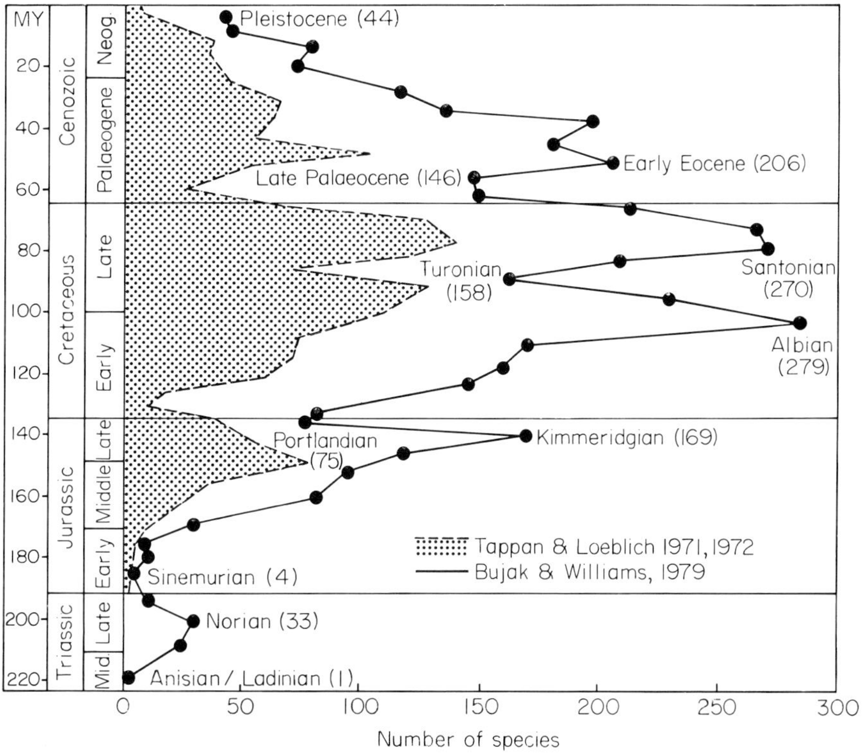

Fig. 15.1. Comparison of the number of described Mesozoic and Cenozoic dinoflagellate cyst species per stage according to Tappan & Loeblich (1971, 1972) and Bujak & Williams (1979). Note similarity of the two curves despite differences in data base. Approximately 100 cyst species are known in modern oceans. This indicates a less severe drop in cyst diversity than would be anticipated from Neogene and Pleistocene data alone; this time interval may be underrepresented in terms of species of the few studies thus far completed relative to older and younger intervals. Middle Triassic portion is based on unpublished data from L. E. Stover and R. Helby (figure modified from Bujak & Williams 1979).

changes in assemblage composition (ultimately tied to the environment) to other independent criteria (e.g. other fossil groups, facies, seismic, depositional sequences, etc.). Distinctive associations or particular taxa can then be used to infer palaeoenvironments in regions for which there are no other data. The associations themselves have palaeoenvironmental significance, and we do not necessarily have to understand every step in the process which converts a living community to a cyst assemblage before interpreting the broad environmental significance of particular associations or of compositional trends through time. Statements made in subsequent sections of this chapter (with the exception of that on 'Evolution') are based on the preserved record of fossil cysts. Therefore, any reference to fossil dinoflagellates is generally meant in the context of 'preservable dinoflagellate cysts.'

3 FOSSIL CYSTS

3.1 Major categories and affinity

The Class Dinophyceae comprises a maximum of eighteen orders (Loeblich 1982; Taylor 1980; Piel & Evitt 1980; Tangen *et al.* 1982), although only six of these are known to have a fossil record (Fig. 15.2). They are the Dinophysiales, Gonyaulacales, Gymnodiniales, Nannoceratopsiales, Peridiniales, and Thoracosphaerales, with the vast majority of fossil remains attributed to the Peridiniales and Gonyaulacales. A significant part of the preserved dinoflagellate fossil record (perhaps mainly in the Palaeozoic and early Mesozoic) may not be recognized as such because it consists of taxa attributed to the Group Acritarcha (for example, nearly all the simplest [primitive?] gonyaulacoids that are known to produce cysts, e.g. *Gessnerium* and *Protogonyaulax*, produce 'acritarchous' cysts which would not be attributed to the dinoflagellates without life cycle data). The acritarchs are a polyphyletic group, by definition of uncertain affinity (see Tappan 1980, for a recent review). The affinity of an appreciable number of fossil dinoflagellates within the taxonomic hierarchy of living forms is not known, and it seems reasonable to assume that many evolutionary lines occur only as fossils. Only one order comprised solely of fossils, the Nannoceratopsiales, has been formally proposed (Piel & Evitt 1980); most of those fossil taxa whose affinities are unknown seem destined to remain in a nebulous systematic state, their phylogenetic relationships largely in the realm of speculation.

Organic-walled cysts are the principal type of fossilizable wall produced by dinoflagellates, but calcareous and siliceous cysts occur in the Peridiniales, and the Thoracosphaerales are characterized by a calcified vegetative (motile stage) cell wall. A number of taxa have internal siliceous skeletal elements, for example *Actiniscus* (Gymnodiniales) and *Pavillardinium* (Gonyaulacales), which have been reported in sediments of Late Oligocene to Recent age; they do not contribute appreciably to the dinoflagellate fossil record and will not be discussed further. A brief discussion of those dinoflagellate orders with known or suspected fossil representatives, consisting primarily of organic-walled (sporopollenin) cysts, follows.

Order Prorocentrales

No fossil representatives belonging to this order are known, and preservable resting cysts are unknown among living forms. The prorocentroids (desmokonts) are generally considered to be among the most primitive dinoflagellates, and the superficial resemblance of some predominately Palaeozoic acritarchs to members of this group seems to support Loeblich's (1976) hypothesis that some acritarchs may be prorocentroid cysts. Secondary loss of the ability to produce preservable resting cysts may account for the lack of cyst producers among

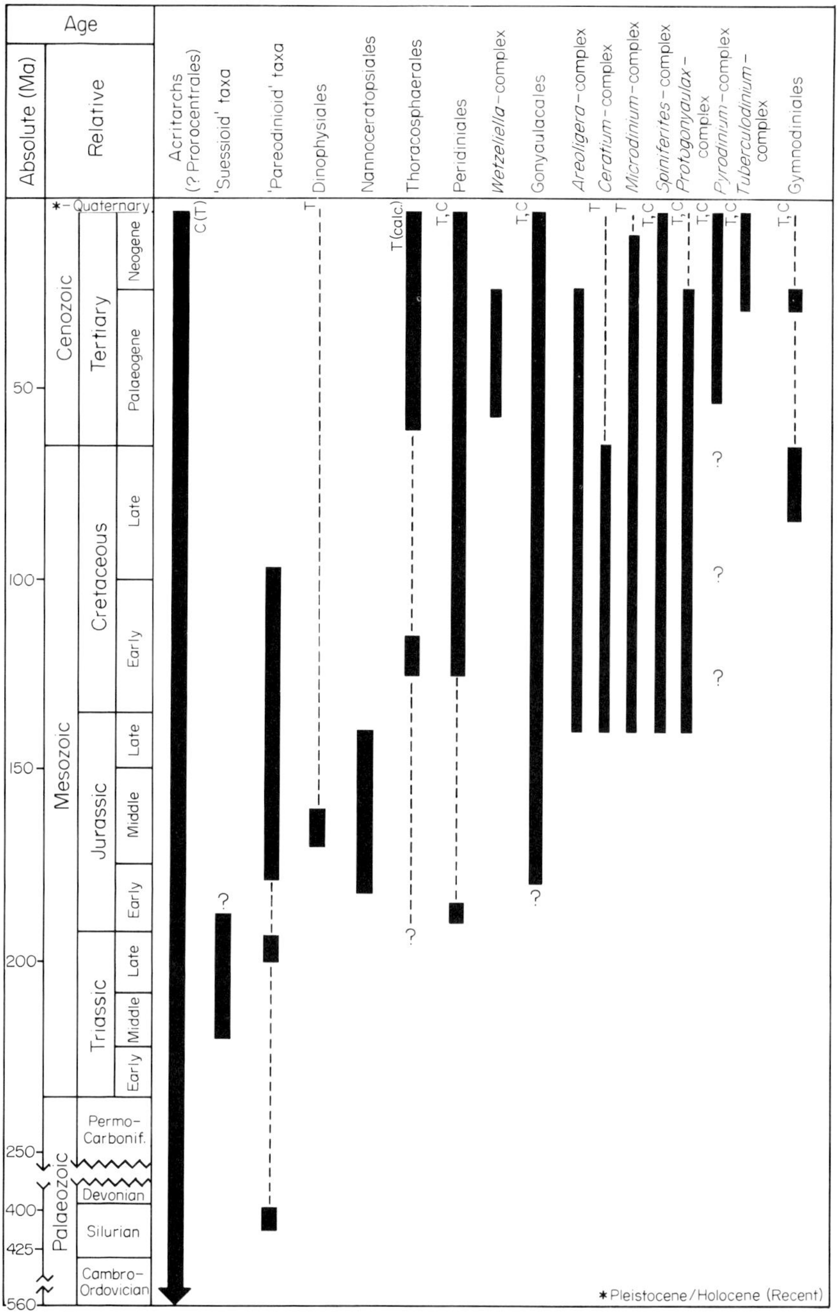

Fig. 15.2. Stratigraphic ranges of the major dinoflagellate lineages and selected species complexes. The acritarchs are a polyphyletic group, some of which may be dinoflagellate cysts. Solid lines are documented occurrences; dashed lines indicate that no known representatives occur for that interval. The Late Silurian genus *Arpylorus* is included in the pareodinioid line, and the Middle Triassic form '*Sahulodinium*' is considered a suessioid cyst. In the Quaternary, a 'T' indicates motile stage (thecate) forms, and a 'C' indicates cysts.

living forms. Dodge (1975) recently discussed phylogenetic relationships among living species of *Prorocentrum*.

Order Dinophysiales

Preservable resting cysts are unknown in living forms, and there is a single possible fossil representative of the group—'*Ternia*' (L. E. Stover and R. Helby, pers. comm.). Known only from the Middle Jurassic of Australia, '*Ternia*' (Fig.

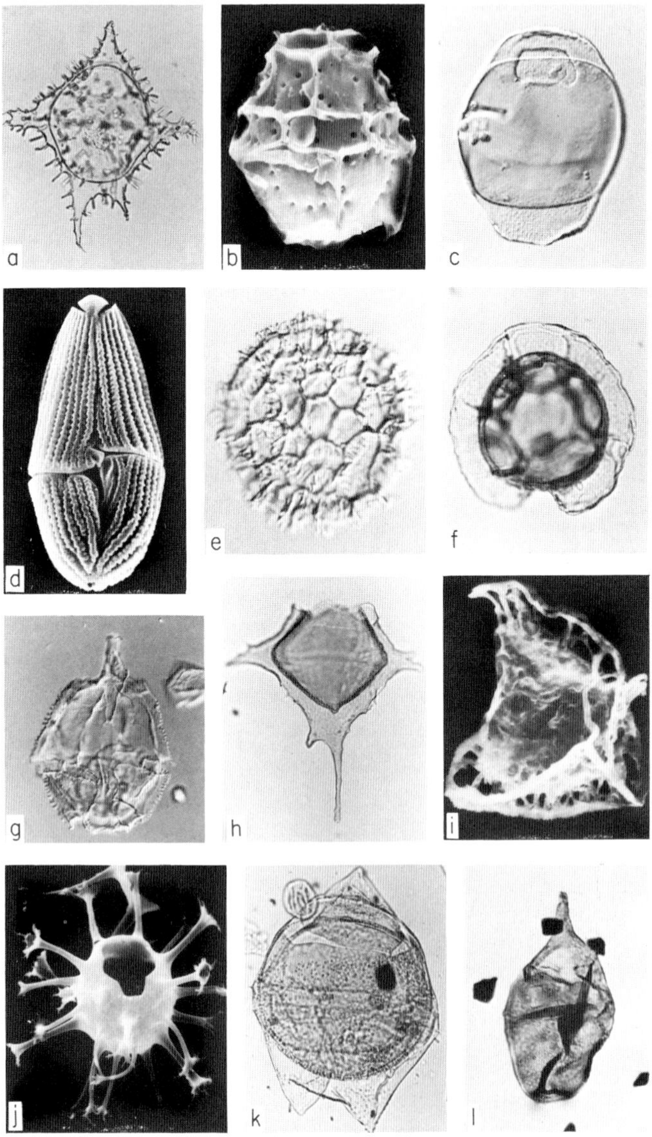

15.3i) exhibits a few morphological features which invite comparison to members of the Dinophysiales, in particular the living genera *Dinophysis* and *Ornithocercus*. The fossil lacks paratabulation, which precludes unequivocal determination of its affinity; however, the arrangement of several trabecular structures on the cyst main body (the shape of which may be considered as suggestively dinophysialean) apparently correspond to various 'lists' characteristic of modern thecae, according to Stover and Helby. It should be noted, however, that a slight resemblance to dinophysoids in terms of a few characters (shape, 'lists,' etc.) cannot be used as unequivocal evidence for the affinity of '*Ternia*.' For example, despite its strikingly dinophysialean shape (even more so than '*Ternia*'), *Nannoceratopsis* is now known to be something quite different (see below). Details of archeopyle morphology remain unclear at this time.

Order Nannoceratopsiales

The order has no known living representatives, and is known only for the single fossil genus *Nannoceratopsis* (Fig. 15.4b), several species of which occur in Jurassic marine sediments. Previously attributed to the Dinophysiales (Evitt 1961a; Gocht 1972), scanning electron microscope studies (Piel & Evitt 1980) of the genus clearly demonstrate the paratabulation to combine features of several lineages—the dinophysoids and the peridinioid/gonyaulacoid complex. The hypocystal and sulcal paratabulation patterns are most similar to those in the Dinophysiales, although the nannoceratopsialean parasulcus is also suggestive of *Peridinium* (as is the dinophysialean pattern). In contrast, the epicystal paratabulation of *Nannoceratopsis* is more similar to that in the gonyaulacoids (and somewhat less so to the peridinioids) than in the dinophysoids in having a more clearly defined longitudinal arrangement of plate series. Finally, the cingular paratabulation is more similar to the peridinioid/gonyaulacoid lines than to the dinophysoids regarding the relative size of the

---

Fig. 15.3. (a) *Wetzeliella lunaris* Gocht. Early Eocene, Maryland, USA. Bright field (BF, × 160). (b) *Cladopyxidium* sp. Late Cretaceous (Maastrichtian), Denmark. Scanning electron photomicrograph (SEM, × 1320). (c) *Isabeladinium cretaceum* (Cookson) Lentin and Williams. Late Cretaceous (Maastrichtian), offshore south-eastern Argentina. Normarski interference contrast (NOM, × 255). (d) *Dinogymnium westralium* (Cookson and Eisenack) Evitt *et al.* Late Cretaceous (Maastrichtian), Denmark (SEM, × 750). (e) *Suessia* sp. Late Triassic, offshore northwestern Australia (NOM, × 510). (f) *Pentadinium laticinctum* Gerlach. Late Eocene, Alabama, USA. Apical view (BF, × 255). (g) *Gonyaulacysta jurassica* (Deflandre) Norris and Sarjeant. Late Jurassic (Oxfordian), Scotland (NOM, × 255). (h) *Muderongia* sp. cf. *M. simplex* Alberti. Early Cretaceous (Barremian), offshore western Australia (BF, × 255). (i) '*Ternia balmei*'. Middle Jurassic, offshore north-western Australia (SEM, × 360). (j) *Hystrichosphaeridium tubiferum* (Ehrenberg) Davey and Williams. Late Cretaceous (Maastrichtian), Denmark (SEM, × 450). (k) *Deflandrea phosphoritica* Eisenack. Early Oligocene, Belgium (BF, × 255). (l) *Pareodinia ceratophora* Deflandre. Late Jurassic, Alaska, USA (BF, × 370).

paraplates and in lacking evidence of a sagittal suture or megacytic zone. In addition, *Nannoceratopsis* is the only cyst known to have a cingular (Type C) archeopyle. The combination of these characters suggests a closer phylogenetic relationship among these lineages (dinophysoids and the gonyaulacoid/peridinioid complex) than had previously been thought. However, the temporal distribution of *Nannoceratopsis* and representatives of the gonyaulacoid and peridinioid lines does not indicate the direction of evolutionary change (in large

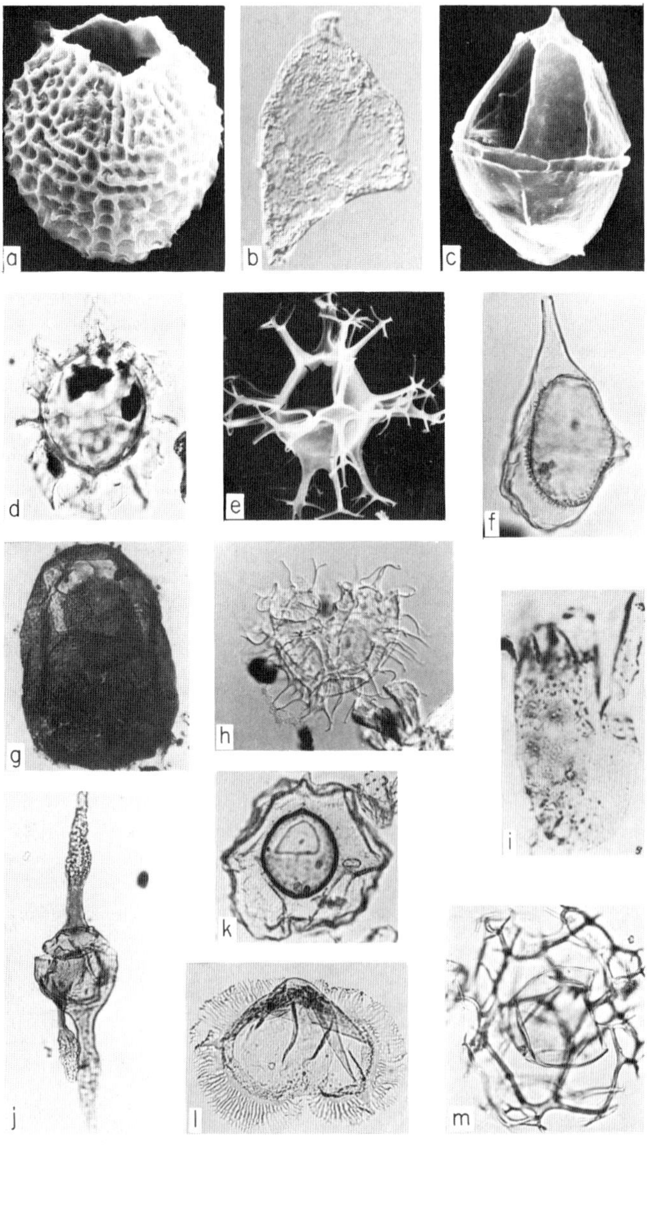

part due to the lack of dinophysialean features in fossils other than in *Nannoceratopsis* itself), i.e. from the dinophysoids to the gonyaulacoids/peridinioids through *Nannoceratopsis*, or in the opposite direction. Nonetheless, reconstructions of dinoflagellate phylogeny (Bujak & Williams 1981) now place *Nannoceratopsis* in a pivotal position along one of the possible evolutionary pathways.

Order Thoracosphaerales

This group occupies a unique position (perhaps along with *Dinogymnium*) among dinoflagellates as it is represented by preservable vegetative (motile) stage cells. The order contains only the Family Thoracosphaeraceae with the single genus *Thoracosphaera*. Until recently (for example, in Tappan 1980) *Thoracosphaera* had been considered as a coccolithophorid genus, although Fütterer (1976, 1977) had previously suggested that several species, including the type species *T. heimii*, be attributed to the Dinophyceae (Keupp 1979, agreed with this proposal). Fütterer placed *Thoracosphaera* in the Peridiniales (Subfamily Thoracosphaeroideae) because incubation experiments (Wall & Dale 1968a; Wall *et al.* 1970) had demonstrated calcareous resting cysts in two modern peridinialean genera (*Ensiculifera* and *Scrippsiella*). Recently, Tangen *et al.* (1982) have shown that the calcareous cell wall in *T. heimii* is not a cyst but rather belongs to the main vegetative stage. Due to this and other differences between *T. heimii* and other dinoflagellates, Tangen *et al.* (1982) placed the species in the new order Thoracosphaerales. Life cycles are unknown for other species of *Thoracosphaera*, and it is possible that some of these are cysts (for example, those with archeopyles) and should be retained in the Peridiniales as calciodinellids. Taylor (1982, pers. comm.) would include the Thoracosphaerales

---

Fig. 15.4. (a) *Eisenackia crassitabulata* Deflandre and Cookson. Early Palaeocene, offshore south-eastern Argentina. Scanning electron photomicrograph (SEM, × 600). (b) *Nannoceratopsis gracilis* Alberti. Early Jurassic, offshore Norway, Nomarski interference contrast (NOM, × 385). (c) *Leptodinium mirabile* Klement. Late Jurassic, Alaska, USA (SEM, × 370). (d) *Danea mutabilis* Morgenroth. Early Palaeocene, Denmark (NOM, × 240). (e) *Spiniferites ramosus* (Ehrenberg) Loeblich and Loeblich. Late Cretaceous (Maastrichtian), Denmark (SEM, × 440). (f) *Dingodinium cerviculum* Cookson and Eisenack. Early Cretaceous (Barremian), offshore western Australia. Bright field (BF) (× 330). (g) *Arpylorus antiquus* Calandra. Late Silurian, Tunisia (BF, × 170) (reproduced from Sarjeant 1978). (h) *Areoligera senonensis* Lejeune-Carpentier. Late Cretaceous (Maastrichtian), The Netherlands (NOM, × 260). (i) '*Sahulodinium*' sp. Middle Triassic (Anisian–Ladinian), offshore northwestern Australia (BF, × 230). (j) *Odontochitina cribropoda* Deflandre and Cookson. Late Cretaceous (Turonian), offshore southeastern Argentina (BF, × 255). (k) *Thalassiphora pelagica* (Eisenack) Eisenack and Gocht. Early Eocene, Maryland, USA (BF, × 95). (l) *Wanaea fimbriata* Sarjeant. Late Jurassic (Oxfordian), Scotland (BF, × 255). (m) *Cannosphaeropsis utinensis* O. Wetzel. Late Cretaceous (Maastrichtian), Denmark (BF, × 190).

sensu Tangen in the Peridiniales, like Fütterer (tabulation and the presence of calcified cysts in the latter), and would not separate the Pyrocystales from the Gonyaulacales. *Thoracosphaera* occurs in the modern plankton of most major oceans, and it is common in Recent bottom sediments (Tangen *et al.* 1982). The oldest reported occurrence of the genus is from the Triassic (Jafar 1979). Calcareous cells attributed to several species of *Thoracosphaera* are present in microfossil assemblages from the Middle Pleistocene of the Central Pacific (Fütterer 1976), Early Palaeocene through Pliocene of the North Atlantic (Fütterer 1976, 1977), and Late Oligocene (Weiler & Sonne 1980) and Early Cretaceous of Germany (Keupp 1979).

Order Peridiniales

Cyst-producing representatives of this group occur in Early Jurassic to Recent sediments. The Peridiniales, along with the Gonyaulacales, comprise the vast majority of fossil dinoflagellates, each having an extensive fossil record. Representative living genera include *Peridinium* (almost exclusively freshwater), and the marine *Protoperidinium, Ensiculifera, Heterocapsa* and *Scrippsiella*; many living species produce preservable resting cysts. Approximately sixty fossil cyst genera with peridinialean affinity are known (see Lentin & Williams 1976, for an extensive treatment of fossil peridinioid cysts). All these fossils can be included in the Family Peridiniaceae (Tappan & Loeblich 1977, attribute the gymnodinioid genus *Dinogymnium* to the peridinialean family Lophodiniaceae; note that some algologists would assign *Lophodinium* to the Gymnodiniales). The generalized paratabulation formula for the group is 4′, 3a, 7″, Xc, 5‴, 2⁗; the paraplate arrangement is characterized by a strong tendency towards bilateral symmetry. Many species have a distinctly peridinioid outline with one apical, two antapical, and sometimes two lateral horns, and they usually have a moderate amount of primary dorso-ventral compression. Preservable resting cysts are predominantly composed of sporopollenin, but calcareous cysts are known in several genera, including extant *Scrippsiella* and *Ensiculifera* (Wall & Dale 1968a; Wall *et al.* 1970) and in the fossils *Calciodinellum, Calciogranellum, Bicarinellum,* and *Pithonella*, among others (Deflandre 1947b, 1948; Gocht 1959; Wall & Dale 1968a; Fütterer 1976, 1977; Keupp 1979, 1980a, b, c), in sediments ranging from Late Jurassic (Oxfordian) to Quaternary in age. A peridinialean paratabulation has been demonstrated in several species of calcareous cysts, either by external features as in *Alasphaera verrucosa* (Keupp 1979) and *Pithonella paratabulata* (Keupp 1980b), or by parasutural markings on the interior surface of the cyst wall as in *P. patricia-creeleyae* (Keupp 1980c). Siliceous cysts such as *Peridinites* are known only from Tertiary sediments and have no known living relatives.

Fossil organic-walled peridiniacean cysts can be grouped into two major subdivisions based on the paratabulation style (Fig. 15.5) on the dorsal epicyst,

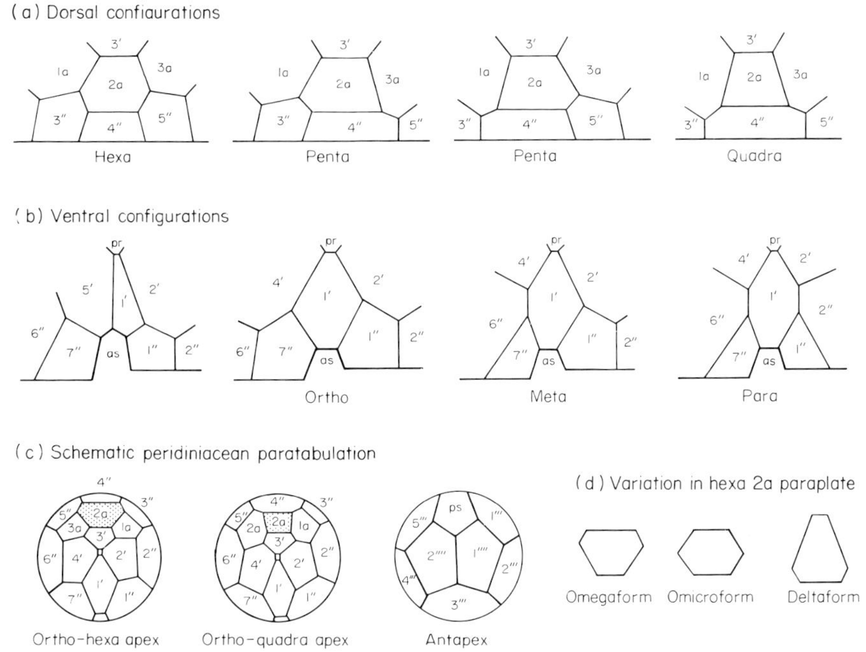

Fig. 15.5. Variation in peridinialean paratabulation. (a) Dorsal configurations: 2a contacts three precingulars in the dominant hexa style, two in the penta (which can occur in either of two modes), and one in the quadra. (b) Ventral configurations: 1′ contacts a single precingular in the unnamed style at the left, two precingulars in the ortho style, three in the meta, and four in the para. The vast majority of fossil peridiniaceans have the ortho ventral style; the others are quite rare. (c) Apical views of the most common styles among fossil cysts (ortho-hexa and ortho-quadra) and standard antapical paratabulation. (d) Schematic variability in the shape of the hexa 2a paraplate for Late Cretaceous and Tertiary cysts.

using parasutural features and/or archeopyle shape. The archeopyle in all peridiniacean cysts involves the mid-dorsal anterior intercalary paraplate (2a), either alone or in combination with other anterior intercalaries, precingulars and/or apicals. In addition, the lower margin of the 2a paraplate is always transverse (straight, not geniculate) in peridiniacean cysts. The majority of these cysts are characterized by the *hexa* style (also common among living dinoflagellates) in which the mid-dorsal anterior intercalary contacts three subjacent precingulars (Fig. 15.5a). This group includes a wide range of morphological types; it ranges from Late Jurassic to Recent, but species are particularly abundant from Late Cretaceous to Early Oligocene time. Representative genera include *Deflandrea* (Fig. 15.3k), *Isabeladinium* (Fig. 15.3c), *Chatangiella, Hexagonifera, Phthanoperidinium, Luxadinium, Palaeocystodinium, Alterbidinium, Lejeunecysta, Liasidinium, Ginginodinium, Nelsoniella,* and *Palaeoperidinium.* These genera appear most closely allied to species of the modern freshwater *Peridinium*. The second principal group is characterized by the *quadra* style of epicystal organization (Fig. 15.5a), in which the mid-dorsal anterior intercalary contacts only one precingular (see Evitt 1978). Members of this group are collectively known as the *Wetzeliella*-complex, which includes *Apectodinium, Gochtodinium, Kisselovia, Rhombodinium, Wetzeliella* (Fig. 15.3a) and *Wilsonidium*. Among living dinoflagellates, the motile stages of some species of *Protoperidinium* seem most similar to this group. Modern cysts with the quadra style are not known; the *Wetzeliella*-complex ranges from the Late Palaeocene through the Oligocene.

A number of fossil cysts are obviously peridiniacean in design, but they cannot be included in the two major groups because they either lack, or have an unusual archeopyle. For example, although the penta style of dorsal paratabulation (Fig. 15.5a) is common among living dinoflagellates, it is exceedingly rare in the fossil record, and only a few species are known from fossil remains (E. H. Davies, pers. comm.). In other genera, e.g. *Palaeohystrichophora* (Harker 1979), details of the dorsal paratabulation cannot be sufficiently resolved to determine affinity, while in still others such as *Subtilisphaera* no clear evidence of an archeopyle has been discerned.

Several paratabulation styles are also developed on the ventral epicyst of peridiniacean cysts (Fig. 15.5b). In the *ortho* style, common among living thecae and the dominant style among fossil cysts, the first apical contacts two precingular paraplates. In the *meta* style, 1′ contacts three precingulars, and in the *para* style, 1′ contacts four precingulars. The meta and para styles are represented by numerous recent species, but for each only one fossil is known. *Phthanoperidinium brooksii* has a meta configuration (Edwards & Bebout 1981), while an undescribed species attributed to *Phthanoperidinium* by Goodman & Ford (1983) is thought to have a para style; both forms are mid-Tertiary in age. Another ventral configuration, as yet lacking a name, is characterized by 1′ having contact with a single precingular; it is present in the Cretaceous cyst *Angustidinium* and in the living genus *Heterocapsa* (Goodman & Evitt 1981).

Order Gonyaulacales

The Order Gonyaulacales was proposed by Taylor (1980) to include a number of taxa formerly attributed to the Peridiniales but referred to informally as 'gonyaulacoid' in the micropalaeontological literature. The group has an abundant fossil record ranging from the Early Jurassic to Recent; many living species produce preservable resting cysts. Modern genera to which a majority of fossil forms seem related are *Gonyaulax, Protogonyaulax, Pyrodinium, Triadinium* (= *Heteraulacus*), *Ceratium, Cladopyxis*, and *Pyrophacus*. Major subdivisions within this group can be recognized using paratabulation (Fig. 15.6). In fact, Taylor (1980) recently proposed a model to derive the major styles of gonyaulacoid thecal tabulation from a hypothetical pattern similar to that of

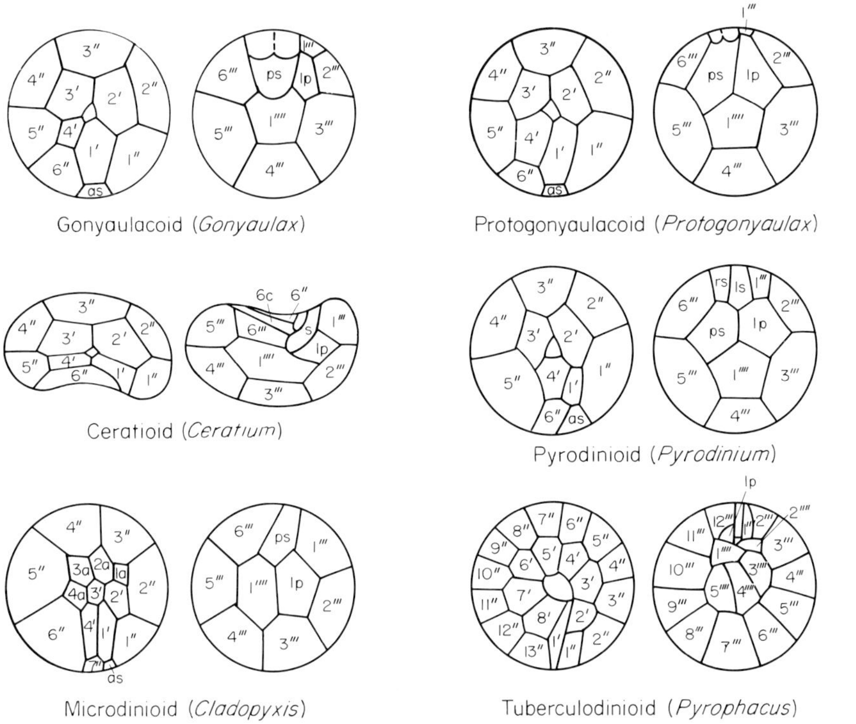

Fig. 15.6. Schematic illustrations of several typical tabulation patterns in the Gonyaulacales. The basic configuration and plate formula remain nearly constant in all styles, with the principal differences being in typological relationship for particular plates, and the presence of anterior intercalary or supernumerary plates in some styles. These styles represent the majority of fossil gonyaulacalean cysts, although many other variations also are known.

Triadinium (see Chapter 2). Using that model, all major styles can be developed by simple distortions and minor topological adjustments of both epi- and hypothecal plates, with few major additions or subtractions of plates or plate series. A greater variation in cyst morphology, ornamentation, archeopyle type, and minor details of paratabulation, is developed in fossil Gonyaulacales, but Taylor's model has proved to be useful in recognizing homologous structures among them (W. R. Evitt, pers. comm. 1985). Application of the model does not require acceptance of an evolutionary direction. A generalized Kofoidean paratabulation for the group is $4'$, $6''$, 6c, $6'''$, 1p, $1''''$. The principal differences between this and the peridinioid pattern are (i) a distinct asymmetry in overall paraplate distribution; (ii) asymmetrical and generally small anterior intercalaries if present; (iii) six rather than seven precingulars; (iv) six rather than five postcingulars; (v) a large posterior intercalary; and (vi) one rather than two antapical paraplates.

Fossil cysts with apparent affinity to *Gonyaulax* are characterized by a six-sided (i.e. one that contacts six superjacent paraplates) antapical paraplate (plate Y of Taylor) on the hypocyst, and on the epicyst by the lack of contact between the second and fourth apical paraplates. This group contains numerous genera, including *Cannosphaeropsis* (Fig. 15.4m), *Ctenidodinium, Gonyaulacysta* (Fig. 15.3g), *Impagidinium, Leptodinium* (Fig. 15.4c), *Meiourogonyaulax, Nematosphaeropsis, Scriniodinium, Spiniferites* (Fig. 15.4e), and possibly *Areosphaeridium, Cordosphaeridium, Hystrichokolpoma, Hystrichosphaeridium* (Fig. 15.3j), *Muratodinium, Oligosphaeridium* and *Thalassiphora* (Fig. 15.4k). Further subdivision in this group can be made on the basis of mode of expression of paratabulation (intratabular processes vs. parasutural features), archeopyle type, and details of ventral paratabulation (S-shaped vs. linear parasulcus). The group ranges from the Jurassic to Recent, and can be accommodated within the taxonomic hierarchy of living dinoflagellates in the Family Gonyaulacaceae.

Two other groups of fossil Gonyaulacales are probably related to the Gonyaulacaceae. One group consists of lenticular cysts with extreme primary dorsoventral compression, an offset sulcal notch, and (coincidentally) an apical archeopyle. Genera include *Areoligera* (Fig. 15.4h), *Canningia, Chiropteridium, Cyclonephelium, Palynodinium, Renidinium, Senoniasphaera*, and others. They range from Late Jurassic to Oligocene in age, and comparable living forms are unknown. The other group is made of those genera with obvious gonyaulacacean characters but which generally lack evidence of paratabulation other than the archeopyle. This heterogeneous group includes *Cleistosphaeridium, Diphyes, Operculodinium, Exochosphaeridium, Tectatodinium, Coronifera*, possibly *Distatodinium*, and many others.

Fossil cysts with a five-sided (contacting five superjacent paraplates) 1'''' (Y) paraplate form another major group related to several extant genera: *Triadinium, Pyrodinium, Protogonyaulax, Alexandrium* and *Gessnerium* (see Taylor 1979), some of which produce fossilizable cysts (*Pyrodinium* has skolochorate cysts; the others are acritarchous—some smooth, with a slit-like archeopyle, some surrounded by a 'calyptra-like' mucoid material). Fossil cysts occur in late Early Cretaceous to Recent assemblages, and include *Alisocysta, Dinopterygium, Eisenackia* (Fig. 15.4a), *Eocladopyxis, Heteraulacacysta, Homotryblium, Tubidermodinium*, and some species currently assigned to *Hystrichokolpoma*. Dodge (1981a) proposed the family Triadiniaceae for species formerly attributed to the modern genera *Heteraulacus* and *Goniodoma*; future allocation to that family of those fossil and living genera listed above is suggested by their shared hypocystal/hypothecal pattern. Further subdivision of fossil genera is possible based on epicystal paratabulation.

A third modern family to which several fossil cysts appear allied is the Cladopyxidaceae, comprised of the modern genera *Cladopyxis, Micracanthodinium*, and *Palaeophalacroma* (= *Sinodinium*). The fossil cysts in this group are typically small, spheroidal, and have relatively large dorsal anterior intercalary paraplates. On the hypocyst, 1p is relatively large and may be subequal in size to $1''''$. In addition, the mid-ventral paratabulation of the group is distinct from that in other fossil Gonyaulacales in the contact relationships of $1'''$, 1p, and the anterior limit of the posterior sulcal paraplate. The fossils, which range from the Early Cretaceous to the Miocene, include *Microdinium, Cladopyxidium* (Fig. 15.3b), *Fibradinium, Histiocysta*, and *Subtilidinium*, and possibly *Druggidium, Gillinia, Phanerodinium* and *Elytrocysta*. Little is known of the group in terms of paratabulation, morphologic variability and stratigraphic distribution (McLean 1972, 1974; Goodman & Stover 1981).

A distinctive group of Mesozoic (Late Jurassic through Late Cretaceous) cysts can be referred to the Family Ceratiaceae (Wall & Evitt 1975). The group is characterized by a modified gonyaulacacean paratabulation and by prominent horns (one apical, one or two antapicals, and one or two postcingulars) which make the general identification as a ceratioid dinoflagellate easy. The family is represented by the single modern genus *Ceratium* which contains numerous marine and a few non-marine species, none of which produce fossilizable cysts except for *C. hirundinella* (some silicific cysts). Wall & Evitt (1975) and R. Helby (in press) give the most thorough treatment of fossil ceratioid dinoflagellates, which include *Aptea, Muderongia* (Fig. 15.3h), *Odontochitina* (Fig. 15.4j), *Phoberocysta, Pseudoceratium*, and *Xenascus*.

The final 'group' within the Gonyaulacales consists of a single genus, *Tuberculodinium*, which is the cyst of the living thecate genus *Pyrophacus* (Wall & Dale 1971). Two living genera, *Fragilidium* and *Pyrophacus*, make up the Family Pyrophacaceae. *Tuberculodinium* cysts occur in sediment of late Oligocene to Recent age (Wall 1967; Drugg 1970; Wall & Dale 1971), and they are characterized by having a higher number of paraplates than more typical Gonyaulacales.

Order Gymnodiniales

Fossils referable to *Dinogymnium* (Fig. 15.3d) may represent the pellicular layer of gymnodinialean motile stage cells (May 1976; Taylor 1980), although an alternative proposal (Tappan & Loeblich 1977) is based on the resemblance of *Dinogymnium* cells to the exospore layer of the resting cysts of *Woloszynskia* and *Lophodinium* (Lophodiniaceae, Peridiniales). The distinctive morphology of the genus (see Evitt *et al.* 1967; May 1977), and its stratigraphic restriction to the Late Cretaceous, seems to indicate a close biological relationship among the several species. Living representatives of the Gymnodiniales include *Gymnodinium, Gyrodinium* and *Amphidinium*.

Cysts of the colonial gymnodinialean dinoflagellate *Polykrikos* (Polykrikaceae) have been reported in Recent and Late Quaternary sediments from several areas in the North Atlantic (see Harland 1981) and in the plankton of the North Atlantic and North Sea (Reid 1978). Ford (1979) recovered the oldest fossil cysts referable to *Polykrikos* from the Late Oligocene of South Carolina. Harland (1981) noted the morphologic similarity between cysts of *P. schwartzii* and the Mesozoic dinoflagellates *Valensiella* (Jurassic) and *Ellipsoidictyum* (Cretaceous), alluding to the possibility of a colonial ancestral form, but recognizing that the similarity may be coincidental.

Cysts of unknown affinity

A relatively small number of fossil cysts have no apparent affinity with living forms. These species are typically among the oldest fossilized dinoflagellates, and they are most common from the Middle Triassic to Early Cretaceous. In most cases, these cysts lack those critical features, in particular a clearly discernible paratabulation and a determinable archeopyle morphology (as in *Heibergella*) necessary to establish biological affinities. Conversely, some forms (e.g. *Suessia* and *Dapcodinium*) have a well defined paratabulation, but the complexity of that pattern is difficult to interpret in an evolutionary sense. For a number of these forms it seems apparent that no other closely allied species exist among fossil and living dinoflagellates, and that they represent a patchy record of several extinct lineages. For others in this group, there are quite surely derivative forms, but relationships cannot be recognized due to insufficient morphologic and stratigraphic information. Because of the incomplete nature of the early fossil record, these cysts have been variously attributed to certain lineages by different paleontologists, each emphasizing a particular feature. For example, Morbey (1975) considered the Late Triassic genus *Suessia* (Fig. 15.3e) to have unknown affinity; Bujak & Williams (1981) believe it to be peridinialean; Dörhöfer & Davies (1980) include it in a separate group of 'suessioid' genera along with *Rhaetogonyaulax, Dapcodinium* and *Noricysta*; and W. R. Evitt (pers. comm.) feels it is possibly a gymnodinialean form (similar to *Woloszynskia* or to those species in the Gymnodiniales with thin plates).

Nonetheless, many early cysts can be grouped into one of several generalized categories based on rudimentary morphologic similarity. One group is characterized by a rather clearly outlined paratabulation (typically indicated by features of low relief) over part or all of the cyst. The archeopyles are generally combination types and involve the apicals, possibly along with some or all of the anterior intercalary and precingular paraplates. This group includes *Suessia, Rhaetogonyaulax, Dapcodinium, Comparodinium, Mancodinium* and *Maturodinium*, and it ranges from the Late Triassic to Early Jurassic. The hypocystal pattern on some of these is suggestive of a gonyaulacacean style, but comparisons are largely speculative. A second group is that comprised of mainly ellipsoidal cysts (often with an apical horn) with intercalary archeopyles involving one or more anterior intercalary paraplates (the posterior margins of which typically are distinctly geniculate). The group ranges from the late Early Jurassic to the mid-Cretaceous, and includes *Pareodinia* (Fig. 15.3l), *Gochteodinia, Kalyptea,*

Broomea (very similar to the modern peridinioid *Podolampas*), *Parvocysta, Facetodinium, Phallocysta, Cantulodinium,* and possibly *Moesiodinium* and *Arpylorus* (Fig. 15.4g). *Batioladinium* may also be included in this group although it has an apical archeopyle. Several other genera, confined mainly to the Middle and Late Triassic, have little apparent affinity with other cysts and are not readily attributed to even simple morphologic groupings. These include types such as *Sverdrupiella, 'Sahulodinium'* (Middle Triassic; see Fig. 15.4i), *Hebecysta, Heibergella, Dodekovia* (Early to Middle Jurassic), and *Noricysta*. The 'pareodinioid' group (see Wiggins 1975) has been considered by some (Bujak & Williams 1981) as an ancestral stock for the gonyaulacoid and peridinioid lineages. Possible phylogenetic associations among some of the Late Triassic to Early Cretaceous genera are given in Dörhöfer & Davies (1980; collectively referred to by those authors as the 'rhaetogonyaulacinean' dinoflagellates), but the proposed relationships are highly speculative.

3.2 Cyst classification

The classification of fossil dinoflagellate cysts is based almost exclusively on cyst morphology (archeopyle, paratabulation, wall structure, shape), whereas modern dinoflagellates are generally classified (Loeblich 1970, 1982) on the basis of thecal morphology (primarily tabulation and shape) and physiology (pigments, ultrastructure, etc.) with little consideration for the cyst. For this reason, there are at present two effectively separate, non-complimentary taxonomies for dinoflagellates. In recent years, incubation experiments (Wall 1965b; Wall & Dale 1966, 1968a, 1969, 1970; Wall *et al.* 1967; Dale 1976, 1977a, b; Dobell & Taylor 1981; and others) have established correlations for approximately fifty cyst-theca pairs among living forms (see Dale 1976; Bradford 1978; Harland 1981). These data have resulted in the beginnings of a unified classification system for cysts and thecae, a system which is supported by some and opposed by others.

Several years ago, Wall & Dale (1968b, 1970) proposed use of a single nomenclature for cysts and thecae at the generic level, although there are some problems implicit to this approach (Evitt 1970). More recently, Dale (1976, 1978) argued for a unified classification system in which both cyst and thecal characteristics are used in species descriptions of living dinoflagellates. Dale (1978) concludes that 'new cyst-based nomenclature should not be created just to "artificially" maintain dual classification,' but that cyst nomenclature be based, insofar as possible, on data both from fossil cysts and living motile and non-motile stages. Using this system, the senior name for an established cyst-theca pair would be the only name applied to both stages. Taylor (1979) distinguished the genus *Protogonyaulax* from *Gonyaulax* both on thecal tabulation and cyst features. *Pyrodinium* is also distinguishable from *Gessnerium* by virtue of differences in cyst morphology (their thecal tabulation patterns being identical). Harland (1982a) has used this approach in a classification system for cysts and thecae of *Protoperidinium*; he indicates that the major difference between cyst and thecal classification is one of taxonomic hierarchy, and that cyst-based taxa regarded as genera by palaeontologists might be thought of as subgenera or sections under the theca-based system of neontologists. The alternative approach (Reid 1974; Bradford 1975, 1978; Harland 1977) to a unified system is to maintain separate nomenclatures and to (i) use the separate names in studies dealing exclusively with either cysts *or* thecae; and (ii) use the senior name in studies dealing with cysts *and* thecae. The tendency in recent years has been to follow the first of these proposed methods (that of Wall and Dale) at the generic and species levels, and apply the senior name in all cases.

Several suprageneric classification systems have been proposed for fossil dinoflagellate cysts. Some of the earlier attempts (Deflandre 1952; Eisenack 1964; Vozzhennikova 1963) attributed several fossil cysts to modern families based on living thecate genera. They included both modern thecae and fossil cysts in the families, partly because these authors thought that the fossils themselves were thecae and not cysts. Other earlier classifications (Eisenack 1961; Vozzhennikova 1965) were more comprehensive in scope, but again were formulated with the supposition that fossil forms were thecate individuals and not resting spores. In these classifications, some families were erected using modern genera as types, while for other families fossil cyst genera were selected. Both prior to and during this time, several authors (Deflandre 1936, 1945, 1947b; Eisenack 1954; Gocht 1957; Neale & Sarjeant 1962) had proposed families based on fossil genera. Several other families have sporadically been proposed in more recent years (Drugg & Loeblich 1967; Wiggins 1972).

A comprehensive classification system based on the knowledge that fossil dinoflagellates are indeed cysts was proposed by Sarjeant & Downie (1966), and this system reappeared some years later (Sarjeant & Downie 1974) in a form redesigned specifically to allow inclusion of modern forms. The most recent suprageneric classification of fossil cysts (Norris 1978a, b) follows the basic approach taken by Sarjeant & Downie (1974); however Norris' scheme, unlike that of his immediate predecessors, is based entirely on fossils and makes little, if any, use of modern dinoflagellate systematics. For this reason, several authors, for example Stover & Evitt (1978), have chosen not to address this problem and simply list genera alphabetically rather than attribute them to one of the existing families because, as Stover & Evitt (1978) state, these more recent attempts at cyst classification lack fundamental information and are therefore phylogenetically misleading. By simply listing genera alphabetically under several main subdivisions based on archeopyle type, no phylogenetic relationships were implied by Stover & Evitt (1978) in their generic analyses.

Therefore, at present a taxonomy of fossil dinoflagellage cysts above the generic level is a predominantly phenetic system based solely on cyst morphology, with little consideration for modern dinoflagellate systematics (e.g. that of Loeblich 1970, 1982). As a result, a complex pattern emerges when comparisons between families in the two systems are made. Despite the apparent incompatibility of the two systems, it is becoming increasingly evident (W. R. Evitt, pers. comm.) that the majority of fossil cysts can be accommodated in only a few existing families within the systematic hierarchy of living dinoflagellates.

Classification systems based solely on a set of morphological features, with no explicit consideration of phylogenetic relationships or stratigraphic distribution, do not result in biologically relevant taxonomic systems (Johnson 1970; Steinbeck & Fleisher 1978). Pheneticists consider morphologic similarity to be a measure of genetic relationship, and therefore, phenetic systems are considered accurate representations of phylogenetic reality. Norris (1978a, p. 302) states that his system of cyst classification (which is fundamentally a phenetic one) removes 'chronistic or cladistic absurdities' and thus 'approaches hopefully, a phylogenetic system of classification'. However, it appears that Norris does not reach his intended goal because of several factors, including (i) no consideration of data on extant motile cell morphology, physiology and systematics; (ii) no attempt to incorporate cyst-theca correlations into his system (which would establish systematic links for at least some groups), and (iii) superficial treatment and overemphasis in family diagnoses of a few morphologic characters (archeopyle type, paratabulation, shape, wall morphology, etc.) where a more detailed, biologically relevant analysis (particularly of paratabulation) is necessary. Typological systems, such as that of Norris, cannot differentiate between morphological similarity caused by common ancestry and that resulting from evolutionary convergence or parallelism (Steineck & Fleisher 1978). An evolutionary classification scheme, which combines motile and non-motile stage morphology (and cellular data), along with available stratigraphic information (interpreted in the context of its selectivity) should provide a more accurate view of dinoflagellate phylogeny. The goal of a comprehensive, phylogenetically reasonable family level classification for fossil and modern dinoflagellates (including both the motile and non-motile states) is being pursued in a joint effort by R. A. Fensome, F. J. R. Taylor, W. A. S. Sarjeant and G. L. Williams (in prep.). Certainly, considerable progress towards a more biologically and evolutionarily sound understanding of certain lineages does seem possible in the near future (see Section 6).

4 GEOLOGIC HISTORY

The post-Palaeozoic fossil record of dinoflagellates, despite its selectivity, is characterized by pervasive trends in the number of (cyst) species (Fig. 15.1), and by the relative distribution of major lineages (Fig. 15.2) and archeopyle types in time. Tappan & Loeblich (1971, 1972) produced the earliest comprehensive species diversity plot for fossil dinoflagellates. The curve derived from the most recent compilation of cyst diversity (Bujak & Williams 1978; based on 1977 data) retains the basic character of the Tappan–Loeblich curve, the principal differences being the marked increase in the number of species and the extension of the curve into the Triassic. The overall similarity of the Bujak–Williams and Tappan–Loeblich curves to cyst diversity plots for certain well documented areas (e.g. offshore eastern Canada), and of these to diversity curves for other plankton groups (Lipps 1970; Tappan & Loeblich 1973; Haq 1973), led Bujak & Williams (1979) to conclude that major trends in dinoflagellate (cyst) species diversity were primarily a function of changing climatic (temperature) regimes and fluctuations in eustatic sea level, with additional influence by salinity, circulation patterns, seasonal flux and nutrient supply. The selectivity of the dinoflagellate fossil record, however, requires a cautious interpretation of these curves in terms of their correlation to other biological and physical events; it does not affect discussion of the major trends themselves. The extent to which the cyst diversity curve is affected by other factors (e.g. availability and accessibility of material due to outcrop area and/or subsurface location, and interest among workers focused on certain time periods; for example, high interest in the Late Jurassic and little in the Neogene) has not been determined in a conclusive manner, although Bujak & Williams (1979) did indicate some potential problems in this area.

The earliest generally accepted (see Bujak & Williams 1981 for discussion) dinoflagellate fossil is *Arpylorus antiquus* from the Late Silurian of northern Africa (Calandra 1964; Sarjeant 1978a), which is approximately 400–410 Ma old. Following that, there is an apparent 200 Ma gap in the record prior to the relatively continuous and widespread occurrence of cysts in Mesozoic and Cenozoic sediments which began in the Middle and Late Triassic. Loeblich (1976), Taylor (1980) and Bujak & Williams (1981) suggest that some Palaeozoic acritarchs may in fact be dinophycean cysts (perhaps produced by prorocentroid thecae), thereby implying that the early fossil history of the group might not be as meagre as it now appears, and that we simply do not know the requisite morphologic criteria necessary to identify dinophyceans among the polyphyletic acritarchs. The general decline of the acritarchs in the early Mesozoic in conjunction with the initial consistent appearance of demonstrably dinophycean cysts at roughly the same time suggests that the dinoflagellates which produced 'acritarchous' cysts might have been replaced by those which produced 'dinophycean' cysts in the Triassic. Alternatively, the cyst producers could have evolved rapidly in the Triassic and Early Jurassic, or the ability to produce fossilizable cysts could have been developed within a relatively short time by one or more existing lineages. At this time, the fossil record neither supports nor refutes any of these possibilities.

Triassic and Early Jurassic assemblages are characterized by low species diversity and by relatively simple cyst morphotypes which typically lack prominent ornamentation (processes, high septa, etc.). The predominant archeopyle among species is a non-peridinioid (geniculate anterior intercalary-precingular boundary) anterior intercalary style (Types I, 2I, 3I), although combination types also occur. The oldest Triassic dinoflagellate is '*Sahulodinium*' (here considered a suessioid form), an ovoidal cyst with an intercalary and/or combination archeopyle (R. Helby & L. E. Stover, pers. comm.). The form is associated with a Middle Triassic (Anisian/Ladinian) palynomorph assemblage from the Bonaparte Gulf Basin, northwest Australia. Late Triassic assemblages commonly contain species of *Rhaetogonyaulax, Suessia, Dapcodinium* and *Comparodinium*. A seemingly provincial Norian (Late Triassic) flora from arctic Canada consists mainly of *Sverdrupiella, Hebecysta, Heibergella* and *Noricysta*, which in large part accounts for the Norian species diversity peak (maximum for Late Triassic/Early Jurassic). Late Triassic assemblages are known in Europe (Sarjeant 1963, and in Davey *et al.* 1966; Morbey & Neves 1974; Harland *et al.* 1975; Morbey 1975, 1978), Spitsbergen (Bjaerke 1977; Bjaerke & Dypvik 1977; Bjaerke & Manum 1977), arctic Canada (Fisher & Bujak 1975; Bujak & Fisher 1976; Fisher & van Helden 1979; W. A. S. Sarjeant (pers. comm.) indicates that the record in arctic Canada now extends back at least to the Middle Triassic, and possibly to the late Early Triassic), Alaska (Wiggins 1973), and Australia (L. E. Stover & R. Helby, pers. comm.).

The trend in low species diversity continues in the Early Jurassic, possibly as a function of an insufficient data base (Bujak & Williams 1981). The few studies on Early Jurassic dinoflagellates include those by Evitt (1961a, b), Gocht (1964), Wall (1965a), Morgenroth (1970), Morbey (1978), and Wille & Gocht (1979). Of the Triassic genera only *Dapcodinium, Rhaetogonyaulax*, and *Comparodinium* range into (and become extinct during) the Early Jurassic. In the Pliensbachian a minor increase in new genera occurs, including (some of which may be early gonyaulacoids) *Mancodinium, Madurodinium, Sciniocassis, Luehndea* (earliest skolochorate cyst), *Valvaeodinium*, and the characteristic *Nannoceratopsis* which continues into the Middle and Late Jurassic. *Pareodinia* first appears in the Toarcian (Sarjeant 1972; Wiggins 1975), as do the suggestively pareodinioid forms *Parvocysta, Phallocysta* and *Facetodinium* (Bjaerke 1980; Dörhöfer & Davies 1980). The earliest peridinioid cyst, *Liasidinium*, is reported (Drugg 1978) from the Sinemurian; it has an intercalary archeopyle with a transverse precingular-intercalary margin, a type which becomes increasingly important in the Late Cretaceous and Tertiary. The cingular archeopyle of *Nannoceratopsis* (Piel & Evitt 1980) is the only other major type to appear in the Early Jurassic.

There is a general increase in species diversity throughout the Middle and Late Jurassic (trend continues into the Aptian), with a sharp decrease in the number of species in sections adjacent to the Jurassic/Cretaceous boundary.

Proximochorate (*Heslertonia*) and skolochorate (*Systematophora, Rigaudella* and *Surculosphaeridium*) cyst types become increasingly abundant, as do distinctly cavate forms (*Dingodinium*, (Fig. 15.4f)) and *Sirmiodinium*. The major evolutionary diversification for this period is in the gonyaulacalean lineage, particularly during the Late Jurassic. The interval is characterized by such gonyaulacoid forms as *Gonyaulacysta, Tubotuberella, Millioudodinium, Leptodinium, Meiourogonyaulax, Scriniodinium, Psaligonyaulax, Ctenidodinium* and *Hystrichogonyaulax*. Among other gonyaulacaleans the earliest ceratioid (*Muderongia*), pyrodinioid (*Glossodinium*) and microdinioid (*Microdinium*) forms occur in the Late Jurassic, as do the first species in the *Spiniferites*-complex and the *Areoligera*-complex. The *Energlynia-Wanaea* (Fig. 15.4l) group (Sarjeant 1976; Fensome 1981) is restricted to the Middle and Late Jurassic. The pareodinioid group (e.g. *Pareodinia, Phallocysta* and *Gochteodinia*) and species of *Nannoceratopsis* are also common elements of the floras during this period. A pronounced increase in archeopyle types is apparently linked to the gonyaulacalean expansion, with most being apical, precingular, combination or panepicystal. Peridinioid cysts, and consequently peridinioid intercalary archeopyles, are extremely rare during the Middle to Late Jurassic. '*Ternia*,' a possible dinophysialean cyst (see Section 3.1), occurs in the Middle Jurassic only.

The trend of increasing species diversity continued through the late Early Cretaceous (Albian), and is followed by a severe drop in the Turonian (evidently related in part to a rise in eustatic sea level) and then by an increase again in the Santonian and Campanian. The major diversification in the Cretaceous is within the peridinioid lineage, in which a wide variety of archeopyle types (all involving at least the mid-dorsal 2a) was developed, particularly in Late Cretaceous forms (Lentin & Williams 1976; Davies 1981). Characteristic peridiniacean genera include *Subtilisphaera, Angustidinium, Ovoidinium* and *Diconodinium* in the Early Cretaceous, and *Isabeladinium, Chatangiella, Nelsoniella, Deflandrea, Trithyrodinium, Palaeoperidinium, Svalbardella, Alterbidinium, Palaeohystrichophora, Amphidiadema* and *Eucladinium* in the Late Cretaceous. Gonyaulacalean cysts, although still common, are no longer as dominant as they were in Late Jurassic sediments. Ceratioid forms are common: *Muderongia, Phoberocysta* and *Pseudoceratium* in the Early Cretaceous; *Endoceratium* and *Aptea* in the mid-Cretaceous; *Odontochitina* and *Xenascus* in the Late Cretaceous. Ceratioid cysts are unknown in sediments younger than Maastrichtian (65 Ma). Microdinioid cysts—*Histiocysta, Microdinium, Gillinia* and *Phanerodinium*—become increasingly varied in the Late Cretaceous. A few pyrodinioid species (in the genus *Dinopterygium*) occur in the Cretaceous, but this line does not become important until the Palaeogene. Finally among the gonyaulacaceans, diversity in the *Areoligera*-complex increases significantly in the Cretaceous with the appearance of such forms as *Areoligera, Canningia, Canninginopsis, Cyclonephelium* and *Senoniasphaera*. The pareodinioid lineage becomes extinct in the early Late Cretaceous, and the gymnodinioid genus *Dinogymnium* is restricted to latest Cretaceous time (Santonian to Maastrichtian).

The Cretaceous–Tertiary boundary is characterized by numerous extinctions of dinoflagellate species if compiled data are used (Bujak & Williams 1979); however, this event is not nearly as dramatic if data from single stratigraphic sections or restricted geographic areas are examined (Benson 1976; Hansen 1977, 1979; McLean 1981; Kjellström & Hansen 1981). Palaeocene species diversity is quite low, followed by a Tertiary maximum in the Early and Middle Eocene and then a return to the pervasive decrease which has continued until the Recent. Palaeogene dinoflagellate assemblages generally follow trends established in the Late Cretaceous. Archeopyle types in the gonyaulacaleans are typically precingular, apical, or panepicystal, and for peridinioids are generally Type I or Type 3I. Peridinioid species continue to be varied and abundant through the Oligocene (*Deflandrea, Ceratiopsis, Spinidinium, Eurydinium, Lejeunecysta, Phthanoperidinium, Vozzhennikovia, Selenopemphix,* and *Palaeocystodinium*), but not as much so as in the Cretaceous. The most noticeable peridiniacean event in the Palaeogene is the development of the highly diverse *Wetzeliella*-complex (*Gochtodinium, Kisselovia, Rhombodinium, Wetzeliella,* and *Wilsonidium*) in the Eocene and Oligocene from an ancestral *Apectodinium* stock (Costa & Downie 1979b; Harland 1979a; Bujak 1979). Gonyaulacalean forms are not nearly as diverse or dominant in the Palaeogene as they were in the Cretaceous. Many Tertiary gonyaulacalean cysts are skolochorate (*Hystrichosphaeridium, Oligosphaeridium, Hystrichokolpoma, Diphyes, Fibrocysta, Areosphaeridium, Systematophora, Eatonicysta*), a cyst morphotype which first became abundant in the Cretaceous. Also, genera in the *Spiniferites*-complex which includes *Spiniferites, Nematosphaeropsis, Impagidinium, Pentadinium* (Fig. 15.3f), *Cannosphaeropsis, Rottnestia, Hystrichostrogylon* and *Achomosphaera*, are common elements in nearly all Tertiary floras. The *Areoligera*-complex, after a Late Cretaceous/Early Tertiary peak (characterized by *Areoligera, Cyclonephelium, Renidinium* and *Palynodinium*), began a progressive decline in the mid-Tertiary (*Chiropteridium* and *Membranophoridium*) and became extinct in the latest Oligocene. Tertiary microdinioid cysts (predominantly *Histiocysta* and *Cladopyxidium*) are common through the Miocene. A major gonyaulacalean evolutionary event is represented by the polymorphic Tertiary pyrodinioid lineage (see Section 6) defined by a peculiar hypocystal paratabulation. A number of new pyrodinioid genera appear in the Palaeocene and Early Eocene (*Alisocysta, Eisenackia, Tubidermodinium, Eocladopyxis*) and persist through the Palaeogene along with later forms (*Homotryblium, Polysphaeridium,* and *Heteraulacacysta*). Another potentially significant Palaeogene gonyaulacalean group is the *Danea*-complex (Damassa 1981), which includes *Danea* (Fig. 15.4d), *Conneximura, Cordosphaeridium, Muratodinium, Lanternosphaeridium, Turbiosphaera, Achilleodinium* and *Thalassiphora*, and ranges from the latest Cretaceous to the Oligocene.

Neogene dinoflagellates are not well known, due to the tendency of micropalaeontologists not to examine such relatively 'young' sediments (foraminifera and coccoliths are more accurate Neogene biostratigraphic indicators and the dinoflagellate assemblages are not particularly interesting to many palaeontologists in comparison with older floras) and of neontologists not to examine such 'old' sediments (with no possibility of incubating Miocene cysts). However, those few Miocene/Pliocene assemblages that have been documented (e.g. Maier 1959; Habib 1971; Baltes 1971; Manum 1976; Barss *et al.* 1979; Costa & Downie 1979a; Harland 1979b; Ballog & Malloy 1981; Piasecki 1980; Bujak & Davies 1981) can be described as relict Oligocene floras. They typically contain lower diversity assemblages of predominantly Oligocene taxa, and perhaps represent the later stages in the progressive decrease of cyst species diversity during the Cenozoic. The apparent drop in cyst species diversity during the Neogene is an anomalous phenomenon because other major planktonic groups (calcareous nannoplankton and foraminifera) experienced an explosive evolutionary diversification in the Miocene in response to more equitable marine climates. Further work on Neogene dinoflagellates may alter this picture by the discovery of additional species, or it may confirm the relatively low number of Neogene cyst species as real by uncovering few new forms.

Quaternary (Pleistocene and Recent) dinoflagellate cyst assemblages have very low diversity, but it is seemingly greater than Neogene diversity based on current figures. Approximately 100 living cyst taxa are known (Bradford 1978) with the majority classified in a few genera: *Spiniferites, Impagidinium, Nematosphaeropsis, Tectatodinium, Operculodinium, Protoperidinium, Lingulodinium, Polysphaeridium* (*Hemicystodinium*), *Tuberculodinium, Polykrikos*, and as unnamed *Peridinium* cysts (several other genera were reduced to sections of *Protoperidinium* by Harland 1982a). Acritarchous cysts and calcareous cysts also occur in Quaternary assemblages. Modern cysts are produced mainly by species of *Gonyaulax*, but also by *Ceratium, Gessnerium, Protogonyaulax, Pyrodinium, Pyrophacus, Peridinium, Protoperidinium, Scrippsiella, Ensiculifera* and *Polykrikos*. The bulk of data on Quaternary cysts comes from the work of Wall, Harland, Reid, Dale and Bradford (see reference list for individual papers). Quaternary cysts are mainly skolochorate (gonyaulacaleans) and have precingular, combination or panepicystal archeopyles. Approximately 2100 living dinoflagellate species are known, but very few of these produce preservable cysts. Dominant lineages in the modern oceans are the dinophysoids (especially tropical; no preservable cysts), ceratioids (no preservable cysts), gymnodinioids (mostly mucoid cysts, few preservable walls), gonyaulacoids (cysts), and the *Protoperidinium* line (cysts) of the peridinioid lineage (*Peridinium* is almost exclusively freshwater). We cannot determine if there was a similar relationship between cyst producers and non-cyst producers among fossil dinoflagellates.

5 EVOLUTION

The dinoflagellates occupy a position near the base of the eukaryotic evolutionary tree (Loeblich 1976; Taylor 1976a, 1978, 1980) because they combine certain primitive characters of the prokaryotes (continuously condensed chromosomes; low levels of chromosomal basic proteins; low molecular weight cytoplasmic ribosomal RNA) along with more advanced or unusual eukaryotic features (high levels of repeated DNA; discrete phase of DNA synthesis; presence of a spindle, etc.) found in most other flagellate groups. They are therefore considered to be among the most primitive of eukaryotic groups (Loeblich 1976; Taylor 1980; Loeblich 1984) despite their great variety in morphology, habitat, cellular organization, behaviour, and in the development of highly specialized organelles in some forms (e.g. ocelloids in the Warnowiaceae which are strikingly similar to metazoan eyes, and sophisticated ejectile bodies in several of the Polykrikaceae and Warnowiaceae; see Chapter 3B and Taylor (1980) for details). The group is thought to have evolved from an early eukaryotic ancestral stock following evolution of repeated DNA (Loeblich 1976). The likely ancestral candidate could have been most similar to anteriorly flagellated desmokonts (Loeblich 1976; Taylor 1980) or like *Oxyrrhis* and the Syndinian parasites (Loeblich 1984).

Thus, dinoflagellates are thought to be a geologically ancient group, perhaps dating back to the Late Pre-Cambrian or earliest Cambrian (550–750 Ma; see Cloud & Glaessner 1982, for a recent review of terminology and biological events during this interval), based on the interpretation (Dodge 1965; Loeblich 1976; Taylor 1980; Loeblich 1984) of cellular characters which would not be preserved in the fossil record. However, the earliest record of a generally accepted (see Section 3) dinophycean cyst is that of *Arpylorus* from the Late Silurian, and this event predates the abundant occurrence of unequivocal fossil cysts in the Middle and Late Triassic by approximately 200 Ma. This apparent disparity in the predicted timing versus actual appearance of the earliest dinoflagellates illustrates the seemingly paradoxical hypotheses generated by interpreting 'hard part' features (identifiable dinophycean cysts) in the fossil record as opposed to 'soft part' features among extant forms to interpolate the early evolutionary history of the group. Several scenarios can be developed to account for the time span between the predicted and the geologically earliest preserved dinoflagellates. The most obvious and widely mentioned of them are: (i) development of preservable cysts did not occur in most groups (the exception being the group containing *Arpylorus*) until the Middle Triassic; (ii) those groups that produce fossilizable dinophycean cysts did not evolve until Late Triassic time; and (iii) preservable cysts were produced by the early dinoflagellates but they can not be recognized as such if current criteria are applied. The diversity, morphologic variation and wide geographic distribution of Late Triassic and Early Jurassic dinoflagellates, coupled with the overall primitive nature of the group and the ability of some modern peridinioid and gonyaulacoid forms to produce acritarchous cysts (see Section 2), suggest that the latter situation may be the essence of this discrepancy, and that the earliest fossil dinoflagellates are in fact represented as acritarchs. The simple thecal organization and flagellar arrangement of the prorocentroids indicate that they may be a primitive lineage among the dinoflagellates (Loeblich 1976; Taylor 1980), and perhaps this group rather than more advanced lines produced the earliest dinophycean acritarchous cysts (see Section 2).

Likewise, several apparently divergent proposals have recently been made to explain the post-Late Triassic evolution of the dinoflagellates, for which there is an abundant (yet selective) fossil record. This situation, like that for the timing of the earliest dinoflagellates, can be attributed in large part to the preferred tendency of workers to base phylogenetic inference on a particular aspect of the group: micropalaeontologists on preserved cysts, and neontologists on structural or biochemical features of living motile stage cells. It is, therefore, not surprising that two contrasting hypotheses have been developed to explain late (post-Late Triassic) dinophycean evolution. In short, neontologists (Loeblich 1976; Taylor 1980) propose an increase in the number of thecal plates, in conjunction with some other criteria, through the whole Phanerozoic (this is essential to the argument and must be contrasted to the palaeontological viewpoint which is restricted to the last 200 Ma); while palaeontologists (Eaton 1980; Dörhöfer & Davies 1980) postulate a general reduction in plate number using the Late Triassic to Recent cyst record. These hypotheses have been termed the plate increase model and plate reduction model, respectively, by Bujak & Williams (1981). The fossil record appears to provide little support for the neontological viewpoint, and in fact there are several inversions in the fossil record of the postulated sequence of evolutionary events based on cellular information, but proponents maintain that the demonstrated inadequacy of the record (e.g. total absence of groups, 'extinction' of ceratioids in the Maastrichtian, etc.) counterweighs this (Taylor 1980). Most recently, Bujak & Williams (1981) have proposed an alternative, the plate fragmentation model, in an attempt to reconcile both palaeontological and neontological factors into a unified evolutionary framework for the dinoflagellates. The major aspects of each of the three alternative models follows.

The conceptual basis of the plate increase model (Fig. 15.7a) derives from the early work of Bergh (1881a, b) and Bütschli (1885). Loeblich (1976) presented biological data which support the model and it has recently been expanded by Taylor (1980). Taylor recognized five fundamental organizational types among living dinoflagellates—prorocentroid, dinophysoid, gonyaulacoid, peridinioid, and gymnodinioid—which he considers to represent the major dinophycean lineages. A sixth type, the fossil nannoceratopsioid lineage, should be included as it combines (Piel & Evitt 1980) features of perhaps three of the above lineages (see Section 5). According to this model, thecate dinoflagellates represent a

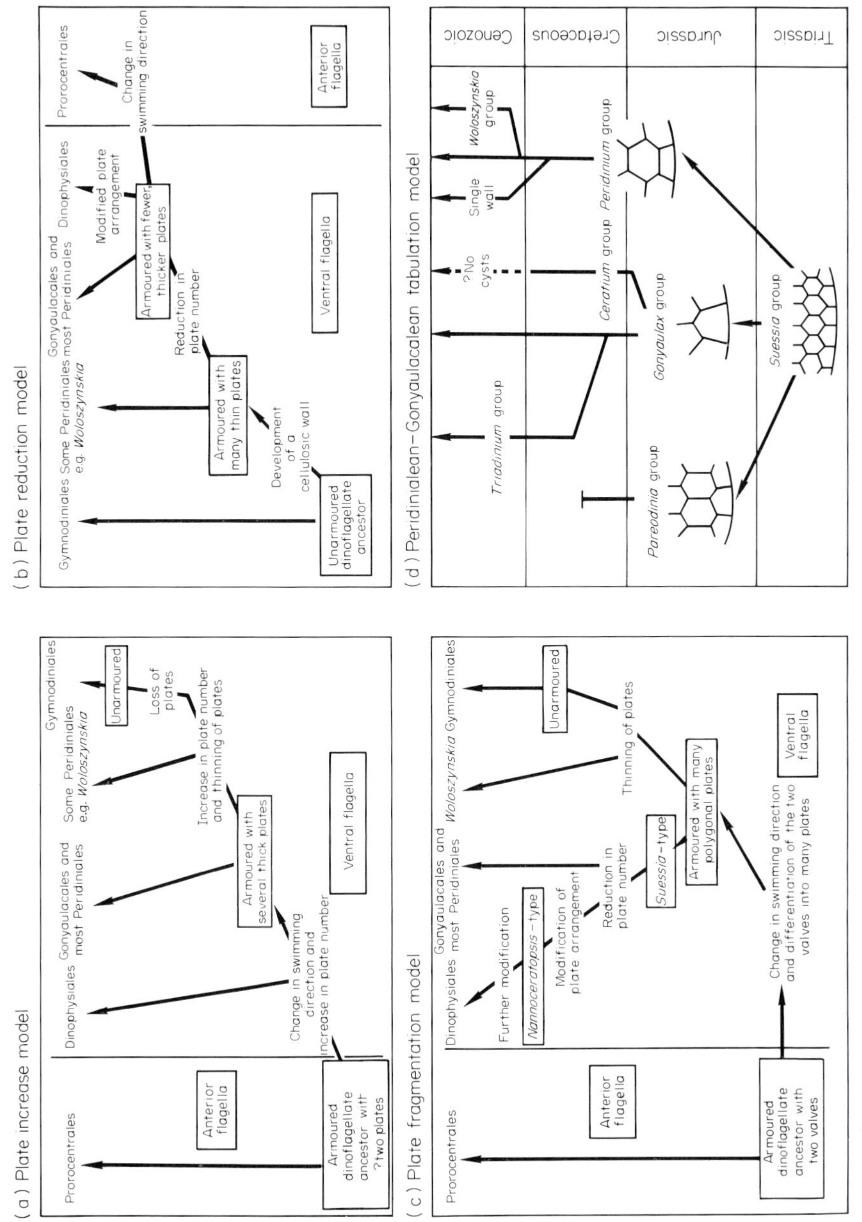

Fig. 15.7. Models for dinoflagellate evolution. (A) Plate increase model based mainly on biological data, (B) Plate reduction model based on palaeontological data. (C) Plate fragmentation model based on an analysis of both fossil and living taxa. (D) Development of peridinialean, gonyaulacalean and pareodinioid tabulation from proposed suessioid ancestral stock (all redrawn from Bujak & Williams 1981).

more primitive condition while most (all?) unarmoured forms are a derived, or more advanced, state (Loeblich 1976). Among thecate forms, the prorocentroids (desmokonts) are considered most primitive due to their simple thecal construction, with no sign of girdle or sulcus, and apically inserted flagella (the commonest arrangement in other flagellates). The prorocentroids presumably gave rise at an early stage to the dinophysoids, thecate dinokonts which appear intermediate between desmokonts and dinokonts in several ways (Chapter 2). Loeblich (1976) suggested that the two prorocentroid valves are equivalent to the epi- and hypothecae of the thecate dinokonts (other than the dinophysoids), but Taylor (see Chapter 2) has argued for the fission line during division as being the fundamental separation between the valve homologues. The peridinioid and gonyaulacoid lineages are considered to be later evolutionary lines than the dinophysoids, possibly evolving through an intermediate form similar to the mid-Jurassic nannoceratopsioid line, which shares features of the dinophysoids, peridinioids and gonyaulacoids. The two derived lineages have more thecal plates than the ancestral dinophysoids, and although the number and arrangement of plates is quite similar in both (Loeblich & Loeblich 1979; Eaton 1980), they can be readily separated by plate symmetry relative to the flagellar insertion (Taylor 1980), and by differences in certain aspects of their life cycles such as predominance of desmoschisis in gonyaulacoids and eleutheroschisis in peridinioids (F. J. R. Taylor, 1982, pers. comm.). In addition, all gonyaulacoids *sensu* Taylor are photosynthetic; many peridinioids are not. The most advanced forms, the gymnodinioids, may have arisen from the peridinioids, or gonyaulacoids, or perhaps along several lines from both. A continuation of the trend towards an increasing number of plates and de-emphasis of thecal development is seen in the modern *Woloszynskia* (an intermediate peridinioid/gymnodinioid form), with the ultimate expression in this model being the development of numerous hexagonal amphiesmal vesicles in the unarmoured gymnodinioids (some *Gymnodinium* species have delicate platelets similar to *Woloszynskia*).

Micropalaeontologists, on the other hand, have independently produced an evolutionary picture of the dinoflagellates by examining the sequence of major morphological events as recorded in the Mesozoic to Tertiary fossil record. The plate reduction, or plate fusion, model (Fig. 15.7b) thus developed, interestingly arrives at a hypothesis diametrically opposed to that of the plate increase model, i.e. that a large number of paraplates seems to precede smaller numbers so that the gymnodinioids are considered the most primitive forms and the prorocentroids the most advanced. While the preserved record of fossil cysts does tend to support this trend, the very selectivity of that record (see Section 2) renders the a priori assumption of the validity of the observed sequence questionable in the absence of supportive data. Proponents (Eaton 1980; Dörhöfer & Davies 1980) of the model infer an unarmoured gymnodinioid ancestral stock which may have produced acritarchous cysts if in fact it produced cysts at all (the enigmatic Silurian genus *Arpylorus* causes difficulty with this assumption because it has what are interpreted as paraplates, thereby inferring a very early armoured form preserved in the fossil record). Most recently, Loeblich (1984) has adopted this type of model as a consequence of his view that the athecate genus *Oxyrrhis* may be the most primitive living dinoflagellate. It produces no cyst.

The development of a cellulosic wall with numerous polygonal plates presumably resulted in the appearance of the Late Triassic suessioid forms, whose affinity is unknown and which have more plate series and also more paraplates per series than either the peridinioids or gonyaulacoids. Several other early fossil cyst genera, notably *Dapcodinium* and *Rhaetogonyaulax*, are similar to *Suessia* in general appearance by having slightly more paraplates than other apparently younger lineages. However, they differ from *Suessia* in having three or more large, geniculate anterior intercalaries; like *Suessia*, their affinity to other major groups is not known (speculative evidence regarding hypocystal paratabulation suggests they may be related to gonyaulacoids, but modern gonyaulacoids lack dorsal anterior intercalaries). The continued reduction in the number of thecal plates, in conjunction with a more simplified arrangement of those plates into a few distinct longitudinal series, eventually led to the peridinioid and gonyaulacoid lines. Modification of the plate arrangement in either the gonyaulacoids or peridinioids gave rise to the dinophysoid lineage. According to the model, the most recent major evolutionary event is represented by the non-cyst-producing prorocentroids, having arisen by a change in the swimming direction towards the area of flagellar insertion and by fusion of the epi- and hypothecal plates into the two large prorocentroid valves. Dörhöfer & Davies (1980) have proposed two mechanisms for decreasing the number of plates in several lineages through time (termed 'unidirectional plate growth' and 'multidirectional plate fusion'), but there appears to be little support for this notion in light of observational evidence from fossil cysts and on recent work (Gocht 1979) correlating growth bands on gonyaulacoid thecae to the morphological expression of those features on cysts (the 'relict' parasutures on which Dörhöfer and Davies base their argument are most likely related to regions of thecal growth, according to Gocht; furthermore, subtle angulations in archeopyle outline, which Dörhöfer and Davies propose are related to triple plate junctions, are difficult to unequivocally establish and seem more related to post-depositional deformation and preservation rather than evolutionary processes).

The plate fragmentation model (Fig. 15.7c) was proposed (Bujak & Williams 1981) to unify palaeontological and neontological data into a single coherent framework, and it represents the most recent comprehensive approach to the problem: Taylor's (1980) lengthy discussion on dinoflagellate evolution, which incorporated fossil data, was the first such synthesis in recent years; unfortunately it was not mentioned by Bujak & Williams (1981). The ancestral forms in this model are similar to the living prorocentroids (anteriorly flagellated, cellulosic theca primarily composed of two valves). Differentiation of the theca into numerous plates and a change in the swimming direction (which resulted in one flagellum becoming longitudinal and the other transverse) gave rise to the gonyaulacoids and peridinioids, and ultimately in one line to the gymnodinioids (and some peridinioids, e.g. *Woloszynskia*) through reduction of the theca to the unarmoured condition. The dinophysoids are believed to have been derived from an ancestral gonyaulacoid/peridinioid stock by further modification of plate arrangement and morphology, perhaps through a *Nannoceratopsis*-like intermediary. Bujak & Williams (1981) also provide a somewhat more detailed account of the evolution of the gonyaulacoid and peridinioid lineages, and of the *Pareodinia* group, all of which they postulate to have arisen from a *Suessia*-like ancestor (Fig. 15.7d). Dörhöfer & Davies (1980) present a slightly modified version of this model, in which a suessioid lineage gives rise to the pareodinioids, which in turn is the ancestral stock for the ceratioid dinoflagellates. Both Taylor (1980) and Bujak & Williams (1981) consider the ceratioids to have been derived from the gonyaulacoids, and not the pareodinioids, based on the interpretation of ceratioid paratabulation as a modified *Gonyaulax* style (Wall & Evitt 1975; Taylor 1980). Additional data (discovery of transitional forms, extending ranges of critical forms, further cyst/theca correlations, etc.) should lead to an increased understanding of dinoflagellate evolution in the near future.

6 PHYLETIC TRENDS IN SELECTED LINEAGES

During the past 10 years or so, there has been an increased effort to document morphological trends within supposed evolutionary lineages among fossil dinoflagellate cysts. Interpretation of morphological trends in an evolutionary context is difficult due to the selective preservation of fossil dinoflagellates and to our limited current understanding of the effects of environmental factors on cyst morphology. In many cases only broad trends can be recognized, but for others relatively detailed relationships among several species in the lineages can be established with reasonable confidence. Recognition, documentation and critically constrained interpretation of cyst lineages provide data for extremely precise biostratigraphic application. Lineage studies also have the potential to indicate the tempo and direction of evolutionary development in the dinoflagellates, and possibly to allow micropalaeontologists to establish the relationship between environmental and genetic expression of particular morphologic characters through time. Few such studies have been completed to date and therefore few, if any, generalizations can be made. Results from selected studies are briefly mentioned to illustrate the potential in this area.

Eaton (1971) published the first detailed lineage study in which he proposed a morphogenetic series for five species of gonyaulacacean dinoflagellates from

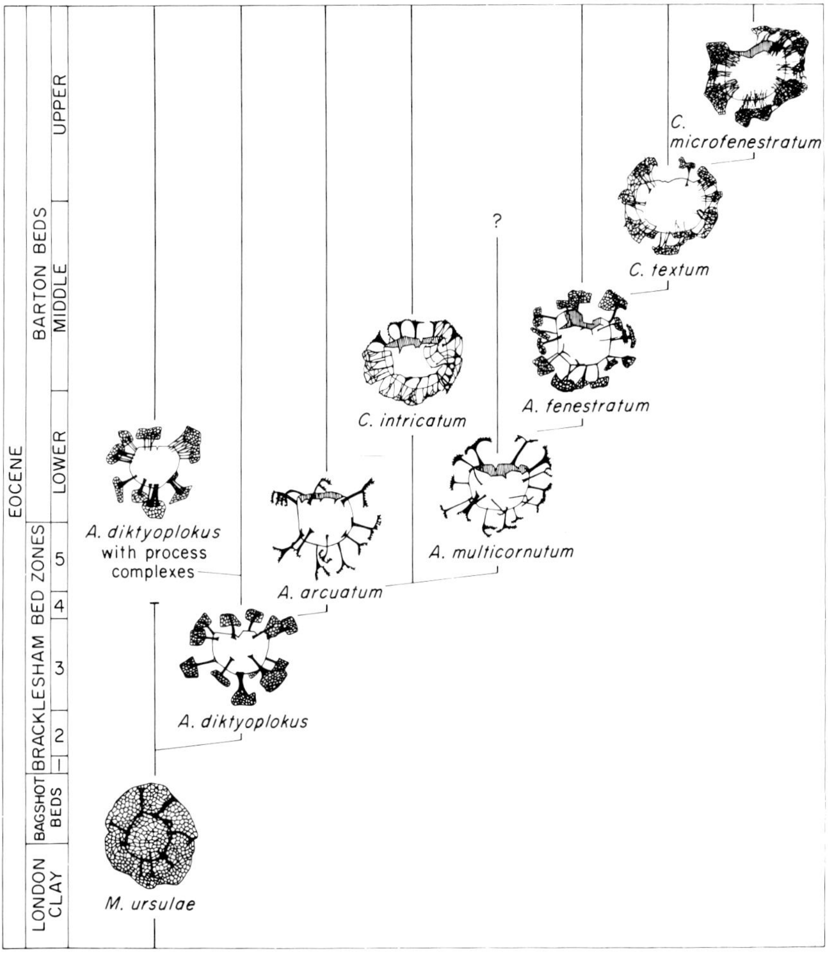

Fig. 15.8. Proposed phylogeny of the *Eatonicysta–Areosphaeridium* lineage for the Early and Middle Eocene in south-eastern England. The ancestral form in the series is *Eatonicysta ursulae* (formerly *Membranilarnacia*). The species is characterized by a process-supported reticulate membrane completely surrounding the main body. *Areosphaeridium diktyoplokus* is thought to have developed through a reduction in the membrane to areas immediately surrounding the process tips, and this line eventually gave rise to forms with process complexes united distally by the trabecular network. Along another morphologic gradient from *A. diktyoplokus*, a continued reduction in process terminations and the development of additional hypocystal processes gave rise to *A. arcuatum*. *Glaphyrocysta intricata* (formerly *Cyclonephelium*) and *A. multicornutum* seem to have evolved from *A. arcuatum* through the development of cingular processes, additional hypocystal processes, and (in the case of *G. intricata*) complex process branching. *Areosphaeridium multicornutum* is proposed to have evolved into *A. fenestratum* by the development of distal process platforms, and *G. texta* (formerly *Cyclonephelium*) were derived from *A. multicornutum* by process branching which eventually formed process complexes. *Glaphyrocysta microfenestrata* (formerly *Cyclonephelium*) evolved from *G. texta* by an increase in the process number and complexity of the distal process platforms (reproduced by Bujak 1976).

the Early and Middle Eocene of southern England (Fig. 15.8), based on analysis of morphology, paratabulation and stratigraphic distribution. This lineage is now referred to as the *Eatonicysta-Areosphaeridium* lineage. Bujak (1976) described the late Middle Eocene continuation of the lineage, which he considered to represent a true evolutionary series. The major morphologic trends among the species include a progressive reduction among early forms in the reticulate membrane which had completely surrounded the oldest species (*Eatonicysta ursulae*), development of cingular processes and of additional hypocystal processes in intermediate forms, and increasingly complex process branching through which process complexes were developed in younger forms (see Fig. 15.8 for a detailed discussion).

Stratigraphic and morphologic relationships among Late Jurassic and Early Cretaceous species of *Muderongia* in Australia (R. Helby, in press) suggest that evolution in this ceratioid genus proceeded along two basic lines: one group of species with a single antapical horn, and another with two antapical horns (one of which can be severely reduced). In the first group, older (Valanginian–Hauterivian) species such as *M. tetracantha* tend to have two nearly symmetrical paracingular horns, while species higher in the section (Hauterivian–Barremian) like *M. imparilis* tend to be distinctly asymmetrical. This morphologic gradient implies a possible ancestor–descendent relationship between the asymmetric *Muderongia* species and early species of *Odontochitina*, particularly *O. operculata*. Among the group with two antapical horns, older forms (Late Jurassic–Barremian) such as *M. simplex* tend to have very short paracingular horns, while the horns on younger species (Hauterivian–Aptian) such as *M. mcwhaei* are typically much longer. Helby does not discuss the evolutionary implications of these intrageneric trends, however he does consider *Muderongia* to have evolved from a *Senoniasphaera* stock.

Among older dinoflagellates, the evolution of some Early Jurassic species of *Comparodinium* (Fig. 15.9) has been carefully documented (Wille & Gocht 1979) for several sections in southern Germany. The oldest species in this lineage, *C. stipulatum*, has an apparently non-tabular ornament distribution. Younger species supposedly derive from this form, and they are characterized by a clearly defined paratabulation which suggests close affinity to *Rhaetogonyaulax*. In another early cyst lineage, the Middle Jurassic genus *Energlynia* was first suggested to be an ancestor of *Wanaea* by Sarjeant (1976) based on an overall morphological similarity and a similar archeopyle type for the two. Recently, Fensome (1981) demonstrated that species in these genera exhibit an evolutionary trend from species of *Energlynia* which lack paracingular septa to species of *Wanaea* which have one or two such septa. Within *Wanaea*, the height and structural complexity of the septa increase from older to younger species, going from the Callovian *W. spectabilis*/*W. digitata* types through intermediate forms to the late Callovian–Oxfordian species *W. fimbriata* and *W. clathrata* (see Stover & Williams 1982, their Figure 3). The ranges of these

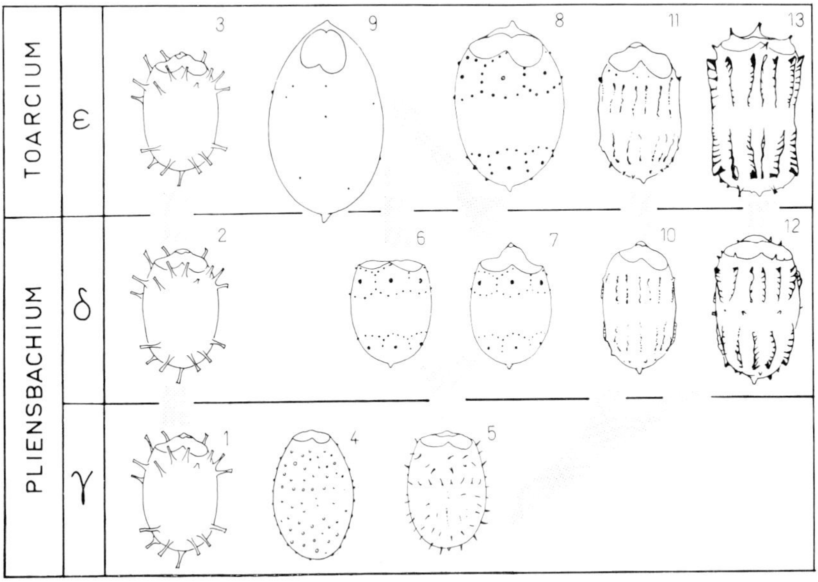

Fig. 15.9. Proposed evolution of Early Jurassic *Comparodinium* species. 1–3: *C.* cf. *koessenium*; 4: *C. perpunctatum*; 5: *C. stipulatum*; 6–8: *C. punctatum*; 9: *C. punctatum* var. *magnum*; 10–11: *C. lineatum*; 12–13: *C. scalatum* (reproduced from Wille & Gocht 1979).

species are poorly known, and subsequent data may alter the interpretation of this morphologic series.

Detailed lineage studies for peridiniacean cysts are relatively rare. Stover (1974) described the stratigraphic distribution of various morphologic characters in several Palaeocene and Eocene species of *Deflandrea* and related genera from Australia, but those taxa were not construed to constitute a lineage (see Section 8.4) for discussion of this paper). Malloy (1972) interpreted a succession of Late Cretaceous and Early Tertiary peridiniacean cysts from offshore Gabon as a lineage representing a 'morphologically transitional evolutionary series.' The general morphological trend among these forms is an elongation of the main body and horns from an ancestral subpentagonal morphotype (*Andalusiella*) to a derived fusiform morphotype (*Palaeocystodinium*); a secondary trend involves the loss of one antapical horn on later forms. The stratigraphic ranges of the species documented in this study do not seem directly related to changing palaeoenvironmental conditions in the section, although fluctuations in their relative abundances may be attributed in part to environmental factors.

The Palaeogene *Wetzeliella*-complex, a suite of six peridiniacean genera containing abour sixty-five species, has been intensely studied for years, principally in terms of its biostratigraphic utility but more recently as the subject of phylogenetic analysis. The complex probably represents one of the better examples of a lineage in the dinoflagellate fossil record. Details of the phylogeny within the group, however, are at present unresolved due to an incomplete appreciation of the biological significance and variability for certain morphological features (archeopyle, distribution of processes, ectophragm), which results in an overly complex generic taxonomy. The ancestral stock for the *Wetzeliella* lineage has not been determined; among living dinoflagellates, species in the complex seem mostly closely related to *Protoperidinium* (Harland 1982a) although *Wetzeliella*-like cysts are unknown for the past 25 Ma. The oldest fossils in the complex occur in the Late Palaeocene and are attributed to *Apectodinium*, a genus which persists until the early Oligocene. *Apectodinium* is considered (Harland 1979a; Costa & Downie 1979b) to represent the ancestral stock for the later genera—*Gochtodinium, Kisselovia, Rhombodinium, Wetzeliella*, and *Wilsonidium*—which appear in the Early Eocene and are common throughout the Eocene and Oligocene. Costa & Downie (1979b) have suggested a phylogeny for all species in the complex based on development and variation in certain features, e.g. horn length, archeopyle morphology, and ectophragm structure. They attach considerable significance to these features and to the succession in which they appear in the stratigraphic record. Bujak (1979) and Harland (1979a) have proposed phylogenetic models restricted to selected groups of taxa in the complex. Harland (1979a) discussed evolution within Palaeocene and Early Eocene species of *Apectodinium*; Bujak (1979) suggested that a soleiform archeopyle (i.e. one with a broadly rounded lower margin) and rounded rhomboidal endocyst outline may distinguish a separate line of evolution within the complex represented by *Rhombodinium* and *Gochtodinium*. Published stratigraphic and morphologic data seem to support these models.

An alternative approach to the interpretation of broad evolutionary gradients in the complex is shown in Fig. 15.10. No explicit taxonomy is required other than a general recognition of an *Apectodinium* line and a *Wetzeliella* line of descent; the figure is simply a schematic presentation illustrating various combinations of three morphologic characters—archeopyle shape, process presence/absence and distribution, and ectophragm development—and their distribution in time. A rough phylogenetic model can thus be constructed from the data without the restrictions imposed by conventional taxonomy. In general, the group appears to have experimented with a number of morphologic characters (including several not shown in Fig. 15.10) but because of the nature of their record only general morphological trends can be recognized with confidence. Phylogenetic or morphogenetic inference to these morphological gradients cannot be made in a substantive manner due to unknown biological, preservational, and environmental factors which contributed to the preserved record.

In most cyst lineages, such as the *Wetzeliella*-complex, overall morphologic similarity among species is the principal criterion used to establish affinity and interpret evolution. In contrast, genera belonging to the pyrodinioid lineage (Goodman, in press) are characterized by a pervasive hypocystal paratabulation pattern that remains constant within an otherwise polymorphic group. These

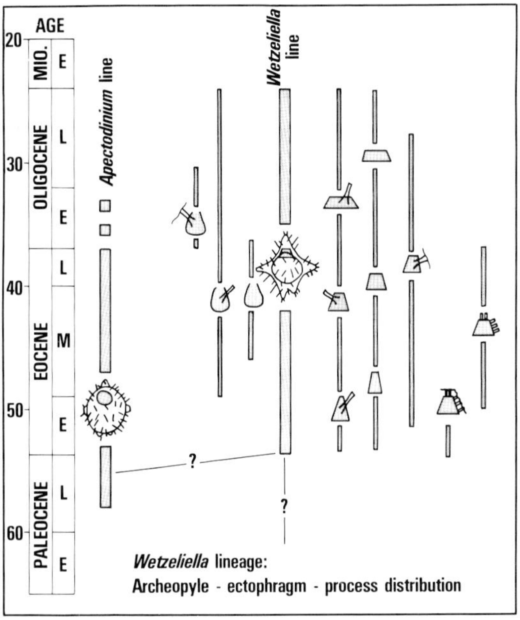

Fig. 15.10. Schematic illustration of possible phylogeny of the *Wetzeliella*-complex. Rather than giving generic ranges, the stratigraphic distribution of morphotypes with known combinations of archeopyle shape, process distribution, and ectophragm development are plotted. Ancestral *Apectodinium* lineage first appeared in the Late Paleocene (Thanetian). Members of the *Wetzeliella* lineage (includes all other described genera) appeared in the Early Eocene and became extinct at the end of the Oligocene. Forms with soleiform archeopyles are on the left side of the main *Wetzeliella* range line. Those with quadrangular archeopyles are on the right. Opercula with a process arising from mid-paraplate signify non-tabular process distribution. Those with processes aligned along their margin indicate a parasutural to penitabular alignment of processes. Wavy line at process tips indicates presence of an ectophragm. Opercula with no processes indicate morphotypes lacking periphragm ornamentation (from Goodman, in press).

forms can be included in the modern dinoflagellate Family Triadiniaceae Dodge, which differs from the Gonyaulacaceae in having an antapical plate that contacts five rather than six superjacent plates (Fig. 15.6). Within the pyrodinioid lineage (Fig. 15.2), two evolutionary lines can be differentiated on the basis of apical (para)plate configuration: one with apparent affinity to the modern *Pyrodinium*, and the other with affinity to the modern *Protogonyaulax* (see Fig. 15.6). Fossil cysts with a pyrodinioid paratabulation include *Polysphaeridinium* (= *Hemicystodinium*), *Heteraulacacysta* and *Homotryblium*; they are most common in Eocene and younger sediments. Fossils with a protogonyaulacoid style are *Alisocysta, Eisenackia, Eocladopyxis, Tubidermodinium*, some species of *Hystrichokolpoma, Dinopterygium*, and possibly *Glossodinium*; this group ranges from the Late Jurassic to Recent, but it is most typical of the Palaeogene. Consistent features of (para)tabulation provide an explicit framework for a more reasonable interpretation of phylogenetic relationships among genera in these lines than has been proposed in the past. The close evolutionary linkage would in all probability not be recognized using traditional morphologic criteria (overall shape, similar archeopyle types, types of ornamentation, etc.). In fact, the genera listed above are classified at present in seven cyst-based families (Norris 1978b), whereas a more reasonable approach in light of similar paratabulation patterns would be to include them all in the Triadiniaceae. This example illustrates the necessity of interpreting the morphology of fossil cysts in relation to modern dinoflagellate morphology and systematics. A number of additional lineages among fossil cysts probably remain unrecognized due to our inability to resolve those characters necessary to demonstrate biological affinity.

Several additional studies address certain aspects of dinoflagellate evolution on a higher level than the examples given above. Wall & Dale (1968a) discussed broad phylogenetic relationships among fossil and living dinoflagellates which would now be classified in the orders of Peridiniales and Gonyaulacales. More recently, Dörhöfer & Davies (1980) outlined their highly conjectural model for early dinoflagellate evolution in the Late Triassic to Early Cretaceous. They propose that an ancestral suessioid stock gave rise to the pareodinioid group in the Middle Jurassic. In the Late Jurassic, the ceratioids evolved from the pareodinioids according to these authors, although an alternative model which considers the ceratioids as a gonyaulacalean lineage (Wall & Evitt 1975) is generally accepted.

Lentin & Williams (1976) discussed in detail the various archeopyles in fossil peridiniacean dinoflagellate cysts, classified them according to shape and type, and presented the distribution and relative importance of the major types from the mid-Cretaceous to the Miocene. Davies (1981) took the next logical step in this approach by further subdividing peridiniacean archeopyles and plotting the major trends in archeopyle shape (without explicit regard to a taxonomy) during the past 100 Ma. Davies recognized three basic categories within the hexa type 2a paraplate (Fig. 15.5d): *deltaform*, in which the length of the anterior transverse margin of 2a is less than the length of the posterior transverse margin; *omicroform*, in which the anterior and posterior margins are equal or nearly so; and *omegaform*, in which the anterior margin is greater in length than the posterior margin. Davies then plotted the stratigraphic distribution of each category and subtype (based on maximum transverse versus longitudinal dimension of 2a), and the apparent trend among subtypes within each category is that the older ones tend to be narrow and the younger ones tend to be broad. Using the deltaform 2a as an example, narrow subtypes are more common in the late Aptian to Cenomanian (they do occur as young as the Eocene), and broad subtypes are more characteristic for the Turonian to Eocene. The hexa 2a style occurs principally in the Late Cretaceous and Early Tertiary; some of the less common peridiniacean types of 2a configuration are uniquely Tertiary and include the penta, quadra and asymmetrical hexa types according to Davies' preliminary charts. General trends in morphologic characters such as the archeopyle undoubtedly have broad evolutionary significance; however, we have just begun to appreciate the insights into dinoflagellate phylogeny provided by the careful documentation of such data.

7 CYSTS IN QUATERNARY SEDIMENTS

7.1 Ecology

The distribution of motile stage dinoflagellates in modern oceans is largely a function of temperature, salinity and distance from the coast (Chapter 11). On a global scale, species tend to occur in broad latitudinal bands forming characteristic low, middle and high latitude assemblages. Within these latitudinal marine climatic regimes, populations vary in composition along a gradient from estuarine/lagoonal to open oceanic environments. Factors influencing the species distribution of marine plankton are numerous; they are summarized in Chapter 11. Although details are now available the interactions are complex and only broad generalizations can be made in most instances. Some information on the distribution of motile marine dinoflagellates can be found in several articles, including those by Graham (1941), Braarud (1962), Wood (1964), and D. B. Williams (1971a), and Taylor (1976b); the most recent reviews of the distribution of living thecae are given by Lentin & Williams (1980) Tappan (1980) and Chapter 11. Absolute abundance, or productivity, of dinoflagellate cells is a more localized phenomenon and is dependent on prevailing and/or seasonal incident and effective radiation, salinity, temperature, nutrient supply, and current conditions (Chapter 11).

The distribution of cysts is dependent on several factors, including distribution of motile stage species, biological and ecological controls over encystment (Chapter 14), and the behaviour of cysts as sedimentary particles in the hydrographic regime. Recognition of latitudinal (tropical, temperate, arctic) cyst assemblages (D. B. Williams 1971a; Reid & Harland 1977; Wall *et al.* 1977) provides a rough empirical correlation between Quaternary cysts and marine climatic zones, which is suggestive of the latitudinal control over thecae (Wall 1971a). Within the latitudinal zones, the cyst-producing species are predominantly neritic so that there is a reasonable probability of the motile cell reaching the euphotic zone after excystment from a benthonic cyst. Thus, most marine cysts occur in sediments deposited in shelf and upper slope environments, and in shallow water estuarine and lagoonal facies; oceanic cysts are relatively rare (Wall *et al.* 1977) and may represent transported assemblages. Cysts of freshwater species are also common, but factors related to their distribution are apparently not as complex as in the marine realm. Wall (1970, 1971a) has discussed various environmental and depositional factors which affect absolute abundances and species composition both in the plankton and in the cyst bottom assemblage. More recently, Reid & Harland (1977) postulated a combination of latitude, water depth, water mass, and sediment type as the primary factors controlling cyst distribution. Wall *et al.* (1977), in the most extensive study of Quaternary cysts to date, suggested that species–water type relationships and prevailing current systems are the principal factors that control cyst distribution in bottom sediments. Because of red tides, the distribution of certain marine thecae are known in considerable detail (see LoCicero 1975; D. L. Taylor & Seliger 1979); these data are summarized in Chapter 11.

Palaeoecological inference using cysts is dependent on several considerations, among which are the biological function of the cyst and the mechanisms involved in cyst dispersion and deposition. Wall (1971b) reviewed the early literature on extant dinoflagellates and listed four biological functions of dinoflagellates cysts—reproduction, protection, propagation and dispersion—which might affect their distribution in sediments. Recent evidence (Dale 1976, 1977b; Anderson & Wall 1978; Pfiester & Anderson, Chapter 14) suggest that preservable (fossilizable) cysts are predominantly thick-walled hypnozygotes formed during a sexual phase of the dinoflagellate life cycle. In the marine realm, Wall *et al.* (1977) presented a complex suite of hydrological systems (Figure 15.11) which have potential significant effects on cyst dispersion in two continental margin models: the 'slope water' model and the 'upwelling-offshore divergence' model (see Chapters 11A and 14).

Documentation of the life cycles for individual taxa probably has considerable potential for providing clearer understanding of the mechanisms controlling phytoplankton distribution in modern oceans, and ultimately distribution patterns of cyst assemblages as well. Information regarding the ecological preferences of individual species of Quaternary dinoflagellate cysts is sparse, but a growing literature now exists which documents such data for several marine species (Chapter 11A). Interpretation of the ecological significance of cyst bottom assemblages in relation to distribution patterns of motile stage cells is still highly speculative. Any interpretation of the bottom thanatocoenosis must be cautiously tempered with the fact that the cyst assemblage does not necessarily reflect accurately either the qualitative or quantitative aspects of the planktonic community due to: (i) taxonomically restricted cyst production (Dale 1976); (ii) variable and largely undocumented cyst/theca abundance ratios (Reid & Harland 1977); (iii) differential preservation (Dale 1976; Reid 1974; Harland 1977); (iv) dispersion by currents (Wall *et al.* 1977); (v) sparse data on encystment triggering mechanisms (e.g. light, temperature; see Chapter 14); (vi) other factors including processing techniques (Reid 1974; Dale 1976). This is even true for regions with unusually favourable sedimentary and hydrographic conditions, and in which extensive plankton records have been kept for several years (Dale 1976). Improved sampling techniques (Anderson *et al.* 1982) may alleviate this situation to some degree.

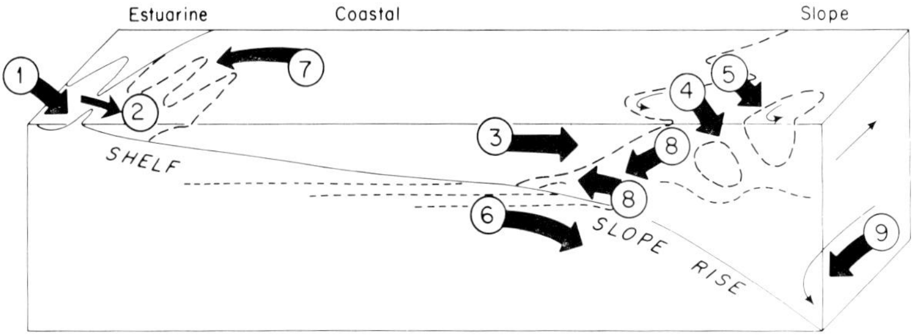

(a) 'Slope water' model for cyst dispersion

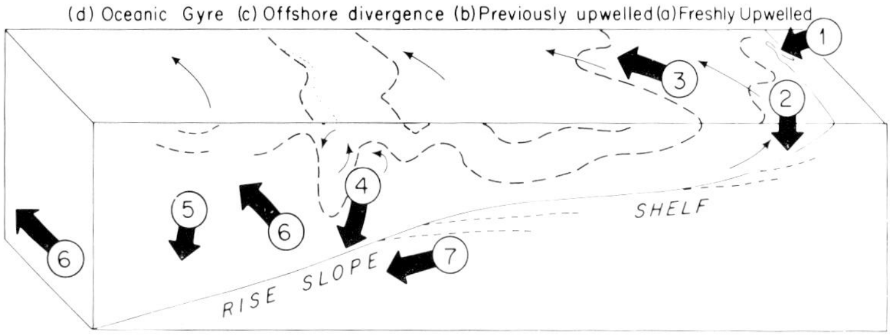

(b) Upwelling-offshore divergence model for cyst dispersion

Fig. 15.11. Models of dynamic motions which contribute to cyst dispersal in the marine environment. *Model A*. 1 = estuarine–lagoonal confinement; 2 = outwash from lagoons; 3 = offsetting coastal waters; 4 = eddies of coastal water migrating seawards into offshore zones; 5 = detached parcels of coastal water containing neritic microfloras; 6 = winnowing of relict shelf deposits; 7 = intrusion of coastal-shelf water into estuaries by mixing or bottom intrusions; 8 = mixing, sinking and bottom indrafting of oceanic water with shelf water across the continental slope and the outer shelf zones; 9 = alongslope subsurface currents of subpolar origin in intermediate waters. *Model B*. 1 = localized inflows of estuarine water; 2 = vertical sinking of the products of nearshore plankton blooms in upwelling areas; 3 = seaward dispersal of nearshore bloom products by wind-driving transport; 4 = vertical sinking of cells from the offshore zone of divergence; 5 = vertical sinking from surface oceanic waters; 6 = alongslope subsurface countercurrents at intermediate depths; 7 = winnowing of outer shelf and slope deposits during submarine erosion (reproduced from Wall *et al*. 1977).

Palaeoecological significance can be attached to distinctive associations of fossil cysts within depositional basins, and particularly to compositional trends in cyst assemblages either through time (in stratigraphic sections) or across an environmental transect (laterally) at a given time. Broad palaeoenvironmental trends can be determined from these data (in conjunction with independent criteria) because the cyst assemblages are ultimately related to environmental factors and thus have palaeoecological meaning in themselves. The palaeontologist therefore is not obligated to reconstruct every step in the process of converting a planktonic dinoflagellate community to a fossil cyst assemblage prior to extracting a coarse palaeoenvironmental interpretation from the assemblage with reasonable confidence.

There are few data on the distribution and ecological significance of Quaternary dinoflagellates in stratigraphic sections. Wall (1970, 1971a) reviewed this subject from studies completed through the early 1970s, and the most recent summaries are given by Reid & Harland (1977) and Harland (1977). In general terms, Quaternary cyst assemblages are low in species diversity, although assemblages containing twenty or more species are not uncommon (Wall 1967; Wall & Warren 1969; Harland 1977). Absolute abundances of cysts can be quite high (Wall & Dale 1968c; Wall & Warren 1969) despite the low diversities. Vertical stratigraphic sections are typically represented by a succession of species groups characterized by one or a few dominant species. Wall & Dale (1968c) recognized five such associations in a borehole through Early Pleistocene sediments in Norfolk, England. Likewise, Wall & Warren (1969) identified six zones in Red Sea piston cores and related them to a Late Pleistocene–Holocene chronostratigraphic scale. Wall & Dale (1973, 1974) described three stratigraphically sequential dinoflagellate associations from Late Quaternary deep-water cores in the Black Sea; the lowest of these associations was almost completely dominated by only two species: *Tectatodinium psilatum* and *Spiniferites cruciformis*. Vertical sequences sometimes show little variation in species content probably due to environmental stability over a very short time interval. For instance, Morzadec-Kerfourn (1966) found little variation in a marine Holocene section from France, as did Harland & Downie (1969) for a Middle Pleistocene sequence of estuarine sediments from England. Recurrent species associations have been demonstrated most conclusively in an Early Pleistocene borehole sequence from Israel (Rossignol 1962) in which one association dominated by *Polysphaeridium* (=*Hemicystodinium*) *zoharyii* alternates three times with another association characterized by species of *Lingulodinium*, *Spiniferites* and *Operculodinium* (all referable to the modern genus *Gonyaulax*). Compositional changes in assemblages are often quite rapid and are attributed by most workers to changes in regional climatic (temperature) or other physical conditions. Rossignol (1962) first recognized this and postulated that the two alternating associations from Israel represented normal versus low salinity conditions. The associations described by Wall & Dale (1968c) and Wall & Warren (1969) were attributed to fluctuations between temperate and cooler glacial periods. Harland (1977) regarded the sequences of 'favourable' and 'unfavourable' zones as defined by dinoflagellate cysts (in terms of 'productivity', or abundance) in Late Quaternary continental shelf sediments around the British Isles to be linked directly to climate. He felt that the zones probably reflect changes in the configuration of water masses. Morzadec-Kerfourn (1975) suggested that directional trends in the relative abundance of certain species are indicative of brackish deposition for a Holocene section in France.

First and last appearance datums for species are relatively rare in Quaternary sections due to the limited amount of time available (c. 2·8 Ma) during that period for evolutionary processes. Therefore, any biostratigraphic application of dinoflagellates to Pleistocene and Holocene stratigraphy must rely primarily on the recognition of characteristic species associations (Reid & Harland 1977). A biostratigraphic framework based on the succession of climatically controlled cyst assemblages has been established as a climatostratigraphic sequence for sections around the British Isles (Harland 1977) and in the North Sea (Harland *et al.* 1978). Assemblage changes in these studies are related to postulated alterations in water mass regimes in the North Atlantic. The biostratigraphy can then be utilized as a climatostratigraphy because the changes in water mass are linked to fluctuations in the positions of the North Atlantic Current and Polar Front. In this context, the species associations can be used for preliminary correlations of sections on a regional scale (Harland 1977).

7.2 Regional distribution patterns

Lateral distribution patterns of Quaternary cysts can be described in terms of two trends: an onshore–offshore ('environmental', or bathymetric) trend and a latitudinal ('climatic') trend (Wall 1971a; Wall *et al.* 1977). There is a relatively limited data base from which to draw conclusions regarding these trends. Early work on environmental trends consists largely of density patterns on sediment surfaces or total cyst content per unit volume of sample rather than relative abundances for individual species. These early data have been reviewed in detail by Wall (1970, 1971a), and include local studies from the Gulf of California (Cross *et al.* 1966), the Orinoco Delta region (Muller 1959), the south-eastern Mediterranean Sea (Rossignol 1961), the Bahamas (Traverse & Ginsburg 1966), and a Pacific atoll in the Caroline Islands (McKee *et al.* 1959). More recent studies present species relative abundance data in addition to absolute abundances across onshore–offshore environmental gradients. The most comprehensive of these studies (Wall *et al.* 1977) suggests that a species richness (diversity) index tends to increase in offshore assemblages along several shelf-to-slope transects in the North and South Atlantic. Absolute abundance of cysts was also shown to increase offshore in some regions, according to Wall *et al.* (1977); however, such trends cannot be extrapolated to include all continental margin configurations because of complexities introduced in cyst dispersion patterns by hydrographic processes (Fig. 15.11). In addition, increasing offshore diversity and abundance may be in part artificial due to contamination by reworked older Quaternary cysts and by offshore transport of neritic and estuarine cysts to deeper depositional settings. Local variations in absolute abundance is indicated in transects along the continental shelf and slope offshore South Africa (Davey 1971; Davey & Rogers 1975).

Wall *et al.* (1977) consider the primary onshore–offshore trend among modern cysts to involve different taxa reaching their peak relative abundances in a succession that parallels changing bottom topography and hydrologic conditions with increasing distance from shore. For example, Wall *et al.* (1977) found species of *Spiniferites*, *Lingulodinium*, *Polysphaeridium* and *Tuberculodinium* are most abundant in estuarine environments, while *Operculodinium*, *Selenopemphix* and *Trinovantedinium* are generally more abundant in shelf sediments. In deeper environments, the peak abundance of *Nematosphaeropsis* is in continental slope and rise sediments, and *Impagidinium* species are restricted to and most abundant in pelagic sediments near and beyond the outer continental shelf. Davey (1971) and Davey & Rogers (1975) found a similar trend between *Spiniferites ramosus* (more abundant inshore) and *Operculodinium centrocarpum* (more abundant offshore), each of which is presumably associated with a particular oceanic current. Davey & Rogers (1975) report species of *Nematosphaeropsis*, cysts of *Protoperidinium* species, and *Impagidinium* in slope samples, which is in general agreement with the observations of Wall *et al.* (1977) for these taxa. Further support for this trend comes from Recent sediments off the coast of France (Morzadec-Kerfourn 1977) in which *Lingulodinium machaerophorum* and some species of *Spiniferites* are most abundant in estuarine and coastal environments, whereas *Impagidinium* and other *Spiniferites* species are characteristic of more offshore zones. Species associations composed of the various taxa also reflect the changes in relative abundance among species on a cumulative basis. Therefore, the distribution of characteristic associations in terms of qualitative and quantitative species content has a recognizable and possibly predictable onshore–offshore trend (which would be extremely useful if a reasonable correlation with information about the thecate stage could be demonstrated).

Distribution data compiled on a larger scale indicate major biogeographic trends in Quaternary cysts which can be correlated to regional water masses and to latitudinal marine climatic (temperature) regimes. A series of studies by Reid (1972, 1974, 1975, 1977) shows a close relationship between water masses and cyst assemblages around the British Isles. This area is one of the most intensely examined on a regional basis for cysts, especially when the distribution data compiled by Reid (op. cit.) and Harland (1974, 1977) are combined with the thecal distribution patterns of Dodge (1981b).

The first study to show a latitudinal (temperature) differentiation of cyst assemblages was an analysis of dinoflagellate cyst biofacies (Fig. 15.12) for the North Atlantic (D. B. Williams 1971b), although the influence of local water mass is also evident from the distribution pattern. Wall *et al.* (1977) recognized this latitudinal trend in the tropical and temperate Atlantic as a gradient in characteristic species associations (in terms of species content and relative abundance) for both nearshore and offshore environments, and as an increase in species diversity towards lower latitudes. Furthermore, a distinctive high latitude cyst assemblage is indicated by pioneering studies of Recent arctic sediments in the Beaufort (Harland *et al.* 1980) and southern Barents (Harland

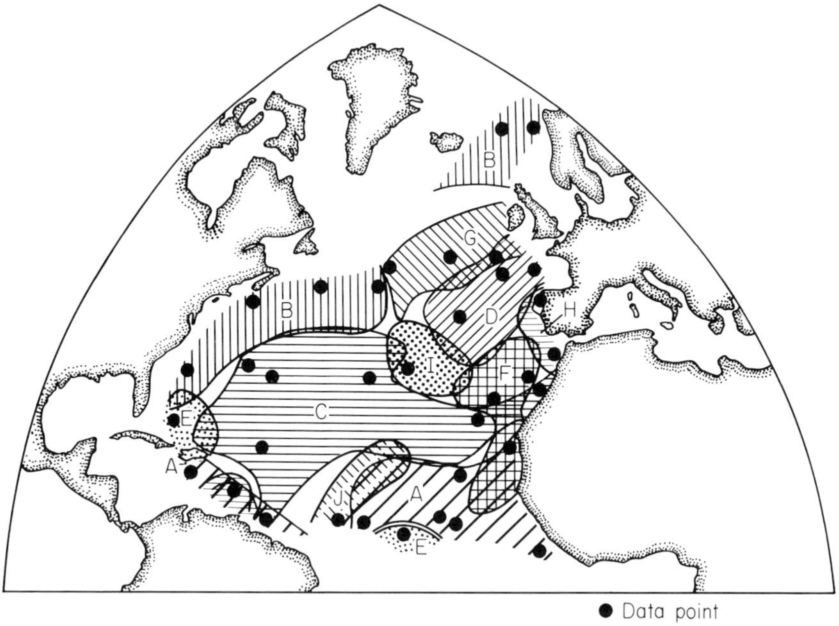

Fig. 15.12. Biofacies map of dinoflagellate cyst assemblages in the North Atlantic, made by superimposing plots of similarity between actual assemblages and ideal assemblages A–J (determined by principal components analysis of assemblage data). Assemblages are characterized by high abundances of particular species (redrawn from D. B. Williams 1971b).

1982b) Seas. The distribution of Quaternary cysts in North Atlantic bottom sediments has been reviewed recently by Reid & Harland (1977), and a comprehensive compilation of data from this region (Harland 1983) is available.

Biogeographic patterns of Quaternary cysts strongly indicate that characteristic assemblages are associated with marine climatic regimes and water masses. Our data base, however, has not increased substantially since Wall (1970, 1971a) last reviewed the status of Quaternary dinoflagellate micropalaeontology, and it remains largely inadequate for areas other than perhaps the North Atlantic (which is poor in information on the distribution of thecate stages; alternatively, there are good distribution data on motile cells from the Indian Ocean (Taylor 1976b), but cyst data are lacking). Some additional information on Quaternary cysts is available for widely separated regions including, among others, New Zealand (Wilson 1973), southern California (Damassa 1980; Ballog & Malloy 1981), Trondheimsfjord, Norway (Dale 1976), the Firth of Forth, Scotland (Harland 1981), the Persian Gulf (Bradford 1975, 1977), the Grand Banks (Harland 1973b), Israel (Rossignol 1961, 1962, 1969), and the Black Sea (Wall & Dale 1973, 1974; Wall *et al.* 1973).

8 DISTRIBUTION OF MESOZOIC AND TERTIARY CYSTS

8.1 Palaeoecology

The application of fossil dinoflagellate cyst information to palaeoecological studies is a relatively recent development. Distinctive associations of cysts were recognized by Scull *et al.* (1966) as being potential indicators of transgressive and regressive facies changes for Lower Tertiary subsurface sections in Texas, based on the size and complexity of processes of certain skolochorate forms. One year earlier Vozzhennikova (1965) had similarly suggested that certain Mesozoic cyst morphotypes could be used to indicate depositional environments, with 'thick-walled' forms more abundant in 'neritic' environments (probably inner to middle neritic) and 'thin-walled' process-bearing forms more characteristic of 'open marine' conditions (probably outer neritic to upper slope). These and several other of the early papers on dinoflagellate palaeoecology were summarized in the first review article on the palaeoenvironmental significance organic-walled microplankton, written by D. B. Williams & Sarjeant (1967). Most recently, Davies *et al.* (1982) reviewed the status of dinoflagellate palaeoecology and classified the various techniques that have been used into four categories. They are: (i) absolute abundance of cysts per unit of sediment or water; (ii) relative abundance of cysts with respect to other palynomorphs or plankton groups; (iii) species diversity and dominance; and (iv) relative abundance of species, groups of related species, or morphotypes within particular assemblages. No generalizations can be drawn with confidence from the studies completed to date because they represent a wide variety of geographic areas and geologic ages, and they tend to ignore information on modern dinoflagellates; therefore, any conclusions made at this time would be premature and undoubtedly would be modified significantly with additional data. However, recent studies strongly indicate that fossil cyst assemblages have a definite palaeoenvironmental character, and that palaeoecologic inference from cyst data is possible within reasonable confidence limits. Most of these studies involved quantitative data of species composition to differentiate associations and recognize trends in time. In many cases, dinoflagellate data must be related to independent criteria in order to attach some type of palaeoecologic meaning to them. The utility of Mesozoic and Tertiary cysts as palaeoenvironmental indicators should be greatly improved by extending ecologic models derived from Quaternary assemblages into older sedimentary deposits, and by correlating cyst data to related factors, e.g. lithology, geochemistry, other microfossil groups, etc. The four categories into which Davies *et al.* (1982) subdivide palaeoenvironmental studies provide a convenient order in which to review our present understanding of this subject for the Mesozoic and Tertiary.

Determination of the absolute abundance of cysts per unit volume of water or sediment is a technique which, to date, has been used almost exclusively for studying Quaternary dinoflagellates (McKee *et al.* 1959; Davey & Rogers 1975; Wall *et al.* 1977; and others). Fauconnier & Slansky (1978, 1980) have used this method to study Cretaceous and Tertiary cysts in north-west Europe and Northern Africa. This method is covered in detail by Davies *et al.* (1982), and will not be discussed further in this section as its application to ancient cyst assemblages has not been fully demonstrated. Relative abundance of cysts with respect to other palynomorphs, microfossil groups, or to other criteria, is a technique that has been applied to Quaternary (Reid 1975) as well as older assemblages. The most commonly used ratio is that between dinoflagellate cysts and spore–pollen. Manum (1976) observed changes in the cyst/pollen ratio and type of palynodebris in upper Eocene through middle Miocene sediments at several DSDP sites from Leg 38, in the Norwegian–Greenland Sea. He postulated that the relative frequencies of the fossil groups reflected changes in environmental conditions probably controlled by shoreline displacements. In this model, high cyst/pollen ratios and low amounts of debris indicate more nearshore conditions or perhaps changes in river transport. In another example, the occurrence of dinoflagellate cysts at particular intervals in well sections was demonstrated (Partridge 1976) to correlate with shallow marine incursions within an overall non-marine depositional sequence in the Lower Tertiary of the Australian Gippsland Basin. Non-marine deposits in the area are characterized by distinctive spore–pollen assemblages which lack dinoflagellate fossils. The utility of this study was enhanced because micropalaeontological changes were tied to seismic stratigraphic sequences and regional application of the palaeoenvironmental data was therefore possible. Mebradu (1978) calculated numerous relative abundance ratios among four palynomorph categories (microplankton; microforaminiferal linings; spores and pollen; tasmanitids) for the Upper Jurassic of Dorset, England, and suggested that changes in the ratios were related to environmental fluctuation. He correlated the high relative abundance of microplankton (predominantly dinoflagellate cysts) to transgressive, or deeper water, depositional phases in the basin. Changes in the relative abundance of dinoflagellate cysts have also been linked to fluctuations in organic carbon and phosphate content in sediments from the mid-Cretaceous of the Paris Basin (Fauconnier & Slansky 1978), and in the Palaeocene and Eocene of the Gafsa Basin, Tunisia (Fauconnier & Slansky 1978, 1980). These results are preliminary, but they have significant implications for the interpretation of possible environmental influence, and additional studies correlating micropalaeontological and geochemical data should confirm similar useful relationships.

Species diversity and dominance are two variables with potential utility in dinoflagellate palaeoecology, but they have seldom been used. Rather than indicating particular environments, these criteria show trends through time which may be related to depositional cycles, nutrient supply, or to other physical and chemical parameters. Wall *et al.* (1977) showed that a species diversity index for Recent cysts tended to increase in offshore assemblages for transects in the Atlantic Ocean. Goodman (1979) used dominance (summed abundance of the two most abundant species divided by the total number of specimens counted per sample) and diversity (total number of species in each sample) trends in a Lower Eocene dinoflagellate assemblage from Maryland (Fig. 15.13a) to suggest possible environmental control for six cyst communities in the sections. The cyst communities, designated A–F, had previously been defined using relative abundance and cluster analysis, and there was a positive correlation between dominance/diversity gradients and major changes in assemblage (species) composition. Using the Quaternary diversity model of Wall *et al.* (1977), Goodman suggested that the relationship between these variables reflected an inshore–offshore distributional trend (Fig. 15.13b), perhaps within an inner neritic to estuarine setting. May (1980) reports similar results for an Upper Cretaceous sequence in New Jersey, in which changes in dominance and diversity trends correspond closely to previously established palaeoenvironments based on invertebrate fossils and lithology. Relatively low to moderate species diversity and dominance by one or two species are characteristic features of assemblages from nearshore gulf (estuarine) sediments. This observation may be roughly analogous to modern 'red tides' (virtually monospecific blooms), which occur typically in coastal embayments or very close to the coast (Chapter 11A), although the two cannot be compared directly because of uncertainties regarding the time represented by most ancient sediment samples. More normal marine shelf assemblages tend to have moderate to high numbers of species with low dominance by any one taxon, in agreement with Goodman (1979). Ford (1979) found a positive correlation between dinoflagellate cyst diversity and the relative abundance of spore–pollen in marine Oligocene sediments from South Carolina, and from this relationship concluded that diversity was related to the influx of a terrestrially-derived nutrient source for that area.

The relative abundance of species or cyst morphotypes is the most widely used measurement to describe dinoflagellate assemblage composition, and to determine palaeoenvironmental trends by changes in species associations through time. Downie *et al.* (1971) used relative abundance to define four cyst associations in the Lower Eocene of southern England (Fig. 15.17). The *Spiniferites* (formerly *Hystrichosphaera*) association is a gonyaulacacean assemblage dominated by species of *Spiniferites*, *Achomosphaera*, *Cordosphaeridium* and *Hystrichosphaeridium*. The *Micrhystridium* association is characterized by high numbers of the acritarch genera *Micrhystridium* and *Comasphaeridium*. The *Areoligera* association, which is gonyaulacacean, contains numerous specimens of *Areoligera* and *Glaphyrocysta* species. The *Wetzeliella* association is dominated by species of the peridiniacean genera *Wetzeliella* and *Deflandrea*. The associations usually occur in particular facies, and therefore they are

(a)

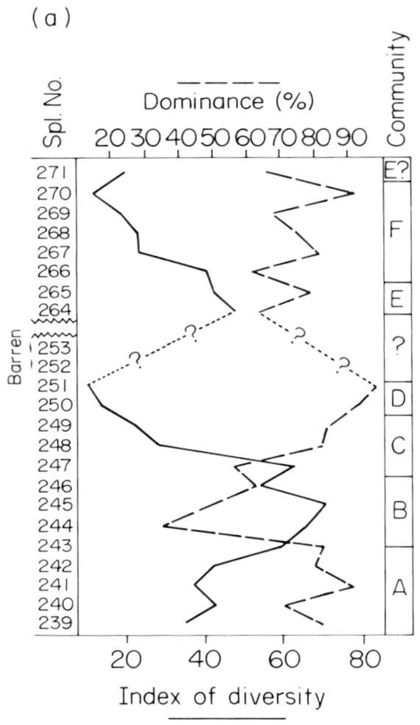

(b)

Cyst community	Relative diversity	Relative dominance	Suggested palaeoenvironmental trend
B	High	Low	More offshore
A/B transition	High	Moderate	More offshore
A	Moderate	Moderate	More offshore
E	Moderate	Moderate	More offshore
F	Low to moderate	Moderate	More inshore
C	Low to moderate	Moderate to high	More inshore
D	Low	High	More inshore

Fig. 15.13. (a) Dominance and species diversity for dinoflagellate cysts in outcrop sections from the Lower Eocene of Maryland (reproduced from Goodman 1979). (b) Relationship between relative dominance, relative species diversity, and suggested palaeoenvironmental trend in Fig. 15.12A (modified after Goodman, 1979).

interpreted as being characteristic of particular palaeoenvironments. The *Areoligera* and *Spiniferites* associations are interpreted to indicate 'open marine' conditions (probably outer neritic/upper slope), the *Micrhystridium* association inner neritic, and the *Wetzeliella* association lagoonal or brackish water environments. These associations thus can be used to recognize transgressive and regressive depositional cycles in the Hampshire and London Basins. Results from a palynological investigation of Sparnacian strata in the Paris Basin (Gruas-Cavagnetto 1967, 1968) generally confirm the findings of Downie *et al.* (1971) because the dinoflagellate associations dominated by species of the *Wetzeliella*-complex (*Apectodinium homomorphum* and *A. parvum*) were recovered from predominately lagoonal facies in the basin. Marine facies contained species of the *Spiniferites* association, and early transgressive intervals were characterized by an acanthomorphid acritarch association similar to that in England.

Quantitative dinoflagellate assemblage data are seldom available in the published literature. However, there are a few localities for which such data are given for Lower Tertiary sections in the Anglo–Paris–Belgian Basin, and in each case one or more of the associations of Downie *et al.* (1971) can be recognized, thus making regional comparisons possible. For example, distinct *Micrhystridium* associations occur in the Lower Eocene of Germany based on the data of Morgenroth (1966), and Gocht (1969) presented relative abundance data which permits recognition of an *Areoligera* association in the Palaeocene and a *Spiniferites* association in the Lower Eocene of Germany (Williams 1977). Abundant quantitative data from the Palaeocene (Schumacker-Lambry 1978) and Lower Eocene (De Coninck 1968, 1975) of Belgium can also be used to recognize each of the four associations defined in southern England. Changes in relative abundance and species composition were also used by Liengjaren *et al.* (1980) to correlate palaeoenvironments to a number of cyst associations for the Upper Eocene and Lower Oligocene of southern England and by Hochuli (1978) to explain assemblages in terms of regional climatic change for the Oligocene and Lower Miocene of the central and western Paratethys.

Two somewhat different approaches to the definition of relative abundance in fossil assemblages have been used by Harland (1973a), Davey (1970) and Chateauneuf (1980). Harland compared the relative abundance of two major groups of cysts by using the ratio of the number of gonyaulacacean cyst species to the number of peridiniacean cyst species, which he called the gonyaulacacean ratio. This ratio tended to be higher in more open marine environments in the Upper Cretaceous Bearpaw Formation of Alberta, which is in agreement with the findings of Downie *et al.* (1971) who found gonyaulacacean associations in similar depositional settings, and with the distributional data of Taylor (1976b) for the Indian Ocean. A second method which may have palaeoecological application is abundance plots for individual species. In a comparison of the relative abundance of two species from the Lower Oligocene of the Paris Basin,

Chateauneuf (1980) reported that *Gerdiocysta conopeum* was more abundant in normal marine environments with relatively high species diversity. In contrast, *Wetzeliella gochtii* was abundant in brackish water, low salinity environments; species diversity in these intervals was low, and dominance high with *W. gochtii* sometimes comprising over 50% of the cyst assemblage. Similarly, the relative distribution of *Palaeohystrichophora infusorioides* and *Cleistosphaeridium huguoniotii* in the Cenomanian of France and England (Davey 1970) may be related to environment, but this has not been demonstrated.

The most recent advance in dinoflagellate palaeoecological analysis has been the application of various statistical techniques to quantitative and binomial (presence–absence) data. Brideaux (1971) applied cluster analysis to binomial data to demonstrate two recurrent species associations of organic microplankton from the Lower Cretaceous of Alberta, and proposed that the associations appeared to be a response to depositional or environmental factors. Cluster analysis has also been applied to presence–absence data in order to substantiate species associations defined by rank abundance analysis (Goodman 1979) from the Eocene of Maryland. Relative abundance (quantitative) data have also been clustered to define associations in the Upper Eocene (Watkins 1979) and Upper Oligocene (Ford 1979, 1981) of South Carolina. Despite the recent trend towards statistical analysis of data and our ever increasing data base on the palaeoecological significance of fossil dinoflagellates, the potential for palaeoenvironmental interpretation and correlation using species associations remains largely unexploited at the present time. The application of dinoflagellates to palaeoecological problems will probably be most reliable within individual depositional basins due to the biological and ecological selectivity of the fossil record which would severely limit detailed comparisons between widely separated areas and ages. Recognition of more general trends, e.g. a tendency for increasing species diversity offshore, may have more widespread application.

8.2 Biogeography and provincialism

The biogeography and provinciality of dinoflagellates during the Mesozoic and Cenozoic is the final stage in understanding cyst palaeoecology. Distribution patterns for species and species association in space and time allow correlation with local/regional palaeoenvironmental conditions or perhaps with characteristic palaeoceanographic events. These data in turn affect the application of dinoflagellates to biostratigraphy, in particular to age determination and biostratigraphic correlation. All species do not occur in all areas simultaneously due to varying environments and a spectrum of environmental tolerances among organisms. These factors also affect such long-term processes as migration, adaptation, phenotypic variation, speciation, extinction, and others, which ultimately translate into a sequence of first and last occurrence data for a suite of species in a stratigraphic section or for a local area. These data must be interpreted in the context of regional and global palaeoceanography (climate, eustatic sea level, oceanic currents, etc.) and of local depositional processes (sedimentation rate, basin subsidence, etc.) to obtain the maximum precision in biochronologic determination. Reliable correlation of biologic events from one sedimentary basin to another, or across an ocean basin, likewise depends on a reasonable understanding of biogeographic species distribution through time. Simply put, the complexities introduced by environment and sedimentary deposition must be considered whenever the distribution of a species is given significance in terms of absolute time. Biogeography provides a generalized framework for this type of analysis, and increases the potential utility of fossil dinoflagellate cysts as palaeoceanographic indices.

Most of the information concerning geographic and stratigraphic distribution of fossil cysts comes from North America, western Europe, Australia, and the surrounding continental margins; somewhat more limited data are available for some areas in the USSR. Data from other areas such as South America are virtually non-existent, or are at best rare and sporadic. Any interpretation of dinoflagellate biogeography is therefore biased by this severely limited data set. Furthermore, there has not been a rigorous effort to define biogeographic trends for those regions with sufficient data until very recently. Thus, the study of dinoflagellate palaeobiogeography is in its formative stages, somewhat behind that for other major microfossil groups, and only a few papers represent our available syntheses to date. These studies generally confirm the findings of Wall *et al.* (1977), who postulate that latitude (or climate) appears to exert the principal influence over the distribution of recent cyst assemblages in terms of large-scale patterns.

The concept of biogeography was first applied to fossil dinoflagellates by Norris (1965), who proposed that provincialism in certain Callovian and Neocomian assemblages was a function of latitudinal differences in palaeoclimate. Norris recognized a low diversity Boreal province occupying the north circumpolar region, and a higher diversity Tethyan province to the south; diversity gradients were attributed to latitudinal factors. A probable southern hemisphere anti-Boreal province was also inferred by Norris, but it could not be demonstrated conclusively due to lack of data. Subsequent studies (Habib 1975, 1977; Fisher & Riley 1980) have generally confirmed these initial findings. In a recent study, Lentin & Williams (1980) recognize three provincial realms for Campanian peridiniacean dinoflagellates (Fig. 15.14). They propose that each of the provincial floras corresponds to a latitudinal climatic regime rather than to local palaeoecological conditions. The Malloy suite is characterized by species of *Andalusiella, Ceratiopsis, Lejeunecysta* and *Senegalinium*, and is apparently restricted to tropical and subtropical palaeolatitudes. The Williams suite is a warm temperature assemblage comprised of species of *Alterbidinium, Chatangiella, Isabeladinium, Spinidinium* and *Trithyrodinium*. The cool tempera-

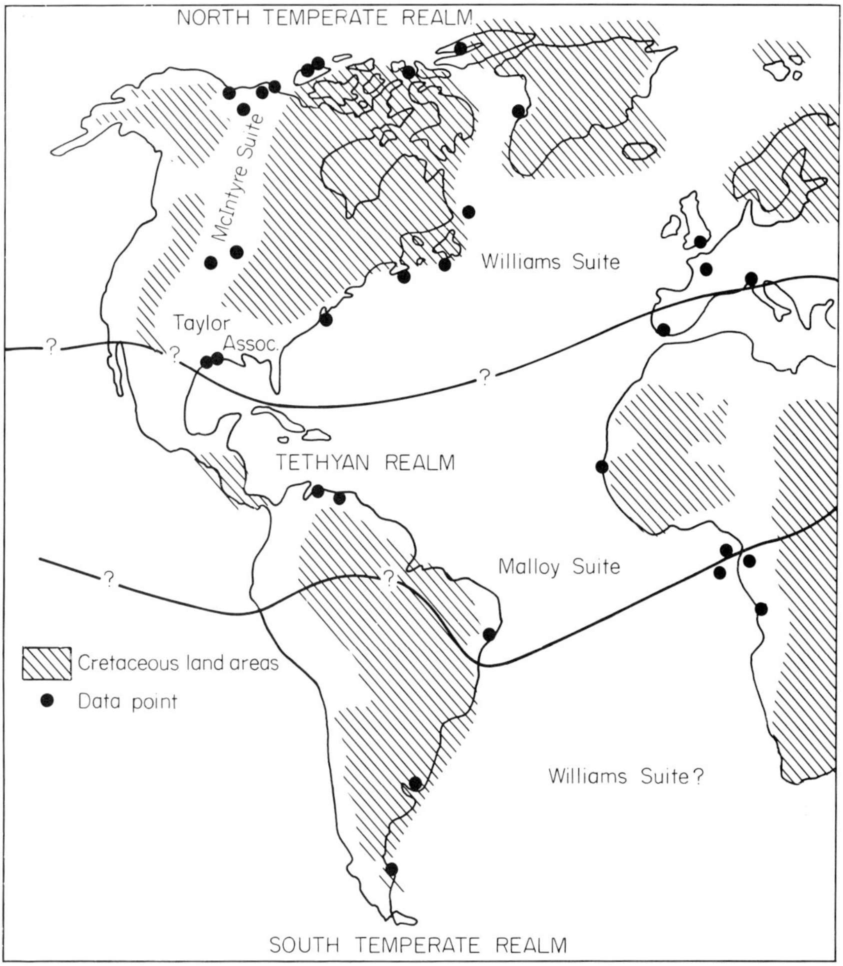

Fig. 15.14. Distribution of peridiniacean provincial suites during the Campanian (Late Cretaceous). Note the latitudinal character of the pattern (redrawn from Lentin & Williams 1980).

ture, or boreal, McIntyre suite contains *Chatangiella* and *Laciniadinium*. Mixed assemblages are reported in transitional regions between the major provinces.

Tertiary biogeography has not been studied in detail. Williams & Bujak (1977a) plot distribution patterns for several Cenozoic species, and suggest a possible ecologic influence for those distributions. This could well be a latitudinal feature similar to that proposed for Cretaceous cyst provinces. One rather well defined Tertiary dinoflagellate flora is the Late Eocene to Early Oligocene transatlantic assemblage (Wrenn & Beckman 1982) characteristic of high southern latitudes. Elements of this assemblage have been reported from the

West Ice Shelf area (Kemp 1972), the northern Ross Sea (Wilson 1968), McMurdo Sound (McIntyre & Wilson 1966; Wilson 1967), Seymour Island (Hall 1977), southern Argentina (Archangelsky 1969), DSDP sites 270 and 274 (Kemp 1975) and 280–283 (Haskell & Wilson 1975) between Australia and Antarctica, and sites 511 and 513 on the Falkland Plateau in the south-west Atlantic (Goodman & Ford 1983).

The *Wetzeliella*-complex, a characteristic Early Tertiary (Fig. 15.2) suite of approximately sixty-five species in six genera, is the only group of fossil cysts for which biogeographic trends have been mapped (Fig. 15.15) through time (Goodman, in press). In general, the distribution of the group appears to be restricted primarily to the middle latitudes, although this pattern may be due in part to limited sample availability in lower latitudes. The Palaeogene biogeographic history of the complex involves an initial appearance in several areas during the Late Palaeocene followed by a pronounced geographic diversification in the Early and Middle Eocene. Beginning in the Late Eocene and continuing to the Late Oligocene, the group underwent a progressive decrease in number of species and retreated from the southern latitudes, possibly in response to a global drop in mean surface water temperature which occurred during this time. The complex went extinct at the end of the Oligocene. The documentation of Mesozoic and Cenozoic dinoflagellate biogeography, and the determination of the palaeoceanographic significance of the distribution patterns, will certainly become the subject of increased activity among micropalaeontologists in future years.

8.3 Distribution in stratigraphic sections: biostratigraphy

Biostratigraphy is the branch of stratigraphy that deals with the spatial distribution and temporal relations of fossils and fossiliferous rocks. It involves age determination, biostratigraphic zonation, and stratigraphic correlation. Several microfossil groups, e.g. the planktonic foraminifera, have been applied to biostratigraphy for many years (see Haq & Boersma 1978); other groups, e.g. radiolarians, calcareous nannofossils, and dinoflagellates have been used as biostratigraphic tools for only the past 15–20 years, due in large part to their utility in providing age determination in regions of hydrocarbon exploration (mainly in the middle and high latitudes). These planktonic microfossils are ideally suited for biostratigraphic application due to their small size, morphologic diversity, and common occurrence in marine sediments around the world.

Fig. 15.15. Biogeographic distribution of the *Wetzeliella*-complex during the (a) Late Palaeocene, (b) Eocene, and (c) Oligocene. The group occurs primarily in a northern middle latitude belt, although a somewhat more weakly delineated southern belt is apparent from the data.

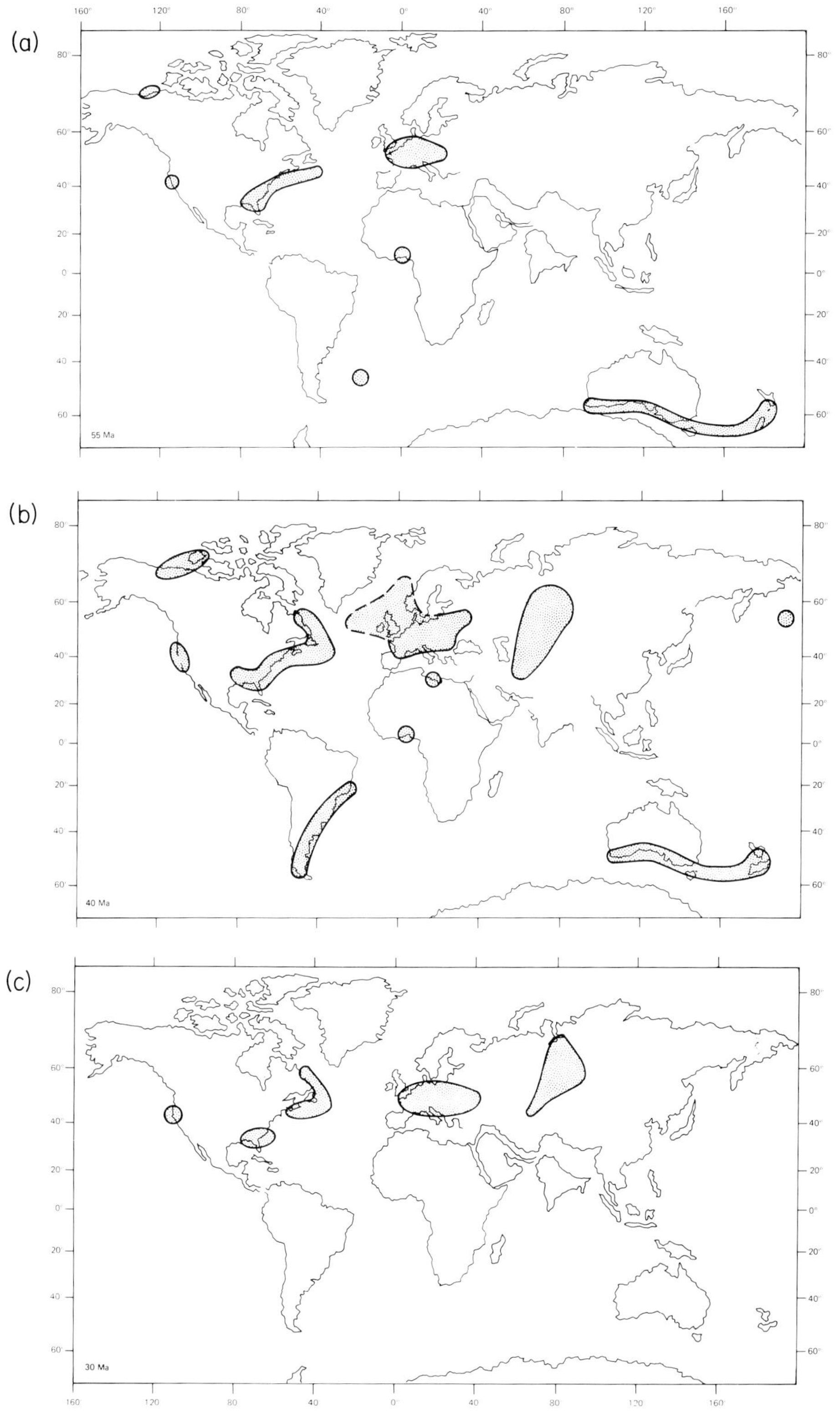
(a)
55 Ma
(b)
40 Ma
(c)
30 Ma

Williams & Bujak (1985) subdivide the use of fossil dinoflagellate cysts in biostratigraphy into four phases. The first is a non-applied descriptive phase which began in the early 1900s and continued until the middle 1950s. The second phase is identified by the initial use of dinoflagellates in applied biostratigraphy as illustrated by the work of Deflandre & Cookson (1955). During the middle 1960s, the development of the first detailed zonations, e.g. that by Clarke & Verdier (1967), marked the start of phase three, which in effect is still in progress. This phase began with the development of predominately European zonal schemes (Fig. 15.16), but the trend has expanded to include many areas of the world at the present time. A major accomplishment of this phase has been the correlation of Cenozoic dinoflagellate zones (Costa & Downie 1976, 1979a; Chateauneuf & Gruas-Cavagnetto 1978) to standard planktonic foraminiferal and calcareous nannofossil zones, which in turn have been correlated to a linear time scale (Hardenbol & Berggren 1978) and to depositional cycles related to eustatic sea level (Vail *et al.* 1977; Vail & Hardenbol 1979). Likewise, the stratigraphic distribution of Mesozoic taxa and zonal schemes have been correlated to ammonite and belemnite zones on a more regular basis (e.g. see Renéville & Raynaud 1981; Fenton & Fisher 1978). Another significant aspect of the third phase is the incorporation of dinoflagellate stratigraphic data with that of other microfossil groups into an integrated palaeontological analysis of major evolutionary and palaeoceanographic events, e.g. documentation of the apparently isochronous fall of eustatic sea level at the end of the Eocene (Van Couvering *et al.* 1981). The fourth phase in the sequence consists of the synthesis of existing dinoflagellate zonations for Mesozoic and Cenozoic sediments which began with Williams' (1977) compilation and is represented at present by the work of Williams & Bujak (1985). The latter paper is a comprehensive review of the nearly fifty published dinoflagellate zonations, including correlation diagrams and illustrations of numerous index species. Williams & Bujak do not propose a single global zonation due to unresolved problems regarding the effects of provincialism on species distribution in time.

Stratigraphic correlation using dinoflagellates is becoming increasingly common as our data base spreads into more regions. For example, Costa & Downie (1979a) and Witmer & Goodman (1980) were able to correlate sections at, respectively, the Rockall Plateau in the North Atlantic and the central Atlantic coastal plain of North America, to surface sections in south-east England using the *Wetzeliella*-based zonation of Costa & Downie (1976). A similiar type of application, biostratigraphic equivalence, has been documented for several Palaeogene sections in south-east England (Downie *et al.* 1971) using characteristic associations of species as a possible indication of similar palaeoenvironmental conditions (Fig. 15.17).

The ranges of taxa need not be the only criterion for biostratigraphic subdivision. For example, Stover (1974) determined the stratigraphic distribution of various combinations of morphologic characters, in particular that of archeopyle style and horn development, for species of *Deflandrea* and related genera in the Lower Tertiary of Australia (Fig. 15.18). Morphotypes having

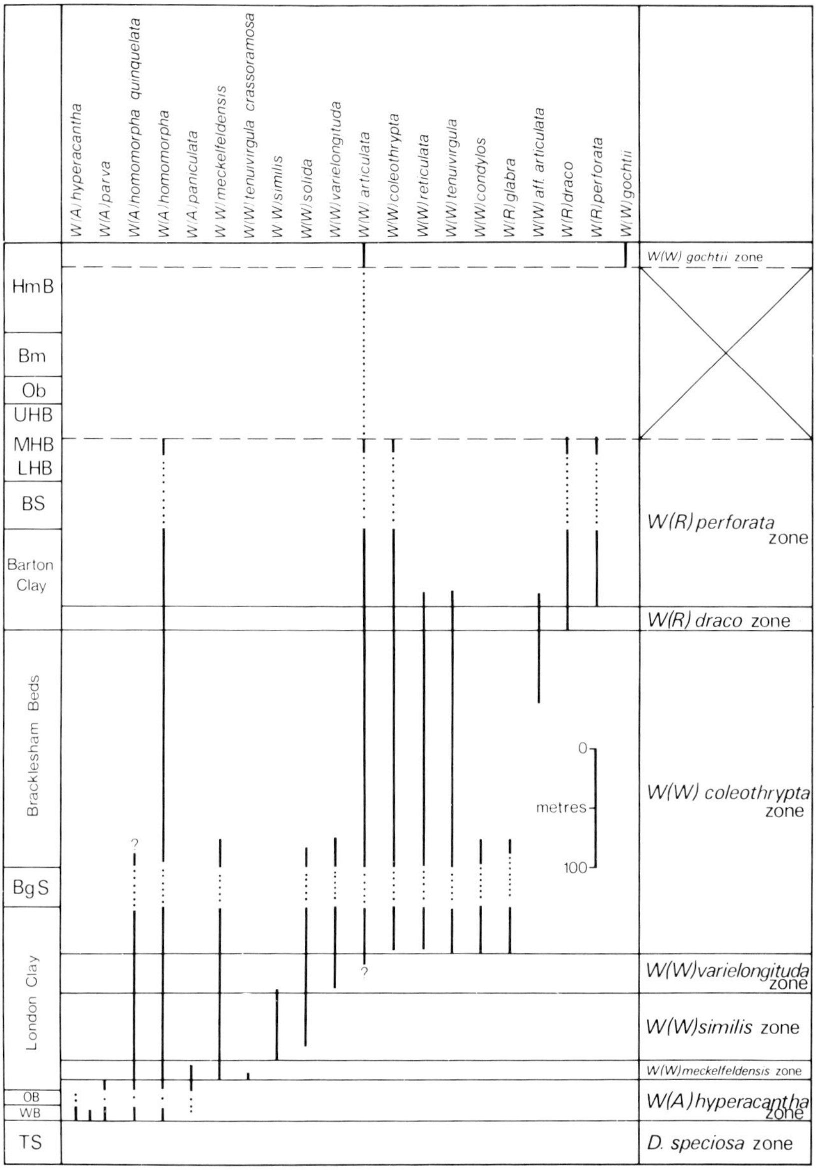

Fig. 15.16. Stratigraphic ranges of selected species in the *Wetzeliella*- complex from south-east England, and a zonal scheme based on first occurrence data of those species (reproduced from Costa & Downie 1976).

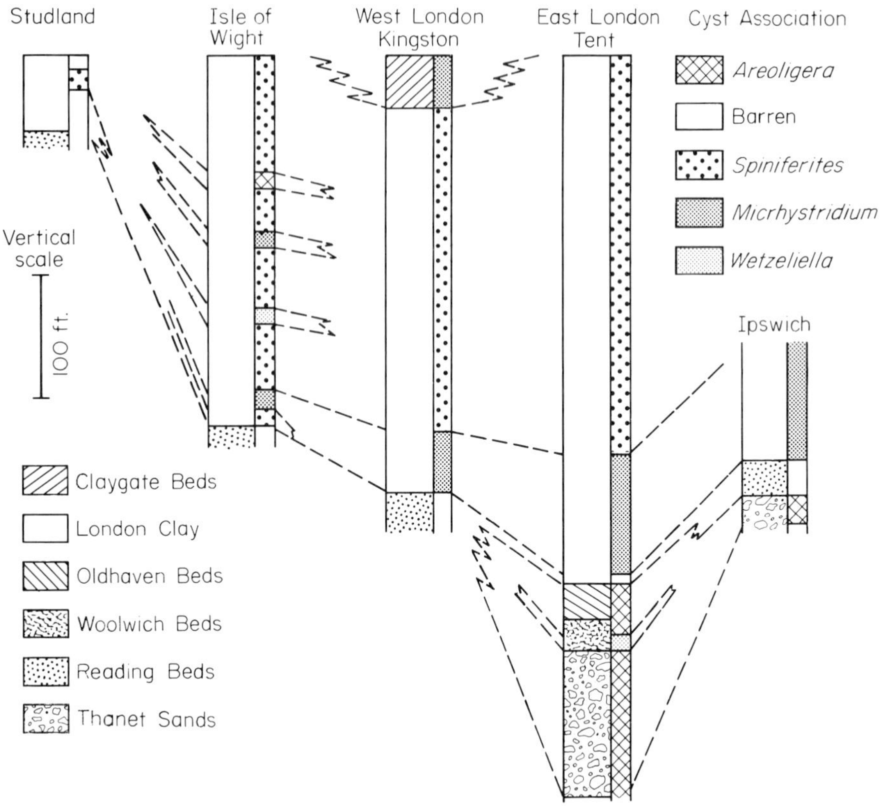

Fig. 15.17. Demonstration of stratigraphic equivalence for sections in the Palaeocene and Eocene of southern England by correlation of cyst associations (determined by relative abundance of species). In each section the lithology is to the left and the association to the right (redrawn from Downie *et al.* 1971).

certain combinations of these features are stratigraphically restricted and can be used as indices for the region. Similarly, the distribution of various morphotypes in the *Wetzeliella*-complex (based on combinations of archeopyle morphology, process distribution and ectophragm development) can be used for preliminary biostratigraphic subdivision in frontier areas without the need to identify individual species (Fig. 15.10).

The final example of dinoflagellate biostratigraphy involves their application to determine the areal extent and temporal magnitude of depositional hiatuses on continental margins. A preliminary study (Kjellström & Hansen 1981) of this type documents dinoflagellate biostratigraphy at the Cretaceous–Tertiary boundary in southern Scandinavia (Fig. 15.19). The results indicate that a more complete depositional sequence is present near the centre of the Danish Embayment, while at the basin margin a greater amount of time is missing

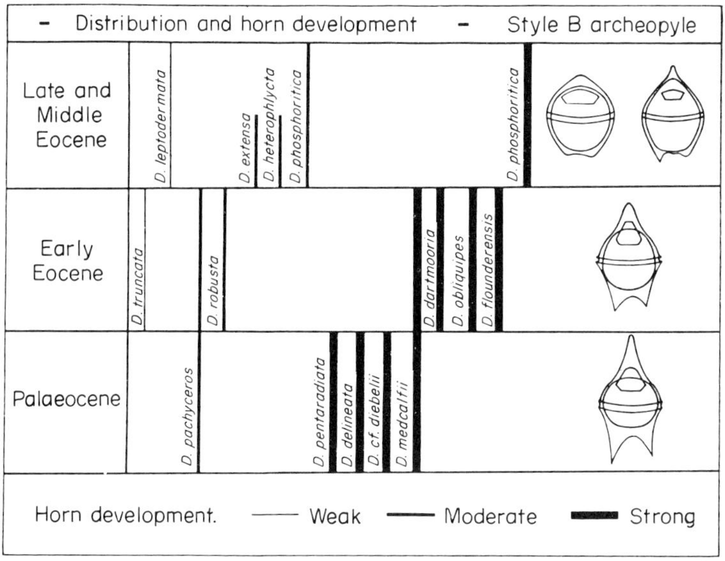

Fig. 15.18. Stratigraphic distribution of selected species of *Deflandrea* and allied genera from the Palaeocene and Eocene of south-east Australia. Taxa are arranged according to the prominence of apical and antapical horns: a Type B archeopyle is characterized by an anterior intercalary (2a) paraplate of the hexa style with the width greater than the height (reproduced from Stover 1974).

between the latest preserved Maastrichtian and the earliest Danian sediments. In effect, this model infers that the Cretaceous–Tertiary 'boundary' is conformable at one end of the shelf-to-basin transect, and unconformable at the other end, with a variable amount of time missing depending on the relative position of a section in the basin. Details of these distribution patterns in relation to a depositional model still need to be resolved. Dinoflagellate cysts may prove to be well suited to the study of continental margin sedimentation due to their common occurrence in those environments.

In any case, the difficulties in making correlations between two sections in this type of setting are obvious because the stratigraphic interval between two biohorizons which define an interval zone (e.g. those of Kjellström & Hansen) is a function of depositional processes controlled by sea-level changes and basin subsidence. In effect, the bounding limits of the zone may be depositional rather than biological, depending on local conditions. Because the horizons may be depositional, their actual timing varies with local conditions and they will not necessarily represent the same time line from basin to basin, or in a basin-to-shelf transect. A serious miscorrelation might arise when an attempt is made to correlate the youngest occurrence of a zonal nominate species which in a basinal setting could represent a low-stand sequence boundary, while in a shelf setting, the top range would represent a high-stand sequence boundary. Correlations

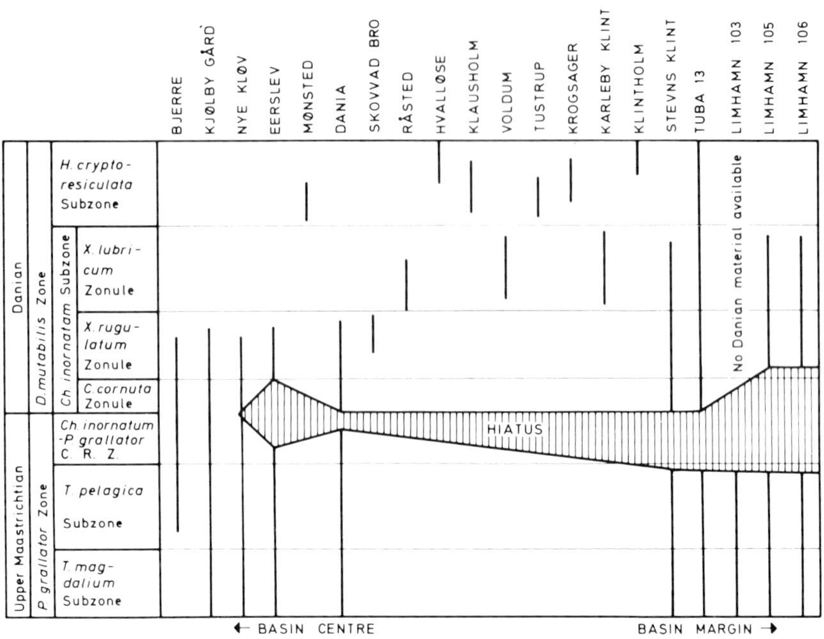

Fig. 15.19. Lateral variation in magnitude of the hiatus at the Cretaceous–Tertiary boundary in southern Scandinavia based on dinoflagellate biostratigraphy. Stratigraphic sections are listed across the top of the figure (reproduced from Kjellström & Hansen 1981).

based on criteria such as this would result in miscorrelation of two obviously non-synchronous data. Further work is required to document the stratigraphic position, magnitude, and precise timing (in a shelf-to-basin transect) of these unconformities, and to determine the biological response to such processes as basin subsidence, sea-level change, and sediment flux. The documentation of these phenomena provide a new dimension to biochronologic and palaeoenvironmental interpretation, and thereby usher in a new era of applied biostratigraphic research in micropalaeontology.

ACKNOWLEDGEMENTS

I wish to thank L. E. Stover for his critical review of this article, and F. J. R. Taylor for his invitation to participate in the project and his seemingly unflappable patience in awaiting the final product. Thanks are also due to those individuals who granted permission to use their published figures. Publication is with the permission of Exxon Production Research Company, Houston, Texas, USA.

9 REFERENCES

ANDERSON D. M., AUBREY D. G., TYLER M. A. & COATS D. W. (1982) Vertical and horizontal distributions of dinoflagellate cysts in sediments. *Limnol. Oceanogr.* **27**, 757–765.

ANDERSON D. M. & WALL D. (1978) Potential importance of benthic cysts of *Gonyaulax tamarensis* and *G. excavata* in initiating toxic dinoflagellate blooms. *J. Phycol.* **14**(2), 224–234.

ARCHANGELSKY S. (1969) Estudio del paleomicroplancton de la Formación Río Turbio (Eoceno), Provincia de Santa Cruz. *Ameghiniana* **6**, 181–218.

ARTZNER D., DAVIES E. H., DORHÖFER G., FASOLA A., NORRIS G. & POPLAWSKI S. (1979) A systematic illustrated guide to fossil organic-walled dinoflagellate genera. *R. Ont. Mus. Life Sci. Misc. Publ.*, 1–119.

BALLOG R. A. & MALLOY R. E. (1981) Neogene palynology from the southern California continental borderland, Site 467, Deep Sea Drilling Project Leg 63. *Init. Rep. Deep Sea Drill Proj.* **63**, 565–576.

BALTES N. (1971) Pliocene dinoflagellata and acritarcha in Romania. *Proc. Second Planktonic Conference*, Rome, 1970, 1–19.

BARSS M. S., BUJAK J. P. & WILLIAMS G. L. (1979) Palynological zonation and correlation of sixty-seven wells, eastern Canada. *Canada Geol. Survey Paper No. 78–42*, 1–118.

BENSON D. G. (1976) Dinoflagellate taxonomy and biostratigraphy at the Cretaceous–Tertiary boundary, Round Bay, Maryland. *Tulane Stud. Geol. Paleontol.* **12**, 169–233.

BERGH R. S. (1881a) Bidrag til Cilioflagellaternes Naturhistorie. *Dansk Vidensk. Medd. Naturhistoriskfor. Kjob.*, Ser. 4, 60–76.

BERGH R. S. (1881b) Der Organismus der Cilioflagellaten. *Morphol. Jahrb.* **7**, 177–288.

BJAERKE T. (1976) Mesozoic palynology of Svalbard II. Palynomorphs from the Mesozoic sequence of Kong Karls Land. *Norsk Polarinst. Arbok* 1976, 83–120.

BJAERKE T. (1980) Mesozoic palynology of Svalbard IV. Toarcian dinoflagellates from Spitsbergen. *Palynology* **4**, 57–77.

BJAERKE T. & DYPVIK H. (1977) Sedimentological and palynological studies of Upper Triassic–Lower Jurassic sediments in Sassenfjorden, Spitsbergen. *Norsk. Polarinst. Arbok* 1976, 131–150.

BJAERKE T. & MANUM S. B. (1977) Mesozoic palynology of Svalbard I. The Rhaetian of Hopen, with a preliminary report on the Rhaetian and Jurassic of Kong Karls Land. *Norsk Polarinst. Skr.* **165**, 1–48.

BRAARUD T. (1962) Species distribution in marine phytoplankton. *J. Oceanogr. Soc. Japan. 20th Anniversary Vol.*, 628–649.

BRADFORD M. R. (1975) New dinoflagellate cyst genera from the recent sediments of the Persian Gulf. *Can. J. Bot.* **53**, 3064–3074.

BRADFORD M. R. (1977) New species attributable to the dinoflagellate cyst genus *Lejeunia* Gerlach, 1961 *Emend.* Lentin and Williams 1975. *Grana* **16**, 45–59.

BRADFORD M. R. (1978) An annotated bibliographic review of Pleistocene and Quaternary dinoflagellate cysts and acritarchs. *Amer. Assoc. Strat. Palynol. Contr. Ser.* **6**, 1–192.

BRIDEAUX W. W. (1971) Recurrent species groupings in fossil microplankton assemblages. *Palaeogeogr. Palaeoclimatol. Palaeoecol.* **9**, 101–122.

BUJAK J. P. (1976) An evolutionary series of Late Eocene dinoflagellate cysts from southern England. *Mar. Micropaleontol.* **1**, 101–117.

BUJAK J. P. (1979) Proposed phylogeny of the dinoflagellates *Rhombodinium* and *Gochtodinium*. *Micropaleontol.* **25**(3), 308–324.

BUJAK J. P. & DAVIES E. H. (1981) Neogene dinoflagellate cysts from the Hunt Dome Kopanoar M-13 well, Beaufort Sea, Canada. *Bull. Can. Pet. Geol.* **29**(3), 420–425.

BUJAK J. P. & FISHER M. J. (1976) Dinoflagellate cysts from the Upper Triassic of arctic Canada. *Micropaleontol.* **22**(1), 44–70.

BUJAK J. P. & WILLIAMS G. L. (1979) Dinoflagellate diversity through time. *Mar. Micropaleontol.* **4**, 1–12.

BUJAK J. P. & WILLIAMS G. L. (1981) The evolution of dinoflagellates. *Can. J. Bot.* **59**(11), 2077–2087.

BÜTSCHLI O. (1885) Dinoflagellata. In *Protozoa (1880–90)*, in Bronn, Klass. u. Ordn. des Thierreichs **1**, 906–1029.

CALANDRA F. (1964) Sur un présumé dinoflagellé, *Arpylorus* nov. gen. du Gothlandien de Tunisie. *C. R. Acad. Sci. Ser. D* **258**, 4112–4114.

CHATEAUNEUF J.-J. (1980) Palynostratigraphie et paléoclimatologie de l'Eocène supérieur et de l'Oligocène du Bassin de Paris (France). *Mém. du B.R.G.M.* **116**, 1–360.

CHATEAUNEUF J.-J. & GRUAS-CAVAGNETTO C. (1978) Les zones de Wetzeliellaceae (Dinophyceae) du bassin de Paris. Comparison et corrélations avec les zones du Paléogène des bassins du nord-ouest de l'Europe. *Bull. B.R.G.M.*, ser. 2, sec. 4, no. 2–1978, 59–93.

CLARKE R. F. A. & VERDIER J.-P. (1967) An investigation of microplankton assemblages from the Chalk of the Isle of Wight, England. *Koninkl. Nederlandse Akad. Wetensch. Verh., Afd. Natuurk., Este Reeks* **24**, 1–96.

CLOUD P. E. & GLAESSNER M. F. (1982) The Ediacarian Period and System: metazoa inherit the earth. *Science* **217** (4562), 783–792.

COSTA L. I. & DOWNIE C. (1976) The distribution of the dinoflagellate *Wetzeliella* in the Palaeogene of north-western Europe. *Palaeontol.* **19** (4), 591–614.

COSTA L. I. & DOWNIE C. (1979a) Cenozoic dinocyst stratigraphy of Sites 403 to 406 (Rockall Plateau), IPOD, Leg 48. *Init. Rep. Deep Sea Drill. Proj.* **48**, 513–529.

COSTA L. I. & DOWNIE C. (1979b) The Wetzeliellaceae; Palaeogene dinoflagellates. *Proc. IV Int. Palynol. Conf.*, Lucknow (1976–77) **2**, 34–43.

CROSS A. T., THOMPSON G. C. & ZAITZEFF J. B. (1966) Source and distribution of palynomorphs in bottom sediments, southern part of Gulf of California. *Mar. Geol.* **4**, 467–524.

DALE B. (1976) Cyst formation, sedimentation, and preservation: factors affecting dinoflagellate assemblages in Recent sediments from Trondheimsfjord, Norway. *Rev. Palaeobot. Palynol.* **22**, 39–60.

DALE B. (1977a) New observations on *Peridinium faeroense* Paulsen (1905), and classification of small orthoperidinioid dinoflagellates. *Br. phycol. J.* **12**, 241–253.

DALE B. (1977b) Cysts of the toxic red-tide dinoflagellate *Gonyaulax excavata* (Braarud) Belech from Oslofjorden, Norway. *Sarsia* **63**, 29–34.

DALE B. (1978) Acritarchous cysts of *Peridinium faeroense* Paulsen: implications for dinoflagellate systematics. *Palynology* **2**, 187–193.

DAMASSA S. P. (1980) Palynomorphs from Holocene sediments of basins in southern California continental borderlands. *Bull. Am. Assoc. Pet. Geol.* **64**, 440 (abstr.).

DAMASSA S. P. (1981) Morphological variability of the dinoflagellate genus *Danea* Morgenroth 1968. *Prog. and Abstr., Am. Assoc. Strat. Palynol. Meeting*, New Orleans, 17–18 (abstr.).

DAVEY R. J. (1970) Non-calcareous microplankton from the Cenomanian of England, northern France and North America, Part II. *British Mus. (Nat. Hist.) Bull., Geology* **18**, 333–397.

DAVEY R. J. (1971) Palynology and paleo-environmental studies, with special reference to the continental shelf sediments of South Africa. *Proc. Second Planktonic Conference*, Roma, 1970, 331–347.

DAVEY R. J. & ROGERS J. (1975) Palynomorph distribution in Recent offshore sediments along two traverse off South West Africa. *Mar. Geol.* **18**, 213–225.

DAVEY R. J., DOWNIE C., SARJEANT W. A. S. & WILLIAMS G. L. (1966) Studies on Mesozoic and Cainozoic dinoflagellate cysts. *British Mus. (Nat. Hist.) Bull., Geology*, Supp. 3, 1–248.

DAVIES E. H. (1981) Archeopyle patterns in Late Cretaceous peridiniaceans. *Palynology* **5**, 234 (abstr.).

DAVIES E. H., BUJAK J. P. & WILLIAMS G. L. (1982) The application of dinoflagellates to paleoenvironmental problems. *Proc. Third North Am. Paleontol. Convention,* **1**, 125–131.

DE CONINCK J. (1968) Dinophyceae et Acritarcha de L'Yprésien du Sondage de Kallo. *Mém. Institut. Royal Sci. Natur. Belg.* **161**, 1–67.

DE CONINCK J. (1975) Microfossiles á paroi organique de L'Yprésien du Bassin Belge. *Admin. Mines, Ser. Géol. Belg. Prof. Paper 12,* 1–151.

DEFLANDRE G. (1936) Microfossiles des silex crétacés I. Généralitiés. Flagellés. *Ann. Paléont.* **25**, 151–191.

DEFLANDRE G. (1945) Microfossiles des calcaires siluriens de la Montagne Noire. *Ann. Paléont.* **31**, 41–76.

DEFLANDRE G. (1947a) Le problème des Hystrichosphères. *Bull. Inst. Océanogr. Monaco* **918**, 1–23.

DEFLANDRE G. (1947b) *Calciodinellum* nov. gen., premier représentant d'une famille nouvelle de Dinoflagellés fossiles à thèque calcaire. *Acad. Sci. (Paris), C.R.* **224**, 1781–1782.

DEFLANDRE G. (1948) Les Calciodinellidés, dinoflagellés fossiles à thèque a calcaire. *Botaniste* **34**, 191–219.

DEFLANDRE G. (1952) Dinoflagellés fossiles. In *Traite de Zoologie. Anatomie, Systématique, Biologie* (ed. P. P. Grassé), pp. 391–406. Masson and Cie., Paris.

DEFLANDRE G. & COOKSON I. (1955) Fossil microplankton from Australian Late Mesozoic and Tertiary sediments. *Austr. J. Mar. Freshwater Res.* **6**, 242–313.

DOBELL P. E. R. & TAYLOR F. J. R. (1981) Viable *Spiniferites* cysts with 2P archeopyles from Recent marine sediments, British Columbia, Canada. *Palynology* **5**, 99–106.

DODGE J. D. (1965) Chromosome structure in the dinoflagellates and the problem of the mesocaryotic cell. *Int. Congr. Ser. Excerpta Med.* **91**, 264–265.

DODGE J. D. (1975) The Prorocentrales (Dinophyceae). II. Revision of the taxonomy within the genus *Prorocentrum. Bot. J. Linn. Soc.* **71**, 103–125.

DODGE J. D. (1981a) Three new generic names in the Dinophyceae: *Herdmania, Sclerodinium* and *Triadinium* to replace *Heteraulacus* and *Goniodoma. Br. phycol. J.* **16**, 273–280.

DODGE J. D. (1981b) *Provisional Atlas of the Marine Dinoflagellates of the British Isles.* Biol. Rec. Centre, Inst. Terr. Ecol., Monks Wood Exp. Sta., Huntington.

DÖRHÖFER G. & DAVIES E. H. (1980) Evolution of archeopyle and tabulation in rhaetogonyaulacinean dinoflagellate cysts. *R. Ont. Mus. Life Sci. Misc. Publ.,* 1–91.

DOWNIE C., HUSSAIN M. A. & WILLIAMS G. L. (1971) Dinoflagellate cyst and acritarch associations in the Paleogene of southeast England. *Geosci. Man.* **3**, 29–35.

DRUGG W. S. (1970) Two new Neogene species of *Tuberculodinium* and one of *Xenikodinium* (Pyrrhophyta). *Proc. Biol. Soc. Wash.* **83**, 115–122.

DRUGG W. S. (1978) Some Jurassic dinoflagellate cysts from England, France and Germany. *Palaeontographica, Abt. B* **168**, 61–79.

DRUGG W. S. & LOEBLICH A. R. Jr (1967) Some Eocene and Oligocene phytoplankton from the Gulf Coast, USA. *Tulane Stud. Geol.* **5**(4), 181–194.

EATON G. L. (1971) A morphogenetic series of dinoflagellate cysts from the Bracklesham Beds of the Isle of Wight, Hampshire, England. *Proc. Second Planktonic Conference, Roma,* 1970, 355–379.

EATON G. L. (1980) Nomenclature and homology in peridinialean dinoflagellate plate patterns. *Palaeontol.* **23**(3), 667–688.

EDWARDS L. E. & BEBOUT J. W. (1981) Emendation of *Phthamoperidinium* Drugg & Loeblich 1967, and a description of *P. brooksii* sp. nov. from the Eocene of the mid-Atlantic Outer Continental Shelf. *Palynology,* **5**, 29–41.

EISENACK A. (1954) Mikofossilien aus Phosphoriten des samlandischen Unteroligozäns und uber die Einheitlichkeit der Hystrichosphaerideen. *Palaeontographica, Abt. A* **105**, 49–95.

EISENACK A. (1961) Einige Erörterungen uber fossile Dinoflagellaten nebst Übersicht uber die zur Zeit bekannten Gattungen. *N. Jb. Geol. Paläontol. Abh.* **112**, 281–324.

EISENACK A. (1964) Ërorterungen über einige Gattungen fossiler Dinoflagellaten ünd uber die Einordnung der Gattungen in das System. *N. Jb. Geol. Paläont. Mon.* **6**, 312–336.

EISENACK A. & KJELLSTRÖM G. (1971a) *Katalog der fossilen Dinoflagellaten, Hystrichosphären und verwandten Mikrofossilien.* Vol. II. *Dinoflagellaten.* E. Schweizerbart'sche Verlagsbuchhandlung, Stuttgart. 1130 pp.

EISENACK A. & KJELLSTRÖM G. (1971b) *Katalog der fossilen Dinoflagellaten, Hystrichosphären und verwandten Mikrofossilien.* Vol. II. *Dinoflagellaten.* 2. *Ergänzungslieferung.* E. Schweizerbart'sche Verlagsbuchhandlung, Stuttgart. 215 pp.

EVITT W. R. (1961a) *Dapcodinium priscum* N. Gen., N. sp., a dinoflagellate from the Lower Lias of Denmark. *J. Paleontol.* **35**, 996–1002.

EVITT W. R. (1961b) The dinoflagellate *Nannoceratopsis* Deflandre: morphology, affinities and infraspecific variability. *Micropaleontol.* **7**, 305–316.

EVITT W. R. (1970) Dinoflagellates—a selective review. *Geosci. Man.* **1**, 29–45.

EVITT W. R. (1975) Proceedings of a Forum on Dinoflagellates, Anaheim, California, 1973. *Am. Assoc. Strat. Palynol. Contr. Ser.* **4**, 1–76.

EVITT W. R. (1978) Special connotations of 'quadra,' 'penta,' and 'hexa' in descriptive terminology of dinoflagellates. *Palynology* **2**, 199–204.

EVITT W. R. (1981) The difference it makes that dinoflagellates did it differently. (unpubl. note). *Int. Comm. Palynol. Newsl.* **4**(1), 6–7.

EVITT W. R. (1985) *Sporopollenin Dinoflagellate Cysts: Their Morphology and Interpretation.* Amer. Assoc. Strat. Palynol. Foundation 333 pp.

EVITT W. R., CLARKE R. F. A. & VERDIER J.-P. (1967) Dinoflagellate studies III. *Dinogymnium acuminatum* n. gen., n. sp. (Maastrichtian) and other fossils formerly referable to *Gymnodinium* Stein. *Stanford Univ. Publ., Geol. Sci.* **10**(4), 1–27.

EVITT W. R., LENTIN J. K., MILLIOUD M. E., STOVER L. E. & WILLIAMS G. L. (1977) Dinoflagellate cyst teminology. *Canada Geol. Survey Paper No. 76–42*, 1–9.

EVITT W. R. & WALL D. (1968) Dinoflagellate studies, IV. Theca and cyst of Recent freshwater *Peridinium limbatum* (Stokes) Lemmermann. *Stanford Univ. Publ., Geol. Sci.* **12**(2), 1–23.

FAUCONNIER D. & SLANSKY M. (1978) Rôle possible des Dinoflagellés dans la sédimentation phosphatée. *Bull. B.R.G.M.* (2), IV, **3**, 191–200.

FAUCONNIER D. & SLANSKY M. (1980) Relations entre le développement des Dinoflagellés et la sédimentation phosphatée du Bassin de Gafsa (Tunisie). *Doc. B.R.G.M.* **24**, 185–204.

FENSOME R. A. (1981) The Jurassic dinoflagellate genera *Wanaea* and *Energlynia*: their morphology and evolution. *N. Jb. Geol. Paläont. Abh.*, **161**(1), 47-61.

FENTON J. P. G. & FISHER M. J. (1978) Regional distribution of marine microplankton in the Bajocian and Bathonian of northwest Europe. *Palinologia*, núm. extraord. **1**, 233–243.

FISHER M. J. & BUJAK J. P. (1975) Upper Triassic palynofloras from arctic Canada. *Geosci. Man* **11**, 87–94.

FISHER M. J. & RILEY L. A. (1980) The stratigraphic distribution of dinoflagellate cysts at the boreal Jurassic–Cretaceous boundary. *Proc. IV. Int. Palynol. Conf.*, Lucknow (1976–77) **2**, 312–329.

FISHER M. J. & VAN HELDEN B. G. T. (1979) Some observations on the fossil dinocyst genus *Rhaetogonyaulax* Sarjeant, 1966. *Palynology* **3**, 265–276.

FORD L. N. Jr (1979) *Dinoflagellates, chlorophytes, and acritarchs from the Cooper Formation (Oligocene) of South Carolina.* M. S. thesis, Va. Poly. Inst. St. Univ.

FORD L. N. Jr (1981) Paleoecological implications of dinoflagellate floras from the Cooper Formation (Chattian) of South Carolina. *Palynology* **5**, 215 (abstr.).

FÜTTERER D. (1976) Kalkige Dinoflagellaten ('Calciodinelloideae') und die systematische Stellung der Thoracosphaeroideae. *N. Jb. Geol. Paläont. Abh.* **151**, 119–141.

FÜTTERER D. (1977) Distribution of calcareous dinoflagellates in Cenozoic sediments of Site 366, eastern North Atlantic. *Init. Rep. Deep Sea Drill. Proj.* **41**, 709–737.

GOCHT H. (1957) Mikroplankton aus dem nordwestdeutschen Neokom (Pt. I). *Paläont. Zeitschr.* **31**, 163–185.

GOCHT H. (1959) Mikroplankton aus dem nordwestdeutschen Neokom (Pt II). *Paläont. Zeitschr.* **33**, 50–89.

GOCHT H. (1964) Planktonische Kleinformen aus dem Lias-Dogger-Grenzbereich Nord und Süddeutschlands. *N. Jb. Geol. Paläont. Abh.* **199**, 113–133.

GOCHT H. (1969) Formengemeinschaften alttertiären Mikroplanktons aus Bohrproben des Erdölfelds Meckelfeld bei Hamburg. *Palaeontographica, Abt. B* **126**, 1–100.

GOCHT H. (1972) Zur Morphologie der Dinoflagellaten–Gattung *Nannoceratopsis* Deflandre. *Lethaia* **5**, 15–29.

GOCHT H. (1979) Korrelation von Überlappungssystem und Wachstum bei fossilen Dinoflagellaten (*Gonyaulax*-Gruppe). *N. Jb. Geol. Paläont. Abh.* **157**(3), 344–365.

GOODMAN D. K. (1979) Dinoflagellate 'communities' from the Lower Eocene Nanjemoy Formation of Maryland, USA. *Palynology* **3**, 169–190.

GOODMAN D. K. (in press) Paleogene biogeography and stratigraphic distribution of *Wetzeliella* and allied dinoflagellate genera. *Amer. Assoc. Strat. Palynol. Contr. Ser.*

GOODMAN D. K. (in press) A revision of the dinoflagellate family Heteraulacacystaceae Drugg and Loeblich, and emendation of *Biconidinium longissimum* Islam. *J. Paleontol.*

GOODMAN D. K. & EVITT W. R. (1981) The dinoflagellate *Angustidinium acribes* (Davey and Verdier) gen. et comb. nov. from the mid-Cretaceous of the northern California Coast Ranges. *Grana* **20**, 43–54.

GOODMAN D. K. & FORD L. N. Jr (1983) Preliminary dinoflagellate biostratigraphy for the middle Eocene to lower Oligocene from the southwest Atlantic Ocean. *Init. Rep. Deep Sea Drill. Proj.* **71**, 859–877.

GOODMAN D. K. & STOVER L. E. (1981) Comparison of *Microdinium* and allied dinoflagellate genera. *Palynology* **5**, 215–216 (abstr.).

GRAHAM H. W. (1941) An oceanographic consideration of the dinoflagellate genus *Ceratium*. *Ecol. Monogr.* **11**, 99–116.

GRUAS-CAVAGNETTO C. (1967) Complexes sporo-poliniques du Sparnacien du bassin de Paris. *Rev. Palaeobot. Palynol.* **5**, 243–261.

GRUAS-CAVAGNETTO C. (1968) Étude palynologique des divers gisements du Sparnacien du bassin de Paris. *Mém. Soc. géol. France (N.S.)* **47**, (110), 1–144.

HABIB D. (1971) Dinoflagellate stratigraphy across the Miocene-Pliocene boundary, Tabiano stratotype section. *Proc. Sec. Int. Plank. Conf.* **1**, 591–598.

HABIB D. (1975) Neocomian dinoflagellate zonation in the western North Atlantic. *Micropaleontol.* **21**, 373–392.

HABIB D. (1977) Comparison of Lower and Middle Cretaceous palynostratigraphic zonations in the western North Atlantic. In *Stratigraphic Micropaleontology of Atlantic Basin and Borderlands* (ed. by F. M. Swain), pp. 341–367. Elsevier, New York.

HALL S. A. (1977) Cretaceous and Tertiary dinoflagellates from Seymour Island, Antarctica. *Nature (Lond.)* **267**, 239–241.

HANSEN J. M. (1977) Dinoflagellate stratigraphy and echinoid distribution in Upper Maastrichtian and Danian deposits from Denmark. *Bull. geol. Soc. Denmark* **26**, 1–26.

HANSEN J. M. (1979) Dinoflagellate zonation around the boundary. In *Symposium Cret.-Tert. Boundary Events. I. The Maastrichtian and Danian of Denmark* (eds. T. Birkelund & R. G. Bromley), pp. 136–143.

HAQ B. U. (1973) Transgression, climatic change and the diversity of calcareous nannoplankton. *Mar. Geol.* **15**, 25–30.

HAQ B. U. & BOERSMA A. (1978) *Introduction to Marine Micropaleontology*. Elsevier, New York.

HARDENBOL J. & BERGGREN W. A. (1978) A new Paleogene numerical time scale. In *Contributions to the Geologic Time Scale* (eds. C. V. Cohee, M. F. Glaessner & H. D. Hedberg), pp. 213–234. *Am. Assoc. Pet. Geol. Stud. Geol. No. 6*.

HARKER S. D. (1979) Archeopyle formation in *Palaeohystrichophora infusorioides* Deflandre, 1935. *N. Jb. Geol. Paläont. Mh.* 1979(6), 369–377.

HARKER S. D. & SARJEANT W. A. S. (1975) The stratigraphic distribution of organic-walled dinoflagellate cysts in the Cretaceous and Tertiary. *Rev. Palaeobot. Palynol.* **20**, 217–315.

HARLAND R. (1973a) Dinoflagellate cysts and acritarchs from the Bearpaw Formation (Upper Campanian) of southern Alberta, Canada. *Palaeontology* **16**(4), 665–706.

HARLAND R. (1973b) Quaternary (Flandrian?) dinoflagellate cysts from the Grand Banks, off Newfoundland, Canada. *Rev. Palaeobot. Palynol.* **16**, 229–242.

HARLAND R. (1974) Quaternary organic-walled microplankton from boreholes 71/9 and 71/10. *Rep. Inst. Geol. Sci.* **73**(14), 37–39.

HARLAND R. (1977) Recent and Late Quaternary (Flandrian and Devensian) dinoflagellate cysts from marine continental shelf sediments around the British Isles. *Paleontographica, Abt. B* **164**, 87–126.

HARLAND R. (1979a) The *Wetzeliella* (*Apectodinium*) *homomorpha* plexus from the Palaeogene/earliest Eocene of Northwest Europe. *Proc. IV Int. Palynol. Conf.*, Lucknow (1976–77) **2**, 59–70.

HARLAND R. (1979b) Dinoflagellate biostratigraphy of Neogene and Quaternary sediments at Holes 400/400A in the Bay of Biscay (Deep Sea Drilling Project Leg 48). *Init. Rep. Deep Sea Drill. Proj.* **48**, 531–545.

HARLAND R. (1981) Cysts of the colonial dinoflagellate *Polykrikos schwartzii* Bütschli 1873, (Gymnodiniales), from Recent sediments, Firth of Forth, Scotland. *Palynology* **5**, 65–79.

HARLAND R. (1982a) A review of Recent and Quaternary organic-walled dinoflagellate cysts of the genus *Protoperidinium*. *Palaeontol.* **25**(2), 369–397.

HARLAND R. (1982b) Recent dinoflagellate cyst assemblages from the southern Barents Sea. *Palynology* **6**, 9–18.

HARLAND R. (1983) Distribution maps of recent dinoflagellate cysts in bottom sediments from the North Atlantic Ocean and adjacent seas. *Palaentol* **26** (2), 321–387.

HARLAND R. & DOWNIE C. (1969) The dinoflagellates of the Interglacial deposits at Kirmington, Lincolnshire. *Proc. York. Geol. Soc.* **37**, 231–237.

HARLAND R., GREGORY D. M., HUGHES M. J. & WILKINSON I. P. (1978) A late Quaternary bio- and climatostratigraphy for marine sediments in the north-central part of the North Sea. *Boreas* **7**, 91–96.

HARLAND R., MORBEY S. J. & SARJEANT W. A. S. (1975) A revision of the Triassic to lowest Jurassic dinoflagellate *Rhaetogonyaulax*. *Palaeontol.* **18**, 847–864.

HARLAND R., REID P. C., DOBELL P. & NORRIS G. (1980) Recent and sub-Recent dinoflagellate cysts from the Beaufort Sea, Canadian Arctic. *Grana* **19**, 211–225.

HARLAND R. & SHARP J. (1980) *Phthanoperidinium obscurum* sp. nov., a non-marine dinoflagellate cyst from the Late Eocene of England. *Rev. Palaeobot. Palynol.* **30**, 287–296.

HASKELL T. R. & WILSON G. J. (1975) Palynology of sites 280–284, off southeastern Australia and western New Zealand. *Init. Rep. Deep Sea Drill. Proj.* **24**, 723–741.

HELBY R. (in press) *Muderongia* and related genera in the Late Jurassic–Early Cretaceous of Australasia. *Alcheringa Monograph Ser.*

HOCHULI P. A. (1978) Palynologische Untersuchungen im Oligozän und Untermiozän der Zentralen und Westlichen Paratethys. *Beitr. Paläont. Österr.* **4**, 1–132.

JAFAR S. A. (1979) Taxonomy, stratigraphy and affinities of calcareous nannoplankton genus *Thoracosphaera* Kamptner. *Proc. IV Int. Palynol. Conf.* Lucknow (1976–77) **2**, 1–21.

JOHNSON L. A. S. (1970) Rainbow's end; the quest for an optimal taxonomy. *Syst. Zool.* **19**, 203–239.

KEMP E. M. (1972) Reworked palynomorphs from the West Ice Shelf area, East Antarctica, and their possible geological and palaeoclimatological significance. *Mar. Geol.* **13**, 145–157.

KEMP E. M. (1975) Palynology of Leg 28 drill sites, Deep Sea Drilling Project. *Init. Rep. Deep Sea Drill. Proj.* **28**, 599–623.

KEUPP H. (1979) Lower Cretaceous Calcisphaerulidea and their relationship to calcareous dinoflagellate cysts. *Bull. Centre Rech. Explor. Prod. Elf-Aquitaine* **3**(2), 651–663.

KEUPP H. (1980a) *Calcigonellum* Deflandre 1948 und *Echinodinella* n. gen. (Kalkige

Dinoflagellaten-Zysten) aus der nordwestdeutschen Unter-Kreide. *Facies* **2**, 123–148.

KEUPP H. (1980b) *Pithonella paratabulata* n. sp., eine unterkretazische Calcisphaere mit äusserer Paratabulation. *Facies* **3**, 239–249.

KEUPP H. (1980c) *Pithonella patriciacreeleyae* Bolli 1974, eine kalkige Dinoflagellaten-Zysten mit interner Paratabulation (Unter-Kreide, Speeton/SE-England). *N. Jb. Geol. Paläont. Mh.* 1980(9), 513–524.

KJELLSTRÖM G. & HANSEN J. M. (1981) Dinoflagellate biostratigraphy of the Cretaceous-Tertiary boundary in southern Scandinavia. *Geol. Fören. Stockholm Förh.* **103**(2), 271–278.

KRUTZSCH W. (1962) Die Mikroflora der Gieseltalbraunkohle. Teil III. Süsswasserdinoflagellaten aus subaquatisch gebildeten Blätterkohlenlagen des mittleren Geiseltales. *Hallesches Jb. Mitteldeutsche Erdgesch.* **4**, 40–45.

LEJEUNE-CARPENTIER M. & SARJEANT W. A. S. (1981) Restudy of some larger dinoflagellate cysts and an acritarch from the Upper Cretaceous of Belgium and Germany. *Ann. Soc. Geol. Belg.* **104**, 1–39.

LENTIN J. K. & WILLIAMS G. L. (1973) Fossil dinoflagellates: index to genera and species. *Canada Geol. Survey Paper No. 73–42*, 1–176.

LENTIN J. K. & WILLIAMS G. L. (1976) A monograph of fossil peridinioid dinoflagellate cysts. *Bedford Inst. Oceanography Rept.* BI-R-75-16, 1–237.

LENTIN J. K. & WILLIAMS G. L. (1977) Fossil dinoflagellates: index to genera and species, 1977 edition. *Bedford Inst. Oceanography Rept.* BI-R-77-8, 1–209.

LENTIN J. K. & WILLIAMS G. L. (1980) Dinoflagellate provincialism with emphasis on Campanian peridiniaceans. *Amer. Assoc. Strat. Palynol. Contr. Ser.* **7**, 1–47.

LENTIN J. K. & WILLIAMS G. L. (1981) Fossil dinoflagellates: index to genera and species, 1981 edition. *Bedford Inst. Oceanography Rept.* BI-R-81-12, 1–345.

LIENGJARERN M., COSTA L. & DOWNIE C. (1980) Dinoflagellate cysts from the Upper Eocene–Lower Oligocene of the Isle of Wight. *Palaeontol.* **23**(3), 475–499.

LIPPS J. H. (1970) Plankton evolution. *Evolution* **24**, 1–22.

LOCICERO V. R. (1975) *Proceedings of the First International Conference on Toxic Dinoflagellate Blooms*. Mass. Sci. Tech. Foundation, Wakefield, Mass.

LOEBLICH A. R. III (1970) The amphiesma or dinoflagellate cell covering *Proc. North Am. Paleontol. Convention*, Chicago, 1969, Part G, 867–929.

LOEBLICH A. R. III (1976) Dinoflagellate evolution: speculation and evidence. *J. Protistol.* **23**(1), 13–28.

LOEBLICH A. R. III (1982) Dinophyceae. In *Synopsis and Classification of Living Organisms*, pp. 101–115. McGraw-Hill Book Company, Inc., New York.

LOEBLICH A. R. III (1984) Dinoflagellate evolution. In *Dinoflagellates* (ed. D. L. Spector) pp. 481–522

LOEBLICH A. R. III & LOEBLICH L. A. (1979) The systematics of *Gonyaulax* with special reference to the toxic species. In *Toxic Dinoflagellate Blooms* (ed. D. L. Taylor & H. H. Seliger), pp. 41–46. Elsevier, New York.

MCINTYRE D. J. & WILSON G. J. (1966) Preliminary palynology of some Antarctic Tertiary erratics. *N.Z. J. Bot.* **4**, 315–321.

MCKEE E. D., CHRONIC J. & LEOPOLD E. B. (1959) Sedimentary belts in lagoon of Kapingamarangi Atoll. *Bull. Am. Assoc. Petrol. Geol.* **43**, 501–562.

MCLEAN D. M. (1972) *Cladopyxidium septatum*, n. gen., n. sp., possible Tertiary ancestor of the modern dinoflagellate *Cladopyxis hemibrachiata* Balech, 1964. *J. Paleontol.* **46**(6), 861–863.

MCLEAN D. M. (1974) Emendation of the dinoflagellate Family Microdiniaceae. *Geosci. Man* **9**, 76 (abstr.).

MCLEAN D. M. (1981) A test of terminal Mesozoic 'catastrophe'. *Earth Planet. Sci. Let.* **53**, 103–108.

MAIER D. (1959) Planktonuntersuchungen in tertiären und quartären marinen Sedimenten.

Ein Beitrag zur Systematik, Stratigraphie und Ökologie der Coccolithophorideen, Dinoflagellaten und Hystrichosphaerideen vom Oligozän bis zum Pleistozän. *N. Jb. Geol. Paläont. Abh.* **107**, 278–340.

MALLOY R. E. (1972) An Upper Cretaceous dinoflagellate cyst lineage from Gabon, West Africa. *Geosci. Man* **4**, 57–65.

MANUM S. B. (1976) Dinocysts in Tertiary Norwegian–Greenland Sea sediments (Deep Sea Drilling Project Leg 38), with observations of palynomorphs and palynodebris in relation to environment. *Init. Rep. Deep Sea Drill. Proj.* **38**, 897–919.

MAY F. E. (1976) Dinoflagellates: fossil motile-stage tests from the Upper Cretaceous of the northern New Jersey coastal plain. *Science* **193**, 1128–1130.

MAY F. E. (1977) Functional morphology, paleoecology, and systematics of *Dinogymnium* tests. *Palynology* **1**, 103–121.

MAY F. E. (1980) Dinoflagellates cysts of the Gymnodiniaceae, Peridiniaceae, and Gonyaulacaceae from the Upper Cretaceous Monmouth Group, Atlantic Highlands, New Jersey. *Palaeontographica, Abt. B* **172**, 10–116.

MEBRADU S. (1978) Stratigraphic palynology of the Upper Jurassic of the Dorset coast, England. Part I. Statistical variables. *Palinologia*, num. extraord. 1, 327–333.

MORBEY S. J. (1975) The palynostratigraphy of the Rhaetian Stage, Upper Triassic in the Kendelbachgraben, Austria. *Palaeontographica, Abt. B* **152**, 1–75.

MORBEY S. J. (1978) Late Triassic and Early Jurassic subsurface palynostratigraphy in northwestern Europe. *Palinologia*, **1**, 355–365.

MORBEY S. J. & NEVES R. (1974) A scheme of palynologically defined concurrent-range zones and subzones for the Triassic Rhaetian Stage (sensu lato). *Rev. Palaeobot. Palynol.* **17**, 161–173.

MORGENROTH P. (1966) Mikrofossilien und Konkretionen nordwesteuropäischen Untereozäns. *Palaeontographica, Abt. B* **119**, 1–53.

MORGENROTH P. (1970) Dinoflagellate cysts from the Lias Delta of Luhnde/Germany. *N. Jb. Geol. Paläont. Abh.* **136**, 345–359.

MORZADEC-KERFOURN M. T. (1966) Étude des Acritarches et Dinoflagellés des sédiments vaseux de la Vallée de la Vilaine aux environs de Redon (Ille-et-Vilaine). *Bull. Soc. géol. min. Bretagne*, 1964–65, N.S., 137–146.

MORZADEC-KERFOURN M. T. (1975) La signification écologique des dinoflagellés et leur intérêt pur l'étude des variations du niveau marin. *Rev. Micropaléontol.* **18**(4), 229–235.

MORZADEC-KERFOURN M. T. (1977) Les kystes de dinoflagellés dans les sédiments Récents le long des côtes Bretonnes. *Rev. Micropaléontol.* **20**, 157–166.

MULLER J. (1959) Palynology of Recent Orinoco delta and shelf sediments. Reports of the Orinoco Shelf Expedition, Vol. 5. *Micropaleontol.* **5**, 1–32.

NEALE J. W. & SARJEANT W. A. S. (1962) Microplankton from the Speeton Clay of Yorkshire. *Geol. Mag.* **99**, 439–458.

NORRIS G. (1965) Provincialism of Callovian–Neocomian dinoflagellate cysts in the northern and southern hemispheres. *Amer. Assoc. Strat. Palynol. Contr. Ser.* **4**, 29–35.

NORRIS G. (1978a) Phylogeny and a revised supra-generic classification for Triassic–Quaternary organic-walled dinoflagellate cysts (Pyrrhophyta). Part I. Cyst terminology and assessment of previous classifications. *N. Jb. Geol. Paläontol. Abh.* **155**, 300–317.

NORRIS G. (1978b) Phylogeny and revised supra-generic classification for Triassic–Quaternary organic-walled dinoflagellate cysts (Pyrrhophyta). Part II. Families and sub-orders of fossil dinoflagellates. *N. Jb. Geol. Paläontol. Abh.* **156**, 1–30.

PARTRIDGE A. D. (1976) The geological expression of eustacy in the Early Tertiary of the Gippsland Basin. *APEA J.* **16**(1), 73–79.

PIASECKI S. (1980) Dinoflagellate cyst stratigraphy of the Miocene Hodde and Gram Formations, Denmark. *Bull. geol. Soc. Denmark* **29**, 53–76.

PIEL K. M. & EVITT W. R. (1980) Paratabulation in the Jurassic dinoflagellate genus *Nannoceratopsis* and a comparison with modern taxa. *Palynology* **4**, 79–104.

POPLAWSKI S. & NORRIS G. (1979) Putative freshwater dinoflagellates in a marine section near the Cretaceous/Tertiary boundary in Alabama, USA. *Symposium Cret.–Tert. Boundary Events, II. Proc.* (eds W. K. Christensen & T. Birkelund), 249 (abstr.).

REID P. C. (1972) Dinoflagellate cyst distribution around the British Isles. *J. mar. biol. Ass. UK.* **52**, 939–944.

REID P. C. (1974) Gonyaulacacean dinoflagellate cysts from the British Isles. *Nova Hedwigia* **25**, 579–637.

REID P. C. (1975) A regional subdivision of dinoflagellate cysts around the British Isles. *New Phytol.* **75**, 589–603.

REID P. C. (1977) Peridiniacean and glenodiniacean dinoflagellate cysts from the British Isles. *Nova Hedwigia* **29**, 429–463.

REID P. C. (1978) Dinoflagellate cysts in the plankton. *New Phytol.* **80**, 219–229.

REID P. C. & HARLAND R. (1977) Studies of Quaternary dinoflagellate cysts from the North Atlantic. *Amer. Assoc. Strat. Palynol. Contr. Ser.* 5A, 147–169.

RENÉVILLE P. DE & RAYNAUD J.-F. (1981) Palynologie du stratotype du Barrémien. *Bull. Centres Rech. Explor.-Prod. Elf-Aquitaine* **5**, 1–29.

RILEY L. A. & SARJEANT W. A. S. (1972) Survey of the stratigraphical distribution of dinoflagellates, acritarchs and tasmanitids in the Jurassic. *Geophytology* **2**, 1–40.

ROSSIGNOL M. (1961) Analyse pollinique de sédiments quaternaires en Israël. I. Sédiments Récents. *Pollen et Spores* **3**, 303–324.

ROSSIGNOL M. (1962) Analyse pollinique de sédiments marins Quaternaires en Israël. II. Sédiments Pleistocénes. *Pollen et Spores* **4**, 121–148.

ROSSIGNOL M. (1969) Sédimentation palynologique dans le domaine marin Quaternaire de Palestine: Étude de paléoenvironment. *Notes Mém. Moyen-Orient. Mus. Nat. d'Hist. Nat.* **10**, 1–272.

SARJEANT W. A. S. (1963) Fossil dinoflagellates from Upper Triassic sediments. *Nature (Lond.)* **199**, 353–354.

SARJEANT W. A. S. (1967) The stratigraphical distribution of fossil dinoflagellates. *Rev. Palaeobot. Palynol.* **1**, 323–343.

SARJEANT W. A. S. (1972) Dinoflagellate cysts and acritarchs from the Upper Vardekløft Formation (Jurassic) of Jameson Land, East Greenland. *Medd. Grønland* **195**, 1–69.

SARJEANT W. A. S. (1974) *Fossil and Living Dinoflagellates.* Academic Press, London.

SARJEANT W. A. S. (1976) *Energlynia*, new genus of dinoflagellate cysts from the Great Oolite Limestone (Middle Jurassic: Bathonian) of Lincolnshire, England. *N. Jb. Geol. Paläont. Mh.* 1976(3), 163–173.

SARJEANT W. A. S. (1978a) *Arpylorus antiquus* Calandra, emend., a dinoflagellate cyst from the Upper Silurian. *Palynology* **2**, 167–179.

SARJEANT W. A. S. (1978b) A guide to the identification of Jurassic dinoflagellate cysts. *La. St. Univ., School. Geosci. Misc. Publ.* 78–1, 1–107.

SARJEANT W. A. S. (1979) Middle and Upper Jurassic dinoflagellate cysts: the world excluding North America. *Am. Assoc. Strat. Palynol. Contr. Ser.* 5B, 133–157.

SARJEANT W. A. S. (1982a) Dinoflagellate cyst terminology: a discussion and proposals. *Can. J. Bot.* **60**(6), 922–945.

SARJEANT W. A. S. (1982b) The dinoflagellate cysts of the *Gonyaulacysta* group: a morphological and taxonomic study. *Am. Assoc. Strat. Palynol. Contr. Ser.* **9**, 1–81.

SARJEANT W. A. S. & DOWNIE C. (1966) The classification of dinoflagellate cysts above the generic level. *Grana Palynol.* **6**(3), 503–527.

SARJEANT W. A. S. & DOWNIE C. (1974) The classification of dinoflagellate cysts above the generic level—a discussion and revisions. *Birbal Sahni Inst. Palaeobotany*, Spec. Publ. No. 3, 9–32.

SCULL B. J., FELIX C. J., MCCALEB S. B. & SHAW W. G. (1966) The inter-discipline approach to paleoenvironmental interpretations. *Trans. Gulf Coast Assoc. Geol. Soc.* **16**, 81–117.

SCHUMACKER-LAMBRY J. (1978) Palynologie du Landenien inférieur (Paléocène) à Gelinden-Overbroek/Belgique. Relations entre les microfossiles et le sédiment. *Lab. Paléobot. Palynol.* (éd.), Univ. Liége.

SPECTOR D. L. (ed.) (1984) *Dinoflagellates*. Academic Press, Orlando. 545 pp.

STEINECK P. L. & FLEISHER R. L. (1978) Towards the classical evolutionary reclassification of Cenozoic Globigerinaceae (Foraminiferida). *Jour. Paleontol.* **52**(3), 618–635.

STOVER L. E. (1974) Palaeocene and Eocene species of *Deflandrea* (Dinophyceae) in Victorian coastal and offshore basins, Australia. *Geol. Soc. Australia Spec. Publ. No. 4*, 167–188.

STOVER L. E. & EVITT W. R. (1978) Analyses of pre-Pleistocene organic-walled dinoflagellates. *Stanford Univ. Publ., Geol. Sci.* **15**, 1–300.

STOVER L. E. & WILLIAMS G. L. (1982) Dinoflagellates. *Proc. Third North Am. Paleontol. Convention*, **2**, 525–533.

TANGEN K., BRAND L. E., BLACKWELDER P. L. & GUILLARD R. R. L. (1982) *Thoracosphaera heimii* (Lohmann) Kamptner is a dinophyte: observations on its morphology and life cycle. *Mar. Micropaleontol.* **7**, 193–212.

TAPPAN H. (1980) Dinoflagellates. In *The Paleobiology of Plant Protists*, pp. 225–462. W. H. Freeman and Co., San Francisco.

TAPPAN H. & LOEBLICH A. R. JR. (1971) Geobiologic implications of fossil phytoplankton evolution and time-space distribution. In *Symp. Palynology of the Late Cretaceous and Early Tertiary* (eds R. M. Kosanke & A. T. Cross), pp. 247–340. Geol. Soc. Am., Spec. Pap. 127.

TAPPAN H. & LOEBLICH A. R. JR (1972) Fluctuating rates of protistan evolution, diversification and extinction. *Int. Geol. Congr. 24*, Montreal, 1972, Sect. 7, *Paleontology*, 205–213.

TAPPAN H. & LOEBLICH A. R. JR (1973) Evolution of the oceanic plankton. *Earth-Sci. Rev.* **9**, 207–240.

TAPPAN H. & LOEBLICH A. R. JR (1977) Peridinialean cyst affinity, rather than gymnodinialean motile stage, of the Late Cretaceous dinoflagellate *Dinogymnium*. *Trans. Amer. Microsc. Soc.* **96**(4), 497–505.

TAYLOR D. L. & SELIGER H. H. (1979) *Toxic Dinoflagellate Blooms*. Elsevier, New York.

TAYLOR F. J. R. (1976a) Flagellate phylogeny: a study on conflicts. *J. Protozool.* **23**, 28–40.

TAYLOR F. J. R. (1976b) Dinoflagellates from the International Indian Ocean Expedition. A report on material collected by the R. V. 'Anton Bruun' 1963–1964. *Bibliotheca Botanica* **132**, 1–234.

TAYLOR F. J. R. (1978) Problems in the development of an explicit hypothetical phylogeny of the lower eukaryotes. *BioSystems* **10**, 67–89.

TAYLOR F. J. R. (1979) The toxigenic gonyaulacoid dinoflagellates. In *Toxic Dinoflagellate Blooms* (eds D. L. Taylor & H. H. Seliger), pp. 47–56. Elsevier, New York.

TAYLOR F. J. R. (1980) On dinoflagellate evolution. *BioSystems* **13**, 65–108.

TRAVERSE A. & GINSBURG R. N. (1966) Palynology of the surface sediments of Great Bahama Bank, as related to water movement and sedimentation. *Mar. Geol.* **4**, 417–459.

TURPIN D. H., DOBELL P. E. R. & TAYLOR F. J. R. (1978) Sexuality and cyst formation in Pacific strains of the toxic dinoflagellate *Gonyaulax tamarensis*. *J. Phycol.* **14**(2), 235–238.

VAIL P. R. & HARDENBOL J. (1979) Sea-level changes during the Tertiary. *Oceanus* **22**, 71–79.

VAIL P. R., MITCHUM R. M. & THOMPSON S. III (1977) Seismic stratigraphy and global changes of sea level. *Am. Assoc. Pet. Geol. Mem.* **26**, 83–97.

VAN COUVERING J. A., AUBRY M. P., BERGGREN W. A., BUJAK J. P., NAESER C. W. & WIESER T. (1981) The terminal Eocene event and the Polish connection. *Palaeogeogr. Palaeoclimatol. Palaeoecol.* **36**, 321–362.

VOZZHENNIKOVA T. F. (1963) *Type Pyrrophyta. Pyrrophytic algae*. Inst. Geol. Geophys., Siberian Branch Acad. Sci. USSR, pp. 171–186. (In Russian.)

VOZZHENNIKOVA T. F. (1965) *Introduction to the study of fossil peridinian algae* (transl. by K. Syers, ed. by W. A. S. Sarjeant). Boston Spa, Yorkshire, England, Natl. Lending Library for Science and Technology, 1967.

VOZZHENNIKOVA T. F. (1967) Fossilized peridinid algae in the Jurassic, Cretaceous and Palaeogene deposits of the USSR (transl. by R. Lees, ed. by W. A. S. Sarjeant). Boston Spa, Yorkshire, England, Natl. Lending Library for Science and Technology, 1971.

WALL D. (1965a) Microplankton, pollen, and spores from the Lower Jurassic in Britain. *Micropaleontol.* **11**, 151–190.

WALL D. (1965b) Modern hystrichosphaeres and dinoflagellate cysts from the Woods Hole region. *Grana Palynol.* **6**, 197–314.

WALL D. (1967) Fossil microplankton in deep-sea cores from the Caribbean Sea. *Palaeontology* **10**, 95–123.

WALL D. (1970) Quaternary dinoflagellate micropaleontology, 1959 to 1969. *Proc. North Am. Paleontol. Convention*, Chicago, 1969, Proc. G, 844–866.

WALL D. (1971a) The lateral and vertical distribution of dinoflagellates in Quaternary sediments. In *Micropaleontology of Oceans* (eds B. M. Funnell & W. R. Reidel), pp. 399–405. Cambridge Univ. Press, Cambridge.

WALL D. (1971b) Biological problems concerning fossilizable dinoflagellates. *Geosci. Man* **3**, 1–15.

WALL D. & DALE B. (1966) 'Living fossils' in western Atlantic plankton. *Nature (Lond.)* **211**, 1025–1026.

WALL D. & DALE B. (1968a) Quaternary calcareous dinoflagellates (Calciodinellideae) and their natural affinities. *J. Paleontol.* **42**, 1395–1408.

WALL D. & DALE B. (1968b) Modern dinoflagellate cysts and the evolution of the Peridiniales. *Micropaleontol.* **14**, 265–304.

WALL D. & DALE B. (1968c) Early Pleistocene dinoflagellates from the Royal Society borehole at Ludham, Norfolk. *New Phytology* **67**, 315–326.

WALL D. & DALE B. (1969) The 'hystrichosphaerid' resting spore of the dinoflagellate *Pyrodinium bahamense* Plate, 1906. *J. Phycol.* **5**, 140–149.

WALL D. & DALE B. (1970) Living hystrichosphaerid dinoflagellate spores from Bermuda and Puerto Rico. *Micropaleontol.* **16**, 47–58.

WALL D. & DALE B. (1971) A reconsideration of living fossil *Pyrophacus* Stein, 1883 (Dinophyceae). *J. Phycol.* **7**(3), 221–235.

WALL D. & DALE B. (1973) Paleosalinity relationships of dinoflagellates in the Late Quaternary of the Black Sea—a summary. *Geosci. Man* **7**, 95–102.

WALL D. & DALE B. (1974) Dinoflagellates in Late Quaternary deep-water sediments of the Black Sea. In *The Black Sea: Its Geology, Chemistry and Biology* (eds E. T. Degens & D. A. Ross), pp. 364–380. Am. Assoc. Pet. Geol. Mem. 20.

WALL D. & DALE B. & HARADA K. (1973) Descriptions of new fossil dinoflagellates from the Late Quaternary of the Black Sea. *Micropaleontol.* **19**, 18–31.

WALL D., DALE B., LOHMANN G. P. & SMITH W. K. (1977) The environmental and climatic distribution of dinoflagellate cysts in modern marine sediments from regions in the North and South Atlantic Oceans and adjacent seas. *Mar. Micropaleontol.* **2**, 121–200.

WALL D. & EVITT, W. R. (1975) A comparison of the modern genus *Ceratium* Schrank, 1793, with certain Cretaceous marine dinoflagellates. *Micropaleontol.* **21**(2), 14–44.

WALL D., GUILLARD R. R. L. & DALE B. (1967) Marine dinoflagellate cultures from resting spore. *Phycologia* **6**, 83–86.

WALL D., GUILLARD R. R. L., DALE B., SWIFT E. & WATABE N. (1970) Calcitic cysts in *Peridinium trochoideum* (Stein) Lemmermann, an autotrophic marine dinoflagellate. *Phycologia* **9**(2), 151–156.

WALL D. & WARREN J. S. (1969) Dinoflagellates in Red Sea piston cores. In *Hot Brines and Recent Heavy Metal Deposits in the Red Sea* (eds E. T. Degens & D. A. Ross), pp. 317–328. Springer-Verlag, New York.

WATKINS D. K. (1979) *Dinoflagellate cyst, chlorophyte, and acritarch associations from the Eocene of South Carolina.* M.S. thesis, Va. Poly. Inst. St. Univ.

WEILER H. & SONNE V. (1980) Ein Fund von kalkigen Dinoflagellaten-Zysten im Unteren Meeressand (Rupel) der Sandgrube am 'Zeilstück' bei Alzey-Weinheim (Mainzer-Becken) *Mainzer geowiss. Mitt.* **8**, 219–221.

WIGGINS V. D. (1972) Two new Lower Cretaceous dinoflagellate genera from southern Alaska. *Rev. Palaeobot. Palynol.* **14**, 297–308.

WIGGINS V. D. (1973) Upper Triassic dinoflagellates from arctic Alaska. *Micropaleontol.* **19**(1), 1–17.

WIGGINS V. D. (1975) The dinoflagellate family Pareodiniaceae: a discussion. *Geosci. Man* **11**, 95–115.

WILLE W. & GOCHT G. (1979) Dinoflagellaten aus dem Lias Südwestdeutschlands. *N. Jb. Geol. Paläont. Abh.* **158**(2), 221–258.

WILLIAMS D. B. (1971a) The distribution of marine dinoflagellates in relation to physical and chemical conditions. In *Micropaleontology of Oceans* (eds B. M. Funnell & W. R. Riedel), pp. 91–95. Cambridge Univ. Press, Cambridge.

WILLIAMS D. B. (1971b) The occurrence of dinoflagellates in marine sediments. In *Micropaleontology of Oceans* (eds B. M. Funnell & W. R. Riedel), pp. 231–243. Cambridge Univ. Press, Cambridge.

WILLIAMS D. B. & SARJEANT W. A. S. (1967) Organic-walled microfossils as depth and shoreline indicators. *Mar. Geol.* **5**, 389–412.

WILLIAMS G. L. (1977) Dinocysts—their classification, biostratigraphy and palaeoecology. In *Oceanic Micropalaeontology* (ed. A. T. S. Ramsay), pp. 1231–1325. Academic Press, London.

WILLIAMS G. L. (1978) Dinoflagellates, acritarchs and tasmanitids. In *Introduction to Marine Micropaleontology* (eds B. U. Haq & A. Boersma), pp. 293–326. Elsevier, New York.

WILLIAMS G. L. & BUJAK J. P. (1977a) Distribution patterns of some North Atlantic Cenozoic dinoflagellate cysts. *Mar. Micropaleontol.* **2**, 223–233.

WILLIAMS G. L. & BUJAK J. P. (1977b) Cenozoic palynostratigraphy of offshore eastern Canada. *Amer. Assoc. Strat. Palynol. Contr. Ser.* **5A**, 14–437.

WILLIAMS G. L. & BUJAK J. P. (1985) Mesozoic and Cenozoic dinoflagellates. In *Plankton Stratigraphy* (eds H. M. Bolli, J. B. Saunders & K. Perch-Nielsen). pp. 847–964. Cambridge University Press, Cambridge.

WILSON G. J. (1967) Some new species of Lower Tertiary dinoflagellates from McMurdo Sound, Antarctica. *N.Z. J. Bot.* **5**, 57–83.

WILSON G. J. (1968) On the occurrence of fossil microspores, pollen grains, and microplankton in bottom sediments of the Ross Sea, Antarctica. *N.Z. J. Mar. Freshwater Res.* **2**, 381–389.

WILSON G. J. (1973) Palynology of the Middle Pleistocene Te Piki Bed, Cape Runaway, New Zealand. *N.Z. J. Geol. Geophys.* **16**, 345–354.

WILSON G. J. & CLOWES C. D. (1980) A concise catalogue of organic-walled fossil dinoflagellate genera. *Rept. N. Zeal. Geol. Survey* **92**, 1–199.

WITMER R. J. & GOODMAN D. K. (1980) Early Tertiary *Wetzeliella* assemblages from the Virginia-Maryland (USA) coastal plain. *Palynology* **4**, 254–255 (abstr.).

WOOD F. E. J. (1964) Studies in microbial ecology of the Australasian region. *Nova Hedwigia* **8**(1/2), 1–20, 35–54: (3/4), 461–527, 548–568.

WRENN J. H. & BECKMAN S. W. (1982) Maceral, total organic carbon, and palynological analyses of Ross Ice Shelf Project Site J9 cores. *Science* **216**, 187–189.

APPENDIX
TAXONOMY AND CLASSIFICATION

F. J. R. TAYLOR

Departments of Oceanography and Botany, University of British Columbia, Vancouver, BC, Canada V6T 1W5

1 INTRODUCTION

In this Appendix some general considerations regarding dinoflagellate taxonomy, particularly those arising from the text, will be briefly considered, followed by an outline classification of the living dinoflagellate genera. Currently dinoflagellate taxonomy is in a greater than usual state of flux, arising partly from the rediscovery and examination, by modern methods (particularly the scanning electron microscope, but also culture and molecular studies), of many taxa not seen since first description early in this century or in the last. Life-cycle features are now taken more into account when circumscribing taxa, particularly in attempting to reconcile the taxonomy of fossils or remains in sediments, where only the cyst stage is usually available, with living taxa in which the cyst stage, if any, may not be known. Differences in the degree of morphological difference required to distinguish genera and the heavy emphasis on tabulation in living forms, makes the task greater. Dale (1983) has aptly described the fossil members as 'over-classified' and the living forms as 'under-classified'. An attempt to produce a detailed, unified classification of living and fossil taxa is currently under way (R. A. Fensome, F. J. R. Taylor, G. Norris, W. A. S. Sarjeant & G. L. Williams, in prep.). The distinction of species is usually considered as objective, based on sexual compatability (the Biological Species Concept) but the information needed to make this judgement is lacking for nearly all species recognized at present and the current picture for *Crypthecodinium cohnii* (Beam & Himes 1984), the only species known well enough to try, suggests that the creation of multiple sibling species within each morphologically determinable species ('morphospecies'—Taylor 1985) may be necessary if the concept is extrapolated directly from the Metazoa. Detailed studies of 'zooxanthellae' also indicate considerable genetic diversity (Trench, Chapter 12, this volume), without information on sexual compatability. It appears that, as more detailed information on the physiology and biochemistry of living morphospecies accumulates, the recognition of the considerable genetically-based diversity that is possible within a given morphotype (Cembella & Taylor 1985, 1986) will alter our view of the properties to be expected of 'species' in dinoflagellates.

2 FORMALITIES

It is unfortunate for dinoflagellate taxonomy that the group, together with other 'phytoflagellates', i.e. flagellates possessing photosynthetic members (euglenoids, cryptomonads, chrysomonads, silicoflagellates, prymnesiomonads, chloromonads, prasinomonads, volvocalean chlorophyceans and, possibly, choanoflagellates) are claimed by both botanists and zoologists, as algae and protozoans, respectively. Consequently their nomenclature has been formally regulated by both the International Code of Botanical Nomenclature (ICBN) and the International Code of Zoological Nomenclature (ICZN). Although these are similar, for the most part, they differ in some key features (see Jeffrey 1973 and Sournia *et al.* 1977, for a summary), most obviously in the requirement for a Latin diagnosis for taxa described under the ICBN, but also in other regulations. These can cause procedural headaches (examples in Sournia *et al.* 1977; Taylor 1985). Furthermore, the codes permit the use of identically-spelled names (homonyms) for genera described in the Plant and Animal Kingdoms. Thus specialists considering themselves botanists are presently permitted to use generic names that zoologists may not, and vice versa. Taylor *et al.* (1986) have put forward proposals to rectify the situation, including the rejection of homonymy in any kingdom, i.e. if a generic name has been used before in any kingdom it should not be accepted for 'phytoflagellates' (predominantly flagellated groups in which both photosynthetic and non-photosynthetic occur, thus including dinoflagellates); Latin should be recommended but not essential in the diagnosis of a new taxon; and names should have priority only in their own rank. The outcome of these proposals depends on decisions of the committees which review and approve/reject modifications to the Codes of both Botanical and Zoological Nomenclature. At the time of writing, these exist as proposals only.

3 TAXONOMIC CRITERIA IN CURRENT USE

While there is not complete uniformity in the weight that different authors have placed on various criteria in dinoflagellate taxonomy, nevertheless there is a rough consensus which has developed over the years. The table below provides a rough indication of the significance of various features used to distinguish living species and genera at present. Note that there are exceptions to nearly all these.

Table 1. Criteria in current use for approximate rank distinctions

Character	Species	Genus	Higher
Size	+		
Form	+	+	+
girdle position	+	+	
displacement	+	+	
Tabulation	+	+	+
girdle plates		+	+
antapicals		+	+
Ornamentation	+		
Plastid presence	+	1	
pyrenoid number		2	
Biochemical	3		
Life cycle			
phase dominance		+	+
cyst morphology	+	+	

1 Usually not used as a generic criterion in dinoflagellates although commonly used at this level in other groups. In *Pheopolykrikos* its distinction from *Polykrikos* is bolstered by the nuclear number.
2 The number of pyrenoids has been used to distinguish the coccoid, symbiotic genera *Symbiodinium* and *Zooxanthella*, although not all authors agree.
3 Features such as toxins and luminescence have sometimes been used as primary characters, e.g. in *Protogonyaulax*, but have been rejected as unreliable or impractical later.

In thecate dinoflagellates, the tabulation pattern has been considered to be pre-eminent in their classification, with variation in the most conservative regions (girdle, sulcus, antapex) of greatest significance. In fossil dinoflagellates, shape and ornamentation have been used as generic or even higher distinguishing features (consequently rendering comparisons between numbers of living and fossil taxa highly deceptive), but this practice has been largely due to the difficulty or impossibility of determining paratabulation patterns. Contemporary palynologists have become highly skilled at determining probable paratabulations (see especially Evitt 1985) and, with this has come a greater emphasis on these underlying patterns in the taxonomy of the fossils.

4 RELATIVELY RECENT NOMENCLATURAL REVISIONS INVOLVING THIS TEXT

In this book an attempt has been made, where possible or acceptable to the contributors, to refer to species by only one name, since it is difficult for those unfamiliar with the group to combine the relevant data when the different names used in the original literature are used.

It is difficult to know how recent one should be in summarizing these changes, since it is evident from the literature that many non-taxonomists are unaware of revisions in place for twenty years or more. The summary below indicates the most major changes of the last twenty years, with very brief indications for the reasons for the changes. It must be emphasized that the issues were sometimes complex and disputatious and so the brief summaries below should be used as rough guidelines only. The references are to sources of further discussion, not necessarily the original proposals.

Balechina Loeblich Jr. & Loeblich III was created when members of the subgenus *Pachydinium* Kofoid & Swezy of the genus *Gymnodinium* Stein were raised to the status of a genus and a genus *Pachydinium* Pavillard was already in existence (see Taylor 1976).

Cachonina Loeblich III has been sunk into *Heterocapsa* Stein because it cannot be adequately separated from it (Morrill & Loeblich 1981; however, see Sournia 1984).

Diplopeltopsis Pavillard was replaced by *Zygabikodinium* Loeblich & Loeblich (1970) because it is pre-occupied by a lichen genus.

Erythropsidinium P. Silva was substituted for *Erythropsis* Hertwig because the latter was pre-occupied by a plant genus (Silva 1960).

Exuviaella Cienkowski was 'sunk' into *Prorocentrum* Ehrenberg because various attempts to distinguish them on the basis of anterior spine development or periflagellar platelet presence, were inadequate (see Taylor 1976).

Glenodinium Ehrenberg has been interpreted in many, irregular ways. However, as pointed out by Loeblich (1980), the features of the type species, *G. cinctum*, are clearly definable. When this is done they correspond to the genus *Sphaerodinium* Woloszynska, which should be sunk into it. Note that *none* of the species attributed to *Glenodinium* by Schiller (1937) would now be placed in this genus!

Goniodoma Stein was replaced with *Triadinium* Dodge (1981), because it is pre-occupied by a genus of butterfly. However, *Triadinium* is also pre-occupied, by a genus of ciliate. If the view is adopted that a generic name should not have been used previously in any kingdom (Taylor *et al.* 1986), then a further new generic name is required.

Massartia Conrad had to be replaced with *Katodinium* Fott because it was pre-occupied by a genus of mushroom (Fott 1957).

Phalacroma Stein was sunk into *Dinophysis* Ehrenberg because of the existence of species with intermediate features (although other genera which intergrade with each other, such as *Gymnodinium* with *Gyrodinium* and *Amphidinium*, have not been combined), this also solving the problem of pre-occupation of the former name by a genus of trilobites (discussed in Taylor 1976). It should be noted that the majority of species formerly attributed to *Phalacroma* are non-photosynthetic, whereas those formerly attributed to *Dinophysis* apparently are all photosynthetic.

Pratjetella Lohmann should be used instead of *Leptodiscus* Hertwig because the latter is pre-occupied.

Porella Schiller had to be renamed *Mesoporos* Lillick because it was pre-occupied in both the Animal and Plant Kingdoms. Although this was done many years ago (Lillick 1937) it still seems to escape some authors' notice. Thus, *Dinoporella* Halim was redundant when created twenty three years later.

Protogonyaulax Taylor (1979) was created for the 'tamarensis group' of *Gonyaulax* Diesing because of differences in tabulation (apical and antapical) and cyst form. Loeblich III & Loeblich (1979) expanded the genus *Gessnerium* Halim to include them. However, if that approach is taken, the genus *Alexandrium* Halim has priority (see Taylor 1984, 1985 and Balech 1985).

Symbiodinium Freudenthal and *Zooxanthella* Brandt both differ from the genera that their motile stages resemble in that they divide exclusively in the non-motile state. They can be distinguished from each other in the coccoid state by the number of pyrenoids, which are readily visible, being one (rarely two) in the former and four to six in the latter. It has also been reported that the motiles differ in that the former is gymnodinoid/woloszynskioid and the latter is amphidinioid.

5 AN OUTLINE CLASSIFICATION OF LIVING DINOFLAGELLATES

The following classification is intended simply as a rapid indication of the apparent affinities of the genera of living dinoflagellates. The characteristics of each suprageneric grouping can be found mostly in Loeblich (1982), although not all the groupings agree with his, and in full detail, combined with those of fossil taxa, in Fensome *et al.* (in prep. probable publication 1988).

Table 2. An outline classification of the living dinoflagellate genera

Order Desmomonadales
- Family 1 Adinomonadaceae
  - Genus *Adinomonas*
- Family 2 Desmomonadaceae
  - Genera *Desmomastix*, *Pleromonas*, *Haplodinium*

Table 2 (continued)

Order Desmocapsales

Family 1 Desmocapsaceae

Genus *Desmocapsa*

Order Prorocentrales

Family Prorocentraceae

Genera *Mesoporus*, *Prorocentrum*

Order Dinophysiales

Family 1 Dinophysiaceae

Genera *Citharistes*, *Dinofurcula*, *Dinophysis*, *Histioneis*, *Histiophysis*, *Latifascia*, *Metadinophysis*, *Metaphalacroma*, *Ornithocercus*, *Parahistioneis*, *Proheteroschisma*, *Pseudophalacroma*, *Sinophysis*, *Thaumatodinium*.

Family 2 Amphisoleniaceae

Genera *Amphisolenia*, *Oxyphysis*, *Triposolenia*

Order Gonyaulacales

Family 1 Ceratiaceae

Genus *Ceratium*

Family 2 Ceratocoryaceae

Genus *Ceratocorys*

Family 3 Cladopyxidaceae

Genera *Cladopyxis*, *Micracanthidinium*, *Palaeophalacroma*, *Sinodinium*

Family 4 Crypthecodiniaceae

Genus *Crypthecodinium*

Family 5 Gonyaulacaceae

Genera *Alexandrium* (= *Gessnerium*), *Amphidoma*, *Gonyaulax*, *Protoceratium*, *Protogonyaulax*, *Pyrodinium*, *Spiraulax*

Family 6 Heterodiniaceae

Genera *Dolichodinium*, *Heterodinium*

Family 7 Ostreopsidaceae

Genera *Coolia*, *Gambierdiscus*, *Ostreopsis*

Family 8 Oxytoxaceae

Genera *Centrodinium*, *Corythodinium* (= *Pavillardinium*), *Oxytoxum*

Family 9 Pyrocystaceae

Genus *Pyrocystis*

Family 10 Pyrophaceae

Genera *Fragilidium*, *Helgolandinium*, *Pyrophacus*

Family 11 Triadiniaceae

Genus *Triadinium* (= *Heteraulacus*)

Order Peridiniales

Family 1 Calciodinellaceae

Genera *Ensiculifera*, *Scrippsiella*, *Thoracosphaera*

Family 2 Glenodiniaceae

Genera *Glenodiniopsis*, *Glenodinium* (= *Sphaerodinium*)

Family 3 Peridiniaceae

Genera *Chalubinskia*, *Dinosphaera*, *Heterocapsa* (incl. *Cachonina*), *Peridiniopsis*, *Peridinium*, *Thompsodinium*

Family 4 Podolampadaceae

Genera *Blepharocysta*, *Podolampas*

Family 5 Protoperidiniaceae

Genera *Apsteinia*, *Boreadinium*, *Diplopsalis*, *Diplopsalopsis*, *Dissodium*, *Gotoius*, *Oblea*, *Protoperidinium*, *Zygabikodinium*

Family 6 Thecadiniaceae
Genera *Amphidiniopsis*, *Roscoffia*, *Thecadinium*
Family 7 Thoracosphaeraceae
Genus *Thoracosphaera*

Order Gymnodiniales
Family 1 Actiniscaceae
Genus *Actiniscus*, *Dicroerisma*, *Plectodinium*
Family 2 Gymnodiniaceae
Genera *Amphidinium*, *Cochlodinium*, *Gymnodinium*, *Gyrodinium*, *Katodinium*, *Lophodinium*, *Torodinium*, *Woloszynskia*
Family 3 Polykrikaceae
Genera *Pheopolykrikos*, *Polykrikos*
Family 4 Warnowiaceae
Genera *Erythropsidinium*, *Greuetodinium* (= *Leucopsis*), *Nematodinium*, *Nematopsides*, *Proterythropsis*, *Protopsis*, *Warnowia*
Family 5 Zooxanthellaceae
Genera *Endodinium*, *Symbiodinium*, *Zooxanthella*

Order Kolkwitziellales
Family 1 Brachydiniaceae
Genera *Asterodinium*, *Brachydinium*
Family 2 Kolkwitziellaceae
Genera *Balechina*, *Berghiella*, *Herdmania*, *Kolkwitziella*, *Ptychodiscus*, *Sclerodinium*

Order Noctilucales
Family 1 Kofoidiniaceae
Genera *Kofoidinium*, *Pomatodinium*, *Spatulodinium*
Family 2 Leptodiscaceae
Genera *Abedinium* (= *Leptophyllus*), *Cachonodinium* (= *Leptodinium*), *Cymbodinium*, *Craspedotella*, *Pratjetella* (= *Leptodiscus*), *Petalodinium*, *Scaphodinium*
Family 3 Noctilucaceae
Genera *Noctiluca*, *Pavillardia*, *Pronoctiluca*, *Protodinifer*

Order Phytodiniales
Family 1 *Gloeodiniaceae*
Genera *Gloeodinium* (= *Hemidinium?*), *Rufusiella*
Family 2 Phytodiniaceae
Genera *Cystodinedria*, *Cystodinium*, (incl. *Dinococcus*?), *Dinastridium* (incl. *Bourrellyella*), *Dinopodiella*, *Hypnodinium*, *Manchudinium*, *Paulsenella*, *Phytodinedria*, *Phytodinium*, *Rhizodinium*, *Stylodinium*, *Tetradinium*

Order Blastodiniales
Family 1 Apodiniaceae
Genus *Apodinium*
Family 2 Blastodiniaceae
Genus *Blastodinium*
Family 3 Cachonellaceae
Genera *Actinodinium*, *Cachonella*
Family 4 Haplozoaceae
Genus *Haplozoon*
Family 5 Oodiniaceae
Genera *Amylodinium*, *Caryotoma*, *Dissodinium*, *Oodinium*
Family 6 Protoodiniaceae
Genera *Crepidoodinium*, *Piscinoodinium*, *Protoodinium*

Table 2 (continued)

Order Chytriodiniales

Family Chytriodiniaceae

Genera *Chytriodinium, Myxodinium, Sporodinium*

Subclass SYNDINIOPHYCIDAE

Order Coccidiniales

Family 1 Amoebophryaceae

Genus *Amoebophrya*

Family 2 Coccidiniaceae

Genus *Coccidinium*

Family 3 Duboscquellaceae

Genus *Dogelodinium, Duboscquella, Keppenodinium*

Family 4 Sphaeriparaceae

Genus *Sphaeripara*

Order Syndiniales

Family Syndiniaceae

Genera *Hematodinium, Ichthyodinium, Merodinium, Solenodinium, Syndinium, Trypanodinium*

Incertae sedis:

Adenoides	*Lissodinium*
Amphilothus	*Oxyrrhis*
Archaeosphaerodiniopsis	*Protaspis*
Ceratoperidinium	*Pseliodinium*
Dinameoba	*Spiromonas*
Filodinium	*Thaurilens*

6 REFERENCES

BALECH E. (1985) The genus *Alexandrium* or *Gonyaulax* of the Tamarensis group. In *Toxic Dinoflagellates* (eds D. M. Anderson, A. W. White & D. G. Baden) pp. 33–38. Elsevier, New York.

BEAM C. A. & HIMES M. (1984) Dinoflagellate genetics. In *Dinoflagellates* (ed. D. L. Spector) pp. 263–298. Academic Press, Orlando.

CEMBELLA A. D. & TAYLOR F. J. R. (1985) Biochemical variability within the *Protogonyaulax tamarensis/catenella* species complex. In *Toxic Dinoflagellates* (eds. D. M. Anderson, A. W. White & D. G. Baden) pp. 55–60. Elsevier, New York.

CEMBELLA A. D. & TAYLOR F. J. R. (1986) Electrophoretic variability within the *Protogonyaulax tamarensis/catenella* species complex: pyridine linked dehydrogenases. *Biochem. System. Ecol.* **14**, 311–323.

DALE B. (1983) Dinoflagellate resting cysts: benthic plankton. In *Survival Strategies of the Algae* (ed. G. A. Fryxell) pp. 69–136. Cambridge Univ. Press, Cambridge.

DODGE J. D. (1981) Three new generic names in the Dinophyceae: *Herdmania, Sclerodinium* and *Triadinium* to replace *Heteraulacus* and *Goniodoma*. *Brit. Phycol. J.* **16**, 273–280.

EVITT W. R. (1985) Sporopollenin dinoflagellate cysts: their morphology and interpretation. *Amer. Assoc. Stratigr. Palynol.* 333pp.

FOTT B. (1957) Taxonomie drobnohledne flory nasich vod. *Preslia* **29**, 278–319.

JEFFREY C. (1973) *Biological Nomenclature*. Edward Arnold, London.

LILLICK L. C. (1937) Seasonal studies of the phytoplankton off Wood's Hole, Massachusetts. *Biol. Bull.* **73**, 488–503.

LOEBLICH A. R. Jr. & LOEBLICH A. R. III (1970) Index to the genera, subgenera and sections of the Pyrrhophyta [V]. *Phycologia* **9**, 199–203.

LOEBLICH A. R. III (1980) Dinoflagellate nomenclature. *Taxon* **29**, 321–324.

LOEBLICH A. R. III & LOEBLICH L. A. (1979) The systematics of *Gonyaulax* with special reference to the toxic species. In *Toxic Dinoflagellate Blooms* (eds. D. L. Taylor & H. H. Seliger) pp. 41–46. Elsevier/North Holland, New York.

MORRILL L. & LOEBLICH A. R. III (1981) A survey for body scales in dinoflagellates and a revision of *Cachonina* and *Heterocapsa* (Pyrrhophyta). *J. Plankt. Res.* **3**, 53–65.

SCHILLER J. (1937) Dinoflagellatae. In *Rabenhorst's Kryptogamenflora* 10 (2). Akadem. Verlagsges., Leipzig. 590pp.

SILVA P. C. (1960) Remarks on algal nomenclature. III. *Taxon* **9**, 18–25.

SOURNIA A. (1984) Classification et nomenclature de divers dinoflagellés marins (Dinophyceae). *Phycologia* **23**, 345–355.

SOURNIA A., CACHON J. & CACHON M. (1975) Catalogue des éspeces et taxons infraspecifiques de Dinoflagellés marins actuels publies depuis la revision de J. Schiller. II Dinoflagellés parasites ou symbiotiques. *Arch. Protistenk.* **117**, 1–19.

TAYLOR F. J. R. (1976) Dinoflagellates from the International Indian Ocean Expedition. *Bibliotheca Botanica* **132**, 1–234.

TAYLOR F. J. R. (1979) The toxigenic gonyaulacoid dinoflagellates. In *Toxic Dinoflagellate Blooms* (eds D. L. Taylor & H. H. Seliger) pp. 47–56.

TAYLOR F. J. R. (1984) Toxic dinoflagellates: taxonomic and biogeographic aspects with emphasis on *Protogonyaulax*. In *Seafood Toxins* (ed. E. P. Ragelis) pp. 77–97. *Amer. Chem Soc.*, *Symp. Ser.* 262. Wash., D.C.

TAYLOR F. J. R. (1985) The taxonomy and relationships of red tide flagellates. In *Toxic Dinoflagellates* (eds. D. M. Anderson, A. W. White & D. G. Baden) pp. 11–26. Elsevier, New York.

TAYLOR F. J. R., SARJEANT W. A. S., FENSOME R. A. & WILLIAMS G. L. (1986) Proposals to standardize the nomenclature in flagellate groups currently treated by both the Botanical and Zoological Codes of Nomenclature. *Taxon* **35**, 890–896.

TAXONOMIC INDEX

(See also the classification outline in the Appendix pp. 727–730 which is not indexed.)

* indicates a non-dinoflagellate taxon.

SUBJECT INDEX

Numbers in italics indicate illustrations